ISBN 978-1-330-16719-9
PIBN 10042729

FB &c Ltd, Dalton House, 60 Windsor Avenue, London, SW19 2RR.
Company number 08720141. Registered in England and Wales.

For support please visit www.forgottenbooks.com

HANDBOOK

FOR

ARCHITECTS AND BUILDERS

PUBLISHED
UNDER THE AUSPICES
OF THE

ILLINOIS SOCIETY OF ARCHITECTS

THIRTY-FIFTH EDITION

1938-39

Emery Stanford Hall, Editor

PREFACE

The thirty-fifth volume of the Handbook for Architects and Builders is again presented to the architectural profession after a long period of suspension. We have produced the work in its usual high standard and no time, effort or expense has been spared to give the architect the information he requires in his daily practice.

The Handbook for architects and builders covers a peculiarly exclusive field and is a recognized reference for everyone in the architectural field.

We have made but few changes from the general plan or arrangement that has proven satisfactory in former editions, although nearly all the subject matter is new or has been brought up to date.

The new Chicago Building Code is published in this volume as passed by the City Council of Chicago and a detailed comprehensive index wherein all sections and chapters can be easily located.

The staff of contributors remains the same with the addition of John Jeffrey Davey, C. W. Berghorn, W. E. Fosket and P. W. Vandenberg, Wayne R. Fuller, Fred Morgan and Deane Perham.

We realize that notwithstanding the care and caution which has been exercised in editing and preparing this volume that inaccuracies may have found their way into this work and for such faults we ask our readers to forward to us their friendly criticism and constructive suggestions in order to improve succeeding editions.

The demand for the Handbook for Architects and Builders is constantly increasing and it has become almost indispensable to architects, engineers contractors, builders and those allied to the building industry.

The classified list of advertisers which appears on page 791 furnishes the architect with a list of those engaged in the manufacture and sale of materials and the contracting business. We have exercised our best judgment in the selection of those represented in our advertising pages.

To the advertisers we extend our appreciation for their patronage and bespeak for them an improved and successful future period.

¶ We call especial attention to the business announcements on these pages. We have accepted only such houses as are absolutely first-class and reliable. In the light of knowledge which we have upon the subject, acquired by experience, we feel that we have used every discretion in the matter of those represented herein

RALPH C. LLEWELLYN
TRES.
ARTHUR WOLTERSDORF
1st VICE PRES.
2nd VICE PRES.
RALPH C. HARRIS
ELMER C. JENSEN
PRESIDENT
HERMAN L. PALMER
FIN. SECY.
SECY.
STANLEY D. FAIRCLOUGH
OFFICERS

WILLIAM P. FOX
EDWARD PROBST
DIRECTORS
JOHN A. HOLABIRD
VICTOR A. MATTESON
PAUL GERHARDT, Jr.
LEO J. WEISSENBORN

IRVING K. POND
JOHN A. HOLABIRD
HERBERT HEWITT
RICHARD E. SCHMIDT
EMERY S. HALL
A. S. ALSCHULER
FREDERICK J. THIELBAR
BOARD OF ARBITRATION

OLSON CONVEYORS
Elevating and conveying systems for handling practically every type of material. For industrial plants, food and beverage plants, department stores. Food and dish handling conveyors for restaurants, hotels and hospitals.
Belt and Gravity Conveyors
Spiral Chutes
Apron Conveyors
Linen and Rubbish Chutes
Coal and Ash Conveyors
Automatic Elevators
Subveyors for handling foods and dishes
Catalog on Request
SAMUEL OLSON MFG. COMPANY, INC.
2418 Bloomingdale Road
Chicago, Illinois

JOSEPH W. ROYER
URBANA
LEO H. PLEINS
SPRINGFIELD
BENJ. A. HORN
ROCK ISLAND
WM. P. FOX
CHAIRMAN · · · CHICAGO
HAROLD B. McELDOWNEY
CHICAGO
COMMITTEE
ON
PUBLIC ACTION
WILLIAM R. McCOY
MOUNT VERNON
HERBERT E. HEWITT
PEORIA
WM. H. SCHULZKE
MOLINE
JOHN E. COYLE
JOLIET
FRANCIS A. FLAKS
CHICAGO

ALFRED S. ALSCHULER
HERBERT HEWITT
L.H. PROVINE
EXAMINING COMMITTEE
of ARCHITECTS
· STATE of ILLINOIS ·
JOHN J. HALLIHAN
DIRECTOR of
REGISTRATION & EDUCATION
EMERY S. HALL
R.C. OSTERGREN

ILLINOIS SOCIETY OF ARCHITECTS
Organized January 12, 1897
Incorporated June 25, 1897

EDITORIALS

By EMERY STANFORD HALL, F. A. I. A.

THE MAN FOR ARCHITECTURE

Obviously practice is essential to the existence of an architect. It should be equally evident that an architect without a knowledge of the actualities of architecture can be nothing but a discredit to his profession. Clients are obtained and held by the establishment of worth.

Professional equipment is generally understood to be made up of natural gifts, educational background and special technical training for architecture. These are qualities which can be measured by comprehensive entrance-to-practice examinations, but do not constitute the complete equipment necessary to competent practice. There are other and seemingly paramount qualities to architecture, which are not measurable by examination. The man for architecture must be a sensible man. He needs to be possessed of that instinctive quality sometimes called horse-sense, the intuition which an automobile does not have, but which keeps the man who rides behind horses out of the ditch. Common sense is particularly important to all learned professions, but is imperatively necessary to one who practices architecture. It enables an architect to lead a temperamental client in the way which he should properly go. Wise common sense guides to the successful mobilation of construction. The same sort of sense generals to the successful completion of a building project.

The architect must know the human animal and be able to give advice and have it kindly received. The diplomacy of the architect is severely tested in dealing with the ever present problem of connivance for special position or dominance in council. Political scheming is not confined alone to affairs of state; it often prevades family and corporate councils. A fine sense of humor has bridged many a papa-mamma-Johnny-Mary eruption; likewise it has unscrambled vice-presidents, general managers and would be's to the saving of the job.

Finally, the man for architecture must be likable, know his stuff, have diplomatic force, but, above all, he must have judicial poise, incontestable sense of what is fair between man and man and real unafraid integrity. Reduced to simplest statement, honesty and competency of the kind that is both correct and comfortable.

ESSENTIAL TO THE JOB?

The degree to which the architect is essential to the building program approximates the extent to which he will be employed. If the job seems to be able to run without him he is classed as a luxury, to be avoided in the interest of building economy. It is on this plea that the architect's chief competitor, the general contractor, steps in and grabs the bacon.

Demand for the right to command must be premised on an intelligent knowledge of the technic and custom incident to the production of the work over which command is asked. The architect who can, sufficient time being allowed, detail and specify every important element of the structure over which he assumes to preside does not have to do so, instead he is able to secure the loyal cooperation of the specialists, who do the work under his direction. The architect who knows trade practice so well that he can correctly segregate the several trades in his specifications and detail and cover them completely, leaving no gaps between and no over-lappings, is in a position to let his work under the separate trade contract system. He does not have to let his work in this manner, but is in a position where he is not subject to the merciful whims of others. As an essential contributor to a building project the architect may properly expect adequate remuneration for his service. As merely an ornament to the job he has to be content with largess.

It is a matter of common knowledge, whether legally or otherwise, that a great many people do employ general contractors to both design and construct their buildings. This practice has grown out of the popular notion that this method means less cost, less trouble and satisfactory results. This practice of employing contractors to displace architects can not be stopped by law. Unless people in general understand the why of a law they will not obey it and law officers dare not enforce it. This statement is bad legal theory, but represents the stone wall of fact. So long as a considerable number of folks are sold to the idea that they can save money by giving the architect the "go-by" they are not going to put an architect in charge of their jobs.

The idea that contractor controlled design and construction is cheaper has been subtly fed to the public by those who would profit thereby. It is unconscionably false when competent architects are employed but, unfortunately, there are incompetent architects in circulation in sufficient numbers to give credence to this inspired propaganda. Irrespective of its truth or falsity, this notion on the part of so many people constitutes a stiff indictment of the prevailing system of training for architects. Translated, it implies that the average of eleven years and large expense required, in order to take a boy from the grades and prepare him for entrance upon architectural practice, is wasted when another boy of no greater talent, with say trade school training of not more than high school rank plus a short service on pay as a trade apprentice can, under the name of contractor, design and build cheaper and better than the boy on whom there has been expended more than twice as much of time and money for architectural training. The grand jury of the people has returned the indictment. It is up to the architectural profession to see that it is quashed. Only they know the falsity of the accusation.

WEAKNESS

In every sphere of life the boy or the man who does only those things he likes to do is not and never will be a contributor. The world does not owe any man a living. It demands that each human being shall contribute his share towards the increase of the world's wealth. A country's strength is measured by the percentage of its citizens who are willing to give. Likewise a pro-

fession's strength is dependent on the largeness of the percentage of its members who are contributors.

Some candidates for architecture like to fiddle with structural problems but detest arduous plan studies and mere aesthetics. Others rate aesthetic consideration as the only important element in building design. Most of such candidates seem to ignore the fact that the prime purpose of building is to render a real human service.

Custom has made the architect the umpire on the job. Most folks like to boss, but few like to pass through the strenuous soul and mind discipline which is necessary to sane command. Picture-making without having designed anything worthy of depicting is a pleasant pastime for too many architectural students. There are even some professors of design who are complacent in the practice of teaching their students to render drawings rather than to plan to meet human need. These skip over problems of practical purpose for grand schemes which are both economically and administratively impossible. Again there are both parents and "profs" who are all for youthful self-expression. These are for all elective courses. Jim wants to study "Mechanics of Materials" and sees no reason why he should climb the ladder of arithmetic, algebra, geometry, trigonometry, calculus and analytical geometry. John wants to boss, but sees no reason for the study of fundamental ethics, the measure of equity, how to gather facts and reason from assembled facts to logically correct conclusion. Bill wants to draw contracts without a knowledge of 'the essence of the contract." The schools have turned out too many candidates for architecture who are misfits, but in no sense are all such. There are strong men fortunate in having had the best of college instruction under the wisest of professors who find themselves out in the world without any practical experience and no opportunity to get any.

THE OLDER MAN'S DUTY

Professional education is just about what the men in practice want it to be. In the past they have demanded draftsmen; they have not wanted competitors. Now they are complaining. Why will the young college man work for contractors, lumber and glass dealers at small wages and with no care for the ideals of their profession? Why will hundreds of these young men go into a single competition the total fees for which would not pay five cents an hour for the combined time put in? The answer is easy to find—these boys for four or five years have been fed with the importance of competitive drawing as having precedent to a knowledge of the real problems of architectural practice. In the vernacular of the street these lads have been gypped. Now they find that the only job for which they are fitted is somebody's hired-man. They have not learned how to handle even a small job from start to finish with satisfaction to their client, justice to the contractor and credit to their profession. When there are no draftsmen's jobs in architects' offices, young architectural graduates often find jobs in contractors' or material dealers' offices. In such positions they acquire the habit of making plans for the purpose of selling the job rather than controlling its construction. Contractor plans do not have to be correct. They are only camouflage to land the job—why follow them? Other graduates, possibly more ambitious, choose to pirate plans with no better moral effect. All of these practices operate to the destruction of honorable and competent architectural practice. Right now there are literally thousands of small jobs which these boys could handle with satisfaction to their clients, pleasure to themselves and credit to their profession, if they had essential practical training.

Medical internship training has made it safe for the young doctor to enter upon practice without jeopardy to the public. In some way, architects in practice must see to it that worthy candidates for architecture are provided with practical training comparable with that made available to candidates for medicine.

Every incompetence, recklessness or indiscretion of one member of a profession is reflected on all other members of his profession. Self preservation, as well as duty, requires that the older, more experienced men shall help the younger men to avoid practical errors.

THE MENTORSHIP PLAN

The Mentorship Plan, approved by The American Institute of Architects, the Association of Collegiate Schools of Architecture, the Beaux Arts Institute of Design and directly under the supervision of the National Council of Architectural Registration Boards offers the organization machinery necessary to enable the architect in practice, to extend a helping hand to the candidate for architecture. The way has been opened; now all that is needed is the cooperation of the individual members of the profession.

President Grant said: "The way to resume specie payment is to resume." Likewise the way to start the mentorship plan is to start. It will not do to wait until every architect in the United States has agreed to do his share. Each architect as soon as he has learned about this plan should volunteer his service and take his assignment in an earnest cooperative spirit. The fact that other men are not doing their share is no excuse for anyone not doing his. Duty is personal. One may be sorry that some other person is not performing his obligation to his profession, but the fact that someone else is neglectful of duty offers him no alibi.

THE MENTOR

The Mentorship Plan introduces two characters, first, the candidate for architecture; second, the mentor. Both are human, each good or bad according to bent. It is important to persuade the candidate who is without natural equipment for architecture to seek other occupation better suited to his natural gifts. It is squarely up to organized architecture to see to it that the worthy candidate gets an appropriate mentor.

Dean Emerson says that the mentor "should be a man of an essentially broad-minded quality, ready to see a thing from more than one standpoint. He should be a man who practices what he preaches, who is not merely a paper performer, who sincerely carries out what he believes. In other words, he is a sort of guide, philosopher and friend, a combination of the man whom I knew as Professor William R. Ware, who devoted himself to his students, whose students were his family.—You might throw into the mixture of which Professor Ware is assumed to be a part, a touch of C. Howard Walker. There would be a certain spice there that would appeal to us all; it would insure an understanding of the human qualities and, at the same time, a readiness to accept any challenge that might come along."

The mentor should have cultural background as well as broad practical experience. The mentor is expected to guide the candidate into an understanding of the every day problems of practice. He is expected to paint clearly to the intelligence of his ward, the limitations and possibilities of the human as well as the machine elements involved in the carrying out of the building program. The candidate must be guided so that he will observe and assemble all of the facts of bearing

on the problem in hand. Having acquired facility in the noting and bringing together of facts he must be trained in the art of evolving a logical program of procedure, premised on the facts observed. Stated in another way, a clearly thought through enumeration of the conditions which must be met in order to solve, in all of its elements, the building problem involved. When the candidate has learned how to diagnose and state a building objective, he must be taught how to proceed to the invention of a solution of the conditions stated in his program of requirements. Having acquired the art of visualizing essential conditions and skill in inventing and depicting a satisfactory solution, the candidate for architecture must be trained to take building conditions as they are and assemble materials, labor, management and finance in such a way as to be able to deliver a building carrying out his correct design in an economical, substantial and attractive manner to the satisfaction of his client. Such skill can only be acquired by the actual doing of these things in a real architect's office, under the guidance of an experienced practitioner. This is the justification for the demand that after graduation the candidate for architecture shall pass through a three-year period of practical training under the guidance of an experienced mentor. It is supported by the same logic as justifies the medical internship period required of candidates for medicine. It would be a fine opportunity for the candidate if he could find employment during at least a portion of the period of his mentorship training in the office of his mentor. The personal daily contact would be valuable. It is essential, however, that he shall find widely diversified experience covering all of the varied practical functions of an architect, and employment in different offices gives diversity of viewpoint. At some time during the period of his mentorship training the candidate must gain experience in preliminary research, design in all of its phases, specifications, contracts, detailing, supervision of construction and arbitrations.

The job mapped out for the mentor is a large one, but it has its compensation. No one ever knows anything about the joy of living until he has been able to help some one else. There is no opportunity to serve another comparable with that given to the older men to help the beginners in their profession. Every time one tries to explain a subject matter to another he understands that material better than he did before. If the person to whom he expounds it is young and keen minded, the teacher has to keep himself mentally alert, which means that he profits along with his pupil.

THE ILLINOIS ARCHITECTURAL ACT

The Illinois Architectural Act is a combination of weakness and strength. It needs amending to clarify its obscurities and to provide effective means of enforcement. Its chief strength lies in two facts as follows: First, that it exclusively and unequivocally guarantees the use of the title "Architect" to individual architects who are registered under the Act (See Sec. 1 and Sec. 16 of Act); and Second, that its provisions for examination and registration are excellent. Corporations being impersonal are constitutionally incapable of being examined to qualify for registration. Furthermore, the law specifically states ("Section 3") "No corporation shall be licensed to practice architecture in this State or be granted a certificate of registration under this Act." Only registered persons may use or advertise the title "Architect" in connection with their names. These conditions thus make it illegal for both corporations and unregistered individuals to use in any way, shape or form the title "Architect". (See Sec. 16 of the Act.)

The law is weak in that it seems to allow unregistered persons and corporations to do what amounts to architectural work provided they do not say or advertise that they are architects. (See Sec. 4 of the Act.) Plans for all residential or farm purposes which are located "outside of the corporate limits of any city or village" are specifically exempted, as is also "any building remodeling or repairing of any building or structure within the corporate limits of any city or village, where the total cost of said building remodeling or repairing does not exceed the sum of Seventy-five hundred dollars."

The sole purpose of registration of architects is to protect the public. The life of a person who lives, works, is amused, instructed or worships in a building that costs less than Seventy-five hundred dollars is just as precious to him as is the life of a person who occupies a more expensive building. The relative cost of a building is no comparable criterion of human jeopardy, thus it follows that this sort of exception is not good law. If the exemption were based on size, height, number of occupants or congestion of position, it might be logically justified. For illustration a poorly planned and constructed reviewing stand of small cost could risk the lives of hundreds of people.

The exception in Section 3 of the Act is misleading although this section does not say that corporations may practice architecture it does grant permission for corporations to do work, falling within the scope of Architectural practice. This particular exception is the cloak for the major violations of the Act.

The real deficiency in the Act lies in its ineffective provisions for enforcement. The act does not provide adequate annual income to defray the expenses of honest competent enforcement service. Professional registration acts can not be successfully enforced unless there is provision for the employment of competent technical inspection and legal service. An inspector in professional cases, to be effective, must know the customary practice of the profession he inspects. Lawyers to try professional cases successfully must know the fundamentals of professional law, understand the underlying technique of the profession concerned, and have had actual experience in trying cases. Inspectors should proceed, as do the G-Men of the U. S. Secret Service by following the case until they have accumulated overwhelmingly adequate evidence to assure conviction. They should be safely secure in their position and adequately remunerated. Politics should be completely divorced from inspection service. If the Illinois Architectural Act were amended to properly clarify its wording, provide a technically competent civil service inspector and require the assignment to architectural cases of an assistant Attorney General, especially qualified to prosecute such cases, the law could be enforced effectively. New York has such officers and does a good job in enforcing their architectural act. To support such a program would require the raising of the annual registration renewal fee from $1.00 to say $12.00, and empounding the fund thus raised to defray the expenses of administering the act.

The architectural examining committee should be changed to a permanent commission requiring the appointment to one member each year, each to serve for a period of five years, thus making it impossible for any governor to interrupt its continuity except on trial and proof of malfeasance in office.

THE ILLINOIS SOCIETY OF ARCHITECTS CANONS OF PROFESSIONAL ETHICS

Preamble.

The architect is engaged in a profession which carries with it grave responsibilities to the public. These duties and responsibilities cannot be met unless the motives, conduct and ability of the members of the profession are such as to command respect and confidence.

The profession of architecture calls for men of the highest integrity, and executive and artistic ability.

The architect is entrusted with financial undertakings where his honesty of purpose must be above suspicion; he acts as professional adviser to his client, and his advice must be absolutely disinterested; he is charged with the exercise of judicial functions as between client and contractor, and must act with entire impartiality, and he has moral responsibilities toward his professional associates and subordinates.

The people of the State of Illinois have a right to expect a high standard of practice and conduct on the part of the architects whom they have licensed to practice. Because an architect is a quasi public official it is imperative that he assume no obligations which shall place official duty and self-interest in conflict.

The Canons of Ethics

No set of rules can be framed which particularize all the duties of the architect in his various relations to the public, to his client, to the building trades and to his professional brethren.

The following canons of ethics cover certain broad principles which should govern the conduct of members of the profession and should serve as a guide in circumstances other than those enumerated:

I.—On Certain Duties to the Public.

The architect's more important work is of a character so permanent and enduring that he owes it to the public to use his best efforts to make it such as may raise the standard of taste in the community and be in itself a public ornament. He should design with due regard to surroundings and should endeavor to check any individualism, whether in himself or his client, that is opposed to the public good. He should take part in those movements for public betterment in which his training and experience enable him to give useful service. He should insist on safe and sanitary construction and he should at all times hold the safeguarding of human life and health as of paramount importance to the interests of client, contractor or self.

II.—On the Architect's Status.

The architect's relation to his client is primarily that of professional advisor. This relation maintains throughout the entire period of his service. When, however, a contract is executed between his client and a builder or other person by the terms of which the architect becomes the official interpreter of its conditions and the judge of its performance, a new relation is created. In respect to the matters under contract, it is incumbent upon the architect to side neither with the client nor contractor, but to endeavor, in so far as his action may determine, that the contract be faithfully carried out according to its true spirit and intent.

It is not proper for the architect to assume to act as the owner's agent unless he has been specifically empowered so to act: by so doing he becomes a party to the contract and in a sense disqualified in his judicial capacity.

The fact that the architect's payment comes through the client does not invalidate his professional obligation to act with impartiality to both parties to the contract. It is essential, however, in order to eliminate the influence of self-interest, that the architect shall not enter into any contract with the client which shall condition his payment upon his decisions or advice.

III.—On Preliminary Drewings and Estimates.

The architect should impress upon his client at the outset the importance of sufficient time for the study and preparation of drawings and specifications. If, on the basis of approved preliminary sketches, the approximate cost of the work has been mutually considered, the architect should endeavor to bring his working drawings to meet such approximate cost, provided that his client has requested no departure from the original basis of estimate. But at the same time he should acquaint his client with the conditional character of preliminary estimates. Complete and final figures can be had only from complete and final drawings and specifications. If an unconditional limit of cost is imposed before such drawings are made and estimated, the architect must be free to make such adjustments as seem necessary to that end.

IV.—On Superintendence and Expert Service.

On all work except the simplest, it is to the interest of the client to employ an inspector or clerk-of-the-works; in many engineering problems and in certain esthetic problems such as sculpture, decorative painting, gardening and the like, it is to the interest of the client to have specialized expert service. The architect should so inform the client and assist him in obtaining such service. In order to secure unified and harmonious working organization, only such persons should be selected by the owner for consulting experts as shall work in harmony with the architect and shall be approved by him.

V.—On the Architect's Charges.

The schedule of charges of the Illinois Society of Architects is recognized as a proper minimum of payment, but where no other architect is affected it is allowable for an architect to make such an arrangement with his client as is mutually satisfactory. He may not reduce his fee below the schedule of charges in an attempt to supplant another architect; it is reasonable and proper to charge higher rates than those of the schedule when his special skill and the quality of his service justify the increase.

A system of compensation based on the actual cost to the architect on a given piece of work plus an agreed professional fee, has much to commend it.

VI.—On Needless Expenditure.

The architect should scrupulously guard cost, and refrain from introducing needless expense or any extravagance in material or construction that may add to cost of building, without compensating gain to the client.

VII.—On Payments for Expert Service.

When retained as an expert, whether in connection with competitions or otherwise, the architect should receive a compensation proportionate to the responsibility and difficulty of the service. No duty of the architect is more exacting than such service, and the honor of the profession is involved in it. Under no circumstances should experts, knowingly, name prices in competition with each other for a given employment. Where governmental regulations prohibit adequate

Association. The use of the initials designating degrees or technical society membership is proper in connection with any professional service and is encouraged as helping to make known the nature of the honor they imply.

XVI.—On Competitions.

In no way does the architect come more conspicuously before the public than through competitions. It is especially desirable that in such circumstances he should conduct himself with self-respect and dignity. To undervalue and cheapen his service or to compete where a just award is not safe guarded is inconsistent with this position. Competitions are undesirable from the standpoint of both the client and the architect and a member of the Association should discourage the holding of same. If a competition becomes inevitible, because of governmental regulations, he should not enter either as a competitor or a professional advisor unless the competition is to be conducted according to the best practice and usage of the profession as formulated from time to time by the American Institute of Architects. Except as an authorized competitor he may not attempt to secure work for which competition has been instituted.

He may not present drawings to secure work for which competition has been closed but not decided.

He may not attempt to influence the award in any competition.

XVII.—On the Expert's Future Status.

An architect may not undertake a further commission on any building or work after having acted in an expert capacity in formulating a program which later is put into effect, or after having acted in an advisory capacity in the matter of awards in competition. Having acted in either or both of such capacities should bar an architect from eligibility to execute commissions upon the work in question.

XVIII.—On Criticising the Work of Others.

An architect may not criticise publicly in the press the work of a fellow architect except over his own signature, or editorially; and he may not intentionally injure, directly or indirectly, the reputation, prospects or business of a fellow architect.

XIX.—On Undertaking the Work of Another.

An architect may not undertake a commission while the just claim of a fellow architect, who had previously undertaken it, remains unsatisfied; nor may he attempt to supplant a fellow architect or to obtain a commission after steps have been taken toward the appointment of another architect.

XX.—On Duties Toward the Student Draughtsman.

It is the duty of the architect to advise and assist those who intend making architecture their career. The intending student should be urged to secure a preparation of broad general culture equivalent to that required for the degree of A. B., concurrently with or followed by a thorough course in a well organized school of architecture.

In cases where such preparation is out of the question and the beginner must get his training in the office of an architect, the latter should assist him to the best of his ability by instruction and advice. An architect, should as far as possible, urge his draughtsmen to avail themselves of educational opportunities. To this end he should give encouragement to all worthy schemes and institutions for architectural education.

Members of the society cannot too strongly insist that a thorough technical preparation for the practice of architecture should rest upon a foundation of general culture.

are Specified by

LEADING ARCHITECTS
THROUGHOUT THE UNITED STATES

Why?

Designs are modern, attractive and efficient.
Specifications are followed accurately.
Service is dependable, based on 52 years of experience. . . . . . .
Construction substantial and time tested.

OUR SPECS
PAGE 21/98
IN
SWEETS

The TABLET & TICKET CO.
1019 W. ADAMS STREET · CHICAGO, ILL.
115 EAST 23rd STREET
NEW YORK CITY
407 SANSOME STREET
SAN FRANCISCO

XXI.—On Duties Toward Building Authorities.

The architect should support all federal, state and municipal officials who have charge of matters relating to building and endeavor to maintain or improve the standards of their departments. His quasi public official capacity requires him to show respect for law by careful and conscientious compliance with all building regulations, and if any such appear to him unwise or unfair, he should endeavor to have such regulations altered, but until so altered he should comply with them. An architect because of his official relation to the state and of his moral obligation should not even under his client's instructions encourage any practices contrary to law or hostile to public interests; for he is not obliged to accept a given piece of work, hence he cannot urge in extenuation and to escape the condemnation attaching to his acts that he has but followed his client's instructions.

XXII.—On Professional Qualifications.

The assumption of the title of architect should be held to mean that the bearer has the professional knowledge, both theoretical and practical, and the natural ability needed for the proper invention, illustration and supervision of all building operations which he may undertake.

XXIII.—On Matters Adjudged Unprofessional.

The following code, based on a report of a special committee of the American Institute of Architects, is adopted by the Illinois Society of Architects as a general guide, yet the enumeration of particular duties should not be construed as the denial of the existence of others equally imperative though not specifically mentioned. It should also be noted that these sections indicate offenses of greatly varying degrees of gravity:

It is unprofessional for an architect—

1. To engage in any of the building trades or to form any trade partnership or agreement with any person or firm engaged therein.
2. To guarantee an estimate or contract by bond or otherwise.
3. To accept a commission or any substantial service or favor from a contractor, or anyone connected with the building trades.
4. To advertise in any form.
5. To enter any competition the terms of which are not in harmony with principles approved by the American Institute, especially if such terms have been specifically condemned by the American Institute or a local chapter thereof.
6. To attempt in any way except as a duly authorized competitor to secure work for which a competition has been instituted.
7. To attempt to influence the award of a competition.
8. To injure intentionally the fair reputation, prospects or business of another architect.
9. To criticise anonymously in the public prints, except editorially, the professional conduct or work of a fellow architect.
10. To undertake a commission while the just claim of another architect who has previously undertaken it remains unsatisfied.
11. To attempt to supplant a fellow architect after definite steps have been taken toward his employment.
12. To offer or perform services at rates lower than those approved as minimum by the Illinois Society of Architects in an attect.
tempt to supplant or underbid another archi-
13. To act in a manner detrimental to the best interests of the profession.

SCHEDULE OF PROPER MINIMUM CHARGES AND PROFESSIONAL PRACTICE OF ARCHITECTS RECOMMENDED BY THE ILLINOIS SOCIETY OF ARCHITECTS

1. The architect's professional services consist of the necessary conferences, the preparation of preliminary studies, working drawings, specifications, large scale and full size detail drawings, and of the general direction and supervision of the work, for which, except as hereinafter mentioned, the minimum charge is six per cent (6%), based upon the total cost of the work complete.

In case of the discontinuance or abandonment of the work, the architect's charge shall be based upon an *estimated* total cost, which estimated total cost may be determined by the architect, by experts, or by the lowest bids of responsible contractors. *Total cost* is to be interpreted as the cost of all materials and labor necessary to complete the work, plus contractors' profits and expenses, as such cost would be if all materials were new and all labor fully paid, at market prices current when the work was ordered.

2. On residential work, on alterations to existing buildings, on monuments, furniture, decorative and cabinet work, and landscape architecture, it is proper to make a higher charge than above indicated.

3. The architect is entitled to compensation for articles purchased under his direction, even though not designed by him.

4. If an operation is conducted under separate contracts, rather than under a general contract, it is proper to charge a special fee in addition to the charges mentioned elsewhere in this schedule.

5. Where the architect is not otherwise retained, consultation fees for professional advice are to be paid in proportion to the importance of the questions involved and services rendered.

6. Where heating, ventilating, mechanical, structural, electrical and sanitary problems are of such a nature as to require the services of a specialist, the owner is to pay for such services in addition to the architect's regular commission. Chemical and mechanical tests and surveys, when required, are to be paid for by the owner.

7. Necessary traveling expenses are to be paid by the owner.

8. If, after a definite scheme has been approved, changes in drawings, specifications or other documents are required by the owner; or if the architect be put to extra labor or expense by the delinquency or insolvency of a contractor, the architect shall be paid for such additional services and expense.

9. The architect's entire fee is itemized and proportionate payments on account are due the architect, as the following items are completed:

Preliminary Studies	.2
General drawings	.3
Specifications	.1
Scale and full size details	.1
General Supervision of the work	.3
Total	1.00

Fee for complete services as agreed, or see paragraphs 1 and 12.

10. Items of service are comprehended as follows:

(a) **Preliminary Studies** consist of the necessary conferences, inspections, studies and sketches modified and remodified to determine the client's problem and illustrate a satisfactory general solution of same, both as to plan and elevation. Illustrative sketches for this purpose need not be to accurate scale, but should be approximately correct as to general dimensions and proportion.

(b) **General Drawings** include figured scale plans of the various stories, elevations of all the fronts, such general vertical sections as may be necessary to elucidate the design, and such details, drawn to still larger scale as, with the assistance of printed notes, and of the accompanying specifications, may make the whole scheme clearly evident to the mind of the competent builder and give him a full and complete comprehension of all the structure conditions as they affect the vital questions of quality and quantity of materials, of character of workmanship, and of cost.

(c) **Specifications** consist of a supplementary statement in words, of at least all those items of information regarding a proposed building which are not set forth in the drawings.

(d) **Detail Drawings** include all the necessary supplementary drawings required for the use of the builders, to enable them to so provide and shape their material that it may be adjusted to its proper place or function in the building with the least delay, and the smallest chance for errors and misfits. If not prepared until after the contract for the building is let they must not impose on the contractor any labor or material which is not called for by the spirit and intent of the "General Drawings" and "Specifications."

(e) The **Supervision** of an architect (as distinguished from the continuous personal superintendence which may be secured by the employment of a clerk-of-the-works or inspector of construction) means such inspection by the architect or his deputy, of work in studios and shops or a building or other work in process of erection, completion or alteration, as he finds necessary to ascertain whether it is being executed in general conformity with his drawings and specifications or directions. He has authority to reject any part of the work which does not so conform and to order its removal and reconstruction. He has authority to act in emergencies that may arise in the course of construction, to order necessary changes, and to define the intent and meaning of the drawings and specifications. On operations where a clerk-of-the-works or inspector of construction is required, the architect shall employ such assistance at the owner's expense.

11. Drawings and specifications, as instruments of service, are the property of the architect.

12. Exceptions.

Dwellings costing less than $10,000	10%
Lofts not requiring special planning for machinery or arrangement	5%
Additions and alterations to dwellings	12%
Additions and alterations to business buildings	10%

N. B.—Above schedule is considered minimum for ordinary and usual professional service. It is not considered fair or reasonable for highly specialized service.

FLOOR COVERING SPECIFICATION *Gives the Architect* A NEW SERVICE OPPORTUNITY

FOR YOUR FILES: Mohawk's Handbook of Rugs and Carpets is filled with information on weaves, grades, wearing qualities and suggestions on selection. It is fully illustrated, and belongs in your files. It is free, and we will be glad to send you a copy. Ask for it, today.

MORE AND MORE, building owners are looking to the architect for the creation and execution of complete and harmonious interiors.

This now includes not only finish and color for walls and ceilings — but also the specification of Floor Coverings — a factor so important that an inappropriate selection can easily mar the entire decorative scheme.

Here is a new service opportunity for you . . . and a new financial opportunity.

Mohawk's experience is available to you in solving floor covering problems for all types of buildings . . . homes, stores, theaters, hotels, public buildings. Mohawk's design department will work with you in producing carpets of special designs and colors . . . giving you faithful interpretation of your own creation, or carrying an idea or sketch through to completion.

Mohawk offers you America's widest selection of weaves and grades . . . because Mohawk alone weaves all types. These include: Chenille, the most luxurious and versatile of carpets; Seamless Broadlooms, universally recognized as supreme; Wiltons; Axminsters; Velvets.

We invite you to take advantage of this Mohawk Service.

Mohawk Carpet Mills

295 FIFTH AVENUE, NEW YORK

REGIONAL SALES OFFICES

BOSTON · CHICAGO · DALLAS · DETROIT · HIGH POINT

LOS ANGELES · PHILADELPHIA · SAN FRANCISCO · ST. LOUIS

THE ILLINOIS ARCHITECTURAL ACT

An Act to provide for the licensing of architects and to regulate the practice of architecture as a profession and to repeal certain Acts therein named, approved June 24, 1919, as amended June 26, 1923, and July 8, 1937. Be it enacted by the people of the State of Illinois, represented in the General assembly:

Section 1. Be it enacted by the People of the State of Illinois, represented in the General Assembly: It shall be unlawful for any person to practice architecture or advertise or put out any sign or card or other device which might indicate to the public that he or she is entitled to practice as an architect, without a certificate of registration as a registered architect, duly issued by the Department of Registration and Education under this Act, and as provided for in the Civil Administrative Code of Illinois.

Sec. 2. Any one, or any combination of the following practices by a person shall constitute the practice of architecture, namely: The planning or supervision of the erection, enlargement or alteration of any building or buildings or of any parts thereof, to be constructed for others. A building is any structure consisting of foundations, floors, walls, columns, girders, beams and roof, or a combination of any number of these parts, with or without other parts.

Sec. 3. No corporation shall be licensed to practice architecture in this State or be granted a certificate of registration under this Act, but it shall be lawful for a stock company or a corporation to prepare drawings, plans and specifications for buildings and structures as defined in this Act which are constructed, erected, built, or their construction supervised by such stock company or corporation, provided that the chief executive officer or managing agent of such stock company or corporation in the State of Illinois shall be a registered architect under this Act, and provided further that the supervision of such buildings and structures shall be under the personal supervision of said registered architect and that such drawings, plans and specifications shall be prepared under the personal direction and supervision of such registered architect and bear the stamp of his official seal.

It shall be lawful, however, for one or more registered architects to enter a partnership with one or more licensed structural engineers, licensed under the laws of this State, for the practice of their professions.

Sec. 4. Nothing contained in this Act shall prevent the draftsmen, students, clerks of works, superintendents and other employes of those lawfully practicing as registered architects under the provisions of this Act, from acting under the instruction, control or supervision of their employers, or to prevent the employment of superintendents of the construction, enlargement or alteration of buildings or any parts thereof, or prevent such superintendents from acting under the immediate personal supervision of the registered architect by whom the plans and specifications of any such building, enlargement or alteration were prepared. Nor shall anything contained in this Act prevent persons, mechanics or builders from making plans, specifications for or supervising the erection, enlargement or alteration of buildings or any parts thereof to be constructed by themselves or their own employes for their own use, provided that the working drawings for such construction are signed by the authors thereof with a true statement thereon of their relation to such construction and that the makers thereof are not architects:

Provided, nothing in this Act contained shall be held or construed to have any application to any building, remodeling or repairing of any building or other structure outside of the corporate limits of any city or village, where such building or structure is to be, or is used for residential or farm purposes, or for the purposes of outbuildings or auxiliary buildings in connection with such residential or farm premises; nor shall said Act apply to any building, remodeling or repairing of any building or structure within the corporate limits of any city or village, where the total cost of said building, remodeling or repairing does not exceed the sum of seventy-five hundred dollars.

Sec. 4-a. The Department of Registration and Education shall exercise, but subject to the provisions of this Act, the following functions, powers and duties:

(1) Conduct examinations to ascertain the qualifications and fitness of applicants for certificates of registration as registered architects, and pass upon the qualifications of applicants for reciprocal licenses, certificates and authorities.

(2) Prescribe rules and regulations for a method of examination of candidates.

(3) Prescribe rules and regulations defining what shall constitute a school, college or university, or department of a university, or other institution, reputable and in good standing, and to determine whether or not a school, college or university, or department of a university, or other institution is reputable and in good standing by reference to a compliance with such rules and regulations, and to terminate the approval of such school, college or university or department of a university or other institution as reputable and in good standing for non-compliance with such rules and regulations; *provided* that no school, college or university, or department of a university or other institution that refuses admittance to applicants, solely on account of race, color or creed shall be considered reputable and in good standing.

(4) Establish a standard of preliminary education deemed requisite to admission to a school, college or university, and to require

atisfactory proof of the enforcement of such tandard by said schools, colleges and universities.

(5) To conduct hearings on proceedings to suspend or revoke or refuse renewal of licenses, certificates or authorities of persons applying for registration or registered under the provisions of this Act and to revoke or refuse to renew such licenses or certificates or authorities.

(6) Formulate rules and regulations when required in any Act to be administered.

None of the foregoing functions or duties enumerated shall be exercised by the Department of Registration and Education except upon the action and report in writing of the Examining Committee which shall be composed of persons designated from time to time by the Director of Registration and Education to take such action and to make such report for the profession involved herein as follows:

Five (5) persons, one of whom shall be a member of the Faculty of the University of Illinois, and the other four of whom shall be architects, residing in this State, who have been engaged in the practice of Architecture at least ten years.

The action or report in writing of a majority of the Committee designated shall be sufficient authority upon which the Director of Registration and Education may act.

In making the designation of persons to act, the Director shall give due consideration to recommendations by members of the profession and by organizations therein.

Whenever the Director is satisfied that substantial justice has not been done in an examination, he may order re-examinations by the some or other examiners. (Added by Act paproved July 8, 1937.)

Sec. 5. Any person who is twenty-one years of age and of good moral character is qualified for an examination for a certificate of registration as a registered architect, provided he or she has graduated from a high school or secondary school, approved by the Department of Registration and Education, or has completed an equivalent course of study as determined by an examination conducted by the Department of Registration and Education, and has subsequently thereto completed such courses in mathematics, history, and language, as may be prescribed by said Department, and has had at least three years' experience in the office or offices of a reputable architect or architects.

Sec. 6. Upon payment of the required fee, an applicant who is an architect, registered or licensed under the laws of another State or territory of the United States, or of a foreign country or province, may, without examination, be granted a certificate of registration as a registered architect by the Department of Registration and Education in its discretion upon the following conditions:

(a) That the applicant is at least twenty-one years of age, of good moral character and temperate habits; and

(b) That the requirements for the registration or licensing of architects in the particular State, territory, country or province were, at the date of the license, substantially equal to the requirements then in force in this State.

Sec. 7. Every person who desires to obtain a certificate of registration shall apply therefor to the Department of Registration and Education in writing, upon blanks prepared and furnished by the Department of Registration and Education. Each application shall be verified by the applicant under oath and shall be accompanied by the required fee.

Sec. 8. The Department of Registration and Education shall hold examinations of applicants for certificates of registration as registered architects at such times and places as it may determine. The examination of applicants for certificates of registration as registered architects, where these applicants have had less than ten years proved architectural practice as a principal in the practice of architecture as a profession, shall consist of written and drafting tests supplemented by verbal examination at the discretion of the Examinating Committee and shall embrace the following subjects:

(a) The planning, designing and construction of buildings.

(b) The strength of building materials.

(c) The principles of sanitation and ventilation as applied to buildings.

(d) The ability of the applicant to make practical application of his knowledge in the ordinary professional work of an architect and in the duties of a supervisor of mechanical work on buildings.

(e) The examinations of applicants for certificates of registration as registered architects, where the applicant shall have had ten or more years proved architectural practice as a principal in the practice of architecture as a profession, shall be by exhibits of preliminary studies, general drawings, specifications and detail drawings, prepared under the personal supervision of the applicant; by photographs of executed work and evidence of authorship, supplemented by a verbal quiz as to reasons for methods used and procedure shown and by proof of honorable practice, or by any or all of these, which in the judgment of the Examining Committee are necessary to determine the applicant's qualifications as an architect, which shall be equivalent to or superior in relative value to the requirements set forth in the preceding paragraphs of this section for an applicant having had less than ten years' experience.

The Department of Registration and Education may by rule prescribe additional subjects for examination. (As amended by Act approved June 26, 1923, in force July 1, 1923.)

Sec. 9. Whenever the provisions of this Act have been complied with by an applicant the Department of Registration and Educa-

Acme
steel kitchen-unit
Combinations

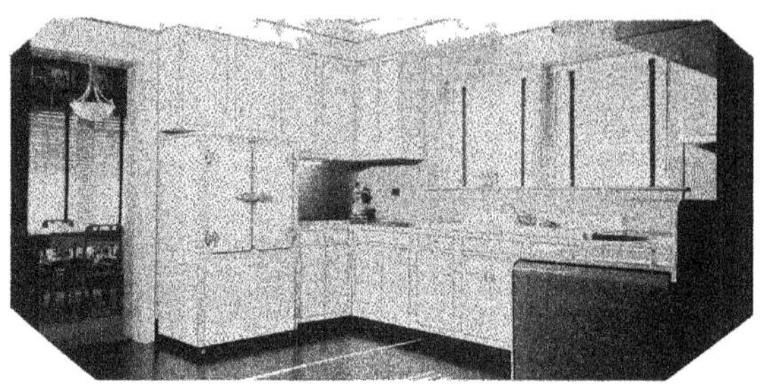

Built to conform to enamelware sink specifications, Acme steel kitchen-units are preferred for arranging such completely modern, step and time-saving kitchens as the one illustrated here.

For, it is around the sink, the most constantly in use of the three "working centers" making up a well ordered kitchen, that the planned kitchen is developed. And, because the base cabinets with their working tops must join the sink and top smoothly and accurately, Acme steel kitchen units are so designed and made.

Acme steel kitchen-units are also made to a wide range of dimensions developed as the result of much research and experimentation and capable of being installed in the greatest variation and forms of wall and floor space.

Acme steel kitchen-units are as adaptable to remodeling jobs as to new construction. Have your plumbing contractor get a Plan from Acme Kitchen Planning Service in accordance with the space you have, or ask for it direct.

Specify "ACME" for all your kitchen jobs.

Acme steel kitchen-units and Combinations are illustrated with full specifications in Sweets Catalog.

ACME METAL PRODUCTS CORP.

1845 West 74th Street -:- CHICAGO

ion shall issue a certificate of registration o the applicant as a registered architect, hich certificate shall have the effect of a icense to the person to whom it is issued o practice architecture in this State, subject o the provisions of this Act.

Any license or certificate of registration ıeretofore issued under the laws of this 'tate authorizing its holder to practice archiecture shall, during the unexpired period for which it was issued, serve the same purpose as the certificate of registration provided for by this Act.

Sec. 10. Any person licensed to practice architecture in this State or registered as an architect under this Act shall be exempt from the provisions of any and all Acts in force in this State regulating the practice of structural engineering.

Sec. 11. Every holder of a certificate of registration as a registered architect shall display it in a conspicuous place in his principal office, place of business or place of employment.

Every registered architect shall have a seal, the impression of which shall contain the name of the architect and the words, "Registered Architect," "State of Illinois." He shall stamp with this seal all working drawings and specifications prepared by him or under his supervision. Any seal heretofore authorized under the laws of this State shall serve the same purpose as the seal provided for by this Act.

Sec. 12. Every registered architect who continues in active practice shall, annually, on or before the first day of July, renew his certificate of registration and pay the required renewal fee. Every license or certificate of registration which has not been renewed during the month of July in any year shall expire on the first day of August in that year. A registered architect whose certificate of registration has expired may have his certificate restored only upon payment of the required restoration fee.

Any architect registered or licensed in this State who has retired from the practice of architecture for a period of not more than five (5) years may have his certificate of registration renewed, at any time within a period of five (5) years after so retiring, upon making application to the Department for such renewal and upon payment of all lapsed annual renewal fees.

Sec. 13. The Department of Registration and Education may refuse to renew, or may suspend, or may revoke, any certificate of registration for any one or any combination of the following causes:

(a) Gross incompetency.

(b) Recklessness in the construction of buildings or their appurtenances.

(c) Dishonest practice.

(d) When the architect has been twice convicted for a violation of any of the provisions of this Act.

(e) A person who has by false or fraudulent representation obtained or sought to obtain a certificate of registration as an architect.

The Department may upon its own motion and shall upon the verified complaint in writing of any person setting forth facts which if proved would constitute grounds for refusal, suspension or revocation as hereinabove set forth, investigate the actions of any person holding or claiming to hold a certificate. The Department shall, before refusing to issue, suspending or revoking any certificate, at least ten (10) days prior to the date set for the hearing, notify in writing the applicant or the holder of such certificate of any charges made and shall afford such accused person an opportunity to be heard in person or by counsel in reference thereto. Such written notice may be served by delivery of the same personally to the accused person, or by mailing the same by registered mail to the place of business last theretofore specified by the accused person in his last notification to the Department. At the time and place fixed in the notice, the Examining Committee designated by the Director of Registration and Education, as provided in this Act, shall proceed to hearing of the charges and both the accused person and the complainant shall be accorded ample opportunity to present in person or by counsel, such statements, testimony, evidence and argument as may be pertinent to the charges or to any defense thereto. The Committee may continue such hearing from time to time. If the Committee shall not be sitting at the time and place fixed in the notice or at the time and place to which hearing shall have been continued, the Department shall continue such hearing for a period not to exceed thirty (30) days. (As amended by Act approved July 8, 1937.)

Sec. 13-a. The Department shall have power to subpoena and bring before it any person in this State and to take testimony either orally or by deposition, or both, with the same fees and mileage and in the same manner as prescribed by law in judicial procedure in civil cases in courts of this State.

The Director, Assistant Director, Superintendent of Registration and any member of the Examining Committee shall each have power to administer oaths to witnesses at any hearing which the Department is authorized by law to conduct, and any other oaths required or authorized in any Act administered by the Department. (Added by Act approved July 8, 1937.)

Sec. 13-b. Any Circuit or Superior Court or any judge thereof, either in term time or vacation, upon the application of the accused person or complainant or of the Department, may by order duly entered, require the attendance of witnesses and the production of relevant books and papers before the Department in any hearing relative to the application for a refusal, recall, suspension or revocation of certificate of registration, and the court or judge may compel obedience to its or his order by proceedings for contempt. (Added by Act approved July 8, 1937.)

cused person is not a resident of this State and the venue is not otherwise fixed herein, such venue shall be in Sangamon County.

Such writ shall be issued by the Clerk of the Court upon praecipe and it shall be served at least ten days before the return day thereof. Service upon the Director, Assistant Director or Superintendent of Registration shall be service on the Department. Such suit shall be commenced within twenty days of the accused person's receipt of notice of the order of refusal, revocation or suspension. The Department shall not be required to certify the record of its proceedings unless the accused person shall first pay to it the sum of five cents per one hundred words of such record. Exhibits shall be certified without cost.

No department order of suspension or revocation shall be set aside or vacated on any ground not specified in the written motion for rehearing provided for in this Act. (Added by Act approved July 8, 1937.)

Sec. 13-g. An order of revocation or suspension or a certified copy thereof, over the seal of the Department and purporting to be signed by the Director shall be prima facie proof that

1. Such signature is the genuine signature of the Director.

2. That such Director is duly appointed and qualified.

3. That the Committee and the members thereof are qualified to act.

Such proof may be rebutted. Such order of revocation or suspension shall be conclusive proof that all precedent and concurrent acts of department officers and of the committee necessary to the validity of such order were pursuant to authority conferred by the Director. (Added by Act approved July 8, 1937.)

Sec. 13-h. Appeals from all final orders and judgments entered by a Circuit or Superior Court in review of an order of the Department may be taken directly to the Supreme Court by either party to the action within sixty days after service of a copy of the order or judgment of the Circuit or Superior Court, and shall be governed by the rules applying to other civil cases appealed to said Supreme Court, except that formal pleadings shall not be required. (Added by Act approved July 8, 1937.)

Sec. 13-i. The pendency of an appeal or writ of certiorari shall not of itself stay or suspend the operation of an order or suspension; but during the pendency of such suit or appeal, the Circuit or Superior Court or the Supreme Court, as the case may be, in its discretion, may stay the operation of such order in whole or in part upon such terms and conditions as the court may prescribe. No such stay shall be granted by the court otherwise than upon ten days' notice to the Department and after a hearing. (Added by Act approved July 8, 1937.)

Sec. 14. The fee to be paid by an applicant for an examination to determine his fitness to receive a certificate of registration as a

registered architect shall be ten dollars ($10).

The fee to be paid by an applicant for a certificate of registration as a registered architect shall be five dollars ($5).

The fee to be paid for the restoration of an expired certificate of registration shall be five dollars ($5).

The fee to be paid upon renewal of a certificate of registration shall be one dollar ($1).

The fee to be paid by an applicant for a certificate of registration who is an architect registered or licensed under the laws of another state or territory of the United States, or of a foreign country or province, shall be fifteen dollars ($15).

Sec. 15. The Department of Registration and Education shall adopt rules and regulations in accordance with the provisions of Section 60 of said Civil Administrative Code, and not inconsistent with this Act, to carry out fully and enforce the provisions of this Act. Sec. 15. (Repealed by Act approved July 8, 1937.)

Sec. 16. Each of the following Acts constitutes a misdemeanor punishable upon conviction by a fine of not less than twenty-five dollars ($25) nor more than two hundred dollars ($200) for each offense:

(a) The practice of architecture by any person or the advertising or putting out of any sign or card or other device which might indicate to the public that he or she is entitled to practice as an architect, without a certificate of registration as a registered architect issued by the Department of Registration and Education of this State.

(b) The making of any wilfully false oath or affirmation in any matter or proceeding where an oath or affirmation is required by this Act.

(c) The affixing of a registered architect's seal to any plans, specifications or drawings which have not been prepared by him or under his immediate personal supervision.

(d) The violation of any provision of Section 11 of this Act.

All fines and penalties shall inure to the Department of Registration and Education of this State.

Sec. 17. The Department of Registration and Education shall keep a record open to public inspection at all reasonable times of its proceedings relating to the issuance, refusal, renewal, suspension or revocation of certificates of registration. This record shall also contain the name, place of business and residence, and the date and number of registration of each registered architect in this State.

Sec. 18. The following Acts are hereby repealed: "An Act to provide for the licensing of architects and regulating the practice of architecture as a profession," approved June 3, 1897, and in force July 1, 1897, and the following Acts amendatory thereof, to-wit: An Act approved April 19, 1899, and in force July 1, 1899. An Act approved May 16, 1905, and in force July 1, 1905, and an Act approved May 26, 1911, and in force July 1, 1911.

Sec. 19. This Act may be known and cited as "The Illinois Architectural Act."

GENERAL RULES GOVERNING EXAMINATIONS.

Section 1. All communications should be addressed to the Superintendent of Registration.

Sec. 2. Lists of successful applicants only will be announced.

Sec. 3. All examinations must be written in the ENGLISH language.

Sec. 4. Application and fee must be on file at least fifteen days before date of examination. All preliminary qualifications must be verified before examination.

Sec. 5. Unmounted photo, 3x5 inches, must accompany application. A duplicate of the photo must be presented at examination, both bearing certificate as to signature and identity by the two persons who signed the letters of recommendation accompanying photo.

Sec. 6. Applicants must present cards of admission on day of examination.

Sec. 7. Examinations will begin promptly at 8:30 A. M.

Sec. 8. Grades assembled at least fifteen days after close of examination.

Sec. 9. Applicants failing in practical work must retake all subjects.

Sec. 10. Applicants failing in written work allowed credit on all grades over 75% at the following examination only.

Sec. 11. Remittances should be made by postal or express money order or bank draft. DO NOT SEND CURRENCY OR PERSONAL CHECK.

REGISTRATION OF ARCHITECTS

STATE OF ILLINOIS

INTERPRETATIONS OF THE ACT.

On the design and construction of a building, as defined in paragraph 2 of the Illinois Architectural Act, the five fundamental functions of an architect are: (a) the preparation of preliminary studies, (b) general working drawings, (c) detail drawings, (d) specifications, and (e) the general supervision of the construction.

The Supreme Court of Illinois, in constructing certain provisions of the Illinois Architectural Act in the case of People v. Rodgers (277 Ill. 151), held that

"The real purpose of the statute is the protection of the public against incompetent architects, from whose services damage might result to the public by reason of dangerous and improperly constructed buildings and by badly ventilated and poorly lighted buildings."

The Attorney General in a letter dated August 31, 1929, to the Department of Registration and Education (File 2164), rendered an opinion containing the following statements:

"Section 4 of the Act, being paragraph 4 of Chapter 10½ Smith-Hurd's Illinois Revised Statutes, 1927, provides in part as follows:

* * * 'Nor shall anything contained in this Act prevent persons, mechanics, or builders from making plans, specifications for, or supervising the erection, enlargement, or alteration of buildings or any parts thereof to be constructed by themselves or their own employees for their own use; provided, that the working drawings for such construction are signed by the authors thereof with a true statement thereon of their relation to such construction and that the makers thereof are not architects;'

"The latter section also provides that the Act shall not apply to any building within the corporate limits of any city or village where the total cost thereof does not exceed the sum of $7,500.

"The words 'for their own use,' as used in Section 4 clearly have reference to buildings of which the owner makes exclusive personal use. In other words, the exemption therein provided is not based upon mere ownership of property but rather its use. To construe the Act to mean that the owner of property may construct any sort of a building irrespective of its use, without the services of a licensed architect, would be to nullify the purpose of the Act."

One reason for the provision in Section 3 of the Act barring the registration of corporations is that corporations have a legal existence apart from their membership and can perform acts for which their members are not personally liable; furthermore, the professional service rendered by an architect, like the service performed by a physician and by a lawyer, is a personal service which corporations, being impersonal, are, actually and theoretically, incapable of rendering. A stock company or corporation may cause architectural services to be rendered others by having on its staff a registered architect, but said architect must be guaranteed sufficiently complete authority in the preparation of plans and specifications and in the supervision of construction so that he can be held as personally liable for his work as though he were an independent practitioner.

To restore an expired certificate of registration (See Section 14), an applicant whose registration has lapsed for more than five years must file a new application involving a $10 fee in addition to paying the restoration fee of $5. He must appear before the Examining Committee to prove himself as well qualified to practice architecture as he was when his Certificate of Registration expired.

General Information.

The first Act licensing architects and regulating the practice of architecture as a profession in the State of Illinois became effective July 1, 1897. An Act providing for the registration of structural engineers became effective in 1915.

The Civil Administrative Code of the State of Illinois, in force July 1, 1917, created nine State departments (now increased to eleven), one of which was the Department of Registration and Education, which department is administered by a director appointed by the Governor. Among the responsibilities placed upon this Department is the exercise of the rights, powers, and duties formerly vested by law in the State Board of Examiners of Architects. The functions and duties enumerated in the Administrative Code (Revised Statutes, Illinois, Hurd, 1931, Chap. 10½, pp. 162-164) pertaining to the conduct of examinations, the determination of examination standards and equivalents, the formulation of rules and regulations governing procedure, and the conduct of hearings on proceedings to revoke or to refuse renewal of licenses, are exercised by the Department of Rgistration and Education only upon the action and report in writing of a committee composed of five architects designated from time to time by the Director, one of whom shall be a member of the faculty of the University of Illinois, and the other four of whom have been engaged in the practice of architecture for at least ten years.

The powers and duties of the professional committee provided by the Code are as follows:

1. To conduct examinations to ascertain the qualifications and fitness of applicants for registration.

2. To prescribe rules and regulations for conducting examinations.

3. To determine the institutions giving academic training in architecture which shall be deemed accredited.

4. To conduct hearings on proceedings to revoke certificates of registration.

The Civil Administrative Code provides the necessary legal machinery for the enforcement of the provisions of the Act.

Rules Governing Examinations.

1. All communications should be addressed to the Department of Registration and Education, Capitol Building, Springfield, Illinois.

2. Application and fee must be on file at least fifteen days before the date of examination. All preliminary qualifications must be verified before the examination.

3. Remittances should be made by postal or express money order or bank draft. DO NOT SEND CURRENCY OR PERSONAL CHECK.

4. An unmounted photograph, 3 x 5 inches, must accompany the application. A duplicate of the photograph must be presented at the examination, both bearing certification as to signature and identity by two persons.

5. Applicants must present cards of admission on the day of examination.

6. Junior examinations will begin promptly at 8:30 A. M. and 1:30 P. M.

7. All examinations must be conducted in the English language.

8. Names of successful applicants only will be announced.

9. The Architectural Examining Committee meets six times a year. Junior examinations are held at the May and November meetings. Senior examinations may be held by appointment at any meeting.

The Examination.

Applicants for registration fall within three general classifications as follows:

1. Junior Applicants (written examination);

2. Senior Applicants (oral and documentary examination);

3. Applicants who are registered in other states, territories, provinces, or countries (documentary or documentary and oral examination).

1. Junior applicants are persons qualifying under Section 5 of the Illinois Architectural Act, who have had less than ten years proved independent legal practice as a principal in the profession of architecture. Years of practice as a principal means years during which the applicant signed all drawings issued from his office as architect, and had complete control or joint control with another architect over all draftsmen, superintendents, and other employees of the office having to do with the designing and supervision of construction; in other words, the full and complete authority and responsibility of a practicing architect. Work as an employee of a contractor, work as a draftsman, superintendent, or manager of an architect's office not in the applicant's own name and under the control of another, shall not be construed as the practice of architecture as a principal.

Examinations of Junior applicants (See Section 8 of the Act) are held twice a year at such times and places as the Department may determine, and embrace the following subjects:

(For Examinations A and B, applicant must bring all necessary drawing instruments, including a 24-inch T-square, triangles, scale, thumb tacks, etc. For Examination A, at least six sheets of medium weight tracing paper 20 x 24 inches, and one backing sheet of heavy detail paper also 20 x 24 inches. The actual trimmed size of all drawings will be 17 x 22 inches. Candidates will not be allowed to leave the room on the first day and should bring lunch.

For Examinations D, E, and F, a filled fountain pen, pencils, scales, and triangles alone will be sufficient. The use of a slide rule for mathematical calculations is permitted but special slide rules for concrete and steel design are not to be used.

The Department will furnish paper for these examinations.)

Examination A. The Science of Planning and the Art of Designing Buildings.

Time—8:30 A. M. to 4:30 P. M. en loge.

No reference material will be permitted.

This examination is held on the first day with a time allowance of eight hours continuous session and consists of:

(a) A test in the science of planning, particularly with reference to practical, logical, and economical arrangement, the securing of comfort, and the safeguarding of the life and health of the occupants of the building. (Grade Value 100.)

(b) A test in the art of designing, particularly with reference to orderly and consistent expression of purpose, logical meeting of conditions, and pleasing composition. (Grade Value 100.)

This examination is not a test in the knowledge of historic styles. The grades will be based solely on the degree of perfection in meeting the elemental requirements of good design. The test requires plans, elevations, sections, and such other drawings as the program may require, for a building the nature of which will be set forth in a program which a well informed owner might be expected to give an architect.

(Total Grade Value 200.)

Examination B. Graphic Analysis and Truss Design.

Time—8:30 A. M. to 12:30 P. M.

Reference material will be permitted.

This examination is held on the morning of the second day with a time allowance of four hours continuous session. It consists of a test in the science of graphic analysis as applied to building problems assuming that the preliminary design of the building is complete and loads determined. (Grade Value 100.)

Examination D. Specifications, Practice and Precedent.

Time—1:30 P. M. to 5:00 P. M.

No reference material will be permitted.

This examination is held the afternoon of the second day with a time allowance of three and one-half hours continuous session. It consists of:

(a) A test in the knowledge of specification writing, and of knowledge of the essence of the contract and of general architectural practice as it pertains to relationship between the public, owner, contractor, and architect. (Grade Value 80.)

(b) A test of general knowledge of the history of architecture and its place in social economy. (Grade Value 20.)

(Total Grade Value 100.)

Examination E. Mechanics of Materials.

Time—8:30 A. M. to 12:30 P. M.

Reference material will be permitted.

This examination is held on the morning of the third day, with a time allowance of four hours continuous session. It consists of a test in the science of determining the strength of materials, and the applicant's knowledge of applied mechanics. The test requires the applicant to design the various parts of a structure and show ability to determine the practical working sizes and shapes of structural parts of a building. Instructions will be furnished the applicant showing conditions and loading. (Grade Value 100.)

Examination F. Reinforced Concrete Design.

Time—1:30 P. M. to 5:00 P. M.

Reference material will be permitted.

This examination is held on the afternoon of the third day, with a time allowance of three and one-half hours continuous session. It consists of a test in the science of computing stresses in reinforced concrete structures and involves exercises to show the applicant's knowledge of the correct design and detailing of reinforced concrete structural parts, such as footings, columns, girders, beams and floor slabs, assuming that the preliminary design of the building is complete. The candidate is required to compute stresses and determine sizes which are safe, practical, and economical for the purposes intended. (Grade Value 100.)

Grading of Examination Papers.

Examination papers are graded as follows:

The maximum grade allowed on Examination A is 200; on Examinations B, D, E and F, each 100. The grade given the applicant on the whole examination is obtained by dividing the sum of the grades for A, B,

D, E, and F by six. To be successful, an applicant must have a general average of not less than 75, with a grade in Examination A of at least 120, and with no other grade below 60.

Applicants failing in written work are allowed credit on all grades of 75 or above, at either of the two succeeding examinations.

2. Senior applicants are persons qualifying under Section 8 (e) of the Illinois Architectural Act, who have had ten or more years of proved, independent, legal practice in the profession of architecture. Under this classification, architects who were registered in Illinois "by exemption" in 1897 are furnished a means for changing their status from "registered by exemption" to "registered by examination," which will aid them in securing registration by reciprocal transfer to other states.

3. Applicants who are Registered in Other States, Territories, and Countries. See Section 6 of the Act.

Investigation and report on the applicant's education and record in practice by the National Council of Architectural Registration Boards, if satisfactory to the Architectural Examining Committee, will be accepted in lieu of required personal appearance.

Rules of the Examining Committee.

1. Failure to pass the written examination for registration shall not disqualify an applicant otherwise eligible for oral examination under Section 8, paragraph (e), of the Illinois Architectural Act; but such failure may be considered in the grading of the oral examination.

2. An applicant for registration by reciprocal transfer of registration credit from another state, who has at some time failed in the Illinois examination, either written or verbal, may be granted such registration only in case at least six months have elapsed between the examination taken in Illinois and the examination taken in another state, and that the examination standards under which the applicant was registered were the equivalent of the requirements in Illinois as of the date of the examination.

3. "Practical experience" shall be construed as meaning working in the office of a legally qualified architect or architects as draftsman, specification writer, superintendent, or clerk of the works. No period of less than one month of continuous full time employment shall be counted toward meeting the practical experience requirement.

4. "College" for the purpose of these rules shall mean an educational institution maintaining a course in architecture with standards of entrance, curricula, and teaching equal to the standards required for membership in the Association of Collegiate Schools of Architecture.

5. Full time teaching experience of two academic college years in an accredited Department of Architecture will be accepted in lieu of the one year of practical experience required for college graduates.

6. An honorary doctor's degree conferred on an architect not having an academic degree shall be considered the equivalent of graduation from college wherever graduation from college is required, provided (a) the college or university conferring the degree is of good standing, and provided (b) the architect has had ten years or more of independent legal practice as a principal prior to the conferring of the honorary degree.

7. "High School graduation or equivalent" shall be construed as meaning either (a) that a person has graduated from a high school or academy, the standards of which are approved by the University of Illinois for admission to its College of Architecture; or (b) that a person has matriculated at an approved college of architecture; (c) has passed the high school examination conducted by the Department of Registration and Education; or (d) has had not less than ten years' employment under a legally qualified architect (or architects) and has attained and held the rank of chief draftsman, foreman, or squad boss for at least two years, plus certification of that fact by his employer or emloyers.

8. Applications for registration, accompanied by an N. C. A. R. Board certified report, may be acted upon between regular meetings of the Architectural Examining Committee by the following procedure:

(a) Upon receipt of the N. C. A. R. Board report and application, the chairman of the Architectural Examining Committee will prepare two findings in quintuplicate,—one for and one against recommending that the applicant be granted registration in Illinois; and he shall record his vote by signing all five copies of the finding with which he agrees. The two findings in quintuplicate shall then be sent consecutively to each member of the committee, the secretary being the last. Each member shall review the record and sign all five copies of the finding of his choice.

(b) The secretary of the committee shall review the ballots of the individual members, and, if he finds they are unanimous for or against the registration of the applicant, he shall send to the Director of the Department of Registration and Education the recommendation of the committee that registration be immediately granted or not granted according as the signed ballots indicate. With the transmission of this report he shall include three (3) copies of the signed finding. A fourth copy of the finding shall be retained for the secretary's record, and the fifth copy shall be sent to the chairman of the committee.

(c) In case of failure of unanimous vote either for or against, action on the application shall be postponed until the next meeting of the committee.

STATES REQUIRING ARCHITECTURAL REGISTRATION

(Date in parentheses indicates year in which law was originally passed.)
*Not Members of the National Council of Architectural Registration Boards.

STATES

Alabama (1931)—Board of Registration of Architects. Harry H. Jones, Sec.-Treas., 315 Shepherd Building, Montgomery.

Arizona (1921)—State Board of Technical Registration for Architects, Engineers, Land Surveyors and Assayers. Lewis S. Neeb, Secretary, P. O. Box 1029, Phoenix.

California (1901)—State Board of Architectural Examiners. Northern District—Frederick H. Reimers, Secretary, 515 Van Ness Avenue, San Francisco. Southern District—A. M. Edelman, Secretary, 907-C State Building, Los Angeles.

Colorado (1909)—State Board of Examiners of Architects. W. Gordon Jamieson, Sec.-Treas., 816 Twelfth Street, Denver.

Connecticut (1933)—Architectural Examining Board. Edward B. Caldwell, Secretary, 1029 Fairfield Ave., Bridgeport.

Delaware (1933)—State Board of Examiners and Registration of Architects. Roscoe Cook Tindall, Secretary, 914 Orange St., Wilmington.

Florida (1915)—State Board of Architecture. Mellen C. Greeley, Sec.-Treas., Suite 925 Barnett National Bank Bldg., Jacksonville.

Georgia (1919)—State Board for the Examination and Registration of Architects. R. C. Coleman, Joint Secretary, State Examining Boards, Department of State, Atlanta.

Idaho (1917)—Department of Law Enforcement. J. L. Balderston, Commissioner of Law Enforcement, State Capitol, Boise.

Illinois (1897)—Department of Registration & Education (Architects' Division). Homer J. Byrd, Supt. of Registration, Springfield; Professor L. H. Provine, University of Illinois, Urbana, Secretary, Illinois Architects Examining Committee.

Indiana (1929)—State Board of Registration for Architects. Leighton Bowers, Secretary, State Capitol, Indianapolis.

Iowa (1927)—Board of Architectural Examiners. William L. Perkins, Sec.-Treas., Chariton.

Kentucky (1930)—State Board of Examiners & Registration of Architects. C. Julian Oberwarth, Secretary, Frankfort.

Louisiana (1910)—State Board of Architectural Examiners. Walter Cook Keenan, Secretary, 4239 St. Charles Avenue, New Orleans.

Maryland (1935)—Board of Examiners for the Registration and Licensing of Architects. Charles Dana Loomis, Executive Secretary, 1309 Lexington Building, Baltimore.

Michigan (1915)—State Board of Examiners for Registration of Architects, Engineers, Surveyors. C. T. Olmsted, Secretary, 306 Transportation Building, Detroit.

Minnesota (1921)—Board of Registration for Architects, Engineers & Land Surveyors. I. M. Clemetson, Executive Secretary, 620 New York Building, St. Paul.

***Mississippi** (1928)—State Board of Architecture. Frank P. Gates, Sec.-Treas., 601 Millsaps, Jackson.

Montana (1917)—State Board of Architectural Examiners. W. R. Plew, Executive Sec.-Treas., Bozeman.

Nebraska (1937)—State Board of Examiners for Professional Engineers and Architects. D. L. Erickson, Secretary, Room 1009, State Capitol, Lincoln.

New Jersey (1902)—State Board of Architects. Gilbert C. Higby, Sec.-Treas., 32 Walnut Street, Newark.

New Mexico (1932)—Board of Examiners for Architects. Willard C. Kruger, Secretary, Box 1587, Santa Fe.

New York (1913)—The State Education Department. Herbert J. Hamilton, Chief Bureau of Professional Examinations, Albany; William P. Bannister, Secretary, State Board of Examiners of Architects, 339 Lexington Avenue, New York City.

North Carolina (1913)—State Board of Architectural Examination and Registration. Ross Shumaker, Secretary, Box 5445, Univ. of N. C., Raleigh.

North Dakota (1913)—State Board of Architecture. E. W. Molander, Acting Secretary, 508 First Avenue Building, Minot

Ohio (1931)—State Board of Examiners of Architects. R. C. Kempton, Executive Secretary, A.I.U. Building, Columbus.

Oklahoma (1925)—State Board of Examiners of Architects. Leonard H. Bailey, Sec.-Treas., 1217 Colcord Building, Oklahoma City.

Oregon (1919)—State Board of Architect Examiners. Margaret Goodin Fritsch, Secretary, 1601 Public Service Building, Portland.

Pennsylvania (1919)—State Board of Examiners of Architects. James W. Minick, Secretary, 375 Education Buiding, Harrisburg.

***Rhode Island** (1936)—Board of Examination and Registration of Architects. Milton MacIntosh, Secretary, 44 Franklin Street, Providence.

South Carolina (1917)—State Board of Architectural Examiners. Professor Walter E. Rowe, Sec.-Treas., 210 Sloan College, Univ. of S. C., Columbia.

South Dakota (1925)—State Board of Engineering and Architectural Examiners. Charles A. Trimmer, Secretary, Madison.

Tennessee (1921)—State Board of Architectural and Engineering Examiners. Joseph W. Holman, Sec.-Treas., 702 Stahlman Bldg., Nashville.

Texas (1937)—Board of Architectural Examiners. Thomas D. Broad, Sec., Burt Building, Dallas.

***Utah** (1911)—Department of Registration. S. W. Golding, Director, State Capitol, Salt Lake City.

Virginia (1920)—State Board for the Examination and Certification of Professional Engineers, Architects and Land Surveyors. C. G. Massie, Secretary, 4030 Fort Avenue, Lynchburg.

***Washington** (1919)—Department of Licenses. Harry C. Huse, Director, Olympia.

West Virginia (1921)—Board of Architects. A. Ford Dickey, Secretary, Professional Building, Huntington.

Wisconsin (1917)—Registration Board of Architects and Professional Engineers. Arthur Peabody, Secretary, State Capitol, Madison.

DISTRICTS

District of Columbia (1924)—Board of Examiners and Registrars of Architects. Robert F. Beresford, Secretary, 810 Eighteenth Street, N.W., Washington.

TERRITORIES

Hawaii (1923)—Board of Registration for Professional Engineers, Architects and Land Surveyors. William C. Furer, Secretary, 506 Hawaiian Trust Building, Honolulu.

***Philippines** (1924)—Boards of Examiners for Architects—Bureau of Civil Service. Jose Gil, Executive Officer, Manila.

***Puerto Rico**—Board of Examiners of Engineers, Architects and Surveyors. Pedro A. De Castro, Secretary-Treasurer, San Juan.

NATIONAL COUNCIL OF ARCHITECTURAL REGISTRATION BOARD

In Cooperation with The American Institute of Architects.
The Association of Collegiate Schools of Architecture.

CONCERNING THE COUNCIL

(Council Circular of Advice No. 14.)

1. Under the Constitution of the United States there can be no national registration in any profession, the authority to register or license citizens to practice a given profession being vested in the several states.

2. The great majority of the States now have laws for the registration of architects. Registration in one State does not qualify an architect to practice in another State, for standards or requirements differ.

3. The National Council of Architectural Registration Boards is an organization made up of, and controlled by the architectural registration authorities of the several States. It is the national quasi-official authority recognized by The American Institute of Architects as qualified to establish a proper standard examination.

4. The requirements for a Standard N.C.A. R.B. Examination are intended to include everything required by any and all of the States; and the standard in every subject is intended to be the highest set by any State.

5. After an architect has successfully complied with the Examination Procedure outlined below, the Council certifies that he has proved qualification for practice and, on request, furnishes its certified data to State registration authorities, recommending that he be granted registration, without further examination, on the basis of evidence submitted and his State examination in accordance with Council standards.

6. The National Council facilitates registration in any State for the holders of its Certificate. While the National Council has no legal competence to guarantee registration in any State to an architect who holds its Certificate, its approval of an applicant is usually accepted in any State, as all State registration authorities are joined together in the Council in establishing the Council's requirements and standards.

7. The National Council, upon the request of a Certificate holder, recommends him for registration in any State without further formality than payment of the Council fee and the State fees.

Examination Procedure.

8. The National Council does not itself conduct any examinations. It does, however, undertake to make arrangements with State examining boards to examine the Candidate for a National Council Certificate in such manner and in such subjects as are necessary to fulfill the Council requirements for the Standard N.C.A.R.B. Examination.

9. The National Council requires all applicants for a Standard N.C.A.R.B. Examination to enroll in writing. It forwards to each applicant an application blank or questionnaire concerning qualifications. If the applicant's qualifications are found to be satisfactory, the National Council accepts him as a Candidate for the Examination for the Certificate.

10. Candidates are divided into two categories:

(a) The Senior, i: e:. architects who have been in practice as principals ten years or more;

(b) The Junior, the men who have been in practice for a shorter period.

11. The requirements for the Junior and Senior Examinations are stated in Circulars of Advice Nos. 16 and 17, respectively, and in detail in Circular of Advice No. 3.

12. The examination of Junior and Senior Candidates is completed when the State examining board authorized to conduct the examination has reported the results to the Council and the Council secretary has attached his certificate of approval thereto.

13. The National Council of Architectural Registration Boards bases its decision on the Candidate's fitness for practice on his training, his character, his record in practice, and his standing in the examination as reported by his examiners.

14. Finding that the Candidate meets the above requirements, the Council issues its Certificate declaring the holder to have passed the Standard N.C.A.R.B. Examination.

Other Council Activities.

1. The Council renders various services stated in brief as follows:

(a) Gives advice concerning essential training and experience prerequisite to admission to a Standard N.C.A.R.B. Examination.

(b) Approves and records Mentors and maintaining advisory contact with them.

(c) Collates and verifies records of preparation for and attainment in architectural practice.

(d) Serves architects who are not holders of the Council Certificate but who are engaged in inter-State practice by assisting them in the proper presentation of their credentials for registration in the various States.

(e) Exercises a unifying influence over examination and registration procedure in the several States.

(f) Acts as a clearing house for matters of interest and importance to State registration authorities, including legal decisions affecting professional registration, examina- subject matter, and administrative problems.

(g) Mantains a file of all registration laws and application forms in use in each State.

16. The National Council of Architectural Registration Boards, which is an Illinois corporation organized not for profit, was founded in 1920. It has since that time developed its procedure, which is recognized by the architectural profession and schools and by The American Institute of Architects as sound and in the interest of the Profession.

INTERSTATE PRACTICE.

(Reprint from Council Circular of Advice Number 4.)

1. Interstate practice may be the privilege and is certainly the ambition of every architect. When the opportunity comes to an architect to secure a commission in a State in which he is not registered, his first thought is how to secure registration in that State as quickly as possible at a minimum of expense. It was for the purpose of facilitating reciprocal registration among the States having laws regulating the practice of architecture, and of securing greater uniformity in the requirements for registration in all States that the National Council of Architectural Registration Boards was organized.

2. Architectural registration is personal. Each individual whose name appears before or above the professional title "Architect" must be registered as an architect in every State in which he does business. Individuals or members of firms practicing without registration in States where there are registration laws are definitely law-breakers, subject to the several penalties prescribed by law.

3. Corporations are impersonal; they have an existence apart from the personnel composing them, and can not be examined as to personal qualifications nor be subject to the pains and penalties of imprisonment or capital punishment. Therefore, corporations are constitutionally incapable of assuming the obligations incident to professional practice. Corporations may act as agents for professional men, but may not be allowed to practice architecture or any other profession, nor lawfully use a registered professional title in connection with a corporate name.

These statements are supported on the grounds of public benefit because the purpose of professional registration is to protect the public by first establishing proof of the competency of the individual, and then placing responsibility upon him in an inescapable way.

By contrast, corporations are designed primarily to relieve the individual from personal responsibility. For this reason, so-called professional corporations are in contravention of the purpose of professional registration.

4. In order to protect the public, it is necessary that State registration boards have first hand information concerning architects from other States who apply to them for registration. Many of the registration boards do not have the funds or other facilities for making a satisfactory investigation. Were they able to do this, it would mean that each State board would have to repeat the process of investigation every time an out-of-State architect applied for registration. Such duplication could not be carried on without unending delay in connection with each separate registration and undue annoyance to an architect's references and the educational institutions in which he received his training.

5. An architect who desires reciprocal registration credit will find it convenient and economical to apply through the National Council for such registration as, in most cases, he will be saved the necessity of personal appearance and an examination before any examining board other than that of his home State. While the Council does not guarantee to an architect that, in employing the services of the Council, he will be granted registration in any State to which he may apply, it does act as a clearing house by furnishing state registration boards with information concerning the qualifications, both educational and professional, of architects who apply for reciprocal registration. This the Council does by a thorough and disinterested investigation, the results of which are called a Council "record."

6. A Council record is a certified transcript, in booklet form, of all replies to inquiries which have been addressed by the National Council to the applicant's references. It should be understood, however, that the mere possession of a Council record does not carry with it the endorsement or recommendation of the National Council of Architectural Registration Boards, nor does it give an architect any prestige. A Council record is simply unbiased authentic evidence concerning an architect's training, practice, and professional standing. With a Council record in hand, the registration authorities of the State in which the applicant desires reciprocal registration are supplied with evidence which is generally considered sufficient to determine whether or not an applicant is eligible for registration in accordance with the law of that State. In other words, if, in the opinion of the examining state board, an applicant's education and practice record, as well as his previous examination in another state as shown by his certified council record, are equivalent or superior to that required in the state to which he is applying for registration he gets registration in that state; if not, he does not.

7. The time required to make the investigation and complete a Council record is usually from four to six weeks. For this reason, an architect who has any prospect whatsoever of securing business in States other than his State of original registration is at an advantage if he has had a Council record prepared in advance and ready for immediate use.

Many states now require a Council record before considering the application of an architect from another State. All seem to prefer it.

8. To employ the services of the Council for the purpose of securing a Council record, request should be made to the Secretary of the National Council of Architectural Registration Boards, Suite 2300, 175 West Jackson Boulevard, Chicago, for the necessary blanks and instructions. This request should be under the applicant's own letterhead and be accompanied by the following:

(a) A money order, check, or bank draft for Twenty-five Dollars ($25.00) plus exchange, if any, made payable to the "N.C.A.R. Boards," to defray the expense of preparing a Council record and transmitting the record to the registration board in the State in which application is being made for registration.

(b) A statement giving the State or States in which the applicant is at present registered, with the date and official number of each registration and, in each case, the manner in which such registration was obtained, whether by "exemption," "reciprocal registration credit," "written examination," or "oral examination." All architects should understand that registration "by exemption" does not constitute legal proof of competency and is therefore not a basis for registration credit.

(c) A statement as to the number of years that the applicant has been engaged in practice as a principal. A person may be considered a principal in the practice of architecture if, under the professional title "architect," he serves only one client, on a salary or other basis, in an office maintained by such client, if it can be established that he is entirely responsible for and in complete control of the architectural work and the management of that office, including the employment and discharge of all persons engaged on the design and supervision of the work.

(d) Except in States which do not have laws regulating the practice of architecture, it is illegal for an architect to practice as a principal unless he is registered. The mere fact that one participates in the profits of an architect's office is not practicing architecture as a principal.

(e) It should be noted: (1) that employment in an architect's office as foreman, draftsman, specification writer, or superintendent is not practice as a principal, or one of the principals, in charge of an architect's office; (2) that employment by a corporation is not practicing architecture as a principal even though the title "architect" is used, where some officer of the corporation, or person other than the architect, has control of the employment and discharge of those engaged in the preparation of plans and specifications or supervision of construction; and (3) that business as a building-contractor, or employment as such contractor's superintendent, draftsman, or estimator, is not the practice of architecture.

General Instructions.

9. The Secretary of the National Council will forward to the applicant an information blank upon receipt of the statements above requested and payment of the fee. Two forms are used in making application for Council records: one for Senior applicants, or architects who have been engaged in practice as a principal for ten or more years; and the other for Junior applicants, or those who have been in practice a shorter time.

10. The information blank is used as a negative for blueprinting and drafting ink must be used for filling in the blank. In case a typewriter is used, the form must be backed with a fresh black carbon. The applicant should retain a blueprint of his information blank in order to provide himself with a copy of the information filed with the National Council. By carefully checking over the blueprint, the applicant is able to determine whether every item, including signatures, is perfectly legible. Every part must be perfectly legible when blue printed.

11. Emphasis is laid on the importance of using great care in filling out the information blank and making the statements as complete and comprehensive as possible, particularly in the following points:

(a) Actual dates of attendance at schools are important, and addresses of schools, secretaries of societies, clients, and architectural references should be accurately given.

(b) Architect references should be given regardless of applicant's classification as Junior or Senior.

(c) Please avoid, if possible giving the names of relatives or partners as either architect or client references.

(d) "Architect" references (page 2) and "architect-employer" references (page 3) on the Junior information blank should not be duplicated.

(e) If possible, the members of examining boards should not be used as architect references, as such architects may be obliged to pass on the record.

(f) If the college or university from which the applicant graduated has a rule that a transcript of grades will not be furnished except at the request of the graduate, the applicant should enclose with his information blank a letter addressed to that institution authorizing it to furnish the Council direct with such a transcript. This authorization will be mailed to the registrar, or other proper officer, with the Council's request for a transcript of grades.

(g) Do not send letters of recommendation, photostats of diplomas or certificates, or transcripts of school records unless requested, as the Council uses only material which has been received in direct response to requests made by the Council.

12. An identification photograph is required and during the time that his record is being prepared, the applicant should arrange to purchase the film negative of an identification photograph, not less than 3"x5", and preferably 5"x7" in size, which he will furnish, with two (2) prints, to the Council office. Before making the prints, the negative shall be signatured by the applicant and his signature attested by two others, in drafting ink, so that all signatures will appear on the face of the prints. Attestations on the back of the prints are of no value as the prints are mounted in the Council record. Glass negatives are not acceptable.

13. A State fee is required and either at the time of sending the Information blank, or at some time during the preparation of his Council record, the applicant will be informed, by the National Council, as to the amount of the preliminary fee which he will be required to pay to the State in which he is applying for registration. Upon receipt of this information, the applicant will forward to the Council the amount of that fee (plus exchange, if any), made payable to the registration board of the State in which he is applying for registration. Only money orders, certified checks, or drafts are acceptable to State registration boards, and under no circumstances is a State fee to be made payable to the National Council. The remittance for the State fee will be held by the Council until the applicant's record is fully completed, when it will be forwarded with the Council record and (if one is required) the application form of the State.

The Council attempts to keep on file, in the Secretary's office, copies of the application forms and the laws of all States, so that it is in a position to prepare a State application for any applicant and forward it to him for his signature. From the State laws on file, it is possible to advise the applicant concerning the amount of State fees and other registration requirements.

Subsequent Applications for Reciprocal Registration.

14. The original Council record is kept on file in the office of the Secretary of the Council. Whenever an applicant wishes to apply for registration in additional States, it is only necessary to advise the Council the name of the State in which application is to be made and enclose the Council's fee of $10.00 to cover the expense of duplicating and transmitting the applicant's record to the registration board in that State. As in the case of first applications for registration credit, the applicant will be furnished the proper State application blank and will be notified as to the amount of the State fee to be paid.

If a long period of time has elapsed between the original investigation and the subsequent application for reciprocal registration, the Council will make further investigation to ascertain whether the applicant has maintained the professional standard indicated by the original record, and the applicant may be requested to furnish additional architect and client references.

The following Circulars of Advice are available on application to the N.C.A.R. Boards:

No. 2—The Theory on which Examination and Registration of Architects Is Based

No. 3—Regulations Concerning the Standard N.C.A.R.B. Examination.

No. 4—Interstate Practice.

No. 14—Concerning the Council.

No. 15 — The Mentor: Concerning His Duties.

No. 16—Concerning the Standard N.C.A.R.B. Junior Examination.

No. 17—Concerning the Standard N.C.A.R.B. Senior Examination.

Documents.

No. 21—Primer on Professional Registration.

No. 22—Registration Objective.

LIST OF REGISTERED ARCHITECTS

Members of the Profession Permitted to Practice in the State of Illinois
Compared with Official Records of November 19, 1938.

CHICAGO

Abel, Lester A., 333 N. Michigan Ave.
Adelman, Gerald D., 540 N. Michigan Ave.
Adler, David, 220 S. Michigan Ave.
Aga, Ole W., 1335 Belmont Ave.
Ahlschlager, Walter W., 700 N. Wabash Ave.
Albano, Joseph F., 5439 S. Woodlawn Ave.
Alderman, William N., 19 E. Pearson St.
Allen, Alfred P., 225 N. Michigan Ave.
Allison, Lyman, J., 8622 Parnell Ave.
Alper, Max, 2753 W. North Ave.
Alschuler, Alfred S., 28 E. Jackson Blvd.
Alschuler, Alfred S., Jr., 28 E. Jackson Blvd.
Anderson, B. D. Andy, 4832 N. Bell Ave.
Anderson, Harold E., 1643 E. 86th Place.
Anderson, Helge A., 4157 Cullom Ave.
Anderson, Helmer Nels, 5948 Midway Park
Anderson, Karl C., 2617 Berwyn Ave.
Anderson, Russell A. M., R. 1144—53 W. Jackson Blvd.
Anderson, Walter G., 4535 N. Bernard St.
Apfelbach, Henry J., 2834 Lyndale Ave.
Archer, Chas. S., 1107 W. Marquette Rd.
Armstrong, John A., 11 S. LaSalle St.
Bacci, Alexander H., 534 Roscoe St.
Bacon, James Earl, 7332 Phillips Ave.
Baggesen, Walter Wm., 8001 Blackstone Ave.
Bailey, Walter T., 4322 Prairie Ave.
Baldwin, Francis M., 1414 Sedgwick St.
Balluff, Louis N., 750 N. Michigan Ave.
Banse, Herbert G., 8 E. Huron St.
Bargman, Ewald F., 1408 Jarvis Ave.
Barrett, Fred Smith, 5016 W. Erie St.
Barrett, Frederick L., 4354 N. Keeler Ave.
Barry, Gerald A., Jr., 5202 W. Chicago Ave.
Baur, Ervin, 842 Waveland Ave.
Beaudry, Ralph L., 9711 Calumet Ave.
Beck, Bereford E., 307 N. Michigan Ave.
Beck, Willis J., 1216 Leland Ave.
Bednarik, Jos., 3700 S. Halsted St.
Behrns, Elmer F., 3429 N. Troy St.
Beidler, Herbert B., 952 N. Michigan Ave.
Bein, Maurice L., 1159 W. Roosevelt Rd.
Beman, Spencer S., 189 W. Madison St.
Bender, Carl R., 501 N. Central Ave.
Benn, William W., 104 S. Michigan Ave.
Bennett, Edward H., 1800—80 E. Jackson
Benson, Arthur E., 520 N. Michigan Ave.
Berbiers, Lee W., 1358 Sedgwick St.
Berger, Bernard, 842 Gunnison St.
Berkson, Aaron, 3221 Douglas Blvd.
Bernham, F. M., 8 S. Michigan Ave.
Bethke, Edward Joseph, 5737 Kenmore Ave.
Bianco, Anthony J. Del, 4050 Crystal St.
Bialles, Theo. P., 5119 N. Marmora Ave.
Bicknell, Alfred H., 3601 N. Hoyne Ave.
Bishop, Thomas R., 35 S. Dearborn St.
Blouke, Pierre, 2600 Tribune Tower
Boehm, Rudolph P., 1830 W. 103rd St.
Bohasseck, Charles, 333 N. Michigan Ave.
Bolen, John C., 1737 N. Meade Ave.
Bollenbacher, J. C., 333 N. Michigan Ave.
Bonnevier, Clarence Julius, 8520 S. May St.
Boothey, Donald D., 105 W. Adams St.
Bouchard, Lewis C., 228 N. LaSalle St., R. 536
Bourke, Robt. E., 1401, 39 S. LaSalle St.
Bowman, Irving H., 1338 Howard St.
Braband, Frank J. E., 901 Wrightwood Ave.
Brabant, Gifford, 2717 N. Kedzie Ave.
Brand, Herbert A., 400 W. Madison St.
Braucher, Ernest N., 6 N. Clark St.
Braun, Donald George, 6336 Kimbark Ave.
Braun, George Jr., 3019 N. Kenneth Ave.
Braun, Isadore H., 123 W. Madison St.
Braun, Martin Henry, 134 N. LaSalle St.
Braun, Wm. T., 228 N. LaSalle St.
Britton, Frank, c/o Libby, McNeil & Libby, Union Stock Yards
Brown, Arthur Robinson, 1059 Glenlake Ave.
Bruns, Benedict J., 4734 N. Hermitage Ave.
Bruns, Herman H. N., 738 N. Albany Ave.
Buchsbaum, Emanuel V., 7810 Crandon Ave.
Buckley, Geo. H., 2237 W. 108th Place.
Buerger, A. J., 4906 Iowa St.
Burger, Walter David, 3740 Lake Shore Drive.
Burgess, Ralph R., 6523 N. Newgard Ave.
Burnham, Daniel H., 134 S. LaSalle St.
Burnham, Hubert, 160 N. LaSalle St.
Burnes, Alfred, 1444 S. St. Louis Ave.
Burns, John J., 1517 Chase Ave.
Buss, Truman C., Jr., 11006 S. Michigan Ave.
Byers, Edwin M., 1358 Sedgwick St.
Cabeen, Richard McP., c/o Holabird & Root, 333 N. Michigan Ave.
Cable, Max Lowell, 1241 N. Ashland Ave.
Camburas, Peter E., 123 W. Madison St.
Capraro, Alexander V., 64 W. Randolph St.
Carlburg, Ralph H., 1408 Carmen Ave.
Carlson, Elmer C., 7936 Cottage Grove Ave.
Carlson, Richard J., 82 W. Washington St.
Carnegie, William G., 400 E. 111th St.
Carr, Charles Alban, 322 W. Chicago Ave.
Carr, George Wallace, 333 N. Michigan Ave.
Carstens, Milton S., 1 N. LaSalle St.
Carter, Thomas A., 6651 N. Newgard Ave.
Cerny, Jerome J., 2341 S. Austin Blvd.
Charles, Walter T., 1126—177 W. Washington
Charn, Victor L., 664 N. Michigan Ave.
Chatten, Melville C., 160 N. LaSalle St.
Cheatle, Edwin, 2308 W. 109th St.
Cheney, Howard Lovewell, 208 S. LaSalle St.
Childs, Frank A., 430 N. Michigan Ave.
Christensen, Eli, 6943 Indiana Ave.
Christensen, Hans C., 616 S. Michigan Ave.
Christensen, John C., 228 N. LaSalle St.
Christiansen, Roy T., 360 N. Michigan Ave.
Chubb, John D., 109 N. Dearborn St.
Clark, Edwin H., 8 E. Huron St.
Cohen, Arthur S., 5490 S. Greenwood Ave.
Cohen, Isadore, 7748 Colfax Ave.
Cohen, Joseph, 3921 W. 19th St.
Colburn, Howard F., 8028 Ingleside Ave.
Coleman, Linza F., 453 W. 63rd St.
Comm, Benjamin Albert, 20 W. Jackson Blvd.
Conners, William Joseph, 6342 N. Maplewood
Cook, Oswald J., 6522 Glenwood Ave.
Cooke, Thomas E., c/o P. J. Clancy, 221 N. Lamon Ave.
Corse, Redmond O., 5458 Magnolia Ave.
Coughlen, Gardner C., 9406 S. Vanderpool

Cowles, Knight C., 30 N. Michigan Ave.
Criz, Albert, 43 E. Ohio St.
Cromelin, John S., 70 Scott St.
Crosby, Wm. S., 6 N. Michigan Ave.
Dahlquist, Clarence L., 720 N. Michigan Ave.
Dalsey, Harry I., 139 N. Clark St.
Davey, John Jeffrey, 29 W. Quincy St.
Davidson, Lawrence Whitney, 154 W. 68th St.
Davis, Zachary T., 228 N. LaSalle St.
Dean, Arthur R., 205 W. Wacker Drive.
De Golyer, Robt. S., 307 N. Michigan Blvd.
De Long, Albert J., 1814 Lincoln Park West
De Money, Frank O., 30 N. LaSalle St.
De Muth, John J., 4552 N. Wolcott Ave.
Del Campo, Scipione, 2955 N. Kilpatrick Ave.
Dewey, Charles, 4342 Drexel Blvd.
Dippold, Albert P., 3948 Cottage Grove Ave.
Dittrich, Francis J., 3100 S. Kedzie Ave.
Doerr, William P., 5487 Hyde Park Blvd.
Doerr, William Phillip, Jr., 5512 Everett Ave.
Doig, Duncan C., 855 Buena Ave.
Doll, Clarence W., 118 E. 26th St.
Dornbusch, Charles H., 333 N. Michigan Ave.
Dougherty, Floyd E., 79 W. Monroe St.
Drainie, John G., 6854 Cornell Ave.
Drielsma, Arthur J., 540 N. Michigan Ave.
Dubin, George H., 127 N. Dearborn St.
Dubin, Henry, 127 N. Dearborn St.
Dunning, N. Max, 306 S. Wabash Ave.
Dubsky, Frank A., 6154 S. California Ave.
Dyer, Scott C., 38 S. Dearborn St.
Ehmann, William H., 104 S. Michigan Ave.
Edmunds, Ralph G., 4541 N. Hamilton St.
Eich, George B., 343 S. Dearborn St.
Eichberg, S. Milton, 208 S. LaSalle St.
Eichenbaum, Edward E., 7558 Crandon Ave.
Ekberg, Arthur B., 7627 Saginaw Ave.
Ekroth, Roy Albert, 4837 N. Hamlin Ave.
Elmslie, Geo. C., 122 S. Michigan Ave.
Elting, Winston, 333 N. Michigan Ave.
Eppenstein, James F., 35 E. Wacker Drive.
Erikson, Carl A., 104 S. Michigan Ave.
Esser, Curt A., 225 N. Michigan Ave.
Esser, Paul F., 6164 N. Hoyne Ave.
Ettinger, Paul, 6800 Sheridan Rd.
Everds, William H., 59 E. Van Buren St.
Eugenidas, Alexander K., 1316 W. 63rd St.
Evans, Floyd, 6901 Merrill Ave.
Ewer, Warren B., 400 W. Madison St.
Fairclough, Stanley D., 819 Exchange St.
Falls, Alexander S., 2045 W. 103rd St.
Faro, Philip Anthony, 9254 Anthony St.
Faro, Robert Vale, 120 S. LaSalle St.
Faulkner, Chas. D. 307 N. Michigan Ave.
Fellows, Wm. K., 814 Tower Court.
Fielder, Fred A., 118 E. 26th St.
Finck, Sidney C., 134 N. LaSalle St.
Fischer, Frederick William, 3018 E. 91st St.
Fischer, John B., 7322 La Fayette Ave.
Feschman, Oscar, 2821 Milwaukee Ave.
Fisher, Albert J., 7128 N. Bell Ave.
Fisher, Howard T., 3031 Pine Grove Ave.
Flesch, Eugene W. P., 1209 Madison Park.
Floto, Julius, 176 W. Adams St.
Forte, Attilio, 664 N. Michigan Ave.
Forrester, Benjamin X., 540 N. Michigan Ave.
Fortin, Joseph T., 600 Blue Island Ave.
Fournier, Lawrence A., 1225 E. 54th St.
Fox, Elmer J., 57 E. Jackson Blvd.
Fox, John J., 549 W. Randolph St.
Fox, William Paul, 737 N. Michigan Ave.
Foy, Wm. D., 540 N. Michigan Ave.
Franklin, Robert L., 9649 S. Seeley Ave.
Frazier, Clarence E., 64 W. Randolph St.
Frazier, Walter S., 540 N. Michigan Ave.
Frederick, Erwin G., 307 N. Michigan Ave.
Freiberg, Frederick A., 7910 Drexel Ave.
Friedman, Raphael N., 28 E. Jackson Blvd.
Frodin, Rube S., 9 E. Superior St.
Frost, Harry Talfourd, 1800 Railway Exch.
Fry, Frank L., 2142 Washington Blvd.
Fugard, John Reed, 520 N. Michigan Ave.
Fuhrer, Eugene, 188 W. Randolph St.
Fullenwider, Arthur E., 7209 Harvard Ave.
Fuller, Ravilo Franklin, 75 E. Wacker Drive.
Furst, Wm. H., 11 S. LaSalle St.
Garden, Hugh M., 104 S. Michigan Ave.
Gaul, Hermann, 228 E. Superior St.
Gaul, Michael Felix, 228 E. Superior St.
Geis, Lester E., 628 Sheridan Rd.
Gerber, Arthur V., 72 W. Adams St.
Gerhardt, Herman Otto, 7552 Champlain Ave.
Gerhardt, Paul, 447 Fullerton Parkway
Gerhardt, Paul, Jr., 1012 City Hall.
Gliatto, Leonard A., 5748 Race Ave.
Glicken, Maurice Jerome, 203 N. Wabash Ave.
Glicken, Maxwell E., 203 N. Wabash Ave.
Goff, Bruce Alonzo, 1515 Howard St.
Goldberg, Bertrand, 1755 E. 55th St.
Golden, Everett, 6828 Lakewood Ave.
Goldman, Joseph, 901 Argyle Ave.
Goldsborough, Robert V., 4167 Drexel Blvd.
Goo, Robert Yau, 360 N. Michigan Ave.
Gormley, James R., 1546 Arthur Ave.
Gould, Albert Allen, 2117 Oakton St.
Graham, William E., 80 E. Jackson Blvd.
Granger, Alfred H., 333 N. Michigan Ave.
Granger, Stewart Stanford, 10140 Rhodes Ave.
Green, Martin, 7538 S. Wabash Ave.
Greenfield, Geo. H., 6023 Kenwood Ave.
Greengard, Bernhard C., 726 Junior Terrace.
Gregori, Raymond, 33 N. LaSalle St.
Griser, Louis Keith, 4823 N. Winchester Ave.
Grosguth, Joseph Jr., 1833 N. Washtenaw
Grossman, Melvin, 550 Arlington Blvd.
Grotz, Chas. J., 3184 Milwaukee Ave.
Grunsfeld, Ernest A., Jr., 435 N. Michigan
Guenzel, Louis, 879 N. State St.
Gustafson, Carl H., 8034 Bishop St.
Gustafson, Virgil E., 5300 N. Kimball Ave.
Gylleck, Elmer A., 10 N. Clark St.
Haagen, Paul T., 155 N. Clark St.
Hagerup, Leonard O., 201 N. Wells St.
Halana, John P., 1615 Farwell Ave.
Hall, Eric E., 123 W. Madison St.
Hall, Emery Stanford, 175 W. Jackson Blvd.
Hallberg, L. G., 221 N. LaSalle St.
Hallman, Edward C., 5443 N. Lincoln Ave.
Halperin, Casriel, 119 N. Waller Ave.
Hamilton, John L., 814 Tower Court.
Hammond, Charles Herrick, 160 N. LaSalle.
Handler, Harry A., 1251 Granville Ave.
Hanson, Herbert Carl, 12013 Eggleston Ave.
Harris, Mandel H., 4940 N. Monticello Ave.
Haselhuhn, Robert G., 1744 Chase Ave.
Hauber, Carl, 25 E. Jackson Blvd.
Hawk, A. T., 806 LaSalle St., R. R. Station.
Hawkins, Mortimer Hill, 75 E. Wacker Dr.
Hecht, Albert S., 420 Surf St.
Heda, Arthur M., 3312 Eastwood Ave.
Hegsted, Martin A., 4630 Altgeld St.
Heimbeck, Walter C., 1736 W. 102nd St.
Heino, Albert F., 11059 Hale Ave.

Henshien, H. Peter, 59 E. Van Buren St.
Herter, John T., 2638 Fargo Ave.
Hertel, Morris Charles, 841 N. Michigan Ave.
Herzog, William T., 842 Grace St.
Hetherington, Murray D., 10153 S. Prospect
Hettinger, Robert John, 4708 Dover St.
Higgins, Edgar J. S., 2950 S. LaSalle St.
Himelblau, A. L., 1535 Touhy Ave.
Hirschfield, Leo S., 65 E. South Water St.
Hodgdon, Charles, 111 W. Monroe St.
Hodgdon, Frederick M., 307 N. Michigan Ave.
Hodgdon, John M., 111 W. Monroe St.
Hodgkins, Howard G., 7357 N. Ashland Ave.
Hofer, Victor George, 6108 Glenwood Ave.
Hoffman, Paul Ernest, 5045 Mango Ave.
Holabird, John Augur, 333 N. Michigan Ave.
Holmes, Harold, 1708 Chase Ave.
Holmes, Kenneth E., 1708 Chase Ave.
Holmes, Morris G., 4815 Drexel Ave.
Holsman, Henry K., 140 S. Dearborn St.
Holsman, John T., 140 S. Dearborn St.
Hooper, Wm. T., 625 N. Laramie Ave.
Horowitz, Wm. H., 28 E. Jackson Blvd.
Horwitz, Max R., 2442 Rosemont Ave.
Hosek, Joseph E., 225 N. Michigan Ave.
Hoskins, Henry J. B., 333 N. Michigan Ave.
Hoskins, John M., 2830 W. Madison St.
Hotton, Bartholomew J., 2680 Eastwood Ave.
Houlihan, Raymond F., 5944 N. Artesian Ave.
Huber, Julius H., 2039 Greenleaf Ave.
Hulsfeld, Joseph J., 4741 N. Kilbourne Ave.
Hull, Denison B., 77 W. Washington St.
Hunt, Melvin B., 221 N. LaSalle St.
Hussander, Arthur F., 730 Oakdale Ave.
Huston, Sanford K., Jr., 1163 E. 54th Place.
Huszagh, Ralph D., 6 N. Michigan Ave.
Hyland, Paul V., 6308 N. Leavitt St.
Irwin, Edgar S., 7335 Kingston Ave.
Isensee, Frederic M., 8201 Harper Ave.
Jacobs, Arthur, 160 N. LaSalle St.
Jacobs, Victor Herbert, 1348½ Estes Ave.
James, Erwin Harry, 4915 Ravenswood Ave.
Janik, Ladislar I., 127 N. Dearborn St.
Jannson, Edward F., 740 Rush St.
Jarvis, John P. D., 7748 Ridgeland Ave.
Jenkins, Austin D., 400 N. Michigan Ave.
Jensen, Clarence A., 3925 N. Keeler Ave.
Jensen, Elmer C., 39 S. LaSalle St.
Jensen, Jens J., 1105 Lawrence Ave.
Johnck, Frederick, 104 S. Michigan Ave.
Johnson, Benj. L., 878 N. Sacramento Ave.
Johnson, Glenn Quincy, 2432 N. Major St.
Johnson, Harry Nels, 4850 N. Winchester
Johnson, Harold T., 2833 Lunt Ave.
Johnson, Reuben Harold, 7001 N. Clark St.
Johnston, William K., 6805 Yale Ave.
Jones, Geo. Maceo, 435 Oakenwald Ave.
Juntenan, 5419 Harper Ave.
Kalischer, Mark Daniel, 520 N. Michigan Ave.
Kallenbach, Henry, Jr., 1311 Newport Ave.
Kane, Robert L., 6521 Stewart Ave.
Kaplan, Leonard H., 8156 S. May St.
Karlin, Irving K., 8124 Crandon Ave.
Kartowicz, Frank Geo., 2242 Lyndale Ave.
Kausal, John F., 333 N. Michigan Ave.
Kay, Fred D., 8904 S. Morgan St.
Keck, George F., 612 N. Michigan Ave.
Keck, William F., 612 N. Michigan Ave.
Kelly, Payton Shaw, 842 Ainslie Ave.
King, John Lord, 70 E. Cedar St.
Kingsley, Donald, Rm. 2306—10 W. 69th St.
Klafter, David Saul, 100 N. LaSalle St.
Klafter, Joseph H., 100 N. LaSalle St.
Klamt, Edward A., 713 Wrightwood Ave.
Klein, William J., 5124 Greenwood Ave.
Kleinhous, George F., 552 Oakdale Ave.
Klekamp, Bernard R., 140 S. Dearborn St.
Klewer, Geo. W., 7631 East Lake Terrace.
Klibanow, David William, 502 E. 88th Place.
Klingensmith, Willey P., 11164 Homewood
Knickerbocker, Earl B., 327 N. Parkside Ave.
Knudson, Harold Nels, 5263 Crystal St.
Kobylanski, Joseph L., 5450 Leland Ave.
Koefner, Uda H., 9908 S. Western Ave.
Koerner, Warren A., 7337 S. Vernon Ave.
Koenigsberg, Nathan, 155 N. Clark St.
Kohfeldt, Walter G., 6846 Clyde Ave.
Komar, Morris L., 64 W. Randolph St.
Kopp, Joseph Diermar, 660 Barry Ave.
Koppel, A. Emanuel, 6705 N. Campbell Ave.
Koppel, Maurice G., 5119 N. Kenmore Ave.
Kopple, William H., 9533 Longwood Ave.
Koster, John L., 4639 Drexel Ave.
Krauss, Bernard, 5009 Sheridan Rd.
Krajewski, Casimir, 844 Rush St.
Kroman, M. Louis, 180 N. Michigan Ave.
Kuehne, Carl Oskar, 1572 N. Halsted St.
Kurzon, Bernard R., 134 N. LaSalle St.
Lamb, Theodore Warren, 20 E. Goethe St.
Lang, Frederick W., 7653 N. Ashland Ave.
Lapasso, Frank J., 6806 Stony Island Ave.
Larsen, Emil L., 4814 N. Damen Ave.
Larson, Godfrey E., 77 W. Washington St.
Larson, Harry Ivan, 6234 S. Karlov Ave.
Lautz, William H., 806 W. 79th St.
Lavell, Haacon S. S., 80 E. Jackson Blvd.
Lawrence, Albin J., 5925 Belle Plaine Ave.
Lawrence, Clark J., 224 S. Michigan Ave.
Layer, Robert, 221 N. LaSalle St.
Leavell, John C., 6606 University Ave.
Ledebuhr, Arthur E., 6447 Rhodes Ave.
Lenske, Charles H., 3912 N. Central Park
Levinson, Mark B., 2019 N. Kedzie Ave.
Levy, Albert D., 2005 N. Kedzie Ave.
Levy, Alexander L., 179 W. Washington St.
Lewin, Edw. P., 134 N. LaSalle St.
Lewis, Jacob, 30 N. Dearborn St.
Lichtmann, Samuel A., 1526 E. 53rd St.
Liebert, Hans T., 5112 N. Kenmore Ave.
Lindquist, Joseph B., 7015 Paxton Ave.
Lindstrom, Robert Seth, 1828 E. 72nd St.
Linkonis, Anthony W., 5454 S. Albany Ave.
Liska, Emil, 9601 Prospect Ave.
Llewellyn, Ralph C., 38 S. Dearborn St.
Loebl, Irving Jerrold, 333 N. Michigan Ave.
Loewenberg, Israel S., 111 W. Monroe St.
Long, Frank B., 333 N. Michigan Ave.
Lorenc, Joseph A., 228 N. LaSalle St.
Lovdall, Geo. F., 1807 Cuyler Ave.
Love, Robert J., 104 S. Michigan Ave.
Lovell, McDonald, 224 E. Ontario St.
Ludgin, Joseph G., 37 S. Wabash Ave.
Lundstrom, John E., 5706 N. Maplewood Ave.
Lurvey, Louis, 3425 W. Monroe St.
Luther, Otto L., 11216 S. Campbell Ave.
Lutz, Harold R., 6108 N. Glenwood Ave.
MacBride, E. Everett, 43 E. Ohio St.
Madden, Thomas A., 5541 Everett Ave.
Maher, Philip Brooks, 157 E. Erie St.
Malter, Jerome Mitchell, 2515 Winnemac Ave.
Manning, Edgar James, 847 Bradley Pl.
Manning, Harry F., 5220 Kenmore Ave.
Marberg, G. Albert, 842 Grace St.
Marks, Edward, 2734 Argyle Ave.

Marley, James J., 9720 Emerald Ave.
Martin, Albert R., Jr., 400 Briar Pl.
Martin, Edgar D., 540 N. Michigan Ave.
Marx, Elmer W., 2734 Mildred Ave.
Marx, Samuel A., 333 N. Michigan Ave.
Matteson, Victor Andree, 20 N. Wacker Dr.
Mayfield, Leonard H., 195 E. Chestnut St.
Mayo, Peter B., 53 W. Jackson Blvd.
McArthur, William James, 2551 Rosemont
McCarthy, Charles A., 6717 East End Ave.
McCarthy, Joe W., 43 E. Ohio St.
McCarthy, Michael John, 6848 Calumet Ave.
McCauley, Willis Joseph, 919 N. Michigan
McClellan, Edward G., 7439 Cottage Grove.
McCurry, Paul D., 9350 S. Hamilton Ave.
McDougall, Walter A., 140 S. Dearborn St.
McDowell, Harold L., 7723 S. Wolcott Ave.
McEldowney, Harold B., 1814 Lincoln Park West.
McGavick, Joseph P., 4507 Lake Park Ave.
McGrath, Paul J., 3260 Lake Shore Dr.
McLaren, Robert J., 5934 Midway Park.
McMurry, Oscar L., 10405 S. Seeley Ave.
McNamara, Louis G., 155 N. Clark, Rm. 1217.
McPherson, John V., 6730 Jeffrey Ave.
McWhorter, Jerome C., 2122 E. 75th St.
Meder, Everett S., 2515 N. Nordica Ave.
Mell, Alfred Lorenz, 30 N. Michigan Ave.
Melstrom, John H., 3163 Cambridge Ave.
Menees, Thomas O., 7222 Harvard Ave.
Merrill, Edward A., 540 N. Michigan Ave.
Merrill, John O., 333 N. Michigan Ave.
Metz, Carl A., 520 N. Michigan Ave.
Meyer, Frederic H., 820 Tower Court.
Meyer, Martin, 139 N. Central Ave.
Michaelsen, Christian S., 3256 Franklin Blvd.
Michel, David D., 3800 Lake Shore Dr.
Miller, John W., 4438 Greenwood Ave.
Miller, Joseph A., 6440 Bosworth Ave.
Milman, Ralph E. 104 S. Michigan Ave.
Mills, Albert B., 228 N. LaSalle St.
Minchin, Sidney C., 4339 Clarendon Ave.
Mink, Emil Joseph, 1537 W. Grace St.
Minkus, Robert L., 3408 Franklin Blvd.
Mittlebresher, Edwin H., 104 S. Michigan Ave.
Mohr, Frederick J., 4345 Kenmore Ave.
Monberg, Lawrence, 232 E. Erie St.
Morphett, Archibald S., 104 S. Michigan Ave.
Morrison, James R. M., 5810 Harper Ave.
Morse, Frederic J., 3950 Pine Grove Ave.
Morse, Harry Leon, 6143 N. Mozart St.
Mrazek, Joseph M., 1239 S. Tripp Ave.
Murison, George W. Jr., 5832 Stony Island
Murison, Richard V., 840 N. Michigan Ave.
Murphy, Charles Frances, 80 E. Jackson Blvd.
Nadherny, Joseph J., 1518 W. Roosevelt Rd.
Naess, Sigurd Edor., 1661-80 E. Jackson Blvd.
Narovec, George, 124 N. Parkside Ave.
Nathan, Myer O., 337 W. Madison St.
Navrated, John A., 2532 S. Millard Ave.
Nedved, Geo. M., 8951 S. Ada St.
Neebe, John, 3713 N. Kedvale Ave.
Neil, Sidney S. 4311 N. Ashland Ave.
Nelson, Joseph A., 2432 Irving Park Blvd.
Nelson, Melvin A., 4633 N. Hermitage Ave.
Nelson, Theodore W., 118 E. 26th St.
Nettenstrom, Joel V., 2130 Bingham St.
Newhouse, Henry L., II, 8 S. Michigan Ave.
Newman, Edgar M., 5745 Newmark Ave.
Nielsen, Carl, 7249 N. Bell Ave.
Nielsen, Elker Rosehill, 3059 Augusta St.
Niemz, Arthur R., 3918 Broadway.
Nimmons, George C., 333 N. Michigan Ave.
Nitsche, Edward A., 2843 N. Kilbourne Ave.
Nock, Samuel A., 5656 School St.
Noonan, Thomas C., 1223 Columbia Ave.
Nordeen, Harvey A., 5633 N. Bernard St.
Norling, Engelbert C., 4850 W. Cortez Ave.
Norstrand, Fabian, 844 Rush St.
Nortman, Harry Rodee, 8020 Eberhart Ave.
Odgers, Philip G., 5835 Nina Ave.
Ohlinger, Earl William, 6735 Bishop St.
Olesky, Walter J., 5543 Drummond Pl.
Oliver, Ralph H., 8 S. Dearborn St.
Olsen, Leif E., 228 N. LaSalle S.
Olsen, Paul Frederick, 404 N. Wells St.
Olson, A. Einar, 115 W. 74th St.
Olson, Benj. Franklin, 19 S. LaSalle St.
Oman, Samuel S., 162 E. Ohio St.
Ostergren, Robert C., 4300 N. Clark St.
Owings, Nathaniel A., 104 S. Michigan Ave.
Pareira, David A., 7716 Luella Ave.
Parsons, Wm. Edward, 80 E. Jackson Blvd.
Patelski, Erich J., 1 West Walton Pl.
Pearce, Harry W., 1853 West Polk St.
Pearson, Gustav E., 4432 Wrightwood Ave.
Pearson, Warner M., 720 N. Michigan Ave.
Peel, Carlyle, 5208 Kimbark Ave.
Pereira, Wm. L., 221 N. LaSalle St.
Perkins, Lawrence B., 225 N. Michigan Ave.
Perkins, Ruth H., Miss, 1528 Fargo Ave.
Perry, Ronald F., 32 W. Randolph St.
Pesetsky, David, 7026 Clyde Ave.
Peterson, Ivan Robert, 1245 Fletcher St.
Peterson, Raymond A., 6506 N. Claremont
Pfendt, Joseph, 2258 N. Avers Ave.
Pingrey, Roy E. 120 S. LaSalle St.
Piontek, Clement L. 1608 Milwaukee Ave.
Pirola, Louis, 4717 Beacon St.
Pistorius, Bernhard H., 700 Melrose Ave.
Phlak, Ray L., 7423 Rogers Ave.
Poe, Morris, 4900 N. Harding Ave.
Polito, Frank F., 75 E. Wacker Dr.
Pond, Irving K., 14 W. Elm St.
Powers, Richard, 6 N. Michigan Ave.
Prather, Fred V., 1791 Howard St.
Pridmore, John E. O., 109 N. Dearborn St.
Prindeville, Chas. H., 58 E. Washington St.
Probst, Edward, 80 E. Jackson Blvd.
Probst, Marvin G., 80 E. Jackson Blvd.
Pruyn, John A., 455 Barry Ave.
Puckey, Francis W., 400 N. Michigan Ave.
Quinn, Everett Francis, 360 N. Michigan Ave.
Quinn, James Edwin, 646 N. Michigan Ave.
Quitsow, Anthony H., 228 N. LaSalle St.
Rabig, Charles E., Klari 6408 Kenwood Ave.
Raftery, John Howard, 540 N. Michigan Ave.
Rapp, George Leslie, 230 N. Michigan Ave.
Rapp, Mason G., 230 N. Michigan Ave.
Rappaport, Benjamin J.. 1545 E. 60th St.
Rawson, Chas. P., 5343 Kenmore Ave.
Rebori, Andrew N., 29th Fl. Tribune Tower.
Recher, George D., 1224 Winona Ave.
Reddersen, Edward H., 703 Bunea Ave.
Reed, Earl Howell, Jr., 431 N. Michigan Ave.
Reichert, Wm., 30 N. LaSalle St.
Reily, S. L., 300 W. Adams St.
Reiner Eugene B., 2553 Logan Blvd.
Reinholdt, Martin T. 122 S. Michigan Ave.
Repp, George W., 1743 W. 100th Pl.
Reynolds, Harold F., 1339 Glenlake Ave.
Rezny, Adrian, 2202 S. Crawford Ave.
Rezny, James B., 2202 S. Crawford Ave.

Rice, Raymond, 7143 S. Greenwood Ave.
Rich, William M., 11 S. LaSalle St.
Richards, Ernest R., 4039 N. Lincoln Ave.
Riddle, Herbert H., 5626 Woodlawn Ave.
Rigoli, Paul A., 1922 Morse Ave.
Rissman, Maurice B., 65 E. South Water St.
Roberts, Eben E., 82 W. Washington St.
Roberts, Elmer C., 82 W. Washington St.
Robertson, David, 5338 Blackstone Ave.
Robinson, Harry F., 7152 Ridgeland Ave.
Rodde, Herbert Leslie, 4129 N. Kenneth Ave.
Rohm, Jean Baptist, 1440 Foster Ave.
Rolm, E. Theodore, 5526 N. Campbell Ave.
Root, John Wellborn, 333 N. Michigan Ave.
Rosen, Alfred Joseph, 6813 Lakewood Ave.
Rosen, Nathan R., 6111 S. Albany Ave.
Roth, Edgar, 7550 Saginaw Ave.
Rummell, Chas. G., 6608 Bosworth Ave.
Rupert, Edward P., 57 W. Monroe St.
Rupinski, Edward W., 2130 N. Kedzie Ave.
Russell, Lewis E., 105 W. Adams St.
Rusy, Anthony F., 1339 S. Avers Ave.
Ruttenberg, Albert, 1319 Estes Ave.
Ryan, William J., 43 E. Ohio St.
Sachtleben, Albert C., 2253 W. 111th St.
Safrank, Irwin Robert, 3542 N. New England Ave.
Sagett, Samuel, 4434 Lawrence Ave.
Sailor, Homer G., 733 W. 64th St.
Sandegren, Andrew H, 5735 Sheridan Rd.
Sandel, Monroe R., 228 N. LaSalle St.
Sanders, Lewis Miles, 35 S. Dearborn St.
Sandquist, Oliver, 5055 N. Keeler Ave.
Sandstedt, Julius S., 5755 S. Wells St.
Sandstrom, Reuben S., 1342 Winnemac Ave.
Saxe, Albert Moore, 43 E. Ohio St.
Schaffner, Arnold S., 1356 S. Komensky Ave.
Schaffner, Daniel J., 72 W. Washington St.
Scheller, Jesse E., 7907 Evans Ave.
Schlossman, Norman J., 333 N. Michigan Ave.
Schmidt, Richard Ernest, 104 S. Michigan.
Schnakenberg, Henry, 3510 N. Oakley Ave.
Scoyoc, Lee Van, 6056 Ingleside Ave.
Schreiber, Arthur Henry, 1401 Sherwin Ave.
Schroeder, Charles W., 6704 S. Union Ave.
Schulze, Carl Elliott, 1511 W. Jackson Blvd.
Schwab, Raymond John, 6224 N. Washtenaw
Schwarz, Julius T., Box 1665.
Seneseney, Geo. T., 600 S. Michigan Ave.
Serpico, Frank J., 5241 N. Sawyer Ave.
Setterberg, Wm. N., 8132 Lafayette Ave.
Sevic, William, 1900 Blue Island Ave.
Shantz, Cedric A., 38 S. Dearborn St.
Shattuck, Walter F., 221 N. LaSalle St.
Shaw, Alfred, 1661—80 E. Jackson Blvd.
Sherrick, John C., 8040 Evans Ave.
Sholes, E. Ray, 330 S. Wells St.
Sieja, Edward M., 4132 Barry Ave.
Sillani, Muzio, 2511 N. Clark St.
Sir, Arthur J., 4859 N. Rockwell St.
Skidmore, Louis, 104 S. Michigan Ave.
Sloan, Fred E., 737 N. Michigan Ave.
Slovinec, John, 1614 S. Ruble St.
Slupkowski, Joseph A., 3024 Haussen Court.
Small, John S., 43 E. Ohio St.
Smith, David Henry, 5648 S. Morgan St.
Smith, George H., 30 N. LaSalle St.
Smith, George S., 8332 Luella Ave.
Smith, Kenneth Alan, 846 Ainslee St.
Smith, Wm. Jones, 430 N. Michigan Ave.
Smith, Z. Erol, 7101 Stony Island Ave.
Sobel, Herbert, 540 N. Michigan Ave.
Sobel, Walter H., 457 Melrose St.
Solomon, Irving, 185 N. Wabash Ave.
Solomon, Louis R., 185 N. Wabash Ave.
Soltker, David Lee, 2307 Cortez St.
Sommer, Charles Bernard, 3730 Magnolia Ave.
Sosna, Ned S., 6747 Chappel Ave.
Spencer, Charles B., 309 W. Jackson Blvd.
Spitz, Alexander H., 220 S. State St.
Spitzer, Herbert I., 205 W. Wacker Dr.
Sponholz, William C., 2249 N. Campbell Ave.
Stahl, Harold A., 3756 N. Leavitt St.
Stanhope, Leon, 410 N. Michigan Ave.
Stanton, Francis Rew, 2457 Orchard St.
Stanton, Frederick C. H., 307 N. Michigan.
Starbuck, Fred L., 224 S. Michigan Ave.
Steif, Benjamin Leo, 919 N. Michigan Ave.
Steen, Benjamin H., 1802 S. Springfield Ave.
Steen, Jack Stewart, 600 S. Michigan Ave.
Steinbach, John G., 5547 Kenmore Ave.
Steinberg, Edward, 205 W. Wacker Dr.
Steinborn, Edward, 330 S. Wells St.
Stenbeck, Eric G., 332 S. LaSalle St.
Stephens, Burett, H., R. 816, 192 N. Clark St.
Stern, Isaac S., 306 S. Wabash Ave.
Stevens, Charles W., 705 Garrick Bldg.
Stewart, Cedric John, 409 N. Ridgeway Ave.
Stockton, Walter J. 307 N. Michigan Ave.
Stoetzel, Ralph E., 612 N. Michigan Ave.
Stoody, Clyde A., 307 N. Michigan Ave.
Strandberg, Carl Bert, 7951 S. Winchester
Strauch, Moriz F., 2216 Giddings St.
Streeter, Andrew G., 5842 S. Wells St.
Strelka, Leo, 205 S. Second Ave.
Stromquist, Victor H., 59 E. Van Buren St.
Sturges, Howard Putnam, 4314 N. California
Sturnfield, Chas. H., 2651 W. 15th St.
Sugarman, Irwin Alvin, 140 N. Parkside Ave.
Sumarkoff, Leonard P., 7433 N. Seeley Ave.
Sullivan, Andrew W., 5930 S. Talman Ave.
Suter, Walter Lindsay, 189 W. Madison St.
Svoboda, Albert F., 155 N. Clark St.
Swanson, Robert C., 5137 N. Troy St.
Swarz, August, 7 S. Dearborn St.
Syversion, George M., 1912 N. Kedvale Ave.
Tallmadge, Thos. E., 19 E. Pearson St.
Taylor, Carl Oder, 140 S. Dearborn St.
Tedeman, Henry P. T., 4647 N. Keating Ave.
Teich, Frederick J., 649 N. Arlington Pl.
Teisen, Axel V., 5332 N. Spaulding Ave.
Temps, Martin, 510 N. Dearborn St.
Terp, George Wm., Jr., 10529 Avenue H
Tesch, George D., 228 N. LaSalle St.
Tessing, Arvid F., 2020 Bradley Pl.
Teutsch, Carl M., 1105 Lawrence Ave.
Thielbar, Frederick J., 520 N. Michigan Ave.
Thomas, Theodore G., 72 W. Adams St.
Tilton, John Neal, Jr., 11 S. LaSalle St.
Tocha, Anton A., 1459 N. Bosworth Ave.
Tourtelot, Edward M. J., 104 S. Michigan
Towsend, John S., Jr., 10527 Hale St.
Traveletti, Rene F., 189 W. Madison St.
Tucker, George E. Loane, 5311 E. Kenmore
Turbyfill, David W. T., 4345 Kenmore Ave.
Turk, Harry E., 43 E. Ohio St.
Turner, John W., 2103 Turner Ave.
Tyson, Herbert, 719 S. Grove Ave.
Uffendell, W. Gibbons, 232 E. Erie St.
Urbain, Jules, Jr., 228 N. LaSalle St.
Urbain, Leon F., 605 W. Washington St.
Urbanek, Chas. A., 1423 S. Avers Ave.

Vade, Louis Henri, 7614 Dorchester Ave.
Varraveto, Patrick Michael, 5471 W. Division
Vesely, William J., 1008 N. Rush St.
Viehe-Naess, Ivar, 5809 Ridge Ave.
Viker, Guttorm A., 1415 Greenleaf Ave.
Viscariello, Vincent, 2931 N. Monticello Ave.
Vitzhum, Karl M., 1 N. LaSalle St.
Voita, Eugene, 837 N. Lorel Ave.
Von Holst, Herman V., 140 S. Dearborn St.
Walker, Frank Chase, 1065 Balmoral Ave.
Walker, Willard C., 307 N. Michigan Ave.
Wall, Richard J., 1530 Elmdale Ave.
Wallace, Maurice R., 5004 N. Springfield Ave.
Walter, George S., 510 N. Dearborn St.
Ward, Robert Allen, 240 E. Delaware Pl.
Wares, James A., 208 W. Washington Blvd.
Waterman, Harry Hale, 3915 Vincennes
Wayman, Leonard, 5107 Strong St.
Weary, Edwin F., 12 S. Michigan Ave.
Weber, Bertram, 820 N. Michigan Ave.
Webster, Maurice H., 225 N. Michigan Ave.
Weisfeld, Leo H., 155 N. Clark St.
Weissenborn, Leo Julius, 435 S. Michigan Ave.
Wenisch, Walter F., 1219 Farwell Ave.
West, Philip D., 520 N. Michigan Ave.
Whalley, Stanley E., 1819 Humboldt Blvd.
Wheeler, Chas. F., 4635 N. Kenton Ave.
Wheeler, Edward Todd, 225 N. Michigan Ave.
Whitney, William Parker, 1917 Lincoln Ave.
Wiener, Lewis E., 1628 Columbia Ave.
Wilkinson, Lawrence E., 520 N. Michigan Ave.
Will, Philip, Jr., 225 N. Michigan Ave.
Wilmanns, August C.. 208 N. Wells St.
Winiarski, Maryan F., 2938 N. Talman Ave.
Woerner, Adolph, Jr., 3166 Lincoln Ave.
Wolff, Richard G., 1755 Albion Ave.
Woltersdorf, Arthur F., 520 N. Michigan Ave.
Work, Robert G., 75 E. Wacker Drive.
Work, William D., 6200 Kenwood Ave.
Worthmann, Henry, 2649 Lawrence Ave.
Wright, Clark C., 333 N. Michigan Ave.
Wright, William Campbell, 140 S. Dearborn
Yeretsky, Norman M., 664 N. Michigan Ave.
Yerkes, Wallace F., 435 N. Michigan Ave.
Zakharoff, Alexis, 5845 Blackstone Ave.
Zaldokas, Mathew E., 1908 W. Division St.
Zimmerman, Ralph W., 43 E. Ohio St.
Zook, R. Harold, 140 S. Dearborn St.
Zucco, Angelo, 936 Lakeside Ave.
Zuckerman, Benj. S., 228 N. LaSalle St.
Zumkeller, Rev. Emil, 816 S. Clark St.

CHICAGO SUBURBS

BARRINGTON

Cerny, Otto Frank, 123 George St.
Maxon, Norman T., 125 E. Hillside Ave.
Luckman, Charles, Irving, R. 2, Brinke Rd.

BERWYN

Clark, Leslie Doane, 6910 Cermak Rd.
Novak, Vladimir James V., 2527 S. Ridgeland Ave.
Vittner, Clement, 6701 W. 19th St.

BLUE ISLAND

Skubic, Leroy Frederick, 2147 W. 121st Pl.
Van Gilder, Arthur, 2211 W. 19th Place.

CICERO

Filas, Thomas M., 5012 W. 25th St.
Moravéc , Bohumil T., 1953 S. 56th St.
Pechota, William A., 2108 S. 49th Ave.
Vedra, Charles, 6135 W. Cermak Rd.
Zelenka, Anthony J., 2331 S. Lombard Ave.

CONGRESS PARK

Busch, Earl, 4442 DuBois Blvd.

CRYSTAL LAKE

Scoville, David B., Country Club Addition.

DES PLAINES

Liedberg, Hugo J., 1300 River Dr.
Saylor, Raymond L., 1650 Mill St.
Winsauer, Louis M., Box 92, Hawthorne Lane.

DOWNERS GROVE

Feller, Anton M., 4927 Main St.
McLaughlin, Daniel F., 4703 Oakwood Ave.

ELMHURST

Brydges, E. Norman, 257 W. St. Charles Rd.
Downton, Herbert E., 118 Pine St.
Hartmann, Theodore A., 374 Addison Ave.
Hanebuth, Edgar C., 146 Willow Rd.
Helma, Kenneth A., 255 Elm Park Rd.
Meredith, Davis D., 251 South St.
Price, Minor Carr, 313 Addison St.

EVANSTON

Anderson, Edwin F., 2735 Hartzell Ave.
Beersman, Charles G., 207 Hamilton St.
Bjork, David T., 624 Asbury Ave.
Black, Gilmer V., 1250 Asbury Ave.
Blake, Edgar O., 604 Davis St.
Bristle, Joseph H., 2505 Park Pl.
Buckelt, Arthur C., 1010 Main St.
Cauley, Frank W., 1519 Hinman Ave.
Cook, Robert S., 2131 Orrington Ave.
Danielson, Philip A., 1625 Hinman Ave.
Dolke, W. Fred, 1841 Asbury Ave.
Edholm, W. L., 126 Callan Ave.
Finlayson, Frank L., 2525 Park Pl.
Gratiaa, Simon D., 2744 Asbury Ave.
Hughes, Everett Hobart, 1244 Elmwood Ave.
Irwin, Howard E., 1006 Main St.
Kegley, Frank T., 1515 Sherman Ave.
Kincaid, James L., 723 Dobson St.
Kraybill, Emmett E., 2638 Central Park Ave.
Lester, Bemis, 1406 Church St.
Maher, Harry E., 1564 Sherman Ave.
Markel, Charles H., 619 Church St.
Marshall, Benjamin H., 1515 Sherman Ave.
McGrew, Kenneth A., 1564 Sherman Ave.
Mueller, Carl O., 2806 Lincoln St.
Mueller, Herbert E., 2806 Lincoln St.
Perkins, Dwight H., 2319 Lincoln St.
Sanford, Trent E., 800 Michigan Ave.
Schmid, Karl Morton, 605 Hinman Ave.
Schmidt, Frederick B., 939 Maple Ave.
Sorgatz, Wm. D., 5551 Judson Ave
Sturm, Meyer J., 708 Church St.
Swensson, Walter F., 2200 Bennett Ave.
Thompson, Louis K., 2505 Prairie Ave.
Walton, Lewis B., 1515 Sherman Ave.
Whitman, Mrs. Bertha, 2656 Lincolnwood Dr.

EVERGREEN PARK

Stott, Roy W., 2957 W. 59th St.

FLOSSMOOR

Kocher, Jacques J., Park Dr.
Pipher, Wesley V., Sterling Ave.

FOREST PARK

Kastrup, Carl Joseph, 7346 Madison St.

GLENCOE

Binford, Wilbur H., 414 Washington St.
Hill, Edgar Arnold, 45 Lakewood Rd.
Schnur, James C., 1017 Forest Ave.
Tomlinson, Herbert W.. 414 Washington St.

GLEN ELLYN

Allen, Lawrence Eugene, 566 N. Newton St.
Mattson, Raymond A., 715 Hill Ave.
Walker, Frederick G., 717 Forest Ave.

GLENVIEW

Maynard, Henry S., Jr., 26 Park Drive.

HARVEY

Beck, H. Frederick, 15 E. 155th St.

HIGHLAND PARK

Abbott, James Fraser, 340 Glenwood Ave.
Bills, William B., 15 N. Sheridan Rd.
Corrough, D. D., 538 Central Ave.
Flinn, Raymond W., 344 Park Ave.
Hutchinson, George Alfred, Jr., 920 S. Sheridan Rd.
Loewenstein, Edward, 192 Vine Ave.
Mann, William D., 362 Park Ave.
Van Bergen, John S., 234 Cedar St.

HINSDALE

Barfield, Norman D., 112 N. Lincoln St.
Bebb, Huberg C., 622 W. North St.
Burt, Paul Gordon, 425 S. Elm St.
Field, Harford, 10 E. Hinsdale Ave.
Flaks, Francis A., 138 E. The Lane.
Sommers, Abner L., 339 N. Garfield Ave.

HOMEWOOD

Kline, Edward L., 1655 Burr Oak.

KENILWORTH

Park, John N., 144 Tudor Pl.

LA GRANGE

Caldwell, Will Carleton, 404 S. Catherine.
Fierbough, John F., 341 S. Stone St.
Fletcher, Robert C., 42 S. Waiola Ave.
Robinson, Argyle E. 425 S. La Grange Rd.
Schimek, Alfred F., 601 N. Spring Ave.
Trow, Robert C., 444 N. Spring Ave.
Valerio, Francis M., R. F. D. Box 71
Wilkins, Samuel Willis, 14 S. 5th Ave.

LAKE BLUFF

Behel, Vernon W., 30 Center Ave.

LAKE FOREST

Anderson, Stanley D., 262 Deerpath Rd.
Cerny, Jerome Robert, Clock Tower.
Ganster, William A., c/o Stanley Anderson.
Hill, Boyd, 137 Westminster Rd.
Jones, F. W., 233 Deerpath Ave.
Lind, Kenneth Nels, Clock Tower.
Ticknor, James H., 262 Deerpath Rd.

LOMBARD

Dillard, Frank G., 353 W. Willow St.
Ericksen, Arndt F., 141 Lombard Ave.

MAYWOOD

Bartlett, Harry C., 1467 S. 16th St.
Chiaro, John A., 12 N. Fifth Ave.
Koenig, Fred. 133 S. 7th Ave.
Van Gunten, Tillman G., 1501 S. 6th Ave.

NILES CENTER

Dockendorf, Joseph A., 7900 Lorel Ave.
Fisher, Joseph G., 5332 Mulford St.
Meldrum, John, 9128 La Crosse Ave.
Raymond, Emerson E., 8324 Keating Ave.

OAK PARK

Ansel, Anton, 1115 Wisconsin Ave.
Barnes, Allen L., 332 S. East Ave.
Benedict, Earl R., Jr., 320 S. Scoville Ave.
Bowes, Frederick W., 729 N. Kenilworth.
Clare, Wm. H., 633 S. Oak Park Ave.
Codd, Henry George, 809 S. Euclid Ave.
Connell, Wilson, Jr., 1134 Paulina Ave.
Dunning, Hugh B., 104 N. Oak Park Ave.
Fyfe, James L., 316 S. Euclid Ave.
Harley, William, Jr., 310 S. Cuyler Ave.
Hookham, Fred H., 825 S. Scoville Ave.
Hotchkiss, Roy James, 127 N. Oak Park.
Hulla, John, 225 Clinton Ave.
Johnson, Stanley T., 1108 Wisconsin Ave.
Yost, Lloyd Morgan, 1150 Wilmette Ave.
Kirby, Weymouth W., 513 S. Cuyler Ave.
Kristen, Charles Andre, 701 N. Lombard.
Kraemer, William F., 624 Clarence Ave.
La Belle, Edward C., 410 Wisconsin Ave.
Maiwurm, Arthur B., 516 S. Euclid Ave.
Maiwurm, Rudolph G., 1047 Pleasant Ave.
Martling, W. Lockwood, 401 Wisconsin Ave.
Peterson, J. Edwin, 204 S. Humphrey Ave.
Poulsen, George F., 643 N. Elmwood Ave.
Pray, Frank M., 634 Harrison St.
Redden, John Stokes, 803 S. Euclid Ave.
Rippel, Fred O., 948 Mapleton Ave.
Roncoli, Louis P., 30 Le Moyne Parkway
Roos, Bernard L., 734 S. Elmwood Ave.
Ullrich, Clarence A., 904 N. Hayes Ave.
Washburne, Fred R., 1012 Chicago Ave.
Young, Gilman B., 809 S. Euclid Ave.

ORLAND PARK

Pomeroy, Jim T., P. O. Box 143.

PALATINE

Klopp, Charles A., 203 S. Rockwell St.

PARIS

Carter, Harry B., 227 E. Court Ave.

PARK RIDGE

Martin, George, 927 Prairie Ave.
McCaughey, William F., Pickwick Bldg.
Pope, Charles F., Jr., 621 Courtland Ave.
Rowe, Charles Barr, 1103 Crescent Ave.
Schiewe, Edward A., 311 Cuttress Pl.

RAVINIA

Manasee, Dewitt J., 1637 S. Green Bay Rd.

RIVER FOREST

Littrell, Donald B., 231 Park Row.
Probst, Edward Eugene, 1037 Forest Ave.

RIVERSIDE

Barnum, George Loyd, 8001 Edgewater Rd.
Halstead, Edward Gray, 143 Bloomingbank Rd.
Keeber, C. Hamilton, 148 Maplewood Rd.
Krieg, Arthur W., 180 Maplewood Ave.
Krieg, William G., 180 Maplewood Ave.
Novy, Joseph J., 160 N. Desplaines Rd.
Skow, Arnold Paul, 40 East Ave.
Smithson, Albert T., 298 Lionel Rd.
Wills, Arthur Dougless, 224 Herrick Rd.

WAUKEGAN

Ekstrand, Henry E., 730 Lenox Ave.
Mullin, Wilbur A., 5 N. Genesee Ave.
Schad, Eugene John, 116 Washington Park

WEST CHICAGO

Fairbanks, Edward Hale, P. O. Box 243.

WESTERN SPRINGS

Harmon, Harry Jones, 4098 Western Ave.
Heimbrodt, Carl Edward, 4110 Lawn Ave.

WHEATON

Garbe, Raymond Wm., Box 224.
Roller, Herman D., 114 W. Indiana St.
Salisbury, Robert N., Liberty Drive.
Schmidt, Hugo, 421 N. Wheaton Ave.

WILMETTE

Arnold, Robert S., 1150 Wilmette Ave.
Eppig, Arthur George, 2800 Blackhawk Rd.
Ellert, Frank J., 623 Gregory Ave.
Harper, Sterling H., 1011 Greenleaf Ave.
Gathercoal, James J., 124 Lockerbie Lane
Mayer, Carl H., 1524 Forest Ave.
Naper, Herbert J., 1127 Lake Ave.
Peterson, Stanley M., 213 17th St.

Rae, Robert, Jr., 431 Greenleaf Ave.
Reinhardt, G. A., 1431 Lake Ave.

WINNETKA

Gibson, Charles H., 1447 Tower Rd.
Lockner, Herman Herbert, 339 Linden Ave.
Otis Samuel S., 516 Walnut Ave.
Windes, Frank A., 598 Birch St.

CITIES OUTSIDE OF CHICAGO AND CHICAGO SUBURBS

ALTON (Madison Co.)

Deeter, Russell O., 3412 Oakwood St.
Maupin, James M., 518 Commercial Bldg.
Pfeiffenberger, Geo. D., 102 W. Third St.
Sheley, Robert N., 1821 Worden Ave.
Wuellner, Walter V., 1821 Liberty St.

AURORA (Kane Co.)

Gray, Frank B., 73 S. LaSalle St.
Malmer, Eugene, 233 W. Park Ave.
Spieler, Herbert E., 211 Calumet Ave.
van der Meer, Wybe J., 70 S. May St.

BATAVIA

Elwood, Franklin G., 173 Illinois Ave.

BELLEVILLE

Chorlton, Wesley W., 501 West C St.
Imbs, Thomas F., 529 W. Main St.
Rubach, Otto W., Belleville Bank and Trust Bldg.
Weisenstein, Lyman J., 234 Lebanon Ave.

BLOOMINGTON

Agle, Charles K., 1121 E. Monroe St.
Gemeny, Blaine B., 703 E. Front St.
Hilfinger, Dean F., 602 Corn Belt Bank Bldg.
Lundeen, Edgar E., Corn Belt Bank Bldg.
Moratz, Arthur M., People's Bank Bldg.
Schaeffer, Archie N., People's Bank Bldg.
Simmons, Aaron T., Williams Bldg.
Wilson, Harrold B., 1609 E. Taylor St.

CARBONDALE

Thompson, Thomas S., 1000 Thompson St.

CHAMPAIGN

Berger, Frederick E., 518 Lincoln Bldg.
Davis, Harold B., 406 Springfield Ave.
Day, Warren W., 612 E. Daniel St.
Kelley, Ralph L., 618 Lincoln Bldg.
Kratz, Elvin V., 315 S. State St.
Lescher, Frank M., 1005 S. First St.
Ramey, George Edwin, Robeson Bldg.
Varney, Ralph W., 1009 W. Green St.
Weidemeyer, William M., 607 N. New St.

CHARLESTON

Spiser, Ralph F., 1441 7th St.

COLLINSVILLE (Madison Co.)

Eberhardt, Henry W., 207 E. Main St.

DANVILLE (Vermillion Co.)

Lewis, Charles M., 204 Franklin St.
Ludwig, George W., Temple Bldg.
Skadden, Harvey F., 210 N. Walnut St.
Strader, George M., 138½ N. Vermillion St.

DECATUR (Macon Co.)

Aschauer, Chas. J., Citizens Bank Bldg.
Bramhall, Arthur E., 254 E. Main St.
Brooks, Barkley S., 254 E. Main St.
Clausen, Swen A., Standard Life Bldg.
Dague, England D., 254 E. Main St.
Harris, Charles, Standard Life Bldg.
Miers, Clayton T., R. R. 7.
Waggoner,, Arthur M., 204 Citizens Bank Bldg.

EAST ST. LOUIS (St. Clair Co.)

Goedde, Bernice R., 2040 Illinois Ave.
Jost, Charles F., 626 N. 33rd St.
Kennedy, John W., 306 First Nat. Bank Bldg.
Saunders, Paul J., 620 N. 10th St.

EDWARDSVILLE (Madison Co.)

Kane, Michael B., Bohm Bldg.

ELGIN (Kane Co.)

Abell, Ralph E., 223 N. Worth Ave.
Morris, George E., Sherwin Bldg.
Schwarzwalder, Clarence F., 920 Prospect
Thompson, LeRoy W., 735 St. John St.

FOX LAKE

Smith, Paul

GALESBURG (Knox Co.)

Aldrich, Harry G., 311 E. Main St.
Payne, Edgar A., 269 Seminary Ave.

GIRARD

Garretson, W., Gordon

GRANITE CITY

Pauly, Edward C., 19th and Cleveland Blvd.

GRAYS LAKE

Mahaffey, David, Box No. 7.

HIGHLAND

Knoebel, Wilbert G.
Pabst, Selmar T.

JACKSONVILLE (Morgan Co.)

Buckingham, Clarence W., 679 S. West St.
Johnson, Raymond G., 360 W. College Ave.

JOLIET

Coyle, John Edward, 215 Hirkimer St.
Kruegal, Arnold J., 373 Webster Ave.
Lindblad, Raymond John, 517 Leach Ave.
Tomlinson, Henry Webster, 611 Morris Bldg.
Vibelius, Fred N., 650 2nd Ave.
Wallace, Chas. L., 409 Campbell St.

KANKAKEE

Shank, John Adam, 219 E. Court St.

KEWANEE

Scribbins, John A., 611 S. Tremont St.

LINCOLN (Logan Co.)

Deal, Joe Mindert, I. O. O. F. Bldg.

LOVINGTON

Holmstrand, Frank E.

MARSAILLES

Vogel, Donald Frank, R. F. D. 1.

METROPOLIS

Daly, Samuel Lester, 708 Market St.

MOLINE (Rock Island Co.)

Beckstrom, M. R., 1518 5th Ave.
Eckerman, Oscar A., 640 19th St.
Keller, Norman Edward, 1215 15th St.
Lundeen, Curt, 321 17th St.
Oldefest, Edward G., 1837 15th Pl.
Parkhurst, Howard M., 1832 17th St.
Schulzke, Wm. H., 5th Ave. Bldg.

MONTICELLO

Gregg, John Wyatt.
Reed, William V., 1012 N. State St.

MOUNT VERNON

McCoy, William R., 1st Nat. Bank Bldg.
Wilson, Dana Clarence, 1sf Nat. Bank Bldg.

MURPHYSBORO

Gill, Rudolph Z., 227 N. 14th St.

OTTAWA (La Salle Co.)

Cook, Norman W., 700 Webster St.
Gerding, Louis H., 708 La Salle St.

PEKIN

Smith, Frank Rupert, 1109 S. 5th St.
Wearda, George P., cor. Capitol & St. Marip St.

PEORIA

Belsterling, Richard G., 104 Belmont Pl. (Peoria Heights).
Caldwell, G. Thompson, 113 Sherman Ave.
Dox, Hamilton B., 502 Lehmann Bldg.
Emerson, Frank N., 1600 Alliance Life Bldg.
Foley, Cletis Roy, 1805 Knoxville Ave.
Gregg, Richard S., 1600 Alliance Life Bldg.
Harrison, Elbert, 1014 Alliance Life Bldg.
Hercules, Jacob W., 212 N. Monroe St.
Hewitt, Carter E., 1600 Alliance Life Bldg.
Hewitt, Herbert, 1600 Alliance Life Bldg.
Hotchkiss, Robert J., 330 Central Life Bldg.
Jameson, Walter Guy, 1012 Alliance Life Bldg.
Kelly, Rudolph L., 1600 Alliance Life Bldg.
Koch, Henry A., 726 Thrush Ave.
Lankton, J. Fletcher, 1805 Knoxville Ave.
Lynge, Morgan J., 409 Wisconsin Ave.
Powers, Gordon E., 209 N. Institute Pl.
Robinson, L. Eugene, 842 Jefferson Bldg.
Sedgwick, Wm. C., 404 E. Melbourne Ave.
Whitmeyer, Mark H., 207 S. Glenwood Ave.
Ziegle, John Nicholas, 1805 Knoxville Ave.

PERU

Wachter, Henry F. A., 504 Grant St.

PLAINFIELD

Cowell, Herbert.

PONTIAC

La Rowe, James Hull, 513 E. Prairie St.

QUINCY

Behrensmeyer, Charles F. A., 425 S. 14th St.
Behrensmeyer, George F., 333 E. Ave.
Geise, Martin J., 109 N. 8th St.
Hafner, Wilbert E., 2110 Grove Ave.
Jokiel, Joseph, 226 N. 14th St.
Wood, Ernest M., 126 N. 8th St.

RANTOUL

Baker, Cecil F., 424 Champaign Ave.

ROCKFORD

Barloga, Jesse A., 222 Rockford News Tower.
Bendus, William Q., 327 N. Church St.
Bradley, Charles W., 414 Brown Bldg.
Bradley, Harold S., 414 Brown Bldg.
Carpenter, Frank A., 749 John St.
Eklund, Herman, 121 7th St.
Hubbard, Willis W., 725 Gas-Electric-Bldg.
Johnson, Gilbert A., 406 Swedish American Bank Bldg.
Orput, Raymond, 3rd Nat. Bank Bldg.
Titus, Armour H., 2323 Melrose St.
Wolfley, Chester E. H., 1115 Talcott Bldg.

ROCK ISLAND

Cervin, Olaf Z., 310 Safety Bldg.
Chapler, Elijah C., 1635 12th St.
Horn, Benjamin A., Rock Island Bank Bldg.
Lerch, Edward, 2211 26th St.
Sandberg, Rudolph C., 501 Rock Island Bank Bldg.
Stuhr, Wm., 310 Safety Bldg.
Swanson, Carl Ernest, 301 Cleveland Bldg.

RUSHVILLE

Thompson, James Arthur, 128 E. Clinton St.

SHELDON

Clark, Charles R.

SPRINGFIELD

Booton, Joseph F., 1017 West Edwards St.
Bullard, Clark W., 317 Unity Bldg.
Bullard, Robert A., 317 Unity Bldg.
Bretscher, Carl Eduard, 1139 Monroe St.
Decker, Daniel W., 1828 S. Spring St.
Dunlap, Russell R., 1018 S. 7th St.
Golabowski, Joseph F., 1813 Noble St.
Hadley, Bryant E., 1005 S. 4th St.
Hanes, Murray S., 205½ S. 6th St.
Hanes, Samuel J., 205½ S. 6th St.
Harris, Ralph C., St. Nicholas Hotel.
Helmle, Henry R., First Nat. Bank Bldg.
Henderson, Russell D., 2124 S. 4th St.
Kelly, Joseph, 125 W. Myrtle St.
Knox, Robert, Div. of Arch. and Eng., Capitol Bldg.
Lindblad, Alfred G., Y. M. C. A.
Macardell, Cornelius W., Div. of Architecture, State Capital.
Meyer, Carl T., 712 Meyers Bldg.
Mills, Kenneth, W., 805 W. Laurel St.
Most, Frederick W., 101 S. Walnut St.
Noble, James B., Div. of Arch. and Eng., State Armory.
Pleins, Leo H., 527 W. Monroe St.
Reiger, Harry J., 403 Security Bldg.
Worthington, Earl C., 822 W. Edwards St.
Zimmer, John E., Div. of Arch. & Eng., Capitol Bldg.

STREATOR

Allen, Alonzo W., 601 W. Bridge St.

URBANA

Danely, Paul, Lincoln Hotel Bldg.
Davis, Arthur Seward, Jr., 810 W. Michigan
Doak, John, 110 Power Plant.
Keith, Granville S., Dept. of Architecture, U. of I.
Maxwell, Wymer W., 512 W. Oregon St.
Provine, Loring H. Prof., 505 Michigan Ave.
Royer, Joseph W., Flat Iron Bldg.
Schieck, William H., 403 E. Washington St.
Smith, Herbert Argo, 101 S. Broadway.
Sterner, William Henry, 256 Administration Bldg.
Stouffer, Ernest L., 256 Administration Bldg.
Sweet, John Elmo, Dept. of Architecture, U. of I.
Toth, Edmund F., 304 Architects Bldg.
Wilkinson, Nathan, Dept. of Architecture, U. of I.

WAUKEGAN

Ganster, William A., 326 N. Utica St.

ARCHITECTS OUTSIDE OF THE STATE OF ILLINOIS

ALABAMA

BIRMINGHAM

McCauley, Charles H., 700 Jackson Bldg.

ARIZONA

PHOENIX

Green, Herbert H., 214 Professional Bldg.

CALIFORNIA

GARDEN GROVE

McDonald, Luther W., R. I. Box 285.

GLENDALE

Vail, Morrison H., 418 Hawthorne St.

HOLLYWOOD

Ayres, Donald Port, 7970 Sunset Blvd.
Hall, Gilbert P., 1354 N. Harper Ave.
Von Mueller, Curt, 4953 Hollycourt Blvd.

LOS ANGELES

Corbey, Leon J., 752½ N. La Fayette Park Pl.
Crow, Ralph M., 237 N. Flower St.
Frankel, Abert B., 981 S. Westmoreland Ave.
Groesse, Paul, 1606 W. 51st Pl.
Lee, Simeon Charles, 1648 Welshire Blvd.
Luekert, Otto, 333 Medio Drive.
Miller, Marcus P., 707 Board of Trade Bldg.
Mueller, Floyd E., 6276 Crestwood Way.
Norton, Francis J., Box 1233.
Stoshitch, Savo Milan, 2627 Monmouth St.

OJAI

Brandt, Berkley, The Hermitage Ranch.

MODESTO

Wach, Edward F., P. O. Box 1396.

OAKLAND

Ferree, Harold C., 435 Valle Vesta Ave.

PASADENA

Hulsebuse, Bernard L., 2416 Las Lunas St.
Purcell, William Gray, R. I. Box 637.

SACRAMENTO

Peterson, Jens C., 812 26th St.

SAN FRANCISCO

Lansburgh, Gustav Albert, 321 Rush St.
Maybeck, Bernard R., 408 Ruso Bldg.

SANTA ANA

Vrydagh, Jupiter G., 505 S. Birch St.

COLORADO

DENVER

Winkel, Benno J., 3985 W. 1st St.

CONNECTICUT

OLD GREENWICK

Braun, Carl Conrad, 50 Center Dr.

DELAWARE

WILMINGTON

Carlson, Walter, 3184 DuPont Bldg.

DISTRICT OF COLUMBIA

WASHINGTON

Arneson, Stephen V., 1739 N St., N. W.
Baumeister, Geo. E., 2701 Courtland Pl., N. W.
Chladek, Arthur Loval, 1445 Otis Pl., N. W.
Colean, Miles L., 15th & K Sts., N. W.
Conner, Geo. D., 5714 Chevy Chase Pkwy.
Davis, Jerome R., 3618 T St., N. W.
Eliel, Arthur G., 4224 16th St., N. W.
Ferrenz, Terrell J., 1323 Ingraham St., N. W.
Holcomb, F. Morse., 1447 Belmont St., N. W.
Karlson, Joseph A., 2356 40th St., N. W.
Knox, Raymond K., 4700 Connecticut, N. W.
Lindeberg, Geo. L., Hotel Powhattan.
Lindstrom, William J., 4892 Conduit Rd., N.W.
Nedved, Rudolph J., 1613 Riggs Pl., N. W.
O'Connor, Wm. J., 2145 C St., N. W.
Ratcliffe, H. E., 1404 Longfellow St., N. W.
Rowe, Lindley P., 1309 17th St., N. W.
Schumacher, August J., 1365 Kennedy St. N. W.
Sjolin, Gosta, 1823 Newton St., N. W.
Speer, Geo. Archibald, 1750 16th St., N. W.
Starr, F. Charles, Cosmos Club.
Twery, Lewis Edw., 605 Roxboro Pl., N. W.
Vigeant, Xavier, 2001 16th St., N. W.

FLORIDA

COCOANUT GROVE

Lang, Albert, 3608 W. Gaudens Rd.

GAINESVILLE

Reeve, Keith G., Dept. of Arch., University of Florida.

MIAMI

Christie, Louis R., Box 1193 Little River St.
Eskridge, Fred A., 218 S. W. 6th Ave.
France, Roy F., 1620 West Ave.

MIAMI BEACH

Bernard, Clifford S., 1604 Michigan Ave.

ORLAND

Spencer, Robert Closson, 302 Ridgewood Ave.

STUART

Morgan, Charles L., P. O. Box 263.

WEST PALM BEACH

Ohlhaber, William, 205 Pilgram Rd.

WINTER PARK

Bellas, Charles.

GEORGIA

AUGUSTA

Irwin, Willis, 620 S. F. C. Bldg.

INDIANA

BLOOMINGTON

Strain, William J., 4 Allen Bldg.

COLUMBUS

Dunlap, Elmer E., 806 Franklin St.

CRAWFORDSVILLE

Beeson, Carroll O., 220 Ben Hur Bldg.

EAST CHICAGO

Palmer, Michael, 4830 Melville Ave.
Norris, Karl D., 206 Calumet Bldg.
Smith, Richard Scott, 3522 Fir St.

EVANSVILLE

Berendes, Edwin A., 121 N. W. Fourth St.
Boyle, Harry E., Furniture Bldg.
Fowler, Frank E., 505 Central Union Bank Bldg.
Legeman, Ralph Earl, 506 Central Union Bldg.
Schlotter, Frank J., 114 N. W. 4th St.
Shopbell, Clifford, 307 Chandler Ave.

FORT WAYNE

Reidel, John M. E., 42 E. State Blvd.
Strauss, Alvin Max, 415 Cal Wayne Bldg.

GARY

Peterson, Harold Emil, 743 Arthur St.
Ransel, Joseph A., 1711 E. 5th Ave.

HAMMOND

Bachman, William J., 6419 Harrison St.
Barnard, Leslie Cosby, 7241 Forest Ave.
Berry, Addison C., 5248 Hohman Ave.
Hutton, J. T., 5231 Hohman Ave.
McComb, Harry J., 5266 Hohman Bldg.
Vaughn, Kenneth R., 7425 Jefferson St.

INDIANAPOLIS

Allen, J. Lloyd, 624 Architects & Builders Bldg.
Bohlen, August C., 1001 Majestic Bldg.
Foltz, Herbert W., 333 N. Penn St.
Henning, Arthur B., 226 E. 12th St.
Honeywell, Albert A., 2404 Broadway.
Jaunra, Robert, 6750 W. 18th St.
Kelley, John Robert, 1013 Architects Bldg.
Nedved, Elizabeth Kimball, R. R. 16, Box 258.
Senefeld, Claude L., 909 Architects Bldg.
Stern, Walter B., 321 E. 50th St.

LA FAYETTE

Marshall, Warren Daniel, 1213 W. L. A. Ave.
Scholer, Walter, 1114 State St.

LOGANSPORT

Canfield, Herbert H., 316 Leath St.
Horn, Carl J., 216½ Fourth St.
Nelson, Lee G., 222 Third St.

LOWELL

Wiley, W. Clifford.

MARTINSVILLE

Branch, James E., 510 E. Washington St.

MICHIGAN CITY

Boonstra, Samuel P., 622 Franklin Blvd.

SOUTH BEND

Austin, Ennis R., 625 J. M. S. Bldg.
Ellwood, W. W., 219 Christman Bldg.
Toth, Andrew A., 1509 N. O'Brien St.

TERRE HAUTE (Vigo Co.)

Miller, Warren D., 222 Van Buren St.
Peddle, Juliet Alice, 2117 W. Tenth St.
Yeager, Ralph O., 215 Van Buren St.

VINCENNES (Knox Co.)

Routt, Lester W., 1012 Perry St.
Schucker, Rudolph W., Bayard Bldg.
Sutton, Byron, 1312 Busseron St.

IOWA

BOONE

Lautz, Rueben S., 614½ Story St.

BURLINGTON

Carswell, Robin B., 307 Iowa State Bank Bldg.

CLINTON

Morrell, Albert H., 410 Howes Bldg.

DAVENPORT (Scott Co.)

Ebeling, Arthur H., 719 Kahl Bldg.
Klein, Carrol A., 111 E. 2nd St.
Krause, Walter O., 910 Kahl Bldg.
Muesse, Howard S., 908 Davenport Bank Bldg.
Parish, William L., 910 Kahl Bldg.
Temple, Seth Justin, 419 Union Bank Bldg.
Whitaker, Raymond C., 1202 Adams St.

DES MOINES

Borg, Elmer Herman, 815 Hubbell Bldg.
Groth, Paul William, 418 Lincoln Ct.
Souers, Clark, 720 Hubbell Bldg.
Woodburn, Chester C., 310 Old Colony Bldg.

DUBUQUE

Heer, Fridolin, Jr., 2535 Trout Terrace.
Horner, George Lewis, 308 Grand Ave.

MASON CITY

Waggoner, Karl M., 11½ S. Federal Ave.

MUSCATINE

Kennison, Herbert A., 617 Iowa St.

OTTUMWA

Kerns, George M., 211 E. 2nd St.

SIOUX CITY

Arnold, Ralph, Insurance Exchange Bldg.

KANSAS

ARKANSAS CITY

Gilbert, Ernest Eugene, R. R. 4.

KENTUCKY

LOUISVILLE

Davis, Brinton B., 508 Washington Bldg.
Gore, William E., 4630 S. Third St.

LOUISIANA

MONROE

Land, Herbert Henry, P. O. Box 744.

MARYLAND

FOREST GLEN

Pritz, Richard H., Box 25.

SILVER SPRING

Janes, Milo F., 753 Silver Spring Ave.
Klaber, Eugene Henry, 1556 West Highway.
Kleinman, Maurice W., 1521 E. Falkland Ave.

MASSACHUSETTS

BOSTON

Collens, Charles, 75 Newbury St.
Robbins, Henry Chandler, 31 St. James St.

MICHIGAN

BANGOR

Martini, Elizabeth A., Route 2.

BERRIEN SPRINGS

Watson, Vernon S., Route 2.

DETROIT

Behel, Vernon W., 98 Gladstone Ave.
Henderson, Ross L., 14456 Coyle Ave.
Kahn, Albert, New Center Bldg.
Kimball, Edgar R., 748 Free Press Bldg.
Maguolo, George John, 608 Fisher Bldg.
Pickens, Buford L., Dept. of Architecture, Wayne University.
Pollmar, F. Carl, 2539 Woodward Ave.
Smith, Fred L., 800 Marquette Bldg.

ESCANABA

Arntzen, Gothard, 820 S. 16th St.

FENVILLE

Meles, Edmund J.

GRAND RAPIDS

Dykema, Claude Dale, R. R. 3 Cascadia.
Happel, Otto G., 110 Fountain St. N. E.
McCall, John Alexander, 2814 Woodcliff Dr.

JACKSON

Kressbach, Carl C. F., 212 W. Michigan Ave.
Olson, Raymond I., 906 Union St.
Schwenkmeyer, Carl Henry, 200 S. Webster St.

MIDLAND

Fraser, Willard E., 201 E. Ellsworth Ave.
Pfeifer, Reuben John, 201 E. Ellsworth Ave.
Hahn, Stanley Worth, Muskegon Bldg.

MUSKEGON

Valentine, Edwin E., 308 Hackley Union Bank Bldg.

SAUGATUCK

Hoerman, Carl, The Chalet Studio.

MINNESOTA

MINNEAPOLIS

Becker, George, 3247 Lyndale Ave.
Dunham, Arthur B., Essex Bldg.
Hills, James B., 1004 Marquette Bldg.
Kaplan, Seeman, 710 McKnight Bldg.
Liebenberg, Jacob J., 710 McKnight Bldg.
Streeter, Serano E., 902 Baker Bldg.

ROCHESTER

Allerton, Oscar J., 112 7th St., N. E.
Crawford, Harold H., 514 8th Ave., S. W.

ST. PAUL

Chase, S. Bruce, 615 Commerce Bldg.
Lagergren, Gustaf P., 1500 Breda St.
Olson, Carl M., 32 Cambridge Ave.

THIEF RIVER FALLS

Eckland, Henry C.

MISSISSIPPI

GULFPORT

Bean, Ralph H., 702 Nat. Bank of Gulfport.

MISSOURI

JEFFERSON CITY

Gloyd, Galen V. R., 421 E. State St.

KANSAS CITY

Boller, Robert O., 7332 Brooklyn Ave.
Carroll, Maurice John, 1320 R A Long Bldg.
Cozad, Fred Paul, 202 Jackson County Court House.
Vigeant, Gregory, Jr., 205 E. Winthrop Rd.

ST. LOUIS

Bradshaw, Preston J., 718 Locust St.
Brockmeyer, Edwin J., 634 N. Grand Mo. Bldg.
Brunson, Elmer A., 1976 Railway Exchange.
Corrubia, Angelo B. M., 812 Olive St.
Farrar, Benedict, 1367 Arcade Bldg.
Fishman, M. Maurice., 220 N. Kingshighway.
Helmuth, George W., Commerce Bldg.
Henderson, Gale E., 907 Wainwright Bldg.
Hoener, Percival John, 3417 S. Kingshighway.
Ittner, Wm. B., Jr., 408 Board of Education Bldg.
Janssen, Ernest C., 3631 Flad Ave.
Klassing, Arthur F., 3237-a Delor
Klingensmith, Paul Q., 4232 West Pine Blvd.
Klutho, Victor J., Syndicate Trust Bldg.
La Beaume, Louis E., 315 N. 7th St.
Levy, Will, 1947 Railway Exchange.
Moritz, Raymond Edward, Chemical Bldg.
Pendleton, Louis D., 1205 Olive St.
Price, Robert Marr, 812 Olive St.
Rathmann, Walter L., 316 N. 8th St.
Rixman, F. E., 1571 Arcade Bldg.
Russell, E. J., 1620 Chemical Bldg.
Schloemann, Carl F., 6329 San Bonita Ave.
Schopp, Lawrence O., 3832 Washington
Stiegemeyer, Oliver Wm., 6332 Southwood
Steinmeyer, Theodore J. W., 136 S. Kirkwood.
Study, Guy, 1367 Arcade Bldg.
Thoresen, Thorgils, 5476 Vernon Ave.
Watson, Jesse N., 1508 Chemical Bldg.
Wedemeyer, William, 105 N. 7th St.
Wells, Willis L., Security Bldg.
Wessbecher, Louis, 406 Market St.
Young, William Ridgeley, Chemical Bldg.

UNIVERSITY CITY

Davis, Neal C., 539 N. & S. Rd.

VALLEY PARK

Bailey, Norman S.

NEBRASKA

OMAHA

Cone, Chester A., 545 S. 31st St.
Scudder, George M., 677 George Blvd.
Wellman, William Thomas, 4724 Davenport

NEW JERSEY

CHATHAM

Doerr, Harold F., 249 Main St.

ENGLISHTOWN

Nachtsheim, Peter, R. F. D. No. 1.

NEWARK

Klein, S. Lawrence, 24 Commerce St.

PATERSON

Hewitt, George, 152 Market St.
Lee, Ellsworth M., 152 Market St.

VERONA

Bonta, Edwin W., 89 Fairview Ave.

NEW MEXICO

ALBUQUERQUE

Lane, Harry L., P. O. Box 1114.

NEW YORK

MOUNT VERNON

Paul, Clarence T., 231 E. 47th St.

NEW YORK CITY

Abramovitz, Max, 45 Rockefeller Plaza.
Brockman, Henry C., 17 Battery Pl.
Byrne, Francis Barry, 64 E. 86th St.
Chayes, Frank, 174 W. 76th St.
Church, Walter S., 449 W. 14th St.
Civkin, Victor, 570 Lexington Ave.
Corbet, Harvey Wiley, 130 W. 42nd St.
Del Gaudio, Mathew W., 545 Fifth Ave.
Eberson, John, 1550 Broadway.
Ennbury, Aymar II, 150 E. 61st St.
Francisco, Ferris Le Roy, 511 Fifth Ave.
Hanna, John Paul, 630 Fifth Ave.
Harrison, Wallace K., 45 Rockefeller Plaza.
Hopkins, J. Edwin, 250 Park Ave.
Jacobus, Robert F., 511 Fifth Ave.
Kahn, Ely Jacques, 2 Park Ave.
Kattelle, Walter R., 195 Broadway.
Kropel, Carl John, 200 11th Ave.
Lindeberg, Harrie T., 2 W. 47th St.
MacKenzie, George C., 119 Broad St.
MacMurray, Wm. H., 130 W. 42nd St.
Math, Earl R., 195 Broadway.
Mitchell, Joseph Orlando, 3208 Woolworth Bldg.
Ogg, John Williams, Briarcliff Manor.
Pfohl, Louis H., 260 11th Ave.
Preis, Carl G., 230 Park Ave.
Rogers, James Gamble, 156 E. 46th St.
Schaeffer, Joseph Carl, 52 Vanderbilt Ave.
Seffert, Edward F., 114 Fifth Ave.
Thomas, Andrew J., 23 W. 47th St.

SEARSDALE

Schneberger, John G., 11 Grand Blvd.
Smith, Benjamin A., 20 Garth Rd.

TUCKAHOE

Johnson, Arthur R., 36 Westview Ave.

WOODSIDE

Ephraim, George, 30 Hobart St.

NORTH DAKOTA

LISBON

Ellis, Charles L., Box 292.

OHIO

AKRON

Long, Chester V., 639 Carpenter St.
Thompson, William Berle, c/o B. F. Goodrich

CINCINNATI

Hale, Robert S., 2251 Harrison Ave.
Salisbury, Ralph D., 1420 Enquirer Bldg.

CLEVELAND

Cummings, Ralph W., 1770 E. 11th St.
Thomas, James William, 3868 Carnegie Ave.

MIDDLETOWN

Weich, Peter John, Cremo Ave.

TOLEDO

Mills, George S., 518 Jefferson Ave.

OKLAHOMA

BARTELSVILLE

Reinhardt, Clarence F., 717 Dewey Ave.

PORTLAND

Howell, Leslie Dillon, 404 U. S. Nat. Bank

TULSA

Thorne, A. Thomson, 1643 S. Florence Pl.

PENNSYLVANIA

PHILADELPHIA

Cret, Paul P., Architect's Bldg.
Gustafson, Carl A., 515 Vernon Rd.
Klauder, Charles Z., 1429 Walnut St.
Zantzinger, C. C., 1602 Architect's Bldg.

PITTSBURGH

Angelis, Michael J., 207 Investment Bldg.
Hogner, Pierre R. I., 537 Perry Highway West View.
McCormick, W. D., 121 127 Negley Ave.
McMullen, Leo A., 900 Renshaw Bldg.
Prack, Bernhard H., 119 Federal St.

SCRANTON

Miller, Charles A., 1522 Wyoming Ave.

UPPER DARBY

Nelson, Edgar Harold, 7 E. Clearfield Rd.

RHODE ISLAND

PROVIDENCE

Sheldon, Arthur W., Hospital Trust Bldg.

SOUTH CAROLINA

ESTELL

Gauger, Raymond Julius.

SOUTH DAKOTA

SIOUX FALLS

Schoening, John A., 102 S. Euclid Ave.
Spitznagel, Harold, 309 Western Surety

TENNESSEE

NASHVILLE

Holman, Joseph W., 701 Stahlman Bldg.

STOP SPRINGS

Wallace, Wellington I. H.

TEXAS

CORPUS CHRISTIE

Johnson, Otis Floyd, 618 Medical Prof. Bldg.

DECATUR

Prendegast, Richard Ward.

EL PASO

Herlin, Geo. W., Box 512.
Wuehrmann, William, 1400 Bassett Bldg.

HOUSTON

Kamrath, Karl F., 2017 W. Gray St.
Mackie, Frederick J., Jr., 2017 W. Gray St.

LEGION

Cummings, Raymond H.

McALLEN

Gatterdam, Fred E., Box 1689.
Pierce, Richard Gordon.

SAN ANTONIO

Riley, Ivan H., 2902 Fredericksburg.

VIRGINIA

ARLINGTON

Kendall, David A., 306 N. Irving St.
Sterner, Carl John, 1311 Arlington Ridge Rd.

ALEXANDRIA

Fox-Thomas, Stafford, 1404 Mount Vernon
Thompson, Magnus, 1500 Russell Rd.

WASHINGTON

SEATTLE

Schulze, William, 2038 N. 78th St.

WISCONSIN

AMERY

Hubbard, Archie H., Box 342.

APPLETON

Kauffmann, Gerhard Max, 215 W. College
Le Vee, Raymond W., 117 E. College St.

EPHRAIM

Bernhard, William.

FORT ATKINSON

Waterman, Arthur E., 425 Adams St.

GREEN BAY

Connell, John Francis, 226 N. Washington
Fogel, Reuben W., Y.M.C.A.
Oppenhamer, William Adam, 110 S. Washington St.

KENOSHA

Lindl, Gordon J., 830 64th St.

LAKE GENEVA

Allen, James Roy, 1109 Wisconsin St.

LA CROSSE

Dockendorff, Bernhard, 404 Linker Bldg.
Parkinson, Albert E., 404 Linker Bldg.

MADISON

Balch, Harold C., 16 N. Carroll St.
Claude, Louis W., 114 N. Carroll Ave.
Kaiser, William Vogt, 905 University Ave.
Kinne, William S., Jr., 1901 Vilos Ave.
Law, Edward John, 111 N. Allen St.
Law, James R., 101 N. Prospect Ave.
Mead, Daniel W., State Journal Bldg.
Potter, Ellis J., Shorewood Hills.
Pugh, Myron E., 2256 Keyes Ave.
Sheldon, Karl H., 8 S. Carroll St.
Teesdale, Lawrence V., c/o Forest Product Laboratories.
Weiler, Joseph J., 315 N. Pickney St.

MILWAUKEE (Milwaukee Co.)

Brielmaier, Joseph, 1st Wis. Nat. Bank Bldg.
Brust, Peter, 135 Wells St.
Byerly, Fred I., 2953 N. Farwell Ave.
Dowling, Edward F., 430 N. 34th St.
Grellinger, Alvin Edward, 5428 W. Lloyd St.
Gruhl, Clarence J., 2857 N. Lake Dr.
Haeuser, Hugo C., 759 Milwaukee St.
Hengels, Henry C., 759 Milwaukee Ave.
Hunt, Leigh, 152 N. Wisconsin Ave.
Johnson, Elmer A., 734 N. Jefferson St.
Kloppenburg, Ralph H., 708 E. Green Tree Rd.
Phillip, Richard, 770 Broadway.

MONROE

Howe, Stanley W., 1518 11th St.

RACINE

Hoffman, Frank J., 503 Janes Bldg.
Jillson, Byron H., 1505 Fleet Ave.
Matson, Justave Mandor, 528 Baker Blk.
Wright, George Ellery, 842 Main St.

TWIN LAKES

Sexton, Frank A.

WEST ALLIS

Lloyd, Albert L., R. 5, Box 410.

Illinois Society of Architects

OFFICERS

1938-1939

ELMER C. JENSEN, President . . .	39 S. LaSalle St.
ARTHUR WOLTERSDORF, 1st Vice-President	520 N. Michigan Ave.
RALPH C. HARRIS, 2nd Vice-President .	Springfield, Ill.
RALPH C. LLEWELLYN, Treasurer . .	38 S. Dearborn St.
STANLEY D. FAIRCLOUGH, Secretary .	819 Exchange Ave.
H. L. PALMER, Financial Secretary . .	134 N. LaSalle St.

DIRECTORS

WILLIAM P. FOX	737 N. Michigan Ave.
PAUL GERHARDT, JR.	1012 City Hall
JOHN A. HOLABIRD	333 N. Michigan Ave.
VICTOR ANDRE MATTESON . . .	20 N. Wacker Dr.
EDWARD PROBST	80 E. Jackson Blvd.
LEO J. WEISSENBORN	Tribune Tower

BOARD OF ARBITRATION

ALFRED S. ALSCHULER	28 E. Jackson Blvd.
EMERY STANFORD HALL . . .	175 W. Jackson Blvd.
HERBERT HEWITT	Alliance Life Bldg., Peoria, Ill.
JOHN A. HOLABIRD	333 N. Michigan Ave.
IRVING K. POND	14 W. Elm St.
RICHARD E. SCHMIDT	104 S. Michigan Ave.
FREDERICK J. THIELBAR . . .	520 N. Michigan Ave.

EDITOR MONTHLY BULLETIN

ARTHUR WOLTERSDORF . . .	520 N. Michigan Ave.

STANDING COMMITTEES

Committee on Education

JOHN LEONARD HAMILTON, Chairman .	814 Tower Court

Entertainment Committee

LEO J. WEISSENBORN, Chairman . .	Tribune Tower

Legislative Committee

PAUL GERHARDT, JR., Chairman . .	1012 City Hall
ROBERT L. FRANKLIN	9647 S. Seeley Ave.
GEORGE E. MORRIS	Sherwin Bldg., Elgin, Ill.
HENRY L. NEWHOUSE, II. . . .	8 S. Michigan Ave.
MARVIN PROBST	80 E. Jackson Blvd.
LEO H. PLEINS	527 W. Monroe, Springfield, Ill.

Materials and Methods Committee

JOHN J. DAVEY, Chairman	29 W. Quincy St.

Membership Committee

H. L. PALMER, Chairman	134 N. LaSalle St.
JOSEPH BOOTON	1017 W. Edwards St., Springfield, Ill.
JOHN J. DAVEY	29 W. Quincy St.
SIDNEY C. FINCK	134 N. LaSalle St.
GILBERT A. JOHNSON	Swedish-American Bank Bldg. Rockford, Ill.

Public Action Committee

WILLIAM PAUL FOX, Chairman	737 N. Michigan Ave.
JOHN E. COYLE	215 Hirkimer St., Joliet, Ill.
FRANCES A. FLAKS	104 S. Michigan Ave.
HERBERT HEWITT	Alliance Life Bldg., Peoria, Ill.
BENJAMIN A. HORN	Central Trust Bldg., Rock Island, Ill.
WILLIAM R. McCOY	419 S. 19th St., Mt. Vernon, Ill.
HAROLD B. McELDOWNY	1814 Lincoln Park West
LEO H. PLEINS	527 W. Monroe St., Springfield, Ill.
JOSEPH W. ROYER	209 Broadway, Urbana, Ill.
WILLIAM H. SCHULZKE	Swedish-American Bank Bldg. Rockford, Ill.

Legal Service Committee

Composed of Entire Board of Directors

State Art Committee

RALPH C. HARRIS	St. Nicholas Hotel, Springfield, Ill.

SPECIAL COMMITTEES

Building Code Committee

ROBERT S. De GOLYER, Chairman . . .	307 N. Michigan Ave.
ROBERT L. FRANKLIN	160 N. LaSalle St.
JOSEPH W. McCARTHY	221 N. LaSalle St.
FREDERICK J. THIELBAR	520 N. Michigan Ave.
HOWARD J. WHITE	80 E. Jackson Blvd.

Sub-Committee on State Building Code

LEO H. PLEINS	527 W. Monroe St., Springfield, Ill.

Building Valuations Committee

EUGENE FUHRER, Chairman	188 W. Randolph St.
DANIEL H. BURNHAM	134 S. LaSalle St.
CHRISTIAN S. MICHAELSEN	3256 Franklin Blvd.

Budget Committee

STANLEY D. FAIRCLOUGH	Secretary
H. L. PALMER	Financial Secretary
RALPH C. LLEWELLYN	Treasurer

Credentials Committee

H. L. PALMER, Chairman	134 N. LaSalle St.
STANLEY D. FAIRCLOUGH . . .	819 W. Exchange Ave.

PAST OFFICERS

1897

John M. Van Osdel, President.
Harry B. Wheelock, 1st Vice-President.
Samuel A. Treat, 2d Vice-President.
Lawrence G. Hallberg, Treasurer.
Charles R. Adams, Secretary.

1898

Harry B. Wheelock, President.
Samuel A. Treat, 1st Vice-President.
Normand S. Patton, 2d Vice-President.
Lawrence G. Hallberg, Treasurer.
Charles R. Adams, Secretary.

1899

Samuel A. Treat, President.
Normand S. Patton, 1st Vice-President.
George Beaumont, 2d Vice-President.
O. H. Postle, Treasurer.
Charles R. Adams, Secretary.

1900

George Beaumont, President.
Charles W. Nothnagel, 1st Vice-President.
Lawrence G. Hallberg, 2d Vice-President.
Samuel A. Treat, Treasurer.
Charles R. Adams, Secretary.

1901

George Beaumont, President.
Emery Stanford Hall, 1st Vice-President.
Edgar M. Newman, 2d Vice-President.
Samuel A. Treat, Treasurer.
Charles R. Adams, Secretary.

1902

Emery Stanford Hall, President.
Edgar M. Newman, 1st Vice-President.
Arthur F. Woltersdorf, 2d Vice-President.
Samuel A. Treat, Treasurer.
Charles R. Adams, Secretary.

1903

Edgar M. Newman, President.
Arthur F. Woltersdorf, 1st Vice-President.
Henry Lord Gay, 2d Vice-President.
Samuel A. Treat, Treasurer.
Charles R. Adams, Secretary.

1904

George L. Pfeiffer, President.
William W. Clay, 1st Vice-President.
S. Milton Eichberg, 2d Vice-President.
Samuel A. Treat, Treasurer.
Charles R. Adams, Secretary.

1905

William C. Clay, President.
Joseph T. Fortin, 1st Vice-President.
Charles J. Furst, 2d Vice-President.
Samuel A. Treat, Treasurer.
Charles R. Adams, Secretary.
H. L. Palmer, Asst. Secretary.

1906

Harry B. Wheelock, President.
Charles J. Furst, 1st Vice-President.
Robert C. Berlin, 2d Vice-President.
Samuel A. Treat, Treasurer.
Emery Stanford Hall, Secretary.
H. L. Palmer, Asst. Secretary.

1907

Normand S. Patton, President.
Arthur F. Woltersdorf, 1st Vice-President.
Irving K. Pond, 2d Vice-President.
Samuel A. Treat, Treasurer.
Emery Stanford Hall, Secretary.
H. L. Palmer, Asst. Secretary.

1908

Irving K. Pond, President.
Richard G. Schmid, 1st Vice-President.
Edmund R. Krause, 2d Vice-President.
Samuel A. Treat, Treasurer.
Emery Stanford Hall, Secretary.
H. L. Palmer, Asst. Secretary.

1909

George Beaumont, President.
Samuel N. Crowen, 1st Vice-President.
Leon E. Stanhope, 2d Vice-President.
Samuel A. Treat, Treasurer.
Emery Stanford Hall, Secretary.
H. L. Palmer, Asst. Secretary.

1910

Arthur F. Woltersdorf, President.
Robert C. Berlin, 1st Vice-President.
Argyle E. Robinson, 2d Vice-President.
Samuel N. Crowen, Treasurer.
Emery Stanford Hall, Secretary.
H. L. Palmer, Asst. Secretary.

1911

Julian Barnes, President.
Argyle E. Robinson, 1st Vice-President.
Peter J. Weber, 2d Vice-President.
Samuel N. Crowen, Treasurer.
Emery Stanford Hall, Secretary.
H. L. Palmer, Asst. Secretary.

1912

Argyle E. Robinson, President.
George W. Maher, 1st Vice-President.
George Beaumont, 2d Vice-President.
Samuel N. Crowen, Treasurer.
Emery Stanford Hall, Secretary.
H. L. Palmer, Asst. Secretary.

1913

Meyer J. Sturm, President.
Arthur F. Woltersdorf, 1st Vice-President
Joseph C. Llewellyn, 2d Vice-President.
Samuel N. Crowen, Treasurer.
Emery Stanford Hall, Secretary.
H. L. Palmer, Asst. Secretary.

1914

Emery Stanford Hall, President.
Frank E. Davidson, 1st Vice-President.
John Devereux York, 2d Vice-President.
Samuel N. Crowen, Treasurer.
John Reed Fugard, Secretary.
H. L. Palmer, Financial Secretary.

1915

Frank E. Davidson, President.
Stafford Fox Thomas, 1st Vice-President.
Robert Seth Lindstrom, 2d Vice-President.
Samuel N. Crowen, Treasurer.
John Reed Fugard, Secretary.
H. L. Palmer, Financial Secretary.

1916

Frank E. Davidson, President.
Stafford Fox Thomas, 1st Vice-President.
William G. Carnegie, 2d Vice-President.
Samuel N. Crowen, Treasurer.
John Reed Fugard, Secretary.
H. L. Palmer, Financial Secretary.

1917

Stafford Fox Thomas, President.
Arthur F. Hussander, 1st Vice-President.
James B. Dibelka, 2d Vice-President.
Samuel N. Crowen, Treasurer.
John Reed Fugard, Secretary.
H. L. Palmer, Financial Secretary.

1918

George W. Maher and Arthur F. Hussander, Presidents.
Arthur F. Hussander, President.
James B. Dibelka, 1st Vice-President.
George W. Maher, 2nd Vice-President.
Samuel N. Crowen, Treasurer.
George A. Knapp, Secretary.
H. L. Palmer, Financial Secretary.

1919

Charles Herrick Hammond, President.
Robert C. Berlin, 1st Vice-President.
N. Max Dunning, 2nd Vice-President.
John A. Armstrong, Treasurer.
Ralph C. Harris, Secretary.
H. L. Palmer, Financial Secretary.

1920

Frank E. Davidson, President.
Herbert E. Hewitt, 1st Vice-President.
John A. Nyden, 2nd Vice-President.
John A. Armstrong, Treasurer.
Ralph C. Harris, Secretary.
H. L. Palmer, Financial Secretary.

1921

Frank E. Davidson, President.
Herbert E. Hewitt, 1st Vice-President.
John A. Nyden, 2nd Vice-President.
John A. Armstrong, Treasurer.
Ralph C. Harris, Secretary.
H. L. Palmer, Financial Secretary.

1922

Frank E. Davidson, President.
Chas. E. Fox, 1st Vice-President.
Herbert E. Hewitt, 2nd Vice-President.
Robert C. Ostergren, Treasurer.
Ralph C. Harris, Secretary.
H. L. Palmer, Financial Secretary.

1923

Charles E. Fox, President.
Byron H. Jillson, 1st Vice-President.
Herbert E. Hewitt, 2nd Vice-President.
Robert C. Ostergren, Treasurer.
Ralph C. Harris, Secretary.
H. L. Palmer, Financial Secretary.

1924

Charles E. Fox, President.
Byron H. Jillson, 1st Vice-President.
Frank A. Carpenter, 2nd Vice-President.
Robert C. Ostergren, Treasurer.
Ralph C. Harris, Secretary.
H. L. Palmer, Financial Secretary.

1925

Charles E. Fox, President.
Byron H. Jillson, 1st Vice-President.
Frank A. Carpenter, 2nd Vice-President
Robert C. Ostergren, Treasurer.
Ralph C. Harris, Secretary.
H. L. Palmer, Financial Secretary.

COOK COUNTY
NURSES' HOME
Chicago, Illinois
45,000 sq. ft.

1926

Leon E. Stanhope, President.
Byron H. Jillson, 1st Vice-President.
George B. Helmle, 2nd Vice-Pesident.
Robert C.. Ostergren, Treasurer.
Ralph C. Harris, Secretary.
H. L. Palmer, Financial Secretary.

1927

Leon E. Stanhope, President.
Howard J. White, 1st Vice-President.
George B. Helmle, 2nd Vice-President.
Robert C. Ostergren, Treasurer.
Walter A. McDougall, Secretary.
H. L. Palmer, Financial Secretary.

1928

Howard J. White, President.
Robert S. De Golyer, 1st Vice-President.
George B. Helmle, 2nd Vice-President.
Robert J. Ostergren, Treasurer.
Walter A. McDougall, Secretary
H. L. Palmer, Financial Secretary.

1929

Alfred Granger, President.
William P. Fox, 1st Vice-President.
George B. Helmle, 2nd Vice-President.
Robert C. Ostergren, Treasurer.
Walter A. McDougall, Secretary.
H. L. Palmer, Financial Secretary.

1930

Robert C. Ostergren, President.
Victor A. Matteson, 1st Vice-President.
George B. Helmle, 2nd Vice-President.
Clarence E. Frazier, Treasurer.
Walter A. McDougall, Secretary.
H. L. Palmer, Financial Secretary.

1931

Robert C. Ostergren, President.
John Reed Fugard, 1st Vice-President.
George B. Helmle, 2nd Vice-President.
W. Gibbons Uffendell, Treasurer.
Carl Hauber, Secretary.
H. L. Palmer, Financial Secretary.

1932

John Reed Fugard, President.
Robert C. Ostergren, 1st Vice-President.
George B. Helmle, 2nd Vice-President.
W. Gibbons Uffendell, Treasurer.
Carl Hauber, Secretary.
H. L. Palmer, Financial Secretary.

1933

John Reed Fugard, President.
Terrell J. Ferrenz, 1st Vice-President.
George B. Helmle, 2nd Vice-President.
Ralph C. Llewellyn, Treasurer.
Carl Hauber, Secretary.
H. L. Palmer, Financial Secretary.

1934

Elmer C. Jensen, President.
John C. Bollenbacher, 1st Vice-President.
George B. Helmle, 2nd Vice-President.
Ralph C. Llewellyn, Treasurer.
Carl Hauber, Secretary.
H. L. Palmer, Financial Secretary.

1935

Elmer C. Jensen, President.
Leo J. Weissenborn, 1st Vice-President.
George B. Helmle, 2nd Vice-President.
Ralph C. Llewellyn, Treasurer.
Harold B. McEldowney, Secretary.
H. L. Palmer, Financial Secretary.

1936

Elmer C. Jensen, President.
Arthur Woltersdorf, 1st Vice-President.
George B. Helmle, 2nd Vice-President.
Ralph C. Llewellyn, Treasurer.
Harold B. McEldowney, Secretary.
H. L. Palmer, Financial Secretary.

1937

Elmer C. Jensen, President.
Arthur Woltersdorf, 1st Vice-President.
Ralph C. Harris, 2nd Vice-President.
Ralph C. Llewellyn, Treasurer.
Stanley D. Fairclough, Secretary.
H. L. Palmer, Financial Secretary.

MEMBERS

Adler, David, 220 S. Michigan Ave.
Alschuler, Alfred S., 28 E. Jackson Blvd.
Anderson, Helmar M., 5948 Midway Park.
Anderson, Stanley D., 262 Deerpath Rd., Lake Forest, Ill.
Ansel, Anton, 1115 Wisconsin Ave., Oak Park, Ill.
Armstrong, John A., 11 S. LaSalle St.
Bean, Ralph H., 702 Nat. Bank of Gulfport, Gulfport, Miss.
Behrensmeyer, Charles F., 425 S. 14th St., Quincy, Ill.
Beidler, Herbert B., 936 N. Michigan Ave.
Bein, Maurice L., 1159 W. Roosevelt Rd.
Bernham, F. M., 8 S. Michigan Ave.
Bishop, Thomas R., 35 S. Dearborn St.

Bjork, David T., 624 Asbury Ave., Evanston, Ill.
Bohasseck, Charles, 333 N. Michigan Ave.
Boller, Robert, 7332 Brooklyn Ave., Kansas City, Mo.
Booton, Joseph F., 1017 W. Edwards St., Springfield, Ill.
Bourke, Robert, 39 S. LaSalle St.
Braband, F. J. E., 901 Wrightwood Ave.
Bradley, Chas. W., Brown Bldg., Rockford, Ill.
Bradley, Harold, Brown Bldg., Rockford, Ill.
Bradshaw, Preston J., 718 Locust St., St. Louis, Mo.
Buchsbaum, Emanuel V., 7810 S. Crandon Ave.
Buckingham, Clarence W., Jacksonville, Ill.
Buerger, Albert J., 6450 S. Bishop St.
Burgess, Ralph R., 6523 Newgard Ave.
Burnham, D. H., 134 S. LaSalle St.
Burnham, Hubert, 160 N. LaSalle St.
Byerly, F. I., 2953 N. Farwell Ave., Milwaukee, Wis.
Capraro, Alexander, 64 W. Randolph St.
Carnegie, Wm. G., 11041 S. Park Ave.
Carpenter, Frank A., 749 John St., Rockford, Ill.
Carr, Chas. Alban, 322 W. Chicago Ave.
Carr, George Wallace, 333 N. Michigan Ave.
Cerny, Jerome J., 2341 S. Austin Blvd.
Charles, Walter T., 77 W. Washington St.
Chatten, Melville C., 160 N. LaSalle St.
Christensen, Chas. Werner, 845 Ainslie St.
Christensen, John C., 228 N. LaSalle St.
Chubb, John D., 109 N. Dearborn St.
Clark, Charles R., Sheldon, Ill.
Clark, Edwin H., 8 E. Huron St.
Coffin, A. S., 6 N. Michigan Ave.
Cohen, Joseph, 3921 W. 19th St.
Cook, Norman W., 810 Columbia St., Ottawa, Ill.
Coughlen, Gardner C., 9406 S. Vanderpoel Ave.
Coyle, John E., 215 Herkimer St., Joliet, Ill.
Crosby, William S., 6 N. Michigan Ave.
Dalsey, Harry I., 10 N. Clark St.
Daly, S. Lester, 412 Girard St., Metropolis, Ill.
Danielson, Phillip A., 1659 Sherman Ave., Evanston, Ill.
Davey, John J., 431 S. Dearborn St.
Davis, George H., 527 Moss St., Peoria, Ill.
Davis, Zachary T., 228 N. LaSalle St.
Day, Warren W., 373 Fourth St., New York, N. Y.
Deal, Joe Mindert, I. O. O. F. Bldg., Lincoln, Ill.
Decker, Daniel W., 1728 S. Spring St., Springfield, Ill.
De Golyer, Robert S., 307 N. Michigan Ave.
De Money, Frank O., 30 N. LaSalle St.
Dippold, Albert P., 3948 Cottage Grove Ave.
Dougherty, Floyd E., 79 W. Monroe St.
Drainie, John G., 6754 S. Cornell Ave.
Drielsma, J. Arthur, 540 N. Michigan Ave.
Drummond, Wm., 547 Edgewood Pl., River Forest, Ill.
Dubin, Geo. H., 127 N. Dearborn St.
Dunning, N. Max, c/o Hotel Powhattan Penn. Ave., 18th and H Sts., N. W., Washington, D. C.
Eichberg, S. M., 208 S. LaSalle St.
Emerson, Frank N., 1600 Peoria Life Bldg., Peoria, Ill.
Eppenstein, James F., 35 E. Wacker Dr.
Esser, Curt A., 225 N. Michigan Ave.
Fairclough, Stanley D., 819 Exchange Ave., U. S. Yards.
Fehlow, Albert C., 717 N. State St.
Ferrenz, Tirrell J., 1323 Ingraham St., N. W., Washington, D. C.
Finck, Sidney C., 134 N. LaSalle St.
Fischer, Frederick William, 3018 E. 91st St.
Fishman, M. Maurice, 205 W. Wacker Dr. (1905).
Flaks, Francis A., 150 The Lane, Hinsdale, Ill.
Flinn, Raymond W., 538 Central Ave., Highland Park, Ill.
Floto, Julius, 176 W. Adams St.
Foltz, Frederick C., 6455 S. Central Ave.
Fortin, Joseph T., 600 Blue Island Ave.
Fox, John J., 549 W. Randolph St.
Fox, Wm. P., 737 N. Michigan Ave.
Fox, Thomas Stafford, 1404 Mt. Vernon Ave., Alexandria, Va.
Franklin, Robert L., 9647 S. Seeley Ave.
Frazier, Clarence F., 64 W. Randolph St.
Friedman, Raphael N., 28 E. Jackson Blvd.
Fry, Frank L., 2111 Maypole Ave.
Fugard, John Reed, 520 N. Michigan Ave.
Fuhrer, Eugene, 188 W. Randolph St.
Furst, Wm. H., 11 S. LaSalle St.
Garretson, W. Gordon, Girard, Ill.
Gatterdam, F. E., 6307 Holbrook Ave.
Gaul, Herman J., 228 E. Superior St.
Gerber, Arthur, 76 W. Adams St.
Gerding, Louis H., 708½ LaSalle St., Ottawa, Ill.
Gerhardt, Paul, 64 W. Randolph St.
Gerhardt, Paul, Jr., 1015 City Hall.
Gill, Rudolph Z., 1328 Walnut St., Murphysboro, Ill.
Golabowski, Joseph F., 1701 S. Pasfield St., Springfield, Ill.
Granger, Alfred, Few Acres, Roxbury, Conn.
Gray, Frank B., 1st Nat. Bk. Bldg., Aurora, Ill.
Green, H. H., 1 E. Country Club St., Phoenix, Arizona.
Guenzel, Louis, 879 N. State St.
Hadley, Bryant E., 1005 S. 4th St., Springfield, Ill.
Hall, Emery Stanford, 175 W. Jackson Blvd.
Hall, Eric E., 123 W. Madison St.
Hamilton, John L., 814 Tower Court.
Hammond, Charles Herrick, 160 N. LaSalle St.
Harris, Ralph C., Central Garage Bldg., 200 E. Ash St., Springfield, Ill.
Hatzfeld, Clarence, 417 S. Dearborn St.
Hauber, Carl, 25 E. Jackson Blvd.

SOME TYPICAL INSTALLATIONS OF

1st National Bank Bldg., St. Paul, Minn.—60,000 Ft. Moultile in Offices
GRAHAM, ANDERSON, PROBST & WHITE, Architects

THOS. MOULDING

MOULTILE

MASTER ASPHALT TILE

FLOORS for Every Purpose

MOULTILE characteristics adapt this floor covering to virtually all conditions. An almost unlimited choice of colors and patterns permit its use with any decorative scheme. It is long-wearing, quiet, resilient and sanitary. Surprisingly, MOULTILE can be installed at lower cost than most floor coverings.

Methods and materials have been developed to make existing wood or concrete floors, no matter what their condition, into suitable foundations for MOULTILE. It is especially well suited for installation over ground floor cement, where the moisture and alkali always present will rot other types of floor covering. Samples and technical data will gladly be furnished on request.

165 West Wacker Drive **Chicago, Illinois**

Wieboldt Dept. Store, River Forest, Ill.—100,000 Ft. Moultile in All Areas
HOLABIRD & ROOT, Architects

C. A. Hemphill Home, Kenilworth, Ill.—Moultile in Basement Billiard Room

Winnetka Congregational Church, Winnetka, Ill.—8,000 Ft. Moultile in Sunday School Areas
AYMAR EMBURY II, Architect
JOHN LEONARD HAMILTON, Assoc. Architect

Hecht, Albert S., 420 W. Surf St.
Heimbrodt, Carl E., 4110 Lawn Ave., Western Springs, Ill.
Hellmuth, George W., Commercial Bldg., St. Louis, Mo.
Helmle, Henry R., First National Bank Bldg., Springfield, Ill.
Henderson, Russell D., 2124 S. 4th St., Springfield, Ill.
Henschien, H. Peter, 59 E. Van Buren St.
Hewitt, Herbert, 1600 Alliance Life Bldg., Peoria, Ill.
Himelblau, A. L., 1535 Touhy Ave.
Hirschfield, Leo S., 65 E. South Water St.
Hodgdon, Fred H., 333 N. Michigan Ave.
Hodgkins, H. G., 7740 Malvern Ave.
Hogner, Perrie R. L., 537 Perry Highway (West View), Pittsburgh, Pa.
Holabird, John A., 333 N. Michigan Ave.
Holsman, Henry K., 307 N. Michigan Ave.
Hooper, William T., 625 N. Laramie St.
Horn, Benjamin A., Central Trust Bldg., Rock Island, Ill.
Hosmer, Clare C., 221 William St., East Orange, N. J.
Hotchkiss, Robert J., Central Nat. Bank Bldg., Peoria, Ill.
Hubbard, Archie H., Box 214, Durand, Wis.
Hull, Denison B., 77 W. Washington St.
Hulsebus, Bernard L., 1001 Jefferson Bldg., Peoria, Ill.
Hunt, Jarvis, 30 N. Michigan Ave.
Hussander, A. T., 8 S. Dearborn St.
Huszagh, Ralph D., 6 N. Michigan Ave.
Hussey, Harry H., No. 1 Ta Soo Chow Huting, Pekin, China.
Hyland, Paul V., 6308 N. Leavitt St.
Jacobs, Arthur, 160 N. LaSalle St.
Jensen, Elmer C., 39 S. LaSalle St.
Jensen, Jens J., 1105 Lawrence Ave.
Jillson, Byron H., 9450 S. Winchester Ave.
Johnson, Gilbert A., Swedish American Bank Bldg., Rockford, Ill.
Johnston, W. K., 6805 Yale Ave.
Joy, Samuel Scott, 342 Emerson St., N. W., Washington, D. C.
Kahn, Albert, 1000 Marquette Bldg., Detroit, Mich.
Kane, Michael B., Bohm Bldg., Edwardsville, Ill.
Karlin, Irving K., 8124 S. Crandon Ave.
Kelly, Joseph, 125 W. Myrtle St., Springfield, Ill.
Kennison, Herbert A., 803 Roshek Bldg., Dubuque, Iowa.
Kingsley, Geo. S., Douglas, Michigan.
Klamt, Edward A., 713 Wrightwood Ave.
Klein, William J., 64 W. Randolph St.
Klekamp, Bernard L., 140 S. Dearborn St.
Knox, Robert, 518 E. Capitol Ave., Springfield, Ill.
Kocher, Jacques J., 506 W. 63rd St.
Koenigsberg, Nathaniel, 155 N. Clark St.
Koerner, Uda H., 9908 S. Western Ave.
Kohfeldt, Walter G., 11207 S. South Park Ave.
Krieg, Wm. G., 180 N. Maplewood Ave., Riverside, Ill.
Kruegal, Arnold J., 417 Western Ave., Joliet, Ill.
LaBeaume, Louis C., 720 Crompton Bldg., St. Louis, Mo.
Laist, Theodore F., Yellow Springs, Ohio.
Lampe, Clarence W., 134 N. LaSalle St.
Lankton, J. Fletcher, 1805 N. Knoxville Ave., Peoria, Ill.
Larson, Godfrey E., 77 W. Washington St.
Lautz, William H., 806 W. 79th St.
Layer, Robert W., 221 N. LaSalle St.
Levy, Alex. L., 179 W. Washington St.
Lewin, Edwin P., 134 N. LaSalle St.
Lewis, Jacob, 30 N. Dearborn St.
Lindquist, Frederick, 9545 S. Hoyne Ave.
Lindblad, Alfred G., 317 S. 7th St., Springfield, Ill.
Lindstrom, William J., 2000 S. St., Washington, D. C.
Liska, Charles O., 216 S. Chester St., Park Ridge, Ill.
Liska, Emil, 9601 Prospect Ave.
Llewellyn, Ralph C., 38 S. Dearborn St.
Loewenberg, Israel S., 111 W. Monroe St.
Lovdall, George F., 1807 Cuyler Ave.
Ludgin, Joseph G., 37 S. Wabash Ave.
Macardell, Cornelius W., 606 S. 4th St., Springfield, Ill.
Mahaffey, David, Box 7, Grays Lake, Ill.
Maher, Harry E., 1564 Sherman Ave., Evanston, Ill.
Maher, Philip Brooks, 157 E. Erie St.
Marshall, Benjamin H., 1515 Sherman Ave., Evanston, Ill.
Martin, Edgar D., 540 N. Michigan Ave.
Martini, Elisabeth A., Route 2, Bangor, Mich.
Marx, Elmer W., 2734 Mildred Ave.
Marx, Samuel A., 333 N. Michigan Ave.
Matteson, Victor Andre, 20 N. Wacker Dr.
Maupin, James M., Comm'l. Bldg., Alton, Ill.
Maynard, Henry S., 26 Park Drive, Glen View, Ill.
McArthur, Albert C., 1414 N. Morningside Ct., Hollywood, Cal.
McCarthy, Joseph W., 43 E. Ohio St.
McCaughey, William F., Jr., Pickwick Bldg., Park Ridge, Ill.
McCoy, William R., 426 S. 19th St., Mt. Vernon, Ill.
McDougall, Walter A., 140 S. Dearborn St.
McEldowney, Harold B., 1814 Lincoln Park West.
McGrew, Kenneth Axtel, 1564 Sherman Ave., Evanston, Ill.
McLaughlin, Daniel F., 4703 Oakwood Ave., Downers Grove, Ill.
Mead, Daniel W., State Journal Bldg., Madison, Wis.
Meder, Everett S., 2515 N. Nordica Ave.
Meldrum, John, Niles Center, Ill.
Meredith, Davis D., 251 South St., Elmhurst, Ill.
Merrill, John C., 134 N. LaSalle St. (18th floor).
Metz, Carl A., 520 N. Michigan Ave.
Meyer, Carl F., 712 Meyers Bldg., Springfield, Ill.
Meyer, Frederic H., 820 Tower Court.
Michaelsen, Christian S., 3256 Franklin Blvd.

Miller, Chas. A., 1522 Wyoming Ave., Scranton, Pa.
Miller, John W., 4438 Greenwood Ave.
Miller, Joseph A., 6440 Bosworth Ave.
Morehouse, Merritt J., Burge Plantation, Mansfield, Ga.
Morgan, Chas. L., 5218 S. Cornell Ave.
Morris, George E., Sherwin Bldg., Elgin, Ill.
Muesse, Howard S., 925 American Bank Bldg., Davenport, Iowa.
Mundie, W. B., 39 S. LaSalle St.
Murphy, Charles, 80 E. Jackson Blvd.
Nadherny, Joseph J., 1518 W. Roosevelt Rd.
Naess, Sigurd E., 80 E. Jackson Blvd.
Neebe, John K., 3713 N. Kedvale Ave.
Nelson, Joseph A., 2432 Irving Park Blvd.
Nelson, Melvin A., 4633 N. Hermitage Ave.
Newhouse, Henry L., II, 8 S. Michigan Ave.
Newman, Edgar M., 107 N. Clark St.
Nicol, Chas. Wheeler, 59 E. Van Buren St.
Nimmons, Geo. C., 333 N. Michigan Ave.
Nitsche, Edward A., 28 E. Huron St.
Noble, James, c/o Div. of Architecture, State Armory, Springfield, Ill.
Nock, Samuel A., 719 N. State St.
Norman, Andrew, 1754 Granville Ave.
Oliver, Ralph H., 8 S. Dearborn St.
Olsen, Leif E., 228 N. LaSalle St.
Olsen, Paul Frederick, 130 N. Wells St.
Olson, Benjamin Franklin, 19 S. LaSalle St.
Oman, Samuel H., 162 E. Ohio St.
Orput, Raymond A., 202 3rd Nat. Bank Bldg., Rockford, Ill.
Orrell, Jay C., 422 E. 81st St.
Ostergren, Robert C., 4300 N. Clark St.
Pearson, Gustav E., 4437 Wrightwood Ave.
Perkins, Dwight H., 2319 Lincoln Rd., Evanston, Ill.
Perry, Ronald F., 32 W. Randolph St.
Peterson, Ivan R., 4826 Elm St., Niles Center, Ill.
Pfeiffenberger, Geo. D., 102 W. 3rd St., Alton, Ill.
Pipher, W. Vernon, Sterling Ave., Flossmoor, Ill.
Pleins, Leo H., 527 W. Monroe St., Springfield, Ill.
Pond, Irving K., 14 W. Elm St.
Prather, Fred V., 1791 Howard Ave.
Pridmore, J. E. O., 38 S. Dearborn St.
Prindeville, Chas. H., 320 Lee St., Evanston, Ill.
Probst, Edward, 80 E. Jackson Blvd.
Probst, Marvin, 80 E. Jackson Blvd.
Quinn, James Edwin, 10 S. LaSalle St.
Quitsow, Anthony H., 228 N. LaSalle St.
Rae, Robert, 431 Greenleaf Ave., Wilmette.
Raeder, Henry, 19 S. Stone St., La Grange, Ill.
Rapp, George L., 230 N. Michigan Ave.
Rapp, Macon G., 230 N. Michigan Ave.
Rathmann, Walter L., 316 N. 8th St., St. Louis, Mo.
Rezny, James B., 2202 S. Crawford Ave.
Rice, Nelson P., 7859 South Shore Dr.
Riddle, Herbert H., 5626 Woodlawn Ave.
Riley, Ivan H., Harlingen, Texas.
Rissman, Maurice B., 65 E. South Water St.
Roberts, E. E., 82 W. Washington St.
Robinson, Argyle E., 35 S. Oak St., Hinsdale, Ill.
Root, John Willborn, 333 N. Michigan Ave.
Rowe, Charles B., 1103 S. Crescent Ave., Park Ridge, Ill.
Royer, Joseph W., 209 Broadway, Urbana.
Rupinski, Edward W., 2130 N. Kedzie Ave.
Rusy, Anthony F., 1339 S. Avers Ave.
Ruttenberg, Albert M., 6800 Sheridan Road.
Ryan, Wm. J., 43 E. Ohio St.
Sandel, Monroe R., 228 N. LaSalle St.
Sanders, Lewis Miles, 35 S. Dearborn St.
Saylor, Raymond Lewis, 1650 Mill St., Des Plaines, Ill.
Shantz, Cedric A., 38 S. Dearborn St.
Schmidt, Hugo, 421 N. Wheaton Ave., Wheaton, Ill.
Schmidt, Richard E., 104 S. Michigan Ave.
Schopp, Lawrence O., 3832 Washington Blvd., St. Louis, Mo.
Schulzke, Wm. H., People's Bank Bldg., Moline, Ill.
Schwartz, Albert A., 163 E. Ohio St.
Schwarz, Julius J.,P. O. Box 1865.
Scribbins, John A., Kewanee, Ill.
Sexton, Frank A., 220 S. W. 50th St., Miami, Fla.
Shattuck, W. F., 221 N. LaSalle St.
Sheldon, Karl H., Shorewood Hills, Madison, Wis.
Slupkowski, Joseph A., 3024 Haussen St.
Smith, Harold, 30 N. LaSalle St.
Sobel, Herbert, 540 N. Michigan Ave.
Spencer, Charles B., 7411 S. Euclid Ave.
Spencer, Nelson S., 1422 E. 65th St.
Spieler, Herbert E., 435 Dwight St., Elgin, Ill.
Spitzer, Herbert, 205 W. Wacker Drive.
Spitzer, Maurice, 205 W. Wacker Drive.
Stanhope, Leon E., 410 N. Michigan Ave.
Stanton, Frederick C. H., 307 N. Michigan.
Starr, F. Charles, Cosmos Club, Washington, D. C.
Steif, B. Leo, 920 N. Michigan Ave.
Stern, Isaac S., 306 S. Wabash Ave.
Stouffer, Ernest Lawrence, 256 Administration Bldg., Champaign, Ill.
Stromquist, Victor H., 59 E. VanBuren St.
Study, Guy, 1363 Arcade Bldg., St. Louis, Mo.
Stuhr, Wm., 310 Safety Bldg., Rock Island.
Sturm, Meyer J., 708 Church St., Evanston, Ill.
Sturnfield, Charles H., 2651 W. 15th St.
Swensson, Walter F., 2200 Bennett Ave., Evanston, Ill.
Swern, Perry W., 19 S. LaSalle St.
Tallmadge, Thomas E., 19 E. Pearson St.
Teesdale, Lawrence V., c./o. Forest Products Laboratory, Madison, Wis.
Teisen, Axel V., 5332 N. Spaulding Ave.
Ticknor, James H., 262 Deerpath Rd., Lake Forest, Ill.
Thielbar, Frederick J., 520 N. Michigan Ave.
Tomlinson, H. Webster, 106 N. Chicago Ave., Joliet, Ill.
Turk, Harry E., 232 E. Erie St.

Urbain, Jules, Jr., 228 N. LaSalle St.
Urbain, Leon F., 605 W. Washington St.
Urbanek, Charles A., 1428 S. Avers Ave.
Vail, Morrison, 1311 Chicago Ave., Evanston, Ill.
Van Bergen, John S., 234 Cedar St., Highland Park, Ill.
Van der Meer, Wybe J., 70 S. May St., Aurora, Ill.
Vitzthum, Karl M., 1 N. LaSalle St.
Wachter, Henry A., 504 Grant St., Peru, Ill.
Wallace, Chas. L., Will County Natl. Bank Bldg., Joliet, Ill.
Wallace, Dwight G., 14 Rockridge Rd., Mt. Vernon, N. Y.
Wallace, Maurice R., 5004 N. Springfield Ave.
Waterman, Harry Hale, 3915 Vincennes Ave.
Watson, Jesse N., Chemical Bldg., St. Louis.
Watson, Vernon S., 643 Fair Oaks Ave., Oak Park, Ill.
Weber, Bertram A., 820 N. Michigan Ave.
Weisfeld, Leo H., 155 N. Clark St.
Weiss, John W., 343 S. Dearborn St.
Weissenborn, Leo, Tribune Tower.
Whitmeyer, Mark H., 207 S. Glenwood Ave., Peoria, Ill.
Wilmanns, August C., 4414 Malden Ave.
Woltersdorf, A. F., 520 N. Michigan Ave.
Worthington, Earl Carlton, 900 S. State St., Springfield, Ill.
Worthmann, Henry, 1440 Roscoe St.
Wright, Clark C., 333 N. Michigan Ave.
Wright, William Campbell, 140 S. Dearborn St.
Yeretsky, Norman M., 664 N. Michigan Ave.
Zelenka, Anthony James, 2331 W. 22nd St., Cicero, Ill.
Zimmer, John E., 1139 W. Lawrence St., Springfield, Ill.
Ziegle, John N., 1805 Knoxville Ave., Peoria, Ill.
Zook, R. Harold, 140 S. Dearborn St.
Zuckerman, Benjamin F., 4821 Keystone Ave.

HONORARY MEMBERS

Goodnow, Charles N., 100 N. LaSalle St.
Huber, Julius H., 2039 Greenleaf Ave.
Palmer, H. L., 134 N. LaSalle St.
Pfeiffer, Geo. L., Miami, Florida.
Polym, Francis J., Niles, Mich.

In Memoriam

1932-1939

CARL M. ALMQUIST
GEORGE S. BANNISTER
JULIUS J. BARTHEL
FRANCIS M. BARTON
ROBERT C. BERLIN
EDWARD A. BLONDIN
FRANK D. CHASE
ROBERT C. CLARK
WILLIAM H. CONWAY
SAMUEL N. CROWEN
RALPH W. ERMELING
JOHN W. GADDIS
HAROLD E. GALLUP
WILLIAM GAUGER
ERNEST R. GRAHAM
ANKER S. GRAVEN
GEORGE PALMER GRAVES
RICHARD GRIESSER
WILLIAM F. GUBBINS
JOHN HANIFEN
GEORGE B. HELMLE
JOHN T. HETHERINGTON
RAYMOND M. HOOD
BERT C. HUBBARD
WILLIAM B. ITTNER
PERCY T. JOHNSTONE
KARL E. JYRCH
OTTO A. KUPFER
MORTON LEVITON
JOHN A. LINDSTRAND
JOSEPH C. LLEWELLYN
ADOLPH LONEK
SIDNEY C. LOVELL
ANDERS G. LUND
JOHN A. MALLINGER
JOHN LAWRENCE MAURAN
JOHN A. NYDEN
JASON F. RICHARDSON
LOUIS E. RITTER
FERDINAND W. C. ROEDDIGER
SIGURD A. ROGNSTAD
RICHARD O. ROSEN
RUDOLPH SCHENCK
HENRY J. SCHLACKS
RICHARD G. SCHMID
FRANK SHOENFELDT
SINCLAIR M. SEATOR
ORLANDO VAN GUNTEN
ALFRED P. WEBER
CHARLES WOODS WEBSTER
JOHN W. WEISS
HARRY B. WHEELOCK
HOWARD J. WHITE
JAMES McLAREN WHITE
JOHN E. YOUNGBERG

THE AMERICAN INSTITUTE OF ARCHITECTS

THE OCTAGON HOUSE, WASHINGTON, D. C.

OFFICERS, 1938-1939

CHARLES D. MAGINNIS, President........................Statler Building, Boston, Mass.
FREDERICK H. MEYER, Vice-President...........1201 Kohl Building, San Francisco, Calif.
CHARLES T. INGHAM, Secretary.................1211 Empire Building, Pittsburgh, Pa.
EDWIN BERGSTRUM, Treasurer........Citizens National Bank Bldg., Los Angeles, Calif.

CHAPTERS AND OFFICERS

LEIGHT HUNT.....State Association Representative, 152 Wisconsin Ave., Milwaukee, Wis.

Presidents (*) and Secretaries (†) Listed as of October 1, 1938

Alabama—* Jack Bass Smith, Steiner Bank Bldg., Birmingham, Ala.; †Marshall E. Van Arman, 1202 Martin Bldg., Birmingham, Ala.

Albany—*Gilbert L. Van Auken, Delmar, N. Y.; †August Lux, 100 State Street, Albany, N. Y.

Arizona—*Leslie J. Mahoney, 1100 Title Bldg., Phoenix, Ariz.; †Fred W. Whittlesey, 900 Security Bldg., Phoenix, Ariz.

Arkansas—*Harry Wanger, 1316 Donaghey Bldg., Little Rock, Ark.; †Lawson L. Delony, 2407 Louisiana St., Little Rock, Ark.

Baltimore—*D. K. Este Fisher, Jr., 1012 N. Calvert St., Baltimore, Md.; † Francis H. Jencks, 113 W. Mulberry St., Baltimore, Md.

Boston—*John T. Whitmore, 50 Congress St., Boston, Mass.; †Stanley E. Davidson, 185 Devonshire St., Boston, Mass.

Brooklyn—*Ralph M. Rice, 655 Fifth Ave., New York, N. Y.; † Henry V. Murphy, 1 Hanson Pl., Brooklyn, N. Y.

Buffalo—*Paul H. Harbach, 505 Franklin St., Buffalo, N. Y.; †Stanley C. Podd, 1376 Amhurst St., Buffalo, N. Y.

Central Illinois—*Bryant E. Hadley, 1005 S. 4th St., Springfield, Ill.; †F. M. Lescher, 304 Architecture Bldg., Urbana, Ill.

Central New York—*Egbert Bagg, 258 Genesee St., Utica, N. Y.; †Clement R. Newkirk, 258 Genesee St., Utica, N. Y.

Chicago—*Elmer C. Roberts, 82 W. Washington St., Chicago, Ill.; †Carl E. Heimbrodt, 20 N. Wacker Drive, Chicago, Ill.

Cincinnati—*Charles R. Strong, 3701 Carew Tower, Cincinnati, Ohio; †John Becker, 15 E. 8th St., Cincinnati, Ohio.

Cleveland—*Joseph L. Weinberg, 1836 Euclid Ave., Cleveland, Ohio; †Francis K. Draz, 13124 Shaker Square, Cleveland, Ohio.

Colorado—*Roland L. Linder, 507 Insurance Bldg., Denver, Colo.; †R. Ewing Stiffler, 1925 Ivanhoe St., Denver, Colo.

Columbus—*Kyle W. Armstrong, 21 E. State St., Columbus, Ohio; †Ralph Chas. Kempton, 50 W. Broad St., Columbus, Ohio.

Connecticut—*Charles Scranton Palmer, 75 Whitney Ave., New Haven, Conn.; †Herbert Gibson, Jr., 904 Main St., Hartford, Conn.

Dayton—*Clifford C. Brown, 1129 Reibold Bldg., Dayton, Ohio; †Geo. T. Neuffer, 437 Ludlow Arcade, Dayton, Ohio.

Delaware—*Reah de B. Robinson, 312 Equitable Bldg., Wilmington, Del.; † Samuel E. Homsey, 611 Industrial Trust Bldg., Wilmingtou, Del.

Detroit—*Richard P. Raseman, 3 Academy Rd., Bloomfield Hills, Mich.; †Talmage C. Hughes, 120 Madison Ave., Detroit, Mich.

Eastern Ohio—*Charles E. Firestone, 1342 Cleveland Ave., N. W., Canton, Ohio; †Charles F. Owsley, 211 N. Champion St., Youngstown, Ohio.

Florida Central—*Franklin O. Adams, 305 Morgan St., Tampa, Fla.; †Norman F. Six, 212 Franklin St., Tampa, Fla.

Florida North—*Jefferson D. Powell, Professional Bldg., Jacksonville, Fla.; †Lee Roy Sheftall, 305 Main St., Jacksonville, Fla.

Florida South—*Vladimir Virrick, 1629 Michigan Ave., Miami Beach, Fla.; †George H. Spohn, 930 Seybold Bldg., Miami, Fla.

Georgia—*Samuel I. Cooper, 827-30 Forsyth Bldg., Atlanta, Ga.; †J. Warren Armistead, Jr., 1330 Candler Bldg., Atlanta, Ga.

Grand Rapids—*Warren L. Rindge, 740 Michigan Trust Bldg., Grand Rapids, Mich.; †John P. Baker, 756 Bristol Ave., N. W., Grand Rapids, Mich.

Hawaii—*Claude Albon Stiehl, P. O. Box 82, Honolulu, T. H.; †Raymond L. Morris, 300 Boston Bldg., Honolulu, T. H.

Indiana—*Edward D. Pierre, 909 Architects & Builders Bldg., Indianapolis, Ind.; †John R. Kelly, 1034 Architects & Builders Bldg., Indianapolis, Ind.

Iowa—*J. Woolson Brooks, 815 Hubbell Bldg., Des Moines, Iowa; †John Normile, 511 Hubbell Bldg., Des Moines, Iowa.

Kansas City—*A. W. Archer, 916 Pioneer Trust Bldg., Kansas City, Mo.; † Homer Neville, 4 W. 13th St., Kansas City, Mo.

Kansas—*Chas. W. Shaver, 823 United Life Bldg., Salina, Kan.; †Thomas Larrick, 1021 Massachusetts St., Lawrence, Kan.

Kentucky—*Thomas J. Nolan, 311 Kentucky Home Life Bldg., Louisville, Ky.; †Ossian P. Ward, 1002 Washington Bldg., Louisville, Ky.

Louisiana—*Arthur Feitel, Carondelet Bldg., New Orleans, La.; †Douglass V. Freret, 620 Audubon Bldg., New Orleans, La.

Madison—*Arthur Peabody, State Capitol, Madison, Wis.; †Edward J. Law, First Central Bldg., Madison, Wis.

Maine—*John P. Thomas, 21 Free St., Portland, Me.; †Josiah T. Tubby, 21 Free St., Portland, Me.

Minnesota—*Louis B. Bersback, 702 Wesley Temple Bldg., Minneapolis, Minn.; †Edwin W. Krafft, 715 Rand Tower, Minneapolis, Minn.

Mississippi—*Eugene D. Drummond, Old Merchants Bank Bldg, Jackson, Miss.; †R. W. Naef, 411 E. Capitol St, Jackson, Miss.

Montana—*Fred A. Brinkman, Kalispell, Mont.; †W. R. Plew, Bozeman, Mont.

Nebraska—*Linus Burr Smith, University of Nebraska, Lincoln, Neb.; †N. R. Brigham, 5404 Western Ave., Omaha, Nebr.

New Jersey—*Arthur B. Holmes, 18 Burnside St., Upper Montclair, N. J.; †Clement W. Fairweather, Metuchen, N. J.

New York—*Arthur Loomis Harmon, 11 E. 44th St., New York, N. Y.; †Robt. B. O'Connor, 101 Park Ave., New York, N. Y.

North Carolina—*George Watts Carr, 111 Corcoran St., Durham, N. C.; †Roy Marvin, 111 Corcoran St., Durham, N. C.

North Louisiana—*Edward F. Neild, Shreveport, La.; †Seymour Van Os (Acting), Ricou Brewster Bldg., Shreveport, La.

North Texas—*Walter C. Sharp, 707 Construction Bldg., Dallas, Texas; †Geo. L. Dahl, Insurance Bldg., Dallas, Texas.

Northern California—*Warren C. Perry, 260 California St., San Francisco, Calif.; †James H. Mitchell, 369 Pine St., San Francisco, Calif.

Northwestern Pennsylvania — *Clement S. Kirby, 4502 Homeland Ave., Erie, Pa.; †J. Howard Hicks, 124 W. 7th St., Erie, Pa.

Oklahoma—*Joseph Edgar Smay, University of Oklahoma, Norman, Okla.; †Leonard H. Bailey, Colcord Bldg., Oklahoma City, Okla.

Oregon—*Leslie D. Howell, 404 U. S. National Bank Bldg., Portland Ore.; †Roi L. Morin, 1603 Public Service Bldg., Portland, Ore.

Philadelphia—*Roy F. Larson, 1700 Architects Bldg., Philadelphia, Pa.; †Thomas Pym Cope, Architects Bldg., Philadelphia, Pa.

Pittsburgh—*Lawrence Wolfe, 119 E. Montgomery Ave., Pittsburgh, Pa.; †Rody Patterson, 2422 Koppers Bldg., Pittsburgh, Pa.
Rhode Island—*Wallis E. Howe, 1014 Turks Head Bldg., Providence, R. I.; †J. Peter Geddes, II, 840 Hospital Trust Bldg., Providence, R. I.
San Diego—*Sam W. Hamill, Bank of America Bldg., San Diego, Calif.; †Louis J. Gill, Sefton Bldg., San Diego, Calif.
Santa Barbara—*Ralph W. Armitage, 235 W. Victoria St., Santa Barbara, Calif.; †Chester L. Carjola, 209 Picacho Lane, Santa Barbara, Calif.
Scranton-Wilkes-Barre — *Arthur P. Coon, Union Bank Bldg., Scranton, Pa.; †Searle H. Von Storch, Union Bank Bldg., Scranton, Pa.
South Carolina—*J. Whitney Cunningham, 2 Pierson Bldg., Sumter, S. C.; †G. Thomas Harmon, III, Valley Road, Hartsville, S. C.
South Georgia—*Morton H. Levy, 3d Floor, Levy Store Bldg., Savannah, Ga.; †Walter P. Marshall, 1108 E. Henry St., Savannah, Ga.
South Texas—*Addison Stayton Nunn, 212 Scanlan Bldg., Houston, Texas; †Theo. F. Keller, 6551 S. Main St., Houston, Texas.
Southern California — *Eugene Weston, Jr., Architects Bldg., Los Angeles, Calif.; †Edgar Bissantz, 816 W. 5th St., Los Angeles, Calif.
Southern Pennsylvania—*James W. Minick, 505 N. 2nd St., Harrisburg, Pa.; †Rolf G. Loddengaard, 306 Park Ave., New Cumberland, Pa.
St. Louis—*Benedict Farrar, 1367 Arcade Bldg., St. Louis, Mo.; †Arthur E. Koelle, 316 N. 8th St., St. Louis, Mo.
St. Paul—*Roy H. Haslund, 2020 Juliet St., St. Paul, Minn.; †Paul M. Havens, First National Bank Bldg., St. Paul, Minn.
Tennessee—Charles I. Barber, 517½ W. Church Ave., Knoxville, Tenn.; †William P. Bealer, 1547 W. Clinch Ave., Knoxville, Tenn.
Toledo—*John N. Richards, 518 Jefferson Ave., Toledo, Ohio; †Mark B. Stophlet, Security Bank Bldg., Toledo, Ohio.
Utah—*Walter E. Ware, 610 Utah Savings & Trust Bldg., Salt Lake City, Utah; †Lloyd W. McClenahan, 818 Garfield Ave., Salt Lake City, Utah.
Virginia—*Walter R. Crowe, 609 Krise Bldg., Lynchburg, Va.; †Louis Philippe Smithey, 112 Kirk Ave., Roanoke, Va.
Washington, D. C.—*Louis Justement, 1223 Conn. Ave. N. W., Washington, D. C.; †Alfred Kastner, 2 Dupont Circle, Washington, D. C.
Washington State—*B. Marcus Priteca, 515 Palomar Bldg., Seattle, Wash.; †Victor N. Jones, 504 Republic Bldg., Seattle, Wash.
West Texas—*Harvey P. Smith, National Bank of Commerce Bldg., San Antonio, Texas; †Glenn C. Wilson, 404 W. Laurel Pl., San Antonio, Texas.
West Virginia—*C. C. Wood (Acting), Lowndes Bldg., Clarksburg, W. Va.; †C. E. Silling, Box 861, Charleston, W. Va.
Westchester—*Kenneth K. Stowell, 28 Tanglewylde Ave., Bronxville, N. Y.; †Robert H. Scannell, 80 Palmer Ave., Bronxville, N. Y.
Wisconsin—*Richard Philipp, 707 N. Broadway, Milwaukee, Wis.; †Alexander H. Bauer, 606 W. Wisconsin Ave., Milwaukee, Wis.

CENTRAL ILLINOIS CHAPTER
AMERICAN INSTITUTE OF ARCHITECTS

OFFICERS

Bryant E. Hadley, President
Walter G. Jameson, 1st Vice-President
Thomas E. O'Donnell, 2nd Vice-President
Frank M. Lescher, Secretary-Treasurer

Directors

One Year	*Two Years*	*Three Years*
Cornelius W. Macardell	Frank N. Emerson	Ernest L. Stouffer

Members

Berger, F. E., 304 Lincoln Bldg., Champaign, Ill.
Booton, Joseph F., Div. of Arch. & Eng., Armory-Office Bldg., Springfield, Ill.
Deam, Arthur F., 103 Architecture Bldg., Urbana, Ill.
Emerson, Frank N., 1600 Alliance Life Bldg., Peoria, Ill.
Foster, William Arthur, Ag. Eng. Bldg., Urbana, Ill.
Gregg, Richard S., 1600 Alliance Life Bldg., Peoria, Ill.
Hadley, Bryant E., Div. of Arch. & Eng., Armory-Office Bldg., Springfield, Ill.
Harrison, Elbert I., 1014 Alliance Life Bldg., Peoria, Ill.
Hooton, Philip R., 710 Peoples Bank Bldg., Bloomington, Ill.
Hotchkiss, Robert J., 201 Fredonia Ave., Peoria, Ill.
Jameson, W. G., 1014 Alliance Life Bldg., Peoria, Ill.
Lescher, Frank M., 304 Architecture Bldg., Urbana, Ill.
Lundeen, Edgar E., Corn Belt Bank Bldg., Bloomington, Ill.
Macardell, Cornelius W., Div. of Arch. & Eng. Armory-Office Bldg., Springfield, Ill.
McCoy, William, Mt. Vernon, Ill.
Newcomb, Rexford, 110 Architecture Bldg., Urbana, Ill.
Noble, James B., Div. of Arch. & Eng. Armory-Office Bldg., Springfield, Ill.
O'Donnell, Thomas E., 119 Architecture Bldg., Urbana, Ill.
Pleins, Leo. H., 527 W. Monroe St., Springfield, Ill.
Provine, Loring H., 104 Architecture Bldg., Urbana, Ill.
Ramey, George E., Robeson Bldg., Champaign, Ill.
Reiger, Harry J., Booth Bldg., Springfield, Ill.
Royer, Joseph W., Urbana-Lincoln Bldg., Urbana, Ill.
Schaeffer, A. N., 710 Peoples Bank Bldg., Bloomington, Ill.
Stouffer, Ernest L., 256 Administration Bldg., Urbana, Ill.
Worthington, Earl C., 900 S. State St., Springfield, Ill.

YEAR 'ROUND PROTECTION

Manufactured by

PROTEX WEATHERSTRIP MFG. CO.

2306-10 West 69th Street

CHICAGO, ILL. PROspect 0940-1

Refer to Sweet's Architectural Catalogue for "The Blue Book of Weatherstrips" to complete specifications and details. University of Wisconsin and R. W. Hunt Company Testing Laboratories for infiltration factors.

PRODUCTS

- **METAL WEATHERSTRIPS**
  Zinc—Brass—Bronze
  Stainless Steel
- **WINTER PANES**
  Double Glazing Units
- **THRESHOLDS**
  Extruded Brass—Aluminum
- **CALKING COMPOUNDS**
  Gun and Knife Types

INSTALLATIONS

U. S. Federal Housing Projects
Evansville, Indiana
Columbia, South Carolina
Oklahoma City, Oklahoma
Cleveland, Ohio
Chicago, Illinois

John Carroll, Univ.	Cleveland, Ohio
Knox College	Galesburg, Ill.
High School	Lake Forest, Ill.
Naval Air Station	Lakehurst, N. J.
Signal Mt. Hotel	Signal Mt., Tenn.
Dept. of Int. Bldgs.	Washington, D.C.

THE LODGE—SUNSET VALLEY WINTER RESORT—IDAHO

CHICAGO CHAPTER
AMERICAN INSTITUTE OF ARCHITECTS

OFFICERS FOR 1938-1939

ELMER C. ROBERTS, President
JOHN H. RAFTERY, 1st Vice-President.
JERROLD LOEBL, 2d Vice-President
LAWRENCE B. PERKINS, Treasurer
CARL E. HELMBRODT, Secretary

Executive Committee

JOHN O. MERRILL
PHILIP B. MAHER
ROBERT S. DEGOLYER
THOMAS E. TALLMADGE

All addresses are Chicago, Ill., except where otherwise noted.

Fellows

Bennett, Edward H., 80 E. Jackson Blvd.
Bollenbacher, John C., 333 N. Michigan Ave.
Burnham, Daniel H., 134 S. La Salle St.
Burnham, Hubert, 160 N. La Salle St.
Chatten, Melville C., 160 N. La Salle St.
DeGolyer, Robert S., 307 N. Michigan Ave.
Dunning, N. Max, Powhatan Hotel, Washington, D. C.
Fellows, William K., 814 Tower Ct.
Garden, Hugh M. G., 104 S. Michigan Ave.
Hall, Emery Stanford, 175 W. Jackson Blvd., Suite 2300
Hamilton, John L., 814 Tower Ct.
Hammond, Charles Herrick, 160 N. La Salle St., Room 1900
Holabird, John A., 333 N. Michigan Ave.
Holsman, Henry K. (M.E.), 140 S. Dearborn
Jensen, Elmer C., 39 S. La Salle St.
Matteson, Victor A., 20 N. Wacker Drive
Mundie, W. B. (M.E.), 39 S. La Salle St.
Nimmons, George C. (M.E.), 333 N. Michigan
Perkins, Dwight H., 2319 Lincoln St., Evanston, Ill.
Pond, Irving K. (M.E.), 14 W. Elm St.
Prindeville, Charles H. (M.E.), 320 Lee Ave., Evanston, Ill.
Root, John W., 333 N. Michigan Ave.
Schmidt, Richard E., 104 S. Michigan Ave.
Smith, William Jones, 430 N. Michigan Ave.
Spencer, Robert C., Jr., 302 Ridgewood Ave., Orlando, Fla.
Tallmadge, Thomas E., 19 E. Pearson St.
Woltersdorf, Arthur, 520 N. Michigan Ave.

Chapter Members

Adler, David, 220 S. Michigan Ave.
Alderman, Wm. N., 19 E. Pearson St.
Alschuler, Alfred S., 28 E. Jackson Blvd.
Anderson, B. D. Andy, 4832 N. Bell Ave.
Anderson, Stanley D., 262 E. Deerpath Ave., Lake Forest, Ill.
Armstrong, John A., 11 S. La Salle St.
Baker, Cecil F., 431 N. 39th St., Omaha, Nebr.
Barry, Gerald A., 5202 W. Chicago Ave.
Baur, Ervin F., 842 W. Waveland Ave.
Beidler, Herbert B., 952 N. Michigan Ave.
Bentley, Harry H., 337 Woodland Road, Highland Park, Ill.
Bieg, Harry K., 720 Clark St., Evanston, Ill.
Blouke, Pierre M., 431 N. Michigan Ave.
Bohasseck, Charles, 333 N. Michigan Ave.
Bourke, Robert E., 39 S. La Salle St.
Brand, Herbert Amery, 400 W. Madison St.
Carr, George W., 333 N. Michigan Ave.
Cervin, Olaf Z., 310 Safety Bldg., Rock Island, Ill.
Chapman, George A., 105 W. Adams St., Room 2524
Cheney, Howard L., 3031 Sedgwick St., N. W. Washington, D. C.
Childs, Frank A., 430 N. Michigan Ave.
Church, Walter S., 449 W. 14th St., New York, N. Y.
Clark Edwin H., 8 E. Huron St.
Clark, William J., 1231 Trestle Glenn Road, Oakland, Calif.
Clas, Angelo Robert, 212 Shoreham Bldg., Washington, D. C.
Colean, Miles L., 1514 44th St., Washington, D. C.
Cowles, Knight C., 30 N. Michigan Ave.
Danielson, Philip A., 1625 Hinman Ave., Evanston, Ill.
Davis, Zachery T., 228 N. La Salle St.
Doerr, William P., 5487 Hyde Park Blvd.
Dubin, Henry, 127 N. Dearborn St.
Dunning, Hugh Baker, 104 N. Oak Park Ave., Oak Park, Ill.
Dyer, Scott S., 38 S. Dearborn St.
Eastman, Albert Reyner, 120 S. Wyman St., Rockford, Ill.
Ehmann, William, 104 S. Michigan Ave.
Elmslie, George E., 122 S. Michigan Ave.
Elting, Winston, 33 N. Michigan Ave.
Erikson, Carl A., 104 S. Michigan Ave.
Fairclough, Stanley D., 809 Exchange St., U. S. Yards
Ferrenz, Tirrell J., 1323 Ingraham St., N. W. Washington, D. C.
Frazier, Walter S., 540 N. Michigan Ave.
Frost, Harry T., 80 E. Jackson Blvd.
Fugard, John Reed, 520 N. Michigan Ave.
Fuhrer, Eugene, 188 W. Randolph St.
Fuller, Revilo F., 6 N. Michigan Ave.
Furst, William H., 11 S. La Salle St.
Ganster, Wm. Allaman, 28 N. Genesee St., Waukegan, Ill.
Gerhardt, Paul, Jr., 1012 City Hall
Gollnick, Louis R., 316 N. Oxford St., Arlington, Va.
Grunsfeld, Ernest A., 435 N. Michigan Ave.
Guenzel, Louis (M.E.), 879 N. State St.
Gylleck, Elmer A., 10 N. Clark St.
Haagen, Paul T., 155 N. Clark St.
Heimbrodt, Carl Edward, 4110 Lawn Ave., Western Springs, Ill.
Henderson, Charles Clinton, 1405 Forest Ave., Wilmette, Ill.
Herter, John T., 2638 Fargo Ave.
Heun, Arthur, 410 S. Michigan Ave.

"Alumilite" ALUMINUM SLAT
★ VENETIAN BLINDS ★
CHICAGO VENETIAN BLIND CO.
EXCLUSIVE MANUFACTURERS
1210 SOUTH MORGAN STREET CHICAGO, ILLINOIS

PRODUCT . . . The new, Chicago "Alumilite" Aluminum Slat Venetian Blinds, with the Chicago Metal-Box Closed Head that eliminates light streaks at the top and conceals all operating hardware, represent the last word in venetian blind design and construction. They introduce a new theme in color and decoration that makes the window an independent unit for beautiful treatment.

WHAT ARE "ALUMILITE" ALUMINUM SLAT VENETIAN BLINDS—They are blinds composed of Aluminum Slats which have been subjected to an electrolytic process which builds up an oxide coating on the surface of the aluminum.

CHARACTERISTICS OF THE FINISH—The "Alumilite" process provides a platinum gray durable finish with a high abrasive resistance, which is also highly resistant to corrosion. It replaces plated metal, eliminating any chipping, flaking, or peeling. The process, in reality, produces a finish which is actually integral with the metal, and which reduces maintenance costs to a minimum.

NEWEST THEME IN COLOR AND DECORATION . . . The platinum gray tone of these new blinds blends harmoniously with every color scheme in every type of surrounding. Captivating color effects are easily achieved and accented by the use of various colors of tapes, as the aluminum slats actually reflect the tape colors.

EASE OF INSTALLATION . . . Chicago "Alumilite" Aluminum Slat Blinds are delivered complete—with all necessary hardware—ready for installation. As each blind is built to fit a particular opening—installation requires but a few minutes.

FINISH—Slats and bottom bar are "Alumilite" finished in accordance with specifications of patents controlling the anodizing process. This finish—in a neutral tone—provides a maximum reflection of daylight without glare and protects against tarnishing, oxidation and pitting.

LADDER TAPE—Best quality imported or domestic. Standard colors.

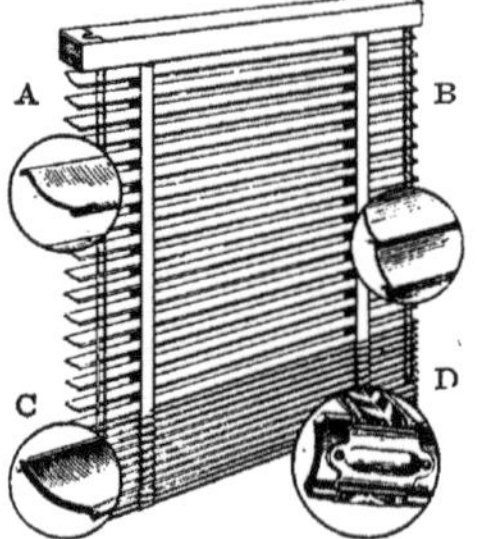

A. "Alumilite" Aluminum Slats are streamlined to facilitate deflection of direct sun rays to produce indirect lighting with elimination of glare and gloom areas inside of room.

B. Slat design permits proper overlap to shut out light when desired.

C. Beautiful design of bottom bar enhances blind appearance.

D. Tapes easily removed from bottom bar with special plates. No tacks or staples used. Every blind may be lengthened or shortened at least one inch without undue effort.

CONSTRUCTION . . . SLATS —Aluminum alloy 2-SH—scientifically formed to a special shape that provides ample rigidity and proper light deflection. Dimensions: 24 gauge by 2 inches wide.

A BLIND FOR EVERY BUILDING . . . the unique features of CHICAGO "ALUMILITE" ALUMINUM SLAT VENETIAN BLINDS make them readily adaptable to every type of building and size of opening to be shaded.

The permanence of the aluminum slats that maintain their finish indefinitely makes these blinds highly desirable for the Monumental or Governmental type of building. In Office and Factory buildings, Schools, Hospitals, and Institutions where light control is vital and ease of cleaning and low maintenance costs are important factors—these new blinds meet every requirement. For the Hotel and Apartment Building their desirability, ease of cleaning and neutral color of the aluminum slats, that harmonizes with every surrounding, make possible immense savings. In the Residence, they breathe an entirely new note into the modern decorative spirit.

FOR COMPLETE INFORMATION SEE "SWEET'S 1938 CATALOG"

Hirschfeld, Leo Saul, 65 E. South Water St.
Hoover, Ira W., 408 S. Oxford Ave., Los Angeles, Calif.
Horn, Benjamin A., Rock Island Bank Bldg., Rock Island, Ill.
Hoskins, John M. (M.E.), 2830 W. Madison
Hull, Denison B., 77 W. Washington St.
Irwin, Howard Emsley, 1006 Main St., Evanston, Ill.
Jackson, Emery B., 5617 Dorchester Ave.
Jansson, Edward F., 740 N. Rush St.
Jensen, Clarence A., 3925 N. Keeler Ave.
Johnck, Frederick, 104 S. Michigan Ave.
Kastrup, Carl J., 7346 Madison St., Forest Park, Ill.
Kegley, Frank T., 1515 Sherman Ave., Evanston, Ill.
Klaber, Eugene H., 1556 E. West Highway, Silver Spring, Md.
Kroman, M. Louis, 180 N. Michigan Ave.
Kurzon, R. Bernard, 134 N. La Salle St.
Lawrence, Clark J., 960 Sheridan Rd., Hubbard Woods, Ill.
Llewellyn, Ralph C., 38 S. Dearborn St.
Loehl, Jerrold, 333 N. Michigan Ave.
Lovell, McDonald, 224 E. Ontario St.
MacBride, Edward E., 43 E. Ohio St.
Maher, Philip B., 157 E. Erie St.
Marshall, Benjamin H., 1515 Sherman Ave., Evanston, Ill.
Martin, Edgar, 540 N. Michigan Ave.
Marx, Sam A., 333 N. Michigan Ave.
Mayo, Peter B., 53 W. Jackson Blvd.
McCarthy, Jos. W., 43 E. Ohio St.
McLaren, Robert J., 5934 Midway Park
Merrill, Edward A., 540 N. Michigan Ave.
Merrill, John O., 134 N. La Salle St. (F.H.A.)
Meyer, Frederic H., 820 N. Michigan Ave.
Mittelbusher, Edwin H., 104 S. Michigan Ave.
Murphy, Charles Francis, 80 E. Jackson Blvd.
Naess, Sigurd Edor, 80 E. Jackson Blvd.
Newhouse, Henry L., II, 8 S. Michigan Ave.
Ogg, John W., 347 Madison Ave., Room 807, New York, N. Y.
Oldefest, Edward G., 301 Moline National Bank Bldg., Moline, Ill.
Oman, Samuel S., 162 E. Ohio St.
Orput, Raymond A., 202 Third National Bank Bldg., Rockford, Ill.
Ostergren, R. C., 4300 N. Clark St.
Owings, Nathaniel A., 104 S. Michigan Ave.
Parsons, William E., 80 E. Jackson Blvd.
Pereira, Hal, 221 N. La Salle St.
Pereira, Wm. L., 221 N. La Salle St.
Perkins, Lawrence B., 225 N. Michigan Ave.
Perola, Louis, 4717 Beacon St.
Pridmore, John E. O., 109 N. Dearborn St.
Puckey, Francis W., 400 N. Michigan Ave.
Raeder, Henry (M.E.), 19 S. Stone Ave., La Grange, Ill.
Raftery, John H., 540 N. Michigan Ave.
Rapp, Mason Gerardi, 230 N. Michigan Ave.
Rebori, A. N., 435 N. Michigan Ave.
Reed, Earl H., Jr., 431 N. Michigan Ave.
Renwick, Edward A., 28 E. Jackson Blvd.
Riddle, Herbert Hugh, 5626 Woodlawn Ave.
Sanford, Trent Elwood, Museum of Science and Industry, 57th St. and Lake Michigan
Rissman, Maurice Barney, 65 E. South Water
Roberts, Elmer C., 82 W. Washington St.
Sandel, Monroe R., 228 N. La Salle St.
Saxe, Albert M., 43 E. Ohio St.
Schlossman, Norman J., 333 N. Michigan Ave.
Schulzke, Wm. H., 830 Fifth Ave. Bldg., Moline, Ill.
Senseney, George Towner, 600 S. Michigan Ave.
Shaw, Alfred, 80 E. Jackson Blvd.
Skidmore, Louis E., 104 S. Michigan Ave.
Small, John S., 43 E. Ohio St.
Sorgatz, Wm. David, 551 Judson Ave., Evanston, Ill.
Starr, Frank Charles, Cosmos Club, Washington, D. C.
Starbuck, Fred L., 224 S. Michigan Ave.
Steif, B. Leo, 919 N. Michigan Ave.
Stockton, Walter T., 307 N. Michigan Ave.
Stoetzel, Ralph L., 612 N. Michigan Ave.
Stromquist, Victor H., 59 E. Van Buren St.
Suter, W. Lindsay, 189 W. Madison St.
Thielbar, Frederick J., 520 N. Michigan Ave.
Tomlinson, H. Webster, 611 Morris Bldg., Joliet, Ill.
Travelletti, Rene Paul, 189 W. Madison St.
Uffendell, W. Gibbons, 232 E. Erie St.
Urbain, Leon F., 605 W. Washington Blvd.
Van Bergen, John S., 234 Cedar Ave., Highland Park, Ill.
Venning, Frank L., 333 N. Michigan Ave.
Walcott, Chester H., 75 E. Wacker Drive
Walker, Frank Chase, 1065 Balmoral Ave.
Walker, Willard, 307 N. Michigan Ave.
Walton, Lewis B., 1515 Sherman Ave., Evanston, Ill.
Waterman, H. H., 3915 Vincennes Ave.
Weber, Bertram A., 820 N. Michigan Ave.
Weissenborn, Leo J., 431 N. Michigan Ave.
Wilkinson, Lawrence E., Hotel Maryland, 900 Rush St.
Work, Robert C., 75 E. Wacker Drive
Wright, Clark G., 333 N. Michigan Ave.
Yerkes, Wallace F., 435 N. Michigan Ave.
Yost, L. Morgan, 1150 Wilmette Ave., Wilmette, Ill.
Zimmerman, Ralph W., 43 E. Ohio St.
Zakharoff, Alexis A., 5845 Blackstone Ave.
Zook, R. Harold, 140 S. Dearborn St.

Associates

Anderson, Edwin F., 2725 Hartzell Ave., Evanston, Ill.
Beckstrom, Melvin R., 2129 16th St., Moline, Ill.
Benedict, Earl R., Jr., 320 S. Scoville Ave., Oak Park, Ill.
DeLong, Albert J., 1814 N. Lincoln Park West
Edmunds, Ralph G., 4541 N. Hamilton Ave.
Gross, Ralph F., 1119 N. Harding Ave.
Mulig, Thomas J., 3036 W. 55th St.
Murison, George W., Jr., 5832 Stony Island Ave.
Peterson, Raymond A., 6506 N. Claremont Ave.
Pickens, Buford L. (moved)
Sobel, Walter H., 540 N. Michigan Ave.
Scoyoc, Lee Van, 6056 Ingleside Ave.
Wayman, Leonard, 3005 Clifton Ave.
Ward, Robert Allen, 850 Lake Shore Drive.
White, Eugene B., 19 S. La Salle St.

OFFICE PRACTICE

By Illinois Society of Architects.

Believing that uniform practice in various architects' offices is desirable for all concerned, this Society recommends that the following conditions prevail in architects' offices of the State of Illinois:

Classification of Employes.

First. That employes be classed as Regular and Special;

Second. Employes classified as "Regular" will be those continually engaged for a period of not less than one year, on a weekly salary basis; it is expected that such employes will assume greater responsibilities to their employers and be granted special privileges, in consideration of faithful service;

Third. Employes classified as "Special" will be those engaged temporarily. It is deemed proper that such employes be paid by the hour for actual service rendered, making no allowance for vacations or holidays, it being considered fair under these circumstances to allow these draughtsmen a slightly higher rate per hour than regular employes who enjoy privileges of vacations and holidays.

Office Hours.

First. It is understood that draughtsmen are expected to be in their respective offices ready to begin actual work at the hours stated, and that they will continue in service at least until the hours fixed for cessation of work;

Second. The regular opening time of offices shall be 8:30 A. M., throughout the year;

Third. Period of service for Monday, Tuesday, Wednesday, Thursday and Friday, in the morning, shall be four hours, extending to 12:30 P M., that the lunch hour shall be one hour, extending from 12:30 to 1:30 P. M.; that the afternoon period shall be four hours, extending from 1:30 to 5:30 P. M.;

Fourth. That the Saturday period of service shall consist of 4½ hours, extending from 8:30 A. M. continously to 1:00 P. M.

Units of Service.

First. One week's service will consist of 44½ hours;

Second. One year's service will consist of 2,180½ hours.

Pay-Day.

First. That pay-day shall be on Monday of every week;

Second. That each pay-day draughtsmen be paid up to the Saturday night preceding.

Holidays and Vacations.

First. We recommend that "Regular' draughtsmen be given the following holidays on full pay: New Year's, Decoration Day, July Fourth, Labor Day, Thanksgiving Christmas;

Second. That all "Regular" draughtsmen having been in the employ of an architect for more than one year be given two weeks' vacation on full pay, at time most convenient for employer;

Third. It should be understood that "Regular" draughtsmen, quitting the employer's service of their own volition, preceding the completion of any year's service, shall not be entitled to vacation allowance;

Fourth. "Regular" employes terminating service at the request of their employer shall be entitled to an allowance in cash proportionate to two weeks' salary allowed for vacation in the same ratio as period of service bears to one year;

Fifth. Vacations and holidays are understood to be granted to employes for rest and recuperation, the employe being understood to be in the service of the employer during vacation and holiday time just to the same extent as when regularly engaged in the office;

Sixth. It is recognized that an average of 44½ hours per week's service is the maximum efficient service that can be continuously rendered without detriment to the health or efficiency of the employe, and that where the employe engages in outside architectural service of any sort for others, he does so at the expense of his employer, and his employer should be credited for corresponding loss of time. The practice of employes of one employer working nights or holidays for another is condemned as detrimental to the best interests of both employer and employe;

Seventh. In case of emergencies of short duration, "Regular" employes are expected to work over-time for the employer without extra remuneration other than a reasonable allowance for the expense of taking meals away from regular lodging place. In such cases, however, the employes will be credited with off time on account of sickness or otherwise, equivalent to the amount of overtime service rendered in cases of emergency;

Eighth. Draughtsmen are encouraged, however, to make use of a portion of their time off for educational improvement.

Illinois Society of Architects,
1906—134 N. La Salle St.

Illinois Society of Architects

Suite 1906, 134 N. La Salle Street, Chicago

The following is a list of the publications of the Society; further information regarding same may be obtained from the Financial Secretary.

FORM NO. 21, "INVITATION TO BID"—Letter size, 8½x11 in., two-page document, in packages of fifty at **75c**, broken packages, two for **5c.**

FORM NO. 22, "PROPOSAL"—Letter size, 8½x11 in., two-page documents, in packages of fifty, at **75c**, broken packages, two for **5c.**

FORM NO. 23, "ARTICLES OF AGREEMENT"—Letter size, 8½x11 in., two-page document, in packages of fifty, at **75c**, broken packages, two for **5c.**

FORM NO. 24, "BOND"—Legal size, 8x13 in., one-page document, put up in packages of twenty-five, at **25c** per package, broken packages, three for **5c.**

FORM NO. 25, "GENERAL CONDITIONS OF THE CONTRACT"—Intended to be bound at the side with the specifications, letter size, 8½x11 in., ten-page document, put up in packages of fifty at **$2.50**, broken packages, three for **25c.**

FORM 26, CONTRACT BETWEEN ARCHITECT AND OWNER. Price, **5c** each, in packages of fifty, **$1.25.**

FORM 1, BLANK CERTIFICATE BOOKS—Carbon copy, from 3¾x8½ in., price, **50c.** Two for **5c.**

FORM 4, CONTRACT BETWEEN THE OWNER AND CONTRACTOR—(Old Form.) Price, two for **5c.** Put up in packages of 50 for **$1.00.**

FORM E, CONTRACTOR'S LONG FORM STATEMENT—As required by lien law. Price, two for **5c.**

FORM 13, CONTRACTOR'S SHORT FORM STATEMENT—Price, **1c** each.

CODES OF PRACTICE AND SCHEDULE OF CHARGES—8½x11 in. Price, five for **10c.**

These documents may be secured at the Financial Secretary's office, suite 1906, 134 N. La Salle St., telephone Cent. 4214. We have no delivery service. The prices quoted above are about the cost of production. An extra charge will be made for mailing or expressing same. Terms strictly cash, in advance, with the order; except that members of the Society may have same charged to their account.

SUGGESTIONS FOR FIRMS ISSUING CATALOGUES AND PRINTED MATTER

Architects are technically educated and are charged with selection on technical merit.

Exact and specific technical detail appeals to an architect because it enables him to judge quickly and correctly.

Drawings to scale of parts or the whole make arrangement or mechanism most quickly clear to the technically educated.

Testimonials from those technically incompetent to judge carry no weight with the competent.

Architects want authentic technical information about all building materials and devices.

Architects do not want to wade through a sea of laudatory verbiage in order to discover an islet of real usable information.

Architects must cover an immense variety and amount of detail in selecting the numerous materials that enter into a building.

Where much detail is handled by a single individual, success is dependent on system.

Information to be immediately available for architects must be classified so that each detail can be considered separately and in order.

Advertisers recognizing these principles and presenting exact technical information under proper classification, free from irrelevant matter and in convenient form for filing, so as to be available when that item is up for consideration, are most likely to secure satisfactory results from their efforts.

It is believed that most architects have their own particular system of filing and classification and would not take kindly to any advertising scheme contemplating the placing of filing cabinets in architects' offices and distribution by those interested in the promotion of advertising scheme. Architects do not take kindly to allowing outsiders access to their private catalogue filing cabinets, and it is impractical to have two filing systems in the same office.

Practical requirements in the preparation of specifications make it necessary for architects to divide their specifications into topics very similar to trade divisions brought about by divisions of labor promulgated by labor authorities, and no single division or chapter of a catalogue should contain matter pertaining to more than one trade; unless the material referred to is used by several trades. It is hoped that eventually the architects may agree on a satisfactory universal building material classification or index. But it is certain that this time has not yet arrived and that no person not actually having had extended experience in the preparation of architects' specifications is capable of preparing such an index that would be practical.

STANDARD SIZES
Requested by Architects

Believing that uniform practice by the various publishers of catalogues and literature for distribution to architects is desii able for all concerned, and wishing to be in accord with the recommendations of the American Institute of Architects, the Illinois Society of Architects advise that all literature for this purpose be prepared to comply as nearly as possible with the conditions set forth, as follows:

First: That 8½"x11" shall be the standard sized page for all general catalogues and bulletins intended for permanent filing by architects; thus making a size convenient for filing in the standard letter-size vertical filing cabinets, such as may be procured from any concern dealing in office filing devices.

Second: That 3¾"x8½" shall be the standard size for post cards and pocket editions intended for the use of architects; thus making a size convenient for filing three to the page, side by side, in standard letter-size vertical filing cabinets; or one to the page, on side, in standard vertical check files; or on end in standard legal document files; also convenient for mailing in standard legal size envelopes.

Third: That all catalogues should be issued in the form of separate bulletins, or chapters separated by a blank page, each treating of but one subject, on both sides of the same sheet, so as to make separation easy for classification purposes.

Fourth: That it is important to have pages cut to exact size; if over size in any particular they may not go into files; if under size, they may be overlooked in running through the files hastily.

Fifth: That these recommendations go into effect January 1, 1915, and that following that date, architects be advised to decline to receive literature for filing which does not comply with standard sizes.

Illinois Society of Architects
134 No. La Salle St.

OFFICIALS, CITY OF CHICAGO

Mayor	EDWARD J. KELLY
Secretary to the Mayor	BESSIE C. O'NEILL
City Sealer	JAMES M. O'KEEFE
City Comptroller	R. B. UPHAM
Deputy City Comptroller	A. J. KEEFE
Commissioner of Public Works	OSCAR E. HEWITT
City Treasurer	GUSTAVE A. BRAND
City Clerk	PETER J. BRADY
Chief Clerk, City Clerk's Office	EDWARD J. PADDEN
Assistant Commissioner of Streets and Electricity	WILLIAM A. JACKSON
Superintendent, Bureau of Central Purchasing	JOHN A. CERVENKA
City Collector	LOUIS RIXMANN
Deputy City Collector	GEORGE F. LORMAN
President, Board of Health	DR. HERMAN N. BUNDESEN
Assistant to the President, Board of Health	JOEL I. CONNOLLY
Chief Medical Inspector	DR. ISAAC D. RAWLINGS
Superintendent of Streets	JOSEPH J. BUTLER
Civil Service Commission	JOSEPH P. GEARY, President JOHN E. BRENNAN WENDELL E. GREEN
Secretary Civil Service Commission	JAMES S. OSBORNE
Building Commissioner	RICHARD E. SCHMIDT
Chief Deputy Building Commissioner	ROBERT KNIGHT
Commissioner of Police	JAMES P. ALLMAN
Chief of Uniformed Force	CAPT. JOHN C. PRENDERGAST
Chief, Detective Bureau	JOHN L. SULLIVAN
Secretary of Police	JAMES McSWEENEY
Corporation Counsel	BARNET HODES
City Attorney	ALEXANDER M. SMIETANKA
City Prosecutor	MICHAEL L. ROSINIA
Map Department	HOWARD C. BRODMAN
City Physician	DR. DAVID J. JONES
Superintendent, Bureau of Water	H. L. MEITES
Commissioner of Public Service	JEFFREY A. O'CONNOR
Board of Examining Engineers	FRANK J. SMITH, President MICHAEL KOUKOLENSKI WILLIAM R. O'TOOLE
Inspector of Steam Boilers, Steam Plants and Smoke Inspection	HARRY KOHL
Superintendent of Sidewalks	HARRY L. BAILEY
City Engineer	LORAN D. GAYTON
Board of Local Improvements	JAMES P. BOYLE, President CHARLES H. WELLER WILLIAM J. CONNORS WILLIAM W. LINK MARTIN J. McNALLY
Secretary of Board of Local Improvements	E. J. GLACKIN
Fire Commissioner	MICHAEL J. CORRIGAN
Superintendent of Sewers	THOMAS D. GARRY
Municipal Librarian	FREDERICK REX
Bureau of Compensation	H. J. WIELAND
Superintendent, House of Correction	EDWARD J. DENEMARK
City Architect	PAUL GERHARDT, JR.
Deputy Smoke Inspector in Charge	FRANK A. CHAMBERS
Board of Examining Plumbers	JOHN A. CASTENS JULIUS NEWMAN (Journeyman)
Board of Examining Mason Contractors	WILLIAM P. CROWE (Chairman) NICHOLAS DIRE, JR.

1935-1939

THE CITY COUNCIL, CHICAGO

HONORABLE EDWARD J. KELLY, Mayor

Alderman James B. Bowler, President Pro Tem.

PETER J. BRADY, City Clerk **EDW. J. PADDEN, Chief Clerk**

Ward	ALDERMEN
1ST WARD	JOHN J. COUGHLIN, Deceased
2ND WARD	WILLIAM L. DAWSON, 3140 S. Indiana Ave..................Calumet 7073
3RD WARD	ROBERT R. JACKSON, 133 E. 47th St..........................Drexel 3742
4TH WARD	(Alderman deceased—no re-election)
5TH WARD	JAMES J. CUSACK, Jr, 11 S. La Salle St., Room 715............Central 0078
6TH WARD	JOHN F. HEALY, 6120 S. Rhodes Ave..........................Normal 4325
7TH WARD	THOMAS J. DALEY, 1920 E. 71st St........................Dorchester 2311
8TH WARD	MICHAEL F. MULCAHY, 758 E. 79th St........................Stewart 7792
9TH WARD	ARTHUR G. LINDELL, 42 E. 115th St.......................Pullman 4700
10TH WARD	WM. A. ROWAN, 9211 S. Ewing Ave..................South Chicago 3400
11TH WARD	HUGH B. CONNELLY, 551 W. 37th St.......................Boulevard 4340
12TH WARD	BRYAN HARTNETT, 3516 S. Washtenaw Ave.............Lafayette 1953
13TH WARD	JOHN E. EGAN, 5908 S. Kedzie Ave..........................Grovehill 0090
14TH WARD	JAMES J. McDERMOTT, 111 W. Washington St., Room 1449...Franklin 5464
15TH WARD	JAMES F. KOVARIK, 5022 S. Marshfield Ave.................Republic 0322
16TH WARD	TERENCE F. MORAN, 5641 S. Loomis St..................Englewood 6593
17TH WARD	WILLIAM T. MURPHY, 33 N. La Salle St......................Franklin 5858
18TH WARD	HARRY E. PERRY, 1035 W. 79th St..........................Stewart 9795
19TH WARD	JOHN J. DUFFY, 1548 W. 95th St.............................Beverly 8766
20TH WARD	WILLIAM V. PACELLI, 767 W. Taylor St.................Haymarket 5586
21ST WARD	JOSEPH F. ROPA, 1700 W. 21st St...............................Canal 7552
22ND WARD	HENRY SONNENSCHEIN, 2406 S. Ridgeway Ave.............Rockwell 1364
23RD WARD	JOSEPH KACENA, Jr., 4144 W. 15th St.....................Lawndale 3131
24TH WARD	J. M. ARVEY, 1 N. La Salle St................Randolph 8000, Extension 404
25TH WARD	JAMES B. BOWLER, 1311 S. California Blvd................Crawford 1345
26TH WARD	FRANK E. KONKOWSKI, 1030 W. Chicago Ave................Monroe 4614
27TH WARD	HARRY L. SAIN, 39 S. La Salle St., Room 1215....................State 2247
28TH WARD	GEORGE D. KELLS, 3146 W. Franklin Blvd...............Haymarket 0812
29TH WARD	THOMAS J. TERRELL, 3549 W. Madison St....................Kedzie 4036
30TH WARD	EDWARD J. UPTON, 5072 W. Monroe St...................Mansfield 1597
31ST WARD	THOMAS P. KEANE, 2935 W. Augusta Blvd................Humboldt 6488
32ND WARD	JOSEPH ROSTENKOWSKI, 1347 N. Noble St...........Brunswick 3306, 3307
33RD WARD	Z. H. KADOW, 160 N. La Salle St., Room 1836.............Dearborn 3235
34TH WARD	MATT PORTEN, 1857 N. Fairfield Ave.......................Humboldt 2490
35TH WARD	WALTER J. ORLIKOSKI, 3158 N. Pulaski Road..............Belmont 2811
36TH WARD	GEORGE W. ROBINSON, 5336 W. North Ave................Merrimac 5400
37TH WARD	ROGER J. KILEY, 111 W. Washington St., Room 1449........Franklin 5464
38TH WARD	P. J. CULLERTON, 134 N. La Salle St., Room 404..............Central 8686
39TH WARD	H. L. BRODY, 4370 N. Elston Ave...........................Palisade 6433
40TH WARD	JOSEPH C. ROSS, 3225 W. Lawrence Ave......................Juniper 0147
41ST WARD	WM. J. COWHEY, 160 N. La Salle St., Room 1635..............Central 6087
42ND WARD	DORSEY R. CROWE, Com. R. 303, City Hall.........Randolph 8000, Ext. 322
43RD WARD	MATHIAS BAULER, 1648 N. Sedgwick St.........Randolph 8000, Ext. 400
44TH WARD	JOHN J. GREALIS, 105 W. Adams St., Room 2312.............Central 0177
45TH WARD	EDWIN F. MEYER, 3744 N. Ashland Ave..................Bittersweet 8685
46TH WARD	JAMES F. YOUNG, 69 W. Washington St.......................Central 8826
47TH WARD	ALBERT F. SCHULZ, 4717 N. Paulina St..................Ravenswood 3070
48TH WARD	JOHN A. MASSEN, 134 N. La Salle St., Room 1424...........Franklin 2892
49TH WARD	FRANK KEENAN, 33 N. La Salle St...........................Central 4451
50TH WARD	JAMES R. QUINN, 111 W. Washington St., Room 1504........Franklin 5937

WILLIAM F. HARRAH, Sergeant-at-Arms.
ALBERT T. JOHNSON, Assistant Sergeant-at-Arms.
JOHN J. DOHNEY, Assistant Sergeant-at-Arms.
JOHN FAHEY, Assistant Sergeant-at-Arms.
ARTHUR X. ELROD, Assistant Sergeant-at-Arms.
CLEMENT J. McDERMOTT, Committee Secretary.

MEMBERSHIP OF COUNCIL COMMITTEES

1937-1939

Finance—ARVEY (Chairman), Kiley (Vice-Chairman), Jackson, Cusack, Healy, Lindell, Rowan, Hartnett, McDermott, Kovarik, Moran, Perry, Sonnenschein, Sain, Kells, Terrell, Keane, Orlikoski, Robinson, Ross, Crowe, Grealis, Massen, Quinn, Bowler (ex officio).

Local Transportation—QUINN (Chairman), Healy (Vice-Chairman), Dawson, Jackson, Daley, Rowan, Hartnett, McDermott, Kovarik, Moran, Pacelli, Konkowski, Sain, Kells, Keane, Rostenkowski, Porten, Orlikoski, Robinson, Crowe, Bauler, Massen, Duffy, Cullerton, Keenan, Bowler (ex officio).

Utilities—McDERMOTT (Chairman), Upton (Vice-Chairman), Coughlin, Mulcahy, Lindell, Connelly, Egan, Moran, Perry, Duffy, Ropa, Sonnenschein, Kacena, Konkowski, Terrell, Rostenkowski, Kadow, Cullerton, Brody, Cowhey, Schulz, Keenan, Bowler (ex officio).

Local Industries, Streets and Alleys—MORAN (Chairman), Robinson (Vice-Chairman), Coughlin, Cusack, Daley, Connelly, Hartnett, Perry, Pacelli, Ropa, Kacena, Konkowski, Sain, Kells, Terrell, Keane, Kadow, Porten, Kiley, Ross, Crowe, Meyer, Bowler (ex officio).

Judiciary and State Legislation—BRODY (Chairman), Keenan (Vice-Chairman), Dawson, Jackson, Cusack, Mulcahy, Murphy, Pacelli, Ropa, Kacena, Konkowski, Upton, Kadow, Kiley, Cullerton, Ross, Grealis, Young, Massen, Bowler (ex officio).

License—KEANE (Chairman), Kells (Vice-Chairman), Coughlin, Dawson, Jackson, Cusack, Healy, Rowan, Connelly, Moran, Pacelli, Ropa, Konkowski, Terrell, Rostenkowski, Porten, Orlikoski, Cowhey, Bauler, Meyer, Young, Bowler (ex officio).

Traffic and Public Safety—MASSEN (Chairman), Lindell (Vice - Chairman), Coughlin, Dawson, Daley, Egan, Murphy, Duffy, Sonnenschein, Kacena, Terrell, Upton, Kadow, Porten, Kiley, Cullerton, Brody, Cowhey, Bauler, Meyer, Keenan, Bowler (ex officio).

Consolidation, Reorganization and Taxation—KONKOWSKI (Chairman), Mulcahy (Vice-Chairman), Cusack, Rowan, Hartnett, Kovarik, Duffy, Upton, Robinson, Kiley, Brody, Cowhey, Schulz, Massen, Keenan, Bowler (ex officio).

Police and Municipal Institutions—ROSTENKOWSKI (Chairman), Porten (Vice-Chairman), Dawson, Daley, Mulcahy, Connelly, Hartnett, McDermott, Kovarik, Murphy, Ropa, Sonnenschein, Kaccna, Sain, Kells, Upton, Keane, Cowhey, Bauler, Meyer, Bowler (ex officio).

Buildings and Zoning—CROWE (Chairman), Coughlin (Vice-Chairman), Dawson, Jackson, Lindell, Egan, McDermott, Perry, Duffy, Pacelli, Kacena, Konkowski, Sain, Kells, Terrell, Kadow, Porten, Cullerton, Brody, Cowhey, Schulz, Keenan, Bowler (ex officio).

Railway Terminals—PERRY (Chairman), Daley (Vice-Chairman), Coughlin, Dawson, Jackson, Cusack, Healy, Mulcahy, Connelly, McDermott, Murphy, Ropa, Sonnenschein, Rostenkowski, Cullerton, Brody, Crowe, Young, Schulz, Massen, Bowler (ex officio).

Recreation and Aviation—HARTNETT (Chairman), Kovarik (Vice-Chairman), Healy, Lindell, Connelly, Egan, Perry, Duffy, Upton, Rostenkowski, Orlikoski, Kiley, Cullerton, Ross, Cowhey, Crowe, Grealis, Young, Schulz, Bowler (ex officio).

Harbors, Wharves and Bridges—GREALIS (Chairman), Jackson (Vice-Chairman), Coughlin, Dawson, Healy, Daley, Mulcahy, Lindell, Rowan, Egan, Kovarik, Murphy, Perry, Duffy, Rostenkowski, Crowe, Young, Bowler (ex officio).

Special Assessments — EGAN (Chairman), Pacelli, (Vice - Chairman), Murphy, Sain, Keane, Kadow, Porten, Robinson, Cullerton, Ross, Bauler, Grealis, Meyer, Quinn, Bowler (ex officio).

Housing—ROWAN (Chairman), Schulz (Vice-Chairman), Coughlin, Dawson, Cusack, Healy, Daley, Hartnett, Egan, Kovarik, Pacelli, Orlikoski, Kiley, Bauler, Grealis, Keenan, Bowler (ex officio).

Health — TERRELL (Chairman), Ross (Vice-Chairman), Mulcahy, Lindell, Egan, Moran, Murphy, Duffy, Kells, Upton, Orlikoski, Robinson, Kiley, Meyer, Young, Schulz, Keenan, Bowler (ex officio).

Schools, Fire and Civil Service—SONNENSCHEIN (Chairman), Meyer (Vice-Chairman), Lindell, Connelly, Kacena, Sain, Kadow, Orlikoski, Brody, Ross, Bauler, Grealis, Keenan, Quinn, Bowler (ex officio).

Tag Days—ORLIKOSKI (Chairman), Dawson (Vice-Chairman), Lindell, Rowan, Sonnenchein, Robinson, Schulz, Bowler (ex officio).

Committees and Rules—ARVEY (Chairman), Lindell, Moran, Keane, Quinn, Bowler (ex officio).

MECHANICS LIEN LAW

State of Illinois

AN ACT

To Revise the Law in Relation to Mechanics' Liens; To Whom, What For and When Lien Is Given; Who Is Contractor; Area Covered by and Extent of Lien; When the Lien Attaches. (Approved May 18, 1903, with amendments in Force July 6, 1937.)

Section 1. **When Lien Given.)** Be it Enacted by the People of the State of Illinois, Represented in the General Assembly: That any person who shall by any contract or contracts, express or implied, or partly expressed or implied, with the owner of a lot or tract of land, or with one whom such owner has authorized or knowingly permitted to contract for the improvement of, or to improve the same, furnish material, fixtures, apparatus or machnery, forms or form work used in the process of construction where cement, concrete or like material is used for the purpose of or in the building, altering, repairing or ornamenting any house or other building, walk or sidewalk, whether such walk or sidewalk be on the land or bordering thereon, driveway, fence or improvement or appurtenances thereto on such lot or tract of land or connected therewith, and upon, over, or under a sidewalk, street or alley adjoining; or fill, sod or excavate such lot or tract of land, or do landscape work thereon or therefor; or raise or lower any house thereon or remove any house thereto; or perform services as an architect or as a structural engineer for any such purpose; or drill any water well thereon; or furnish or perform labor or services as superintendent, timekeeper, mechanic, laborer or otherwise, in the building, altering, repairing or ornamenting of the same; or furnish material, fixtures, apparatus, machinery, labor or services, forms or form work used in the process of construction where concrete, cement or like material is used, or drill any water well on the order of his agent, architect, structural engineer or superintendent having charge of the improvement, building, altering, repairing or ornamenting the same, shall be known under this Act as a contractor, and shall have a lien upon the whole of such lot or tract of land and upon the adjoining or adjacent lots or tracts of land of such owner constituting the same premises and occupied or used in connection with such lot or tract of land as a place of residence or business; and in case the contract relates to two or more buildings, on two or more lots or tracts of land, upon all such lots and tracts of land and improvements thereon for the amount due to him for such material, fixtures, apparatus, machinery, services or labor, and interest from the date the same is due. This lien shall extend to an estate in fee, for life, for years, or any other estate or any right of redemption, or other interest which such owner may have in the lot or tract of land at the time of making such contract or may subsequently acquire therein, and shall be superior to any right of dower of husband or wife in said premises: **Provided,** the owner of such dower interest had knowledge of such improvement and did not give written notice of his or her objection to such improvement before the making thereof; nor shall the taking of additional security by the contractor or sub-contractor be a waiver of any right of lien which he may have by virtue of this Act, unless made a waiver by express agreement of the parties; and this lien shall attach as of the date of the contract. Approved May 27, 1937.

Section 2. **Liens for Work or Materials by Mistake Put Upon Land Other Than the Contracting Parties.)** Any person furnishing services, labor or material for the erection of a building, or structure, or improvement by mistake upon land owned by another than the party contracting as owner, shall have a lien for such services, labor or material upon such building, or structure or improvement, and the court in the enforcement of such lien, shall order and direct such building, structure or improvement to be separately sold under its decree, and the purchaser may remove the same within such reasonable time as the court may fix.

Section 3. **Liens for Work or Materials Under Contract with Husband on Land of Wife.)** If any such services or labor are performed upon or materials are furnished for lands belonging to any married woman with her knowledge and not against her protest in writing as provided in section 1 of this Act, in pursuance of a contract with the

When You Build--Build With

COMMON BRICK

for

BEAUTY - ECONOMY
ENDURANCE - FIRE SAFETY

Building With Common Brick Insures

LOW Maintenance - LOW Depreciation - HIGH Re-sale Value

For Quality Common Brick Phone

CHICAGO BRICK CO.

(Subsidiary of Wedron Silica Co.)

RANDOLPH 2780

38 S. Dearborn Street Chicago, Illinois

husband of such married woman, the person furnishing such labor or materials shall have a lien upon such property, the same as if such contract had been made with (the) married woman, and in case the title to such lands upon which improvements are made is held by husband and wife jointly, the lien given by this Act shall attach to such lands and improvements, if the improvements be made in pursuance of a contract with both of them, or in pursuance of a contract with either of them, and, in all such cases, no claim or homestead right set up by a husband or wife shall defeat the lien given by this Act.

Section 4. **Breach of Contract by Owner—Recovery for Material—Partial Performance—Quantum Merit—Right to Reclaim—Unused Material.)** When the owner of the land shall fail to pay the contractor moneys justly due him under the contract at the time when the same should be paid, or fails to perform his part of the contract in any other manner, the contractor may discontinue work, and the contractor shall not be held liable for any delay on his part during the period of, or caused by such breach of contract on the part of the owner, and if after such breach for the period of ten days the owner shall fail to comply with his contract, the contractor may abandon the work and in such case the contractor shall be entitled to enforce his lien for the value of what has been done, and the court shall adjust his claim and allow him a lien accordingly. In such cases all persons furnishing material which has not been incorporated in the improvement shall have the right to take possession of and remove the same if he so elects.

Section 5. **Contractors to Notify Owners of Sub-Contracts and Amounts of Their Claims—Owner's Duty with Regard Thereto and Rights in Case of Default—Contractor's Liability for Failure to Give Statement—Contractors to Whom This Section Does Not Apply.)** It shall be the duty of the contractor to give to the owner, and the duty of the owner to require of the contractor before the owner or his agent, architect or superintendent, shall pay or cause to be paid to said contractor or to his order any moneys or other consideration due to or become due such contractor or make or cause to be made to such contractor any advancement of any moneys or any other consideration, a statement in writing under oath or verified by affidavit, of the names of all parties furnishing materials and labor, and of the amounts due or to become due each. Merchants and dealers in materials only shall not be required to make statements herein provided for.

Section 6. **Time for Completion of Contract.)** In no event shall it be necessary to fix or stipulate in any contract a time for the completion or a time for payment in order to obtain a lien under this Act: **Provided,** that the work is done or material furnished within three years from the commencement of said work or the commencement of furnishing said materials.

Section 7. **Limitations as Against Third Parties—Claim for Lien—What shall Consist of—When Claim May Be Filed and When Amended—As to Errors in—Proof of Delivery of Material, Not Used, Sufficient—Delivery of Material at one Building Good for All Buildings.)** No contractor shall be allowed to enforce such lien against or to the prejudice of any other creditor or incumbrancer or purchaser, unless within four months after completion, or if extra or additional work is done or material is delivered therefor within four months after the completion of such extra or additional work or the final delivery of such extra or additional material, he shall either bring suit to enforce his lien therefor or shall file with the clerk of the Circuit Court in the county in which the building, erection or other improvement to be charged with the lien is situated, a claim for lien, verified by the affidavit of himself, or his agent or employee, which shall consist of a brief statement of the contract, the balance due after allowing all credits, and a sufficiently correct description of the lot, lots or tracts of land to identify the same. Such claim for lien may be filed at any time after the contract is made, and as to the owner may be filed at any time after the contract is made and within two years after the completion of said contract, or the completion of any extra work or the furnishing of any extra material thereunder, and as to such owner may be amended at any time before the final decree. No such lien shall be defeated to the proper amount thereof because of an error or overcharging on the part of any person claiming a lien therefor under this Act, unless it shall be shown that such error or overcharge is made with intent to defraud; nor shall any such lien for material be defeated because of lack of proof that the material after the delivery thereof, actually entered into the construction of such building or improvement, although it be shown that such material was not actually used in the construction of any such building or improvement: **Provided,** it is shown that such material was delivered either to said owner or his agent for such building or improvement, to be used in said building or improvement, or at the place where said building or improvement was being constructed, for the purpose of being used in construction or for the purpose of being employed in the process of construction as a means for assisting in the erection of the building or improvement in what is commouly termed forms or form work where concrete, cement or like material is used, in whole or in part. **And, provided, further,** that in case of the construction of a number of buildings under contract between the same parties, it shall be sufficient in order to establish such lien for material, if it be shown that such material was in good faith delivered at one of the said buildings for the purpose of being used in the construction of any one or all of such buildings, or delivered to the owner or his agent for such buildings, to be used therein; and such lien for such material shall attach to all of said buildings, together with the land upon which the same are being constructed, the same as in a single building or improvement. **And, provided, further,** that in the event the contract relates to two or more buildings on two or more lots or tracts of land, then all of said buildings and lots or tracts of land may be included in one statement of claim for a lien. As amended by Act approved June 16, 1913, in force July 1, 1913.

Section 8. **Assignability of Liens or Claims for Liens—Rights of Assignee.)** All liens or claims for liens which may arise or accrue under the terms of this Act shall be assignable, and proceedings to enforce such liens or claims for lien may be maintained by and in the name of the assignee, who shall have as full and complete power to enforce the same as if such proceedings were taken under the provisions of this Act and by and in the name of the lien claimant.

Section 9. **When, How and in What Court Suit May be Brought—Two or More Lien Holders May Join in Bringing Suit—Answers Stand as Cross Bills—Original Bill Cannot be Dismissed Without Consent of Parties—Lien Claimants May Contest Each Other's Claims Without Formal Issues of Record—Rights of in Case of Surprise—Limitation.)** If payment shall not be made to the contractor having a lien by virtue of this Act of any amount due when the same becomes due, then such contractor may bring suit to enforce his lien by complaint or petition in any court of competent jurisdiction in the county where the improvement is located, and in the event that the contract relates to two or more buildings or two or more lots or tracts of land, then all of said buildings and lots or tracts of land may be included in one complaint or petition. Any two or more persons having liens on the same property may join in bringing such suit, setting forth their respective rights in

their complaint or petition; all lien claimants not made parties thereto may upon application become defendants and enforce their liens by answer to the complaint or petition in the nature of an intervening petition, and the same shall be taken as a counterclaim against all the parties to such suit; and the said complaint or petition shall not thereafter be dismissed as to any such lien claimant, or as to the owner or owners of the premises without the consent of such lien claimant. The plaintiff or petitioner, and all defendants to such complaint or petition may contest each other's right without any formal issue of record made up between them other than that shown upon the original complaint or petition, as well with respect to the amount due as to the right to the benefit of the lien claimed: **Provided,** that if by such contest by co-defendants any lien claimants be taken by surprise, the court may, in its discretion, as to such claim, grant a continuance. The court may render judgment against any party summoned and failing to appear, as in other cases of default. Such suit shall be commenced or answer filed within two years after the completion of the contract, or completion of the extra or additional work, or furnishing of extra or additional material thereunder. (Amended by Act approved June 28, 1935.)

Section 10. **Personal Representatives—Death of Parties in Interest.)** Suits may be instituted under the provisions of this Act in favor of administrators or executors, and may be maintained against the representatives in the interest of those against whom the cause of action accrued and in suits instituted under the provisions of this Act the representatives of any party who may die pending the suit shall be made parties.

Section 11. **Who Are Parties in Interest—How and When Made—Or May Become Parties to Suit—Publication, Service of Process on Non-Resident—Claims Not Due, Etc.—Pleading, Requisites of Bill or Petition—Diligence Required in Prosecuting Claim—When and How Party Bringing Suit May Dismiss Same.)** The complaint or petition shall contain a brief statement of the contract or contracts on which it is founded, the date, when made, and when completed, if not completed, why, and it shall also set forth the amount due and unpaid, a description of the premises which are subject to the lien, and such other facts as may be necessary to a full understanding of the rights of the parties. Where plans and specifications are by reference made a part of the contract it shall not be necessary to set the same out in the pleadings or as exhibits, but the same may be produced on the trial of the suit. The plaintiff or petitioner shall make all parties interested, of whose interest he is notified or has knowledge, parties defendant, and summons shall issue and service thereof be had as in other civil actions; and when any defendant resides or has gone out of the State, or on inquiry cannot be found, or is concealed within this State, so that process cannot be served on him, the plaintiff or petitioner shall cause a notice to be given to him, or cause a copy of the complaint or petition to be served upon him, in like manner and upon the same conditions as is provided in other civil actions, and his failure to so act with regard to summons or notice shall be ground for judgment or decree against him as upon the merits. The same rule shall prevail with cross-petitioners with regard to any person of whose interest they have knowledge, and who are not already parties to the suit or action. Parties in interest, within the meaning of this Act, shall include persons entitled to liens thereunder whose claims are not, as well as are, due at the time of the commencement of suit, and such claim shall be allowed subject to a reduction of interest from the date of judgment to the time the claim is due; also all persons who may have any legal or equitable claim to the whole or any part of the premises upon which a lien may be attempted to be enforced under the provisions thereof, or who are interested in the subject matter of the suit. Any such persons may, on application to the court wherein the suit is pending, be made or become parties at any time before final judgment. No action or suit under the provisions of this Act shall be voluntarily dismissed by the party bringing the same without due notice to all parties before the court and leave of court upon good cause shown and upon terms named by the court. (Amended by Act approved June 28, 1935.)

Section 12. **Practice—Power of Courts—When Receivers May Be Appointed.)** The court shall permit amendments to any part of the pleadings, and may issue process, make all orders, requiring parties to appear, and requiring notice to be given, that are or may be authorized in other civil actions and shall have the same power and jurisdiction of the parties and subject matter, and the rules of practice and proceedings in such cases shall be the same as in other civil cases, except as is otherwise provided in this Act. The court shall have power to appoint receivers for property on which liens are sought to be enforced in the same manner for the same causes and for the same purposes as in cases of foreclosure of mortgages, as well as to complete any unfinished building where the same is deemed to be to the best interest of all the parties interested. (Amended by Act approved June 28, 1935.)

Section 13. **Practice—Answer—Defense—Right to Recover on Counter Claim.)** Defendant shall answer as in other civil actions. The owner shall be entitled to make any defense against the contractor by way of counter claim that he could in any action of law, and shall be entitled to the same right of recovery on proof of such in excess of the claim of the contractor against the contractor only, but for matters not growing out of the contract such recovery shall be without prejudice to the rights of the sub-contractors thereunder for payment out of the contract price or fund; and in event that the court shall find, in any proceeding in chancery, that no right to a lien exists, the contractor shall be entitled to recover against the owner as at law, and the court shall render judgment as at law for the amount which the contractor is entitled to, together with costs in the discretion of the court. In any proceedings to enforce a lien it shall only be necessary for all persons seeking a lien on account of wages due for labor to file in such proceedings an affidavit giving the amount due, between what dates the same was performed and the kind of labor performed, and the court shall direct the amount due for wages as therein specified to be paid within a short day to be fixed by the court, unless within ten days after the filing of said claim for wages the amount claimed is contested by the owner or some other party to the suit, and in order to contest the amount due for wages it shall be necessary for the party making such contest to file an affidavit in which he shall state the defense he has to the allowance of such claim, and the court shall proceed at once to hear such evidence as the parties may adduce, and determine the merits as to the allowance of such claim for wages, and in the event that the allowance for wages is not paid within the time fixed by the court, then the court shall order the premises sold to pay such amount in such manner as the court shall direct. (Amended by Act approved June 28, 1935.)

Section 14. **Trials—Parties Ready Not to Be Delayed—When Court May Delay Order for Sale or Distribution.)** In no case shall the want of preparation for trial of one claim delay the trial in respect to others, but trial shall be had upon issues between such parties as are prepared without references to issues between other parties; and when one creditor shall have obtained a decree or judgment for the amount due, the court may order a sale of the premises on which the lien operates,

or a part thereof, so as to satisfy the decree or judgment: **Provided,** that the court may, for good cause shown, delay making any order for sale or distribution until the rights of all the parties in interest are ascertained and settled by the court.

Section 15. **Preference to Laborers — No Preference to First Contractors.)** Upon all questions arising between different contractors having liens under this Act, no preference shall be given to him whose contract was made first, except the claim of any person for wages by him personally performed, shall be a preferred lien.

Section 16. **Incumbrances—Apportionment —on Improvements Made After Record of Incumbrance—Lien Holders Have Pro Rata Benefit in What Owner Pays for—Fraudulent Incumbrances—Disposition of.)** No incumbrance upon land, created before or after the making of the contract under the provisions of this Act, shall operate upon the building erected, or materials furnished until a lien in favor of the persons having done work or furnished material, shall have been satisfied, and upon questions arising between incumbrances and lien creditors, all previous incumbrances shall be preferred to the extent of the value of the land at the time of making of the contract, and the lien creditor shall be preferred to the value of the improvements erected on said premises, and the court shall ascertain by jury or otherwise, as the case may require, what proportion of the proceeds of any sale shall be paid to the several parties in interest. All incumbrances, whether by mortgage, judgment or otherwise charged and shown to be fraudulent, in respect to creditors, may be set aside by the court, and the premises freed and discharged from such fraudulent incumbrance.

Section 17. **Costs—How Taxed—Attorney's Fees.)** The costs of proceedings, as between all parties to the suit, shall be taxed equitably against the losing parties, and where taxed against more than one party, shall be so taxed against all in favor of the proper party but equitably as between themselves; and the costs, as between creditors aforesaid in contests relative to each other's claims, shall be subject to the order of the court, and the same rule shall prevail in respect to costs growing out of the proceedings against and between incumbrances. In all cases where liens are enforced, the court shall in its discretion, order a reasonable attorney's fee taxed as a part of the costs in favor of the lien creditor.

Section 18. **What Estate to be Sold—Manner of Making Sales When Part May be Sold.)** Whatever right or estate such owner had in the land at the time of making the contract may be sold in the same manner as other sales of real estate are made under decrees in chancery. If any part of the premises can be separated from the residue, and sold without damage to the whole, and if the value thereof is sufficient to satisfy all the claims proved in the cause, the court may order a sale of that part.

Section 19. **Proceeds of Sale—Application of Pro Rata—Labor Claims Preferred—Deficiency Decrees—Excess, to Whom Paid.)** The court shall ascertain the amount due each lien creditor and shall direct the application of the proceeds of sale to be made to each in proportion to their several amounts, according to the provisions of this Act, but the claims of all persons for labor as provided in section fifteen (15) shall be first paid. If, upon making sale under this Act of any or all premises, the proceeds of such sale shall not be sufficient to pay all claims of all parties, according to their rights, the decree shall be credited by the amount of said sale and execution may issue in favor of any creditor whose claim is not satisfied for the balance due as upon a deficiency decree in the foreclosure of a mortgage in chancery and such deficiency decree shall be a lien upon all real estate and other property of the party against whom it is entered to the same extent and under the same limitations as a judgment at law; and in cases of excess of sales over the amount of the decree, such excess be paid to the owner of the land, or to the person who may be entitled to the same, under the direction of the court.

Section 20. **Redemption.)** Upon all sales under this Act, the right of redemption shall exist in favor of the same persons, and may be made in the same manner as is or may be provided for redemption of real estate from sales under judgments and executions at law.

Section 21. **Sub-Contractors—Liens of Sub-Contractors—Who Are—Extent of Their Liens Superior to Creditors or Contractors on Money Due Contractors—Duty of Owner and Contractor to File Notice of Waiver of Lien—Limit of Owner's Liability—Owner Liable for Sub-Contracts Performed After Notice Thereof—Rights of In Case Contractor Default May Complete, If Contractor Abandons.)** Every mechanic, workman or other person who shall furnish any materials, apparatus, machinery or fixtures, or furnish or perform services or labor for the contractor, or shall furnish any material to be employed in the process of construction as a means for assisting in the erection of the building or improvement in what is commonly termed form or form work where concrete, cement or like material is used in whole or in part, shall be known under this Act as a sub-contractor, and shall have a lien for the value thereof, with interest on such amount from the date the same is due, from the same time, on the same property as provided for the contractor, and, also, as against the creditors and assignees, and person and legal representatives of the contractor, on the material, fixtures, apparatus or machinery furnished, and on the moneys or other considerations due or to become due from the owner under the original contract. If the legal effect of any contract between the owner and contractor is that no lien or claim may be filed or maintained by any one, such provision shall be binding; but the only admissible evidence thereof as against a sub-contractor or material man, shall be proof of actual notice thereof to him before any labor or material is furnished by him; or proof that a duly written and signed stipulation or agreement to that effect has been filed in the office of the recorder of deeds of the county or counties where the house, building or other improvement is situated, prior to the commencement of the work upon such house, building or other improvement, or within ten days after the execution of the principal contract or not less than ten days prior to the contract of the sub-contractor or material man. And the recorder of deeds shall record the same at length in the order of time of its reception in books provided by him for that purpose, and the recorder of deeds shall index the same, in the name of the contractor and in the name of the owner, in books kept for that purpose, and also in the tract or abstract book of the tract, lot, or parcel of land, upon which said house, building or other improvement is located, and said recorder of deeds shall receive therefor a fee, such as is provided for the recording of instruments in his office.

In no case, except as hereinafter provided, shall the owner be compelled to pay a greater sum for or on account of the completion of such house, building or other improvement than the price or sum stipulated in said original contract or agreement, unless payment be made to the contractor or to his order, in violation of the rights and interests of the persons intended to be benefited by this Act: **Provided,** if it shall appear to the court that the owner and contractor fraudulently, and for the purpose of defrauding sub-contractors fixed an unreasonably low price in their original contract for the erection or repairing of such house, building or other improvement, then the court shall ascertain how much of a difference exists between a fair price for

labor and material used in said house, building or other improvement, and the sum named in said original contract, and said difference shall be considered a part of the contract and be subject to a lien. But where the contractor's statement, made as provided in section five (5), shows the amount to be paid to the sub-contractor, or party furnishing material, or the sub-contractor's statement, made pursuant to section twenty-two (22), shows the amount to become due for material; or notice is given to the owner, as provided in sections twenty-four (24) and twenty-five (25), and thereafter such sub-contract shall be performed, or material to the value of the amount named in such statement or notice, shall be prepared for use and delivery, or delivered without written protest on the part of the owner previous to such performance or delivery, or preparation for delivery, then, and in any of such cases, such sub-contractor or party furnishing or preparing material, regardless of the price named in the original contracts, shall have a lien therefor to the extent of the amount named in such statements or notice. Also in case of default or abandonment by the contractor, the sub-contractor or party furnishing material, shall have and may enforce his lien to the same extent and in the same manner that the contractor may under conditions that arise as provided for in section four (4) of this Act, and shall have and may exercise the same rights as are therein provided for the contractor. (As amended by Act approved June 16, 1913, in force July 1, 1913.)

Section 22. **Where Partners Taken in After Contract—Lien for Material Furnished to Sub-Contractor — Lien of Sub-Contractor — Statement of Sub-Contractor to Owner or Contractor — Penalty for Failure to Give Statement.)** Whenever, after a contract has been made, the contractor shall associate one or more persons as partners or joint contractors, in carrying out the same, or any part thereof, the lien for materials or labor furnished by a sub-contractor to such contractor and his partners or associates, as originally agreed upon, shall continue the same as if the sub-contract had been made with all of said partners. When the contractor shall sub-let his contract, or a specified portion thereof, to a sub-contractor, the party furnishing material to or performing labor for such sub-contractor shall have a lien therefor, and may enforce his lien in the same manner as is herein provided for the enforcement of liens by sub-contractors. Any sub-contractor shall, as often as requested in writing by the owner, or contractor, or the agent of either, make out and give to such owner, contractor or agent, a statement of the persons furnishing material and labor, giving their names and how much, if anything, is due or to become due to each of them, and which statement shall be made under oath if required. If any sub-contractor shall fail to furnish such statement within five (5) days after such demand, he shall forfeit to such owner or contractor the sum of fifty (50) dollars for every offense, which may be recovered in an action of debt before a justice of the peace, and shall have no right of action against either owner or contractor until he shall furnish such statement, and the lien of such sub-contractor shall be subject to the liens of all other creditors.

Section 23. **Lien Against Fund Due or to Become Due—Contractors for Public Improvements, Notice—Duty and Liability of Officer Notified.)** Any person who shall furnish material, apparatus, fixtures, machinery or labor to any contractor having a contract for public improvement for any county, township, school district, city or municipality in this State, shall have a lien on the money, bonds or warrants due or to become due such contractor under such contract: **Provided,** such person shall, before payment or delivery thereof is made to such contractor, notify the official or officials of the county, township, school district, city or municipality whose duty it is to pay such contractor of his claim by a written notice; **and, provided further,** that such lien shall attach only to that portion of such money, bonds or warrants against which no voucher or other evidence of indebtedness has been issued and delivered to the contractor by or on behalf of the county, township, school district, city or municipality as the case may be at the time of such notice. It shall be the duty of any such official so notified to withhold a sufficient amount to pay such claim until the same is admitted by the contractor, or adjusted by the agreement of the parties, or there has been an adjudication of the same in a court of competent jurisdiction, and thereupon to pay the amount so determined to be due such claimant, if any, and to that end the said county, township, school district, city or municipality or any of the other parties interested may institute suit in the same manner as is provided herein in case of privately owned real estate to determine the rights of the parties when such claim is filed.

Any person who shall furnish material, apparatus, fixtures, machinery or labor to any contractor having a contract for public improvement for the State, may have a lien on the money, bonds or warrants due or about to become due such contractor under the contract, by filing with the director of the department whose duty it is to let such contract a sworn statement of the claim showing with particularity the several items and the amount claimed to be due on each; but the lien shall attach only to that portion of the money, bonds or warrants against which no voucher has been issued and delivered out of such department.

The person so claiming a lien shall, within thirty (30) days after filing notice with the State official, commence proceedings by complaint for an accounting, making the contractor to whom such material, apparatus, fixtures, machinery or labor was furnished, party defendant, and shall, within the same period notify the official of the State of the commencement of such suit by delivering to him a certified copy of the complaint filed; **provided,** that suit shall be commenced and a copy of the complaint served upon the State official not less than fifteen (15) days before the date when the appropriation from which such money is to be paid, will lapse. It shall be the duty of the State official after the sworn statement has been filed with him, to withhold payment of a sum sufficient to pay the amount of such claim, for the period limited for the filing of suit, unless otherwise notified by the person claiming the lien.

Upon the expiration of this period the money, bonds or warrants so withheld shall be released for payment to the contractor unless the person claiming the lien shall have instituted proceedings and served the official of the State with the certified copy of the complaint as herein provided, in which case the amount claimed shall be withheld until the final adjudication of the suit is had: **Provided,** the State official may pay over to the clerk of the court in which such suit is pending a sum sufficient to pay the amount claimed to abide the result of such suit and be distributed by the clerk according to the decree rendered.

Any payment so made to such claimant or to the clerk of the court shall be a credit on the contract price to be paid to such contractor. Any officer violating the duty hereby imposed upon him shall be liable on his official bond to the claimant serving such notice for the damages resulting from such violation, which may be recovered in an action at law in any court of competent jurisdiction. There shall be no preference between the persons serving such notice, but all shall be paid **pro rata** in proportion to the amount due under their respective contracts. (Amended by Act approved June 28, 1935.)

Section 23. For the purpose of this section "contractor" includes any sub-contractor. Any person who shall furnish material, apparatus, fixtures, machinery or labor to any contractor having a contract for public improvement for any county, township, school district, city or municipality in this State, shall have a lien on the money, bonds, or warrants due or to become due the contractor having a contract with such county, township, school district or municipality in this State under such contract: **Provided,** such person shall, before payment or delivery thereof is made to such contractor, notify the official or officials of the county, township, school district, city or municipality whose duty it is to pay such contractor of his claim by a written notice; and furnish a copy of said notice at once to said contractor **and, provided further,** that such lien shall attach only to that portion of such money, bonds, or warrants against which no voucher or other evidence of indebtedness has been issued and delivered to the contractor by or on behalf of the county, township, school district, city or municipality as the case may be at the time of such notice. The person so claiming a lien shall, within sixty (60) days after filing such notice, commence proceedings by complaint for an accounting, making the contractor having a contract with the county, township, school district, city or municipality and the contractor to whom such material, apparatus, fixtures, machinery or labor was furnished, parties defendent, and shall within the same period notify the official or officials of the county, township, school district, city or municipality of the commencement of such suit by delivering to him or them a certified copy of the complaint filed. It shall be the duty of any such official so notified upon receipt of the first notice herein provided for to withhold a sufficient amount to pay such claim for the period limited for the filing of suit, unless otherwise notified by the person claiming the lien. Upon the expiration of this period the money, bonds or warrants so withheld shall be released for payment to the contractor unless the person claiming the lien shall have instituted proceedings and served the official of the county, township, school district, city or municipality with the certified copy of the complaint as herein provided, in which case, the amount claimed shall be withheld until the final adjudication of the suit is had; provided, that the official so notified may pay over to the clerk of the court in which such suit is pending a sum sufficient to pay the amount claimed to abide the result of such suit and be distributed by the clerk according to the decree rendered. Any payment so made to such claimant or to the clerk of the court shall be a credit on the contract price to be paid to such contractor until the same is admitted by the contractor, or adjusted by the agreement of the parties, or there has been an adjudication of the same in a court of competent jurisdiction, and thereupon to pay the amount so determined to be due such claimant, if any, and to that end the said county, township, school district, city or municipality or any of the other parties interested may institute suit in the same manner as is provided herein in case of privately owned real estate to determine the rights of the parties when such claim is filed.

Any person who shall furnish material, apparatus, fixtures, machinery or labor to any contractor having a contract for public improvement for the State, may have a lien on the money, bonds or warrants due or about to become due the contractor having a contract with the State under the contract, by filing with the Director of the Department whose duty it is to let such contract a sworn statement of the claim showing with particularity the several items and the amount claimed to be due on each; but the lien shall attach to only that portion of the money, bonds or warrants against which no voucher has been issued and delivered out of such department.

The person so claiming a lien shall, within sixty (60) days after filing notice with the State official, commence proceedings by bill in equity for an accounting, making the contractor having a contract with the State and the contractor to whom such material, apparatus, fixtures, machinery or labor was furnished, parties defendant, and shall, within the same period notify the official of the State of the commencement of such suit by delivery to him a certified copy of the complaint filed; **provided,** that suit shall be commenced and a copy of the complaint served upon the State official not less than fifteen (15) days before the date when the appropriation from which such money is to be paid, will lapse. It shall be the duty of the State official after the sworn statement has been filed with him, to withhold payment of a sum sufficient to pay the amount of such claim, for the period limited for the filing of suit, unless otherwise notified by the person claiming the lien.

Upon the expiration of this period the money, bonds, or warrants so withheld shall be released for payment to the contractor unless the person claiming the lien shall have instituted proceedings and served the official of the State with the certified copy of the complaint as herein provided, in which case, the amount claimed shall be withheld until the final adjudication of the suit is had: **Provided,** the State official may pay over to the clerk of the court in which such suit is pending, a sum sufficient to pay the amount claimed to abide the result of such suit and be distributed by the clerk according to the decree rendered.

Any payment so made to such claimant or to the clerk of the court shall be a credit on the contract price to be paid to such contractor. Any officer of the State, county, township, school district, city or muncipality violating the duty hereby imposed upon him shall be liable on his official bond to the claimant serving such notice for the damages resulting from such violation, which may be recovered in a civil action at law in any court of competent jurisdiction. There shall be no preference between the persons serving such notice, but all shall be paid **pro rata** in proportion to the amount due under their respective contracts. (Amended by Act approved July 2, 1935.)

NOTE—Section 23 was amended by two Acts of the Fifty-ninth General Assembly as set out above. Amended by Act approved July 6, 1937.

Section 24. **Notice to the Owner by Sub-Contractor—Limitation for Service of—May be Served on Owner, Agent, Architect or Superintendent in Charge—Duties and Liabilities of Agents, Architects and Superintendent Notified—Excuse of Notice—Sub-Contractors Protected to Amount Named in—Form of.)** Sub-contractors, or party furnishing labor or materials may at any time after making his contract with the contractor, and shall within sixty (60) days after the completion thereof, or, if extra or additional work or material delivered thereafter, within sixty (60) days after the date of completion of such extra or additional work or final delivery of such extra or additional material, cause a written notice of his claim and the amount due or to become due thereunder, to be personally served on the owner or his agent or architect, or the superintendent having charge of the building or improvement, **provided, however,** that if the lot or lots and tract or tracts of land in question are registered under the provisions of "An Act concerning land titles," approved May 1, 1897, as amended, said notice shall not be served as aforesaid, but shall be filed in the office of the registrar of titles of the county in which such lot or lots and tract or tracts of land are situated, **and, provided, further,** such notice shall not be necessary when the sworn statement of the contractor or sub-contractor provided for herein shall

serve to give the owner notice of the amount due and to whom due, but where such statement is incorrect as to the amount, the sub-contractor or material man named shall be protected to the extent of the amount named therein as due or to become due to him.

The form of such notice may be as follows:

To (name of owner): You are hereby notified that I have been employed by (the name of contractor) to (state here what was the contract or what was done, or to be done, or what the claim is for) under his contract with you, on your property at (here give substantial description of the property) and that there was due to me, or is to become due (as the case may be) therefor, the sum of......

Dated at ....this......
day of......................A. D. ....

(Signature).......................

(Amended by Act approved June 30, 1923.)

Section 25. **Notice to Non-Resident Owner by Filing Claim with Circuit Court, What Claim Shall Consist of—When Itemized Account Not Necessary.)** In all cases where the owner, agent, architect or superintendent can not upon reasonable diligence, be found in the county in which said improvement is made, or shall not reside therein, the sub-contractor or person furnishing materials, fixtures, apparatus, machinery, labor or services may give notice by filing in the office of the clerk of the Circuit Court against the person making the contract and the owner a claim for lien verified by the affidavit of himself, agent, or employe, which shall consist of a brief statement of his contract or demand, and the balance due after allowing all credits and a sufficient correct description of the lot, lots, or tract of land to identify the same. An itemized account shall not be necessary.

Section 26. **Lien of Laborer's Preferred—Limitation As to Laborer's Notice.)** The claim of any person for wages as a laborer under sections fifteen, twenty-one and twenty-two of this Act shall be a preferred lien.

Section 27. **Owner's Duty to Retain and Pay Money After Notice—Preference to Laborers—Manner in Which He Shall Make Payment—Liability of Owners.)** When the owner or his agent is notified as provided in this Act, he shall retain from any money due or to become due the contractor, an amount sufficient to pay all demands that are or will become due such sub-contractor, tradesman, materialmen, mechanic, or workmen of whose claim he is notified, and shall pay over the same to the parties entitled thereto.

Such payments shall be as follows:

First—All claims for wages shall be paid in full.

Second—The claims of tradesmen, materialmen and sub-contractors, who are entitled to liens **pro rata,** in proportion to the amount due them respectively. All payments made as directed shall, as between such owner and contractor, be considered the same as if paid to such contractor. Any payment made by the owner to the contractor after such notice, without retaining sufficient money to pay such claims, shall be considered illegal and made in violation of the rights of the laborers and sub-contractors, and the rights of such laborers and sub-contractors to a lien shall not be affected thereby, but the owner shall not be held liable to any laborer and sub-contractor or other person whose name is omitted from the statement provided for in sections five (5) and twenty-two of this Act, nor for any larger amount than the sum therein named as due such person (provided such omission is not made with the knowledge or collusion of the owner), unless previous thereto or to his payment to his contractor, he shall be notified, as herein provided, by such person of their claim and the true amount thereof.

Third—The balance, if any, to the contractor.

Section 28. **Suits to Enforce Lien by Sub-Contractors—When Can Be Brought, Pleadings, Action at Law Against Owner and Contractor — Proceedings, Extent of Owner's Liability.)** If any money due to the laborers or sub-contractors be not paid within ten (10) days after his notice is served as provided in sections five (5), twenty-four (24), twenty-five (25), and twenty-seven (27), then such person may either file his petition and enforce his lien as hereinbefore provided for the contractor in sections nine (9) to twenty (20) inclusive, of this Act, except as to the time within which suit shall be brought or he may sue the owner and contractor jointly for the amount due him in any court having jurisdiction of the amount claimed to be due, and a personal judgment may be rendered therein, as in other cases. In such actions at law, as in suits to enforce the lien, the owners shall be liable to the plaintiff for no more than the pro rata share that such person would be entitled to with other sub-contractors out of the funds due to the contractor from the owner under the contract between them, except as hereinbefore provided for laborers, and such action at law shall be maintained against the owner only in case the plaintiff establishes his right to the lien. All suits and actions by sub-contractors shall be against both contractor and owner jointly, and no decree or judgment shall be rendered therein until both are duly brought before the court by process or publication, and in all courts including actions before a justice of the peace and police magistrates, such process may be served and publication made as to all persons except the owners as in other civil actions. All such judgments, where the lien is established shall be against both jointly, but shall be enforced against the owner only to the extent that he is liable under his contract as by this Act provided, and shall recite the date from which the lien thereof attached according to the provisions of sections one (1) to twenty (20) of this Act; but this shall not preclude a judgment against the contractor, personally, where the lien is defeated. (Amended by Act approved June 28, 1935.)

Section 29. **Judgment Before Justice of the Peace—When Transcript of May be Filed—Executed Thereon—Liens Thereof.)** If the execution issued on a judgment obtained before a justice of the peace or police magistrate shall be returned not satisfied a transcript of such judgment may be taken to the Circuit Court and spread upon the records thereof, and execution issued thereon as in other cases except that the lien of the same shall be preserved as a preferred lien on the property improved from the date recited in the judgment and enforced thereon the same as if a decree had been rendered by the Circuit Court in a suit to enforce such lien under the provisions of this Act.

Section 30. **Proceedings for General Settlement—Interpleader—How Liens and Claims Cut Off and Judgments Thereon Stayed in Such Proceedings.)** If there are several liens under sections twenty-one (21) and twenty-two (22) upon the same premises, and the owner or any person having such a lien shall fear that there is not a sufficient amount coming to the contractor to pay all such liens, such owner or any one or more persons having such lien may file his or their complaint or petition in the circuit court of the proper county, stating such fact and such other facts as may be sufficient to a full understanding of the rights of the parties. The contractor and all persons having liens upon or who are interested in the premises, so far as the same are known to or can be ascertained by the claimant or petitioner, upon diligent inquiry shall be made parties. Upon the hearing the court shall find the amount coming from the owner to the contractor, and the amount due to each of the persons having liens, and in case the amount found to be coming to the contractor shall be insufficient to discharge all the liens in full, the amount so found in favor of the contractor shall be divided between the persons entitled to such liens **pro**

rata after the payment of all claims for wages in proportion to the amounts so found to be due them respectively. If the amount so found to be coming to the contractor shall be sufficient to pay the liens in full, the same shall be so ordered. The premises may be sold as in other cases under this Act. The parties to such suit shall prosecute the same under like requirements as are directed in section eleven (11) of this Act, and all persons who shall be duly notified of such proceedings, and who shall fail to prove their claims, whether the same be in judgment against the owner or not, shall forever lose the benefit of and be precluded from their liens and all claims against the owner. Upon the filing of such complaint or petition the court may, on the motion of any person interested, and shall, upon final decree, stay further proceedings upon any suit against the owner on account of such liens, and costs in such cases shall be adjusted as provided for in section seventeen (17). (Amended by Act approved June 28, 1935.)

Section 31. **Failure to Complete Contract by Contractors—Requisites and Manner of Sub-Contractor's Suit in Case of — Owner's Liability in Case of.)** Should the contractor, for any cause, fail to complete his contract, any person entitled to a lien as aforesaid may file his petition in any court of record against the owner and contractor, setting forth the nature of his claim, the amount due, as near as may be, and the names of the parties employed on such house or other improvements subject to liens; and a notice of such suit shall be served on the persons therein named, and such as shall appear shall have their claim adjudicated. The premises may be sold as in other cases under this Act. The parties to such suit shall prosecute the same under like requirements as are directed in section eleven (11) of this Act.

Section 32. **Payments of Owner to Contractor—When Wrongful.)** No payments to the contractor or to his order of any money or other considerations due or to become due to the contractor shall be regarded as rightfully made, as against the sub-contractor, laborer, or party furnishing labor or materials, if made by the owner without exercising and enforcing the rights and powers conferred upon him in sections five (5) and twenty-two (22) of this Act.

Section 33. **Limitation As to Suit of Sub-Contractors to Enforce Lien.)** Petition shall be filed or suit commenced to enforce the lien created by sections twenty-one (21) and twenty-two (22) of this Act within four months after the time that the final payment is due the sub-contractor, laborer or party furnishing material.

Section 34. **General Provisions—Suit to Be Commenced or Answer Filed by Lien Claimants, and Within Thirty (30) Days on Demand of Owner, Liener or Interested Party.)** Upon written demand of the owner, liener, or any person interested in the real estate, or their agent or attorney, served on the person claiming the lien, or his agent or attorney, requiring suit to be commenced to enforce the lien, or answer to be filed in a pending suit, suit shall be commenced or answer filed within thirty days thereafter, or the lien shall be forfeited and same released, if a claim for a lien has been filed with the clerk of the Circuit Court.

Section 35. **Neglect to Satisfy Lien Paid or to Release Where Not Sued on Time—Penalty.)** Whenever a claim for lien has been filed with the clerk of the Circuit Court, either by the contractor or sub-contractor, and is afterward paid, with cost of filing same, or where there is a failure to institute suit to enforce the same after demand, as provided in the preceding section, within the time by this Act limited, the person filing the same or some one by him duly authorized in writing so to do, shall acknowledge satisfaction or release thereof in the proper book in such office, in writing on written demand of the owner, and on neglect to do so for ten days after such written demand, he shall forfeit to the owner the sum of twenty-five (25) dollars, which may be recovered in an action of debt before a justice of the peace.

Section 36. **Penalty for Wrongful Sale, Use or Removal of Materials.)** Any owner, contractor, sub-contractor or other person who shall purchase materials on credit and represent at the time of purchase that the same are to be used in a designated building or buildings, or other improvement, and shall thereafter sell, use or cause to be used, the said materials in the construction of or remove the same to any building or improvement other than that designated, or dispose of the same for any purpose, without the written consent of the person of whom the materials were purchased, with intent to defraud such person, shall be deemed guilty of a misdemeanor, and, on conviction, shall be punished by a fine not exceeding five hundred dollars ($500), or confined in the county jail not exceeding one year, or both so fined and imprisoned.

Section 37. **Liens Against Boats, Barges and Water Craft.)** Any architect, contractor, sub-contractor, materialman, or other person furnishing services, labor or material for the purpose of or in constructing, building, altering, repairing or ornamenting a boat, barge or other water craft, shall have a lien on such boat, barge or other water craft, for the value of such services, labor or material in the same manner as in this Act provided for services, labor or material furnished by such parties for the purpose of building, altering, repairing or ornamenting a house or other building. And such lien may be established and enforced in the same manner as liens are established and enforced under this Act, and the parties shall be held to the same obligations, duties and liabilities as in case of a contract for building, altering, repairing or ornamenting a house or other building.

Section 38. **Circuit Court Clerk's Duties with Regard to Claims Filed—Abstract Fee.)** When claims for liens are filed pursuant to the provisions of sections seven (7) and twenty-five (25), the clerk of the Circuit Court shall endorse thereon the date of filing and make an abstract thereof in a book kept for that purpose and properly indexed, containing the name of the person filing the lien, the amount of the lien, the date of filing, the name of the person against whom the lien is filed, and a description of the property charged with the lien, for which the person filing the lien shall pay one dollar ($1) to the clerk.

Section 39. This Act is and shall be liberally construed as a remedial Act.

Section 40. An Act entitled, "An Act to revise the law in relation to mechanic's liens," approved and in force June 26, 1895, and all other Acts and parts of Acts inconsistent with this Act are hereby repealed, **provided** that this section shall not be construed as to effect any rights existing or actions pending at the time this Act shall take effect. (Approved May 18, 1903.)

UNIVERSAL FORM CLAMP COMPANY

Manufacturers of Accessories for All Types of Concrete Construction

972 MONTANA STREET, CHICAGO, ILL.

UNIVERSAL'S FORM CLAMPING AND TYING DEVICES

FORM CLAMPING ASSEMBLIES

For watertight and stainproof walls and wherever ties must be back from the concrete face.

All the assemblies illustrated below, are drilled and tapped for an inside threaded rod. These assemblies have the following advantages:

The inside threaded rods are kept the required distance back from the concrete face.

The holes left by the rods are small and are easily grouted.

The assemblies are easily inserted through holes drilled in the sheathing of the wall forms.

Range of adjustment to accommodate any dimension of form lumber without removing the clamp.

Outer rods are removed without stripping the forms or removing the wales.

UNIVERSAL'S "SPI-RO-LOC" ASSEMBLY

Patented

This assembly provides more speed than a wedge assembly with the fine adjustment of square cut threads and a clamp that will slide over the threaded rod into proper position against the wale and tighten with less than one complete turn.

UNIVERSAL'S "C-P" CLAMP ASSEMBLY

Patent Pending

This assembly enables the contractor to entirely remove the tie rods from the wall. Nailholes will be provided in this assembly to give spreader action when required.

THE FORM CLAMP

Patented

For all types of concrete construction.

The features of the Form Clamp:

Rods can be entirely removed from the concrete.

Holes left by the rods are small and are easily grouted.

Two Form Clamps and a length of mild steel rod comprise a form tie.

A single unit, exceedingly simple in principle and application.

The useful life of the Form Clamp is unlimited.

The *sure grip principle* wherein the setscrew does not carry the load but merely holds the rod in the depression, throwing the load on the shoulder of the clamp. The depression of the rod between the shoulders insures the Clamp against slipping or moving on the rod.

FORM CLAMPS WITH SPREADER CONES

Patented

An inexpensive and efficient spreader for use with Form Clamps and plain round rods. The clamps, rods and cones are removed and re-used, the paper tubing being the only item lost.

FORM CLAMPS WITH CONE NUTS

Cone Nuts with inside tie rods, stud rods and Form Clamps provide a combined tie and positive spreader. The clamps, stud rods and cones are removed and re-used, the tie rods being the only item lost.

TWISTYES

Patented

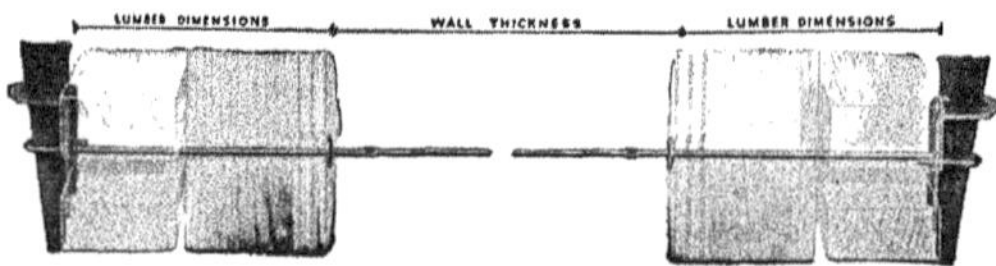

For ordinary, finished and watertight walls.

A strong spreader tie that can be twisted off an inch back from the wall face without removing the forms—combined with an improved clamp that has ample take-up and sufficient bearing surface.

UNIVERSAL FORM CLAMP COMPANY

Manufacturers of Accessories for All Types of Concrete Construction

972 MONTANA STREET, CHICAGO, ILL.

UNIVERSAL'S BAR SPACERS AND CHAIRS

For Reinforced Concrete Construction

Universal's Bar Spacers and Chairs meet requirements where reinforcing steel must be accurately located in the forms and firmly held in place, before and during pouring. They insure against displacement during construction and keep the steel at a proper distance from the forms.

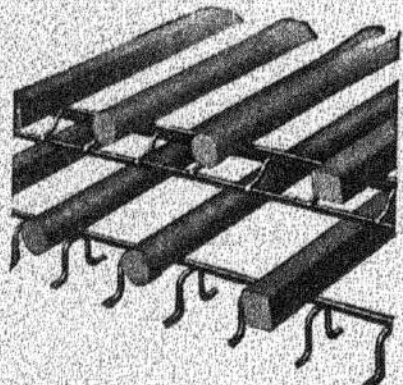

Beam Bar Bolster and Spacer

Bolster supports bars regardless of spacing and can be conveniently cut to beam widths on the job. Spacers provide deep seat for each bar with a leg under each bar.

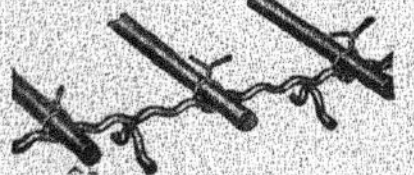

Slab Bar Bolster and Spacer

Bolster is fabricated with a top wire having corrugations 1 in. apart that act as a spacing guide when setting steel and prevent sliding of bars. Spacer is furnished with straight top wire and a leg under each bar.

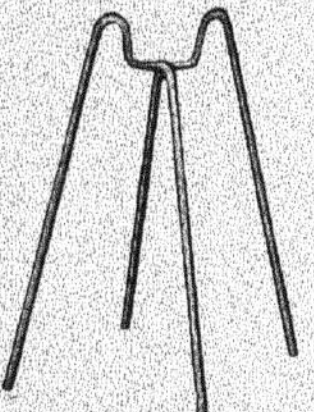

Individual High Chair

These chairs are used for supporting top layer of steel in combination with a support bar. They are firm non-collapsible chairs, providing bar seats with unusual strength.

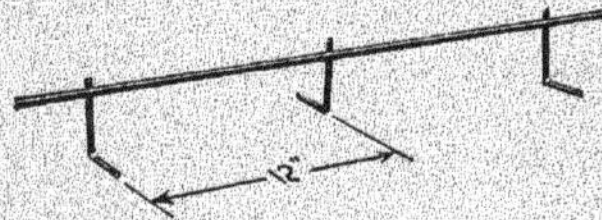

Continuous High Chair

Fabricated with a plain or corrugated top wire, is advantageously used for supporting and securely holding bent bars in position at the top of the slab.

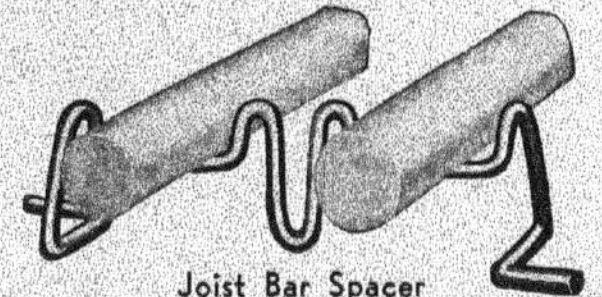

Joist Bar Spacer

For supporting joist reinforcing steel. Legs are so formed that they prevent spacers from slipping between pan and soffit.

UNIVERSAL'S ADJUSTABLE INSERTS—MALLEABLE

For hanging shelf angles, shaft hangers, sprinkler systems, plumbing and heating systems, etc. This method saves considerable time and labor, as the slot into which the nut or bolt head is inserted allows adjustment the full length of the slot.

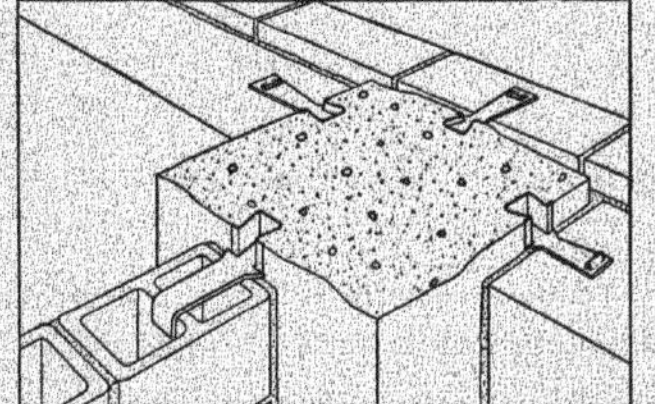

ANCHORS IN COLUMNS WITH BRICK FACING AND TILE PARTITION
Showing the versatility of Universal Dovetail Anchors and Slot

UNIVERSAL'S DOVETAIL ANCHOR SLOT AND ANCHORS

Universal's Anchor Slot with the Felt Filler prevents seepage of cement grout.

Slot and Filler are very easily installed and because seepage is prevented by use of the Filler, cutting, chipping, digging, cleaning and other difficulties are eliminated—leaving the slot entirely clean and immediately ready for installation of anchors at any desired spacing.

Dovetail Anchor Slot with Dovetail Anchors provide an unexcelled and economical method of anchoring masonry to concrete walls, beams and columns. The Slot is unique and superior because of its wide flange which adds so much strength in section that it is impossible for concrete to distort the slot, making it always easy to insert anchors. Any possibility of loosening or removing the anchor slot is eliminated by the greater dovetail or flare.

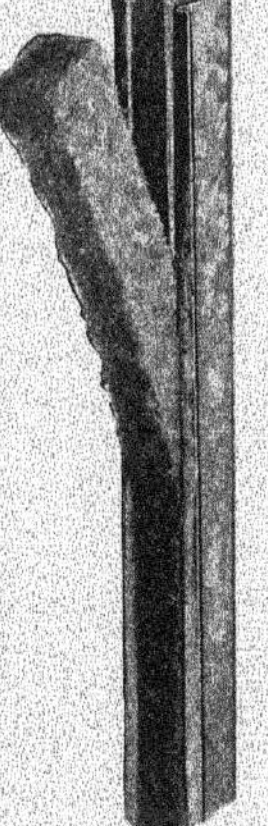

ANCHOR SLOT WITH FILLER

Colonial Damper

Cast Iron Stands the Gaff

There is no fireplace fire control comparable to the Colonial Damper for over thirty years the standard with architects and builders.

Because cast iron can take it the Colonial head throat and damper delivers life long service. Patented features give perfect fire control. Colonial dampers can be set two, three or more courses above the opening where a fireplace damper functions best.

Specify the Colonial Damper in all your designs.

Three styles of operation. 12 sizes to meet every condition.

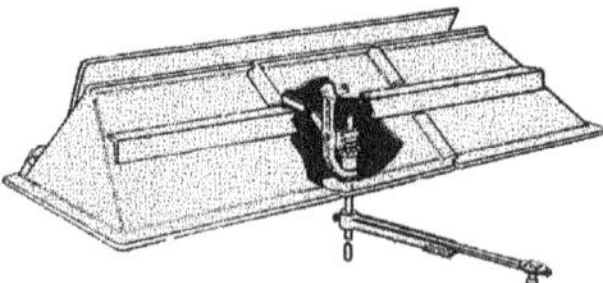

Style "G" extension operating device enables the Colonial Damper to be set up high in the throat where only the best results of draft control can be obtained.

Service to Architects

★ As fireplace experts for over 30 years we offer architects and builders free counsel in meeting specific problems. Details and information free. Rely on fireplace headquarters for dampers, ash traps, ashpit doors, fenders, screens, andirons, firesets and all fireplace furnishings. Wood Mantels—Catalog 132—Copy free.

DISTRIBUTORS FOR CIRCULATING FIREPLACE UNITS

Write for Our Complete Catalog and Fireplace Construction Blue Print

COLONIAL FIREPLACE COMPANY

4626 W. ROOSEVELT ROAD CHICAGO, ILLINOIS

INDEX TO BUILDING ORDINANCE

The items indexed herein are according to Chapters and section numbers, and the language used is as it appears in the Ordinance.

Explanatory Note to Index.

Bold face lines indicate Chapter Titles or Initial Use of a word. The first Light Face line under a Bold Face line indicates either a subhead in a chapter outline or the omission of the first word.

Indentations are used following the first Light Face line to show further Subheads under Chapter Classifications or to show the omission of additional words.

CHAPTERS OF THE BUILDING ORDINANCE PASSED BY THE CITY COUNCIL

Chapter No.	Title of Chapter	Date of Passage
1	Title and Scope	October 13, 1937
2	Administration	July 13, 1938
3	Arbitration, Standards, Registration and Penalties	July 13, 1938
4	Permits, Plans, Fees, Inspections and Certificates	July 13, 1938
5	Definitions	June 3, 1938
6	General Occupancy Provisions	June 3, 1938
7	Single Dwellings	June 3, 1938
8	Multiple Dwellings	June 10, 1938
9	Institutional Buildings	June 10, 1938
10	Business Units	July 6, 1938
12	Garages	August 3, 1938
13	Theatres	July 13, 1938
14	Open Air Assembly Units	July 13, 1938
15	Public Assembly Units	July 13, 1938
16	Churches	July 6, 1938
17	Schools	July 6, 1938
18	Other Buildings and Structures—Billboards and Signboards, Fences, Hangars, Ice Houses, Illuminated and other Roof Signs, Incinerators, Jails, Police Stations, Prisons and Buildings of Detention, Roundhouses and Carbarns, Sheds and Shelter Sheds, Stables, Tanks and Exposed Metal Structures	July 13, 1938
19	Frontage Consents	August 3, 1938
20	Existing Buildings and Structures	July 6, 1938
21	General Construction Provisions	July 6, 1938
24	Size and Location of Rooms	July 13, 1938
25	Multiple Use Buildings	July 13, 1938
26	Means of Exit	June 20, 1938
27	Windows, Skylights and Ventilation	July 13, 1938
30	Artificial Lighting and Exit Signs	July 13, 1938
31	Construction Type Requirements	May 18, 1938
32	Fire Resistive Standards	May 18, 1938
33	All Types and Kinds of Fire Doors	July 13, 1938
34	Structural Provisions	October 13, 1937
35	Foundations	October 13, 1937
36	Wood Construction	October 13, 1937
38	Masonry Construction	November 3, 1937
39	Reinforced Concrete Construction	October 13, 1937
40	Steel and Metal Construction	November 3, 1937
41	Chimneys	November 3, 1937
42	Floor and Roof Construction	May 18, 1938
43	Trussed Steel Joists and Other Construction	May 18, 1938
44	Elevators, Dumbwaiters, Escalators and Mechanical Amusement Devices	July 20, 1938
46	Heating Provisions	July 20, 1938
47	Ventilation	July 20, 1938
47A	Administration of Ventilating and Plumbing Provisions	July 20, 1938
48	Plumbing Provisions	August 3, 1938
49	Water Supply and Distribution Systems	August 3, 1938
51	Administration of Ordinances Regulating Steam Boilers, Unfired Pressure Vessels and Mechanical Refrigerating Systems	August 3, 1938
52	Steam Boilers and Unfired Pressure Vessels	August 3, 1938
53	Mechanical Refrigeration	August 3, 1938

N O T I C E

The thirty-fifth edition of the Handbook for Architects and Builders has been completed for some time, but owing to the fact that we were advised that amendments to various chapters of the new Chicago Building Ordinance were to be made, we delayed the issuance in order to include same.

Amendments passed January 26, 1939 are on the next page.

Map of the new Fire limits of the City of Chicago passed January 26, 1939 is on pages 801 and 802.

AMENDMENTS OF SECTIONS OF THE BUILDING ORDINANCE OF THE CITY OF CHICAGO

Passed January 26, 1939.

That the ordinance entitled "An ordinance to Revise the Building Code of the City of Chicago," passed October 13, 1937, as amended, be and the same is hereby further amended.

Section 405.04.

By striking out of Section 405.04 of Chapter 4 the words "and the Commissioner of Public Works".

Section 1208.01.

By striking out from Section 1208.01 Subdivision c), and by inserting in lieu thereof the following:

"c) **Gas Unit Heaters.** Gas-fired unit heaters, subject to the following provisions, shall be permitted in garages:

(1) The flame shall be protected from drafts of air and contact with combustible materials.

(2) All gas unit heaters installed in garages shall be installed so that the bottom of the heater shall be not less than eight (8) feet above the floor. They shall be constructed and arranged to conform to the "American Standard Approval Requirements for Gas Unit Heaters," effective January 1, 1936; Approved American Standards Association, August 13, 1934.

(3) Nothing in the provisions of this ordinance shall be interpreted or constructed as prohibiting or preventing the use in Class 2 garages or in office space of Class 1 garages where such spaces are separated from the garage proper by a wall of not less than two (2) hour fire-resistive value of a gas-fired heater constructed and arranged to conform to the requirements for "Private Garage Heaters," approved American Standards Association, December 28, 1932.

d) **Electric Unit Heaters.** Electric unit heaters constructed and installed to conform to the Underwriters Laboratories, Inc., "Standard for Electric Heating Appliances" Third Edition, December, 1936, shall be permitted in garages."

Subdivisions "a" and "b" of Section 1809.03.

(a) by striking out of subdivision "a" of Section 1809.03 the comma after the word "signboards" and the words "within the fire limits and provisional fire limits,"

(b) by striking out the period at the end of subdivision "b" of Section 1809.03 and adding the following language:

"and if erected on the ground shall be located not less than ten (10) feet from any building or other structure or from any thoroughfare or other public space."

Section 4702.02.

By striking out the following from the table of ventilating requirements provided in Section 4702.02:

Subdivision "c" of Section 1809.03.

By adding at the end of Section 1809.03 the following:

"c) The height of the top of any billboard or signboard shall not exceed fifteen (15) feet six (6) inches above the level of the adjoining street level or above the adjoining ground level if such ground level is above the street level."

Section 2501.03.

By changing Subdivisions (b) and (e) of Section 2501.03 to read as follows:

"b) **Single Dwellings.** Four (4) hour fire-resistive separation from any theatre, open air assembly unit, Class 1 garage or school. One (1) hour fire-resistive walls and ceiling separation from any Class 2 garage. Two (2) hour fire-resistive separation from any church, multiple dwelling or public assembly unit, and in buildings more than two (2) stories high, from any business unit. One (1) hour fire-resistive separation from any business unit in buildings two (2) stories or less in height.

"e) **Business Units.** Four (4) hour fire-resistive separation from any institutional building, Class 1 garage, theatre or open air assembly unit. Two (2) hour fire-resistive separation from any multiple dwelling, public assembly unit, church or school, and in buildings more than two (2) stories high, from any single dwelling. One (1) hour fire-resistive separation from any single dwelling two (2) stories or less in height.' '

Section 2605.03.

By striking out of paragraph i) of Section 2605.03 the word and figures "twelve (12)", and by inserting in lieu thereof the word and fiures "fourteen (14)".

Elimination of Section 3401.08.

By striking out all of Section 3401.08.

Section 4803.03.

By changing Subdivision h) of Section 4803.03:

"h) **Hard Drawn Copper Tubing.** All copper tubing shall conform to the A.S.T.M. "Standard Specifications for Copper Water Tube" (serial designation BSS-33). All sizes two (2) inch and smaller shall be Class K or L; sizes over two (2) inch shall be Classes K, L or M."

Section 4806.01.

By inserting in Section 4806.01 the word "lead" between the words "iron" and "brass," as the same appear in the last line of said section.

"Sales Rooms (except Department Stores)	5	Stories below that nearest to grade	S 1.2 and E 1.2
	5	Stories nearest to grade and above	No Requirements
			S .5"

and by inserting in lieu thereof the following:

"Sales Rooms (except Department Stores)	5	Stories below that nearest to grade	S 1.2 and E 1.2
	5	Stories nearest to grade and above	No Requirements
			S .5 or E .5"

NOTE: It is suggested that a notation be made at the proper place in the code referring to these amendments.

BUILDING ORDINANCE
OF THE CITY OF CHICAGO

List of Chapters with date of passage is on Page 170.
Index to Building Ordinance starts on Page 139.

SUGGESTED USE OF THE ORDINANCES RELATING TO BUILDINGS.

In order to ascertain the requirements for any proposed building, it is first essential to determine the zoning restrictions applicable to the location. Refer to the Zoning Ordinance and zoning maps. Note the use district in which the property in question is located, and from the corresponding section in the text, determine whether the intended use of said property is permitted. Note also the volume district number and consult the text for restrictions upon height and volume, area, lot occupancy and requirements for yards, open spaces and set-backs.

From the Fire Prevention Ordinance, learn whether the property is inside or outside of the fire limits, as this may determine the type of construction to be used.

Refer next to the Building Ordinance. Chapter 19 of the Building Ordinance adopts the requirements for frontage consents of the previous ordinance. These should be consulted to see whether the consent of owners of adjacent properties is required.

Chapter 6 lists the uses embraced by each of the occupancy classes. Chapters 7 to 18 prescribe the requirements peculiar to each class of occupancy.

In the case of a building devoted to the uses of more than one occupancy class, additional provisions are contained in "Chapter 25—Multiple Use Buildings." Alteration of existing buildings is regulated by Chapter 20. Provisions common to two or more occupancies are described in Chapters 21 to 33.

The types of construction required by the occupancy chapters are described in detail in Chapter 31 and also in the chapter describing the occupancy.

The fire resistive values for different materials of given thicknesses are given in Chapter 32.

The structural provisions, Chapters 34 to 43 are the regulations of engineering design and construction, and are applicable to all occupancies and all types of construction.

The mechanical provisions, Chapters 44 to 53, like the structural provisions, are general in their application and detail the requirements for the mechanical and sanitary equipment of buildings.

Sprinkler, standpipe and other fire prevention regulations are contained in the Fire Prevention Ordinance.

Requirements for Permits, Plans, Fees, Inspections and Certificates are stated in Chapter 4. (See especially Sec. 401.03b.)

It should be understood, in order to expedite cross references, that the Chapters of the Building Ordinance are divided into Articles, designated by whole numbers consisting of the chapter number with two integers appended, as Article 3101 of Chapter 31; that the Articles are divided into Sections designated by decimal numbers appended to the number of the Article of which they are a part, as Section 3101.01; that the Sections are sometimes divided into Paragraphs which are designated alphabetically; and that the term "Item", followed by a numeral, refers to a subdivision of a Paragraph.

Charles G. Brookes, Editor,
The Commission for Revising the Building Code of the City of Chicago.

CHAPTER 1
Title, Purpose and Scope

101.01 Title: This ordinance shall be known as the "Building Ordinance of the City of Chicago" and may be cited as such.

101.02 Purpose and Scope: The purpose of this ordinance is to provide minimum standards, provisions and requirements to insure safe and stable design, methods of construction and use of material in buildings or structures hereafter erected, constructed, enlarged, altered repaired, moved, converted to other uses or demolished and to regulate the equipment, maintenance, use and occupancy of all buildings, structures and premises. In interpreting and applying the provisions of this ordinance such provisions shall in every instance be held to be the minimum requirements adopted for the promotion of the public health, safety, comfort or welfare.

101.03 Validity of Ordinance: If any section, paragraph, subdivision, clause, sentence or provision of this ordinance shall be adjudged by any court of competent jurisdiction to be invalid, such judgment shall not affect, impair, invalidate or nullify the remainder of this ordinance but the effect thereof shall be confined to the section, paragraph, subdivision, clause, sentence or provision immediately involved in the controversy in which such judgment or decree shall be rendered.

CHAPTER 2
Administration

ARTICLE 201

Officers, Powers and Duties

201.01 Department of Buildings—Officers: There is hereby established an executive department of the municipal government of the City of Chicago which shall be known as the Department of Buildings, and which shall embrace a Commissioner of Buildings, a Deputy Commissioner of Buildings, a Building Inspector in Charge, an Elevator Inspector in Charge, a Secretary of the Department of Buildings, a Private Secretary to the Commissioner of Buildings and such number of Assistant Building Inspectors in Charge, Building Inspectors, Elevator Inspectors and Fire Escape Inspectors and other officers, assistants and employees as may from time to time be provided for in the annual appropriation ordinance. The offices enumerated above are hereby created, and the incumbents of these offices shall be known and designated by their respective titles as herein set forth.

201.02 Commissioner of Buildings—Appointment: The Commissioner of Buildings shall be the head of said Department of Buildings and shall be a registered architect, or a registered structural engineer and shall have been engaged as an architect, or a structural engineer for a period of not less than ten (10) years prior to his appointment. During his term of office as Commissioner of Buildings he shall not be engaged in any other business. He shall be appointed by the Mayor, by and with the advice and consent of the City Council.

201.03 Bond: The Commissioner of Buildings, before entering upon the duties of his office, shall execute a bond to the city in the sum of twenty-five thousand dollars ($25,000.00), with such sureties as the City Council shall approve, conditioned for the faithful performance of his duties as Commissioner of Buildings.

201.04 Appointment of Subordinates: The Commissioner of Buildings shall have the management and control of all matters and things pertaining to the Department of Buildings, and shall appoint and may remove, according to law, all subordinate officers and assistants in his department. All subordinate officers, assistants, clerks and employees in said department shall be subject to such rules and regulations as shall be prescribed from time to time by said Commissioner.

201.05 Duties of Commissioner:

(a) **General.** The Commissioner of Buildings shall institute such measures and prescribe such rules and regulations for the control and guidance of his subordinate officers and employees as shall secure the careful inspection of all buildings while in process of construction, alteration, repair or removal and the strict enforcement of the several provisions of this ordinance. It shall be the duty of the Commissioner of Buildings to administer and enforce the provisions of this ordinance and to administer and enforce the provisions of all general ordinances, now in force or hereafter to be passed by the City Council that relate to the erection, construction, alteration, repair, removal and safety of buildings and the use of buildings and premises with the exception of those ordinances which by their terms are to be under the direct and immediate supervision of the Board of Health or of the Bureau of Fire Prevention, or of such other departments or officers of the City of Chicago designated by such ordinances.

(b) **Personal Liability.** In all cases where any action is taken by the Commissioner of Buildings to enforce the provisions of any of the sections contained in this ordinance or to enforce the provisions of any of the general ordinances of the city now or at any time hereafter in force, whether such action is taken in pursuance of the express provisions of such sections or ordinances or in a case where discretionary power is given by the ordinances of said city to the Commissioner of Buildings, such acts shall be done in the name of and on behalf of the City of Chicago, and the said Commissioner of Buildings in so acting for the city shall not render himself liable personally, and he is hereby relieved from all personal liability for any damage that may accrue to persons or property as a result of any such act committed in good faith in the discharge of his duties, and any suit brought against the said Commissioner of Buildings by reason thereof shall be defended by the Department of Law of said city until the final termination of the proceedings therein.

(c) **Power to Pass on Ordinances.** The Commissioner of Buildings shall have full power to pass upon any question arising under the provisions of this ordinance or under any of the provisions relating to buildings contained in all general ordinances of the city, subject to the conditions, modifications and limitations contained therein.

(d) **Power to Make Rules.** The Commissioner of Buildings may adopt reasonable rules not inconsistent with the ordinances, to determine the quality of materials and workmanship in building construction and equipment and in the repair thereof, provided that any standards adopted by such rules shall be according to the practice, custom and usage prevailing in the industry involved. Copies of said rules shall be published and shall be kept always on file in the office of the Commissioner of Buildings. Copies of all such rules so adopted shall be transmitted to the City Council at the first regular meeting held after adoption of same.

(e) **Inspection of Buildings or Structures Where Complaint Is Made.** It shall be the duty of the Commissioner of Buildings where any citizen represents that any building or structure, or part thereof, is in an unsafe or dangerous condition; or that the stairways, corridors, exits or fire escapes in any factory or workshop or other place of employment are insufficient for the escape of employees in case of fire, panic or accident; or that the stairways, exits and fire escapes of any building or structure in the city do not comply with the requirements of this ordinance or the requirements contained in other general ordinances of the city; to make an examination of such building or structure; and if such representation is found to be true the said Commissioner shall give notice in writing to the owner, occupant, lessee or person in possession, charge or control of such building or structure to make such changes, alterations or repairs as safety, or the ordinances of the city may require. It shall be unlawful to continue the use of such buildings until the changes, alterations or repairs found necessary by the Commissioner of Buildings to make such building or structure, or part thereof, safe or to bring it into compliance with this ordinance and with the provisions of all other ordinances of the city shall have been made. Upon failure of parties so notified to comply with the requirements of said notice the matter shall be placed in the Department of Law of the City of Chicago for prosecution.

(f) **Examination and Approval of Plans.** The Commissioner of Buildings and his assistants shall pass upon all questions relating to the strength and durability of buildings or structures and shall examine and approve all plans before a permit is issued for the construction of any building or structure.

(g) **Certificates — Notices.** The Commissioner of Buildings shall sign, or cause to be signed, all certificates and notices required to be issued from the Department of Buildings and shall keep a record of the same, and shall issue, or cause to be issued, all permits authorized by this ordinance or as authorized by the provisions of any other ordinance of the city.

(h) **Records.** The Commissioner of Buildings shall also keep a proper record of all transactions and operations of the department and such record shall be at all times open to the inspection of the Mayor, Comptroller, Commissioner of Police, Chief Fire Marshal and Members of the City Council. The Commissioner of Buildings shall keep in proper books for that purpose an accurate account of all fees charged, giving the name of the person to whom same is charged, the date on which said charge is made and the amount of each such fee. The Commissioner of Buildings shall cause a complete record to be kept showing the location and character of every building or other structure for which a permit is issued, and shall cause to be filed every report of inspection, which reports shall bear the signatures of the inspectors making such inspections. He shall cause a record to be kept of all complaints of violations of the building ordinances and shall cause all such complaints to be investigated.

(i) **Annual Reports and Estimates.** The Commissioner of Buildings shall annually on or before the first day of March in each year, prepare and present to the City Council, a report showing the receipts and expenditures and entire work of the Department of Buildings during the previous fiscal year, and he shall on or before November first of each year, prepare and submit to the Comptroller, an estimate of the whole cost and expense of providing for and maintaining his office during the ensuing fiscal year.

(j) **Daily Report of Permits.** The Commissioner of Buildings shall prepare each day a report of the permits issued on the previous day, giving all the necessary information contained in same, including the legal description, when given, and the value of the building to be constructed or altered, and to file one (1) copy of this report with the Cook County Assessor; one (1) copy with the Board of Appeals of Cook County; and one (1) copy with the Corporation Counsel of the City of Chicago.

201.06 Buildings in Unsafe Condition:

(a) **Authority of Commissioner.** Whenever the Commissioner of Buildings shall find any building or structure, or part thereof in the city in such an unsafe condition as to endanger life, but in such condition that by the immediate application of precautionary measures such danger may be averted, he shall have authority, and it shall be his duty, to forthwith notify in writing, the owner, agent or person in possession, charge or control of such building or structure, or part thereof, to adopt and put into effect such precautionary measures as may be necessary or advisable in order to place such building or structure, or part thereof, in a safe condition; such notice shall state briefly the nature of the work required to be done and shall specify the time within which the work so required to be done shall be completed by the person, firm or corporation notified, which shall be fixed by said Commissioner of Buildings, upon taking into consideration, the condition of such building or structure, or part thereof, and the danger to life or property which may result from its unsafe condition.

(b) **Notices When Owner Not Found.** Whenever the Commissioner of Buildings shall be unable to find the owner of such building or structure, or part thereof, or any agent or person in possession, charge or control thereof, upon whom such notice may be served, he shall address, stamp and mail such notice to such person or persons at his or their last known address, and in addition thereto shall place or cause to be placed the notice herein provided for upon such building or structure at or near its principal entrance and shall also post or cause to be posted in a conspicuous place at each entrance to such building in letters not less than two (2) inches high, a notice as follows:

"THIS BUILDING IS IN DANGEROUS CONDITION AND ITS USE OR OCCUPANCY HAS BEEN PROHIBITED BY THE COMMISSIONER OF BUILDINGS."

It shall be unlawful for any person, firm or corporation to remove said notice or notices without written permission from the Commissioner of Buildings.

(c) **Commissioner's Power to Tear Down Unsafe Structures — May Direct Commissioner of Public Works to Remove—Continued Use Forbidden.** If, at the expiration of the time specified in such notice for the completion of the work required to be done by the terms of such notice in order to render the building or structure safe, said notice shall not have been complied with, and said building or structure is in such an unsafe condition as to endanger life or property, it shall be the duty of the Commissioner of Buildings to direct the Commissioner of Public Works to proceed forthwith to tear down or destroy that part of said building or structure that is in such unsafe condition as to endanger life or property, and in cases where an unsafe building or structure cannot be repaired or rendered safe by the application of precautionary measures, such building or structure, or the dangerous parts thereof, shall, on the order of the said Commissioner of Buildings, be torn down by the Commissioner of Public Works, and the expense of tearing down any part of such building or structure shall be charged to the person owning or in possession, charge or control of such building or structure, or part thereof, and the said Commissioner of Buildings shall recover or cause to be recovered from such owner or person in possession, charge or control thereof the cost of doing such work, by legal proceedings prosecuted by the Department of Law. If the owner, agent or person in possession, charge or control of such building or structure, or part thereof, when so notified shall fail, neglect or refuse to place such building or structure, or part thereof, in a safe condition, and to adopt such precautionary measures as shall have been specified by said Commissioner within the time specified in such notice, in such case, at the expiration of such time it shall be unlawful for any person, firm or corporation to occupy or use said building or structure, or part thereof, until said building or structure, or part thereof, is placed in a safe condition; and where a building or structure, or part thereof, is in a dangerous or unsafe condition and has not been placed in a safe condition within the time specified in the notice of the Commissioner of Buildings, such building or structure, or such part thereof, shall be forthwith vacated, and it shall be unlawful for any person or persons to enter same except for the purpose of making repairs required by the Commissioner of Buildings and the ordinances of the City of Chicago.

(d) **In Urgent Cases — Commissioner's Power Final.** Whenever the decision of the Commissioner of Buildings upon the safety of any building, or structure, or part thereof, is made in a case which in his opinion is so urgent that failure to properly carry out his orders to demolish or strengthen such building or structure, or part thereof, may endanger life and limb, the decision and order of the Commissioner of Buildings shall be absolute and final.

201.07 Building or Part of Building Constructed in Violation of Ordinances:

(a) **General.** Whenever it shall be found that any building or structure, or part thereof, is being, or shall have been constructed or built in violation of any of the provisions of this ordinance or any of the provisions of the general ordinances of the city, which it is his duty to enforce, the Commissioner of Buildings shall forthwith notify the owner, agent, architect and the contractor engaged in erecting such building or structure, or part thereof, of the fact that such building or structure, or part thereof, has been, or is being constructed or erected contrary to the provisions of such ordinances, and shall specify briefly in such notice in what manner the provisions of such ordinances have been violated, and shall require the person so notified to forthwith make such building or structure, or part thereof, conform to and comply with the provisions of such ordinances, specifying in such notice the time within which such work shall be done.

(b) **Authority of Commissioner to Tear Down—May Direct Commissioner of Public Works to Remove.** If, at the expiration of the time set forth in such notice, the person so notified shall have refused, neglected or failed to comply with the requirements made in such notice to have such building or structure, or part thereof, concerning which notice was sent, changed so as to conform to and comply with the provisions of the building ordinance of the city, the Commissioner of Buildings shall have the authority, and it shall be his duty, to direct the Commissioner of Public Works to proceed forthwith to tear down such building or structure, or such part thereof, as shall be or may have been erected and constructed in violation of any of the provisions of such ordinance, and the cost of such work shall be charged to and be recovered from the owner of such building or from the person for whom such building or structure is being erected, in legal proceedings prosecuted by the Department of Law.

(c) **May Direct Commissioner of Public Works to Remove.** The Commissioner of Buildings shall have authority to direct the Commissioner of Public Works to tear down any defective or dangerous wall or structure or any building or structure, or part thereof, which may be constructed in violation of the terms of this article or of the provisions contained in any of the building ordinances of the city, after written notice has been served upon the owner, lessee, occupant, agent or person in possession, charge or control, directing them or him to tear down or remove any defective wall, building or structure, or any part thereof, which is in a dangerous condition, or which has been or is being constructed or maintained in violation of the terms of such ordinances. In case of the destruction or partial destruction of buildings by fire, decay or otherwise, when any department of the city government, pursuant to the ordinances of the city, shall make an outlay of money or incur any liability for the payment of any expense on behalf of the city in an effort to preserve or prevent the destruction of such building or buildings, or structure, or for the preservation of life of its citizens, it shall be the duty of the Commissioner of Buildings to ascertain the amount of such outlay or expenditure and present a bill therefor to the owner or owners of any such building or buildings, or its or their agent or agents, and it shall be the duty of the said Commissioner of Buildings to refuse to issue a permit for the construction, reconstruction, alteration or repair of any building or buildings or structure by any such owner or owners, lessee, occupant, agent or person in possession, charge or control thereof until such outlay or expenditure shall be repaid to the city by the owner, lessee, occupant, agent or person in possession, charge or control of such building or buildings thus totally or partially destroyed in the manner aforesaid. Said Commissioner shall also proceed forthwith to collect the amount of such bill from such owner or owners by legal proceedings prosecuted by the Department of Law.

(d) **May Stop Construction of Buildings.** The Commissioner of Buildings shall have power to stop the construction of any building, or the making of any alterations or repairs of any building within the city when the same is being done in a reckless or careless manner, or with defective material, or if incompetent workmen are employed thereon and performing the work in a manner not in conformity with the practice, custom and usage prevailing in the various branches of the building industry, such as failing to properly and securely assemble, join and erect any materials used in the construction, alteration or repair of any building, or in violation of any ordinance, and to order, in writing or by word of mouth, any and all persons in any way or manner whatever engaged in so constructing, altering or repairing any such building, to stop and desist therefrom.

(e) **May Stop Wrecking of Buildings.** The Commissioner of Buildings shall also have power to stop the wrecking or tearing down of any building or structure, or part thereof, within the city when the same is being done in a reckless or careless manner or in violation of any ordinance or in such a manner as to endanger life or property, and to order any and all persons engaged in said work to stop and desist therefrom. When such work has been stopped by the order of said Commissioner, it shall not be resumed until said Commissioner shall be satisfied that adequate precautions will be taken for the protection of life and property, and that said work will be prosecuted carefully and in conformity with the ordinances of the city.

201.08 Deputy Commissioner of Buildings:

(a) **Appointment and Qualifications:** The Deputy Commissioner of Buildings shall be appointed by the Commissioner of Buildings according to law. The person certified to fill this office shall be either a civil, structural or architectural engineer or an architect, an experienced building contractor or an efficient building mechanic with at least five (5) years' experience and training.

(b) **Duties.** The Deputy Commissioner of Buildings shall act as Commissioner of Buildings in the absence of the Commissioner of Buildings from his office and while so acting shall discharge all the duties and possess all the powers imposed upon or vested in the Commissioner of Buildings. The Deputy Commissioner of Buildings shall, under the direction of the Commissioner of Buildings, have general control of all matters and things pertaining to the work of the Department of Buildings and shall perform such other duties as may be required of him by the Commissioner of Buildings.

201.09 Secretaries:

(a) **Appointment.** The Secretary of the Department of Buildings and the Private Secretary to the Commissioner of Buildings shall be appointed by the Commissioner of Buildings according to law.

(b) **Duties.** The Secretary of the Department of Buildings shall, under the supervision and direction of the Commissioner of Buildings, preserve and keep all books, records and papers belonging to the office of the Department of Buildings, or which are required by law to be filed therein. He shall perform such other duties as may be required of him by the Commissioner of Buildings.

(c) The Private Secretary to the Commissioner of Buildings shall have charge of the correspondence of the office of the Commissioner of Buildings and shall perform such other duties as may be required by the Commissioner of Buildings.

201.10 Engineering Staff:

(a) **Appointment and Qualifications.** The Commissioner of Buildings shall appoint, according to law, one (1) or more Structural Engineers and such other engineers and assistants as the City Council may by ordinance provide for service on the engineering staff of the Department of Buildings. Every person certified to fill the position of Structural Engineer, shall be a civil, structural or architectural engineer of at least five (5) years' training and experience.

(b) **Duties.** The Structural Engineers, shall under the direction of the Commissioner of Buildings, examine all plans submitted for the purpose of obtaining a permit. They shall also examine and verify the figures on all applications for and on all floor load placards before such placards are approved for posting. They shall, in addition thereto, perform such other duties as may be required of them by the Commissioner of Buildings.

201.11 Building Inspector in Charge:

(a) **Appointment and Qualifications.** The Building Inspector in Charge shall be appointed by the Commissioner of Buildings according to law. The person certified to fill this position shall be a civil, structural, architectural or fire protection engineer, or an architect or building superintendent or a building mechanic, with at least five (5) years' experience in general building construction.

(b) **Duties.** In the absence of the Commissioner of Buildings and the Deputy Commissioner of Buildings from their offices, the Building Inspector in Charge shall act as Commissioner of Buildings and while so acting he shall discharge all of the duties and possess all of the powers imposed upon or vested in the Commissioner of Buildings. He also shall have immediate charge of the investigation of complaints and of the periodical inspection of buildings and structures being erected, enlarged, altered or repaired, excepting only such inspection as is expressly assigned to the elevator or fire escape inspectors or as by law assigned to some other department of the city government and shall perform such other duties as may be required of him by the Commissioner of Buildings.

201.12 Assistant Building Inspectors in Charge:

(a) **Appointment and Qualifications:** The Commissioner of Buildings shall appoint, according to law, at least four (4) Assistant Building Inspectors in Charge. Every person certified to fill the position of Assistant Building Inspector in Charge shall be a civil, structural, architectural or fire protection engineer, or an architect, a building superintendent, or a building mechanic, with at last five (5) years' experience in general building construction.

(b) **Duties.** The Assistant Building Inspectors in Charge shall have immediate charge of the several districts assigned to them by the Commissioner of Buildings and shall perform such other duties as may be required of them by the Commissioner of Buildings.

201.13 Building Inspectors:

(a) **Appointment and Qualifications.** The Commissioner of Buildings shall appoint, according to law, such Building Inspectors as may be provided for by the City Council. Every person certified to fill the position of Building Inspector shall be a civil, structural, architectural or fire protection engineer, or an architect, or a building superintendent or a building mechanic, with at least five (5) years' experience in general building construction.

(b) **Duties.** The Building Inspectors shall, under the direction of the Building Inspector in Charge, examine all buildings and structures in the course of erection, enlargement, alteration, repair or removal, as often as is required for efficient supervision, and shall make such periodical examinations of existing structures as shall be assigned to them. They shall examine all buildings, structures and walls reported to be in dangerous condition. They shall examine all buildings and other structures for the enlargement, alteration, raising or removing of which application for permit shall be made.

(c) **Reports.** Every Building Inspector shall make written reports daily to the Commissioner of Buildings as to the condition in which he found each building examined and as to violations, if any, of the ordinances which the Commissioner of Buildings is required to enforce, together with the street and number of the premises where such violations, if any, were found, the names of the owner, agent, lessee and occupant thereof, and of the architect and contractor engaged in and about the work in question. The Building Inspectors shall perform such other duties as may be required of them by the Commissioner of Buildings.

201.14 Elevator Inspectors:

(a) **Appointment and Qualifications.** The Commissioner of Buildings shall appoint, according to law, an Elevator Inspector in Charge, an Assistant Elevator Inspector in Charge and such other Elevator Inspectors as may be provided for by the City Council. The person certified to fill the position of Elevator Inspector in Charge shall be a graduate in engineering from a recognized technical school, shall be versed in the principles of both mechanical and electrical engineering, and shall have had at least five (5) years' experience in design or construction work. The person certified to fill the position of Assistant Elevator Inspector in Charge shall be well versed in the principles of both mechanical and electrical engineering and shall have had at least five (5) years' experience in design, construction or inspection work. Every person certified to fill the position of Elevator Inspector shall be a mechanical engineer, machinist or elevator mechanic, well versed in the principles of mechanical and electrical construction.

(b) **Duties.** The Elevator Inspector in Charge, under the direction of the Commissioner of Buildings, shall examine and approve all drawings for the installation of all elevators, mechanical equipment used for the raising or lowering of any proscenium fire curtain, or stage orchestra floor, or any platform, dumbwaiter, escalator and mechanical amusement device. No such elevator, dumbwaiter, escalator, mechanical equipment, device or apparatus shall be installed, nor shall any such equipment be operated without the approval of the Elevator Inspector in Charge. The Elevator Inspector in Charge shall cause such inspection to be made of all new installations as may be necessary to insure construction in accordance with the drawings approved by him. He shall cause such periodic inspections to be made of existing installations of such equipment as may be necessary to insure safe maintenance of this equipment and as required by this ordinance. He shall also perform such other duties as may be required of him by the Commissioner of Buildings. The Assistant Elevator Inspector in Charge shall perform all the duties of the Elevator Inspector in Charge during his absence, and such other duties as are assigned to him by the Elevator Inspector in Charge. The Elevator Inspectors shall inspect all elevators, mechanical equipment used for the raising or lowering of any proscenium fire curtain, or stage, orchestra floor, or any platform, dumbwaiter, escalator and mechanical amusement device as shall be assigned to them by the Elevator Inspector in Charge.

(c) **Reports.** They shall make written reports daily to the Commissioner of Buildings as to the condition in which they find the proscenium fire curtains, elevators, dumbwaiters, escalators, mechanical equipment, devices and apparatus, inspected by them, and of any violations of the requirements of this ordinance pertaining to such matters, together with the street and number of the premises where such violations occur, the names of the owner, agent, lessee and occupant thereof, and if a new installation, the names of the architect, or engineer and contractor engaged in or about the construction and installation of such equipment. The inspectors shall also perform such other duties as may be required of them by the Commissioner of Buildings.

201.15 Clerical Assistants: The Commissioner of Buildings shall appoint, according to law, such clerical assistants, stenographers, messengers and other employees as may be provided for by the City Council; and they shall perform such duties as may be required of them by the Commissioner of Buildings.

201.16 Bonds: The Deputy Commissioner of Buildings, the Building Inspector in Charge, the Assistant Building Inspector in Charge, the Elevator Inspector in Charge and the Structural Engineers, before entering upon the duties of their offices or positions, shall each execute a bond running to the City of Chicago, conditioned for the faithful performance of his duties, with such sureties as the City Council shall approve in the following sums: The Deputy Commissioner of Buildings, ten thousand dollars ($10,000.00); the Building Inspector in Charge, the Assistant Building Inspectors in Charge, the Elevator Inspector in Charge, and the Structural Engineers, five thousand dollars ($5,000.00) each.

201.17 Employees Not to Engage in Another Business: Every employee in the Department of Buildings shall devote his entire time to such employment and shall not be engaged in any other business or vocation.

201.18 Duty of Police to Assist Commissioner in Enforcing Provisions of This Ordinance: Whenever it shall be necessary, in the opinion of the Commissioner of Buildings, to call upon the Department of Police for aid or assistance in carrying out or enforcing any of the provisions of this ordinance and of the provisions of any other ordinance of the city, which it is his duty to enforce, he shall have the authority to do so and it shall be the duty of the Commissioner of Police, or of any member of the Department of Police, when called upon by said Commissioner of Buildings to act according to the instructions of, and to perform such duties as may be required by said Commissioner of Buildings in order to enforce or put into effect the said provisions of this ordinance and of such other ordinances as it is his duty to enforce.

201.19 Power of Entry:

(a) **Buildings Under Construction.** The Commissioner of Buildings and his assistants are empowered to enter any building, structure or premises, whether completed or in process of erection, for the purpose of determining whether the same has been or is being constructed, maintained and used in accordance with the provisions of this ordinance and the provisions contained in other general ordinances of the city and it shall be unlawful to exclude them from any such building, structure or premises.

(b) **Occupied Buildings.** The Commissioner of Buildings, President of the Board of Health, Fire Commissioner, Commissioner of Public Works, Commissioner of Streets and Electricity, Commissioner of Police, or any of them and their respective assistants, shall have the right to enter any building, or premises, and any and all parts thereof, at any reasonable time, and at any time when occupied by the public in order to examine such buildings, or premises to judge of the condition of the same and to discharge their respective duties, and it shall be unlawful for any person to interfere with them, or any of them, in the performance of their duties.

201.20 City Officers Empowered to Close: The Commissioner of Buildings, President of the Board of Health, Fire Commissioner, Commissioner of Public Works, Commissioner of Streets and Electricity, Commissioner of Police, or any one of them, shall have the power, and it shall be their joint and several duty, to order any building or premises closed, where it is discovered that there is any violation of any of the provisions of this ordinance, and to keep same closed until such provisions are complied with.

201.21 License—Mayor Shall Revoke: Upon a report to the Mayor by the Commissioner of Buildings, President of the Board of Health, Fire Commissioner, Commissioner of Public Works, Commissioner of Streets and Electricity, or the Commissioner of Police that any requirements of this ordinance, or that any order given by them, or any of them in regard thereto, has been violated, or not complied with, the Mayor shall revoke the license of any such business, or occupation of any such building so reported and cause the same to be closed.

201.22 Electrical Code: The installation of all electric wiring and electric equipment provided for in this ordinance shall be done in strict accordance with the ELECTRICAL CODE of the City of Chicago.

201.23 Plumbing and Drainage: The installation of all plumbing and drainage provided for in this ordinance shall be done in conformity with the provisions of Chapters 47-A and 48.

201.24 Ventilation: The installation of all ventilating systems provided for in this ordinance shall be done in conformity with the provisions of Chapters 47 and 47-A.

CHAPTER 3
Arbitration, Standards, Registration and Penalties

ARTICLE 301
Arbitration

301.01 General: In all cases where discretionary power is given to the Commissioner of Buildings to estimate damage to buildings, as also in questions relating to the security of any building or buildings or structures, or parts thereof, and in all other cases where discretionary powers are given by ordinance to the Commissioner of Buildings, any party or parties believing themselves injured or wronged by any decision of the Commissioner of Buildings, may, before instituting any suit, make an appeal for arbitration as provided by this article.

301.02 Time for Making Appeal: Any person wishing to make an appeal shall do so within five (5) days' time after written notice of the decision or order of the Commissioner of Buildings has been given. An appeal made later than five (5) days after the serving of the notice of the Commissioner of Buildings shall not entitle the appellant to any arbitration. The request for arbitration shall be in writing and shall state the object of the proposed arbitration and the name of the person who is to represent the appellant as arbitrator.

301.03 Cost Apportionment of Arbitrators: The Commissioner of Buildings shall thereupon inform the appellant of the cost of such arbitration and such appellant shall, within twenty-four (24) hours from the receipt of such information, deposit with the Commissioner of Buildings the sum of money requested for defraying the expense of the same, which sum shall be fixed in each case by said Commissioner in proportion to the time it will take and the difficulty and importance of the case, but shall in no case be more than the cost of similar service in the course of ordinary business of private individuals or corporations. As soon as such sum of money shall have been deposited with him, the Commissioner of Buildings shall appoint an arbitrator to represent the city and the two (2) arbitrators thus chosen shall, if they cannot agree, select a third arbitrator, and the decision of any two of these arbitrators, after investigation and consideration of the matter in question, shall be final and binding upon the appellant as well as the city unless an appeal is taken therefrom as provided in case of an appeal under a statutory arbitration, within five (5) days thereafter.

301.04 Arbitrators to Take Oath — Power to Examine Witnesses: The arbitrators shall themselves, before entering upon the discharge of their duties, be placed under oath by the City Clerk, to the effect that they are unprejudiced as to the matter in question and that they will faithfully discharge the duties of their position. They shall have the power to call witnesses and place them under oath, and their decision or award shall be rendered in writing both to the Commissioner of Buildings and to the appellant. The fee deposited by the appellant with the Commissioner of Buildings shall be paid by the Commissioner of Buildings to the arbitrators upon the rendering of their report and shall be in full of all costs incident to the arbitration; but should the decision of said Board of Arbitration be rendered against the Commissioner of Buildings, then the money deposited by the aforesaid appellant shall be returned to him and the entire cost of such arbitration shall be paid by the city.

ARTICLE 302
Standards and Tests

302.01 General: The Committee on Standards and Tests, standard of quality of materials and construction, and standard tests required by this ordinance shall be in accordance with the provisions of this article.

302.02 Committee on Standards and Tests:

(a) **Appointment.** For the purpose of insuring public safety and for the purpose of investigating or testing new materials, methods, or systems of construction, or new arrangements of materials, not permitted by, or varying from the requirements established by this ordinance but which are claimed to be equally as good or superior to those permitted thereunder the Mayor shall appoint a Committee on Standards and Tests consisting of seven (7) members, of which Committee one member shall be the Commissioner of Buildings, three members shall be either architects or engineers licensed by the State of Illinois, who shall have been residents of the County of Cook for a period of at least one year preceding the date of appointment, and three members shall be members of the Chicago City Council.

(b) **Application for Test.** Any person desiring to submit any new building materials, methods or systems of construction, or new arrangements of materials not provided for in this ordinance for determining or testing the adaptability or safety of such materials, methods, systems or arrangements for building purposes under this ordinance or to establish the safety qualifications of any substance for occupancy purposes, shall make application in writing to the City Clerk, setting forth the merits claimed and the purposes desired, and shall deposit with the City Treasurer a sum sufficient to cover the expenses of making said test. The Committee on Standards and Tests shall also determine the classification of fume and explosion hazard gases or liquids in all cases where no classification of such gases or liquids has been made in this ordinance.

(c) **Methods.** The methods of the Committee on Standards and Tests in making such tests shall follow the published standard specifications of the U. S. Bureau of Standards, or of the American Society for Testing Materials, insofar as such standard specifications shall govern and also conform to Section 302.03, for standard specifications for materials and tests. If, after such testing or investigation, the materials, methods, systems or arrangements of materials for construction shall be found satisfactory to the majority of all members of such Committee and they shall approve its use in buildings erected under this ordinance, they shall submit their recommendation, if any, to the City Council.

302.03 Standard Specifications for Materials and Tests:

(a) **Documents.** Wherever in this ordinance materials used in construction, methods of installation, fire tests of materials, or other building operations are required to conform to or be at least equal to the standard specifications of materials or of practice, standard tests of materials, fire tests of construction or of materials, or any other standard specification requirement recognized and accepted by a federal or state bureau, national technical organization, or national board of fire underwriters, and the specifications or test of the bureau or society has been identified in this ordinance by the designation of the said bureau or society and the date of its publication, the said documents, or so much of them as are adopted by this ordinance, are hereby declared to be a part of this ordinance the same as if the specifications, regulations or tests were herein fully set out; and it is further provided that authenticated copies of said documents shall be kept on file in the office of the City Clerk and the Commissioner of Buildings, from and after the date of the adoption of this ordinance.

(b) **Standard Quality of Materials.** All building materials shall be of a quality to meet the intent of this ordinance, and shall conform to requirements promulgated as rules by the Commissioner of Buildings in accordance with the provisions of Chapter 2, relating to Administration.

(c) **Standard Quality of Workmanship.** Workmanship in the fabrication, preparation and installation of building materials shall meet the intent of this ordinance, and shall conform to generally accepted standard practice or requirements promulgated as rules by the Commissioner of Buildings in accordance with the provisions of Chapter 2 relating to Administration.

ARTICLE 303

Registration

303.01 Registry With Department of Buildings: Every person, firm or corporation engaged in the business of constructing, repairing, removing or demolishing the whole or any part of buildings or structures, or the appurtenances thereto in the city of Chicago, shall before undertaking the erection, enlargement, alteration, repair, removal or demolition of any building or structure, for which permits are required by the ordinances of the City, register the name and address of such person, firm or corporation in a book kept by the Commissioner of Buildings and used for this purpose. No permit shall be granted for the erection, enlargement, alteration, repair, removal or demolition of any building or structure unless the name and address of the person, firm or corporation that is about to undertake such work is contained in the registration book kept for that purpose.

303.02 Excavating Work: When application is made for a permit for excavating work only, the provisions of Section 303.01, shall not apply to any person, firm or corporation licensed as an excavator as provided in and by the ordinances of the city.

303.03 Where Masonry Work Only Is Required: When application is made for a permit, and the work of construction involves masonry construction only, the provisions of Section 303.01 shall not apply to any person, firm or corporation licensed as a mason contractor or employing mason, as provided in and by the ordinances of this city. Where the work of construction, for which a permit is sought, involves construction other than masonry construction, any mason contractor or employing mason, licensed as aforesaid, engaged in or undertaking the work of such construction other than masonry construction, must register his, their or its name or names, and comply with the requirements of this article before a permit for such work is issued.

303.04 Warm Air Furnace Work: Any person, firm or corporation desiring to engage in the business of constructing, replacing or installing warm air furnaces shall apply for registration to the Commissioner of Buildings. Such applicant shall register the name and address of such person, firm or corporation in a book kept by the Commissioner of Buildings for this purpose. The fee for such initial registration for the first calendar year shall be twenty-five dollars ($25.00) which shall be paid to the City Collector upon the filing of such application, and which shall expire on the thirty-first day of December after such initial registration. Any registrant may renew such registration annually upon the payment of an annual renewal fee of ten dollars ($10.00) payable on January first of each year.

303.05 Liability for Violations: If any person, firm or corporation registered as provided by this article, shall fail in the execution of any work for which a permit was issued, to comply with the ordinances of the city relative to the erection, enlargement, alteration, repair, removal or demolition of any building, or part thereof, either the Commissioner of Buildings or the President of the Board of Health may bring suit and prosecute such person, firm or corporation for such failure or violation, and in case of conviction, his, their or its name or names shall be stricken from the said registration book and shall not be re-entered or reinstated during such time as any violation exists or any judgment remains unsatisfied with regard to said conviction.

303.06 Reinstatement: Any person, firm or corporation that shall have been convicted under the preceding section and shall have had his, their or its name or names stricken from such registration book may have such name or names re-entered therein on filing with the Commissioner of Buildings a certificate signed by the City Prosecutor, the Commissioner of Buildings and the president of the Board of Health to the effect that all violations of ordinances with reference to which conviction was secured have been correoted and are non-existent and that all claims and judgments arising from such convictions have been paid.

303.07 Waiver of Fees: Any person, firm or corporation who is required by the provisions of this Article to register his or their names with the Commissioner of Buildings, or who is required by this or other ordinances of the city to pay a fee for such registration, or a license fee for the business of excavating work or for masonry or mason work shall not be required to pay any such fee if such person, firm or corporation is already registered and licensed for such business for the current year in another city or village within the State of Illinois; provided however, that any such person, firm or corporation shall register his or their names as required by this article and, before any permit is issued, shall pay to the City of Chicago the difference in amount that such fee as required by the City of Chicago exceeds the fee paid at the place of registration or licensing.

ARTICLE 304
Penalties

304.01 General: Any person, firm or corporation violating, neglecting or refusing to comply, with or resisting or opposing the enforcement of any of the provisions of this ordinance, where no other penalty is provided, shall, upon conviction, be fined not less than twenty-five dollars ($25.00) nor more than two hundred dollars ($200.00) for each offense and every such person, firm or corporation shall be deemed guilty of a separate offense for every day on which such violation, neglect or refusal shall continue; and any builder or contractor who shall construct any building in violation of the provisions of this ordinance, and any architect designing, drawing plans for, or having supervision of such building, or who shall permit it to be constructed, shall be liable for the penalties provided and imposed by this section.

304.02 Nuisance: Every building or structure constructed or maintained in violation of any of the provisions of this ordinance, or which is in an unsanitary condition, or in an unsafe or dangerous condition or which in any manner endangers the health or safety of any person or persons, is hereby declared to be a public nuisance. Every building or part thereof which is in an unsanitary condition by reason of the basement or cellar being damp or wet, or by reason of the floor of such basement or cellar being covered with stagnant water, or by reason of the presence of sewer gas, or by reason of any portion of a building being infected with disease or being unfit for human habitation, or which, by reason of any other unsanitary condition, is a source of sickness, or which endangers the public health, is hereby declared to be a public nuisance.

304.03 Building Operations at Night in Residential Districts Prohibited—Penalty: It shall be unlawful for any person, firm or corporation, in conducting any building operations between the hours of ten o'clock in the evening and six o'clock in the morning to operate or use any pile drivers, steam shovels, pneumatic hammers, derricks, steam or electric hoists or other apparatus, in any block in which more than half of the buildings on either side of the street are used exclusively for dwelling purposes. Any person, firm or corporation violating any of the provisions of this section shall upon conviction be fined not less than five dollars ($5.00) nor more than one hundred dollars ($100.00) for each offense, and each day's violation of same shall be considered a separate and distinct offense.

CHAPTER 4
Permits, Plans, Fees, Inspections and Certificates

ARTICLE 401
Permits

401.01 General:

(a) **Permit Required.** Before proceeding with the erection, enlargement, alteration, repair, removal or demolition of any building or structure in the city, a permit for such erection, enlargement, alteration, repair, removal or demolition shall first be obtained by the owner or his agent from the Commissioner of Buildings, and it shall be unlawful to proceed with the erection, enlargement, alteration, repair, removal or demolition of any building, structure or structural part thereof within the city unless this permit shall have first been obtained from the Commissioner of Buildings as provided in this section.

(b) **Applications.** Applications for building permits shall be in such form as shall be prescribed by the Commissioner of Buildings. Every such application for a permit shall be accompanied by a copy of every recorded easement on the lot on which the building is to be erected, and on the immediately adjoining lots, showing the use or benefit resulting from such easement. All such applications shall be accompanied by drawings, plans and specifications in conformity with the provisions of this Chapter. Where alterations or repairs in buildings are made necessary by reason of damage by fire, that fact shall be stated in the application for a permit. In such cases, before a permit shall issue, the Commissioner of Buildings shall case a thorough inspection to be made of the damaged premises with the view of testing the structural integrity of the damaged parts.

(c) **Exceptions, Permit Not Required.** A permit shall not be required for any minor repairs, as may be necessary to maintain existing parts of buildings, but such work or operations shall not involve the replacement or repair of any structural load-bearing members, nor reduce the means of exit, affect the light or ventilation, room size requirements, sanitary or fire-resistive requirements, use of materials not permitted by this ordinance, nor increase the height, area or capacity of the building.

401.02 Approval by Other Departments:

(a) **General.** All drawings and plans for the construction erection, addition to or alterations of any building or other structure, for which a permit is required shall first be presented to the Commissioner of Buildings for examination and approval as to proper use of building and premises and as to compliance in all other respects with the Chicago Zoning Ordinance and shall therefore thereafter be presented to the Board of Health, the Department of Smoke Inspection, Fire Department, Department of Boiler Inspection and Department of Public Works for submission to the proper official of these departments and bureaus for his examination and approval with regard to such ordinances as are within the duty of such office to enforce, and after said drawings and plans have been examined and passed upon, the same shall be returned to the Commissioner of Buildings where they shall be taken up for examination and approval by the Commissioner of Buildings.

(b) **Indemnity Bonds.** Before any building permit is issued, the applicant shall produce evidence that he has filed with and had approved by the Commissioner of Public

Works of the city, an indemnifying bond protecting the city against any and all damages that may arise to the streets or alleys upon which such building abuts, and to the city, and to any person, in consequence, or by reason of, the proposed operations to be authorized by such permit, or by reason of any obstruction or occupation of any streets or sidewalks in and about such building operations.

401.03 Issuance of Permits:

(a) **General.** At the proper time notice shall be given by the Commissioner of Buildings to the applicant that his plans have been examined and are ready to be returned to him, and if such plans have been approved as submitted to the various departments and bureaus as aforesaid, the Commissioner of Buildings shall, according to ordinance, issue a permit for the construction, erection, repair or alteration of such building or structure; and shall file such application, and shall apply to such plans a final official stamp, stating that the drawings to which the same has been applied comply with the provisions contained in this ordinance. The plans so stamped stall then be returned to such applicant.

(b) **Driveways — Permit Required.** No permit shall issue for the construction, erection, repair or alteration of any building or structure designed or intended for use as a garage or any other business, the operation of which will require a driveway across a public sidewalk, until the applicant therefor has first obtained from the Commissioner of Public Works a permit for driveway or driveways as prescribed by Article II of Chapter 14 of the Revised Chicago Code of 1931.

401.04 Daily Report of Permits: The Commissioner of Buildings shall prepare each day, a report of the permits issued on the previous day as required by Section 201.05, paragraph (j).

401.05 Time Limits: If, after a building or other required permit shall have been granted, the operations called for by such permit shall not be begun within six (6) months after the date thereof, such permit shall be void and no operations thereunder shall be begun or completed until an extended permit shall be taken out by the owner or his agent and a fee of twenty-five (25) per cent of the original cast of permit shall be charged for such extended permit; provided however, that in no case shall a permit be issued, or renewed, for a less fee than two dollars ($2.00). An extended permit shall be valid for six (6) months following the date of expiration of the original permit and must be applied for within ten (10) days of expiration of the original permit. Two (2) extensions only shall be granted and if work is not begun within eighteen (18) months after date of issuance of original permit, all rights under such permit shall thereupon terminate by limitation. Where, under authority of a permit, or extended permit, work has begun and has been abandoned for a continuous or cumulative period of twelve (12) months, all rights under such permit shall thereupon terminate by limitation.

401.06 Construction Contrary to Permit: It shall be unlawful for any owner, agent, architect, structural engineer, contractor or builder engaged in erecting, altering or repairing any building to make any departure from the drawings or plans, as approved by the Commissioner of Buildings, of a nature which involves any violation of the ordinances on which the permit has been issued. Any such departure from the approved drawings and plans involving a violation of requirements, shall operate to annul the permit which has been issued for such work and shall render the same void.

401.07 Power to Stop Work: No person shall begin work on any building or structure, for which a permit is required until such permit shall have been secured. In case any work is begun on the erection, alteration, repair or removal of any building or structure without the issuance of a permit authorizing same, the Commissioner of Buildings shall have power to stop such work at once and to order all persons engaged thereon to stop and desist therefrom until the proper permit is secured. Any case wherein any work is done under a permit anthorizing erection, alteration or repair of a building or structure, which work is being done contrary to the approved drawings and plans, the Commissioner of Buildings or the President of the Board of Health shall have power to at once stop such work and to order all persons engaged thereon to stop and desist therefrom, which work shall not be resumed until satisfactory assurance has been given to the Commissioner of Buildings or the President of the Board of Health, as the case may be, that it will be according to the approved drawings and plans. Nothing in this paragraph shall be construed to prevent minor changes in arrangement or decoration which do not affect the requirements of any ordinance.

401.08 Street Obstructions — Permits — Bonds—Fees: Permits for the obstruction of streets shall be issued by the Commissioner of Public Works and shall be paid in proportion to the street frontage occupied at the rate of five dollars ($5.00) per month for every twenty-five (25) feet or fractional part thereof, of frontage so occupied; and before any permit shall be granted to any person, firm or corporation for the obstruction of any street or streets, or sidewalk, an estimate of the cost of restoring said street and sidewalk, to a condition equally as good as before it was obstructed with a fair additional margin for contingent damages, shall be made by the Commissioner of Public Works. Such estimate shall be in no case less than two dollars ($2.00) per foot, or fractional part thereof, for the frontage of the portion of the street to be obstructed, and a deposit shall be required of the person, firm or corporation desiring to obstruct such street or sidewalk. Such deposit, less the charge of five dollars ($5.00) per month for each twenty-five (25) feet of frontage used, shall be returned upon the restoration of the said street and sidewalk to a condition equally as good as before it was obstructed. When the Commissioner of Public Works shall receive satisfactory proof that said street and sidewalk have been restored to a condition equally as good as before it was obstructed, he shall issue a certificate to the Comptroller, certifying to said fact and the Comptroller shall thereupon forthwith issue a warrant on the City Treasurer for the amount of money thus deposited for costs which it, the city, may suffer or be put to, or which may be recovered from it from, or by reason of the issuance of such permit, or by reason of any act or thing done or neglected to be done under or by virtue of the authority given in such permit and the requirements of the city ordinance. Any permit issued pursuant to the terms of this section may be revoked for cause by the Commissioner of Public Works at any time.

401.09 Canopy, Plans—Permits: It shall be unlawful for any person, firm or corporation to erect or construct any canopy attached to a building or structure under any general or special ordinance now in force, or which shall or may hereafter be adopted, any part of which canopy shall project over a public street, alley or other public space, without first submitting the plans of such canopy, and also of the part of the building or other structure to which it is to be attached, to the Commissioner of Buildings for his approval. No permit shall be issued by the Department of Public Works unless the plans of such canopy shall have been approved by the Department of Buildings and a permit to attach said canopy to the building from which it is intended to project shall have been obtained from the Commissioner of Buildings. No canopy that has been or may hereafter be authorized by any general or special ordinance shall at any time be inclosed by canvas or other cloth or material in whole or in part so as to obstruct free passage underneath same, or so as to obstruct or reduce any required exit width.

401.10 Boilers, Approval of Plans: The size, number and location of power or heating boilers to be installed in any building shall be marked on the plans, and, except in single dwellings, shall be approved by the Department for the Inspection of Steam Boilers, Unfired Pressure Vessels and Cooling Plants and by the Department of Smoke Inspection and Abatement, before a permit is issued by the Department of Buildings for the erection of such building.

401.11 Elevator Construction or Alterations:

(a) **Permit Required.** Before proceeding with the construction, installation or alteration of any elevator or mechanical equipment used for the raising and lowering of any curtain, or stage, or orchestra floor, or any platform, dumbwaiter, escalator, or mechanical amusement device or apparatus, application for a permit for such construction, installation or alteration, shall be submitted to the Commissioner of Buildings either by the owner or agent of the building, or of the premises on which such equipment is to be installed. A permit shall be obtained for any alteration in such elevator equipment except that this requirement shall not apply to the replacement of existing parts with other parts which are identical with those which are replaced.

(b) **Application.** The application for permit shall specify the number and kind of equipment which it is desired to install, or the nature of the alteration to be made and the location of the building or structure or premises, and shall be accompanied by such drawings and specifications as shall be necessary to inform said Commissioner of the plan of construction, type of elevator, dumbwaiter, escalator or mechanical amusement device; method of alteration and the location thereof. Every application for a permit for a mechanical amusement device shall also be accompanied by a detailed drawing and description of the construction proposed, with a certificate signed by an architect or engineer certifying to the strength and safety of such device. If such drawings and specifications show that the equipment is to be installed or altered in conformity with the provisions of this ordinance, the Commissioner of Buildings shall approve the same and shall issue a permit to such applicant upon the payment by such applicant of the permit fee hereinafter named. It shall be unlawful for any owner, agent or contractor to permit or allow the installation or alteration of any such equipment until a permit has been obtained, and the permit fee paid.

401.12 Fences: It shall be unlawful for any person, firm or corporation to erect or construct any fence more than five (5) feet in height within the city limits without first obtaining a permit from the Commissioner of Buildings.

401.13 Permit for Moving Frame Buildings:

(a) **Permit and Frontage Consent.** No person, firm or corporation shall be permitted to move any building which has been damaged to an extent greater than fifty (50) per cent of its value by fire, decay or otherwise; nor shall it be permissible to move any frame building of such character as it prohibited to be constructed within the fire limits from any point outside the fire limits to any point within the fire limts; nor shall it be permissible to move any building to a location at which the use for which such building is designed are prohibited by ordinance. Permits for the moving of frame buildings, other than those the moving of which is herein prohibited, shall be granted upon the payment of a fee of ten (10) cents for each one thousand (1000) cubic feet of volume, or fractional part thereof of such building, and upon securing and filing the written consent of two-thirds (⅔) of the property owners according to frontage on both sides of the street in the block in which such building is to be moved. No permit shall be issued to move any building used or designed to be used for purposes for which frontage consents are required until frontage consents in the block to which such building is to be moved have also been secured and filed as required by the ordinances relating to such use.

(b) **Lot Occupancy.** No building used for residence or multiple dwelling purposes shall be moved from one lot to another or from one location to another upon the same lot unless the space to be occupied on such lot shall comply with the provisions of Chapter 8—MULTIPLE DWELLINGS, of this ordinance.

401.14 Tanks for Flammable Liquids:

(a) **Permit Required.** Any person, firm or corporation desiring to install a tank for the storage of any flammable liquids, as provided in Chapter 11—HAZARDOUS USE UNITS, shall first obtain a permit from the Commissioner of Buildings; provided however, that no permit shall be required for an aggregate capacity of tanks of one hundred twenty (120) gallons or less for Class I and Class II liquids, nor for an aggregate capacity of tanks of five hundred fifty (550) gallons or less for Class III and Class IV liquids. The application for permit shall be made by the owner or his agent as required by this chapter. Before issuing such permit the Commissioner of Buildings shall first cause to be inspected, the location or site where such tank is to be installed, and if the site is satisfactory, such permit shall be issued upon the payment of fees hereinafter provided. Such permit shall not be assigned nor shall any right or privilege thereunder be transferred or assigned except by written consent of the Commissioner of Buildings.

(b) **Application.** Every application for a permit for any such tank shall be in writing, stating the location, the space desired to be used, the length, breadth and depth, together with the measurement in feet from the surface of the ground to the top of such tank, and shall contain the plans and specifications for the construction of said tank, its connections, fittings, openings and safety applianoes, all as required by Chapter 11—HAZARDOUS USE UNITS. No such tank or equipment shall be covered or used until the installation, material and workmanship have been finally inspected, approved and certified by the Department of Buildings.

(c) **Permit Not Required.** Nothing in this section shall be construed as requiring any owner or occupant of a building, or his agent, to obtain a permit for the use, nor to prohibit the use by him of oils, paints, varnishes or similar flammable mixtures unless the storage is to be maintained longer than thirty (30) days, or the aggregate quantities are in excess of one hundred twenty (120) gallons for Class I and Class II liquids, or in excess of five hundred fifty (550) gallons for Class III or Class IV liquids.

401.15 Wrecking Buildings:

(a) **Permit.** Before proceeding with the wrecking or tearing down of any building or other structure more than one (1) story in height, a permit for such wrecking or tearing down shall first be obtained by the owner or his agent from the Commissioner of Buildings, and it shall be unlawful to proceed with the wrecking or tearing down of any building or structure or any structural part of such building or structure unless such permit shall first have been obtained. Application for such permit shall be made by such owner, or his agent, to the Commissioner of Buildings, who shall issue such permit upon such application and the payment of the fee herein provided for.

(b) **Application.** Every application shall state the location and describe the building which it is proposed to wreck or tear down. The fee for such permit shall be five dollars ($5.00) for every twenty-five (25) feet, or fractional part thereof, of frontage. Upon the issuance of such permit, such building may be wrecked or torn down, provided that all the work done thereunder shall be subject to the supervision of the Commissioner of Buildings and to such reasonable restrictions as he may impose in regard to elements of safety and health; and provided further, that the work shall be kept sprinkled and sufficient scaffolding be provided to insure safety to human life, and to comply with the Provisions of the Act of the General Assembly, passed June 3, 1907, in force July 1, 1907, providing for the safety of workmen in and about the construction and removal of buildings.

(c) **Bond.** Before any permit is issued granting authority to wreck a building or structure for which such permit is required, the person, firm or corporation engaged in the work of wrecking same shall file with the City Clerk a bond with sureties to be approved by the City Comptroller to indemnify, keep and save harmless the city against any loss, cost, damage, expense, judgment or liability of any kind whatsoever which the City may suffer, or which may accrue against, be charged to or be recovered from said City, or any of its officials from or by reason or on account of accidents to persons or property during any such wrecking operations, and from or by reason or on account of anything done under or by virtue of any permit granted for any such wrecking operations. Such bond in each case shall extend to and cover all such wrecking operations carried on through permits obtained thereunder by such person, firm or corporation during any fiscal year beginning January first and ending December thirty-first, and no permit shall be issued for any wrecking work, except as hereinbefore otherwise provided during such fiscal year until such bond is filed. Said bond shall be in the penal sum of twenty thousand dollars ($20,000) for all wrecking operations on such buildings and other structures not more than three stories in height, and there shall be an additional bond filed in the penal sum of twenty thousand dollars ($20,000) or a bond in the penal sum of forty thousand dollars ($40,000) shall be filed in the first instance in case of wrecking operations on buildings and other structures four (4) or more stories in height. Upon the filing of such bond or bonds, the person, firm or corporation engaged in the work of wrecking such buildings and other structures may obtain permits for such wrecking operations as are authorized under the said bond or bonds as hereinabove provided for, during the fiscal year in which the same is or are filed; provided however, that in case of accident or casualty in the progress of any wrecking operations carried on under any permit so issued, or the happening of any circumstance which might, in the opinion of the Commissioner of Buildings render such bond or bonds inadequate, the said Commissioner may, in his discretion, require such additional bond as he may deem necessary to fully protect the city from loss resulting from the issuance of such permits before he allows the work to proceed or before any additional permits are issued by him.

(d) **United States and City Authorities.** The Administrator of Public Works of the United States or such other authority as may be created by acts of Congress with power to cooperate with the city in the making of public improvements, the Department of Public Works, the Department of Streets and Electricity and the Fire Department may engage in the work of wrecking of buildings and structures, and in such cases where any of these agencies make application for a permit to wreck buildings or structures, the Commissioner of Buildings shall issue such permit without the fee provided herein and shall not require the filing of a bond with sureties as provided heretofore in this section.

401.16 Warm Air Furnaces: It shall be unlawful for any person, firm or corporation to construct, replace or install any warm air heating furace, with appurtenances, ducts or registers, without first obtaining a permit from the Commissioner of Buildings for such work, as provided by this chapter.

401.17 Amusement Devices, Roller Coasters, Scenic Railways: Before any mechanical amusement device, roller coaster, scenic railway, water chute, or other mechanical riding, sailing, sliding, or swinging device is erected, either in existing or new amusement parks, or places or sites where such devices are operated under carnival, fair of similar auspices, a detailed plan shall be submitted to the Commissioner of Buildings for his approval or rejection, and if approved, a permit shall be procured by the person, firm or corporation desiring to erect such device.

401.18 Calcium Carbide Storage:

(a) **Permit.** It shall be unlawful for any person, firm or corporation to keep or store calcium carbide in excess of six hundred (600) pounds within the City of Chicago without first obtaining a permit, as hereinafter provided, for each location where such quantity of calcium carbide is to be kept or stored.

(b) **Application.** Application for such permit shall be made in writing and shall conform to the provisions of this ordinance, relating to applications for building permits and set forth the location, plans and description of the building in which such calcium carbide, together with the quantity is to be kept or stored. Every such application shall be approved by the Commissioner of Buildings before a permit is issued. If it shall appear from the application so filed and approved that the premises in which applicant proposed to keep or store such calcium carbide conform to the requirements of this ordinance, then, upon the payment by the applicant to the City Collector of the permit fee hereinafter provided for, the Commissioner of Buildings shall issue or cause to be issued to such applicant a permit authorizing such applicant to keep and store calcium carbide in the place designated in the permit for the period therein stated.

401.19 Revocation of Permit. If the work in, upon or about any building or structure shall be conducted in violation of any of the provisions contained in this ordinance, it shall be the duty of the Commissioner of Buildings to revoke the permit for the building or wrecking operations in connection with which such violation shall have taken place. It shall be unlawful, after the revocation of such permit, to proceed with such building or wrecking operations unless such permit shall first have been reinstated or re-issued by the Commissioner of Buildings. Before a permit so revoked may be lawfully re-issued or reinstated, the entire building and building site shall first be put into condition corresponding with the requirements contained in this ordinance, and any work or material applied to the same in violation of any of the provisions shall be first removed from such building, and all material not in compliance with the provisions of this ordinance shall have been removed from the premises.

ARTICLE 402

Plans

402.01 General: All plans and drawings for buildings or for structures other than buildings, shall be presented to the Commissioner of Buildings for his approval, and each set of plans presented, shall be approved by the Commissioner of Buildings before a permit will be granted. All such plans and drawings shall be drawn to a scale of not less than one-eighth (⅛) of an inch to the foot, on paper of cloth, in ink, or by some process that will not fade or obliterate. All distances and dimensions shall be accurately figured, and drawings made explicit and complete, showing the lot lines and the entire sewerage and drain pipes and the location of all plumbing fixtures within such building or structure.

402.02 Architect or Engineer to Seal: No plans shall be approved for permits unless such plans are signed and sealed either by an architect licensed to practice architecture, as provided by "The Illinois Architectural Act," or by a structural engineer licensed to practice structural engineering, as provided by "The Illinois Structural Engineering Act."

402.03 Architect or Engineer Must Certify: It shall be unlawful for any architect or structural engineer or other person permitted under the laws of the State to make drawings and plans, to prepare or submit to the Commissioner of Buildings, for his approval, any final drawings or plans for any building or structure, which do not comply with the requirements contained in this ordinance. It shall be the duty of the Commissioner of Buildings to require that all drawings and plans submitted to him for approval, for any building or structure, shall be accompanied by a certificate of such architect or structural engineer preparing such drawings and plans, that said drawings and plans comply with the requirements contained in this ordinance.

402.04 Approval of Preliminary Drawings and Plans: The Commissioner of Buildings may in his discretion, issue a permit for the construction of a building or structure, or part thereof, upon the approval of preliminary drawings and plans, a true copy of which shall be filed with the Commissioner of Buildings, and before the entire working drawings and plans and detailed statements of said building have been completed and submitted for approval, if such preliminary drawings and plans and detailed statements shall be of sufficient clarity to indicate the nature and character of the work proposed and to show their compliance with this ordinance; provided however, that the complete working drawings and plans shall be submitted to the Commissioner of Buildings before construction is permitted. The Commissioner of Buildings shall check the completed drawings and plans and if approved shall stamp his official approval thereon as provided by Section 401.03.

402.05 Encroachment on Public Domain:

(a) **General.** The Commissioner of Buildings shall not issue any permit authorizing the construction, erection, repair or alteration of any building or structure unless the drawings and plans submitted for his approval clearly show that such building or structure with all its appurtenances, foundations and parts can be erected entirely within the limits of the lot or tract of land upon which it is proposed to erect such building or structure, except as hereinafter provided, and except as otherwise provided by the ordinances of the City of Chicago, and no permit to erect, repair or alter any building or structure shall authorize the use of or encroachment upon any part of any public highway or other public ground for the construction of or maintenance of such building or structure, except as hereinafter provided, and except as otherwise provided by the ordinances of the City of Chicago, nor shall any permit be issued for the construction or maintenance of any balcony or canopy extending over any public highway or other public ground unless permits therefor have been obtained from the Department of Public Works pursuant to an ordinance specifically authorizing the same. The drawings and plans of every building or structure which show that any part of said building or structure, or any of its appurtenances, or attachments thereto, extend over any part of any public highway or other public ground than hereinafter provided for shall, previous to being submitted to the Commissioner of Buildings, be submitted to the Commissioner of Public Works and notice thereby given to him of the proposed encroachment upon any public highway or other public ground. Proof of such notice to the Commissioner of Public Works shall accompany drawings and plans when same are presented to the Commissioner of Buildings.

(b) **Driveways.** No permit shall issue for the construction, erection, repair or alteration of any building or structure designed or intended for use as a garage or any other business, the operation of which will require a driveway across a public sidewalk, until the applicant therefor has first obtained from the Commissioner of Public Works, a permit for driveway or driveways as prescribed by Article II of Chapter 14 of the Revised Chicago Code of 1931.

402.06 Submission of Plans to City Engineer and Harbor Master:

(a) **Foundations Below Datum.** When the plans for new buildings or structures involve foundations, piles, piers, or footings extending lower than twenty-five (25) feet below City Datum, the plans for such foundations shall be submitted to the City Engineer for advice as to the presence on or adjacent to the property of any public utility. Whenever application is made for a permit to erect any building or structure with a foundation, or part thereof, designed to extend to a depth of forty (40) feet, or more, below City Datum, the plans of said building or structure shall be submitted to the City Engineer and his approval secured before a permit is issued by the Commissioner of Buildings for the erection of such building or structure.

(b) **Harbor Structures.** No building or structure shall be erected within forty (40) feet of any part of the harbor of the city without first obtaining a permit in writing from the Commissioner of Public Works as provided in Section 2873 and paying the fees provided for in Section 2874 of the Revised Chicago Code of 1931.

402.07 Drawings and Plans Filed with Department of Buildings: True copies of the drawings and plans bearing the approval stamp of the Commissioner of Buildings shall be filed with the Department of Buildings and shall remain on file in that office for a period of six (6) months after the occupation of such building, after which upon demand, such drawings and plans shall be returned by the Commissioner of Buildings to the person by whom they have been deposited. It shall not be obligatory upon the Commissioner of Buildings to retain such drawings and plans in his custody for more than six (6) months after the occupation of the building to which they relate.

402.08 Plans to Be Kept at the Building: In all construction work for which a permit is required, the approved and stamped drawings, plans and permit shall be kept on file at the construction site while the work is in progress.

402.09 Alterations Upon Stamped Plans Not Permitted: It shall be unlawful to erase, alter or modify any lines, figures or coloring contained upon drawings or plans bearing the approval stamp of the Commissioner of Buildings or filed with him for reference. If during the progress of the execution of such work, it is desired to deviate in any manner affecting the construction, or other essentials of the building from the terms of the application or drawing, notice of such intention to alter or deviate shall be given to the Commissioner of Buildings and an amended plan showing such alteration or deviation shall be submitted for his approval, and his written assent shall first be obtained before such alteration or deviation shall be made.

402.10 Plat to Be Filed: At the time of applying for a permit for the erection of, alteration of, addition to, or moving of any building or structure, the applicant shall submit to the Commissioner of Buildings, a plat of the lot, showing the dimensions of the same and the position to be occupied by the proposed building, or by the building to be altered or added to or by the buildings to be moved thereon, and the position of any other building or buildings that may be on the lot. The measurements shall in all cases be taken at the top of the first story and shall not include any portion of any street or alley, or other public ground. Each application for a building permit shall be accompanied by a plat in duplicate, drawn to scale and in such form as may be prescribed by the Commissioner of Buildings, showing the actual dimensions of the lot to be built upon, the size of the building to be erected, its position on the lot, and such other information as may be necessary to provide for the enforcement of the regulations contained in this ordinance. A careful record of such applications and plats shall be kept in the office of the Commissioner of Buildings.

402.11 Fire Escapes and Exterior Stairways: No permit for a Type T5 stairway, or a fire escape more than twenty-four (24) inches in width shall be granted unless a detailed plan for such stairway or fire escape, approved by an architect or a structural engineer, shall be submitted to the Commissioner of Buildings, and a copy of such plans shall be left on file with said Commissioner. No change in the position of any existing fire escape or stairway shall be made, nor shall any change in the position of any stairway or fire escape, as shown on approved plans be permitted, unless the written consent of the Commissioner of Buildings shall first have been obtained.

402.12 Tanks for Flammable Liquids:

(a) **General.** Every application for a permit to install a tank or tanks for flammable liquids shall be made to the Commissioner of Buildings and shall be accompanied by a plat of survey showing the location and dimensions of all the property coming within the frontage area, the name and address of the owner or owners of each parcel of ground coming within such area, including the filling station site, and the total frontage in feet, with the consents of the required majority of such frontage as provided in Chapter 19—FRONTAGE CONSENTS.

(b) **Driveways—Permit Required.** In any location where a driveway or driveways across a public sidewalk are required, permit shall not be issued until the applicant therefor has first obtained from the Commissioner of Public Works a permit for driveway or driveways, as prescribed by Article II of Chapter 14 of the Revised Chicago Code of 1931.

402.13 Warm Air Furnaces: Every application for a permit to construct, replace or install any warm air heating furnace shall be made to the Commissioner of Buildings and shall be accompanied by drawings or plans, and such specifications or statements as shall be required to show all details of construction and mechanical devices for approval of the Commissioner of Buildings before the issuance of such permit.

ARTICLE 403

Fees

403.01 General: Fees for the issuance of permits for new buildings, alterations and other structures and other permits shall be payable to the City Collector when such permits are issued, as required by this article.

403.02 Use of Water:

(a) **Deposit with Bureau of Water Indemnity Bonds.** Before the Commissioner of Buildings issues a permit, as provided herein, he shall require evidence from the applicant that payment has been made to the Bureau of Water of the city for the water to be used, or for a water meter for measuring all the water to be used, in the construction of such building, in accordance with the regulations of the Bureau of Water. Such applicant shall produce evidence that he has filed with and had approved by the Commissioner of Public Works of the city, an indemnifying bond protecting the city against any and all damages that may arise to the streets or alleys upon which such building abuts, and to the city, and to any person, in consequence, or by reason of the proposed operations to be authorized by such permit, or by reason of any obstruction or occupation of any streets or sidewalks in and about such building operations.

(b) **Fees.** The fees to be paid for water used in connection with the erection of buildings shall be as follows, to-wit: At the rate of five cents ($.05) for every one thousand (1,000) bricks, wall measure, used in connection therewith; six cents ($.06) for every one hundred (100) cubic feet of rubble stone used in connection therewith; eight cents ($.08) for every one hundred (100) cubic feet of concrete used in connection therewith; fifteen cents ($.15) for every one hundred (100) yards of plastering used in connection therewith; five cents ($.05) for every one hundred (100) cubic feet of hollow tile arch, partition, or fireproof covering used in connection therewith.

403.03 Fees for New Buildings:

(a) **Sheds.** For sheds not exceeding three hundred (300) square feet in area three dollars ($3.00); for shelter sheds, at the rate of two dollars ($2.00) for each one thousand (1,000) cubic feet or fractional part thereof.

(b) **Other Buildings.** For all buildings or other structures other than sheds and shelter sheds, the fee for the permit shall be at the rate of thirty cents ($.30) for every one thousand (1,000) cubic feet or fractional part thereof for buildings containing not to exceed two hundred thousand (200,000) cubic feet of volume; for buildings exceeding two hundred thousand (200,000) cubic feet in volume thirty cents ($.30) per thousand (1,000) cubic feet for the first two hundred thousand (200,000) cubic feet and sixty cents ($.60) per thousand (1,000) cubic feet for each additional one thousand (1,000) cubic feet of volume or fractional part, the cubic contents being measured to include every part of the building from the basement to the highest point of the roof, and to include all bay windows and other projections.

(c) **Minimum Permit Fee.** In no case shall any such permit be issued for a lesser fee than two dollars ($2.00).

403.04 Alterations: The fee to be charged for permits issued for alterations and repairs in or to any building or other structure shall be based on the cost of such alterations and repairs, and shall be at the rate of two dollars ($2.00) for the first one thousand dollars ($1,000.00) or part thereof, and one dollar ($1.00) additional for each additional one thousand dollars ($1,000.00) or part thereof, to be expended therefor. The fee for a permit to raise any building other than a frame building, shall be for shoring up, raising, underpinning or moving any building other than a frame building, twenty cents ($.20) per one thousand (1,000) cubic feet of volume, or fractional part thereof; provided however, that in no case shall such permit be issued for a lesser fee than five dollars ($5.00).

403.05 Other Permit Fees:

(a) **General.** In addition to the permit fees for buildings, permit and inspection fees shall be charged as follows:

(b) **Fire Escapes.** For the erection of a fire escape or Type T5 stairway, five dollars ($5.00) for each fire escape or Type T5 stairway, four (4) stories or less in height; and seventy-five cents ($.75) additional for each story above four (4) stories in height.

(c) **Canopy, Marquee or Chimney.** For the erection or alteration of a canopy or marquee, ten dollars ($10.00). For the erection of isolated chimneys or for chimneys extending over fifty (50) feet above the roof of any building—ten dollars ($10.00).

(d) **Roof Tanks.** For tank above roof or tower more than four hundred (400) gallons capacity, ten dollars ($10.00); and for tank above roof or tower of four hundred gallons (400) capacity or less five dollars

($5.00). For structural supports for tank above or upon a roof, for any tank exceeding two hundred fifty (250) gallons capacity, ten dollars ($10.00); provided however, that a permit for any new building may include such structural supports as provided in Section 403.03, paragraph (b) for new building.

(e) **Roofs.** For recoating or recovering the roof of any building, two dollars ($2.00).

(f) **Elevators.** For each elevator or mechanical equipment used for the raising or lowering of any proscenium fire curtain, stage or orchestra floor, or any platform, dumbwaiter, or escalator, constructed, installed, or altered, shall be five dollars ($5.00).

(g) **Amusement Devices.** The permit fee, including the initial inspection, for each portable mechanical amusement device, shall be twenty-five dollars ($25.00) for each assembly or installation on premises; and for each amusement device or other mechanical riding, sliding, sailing, or swinging device built on the premises, installed or altered—fifty dollars ($50.00).

(h) **Ventilating Systems.** For every mechanical ventilating supply system and mechanical ventilating exhaust system, as required by Chapter 43—VENTILATION, of this ordinance, which charge shall include the inspection and testing of the system prior to its approval—five dollars ($5.00) for a system handling five thousand (5,000) cubic feet of air per minute or less, plus one dollar ($1.00) for each additional one thousand (1,000) cubic feet of air per minute. The permit fee for every increase in the total capacity of an existing equipment shall be one dollar ($1.00) for every one thousand (1,000) cubic feet of air per minute additional capacity. The sum of the capacity of both supply and exhaust systems shall be the basis of computing capacity as required by this section.

(i) **Fences.** For the erection of a permanent fence two dollars ($2.00) for one hundred (100) lineal feet or less, and one dollar ($1.00) for each additional one hundred (100) lineal feet, or part thereof, for fences more than five (5) feet in height.

(j) **Standpipes.** For the installation and initial inspection test of a standpipe system in any building five (5) stories or more in height, ten dollars ($10.00) for the first standpipe and five dollars ($5.00) for each additional standpipe, and five dollars ($5.00) for each unit of pumping capacity of fifty (50) gallons per minute or fraction thereof.

(k) **Sprinkler Systems.** For the installation of a standard sprinkler system—ten dollars ($10.00) for one hundred (100) sprinkler heads or less, and five dollars ($5.00) additional for each additional one hundred (100) sprinkler heads, or fraction thereof; and for a fire pump used in connection with a sprinkler system, five dollars ($5.00) for each fifty (50) gallons pumping capacity, per minute or fraction thereof.

(l) **Street Obstructions.** For street obstructions see Section 401.08.

(m) **Tanks for Flammable Liquids.** For the installation of tanks for flammable liquids of Classes I and II, ten dollars ($10.00) for an aggregate capacity of tanks from one hundred twenty-one (121) gallons to five hundred (500) gallons, and twenty-five cents ($.25) for each additional one hundred (100) gallons capacity, or fraction thereof; and for liquids of Classes III and IV, ten dollars ($10.00) for an aggregate capacity of tanks from five hundred fifty-one (551) gallons to one thousand (1,000) gallons and twenty-five cents ($.25) for each additional one hundred (100) gallons capacity, or fraction thereof. Such permit fee of ten dollars ($10.00) shall be paid at the time the written application for a tank permit is filed and shall be forfeited to the city in case the site is not approved for permit.

(n) **Warm Air Furnaces.** For the construction, replacement or installation of any warm air heating furnace of either gravity or forced air type, five dollars ($5.00) for each such furnace.

403.06 Billboards, Signs and Signboards: The fees to be charged for permits issued for the erection or construction of billboards or signboards or for the alteration thereof shall be as follows:

Square Feet in Area	Fees	
Up to 150......	$2.00	
151 to 225......	3.00	
226 to 375......	5.00	
Over 375	5.00	for each 375 square feet or of fractional part thereof;

provided however, that where such signboard or billboard does not exceed sixty-five (65) square feet in area and is attached to the surface of a permanent building, in accordance with the provisions of Article 1809 and is designed to give publicity to the business carried on within such building, such as the name and address of owner and the nature of business, but in no event to advertise any article manufactured by any other person, and no part of said sign is more than eighteen (18) feet above the average inside grade at the front of the building, no fees for erection shall be charged, but not more than one (1) sign of sixty-five (65) square feet shall be allowed for each twenty-five (25) lineal feet of frontage, unless the fees for erection are paid as herein provided; and provided further, that where such signboard or billboard does not exceed twenty (20) square feet in area and is attached to the surface of a permanent building in accordance with the provisions of Article 1809 and is designed to give publicity to some article sold on the premises, and no part of said sign is more than eighteen (18) feet above the average inside grade at the front of the building, no fees for the erection shall be charged; and provided also, that wheer such signboard or billboard does not exceed twenty-four (24) square feet in area, when attached to the front, sides or rear walls of any building, so that flat surface of the same is against the building, or when erected on the ground, if not erected nearer than ten (10) feet to any building, structure, other signboard or public sidewalk, which is used to advertise the sale or lease of the property upon which it shall be charged.

403.07 Calcium Carbide Storage: The permit fee for the storage of calcium carbide in quantities exceeding six hundred (600) pounds is hereby fixed at fifty dollars ($50.00) per annum. Every such permit shall expire on the 31st day of December after its issuance. If at the time of the application for a permit, under the provisions of this section, less than three months of the current permit year shall have expired, the applicant for a permit shall be required to pay the full annual permit fee hereinbefore fixed. If three (3) months or more, but less than six (6) months, shall have expired, he shall be required to pay three-fourths (¾) of the annual permit fee. If six (6) months or more, but less than nine (9) months shall have expired, the applicant shall be required to pay one-half (½) the annual permit fee. If nine (9) months or more shall have expired, the applicant shall be required to pay one-fourth (¼) of the annual permit fee hereinbefore specified. No permit granted under the provisions of this section shall be assigned or transferred to any other person, firm or corporation. Such permit shall be posted in a conspicuous place in the building where the calcium carbide is stored.

403.08 Illuminated Roof Signs: Any person, firm or corporation desiring to erect or maintain an illuminated roof sign, as described in this ordinance, shall pay to the city, to cover the cost of inspection and approval by the Commissioner of Buildings, of the plans and specifications of such sign, when erected, a fee of fifty dollars ($50.00) for the first five hundred (500) square feet of superficial area of such sign or fractional part thereof, and five ($.05) cents for each additional square foot.

403.09 Floor Load Placards: Fees for the approval of floor load placards required by Chapter 21 — GENERAL, of this ordinance, shall be charged as follows: it shall be the duty of the owner, agent or lessee to pay to the City Collector, a fee amounting to five dollars ($5.00) for each ten thousand (10,000) square feet of floor area or less; for more than ten thousand (10,000) square feet of floor area and not to exceed fifty thousand (50,000) square feet of floor area, ten dollars ($10.00); for each additional fifty thousand (50,000) square feet of floor area in excess of the first fifty thousand (50,000) square feet of floor area, ten dollars ($10.00) additional; and for issuing new placards in place of lost placards, the fee shall be: for ten thousand (10,000) square feet or less, two dollars ($2.00); for more than ten thousand (10,000) square feet, five dollars ($5.00). For the purpose of determining the amount of the fee herein required to be paid, every part of a structure separated by dividing walls as required by Chapter 22—SIZE AND LOCATION OF ROOMS, shall be considered as a separate building.

403.12 Plumbing: A fee of two dollars and fifty cents ($2.50) shall be paid to the City Collector for the approval of plans and inspection and test of any plumbing within any building containing not more than five (5) plumbing fixtures. An additional fee of fifty cents ($.50) shall be paid for every plumbing fixture in excess of five (5) within such building.

403.13 Furnaces and Other Fuel Burning Apparatus: The fees for the inspection of plans and issuing of permits and for the inspection of furnaces or other fuel burning equipment or devices, and issuing of certificates shall be as follows:

For inspecting plans of new plants and of plants about to be reconstructed	$2.00
For inspecting plans for repairs and alterations	1.00
For permits for the erection, installation, reconstruction, repair or alteration of any furnace or other fuel burning apparatus, smoke-prevention device or chimney each unit, or single apparatus	5.00
For examining or inspecting any new or reconstructed furnace connected to a high pressure boiler after its erection or reconstruction and before its operation and maintenance, first unit or single apparatus	5.00
Each additional unit or single apparatus	3.00
For examining or inspecting any new or reconstructed furnace connected to a low pressure boiler or any other fuel burning equipment, or any smoke prevention device, after its erection or reconstruction and before its operation and maintenance, each unit or single apparatus	3.00

provided however, that this section shall not apply to furnaces or other fuel burning apparatus or devices installed or used to heat private residences, tenements or buildings consisting of two (2) apartments or less. The aforesaid fees shall be paid to the City Collector prior to the approval of plans for such installations by the Deputy Smoke Inspector in Charge. The fee for the examination or inspection shall include the issuance of a certificate for operation in case such certificate for operation is granted and shall be paid at the time the permit is issued.

The Deputy Smoke Inspector in Charge may, and he is hereby directed and instructed to remit all inspection or examination fees charged against any and all charitable, religious and educational institutions, when the furnace or other device or apparatus inspected is located in or upon premises used and occupied exclusively by such charitable, religious or educational institution; provided however, that such charitable, religious or educational institution is not connected or carried on for private gain or profit; and provided further that the Deputy Smoke Inspector in Charge may require every application for the remission of such fees to be verified by the affidavit of one (1) or more taxpayers of the city.

ARTICLE 404

Inspections and Inspection Fees

404.01 General: The Commissioner of Buildings shall cause to be inspected annually or semi-annually, or otherwise, such buildings, structures, equipment, sites or parts thereof as shall be provided by this article, except as otherwise required by this chapter under permits for new buildings or alterations.

404.02 Annual Inspection of Buildings—Stairways and Means of Egress:

(a) **Inspection Required.** The Commissioner of Buildings and his assistants shall make an annual inspection of all theatres, churches, schools and public assembly units and open air assembly units; and also all buildings over one (1) story in height, except single dwellings, multiple use buildings consisting of business and dwelling units two (2) stories or less in height and except multiple dwellings three (3) stories or less in height. It shall be the duty of every owner, agent, lessee or occupant of any such building as is referred to in this section, and of the person in charge or control of same, to permit the making of such annual inspection by the Commissioner of Buildings or by a duly authorized Building Inspector at any time upon demand being duly made.

(b) **Certificate of Compliance.** Whenever any such inspection shows the building to be in compliance with the requirements of this chapter with respect to stairways, means of egress and all other respects, it shall be the duty of the Commissioner of Buildings to issue, or cause to be issued, a certificate setting forth the result of such inspection, containing the date thereof, and a statement to the effect that such building complies in all respects with the provisions of this chapter or complies with all provisions of the ordinances under which such building was constructed, upon the payment of the inspection fee herein required; provided however, that such certificate shall be issued for charitable, religious or educational institutions as otherwise provided by Section 404.04. It shall be the joint and several duty of the owner, agent, lessee or occupant of the building so inspected, and of each and every person in charge and control of the same to frame the said certificate and place it in a conspicuous place near the main entrance of such building.

(c) **Building Plan.** It shall be the joint and several duty of the owner, agent, lessee or occupant of every building described in this section to provide a typical floor plan of such building reproduced on a sheet eight and one-half by eleven (8½x11) inches in size. Said plan shall be drawn on as large a scale as will be practicable on such sheet, and said sheet shall also state the street address of such building and shall give the class of the building, the kind of construction used therein, the height and number of stories contained therein, and the nature of the occupancy. It shall also be the joint and several duty of such owner, agent, lessee or occupant to deliver a copy of said sheet and place to the Commissioner of Buildings and to frame a copy of said sheet and place the same near the framed certificate hereinabove required. It shall also the joint and several duty of the said owner, agent, lessee or occupant to substitute a new sheet for the sheet on file with the Commissioner of Buildings, and also the sheet framed as above required, whenever such changes or alterations are made in such building as will affect the substantial accuracy of the sheet previously furnished said Commissioner and framed as above required.

(d) **Notice of Non-Compliance.** Where the result of such inspection shall show that such building fails in any respect to comply with the requirements of this ordinance, it shall be the duty of the Commissioner of Buildings to notify the owner, agent, lessee or occupant of such building to this effect and to specify wherein such building fails to comply with the requirements of this ordinance; and it shall thereupon become the joint and several duty of such owner, agent, lessee or occupant to proceed forthwith to make whatever changes or alterations may be necessary to make such building comply in all respects with the requirements of this ordinance, and to complete such changes and alterations within thirty (30) days after the receipt of such notice.

404.03 Annual Inspection Fees: For every such annual inspection, it shall be the duty of the owner to pay to the City Collector an annual inspection fee for same amounting to five dollars ($5.00) where the said building contains not to exceed twenty-five thousand (25,000) square feet of floor area. Where the building has a floor area in excess of twenty-five thousand (25,000) square feet, an annual inspection fee of five dollars ($5.00) shall be paid for the first twenty-five thousand (25,000) square feet of floor area and for each additional twenty-five thousand (25,000) square feet of floor area, or fractional part thereof, an additional fee of three dollars ($3.00) shall be paid, except as otherwise provided for religious, charitable or educational institutions. For the purpose of determining the amount of the fee herein required to be paid, every part of a building or structure separated by dividing walls as required by the provisions of Chapter 22—SIZE AND LOCATION OF ROOMS, of this ordinances, shall be considered as a separate building.

404.04 Exemptions — Charitable, Religious and Educational Institutions: The Commissioner of Buildings and the City Collector shall remit all inspection fees charged, or that may hereafter be charged, against any and all charitable, religious and educational institutions, when the building, or part thereof, so inspected, is located in or upon premises used or occupied exclusively and owned by such charitable, religious or educational institution; provided however, that such charitable, religious or educational institution is not conducted or carried on for private gain or profit; and provided further, that every application for the remission of such fees shall be supported by the affidavit of one (1) or more taxpayers of the city as to such facts.

404.05 Amusement Parks and Devices:
(a) **Buildings.** The Commissioner of Buildings shall inspect, or cause to be inspected, all buildings to be used for purposes of exhibition, amusement or entertainment, which are attended by the public, that are within or connected with an amusement park, each year before said buildings are open to the public, for the purpose of ascertaining whether they comply with the city ordinances and the rules and regulations of the Department of Buildings. The fee for such annual inspection shall be five dollars ($5.00) for each building so inspected.

(b) **Devices.** The Commissioner of Buildings shall inspect, or cause to be inspected annually, all amusement devices, mechanisms and structures, other than riding devices and other than buildings, within an amusement park, for the purpose of ascertaining whether they comply with the city ordinances and the rules and regulations of the Department of Buildings; and the fee for such annual inspection shall be ten dollars ($10.00) for each device, mechanism and structure so inspected. The Commissioner of Buildings shall inspect annually, or cause to be inspected, all amusement devices operated by animals or by other motive power and all other riding, sliding, sailing, swinging or rolling devices situated on any lot or tract of land outside of any amusement park before said devices are opened to the public. Where said devices are taken down, removed and re-assembled or re-erected in another location, the Commissioner of Buildings shall inspect or cause said devices to be reinspected after each removal and before said devices are opened to the public, for the purpose of ascertaining whether they comply with the city ordinances and the rules and regulations of the Department of Buildings. A fee, as provided in Section 403.05, paragraph (g) shall be paid for every such annual inspection or reinspection.

404.06 Elevators, Dumbwaiters, Escalators:
(a) **Inspection.** Every elevator or movable stage or orchestra floor or platform, or dumbwaiter, or escalator now in operation, or which may hereafter be installed, together with the hoistway and all equipment thereof, shall be inspected under and by the authority of the Commissioner of Buildings at least once every six (6) months, and in no case shall any new equipment be placed in operation until an inspection of the same has been made. It shall be the duty of every owner, or agent, lessee or occupant of any building wherein any elevator, dumbwaiter or escalator is installed, and of the person in charge or control of any such equipment to permit the making of a test and inspection of such elevator, dumbwaiter or escalator, and all devices used in connection therewith upon demand being made by the Commissioner of Buildings or by his authorized Elevator Inspector within five (5) days after such demand has been made.

(b) **Certificate of Compliance.** Whenever any such equipment has been inspected and the tests herein required shall have been made of all safety devices with which such elevator, dumbwaiter or escalator is required to be equipped and the result of such inspection and tests shows such equipment to be in good condition, and that such safety devices are in good working condition and in good repair, it shall be the duty of the Commissioner of Buildings to issue, or cause to be issued, a certificate setting forth the result of such inspection and tests and containing the date of inspection; the weight which the elevator, dumbwaiter or escalator will safely carry and a statement to the effect that the shaft doors, hoistway and all equipment, including safety devices, are constructed in accordance with the provisions of this ordinance, upon the payment of the inspection fee required by this ordinance. It shall be the joint and several duty of the owner, agent, lessee or occupant of the building in which such equipment is located and of each person in charge or control of such equipment to frame the certificate and place same in a conspicuous place in each elevator, and near such dumbwaiter or escalator. The words "safe condition" in this section shall mean that it is safe for any load up to the approved weight named in such certificate.

(c) **Repairs Required.** Where the result of such inspection or tests shall show such elevator, dumbwaiter or escalator to be in an unsafe condition or in bad repair, or shall show that the safety devices or any of them, which are required by this ordinance, have not been installed, or if installed, are not in good working order or not in good repair, such certificate shall not be issued until such elevator, its hoistway and its equipment, or such dumbwaiter or escalator, or such device or devices shall have been put in good working order. The inspection fees herein required shall be paid either at the time application is made for inspection or upon the completion of such inspection and tests.

(d) **Power of Commissioner to Stop Operation.** Whenever any elevator inspector finds any elevator or dumbwaiter and its equipment and hatchway, including doors, or any escalator or other mechanical equipment mentioned in this section, in an unsafe condition he shall immediately report the same to the Elevator Inspector in Charge, who shall report it to the Commissioner of Buildings, together with a statement of all facts relating to the condition of such equipment. It shall be the duty of the Commissioner of

Buildings, upon receiving from the Elevator Inspector in Charge a report of the unsafe condition of such equipment and hatchway, including doors, to order the operation of such equipment to be stopped and to remain inoperative until it has been placed in a safe condition, and it shall be unlawful for any agent, owner, lessee or occupant of any building, wherein any such equipment is located, to permit or allow same to be used after the receipt of a notice from the Commissioner of Buildings which notice is in writing, that such equipment is in unsafe condition and until it has been restored to a safe and proper condition as required by the provisions of this ordinance.

404.07 Grand Stands: The Commissioner of Buildings shall inspect, or cause to be inspected, all tiers of seats and grand stands each year before same are opened to the public for the purpose of ascertaining whether they comply with the city ordinances and the rules and regulations of the Department of Buildings. A fee shall be charged for such annual inspection as follows: where the seating capacity is five thousand (5000) or less, ten dollars ($10.00); where the seating capacity is more than five thousand (5000), twenty-five dollars ($25.00).

404.08 Other Inspection Fees: For semi-annual inspection of elevator or movable stage or orchestra platform, dumbwaiter, or escalator, five dollars ($5.00); for semi-annual inspection of iron or steel curtain, fifteen dollars ($15.00); for semi-annual inspection of asbestos curtain, five dollars ($5.00).

404.09 Billboards Signs and Signboards:

(a) **Inspection.** It shall be the duty of the Commissioner of Buildings to exercise supervision over all billboards and signboards erected or being maintained under the provisions of this ordinance, and to cause inspection by inspectors in his department of all such billboards and signboards to be made once each year and oftener where the condition of such boards so requires; whenever it shall appear to said Commissioner that any such billboard or signboard has been erected in violation of this ordinance or is in an unsafe condition or has become unstable or insecure, or is in such a condition as to be a menace to the safety or health of the public, he shall thereupon issue, or cause to be issued, a notice in writing to the owner of such billboard or signboard, or person in charge, possession or control thereof, if the whereabouts of such person is known, informing such person, firm or corporation of the violation of this ordinance and the dangerous condition of such billboard or signboard and directing him to make such alterations or repairs thereto, as shall be necessary or advisable to place such billboard or signboard in a safe, substantial and secure condition and to make the same comply with the requirements of this ordinance within such reasonable time as may be stated in said notice. If the owner or person in charge, possession or control of any billboard or signboard, when so notified, shall refuse, fail or neglect to comply with and conform to the requirements, of such notice, said Commissioner shall, upon the expiration of the time therein mentioned, tear down or cause to be torn down such part of such billboard or signboard as is constructed and maintained in violation of this ordinance and shall charge the expense to the owner or person, in possession, charge or control of such billboard or signboard and the same shall be recovered from such owner or person by appropriate legal proceedings.

(b) **Where Owner Cannot Be Found.** If the owner of such billboard or signboard, or the person, in charge, possession or control thereof, cannot be found or his or their whereabouts cannot be ascertained, the Commissioner shall attach, or cause to be attached to said billboard or signboard, a notice of the same import as that required to be sent to the owner or person in charge, possession or control thereof, where the owner is known; and if such billboard or signboard shall not have been made to conform to this ordinance and placed in a secure, safe and substantial condition, in accordance with the requirements of such notice, within thirty (30) days after such notice shall have been attached to such billboard or signboard, it shall be the duty of the Commissioner of Buildings to thereupon cause such billboard or signboard or such portion thereof as is constructed and maintained in violation of this ordinance to be torn down; provided however, that nothing herein contained shall prevent the Commissioner of Buildings from adopting such precautionary measures as may be necessary or advisable in case of imminent danger in order to place such billboard or signboard in a safe condition, the expense of which shall be charged to and recovered from the owner of such billboard or signboard, or person in charge, possession or control thereof in any appropriate proceedings therefor.

(c) **Bond.** Every person, firm or corporation constructing and erecting billboards or signboards shall file with the City Clerk a bond, with sureties to be approved by the Commissioner of Buildings, in the penal sum of twenty-five thousand dollars ($25,000.00), conditioned that such person, firm or corporation shall faithfully comply with all the provisions and requirements of this article with respect to the construction, alteration, location and safety of billboard or signboards and for the payment of the inspection fees required by this article; and conditioned further to indemnify, save and keep harmless said City of Chicago and its Officials, from any and all claims, damages, liabilities, losses, actions, suits or judgments which may be presented, sustained, brought or secured against the City of Chicago or any of its Officials on account of the construction, maintenance, alteration or removal of any of said billboards or signboards, or by reason of any accidents caused by or resulting therefrom.

(d) **Fees.** The annual inspection fees to be charged for the inspection of billboards or signboards shall be as follows: For signs up to fifty (50) square feet in area, fifty cents ($.50); for signs from fifty (50) square feet to three hundred seventy-five (375) square feet in area, one dollar ($1.00); for signs in excess of three hundred seventy-five (375) square feet in area, one dollar ($1.00) for each three hundred seventy-five (375) square feet, or fractional part thereof: provided however, that where such billboard or signboard does not exceed sixty-five (65) square feet in area and is attached to the surface of a permanent building in accordance with the provisions of Article 1809 and is designed to give publicity to the business carried on within such building, such as the name and address of owner and the nature of business, but in no event to advertise any article manufactured by any other person, and no part of said sign is more than eighteen (18) feet above the average inside grade at the front of the building, no fees for inspection shall be charged, but not more than one (1) sign of sixty-five (65) square feet shall be allowed for each twenty-five (25) lineal feet of frontage, unless the fees for inspection are paid as herein provided: and provided further, that where such billboard or signboard does not exceed twenty (20) square feet in area and is attached to the surface of a permanent building in accordance with the provisions of Article 1809 and is designed to give publicity to some article sold on the premises, and no part of said sign is more than eighteen (18) feet above the average inside grade at the front of the building, no fees for inspection shall be charged; and provided further, that where such billboard or signboard does not exceed twenty-four (24) square feet in area, when attached to the front, sides, or rear walls of any building, so that the flat surface of the same is against the building, or when erected on the ground if not erected nearer than ten (10) feet to any building, structure, other signboard or public sidewalk, which is used to advertise the sale or lease of the property upon which it shall be erected, no fees for inspection shall be charged.

(e) **Penalty.** Any person, firm or corporation owning, operating, maintaining or in charge, posession or control of any building, structure, billboard or signboard within the city, that shall neglect or refuse to comply with the provisions of this ordinance, or that erects, constructs or maintains any billboard or signboard that does not comply with the provisions of this ordinance, in all cases where no specific penalty is fixed herein, shall be fined not less than twenty-five dollars ($25.00) nor more than two hundred dollars ($200.00) for each offense; and each day on which any such person shall permit or allow any billboard or signboard owned, operated, maintained or controlled by him, to be erected, constructed or maintained, in violation of any of the provisions of this ordinance, shall constitute a separate and distinct offense.

404.10 Illuminated Roof Signs:

(a) **Inspection.** It shall be the duty of the Commissioner of Buildings to cause the Building Inspectors to make an inspection annually of each illuminated roof sign erected or constructed or being maintained under the provisions of this ordinance, for the purpose of ascertaining whether such sign is safely and securely constructed and so anchored, supported and fastened to the building or structure; provided however, that the provisions of this section shall not apply to the erection, construction and maintenance of billboards and signboards as regulated by the ordinances of the City of Chicago.

(b) **Fees.** For each annual inspection of any illuminated roof sign by the Commissioner of Buildings, subsequent to the first inspection, there shall be paid a fee of fifty dollars ($50.00) for the first five hundred (500) square feet, or fractional part thereof; and five cents ($.05) additional for each additional square foot area over five hundred (500) square feet. In addition to the fees herein required to be paid for inspection, there shall be paid by the owner or person having charge or control of any illuminated roof sign, as herein described, an annual inspection fee to cover the cost of such inspection, which shall be made by the Commissioner of Streets and Electricity, whose duty it shall be to cause such annual inspection to be made and such fee shall be at the rate provided by the ordinances of the city.

(c) **Bond.** No person, firm or corporation shall be permitted to erect or maintain an illuminated roof sign, or any structurally erected sign, unless he shall execute and file with the City Clerk of Chicago, with sureties to be approved by the Commissioner of Buildings, a bond to the City of Chicago in the penal sum of fifteen thousand dollars ($15,000.00), conditioned to indemnify, save and keep harmless the City of Chicago, and its officers and agents, from any damage which it, the said city, or any of said officers may suffer, or from any costs, liability or expense of any kind, whatsoever which it the said city, or any of its officers may be put to, or which may be recovered against the said city, or any of its officers, from or by reason of the construction, erection and maintenance of such sign and conditioned further to faithfully observe and perform all the provisions and conditions of this Chapter and of any ordinance now in force or which may hereafter be passed by the City Council of the City of Chicago, relating to or governing the erection, maintenance, use or inspection of illuminated roof signs.

(d) **Revocation and Removal.** The permission and authority granted under this chapter, shal be removed at the expense of the discretion of the Mayor. In case of the termination of the privileges herein granted by the exercise of the Mayor's discretion, as aforesaid, all such electrical signs erected by virtue of the authority conferred by this chapter, shall be removed at the exuense of the owner or owners of the building, or the person, firm or corporation or individual who are then maintaining same without any cost or expense of any kind whatsoever to the City of Chicago; provided howerer, that in the event of the failure, neglect or refusal on the part of the owner of the building or structure upon which said illuminated electric sign is constructed or the person, firm, corporation or individual operating and maintaining said electric sign to remove said electric sign upon the revocation of the permit by the Mayor as herein provided, the Commissioner of Buildings may proceed to remove same and charge the expense thereof to the owner of the building or structure upon which said illuminated electric sign is constructed or to the person, firm, corporation or individual operating or maintaining same.

(e) **Penalty.** Any person, firm or corporation who shall erect, construct or maintain an illuminated roof sign in violation of any of the provisions of this section shall be fined not less than fifty dollars ($50.00) nor more than two hundred dollars ($200.00) for each offense; and each day on which any such person shall permit or allow any illuminated roof sign, owned, operated, maintained or controlled by him, to be erected, constructed or maintained in violation of any of the provisions of this ordinance, shall constitute a separate and distinct offense.

404.11 Canopies and Marquees:

(a) **Inspection.** The Commissioner of Buildings shall make an annual inspection of canopies and marquees attached to buildings or other structures which shall extend into or over any street, alley or any public place.

(b) **Fees.** The annual inspection fee to be charged for the inspection of canopies and marquees shall be as follows: where the horizontal projection of the canopy or marquee does not exceed two hundred (200) square feet in area, five dollars ($5.00); and where the horizontal projection of the canopy or marquee exceeds two hundred (200) square feet in area, five dollars ($5.00) for the first two hundred square feet, and one dollars ($1.00) additional for each additional fifty (50) square feet in the area of such canopy or marquee.

404.12 Mechanical Refrigerating Systems: The fee for household multiple and remote systems shall be as follows:

Class B: $10.00 for each compressor unit or generator.

Class C: $5.00 for each compressor unit or generator.

Class D: $3.00 for each compressor unit or generator.

Class E: $3.00 for each compressor unit or generator.

The fee for annual inspection for commera and industrial systems shall be as follows:

For each compressor or generator of 5 tons or less capacity	$ 3.00
For each compressor or generator over 5 tons and not over 5 tons capacity	5.00
For each compressor or generator over 35 tons and not over 100 tons capacity	10.00
For each compressor or generator over 100 tons capacity	12.00

The first inspection fee shall be paid for at the time the permit is issued. All fees required hereunder shall be paid to the City Collector.

Compressor capacity of rating shall be based upon five (5) degrees Fahrenheit evaporator temperature and eighty-six (86) degrees Fahrenehit condenser temperature. Compressor displacement shall conform to the following table for the refrigerants named:

DISPLACEMENT PER MINUTE PER TON

		Cu. In.
Carbon Dioxide	CO_2	1,625
Ammonia	NH_3	6,912
Dichlorodifluoronethane	CCL_2F_2	12,528
Methyl Chloride	CH_3CL	13,824
Sulphur Dioxide	SO_2	20,736

404.13 Revolving Doors: Every revolving door now in operation, or which may hereafter be installed, together with all the equipment and mechanism thereof shall be inspected annually under the authority of the Commissioner of Buildings. Whenever such inspection shows a revolving door to be in good working order and in compliance with this ordinance, the Commissioner of Buildings shall issue, or cause to be issued, a certificate to that effect, and for each such inspection and certificate a fee of three dollars ($3.00) shall be charged.

404.14 Standard Sprinkler System:

(a) **Inspection.** Every standard sprinkler system now existent, or which may hereafter be installed, shall be inspected semi-annually under the authority of the Division Fire Marshal in charge of Fire Prevention.

(b) **Certificate and Fee.** Whenever such inspection shows the standard sprinkler system to be in good order and in compliance with this ordinance the said Division Fire Marshal shall issue a certificate to that effect and for each such inspection and certificate a fee of five dollars ($5.00) shall be charged.

(c) **Repairs Required.** In case such inspection discloses any violation of or variation from the requirements of this ordinance or any change in the construction or occupancy of the building, or any condition, such as defective parts, frozen tanks, closed valves, obstructed heads, which, in the opinion of the Division Fire Marshall in charge of Fire Prevention, would seriously handicap the operation of the sprinkler equipment, notice shall immediately be sent to the owner or owners, agent or person in control of the building containing such sprinkler systems, to remove or correct the defective condition as set forth in said notice within such time as may be specified by the said Division Fire Marshal, in said notice.

(d) **Penalty.** Upon the failure of said owner or owners of such building to make the required corrections as set forth in said notice, he or they shall, upon conviction, be fined not less than twenty-five dollars ($25.00), nor more than two hundred dollars ($200.00) for each offense, and each and every day that said building is occupied after the expiration of the time set forth in said notice without having the proper corrections made as set forth in said notice shall be considered a separate and distinct offense.

404.15 Standpipe System:

(a) **Inspection.** It shall be the duty of the Division Fire Marshal in charge of Fire Prevention to cause an inspection to be made of all standpipe systems at least once every six (6) months.

(b) **Fee.** For every such semi-annual inspection, it shall be the duty of the owner to pay to the City Collector an inspection fee of five dollars ($5.00).

(c) **Repairs Required.** In case such inspection discloses any violations of or variation from the requirements of this ordinance or any defective conditions, which would handicap the operation of the standpipe system, notice shall be sent to the owner or agent in control of the building containing such standpipe system, to remove or correct such defective conditions within such time as shall be set forth by the said notice.

404.16 Tanks for Flammable Liquids:

(a) **Inspection.** Every tank for flammable liquids of a capacity of one thousand (1000) gallons or more, either above ground or within buildings, shall be inspected annually under the authority of the Commissioner of Buildings.

(b) **Fees.** For every such annual inspection, with certificate of compliance, a fee of five dollars ($5.00) shall be charged for each such tank.

(c) **Repairs Required.** In case such inspection discloses any violations of this ordinance, or any defective conditions, notice shall be sent to the owner or agent in control of such tanks, to remove or correct such defective conditions within such time as shall be set forth in said notice.

404.17 Mechanical Ventilating System: Every mechanical ventilating system shall be inspected annually by the Board of Health. The fee for such annual inspection shall be fifty cents ($.50) per one thousand (1000) cubic feet of air per minute or fractional part thereof of air required by this ordinance to be circulated for ventilating purposes, including the sum of mechanical supply and exhaust systems; provided however, that no such charge shall be less than two dollars ($2.00). Such annual fee shall be due and payable to the City Collector on the first day of March of each year.

404.18 Gas Holders: Every tank or gas holder containing more than twenty-five hundred (2500) cubic feet of explosion hazard gases within the city shall be inspected at least once every five (5) years, as required by Section 1103.07 for HAZARDOUS USE UNITS, and under the direction of a building inspector appointed by the Commissioner of Buildings. For every such five year inspection, with a certificate of compliance, a fee of ten dollars ($10.00) shall be charged for each tank or gas holder.

404.19 Other Tanks: Every tank having a capacity of more than two hundred fifty (250) gallons, and located above the roof or above the floors of any building, or on any other structure, except as provided by Sections 404.16 and 404.18 shall be inspected annually by the Commissioner of Buildings as provided by this article. For every such annual inspection, it shall be the duty of the owner to pay an inspection fee of five dollars ($5.00) for each such tank; provided however, that for any building required to be inspected annually, the inspection fee for such tank therein or thereon shall be computed as required by Section 404.03 by floor area.

404.20 Boilers and Unfired Pressure Vessels: Annual Inspection Fees for Boilers and Unfired Pressure Vessels shall be as follows:

For each low pressure boiler	$3.00
For each miniature boiler	3.00
For each high pressure boiler containing not more than 250 square feet of heating surface	5.00
For each high pressure boiler containing more than 250 square feet and not more than 1500 square feet of heating surface	6.00
For each high pressure boiler containing more than 1500 square feet and not more than 5000 square feet of heating surface	7.00
For each high pressure boiler containing more than 5000 square feet of heating surface	8.00
For each unfired pressure vessel carrying 15 pounds pressure or less	3.00
For each unfired pressure vessel carrying more than 15 pounds pressure	5.00

The fee for inspection of boilers and other apparatus above provided for, shall be double the respective amounts above specified when an inspection is made on Sunday or Legal Holidays at the request of the person or corporation owning or operating said boiler or other apparatus.

404.21 Electrical Equipment of Buildings:

(a) **Wiring only for Lighting Circuits not including Fixtures, Sockets or Receptacles.** For the inspection of each complete branch lighting circuit of one thousand (1000) watts or less:

For one circuit	$1.50
For each of the next four circuits	1.20
For each of the next five circuits	1.00
For each of the next five circuits	.85
For each of the next five circuits	.75
For each of the next five circuits	.65
For each succeeding circuit	.60

For the inspection of each complete branch lighting circuit of larger capacity than one thousand (1000) watts, and not more than two thousand (2000) watts, the charge shall be double the fee of an equal number of one thousand (1000) watt circuits.

(b) **Additional Outlets.** For the inspection of additional outlets on existing circuits, ten cents ($.10) for each outlet on which a socket, receptacle or fixture will be attached.

(c) **Electrical Fixtures, Sockets and Receptacles, not including the Circuit Feeding Same.** For the inspection of fixtures, sockets or receptacles for lamps of nominal fifty (50) watts capacity:

1 to 15	lamps	$0.50
16 to 20	lamps	.75
21 to 25	lamps	1.00
26 to 30	lamps	1.25
31 to 40	lamps	1.50
41 to 50	lamps	1.75
51 to 60	lamps	2.00
61 to 70	lamps	2.25
71 to 80	lamps	2.50
81 to 90	lamps	2.75
91 to 100	lamps	3.00
101 to 110	lamps	3.20
111 to 120	lamps	3.40
121 to 130	lamps	3.60
131 to 140	lamps	3.80
141 to 150	lamps	4.00
151 to 160	lamps	4.20
161 to 170	lamps	4.40
171 to 180	lamps	4.60
181 to 190	lamps	4.80
191 to 200	lamps	5.00

above two hundred (200) lamps, five dollars ($5.00) for the first two hundred (200) lamps and twenty-five cents ($.25) for each group of twenty-five (25) lamps or less. For lamps of capacity greater than fifty (50) watts, the charge shall be in proportion to the wattage of the lamp.

(d) **Wiring and Fixtures.** For the inspection of both circuit wiring and fixtures, sockets or receptacles; the aggregate sum of the fees as shown above for wiring and for electrical fixtures.

(e) **Motors and Other Forms of Power.** For the inspection of each electrical horsepower of seven hundred forty-six (746) watts used for mechanical or other purposes than above mentioned, one (1) motor two dollars ($2.00) plus ten cents ($.10) per horse power; additional motors fifty cents ($.50) plus ten cents ($.10) per horse power. This fee to be applied to all motors over one-fourth (¼) horse power; motors of one-fourth (¼) horse power or under to be c a ge on an equivalent incandescent lamp basis. d

(f) **Temporary Work, Outside Work, etc.** Inspections of electric lights, other than electric signs, as herein defined, placed on a public street or alley, for the purpose of illuminating the same; temporary installations for show window exhibitions, conventions and the like; underground or overhead wires and apparatus and all other inspections not specifically provided for herein shall be charged for according to the time required for such inspections at the rate of two dollars ($2.00) per hour.

(g) **Reinspections.** Each reinspection of any overhead, underground or interior wires or apparatus, altered, changed or repaired and where a permit is required, shall be charged for according to the time required for such reinspection at the rate of two dollars ($2.00) per hour.

(h) **Extra Inspections.** Where extra inspections are made on account of any of the following reasons, a charge of one dollar and fifty cents ($1.50) shall be made for each such inspection:

Inaccurate or incorrect information
Failure to make necessary repairs
Faulty construction.

(i) **Minimum Fee.** No inspection shall be made for a less amount than one dollar and fifty cents ($1.50).

ARTICLE 405

Certificates

405.01 General: Certificates of occupancy and other certificates shall be required as provided in this article, except as otherwise required by this chapter for certificates of compliance.

405.03 Capacity—Certification for License: The Commissioner of Buildings shall determine the number of persons which every building or room used for public purposes may acommodate according to the provisions of Chapter 22—SIZE AND LOCATION OF ROOMS—of this ordinance, and shall certify the same to the Bureau of Fire Prevention and City Clerk. No more than the number so certified shall be allowed in such room at any one time, in any building used for a hospital, business unit, theatre, open air assembly unit, public assembly unit, church or school.

405.04 No Amusement License to Issue Without Certificate: No license shall be issued to any person, firm or corporation to produce, present, conduct, operate or offer for gain or profit, any theatricals, shows or amusements until the Commissioner of Buildings, the Board of Health, the Fire Commissioner and the Commissioner of Streets and Electricity and the Commissioner of Public Works shall have certified in writing that the room or place where it is proposed to produce, present, conduct, operate or offer such theatricals, shows or amusements complies in every respect with the ordinances of the City of Chicago relating to their respective departments.

405.05 Multiple Dwelling Occupancy:

(a) **Certificate.** No multiple dwelling hereafter erected, shall be occupied in whole or in part for human habitation until the issuance of a certificate by the Board of Health that said building conforms to the requirements relative to plumbing and drainage applicable to such buildings, nor until the issuance by the Commissioner of Buildings of a certificate that the said building conforms to the requirements of this ordinance relative to all general, structural and special provisions applicable to new multiple dwellings. Within five (5) days from date of application for any such certificate, a certificate of occupancy shall be issued or the official concerned shall state in writing his reasons for his refusal to issue said certificate.

(b) **Advance Occupancy.** The certificate above referred to may be issued in the case of a new building comprising more than three (3) apartments so as to allow the occupancy of any completed section of the building extending from the basement to the roof in advance of the completion of the other portions of the building, when such portion is completely cut off from other parts of the building by a standard fire separation and all provisions for exits required by this ordinance have been complied with.

(c) **Inspection.** When the outer walls of a multiple dwelling have been erected so as to outline the position of the courts and shafts required for the lighting and ventilation of habitable rooms, the owner of the building, or his representative, shall be entitled upon application in writing, to an inspection of the same by the Commissioner of Buildings, and if the work to that point is in compliance with the provisions regarding the size of the shafts and the location of the building, to a certificate setting forth those facts. When the work of constructing partitions has advanced to such a degree, on any floor, that the rooms on that floor are determined in their dimensions, the owner, or his representative shall be entitled to an inspection from the Commissioner of Buildings, and, if the rooms thus outlined, conform in their dimensions to the plans filed and to the requirements of this ordinance, he shall be entitled to a certificate stating that fact.

(d) **Violation.** If a multiple dwelling hereafter erected is occupied as a place of habitation in any of its parts in violation of this section, it shall forthwith be subject to notice from the Commissioner of Buildings, and shall be vacated upon such notice, and shall not again be occupied until made to conform with the provisions of this ordinance, nor until after the issuance of the certificate required in this section.

CHAPTER 5

ARTICLE 501

Definitions

501.01 General: For the purposes of this ordinance, wherever the following words, terms or phrases occur either in this ordinance, or in any application, certificate, drawing, permit, plan or specification submitted for the purpose of obtaining a building permit, the definition herein given shall be deemed to be the meaning of such words, terms or phrases, except where otherwise expressly stated.

Addition. Any construction which increases the area or cubic contents of a building or structure.

Aisle. A passageway between rows of seats, or between rows of seats and a wall in an assembly room or auditorium, or between desks, tables, counters or other materials, or between such articles or materials and a wall in other rooms or spaces. 1. **Longitudinal Aisle.** In an assembly room or auditorium, an aisle approximately perpendicular to the rows of seats served thereby. 2. **Transverse Aisle.** In an assembly room or auditorium, an aisle approximately parallel to the rows of seats.

Alcove. A recess adjoining and connected with a larger room, with an unobstructed opening into such room equal in area to not less than twenty (20) per cent of the entire wall area of the alcove.

Alley. A narrow thoroughfare upon which abut generally the rear of premises, or upon which service entrances of buildings abut, and not generally used as a thoroughfare by both pedestrians and vehicles, or which is not used for general traffic circulation, or which is not in excess of thirty (30) feet wide at its intersection with a street.

Alteration. 1. Any change or modification of construction or space arrangement in an existing building or structure not increasing the area or cubic content thereof. 2. Any change which decreases the area or cubic content of a building.

Amphitheatre, Teaching. (See Teaching Amphitheatre.)

Apartment. A dwelling unit containing a kitchen or kitchenette or any equipment or facilities for the heating, cooling, preserving or serving of food in a multiple dwelling or in a multiple use building.

Apartment House. A multiple dwelling or that part of a building containing two or more apartments having a common entrance.

Area. A particular extent of any surface, except as otherwise defined or qualified. 1. **Building Area.** The maximum horizontal projected area of the building at grade. 2. **Floor Area.** The area of a floor within the inclosing walls of a building excepting space occupied by walls, columns, stairs, elevator shafts, well holes, chimneys, courts, and pipe, wire or vent shafts. 3. **Net Floor Area.** A floor area, as defined above, less also the area of lobbies, inclosed corridors and other inclosed means of exit, toilet rooms, janitors' closets and vaults.

Assembly Room. A room arranged, used or intended to be used by a group of persons assembled for any purpose other than for a regular business or dwelling.

A.S.T.M. Abbreviation for American Society for Testing Materials.

Basement. A story below the first story in a building having no ground story, or below the ground story in a building in which a ground story is included. In a building having more than one (1) basement, the story immediately below the first story or ground story shall be defined as "Basement," the story next lower "Basement B," and so on for the full number of basements which the building contains.

Bathroom. A room containing a tub, shower compartment or other facilities for bathing.

Building. When separated by standard fire division walls each unit so separated shall be deemed a separate building.

Ceiling. The undersurface of the overhead covering of a room.

Cellar. A basement story more than one-half (½) below the grade.

Class Room. A room used for the instruction of or recitation by students in groups.

Closet. A non-habitable room used for storage.

Convent. A building occupied by nuns as a place of residence.

Converted. Changed from one (1) class of occupancy to another class.

Corner Lot. (See Lot.)

Court. An open, unobstructed and uncovered space other than a yard, on the same lot with a building. 1. **Lot Line Court.** A court bounded on one side by a lot line and on its remaining sides by the walls of a building or another lot line or lot lines. 2. **Inner Court.** A court entirely surrounded by the walls of a building. 3. **Open Lot Line Court.** A lot line court open at one end, for the required dimensions of the court, to a street, alley or yard. 4. **Outer Court.** A court bounded by the walls of a building on all sides but one which shall be open, for the required dimensions of the court, to a street, alley, yard or through court. 5. **Through Court.** A court between a lot line and a building, extending from a street at one end to a street, alley or yard at the other end.

Curb Grade. Elevation of the curb as fixed by city ordinances.

Face. A wall surface of any other plane shall be considered to face a line or another plane when it makes an angle of less than forty-five (45) degrees.

Family. One or more persons living, sleeping, cooking and eating on the premises as a single housekeeping unit.

First Story. The lowest story having a floor, every point of which shall be not more than one (1) foot below the established inside grade, except as otherwise provided in Chapters 7 to 18, inclusive.

Grade. The finished grade of improved premises is the elevation of the surface of the ground adjoining the building. The established grade of premises, whether vacant or improved is the elevation of the sidewalk at the property line as fixed by city ordinance. On a two level thoroughfare, grade shall be taken at the upper level except as otherwise expressly provided.

Grand Stand. A structure containing tiers of seats, with or without roof or walls, used for viewing open air games or spectacles.

Ground Story. A story immediately below the first story on premises abutting an existing or proposed and authorized two level thoroughfare, the lowest point of the floor of which shall be not more than one (1) foot below the lowest established grade unless otherwise provided.

Habitable Room. Any room complying with the provisions of this ordinance, in which persons sleep, eat or carry on their usual business, domestic or social vocations, except private laundries, bathrooms, toilet rooms, pantries, closets, corridors, rooms for mechanical equipment and spaces used for service and utilities.

Hotel. A building or structure kept, used or maintained as, or advertised or held out to the public to be an inn, family hotel, apartment hotel, lodging house, dormitory or place where sleeping or rooming acommodations are furnished for hire or rent, whether with or without meals, in which five (5) or more sleeping rooms are used and maintained for the accommodation of guests, lodgers or roomers.

Kitchen. A space where food is prepared.

Locker Room. A room designed, intended or used for the accommodation of lockers for the storage of goods or wearing apparel, with or without provisions for changing apparel.

Lot. A parcel of land or premises occupied or which it is intended shall be occupied by one building with its usual auxiliary buildings or uses customarily incident thereto, including such open spaces as are required by this ordinance and such open spaces as are arranged and designed to be used in connection with such building. 1. **Corner**

Lot. A lot which abuts upon two streets making an angle on the lot side, at their intersection, of one hundred twenty (120) degrees or less. 2. **Front of Lot.** The shortest street frontage. 3. **Rear of Lot.** The boundary opposite the front of the lot as herein defined.

Lot Line. A boundary line between a lot and an adjoining lot.

Mezzanine. A partial floor used for the same purposes as the floor below, having at least one side open into the story containing such mezzanine, and having an area not to exceed twenty (20) per cent of the area of the floor of the containing story.

Monastery. A building occupied by monks as a place of residence.

Ordinance. The term "this ordinance" wherever contained herein shall be construed as the Building Ordinance, Chapters 1 to 46 inclusive. The term "other ordinances of the City of Chicago" shall be construed as ordinances other than the building ordinance.

Owner. The term "owner" for the purposes of this ordinance shall mean the owner or owners of the freehold of the premises or of a lesser estate therein, a vendee in possession, or the lessee or joint lessees of the whole thereof.

Pantry. A space accessory to a dining room or kitchen, for the storage of food, dishes and utensils.

Partition. A vertical separating construction extending from floor to ceiling.

Principal Entrance. That entrance from a street, alley or other open space serving as a way of approach to a building, to which is apportioned the greater aggregate of the required outside exit doorway width.

Proscenium. The vertical plane of separation between an auditorium and a stage.

Public Hall. A hall, corridor or passageway within a building for the common use of its occupants.

Purlin. A roof beam.

Repair. The renewal or reconstruction of any part of a building or other structure for the purpose of its maintenance, which does not increase the area or cubic content of a building or other structure.

Row of Seats. A group of adjoining seats arranged side by side.

Shall. As used in this ordinance is mandatory.

Shaft. A space inclosed with side walls and extending through one (1) or more stories.

Shed. A structure not exceeding three hundred (300) square feet ground area, and not to exceed fourteen (14) feet in height above grade to the highest point thereof, with a roof covering of fire-retarding material, and inclosing walls of combustible materials, no part of which is located on the front half of any lot. 1. **Shelter Shed.** A structure having a roof and with more than fifty (50) per cent of the area of its sides open.

Skeleton Construction. A construction type wherein all external and internal loads and stresses are transmitted cumulatively from the top of a building to the foundation by a skeleton or frame-work.

Spandrel. The wall construction extending from the top of an opening to the sill of another opening in the next story above and contained between vertical lines produced from the jambs of the openings.

Square. When the word "square" is used it shall be held to embrace a plot of ground surrounded and bounded and inclosed by public streets, railway right of water, waterway, public place or parks.

Stage. A structure or platform upon which an act or performance is presented.

Stair Hall. That portion of a building or other structure including the stairways, and those portions of the public halls through which it is necessary to pass in the direction of exit by means of a stairway to an outside exit doorway.

Story. That part of a building or other structure included between the top of any floor and the top of the floor next above, except that the topmost story shall be that portion of a building or other structure included between the top of the floor and the ceiling above.

Street. A thoroughfare used for public, foot and vehicular traffic, other than an alley.

Teaching Amphitheatre. A room in which students stand or sit in banks for the purpose of witnessing surgical operations, X-ray treatments, or other medical, surgical or dental practices, anatomical exhibitions and demonstrations and the practice of section or dissection or other scientific demonstrations.

Terra Cotta. A burned clay product, sometimes used for the facing of walls or exteriors, as distinguished from structural clay tile or brick.

Toilet Room. A room containing one (1) or more water closets or urinals.

Veneer. When referring to a building or other structure shall mean an outer facing, not an integral part of the wall.

Vomitory. A passageway through the floor of any seating level by means of a level or ramped floor or steps of any combination thereof, and used as a means of ingress to or egress from the floor through which it passes.

Well. An opening through a floor, roof or ceiling.

Yard. An open, unobstructed and uncovered space at the rear of a lot, other than a court, extending the entire width of the lot, and whose depth shall be deemed to be the distance from the rear wall of a building to the rear of the lot.

CHAPTER 6

General Occupancy Provisions

601.01 Classification of Buildings: All buildings and structures existing at the time of the enactment of this ordinance, or hereafter designed, erected, altered or converted, shall be classified according to occupancy and use in accordance with the provisions of this chapter. The classification of buildings and structures provided in this ordinance shall be applicable to the whole of the Revised Chicago Code of 1931, and wherever a reference is made to a building or structure of a certain class, it shall be construed as referring to the classification of use hereinafter described, unless it plainly appears that some other classification is intended. Any requirements for buildings not herein classified shall be determined by the Commissioner of Buildings subject to arbitration as provided in this ordinance.

601.02 Single Dwellings: A single dwelling, for the purposes of this ordinance, is hereby defined as a building, or part of a building, other than a multiple dwelling, designed as, intended for, or used as a residence for a single family, or for a group of persons other than a single family, when such group does not exceed ten (10) in number. Dwellings which are completely separated from each other from the ground to and through the roof by fire-resistive construction of the value required under Section 2110.02, without any communicating opening and which are without any means of exit in common, shall be classified as Single Dwellings.

601.03 Multiple Dwellings: A multiple dwelling, for the purposes of this ordinance, is hereby defined as a building, or part of a building, designed, intended or used as an apartment house, apartment hotel, tenement house, hotel, rooming house, lodging house, clubhouse with sleeping quarters, convent, monastery, ecclesiastical home, nurses' home, or other use in which there is more than one (1) dwelling.

601.04 Institutional Buildings: An Institutional building is hereby defined as a building, or part of a building designed, intended or used as a hospital, sanitarium, medical unit, infirmary or other similar unit for the care or treatment of men, women or children; a home or asylum for bedridden or decrepit persons, the blind, the aged, children, or the insane; or a nursery or day nursery for infants.

units as are otherwise classified under this ordinance. The displaying of motion pictures shall include the displaying of sound pictures and motion pictures and the presenting of theatrical performances shall include the presenting of dramatic, operatic, comic, pantomime or vaudeville performances, also the presenting of other amateur or professional theatricals, productions or spectacles.

601.09 Open Air Assembly Unit: An open air assembly unit, for the purposes of this ordinance, is hereby defined as a structure, or part of a structure, or group of structures, or inclosure, designed, intended or used for the purpose of seating, sheltering or assembling a group of persons in the open air, together with such other buildings and occupancies as are necessary to the operation of open air assembly units.

601.10 Public Assembly Unit: A public assembly unit, for the purposes of this ordinance, is hereby defined as a building, or part of a building, designed, intended or used for the congregating, gathering or assembling for any purpose, of a group of more than fifty (50) persons in one (1) or more rooms, wether such congregation, gathering or assemblage be of a public, restricted or private nature, except theatres, open air assembly units, churches and schools, treated in Chapters 13, 14, 16 and 17 and except such other assembly uses as are otherwise classified under this ordinance.

601.11 Church Building: A church building, for the purposes of this ordinance, is hereby defined as a building, or part of a building, designed, intended or used as a place of worship and for other religious purposes.

601.12 School Building: A school building, for the purposes of this ordinance, is hereby defined as a building, or part of a building, designed, intended or used as a place of public or private instruction, but not including such occupancies as are otherwise classified under this ordinance.

601.13 Miscellaneous Buildings and Structures: Miscellaneous buildings and structures, for the purposes of this ordinance, shall be defined as all buildings and structures not included in Chapter 7—SINGLE DWELLINGS to Chapter 17—SCHOOL BUILDINGS, inclusive, of this ordinance, including billboards, signboards, fences, hangars, ice houses, illuminated signs, incinerators, crematories, jails, police stations, fire stations, prisons, certain railroad structures, sheds, tables and miscellaneous buildings and structures not otherwise provided for in this ordinance.

CHAPTER 7
Single Dwellings

ARTICLE 701
General

701.01 Single Dwellings Defined: A single dwelling, for the purposes of this ordinance, is hereby defined as a building, or part of a building, other than a multiple dwelling, designed as, intended for, or used as a residence for a single family, or for a group of persons other than a single family, when such group does not exceed ten (10) in number; or for any of the purposes of a hospital specified in Section 901.01, paragraph (a) when the total number of patients, or inmates, does not exceed ten (10) persons. Dwellings which are completely separated from each other, from the ground to and through the roof, by fire-resistive construction of the value required under Chapter 21, without any communicating opening, and which are without any means of exit in common, shall be classified each as a single dwelling.

701.02 Requirements for Single Dwellings: Every building or part thereof, hereafter designed, erected, altered, converted or used as a single dwelling, shall comply with the General Provisions of this ordinance and in addition thereto, shall comply with the special provisions set forth in this chapter.

ARTICLE 702

Lot Occupancy

702.01 Location on Lot:

(a) **General.** The minimum distance from property dividing lot lines to door or window openings through exterior walls, shall be as follows: In walls built at angles of less than forty-five (45) degrees from such lot lines the minimum distance from such lot lines to closest jambs of door or window openings shall be three (3) feet. In walls built at angles of forty-five (45) to eighty-nine (89) degrees inclusive, from such lot lines, the minimum distance from such lot lines to closest jambs of door or window openings shall be one and one-half (1½) feet. In walls built at angles of ninety (90) degrees or more from such lot lines, the minimum distance from such lot lines to closest jambs of door or window openings shall be one (1) foot.

(b) **Dwellings of Wood Frame Construction.** Exterior inclosing walls of dwellings of wood frame construction shall not be erected closer than three (3) feet to property dividing lot lines.

ARTICLE 703

Construction

703.01 Wood Frame Construction:

(a) **Within the Provisional Fire Limits.** A building of wood frame construction as provided in Article 3108, to be occupied exclusively as a single family dwelling, not exceeding two (2) stories or thirty (30) feet in height and having no one floor with an area of more than fifteen hundred (1500) square feet above or below the first floor may be erected on any vacant lot within the provisional fire limits of the City of Chicago, except in any block in which fifty-one (51%) per cent or more of the lots in such block are occupied by buildings of other than wood frame construction. The word "block" as used herein shall be deemed to be that property abutting on both sides of a street and lying between the two nearest intersecting streets.

(b) **Outside the Fire Limits and Provisional Fire Limits.** Wood frame construction may be used outside the fire limits and the provisional fire limits of the City of Chicago in any building to be occupied as a single family dwelling not exceeding thirty (30) feet in height, and having no one (1) floor with an area of more than fifteen hundred (1500) square feet above or below the first floor.

703.02 Ordinary or Heavy Timber Construction: Ordinary Construction or heavy timber construction, or a superior type of construction shall be used in any single dwelling not exceeding three (3) stories and basement or forty-four (44) feet in height and having no one (1) floor with an area of more than twenty-five hundred (2500) square feet above or below the first floor.

703.03 Semi-Fireproof and Fireproof Construction: Semi-fireproof or fireproof construction shall be used in any single dwelling exceeding three (3) stories and basement, or forty-four (44) feet in height and in any single dwelling having any one (1) floor with an area of more than twenty-five hundred (2500) square feet either above or below the first floor.

ARTICLE 704

Size and Location of Rooms

704.01 Size of Habitable Rooms: Every habitable room in a single dwelling shall have a minimum floor area of seventy (70) square feet and at least seventy-five (75) per cent of the floor area shall have a height of not less than eight (8) feet, measured from finished floor line to finished ceiling line and must contain not less than five hundred sixty (560) cubic feet.

704.02 Habitable Rooms in Basements: Habitable rooms may be located in a basement of a single dwelling; provided however, that the floor of such basement is not more than four (4) feet below grade.

704.03 Kitchens, etc.:

(a) **Limitation Based on Area.** Every dwelling containing a cooking stove, sink or other equipment to prepare food by use of heat from any source shall contain at least one (1) room with a floor area of not less than one hundred twenty (120) square feet.

(b) **Dining Kitchen.** In a dwelling containing at least one (1) other habitable room, a kitchen and dining room together may be considered as one (1) habitable room; provided however, that the combined area of the vertical surfaces of any partitions, pilasters, columns or other separation between kitchen and dining room spaces shall not exceed, in square feet, the product of three (3) feet multiplied by the clear ceiling height of the room, all as measured on the kitchen side of the separation; and provided further, that the opening between the kitchen and the dining spaces is not less than three (3) feet in clear width and extends to within six (6) inches of the finished ceiling on the kitchen side. The combined floor area of the kitchen and the dining spaces and the combined natural ventilation of such spaces, shall meet the requirements of paragraph (a) of this section. Any opening between such a dining kitchen and any other part of the dwelling may be either with or without a door.

ARTICLE 705

Multiple Use Buildings

705.01 Other Occupancies Permitted and Prohibited: Subject to the provisions of this article, a single dwelling may be located in the same building with any other occupancy, except a garage building of an area greater than eight hundred (800) square feet, and except buildings for hazardous use occupancy prohibited in a single dwelling under Section 1105.03 and except in buildings as otherwise provided in Chapters 6 to 19, inclusive of this ordinance.

705.02 Separating Construction: Every single dwelling in a multiple occupancy building of any construction other than fireproof or semi-fireproof shall be separated from every other occupancy in the same building by floors, ceiling and walls of fireproof construction of two (2) hour fire-resistive value and shall be provided with a door giving direct exit to the outside of the building or by an exit corridor of fireproof construction giving direct exit to the outside of the building.

ARTICLE 706

Means of Exit

706.01 Stairways:

(a) **General.** In any single dwelling more than one (1) story in height, and less than three (3) stories in height, there shall be one (1) stairway not less than three (3) feet wide, extending from the topmost habitable story to the first floor and there shall be a front and rear exit from the first floor. There shall be one (1) stairway not less than two (2) feet six (6) inches in width extending from the first floor to any basement and a direct exit from the basement. If any floor above the first floor exceeds twenty-five hundred (2500) square feet in area, or if the floor of any habitable story is more than sixteen (16) feet above the average finished level of the ground adjoining the building, there shall be one (1) additional stairway from such floor to the first floor. The clear width of such additional stairway shall be not less than two (2) feet, six (6) inches. Where two (2) stairways are required, the basement stairway shall be inclosed in the basement or in the first story with not less than one (1) hour fire-resistive construction with one and three-fourths (1¾) inch thick wood slabs or more fire-resistive doors. Where two (2) stairways are required, they shall be separated by a distance of not less than fifteen (15) feet at the top and at the bottom such distances being measured from doorway to doorway if both stairs are inclosed, from top rise to top rise and from bottom rise to bottom rise, if both stairways are open, and from top rise to doorway and from bottom rise to doorway, if one (1) stairway is inclosed and the other stairway is open.

(b) **Buildings Used for Hospital Purposes.** A single dwelling used for any of the purposes of a hospital and having any sleeping accommodations above the first story, shall be provided with at least two (2) stairways, each not less than three (3) feet wide. A separate door exit shall be provided for each such stairway and from the basement to the outside of the building. Any single dwelling more than two (2) stories high used for hospital purposes shall be of fireproof or semi-fireproof construction throughout.

706.02 Exits from Rooms: Every habitable room in a single dwelling shall have at least one (1) doorway which will lead to an exit from the building without the necessity of passing through a bedroom, a bathroom or toilet room.

706.03 Basement Exits: In a single dwelling, every basement with a floor area of twenty-five hundred (2500) square feet, or less, shall be provided with an interior or exterior stairway not less than two (2) feet, six (6) inches wide to grade with a direct exit to the outside of the building. Every basement with a floor area of more than twenty-five hundred (2500) square feet shall be provided with a stairway not less than three (3) feet wide to grade and with one (1) other completely separated means of exit. Said other means of exit may be either a stairway not less than two (2) feet six (6) inches wide to grade or to a window measuring not less than two (2) feet, six (6) inches by three (3) feet, having its sill not more than four (4) feet, six (6) inches above the basement floor and opening to a street, a public alley, a private alley, an easement, a yard, an open lot line court, a through court or an outer court. Such opening may be through an open and unobstructed areaway; provided however, that the top of said areaway is not more than three (3) feet above the bottom of said areaway.

ARTICLE 707

Windows and Ventilation

707.01 Habitable Rooms: Every habitable room in a single dwelling shall have windows and ventilation, as required by Section 807.01 for multiple dwellings.

707.02 Non-Habitable Rooms and Spaces: Every kitchen, bathroom or toilet room and every compartment containing any water closet, or urinal fixture shall meet the requirements of Section 807.02 for multiple dwellings.

707.03 Openings on Open Porches: A window or doorway, opening upon an open porch shall meet the requirements of Section 807.05 for multiple dwellings.

707.04 Inclosure of Porches: Every porch inclosure shall meet the requirements of Section 807.06 for multiple dwellings.

CHAPTER 8
Multiple Dwellings

ARTICLE 801

General

801.01 Multiple Dwelling: A multiple dwelling, for the purposes of this ordinance, is hereby defined as a building, or part of a building, designed, intended or used as an apartment house, apartment hotel, tenement house, hotel, rooming house, lodging house, club house with sleeping quarters, convent, monastery, ecclesiastical home, nurses' home, or other use in which there is more than one (1) dwelling.

801.02 Other Terms Defined:

(a) **General.** Certain terms in this chapter are defined for the purposes thereof as follows:

(b) **Habitable Room.** Any room in which persons sleep, eat or carry on their usual domestic, or social vocations, or avocations, but shall not include kitchens, laundries, bathrooms, water closet compartments, serving and storage pantries, storage rooms and closets less than forty (40) square feet in floor area, mechanical equipment rooms, cellars, corridors and similar spaces, used neither frequently, nor during extended periods.

(c) **Corridor Type Multiple Dwelling.** A multiple dwelling in which all vertical means of exit from any one (1) dwelling can be reached through a common corridor and in no other way.

(d) **Kitchen.** A space where food is prepared by the use of heat.

(e) **Dining Kitchen.** A room used as both kitchen and dining room but not used for other purposes.

(f) **Kitchen Alcove.** An alcove used as a kitchen.

(g) **Closet Kitchen.** A space containing kitchen equipment, provided with a door, or doors, whereby it may be closed off from the rest of the room in which it is located and measuring not more than three (3) feet from its back wall to the inside face of the door, or doors, when closed.

(h) **Rectangular Court.** A court, every angle of which measures approximately ninety (90) degrees or two hundred seventy (270) degrees.

(i) **Non-Rectangular Court.** Any court other than a rectangular court.

(j) **Yard.** An unoccupied space on the same lot with a building separating every part of every building on the lot from the rear line of the lot.

(k) **Court.** An open unoccupied unobstructed space other than a yard on the same lot with a building.

(l) **Closed Lot Line Court.** A court bounded on one (1) side by a lot line and on its remaining sides either by the walls of a building, or another lot line or lot lines.

(m) **Inner Court.** A court entirely surrounded by the walls of a building.

(n) **Open Lot Line Court.** A court between a lot line and a building open at one end for the required dimensions of the court to a street, alley or rear yard.

(o) **Outer Court.** A court bounded by walls of a building on all sides except one (1) and open on that side for the required dimension of the court to a street, alley, rear yard or through court.

(p) **Through Court.** A court between a lot line and a building extending from a street at one (1) end to a street alley or rear yard at the other end.

801.03 Requirements for Multiple Dwellings: Every building, or part thereof, hereafter designed, erected, altered or converted for use as a multiple dwelling, shall comply with the General Provisions of this ordinance and in addition thereto shall comply with the special provisions set forth in this chapter; in case such provisions conflict the special provisions of this chapter shall govern.

ARTICLE 802

Lot Occupancy and Court Requirements

802.01 Rear Yard Requirements:

(a) **General.** On a lot having only one (1) street frontage, when the rear line of said lot does not abut upon a public alley, a public park, or a public waterway, there shall be a rear yard abutting the entire length of said rear line and measuring not less than eight (8) feet at a right angle to said rear line at any point, except as otherwise provided in paragraphs (b), (c), (d) and (e) of this section. Where the rear line of said lot does abut upon an alley, the distance from any part of the building on such lot to the center line of such alley shall be not less than the requirements of the Chicago Zoning Ordinance.

(b) **Districts of One Hundred (100) Per Cent Occupancy to Thirty (30) Foot Level.** In a Volume District in which one hundred (100) per cent of occupancy is allowed up to thirty (30) feet above grade, a rear yard on a lot, either with or without alley, may begin at such thirty (30) foot level, but the width of any yard above that level shall be established as from grade.

(c) **Exemption in Certain Districts.** In a Fourth or Fifth Volume District, which is also located either in a Commercial or Manufacturing District, as defined in the Chicago Zoning Ordinance, a rear yard shall be required only as provided by said Zoning Ordinance.

(d) **Exemption on Corner Lots.** The owner of a corner lot may select and designate which side of the lot shall be considered to be the rear. Any side, other than a street side, may be so chosen; provided however, that for a lot bounded by three (3) or more streets, any side may be named as the rear; provided further, that the building line requirements of the Chicago Zoning Ordinance shall be complied with for the true frontage of the lot as defined in said Zoning Ordinance without regard to the street upon which the building faces.

(e) **Exemption on Certain Lots Extending from Street to Street.** Any interior lot which extends between and abuts upon two (2) approximately parallel streets, which are separated by an average distance of two hundred (200) feet or less, is exempt from the provisions of this section for a rear yard.

(f) **Requirements on Certain Lots Extending from Street to Street.** Where an interior lot extends through so as to abut upon two (2) approximately parallel streets and where the average distance between such streets is more than two hundred (200) feet, there shall be a rear yard with a minimum dimension of sixteen (16) feet, measured approximately at right angles to said streets, such rear yard extending the entire distance across said lot and being so located that its center line shall be not more than eight (8) feet distant from a line midway between the two (2) said streets; provided however, that in lieu of such rear yard there may be an open lot line court on each side of the lot, with a minimum dimension of eight (8) feet, measured at a right angle to the lot line and extending from either street back to a point eight (8) feet beyond the line midway between said streets; and provided further that where a sixteen (16) foot rear yard exceeds the minimum requirements for dimensions and areas at all floors by not less than one hundred (100) per cent, said rear yard may be located at any place on the lot without requiring an open lot line court. Such rear yards and such open lot line courts shall comply with all regulations of this Section and Section 802.02 respectively, as to increase of dimensions in proportion to height.

(g) **Zoning Ordinance Provisions.** Nothing in this section shall be construed to require less than the minimum yard, court, building line or setback requirements of the Chicago Zoning Ordinance.

802.02 Court Requirements:

(a) **General.** The provisions of this section shall apply to courts which are required for purposes of lighting or ventilation and nothing in this section shall be construed to prevent the construction or use of any court, regardless of size or dimensions, so long as such court is not required for purposes of lighting or ventilation.

(b) **Areas and Dimensions of Non-Rectangular Courts.** At any height in a non-rectangular court, the area shall be not less than the minimum area and the axial dimensions shall be not less than the minimum dimensions required at the same height in a rectangular court. All space within a non-rectangular court that is included between the lot line and a court wall forming an angle with the lot line or between two (2) angling walls of a court where the included angle is less than forty-five (45) degrees shall not be considered as court area for the purpose of light or ventilation where the minimum distance across the said angle is less than one-half (½) the required minimum dimensions of a rectangular court measured in the same direction.

(c) **Window and Door Locations on Lot Line Courts.** The jamb of any window, or door, in a wall of a court forming an angle with an abutting lot line, shall be not nearer than one and one-half (1½) feet to such lot line.

(d) **Porches, Steps, Fire Escapes and Other Obstructions in Yards and Courts.** The required dimensions, and area of any yard or court shall be exclusive of the space occupied by any porch, stairway, fire escape or other obstruction, except as otherwise provided in the following Items, numbered 1, 2 and 3.

Item 1. Any required fire escape or an unenclosed outside fireproof stairway or a solid floor balcony to an interior fire shield stairway, built as required by this ordinance, if projected not more than four (4) feet into a yard or court, or the projections of window sills, belt courses, cornices projecting not more than six (6) inches from the face of the building at any height, shall be exempt from the restrictions provided by this paragraph.

Item 2. An entrance canopy, or similar roofed space, steps and landings serving main exit level and their balustrades, open fences or railings, or similar structures conforming to the areas, dimensions and other requirements of the Chicago Zoning Ordinance, shall be exempt from the restrictions provided by this paragraph.

Item 3. A masonry stack is permitted in a court or yard; provided however, that it adjoins a wall which is vertical and without any offset throughout its height; that it does not reduce the open area of the yard or court below the required minimum; that it does not project into a court in such a manner as to reduce the minimum dimension at such place more than fifty (50) per cent and that it does not reduce or obstruct any required exit passage. A metal stack may be erected with the same restrictions, except that instead of adjoining the building it may be separated from a non-offset wall by a clear distance of not more than twelve (12) inches at any point and except that it may be erected beside a vertically offset wall if it is so arranged as to be not more than twelve (12) inches distance from the face of the wall at one (1) point at each offset.

(e) **Alley Side Line Courts.** Where the side of an interior lot is bounded by a public alley, the minimum required width and area of any court adjoining such alley shall be computed from the center line of the alley.

(f) **Through Courts.** The minimum horizontal dimensions of a through court measured at a right angle to the lot line, shall be three (3) feet. Beginning at a height of ten (10) feet above the bottom of the court, said minimum horizontal dimension shall be increased at the rate of not less than one (1) foot for each ten (10) feet of additional height, until a total horizontal court dimension of fifteen feet is attained. Above such point, the wall of the court may be carried up without further increase in said court dimension except as otherwise required by the Chicago Zoning Ordinance for setbacks from side lot lines.

(g) **Rectangular Open Lot Line Courts.** The minimum horizontal dimension of a rectangular open lot line court, measured at a right angle to the lot line shall be four (4) feet for a distance of fifty (50) feet measured from the closed end of the court. For the next succeeding fifty (50) feet, or fraction thereof, the minimum dimension shall be five (5) feet and for each succeeding fifty (50) feet or fraction thereof, thereafter, said minimum dimension shall be increased one (1) foot. Beginning at a height of ten (10) feet above the bottom of an open lot line court, the required horizontal dimension or dimensions, measured at a right angle to the lot line shall be increased at the rate of one (1) foot for each ten (10) feet of additional height, until a total horizontal court dimension of fifteen (15) feet is attained. Above such point, the court may be carried up without further increase in said court dimension, except as otherwise required by the Chicago Zoning Ordinance.

(h) **Rectangular Outer Courts.** The minimum horizontal dimension of a rectangular outer court, measured parallel to the open side of such court, shall be eight (8) feet for a distance of fifty (50) feet, measured from the inner, or closed end, of the court. For the next succeeding fifty (50) feet or fraction thereof, the minimum dimension shall be ten (10) feet and for each succeeding fifty (50) feet or fraction thereof, thereafter, said minimum dimension shall be increased two (2) feet. Beginning at a height of ten (10) feet above the bottom of an outer court, the required horizontal dimension or dimensions, measured parallel to the open side of the court shall be increased at the rate of two (2) feet for each ten (10) feet of additional height until a total horizontal court dimension of thirty (30) feet is attained. Above such point, the court may be carried up without further increase in said court dimension, except as otherwise required by the Chicago Zoning Ordinance. The increase in horizontal dimensions, required by this paragraph, may be obtained by stepping back either or both of the side walls.

(i) **Rectangular Closed Lot Line Courts.** The minimum horizontal dimension of a rectangular closed lot line court, measured at a right angle to the lot line, shall be five (5) feet. Beginning at a height of ten (10) feet above the bottom of the court, the required horizontal dimension, measured at right angle to the lot line, shall be increased at the rate of one (1) foot for each ten (10) feet of additional height, until a total horizontal court dimension of fifteen (15) feet is attained. Above such point, the court may be carried up without further increase in said court dimension, except as otherwise required by the Chicago Zoning Ordinance. The minimum horizontal dimension of a rectangular closed lot line court, measured parallel to the lot line at any height, shall be not less than twice the required minimum dimension measured at a right angle to the lot line at the same height. The increase in horizontal dimension parallel to the lot line may be obtained by stepping back either, or both, of the end walls. The minimum area of a rectangular closed lot line court, at any height, shall be determined from Table 802.02 (i). For any height not given directly in the table, said minimum area shall be determined by interpolation between the two (2) nearest heights.

Table 802.02 (i)

Areas of Closed Lot Line Courts

Height Above Bottom of Court Feet	Minimum Allowable Area Square Feet
0	100
10	100
20	132
30	168
40	208
50	252
60	300
70	352
80	408
90	468
100	532
110	600
More than 110	600

(j) **Rectangular Inner Courts.** The minimum dimension of a rectangular inner court shall be ten (10) feet. Beginning at a height of ten (10) feet above the bottom of the court, said minimum dimension shall be increased at the rate of two (2) feet for each ten (10) feet of additional height, until a total dimension of thirty (30) feet is attained. Above such point, the court may be carried up without further increase in any court dimension; provided however, that the areas required by Table 802.02 (j) are maintained. The increase in horizontal dimensions required by this paragraph may be obtained by stepping back either or both of the walls which face each other across the court. The minimum area of a rectangular inner court at any height shall be determined from Table 802.02 (j). For any height not given directly in the table, said minimum area shall be determined by interpolation between two (2) nearest heights.

Table 802.02 (j)

Areas of Inner Courts

Height Above Bottom of Court Feet	Minimum Allowable Area Square Feet
0	200
10	200
20	264
30	336
40	416
50	504
60	600
70	704
80	816
90	936
100	1064
110	1200
More than 110	1200

(k) **Offsetting Stories.** The bottom of a yard or court may be established of minimum dimensions in any story above grade, with offsets above such story, or stories, in conformity with the requirements of this section, so that at any story the required minimum dimension shall be maintained. In offsetting, one (1) or more walls of any court may be vertical throughout the entire height so long as minimum dimensions and areas as established by this section are maintained.

(l) **Exceptions for Buildings of Not More Than (3) Stories.** The minimum dimension and area through the entire height of any court, in a building not more than three (3) stories and basement, need not exceed the dimension and area required by paragraphs (a) to (k) inclusive of this section for a court of ten (10) feet high; provided, however, that in every dwelling there shall be at least one (1) habitable room with a door or window opening on a street, alley, casement, yard as provided in Section 802.01 or court of the full dimensions and area required by paragraphs (a) to (k) inclusive of this section.

(n) **Ventilation of Courts.** Every closed lot line court and every inner court shall be ventilated from the lowest level of the court by a tunnel or passage, having a minimum cross-sectional area of thirty (30) square feet and extending from the court to a street, alley, rear yard, side yard, public waterway, or public park. Such tunnel or passage, shall not be closed or obstructed wholly or partially by anything that will affect ventilation and if it serves as an exit, it shall conform to the requirements of Section 806.07. Two (2) or more tunnels may be used if their combined openings are equivalent to the area herein required, but in no case shall the least dimension of such tunnel be less than two (2) feet.

ARTICLE 803

Construction

803.01 Fireproof Construction:

(a) **General.** Fireproof construction may be used in any multiple dwelling and shall be used in any multiple dwelling building exceeding eighty (80) feet in height.

(b) **Special Requirement for Partitions and Public Corridors.** In any multiple dwelling required to be of fireproof construction, every dwelling and every public corridor which serves as the only means of exit from one (1) or more dwellings, shall have inclosing walls as required in Section 803.02 paragraphs (b) and (c) for semi-fireproof construction.

803.02 Semi-Fireproof Construction:

(a) **General.** Semi-fireproof construction may be used in any multiple dwelling not exceeding eighty (80) feet in height and semi-fireproof or fireproof construction shall be used in any multiple dwelling of greater height than four (4) stories and basement or fifty (50) feet in height.

(b) **Special Requirements for Inclosing Partitions.** In any multiple dwelling required to be of semi-fireproof construction, every dwelling shall have inclosing walls or partitions of one (1) hour fire-resistive construction not less than three (3) inches thick, exclusive of plaster or other finish and every such wall shall extend from the top of the fire-resistive floor slab to the underside of the fire-resistive slab at the ceiling. All other partitions in a multiple dwelling of semi-fireproof construction shall be of one (1) hour fire-resistive construction, and no combustible studding or combustible lath shall be permitted in any partition in a building of semi-fireproof construction.

(c) **Inclosure for Public Corridors.** In a multiple dwelling required to be of semi-fireproof construction, every public corridor which serves as the only means of exit from one (1) or more dwellings shall have walls of two (2) hour fire-resistive construction, not less than three (3) inches thick, exclusive of plaster or other finish, and every such wall shall extend from the top of the fire-resistive floor slab to the underside of the fire-resistive slab above.

803.03 Ordinary and Heavy Timber Construction:

(a) **General.** Ordinary or heavy timber construction may be used for multiple dwelling buildings subject to the provisions of this section in the following paragraphs (b) to (h) inclusive. The height in stories hereinafter stated shall be the number of stories exclusive of the height of any basement.

(b) **One (1) Story Buildings.** Multiple dwellings, not exceeding one (1) story in height, may be of ordinary or heavy timber construction, and if there are no living rooms in basement shall be exempt from the special provisions of paragraphs (c), (d) and (e) of this section.

(c) **Special Provisions for Buildings More Than One (1) Story High.** In a building more than one (1) story high and less than four (4) stories high, all structural metal or concrete reinforcement in any column or other member supporting a required division wall between dwellings or any part of such a wall shall have a two (2) hour fire-resistive covering.

(d) **Special Provisions for Buildings More Than Two (2) Stories High.** In a building more than two (2) stories high and less than four (4) stories high, there shall be a ceiling of metal lath and plaster between the first story and the entire basement. The provisions of paragraph (c) shall also apply to such a building.

(e) **Special Provisions for Buildings More Than Three (3) Stories, or Forty-four (44) Feet High.** In a building more than three (3) stories and basement, or forty-four (44) feet high, every wall, partition and ceiling shall be covered with metal lath and plaster meeting the requirements of Chapter 32—FIRE RESISTIVE STANDARDS of this ordinance. Any metal column, or other metal member supporting any floor in such a building shall have a two (2) hour fire-resistive covering. The provisions of paragraphs (c) and (d) of this section shall also apply to such a building.

(f) **Special Provisions for Buildings Four (4) Stories and Basement But Not More Than Fifty (50) Feet High.** Any multiple dwelling may be built of ordinary or heavy timber construction not exceeding four (4) stories high but not to exceed fifty (50) feet in height; provided however, that with the exception of required fire-resistive walls between dwellings and for enclosing walls of stairways, every wall partition and ceiling shall be protected with a covering that shall be at least as fire-resistive as metal lath and plaster. Where there are living rooms in the basement there shall be a floor of two (2) hour fire-resistive construction between the first story and the entire basement. Where there are no living rooms in the basement, there shall be either a ceiling of metal lath and plaster between the first story and the entire basement or a standard system of automatic sprinklers over the entire basement. Any metal column or other metal members supporting any floors in such a building shall have a two (2) hour fire-resistive covering. The provisions of paragraph (c) of this section shall also apply to such building.

(g) **Separation of Dwellings.** In a multiple dwelling required to be of ordinary or heavy timber construction, every dwelling shall be vertically separated from all other parts of the building by walls of two (2) hour fire-resistive, non-combustible construction; provided however, that where two (2) or more dwellings are contained within a floor area not exceeding eight hundred fifty (850) square feet, such vertical separation walls shall be required only to include such area and not between dwellings within such area. In a hotel, rooming house, lodging house, club house with sleeping quarters, convent, monastery, ecclesiastical home, nurses' home or other use in which there are sleeping quarters for more than ten (10) persons, there shall be for every eight (8) rooms on any one (1) floor, dividing walls of two (2) hour fire-resistive construction separating such eight (8) rooms from the contiguous space. Nothing in this paragraph shall be construed to require a fire separation within a dwelling of any area whatever if such area is obviously intended for one (1) apartment for the use of only one (1) family. Every required separation wall between dwellings shall extend from the basement floor, or from a floor of two (2) hour fire-resistive construction directly over the entire basement, to the underside of the roof sheathing; provided however, that any such wall may be offset at any floor if the construction between such offsets is of two (2) hour fire-resistive construction and all structural supports for the offset portion of the wall are fireproofed as provided in this ordinance. Every doorway in a required vertical separation wall between dwellings shall be equipped with a standard forty-five (45) minute fire-resistive door. No fire-resistive, or non-combustible door shall be required in any wall between a dwelling and a corridor, or stair, except in a standard fire division wall and as otherwise required by this chapter. Any dwellings in a basement which is required by this ordinance to be covered in whole, or in part, by a floor of two (2) hour fire-resistive construction, shall be separated from one another by walls of two (2) hour fire-resistive construction.

(h) **Inclosure for Public Corridors.** In a corridor type multiple dwelling, required to be of ordinary or heavy timber construction, every public corridor shall have walls, floors and ceilings of two (2) hour fire-resistive non-combustible construction. Every doorway from a dwelling into such a corridor shall be equipped with a door not less fire-resistive than a wood slab door one and three-fourths (1¾) inch thick.

803.04 Wood Frame Construction: Wood frame construction may be used in any multiple dwelling outside the fire limits, or the provisional fire limits of the City of Chicago; but shall not be used within such limits. Wood frame construction shall not be used in any corridor type multiple dwelling. No multiple dwelling of wood frame construction shall contain more than four (4) dwellings, or exceed twenty-five (25) feet in height.

ARTICLE 804

Size and Location of Rooms

804.01 Size of Habitable Rooms: Every habitable room in a multiple dwelling shall have a minimum floor area of seventy (70) square feet and at least seventy (75) per cent of the floor area shall have a height of not less than eight (8) feet measured from finished floor line to finished ceiling line and must contain not less than five hundred sixty (560) cubic feet. Every room used as a sleeping room for more than one (1) person shall contain not less than four hundred (400) cubic feet for each adult person.

804.02 Habitable Rooms in Basements: Habitable rooms may be located in a basement of a multiple dwelling; provided however, that the floor of such basement is not more than two (2) feet below inside sidewalk grade. No habitable rooms shall be located in any basement having a lower floor level.

804.03 Floor Areas and Wall Openings in Alcove or Alcove Rooms: For the determination of ventilation and lighting requirements, the floor area of an alcove shall be considered as part of the floor of the room to which it is connected, up to thirty (30) per cent of the floor area of the room to which it is connected, but it shall not be considered as a part of the seventy (70) square feet of floor area required by Section 804.01. Any alcove containing a floor area greater than thirty (30) per cent of the floor area of the room to which it is connected shall comply in all respects with the requirements for a separate room. The opening between an alcove and a habitable room shall be unobstructed and shall be equal in area to not less than twenty (20) per cent of the entire wall area of the alcove.

804.04 Kitchens:

(a) **Limitations Based on Area.** Every dwelling containing a cooking stove, kitchen sink, or other kitchen equipment shall contain at least one (1) room with a floor area of not less than one hundred twenty (120) square feet.

(b) **Dining-Kitchen.** In a dwelling containing at least one (1) other habitable room, a kitchen and dining room together may be considered as one (1) habitable room; provided however, that the combined vertical surfaces of any partitions, pilasters, columns, or other separation between kitchen and dining room spaces shall not exceed in square feet, the product of three (3) feet multiplied by the clear ceiling height of the room, all as measured on the kitchen side of the separation; and provided further, that the opening between the kitchen and dining spaces is not less than three (3) feet in clear width and extends to within six (6) inches of the finished ceiling on the kitchen side. The combined floor area of the kitchen and dining spaces and the combined natural ventilation of such spaces, shall meet the requirements for a habitable room. Any opening between such a dining kitchen and any other part of the dwelling may be either with or without a door.

(c) **Kitchen Alcove.** A kitchen alcove shall comply with all the requirements of Section 804.03 for an alcove room.

(d) **Closet Kitchen.** A closet kitchen shall be ventilated as provided in Section 807.02, Item 3, but shall not be subject to any other requirement of Article 807, nor shall it be considered as a room for any purpose of this ordinance. In all of its effects upon required light and ventilation, the floor space occupied by a closet kitchen shall be considered as a part of the floor area of the room in which said closet kitchen is located, but the floor space occupied by the closet kitchen shall not be considered as a part of the minimum floor area required for such room.

804.05 Sleeping Stalls In Rooms: Sleeping stalls shall not be constructed or used in any room in any building, now existing or hereafter erected and devoted in whole or in part to the purposes of a lodging or rooming house, unless such room has a ceiling height of not less than nine (9) feet and unless such room has two (2) or more windows which face directly upon a street, alley, yard, court, public waterway or public park and which have a combined glass area and ventilating area as required for habitable rooms by Section 807.01 of this ordinance; nor unless the semi-partitions forming such stalls are so constructed that there is a clear and unobstructed interval of at least thirty(30) inches between the top of each semi-partition and the ceiling of the room; nor unless each such stall shall open directly into an aisle or passageway, leading directly to a stairway, or stairway fire escape, the location of which is indicated by a red sign and at night by a red light also. Such sleeping stalls shall not be installed in any room in such numbers that there shall be less than five hundred (500) cubic feet of air per person. The semi-partitions forming such stalls hereafter constructed shall be of non-combustible material.

804.06 Porches and Exterior Stairways:

(a) **General.** Every porch or exterior stairway of a multiple dwelling more than three (3) stories and basement in height shall be constructed wholly of noncombustible material, except as provided in paragraph (b) of this section. Every such porch or exterior stairway, more than two (2) feet above finished ground level shall be provided with a guard, fence, balustrade, or wall on every open side, constructed as required by Chapter 21 of the General Provisions.

(b) **Wood Construction.** Roofs, floors, joists, risers, treads and stringers of wood may be used in porches and exterior stairways of multiple dwellings four (4) stories or less in height, if wholly supported on masonry, or reinforced concrete walls or piers, as follows:

Corner piers of brick masonry supporting a stairway or porch which projects beyond the outer wall of the building shall be not less than sixteen (16) inches square; or if angle-shaped, shall have outer faces measuring not less than twenty (20) inches with a thickness not less than twelve (12) inches.

Intermediate piers of brick masonry, shall measure not less than twelve (12) inches in least dimension and shall have a cross-sectional area of not less than two hundred (200) square inches. Every opening between piers shall be filled in with a guard wall of masonry, not less than nine (9) inches thick or of reinforced concrete or metal and not less less than three (3) feet, six (6) inches high, measured vertically from the platform or stair casing and supported on a steel or reinforced beam or beams.

ARTICLE 805

Multiple Use Buildings

805.01 Other Occupancies Permitted and Prohibited: Subject to the provisions of this Article, a multiple dwelling may be located in the same building with any other occupancy, except hazardous use occupancies prohibited in a multiple dwelling under Section 1105.03 and except as provided in Chapters 6 to 19, inclusive of this ordinance.

805.02 Separating Construction: Every part of a building used as a multiple dwelling in a building of multiple occupancy, shall be separated from every other occupancy in the same building by construction of the fire-resistive value required by Chapter 25.

ARTICLE 806

Means of Exit

806.01 Exit From Rooms: Every habitable room in a multiple dwelling shall have one (1) doorway which will lead to an exit from the dewlling without the necessity of passage through a bedroom or bathroom or toilet room.

806.02 Exit From Dwellings:

(a) **General.** Every dwelling above the first story shall be provided with an exit, or exits, as set forth in the following paragraph (b). No part of any such exit shall be through another dwelling.

(b) **Number of Exits.** Each dwelling unit containing more than four (4) habitable rooms shall have two (2) separate means of exit to a public corridor. Each dwelling unit shall have one (1) exit doorway so located that it may be reached from any habitable room in the dwelling unit without passing through more than one (1) other habitable room; provided however, that any dwelling unit with a single means of exit to a public corridor shall be in a building of ordinary, heavy timber semi-fireproof or fireproof construction.

(c) **Exit Connections.** Every required exit from a dwelling above the first story shall lead either directly to two (2) vertical means of exit or into a public corridor lobby or other space leading to two (2) vertical means of exit. Where two (2) means of exit are required, each exit from the dwelling shall be from different rooms, or other spaces, but both such exists may open into the same public corridor, lobby or other space, if such space is provided with two (2) separate vertical means of exit, meeting the requirements of Section 806.03. In paragraph (b) of this Section, where a single doorway is permitted as the only means of exit from a dwelling above the main exit level, such doorway shall open into a public corridor, lobby or other space, from which there are two (2) separate vertical means of exit meeting the requirements of Section 806.03.

806.03 Public Vertical Means of Exit:

(a) **Number.** In any multiple dwelling there shall be at least two (2) public vertical means of exit leading to the main exit level of the building, accesible from each dwelling.

(b) **Width.** The minimum required width of stairways in multiple dwelling shall be as required in Chapter 26—MEANS OF EXIT.

(c) **Location.** Means of exit required by paragraph (a) of this section, shall be so located that no exit door from any dwelling above the main exit level shall be distant more than thirty (30) feet, in a building of wood frame construction, or more than fifty (50) feet in a building of ordniary or heavy timber construction, except as hereinafter provided, or more than one hundred (100) feet in a building of semi-fireproof or fireproof construction from a doorway into the inclosure of such a means of exit or from a doorway or window opening to a fire escape or to an outside stairway. Such distance shall be measured as a horizontal radius from the center of said doorway, or window opening. In any multiple dwelling, of the corridor type, there shall be a vertical means of exit within twenty-five (25) feet of every end of a corridor, if the building is of ordinary or heavy timber construction and within fifty (50) feet of every end of a corridor if the building is of semi-fireproof construction. Access from any dwelling to the two (2) required vertical means of exit may be by way of a single corridor and it shall be permissible to locate such vertical means of exit in such a manner that it is necessary to pass by one of them in order to reach the other, if both are inclosed with two (2) hour fire-resistive inclosure. It shall not, however, be permissible to pass through any portion of the inclosure of one (1) vertical means of exit to reach another vertical means of exit.

806.04 Exits From Buildings: Every required vertical means of exit shall give egress directly to the outer air at grade, or at the main exit level of the building, or it shall terminate in a foyer, lobby or other public room having the required number and width of exits to the outer air at the main exit level of the building but such public space at main exit level shall have at least two (2) exits. Vertical means of exit, consisting of inclosed stairways above the main exit story and continuing as open stairways through that story, shall be considered as meeting the requirements of this section.

806.05 Shaft and Corridor Doors: Every opening into a stairway or other vertical shaft in a multiple dwelling more than four (4) stories and basement in height, and every opening into a fireshield or two-way stairway of any height shall be provided with a forty-five (45) minute fire-resistive door. Every such opening in any multiple dwelling of four (4) stories or less in height, except a fire shield or two-way stairway, and except a stairway in a two (2) story building containing only two (2) dwellings, shall be provided with a door not less fire-resistive than two and one-fourth (2¼) inch wood slab door. In any corridor type multiple dwelling of ordinary or heavy timber construction, more than one (1) story and basement in height, every doorway from a dwelling into a corridor shall be equipped with a door not less fire-resistive than a one and three-fourths (1¾) inch wood slab door.

806.06 Non-required Stairways: A non-required stairway located entirely within the inclosing walls of one (1) dwelling is not required to be further inclosed. Such a stairway may be of any material not less fire-resistive than wood and may be of any dimensions and proportions. Every non-required stairway which is open to use by occupants of more than one (1) dwelling shall conform to all the requirements of a required stairway.

806.07 Width of Public Exits: A public doorway, corridor, stairway, ramp or other part of a public exit in a multiple dwelling shall be not less than three (3) feet in clear width at any point; provided however, that handrails in stairways shall be considered as obstructions; and provided also that the opening of a revolving door may be of lesser width as provided in Chapter 26. The aggregate width of doorways forming exits from the building shall be not less than the aggregate width of required stairways; provided however, that where the aggregate width of exits leading to such outside exit is greater than three (3) feet, six (6) inches, the exit doorway may be not less than eighty-seven and one-half (87½) per cent of the aggregate width required by this Section.

ARTICLE 807

Windows and Ventilation

807.01 Habitable Rooms: Every habitable room in a multiple dwelling, except as especially provided in Section 807.03 for certain public rooms, shall have one (1) or more windows facing on an opening upon a street, a public alley, a private alley on the same premises, an easement, a public park, a public waterway, or upon a yard, or court meeting the requirements of Section 802.01 or Section 802.02. The combined glass area in such windows in such room shall be not less than one-tenth (1/10) of the floor area of the room and in no case less than ten (10) square feet. The total width of glass in any required sash shall be not less than one (1) foot. The top of every window sash containing any required glass shall be at least six (6) feet, six (6) inches above the floor. Every window shall be so constructed as to permit its being opened for not less than forty-five (45) per cent of its full area. The combined area of ventilating openings in any habitable room shall be not less than five (5) per cent of the floor area of that room.

See illustration 2702.04.

807.02 Certain Non-Habitable Rooms and Spaces: Every kitchen or toilet room and every compartment containing any water closet or urinal fixture and every storage room or room or closet having a floor area greater than forty (40) square feet shall be ventilated as required by Section 807.01 with the following excepted items:

Item 1. The smallest allowable combined glass area shall be six (6) square feet, instead of ten (10) square feet.

Item 2. A system of mechanical ventilation meeting the requirements of Chapter 47—VENTILATION may be used in lieu of the natural ventilation and light otherwise required in every kitchen or toilet room and every compartment containing any water closet or urinal fixture.

Item 3. Every closet kitchen shall be provided with a flue, or duct, for exhaust ventilation with a free cross-sectional area of not less than thirty-two (32) square inches for each closet kitchen served thereby and such flue or duct shall lead to the atmosphere approximately in a vertical direction, or as required by Item 2 of this section.

807.03 Certain Public Rooms: Every public dining room, banquet room, ballroom, or similar public room, shall have windows, doors, transoms and skylights, or either of them, as required by Section 807.01 for a habitable room, or shall have a system of mechanical ventilation meeting the requirements of Section 807.01 and Chapter 47—VENTILATION; provided, however, that no window or other ventilation shall be required for a public lobby or public space having a volume of less than six thousand (6000) cubic feet.

807.04 Rooms in Which Persons Are Employed: Every room in which any person is employed shall have natural ventilation, as required by Section 807.01 for a habitable room, or shall have a system of mechanical ventilation, meeting the requirements of Section 807.01 and Chapter 47—VENTILATION. Nothing in this section shall be construed as permitting any violation of the Chicago Zoning Ordinance or to require the ventilation of any public room which is exempted from such requirement by Section 807.03.

807.05 Openings on Open Porches: A window opening on an open porch shall be considered as meeting the requirements of Sections 807.01 to 807.04 inclusive, if said porch faces upon a street, public alley, private alley on the same premises, casement, public park, public waterway or a yard, or court, meeting the requirements of Section 802.01 or Section 802.02. No point of such window shall be distant more than eight (8) feet horizontally from the nearest outside line of said porch. The inclusure of any porch shall be only as provided in Section 807.06.

807.06 Inclosure of Porches: Every inclosure shall have a glass area not less than twice the glass area in all required windows, doors and transom openings into said porch, from any room or rooms in the case of a street front porch and not less than three (3) times such area in the case of a porch not on a street front. Inclosing sash and frame may be of either wood or metal. The top of every inclosing sash shall be not less than six (6) inches higher than the top of any required window, door, or transom opening into said porch. At least forty (40) per cent of the required glass area of the inclosure shall be arranged to open.

807.07 Skylights: A skylight or skylights meeting the requirements of Chapter 27 may be used in lieu of a window, or windows in any habitable, or non-habitable room, in the story beneath a roof, but not in a room in any other location. The glass area and amount of opening of such a skylight shall be computed the same as for a window.

ARTICLE 808

Equipment

808.01 Lavatories, Baths and Water Closets: Every dwelling shall have access to at least one (1) lavatory, one (1) bath tub or shower, and one (1) water closet. Every family dwelling within an apartment house, apartment hotel or tenement house hereafter erected shall contain within the inclosing walls of such dwelling not less than one (1) lavatory or sink, one (1) bath tub or shower, and one (1) water closet. Public lavatories, bath tubs, or showers, and water closets shall be provided for the occupants of dwellings which do not have private fixtures of these respective kinds. In every story of a multiple dwelling there shall be at least one (1) public lavatory, one (1) public bath tub or shower, one (1) public water closet, for each ten (10) sleeping rooms, or fraction thereof, not included in dwellings which have private fixtures of the same respective kinds or there shall be at least one (1) each of said fixtures for each twenty (20) persons or fraction thereof, for whom sleeping accommodations are provided in rooms not included in dwellings which have private fixtures of the same respective kinds, if such calculation, based on the number of persons, gives a greater number of fixtures than does the calculation based on the number of rooms. No lavatory, bath tub, shower or water closet shall be considered as both a public and private fixture, nor as a private fixture for more than one (1) dwelling. There shall be at least one (1) lavatory in every public toilet room.

808.02 Communication Between Bedrooms and Water Closets: Every room used exclusively as a bedroom shall have access to at least one (1) water closet compartment without passing through another room used exclusively as a bedroom.

ARTICLE 809

Artificial Lighting and Exit Signs

809.01 Normal Illumination of Exits: Every corridor, stairway or ramp which serves as a required means of exit from more than one (1) dwelling shall be provided with a system of electric lighting which will produce an illumination of five-tenths (0.5) foot candle on the floor at every corridor angle, corridor intersection, stair or ramp landing, stair or ramp platform or doorway and an illumination of two-tenths (0.2) foot candle at every other point on the floor of such corridor, stairway or ramp.

809.02 Exit Signs: In a multiple dwelling of the corridor type, there shall be a standard exit sign over every doorway which leads from a public corridor to a vertical means of exit. In a multiple dwelling of any type, there shall be a standard exit sign over every doorway which serves as a required exit from a lobby, dining room or other assembly room which has a floor area or more than five hundred (500) square feet.

CHAPTER 9

Institutional Buildings

ARTICLE 901

General

901.01 Institutional Building Defined: An institutional building, for the purpose of this ordinance, is hereby defined as a building, or part of a building, designed, intended or used for a place in which sick or injured are given medical and surgical treatment and care, including a sanitarium, medical unit, hospital, infirmary or other similar unit for the care and treatment of men and women or children, a home or asylum for bedridden and decrepit persons, the blind, the aged or for children, or the insane, or a nursery or day nursery for infants and in which there are sleeping accommodations for more than ten (10) individuals.

901.02 Requirements for Institutional Buildings: Every building, or part of a building, designed, erected, altered or converted for use as an institutional building, shall comply with the General Provisions of this ordinance and in addition shall comply with the special provisions set forth in this chapter.

901.03 Capacity of an Institutional Building:

(a) **General.** The capacity of an institutional building shall be based on the sum of the maximum capacity of all rooms used as sleeping rooms by patients or inmates and by personnel, employees and others as provided in Section 904.01.

(b) **Assembly Rooms.** Any room with a floor area of more than six hundred (600) square feet, and designed or used as a place of assembly shall be classed as an assembly room and shall have a rated capacity as computed for assembly rooms.

ARTICLE 902

Lot Occupancy

902.01 Frontage Required: Every institutional building shall be located on a lot or tract of land having not less than one (1) frontage on a street.

902.02 Court Requirements: Any court required for purposes of light and ventilation shall conform to the minimum requirements for courts and multiple dwellings as given in Section 802.02 except as otherwise provided in this chapter.

ARTICLE 903
Construction

903.01 Fireproof Construction: Fireproof construction shall be used in any institutional building more than seven (7) stories and basement, or eighty (80) feet in height.

903.02 Semi-Fireproof Construction: Semi-fireproof, or a superior type of construction, shall be used in any institutional building more than one (1) story and basement in height and not more than seven (7) stores or eighty (80) feet in height.

903.03 Ordinary or Heavy Timber Construction: Ordinary or heavy timber construction, or a superior type of construction, shall be used for every institutional building not more than one (1) story above grade in height; provided, however, that the level of the floor shall be at or above grade and located on a fill of non-combustible material. If there is a basement or other open space beneath the first floor, then the floor construction above such space including also the walls, partitions and stairways, or other construction of the basement and inclosing walls and partitions inclosing the stairways shall be of three (3) hour fire-resistive construction.

903.04 Assembly Rooms and Other Rooms: Any room used for assembly, social, educational, devotional or residential purposes located within an institutional building shall be of fire-resistive construction equal to the requirements for the construction of such building.

903.05 Porches and Exterior Stairways: Every porch or exterior stairway more than one (1) story and basement in height shall be constructed of non-combustible material. Every exterior stairway above the main exit level of an institutional building shall be provided with a roof, except the highest flight which shall then have ice-proof treads.

ARTICLE 904
Size and Location of Rooms

904.01 Sleeping Rooms:

(a) **Rooms to be Considered as Sleeping Rooms.** Any room or space designed, intended or used for a sleeping room, ward or dormitory, either for patients and inmates, personnel or employees, or other occupants, shall be considered as sleeping rooms.

(b) **Spaces Not to be Considered as Sleeping Rooms.** Corridors, halls, operating rooms, laboratories, treatment rooms, utility rooms, floor pantries, serving rooms, tea kitchens, linen rooms, janitors' closets, medicine rooms, chart rooms, toilet rooms, bathrooms, dressing rooms, morgues, autopsy rooms or any other kind of rooms used for any service, including medical, surgical, therapeutic and nursing, but not used as sleeping quarters, shall not be factors in the determination of sleeping room capacity.

(c) **Dimensions and Volumes Required.** Every room used or designed to be used for a sleeping room, ward or dormitory, for patients and inmates, personnel, employees or others, shall have a floor area of at least one hundred (100) square feet, and a ceiling height of at least nine (9) feet six (6) inches; provided, however, that a nurses' home, which is built either above or below any space in a building used as an institutional building shall conform to the requirements of this paragraph. The number of individuals for whom sleeping accommodations may be provided in any room, ward or dormitory, shall not exceed the number for whom both floor space and room volume will be provided for in accordance with the following table:

Table 904-01 (c)

	Minimum Floor Area per Person Sq. Ft.	Minimum Volume of Room per Person Cu. Ft.
Adults	80	800
Children, 2 to 14 years of age	50	500
Children under 2 years of age	20	200

904.02 Flammable Films: All flammable photographic X-ray and other films shall be stored in a standard fireproof vault or fireproof cabinet constructed as required by Section 1103.25.

ARTICLE 905
Multiple Use Buildings

905.01 Other Occupancies Permitted and Prohibited: Subject to the provisions of this article, an institutional building may be located in the same building with other occupanies, as provided by Chapter 25. A hazardous use unit, a Class 1 garage, a school or theatre are prohibited in an institutional building, except as otherwise provided in Chapters 7 to 18 inclusive of this ordinance.

905.02 Separating Construction: Every required separation between an institutional building and a nurses' home, or nurses' training school, or a medical school, shall have the fire-resistive value required by Chapter 25; provided however, that any space used for a nurses' home, or nurses' training school, or a medical school, included within the walls or under the same roof as an institutional building shall conform to the requirements of this chapter for construction, size and location of rooms as well as exits.

905.03 Communicating Openings: Every doorway in a wall separating an institutional building from a nurses' home or a medical school, shall be a forty-five (45) minute fire-resistive door.

ARTICLE 906
Means of Exit

906.01 Exit Width and Height:

(a) **Width Based on Capacity.** The width of all doorways, passages, stairways and ramps in an institutional building, and the number and width of stairways from any story containing sleeping rooms shall be in accordance with paragraph (b) of this section and paragraphs (a) and (b) of Section 906.06. The width of doorways, passages, stairways and ramps from any story containing assembly rooms, lecture rooms, outpatients' waiting rooms or similar rooms for the congregating of the public or employees in an institutional building shall be at a rate of not less than two (2) feet six (6) inches per one hundred (100) persons capacity of such rooms, which width shall be maintained throughout the building from said rooms to the outside exit doorways; provided however, that if the stairways provided for sleeping rooms have a total combined width sufficient to meet the requirements of this section, additional width of stairs shall not be required.

(b) **Minimum Width and Height.** Every required passage, corridor, stairway, ramp or other part of a required exit in an institutional building shall be at least three (3) feet eight (8) inches in clear width, unobstructed by any radiator, lighting fixture or other object. Such clear and unobstructed width shall be maintained from the floor to a height of at least seven (7) feet above the floor. Any corridor leading from patients' rooms or wards to other means of exit shall be not less than six (6) feet six (6) inches wide.

906.02 Sleeping Room Doorway Width Every doorway to any patients' or inmates' sleeping room, ward or dormitory shall be at least three (3) feet eight (8) inches wide between jambs.

906.03 Construction of Doors: Every door between a corridor and a store room, a room containing X-ray apparatus, anaesthetizing room, operating room, delivery room, record file room or other similar room, shall, if of combustible material, be not less than one and three-fourths (1¾) inches thick at any point either in a single thickness or in laminated layers.

906.04 Transoms: No transom shall be installed except over doorways to the outer air, vestibules or lobbies, or over doorways connecting corridors. Such transoms shall be at least as fire-resistive as the required doors above which they occur.

906.05 Teaching Amphitheatres: All the means of exit from a teaching amphitheatre shall have a width computed at the rate of one (1) foot eight (8) inches per one hundred (100) persons capacity; provided however, that no aisle or stair shall be less than two (2) feet six (6) inches wide at any point. There shall be not more than ten (10) seats in a row between aisles and not more than five (5) seats in a row between nay aisle and a wall.

906.06 Required Stairways:

(a) **Number.** In every institutional building there shall be at least two (2) stairways between the main exit floor and every habitable floor.

(b) **Location of Stairways.** Stairways shall be so located that no point on any floor above the first story shall be more than seventy-five (75) feet from the center line of the entrance doorway to such stairways as measured on a straight line. Every stairway shall lead, either directly or by way of a public corridor or other public room to an outside exit doorway. The horizontal distance in a straight line between the center line of the entrance doorways to basement exit stairways, shall be not less than one-half (½) the greatest horizontal dimensions of the entire basement or sub-basement measured parallel to an inclosing wall thereof. Basement stairways shall be so located that no point on the basement floor shall be more than one hundred (100) feet from a stairway or from an outside exit doorway.

(c) **Rise and Tread.** Every required stairway, except for a teaching amphitheatre, shall have rises of not more than seven and one-half (7½) inches and treads of not less than ten (10) inches.

(d) **Types of Stairways.** Every required stairway in an institutional building shall conform to the requirements of Chapter 26 —MEANS OF EXIT.

(e) **Construction of Ramps.** Every ramp serving as a required means of exit shall have a slope not greater than one (1) in ten (10).

906.07 Open Stairways: An open stairway may be built from the first or entrance floor of an institutional building to a height of not more than thirty (30) feet above the average level of the sidewalk or ground adjoining the front of the building. Any such stairway shall be wholly in addition to the stairways required in Section 906.06. There shall be a floor and wall of two (2) hour fire-resistive construction between any stairway permitted by this section and every portion of the building containing any ward, dormitory or patients' or inmates' sleeping room. Every doorway in such partition shall be equipped with a forty-five (45) minute fire-resistive door.

906.08 Outside Exit Doorways and Doors:

(a) **Number and Location.** There shall be not less than two (2) outside exit doorways in any institutional building. Outside exit doorways shall be so located that no point on the main exit floor shall be more than ninety (90) feet from the center of such a doorway. Every required outside exit doorway shall open directly to a street or alley, or to a yard or court having direct connection to a street or alley, unobstructed by any gate or door, or to a fireproof roof of a fireproof building or uncovered terrace not less than ten (10) feet wide, nor more than fifteen (15) feet above the average level of the adjoining ground, and having a stairway or ramp not less than three (3) feet six (6) inches wide leading directly to a street or alley or to a yard or court having a direct connection to a street or alley, unobstructed by any gate or door.

(b) **Construction of Outside Exit Doors.** Every outside exit door shall open outward, and shall be without any lock which requires the use of a key to open from the inside, except that where forcible detention is necessary, other locks may be used. Revolving doors may be used in institutional buildings, but if used, they must be wholly in addition to the doors required by this ordinance. No single door nor leaf of a double door of any exit shall be more than four (4) feet wide.

ARTICLE 907

Windows and Ventilation

907.01 Window Facing: Every required window in an institutional building shall face upon a street, public alley, yard or court not less than ten (10) feet in width.

907.02 Windows: Every habitable room, including sleeping rooms, in an institutional building, shall have a window or windows with a glass area of not less than ten (10) per cent of the floor area of the room with closures of the area required by Chapter 47—VENTILATION. All other rooms in an institutional building shall have either windows or a mechanical ventilating system as required by Chapter 47—VENTILATION.

907.03 Skylights: Skylights may be used in lieu of windows as provided in Chapter 27—for any room in an institutional building except sleeping rooms.

ARTICLE 908

Equipment

908.01 Toilets, Lavatories and Drinking Fountains:

(a) **General.** The minimum number of plumbing fixtures on each floor of an institutional building, for the general use of patients or inmates having sleeping accommodations on such floor, except for deductions as provided in paragraphs (b) and (c) of this section shall be as follows:

For each twenty (20) patients or inmates or fraction thereof—1 Lavatory.

For each twenty (20) male patients or inmates or fraction thereof — 1 Water Closet.

For each twenty (20) female patients or inmates or fraction thereof—1 Water Closet.

For each forty (40) patients or inmates or fraction thereof—1 Bathtub, shower bath or spray bath.

For each fifty (50) patients or inmates or fraction thereof—1 Drinking Fountain.

(b) **Deductions for Fixtures in Rooms.** If plumbing fixtures of any required kind are provided for the exclusive use of patients or inmates occupying any sleeping room or ward, the number of patients or inmates so provided for need not be considered in computing the number of fixtures of the same kind which are required for general use; provided however, that the number of patients or inmates having such exclusive use of any such fixture shall not exceed the number permitted in paragraph (a) of this section for a fixture of the same kind.

(c) **Deductions for Nurslings.** Plumbing fixtures, as provided in paragraph (a) of this section are not required for nurslings.

908.02 Cabinets for Storage of Flammable Films: Cabinets for Storage of flammable photographic, X-ray and other films shall be as provided in Section 1103.25 of Chapter 11 —HAZARDOUS USE UNITS.

908.03 Anaesthetizing Rooms and Equipment.

(a) **General.** In every existing institutional building, within one (1) year from the date of the passage of this ordinance, and in every institutional building hereafter designed, erected, altered or converted, every room used for the storage or application of anaesthetics consisting of flammable or explosive gases or mixtures, including cyclepropane, ether, ethyl chloride, ehtylene, propylene or any flammable liquids, shall have safeguards for installation and operation of such rooms and equipment as required by this section.

(b) **Cylinders.** Any cylinders containing anaesthetizing gases or liquids shall be plainly marked with the name of the substance which they contain, and shall comply with the requirements of the Interstate Commerce Commission for such containers.

Such cylinders or containers shall be stored in a room having either natural or mechanical ventilation and shall not be stored in any operating room. Approved regulators or gas flow devices shall be provided for any such substances except low pressure oxygen containers. No such regulators or gas flow devices shall permit the intermixing of gases by any error of manipulation.

(c) **Electric Connections and Grounding.** All anaesthetizing equipment using combustible anaesthetics shall be grounded to the water supply system to reduce the possibility of static electric sparks. Metal door sills, trim, hardware and metal cases shall be grounded. Any motor used in such room shall be explosion proof as required by Section 1108.03 for Hazardous Use Units. Switches controlling any such motor or electrical apparatus or lighting circuits and receptacles in a room where flammable vapors are present, shall not be permitted within the room unless such equipment is protected and approved for use with explosive gases. All the electric lights in such a room shall be inclosed by vapor proof globes. Telephones, telephone ringing apparatus, and electric bells or annunciators shall not be permitted within such a room.

(d) **Warning Signs.** In every room where flammable or explosive gases or materials are stored, signs shall be posted in a permanent location having clear and legible letters as follows: "NO SMOKING"—"NO OPEN FLAMES"—"NO LIVE CAUTERY".

ARTICLE 909

Fire Protection

909.01 Standpipes, Hose and Connections: Every institutional building more than three (3) stories and basement in height shall be provided with a standpipe or standpipes meeting the requirements of the Fire Prevention Ordinance.

909.02 Automatic Sprinklers Required in Certain Locations: A standard system of automatic sprinklers, meeting the requirements of the Fire Prevention Ordinance, shall be provided in every room of an institutional building, used as a paint shop, carpenter or other work shop, a standard fireproof vault for films or waste paper storage room. Every cabinet used for the storage of flammable photographic, X-ray and other films shall be provided with at least one (1) automatic sprinkler head connected with a water supply having a pressure of ten (10) pounds per square inch.

ARTICLE 910

Lighting

910.01 Normal Illumination of Exits: Every means of exit in an institutional building shall have a system of electric lighting as required by Chapter 30 for the normal illumination of exits.

910.02 Additional Illumination: Every institutional building two (2) stories or more in height shall have additional illumination in all exit connections as required by Chapter 30 for additional illumination of exits.

910.03 Lighting Systems in Multiple Use Buildings: The entire lighting system or systems of an institutional building shall be independent of the lighting system of any part of the building used for other than institutional building purposes and shall be on a separate meter and separate service connection from any such other lighting system.

910.04 Exit Signs: In every institutional building, two (2) stories or more in height, standard exit signs, standard directional signs and standard fire escape signs shall be provided to mark the ways of egress as required by Chapter 30 for exit signs.

910.05 Duplicate Lighting Systems in Operating Rooms: Every operating room or delivery room shall be provided with an emergency lighting system in addition to the regular lighting system. Such emergency lighting system shall be supplied by a source of electric energy entirely independent of the normal electric illumination source of the building or shall consist of thirty-two (32) volt electric lights supplied with current from a storage battery which shall be kept continuously and automatically charged from a motor generator, and shall have a twenty-five (25) watt lamp continuously on the control board and also a warning bell to indicate reduction in battery strength below seventy-five (75) per cent. The wiring of the emergency system shall be separate from all other wiring and shall be run in independent conduits.

CHAPTER 10
Business Units

ARTICLE 1001

General

1001.01 Business Units Defined:

(a) **General.** A business unit shall, for the purposes of this ordinance, be defined as any building or part of a building, designed, intended or used for one (1) of the business purposes described under this section.

(b) **Office Unit.** A building, or part of a building, designed, intended or used for office occupancy shall be defined as an office unit, and such units shall include, among others, office buildings or buildings in which any story above the second story is occupied for the purposes of both office and sales use, administrative buildings and laboratories not otherwise provided for.

(c) **Financial Unit.** A building, or part of a building, designed, intended or used for the purposes of banks, trust companies, board of trade rooms, stock exchange rooms and trading rooms having grain pits or trading posts.

(d) **Sales Unit.** A building, or part of a building, designed, intended or used for sales purposes shall be defined as a sales unit and such units shall include, among others, retail stores, wholesale stores, market buildings, merchandise shops and salesrooms not otherwise provided for and including:

Item 1. Specialty Store. A building, or part of a building, designed, intended or used for wholesale or retail dealing in one (1) commodity or in one (1) line of allied merchandise.

Item 2. General Store. A building, or part of a building, designed, intended or used for retail sales or display of merchandise of a diversified character where the net floor area is more than two thousand (2,000) square feet.

(e) **Storage Unit.** A building, or part of a building, designed, intended or used for storage purposes shall be defined as a storage unit and such units shall include the storage of goods, wares, other articles of merchandise or commerce, freight houses, and all other storage uses not otherwise classified under this ordinance.

(f) **Manufacturing Unit.** A building, or part of a building, designed, intended or used for manufacturing purposes shall be defined as a manufacturing unit and such units shall include among others, any operation or process incident to the producing, fabricating, assembling, preparing or adapting for use, repair, servicing or refinishing of any goods, wares or other articles; mills, factories, greenhouses and workshops; as well as buildings used for the production of steam, gas or electricity; or used as telephone exchanges.

ARTICLE 1002

Lot Occupancy and Court Requirements

1002.01 Court Requirements:

(a) **General.** Every court which is necessary to obtain windows or other openings for required natural lighting and ventilation, shall be open to the sky, and shall meet the requirements of this section.

(b) **Minimum Dimensions and Areas of Inner Courts.** In an inner court of any business unit, the minimum horizontal dimension, measured at a right angle to any wall line at any point, and the minimum area easured at any height, shall conform with Table MDAC 1002.03 (b).

Table MDAC 1002.03 (b) Minimum Dimensions and Areas of Inner Courts

Height of Court Stories	Dimension in Feet	Area in Square Feet
1	6	100
2	6	120
3	8	160
4	8	160
5	12	260
6	16	400
7	20	625
8 or more	24	840

(c) **Minimum Dimensions and Areas of Outer Courts and Lot Line Courts.** In a lot line court of any business unit, the minimum horizontal dimension, measured at a right angle to any wall line, street line, alley line or yard line, shall be equal to one-half (½) of the minimum horizontal dimension required for an inner court under Table MDAC 1002.03 (b) and in a lot line court of any business unit, the minimum area measured at any height shall be equal to one-half (½) the minimum area required for an inner court under Table MDAC 1002.03 (b); provided however, that the minimum horizontal dimension and the minimum area required for a lot line court shall obtain irrespective of the presence of or dimensions of courts on other premises bound by the same lot line.

ARTICLE 1003

Construction

1003.01 Fireproof Construction:

(a) **General.** Fireproof construction shall be used for any business unit exceeding eighty (80) feet in height, except as otherwise provided in Section 1003.03 for heavy timber construction.

(b) **Omission of Fireproofing.** Non-combustible structural members supporting a roof only, and located more than twenty (20) feet above the floor directly beneath, need not be fireproofed. The roof slabs, plates or arches shall have a fire-resistive value of one (1) hour.

1003.02 Semi-Fireproof Construction: Semi-fireproof construction may be used for any business unit not exceeding eighty (80) feet in height.

1003.03 Heavy Timber Construction: Heavy timber construction, or a superior type of construction, shall be used for every business unit more than fifty (50) feet in height and not exceeding one hundred (100) feet in height, excepting sales units, which shall not exceed eighty (80) feet in height if built of heavy timber construction; and except as otherwise provided by Section 1003.04. If such unit is equipped throughout with a standard system of automatic sprinklers, the roof sheathing may be cne and five-eighths (1⅝) inches thick in one (1) layer.

1003.04 Ordinary Construction:

(a) **General.** Ordinary construction, or a superior type of construction, shall be used for any business unit fifty (50) feet or less in height; provided however, that in every business unit more than two stories high, if of ordinary construction, ceilings and partitions shall be covered with metal lath and plaster.

1003.05 Special Construction Types for Certain Uses:

(a) **General.** A type of construction in which all exterior walls are composed of a metal frame and sheet metal inclosure, in which all parts that carry loads or resist strains are constructed of metal and in accordance with the structural provisions of this ordinance may be employed in the construction of buildings for the specific uses and in the manner specifically permitted by the following paragraphs of this section.

(b) **Storage and Manufacturing Units.** Any storage or manufacturing unit, not exceeding one (1) story in height, no part of which shall extend nearer than fifty (50) feet to any lot line, alley line or street line, may be constructed as permitted by paragraph (a) of this section. If the exterior walls are constructed of non-combustible studs, metal lath and cement plaster, not less than one and one-half (1½) inches thick, such a building one (1) story high, may be located not less than twenty (20) feet from any such lot line, alley line or street line. If the exterior walls are constructed as provided by this paragraph, such a building may be constructed two (2) stories in height; provided however, that no part of such a building shall extend nearer than fifty (50) feet to any lot line, alley line or street line.

1003.06 Wood Frame Construction: Wood frame construction may be used for any business unit twenty-five (25) feet or less in height and 1,250 square feet or less in area which is also outside the fire limits and provisional fire limits established by other ordinances, except as permitted within the provisional fire limits by other ordinances, and except temporary buildings as otherwise permitted by this ordinance.

1003.07 Mezzanines:

(a) **General.** Every mezzanine in a business unit shall be built to conform to the construction required for the unit, except as provided by this section.

(b) **Fireproof or Semi-fireproof Construction.** In any business unit of fireproof or semi-fireproof construction, except storage units, any mezzanine may be built of non-combustible materials without fireproofing; provided however, that any such mezzanine shall not exceed five hundred (500) square feet, nor shall the aggregate area of mezzanines exceed ten (10) per cent of the floor area or fire area in the first story, nor five (5) per cent of such area in any other story.

(c) **Storage Units.** In any storage unit of fireproof or semi-fireproof construction, any mezzanine may be built of non-combustible materials if the area of such mezzanine does not exceed ten (10) per cent of the floor area or fire area in the first story, or five (5) per cent of the floor area in any other story.

1003.09 Electricity, Gas or Steam Manufacturing Units: A manufacturing unit, used exclusively for the production and distribution of electricity, gas or steam, shall be exempt from all limitations of fire areas, height or mezzanine areas provided in this ordinance; provided however, that such a unit shall be constructed entirely of non-combustible materials with interior framing members either with or without fireproof covering and separated from all other units or other uses with standard fire separation; and provided further, that any such unit shall not be located either above or below any occupancy other than that used for the production or distribution of electricity, gas or steam. Any tank used for a gas reservoir shall be located as required by Chapter 11—HAZARDOUS USE UNITS.

1003.10 Basement Construction: Every ceiling of a basement in any business unit of ordinary construction shall be of one (1) hour fire-resistive construction, and every wall and partition in such a basement shall be of not less than one (1) hour fire-resistive construction; provided however, that the basement of every such sales unit shall be either plastered or sprinkled.

ARTICLE 1004

Size and Location of Rooms

1004.01 Fire Areas:

(a) **General.** Standard fire division walls built as required by the General Provisions of this ordinance shall be provided in business units as required by Article 2402. In addition to subdivision into areas as required by Article 2402 business units shall be subdivided by standard fire division walls running through the full depth of the building from the street frontage wall as required by this section.

(b) **Fireproof Construction.** Any building of fireproof construction which is used for storage or manufacturing purposes and not over two (2) stories in height may have undivided fire areas of any area, if occupied by not more than one person, firm or business enterprise.

(c) **Semi-Fireproof Construction.** Any one (1) story building of semi-fireproof construction which is used for storage or manufacturing purposes may have undivided fire areas of any area, if occupied by not more than one (1) person, firm or business enterprise.

(d) **Heavy Timber Construction.** Any one (1) story business unit of heavy timber construction which is occupied by more than one (1) person, firm or business enterprise, shall have one (1) such wall for each eighty (80) feet of street frontage or part thereof. Any one (1) story building of heavy timber construction used for storage or manufacturing purposes which is occupied by one (1) person, firm or business enterprise, may have undivided fire areas of any area.

(e) **Ordinary Construction.** Any one (1) story business unit of ordinary construction, which is occupied by more than one (1) person, firm or business enterprise, shall have one (1) such wall for each fifty (50) feet of street frontage or part thereof. Any one (1) story building of ordinary construction used for storage or manufacturing purposes which is occupied by only one (1) person, firm or business enterprise, may have undivided fire areas of any area; provided however, if any such structure contains a basement story, such basement story shall be subdivided by standard fire division walls into areas not exceeding nine thousand (9,000) square feet each.

(f) **Partitions.** Any one (1) story business unit of ordinary or heavy timber construction shall have, in addition to the walls required by paragraphs (d) and (e), partitions of one (1) hour fire-resistive construction from floor to ceiling between the different persons, firms or business enterprises in such units.

(g) **Floors Below Grade.** Every story below grade shall be subdivided as required by this section, provided however, that any basement areas may be increased above the required limits by any area under public sidewalks wholly within the curb lines and lot lines produced across sidewalks.

1004.02 Rooms in Basements: No part of any floor in a business unit lower than a basement A story shall be designed, constructed, converted or used for business purposes unless such unit is of fireproof construction. No part of any floor lower than a basement B story in any sales unit shall be used as a salesroom.

1004.03 Public Rooms: Any assembly room in a business unit shall meet the requirements of Chapter 15—PUBLIC ASSEMBLY UNITS, except a banking room, stock exchange, trading room having a grain pit or trading post therein, lobby, loggia or rotunda, which shall meet the requirements of this chapter.

1004.04 Number of Mezzanines:

(a) **General.** Any room in a business unit may have mezzanines not in excess of the following provisions:

(b) **Office and Financial Units.** Not more than one (1) tier permitted unless the building is of fireproof or semi-fireproof construction.

(c) **Sales Units.** Only one (1) mezzanine story in any story and the combined area of mezzanines in any one (1) story or fire area shall not exceed ten (10) per cent of the floor area or fire area in such story; provided however, that a mezzanine in the first story may be not more than twenty (20) per cent of the floor area or fire area.

(d) **Storage Units.** More than one (1) tier of mezzanines permitted.

(e) **Manufacturing Units.** More than one (1) tier of mezzanines permitted; provided however, that the combined area of mezzanines in any one (1) story or fire area shall not exceed twenty (20) per cent of the floor area or fire area in such story.

1004.05 Ceiling Heights: Every room or space in a business unit, except a storage unit, and except any space above or beneath a mezzanine, having an area of one thousand (1,000) square feet or less, shall have a height of not less than eight (8) feet; provided however, that beams or ducts may project not more than twelve (12) inches below such required height.

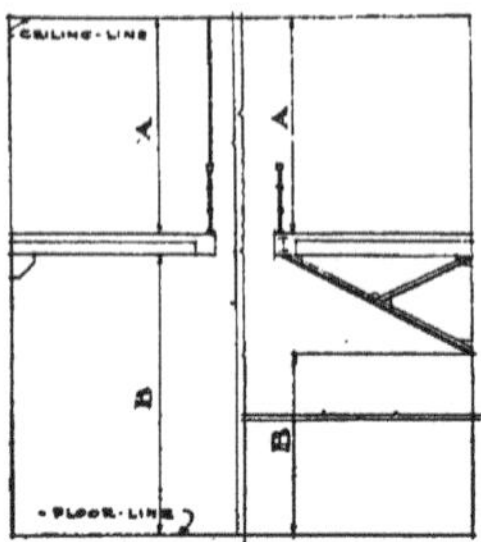

SECTION 1004.05.

A—height from floor of any gallery, mezzanine or intermediate floor to ceiling over same.
B—space between the bottom of such gallery, mezzanine or intermediate floor and the floor of the story in which such gallery, etc., is placed.

Explanation:
A—shall not be less than 7'0".
B—shall not be less than 7'0".

1004.06 Bakeries, Ice Cream Factories: Any bakery or food baking room, or any ice cream factory, in any business unit, except a bakery or ice cream making room which is part of a kitchen for a restaurant or multiple dwelling, shall be so located that the floor shall be at or above a level five (5) feet below the established sidewalk or alley grade adjacent to the building. Such floor, if below grade, shall be constructed of impervious non-combustible material, but may have a hardwood finish surface. All intersections of floors and walls shall be made rat proof. All walls and ceilings, except where heavy timber construction is permitted, shall be of smooth non-combustible material.

ARTICLE 1005
Multiple Use Buildings

1005.01 Other Occupancies Permitted and Prohibited: Subject to the provisions of this article, a business unit may be located in the same building with any other occupancy, except a Hazardous Use Unit and, except also, as otherwise provided in Chapters 6 to 19, inclusive.

1005.02 Separating Construction: Every business unit shall be separated from every other occupancy in the same building, as provided for in Chapter 25 of the General Provisions; provided however, that no separation is required between business units.

ARTICLE 1006
Means of Exit

1006.01 Exits—Number: Every room having an occupancy of more than seventy-five (75) persons, shall have at least two (2) doorways, remote from each other, leading to an exit or exits.

1006.02 Horizontal Exit Connections:

(a) **Where Required.** In each story which is occupied by more than one (1) occupant in any business unit, there shall be a corridor, lobby, hall, passage or similar inclosure, to permit public ingress to and egress from a stairway or other vertical means of exit. In any story which is used by only one (1) occupant, in any business unit, no such corridor, hall, lobby, passage or similar inclosure shall be required and it shall be permissible to enter any such space directly from a stairway or other vertical means of exit; provided however, that on the street exit floor the means of exit from any required inclosed stairway shall be through a passage, corridor, lobby or vestibule, or shall open direct to the outside.

(b) **Width.** The clear width of every corridor, lobby, hall or passageway leading to an exit shall be not less than three (3) feet, eight (8) inches for the first fifty (50) persons to be accommodated thereby, and six (6) inches additional for each additional fifty (50) persons or fraction thereof. Main public corridors shall be not less than five (5) feet in clear width; provided however, that branch public corridors not more than twenty-five (25) feet in length may be made not less than three (3) feet eight (8) inches in clear width. In no case shall a corridor, lobby, hall, passage or similar inclosure in the main exit story have a clear width which is less at any point than the aggregate required width of the vertical means of exit served thereby at such point; provided however, that a basement stairway used wholly by employees engaged in operation or maintenance of a building, and a vertical means of exit having a direct outside exit doorway may be entirely disregarded in computing the said width of such corridor, lobby, hall, passage or similar inclosure.

(c) **Distance to Horizontal Exit Connections.** In any one (1) story building, having combustible contents, which by the provisions of Section 1004.01, may have an undivided fire area of thirty thousand (30,000) square feet or less, the maximum distance of travel from any point to one (1) horizontal exit connection from such fire area shall not exceed one hundred (100) feet. In any one (1) story building having non-combustible contents which, by the provisions of Section 1004.01 may have an undivided fire area, the maximum distance of travel from any point to one (1) horizontal exit connection from such fire area shall not exceed one hundred fifty (150) feet.

1006.03 Horizontal Exits in Lieu of Other Exits: No horizontal exit shall be considered in lieu of other exits under this section unless the fire area on either side of such horizontal exit is sufficient to hold the joint occupancy of both fire areas, allowing not less than three and one-half (3½) square feet of clear floor space per person, and unless the vertical exits for each fire area meet the requirements of this ordinance; provided however, that for a floor area of four thousand (4,000) square feet or less having horizontal exits on both sides connecting with two (2) fire areas, no interior stairway, fire-shield stairway, nor stairway fire escape shall be required if horizontal exits are of same width as required stairways. When vestibules or balconies are used they shall conform to the requirements for vestibules or balconies for fire-shield stairways.

1006.04 Doorways, Width:

(a) **General.** In every business unit, the aggregate clear width of doorways serving as an exit from any room or floor area to a stairway or other means of exit shall be not less than thirty-six (36) inches for the first fifty (50) persons to be accommodated thereby, and six (6) inches additional for each additional fifty (50) persons or fraction thereof. The aggregate clear width of doorways at main exit level serving as an exit from any stairway, hall or passageway, shall be not less than the required width for such stairway, hall or passageway, except that for any stairway from a basement which is used wholly by employees engaged in operation or maintenance of building, such stairway shall not be considered in computing the required width of doorways. No single exit door when used as a unit in a group of doors shall have a clear width of less than two (2) feet, six (6) inches.

(b) **General Stores.** In every general store, the doorways at street exit floor shall equal the aggregate width of stairways, escalators and other vertical exits required as a means of egress from the basement, plus twenty-two (22) inches for each three thousand (3,000) square feet or fraction thereof of sales area on street level floor; provided however, that the aggregate width of doors thus computed shall be equivalent to the aggregate width of stairways, escalators or other vertical exits required as a means of egress from the upper floors, plus twenty-two (22) inches for each three thousand (3,000) square feet or fraction thereof of sales area on street level floor.

1006.05 Vertical Means of Exit:

(a) **General.** Vertical means of exit shall be so located that no point in any floor area shall be more distant from a vertical means of exit than one hundred fifty (150) feet in fireproof units and one hundred (100) feet in other than fireproof units, measured along the line of travel from an exit; provided however, that when any floor area is subdivided into smaller areas, such as rooms in office units, the distance from the corridor entrance door of any such room, along an unobstructed hall to an exit shall be not more than one hundred fifty (150) feet in fireproof units and not more than one hundred (100) feet in other than fireproof units.

(b) **Storage Units.** In storage units of either fireproof or other than fireproof construction, the location of exits required by this paragraph shall be not more than one hundred fifty (150) feet from any point in a floor area.

1006.06 Stairways:

(a) **Number Required.** Every habitable story of a business unit shall have stairways leading to the main exit level of the type required by Chapter 26—MEANS OF EXIT and the number of stairways shall meet the requirements of Section 1006.05 for location and of this section, as follows:

Item 1. 3000 Square Feet or Less. Units three thousand (3000) square feet or less in any net floor area above the main exit level, one (1) fire-shield stairway or one (1) interior stairway and one (1) exterior fire escape stairway; provided however, that any such building of other than fireproof or semi-fireproof construction shall have two (2) stairways and every basement shall have two (2) stairways.

Item 2. 3000 to 7000 Square Feet. Any business unit more than three thousand (3000) square feet and less than seven thousand (7000) square feet in any net floor area, shall have one (1) fire-shield stairway and one (1) interior stairway, or one (1) interior stairway and one (1) exterior fire-escape stairway, except as otherwise provided for sales unit, which shall have one (1) fire-shield stairway and one (1) interior stairway; provided however, that any such business unit three (3) stories or less in height may have two (2) interior stairways in lieu of the above requirements.

Item 3. 7000 Square Feet or More. Any business unit having seven thousand (7000) square feet or more in any net floor area, and not exceeding two hundred sixty-four (264) feet above grade, shall have one (1) fire-shield stairway and one (1) interior stairway, or two (2) interior stairways and one (1) exterior fire-escape stairway, except sales units, which shall have one (1) fire-shield stairway and one (1) interior stairway; provided however, that any such unit which is three (3) stories or less in height may have two (2) interior stairways in lieu of the above requirements.

Item 4. Over 264 Feet High. Units exceeding two hundred sixty-four (264) feet above grade, shall have stairways as determined by this paragraph for floor areas, but at least one (1) of such required stairways shall be a fire-shield stairway extending from the highest roof level to grade.

Item 5. Exterior Fire-Escape Stairway. No stairway shall extend more than two hundred sixty-four (264) feet above grade. Exterior fire-escape stairways shall not be used in lieu of any other required stairway for a sales unit; provided however, that for the purpose of reducing the width of required stairways in any business unit, it shall be permitted to substitute one (1) unit of thirty-six (36) inches of width of an exterior fire-escape stairway in lieu of each unit of twenty-two (22) inches of required stairway width, subject to other requirements of this paragraph for the minimum number and minimum width of stairways.

Item 6. Fire-Shield Stairway. In any business unit more than three (3) stories in height which is required by this article to have four (4) or more vertical means of exit, two (2) of such stairways shall be fire-shield stairways.

(b) **Width.** The unit of stair width shall be twenty-two (22) inches. Fraction of a unit shall not be counted. No stairway required by this article as an exit shall have an unobstructed width of less than three (3) feet, eight (8) inches throughout its length. The aggregate width of stairs in any story of the building shall be such that the stairways can accommodate the total number of persons ordinarily occupying or permitted to occupy the largest floor area served by such stairways above the flight or flights of stairways under consideration, on the basis of forty (40) persons for each unit of stair width; provided however, that the number of persons to be accommodated, as herein provided, may be sixty (60) persons for each such unit of such total number when a horizontal exit is provided in accordance with this article and eighty (80) persons for each unit of such total number when the building is equipped with automatic sprinklers and a horizontal exit is provided.

(c) **Types.** Every required stairway in a department store shall be non-combustible and every required stairway in other business units shall be as required by Chapter 26—MEANS OF EXIT.

1006.07 Height of Flight: Any flight of stairs in a sales unit shall have a vertical rise of not more than nine (9) feet without an intermediate platform or landing and not more than twelve (12) feet in any other business unit.

1006.08 Winding and Spiral Stairways: Any stairway leading to a space or roof, which is used only occasionally by employees engaged in the operation or maintenance of the building, or any stairway, which is for the sole use of employees of an occupant, may be a winding or spiral stairway and the width of treads and risers may be of any dimension.

1006.09 Open Stairways in Basements: Every sales unit having a basement used for sales purposes shall have inclosed stairways from such basement equal in number and width to the required number and width of inclosed stairways above the main exit level, and if a greater number or width of stairways is required from such basement, such additional stairways may be either open or inclosed.

1006.10 Exception for Certain Manufacturing or Distributing Units: Manufacturing units used exclusively for the manufacture or distribution of electricity, gas or steam, shall have means of exit provided for a capacity in persons equivalent to one (1) person for each four hundred (400) square feet of floor area. A minimum of two (2) stairways shall be provided for each thirty thousand (30,000) square feet of floor area on one (1) floor above or below the main exit level.

ARTICLE 1007

Windows and Ventilation

1007.01 Light: Every room in a business unit open to the public or employees shall be lighted by means of windows facing upon a street, public alley, yard, court or other open space, or by means of skylights in the ceiling and roof immediately above the story to be lighted, or by artificial illumination as otherwise required in this ordinance.

1007.02 Ventilation: Ventilation shall be required in all habitable spaces in business units, by means of openings in windows or skylights, with closures of the required area, or a mechanical ventilating system, as provided in Chapter 47—VENTILATION; provided however, that windows or ventilating systems shall not be required for rooms used for cold storage or refrigeration purposes.

1007.03 Windows and Skylights: All windows and skylights for the purposes of this article shall be as provided in Chapter 27 of the General Provisions of this ordinance. Any window for a food baking room, kitchen, or room used for the preparation of food, shall have its sill not less than three (3) feet above grade.

ARTICLE 1008

Equipment

1008.01 Elevators and Dumbwaiters: Every elevator or dumbwaiter in a business unit shall meet the requirements of Chapter 44 of this ordinance.

1008.02 Escalators: An escalator normally operating in the direction of exit may be considered as a means of exit for any business unit to the extent premitted by Chapter 26—MEANS OF EXIT; provided however, that such escalator shall be enclosed and shall be of the horizontal tread type, and without combustible material except where the required stairway construction may be combustible; and provided further that such escalator shall meet the requirements of Chapter 44 of this ordinance. Wherever escalators shall be installed as a means of ingress there shall be provided an equal number of escalators operating normally in the direction of egress. In no case shall the point of entry to such an escalator operating in the direction of egress be more than fifty (50) feet distant from the discharge point of the corresponding escalator providing a means of ingress.

1008.03 Toilet Equipment:

(a) **General.** Every business unit shall be provided with toilet fixtures in accordance with maximum capacity of the unit as follows:

(b) **Water Closets.** Not less than one (1) water closet for twenty-five (25) males in all units and not less than one (1) water closet for twenty-five (25) females in all units, such capacity referring to maximum number of employees; provided however, that for any business unit having not more than five (5) employees, one (1) water closet for both sexes.

(c) **Urinals.** Not less than one (1) urinal for every three (3) water closets for males.

(d) **Lavatories.** Not less than one (1) lavatory in every toilet room containing any water closet.

ARTICLE 1009

Fire Protection

1009.01 Automatic Sprinklers: A standard system of automatic sprinklers meeting the requirements of the Fire Prevention Ordinance shall be provided in every room used for the storage or baling of waste paper in every business unit; for every space used as a carpenter or paint shop, or store room for lumber or furniture in an office or financial unit; for all basement spaces in any department store and throughout all department stores two (2) stories or more in height.

1009.02 Standard Standpipe Systems: A standard inside standpipe system meeting the requirements of the Fire Prevention Ordinance shall be provided in every business unit which is more than eighty (80) feet in height; provided however, that such a standpipe system shall not be required in a sales unit having a standard system of automatic sprinklers.

ARTICLE 1010

Artificial Lighting and Exit Signs

1010.01 Normal Illumination of Exits: Every means of exit in a business unit shall have a system of electric lighting as required by Chapter 30 for the normal illumination of exits.

1010.02 Additional Illumination: Every business unit more than two (2) stories in height shall have additional illumination in property dividing lot line or to any building other than a dwelling on the same lot.

1202.02 Basement Garages: Every Class 1 garage located in a basement shall have frontages on at least two (2) open spaces, one (1) of which shall be a street and the other of which shall be a street, alley not less than ten (10) feet wide leading to a street, or a court or other open space not less than ten (10) feet wide leading to a street or public alley of not less width than ten (10) feet.

ARTICLE 1203

Construction

1203.01 Construction Type Requirements:

(a) **General.** Wherever the construction is required or permitted to be of a specified type, such construction shall meet the requirements of Chapter 31—CONSTRUCTION TYPE PROVISIONS.

(b) **Height of Garages—Determining.** For the purpose of determining the required construction type, the height of a garage shall be deemed to be the height above grade of the topmost floor used for garage purposes in any building. Where the required exit doorways are provided at only one (1) level of a two (2) level street, the story in which such doorways are located shall be considered the first story.

(c) **Basement Garages.** Every basement garage shall be of fireproof construction.

(d) **Class 1 Garages.** Every Class 1 Garage, one (1) or two (2) stories in height shall be of fireproof, semi-fireproof, heavy timber or ordinary construction; provided however, that fireproof construction shall be used for all floor construction other than mezzanines not used for garage purposes and for all columns other than columns supporting such mezzanines, or roofs. Every Class 1 Garage, three (3) stories in height, shall be of fireproof or semi-fireproof construction; provided however, that fireproof construction shall be used for all floors other than mezzanines not used for garage purposes and for all columns other than columns supporting such mezzanines, or roofs. Every Class 1 Garage more than three (3) stories in height shall be of fireproof construction.

(e) **Class 2 Garages.** Class 2 Garages, not more than four hundred (400) square feet in area may be of wood frame or more fire-resistive type of construction, except that the floor thereof shall be of non-combustible material. Class 2 Garages, more than four hundred (400) square feet in area shall be of ordinary construction or a more fire-resistive type of construction, except that the floor thereof shall be of non-combustible material.

(f) **Parking Lots and Structures.** A parking lot consisting of one level only for the accommodation of motor vehicles may be maintained without enclosing walls, but any building or part of a building or any structure or any part of a structure or any mechanism used for any purposes of a garage shall, if more than one story and basement high, be enclosed above the first story with walls of masonry constructed in accordance with the provisions of Chapter 38—MASONRY CONSTRUCTION. In any mechanism used for garage purposes, the lower ten (10) feet immediately above the grade or ground level shall be considered the first story thereof.

ARTICLE 1204

Size and Location of Rooms

1204.01 Fire Division Areas:

(a) **General.** For the purpose of computing fire division areas in garage buildings there shall be included, with the floor area of the main room or rooms, the area of all balconies. There shall also be included the area of all connecting roofed spaces which are to contain any fuel or oil servicing equipment. A fire division area so computed shall not include the area of any ramp, elevator or passage used solely for the ingress or egress of persons or vehicles, if such spaces are inclosed as required for vertical shafts. Mezzanines, as defined in this ordinance not used for garage purposes shall not be included in the computations of fire division areas.

(b) **Maximum Fire Division Areas.** No fire division area in a Class 1 Garage shall exceed the area allowed for such garage by Article 2402.

(c) **Staggered Floor Levels.** Where connecting levels are located within a vertical distance of six (6) feet from each other, such levels shall be considered as parts of the same fire area; provided however, that the combined area of such different levels, plus the area of all connecting ramps shall not exceed the limits allowed under Article 2402.

1204.02 Roof Garages: No garage shall be located on the roof of a building of other than fireproof construction. Any roof area of a building of fireproof construction may be used as a Class 1 Garage, provided:

1. That such roof shall be designed and constructed to support a live load equal to the sum of the live load prescribed for a roof and the live load prescribed for a garage floor.

2. That exit facilities, the same as required for a garage located in a story above the first story shall be provided from such roof.

3. That there shall be parapet walls of masonry or reinforced concrete at least four (4) feet high, meeting the requirements of the Structural Provisions and General Provisions of this ordinance completely surrounding such roof area.

4. That there shall be a continuous wheel guard of masonry or reinforced concrete at least twelve (12) inches high; at least twelve (12) inches thick, located at least three (3) feet from said parapet and extending completely around such roof area.

1204.03 Separation from Heating Plants: Every heating plant and its fuel storage, except for Class 2 garages attached to single dwellings, shall be separated from all garage space by standard fire division walls and standard fire separations. Communication, if any, between any garage space and a heating plant or fuel storage space shall be through a vestibule as provided by this chapter for communicating vestibules in multiple use buildings. There shall be no communication inside the building between any Class 1 garage space and any space containing a heating plant having a combustion chamber.

ARTICLE 1205

Multiple Use Buildings

1205.01 General: Subject to the provisions for multiple uses contained in other chapters of the Occupancy Provisions of this ordinance, a garage may be located in the same building with another occupancy.

1205.02 Construction: Every garage, located in the same building with any other occupancy, shall meet the requirements of Article 1203 of this chapter; provided however, that no garage of other than fireproof construction shall be located in any building containing a dwelling in which the floor of any habitable room is more than twenty-four (24) feet above grade.

1205.03 Separation From Other Occupancies:

(a) **Class 1 Garages.** Every Class 1 garage shall be separated from every space used for other than garage purposes by standard fire division walls and standard fire separations. Openings for vehicular passage, protected as required for openings in standard fire division walls shall be permitted in the first story only, to communicate with spaces in fireproof buildings devoted to related occupancies as defined in Section 1201.01, paragraph (b). Such openings shall not be considered a required means of exit.

(b) **Pedestrian Communication.** Except as otherwise provided in other chapters of the Occupancy Provisions of this ordinance, pedestrian communication between a garage and any other occupancy shall be permitted in the first story; provided however, that such communication shall be through a vestibule; completely inclosed in walls, floors and ceiling of a four (4) hour fire-resistive value, not less than five (5) feet in length of travel with a standard fire door in each opening thereto. Such doors shall be hinged to open in the direction of egress from the garage and may be considered as one (1) of the required means of exit. Such doors shall be operable from the garage side without the use of a key and shall be self-closing.

(c) **Class 2 Garages.** Every Class 2 garage shall be separated from other occupancy spaces by construction having not less than a three (3) hour fire-resistive value; except that a garage for not more than two (2) motor vehicles forming a part of or attached to any single dwelling shall have all separating construction of one hour fire-resistive construction.

ARTICLE 1206

Means of Exit

1206.01 Doorways—General: Every doorway through an exterior wall, stair inclosure or fire division wall shall have its sill located at least two (2) inches above the normal adjoining floor levels. The approach to such doorways shall be ramped at a rate not to exceed one-eighth (⅛) inch per foot.

1206.02 Vehicle Exit Doorways and Doors:

(a) **Number.** There shall be at least two (2) vehicle exit doorways from every garage having a fire division area of more than sixteen thousand (16,000) square feet in any one (1) story.

(b) **Location.** Every required vehicle exit doorway shall open with facilities for the exit of cars into a street, public alley or other open space not less than ten (10) feet wide leading to a street, or into a tunnel or passage of not less than four (4) hour fire-resistive construction, at least ten (10) feet in width and at least nine (9) feet in height leading to a street, public alley or other open space.

(c) **Types of Doors.** Every vehicle exit door, located in a required fire division wall or required fire separation, shall have a fire-resistive value at least equal to the value required for the protection of doorways in such locations by Chapter 33—ALL TYPES AND KINDS OF FIRE DOORS. Every such door may be operated manually or mechanically but shall be arranged to close automatically in case of fire when located in openings requiring standard fire doors.

1206.03 Pedestrian Exit Doorways and Doors:

(a) **Number.** In every garage there shall be at least one (1) pedestrian exit doorway in addition to all vehicle exit doorways. In a garage having a fire area of more than six thousand (6,000) square feet, there shall be at least two (2) pedestrian exit doorways in addition to all vehicle exit doorways; provided however, that in lieu of one (1) such pedestrian exit doorway there may be an opening through a vehicle exit door as hereinafter provided. There shall be at least one (1) pedestrian exit doorway from the first story to every garage; provided, however, that such pedestrian exit from a Class 2 Garage may be an opening through a vehicle exit door.

(b) **Location.** In a Class 1 garage where only one (1) pedestrian exit doorway is required, the distance on a straight line between the center of such pedestrian exit doorway and the center of at least one (1) vehicle exit doorway shall be not less than fifty (50) per cent of the greatest horizontal dimension of the floor. Where two (2) pedestrian exit doorways are required, they shall be separated by a distance at least equal to seventy-five (75) per cent of the greatest horizontal dimension of the floor. Every required pedestrian exit doorway shall open upon a street, alley or other open space not less than three (3) feet wide leading to a street or to an alley leading to a street. A doorway through a standard fire division wall, located as required by this paragraph shall be accepted as a required pedestrian exit doorway. No exterior doorway of a garage shall be located within six (6) feet of any exterior doorway of any other building on the same lot.

(c) **Size and Construction.** Every pedestrian exit doorway shall be at least two (2) feet, six (6) inches wide and seven (7) feet high and not more than three (3) feet, six (6) inches wide and eight (8) feet high. Every required pedestrian exit door shall swing outward in the direction of egress and shall be operable from the inside at all times without the use of a key.

(d) **Pedestrian Exits Through Vehicle Doors.** A pedestrian opening through a vehicle door shall be not less than two (2) feet wide and five (5) feet high, located not more than one (1) foot, six (6) inches above the floor and hinged to swing outward in the direction of egress.

1206.04 Stairways:

(a) **Number.** There shall be at least one (1) stairway or ramp for pedestrians only, connecting the first floor with every floor above or below the first floor. There shall be not less than two (2) such stairways or ramps from any floor exceeding three thousand (3,000) square feet in area; provided however, that a vehicle ramp or an opening through a fire division wall, located as required by Section 1206.03, paragraph (b) shall be accepted as one (1) of such stairways; and provided further, that from every garage located in a basement, except for garages of less than eight hundred (800) square feet in area and attached to a single dwelling, there shall be not less than two (2) stairways leading directly to the outside of the building at grade.

(b) **Location.** At least one (1) required stairway or pedestrian ramp shall lead directly to an outside exit doorway at grade into a street, alley or other open space not less than three (3) feet wide leading to a street or to an alley leading to a street. Entrance doorways to such required stairways or pedestrian ramps shall be located as provided in Section 1206.03, paragraph (b).

(c) **Width.** Every required stairway or pedestrian ramp shall be not less than three (3) feet in clear and unobstructed width.

(d) **Construction and Inclosure.** Every stairway and pedestrian ramp shall be constructed of non-combustible materials, except for handrails and shall be inclosed as required for vertical shafts.

1206.05 Vehicle Ramps and Elevators:

(a) **Number and Type.** Every garage more than two (2) stories in height and every basement garage more than two thousand (2,000) square feet in area and every second floor garage more than two thousand (2,000) square feet in area shall have not less than two (2) single ramps so arranged or marked as to allow only one (1) way traffic or one (1) double ramp for vehicles connecting each level with the first floor or grade level. An elevator or elevators may be provided in addition to vehicle ramps but such provision shall in no way modify the requirements of this section as to number, type and construction of ramps; provided however, that an elevator or elevators may be used in lieu of all required ramps. The number of such elevators is not regulated by this ordinance.

(b) **Width and Construction of Ramps.** Every vehicle ramp shall be of not less than three (3) hour fire-resistive construction. The slope of ramps connecting levels separated more than six (6) feet vertically, shall not exceed twelve and one-half (12½) per cent. The slope of ramps connecting levels separated by a vertical distance of six (6) feet or less shall be not more than sixteen and two-thirds (16⅔) per cent. Single ramps shall be not less than thirteen (13) feet in clear width. The clear width of ramps shall be measured one (1) foot above the floor thereof. A level run not less than twenty (20) feet long shall be provided at each end of every ramp. Double ramps shall be not less than twenty-four (24) feet in clear width.

ARTICLE 1207

Windows and Ventilation

1207.01 Windows: Every window affording required ventilation shall open upon a street, alley, court or other open space not less than three (3) feet wide: In Class 1 garages every window facing upon any open space less than sixteen (16) feet wide, shall be a Type A window and every skylight located within a distance of less than sixteen (16) feet from a property dividing lot line or from the opposite side of any open space shall be a Type A skylight. Windows and skylights, otherwise shall meet the requirements of Chapter 27 — WINDOWS AND VENTILATION and of the General Provisions of this ordinance.

See illustration 2702.04.

1207.02 Ventilation:

(a) **General.** Every room or space in every garage shall be provided with openings to the outside air to provide natural ventilation at all times, or shall be equipped with a mechanical system of ventilation meeting the requirements of Chapter 47 — VENTILATION. Such openings shall, in combined area, equal two (2) per cent of the floor area. Every Class 1 garage shall be ventilated as required by Chapter 47—VENTILATION. No outlet opening for any mechanical ventilating system in a garage shall be located within sixteen (16) feet from any property dividing lot line or within sixteen (16) feet from the opposite side of an alley or other open space.

(b) **Basement Garages.** Every basement garage shall have a system of mechanical ventilation, meeting the requirements of Chapter 47—VENTILATION. In addition to such mechanical ventilation, there shall be window, sidewalk, or vent shaft openings direct to the open air. Said window, sidewalk or vent shaft area shall be equal in area to not less than one (1) per cent of the largest floor area served thereby. The required vent shaft area shall be provided by not less than two (2) vent shafts located immediately adjacent to the stairways or ramps. Each such shaft shall be provided with a damper or dampers. Said dampers, when opened, shall provide the required area and shall not obstruct the required area of the shaft. Said dampers shall normally be closed, but shall be so arranged that they may be opened, manually, from the stairways or ramps.

ARTICLE 1208

Equipment

1208.01 Heating:

(a) **Permitted Heating Systems.** Except as hereinafter provided, garage heating shall be by means of radiation or convection from hot water, vapor or steam heating systems of which the boilers or devices containing combustion chambers and the fuel storage spaces shall be separated from every garage space as required by Section 1204.03 of this chapter.

(b) **Prohibited Heating Systems.** No stove, hot air furnace or similar device containing a combustion chamber shall be installed in any garage and no system of air circulation shall be permitted to penetrate any required fire-resistive separation or inclosure.

(c) **Gas or Electric Unit Heaters.** Electric unit heaters and gas fired unit heaters in which the flame is protected from draughts of air and contact with combustible materials, shall be permitted in Class 2 garages.

ARTICLE 1209

Fire Protection

1209.01 Automatic Sprinklers: Every Class 1 garage located in a basement, every Class 1 garage three (3) stories or more in height and every Class 1 garage having a fire division area in excess of the area permitted under Section 2402.02, shall be equipped throughout with a standard system of automatic sprinklers; *"provided, however, that this provision shall not apply where a garage consists only of an unheated basement the roof over which is used as an open-air garage."* Amended Oct. 4, 1938.

ARTICLE 1210

Artificial Lighting

1210.01 General: Artificial lighting shall be by means of incandescent electric lamps.

1210.02 Illumination of Exits and Exit Signs:

(a) **Normal Illumination.** All means of exit in every Class 1 garage shall be provided with normal illumination meeting the requirements of Chapter 30—ARTIFICIAL LIGHTING AND EXIT SIGNS, except that the minimum intensity of illumination shall be five-tenths (.5) foot-candle at the floor.

(b) **Additional Illumination.** Additional illumination shall be provided in every Class 1 garage more than one (1) story in height and in every garage located in a basement meeting the requirements of Chapter 30—ARTIFICIAL LIGHTING AND EXIT SIGNS.

(c) **Exit Signs.** In every Class 1 garage standard exit signs, standard directional signs and standard fire escape signs shall be provided as required under Chapter 30—ARTIFICIAL LIGHTING AND EXIT SIGNS.

CHAPTER 13

Theatres

ARTICLE 1301

General

1301.01 Theatre Defined: A theatre, for the purposes of this ordinance, is hereby defined as a building, or part of a building, designed, erected, altered, converted or used for the purposes of displaying motion pictures or presenting theatrical performances before an assemblage of persons, whether such assemblage be of a public, restricted or private nature, except such units as are otherwise classified under this ordinance. The displaying of motion pictures shall include the displaying of still pictures, sound pictures and motion pictures and the presenting of theatrical performances shall include the presenting of dramatic, operatic, comic, pantomime or vaudeville performances also the presenting of other amateur or professional theatricals, productions or spectacles.

1301.02 Other Terms Defined:

(a) **General.** Certain terms in this chapter are defined for the purposes thereof, as follows:

(b) **Auditorium.** See general definition of term under Chapter 5. That portion of a theatre in which are located the seats for the audience. The seats may be on one (1) or more levels. Any such level is herein referred to as a "seating level". The main floor is the lowest seating level of such an auditorium. Such portion of an auditorium located between the rearmost transverse aisle of the main floor and the rear of the auditorium shall be governed by the requirements for mezzanines or balconies as defined in this section.

(c) **Balcony.** See general definition of term under Chapter 5. A balcony in a theatre auditorium is a partial floor seating level superimposed above the main floor or a mezzanine or another balcony in the auditorium and having more than ten (10) rows of seats extending across the auditorium, exclusive of any row having less than six (6) seats and exclusive of any box at the sides. When there is more than one (1) balcony, the lowest balcony shall be designated as the "first balcony"; the next higher balcony as the "second balcony" and so on, for the purposes of this ordinance.

(d) **Bank of Seats.** A row of seats at a different level from the row immediately forward, and located between adjacent aisles or between an aisle and a wall, buttress, railing, guard or other abutment for a row of seats.

(e) **Capacity.** The number of persons for whom seats are provided in the auditorium.

(f) **Dressing Room.** A room used or intended to be used by a performer or performers for dressing or changing of clothing.

(g) **Employee.** Any person engaged in the performance of labor or service having to do with the management of a theatre, with the preparation of it for a picture display or theatrical performance, with the conduct or execution of any display, act, or performance therein, with the maintenance of it, or with any other function involved in its operation.

(h) **Entrance.** A means or route of ingress to a building or room, including a doorway passage, passageway, corridor, ramp or stairway. A means of exit shall be considered a means of entrance unless prohibited from being used for such purpose under this ordinance.

(i) **Foyer.** A room adjoining the auditorium and serving as the principal entrance to any seating level thereof.

(j) **Group of Seats.** Two (2) or more seats in one (1) or more rows in any part of the auditorium, arranged together and separated from adjacent seats.

(k) **Lobby.** A room, other than a foyer, adjoining any public room or public space in a theatre and serving as an entrance to such public room or public space.

(l) **Main Exit Floor and Main Exit Level.** A floor, or part of a floor, at or near grade or ground level, on which any exit to grade or ground level is located, other than a ground floor.

(m) **Mezzanine.** A partial floor seating level superimposed above the main floor or another mezzanine in the auditorium and having not more than ten (10) rows of seats extending across the auditorium, exclusive of any row having less than six (6) seats, and exclusive of any box at the sides. When there is more than one (1) mezzanine, the lowest mezzanine shall be designated as the "first mezzanine", the next higher mezzanine as the "second mezzanine."

(n) **Principal Entrance.** That public entrance, from a way of approach for the public to which is apportioned the greater part of so much of the public normal exit width as is, under paragraph (b) of Section 1305.04 required to lead to an entrance doorway or doorways from such way of approach. Where such part of so much of such exit width is so apportioned that there is an equal width leading to two (2) or more public entrances from such a way of approach the owner may select and designate one (1) such entrance as the principal entrance.

(o) **Projection Block.** That portion of a theatre containing a projection room alone or in combination with other rooms appurtenant to the operation thereof.

(p) **Projection Room.** A room in which there is located any projector for motion pictures, any projector for still pictures or any other light projection device, in which flammable film is used.

(q) **Property Room.** A room for the storage of any adjunct of a theatrical act or performance, except scenery, commonly known and described as stage properties.

(r) **Secondary Entrance.** Any entrance for the public, other than the principal entrance.

(s) **Stadium Type Auditorium.** An auditorium in which rows of seats raised above other seats but not superimposed above any part of the main floor are located between the rearmost transverse aisle of the main floor and the rear of the auditorium.

(t) **Stage Block.** That portion of a theatre containing a stage alone or in combination with dressing rooms, storage and property rooms, work shops and other rooms appurtenant to the operation thereof. No space open to the public shall be included in any stage block.

(u) **Stage Workshop.** Any shop or room in which carpentry, electrical work, painting or any other work incidental to the preparation, operation or maintenance of any stage is done.

(v) **Storage Room.** A room, other than a property room used for storage purposes.

(w) **Trap Space.** A space beneath the floor of a stage, used for the handling of stage properties for placement upon and removal from the stage through trap doors in the stage floor, and in which the lifts and other devices incident to such handling of stage properties are located.

1301.03 Existing Theatres:

(a) **General.** The following provisions shall apply to theatres in existence at the time of the passage of this ordinance:

(b) **Repair and Replacement of Parts.** Such repair and replacement of parts of theatres or their equipment as is necessary to maintain an existing theatre in the condition required under the ordinance in effect when such buildings were constructed, altered or converted for the purposes of a theatre, shall be permitted in the forms required under the then existing ordinances; provided however, that no such repair or replacement shall increase the existing capacity of a theatre or reduce the existing width of exits.

(c) **Alterations and Remodeling.**

Item 1. Shall Comply with the Provisions of This Ordinance. Any alteration of or any addition to any theatre or its parts, except as described in paragraph (b) of this section, shall be made to comply with the provisions of this chapter and of the General Provisions of this ordinance.

Item 2. Stage and Projection Blocks. Any alteration or remodeling of any portion of a stage block or projection block shall be made to comply with the provisions of this ordinance and any stage or projection block hereafter altered or remodeled shall be made to comply in its entirety with the provisions of this chapter.

Item 3. Seating and Exits. Any alteration of the seating arrangement or of any means of exit shall be made to comply with the provisions of this chapter and any such alteration of seating shall require the compliance of all means of exit with the provisions of this ordinance.

Item 4. Equipment. No alteration or extension of any existing system of heating or refrigeration or any system of piping or machinery in which hazardous gases or liquids are contained, or which are prohibited in a theatre under this ordinance, shall be made, unless all parts of such system and the rooms or spaces containing them shall be made to comply with the provisions of this ordinance.

Item 5. Rooms Converted to Other Uses. Any room in a theatre hereafter converted for any use requiring special protection under this ordinance shall be made to comply with the provisions of this ordinance for such special protection.

1301.04 Requirements for Theatres: Every building, or part of a building, hereafter designed, erected, altered, converted or used initially for the purposes of a theatre, shall comply with the general provisions of this ordinance in addition thereto shall comply with the special provisions set forth in this chapter.

1301.05 Standing Prohibited:

(a) **Prohibited in Auditorium.** No space shall be arranged to provide standing room for the public in any auditorium nor in any appurtenant space thereto which may be open to an auditorium.

1301.06 Plan Requirements: The working drawings for every theatre shall be submitted to the Department of Buildings for approval and for the granting of a building permit, and shall clearly indicate the exact location of every fixed and portable seat and the arrangement and clear width of every means of exit required under this chapter. On the floor plans of the main floor, any mezzanine and any balcony in the auditorium, the number of seats, both fixed and portable, on such floor shall be stated in conspicuous figures.

ARTICLE 1302

Construction

1302.01 Fireproof Construction:

(a) **General.** Every theatre having a capacity of more than six hundred (600) shall be of fireproof construction.

(b) **Omission of Fireproofing.** Non-combustible structural members, supporting a roof only, and located more than twenty (20) feet above the floor directly beneath, need not be fireproofed. The roof slabs, plates or arches shall have a fire-resistive value of one (1) hour.

1302.02 Semi-Fireproof Construction: Every theatre having a capacity of six hundred (600) persons or less shall be not less fire-resistive than semi-fireproof construction.

1302.03 Auditorium Separations: The auditorium of any theatre shall be separated from every other part of the theatre, excepting the stage block, by not less than one (1) hour fire-resistive construction.

1302.04 Stage Block Separations:

(a) **General.** The inclosing walls, the roof or floor forming the top separation and the floor forming the bottom separation between the stage block and other parts of a theatre or other occupancy shall be of solid construction.

(b) **Fire-Resistive Values.** Except as provided for separations between occupancies in multiple use buildings, the inclosing walls, floors and roofs of any stage block shall be of four (4) hour fire-resistive construction; provided however, that such inclosure for a Type 1 stage shall be of not less than three (3) hour fire-resistive value.

(c) **Parapet Required.** Where no superstructure occurs the inclosing walls of the stage block shall be extended through and above the roof forming a parapet.

1302.05 Stage Floors:

(a) **Finish Floors.** It shall be permissible to cover the structural floor of any stage with a wood finish floor.

(b) **Traps.** Every trap door permitted in the floor of a type 2 stage may be of wood; but shall be not less than two and one-half (2½) inches thick of solid, not laminated material and shall be supported by structural members of non-combustible material.

1302.06 Dressing Room Separations: That part of every theatre in which any dressing room and the exit connections for such rooms are located shall be separated from other parts of the theatre by two (2) hour fire-resistive construction; provided however, that so much of any such separation as is a part of the separating construction required between the stage block and other parts of the theatre shall meet the requirements of Section 1302.04. Dressing rooms shall be separated from each other and from the exit connections thereof by walls of two (2) hour fire-resistive construction. Where the dressing room section is not located within the stage block, any doorway between dressing rooms, or between a dressing room and a horizontal exit connection, may be provided with a combustible door.

1302.07 Workshop, Storage and Property Room Separations: That part of every theatre in which any workshop, storage room or property room is located shall be separated from other parts of the theatre by two (2) hour fire-resistive construction; provided however, that so much of any such separation as is a part of the separating construction required between the stage block and other parts of the theatre shall meet the requirements of Section 1302.04. The wall separation between such rooms shall be of one (1) hour fire-resistive construction and every doorway in such a separating wall shall be provided with a self-closing door.

1302.08 Projection Block Separations: The construction of an inclosure for any projection block shall have a fire-resistive value of not less than two (2) hours. Every window in a projection block inclosure shall be a Type A window.

1302.09 Rooms for Mechanical Equipment:

(a) **Rooms for Cooling and Refrigerating Equipment.** It shall be permissible in any theatre, to install and use not more than five (5) fixed cooling or refrigerating units, if not more than four (4) pounds of any poisonous, toxic, irritant, corrosive, flammable or explosive gas is contained in any of such units.

(b) **Other Rooms for Mechanical Equipment.** In any theatre, every room, other than the rooms otherwise provided for under paragraph (a) of this section, containing any fan, blower, pump, ejector, engine, motor, generator or other such piece of mechanical equipment, shall have an inclosure, with any door, or any window therein of the same construction as is required for such items under the General Provisions of this ordinance; provided however, that in no case shall any doorway in such an inclosure be a combustible door; and provided further that Type A windows shall not be required in such a room, unless connected with a shaft or exit court.

1302.10 Character of Doors and Finish In Dressing Room Sections; Stage and Projection Blocks:

(a) **Doors.**

Item 1. Every door in any doorway in the stage block in storage and property rooms and in the projection block, shall be a non-combustible door, unless required to be a sixty (60) minute fire-resistive door or a forty-five (45) minute fire-resistive door.

Item 2. Every doorway leading from the stage block to another part of a theatre shall be closed with a sixty (60) minute fire-resistive door; provided however, that for Type 1 stages, a forty-five (45) minute fire-resistive door may be used.

Item 3. Every outside doorway from the stage shall be vestibuled and the inner doorway shall have a self-closing non-combustible door.

Item 4. Every doorway leading from a trap space below a stage floor shall be provided with a sixty (60) minute fire-resistive door.

Item 5. Every doorway from the gridiron level to adjacent roof levels shall be provided with a forty-five (45) minute fire-resistive door. Every door located in a required two (2) hour fire-resistive inclosure, shall be a forty-five (45) minute fire-resistive door and every door located in a required four (4) hour fire-resistive inclosure shall be a sixty (60) minute fire-resistive door.

(b) **Finish and Fixtures.** In any dressing room section, not located within any stage block, any base, chair rail, wainscot cap, door trim, window trim, picture moulding, other trim, wainscot or other such finish, and any shelving, make-up table, or other built-in furniture or fixture may be of material not more combustible than wood; provided however, that any such item located within an inclosure for a vertical means of exit, except a handrail, shall be of non-combustible material. In any dressing room section located within any stage block, every such aforesaid item, except a handrail in an inclosure for a vertical means of exit, shall be of non-combustible material. In every stage block, any stage basement, any stage workshop and every projection block, any such aforesaid item, except a handrail in an inclosure for a vertical means of exit, shall be of non-combustible material.

ARTICLE 1303

Size and Location of Rooms

1303.01 Auditoriums: Not more than one (1) auditorium shall be permitted in any theatre. The auditorium in every theatre shall have a clear ceiling height of not less than twelve (12) feet; provided however, that the average clear ceiling height beneath or above any mezzanine or balcony shall not be less than ten (10) feet; provided further, that the lowest point on the soffit of a ceiling or beam, girder or truss projecting below a ceiling shall be not less than eight (8) feet above the floor directly beneath it.

1303.02 Main Floor Levels:

(a) **Highest Aisle Level.** The highest point of any main floor aisle in any auditorium shall be at such elevation that on any route of normal egress, between such point and grade at the principal entrance to the theatre, there shall be no ascending or descending slope of floor or pavement in excess of eight and one-third (8⅓) per cent, one (1) rise in twelve (12) horizontal. On any route of normal egress between such points there shall be no steps. There shall be no limitation of height for any part of any aisle back of the rearmost transverse aisle of the main floor in any stadium type auditorium.

(b) **Lowest Floor Level.** The lowest point of the main floor shall be not more than seven (7) feet below the level of any street, alley, exit court, or other space to which the exit nearest the proscenium leads.

(c) **Slope of Main Floor.** The main floor of an auditorium may be sloped within the limitations prescribed for the slope of aisles. The main floor, except the floor of the longitudinal aisles, may be constructed as a series of level terraces the width of which shall be one (1) row or space or that portion of a row space not otherwise required for the seat legs or supports; provided however, that where such terraces abut an aisle the floor thereof shall be made to meet the aisle without a step or riser by means of ramps, the slope of which shall not exceed the slope of the aisle, unless an aisle at least two (2) feet wide with steps the full width descending in the direction of the proscenium for a distance not to exceed three (3) seat row spaces, be provided from the highest seat level to the level of the main required longitudinal aisle. Such aisle and steps shall be separated from the main aisle by a handrail, except for the outlet thereof.

1303.03 Types of Stages and Limits of Stage Blocks:

(a) **General Requirements.** The stage in every theatre shall be one (1) of two (2) types which, for the purposes of this ordinance shall be designated as Type 1 and Type 2 stages. Each of said types shall conform to the requirements of this section and the stage block in which a Type 1 stage is located shall be limited in size as prescribed, but there shall be no limit of size for the stage block in which a Type 2 stage is located.

(b) **Type 1 Stages.** A type 1 stage shall be used only as a sound projection room. Such a stage shall have no provision whatever made for the use of any movable or combustible scenery. No opening for any trap door shall be permitted in the floor of such a stage. Every Type 1 stage shall have a stage vent meeting the requirements of Section 1306.03; but a fire curtain shall not be required for the proscenium opening of any such stage. The width of a stage block containing a Type 1 Stage measured at a right angle to the center line of the auditorium between the inner faces of the inclosing walls, shall not exceed the clear width of the proscenium opening at any point. The depth of such stage block, measured parallel to the center line of the auditorium between the stage face of the proscenium wall and the inner face of the rear inclosing wall, shall not exceed three (3) feet; provided however, that one hundred eighty (180) square feet of floor area may be added at the rear of the stage, to provide space for sound equipment, if the depth, measured parallel to the center line of the auditorium between the stage face of the proscenium wall and the inner face of the rear inclosing wall of such space for sound equipment, shall not exceed nine (9) feet at any point. The highest point of the underside of the roof or ceiling of any such stage block shall be not more than four (4) feet above the highest point of the proscenium opening and in no case shall such height exceed thirty-five (35) feet above the stage floor. No footlights shall be permitted on a Type 1 Stage.

(c) **Type 2 Stages.** A Type 2 stage shall have provisions made for the use of movable and combustible scenery. It shall be permissible to install removable trap doors in the stage floor and a trap space, meeting the requirements of Section 1303.04 shall be permitted below the floor of such a stage. Every Type 2 stage shall have a stage vent, meeting the requirements of Section 1306.03. The proscenium opening shall be equipped with a fire curtain meeting the requirements of Section 1307.03. It shall be permissible to form a switchboard pocket in the proscenium wall, but the height of said pocket shall not exceed twelve (12) feet and shall be so constructed that the fire-resistive value of the proscenium wall shall not in any way be impaired.

1303.04 Trap Spaces: Any trap space for a Type 2 stage shall be located within an area bounded at the front by a line parallel with, and one (1) foot back of the stage face of the proscenium wall at the sides by lines parallel with the center line of the auditorium and not more distant from such center line than one-half (½) the clear width of the proscenium opening, and at the rear by a line coinciding with the stage face of the rear inclosing wall of the stage block.

1303.05 Projection Rooms and Projection Blocks: In a projection room there may be housed one (1) or more motion picture projectors in which nitrocellulose or equally hazardous films are used and in such room may also be housed one (1) or more still picture projectors, one (1) or more spotlights or one (1) or more other light projection devices in which nitrocellulose or equally hazardous films are used. Every such room shall have a floor area of not less than fifty (50) square feet for each projector, spotlight or other light projection device for which provisions are made therein, and a clear ceiling height of not less than eight (8) feet. Any door, trim, shelf, furniture or fixture in such room and in any appurtenant room within the projection block shall be constructed of or covered with non-combustible material. Every opening in such a projection room for the purpose of picture screen or stage observation, or for the operation of any projector, spotlight or other light projection device, shall be equipped with a sliding shutter which shall close automatically by gravity in case of fire. Every such shutter shall be constructed of sheet iron or sheet steel not less than one hundred forty thousandths (.140) inch in thickness, shall overlap the opening at least one (1) inch all the way round, shall slide in a grooved frame of same material installed inside the projection room and securely anchored to the wall. Every such shutter shall be so made and arranged that it shall close easily when operated manually, and drop freely when released automatically. All of such shutters shall be held normally open by a fine, combustible cord, having one hundred sixty (160) degree heat Fahrenheit actuating device suspended above each projector, spotlight or other light projection device provided for, and so arranged that the burning of the cord at any point, the operation of any heat actuating device, or the manual release of the cord from a point near the exit doorway of the room shall cause all shutters to close quickly and simultaneously. Such cord shall be arranged to offset a constant pull of not less than eight (8) pounds. No provision shall be made in any projection block for storing materials of a combustible nature, except films for one (1) day's operation. Provision shall be made in the projection room or some space within the projection block for the rewinding and storage of films for one (1) day's operation. Every such projection room shall be ventilated as provided under Section 1306.04, and shall have a toilet room as provided under Section 1303.07, paragraph (d).

1303.06 High Pressure Heating Plants: No boiler room or other room in which there is a steam heating plant having a working pressure greater than ten (10) pounds per square inch, shall be located under, or within twenty (20) feet laterally of an auditorium in any theatre, but such a plant shall be permitted elsewhere in any theatre, if separated from other parts of the building by four (4) hour solid fire-resistive construction; with every doorway in such inclosure having a sixty (60) minute fire-resistive door. No doorway to any such room shall be located in the auditorium, in any foyer or lobby, or in any corridor or passage directly connected with any auditorium, foyer or lobby.

1303.07 Toilet Rooms:

(a) **General Requirements.** Separate toilet rooms shall be provided for males and females in every theatre. The number of plumbing fixtures required is stipulated under Section 1307.08.

(b) **Door and Signs.** Entrance doors to public toilet rooms for males and females shall be separated by a distance of not less than ten (10) feet. The entrance door to every toilet room for the public shall be plainly marked for males "MEN" and for females "WOMEN." Same shall apply to toilets for employees.

(c) **Public Toilet Rooms.** Toilet rooms for the public containing the respective plumbing fixtures required for both sexes shall be located either on each seating level, or on the floor next above or below the seating level for which such respective plumbing fixtures are installed; provided however, that general toilet rooms for the public containing all of the plumbing fixtures required in the theatre for either or both sexes, may be located on the main floor or on the floor next above or below the main floor of the auditorium; and provided further that auxiliary toilet rooms for both sexes shall also be installed on each seating level or on the floor next above or below each seating level which is more than one (1) floor removed from the general toilet rooms; and provided also, that every such auxiliary toilet room shall contain at least two (2) water closets, or one (1) water closet and one (1) urinal, and at least one (1) lavatory.

(d) **Employees' Toilet Rooms.**

Item 1. For Projection Rooms. For every projection room designed to house more than one (1) motion picture projector there shall be one (1) toilet room located within the projection block, immediately adjoining and directly accessible from the projection room.

Item 2. For Dressing Room Sections. Toilet rooms shall be provided for the dressing rooms of every theatre, which shall be conveniently located within the dressing room section and shall contain all of the plumbing fixtures required for the number of persons of either sex for whom dressing accommodations are provided.

ARTICLE 1304

Multiple Use Buildings

1304.01 Other Occupancies Permitted and Prohibited: Subject to the provisions of this article, a theatre may be located as a separate unit in the same building with any other occupancy, except a Type One school or a hazardous use unit, prohibited in a theatre under Section 1105.02 and except also as otherwise provided under Chapters 6 to 10 inclusive.

1304.02 Separating Construction: Every theatre shall be separated from every other occupancy in the same building by a standard fire division wall or a standard fire separation.

1304.03 Communicating Openings:

(a) **Communication Between Theatres and Other Occupancies.** Communication between any theatre and another permissible occupancy in the same building shall be permitted in a lobby on the main exit floor of the theatre and as provided elsewhere in such multiple use building; provided however, that additional communication shall be permitted on the respective levels of any existing or proposed and authorized two (2) level street upon which a Multiple Use building abuts.

(b) **Communication Between Adjoining and Adjacent Buildings.** There shall be no communication between a theatre and an adjoining or adjacent building, except in the first story; provided however, that adjoining and adjacent buildings may have communication on any ground floor, subject to the General Provisions; and provided further, that there shall be no opening for communication on any floor between any theatre and any adjoining or adjacent building which is an institution building, a Type One school or hazardous use unit prohibited in a theatre under Section 1105.02.

(c) **Communicating Doorways.** Every doorway in a wall separating a theatre from any part of a building used for other than theatre purposes shall be provided with a double standard fire door. Except for the purposes of Section 1304.05, such doorways shall not exceed sixty (60) square feet in aggregate area. In no case shall any two (2) of such doorways be located closer together than twenty (20) feet, measured horizontally between the nearest jambs. In no case shall any such doorway be considered as a means of exit required under Article 1305.

1304.04 Means of Exit for Theatres in Multiple Use Buildings: Every means of exit required for a theatre under Article 1305 shall be for theatre use only and shall afford no communication with any part of a building used for other than theatre purposes; provided however, that a fire escape serving another permitted occupancy in the same building may be combined for common use with a fire escape other than a fire escape in an exit court, serving a theatre, if the width thereof below the point of junction is made equal to the aggregate width required for the fire escapes so combined. No counterbalanced fire escape, or other fire escape serving another occupancy shall in any way obstruct or reduce any required means of exit for a theatre.

1304.05 Garages:

(a) **General.** A garage shall not be permitted above or below a theatre in any multiple use building, but a garage shall be permitted elsewhere in any multiple use building in which there is a theatre; provided however, that it shall be permissible to design, construct and use below a theatre in any building hereafter erected, whether a multiple use building or not, a passage to a garage, if such passage be separated from all other parts of the building by solid construction of four (4) hour fire-resistive value, and if such passage be ventilated in accordance with the provisions of Chapter 47 — VENTILATION, and Chapter 12 — GARAGES; provided further, that such a passage may have a driveway not more than eighteen (18) feet in width for motor cars and platforms for loading and unloading passengers, but no space for the storage or parking of motor cars shall be included in any such passage, and no flammable liquid or other hazardous material shall at any time be stored therein; and provided also, that such a passage shall be provided with double standard fire doors at the garage and with a sixty (60) minute fire-resistive door at the other end.

(b) **Means of Communication.** Not more than one (1) means of communication between any theatre and any garage shall be permitted; provided however, that if there is a passage with a driveway to a garage, as before described, one (1) additional such means of communication through such passage shall be permitted. Such communication between a theatre and a garage and between a theatre and such passage to a garage shall be through a vestibule constructed in accordance with Section 1205.03, paragraph (b). No such means of communication shall be considered as a means of exit required under Article 1305.

ARTICLE 1305

Means of Exit

1305.01 General: Every room or space open to the public in every theatre shall have, in addition to the entrance thereto, an exit arranged to permit passage through and from the room without returning to the place of entrance, except as otherwise permitted under this Article.

1305.02 Classification of Means of Exit:

(a) **General.** The means of exit required for every theatre shall be of three (3) classes, designated for the purpose of this ordinance, as "Means of Public Normal Exit," "Means of Public Emergency Exit" and "Means of Employees' Exit."

(b) **Means of Public Normal Exit Defined.**

Item 1. The means of public normal exit shall include all of the means normally comprising a route of ingress intervening between a way of approach for the public and the auditorium and providing the usual means of entrances for the public; provided however, that no means of exit hereinafter defined as a means of public emergency exit shall be embraced by any of the items herein described.

Item 2. Every doorway, foyer, lobby, passageway and aisle intervening between the auditorium and a way of approach for the public shall be designated a public horizontal normal exit connection.

Item 3. Every stairway and ramp and the inclosed spaces thereof, leading directly or indirectly to a public horizontal normal exit connection on the main exit floor shall be designated a public vertical means of normal exit.

(c) **Means of Public Emergency Exit Defined.**

Item 1. The means of public emergency exit shall include all of the means of egress provided solely for the use of the public in an emergency intervening between the auditorium and a way of departure for the public; provided however, that none of the usual means of entrance described in paragraph (b) of this section, as means of public normal exit, shall be included in the means of public emergency exit.

Item 2. A direct outside doorway at grade or ground level provided solely for the use of the public in an emergency shall be designated as a public outside emergency exit at grade.

Item 3. An interior stairway or ramp provided solely for the use of the public in an emergency, leading to a public outside emergency exit at grade, and an outside stairway or stairway fire escape or an outside ramp, leading from an outside emergency exit to the grade, shall be designated as a public vertical means of emergency exit.

Item 4. A court at grade or ground level provided solely for the use of the public in an emergency leading directly or through a fireproof passage to a way of departure for the public shall be designated as an exit court.

(d) **Means of Employees' Exit Defined.** Any means of public normal exit and any means of public emergency exit made available for the joint use of the public and employees, shall be considered a joint public and employees' exit but shall be designated respectively by the term used therefor in paragraphs (b) and (c) of this section; provided however, that no means of public emergency exit, except an exit court, shall be designed or arranged to provide a normal exit for employees. Any means of exit provided solely for the use of employees shall be designated respectively as an employees' normal exit or an employees emergency exit.

1305.03 Exit Width and Height:

(a) **Width Based on Capacity.** The width of every required means of public exit in any theatre shall, except as otherwise provided, be computed on the basis of the seating capacity of the auditorium, in accordance with the rules hereinafter given. Other exit width shall be of the specific minimum dimensions hereinafter given.

(b) **Minimum Width and Height.** No doorway, which is a required exit shall be less than three (3) feet in clear width, or less than seven (7) feet in clear height; provided first, that adjacent doorways required for normal exit, which are separated by mullions or by other divisions not more than ten (10) inches wide overall, may be not less than two (2) feet, ten (10) inches in clear width; provided second, that any doorway to a stairway for the sole use of employees may be two (2) inches less in width than the required width for such stairway; provided third, that an exit doorway from an orchestra pit may be reduced in clear height to four (4) feet six (6) inches. No required vertical means of exit shall be less than three (3) feet in clear nected with a main floor longitudinal aisle shall be connected with such an aisle by means of an aisle not less than two (2) feet eight (8) inches wide.

Item 4. Width of Transverse Aisles. The clear width of any transverse aisle required or permitted in any theatre, shall be equal to one (1) foot eight (8) inches per one hundred (100) persons served thereby. The minimum clear width of any transverse aisle, except as permitted in Item 3 of this paragraph, shall be four (4) feet. The clear width of a transverse aisle shall be measured horizontally between vertical planes which exclude evry encroachment thereon, which clear width shall be preserved to a height of at least seven (7) feet from the floor of the aisle. A transverse aisle in a balcony and a transverse aisle at the rear of a main floor in a stadium type auditorium may serve both as a means of public normal exit and as a means of public emergency exit in which case application of the required widths of such means shall be coincidental and not cumulative. Transverse aisles above the main floor may be inclined in the transverse direction to the direction of the aisle, at a rate not to exceed one (1) unit of rise to twelve (12) units of horizontal projection.

Item 5. Steps and Ramps in Aisles. No riser or step shall be permitted in any main floor aisle. No ramp in any such aisle of the main floor proper shall exceed a slope of sixteen and two-thirds (16⅔) per cent, one (1) rise in six (6) horizontal. Where steps are necessary in any mezzanine longitudinal aisle, or in any balcony longitudinal aisle, such steps shall be made the full width of the aisle in every case, with the rise of each step not more than eight (8) inches, with the cut thereof not less than ten (10) inches, and with all treads of equal width and all risers of equal height; provided however, that a change may occur in the pitch of the steps adjoining a transverse aisle, if no more than eight (8) steps are involved and if such steps are of uniform pitch. No riser or step shall be permitted in the floor of any transverse aisle. No riser or step to any emergency doorway and no other riser or step shall be permitted to encroach upon the clear width required for any transverse aisle.

Item 6. Distance of Travel to Exit Doorways. The distance of any seat from a public normal exit doorway from the auditorium, measured along the line of travel shall be not greater than one hundred fifty (150) feet and if more than one hundred (100) feet it shall have not more than one (1) angle or turn.

(b) **Inside Normal Exit Doorways.**

Item 1. From Main Floors and Mezzanines. Every main floor longitudinal aisle, and every mezzanine longitudinal aisle in any auditorium, shall lead directly to an inside normal exit doorway, or, in stadium type auditoriums, to a transverse aisle at the rear of the main floor. The rearmost transverse aisle of the main floor in a stadium type auditorium shall have an inside normal exit doorway, or a vomitory leading to such a doorway, for each side of the auditorium. The clear width of every such doorway shall be at the rate of one (1) foot eight (8) inches per one hundred (100) persons for which it provides exit; provided however, that no such doorway shall be less in clear width than four (4) feet in any case.

Item 2. From Balconies. Every balcony transverse aisle shall have an inside normal exit doorway, or a vomitory leading to such a doorway for each side of the auditorium. The clear width of every such doorway shall be at the rate of one (1) foot eight (8) inches per one hundred (100) persons for which it provides exit: provided however, that no such doorway shall have a clear width of less than four (4) feet.

(c) **Vomitories.** The clear width of every vomitory shall be at least one (1) foot more than the doorway to which it leads, but not in any case less than five (5) feet. No vomitory shall have a floor with a slope greater than sixteen and two-thirds (16⅔) per cent one (1) rise in six (6) horizontal. No doors shall be required between any vomitory and the auditorium or between any vomitory and any foyer above the main floor. Every vomitory shall be so arranged as to enter upon a public horizontal normal exit connection in the direction of or at an angle not more than (90) degrees from the direction of exit.

(d) **Vertical Means of Normal Exit.**

Item 1. From Mezzanines. Not less than two (2) vertical means of public normal exit shall be provided to serve every mezzanine. Every such means of exit provided shall lead to a public horizontal exit connection on the main exit floor. Entrance to such means of exit shall be direct from the mezzanine, or direct from a foyer or lobby with its floor at the same level as the highest level of a mezzanine longitudinal aisle and in either case each of such means of exit shall be from opposite sides of the auditorium. Such means of exit shall be so arranged as to enter upon a public horizontal normal exit connection in the direction of or at an angle not more than ninety (90) degrees from the direction of exit.

Item 2. From Proscenium Boxes. Not less than one (1) vertical means of public normal exit shall be provided to serve any proscenium box or each tier of proscenium boxes, which shall lead to the main floor of the auditorium. Entrance to such means of exit shall be from a doorway, aisle or passage directly connected with every box served by such means of exit, and shall have a clear width of not less than three (3) feet at any point.

Item 3. From Balconies. Not less than two (2) vertical means of public normal exit shall be provided to serve every balcony. Every such means of exit provided shall lead to a public horizontal exit connection on the main exit floor. Entrance to such means of exit shall be direct from the balcony, or direct from a foyer or lobby with its floor at the same level as the highest level of a balcony longitudinal aisle and in either case each of said means of exit shall be from opposite sides of the auditorium. Such means of exit shall be so arranged as to enter upon a public horizontal normal exit connectoin in the direction of or at an angle not more than ninety (90) degrees from the direction of exit.

Item 4. Width. The width of every such means of exit shall be at the rate of one (1) foot eight (8) inches per one hundred (100) persons served thereby; provided however, that no such means of exit from any mezzanine or balcony, shall in any case be less than four (4) feet in width.

Item 5. Distribution of Width. The total required width of Vertical Means of Normal Exit may be divided equally between the required means but in no case shall less than thirty-seven and one-half (37½) per cent of such total be provided from either one-half (½) of any seating level.

Item 6. Stairways. If such means of exit be a stairway with a direct entrance from a mezzanine or balcony, it shall be separated from the auditorium by construction of one (1) hour fire-resistive value; provided however, that no such separation shall be required between the auditorium and any flight of steps in a vomitory. If such means of exit be a stairway with entrance from a mezzanine foyer or lobby, or with entrance from a balcony foyer or lobby, so much of it as is located in any foyer or lobby may be an open stairway, but such parts of it as are not located in any foyer or lobby shall be separated from the adjoining space by construction of two (2) hour fire-resistive value.

1305.06 Public Normal Exit from Rooms Other Than Auditoriums: Every room open to the public and not otherwise provided for under this chapter shall have at least one (1) public inside normal exit doorway opening upon a public horizontal normal exit connection or a public vertical means of normal exit which shall have a width of not less than four (4) feet.

1305.07 Public Emergency Exit from Auditorium:

(a) **General.** In addition to the means of public normal exit required from an auditorium, every theatre auditorium shall have means of public emergency exit to a way of departure from the theatre as required under this section.

(b) **From Main Floors.**

Item 1. Doorways. On the main floor of the auditorium, such means of exit shall consist of public outside emergency exit doorways at grade, leading directly to ground level or directly to the pavement level of a street, sidewalk, public alley, exit court or other open space serving as a way of departure for the public from the theatre, and otherwise meeting the requirements of this section.

Item 2. Aisles. The number of banks of seats on the main floor shall not exceed fifteen (15) unless an intervening or cross aisle is provided between each fifteen (15) banks of seats or a direct exit is provided for each aisle.

(c) **From Mezzanines and Balconies.** On every mezzanine, except as otherwise permitted under paragraph (e) of this section and on every balcony such means of exit shall consist of public emergency exits leading to the ground level, or to the pavement level of a street, sidewalk, public alley, exit court, or other open space serving as a way of departure for the public and otherwise meeting the requirements of this section.

(d) **Required Public Emergency Exit Width.** The clear width of public emergency exits from every seating level of the auditorium in any theatre, shall be at the rate of one (1) foot eight (8) inches per one hundred (100) persons served thereby.

(e) **Distribution of Width.** The total required public emergency exit doorway width shall serve both halves of the seating level for which required. The division of such total may be equal but in no case shall less than thirty-seven and one-half (37½) per cent of the total be provided for either one-half (½) of any seating level; provided however, that any mezzanine having less than ten (10) rows of seats and also having at the end of each longitudinal aisle thereof a public inside normal exit doorway leading to a foyer or lobby with its floor at the same level as the highest level of such aisle may have no means of direct public emergency exit; and provided further, that the required width of public inside normal exit doorway at the end of each longitudinal aisle of such mezzanine shall be increased at the rate of ten (10) inches for each one hundred (100) persons served thereby, and at each side of the auditorium in the foyer or lobby to which such public inside normal exit doorways lead, there shall be provided one (1) public emergency exit to grade, and the combined width of the doorways of such emergency exits shall be computed at the rate of one (1) foot and eight (8) inches per one hundred (100) persons seated on the mezzanine.

(f) **Location on Main Floors.**

Item 1. Where Capacity Is Six Hundred (600) or Less. On the main floor of every auditorium, with a capacity of six hundred (600) persons, or less, the required public emergency exit doorways shall all be located between the proscenium and a point which is not more distant from said proscenium than one-half (½) the distance between the first and last rows of main floor seats; provided however, that the rearmost transverse aisle of the main floor in a stadium type auditorium shall have a public outside emergency exit doorway at each end thereof.

Item 2. Where Capacity Is More Than Six Hundred (600). On the main floor of every auditorium with a capacity of more than six hundred (600) persons, at least fifty (50) per cent and not more than sixty-six and two-thirds (66⅔) per cent of the required public emergency exit doorway width shall be located between the proscenium and a point not more distant from said proscenium than one-half (½) the distance between the first and last 'ows of main floor seats. The rearmost ransverse aisle of the main floor in a stadium ype auditorium shall have a public outside mergency exit doorway at each end thereof.

(g) **Location on Balconies.** On every balcony there shall be one (1) public emergency exit doorway located at each end of every required transverse aisle.

(h) **Steps at Public Emergency Exit Doorways.** At least thirty (30) per cent of the required emergency exit doorways from the main floor shall be reached without steps. No steps ascending in the direction of egress shall be located outside of the walls of the building, or within an exist court or otherwise be exposed to the weather. Not more than five (5) steps ascending in the direction of egress may be located on any level above the main floor inside of and leading to any public emergency exit doorway.

(i) **Vertical Means of Public Emergency Exit, Stairways, Fire Escapes, Ramps.**

Item 1. From Mezzanines. For the public emergency exit doorways required at each side of the auditorium on any mezzanine, or mezzanine foyer, there shall be provided at least one (1) vertical means of exit descending from the doorways to the required level at grade.

Item 2. From Proscenium Boxes. For each tier of proscenium boxes there shall be a means of public emergency exit to grade; provided however, that such means of public emergency exit shall not be required for any tier of proscenium boxes for which two (2) ways of direct escape by the means of public normal exits are provided.

Item 3. From Balconies. For the public emergency exit doorways required on each side of any balcony, there shall be provided at least one (1) vertical means of exit descending from the doorways to the required level at grade.

Item 4. Width. The width of every such vertical means of exit shall be at the rate of one (1) foot, eight (8) inches per one hundred (100) persons served thereby, and this width shall be maintained, from the first doorway thereto through the length of such means of exit by a cumulative increase toward grade; provided however, that no such means of exit shall be less than three (3) feet wide in any case.

Item 5. Shall Serve Only One Side of Auditorium. In no case shall the means of emergency exit from opposite sides of the auditorium be combined into one (1) vertical means of exit.

Item 6. Public Emergency Exit Stairways. If such means of exit be an inside stairway, it shall be contained within an inclosure of two (2) hour fire-resistive construction with every doorway having a forty-five (45) minute fire-resistive door. Such a stairway, its inclosure and doorways shall in no case be less in width than is required for the emergency exit doorway it serves. If such means of exit be an outside stairway, no part thereof shall encroach upon the required width of or otherwise obstruct an exit court below a plane eight (8) feet above the pavement of such court.

Item 7. Public Emergency Exit Fire Escapes. If such means of exit be a stairway fire escape leading to an exit court, no part thereof shall be counterbalanced, and no part thereof shall encroach upon the required width of or otherwise obstruct such a court, below a plane eight (8) feet above the pavement thereof. If such means of exit be a stairway fire escape leading to a street, or alley, it shall be permissible to counterbalance the lower flight thereof but no part thereof shall encroach normally upon a public sidewalk below a plane twelve (12) feet above the pavement of such sidewalk, no part thereof shall encroach normally upon a public alley below a plane fourteen (14) feet above the pavement of such alley. Every such fire escape shall have a ladder extending from its topmost landing to the roof of the auditorium, except where other stories are superimposed thereon or unless located in an exit court having ceiling. It shall be permissible to install any such fire escape on any street, except where prohibited by other ordinances or laws.

Item 8. Public Emergency Exit Ramps. If such means of exit be a ramp, it shall not exceed a slope of sixteen and two-thirds (16⅔) per cent, one (1) rise in six (6) horizontal, and shall conform with such of the requirements of a stairway for such purposes as are applicable.

Item 9. Landing at Public Emergency Exit Doorways. Whether such means of exit be a stairway, fire escape or ramp, there shall be a level landing at every public emergency exit doorway leading thereto, which shall be not less than four (4) feet wide and at least two (2) feet longer than the width of the doorway.

Item 10. Landings in Exit Courts. Whether such means of exit leading to an exit court be a stairway, fire escape or ramp, it shall issue upon the pavement of the court in the direction of egress therefrom.

(j) **Exit Courts.**

Item 1. General Requirements. Every required emergency exit doorway which does not open directly upon a street, or a public alley, or upon another space serving as a way of departure from the theatre, shall open directly upon an exit court leading directly or through a fireproof passage to a street, or a public alley or another open space serving as a way of departure for the public.

Item 2. Pavement of Exit Courts. The floor or pavement of every exit court and passage shall be generally level or ramped to meet the level of the street, sidewalk, alley or other space to which the court leads. If ramped, such pavement shall have a slope not greater than sixteen and two-thirds (16⅔) per cent, one (1) rise in six (6) horizontal. No step, areaway, trap door, coal hole or other opening, shall be constructed in such pavement anywhere, or in the pavement of the sidewalk, alley or other space opposite the exit from such a court or passage. If there is any space below the floor or pavement of such a court or passage, it shall be separated from the court or passage by four (4) hour fire-resistive solid construction. Every such court and passage shall be arranged to drain so that it shall not hold water.

Item 3. Exit Court Walls and Openings. The walls of every such court and passage shall be of four (4) hour fire-resistive solid construction. No doorway shall be permitted in any wall of any such court, except a doorway from a theatre. Every door in any doorway permitted in such a court shall be a forty-five (45) minute fire-resistive door. No window shall be permitted in any wall of an exit court having a ceiling of such nature as is permitted under Item 4 of this paragraph. Windows shall be permitted in the walls of any other exit court. Every such window and every window which is located above an exit court and below a level thirty (30) feet above the top of the highest emergency exit doorway in such court shall be a Type A window.

Item 4. Exposure to Sky or Alternative Natural Ventilation of Exit Courts. Every such court shall be open and unobstructed to the sky; provided however, that outside stairways or fire escapes permitted under this Article shall not, for the purposes of this item be considered as obstructions; and provided further, that any such court may be covered with a ceiling of four (4) hour fire-resistive solid construction, if natural ventilation is provided at the top of the court by one (1) of the following means:

Means 1. By an opening unobstructed to the sky produced by offsetting the top of the court a horizontal distance not more than two and one-half (2½) times the maximum width of the court, said opening being no less in area and said offset being no less in cross-sectional area than seventy-five (75) per cent of the floor area of the court.

Means 2. By an opening or openings to the outside air in a wall or walls the ceiling of the court, said opening or openings aggregating in area seventy-five (75) per cent of the floor area of the court.

Means 3. By an opening or openings such as described under Means 2, into a tunnel of no lesser area than such openings and not longer than two and one-half (2½) times the maximum width of the court and leading directly to the outside air. If any ceiling is provided in any such court, no part of it shall be lower than one (1) foot above the top of the highest public emergency exit doorway leading to such court. If there is a passage such as described in connection with such a court, such passage shall have a ceiling of four (4) hour fire-resistive solid construction and the clear ceiling height of such a passage shall be not less than eight (8) feet.

Item 5. Required Width of Exit Courts. Every exit court shall have a clear width based upon the aggregate number of persons served by the public exit doorways opening thereon in accordance with (Table 1305.07).

(Table 1305.7)

Required Width of Exit Courts

Maximum Number of Persons Served	Minimum Width in Feet
300	5
400	6
500	7
600	8
700	9
800	10
For each additional 250 or fraction thereof	1 foot additional

Such a court may be of the maximum required width for its entire length, or may be of the minimum width permissible under the aforesaid rules, at the doorway which is most remote from the outlet thereof and be increased in width in accordance with the aforesaid rules, in the direction of egress. The width of any passage such as described in connection with such a court shall be not less than the maximum width required for such court.

Item 6. Width of Alleys and Other Spaces Serving as Ways of Departure from Exit Courts. Every public alley, or other space, serving as a way of departure from such court or passage, shall have a clear width of not less than ten (10) feet; provided however, that such alley or other space shall provide an outlet or outlets equal in width to the requirements under this article for an exit court serving the number of persons for whom emergency exit is provided combined with the capacity in number of persons of all normal exit doorways discharging therein.

Item 7. Permitted Uses of Exit Courts. No exit court or passage shall be designed, erected, altered, converted or used for any other purpose than that of exit of the public from a theatre; provided however, that an exit court may also serve as a means of normal or as a means of emergency exit for employees of the theatre. Where two (2) theatres are erected on one (1) lot as defined in Chapter 5—DEFINITIONS, one (1) exit court may be used in common by both theatres; provided however, that such court shall meet the requirements for an exit court serving that theatre which has the greatest capacity.

Item 8. Obstructions Prohibited in Exit Courts. No object shall be permitted to obstruct or reduce the required dimensions of such court or passage other than permitted under this paragraph; and except as the outside opening of such a court or passage, where the clear width may be reduced one (1) foot and the clear height may be reduced to seven (7) feet to allow for gates or doors. Any gate or door provided at such an opening shall be non-combustible material. Rny radiator, pipe, or other object located in such a court or passage shall be recessed to avoid obstruction.

1305.08 Public Emergency Exit from Rooms Other Than Auditorium: Every room open to the public in any theatre having a floor area greater than one hundred (100) square feet and every group of adjoining connecting

(d) **Locks, Panic Hardware.**

Item 1. Locks and Fastenings. Any lock or other fastening device used on any door or other closure for an opening of a required exit, in any theatre, shall be of a type which may be released by any person, at any time, seeking egress from the theatre without the use of a key or other detached opening devices.

Item 2. Panic Hardware. No locking device shall be installed on any door of any public exit or on any door of any employees' outside exit doorway; provided however, that any such door may be fitted with panic hardware; and provided further, that for any door at the principal entrance it shall be permissible to use any locking device.

1305.15 Concealing of Exits, False Doors and Windows: No drapery or other device shall be permitted to conceal a required exit nor to obstruct the required opening thereof. No mirror or other device shall be so arranged or located as to represent or be mistaken for a means of public exit.

1305.16 Roof Promenades Prohibited: No roof of any building in which there is a theatre shall be designed, erected, altered, converted or used for the purposes of a promenade for persons attending any display or performance held in the theatre.

ARTICLE 1306

Windows and Ventilation

1306.01 Windows:

(a) **Window Grilles.** No window opening in any theatre shall have installed therein any grillage of bars, any other grillage or grillework, or any other object which would prevent escape through the opening in any emergency; provided however, that this requirement shall not apply to windows which occur in spaces not open to the public.

(b) **Window in Stage Walls.** No windows shall be required in the walls of any Type 1 stage and no windows shall be permited in the walls of any Type 2 stage.

1306.02 Glass in Foyer and Auditorium Separating Walls: In a wall separating any foyer or lobby from the auditorium of any theatre, it shall be permissible to install glazed openings, other than the required doorways, in the main exit story only. Such openings shall not exceed in area forty (40) per cent of the area of such wall, exclusive of all doorways therein. The area of such wall shall be measured on that side of the wall which has the least area. Every such opening shall be glazed with wired glass. No pane of glass in any such opening shall exceed seven hundred twenty (720) square inches in area, nor forty-eight (48) inches in its greatest dimension. A frame shall be provided for every such opening, which shall be of metal, or kalamein construction, and shall be securely anchored to the wall construction. Every glass stop, any sash, and every division between panes of glass shall be of metal or kalamein construction. No such panes or sash shall be made to open.

1306.03 Stage Vents:

(a) **Number Required.** Every stage in any building hereafter designed, erected, altered, converted or used for the purposes of a theatre, shall be provided with a means of natural ventilation, so arranged and equipped as to function in case of fire. Such means of ventilation shall consist of not less than one (1) vent nor more than three (3) vents to the outer air, meeting the requirements of this section.

(b) **Location and Extent.** The stage opening to every stage vent shall be located in the soffit of the ceiling or roof of the stage, the center of such opening shall be within a radius of twenty (20) feet of a point located on the center line of the proscenium opening and half way between the proscenium wall and the rear wall of the stage, if there be but one (1) such vent; or the center of such opening shall be within a radius of five (5) feet of the center point of its respective zone of stage area if there be more than one (1) such vent. From such opening a shaft shall be extended through the ceiling and roof to a point not less than fifteen (15) feet above the roof, or above any other roof which is within a horizontal distance of ten (10) feet from the shaft at any point; provided however, that in no case shall any such shaft serve more than one (1) such opening; and provided further, that in lieu of being vertical all or any part of such shaft may form an angle of not more than thirty (30) degrees with the vertical; and provided also, that the sum of the changes in direction of any such shaft shall not in any case exceed one hundred eighty (180) degrees.

(c) **Area Required.** The area of the stage opening of any stage vent, or the aggregate area of such openings, shall be not less than five (5) per cent of the floor area of the stage. The cross-sectional area of the shaft connected with any such opening shall in every case be not less than the area of the opening.

(d) **Construction.** The inclosure for the shaft of any stage vent shall be of not less than two (2) hour fire-resistive construction; provided however, that so much of any such shaft as is passed through any part of a building used for other than theatre purposes, shall be of solid construction having a not less than four (4) hour fire-resistive value.

(e) **Dampers.** The stage vent shall be provided with a damper or dampers, which shall tightly close the opening and shall be formed of sheet iron or steel not less than one hundred twenty-five thousands (.125) of an inch in thickness.

(f) **Operation and Control of Dampers.** The damper or dampers of every stage vent shall be equipped with a mechanical device which when operated manually or automatically, shall so function as to cause the damper or dampers to fully open. The use of an electrical device for such purpose shall not be permitted. Such damper opening device shall be so made, arranged and equipped that the damper or dampers may be opened manually and instantly from the stage electrician's station. Such damper opening device shall also be made, arranged and equipped that, in case of fire, it shall be operated automatically by the actuating of a thermostatic device located under each stage vent. Such thermostatic devices shall be actuated as a result of a difference in the rate of rise in temperature, rather than as a result of an increase in temperature. At the stage electrician's station there shall be a conspicuous-non-combustible sign on which plain directions, as to how to open the stage vent damper or dampers, shall be given in letters at least four (4) inches high. Such lettering shall be in white on a black field.

(g) **Hoods.** The top of the shaft of every stage vent shall have a hood constructed entirely of non-combustible material and the eaves thereof shall project not less than one (1) foot. Openings in all walls of the shaft shall be provided below such hood, which opening shall aggregate in the clear not less than the cross-sectional area of the shaft. The curbs of such openings shall stand not less than fifteen (15) feet above the roof.

(h) **Proximity to Property Lines.** No opening in any stage vent shall be located within a horizontal distance of ten (10) feet from any property dividing lot line, measured at any point.

(i) **Proximity of Other Openings.** No window, doorway or other wall openings on the same premises shall be located within a horizontal distance of twenty (20) feet from any opening in a stage vent, measured at any point.

(j) **Approval of Installations.** No building hereafter designed, erected, altered, converted or used for the purposes of a theatre, which is required to have a stage vent, shall be opened to the public for any purpose whatever, unless and until each stage vent and all of its appurtenances have been subjected to and have satisfactorily passed the operating tests prescribed in paragraph (k) of this section, and have been made to fully comply with the provisions of this ordinance. When each stage vent and all of its appurtenances have passed said initial operating tests and have been found by examination to fully comply with the provisions of this ordinance, a certificate to that effect shall be caused to be issued to the owner, manager or other person in control of the theatre by the Commissioner of Buildings and approval of the installation shall be given by the said Commissioner. If any stage vent and all of its appurtenances have not passed said initial operating tests, or have been found by examination not fully to comply with the provisions of this ordinance, no such certificate shall be issued and no such approval shall be given by the said Commissioner, and he shall notify the owner, manager or other person in control of the theatre, in writing, of the reasons for his action. Such certificate shall be posted under glass in a conspicuous place on the stage for a period of thirty (30) days after the date thereof, and during such period shall serve the purposes of the certificate of inspection required under other provisions of this ordinance.

(k) **Operating Tests.** Upon completion of the installation, each stage vent and all of its appurtenances shall pass a test by the Department of Buildings, which test shall demonstrate that each and all of the devices for operation and control, required under this section, are properly operative. Such tests shall be made by the Department of Buildings at the same time that the initial curtain operating tests, required under Section 1307.03 are made. Should the damper, or dampers, fail properly to operate according to the true meaning and intent of this ordinance, or should any part of the construction of any stage vent be found not to comply with this ordinance, the cause of such failure shall be removed and the part not in conformity with this ordinance, made to conform by the owner, manager or other person in control of the theatre, and the required operating tests repeated by the department until operation of the damper or dampers and the construction of every stage vent are found to comply with this ordinance.

1306.04 Mechanical Ventilation for Stages: No system of mechanical or forced ventilation shall be required for any stage.

ARTICLE 1307

Equipment

1307.01 Seats:

(a) **Number of Seats in Rows and Between Aisles.** No row of seats shall have more than twelve (12) seats between aisles or more than six (6) seats abutting an aisle at one (1) end only; provided however, that on the main floor, if the distance between rows is increased above that required by paragraph (b) of this section, the number of seats between aisles may be increased by one (1) seat for each one (1) inch by which the distance between rows exceeds that required by paragraph (b) of this section; and provided further, that if such distance is increased by eight (8) inches, rows of forty-eight (48) seats or less and eighty (80) feet or less in length shall be permitted between aisles; and provided also, that if more than nineteen (19) seats are placed between aisles then such aisles shall be increased in width where necessary to provide a clear width equal to twenty-five (25) inches per one hundred (100) persons served at every point. In no case shall a row of seats abutting an aisle at one (1) end only have more than ten (10) seats. No curved row of seats shall describe an arc of e than ninety (90) degrees between aislesor

(b) **Spacing of Rows.** In every theatre auditorium rows of fixed seats shall be spaced at least two (2) feet and ten (10) inches apart, measured horizontally between the rearmost points of seats in adjacent rows; and the minimum distance, measured horizontally between parallel vertical planes touching the rearmost point of the back of any seat and the foremost point of an arm of any seat in the row next to the rear, shall be one (1) foot, four (4) inches; except as otherwise provided under paragraph (e) of this section.

(c) **Width of Seats.** Arms or other divisions between sittings shall be provided for all fixed seats. The minimum width of ninety (90) per cent of the fixed seats in any auditorium shall be twenty (20) inches, measured between the centers of the arms thereof, or of other divisions between sittings. The minimum width of the remaining ten (10) per cent of such seats shall be nineteen (19) inches, measured in the same manner.

(d) **Fixed and Portable Seats.** Every seat in any theatre auditorium shall be securely fastened in place; provided however, that in any box one (1) portable seat or chair shall be allowed for each five (5) square feet of floor area of such box; provided further that in no case shall any box have more than tdenty (20) portable seats or chairs; and provided also, that no portable seat or chair shall be permitted on any sloping floor.

(e) **Self-Raising Seats.** On the main floor of every auditorium, at least two (2) complete rows of fixed seats, opposite every public emergency exit doorway, shall be of a self-raising type. The minimum clearance between a rearmost point of back and the foremost point of an arm of the seat when raised shall be one (1) foot, six (6) inches, measured as stated under paragraph (b) of this section.

1307.02 Proscenium Fire Curtains for Type 1 Stages: No fire curtain shall be required for the proscenium opening of any Type 1 stage.

1307.03 Proscenium Fire Curtains for Type 2 Stages:

(a) **General.** The proscenium opening between any auditorium and a Type 2 stage shall have a vertically operated curtain. The curtain shall be so constructed and installed as to intercept hot gas, flame and smoke. The period of protection shall be taken as not less than thirty (30) minutes and the fire-resistance of the curtain during such period shall be such as to meet the test prescribed under paragraph (o) of this section; provided however, that the subjection of any curtain to such test shall be required only where stipulated under said paragraph (o). The full closing of the curtain in case of fire shall be effected in less than one (1) minute from the full open position, but the last five (5) feet of travel shall require not less than five (5) seconds, nor more than fifteen (15) seconds. The curtain shall be so arranged and maintained that it will at all times descend safely and close completely the proscenium opening. No paint in which oil is used and no combustible material shall be applied or attached to the curtain, unless sprinkler heads, attached to a standard system of automatic sprinklers are installed in the auditorium, immediately over the proscenium opening in accordance with the provisions of Section 1309.01, paragraph (b), in which case such paint material may be used on the auditorium side of the curtain.

(b) **General Design of Curtain Structures.** The curtain shall have sufficient strength to resist a lateral pressure of ten (10) pounds per square foot of its area when in a closed position, with a factor of safety of not less than two (2) on the ultimate strength of the construction. The mounting and details shall be such as to insure ready and positive closure of the curtain while subjected to a lateral pressure of five (5) pounds per square foot, and ability to retain contact with the stage floor when closed while under such pressure. The design strength of every tension member shall be based on a center deflection of not to exceed one-tenth (1/10) of the span under ordinary temperature. In no case shall the maximum deflection of any part cause a permanent set or bend in the curtain structure, nor shall it cause the curtain to chafe against the wall at the edges of the proscenium opening, nor shall it be such as to otherwise interfere with the proper operation of the curtain. Provisions shall be made to prevent the curtain from getting away from the guide rails, when subjected to lateral pressure due to suction from the stage side. To accomplish this, it shall be permissible to use a curved curtain, so installed that the concave side will be toward the stage; provided however, that the middle ordinate of the curve in any such curtain shall not exceed two (2) inches. Provisions shall be made for the expansion of the curtain and guides due to changes in temperature. An allowance of one-sixteenth (1/16) per inch per foot of length of steel members shall be made for such purpose. The calculations for the strength of the curtain, curtain mountings and all the details thereof shall conform to generally accepted engineering methods and practice, unless hereinafter required to conform with other specific rules. The stresses in materials shall not exceed those prescribed in the structural provisions of this ordinance. No part of any curtain shall be supported by or fastened to any combustible material.

(c) **Thickness of Curtains.** The thickness of the curtain shall be not less than one one-hundred-twentieth (1/120) of the width of the proscenium opening in any case. The arrangement of the structural members of the curtain framework shall be such as to secure a positive separation of the coverings on the two (2) sides with an air space between, equal to the required thickness of the curtain less the aggregate thickness of the coverings, but not in any case less than five (5) inches.

(d) **Curtain Frames.**

Item 1. Framework. The curtain shall have a rigid structural framework of steel, covered either with sheet metal and insulation or with asbestos cloth, as provided under paragraph (e) of this section. Members of the framework shall consist of tubes, structural shapes or bars, horizontal members being spaced not more than eight (8) feet apart, and vertical members not more than ten (10) feet apart.

Item 2. Yielding Pad. The bottom of the curtain shall have securely fastened thereto a yielding pad entirely of non-combustible material, not less than five (5) inches in thickness and height, to form a seal against the stage floor.

Item 3. Guide Shoes. For engagement with the guides, the sides of the curtain shall be equipped with sliding metal shoes which for a curtain twenty-eight (28) feet or less in width may be of cast iron, and for a curtain over twenty-eight (28) feet in width shall be of cast or forged steel smoothly finished; provided however, that in lieu of sliding shoes the curtain may be equipped with wheeled trolleys.

(e) **Curtain Coverings.**

Item 1. Metal Curtains. A curtain covered with sheet metal and insulation, for the purposes of this ordinance, shall be designated as a "metal curtain." Each metal curtain shall have its front or auditorium side covered with well fitted steel sheets, not less than three hundred seventy-five ten-thousandths (.0375) inch in thickness. Such sheets shall be applied directly to the curtain frame and well secured in place, all joints between sheets being properly sealed with fire-resistive cement. The back or stage side shall be covered with insulation consisting of two (2) inch thick fused asbestos air-cell boards, weighing not less than three and six-tenths (3.6) pounds per square foot, made of alternate flat and corrugated asbestos sheets cemented together and fused under a temperature of not less than one thousand

(1,000) degrees Fahrenheit; provided however, that in lieu of such fused asbestos air-cell boards, asbestos insulation boards, one (1) inch thick, weighing not less than four and six-tenths (4.6) pounds per square foot, having both surfaces formed of solid asbestos mill board, not less than one-fourth (¼) inch thick, with a core formed of two (2) thicknesses of corrugated asbestos, separated by a single thickness of flat asbestos, may be used. The insulation shall be applied against the curtain framework and secured thereto, by the means of suitable metal fastening devices, and the joints between boards shall be solidly filled with fire-resistive cement. The metal marginal members of the curtain frame on the stage side shall be insulated in the same manner or in other manner fully equivalent thereto.

Item 2. Asbestos Cloth Curtains. A curtain covered with asbestos cloth, shall be designated as an "asbestos cloth curtain." Such asbestos cloth curtain shall have its framework covered on both sides with asbestos cloth applied in full width strips continuous from top to bottom of the curtain. The metal marginal members of the curtain frame shall have the covering cloth extended around them to form insulation of at least two (2) thicknesses for the edges of the curtain. The covering cloth shall be metal-reinforced close-woven asbestos cloth weighing not less than three and one-fourth (3¼) pounds per square yard. The metal reinforcement for the covering cloth shall be incorporated in the yarn before weaving and shall consist of wire, of either monel metal, nickel, brass, chromel, nichrome or other metal or alloy having not less strength at a temperature of one thousand seven hundred (1,700) degrees Fahrenheit and not less resistance to corrosion at ordinary temperature than the least of said metals or alloys. The wire may be either single or double, but the strength in tension of the wire in each strand of yarn shall be not less than seven (7) pounds and the strength in tension of the yarn with the wire shall be not less than twelve (12) pounds when determined on a four (4) inch length between the gripping jaws of the testing machine at ordinary temperatures. The strength of the cloth in tension shall be not less than one hundred eighty (180) pounds per inch of width of warp and eighty-five (85) pounds per inch of width of filling, when tested in strips one (1) inch wide with a four (4) inch length between the jaws of the testing machine at ordinary temperatures. The head of the asbestos fibre of the yarn may contain cotton or other combustible fibre in an amount not to exceed four (4) per cent of the weight of the asbestos. The total carbon content of the cloth shall not exceed two and one-half (2½) per cent of the total weight of fibre. The strips of cloth shall be lapped not less than one and one-half (1½) inch and sewed with two (2) lines of stitching of asbestos thread, having a wire core, which shall be of the same or greater strength than the yarn of the cloth. The cloth shall be secured to the front and back of the curtain by clamping strips of one and one-fourth (1¼) by five-sixteenths (5/16) inch half oval steel or steel strips of other shape and equivalent section, extended entirely around the curtain at all edges and across the curtain, space at intervals of eight (8) inches, measured vertically from center to center, and secured to the structural members of the curtain framework by means of one-fourth (¼) inch bolts or tap screws spaced eight (8) inches on centers. After erection, the asbestos cloth on both sides of the curtain and on the marginal members of the curtain frame shall be filled with a mineral paint having a silicate of soda binder, to which may be added casein in the proportion of not more than four (4) parts casein to ten (10) parts of concentrated solution of sodium silicate. Such paint shall be applied hot and brushed well into the cloth, so as to make it practically gas and smoke tight.

(f) **Curtain Guides and Smoke Pockets.** The curtain shall be guided throughout its travel by rigid steel guides and shall have a smoke pocket at each side. Each guide and smoke pocket shall be designed to form a stop between the curtain and the wall to intercept gas, flame and smoke. Each guide shall be built integral with its respective smoke pocket and shall extend from the stage floor to a height of at least twice the height of the proscenium opening plus five (5) feet. The surfaces of each guide which contact with the guide shoe, or the trolley wheels shall be planed smooth. Each smoke pocket shall consist of not less than a nine (9) inch steel channel guide post and a fourteen by one-fourth (14x¼) inch steel guard plate or shall consist of not less than equivalent sections, exclusive of such other members as may be necessary to form the required stop between the curtain and the wall. Such guides and smoke pockets shall be securely bolted at least every four (4) feet to suitable steel anchors built into the wall, or shall be secured in place by not less than three-fourth (¾) inch bolts so spaced and extended through the wall, or shall be riveted to structural steel framing of the building. Where the main curtain members carry the stresses from lateral pressure on the curtain as suspension tension members, the guides and their attachment to the building shall have adequate strength to safely carry the reactions from such tension members. The top of the curtain shall have a smoke stop fitted to intercept gas, flame and smoke. Such smoke stop shall be of the sand box type conforming to the following description:

(g) **Description of Smoke Stop.** The smoke stop shall consist of two (2) channels formed of sheet steel, not less than six hundred twenty-five thousandths (.0625) inch thick set level and made continuous for the entire width of the curtain. One (1) channel shall be secured to the proscenium wall, being set with its flanges extending upward so that its web will form a bottom, having its free ends closed and being filled with clean, fine sand to a depth of at least three (3) inches. The other channel shall be secured to the top of the curtain, being set with its flanges extending downward so that its web will form a top and with one (1) flange located midway between the flanges and extending down to within one (1) inch of the web of the other channel. The stop shall be otherwise so made and installed that whenever the curtain is closed the channels will engage in such manner that the sand will form a positive seal between them; provided however, that a smoke stop of other character than the sand box type may be used for the top of the curtain, if the type proposed is proven to be equally effective under all circumstances and conditions that may arise in the operation of the curtain.

(h) **Types of Curtains.** The curtain required in any theatre with a Type 2 stage shall be of the one (1) piece type or a two (2) piece sectional type curtain; provided however, that a curtain of other design may be used subject to the provisions of paragraphs (p) and (r) of this section.

(i) **Sectional Type Curtains.** A two (2) piece sectional type curtain may be used for the proscenium opening; if, first, such curtain is a metal curtain meeting the requirements of paragraph (e) of this section; if, second, the sections of the curtain are so made, arranged and installed that when operated in any required manner both sections shall open or close simultaneously, the lower section being made to travel at a rate of speed which is twice that of the upper section; if, third, the joint between sections of the curtain is provided with a smoke stop of the same character as that required for the top of the curtain; and if, fourth, such curtain is otherwise made to comply with so many of the requirements stipulated under this section for a one (1) piece type curtain as are in any way applicable to a sectional type curtain.

(j) **Cables and Rigging and Limit of Width for Drum Type Curtains.**

Item 1. Cables and Drums. Every drum type curtain shall be hung on steel cables, one (1) at each intermediate point not over ten (10) feet apart, there ebing not less than four (4) cables in any case. Each cable shall be not less than one-half (½) inch in diameter, shall be attached to the top of the curtain and also attached to a drum, not less than twenty (20) inches in diameter, all drums being keyed fast to a heavy rolled steel operating shaft. The operating shaft shall run in machined and self-lubricating journals, one (1) on each side of each drum and one (1) at each end. The journals shall rest on substantial non-combustible supports on which no fireproofing shall be required.

Item 2. Counterweights. Such type of curtain shall be balanced by two (2) sets of counterweights, only to such extent as provided under Item 1 of paragraph (l) of this section, one (1) set of counterweights being located at each side of the curtain. Each set of counterweights shall be supported by two (2) cables not less than one-half (½) inch in diameter, each of which shall be attached to a not less than twenty (20) inch diameter drum keyed fast to the operating shaft. The counterweights in each set shall be connected together with one (1) inch tie rods. Each set of counterweights shall operate on not less than two and one-half by two and one-half by three-eighths (2½x2½x⅜) inch steel tee bar guides, or steel guides of other equivalent shapes, secured to the proscenium wall by not less than two and one-half by three-eighths (2½x⅜) inch steel "U" brackets or steel brackets of other equivalent design. Each bracket shall be secured by two (2) bolts built into the wall. The counterweight guides shall be set on the stage floor and continue for the full travel of the counterweights plus at least five (5) feet. The counterweight guides and counterweight hoistway shall be inclosed with a sheet steel guard, not less than six hundred twenty-five ten-thousandths (.0625) inch in thickness, to a height of eight (8) feet above the stage floor.

Item 3. Limit of Width. A drum type curtain shall not be used for any proscenium opening which is more than fifty (50) feet in width.

(k) **Cables and Rigging for Sheave-Type Curtains.**

Item 1. Cables and Sheaves. Every sheave-type curtain shall be hung on steel cables, one (1) at each end and one (1) at each intermediate point not over ten (10) feet apart, there being not less than four (4) cables in any case. Each cable shall be not less than one-half (½) inch in diameter, shall be attached to the top of the curtain and pass over a not less than twenty (20) inch diameter heavy pattern sheave and over a not less than twenty-four (24) inch diameter heavy pattern headblock to the counterweights. The grooves of all sheaves and head blocks shall be machine turned. The sheaves and head blocks shall be mounted on channels, made continuous and parallel to the proscenium wall, on which no fireproofing shall be required. The channels shall rest on substantial non-combustible supports, also on which no fireproofing shall be required.

Item 2. Counterweights. Such type of curtain shall be balanced by counterweights, only to such an extent as provided under Item 1 of paragraph (l) of this section. The counterweights, counterweight guides and counterweight hoistway guard for this type of rigging shall be the same as is required under paragraph (j) for a drum type curtain; provided however, that where six (6) cables or less are required for suspending a sheave-type curtain, there may be but one (1) set of counterweights, and all of the cables shall lead thereto; and provided further, that where more than six (6) cables are required for suspending a sheave-type curtain, there shall be two (2) sets of counterweights, and half of the cables shall lead to one (1) set of counterweights and half to the other.

(l) **Operating Equipment and Control Stations.**

Item 1. Method of Operating Curtains. The opening of any curtain shall be either by hand, hydraulic or electric power machinery; and closing for emergency or automatic operation shall be the same as for normal operation and shall be by gravity, obtained by underbalancing the curtain with reference to the counterweights by not less than one (1) pound per square foot of curtain; provided however, that when the curtain is equipped with sliding shoes of the anti-friction type and smooth, lubricated guides, the underbalancing with reference to the counterweights may be not less than six-tenths (.6) pounds per square foot of curtain; and provided further, that when the curtain is equipped with wheeled trolleys, the underbalancing with reference to the counterweights may be not less than two-tenths (.2) pounds per square foot of curtain. The counterweights shall be adjusted to overcome the frictional resistance resulting from the type of mounting used and to obtain the required closing speed, but the aforesaid underbalances shall not be lightened in any case.

Item 2. Machinery, Apparatus and Equipment. The machinery, apparatus and equipment for operating the curtain shall be of simple design, suitable for the purpose employed and positive in operation. The raising and lowering machinery, apparatus and equipment shall be designed in accordance with rules and practices at least equal to the rules and practices recommended by the American Standards Association for the construction and operation of passenger elevator machinery, apparatus and equipment, which are applicable to such curtain machinery, apparatus and equipment and not in conflict with this ordinance, set forth in the American Standard Safety Code for Elevators, Dumbwaiters and Escalators, dated July, 1931. Proper provisions shall be made for top overtravel of the curtain, the length of which shall be not less than heretofore stipulated. There shall be provided upper terminal stopping devices, arranged to automatically stop the curtain from any speed attained in operation, within the overtravel, independently of the operating device or any other stopping device. All gears and worms shall be inclosed in self-lubricating housings.

Item 3. Hydraulic Power Raising Machinery. For hydraulic machinery used for the raising of the curtain, the water supply shall be taken from the house tank, from another elevated tank or from a suitable accumulator. Subject to the provisions of other chapters of this ordinance, such water supply may be taken from the water supply tank for a standard system of automatic sprinklers.

Item 4. Electric Power Raising Machinery. For electric machinery used for the raising of the curtain, the current supply shall be fused independently of the house supply. All parts of the electric power wiring and equipment shall be otherwise made and installed in conformity with the provisions of other ordinances of the City of Chicago.

Item 5. Opening Control Stations. The opening of the curtain shall be controlled from the stage electrician's station, which shall be located at either side of the stage.

Item 6. Closing Devices and Control Stations. The curtain shall be equipped with a mechanical device which when operated, manually or automatically, shall so function as to cause the curtain to close. The use of an electrical device for such purpose shall not be permitted. Such curtain closing device shall be so made, arranged and equipped that it can be operated manually and instantly from three (3) points as follows:

First. Normally, from the stage electrician's station.

Second. In an emergency, from the stage fireman's station located at the opposite side of the stage from the stage electrician's station.

Third. In an emergency, from a point on the stage not more than five (5) feet from an outside exit doorway. Such curtain closing device shall also be so made, arranged and equipped, that, in case of fire, it shall be operated automatically and instantly as follows: Near the ceiling of the stage, a cord attached to suitable thermostatic devices shall be suspended and the burning of the cord, or the actuating of any of the thermostatic devices, shall in any case operate the closing device of the curtain. Such cord shall, for the purposes of this ordinance, be designated as "the curtain safety cord," and shall be a tarred, hempen cord, located above the gridiron, extended across the entire width of the stage running parallel with the curtain, and passing beneath the stage vent or each such vent at its center. Such thermostatic devices shall be actuated as the result of a difference in the rate of rise in temperature, rather than as a result of an increase in temperature. If there is but one (1) stage vent, there shall be one (1) such thermostatic device located under such vent at its center, and one (1) such thermostatic device located on each side of the stage vent at points half way between the sides of the stage and the stage vent, in such case there being in all not less than three (3) such thermostatic devices. If there is more than one (1) stage vent, there shall be one (1) such thermostatic device located under each such vent at its center, and at least two (2) other such thermostatic devices, each of which shall be located on different sides of the stage at points halfway between the side of the stage and a stage vent, in such case there being in all at least two (2) more such thermostatic devices than there are stage vents. At each of the three (3) points from which the curtain closing device is required to be manually operative, there shall be a conspicuous non-combustible sign on which plain directions as to how to lower the curtain shall be given in letters at least four (4) inches high. Such lettering shall be in white on a black field.

Item 7. Checking Devices. For a curtain for which hand power or electric power raising machinery is employed, there shall be provided for each set of counterweights, one (1) liquid dash pot, having a smooth bored cylinder fitted with plunger and rod, with not less than a sixteen (16) inch sheave on top of rod and necessary cable, or an equivalent automatic device for checking the lowering of the curtain and to cause it to settle on the stage floor without shock. Such checking device shall be located within the inclosure formed by the guard required for the hoistway of the curtain counterweights. For a curtain for which hydraulic power raising machinery is employed, there shall be provisions made for automatically checking the lowering of the curtain and to cause it to settle on the stage floor without shock, which may be accomplished through the lifting machinery, or by such a checking device as before described.

(m) **Character of Curtain to Be Used for Type 2 Stages.** The proscenium opening of every Type 2 stage shall have a curtain meeting the requirements of paragraphs (a) to (l) inclusive of this section; provided however, that an asbestos cloth curtain shall not be used where the proscenium opening exceeds twenty-eight (28) feet in width, or twenty-two (22) feet in height, or the seating capacity of the auditorium exceeds six hundred (600).

(n) **Approval of Installations.** No building hereafter designed, erected, altered, converted or used for the purpose of a theatre, having a Type 2 stage shall be opened to the public for any picture display or theatrical performance, or for any other purpose whatever, unless and until such curtain and all of its machinery, apparatus, equipment and appurtenances have been subjected to and have satisfactorily passed initially the operating tests prescribed in paragraph (o) of this section and have been made to fully comply with the provisions of this ordinance. When such curtain and all of its machinery, apparatus, equipment and appurtenances have passed said initial operating tests and have been found by examination to fully comply with the provisions of this ordinance, approval of the installation shall be given by the Commissioner of Buildings, and a certificate of inspection to that effect shall be caused to be issued to the owner, manager or other person in control of the theatre by said Commissioner. Such certificate shall be posted under glass in a conspicuous place on the stage. Such certificate shall expire in six (6) months from the date of issuance thereof, and it shall be unlawful to operate initially any such theatre within the City of Chicago without such certificate or for more than ten (10) days after expiration thereof. If such curtain and all of its machinery, apparatus, equipment and appurtenances have not passed said initial operating tests, or have been found not to fully comply with the provisions of this ordinance, no such certificate shall be issued and no such approval shall be given by said Commissioner and he shall advise the architect, owner, manager or other person in control of the theatre, in writing, of the reasons for his action.

(o) **Operating Tests.** Upon completion of the installation, the curtain and all of its machinery, apparatus, equipment and appurtenances shall pass a test by the Department of Buildings, which test shall demonstrate that each and all of the devices for operation and control required under this section are properly operative. Such tests shall be made by the Department of Buildings not more than one (1) week preceding the time set for the initial opening of the theatre to the public. Should the curtain, or any part of its machinery, apparatus or equipment, when tested as aforesaid, fail properly to operate, within the true meaning and intent of this ordinance, the cause of such failure shall be removed by the owner, manager or other person in control of the theatre and the required operating tests repeated by the Department until operation of the curtain is found to fully comply with this ordinance.

1307.04 Motion Picture Equipment:

(a) **Projectors.** Every motion picture projector in which nitrocellulose film, or other flammable film is used, together with any electrical device and any other device accessory to the operation or repair of such projector, shall be located within a projection block, meeting the requirements of Section 1302.08 and every such projector shall be contained within a projection room in such block, meeting the requirements of Section 1303.05. Every motion picture projector provided in any theatre shall, except for purposes of replacement or repair, be permanently fastened in place. No projection room or projection block shall be required for any motion picture projector in which acetate cellulose film, or other slow-burning or non-flammable film is used.

(b) **Shelving, Furniture and Fixtures.** All shelving, furniture and fixtures located in any projection block shall be of metal or other non-combustible material.

(c) **Storage of Film on Hand.** Provision for the storage of the permitted amount of flammable film shall be made in metal containers located within the projection block and meeting the requirements of the Fire Prevention Ordinance.

1307.05 Still Picture and Other Light Projection Equipment: No projection room or projection block shall be required for any still picture projector, spotlight, or other light projection device in which acetatecellulose film or other slow-burning or non-flammable film, or glass mounted slides are used.

1307.06 Cooling and Refrigerating Equipment: No cooling equipment, refrigerating equipment, ice making machinery or equipment or other equipment, machinery or piping, for or in which any toxic, irritant, corrosive, flammable or explosive gas is used as the refrigerant or cooling medium shall be installed in any part of any theatre; excepting such small units as permitted under Section 1302.09.

1307.07 Transformers and Capacitors: Every electrical transformer or capacitor, containing oil, installed in any building hereafter designed, erected, altered or converted for the purposes of a theatre and the inclosure therefor, shall conform to the requirements of Article 2112 of the General Provisions; provided however, that no such room shall be located directly under or within twenty (20) feet laterally of the auditorium in any theatre, but such a room shall be permitted elsewhere in any theatre. No doorway to any such room shall be located in the auditorium, in any foyer or lobby or in any corridor or passage directly connected with the auditorium, or any foyer or lobby.

1307.08 Toilet Equipment:

(a) **General.** The number of plumbing fixtures which shall be provided in any theatre shall meet the requirements of this section; provided however, that an additional number of plumbing fixtures shall be provided in auxiliary toilet rooms as required under Section 1303.07, paragraph (c).

(b) **For the Public.** At least one (1) water closet for males and at least one (1) water closet for females shall be provided for each three hundred (300) persons, or fraction thereof, comprising the capacity of the auditorium in every theatre; provided however, that urinals may be substituted for water closets for males to the extent of not more than sixty (60) per cent of the number of such water closets required. There shall be at least one (1) lavatory in every public toilet room.

(c) **For Employees.**

Item 1. For Employees Engaged in Projection Rooms. In the toilet room required in connection with every Type 1 projection room there shall be at least one (1) water closet and at least one (1) lavatory.

Item 2. For Employees Using Stage Dressing Rooms. In the toilet room for males, required in connection with every dressing room section, there shall be at least one (1) water closet provided for each fifty (50) males or fraction thereof; provided however, that urinals may be substituted for water closets to the extent of not more than fifty (50) per cent of the number of water closets required. In the toilet room for females required in connection with every dressing room section there shall be at least one (1) water closet provided for each fifty (50) females, or fraction thereof. In every such toilet room for males and for females there shall be at least one (1) lavatory, unless there is a lavatory provided in every dressing room.

Item 3. For Employees in Other Parts of Theatres. For employees, other than those engaged in the projection room and those using stage dressing rooms, at least one (1) water closet for males and at least one (1) water closet for females shall be provided for each fifty (50) such employees, or fraction thereof; provided however, that urinals may be substituted for water closets for males to the extent of not more than fifty (50) per cent of the number of such water closets required. In every toilet room for employees there shall be at least one (1) lavatory.

Item 4. Employees' Toilets, Where Not Required. Where the number of employees, other than those engaged in the projection block, stage block and dressing room sections, will be twenty (20) or less, the requirements of Item 3 above, shall not apply and no toilet equipment shall be required for this number of such employees; provided however, that such employees shall have access to the toilets for the public.

1307.09 Drinking Fountains: In every theatre there shall be a drinking fountain installed in such location that any person in the auditorium may, without traveling more than one hundred fifty (150) feet horizontally, have access to a drinking fountain located on the same level as such person is seated, or on not more than one (1) floor above or below such level. No drinking fountain shall be installed in any toilet room.

1307.10 Equipment, Acoustical Treatment and Decorations Used Elsewhere Than on Stage: Permanent and temporary furniture, fixtures and other equipment, not otherwise provided for, scenery, draperies, hangings and other decorations used in front of the proscenium wall in the auditorium, or elsewhere than on the stage in any theatre, shall be made to conform with the provisions of the Fire Prevention Ordinance. Acoustical treatment used in such parts of any theatre shall also meet the provisions of Chapter 21—GENERAL CONSTRUCTION PROVISIONS.

1307.11 Lockers: Every locker used for the storage of wearing apparel in any theatre, shall be entirely of non-combustible material.

ARTICLE 1308

Fire Protection

1308.01 Automatic Sprinklers:

(a) **General.** Every Type 2 stage block in any theatre shall be equipped throughout with a standard system of automatic sprinklers meeting the requirements of the Fire Prevention Ordinance. On every such stage there shall be standard sprinkler heads beneath each fly gallery and beneath the stage ceiling or roof; but no sprinkler head shall be installed in or beneath any stage vent. Any basement under the stage, every stage workshop, every storage room, every property room, every other room or space which is appurtenant to the stage and in which combustible material is handled or stored and every dressing room section which is located within the stage block, including every interior vertical means of exit and every horizontal exit connection for every such section, room or space, shall also be equipped throughout with a standard system of automatic sprinklers.

(b) **Sprinkler Heads Over Proscenium Openings.** Wherever any drape, any hanging or any other decoration of combustible material is used or provided for above or alongside of the proscenium opening in the auditorium of any theatre, there shall be provided in the auditorium immediately over the proscenium opening in a sufficient number of automatic sprinkler heads, spaced not more than six (6) feet apart, attached to a standard system of automatic sprinklers, to properly safeguard such drape, hanging or other decoration, irrespective of the fire-retarding treatment thereof.

1308.02 Standpipes: A standard inside standpipe system shall be installed in the stage block of every theatre hereafter erected, except a theatre having a Type 1 Stage. Such standard inside standpipe system shall meet the provisions of the Fire Prevention Ordinance; provided however, that no two and one-half (2½) inch hose connection shall be required; provided further that location of the standpipe risers and the connections on them within a stairway inclosure shall not be required; and provided also as otherwise required under this section. When not contained within the stage block every entire dressing room section, any basement under the stage, every stage workshop, every storage room, every property room, every other room or space which is appurtenant to the stage and in which combustible material is handled or stored and every other basement room or space which is connected with a basement under the stage shall be served by such system of standpipe fire lines. The standpipe risers shall be not less than two (2) inches in diameter; provided however, that in lieu of such a riser any hose connection may be made direct to any riser of the required sprinkler system through a proper valve. Such system of standpipe fire lines shall have standard semi-automatic hose racks with one and one-half (1½) inch hose, so located on each level of the stage block, except the gridiron level, that every part of any such level may be reached by hose of the maximum length permissible. There shall be at least two (2) such hose racks and hose in any basement under the stage, at least two (2) such hose racks and hose on the stage floor level, and at least one (1) such hose rack and hose on each fly gallery if any. On each of said levels, except any fly gallery level, one (1) of the required hose racks and hose shall be located on each side of the proscenium opening, at any point between the side of such opening and the side of the stage.

ARTICLE 1309

Artificial Lighting and Exit Signs

1309.01 Normal Illumination of Exits

(a) **General.** In every theatre, every vertical means of exit, every horizontal exit connection and every exit court, also every passage, passageway, or alley leading from an exit court or from an outside exit doorway, shall be illuminated to facilitate egress, in accordance with the provisions of this section. Every foyer, lobby, corridor, passage, passageway, aisle or other such interior space for use of the public, or for common use of employees, and which is appurtenant to any vertical means of exit, or which is appurtenant to any horizontal exit connection shall also likewise be illuminated. Such illumination for every public exit shall be provided for continuous use during the entire time that the theatre is open to the public and until the theatre has been entirely vacated by the public; provided however, that such exterior illumination as is required need not be provided to operate before sunset. Electric lighting shall be employed at such places within every such aforesaid exit and appurtenant space to maintain the illumination at least to the full intensities hereinafter required. Every such lighting system shall comply with the provisions of the Electrical Code of the City of Chicago.

(b) **Minimum Intensities of Illumination.**

Item 1. Required Means of Public Exit. Every stairway, ramp, fire escape, corridor, passage, passageway, or other such space, serving as a required means of public normal exit, or as such a means of public emergency exit from any theatre, shall be provided with a system of lighting which shall produce intensities of illumination of not less than one (1) foot-candle on the floor at every stair landing and platform, ramp landing and platform, fire-escape landing and platform, corridor angle and intersection, passage or passageway angle and intersection, inside exit doorway leading from any such space in the direction of exit, outside exit doorway, doorway giving access to a fire escape, and like principal points in any other such exit space. Such system of lighting shall also produce intensities of illumination not less than five-tenths (.5) of a foot-candle at every other point on the floor of such stairway, ramp, fire escape, corridor, passage, passageway or other such exit space. Every foyer and lobby, every aisle on the main floor, any mezzanine, or any balcony of the auditorium and every step in any such aisle or elsewhere in the auditorium, in any theatre, shall be provided with a system of lighting which shall produce intensities of illumination of not less than one (1) foot-candle on the floor or tread over its entire area. Every exit court, every passage or passageway leading away from such a court or from an outside exit doorway, and every alley or other space leading away from such a court or passage or passageway or doorway shall be provided with a system of lighting which shall produce intensities of illumination of not less than two hundredths (.02) of a foot-candle on the pavement for a distance of at least fifty (50) feet from every outside exit doorway in such court, passage, passageway, alley or space.

Item 2. Required Means of **Employees' Exit.** Every stairway, ramp, fire escape, corridor, passage, passageway or other such space, serving as a required means of employees' normal exit, or as such a means of employees' emergency exit, from any theatre shall be provided with a system of lighting which shall produce intensities of illumination of not less than five-tenths (.5) of a foot-candle on the floor at every stair landing and platform, ramp landing and platform, fire escape landing and platform, corridor angle and intersection, inside exit doorway leading from any such space in the direction of exit, outside exit doorway, doorway giving access to a fire escape and like principal points in any other such exit space; and such system of lighting shall also produce intensities of illumination of not less than two-tenths (.2) of a foot-candle at every other point on the floor of such stairway, ramp, fire escape, corridor, passage, passageway or other such exit space.

Item 3. Non-Required Means of Exit. Every non-required stairway, ramp or fire escape in any theatre, which by reason of its location and arrangement is in fact a means of exit or part of a means of exit to any way of departure from the theatre, or to any exit court, either for the public or soley for employees, shall be lighted in the same manner and to the same intensities as is prescribed for such respective required exits under Items 1 and 2 of this paragraph.

Item 4. Reduction in Minimum Intensities Permited in Certain Auditoriums. In any auditorium where still pictures, motion pictures or other projections are made by means of directed light, the illumination on the floor of every aisle and at every exit doorway may be reduced to intensities of not less than one-fifth (1/5) of those required under Item 1 of this paragraph. In any auditorium where any theatrical act or performance for which reduced illumination is necessary is presented, the said floor illumination may be likewise reduced. In other parts of every theatre the full intensity of illumination shall be as prescribed under said Item 1.

Item 5. Additional Illumination. Additional illumination of intensity equal to one-fifth (1/5) of that required by Item 1 of this paragraph shall be provided in all normal exit connections, which shall be supplied from the general lighting source, but controlled separately from the required lights and connected with such source ahead or outside of the main circuit breaker which controls the required general lighting circuits.

1309.02 Emergency Illumination of Exits:

(a) **General.** For every emergency exit from the auditorium and for every exit court there shall be provided a separate system of emergency exit lighting arranged to assure continued illumination in cases of emergency caused by the failure of the principal lighting of the building. Every such emergency lighting system shall be a thirty-two (32) volt or a sixty-four (64) volt electrci lighting system, meeting the requirements of Sub-Section 42 of the Electrical Code of the City of Chicago. Such emergency exit lighting system shall be of sufficient capacity to maintain the required illumination of exit signs continuously during the time required under paragraph (b) of Section 1309.03 under normal conditions and in any emergency, and shall be arranged to assume automatically, in the case of failure of the principal lighting source the illumination of every emergency exit from the auditorium and for every exit court and to maintain such illumination for a period of not less than thirty (30) minutes in any emergency. It shall be optional with the owner to supply the additional illumination called for in Section 1309.01, paragraph (b) Item 5, from this generator battery set in lieu of a separate connection with the principal source of current. The lamp sockets used in connection with every such emergency exit lighting system shall be of the standard intermediate base type, to render difficult the substitution of standard medium base type lamps therein. No lamp socket in connection with any such emergency exit lighting system shall be used in combination with a lamp socket of any other lighting system in any lighting fixture or elsewhere.

(b) **Minimum Intensities of Illumination.** Such emergency exit lighting system shall produce intensities of illumination not less than one-fifth (1/5) of what is required under paragraph (b) of 1309.01 on the floor at the principal points and at every other point on the floor of the exits described in said paragraph (b).

1309.03 Exit Signs:

(a) **General.** Signs shall be provided in every building hereafter designed, erected, altered, converted or used for the purposes of a theatre, to mark the way of egress, in conformity with the requirements of this section. Such signs shall be of three (3) kinds, designated for the purposes of this ordinance as "Standard Exit Signs," "Standard Directional Signs" and "Standard Fire Escape Signs". Every exit sign, directional sign and fire escape sign on any building hereafter erected, altered, converted or used for the purposes of a theatre, shall be made to conform with the provisions of Chapter 30—ARTIFICIAL LIGHTING AND EXIT SIGNS.

(b) **Illumination of Signs.** Every standard exit sign, standard directional sign and standard fire escape sign shall be suitably illuminated by an electric source giving an intensity of not less than five (5) foot-candles on the illuminated surface. Such illumination shall be from lights supplied from circuits of the wiring system for the emergency exit lighting required in exits under Section 1309.02 but controlled separately from such lighting. Such illumination of signs for every public exit shall be provided for continuous use during the entire time the theatre is open to the public and until the theatre has been entirely vacated by the public.

(c) **Standard Exit Signs Required.** In every theatre there shall be standard internally illuminated exit signs in the following locations:

Item 1. Public Exit Doorways. At every doorway through which it is necessary to pass along any route of normal or emergency egress from any seat in the auditorium, or from any room or space for the use or accommodation of the public; provided however, that no exit sign shall be required for any opening having no door or other closure therefor, through which it is necessary to pass along any route of normal or emergency egress from any seat in the auditorium, if a standard exit sign or a standard directional sign is so located that it will be visible through such opening from the approach to the opening at a distance of twenty-five (25) feet or more from the opening; and provided further, that not more than one (1) exit sign shall be required for a group of doorways, if the distance between any two (2) adjacent doorways in the group is not more than ten (10) inches measured horizontally between the nearest jambs, and if all the doorways in the group are so located that they will be visible from the approach to the sign at a distance of twenty-five (25) feet, or more, from the sign.

Item 2. Employees' Exit Doorways. At every doorway through which it is necessary to pass along any route of egress from any room or space for the sole use of employees; provided however, that no exit sign shall be required at any doorway leading from any such room or space directly into a corridor, passage or other horizontal exit connection, if a standard exit sign or a standard directional sign is so located in the corridor, passage or other horizontal exit connection that it will be visible from a point therein opposite the doorway; and provided further that no exit sign shall be required for any doorway for which a standard fire escape sign is required.

Item 3. Exit Courts. In every exit court at any doorway or entrance to a passage leading from such a court in the direction of egress.

(d) **Standard Directional Signs Required.** In every theatre there shall be standard directional signs in the following locations:

Item 1. Horizontal Exit Connection. In every foyer, lobby, corridor and passage for the use or accommodation of the public, at every change in the direction of egress; provided however, that no directional sign shall be required at any such point, if at such point a standard exit sign is visible in the direction of egress; provided further, that no directional sign shall be required at any such point in any corridor or passage, if beyond such point in the direction of egress there is no way out of the corridor or passage except through an exit; and provided also that no directional sign shall be required at any such point in any vomitory

Item 2. Vertical Means of Exit. At the point of entrance in the direction of egress to every stairway or ramp for public use, where no doorway occurs at such point; provided however, that no directional sign shall be required at the head of any stairway or ramp in a vomitory.

Item 3. Exit Courts. In every exit court and in any passage leading from such court where, because of any intersection, interception, change in direction or other peculiarity of arrangement, it is reasonably necessary to indicate the proper direction of egress.

(e) **Standard Fire Escape Signs Required.** At every doorway giving direct access to any fire escape, except a stairway fire escape serving as a required vertical means of public emergency exit.

1309.04 Red Lights Prohibited, Required or Permitted:

(a) **Red Lights.** No red light shall be installed or used anywhere in any building hereafter designed, erected, altered or converted for the purpose of a theatre, or for more than one (1) year after the passage of this ordinance which is designed, arranged, maintained or used for the purposes of a theatre; unless first, it be in connection with a required sign for which a red light is called for under this ordinance; or unless second, it be in connection with a non-required sign of same character as a required type; or unless third, it be in connection with a non-required directional sign of different character than the required type; or unless fourth, it be used on the screen in connection with a picture display; unless, fifth, it be used on the stage in connection with a theatrical act or performance, or unless, sixth, it be in such location and so arranged and maintained that it cannot be mistaken for an exit light, or an exit sign as defined under Section 1309.03.

CHAPTER 14

Open Air Assembly Units

ARTICLE 1401

General

1401.01 Open Air Assembly Unit Defined: An open air assembly unit, for the purposes of this ordinance, is hereby defined as a structure, or part of a structure, or group of structures, designed, intended or used for the purpose of assembling a group of persons in the open air.

1401.02 Types of Occupancies: An open air assembly unit shall, for the purposes of this ordinance, include among others, structures for such occupancies as:

Amusement Parks	Bowls
Athletic Fields	Concession Booths
Automobile Speedways	Grand Stands
Aviation Fields	Observation Platforms
Band Stands	Race Tracks
Baseball Parks	Reviewing Stands
Beach Inclosures	Stadiums
Bleachers	Swimming Pools

Tents for Circuses or Public Meetings together with other buildings and occupancies as are necessary to the operation of open air assembly units.

1401.03 Open Air Assembly Units for Fifty (50) Persons or Less: Any structure or part of a structure, designed, intended or used for the seating of a group of fifty (50) persons or less in the open air, but which is not a part of an open air assembly unit, shall meet the requirements of this ordinance for wood frame construction.

1401.04 Open Air Defined: The term "open air" as applied to an assembly unit shall mean that the structure, or group of structures, or part of a structure in or upon which persons assemble to view or to hear an act, speech, contest, performance, display or event shall be open to the air on one (1) or more sides for a horizontal distance equal to not less than thirty-three and one-third (33⅓) per cent of its perimeter, and for a height of not less than eight (8) feet in each seating level or story, except as hereinafter provided for stages.

1401.05 Requirements for Open Air Assembly Units: Every structure, or part of a structure, hereafter designed, erected, altered or converted for the purposes of an open air assembly unit, shall comply with the General Provisions of this ordinance, and in addition thereto shall comply with the special provisions set forth in this chapter.

Any building or structure or part of a building or structure included in an open air assembly unit or as part of an open air assembly unit that houses or encloses both the persons assembled to view or to hear an act, speech, contest, performance, display or event, together with the said act, speech, contest, performance, display or event, shall comply with all the requirements of Chapter 15—PUBLIC ASSEMBLY UNITS, or Chapter 13—THEATRES, as the case mav be, or as such occupation may be classified therewith.

1401.06 Capacity:

(a) **General.** The capacity of an open air assembly unit and every part thereof shall be the number of fixed seats for spectators plus an allowance of one (1) person for each five (5) square feet of floor or ground area designed or used as standing space or for portable 'seats for spectators. The number of fixed seats shall be computed in accordance with the provisions of Section 1408.01, paragraph (b). Floor area of stairways, ramps, aisles, passages or spaces used for access or circulation, shall not be considered in computing the capacity of an open air assembly unit, and shall not be used for portable seats or for standing room.

(b) **Separate Units and Seating Levels.** Where the space occupied by spectators is divided into separate units, such as grand stands or bleachers on opposite sides of a field, or grand stands or bleachers located side by side but not connected by common means of entrance or exit, the capacity of each such separate unit or space shall be computed in the same manner as in Paragraph (a) of this section.

ARTICLE 1402

Lot Occupancy and Court Requirements

1402.01 Frontages Required: Every open air assembly unit as defined and included herein shall have frontages upon one (1) or more open spaces consisting of streets, public alleys not less than ten (10) feet wide leading to a street, or courts or other open spaces not less than ten (10) feet wide which lead directlv to a street or to a public alley leading to a street, and according to capacity as follows:

Capacity	Frontages
800 Persons or less	1 Street
More than 800 and not more than 2500 persons	1 Street and 1 other open space
More than 2500 and not more than 5000 persons	2 Streets or 1 street and 2 other open spaces
More than 5000 and not more than 10,-000 persons	2 Streets and 1 other open space
More than 10,000 persons	1 Street or 1 open space on every side but at least 2 frontages shall be upon streets. One and only one fireproof passageway built as required by Section 1305.07 for exit courts for theatres, but at least ten (10) feet wide may be used in place of one (1) required open space other than a street.

1402.02 Period of Occupancy for Temporary Structures: Any permit issued for the construction or erection of a temporary open air assembly unit shall not entitle its holder to use such temporary open air assembly unit for any period exceeding fourteen (14) consecutive days.

1402.03 Location of Wood Frame Structures: Any permanent open air assembly unit of wood frame construction more than eight hundred (800) square feet in area shall be located not less than sixty (60) feet from any building or property dividing lot line, or the opposite side of any street or alley. Any temporary open air assembly unit of wood frame construction shall be located not less than thirty (30) feet from any building, except as otherwise provided.

1402.04 Location of Other Structures: Any club house, garage, hangar, paddock or stable which is part of an open air assembly unit shall be located not less than sixty (60) feet from any grand stand or bleachers; provided however, that connecting passageways may be erected as provided by Section 1403.08.

ARTICLE 1403

Construction

1403.01 Superior Types of Construction:

(a) **General.** Any open air assembly unit, structure for seating purposes only, having a capacity of more than ten thousand (10,000) persons, or more than one (1) story in height, or having more than one (1) tier of seats, shall be built of non-combustible materials throughout or of a type of construction superior to heavy timber construction.

(b) **Omission of Fireproofing.** Non-combustible structural members, supporting a roof only, and located more than twenty (20) feet above the floor directly beneath, need not be fireproofed. The roof slabs, plates or arches shall have a fire-resistive value of one (1) hour. Metal columns, beams, girders, trusses, purlins and other framing members need not be fireproofed when located not less than twenty (20) feet from any inside lot line, or any other structure. Roof purlins and roof sheathing may be as required for heavy timber construction when no part of the roof trusses or roof framing is less than ten (10) feet above any floor surface and such a structure need not have inclosing walls if located not less than twenty (20) feet from any inside lot line. Where an inclosing wall as required for heavy timber construction or a superior type of construction is provided in any location twenty (20) feet or less from an inside lot line, then metal members protected by such a wall need not be fireproofed.

1403.02 Heavy Timber Construction: Heavy timber construction, or a superior type of construction shall be used for any such open air assembly until not more than one (1) tier of seats in height, having a capacity of more than five thousand (5000) persons if outside the fire limits or more than twenty-five hundred (2500) persons if inside the fire limits, and not exceeding ten thousand (10,000) persons, provided however, that such a structure need not have luclosing walls if located not less than twenty (20) feet from any lot line or any other structure.

1403.03 Ordinary Construction: Ordinary construction, or a superior type of construction shall be used for any open air assembly unit•for seating purposes only not more than one (1) story or one (1) tier in height, having a capacity of twenty-five hundred (2500) persons or less if inside of the fire limits, or a capacity of more than twenty-five hundred (2500) persons but not exceeding five thousand (5000) persons if outside the fire limits, provided however, that such a structure need not have inclosing walls if located not less than twenty (20) feet from any lot line and if all columns and framing members under the seating floor are of the sizes required by this ordinance for heavy timber construction; and provided further, that any interior ceilings, partitions or inclosing guards and fences beneath the seating level, shall be protected by a covering or inclosure of non-combustible materials.

1403.04 Wood Frame Construction: Wood frame construction may be used for any open air assembly unit for seating purposes only having a capacity of twenty-five hundred (2500) persons or less, if outside the fire limits, subject to the provisions of other ordinances relating to frontage consents and zoning.

1403.05 Observation Towers and Amusement Devices: Any observation tower in an open air assembly unit, shall be used exclusively as an observation tower, and shall not be a part of any building or other structure. A roller coaster, scenic railway, water chute, spiral, circular, spherical, riding, sailing, sliding or swinging mechanical device, and which stands not less than thirty (30) feet from any other structure at any point, may be built of unprotected metal, or a superior type of construction of any height; provided however, that such a structure shall have non-combustible floors, and shall have exits as required by Article 1406 of this chapter.

1403.06 Construction of Stages and Platforms:

(a) **Stages in Permanent Open Air Assembly Units.** Every isolated stage block structure in a permanent open air assembly unit shall have inclosing walls of not less than two (2) hour fire-resistive construction. Any trap space shall be included within such walls, and shall be separated from the assembly space and any space underneath the assembly space by a wall of not less than three (3) hour fire-resistive construction. The stage floor may be entirely of wood construction if the stage does not exceed one thousand (1000) square feet in area, or if there is no trap space; otherwise it shall be of not less than one (1) hour fire-resistive construction but may have a wood finish as provided by Chapter 13—THEATRES for a stage in a theatre.

(b) **Platforms.** Isolated platforms of combustible construction may be erected not more than six (6) feet above the grade in any open air assembly unit, but the aggregate floor area of any such platform shall not exceed ten thousand (10,000) square feet.

1403.07 Projection Booths: Rooms or booths for the housing of motion picture projectors using nitrocellulose or equally hazardous films may be either the fixed or the portable type; but every such room or booth shall afford a complete inclosure of non-combustible material of not less than one (1) hour fire-resistive construction, and every doorway to such a room or booth shall have a non-combustible door.

1403.08 Miscellaneous Rooms: Every dressing room, property room, storage room, workshop or similar room, and every canopy or covered structure in an open air assembly unit shall conform to the requirements for the type of construction of the unit in which it is located or to which it is connected.

1403.09 Parking Space: Every parking space for automobiles in any open air assembly unit shall be entirely outside of any structure except structures of semi-fireproof or fireproof construction, and shall be separated from such a structure and its exits by not less than two (2) hour fire-resistive construction, without openings through such construction.

1403.10 Fences and Inclosing Walls: Any fence or inclosing wall of an open air assembly unit which is not a part of a structure, shall comply with the provisions of Chapter 18 for billboards, signboards and fences.

ARTICLE 1404

Size and Location of Rooms

1404.01 Minimum Ceiling Height: The clear height of any space for the use of spectators shall be not less than eight (8) feet at any point.

1404.02 Stages and Platforms:

(a) **Stages for Use With Scenery.** Any stage in a permanent open air assembly unit shall conform to the requirements for a stage of like character and dimensions in a theatre, with the following exceptions:

Item 1. Ceiling. If a ceiling is used over a stage or stage block, such ceiling shall conform to the provisions of Section 1303.03 for stages in a theatre, unless the space in front of the stage opening is open to the sky for a distance of not less than sixty (60) feet from the front of such stage shell or cover.

Item 2. Ventilator. Any stage having a ceiling area of fifty (50) per cent or less of its floor area, or any stage shell or cover having an open space in front of the stage opening as provided in Item 1, shall not be required to have a stage vent.

Item 3. Fire Curtain. A proscenium fire curtain shall be required only when the assembly seating space in front of such stage is inclosed by walls along more than fifty (50) per cent of the perimeter of such assembly space; and if such space is less than fifty (50) per cent inclosed, the openings shall be so arranged that not less than three (3) sides shall have openings therein at the ground level for a height of not less than eight (8) feet and the openings in any one (1) side shall not exceed twenty-five (25) per cent of the perimeter of such space.

(b) **Trap Space.** Any trap space for a stage in an open air assembly unit shall be governed by the provisions of Chapter 13 for such trap space in theatres.

1404.03 Projection Rooms and Booths: All the provisions of Chapter 13 for projection rooms and projection blocks in theatres shall apply to projection rooms and booths in open air assembly units, except that a non-combustible door shall be permitted in lieu of the sixty (60) minute fire-resistive door, and no toilet room shall be required in a projection room and exits may be as provided in Article 1406.

1404.04 Toilet Rooms: Every permanent open air assembly unit shall be provided with separate toilet rooms for males and females, having entrance doors separated by a distance of not less than twenty (20) feet. The horizontal distance necessary to be traveled by any spectator in reaching a toilet room for either sex shall not exceed four hundred (400) feet in any case. In an open air assembly unit to which admission is charged, all required public toilet rooms shall be located inside the ticket gates. Where dressing rooms are provided for actors, athletes or other performers, separate toilet rooms for males and females shall be provided at convenient locations and shall be designated as "Employees Toilet Rooms."

1404.05 Protective Railings: Every floor level, upper deck, balcony, ramp or stairway, which is more than two (2) feet above grade or ground level, or at any height above another open air floor occupied by the public shall have in all wall openings or open sides thereof a solid curb not less than four (4) inches high and also a protective railing not less than three (3) feet, six (6) inches above the floor, of solid non-combustible material, or in lieu thereof, a wire mesh inclosure with any openings therein one and one-half (1½) inches or less in any dimension, and of the strength required by this ordinance. Every space occupied by the public adjoining any water or other vertical offset having a depth of more than three (3) feet shall have a protective railing at each such offset not less than three (3) feet high with openings not to exceed ten and one-half (10½) inches in one (1) dimension and of the strength required by this ordinance.

1404.06 Protective Floor Guards: Any floor or seating space of an open air assembly unit which has openings therein exceeding one-half (½) inch in any dimension and which is above another floor occupied by the public, shall have thereunder a protective guard of wire mesh or other non-combustible material with any openings therein one-half (½) inch or less in any dimension.

1404.07 Protective Devices: Every elevator, observation tower, roller coaster, scenic railway, water chute, spiral, circular, spherical, riding, sailing, sliding or swinging mechanical device or other amusement device, shall be equipped with public safety appliances and protective handrails, guards or inclosures of such character as shall be approved by the Commissioner of Buildings.

ARTICLE 1405

Multiple Use Buildings

1405.01 Other Occupancies Permitted and Prohibited: Subject to the provisions of Article 1402 and this article, an open air assembly unit may adjoin or be connected to a building with any other occupancy, except a garage of an area greater than one thousand (1000) square feet, and except hazardous use occupancies, and except as otherwise provided in Chapters 6 to 19 inclusive of this ordinance.

1405.02 Separating Construction: Any room, space or structure joined to or forming a part of an open air assembly unit, but used for other purposes, shall meet the requirements for the occupancy which includes such other purposes. Any garage or parking space shall be separated from all other parts of the structure by not less than two (2) hour fire-resistive construction without any doorway or other opening. Every open air assembly unit in multiple occupancy shall be separated from every other occupancy in the same structure by construction of the fire-resistive value required by Section 2110.02.

ARTICLE 1406

Means of Exit

1406.01 General: Every room or space open to the public in an open air assembly unit shall have in addition to the entrance thereto, not less than one (1) exit arranged to permit passage through and from the room toward an outside doorway without returning to the place of entrance; provided however, that any room or space without seating arrangements having a floor area of one thousand (1000) square feet or less may have a single exit doorway.

1406.02 Classification of Exits:

(a) **General.** The means of exit required for every open air assembly unit shall, for the purposes of this ordinance, be designated as either a normal exit, emergency exit or employees' exit as hereinafter provided and as defined in Section 1506.02 for such exits in public assembly units.

(b) **Requirements for Stage and Combustible Scenery.** An assembly room in an open air assembly unit having a stage designed for or used with combustible scenery shall conform to the requirements of Chapter 13 for theatres and for exits in theatres.

1406.03 Width of Normal Exits:

(a) **Width Based on Capacity.** The width of all normal exit doorways, gateways, passages, stairways and ramps in an open air assembly unit shall be computed on the basis of capacity determined in accordance with Section 1401.05 of this ordinance. The width of normal exit doorways, corridors, passageways and horizontal normal exit connections, other than aisles, shall be at a rate of not less than eight (8) inches per one hundred (100) persons served thereby. This minimum width shall be maintained throughout the structure by a cumulative increase toward the outside exit doorways or gateways, so that the width of exit at any point shall be not less than the total required width of exits served thereby. Any building which is a part of a group of open air assembly units but not joined thereto, and which is inclosed with walls, shall have normal exits therefrom as required for its class of occupancy, which shall be maintained toward the outside exit doorways or gateways; provided however, that any building located as required by Section 1402.04 may have outside exits into any open space.

(b) **Minimum Widths of Normal Exits.** Any doorway or gateway used as an exit or for any public purpose in an open air assembly unit shall be not less than three (3) feet in clear width; any horizontal passage or ramp shall be not less than five (5) feet in width; and any stairway or other vertical means of exit shall be not less than four (4) feet in width, except as otherwise provided. Any balcony having a capacity of two hundred fifty (250) persons or less may have a stairway or ramp leading thereto of a clear width of not less than three (3) feet.

(c) **Grand Stands and Other Seating Spaces.**

Item 1. Number and Location of Normal Exits. Every grand stand or other open air assembly unit having seating arrangements shall have normal exit doorways according to capacity as follows:

1000 persons or less	Not less than 2 exit doorways
More than 1000 and not more than 4000 persons	Not less than 3 exit doorways
More than 400 persons	Not less than one additional exit doorway for every four thousand (4000) persons capacity or fraction thereof exceeding 4000.

Item 2. Aisles. In any seating space, the aisles shall be so spaced that there shall be not more than twenty (20) seats in a row between adjacent aisles in a structure of wood frame or ordinary construction, nor more than thirty (30) such seats in a structure of heavy timber or superior type of construction. The number of seats between any aisle and a railing or wall shall be not greater than one-half (½) the allowable number of seats between aisles. The width of aisles at any point shall be computed at a rate of not less than seven (7) inches per one hundred (100) seats for which they provide exit, if such seats have backs; and shall be computed at a rate of not less than six (6) inches per one hundred (100) seats if such seats are without backs. Such aisles shall be not less than three (3) feet wide, and shall be increased in width where necessary in the direction toward exits; provided however, that any aisle having seats on one (1) side only and any aisle leading to box seats may be not less than two (2) feet six (6) inches wide, but the minimum width shall conform to the rate required by this paragraph. There shall be not more than twenty-four (24) consecutive rows of seats between transverse aisles.

Item 3. Seat Spacing. In computing the seating capacity, the width of seats shall be as required by Section 1408.01.

Item 4. The rise of any bank of seats shall not exceed two (2) feet.

1406.04 Horizontal Normal Exit Connections:

(a) **General.** Any longitudinal aisle, passageway, corridor, foyer or lobby which also serves as a way of ingress in an open air assembly unit, shall, for the purposes of this ordinance, be defined as a horizontal normal exit connection and in addition to other provisions of this chapter applying to such spaces, shall meet the following requirements:

(b) **Steps.** Wherever steps are used in a horizontal exit connection, such steps shall extend the full width of the exit connection, and shall conform to the requirements of Chapter 26 of the General Provisions.

(c) **Ramps.** Wherever ramps are used in a horizontal exit connection, the pitch shall be not greater than one and three-fourths (1¾) inches rise per foot of length.

(d) **Obstructions.** There shall be no radiators, doors or other obstructions to exit travel in a horizontal normal exit connection other than handrails having a projection of four (4) inches or less.

(e) **Inside Exit Doorways.** Every horizontal exit connection shall lead to an inside exit doorway or gateway at each end thereof, or to an outside exit doorway or gateway or to a field or other open space leading to an exit gateway.

(f) **Exit Doors.** All normal exit doors shall open outward. Normal exit gateways may have sliding gates.

(g) **Turnstiles.** Any turnstiles or similar devices shall not be considered as a required means of exit.

1406.05 Vertical Means of Normal Exit:

(a) **General.** Any stairway or ramp, or a vomitory with a stairway or ramp therein, which also serves as a way of ingress in an open air assembly unit, shall be considered as a vertical means of exit and in addition to other provisions of this ordinance applying to such spaces shall meet the following requirements:

(b) **Number and Width.** The number of vertical means of normal exit shall be sufficient to meet the requirements for location and the width thereof shall be as required by Section 1406.03 for width of normal exits.

(c) **Location.** Every required vertical means of normal exit shall lead to a horizontal normal connection on the main exit level, and shall be so arranged that the direction of exit travel of the flight nearest the main exit level shall be faced toward an outside normal exit doorway. Every two (2) or more required vertical means of normal exit shall have two (2) most remote means separated at the top thereof by a distance not less than the distance between the longitudinal aisles most remote from each other, as measured along the center line of exit travel toward such means of exit; and not more than fifty (50) per cent of the aggregate required width of vertical means of normal exit shall be located at one (1) such unit.

(d) **Rise and Cut.** In a required stairway, any rise shall be not more than eight (8) inches high any any cut shall be not less than ten and one-half (10½) inches wide.

(e) **Types of Stairways.** Any stairway in an open air assembly unit shall be as required by Article 2601 for Means of Exit.

(f) **Length of Flights.** Any flight of stairs or steps in a means of normal exit except steps in a longitudinal aisle shall have not less than three (3) rises; and any flight shall not exceed eleven (11) feet six (6) inches in height without a landing or platform.

(g) **Ramps and Vomitories.** Any ramp or vomitory used as a vertical means of normal exit shall meet the requirements of Article 2601 for means of exit. Any ramp shall not exceed a rise of ten (10) feet between level landings.

(h) **Towers and Amusement Devices.** Any observation tower or amusement device, built as required by Section 1403.05. may have one (1) or more stairways from the highest level to grade; provided however, that if only one (1) stairway is used, there shall be one (1) other means of exit by elevator, or other mechanical device, or ramp or track, used in the ordinary operation of such structure; and provided further, that the stairway shall be not less than four (4) feet wide of non-combustible material throughout and inclosed for its full height from tread to soffit with wire mesh of not less than number ten (10) gauge, with openings two (2) inches or less in any dimension; or in lieu thereof, a solid wall of non-combustible material with windows as required.

1406.06 Outside Normal Exit Doorways:

(a) **Number.** In every open air assembly unit there shall be not less than two (2) doorways or gateways affording normal exit from the structure directly to a street, or to a public alley or to an open space as provided by Section 1402.01 for frontages. The number of required outside normal exit doorways shall be sufficient to meet the requirements of Section 1406.03 paragraph (c) for number and location of normal exits.

(b) **Location.** The location of outside normal exit doorways or gateways in an open air assembly unit shall be as required by Section 1506.06 for such doorways in a public assembly unit.

1406.07 Emergency Exits: Every roller coaster, scenic railway or other riding, rolling, sailing, sliding or swinging device shall have, in addition to the normal exits therefrom, a means of emergency exit from the main track or slide rails, at intervals of not more than one hundred (100) feet measured along such track or rails; and such means of exit shall be by doorways or stairways not less than two (2) feet six (6) inches in width, or by ladders to grade, with a runway to such ladder not less than one (1) foot six (6) inches wide. The construction of emergency exits required by this paragraph shall be equivalent in fire-resistive value to the construction of required normal exits for such a structure.

1406.08 Employees' Exit:

(a) **General.** Not less than one (1) means of exit shall be provided for every room or space which is occupied by employees but is not regularly open to the public; and at least two (2) means of exit shall be provided for every such room of space which adjoins a stage on which combustible scenery is used, or which is located within a stage block, or which is directly involved in the conduct or execution of any display, act or performance in which combustible scenery is used. The length of any employees' exit shall not exceed three hundred (300) feet measured on the most direct route of travel from said room or space to a point wholly outside the limits of the structure, except as otherwise provided for towers. The width of any employees' exit other than a dressing room or basement shall be not les sthan two (2) feet six (6) inches. Any employees' exit from a dressing room or from a basement space shall be not less than three (3) feet wide.

(b) **Spiral and Winding Stairways.** No spiral or winding stairway shall be permitted for employees' use, except as especially provided by Chapter 13—THEATRES for gridiron stairways in stage blocks.

(c) **Observation Towers.** Any observation tower, built as required by Section 1403.05 shall have in addition to the normal exits provided by Section 1406.05, a means of employees' exit which, if a ladder, shall be not less than one (1) foot six (6) inches wide and inclosed within a shaft as required by Section 1406.05, paragraph (h) for stairway inclosure and if a stairway, shall be non-combustible and not less than two (2) feet wide; provided however, that such a means of employees' exit need not be inclosed except by railings if the height above grade is fifty-five (55) feet or less.

(d) **Projection Booths.** Any projection booth in a tower, separated from any structure by a distance of not less than ten (10) feet may have one (1) ladder exit inclosed with non-combustible material to grade.

ARTICLE 1407

Windows and Ventilation

1407.01 General. Every inclosed room in an open air assembly unit, having a smaller opening to the outer air than that required by Section 1401.03, shall have windows or ventilation as required by Chapter 47—VENTILATION of this ordinance.

1407.02 Projection Booths:

(a) **General.** Every projection room for use with nitrocellulose film in an open air assembly unit, shall be ventilated as required by Chapter 47—VENTILATION, except as otherwise provided by Section 1404.02 and except as follows:

(b) **Exception.** A projection room in an open air assembly space may be ventilated by a gravity ventilating system having an air outlet of not less than two hundred (200) square inches in the ceiling above each projection machine, with a non-combustible duct of not less than two hundred (200) square inches cross-section, extending to a height of at least twenty (20) feet above any part of any floor, passage or walk for the use of spectators. There shall be no damper or other air controlling device in such duct unless it be counter-weighted and so arranged that in case of fire it will be automatically released and opened simultaneously with the shutter or shutters, required in the projection room inclosure for the operation of any light projection device.

ARTICLE 1408
Equipment

1408.01 Seats:
(a) **Number and Spacing** of **Seats:** Seat spacings and the number of seats in a row shall conform to the provisions of Article 1406.

(b) **Width** of **Fixed Seats.** The minimum width of fixed seats in a row in which there are arms or other divisions between sittings, shall be one (1) foot eight (8) inches for eighty-five (85) per cent of such seats and one (1) foot seven (7) inches for the remaining fifteen (15) per cent, such width being measured in each case from center to center of arms or other divisions between sittings. For fixed seats without arms or other divisions between sittings, the minimum width of a sitting shall be computed at one (1) foot six (6) inches.

(c) **Fixed and Portable Seats.** No portable seat or chair shall encroach on any space allotted to fixed seats, or which forms a part of any required means of exit. Every seat in an open air assembly unit shall be securely fastened in place with the following exceptions:

Exception 1. A box or loge may contain twenty (20) or fewer portable seats or chairs; provided however, that the floor area of such box or loge is equal to not less than five (5) square feet for each such seat; and provided further, that no portable seat or chair shall be permitted on any sloping floor having more than one (1) inch pitch to the width of the seat tread.

Exception 2. Portable seats or chairs for spectators may be used on an exhibition or playing field which is overlooked by an open air assembly unit, if there are means of exit meeting the requirements of Article 1406 for such portable seats or chairs in addition to the width required for exits from fixed seats.

1408.02 Stage and Platform Curtains: Any stage or platform in an open air assembly unit may be equipped with a combustible proscenium curtain of the draw, roll or lift type, except as otherwise provided by Section 1404.02. Any proscenium fire curtain required by Section 1404.02 shall conform to the requirements of Chapter 13—THEATRES for such fire curtain in theatres.

1408.03 Electric Wiring and Equipment: All the provisions of Section 1307.07 for electric wiring systems and equipment in a theatre shall apply in an open air assembly unit.

1408.04 Toilet Equipment: Water closets and urinals shall be provided in every permanent open air assembly unit according to capacity, as follows:

Table 1408.04

Type of Occupancy	Water Closets for Women	Water Closets for Men	Urinals for Men
Athletic Fields Automobile Speedways Aviation Fields Baseball Parks Race Tracks	One for each 800 capacity or less	One for each 675 capacity or less	One for each 200 capacity or less
Athletic Bowls and Stadia	One for each 600 capacity or less	One for each 750 capacity or less	One for each 225 capacity or less
Amusement Parks Band Stands Dancing Pavilions Open Air Theatres and all other Open Air Units	One for each 400 capacity or less	capacity or less One for each 1000	One for each 300 capacity or less

The term "capacity" in this paragraph shall include all spaces provided for seating and standing room, as required by Section 1401.05. There shall be at least one (1) lavatory in every toilet room. At least one (1) water closet for males and at least one (1) water closet for females shall be provided for each fifty (50) or fraction of fifty (50) actors, athletes or other performers or employees; provided however, that urinals may be substituted for water closets for males to the extent of not more than fifty (50) per cent of the number of such water closets required.

1408.05 Drinking Fountains: Every permanent open air assembly unit shall be provided with public drinking fountains, which shall be so located that the horizontal distance necessary to be traveled by any spectator in reaching a fountain shall not exceed four hundred (400) feet. The total number of public drinking fountains in any case shall be not less than one (1) bubbler for each two thousand (2000) spectators, or fraction thereof. No drinking fountain shall be installed in any toilet room.

ARTICLE 1409
Fire Protection

1409.01 Automatic Sprinklers for Stages: Every isolated stage block which is designed for, or used with combustible scenery, and is not fully open to the sky, shall be equipped with a standard system of automatic sprinklers.

ARTICLE 1410
Artificial Lighting and Exit Signs

1410.01 Normal Illumination of Exits: Every means of exit in an open air assembly unit, which is within a structure or within twenty (20) feet of any structure, fence, doorway or gateway, shall have a system of electric lighting as required by Chapter 30 for the normal illumination of exits.

1410.02 Exit Signs: In every open air assembly unit having an aggregate capacity of more than eight hundred (800) persons, standard exit signs, standard directional signs, and standard fire escape signs, illuminated by electricity, shall be provided to mark the ways of egress, meeting the provisions of Chapter 30; provided however, that standard exit signs and standard directional signs shall be required only within or upon the structure.

CHAPTER 15
Public Assembly Units

ARTICLE 1501
General

1501.01 Public Assembly Unit Defined: A public assembly unit, for the purposes of this ordinance, is defined as a building, or part of a building, designed, intended or used for the congregating, ga e g or assembling for any purpose of atgroup of persons in one (1) or more rooms, whether such congregation, gathering or assemblage be of a public, restricted or private nature, except theatres, open air assembly units, churches and schools treated in Chapters 13, 14, 16 and 17 and except such other assembly uses as are otherwise classified under this ordinance. Public assembly units used regularly for the exhibition of motion pictures for which admission fees are charged shall conform to the requirements of Chapter 13—THEATRES.

1501.02 Types of Occupancy: A public assembly unit shall, for the purposes of this ordinance, include among others, such occupancies as:

Aquariums	Exhibitions
Armories	Funeral Homes
Art Galleries	Gymnasiums
Athletic Clubs or Rooms	Game Rooms
Auditoriums	Lecture Rooms
Banquet Rooms	Legislative Halls
Bath Houses	Libraries
Boards of Trade	Lodges
Broadcasting Studios	Museums
Cafes	Music Halls
Cafeterias	Natatoriums
Circuses	Passenger Stations
Community or Field Houses	Planetariums
Concert Halls	Reading Rooms
Convention Halls	Restaurants
Council Chambers	Riding Academies
Court Rooms	Rinks
Dance Halls	Roof Gardens
Dining Rooms	Smoking Rooms
	Taverns
	Trading Rooms

except as provided for such occupancies as are otherwise classified under this ordinance.

1501.03 Public Assembly Unit for Ten (10) Persons or Less: Any building, or part of a building, designed, intended or used as a public assembly unit for a group of ten (10) persons or less, shall meet the requirements of this ordinance for ordinary construction.

1501.04 Public Assembly Unit for More Than Ten (10) and Not More Than One Hundred (100) Persons:

(a) **General.** Any building, or part of a building, designed, intended or used as a public assembly unit for a group of more than ten (10) and not more than one hundred (100) persons, shall meet the requirements of this ordinance for schools, except as follows:

(b) **In Buildings of Other Occupancies.** A public assembly unit which is also in an institutional building, theatre, open air assembly unit, church or school, shall meet the requirements of the chapters of this ordinance covering such occupancies.

(c) **Trading Rooms.** A Board of Trade room, Broker's Board room, stock exchange, trading room or other room having a grain pit or trading post therein, shall meet the requirements of Chapter 10—for financial units.

1501.05 Other Terms Defined:

(a) **General.** Other terms used in this chapter are, for the purposes of this ordinance, defined as follows:

(b) **Armory or Rink.** A building, or part of a building, with a floor designed, intended or used as a drilling place, or as place for skating on ice or roller skates, or as a place of playing curling, hockey or similar games on such a floor.

(c) **Athletic Club.** A building, or part of a building, designed, intended or used for athletic or exercising purposes, or for the playing of athletic games; or for such occupancies as athletic clubs, billiard rooms, bowling alleys, exercise rooms, field houses, game rooms, gymnasiums, pool rooms, and turnvereins; except such occupancies as are herein embraced under other definitions.

(d) **Circus.** A building, or part of a building, designed, intended or used for the purpose of exhibition of live stock, or feats of horsemanship, or of performances of animals, or of animal races, or riding academies, with or without acrobatic displays in conjunction with such uses.

(e) **Dance Hall.** A building, or part of a building, designed, intended or used as a place for dancing, including such occupancies as ballrooms, cabarets, casinos, roof gardens or other places having entertainment and eating facilities, or either of them, for the audience or assmblage in conjunction with dancing.

(f) **Eating Place.** A building, or part of a building, designed, intended or used for the purposes of serving food and drink, or either of them, to persons for consumption on the premises, including such occupancies as banquet rooms, cafes, cafeterias, dining rooms, grills, lunch rooms, restaurants, and taverns, except as embraced under other definitions.

(g) **Locker Rooms.** A room, or part of a building, designed, intended or used for the purposes of storage in lockers of wearing apparel or other articles in use by persons, with or without provisions for dressing in conjunction with such lockers.

(h) **Museum or Exhibition Place.** A building, or part of a building, designed, intended or used as a repository for natural, scientific or literary objects of interest, or works of art, or industrial productions, or other productions, specimens, or exhibits arranged for inspection or view, and including such occupancies as aquariums, exhibition buildings, expositions, fairs, galleries, museums and planetariums.

(i) **Promenade.** Any part of a building, designed, intended or used as a place for walking for pleasure or exercise, including arcades, ambulatories, cloisters, colonnades, roof promenades and gardens.

(j) **Reading Place.** A building, or part of a building, designed, intended or used for rea g and library purposes or either of them

(k) **Smoking or Waiting Room.** A building, or part of a building, designed, intended or used for smoking and waiting purposes or either of them, including passenger depot concourses.

(l) **Swimming Bath.** A building, or part of a building, designed, intended or used for swimming purposes, including natatoriums and bath houses.

(m) **Trading Room.** A building, or part of a building, designed, intended or used for the purchase, sale or exchange of stocks or other securities for delivery on the premises or elsewhere; and for the purchase, sale or exchange of grain or other articles of commerce from stock not located on the premises and for delivery elsewhere, including boards of trade, brokers' board rooms, stock exchanges and rooms having a grain pit or trading post.

(n) **Other Terms.** Other terms used in this chapter but not defined herein, shall have the same meaning as defined for general purposes under Chapter 5.

1501.06 Requirements for Public Assembly Units: Every building, or part of a building, hereafter designed, erected, altered or converted for the purposes of a public assembly unit, or a place of public assembly shall comply with the General Provisions of this ordinance and also with the special provisions set forth in this chapter.

1501.07 Capacity:

(a) **Capacity of a Public Assembly Unit.** The capacity of a public assembly unit and of its several parts and rooms, shall be computed as required by Section 2401.02 of the General Provisions of this Ordinance.

(b) **Rooms and Spaces Excepted.** In computing the total capacity of the building, the following rooms or spaces shall not be included: entrance lobbies, foyers and corridors used for access and communication only, kitchens, boiler or furnace rooms, storage rooms, toilet rooms, coat rooms, private offices and similar rooms of an administrative or service use in which persons do not congregate; such portions of aquariums, aviaries, conservatories, exhibitions, stables, zoos, and similar spaces used for the care of animals or exhibits in public assembly units, but not open to the public; bus stations, train sheds, taxicab loading platforms or similar spaces which are open at grade on at least one (1) side, or having an opening or openings without doors equivalent to at least twenty (20) per cent of the perimeter of such space, opening upon an open space at grade and such spaces as may be classified under other occupancy chapters of this ordinance; provided however, that such excepted rooms and spaces shall have exits computed on the basis of their capacity as required by Article 1506 of this chapter.

ARTICLE 1502

Lot Occupancy and Court Requirements

1502.01 Frontages Required:

(a) **Frontages Upon Open Spaces.** Every public assembly unit shall have frontages upon one (1) or more open spaces consisting of streets, public alleys not less than ten (10) feet wide leading to a street, or courts or other open spaces not less than ten (10) feet wide which lead directly to a street or to a public alley leading to a street and according to capacity as follows:

Capacity	Space
200 persons or less....	1 street
More than 200 and not more than 800 persons	1 street and 1 other open space
More than 800 persons.	1 street and 2 other open spaces

(b) **Fireproof Passageway Permitted.** One (1) and only one (1) fireproof passageway, built as required by Section 2601.02 may be used in place of one (1) required open space other than a street.

1502.02 Court Requirements: Every court which is necessary to obtain windows or other openings for required natural lighting and ventilation shall be made to conform with the provisions of Chapter 10—BUSINESS UNITS.

ARTICLE 1503

Construction

1503.01 Fireproof Construction:

(a) **General.** Fireproof construction shall be used for every public assembly unit more than four (4) stories in height, or more than fifty (50) feet above grade in height to the highest floor level, and for every public assembly unit having an aggregate capacity of more than one thousand eight hundred (1,800) persons, except as otherwise provided under this chapter.

(b) **Omission of Fireproofing.** Non-combustible structural members supporting a roof only, and located more than twenty (20) feet above the floor directly beneath, need not be fireproofed. The roof slabs, plates or arches shall have a fire-resistive value of one (1) hour.

1503.02 Semi-Fireproof Construction: Semi-fireproof construction, or a superior type of construction, shall be used for every public assembly unit which is more than two (2) stories in height, and which is not more than fifty (50) feet above grdae in height measured to the highest floor level, and for every public assembly unit having an aggregate capacity of more than eight hundred (800) persons but not more than one thousand eight hundred (1,800) persons, except as otherwise provided under this chapter.

1503.03 Ordinary or Heavy Timber Construction: Ordinary or heavy timber construction, or a superior type of construction, shall be used for every public assembly unit not more than two (2) stories in height and for every public assembly unit having an aggregate capacity of eight hundred (800) persons or less, except as otherwise provided under this chapter.

1503.04 Floor Construction Above Basements: In every public assembly unit of ordinary, heavy timber, or semi-fireproof construction, more than one (1) story and basement in height, and also every assembly room in a public assembly unit of not more than one (1) story in height, the basement and the floor immediately above the basement, if any, shall be of three (3) hour fire-resistive construction.

1503.05 Roof Construction for Certain Annexes: Any public assembly unit may have appended to it a one (1) story or one (1) story and basement structure, used only for assembly purposes, having a roof and its supporting members including columns, unprotected with fire-resistive material; provided however, that such structure shall comply in all respects with the requirements of Chapter 21 of the General Provisions of this ordinance.

1503.06 Assembly Room Separations: Every assembly room in a public assembly unit of ordinary construction, shall be separated from every other part of the unit by partitions not less fire-resistive than wood studding with metal lath and plaster on both sides and by a floor of the character required for such construction, with a metal lath and plaster ceiling thereunder; and every assembly room in a public assembly unit of heavy timber construction shall be separated from every other part of the unit by partitions of one (1) hour fire-resistive construction.

ARTICLE 1504

Size and Location of Rooms

1504.01 General:

(a) **Measurements of Ceiling Height.** For the purposes of this article, ceiling height shall be measured from the floor line of a level floor or from the highest point of a sloping floor to the lowest point of the ceiling; provided however, that beams, girders and ducts may project twelve (12) inches below the required ceiling height.

(b) **Minimum Ceiling Height.** The minimum height of any room in a public assembly unit shall be ten (10) feet, except toilet rooms and spaces below or above a balcony, which shall have a minimum height of eight (8) feet, and except storage rooms, closets, rooms for mechanical equipment or similar service rooms, which shall have a minimum height of seven (7) feet, and except as otherwise provided in Section 1504.02 for assembly rooms.

1504.02 Assembly Rooms and Gymnasiums:

(a) **Ceiling Height.** Every assembly room or gymnasium in a public assembly unit of rangements in a public assembly unit shall have a clear ceiling height of at least twelve (12) feet; except that the clear height need not exceed eight (8) feet either above or below a balcony for the audience.

(b) **Floor Levels in Fireproof Construction.** The floor of an assembly room in any public assembly unit of fireproof construction may be at any height above grade, and shall be not more than fourteen (14) feet below grade at any point.

(c) **Floor Levels in Other Than Fireproof Construction.** The floor of an assembly room or gymnasium in a public assembly unit of other than fireproof construction shall be not more than fifteen (15) feet above grade at any point, except as provided for balconies under Paragraph (d). The lowest point of the main floor of any assembly room or gymnasium thereon shall be not more than seven (7) feet below grade.

(d) **Balconies.** There shall be no spectators' balcony in an assembly room or gymnasium of a public assembly unit of other than fireproof or semi-fireproof construction. In a public assembly unit of semi-fireproof construction there shall be no spectators' balcony located in whole or in part, above any other spectators' balcony; provided however, that gymnasiums, athletic or game rooms, having a combined total capacity on the balcony or balconies of not to exceed fifty (50) persons shall be exempt from the provisions of this paragraph.

(e) **Requirements Where Combustible Scenery Is Provided For or Used.** An assembly room in a public assembly unit having a stage designed for or used with combustible scenery, shall conform to the requirements of Chapter 13—THEATRES.

(f) **Motion Picture Machine Booths.** All motion picture machines for use with nitrocellulose film in a public assembly unit shall be inclosed in a booth or booths, as required by Chapter 17—SCHOOLS.

1504.03 Toilet Rooms: Separate toilet rooms shall be provided for males and females in every public assembly unit, having entrance doors separated by a distance of not less than twenty (20) feet. The entrance door to every toilet room for the assemblage shall be plainly marked for males "MEN" and for females "WOMEN."

1504.04 Storage Rooms and Closets: No storage room or storage closet shall be located under any stairway in any public assembly unit of other than fireproof construction.

1504.05 Rinks: The floor of any room designed, intended or used as a public skating rink by a group of more than twenty-five (25) persons shall be located not more than two (2) feet above grade or the main exit level, or not more than two (2) feet below grade or the main exit level.

ARTICLE 1505

Multiple Use Buildings

1505.01 Other Occupancies Permitted and Prohibited: Subject to the provisions of this article, a public assembly unit may be located in the same building with any other occupancy, except a hazardous use unit prohibited in a public assembly unit under Section 1105.02, and except as otherwise provided in Chapters 6 to 19 inclusive of this ordinance.

1505.02 Separating Construction: Every assembly room and gymnasium in the same building shall be separated from other assembly rooms or gymnasiums and every public assembly unit in multiple occupancy shall be separated from every other occupancy in the same building by construction of the fire-resistive value required in Section 2110.02.

1505.03 Electric Wiring System: The electric wiring system of every public assembly unit shall be separate and distinct from the electric wiring system for parts of a building used for other occupancy; provided however, that current for the entire building may be taken through a common service connection if the main service and such systems are each equipped with an automatic circuit breaker, or with another automatic switching device which shall function likewise.

ARTICLE 1506

Means of Exit

1506.01 General: Every room or space open to the public in a public assembly unit shall have in addition to the entrance thereto, not less than one (1) exit arranged to permit passage through and from the room toward an outside doorway without returning to the place of entrance; provided however, that any proscenium box with twenty (20) seats or less, or any room without seating arrangements having a floor area of one thousand (1000) square feet or less may have a single exit doorway.

1506.02 Classification of Exits:

(a) **General.** The means of exit required for every public assembly unit shall, for the purposes of this ordinance, be designated as either a normal exit, emergency exit or employees' exit as hereinafter provided.

(b) **Normal Exit Defined.** Any means of normal exit shall include either a horizontal normal exit connection, or a vertical means of normal exit, or an outside normal exit doorway as provided by this article, but shall not include any means of exit which is hereinafter defined as a sole emergency exit.

(c) **Emergency Exit Defined.** Any means of emergency exit shall include either a horizontal emergency exit connection, or a vertical means of emergency exit, or an outside emergency exit doorway as provided by this article.

(d) **Employees' Exit Defined.** Any means of employees' exit shall include either a normal or emergency exit which is available for the joint use of the public and employees or any normal or emergency exit which is provided for the sole use of employees.

(e) **Requirements for Stage and Combustible Scenery.** An assembly room in a public assembly unit having a stage designed for or used with combustible scenery shall conform to the requirements of Chapter 13 for exits in theatres.

1506.03 Width of Normal Exits:

(a) **Width Based on Capacity.** The width of all normal exit doorways, passages, stairways and ramps in a public assembly unit shall be computed on the basis of capacity determined in accordance with Section 2401.02 of the General Provisions of this ordinance. The width of normal exit doorways and corridors from any room other than an assembly room, gymnasium, or room having seating arrangements, as provided in paragraph (c) shall be at a rate of not less than one (1) foot, six (6) inches per one hundred (100) persons capacity if on the first floor; and at a rate of not less than one (1) foot ten (10) inches per one hundred (100) persons capacity if on any floor or balcony other than the first floor. This minimum width shall be maintained throughout the building by a cumulative increase toward the outside exit doorways, so that the exit at any point shall be equivalent to the total required width of exits from rooms served by such doorways; provided however, that the total combined width of required stairways or ramps shall be computed according to the greatest capacity of any floor for which they serve as exits and need not be increased for additional floors served thereby, except as required by paragraph (c) for rooms having seating arrangements.

(b) **Minimum Widths of Required Doorways, Passages and Stairways.** Any doorway used as an exit or for any public purpose in a public assembly unit, shall be not less than three (3) feet in clear width; and any passage, stairway or ramp serving as an exit shall be not less than four (4) feet in clear width; provided however, that a balcony having a capacity of two hundred fifty (250) or less may have a stairway and passage leading thereto of a clear width of not less than three (3) feet.

(c) **Assembly Rooms, Gymnasiums and Rooms Having Seating Arrangements.** All doorways, passages, stairways and ramps affording normal exit from any assembly room, gymnasium or other room having seating arrangements in a public assembly unit shall meet the following requirements:

Item 1. Width Based on Capacity. The width of doorways, passages, stairways and ramps required for normal exits from such rooms shall be computed at a rate of not less than one (1) foot, six (6) inches per one hundred (100) persons capacity for all assembly rooms, gymnasiums or other rooms having seating arrangements, and this minimum width shall be maintained by a cumulative increase through all stories toward the outside exit doors.

Item 2. Doorways. Every assembly room, gymnasium, or other room having seating arrangements, and every floor, balcony or tier thereof, shall have normal exit doorways according to capacity as follows:

CAPACITY	DOORWAYS
500 persons or less	Not less than two (2) exit doorways
More than 500 persons and not more than 1,000 persons	Not less than four (4) exit doorways
More than 1,000 persons	Not less than two (2) additional exit doorways for every 1,000 persons capacity or fraction thereof exceeding one thousand (1,000).

Such required normal exit doorways shall be so located that the line of travel from any seat to a normal exit doorway shall be not greater than one hundred fifty (150) feet, and if more than one hundred (100) feet such line of travel shall have not more than one (1) angle or turn.

Item 3. Aisles. In any assembly room, gymnasium or other room having seating arrangements in a public assembly unit of fireproof construction the aisles shall be so spaced that there will be not more than twenty (20) seats in a row between adjacent aisles, and not more than ten (10) seats in a row between an aisle and a wall, if such seats are fixed seats and in a building of other than fireproof construction, or where seats are not fixed seats, there shall be not more than fourteen (14) seats in a row between adjacent aisles and not more than seven (7) seats in a row between an aisle and a wall. The width of the aisles at any point shall be computed at a rate of not less than one (1) foot, ten (10) inches per one hundred (100) persons served, but no aisle shall be less than three (3) feet wide at any point.

Item 4. Seat Spacing. In computing the seating capacity for seat boards without divisions between seats, each one (1) foot, six (6) inches in length of seat boards shall be counted as one (1) seat. Rows of seats with backs shall be not less than two (2) feet, ten (10) inches apart measured from back to back and rows of seats without backs shall be not less than two (2) feet, two (2) inches apart measured horizontally between vertical planes through the rearmost edges of seats in successive rows.

Item 5. Rise of Seats. The rise of any bank of seats shall not exceed one (1) foot, nine (9) inches.

1506.04 Horizontal Normal Exit Connections:

(a) **General.** Any longitudinal aisle, passageway, corridor, foyer or lobby which also serves as a way of ingress in a public assembly unit, shall for the purposes of this ordinance, be defined as a horizontal normal exit connection, and in addition to other provisions of this chapter applying to such spaces shall meet the following requirements:

(b) **Steps.** Wherever steps are used in a horizontal exit connection, they shall extend the full width of such aisle, passageway or other space and the rise and cut of each step shall conform to the requirements of Section 1506.05 for stairways; and the location of any step shall be not nearer than the width of the horizontal exit connection from any end thereof or from any other stairway or exit doorway.

(c) **Ramps.** Wherever ramps are used in a horizontal exit connection, the pitch shall be not greater than one (1) unit rise in five (5) units horizontal, if in an aisle, and not greater than one (1) rise in six (6) horizontal if in any other horizontal exit connection.

(d) **Inside Exit Doorways.** Every horizontal exit connection shall lead to an inside exit doorway at each end unless there is an outside exit doorway at the end thereof; provided however, that the end of any longitudinal aisle nearest the stage may lead to a transverse aisle leading to an emergency exit doorway.

(e) **Exit Doors.** All normal exit doors shall open outward or toward the direction of travel, and shall not project into any means of normal exit; provided however, that normal exit doors may swing outward into any corridor or passageway which is more than seven (7) feet in clear width but such doors shall swing through an arc of not less than one hundred eighty (180) degrees against the wall, and shall not obstruct the clear required width when in such an open position.

1506.05 Vertical Means of Normal Exit:

(a) **General.** Any stairway or ramp, or a vomitory with a stairway or ramp therein, which also serves as a way of ingress in a public assembly unit, shall be considered as a vertical means of exit, and in addition to other provisions of this ordinance applying to such spaces, shall meet the following requirements:

(b) **Number and Width.** Every floor level or balcony, or space occupied by the public, shall have not less than two (2) vertical means of normal exit, which shall be either stairways, as provided by Article 2601, or ramps, as provided by Article 2601, or vomitories as required by this section. The number of vertical means of normal exit required shall be sufficient to meet the requirements for location as given in paragraph (d) and the combined width of such exits shall be as required by Section 1506.03 for width of normal exits.

(c) **Height.** Every stairway or other required vertical means of normal exit shall start at the highest story or level occupied by the public or at the highest level in the part of the building in which it is located.

(d) **Location.** Every required vertical means of normal exit shall lead to a horizontal normal exit connection on the main exit floor level, and shall be so arranged that the direction of exit travel of the flight nearest the main exit floor shall be faced toward an outside normal exit doorway. Any two (2) or more required vertical means of normal exit from a mezzanine or balcony in any assembly room, shall have the two (2) most remote means separated by a distance measured along the center line of exit travel, of not less than one-half (½) the distance between the center lines of the longitudinal aisles most remote from each other as measured along the perimeter of the mezzanine or balcony. The required vertical means of exit from any floor or balcony shall be so arranged that not more than seventy (70) per cent of the aggregate required width of normal exits shall be located on any one (1) side of a room having seating arrangements.

(e) **Rise and Cut.** In a required stairway any rise shall be not more than seven and one-half (7½) inches high, and any cut shall be not less than ten (10) inches wide.

(f) **Length of Flights.** Any flight of stairs or steps in a means of normal exit except steps in a longitudinal aisle shall have not less than three (3) rises; and any flight shall not exceed eleven (11) feet, six (6) inches in height without a landing or platform.

(g) **Vomitories.** Any vomitory used as a vertical means of normal exit, or in lieu of a required stairway shall meet the requirements of this section for required stairways and ramps with respect to number, width, location, inclosure and construction. If any steps are used in such vomitory, there shall be not less than three (3) steps in a flight, and if used in combination with a ramp, such flight of steps shall be separated from any ramp by a level landing not less than four (4) feet long.

1506.06 Outside Normal Exit Doorways—Number: In every public assembly unit there shall be not less than two (2) doorways affording normal exit from the building directly to a street, or to a public alley, not less than ten (10) feet wide, leading to a street, or to a court or other open space not less than ten (10) feet wide leading directly or through a public alley not less than ten (10) feet wide to a street; and if the capacity of the public assembly unit or of the building, exceeds eight hundred (800) persons, there shall be not less than three (3) such doorways; or if the capacity exceeds two thousand (2,000) persons and is not greater than five thousand (5,000) persons there shall be not less than four (4) such doorways and one (1) additional outside normal exit doorway shall be required for each increase of two thousand (2,000) persons capacity.

1506.07 Width of Emergency Exits:

(a) **General.** Any room or story in a public assembly unit, having a capacity of more than two hundred fifty (250) persons, shall have in addition to the normal exits required by this chapter, sole emergency exits as follows:

(b) **Width Based on Capacity.** The width of all emergency exit doorways, passages, stairways and ramps shall be computed on the basis of capacity determined in accordance with Section 2401.02 of the General Provisions of this ordinance. The width of emergency exit doorways, passages, stairways, ramps or other means of emergency exit shall be computed at a rate not less than fifty (50) per cent of the required width of normal exits. This minimum width shall be maintained throughout the building by a cumulative increase toward the outside exit doorways so that the exit at any point shall be equivalent to the total required width of emergency exits; provided however, that the total combined width of required emergency stairways or ramps for any public assembly unit other than assembly rooms or other rooms having seating arrangements, shall be computed according to the greatest capacity of any floor for which they serve as exits and need not be increased for additional floors served thereby, except as required for rooms having seating arrangements.

(c) **Minimum Widths of Required Emergency Exits.** Any required emergency exit shall be not less than the minimum width required by Section 1506.03, paragraph (b) for any exit in a public assembly unit.

(d) **Assembly Rooms, Gymnasiums and Rooms Having Seating Arrangements.** All doorways, passages, stairways and ramps affording emergency exit from any assembly room, gymnasium or other room having seating arrangements shall meet the requirements of Section 1506.03, paragraph (c), except as follows:

Item 1. Width Based on Capacity. The width of emergency exits shall be computed as required by paragraph (b) of this section.

Item 2. Doorways. Every assembly room, gymnasium or other room having seating arrangements, and every floor, balcony or tier thereof shall have emergency exit doorways according to capacity as follows:

CAPACITY	DOORWAYS
More than 250 and not more than 600 persons	Not less than 1 emergency doorway
More than 600 persons	Not less than 2 emergency doorways; provided however, that the number of emergency exit doorways shall be sufficient to meet the requirements for location.

Item 3. Location. Any required emergency exit doorways shall be so located that the horizontal line of travel from any seat to any emergency exit doorway shall be not greater than one hundred fifty (150) feet. Any transverse aisle in an assembly room shall have at each end thereof an emergency exit doorway having a clear width at least equal to the required width of the transverse aisle. Any assembly room having a stage or platform shall have not less than two (2) emergency exit doorways located on opposite sides of the assembly room and between the front of stage or platform and a point which is not more distant from the stage or platform than one-half (½) the distance between the first and last rows of seats.

1506.08 Horizontal Emergency Exit Connections:

(a) **General.** Any transverse aisle, passageway, corridor, foyer, lobby, or outside non-combustible platform, which does not serve as a way of ingress in a public assembly unit, shall, for the purposes of this ordinance, be defined as a horizontal emergency exit connection, and in addition to other provisions of this chapter applying to such spaces, shall meet the following requirements:

(b) **Steps, Ramps and Obstructions.** The provisions for steps, ramps and obstructions in an emergency exit connection shall be as required by Section 1506.04 for normal exit connections.

(c) **Exit Doors.** All emergency exit doors shall open outward as required by Section 1506.04, paragraph (c) for normal exit doors.

(d) **Transverse Aisles.** A transverse aisle in a balcony and a transverse aisle at the rear of a main floor in a stadium type auditorium may serve both as a means of public normal exit and as a means of public emergency exit in which case application of the required widths of such means shall be coincidental and not cumulative.

1506.09 Vertical Means of Emergency Exit:

(a) **General.** Any stairway or ramp which does not serve as a way of ingress in a public assembly unit may be considered as a vertical means of exit, and in addition to other provisions of this ordinance applying to such spaces shall ceet the following requirements:

(b) **Number and Width.** Every floor level or balcony, or other space which is above or below the main exit level, shall have vertical means of emergency exit which shall be sufficient in number and combined width to meet the requirements for all emergency exits leading to such stairway or ramp.

(c) **Height.** Every required vertical means of emergency exit shall be continuous from the story or level which it serves to the main exit level or grade.

(d) **Location.** Every required vertical means of emergncy exit shall lead to a horizontal emergency exit connection or an outside emergency exit doorway, or a court or other open space at grade and shall be so arranged that the direction of exit travel of the flight nearest the main exit level shall be faced toward an outside emergency exit doorway or open space at grade. Any two (2) or more required vertical means of emergency exit from a mezzanine or balcony in an assembly room shall have the two (2) most remote means separated by a distance not less than the width of the assembly room, as measured along the front wall thereof; and not more than seventy (70) per cent of the aggregate required width of emergency exits shall be located on any one side of a room having seating arrangements.

(e) **Rise and Cut.** In a required emergency stairway, any rise shall be not more than seven and one-half (7½) inches high and any cut shall be not less than ten (10) inches wide.

(f) **Length of Flights.** Any flight of stairs or steps in a means of emergency exit shall have not less than three (3) rises and any flight shall not exceed eleven (11) feet six (6) inches in height without a landing or platform.

1506.10 Outside Emergency Exit Doorways:

(a) **Number.** In every public assembly unit required by this chapter to have emergency exits, there shall be not less than one (1) outside doorway for each required means of emergency exit affording exit from the building directly to a street or to a public alley not less than ten (10) feet wide leading to a street, or to a court or other open space not less than ten (10) feet wide leading directly or through a public alley not less than ten (10) feet wide to a street.

(b) **Location.** In a public assembly unit having only one (1) emergency exit doorway, such doorway shall be located on a side other than the side having the principal entrance to the building and if two (2) emergency exit doorways are required, the center lines of such doorways shall be separated by a distance not less than twenty (20) per cent of the perimeter of the building measured on said perimeter; and if more than two (2) emergency exit doorways are required, any two (2) of such doorways shall be separated by a distance not less than fifteen (15) per cent of the perimeter of the building; provided however, that the combined width of all required normal and emergency exits opening into spaces other than streets, shall not exceed the combined width of such open spaces at any point of egress therefrom.

(c) **Landings.** Every outside emergency exit doorway where the means of exit is not a level floor, shall have a level landing at each side thereof not less than four (4) feet long, and extending beyond the door jambs not less than one (1) foot at each jamb unless confined by walls.

1506.11 Employees' Exit:

(a) **Employees' Normal Exit.** Every room or space in a public assembly unit, which is occupied by employees and not regularly open to the public, shall have not less than one (1) means of employees' normal exit which may be any means of normal exit as required by Section 1506.03 or a means of exit for the sole use of employees, or any combination of the two. Such means of exit shall be so located that the vertical means of exit or outside exit doorway shall be not more than one hundred (100) feet from any point in the floor or story served thereby. Every means of employees' normal exit shall be not less than three (3) feet wide. The construction of all such means of exit shall be as required by Sections 1506.04, 1506.05 and 1506.06 for horizontal normal exit connections, vertical means of normal exit and outside normal exit doorways.

(b) **Employees' Emergency Exit.** Every public assembly unit having a stage for combustible scenery, and every basement room occupied by employees shall have means of employee's emergency exit from every part thereof in addition to the employees' normal exit required by this section and such emergency exits shall be as required by Chapter 13—THEATRES for such spaces in theatres.

1506.12 Non-Required Stairways: Every non-required stairway or other vetical means of exit in a public assembly unit, which is either for use of the public or employees, shall conform to the requirements of this chapter for similar required vertical means of exit.

1506.13 Doors in Exits:

(a) **General.** All exit doors for a public assembly unit shall, in addition to other provisions of this chapter, meet the following requirements:

(b) **Doors.** Every outside normal or emergency exit door shall be so arranged that it will not encroach upon a public sidewalk, street, alley or other public space when opened. All exit doors shall conform to the requirements for fire-resistive construction having the value of the required type of construction for the inclosing walls of the rooms or spaces from which such doors open; provided, however, that outside normal exit doors may be of wood.

(c) **Hardware.** Any lock, latch or other fastening device on any required exit door shall be so made that when locked or fastened it may easily be opened by any person on the way out without a key or other opening device; and every outside normal and emergency exit door except the doors of one (1) principal entrance, shall be provided with standard panic hardware; provided however, that exit doors for the sole use of employees are exempt from the provisions of this paragraph.

ARTICLE 1507

Windows and Ventilation

1507.01 Window Facing: Every required window in a public assembly unit shall face upon a street, a public alley, a public park or a yard or court meeting the requirements of Section 1502.02.

1507.02 Rooms Without Seating Arrangements: Every room in a public assembly unit, other than a kitchen, assembly room, gymnasium or other room having seating arrangements, and except entrance lobbies, corridors used for access and communication only, closets and storage rooms, shall have either windows, or skylights, or both, with closures arranged to open so as to provide a total net area of openings for natural ventilation as required by Chapter 47—VENTILATION; or if such openings are not provided there shall be a system of mechanical ventilation which shall meet the requirements of Chapter 47—VENTILATION, applicable to the mechanical ventilation of such a room.

1507.03 Kitchens, Assembly Rooms, Gymnasiums and Rooms Having Seating Arrangements: Every kitchen, assembly room, gymnasium and other room having seating arrangements in a public assembly unit, shall have either natural ventilation or a system of mechanical ventilation as required by Chapter 47—VENTILATION.

1507.04 Projection Rooms: Every projection room for use with nitrocellulose film in a public assembly unit, shall be ventilated as required by Chapter 47—VENTILATION, of this ordinance.

ARTICLE 1508

Equipment

1508.01 Seats:

(a) **Aisle Widths, Spacing of Seats and Spacing of Rows.** The width of aisles, seat spacing and spacing of rows for public assembly units having fixed seats shall be as required by Section 1506.03, paragraph (c).

(b) **Fixed and Portable Seats.** Every seat on any floor of a public assembly unit which is not a level floor shall be a fixed seat, except as otherwise provided under this section. Portable seats may be used on any level floor, level balcony, or bank of seats of an assembly room, or other room, when fastened together side by side in groups of not less than five (5) seats; provided however, that not more than twenty (20) portable chairs may be used in any box or other railed-in inclosure with a level floor and a proper aisle or other means of exit.

1508.02 Stage Curtains: Every opening between a stage and an assembly room, gymnasium or other room having seating arrangements in a public assembly unit, shall be equipped with a fire curtain, as required by Chapter 13—THEATRES, for a proscenium fire curtain in a theatre; provided however, that it shall be permissible to use a draw curtain for a stage or platform which is neither designed for nor used with combustible scenery.

1508.03 Toilet Equipment: The number and location of plumbing fixtures which shall be provided in any public assembly unit shall meet the requirements of Section 1307.08 for theatres.

1508.04 Drinking Fountains: In every public assembly unit, drinking fountains shall be installed in such locations that any person may, without traveling more than one hundred fifty (150) feet horizontally, have access to a drinking fountain located either on the same level or the floor next above or below such level; provided however, that at least one (1) drinking fountain shall be installed for each assembly room, and that no drinking fountain shall be installed in any toilet room. Every drinking fountain shall meet the requirements of Chapter 48—PLUMBING PROVISIONS of this ordinance.

1508.05 Cooling and Refrigerating Equipment: Any cooling equipment, ice-making machinery or equipment, together with piping in connection therewith, installed in any public assembly unit, shall meet the requirements of Chapter 45—MECHANICAL REFRIGERATION, of the mechanical provisions of this ordinance.

ARTICLE 1509

Fire Protection

1509.01 General: Every public assembly unit shall comply with the requirements of the Fire Prevention Ordinance, except as otherwise provided under this chapter.

1509.02 Automatic Sprinklers: Every basement exceeding four thousand (4,000) square feet in aggregate area, if such basement is used in whole or in part for the storage of combustible material, excepting space used solely for the storage of fuel for the heating plant in any public assembly unit, shall be equipped with a system of automatic sprinklers meeting the provisions of the Fire Prevention Ordinance.

1509.03 Standpipes: Any stage or basement, which is required by Section 1509.02 to have automatic sprinklers, shall also be equipped with a standard system of inside standpipes meeting the provisions of the Fire Prevention Ordinance.

ARTICLE 1510

Artificial Lighting and Exit Signs

1510.01 Normal Illumination of Exits: Every means of exit in a public assembly unit shall have a system of electric lighting as required by Chapter 30 for the normal illumination of exits.

1510.02 Exit Signs: In every public assembly unit standard exit signs, standard directional signs and standard fire escape signs illuminated by electricity shall be provided to mark the ways of egress, meeting the provisions of Chapter 30; provided however, that standard exit signs and standard directional signs shall be required only for a public assembly unit having an aggregate capacity in all rooms of more than two hundred fifty (250) persons.

CHAPTER 16.

Churches.

ARTICLE 1601.

General.

1601.01 Church Defined: A church for the purposes of this ordinance, is hereby defined as a building, or part of a building, designed, intended or used as a place of worship and for other religious purposes.

1601.02 Churches for One Hundred (100) or Less: Any building, or part of a building, designed, intended or used as a church and for other religious purposes by a group of one hundred (100) persons or less, shall meet the requirements of this ordinance for ordinary construction; provided, however, that any room used for such purposes shall not be at a higher level than four (4) feet above grade.

1601.03 Requirements for Churches: Every building, or part thereof, hereafter designed, erected, altered, converted or used as a church, shall comply with the General Provisions of this ordinance and with the special provisions set forth in this chapter.

1601.04 Capacity: The capacity of a church and of its several parts and rooms, shall be computed in the manner required by Section 2401.02 of the General Provisions of this ordinance. In computing the total capacity of the building, there shall be excluded the entrance lobbies, foyers and corridors used for access and communication only, offices, private studios, vesting rooms, toilet rooms, kitchens, boiler or furnace rooms, storage rooms and similar rooms of an administrative or service character in which persons do not congregate or in which they congregate only when absent from some other place of assembly in the building.

ARTICLE 1602.

Lot Occupancy and Court Requirements.

1602.01 Frontages Required:

(a) **General.** Every church, with a capacity of six hundred (600) persons or less, shall have a frontage on at least two (2) open spaces, one (1) of which shall be a street, and the other of which shall be a street, a public alley not less than ten (10) feet wide leading to a street, or a court or other open space, not less than ten (10) feet wide, leading directly or through a public alley not less than ten (10) feet wide to a street. Every church with a capacity of more than six hundred (600) persons shall have frontages on at least three (3) open spaces, one (1) of which shall be a street, and the other two (2) of which shall be streets, public alleys, not less than ten (10) feet wide leading to streets, or courts or other open spaces not less than ten (10) feet wide leading directly or through public alleys not less than ten (10) feet wide to streets.

(b) **Fireproof Passageway Permitted.** One (1) and only one (1) fireproof passageway, built as required by Section 2601.02 may be used in place of one (1) required open space other than a street.

ARTICLE 1603.

Construction.

1603.01 Fireproof Construction:

(a) **General.** Fireproof construction shall be used for the basement and the floor immediately above the basement in any church having a total capacity greater than eight hundred (800) persons. Fireproof construction shall be used throughout for every church having a capacity, above the basement, greater than one thousand eight hundred (1800) or in which the highest point of the highest usable floor, balcony or gallery is more tan thirty (30) feet above average grade.

(b) **Omission of Fireproofing.** Fireproofing may be omitted from roof trusses which are located entirely at a height of thirty (30) feet or more above the main floor and which support no construction above the roof other than a spire, fleche or similar structure containing no usable space; provided also that fireproofing may be omitted from roof trusses if there is a ceiling of metal lath and plaster or other not less fire-resistive construction immediately beneath them and they support no structure containing usable space. Noncombustible structural members supporting a roof only and located more than twenty (20) feet above the floor directly beneath need not be fireproofed. The roof slabs, plates or arches shall have a fire-resistive value of one (1) hour.

1603.02 Semi-Fireproof Construction: Except in cases where ordinary or heavy timber construction is used under the provisions of Section 1603.03, semi-fireproof construction, or a superior type of construction shall be used, above the basement for every church, having a capacity above the basement of one thousand eight hundred (1800) or less and in which the highest point of the highest usable floor or balcony is not more than thirty (30) feet above average grade; provided, however, that the basement, if any, and the floor immediately above it shall be of fireproof construction; provided further that unprotected metal roof trusses or wood roof trusses meeting the requirements of heavy timber construction may be used, if located entirely at a height twenty (20) feet or more above the main floor and which support no construction above the roof other than a spire, fleche or similar structure containing no usable space, or if there is a ceiling of metal lath and plaster or other not less fire-resistive construction immediately beneath them and they support no structure containing usable space; and provided also that wood may be used for all roof framing and roof covering supported by such trusses.

1603.03 Ordinary and Heavy Timber Construction: Ordinary or heavy timber construction, or a superior type of construction, shall be used for every church not more than two (2) stories above grade in height, having a total capacity of eight hundred (800) or less.

ARTICLE 1604.

Size and Location of Rooms.

1604.01 General:

(a) **Measurement of Ceiling Height.** For the purposes of this ordinance, ceiling height shall be measured from the floor line of a level floor, or from the highest point of a sloping floor to the lowest point of the ceiling, except that beams and girders may project twelve (12) inches below the required ceiling height.

(b) **Minimum Ceiling Height.** The minimum ceiling height of any room in a church shall be ten (10) feet, except toilet rooms, which shall have a minimum ceiling height of eight (8) feet, except storage and other similar rooms which shall have a minimum ceiling height of seven (7) feet and except as otherwise provided in Section 1604.02, paragraph (a) for assembly rooms.

1604.02 Assembly Rooms:

(a) **Ceiling Height.** Every assembly room in a church shall have a clear ceiling height of at least twelve (12) feet; provided however, that the clear height need not exceed eight (8) feet either above or below a balcony for the audience or congregation; provided further that the provisions of Section 1604.01, paragraph (a) for the projection of beams and girders below the required ceiling height shall not apply to the eight (8) foot clearance required above and below such a balcony.

(b) **Floor Levels in Fireproof Construction.** The floor of an assembly room, in a church of fireproof construction, may be at any height above grade and shall be not more than twenty (20) feet below grade at any point. If such floor is more than twenty (20) feet above grade, or more than ten (10) feet below grade at any point, the exits shall be increased as required in Section 1606.04, paragraph (b).

(c) **Floor Levels in Non-Fireproof Construction.** The main floor of any assembly room, in a church of other than fireproof construction, shall be not more than ten (10) feet above grade at any point, nor more than six (6) feet below grade at any point.

(d) **Floor Levels of Stages and Platforms.** In either fireproof or other than fireproof construction, stages or platforms not intended for occupancy by the audience or congregation may be raised above the level of the main floor of the assembly room.

(e) **Balconies.** There shall be no balcony for the audience or congregation in any church of ordinary or heavy timber construction. In a church of ordinary or heavy timber construction there may be balconies for the use of the organist, the choir or the person, or persons officiating at the service. No such balcony shall have a floor area of more than three hundred (300) square feet. In a church of semi-fireproof construction there may be one (1) balcony for audience or congregation; portions of such balcony may be at different levels and may have exits to more than one (1) floor of the building, but in no case, shall the different levels overlap so that there will be a gallery above a balcony. In a church of fireproof construction there may be a balcony and one (1) or more galleries, one (1) located above another. The highest point of the floor of any balcony shall be not more than thirty (30) feet above average grade in a church of semi-fireproof construction, nor more than twenty (20) feet above grade in a church of ordinary or heavy timber construction.

(f) **Stages With Combustible Scenery Prohibited.** No stage or platform designed for or used with combustible scenery shall be permitted in a church.

(g) **Stages and Platforms Without Combustible Scenery.** A stage or platform, which is neither designed for, nor used with, combustible scenery shall conform to the requirements as to type of construction of the building in which it is located but is not required to meet any other special requirements. Such a stage or platform may be equipped with a draw curtain or curtains.

(h) **Motion Picture Machine Booths.** All motion picture machines for use with flammable film in a church shall be inclosed in a booth, or booths, as required by Chapter 21 of the General Provisions of this ordinance.

ARTICLE 1605.

Multiple Use Buildings.

1605.01 Other Occupancies Permitted and Prohibited: Subject to the provisions of this article, a church may be located in the same building with any other occupancy, except a garage of an area greater than six hundred (600) square feet, except hazardous use occupancies prohibited in a church under Section 1105.02 and except as otherwise provided in Chapters 6 to 19 inclusive of this ordinance.

1605.02 Separating Construction: Every church in multiple occupancy shall be separated from every other occupancy in the same building by construction of the fire-resistive value required by Section 2110.02.

ARTICLE 1606.

Means of Exit.

1606.01 Exit Width:

(a) **Width Based on Capacity.** The width of all doorways, passages, stairways and ramps in any church shall be computed on the basis of capacity determined in accordance with Section 2401.02 of the General Provisions of this ordinance. The width of exit doorways and corridors from any room other than assembly room, as provided in Section 1606.04 shall be at a rate not less than one (1) foot, six (6) inches per one hundred (100) persons capacity; and this minimum width shall be maintained throughout the building by a cumulative increase toward the outer exit doorways so that the width at any point in any exit shall be at a rate not less than one (1) foot, six (6) inches, per one hundred (100) persons served by such doorway or corridor. The width of stairways or ramps from any room other than an assembly room, as provided in Section 1606.04 shall be at a rate not less than one (1) foot three (3) inches per one hundred (100) persons capacity served.

(b) **Minimum Widths at Doorways, Passages and Stairway.** Any doorway used as an exit, or for any public purpose in a church, shall be not less than three (3) feet in clear width; and no other doorway in a church shall be less than two (2) feet, six (6) inches in clear width. No passage or stairway serving as an exit shall be less than four (4) feet in clear width; provided however, that a balcony of a capacity of two hundred fifty (250) persons or less may have a stairway and passage leading thereto of a clear width of not less than three (3) feet.

1606.02 Outside Exit Doorways and Doors:

(a) **Number and Width.** In every church there shall be not less than two (2) doorways affording exit from the building directly to a street, to a public alley, not less than ten (10) feet wide leading to a street or to a court, or other open space, at least ten (10) feet wide, leading directly or through a public alley not less than ten (10) feet wide to a street; and if the capacity of the church exceeds eight hundred (800) persons, there shall be not less than three (3) such doorways. The combined width of such outside exit doorways shall be not less than the aggregate of the required widths of exit doors and corridors from the several rooms within the building, computed at the rates provided in Section 1606.01, paragraph (a) and Section 1606.04, paragrah (b) of this article.

(b) **Location.** In a church having only two (2) required outside exit doorways, the center lines of such doorways shall be separated by a distance not less than twenty (20) per cent of the perimeter of the church, measured on said perimeter. In a church having three (3) required outside exit doorways, the center lines of any two (2) such doorways shall be separated by a distance of not less than fifteen (15) per cent of the perimeter of the church measured on said perimeter.

1606.03 Stairways:

(a) **Number Required.** There shall be not less than two (2) stairways from every story above the first story in a church and not less than two (2) stairways from every story below the first story. If any story above or below the first story in a church, contains an assembly room with a capacity of more than three hundred (300) persons, there shall be not less than three (3) stairways from such story to the first story.

(b) **Location.** Every stairway shall be so located as to discharge directly or through passageways and lobbies to an outside exit doorway located as required in Section 1606.02, paragraph (b). The required exit stairways serving any assembly room shall be so located that one (1) such stairway is directly or through passageway or lobby accessible from each exit doorway of such room required by Section 1606.04, paragraph (a).

1606.04 Exits from Assembly Rooms:

(a) **Doorways and Doors.** There shall be not less than two (2) exit doorways from every assembly room of a capacity of eight hundred (800) persons or less and there shall be at least two (2) normal and two (2) emergency exit doorways from every assembly room which has a capacity of more than eight hundred (800) persons. Every exit doorway from an assembly room in a church shall open through a foyer, or public passage, or directly to a street or public alley, not less than ten (10) feet wide, or to a court or open space not less than ten (10) feet wide, connecting directly or through a public alley not less than ten (10) feet wide with a street. In an assembly room having a capacity of eight hundred (800) persons or less, the required exit doorways shall be located at opposite sides or at opposite ends of the room and in no case adjacent. In an assembly room having a capacity of more than eight hundred (800) persons, the required exit doorways shall be arranged to serve opposite sides of the room in such manner that not more than sixty (60) per cent of the required aggregate width serves one (1) side of the room and not less than forty (40) per cent serves the other. Individual doors shall be as widely separated as practicable.

(b) **Width of Exits.** The combined width of exit doorways and passages from an assembly room, with a capacity of eight hundred (800) persons or less, shall be at a rate not less than two (2) feet, three (3) inches per one hundred (100) persons capacity. The combined width of normal exit doorways and passages from an assembly room, with a capacity of more than eight hundred (800) persons, shall be at a rate not less than one (1) foot, six (6) inches, per one hundred (100) persons capacity and the combined width of emergency exit doorways and passages from such assembly room shall be at a rate not less than nine (9) inches, per one hundred (100) persons capacity. The combined width of stairways and ramps affording exit from an assembly room, with a capacity of eight hundred (800) persons or less, shall be at a rate of not less than one (1) foot, ten and one-half (10½) inches per one hundred (100) persons capacity. The combined width of stairways and ramps affording normal exit from an assembly room with a capacity of more than eight hundred (800) persons shall be at a rate not less than one (1) foot, three (3) inches per one hundred (100) persons capacity and the combined width of stairways and ramps afforing emergency exit from such assembly room shall be at a rate not less than seven and one-half (7½) inches per one hundred (100) persons capacity. If the floor of an assembly room is more than twenty (20) feet above average grade, the aggregate required width of stairways and other exits therefrom shall be increased ten (10) per cent for each five (5) feet or fraction thereof, by which the level of said floor exceeds twenty (20) feet above or ten (10) feet below average grade.

(c) **Aisles.** In any assembly room in a church the aisles shall be so spaced that there shall be not more than fourteen (14) seats in a row between adjacent aisles and not more than seven (7) seats in a row between an aisle and a wall. The width of the aisles at any point shall be computed at the rate of one (1) foot, six (6) inches per one hundred (100) persons served, but no aisle shall be less than two (2) feet, six (6) inches wide at any point.

(e) **Steps in Aisles.** There shall be no steps in aisles, except when necessary, from bank to bank, or from aisle to space between seats. The minimum width of cut of any step in an aisle shall be ten (10) inches and the maximum height of rise of any step in an aisle shall be eight (8) inches.

(f) **Seat Spacing.** Pews shall be spaced not less than two (2) feet, eight (8) inches, measured from back to back. Seats of other type shall be spaced as provided in Section 1506.03.

(g) **Rise of Seats.** The rise from bank to bank of seats shall not exceed one (1) foot, nine (9) inches.

ARTICLE 1607.

Windows and Ventilation.

1607.01 Window Facing: Every required window in a church shall face upon a street, a public alley, a public park, a public waterway, or a court or open space not less than ten (10) feet wide of the character required by Section 1602.01 of this chapter.

1607.02 Ventilation of Rooms Used for Worship: Every room used as a place of worship in a church shall have either windows or skylights, or both, having a combined glass area of not less than five (5) per cent of the floor area, arranged to open for a total area not less than three (3) per cent of the floor area of the room, or a mechanical ventilating system, all as required by Chapter 47—VENTILATION, of this ordinance.

1607.03 Ventilation of Assembly Rooms: Every assembly room in a church, other than a room used as a place of worship, shall have either windows or skylights, or both, having a combined glass area of not less than five (5) per cent of the floor area, arranged to open for a total area not less than the requirements of Chapter 47—VENTILATION of this ordinance.

ARTICLE 1608.

Equipment.

1608.01 Width of Seats: The minimum width of any seat in a bank, in which there are arms or other divisions between sittings, shall be one (1) foot, eight (8) inches, measured from the center of arms or other divisions. Where there are no arms or other separations between sittings, as in a pew, or on a bench, the width of a sitting shall be computed at not more than one (1) foot, six (6) inches.

1608.02 Stage Curtains: In a church, a stage or platform for use without combustible scenery and having a proscenium wall, may be equipped with a draw curtain, provided however, that such curtain is treated with a fire-retarding solution in accordance with the Fire Prevention Ordinance.

ARTICLE 1609.

Fire Protection.

1609.01 Standpipes: A standard system of standpipe fire lines shall be provided in every church which is more than eighty (80) feet in height.

ARTICLE 1610.

Artificial Lighting and Exit Signs.

1610.01 Normal Illumination of Exits: In every church, every required means of exit shall be provided with a system of electric lighting in accordance with Chapter 30 for illumination of exits.

1610.02 Emergency Illumination of Certain Exits: In each assembly room having a capacity of one thousand, eight hundred (1800) or more persons, and throughout the required paths of exit travel from every such room, there shall be provided a separate system of additional electric lighting to facilitate egress in an emergency in accordance with Chapter 30 for illumination of exits.

1610.03 Exit Signs: In every church, standard exit signs, standard directional signs and standard fire escape signs illuminated by electricity shall be provided to mark the ways of egress in accordance with Chapter 30 for illumination of exits.

CHAPTER 17.

Schools.

Article 1701.

General.

1701.01 School Defined: A school, for the purposes of this ordinance, is hereby defined as a building, or part of a building, designed, intended or used as a place of public or private instruction, but not including such occupancies as are otherwise classified under this ordinance.

1701.02 Schools for Ten (10) Persons or Less: Any building, or part of a building, designed, intended or used as a place of public or private instruction, in which the total number of teachers and students for whom seating accommodations are provided in all rooms does not exceed ten (10) shall meet the requirements of this ordinance for a single dwelling; provided however, that any floor above the main exit story shall have exits meeting the requirements of Article 1706.

1701.03 Other Terms Defined:

(a) **General.** Other terms used in this chapter are, for the purposes of this chapter, defined as follows:

(b) **Basement Room.** Any room with a floor level more than two (2) feet below average grade or with a ceiling eight (8) feet or less above average grade.

(c) **Play Room or Recreation Room.** Any room designed, erected, altered, converted or used for no other purpose than that of play or recreation.

1701.04 Classification of Schools:

(a) **General.** Every school shall be classified as a Class One or a Class Two school. Wherever in this chapter there is any provision not stated specifically as applying to one or the other of those two classes, it shall apply to both.

(b) **Class One School.** Class One shall include all rooms or units in schools for the lower grades and high school grades.

(c) **Class Two School.** Class Two shall include all rooms, or units in colleges, or schools for grades higher than high school grades and all commercial and vocational schools or schools for the teaching of the fine arts.

1701.05 Requirements for Schools: Every building, or part of a building, hereafter designed, erected, altered, converted or used for the purposes of a school, shall comply with the General Provisions of this ordinance and also to the special provisions set forth in this chapter. Every building, or part of a building, containing both a Class One and Class Two school shall conform to the requirements for a Class One School, unless the two (2) classes are separated by construction of the fire-resistive value required in Section 2110.02. Where any school room is hereafter designed, erected, altered, converted, or used for the purposes of more than one (1) occupancy, such room shall be made to comply with the requirements of this chapter in respect to each such occupancy and the most restrictive requirement shall govern in any case.

1701.06 Capacity:

(a) **Capacity of a School.** The capacity of a school is the sum of the number of fixed seats, plus an allowance as provided in Chapter 24 of the General Provisions of this ordinance, for any floor area not occupied by fixed seats and their aisles. The following rooms and spaces shall not be included in computing the capacity of a school:

- Assembly Rooms, Gymnasiums and other places of assembly of types suitable only for the use of students when seats are provided elsewhere in the building for all students.
- Bicycle Rooms.
- Cafeterias and other rooms for serving food; also rooms in which provision is made for students to eat lunches brought from the outside.
- Corridors and Offices not used for instruction.
- Boiler and Furnace Rooms,
- Fuel Storage Rooms,
- Locker Rooms,
- Storage Rooms and Toilet Rooms.
- Any other rooms not listed above into which students do not enter;

provided however, that such excepted rooms and spaces shall have exits computed on the basis of their capacity in accordance with the requirements of Article 1706 which are applicable to such exits.

(b) **Capacity of a Room.** The capacity of any room is the sum of the number of fixed seats, plus an allowance based on the provisions of Chapter 24 for any floor area not occupied by fixed seats and their aisles.

(c) **Rooms Classed as Assembly Rooms.** Any room with a floor area of more than one thousand five hundred (1,500) square feet, and used for study or assembly purposes, shall be classed as an assembly room. The capacity of an assembly room is the aggregate capacity of the main floor, balconies, galleries, stages and platforms.

ARTICLE 1702.

Lot Occupancy and Court Requirements.

1702.01 Frontage Required:

(a) **Class One Schools.** Every Class One school shall have frontage on at least two (2) open spaces, one (1) of which shall be a street and the other of which shall be a street, a public alley not less than ten (10) feet wide leading to a street, or a court not less than ten (10) feet wide leading directly or through a public alley not less than ten (10) feet wide to a street.

(b) **Class Two Schools.** Every Class Two school shall have at least one (1) frontage on a street.

1702.02 Location Relative to Lot Lines:

(a) **General.** No required window or doorway of a class room, or study room, shall be nearer than eight (8) feet to a lot line in the case of a Class One school, or nearer than four (4) feet to a lot line in the case of a Class Two school; provided however, that these restrictions shall not apply to any window or doorway which faces a street, public alley, public park or other open public space more than ten (10) feet wide.

(b) **Increase of Width with Height.** The distance from lot lines required by paragraph (a) of this section shall be maintained for the first four (4) stories. Above the fourth story, each such distance shall be increased one (1) foot for each additional story in a Class One school and one-half (½) foot for each additional story in a Class Two school.

ARTICLE 1703.

Construction.

1703.01 Fireproof Construction:

(a) **General.** Fireproof construction shall be used for every school more than four (4) stories above grade in height, or more than fifty (50) feet above grade in height to the fourth floor, except as provided under Section 1703.05.

(b) **Omission of Fireproofing.** Non-combustible structural members supporting a roof only, and located more than twenty (20) feet above the floor directly beneath need not be fireproofed. The roof slabs, plates or arches shall have a fire-resistive value of one (1) hour.

1703.02 Semi-Fireproof Construction: Semi-fireproof construction, or a superior type of construction, shall be used for every school one (1) story above grade in height having a capacity of more than six hundred (600) persons and for every school four (4) stories or less in height and not more than fifty (50) feet above grade in height to the fourth floor; except as otherwise provided under Sections 1703.03 and 1703.05.

1703.03 Ordinary and Heavy Timber Construction: Ordinary or heavy timber construction, or a superior type of construction, shall be used for every school not more than one (1) story above grade in height, having a capacity of not more than six hundred (600) persons, and for every school not more than two (2) stories above grade in height, having a capacity of not more than two hundred (200) persons; except as otherwise provided under Sections 1703.04, 1703.05 and 1703.06.

1703.04 Floor Construction Above Certain Basements: In every school of ordinary, heavy timber or semi-fireproof construction, the floor immediately above the basement, if any, shall be of three (3) hour fire-resistive construction.

1703.05 Roof Construction for Certain Annexes: Any school may have appended to it a one (1) story or one (1) story and basement structure used only for assembly, gymnasium, natatorium or similar purposes having a roof and roof-supporting members, including columns unprotected with fire-resistive material; provided however that such structure shall comply with Chapter 21 of the General Provisions of this ordinance.

1703.06 Portable Buildings:

(a) **General.** Portable buildings of non-combustible frame construction in units not larger than twenty-eight (28) feet by thirty-five (35) feet, nor more than one (1) story high and used solely for school purposes may be erected; provided however, that the roof of each unit shall be covered with metal or other non-combustible material; and provided further that each such building shall have two (2) exit doorways separated by a distance not less than the smallest horizontal dimension of the building and provided also, that each such unit shall in no case be nearer than ten (10) feet to any other building at any point.

(b) **Two (2) Year Limit on Use.** No such unit shall be kept on any lot or block for a period of more than two (2) years after the date of issue of the original permit.

ARTICLE 1704.

Size and Location of Rooms.

1704.01 Measurement of Ceiling Height: For the purpose of this article, ceiling height shall be measured from floor line to lowest ceiling line.. Beams, girders and ducts may project twelve (12) inches below the required ceiling height; provided however, that this provision shall not apply to the eight (8) foot clearance required above or below a balcony.

1704.02 Assembly Rooms and Gymnasiums:

(a) **Ceiling Height.** Every assembly room and gymnasium in a school shall have a clear ceiling height of at least twelve (12) feet; provided however, that such clear height need not exceed eight (8) feet either above or below a balcony.

(b) **Floor Levels in Fireproof Construction.** The floor of an assembly room or gymnasium, in a school of fireproof construction, may be at any height above grade, but shall be not more than six (6) feet below grade at any point. If such a floor is more than twenty (20) feet above grade at any point, the exits shall be increased as required elsewhere in this ordinance.

(c) **Floor Levels and Capacity in Other Than Fireproof Construction.** The floor of an assembly room of gymnasium, in a school of other than fireproof construction, shall be not more than ten (10) feet above grade at any point, nor more than six (6) feet below grade at any point. An assembly room or gymnasium in a school of semi-fireproof construction shall have a capacity of not more than one thousand five hundred (1,500) persons. An assembly room or gymnasium in a school of heavy timber or ordinary construction shall have a capacity of not more than six hundred (600) persons; provided however, that an annex meeting the requirements of Section 1703.05 may have a capacity of not more than one thousand five hundred (1,500) persons.

(d) **Balconies.** There shall be no spectators' balcony in an assembly room or gymnasium of a school of other than fireproof or semi-fireproof construction. In a school of semi-fireproof construction there shall be no spectators' balcony located in whole or in part above any other spectators' balcony; provided however, that a single balcony may have exits to more than one (1) floor of the building.

(e) **Requirements Where Combustible Scenery is Provided For or Used.** In an assembly room, having a stage designed for or used with combustible scenery, the stage, the assembly room and all necessary rooms or spaces, shall conform to the requirements for such rooms or spaces in a theatre as set forth in Chapter 13—THEATERS, it being understood that such an assembly room shall be made to comply with the requirements of a theatre auditorium which are applicable to such a room except as otherwise provided in this chapter. Such a stage or assembly room and accessory rooms or spaces, shall not be located in any building of other than fireproof or semi-fireproof construction.

(f) **Motion Picture Machine Booths.** All motion picture machines for use with nitrocellulose film in a school shall be inclosed in a booth or booths meeting the requirements of Chapter 21 of the General Provisions with the following exceptions:

Exception 1. No inclosure of more than one (1) hour fire-resistive construction shall be required where only one (1) motion picture machine is provided for or installed, unless it be a separation between occupancies in a multiple use building.

Exception 2. No toilet room or water closet shall be required.

Exception 3. Where only one (1) motion picture machine is provided for, or is installed in a booth, a floor area of fifty (50) square feet shall be required. Where two (2) motion picture machines are provided for or installed, an area of seventy-five (75) square feet shall be required. Where more than two (2) motion picture machines are provided for or installed, the entire floor area of the booth shall be the same as required for a theatre. There is no floor area requirement in a school for any machine or projector other than a motion picture projector for use with flammable film.

1704.03 Class Rooms, Study Rooms, Laboratories, Etc.:

(a) **General.** In a Class One school, every class room or study room with a floor area of more than two hundred (200) square feet, shall have a ceiling height of at least twelve (12) feet, except as provided elsewhere and every class room or study room with a floor area of not more than two hundred (200) square feet shall have a ceiling height of at least ten (10) feet. No basement room shall be designed, erected, altered, converted or used for the purposes of study or instruction.

(b) **Commercial Training Rooms.** Office type partitions seven (7) feet six (6) inches or more in height may be installed in any room used exclusively for commercial training; provided however, that such partitions are glazed for their entire height above a level three (3) feet above the floor; and provided further, that no such glazed inclosure has a floor area of more than two hundred fifty (250) square feet; and provided also, that the ventilation of the room which is subdivided by such partitions is equal to the ventilation required for a class room.

1704.04 Cafeterias and Other Rooms for Serving Food: Every cafeteria and other room for serving food in a school, shall have a ceiling height of at least twelve (12) feet, unless it is equipped with a mechanical ventilating system as provided in Chapter 47—VENTILATION, in which case it may have a ceiling height of not less than ten (10) feet, six (6) inches; provided however, that these requirements shall not apply to a room in which provision is made for students to eat lunches brought from the outside.

1704.05 Play Rooms and Recreation Rooms. Every play room or recreation room shall have a ceiling height of at least ten (10) feet.

1704.06 Toilet Rooms: Every toilet room shall have a ceiling height of at least nine (9) feet.

1704.07 Storage Rooms, Closets and Miscellaneous Rooms: No storage room or storage closet shall be located under any stairway in any school of other than fireproof construction.

ARTICLE 1705.

Multiple Use Buildings.

1705.01 Other Occupancies Permitted and Prohibited: Subject to the provisions of this article, a school may be located in the same building with any other occupancy, except a hazardous use unit, prohibited in a school under Section 1105.02, a garage or a theatre and except also as otherwise provided under Chapters 6 to 19 inclusive. No Class One school shall be located above the sixth story in any building, any part of which is used for other than school purposes.

1705.02 Separating Construction: Every Class One school and every Class Two school, in the same building, shall be separated from each other and every school in multiple occupancy shall be separated from every other occupancy in the same building by construction of the fire-resistive value required in Section 2110.02.

ARTICLE 1706.

Means of Exit.

1706.01 Exit Width:

(a) **Width Based on Capacity.** The width of all doorways, passages, stairways and ramps in a school shall be computed on the basis of capacity determined in accordance with Section 1701.06. The width of exit doorways and corridors from any room, other than an assembly room or gymnasium, as provided in Section 1706.03, shall be at a rate of not less than one (1) foot, six (6) inches per one hundred (100) persons capacity. The width of stairways or other vertical means of exit, other than for an assembly room or gymnasium, as provided in Section 1706.03, shall be at a rate of not less than one (1) foot, three (3) inches per one hundred (100) persons aggregate capacity served by such means of exit; provided however, that for Class Two schools, having stairways inclosed, as required elsewhere in this ordinance, the combined width of required stairways shall be computed according to the greatest capacity of any floor for which they serve as exits, and need not be increased for additional floors served thereby, except as required by Section 1706.03.

(b) **Minimum Widths of Required Doorways, Corridors and Stairways.** No doorway used as an exit in a school shall be less than two (2) feet, eight (8) inches in clear width. No main corridor shall be less than six (6) feet in clear and unobstructed width. No stairway shall be less than three (3) feet, eight (8) inches in width and no basement stairway shall be less than two (2) feet, six (6) inches in width, except as otherwise provided in Section 1706.03 for teaching amphitheatres.

1706.02 Corridors:

(a) **Classification of Corridors.** Corridors in a school shall be classified as main and secondary. A main corridor is a corridor connecting any assembly room, gymnasium, class room or other room used for purposes of assembly or instruction with an exit stairway or an outside exit doorway. A secondary corridor is any corridor other than a main corridor.

(b) **Termination of Main Corridors at Stairways.** In Class One schools every main corridor shall terminate, either directly or at a right angle, in access to a stairway.

(c) **Lockers in Corridors.** If lockers are placed in any corridor in a school, the width of such corridor shall be increased so as to maintain the full required clear width when locker doors are open in any position.

(d) **Obstruction of Main Corridors.** No door, or other obstruction, which would interfere with exit travel, shall be placed in any main corridor in a school, but this requirement shall not prohibit the installation of doors, gates or other devices which can be used to shut off a section, or sections of the building for special use; provided however, that each section so formed has all the exit facilities, which under this chapter would be required if the separation were permanent; and provided further, that such door, gate or other device is of a type which, when fully open, is entirely free of the corridor.

1706.03 Exits from Assembly Rooms, Gymnasiums, Etc.:

(a) **Doorways and Doors.** There shall be at least two (2) exit doorways from every assembly room or gymnasium in a school; and there shall be at least three (3) exit doorways from every such assembly room or gymnasium which has a capacity of more than three hundred (300) persons. Every exit doorway from an assembly room or gymnasium in a school shall open into a foyer or public passage, or directly to a street, a public alley not less than ten (10) feet wide leading to a street, or to a court or other open space not less than ten (10) feet wide leading directly or through a public alley not less than ten (10) feet wide to a street. In an assembly room or gymnasium having only two (2) required exit doorways, the vertical center lines of such doorways shall be separated by a distance not less than forty (40) per cent of the perimeter of the room measured on such perimeter. Where three (3) exit doorways are required, they shall be so located that their vertical center lines divide the perimeter of the room into three (3) portions, one (1) of which shall be at least forty (40) per cent of such perimeter, while each of the other two (2) portions shall be at least fifteen (15) per cent of such perimeter.

(b) **Width of Exits.** The combined width of doorways and the combined width of corridors affording exit from any assembly room, gymnasium, balcony or other assembly space with a capacity of more than eight hundred (800) persons, shall be not less than two (2) feet, three (3) inches per one hundred (100) persons capacity. The combined width of stairways affording exit from any such assembly space shall be not less than one (1) foot, ten and one-half (10½) inches per one hundred (100) persons capacity. If the floor of an assembly room is more than twenty (20) feet above average grade, the aggregate required width of stairs and other exits therefrom shall be increased ten (10) per cent for each five (5) feet or fraction thereof, by which the level of said floor exceeds twenty (20) feet above average grade.

(c) **Aisles.** In any assembly room or gymnasium in a school, except in a teaching amphitheatre, the aisles shall be so spaced that there will be no more than fourteen (14) seats in a row between adjacent aisles and not more than seven (7) seats in a row between an aisle and a wall. The width of the aisle at any point shall be computed at the rate of one (1) foot, six (6) inches per one hundred (100) persons served, but no aisle shall be less than two (2) feet, six (6) inches in width at any point. There shall be no steps in any aisle, except where necessary from bank to bank, or from aisle to spaces between seats and except in teaching amphitheatres.

(d) **Seat Spacing.** Rows of seats with backs shall be not less than two (2) feet, eight (8) inches apart, measured horizontally from back to back and rows of seats without backs shall be not less than two (2) feet, two (2) inches apart, measured horizontally between vertical planes through the rearmost edges of seats in successive rows, provided however, that these restrictions shall not apply to teaching amphitheatres, requirements for which are given in Section 1706.04.

(e) **Rise of Seats.** The rise of any bank of seats shall not exceed one (1) foot, nine (9) inches, except in teaching amphitheatres.

1706.04 Aisles, Banks and Seats in Teaching Amphitheatres: The width of any aisle or stair in a teaching amphitheatre, shall be computed at the rate of one (1) foot, six (6) inches per one hundred (100) seats for which it provides exit; provided however, that no aisle or stair shall be less than two (2) feet, six (6) inches wide at any point. There shall be not more than ten (10) seats in a row between aisles and not more than five (5) seats in a row between an aisle and a wall. In a teaching amphitheatre with a seating or standing capacity of not more than twenty (20) persons, rises and cuts of stairs may be of any desired dimension and arrangement. In a teaching amphitheatre of more than twenty (20) persons capacity, rises shall not exceed nine (9) inches in height and cuts shall be not less than nine (9) inches in width and both rises and cuts shall be uniform throughout any stair.

1706.05 Exits from Class Rooms and Study Rooms:

(a) **Doorways and Doors.** Every class room or study room shall have not less than one (1) exit doorway opening to a main corridor or directly to the outer air. If such class room, or study room, has a floor area of more than eight hundred (800) square feet, it shall have not less than two (2) exit doorways, one (1) of which shall open to a main corridor or directly to the outer air and the other of which shall open in like manner or into another class room or study room.

(b) **Aisles.** There shall be an aisle not less than one (1) foot, six (6) inches wide along each wall of every class room or study room in a school. No intermediate aisle shall be less than one (1) foot, four (4) inches wide and no main aisle or cross aisle shall be less than two (2) feet, six (6) inches wide. There shall be not more than fourteen (14) seats in a row between adjacent aisles.

1706.06 Exits from Cafeterias and Other Rooms for Serving Food: Every cafeteria and other room for serving food with a floor area of more than eight hundred (800) square feet shall have two exits spaced as required in Section 1706.03 for exits from an assembly room. The combined width of required exits from such a room shall be not less than two (2) feet, three (3) inches per one hundred (100) persons capacity of said room.

1706.07 Stairways:

(a) **General.**

Item 1. Number. The number of stairways in any school shall be sufficient to meet the requirements for location as given in Item 3 of this paragraph and the combined width of such stairways shall be sufficient to meet the requirements of one (1) foot, three (3) inches per one hundred (100) persons served, as provided in Section 1706.01, paragraph (a); provided however, that there shall be not less than two (2) stairways in any school having a class room, study room or other room in which students congregate, above the main exit story, nor less than two (2) stairways in any school having such a room below the main exit story.

Item 2. Height or Extent. Every stairway shall start at the highest story in a school; or, if the school is not of uniform height, shall start at the highest story in the wing or part of the school in which it is located.

Item 3. Location of Stairways and Stairway Entrances and Exits. Every required exit stairway in a Class One school shall be built against an exterior wall of the building. In any school, the required stairways shall be so located that in any story either above or below the main exit story, the distance from any exit doorway from a class room or study room, to a required stairway entrance, shall not exceed sixty (60) feet where the doorway is located between the dead end of a corridor and the stairway, nor ninety (90) feet in any other case; provided however, that any exit from a basement to a street, to a public alley not less than ten (10) feet wide leading to a street, or to a court or other open space not less than ten (10) feet wide leading directly or through a public alley not less than ten (10) feet wide to a street, shall be considered a stairway if it has the location and width required for a stairway. The distance from a class room or study room exit doorway to a stairway entrance shall be measured on the most direct route of travel from the center line of the exit doorway to the center line of the first riser of the exit flight. Every required or non-required inclosure of a stairway shall have an exit doorway to the outer air at the main exit level of the building.

Item 4. Landings and Platforms. Every stairway in a Class One school shall have a landing at each habitable story.

(b) **Basement Stairways.** Every basement stairway in a school shall lead directly to a room, corridor or other part of the building, from which there is an exit to the outer air, either directly or by way of a corridor, stairway or a combination of a corridor and a stairway or such basement stairway may itself lead directly to the outer air. No basement stairway shall have a cut of less than nine (9) inches nor a rise of more than eight (8) inches, except as provided in Section 1706.04 for teaching amphitheatres.

(c) **Landings for Outside Stairways or Steps.** Where an exit from a school is by way of a stairway or steps located outside the building, there shall be a level landing not less than three (3) feet long between the doorway and said stairway or steps and such landing shall extend inside the building for a further distance of three (3) feet before the commencement of any interior stairway or incline.

(d) **Length and Height of Flights for Class One Schools.** No flight of stairs or steps shall have less than three (3) rises. No flight shall have a greater height than eleven (11) feet, six (6) inches without a level landing or an intermediate platform.

1706.08 Elevators in Schools for Crippled Children: Every floor, above the second floor, in a school for crippled children shall be served by an elevator, or elevators, with a floor area of at least four (4) square feet per person, computed on the basis of a carrying capacity of all elevators sufficient to empty the school, above the second floor, in twenty (20) minutes.

1706.09 Outside Exit Doorways and Doors:

(a) **Number.** In every school, other than a portable school, there shall be two (2) doorways affording exit from the building directly to a street, to a public alley not less than ten (10) feet wide leading to a street, or to a court or other open space not less than ten (10) feet wide leading directly or through a public alley not less than ten (10) feet wide to a street.

(b) **Location.** In a Class One school, having stairways as provided in Section 1706.07, paragraph (a), outside exit doorways shall be located in accordance with Item 3 of said paragraph. In either a Class One or a Class Two school, two (2) outside exit doorways shall be separated by a distance not less than thirty (30) per cent of the perimeter of the school, measured on said perimeter between the door centers.

(c) **Revolving Doors.** No revolving door shall be placed in any Class One school. Revolving doors may be used in Class Two schools; provided however, that they are wholly in addition to the doors required by this ordinance.

1706.10 Exits from Roofs Used for Assembly, Recreation or Instruction: Any roof, or part of roof, used for assembly, recreation or instruction, shall have exits conforming to the requirements for exits from an assembly room as given in Section 1706.03.

1706.11 Exits from Courts: In any school, every inner court, or closed lot line court, shall have an opening at the bottom, not less than three (3) feet in clear width and six (6) feet, six (6) inches in clear height, affording an exit either directly at grade or through a passage leading directly to a main stairway, or a main corridor. Any door or gate, in such an exit shall be without any fastening device which can prevent its opening from the court side by any person at any time.

ARTICLE 1707.

Windows and Ventilation.

1707.01 Window Facing: Every required window in a school shall face upon a street, a public alley, a public park or a yard or court meeting the requirements of Section 1702.02.

1707.02 Class Room and Study Room Windows, Skylights and Ventilation:

(a) **Window and Skylight Area.** In a Class One school, every class room, study room or other room used for purposes of instruction, shall have windows or skylights, or both windows and skylights with a total net glass area equal to at least twenty (20) per cent of the floor area of the room, except in the laboratories, demonstration rooms or teaching amphitheatres in which teaching or demonstration by such lighting is impractical.

(b) **Window Location.** Windows to be considered in the requirements of paragraph (a) of this section must be located wholly on one (1) side of the room.

(c) **Height of Windows.** In any class room or study room of a Class One school, the top of every required window shall be at a height above the floor equal to not less than one-half (½) the width of the room; said height being measured to the top of the window opening at the exterior face of the wall.

(d) **Ceiling Lights.** If a ceiling light is used under a skylight it shall be at least as large as the area of the skylight well and both ceiling light and skylight shall be twenty-five (25) per cent greater in area than is required when a skylight alone is used.

(e) **Ventilation.** Every class room, study room, or other room used for purposes of instruction, shall be ventilated according to one (1) or the other of the two (2) following cases:

Case 1. There shall be windows, skylights or other openings with closures arranged to open, so as to provide a total net area of openings for natural ventilation equal to not less than eight (8) per cent of the floor area of the room.

Case 2. There shall be a system of mechanical ventilation which shall meet the requirements of Chapter 47, which are applicable to the mechanical ventilation of such a room.

1707.03 Stairway Windows: There shall be a window, with not less than twelve (12) square feet of glass area, at every landing or intermediate platform of a stairway in a Class One school. Every window in a required or non-required inclosure of such a stairway shall be a Type A window.

1707.04 Ventilation of Assembly Rooms: Every assembly room in a school shall have windows or skylights, or other openings with closures, or all of them, arranged to open so as to provide a total area for natural ventilation of not less than the percentage of floor area required by Chapter 47—VENTILATION of this ordinance; or shall have a mechanical ventilating system as required by Capter 47—VENTILATION.

1707.05 Cafeteria and Lunch Room Lighting and Ventilation: Every cafeteria and other room for serving food and every room in which provision is made for students to eat lunches brought from home in any school, shall be provided with natural lighting from either windows, skylights or both, having a total net glass area not less than ten (10) per cent of the floor area of the room. Every such room shall have windows, skylights or other openings with closure, or all of them, arranged to open so as to provide a total area for natural ventilation of not less than five (5) per cent of the floor area of the room. If the room contains any equipment for cooking or warming food, there shall also be a mechanical ventilating system which shall meet the requirements of Chapter 47—VENTILATION, which are applicable to such a system.

ARTICLE 1708.

Equipment.

1708.01 Seats:

(a) **Width.** The requirements of this paragraph shall apply to all fixed seats, except in a teaching amphitheatre, the requirements for which are given in Section 1706.04. The minimum width of a seat in a row in which there are arms or other divisions between sittings shall be one (1) foot, eight (8) inches measured from center to center of arms or other divisions. Where seats are arranged without arms or other divisions between sittings, the width of a sitting shall be computed at one (1) foot, six (6) inches.

(b) **Seat Spacing and Aisles.** In any assembly room or gymnasium the spacing of seats, the rise of banks and the width and location of aisles shall be as required in Section 1706.03, paragraphs (c), (d) and (e).

1708.02 Proscenium Fire Curtains: Every opening between a stage and an assembly room, or other room, in any school, shall be equipped with a fire curtain, meeting the requirements of Article 1307 for a proscenium fire curtain in a theatre: provided however, that in any school, it shall be permissible to use a draw curtain for a stage or platform which is neither designed for nor used with combustible scenery.

1708.03 Stage Scenery, Paraphernalia and Equipment:

(a) **General.** Drop curtains, draw curtains, borders, wings, draperies, hangings, stage sets and other scenery, stage paraphernalia of every sort, sheaves, pulleys, cables, their supports and other equipment used back of the curtain in any school shall be made to conform with the requirements of the Fire Prevention Ordinance. Acoustical treatment used on any stage shall meet the provisions of Chapter 21 of the General Provisions.

(b) **Scenery in Basements.** No scenery shall be installed or provided for in any basement of any school.

1708.04 Toilets, Lavatories and Drinking Fountains:

(a) **Doors and Signs.** Entrance doors to toilet rooms for males and females in any school shall be separated by a distance of not less than twenty (20) feet, and shall be plainly market for males "Boys" or "Men" and for females "Girls" or "Women".

(b) **Water Closets and Urinals.** The proportion of sexes in a school shall be assumed as one-half (½) male and one-half (½) female, unless the school is exclusively for students for one (1) sex. Water closets and urinals shall be provided to serve each story on the basis of the combined capacity of all the rooms in that story computed in accordance with Section 1701.06 and shall be located either in that story or in the first story above or below it; provided however, that the capacity of assembly rooms and gymnasiums shall not be considered in determining the number of water closets and urinals to be provided. The number of water closets and urinals required shall be as follows:

For each fifty (50) males or fraction thereof—1 Water Closet.

For each twenty-five (25) females or fraction thereof—1 Water Closet.

For each twenty-five (25) males or fraction thereof—1 Urinal.

Water closets and urinals for use in connection with rooms used exclusively by regular occupants of the school, such as rest rooms or gymnasium locker rooms, may be included in the total number of such fixtures required.

(c) **Lavatories.** There shall be at least one (1) lavatory in every toilet room.

(d) **Drinking Fountains.** There shall be at least one (1) drinking fountain for each two hundred fifty (250) students; and at least one (1) such fountain in each story. No drinking fountain shall be located in any toilet room.

1708.05 Cooling and Refrigerating Equipment: No cooling equipment, refrigerating equipment, ice making machinery or equipment, piping in connection with such equipment or machinery, or other equipment, machinery, or piping, for or in which any toxic, irritant, corrosive, flammable or explosive gas is used as the refrigerant or cooling medium, shall be installed in any part of any school unless such equipment, machinery or piping comprises a small refrigerating or cooling unit meeting the requirements of Chapter 45—MECHANICAL REFRIGERATION of this ordinance.

ARTICLE 1709.

Fire Protection.

1709.01 Fire Alarms:

(a) **General Requirements.** Every school shall be equipped with a fire alarm system meeting the requirements of the Fire Prevention Ordinance.

(b) **Schools for the Deaf.** In every school for the deaf, there shall be an electric light with a red globe near each teacher's desk. Every such light shall be arranged to operate in addition to, and in connection with, the sounding devices required in connection with the fire alarm system.

ARTICLE 1710.

Artificial Lighting and Exit Signs.

1710.01 Room Illumination: Equipment for electric lighting shall be provided in all rooms in a school in accordance with the following table:

TABLE 1710.01.

Illumination.

Room	Intensity of Illumination Required (Foot Candles)
Sewing rooms, draughting rooms, art rooms and other rooms where fine detail work is done	8
Class rooms, study halls, libraries, laboratories, manual training rooms and offices	5
Gymnasiums, play rooms and sports rooms	3
Assembly rooms, cafeterias and other rooms where food is served, rooms in which provision is made for students to eat lunches brought from home and other rooms in which students congregate for extended periods but not for study	2
Locker rooms, toilet rooms and recreation areas........	1

In sewing rooms, draughting rooms, art rooms and other rooms where fine detail work is done, the required illumination shall be on the plane of the work; in all other rooms it shall be on a plane thirty (30) inches above the floor and shall cover every part of such plane.

1710.02 Normal Illumination of Exits: In every school every required means of exit shall be provided with a system of electric lighting in accordance with Chapter 30 of the General Provisions of this ordinance.

1710.03 Emergency Illumination of Certain Exits: In each assembly room having a capacity of one thousand eight hundred (1800) or more persons, and throughout the required path of exit travel from every such room, there shall be provided a separate system of additional electric lighting to facilitate egress in an emergency, in accordance with the provisions of Chapter 30 of this ordinance.

1710.04 Exit Signs: In every school, standard exit signs, standard directional signs and standard fire escape signs, illuminated by electricity, shall be provided to mark the ways of egress, in accordance with Chapter 30 of the General Provisions of this ordinance.

CHAPTER 18.

Other Buildings and Structures.

ARTICLE 1801.

General.

1801.01 Other Buildings and Structures Defined: Other buildings and structures, for the purposes of this ordinance, shall be defined as all buildings and structures not included in Chapters 6 to 17 inclusive of this ordinance, including:

Billboards and Signboards
Car Barns
Fences
Hangars
Ice Houses
Illuminated Signs
Incinerators
Jails
Police Stations
Prisons
Certain Railroad Structures
Sheds
Stables

and other buildings and structures not otherwise provided for in this ordinance.

1801.02 Requirements: All buildings and structures included in this chapter shall comply with the requirements of the General Provisions and of the Structural Provisions of this ordinance and in addition shall comply with the special provisions of this chapter.

ARTICLE 1802.

Hangars.

1802.01 Hangars Defined: A hangar, for the purposes of this ordinance, is hereby defined as a building, or part of a building, designed, intended or used as a place for the housing, shelter or support of one (1) or more aircraft for the purposes of storage, display, sale, demonstration, servicing, repair, testing, remodeling, adjustment, painting or any other purpose; provided, however, that a building or space in a building used only for the manufacture or repair of aircraft without gasoline or other motor fuel, and in which building no flammable solution of cellulose nitrate, cellulose acetate or other similar dope is used, or a room or space for the storage, display or sale of motor vehicles which are entirely without gasoline or other motor fuel, or rooms used as waiting rooms, dining rooms, dressing rooms, toilets or other utility rooms, shall not be deemed to be a hangar if such room or space is separated from all hangar uses by a standard fire division wall.

1802.02 Requirements:

(a) **Floors.** Floors of hangars shall be not less than three (3) inches above grade at the vehicle entrance doors; shall be of non-combustible material and shall be pitcheu not less than one-eighth (⅛) inch in each foot of horizontal distance to drain toward the vehicle entrance doors or to scuppers in the walls.

(b) **Basements Prohibited.** No basement, cellar, pit, duct or other opening may be built beneath any hangar; provided however, that the heating plant for a hangar may be placed immediately below the grade floor of a hangar, subject to the provisions of this ordinance for heating plants.

1802.03 Hangars on Roofs: Except as otherwise provided in this ordinance, hangars may be constructed on the roofs of fireproof buildings which are located outside of the fire limits.

1802.04 Construction:

(a) **General.** Hangars within the fire limits or provisional fire limits, shall be built of, ordinary construction or of more fire-resistive construction, except as otherwise provided in this section.

(b) **Wood Frame Construction.** Hangars built outside the fire limits of wood frame construction shall be located not less than fifty (50) feet from any other building, or property dividing lot line or the opposite side of the street or alley; provided however, that a hangar, housing a single airplane, and having an area of not more than twenty-five hundred (2500) square feet shall be located not less than fifteen (15) feet from any other building or any property dividing lot line. Any hangar built of wood frame construction or of less fire-resistive construction shall not have an area of more than ten thousand (10,000) square feet.

(c) **Hangars on Roofs.** Any hangar built on the roof of a fireproof structure shall be of fireproof or semi-fireproof construction and shall be equipped with a standard system of automatic sprinklers.

(d) **Fireproof Construction.** Any hangar more than one (1) story high above grade shall be of fireproof construction.

1802.05 Fire Areas:

(a) **General.** In computing any fire area in a hangar, the combined area of all balconies and mezzanines shall be included; provided however, that a head house having no aircraft therein, when separated from other parts of the hangar by four (4) hour fire-resistive construction, shall be considered as a business unit and shall comply with the special provisions of Chapter 10 for business units.

(b) **Maximum Allowable Fire Areas.** Any fire area in a hangar shall not exceed the areas given in the following table:

TABLE 1802.05 (b)

Maximum Allowable Fire Area in Hangar Not More than One Story High.

	Hangar not equipped with a standard system of automatic sprinklers (Sq. Ft.)	Hangar equipped with a standard system of automatic sprinklers (Sq. Ft.)
Fireproof or Semi-Fireproof Construction.	30,000	60,000
Heavy Timber Construction..............	25,000	50,000
Ordinary Construction	20,000	40,000
Wood Frame Construction.................	10,000	20,000
Maximum Allowable Fire Area in Hangar More than One Story High.		
Fireproof Construction	30,000	60,000
Semi-Fireproof Construction	25,000	50,000

1802.06 Separation From Heating Plants: Every heating plant and its fuel storage in a hangar shall be completely separated from all hangar uses in the building in which it is located by two (2) hour fire-resistive construction. Entrance to such heating plant shall be from the outside only.

1802.07 Multiple Use Buildings:

(a) **General.** Any hangar in the same building with any other occupancy shall be subject to the requirements of this article and to the requirements of the sections on multiple uses in Chapters 6 to 17, inclusive.

(b) **Types of Construction.**

Item 1. Ordinary, Heavy Timber of Semi-Fireproof Construction. Any hangar of ordinary, heavy timber or semi-fireproof construction shall not be located in any building, designed, erected, altered, converted, used or intended to be used as a multiple dwelling, hospital, school or theatre or in any building containing a single dwelling in which the floor of any habitable room is more than eighteen (18) feet above grade.

Item 2. Wood Frame Construction. Any hangar of wood frame construction shall not be located in any building designed, used or intended to be used for other than hangar purposes; provided however, that the rooms comprising the workshop, locker rooms, dressing rooms and other utility rooms required in connection with such hangar may adjoin the hangar when separated therefrom by four (4) hour fire-resistive construction, having each opening equipped with a double standard fire door.

(c) **Separation.** Where a hangar is located in a building any part of which is used for other than hangar purposes, it shall be separated from such other use by not less than four (4) hour fire-resistive construction.

(d) **Communicating Openings.** Every doorway in a wall separating a hangar from any part of a building used for other than hangar purposes shall be protected with double standard fire doors.

1802.08 Means of Exit: In every hangar with an area of more than ten thousand (10,000) square feet there shall be two (2) or more pedestrian exit doorways in addition to the vehicle exit doorway. Two (2) of these exit doorways shall be located at opposite sides or ends of the building. No point in any hangar shall be more than one hundred fifty (150) feet from any such exit door, measured along the line of travel. Where hangars are built on roofs of fireproof buildings there shall be two (2) or more inclosed stairways from the floor of the hangar to the ground which stairways shall be located each within twenty-five (25) feet of a required exit doorway; provided however, that where only two stairways are required one (1) of them may be a fire escape stairway.

1802.09 Artificial Lighting and Exit Signs:

(a) **Lighting.** Artificial lighting shall be restricted to electricity.

(b) **Exit Signs.** In every hangar standard exit signs, standard directional signs and standard fire escape signs, illuminated by electricity shall be provided to mark the ways of egress, meeting the requirements of Chapter 30.

ARTICLE 1803.

Prisons and Similar Buildings of Detention.

1803.01 General: For the purposes of this ordinance, there shall be included in this occupancy classification, jails, prisons, police stations, reformatories, houses of correction and other similar buildings, where persons are or may be forcibly restrained, excepting institutional buildings.

1803.02 Construction: Every building included in this classification shall be of fireproof construction; provided however, that buildings of not more than two (2) stories in height above the ground and of not more than five thousand (5,000) square feet in area may be of ordinary or of more fire-resistive construction; and provided further that in every such building, that part containing the cell-block or lockup or the patrol wagon quarters shall be of fireproof construction, and this part shall be separated from all other parts of the building by a standard fire-division wall.

1803.03 Size and Location of Rooms:

(a) **Court Rooms.** In all court rooms in buildings referred to in this article, the ceiling height shall be not less than eleven (11) feet.

(b) **Sleeping Rooms Above Grade.** No sleeping rooms nor cell blocks shall be constructed below the first floor level in any building included in this classification.

(c) **Police Stations.** Every room in a police station, other than the cell-block shall have a ceiling height of not less than ten (10) feet.

(d) **Offices.** Offices and other similar rooms not including toilets, closets and storerooms, shall have a floor area of not less than seventy (70) square feet and a ceiling height of not less than eight (8) feet.

(e) **Cell Blocks.** Every cell block shall have three (3) or more walls facing upon a street, alley, open yard or court.

1803.04 Means of Exit:

(a) **Number.** Every floor in every building included under this article shall have two (2) or more separate and distinct exits from each floor. Every room or space open to the public and every room or space occupied by inmates or prisoners or guards, attendants or other employees in any building referred to in this article, which room or other space has a seating capacity of more than one hundred (100) persons, or which has a floor area of more than one thousand (1,000) square feet, shall have, in addition to the entrance thereto, not less than one (1) exit arranged to permit passage through and from an outside doorway without returning to the place of entrance.

(b) **Capacity of Rooms.** Capacity of various rooms shall be computed as determined by Table 2401.02 MOCC, Chapter 24; provided however, that the capacity of cell blocks shall be computed on the basis of one (1) person for each twenty-five (25) square feet of floor area; and provided further that when such cell block shall have provision for persons, including all inmates or prisoners and all guards, attendants and other employees, more than the number computed by floor area, then exits shall be provided for the actual number of persons in such cell blocks.

(c) **Widths.**

Item 1. Normal Exits. The widths of doorways, passages, stairways, ramps or other required means of normal exit from rooms in all buildings referred to in this article shall be computed at a rate of not less than one (1) foot, six (6) inches per one hundred persons capacity and this minimum width shall be maintained by a cumulative increase toward the outside entrance doors.

Item 2. Emergency Exits. In addition to the normal exits required, there shall be provided also, emergency exits from every room or story having a capacity of more than two hundred fifty (250) persons. The widths of doorways, passages, stairways, ramps or other required means of emergency exit from rooms referred to in this article, shall be computed at the rate of not less than nine (9) inches per one hundred (100) persons capacity and this minimum width shall be maintained by a cumulative increase toward the outside exit doors.

(d) **Remote Control of Locks.** In every cell room and in every other room where inmates or prisoners are confined by locked doors, the doors of all such cells or rooms shall be provided with remote control of locks so that they may be promptly and effectively opened by an attendant in case of fire or other emergency.

* (e) **Stairways.** All stairways in buildings included in this article shall meet the requirements of Chapter 26—MEANS OF EXIT. No stairway shall be less than three (3) feet, eight (8) inches in width; provided however, that stairways used only by guards, attendants or other employees may be not less than (3) feet in width. Stairways shall be so located that on any floor the top riser of such stairway, if an open stairway, or the door to such stairway, if a closed stairway, shall be not more than seventy-five (75) feet from the door to any room or cell.

1803.05 Windows: Except as otherwise provided, every habitable room included in this article shall have one (1) or more windows opening directly upon a street, alley, yard or court. The total glass area of such window or windows shall be not less than ten (10) per cent of the floor area of such room. The top of such windows shall be not less than seven (7) feet above the floor; provided however, that where cell blocks have windows on three (3) sides of same with a total glass area equal to twenty-five (25) per cent of the floor area of such block, it shall not be required that each cell have a window.

1803.06 Artificial Lighting and Exit Signs:

(a) **Lighting.** Every means of exit shall have a system of electric lighting as required by Chapter 30 for the normal illumination of exits.

(b) **Exit Signs.** Standard exit signs, standard directional signs and standard fire escape signs shall be provided in every building included in this article to mark the ways of egress as required by Chapter 30 for exit signs.

ARTICLE 1804.

Stables.

1804.01 Stables Defined: A stable, for the purposes of this ordinance, is hereby defined as a building, or part of a building, designed, intended or used for the housing or shelter of horses, cattle or other animals.

1804.02 Lot Occupancy: Every stable housing more than eight (8) animals and every stable with a floor area of more than five hundred (500) square feet shall be located not less than four hundred (400) feet from any school, church, hospital, public park or public playground.

1804.03 Construction: Stables shall be constructed of any of the types of construction described in Chapter 31—CONSTRUCTION TYPE PROVISIONS, of this ordinance, subject to the restrictions as to area and story height provided in Chapter 24—SIZE AND LOCATION OF ROOMS, of this ordinance. Every stable housing more than two (2) animals, on any lot abutting on a street or alley in which a public sewer is constructed, shall have an impervious floor properly drained to such a sewer.

1804.04 Storage Room: Every stable having a floor area of more than forty-five hundred (4500) square feet, shall have an inclosure of not less than one (1) hour fire-resistive construction for any space used for the storage of hay, straw or combustible bedding.

1804.05 Means of Exit: In every stable with an area of more than nine thousand (9,000) square feet, there shall be two (2) or more pedestrian exit doorways. Exit doorways shall be at opposite sides or ends of the building and no point in any stable shall be more distant from any exit doorway than one hundred fifty (150) feet measured along the line of travel. Stables with stories above or below the first story shall have two (2) or more ramps or stairways to the ground leading directly to each required exit doorway; provided however that one (1) such exit stairway may be a fire escape stairway.

1804.06 Artificial Lighting: Artificial lighting shall be restricted to electricity in every stable inside the fire limits.

ARTICLE 1805.
Fire Stations.

1805.01 Fire Stations Defined: A fire station, for the purposes of this ordinance, is hereby defined as a building, or part of a building, designed, intended or used as a place for the housing of one (1) or more pieces of fire-fighting or salvaging equipment together with the sleeping quarters, locker rooms, toilet and bath rooms, heating plant and such other rooms or spaces as are required by the firemen or the equipment.

1805.02 Construction: Fire stations shall be built of ordinary construction or of more fire-resistive construction; provided however, that the ground floor shall be of noncombustible material, and where there is a basement, such story and the floor above the basement shall be of fireproof construction. Fire stations more than three stories in height shall be of fireproof construction.

1805.03 Multiple Use Buildings: A fire station may occupy a part of any building used as a Business Unit or as a Garage, as described in Chapters 10 and 12 respectively, or any building used as a police station; provided however, that such fire station shall be separated from every other part of such building by two (2) hour fire-resistive construction.

1805.04 Means of Exit: Every fire station shall have not less than one (1) pedestrian exit doorway in addition to the vehicle exit doorway. Every fire station, more than one (1) story high shall have one or more stairways from the top floor to the ground.

1805.05 Windows and Ventilation. Every habitable room used for living or sleeping purposes or dining purposes in a fire station shall have a window or windows with ventilating openings having an area of not less than five (5) per cent of the floor area of such room.

ARTICLE 1806.
Incinerators.

1806.01 Incinerator Defined: An incinerator is hereby defined for the purposes of this ordinance as a furnace for the burning of garbage, rubbish of other waste substance, with or without fuel.

1806.02 Restrictions: Incinerators shall not be built within any hazardous use unit having flammable liquids or vapors or flammable dust.

1806.03 Classification: Incinerators shall be divided into two (2) types as hereinafter provided.

1806.04 Non-Fuel-Fired Incinerators:

(a) **General.** In this type shall be included all incinerators which use no fuel for combustion other than the normal refuse deposited; provided however, that a gas flame or other means may be provided to accomplish primary ignition and the incinerator flue and the smoke flue shall be one.

(b) **Construction.** Where the horizontal grate area does not exceed nine (9) square feet, the walls of the combustion chamber shall be constructed of common brickwork not less than three and three-fourths (3¾) inches thick with a lining of fire brick not less than four and one-half (4½) inches thick. Where the horizontal grate area exceeds nine (9) square feet, the walls of the combustion chamber shall be constructed of common brickwork not less than eight (8) inches thick with a lining of fire brick not less than four and one-half (4½) inches thick. Fire brick shall be laid in fire clay mortar.

(c) **Chimney.** The chimney shall be built as provided by Chapter 41—CHIMNEYS of this ordinance. For incinerators having not more than one (1) service opening, the flues shall have a diameter or side of not less than eleven and one-fourth (11¼) inches. For incinerators in buildings not more than six (6) stories in height, with not more than twelve (12) service openings, the flues shall have a diameter or side of not less than fifteen and three-fourths (15¾) inches. For incinerators in buildings more than six (6) stories in height, the flues shall have a diameter or side of not less than seventeen and one-fourth (17¼) inches. The top of every incinerator chimney shall be provided with a spark arrester having a heavy wire mesh not exceeding one (1) inch mesh.

(d) **Service Openings.** The area in the clear of any one (1) service opening shall not exceed one-third (⅓) of the cross-sectional area of the flue. Every service opening into an incinerator flue shall be provided with a charging device firmly built into the masonry in such manner that no part shall project beyond the inner surface of the flue. Such charging device shall be self-closing and shall be so designed that the opening to flue interior shall be closed off during the charging operation.

1806.05 Fuel-Fired Incinerators:

(a) **General.** In this type shall be included all incinerators which are equipped with fuel combustion chambers.

(b) **Location.** Incinerators of this type shall be located in separate rooms, except that refuse material bins or containers may also be kept in such room; provided however, that any material bill which is the terminal for a chute shall not be in the same room with the incinerator; and provided further, that an incinerator may be located in the same room with boilers or other equipment comprising the building heating plant. The incinerator room or the boiler room in which an incinerator is located, shall be separated from the rest of the building by not less than two (2) hour fire-resistive construction.

(c) **Construction.** Every fuel-fired incinerator shall be designed with special regard for the stresses set up by high temperatures. Incinerator combustion chambers, designed to burn not more than two hundred fifty (250) pounds of refuse per hour, shall be constructed of common brickwork not less than eight (8) inches thick with a lining of fire brick not less than four and one-half (4½) inches thick; provided however, that a steel plate casing not less than three-sixteenths (3/16) inch thick may be substituted for four (4) inches of common brickwork. Incinerator combustion chambers designed to burn more than two hundred fifty (250) pounds of refuse per hour, shall be constructed of common brickwork not less than eight (8) inches thick, with a lining of fire brick not less than nine (9) inches thick; provided however, that a steel plate casing not less than three-sixteenths (3/16) inch thick may be substituted for four (4) inches of common brickwork. Fire brick shall be laid in fire clay mortar.

(d) **Chimney.** The chimney of a fuel-fired incinerator shall be built as provided by Chapter 41—CHIMNEYS of this ordinance. An incinerator breeching may be connected to a heating boiler chimney or other chimney if the area of the chimney is not less than four (4) times the area of the incinerator breeching.

(e) **Refuse Chutes.** Refuse chutes shall rest upon substantial brick or concrete foundations. Walls of chutes shall be of solid brickwork not less than eight (8) inches thick or of reinforced concrete not less than six (6) inches thick. Chutes shall extend through the roof and shall be covered with metal skylights glazed with unwired glass, except where a metal ventilator is provided. Each service opening in a refuse chute shall be protected with a self-closing forty-five (45) minute fire-resistive door. Every such service opening shall be located in a separate room or compartment with walls of one (1) hour fire-resistive construction. Chutes shall terminate or discharge into a room other than the room in which the incinerator is located. Such terminal room shall have an inclosure of not less than two (2) hour fire-resistive construction with openings protected with self-closing forty-five (45) minute fire-resistive doors.

ARTICLE 1807.

Sheds and Shelter Sheds.

1807.01 Shelter Sheds Defined: A shelter shed, for the purposes of this ordinance, shall be defined as a structure having a roof, and with more than fifty (50) per cent of the area of its sides open.

1807.02 Sheds: A shed, for the purposes of this ordinance, shall be defined as a structure not exceeding three hundred (300) square feet ground area, and not to exceed fourteen (14) feet in height above grade to the highest point thereof, with a roof covering of fire-retarding material and inclosing walls of combustible material, no part of which is located on the front half of any lot. No part of such shed shall be used as a dwelling, or as an addition to any dwelling, or for any business purpose.

1807.03 Within the Fire Limits: Except as otherwise provided, shelter sheds of wood frame construction, eight hundred (800) square feet or less in area, may be erected within the fire limits or provisional fire limits; provided however, they shall have roofing of non-combustible materials, and no part of the building shall be more than fifteen (15) feet above grade, and the floors shall be of non-combustible material or of planks not less than one and five-eighths (1⅝) inch thick laid directly on the ground. Any shelter shed having an area of more than eight hundred (800) square feet and not exceeding sixteen hundred (1600) square feet, if within the fire limits, shall be located not less than twenty-five (25) feet from any other structure on the same lot, or any lot line and inclosing walls, if any, shall be of non-combustible material. Shelter sheds shall be erected only on the rear of the lot and not more than one (1) such shelter shed 800 square feet or less in area shall be erected on any lot of thirty (30) feet or less in width.

1807.04 Outside the Fire Limits: Except as otherwise provided, shelter sheds of wood frame construction, built as provided in Section 1807.03, and twenty-eight hundred (2800) square feet or less in area, may be erected outside the fire limits or provisional fire limits. Any shelter shed having an area of more than twenty-eight hundred (2800) square feet, and not exceeding four thousand (4,000) square feet in area, shall be located not less than twenty (20) feet from any lot line or other structure or building, and inclosing walls, if any, shall be of non-combustible material.

1807.05 Adjacent to Railways and Waterways: Except as otherwise provided, shelter sheds of wood frame construction, to be used for the storage or handling of coal, brick, lumber, stone, cement, salt or similar commodities which are non-combustible, or for the icing of cars, may be erected within or without the fire limits, upon, along or adjacent to any railroad tracks or navigable waters. Such shelter sheds shall not exceed nine thousand (9,000) square feet in area. If within the fire limits or provisional fire limits, such shelter sheds of wood frame construction shall not exceed thirty-five (35) feet in height, and if outside the fire limits shall not exceed forty-five (45) feet in height. Such shelter sheds shall be located not less than twenty-five (25) feet from any property dividing lot lines and from any other building.

ARTICLE 1808.

Ice Houses.

1808.01 Ice Houses Defined: An ice house, for the purposes of this ordinance, shall be defined as a building used exclusively for the storage of ice.

1808.02 Construction:

(a) **Within the Fire Limits.** Except as otherwise provided, ice houses of wood frame construction, having fire-retarding roof covering, and not exceeding nine thousad (9,000) square feet in area, may be erected within the fire limits or provisional fire limits. Such ice houses shall be not less than twenty-five (25) feet from any property dividing lot lines and from any other building. Such ice houses shall have walls inclosed with an eight (8) inch thick envelope of brick, hollow tile, concrete or other non-combustible material on foundations extending to soid ground and not less han four (4) feet below the surface of the ground. Such ice houses shall be not more than forty-five (45) feet in height.

(b) **Outside the Fire Limits.** Except as otherwise provided, ice houses of wood frame construction having fire-retarding roof covering, and not exceeding twenty thousand (20,000) square feet in area, may be erected outside the fire limits or provisional fire limits, upon, along or adjacent to any railroad tracks or navigable waters. Such ice houses shall be located not less than one hundred (100) feet from any property dividing lot lines and from any other building, and shall be not more than forty-five (45) feet in height. Ice houses larger than twenty thousand (20,000) square feet may be built in accordance with this paragraph; provided however, that such building shall be divided by fire division walls into units of area not greater than twenty thousand (20,000) square feet. Except as otherwise provided, ice houses of wood frame construction, and having fire-retarding roof coverings, and not exceeding eighty thousand (80,000) square feet in area, may be erected outside the fire limits or provisional fire limits contiguous to any lake or other body of water where ice may be harvested. Such ice houses and such other building or buildings as are used in connection with the ice house, shall be not less than six hundred (600) feet from any property dividing lot lines and from any other building. Such ice houses shall be not more than forty-five (45) feet in height. Ice houses larger than eighty thousand (80,000) square feet in area may be built in accordance with this paragraph; provided however, that such building shall be divided by solid walls of masonry into units of area not greater than eighty thousand (80,000) square feet.

ARTICLE 1809.

Billboards and Signboards.

1809.01 Billboards and Signboards Defined:

(a) **Billboard.** A billboard, for the purposes of this ordinance, is hereby defined as a structure with a vertical or nearly vertical surface, erected for the outdoor display of posters.

(b) **Signboard.** A signboard, for the purposes of this ordinance, is hereby defined as a structure with a vertical or nearly vertical surface, erected for the display of notices of any kind, other than posters.

(c) **Exceptions.** The faces of the walls or the flat surface of the roof of a building or other structure, used for the display of posters or other notices, shall not be interpreted as a billboard or signboard within the meaning of this ordinance.

1809.02 Construction:

(a) **Wind Pressure.** Billboards and signboards together with all supports and connections, shall be constructed and maintained to withstand a wind pressure of not less than twenty-five (25) pounds per square foot of surface area without stressing the supporting members beyond the safe limit of stresses provided in the structural provisions of this ordinance.

(b) **Materials.** Billboards and signboards shall be built entirely of non-combustible material, except as otherwise provided by Section 1809.04; provided however, that billboards and signboards more than ten (10 feet from the nearest building or other structure, or from any thoroughfare or other public space may have stringers, uprights and braces of wood.

1809.03 Size and Location:

(a) **Height.** Except as otherwise provided, the faces of billboards and signboards, within the fire limits and provisional fire limits, shall not exceed twelve (12) feet in height.

(b) **Location.** The bottom of the face of any billboard or signboard shall have an open unobstructed space thereunder not less than three (3) feet six (6) inches above the level of the adjoining street level or above the adjoining ground level if such ground level is above the street level.

1809.04 Wood Billboards and Signboards: Billboards or signboards used exclusively to advertise the sale or lease of the property upon which they are erected, and having an area not exceeding twenty-four (24) square feet, may be built entirely of wood or other combustible material; provided however, that if attached to a building or other structure, they shall be fastened flat against the wall surface and if erected on the ground shall be located not less than ten (10) feet from any building or other structure or from any thoroughfare or other public space.

1809.05 Alteration, Repair and Removal: No material alterations of any billboard or signboard, nor removal from one (1) location to another shall be made except upon a written permit issued by the Commissioner of Buildings authorizing such alteration or removal; and such permit shall be issued upon application in writing made to such Commissioner by the owner of such billboard or signboard or by the person in charge, possession or control thereof, accompanied by a plan of the proposed alterations or repairs to be made and a written statement covering the proposed removal from one (1) location to another and its reconstruction in the new location, which said alteration and repairs or removal shall be made in accordance with the provisions of this article and the ordinances of the City of Chicago. Where such plans, specifications and location are in compliance with the requirements of this article, such Commissioner shall issue a permit upon the payment of a fee therefor as herein fixed; but such alteration shall not be construed to apply to the changing of any advertising matter of any billboard or signboard, nor the refacing of the framework supporting same.

1809.06 Owner's Name: Every billboard and signboard shall bear on the top of such structure, the name of the owner.

ARTICLE 1810.

Illuminated and Other Roof Signs.

1810.01 Illuminated and Other Roof Signs Defined: An illuminated or other roof sign, for the purposes of this ordinance, is defined as any structure erected upon or above the roof of any building or other structure for the display of notices of any kind, which may have all or any part of the letters of such notices flush or raised or in outline of incandescent lamps or other lights, or which may have a border of light, and all signs erected upon or above the roof of any building or structure shall comply with the requirements of this article and Article 1813 of this chapter.

1810.02 Construction: Structures included under this article shall be skeleton construction of steel, or a superior type of construction, designed and constructed to withstand a wind pressure of not less than thirty (30) pounds per square foot. Steel or other metal shall be used for the face of the sign and shall be so designed that the area of any surface presented to the wind shall not exceed fifty (50) per cent of the aggregate area of the surface on which the notice is placed. No material shall be used in any sign included under this article which is more combustible than metal, except such material as may be required for insulating wires and conductors, and except that wood structural members, with a cross-sectional area of not less than fifty-two (52) square inches, may be used at the roof or wall surfaces for purposes of securing the steel framework.

1810.03 Height: The total height of any structure included under this article shall not exceed sixty (60) feet above the roof.

1810.04 Location: The bottom of the face of any structure included under this article shall have a clear unobstructed space of not less than five (5) feet between such structure and any roof. The face of any structure included under this article shall be located not less than six (6) feet back of the face of any street front wall.

1810.05 Owner's Name: Every structure included under this article shall bear on the top of such structure, the name of the owner.

1810.06 Revocation: Permission and authority granted under this article may be revoked for cause by the Commissioner of Buildings. In case of such revocation, all such electrical signs shall be removed at the expense of the owner of the building, or the person, firm, corporation or individual who are then maintaining same, without any cost or expense of any kind whatsoever to the City of Chicago; provided however, that in the event of failure, neglect or refusal on the part of the owner of the building or structure upon which said illuminated electric sign is constructed, or the person, firm, corporation or individual operating and maintaining said electric sign, to remove said electric sign, upon the revocation of the permit by the Commissioner of Buildings, as herein provided, the Commissioner of Buildings may proceed to remove same and charge the expense thereof to the owner of the building or structure upon which said illuminated electric sign is constructed, or to the person, firm, corporation or individual operating or maintaining same.

ARTICLE 1811.

Fences.

1811.01 Fence Defined: A fence, for the purpose of this ordinance, is hereby defined as a structure forming a barrier, which is not otherwise a part of any building or structure which may be lawfully erected.

1811.02 Construction: Every fence shall be designed and constructed to resist a horizontai wind pressure of not less than twice that required for buildings. Where the distance from the lot line to the face of any fence is less than the height of the fence, such fence shall have lateral supports on one (1) side with structural braces extending to the top of the fence; provided however, that any free standing masonry wall or masonry sections in any fence shall comply with the provisions of Chapter 38—MASONRY CONSTRUCTION of the structural provisions.

1811.03 Height: A fence of any material, parallel to a street lot line, alley lot line, or any dividing line, and within eight (8) feet of such lot line, shall not exceed eight (8) feet in height above the established grade or above the ground where no grade has been established; provided however, that any such fence which does not obstruct vision to an extent greater than fifty (50) per cent of the aggregate vertical surface area of such fence, including openings, shall be exempt from the provisions of this section, except that any fence constructed in whole or in part of wood shall not exceed eight (8) feet in height.

ARTICLE 1812.

Roundhouses and Carbarns.

1812.01 Roundhouses Defined: A roundhouse, for the purposes of this ordinance, is hereby defined as a structure for the storing and repairing of locomotives using any fuel other than a volatile flammable liquid.

1812.02 Carbarns Defined: A carbarn, for the purposes of this ordinance, is hereby defined as a structure for the storing and repairing of electric street cars and other electric conveyances.

1812.04 Construction: Every roundhouse and carbarn shall have all inclosing walls and all floors and pits of non-combustible materials. Roofs and columns supporting roofs shall be of heavy timber construction or of more fire-resistive construction. Roof coverings shall be of fire-retarding material.

1812.05 Fire Areas: Every roundhouse shall be divided by standard fire division walls from any space used for the storage of combustible fuel or materials. Every carbarn shall be divided by standard fire division walls into fire areas not exceeding thirty thousand (30,000) square feet; provided however, that fireproof carbarns not more than one (1) story in height may be of any area.

1812.06 Means of Exit: Every roundhouse and every carbarn shall have pedestrian exit doorways so located that any point on any floor shall be not more than one hundred (100) feet from one (1) such exit doorway.

ARTICLE 1813.

Tanks and Exposed Metal Structures.

1813.01 Tanks Above Roofs: It shall be unlawful to construct, maintain or allow or permit to remain in, upon or above the roof of any building, any tank of a larger capacity than two hundred fifty (250) gallons unless such tank shall be supported upon a foundation or structure of masonry, reinforced concrete or structural steel or any combination thereof, which shall be designed to meet the requirements of the Structural Provisions of this ordinance.

1813.02 Exposed Metal Structures:

(a) **Maintenance.** It shall be the duty of the owner, person or corporation in charge, possession or control of any building, upon or above which any unprotected metal structure or structures including tank and sign structures, canopies and smoke stacks, are now located or shall be erected, to maintain such structures in good and safe condition. Any exposed tank of larger capacity than two hundred fifty (250) gallons shall have a rustproof plate or tag attached to such tank or its supports, which shall show in letters or figures at least two (2) inches high, the month and year when such tank was installed.

(b) **Inspections.** Within two (2) years after the erection of any exposed metal structure, permitted by this ordinance upon or above the roof of any building, and within two (2) years following the passage of this ordinance, and at least once every five (5) years thereafter, every such exposed metal structure now existing or hereafter erected shall be subjected to a critical examination by a licensed architect or a licensed structural engineer, whom the owner shall hire, and a report of its condition, by said architect or engineer, made to the owner, shall be submitted in duplicate by the owner to the Commissioner of Buildings. One (1) copy of such report, if satisfactory to the Commissioner of Buildings, shall be returned to the owner, bearing the approval stamp of the Commissioner of Buildings.

(c) **Repairs, Reinforcement or Removal.** Every such structure found to be in an unsafe condition or in need of repair or reinforcement, shall be subject to notice by the Commissioner of Buildings to the owner or person or corporation in charge, possession or control of the building whereon such structure is located, immediately to effect such repairs, reinforcement or precautionary measure as will bring the premises in a good and safe condition and further, without delay, to begin and complete the work of permanent repair, reinforcement or removal, which shall be required to make such structure conform to the provisions of this ordinance.

(d) **Paint or Metal Coatings Not Protection.** For the purposes of this section, no form of paint, galvanizing, or similar coating shall be considered as a protection exempting otherwise unprotected metal structures from the provisions of this section.

CHAPTER 19.

Frontage Consents.

ARTICLE 1901.

General.

1901.01 Existing Provisions Adopted by Reference: All ordinances and parts of ordinances of the Revised Chicago Code of 1931 as amended, which require frontage consents for the erection, construction, alteration, enlargement or maintenance of any building or structure or for a license for the use or occupancy of any building or premises before a permit or license is granted therefor, are hereby declared to be a part of this ordinance the same as if the ordinance and parts of ordinances were herein fully set out.

The following paragraphs with Section Numbers are from the Revised Code of 1931 and are included for the convenience of the Architect.

Frontage Consents.

1690 Frontage Consents—General Requirements: Whenever frontage consents are required, for the construction of a building or for any occupation for which a building is about to be constructed or altered, under any section of this ordinance or under any other ordinance of the city, such frontage consents shall be presented to the commissioner of buildings before the issuance of a permit for the erection or alteration of a building for such purpose. Unless otherwise specified the provisions of this article in such case shall apply as to the definition of the word "block," whenever such word is used in any section or ordinance requiring frontage consents.

1698 Frontage Adjacent—How Occupied for Building Purposes: If the written consent of and a waiver of claims for damages against the city by the owners of properties adjoining the site of any proposed building is first obtained and filed with the Commissioner of Public Works, the permission to occupy the roadway and the sidewalk may be extended beyond the limits of such building in front of the property for which the consent of the owner or lessee thereof has been secured upon the same terms and conditions as those herein fixed for the occupatiou of sidewalk and street in front of the building site.

1677 Definition of Word "Block": Whenever a provision is made in this ordinance that frontage consents shall be obtained for the erection, construction, alteration, enlargement or maintenance of any building or structure in any block, the word "block," so used, shall not to be held to mean a square, but shall be held to embrace only that part of a street bounding the square which lies between the two nearest intersecting streets one on either side of the point at which such building or structure is to be erected, constructed, altered, enlarged or maintained unless it shall be otherwise specially provided.

1678 Frontage Consents—Where Required—Uses of Property for Required—Consent in Writing: It shall be unlawful for any person, firm or corporation to locate, build, construct or maintain on any lot fronting on any street or alley in the City in any block in which one-half of the buildings on both sides of the street are used exclusively for residence purposes, or within fifty feet of any such street, any building, structure or place used for a gas reservoir, manufacture of gas, stock yards, slaughter house, packing house, smoke house or place where fish or meats are smoked or cured, soap factory, glue factory, size or gelatine manufactory, renderies, fertilizer manufactory, tannery, storing or scraping of raw hides or skins, lime kiln, cement or plaster of Paris manufactory, oil cloth or linoleum manufactory, rubber manufacture from the crude material, saw or planing mill, wood working establishment, starch factory, glucose or dextrine manufactory, textile factory, laundry run by machinery, factory combined with a foundry, iron or steel works, brass or copper works, sheet metal works, blacksmithing or horseshoeing shop, boiler making, foundry, smelter, metal refinery, machine shop, stone or monument works run by machinery, asphalt manufacture or refining, paint and varnish factory, oil or turpentine factory, printing ink factory, tar distillation or manufacture, tar roofing, tar paper or

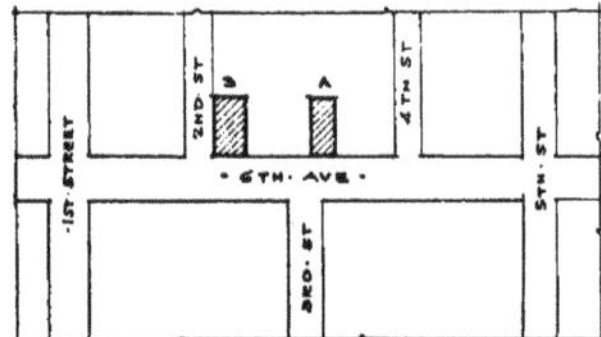

tarred fabric manufactory, ammonia or chlorine or bleaching powder factory, celluloid manufactory, place for the distillation of wood or bones, lamp black factory, sulphurous acid, sulphuric acid, nitric or hydrochloric acid manufacture, factories or other manufacturing establishments using machinery or emitting offensive or noxious fumes, odors or noises, storage warehouses storing or baling of junk or scrap paper or rags, shoddy manufacture or wool scouring, second-hand store or yard, incineration or reduction of garbage or offal, dead animals or refuse, stable for more than five horses, medical dispensary, livery stable, sale stable, boarding stable, without the written consent of a majority of the property owners according to frontage on both sides of such street or alley. Such written consent shall be obtained and filed with the Commissioner of Buildings before a permit is issued for the construction or alteration of any building, structure or place for any of the above purposes: provided, that in determining whether one-half of the buildings on both sides of the street are used exclusively for residence purposes any building fronting upon another street located upon a corner lot shall not be considered.

1679 Reformatories — Sheltering Institutions: It shall be unlawful for any person, firm or corporation to build, construct, maintain, conduct or manage any reformatory, rescue or sheltering institution in any block or square in which one-half of the buildings on both sides of the street or streets on which the proposed reformatory, rescue or sheltering institution or the grounds thereof may have frontage, are used exclusively for residence purposes without the written consent of a majority of the property owners, according to frontage on both sides of the streets bounding such square. Such written consent shall be obtained and filed with the Commissioner of Buildings before a permit is issued for the construction, alteration, or maintenance of such building. Provided, that in determining whether one-half of the buildings on both sides of the street are used exclusively for residence purposes, any building fronting upon another street and located upon a corner lot shall not be considered.

1680 Permit For Moving Frame Buildings — Requirements — Written Consents — Space Occupied on lot: (a) No person, firm or corporation shall be permitted to move any building which has been damaged to an extent greater than 50 per cent of its value by fire, decay or otherwise; nor shall be permitted to move any frame building of such character as is prohibited to be constructed within the fire limits from any point outside the fire limits to any point within the fire limits; nor shall it be permissible to move any building to a location at which the uses for which such building is designed are prohibited by ordinance. Permits for the moving of frame buildings other than those the moving of which is herein prohibited, shall be granted upon the payment of a fee of ten cents for each one thousand cubic feet of volume or fractional part thereof of such building, and upon securing and filing the written consent of two-thirds of the property owners according to frontage on both sides of the street in the block in which such building is to be moved. No permit shall be issued to move any building used or designed to be used for purposes for which frontage consents are required until frontage consents in the block to which such building is to be moved have also been secured and filed as required by the ordinances relating to such use.

(c) No frontage consent shall be required of any person, firm or corporation for removing a building upon his own premises and not going upon the premises of any other person, or upon any street, alley or other public place, in making such removal.

1681 Amusements—Frontage Consents Required: It shall be unlawful for any person, firm or corporation to construct or erect any building or structure designed or intended to be used for the purpose of presenting or carrying on therein any entertainment for which a license is required by the ordinances of the City of Chicago or to devote any grounds or place to such purposes without first obtaining the written consent of the property owners as required by the City ordinances.

1682 Buildings for the Storage of Shavings, Sawdust and Excelsior—Frontage Consents: It shall be unlawful for any person, firm or corporation to construct or erect any building designed or intended to be used for the purpose of storing shavings, sawdust or excelsior therein within the city without first obtaining the written consent of the property owners as required by the City ordinances.

1683 Frontage Consents—Business of Selling Provisions, Etc., in Residence Districts: It shall be unlawful for any person, firm or corporation to carry on the business of selling meats, poultry, fish, butter, cheese, lard, vegetables or any other provisions from any place of business located in any block in which all the other buildings are used exclusively for residence purposes, without first securing and filing with the City Collector the written consent of three-fourths of the property owners according to frontage on both sides of the street in the block in which the building to be thus used is located, provided in determining whether all the buildings in said block are used exclusively for residence purposes, any building fronting on another street and located upon a corner shall not be considered. In case a permit for building a store for such purposes in such block, or converting a building to store purposes in such block is applied for, the frontage consents required by this section shall be filed with the Commissioner of Buildings.

1684 Business of a Store—Requirements as to a Permit for Erection: No permit shall be issued for the erection or remodeling of any building in any block in which the use of buildings is restricted or regulated by ordinance if such building is designed to be used for conducting therein any business or store, without first requiring the applicant for such permit to file with the Commissioner of Buildings a plat showing the use to which all the property in such block is devoted.

1685 Withholding of Building Permit—Protest of Property Owners—Public Hearing: In all cases where an application for a permit is made for the erection of a new building in any square in which a majority of the buildings are used exclusively for residence purposes, or in a square on the opposite side of the street from such square so used for residential purposes; if there shall be filed with the Commissioner of Buildings a protest signed by not less than ten owners of property in such square so used for residential purposes, or in case the ownership of the frontage is in less than twenty persons then by a majority of the owners according to frontage, the Commissioner of Buildings shall withhold the issuance of the permit until the City Council shall have ordered a public hearing similar to that required in an act of the general assembly entitled "An Act to confer certain additional powers upon city councils in cities and presidents and boards of trustees in villages and incorporated towns concerning buildings and structures, the intensity of use of lot areas, the classification of trades, industries, buildings and structures with respect to location and regulations, the creation of districts of different classes, and the establishment of regulations and restrictions applicable thereto," in force June 28, 1921. For the purposes of this section a square shall be understood to be a plot of ground containing city lots surrounded by public streets, railway right of way, natural boundaries, or public places or thoroughfares.

1686 Garages—Frontage Consents Required: No person, firm or corporation shall keep, conduct or operate a garage in this city without first obtaining a license so to do in the manner provided for in this ordinance; and it shall not be lawful for any person, firm or corporation to locate, build, construct or maintain any garage within the territory bounded by the Chicago River and the south branch thereof on the north and west, by Lake Michigan on the east and by Van Buren Street on the south, any part of which is within eighty feet, or the entrance or exit to or from which, for the use of automobiles, is within one hundred and sixty feet of any portion of the street front of any building used as and for a hospital, church or public or parochial school, or such entrance or exit of which is upon a street containing street car tracks and within one block of the entrance of a street railway tunnel, or which shall house within said distance of one hundred and sixty feet of such street front, more than seventy-five cars. It shall not be lawful to locate, build, construct or maintain any garage within two hundred feet of any building used as and for a hospital, church or public or parochial school or the grounds thereof, in any portion of the City of Chicago outside of the territory above named, nor shall any person, firm or corporation hereafter locate, build, construct, establish any garage in the city, on any lot in any block in which dwelling houses, apartment houses and hotels constitute one-half or more of the buildings on both sides of the street in the block, or within one hundred feet of any such street in any such block without the written consent of a majority of the property owners according to frontage on both sides of the street; provided, that all lots which abut only on a public alley or court shall be considered as fronting on the street to which such alley or court leads. It shall not be deemed inconsistent with the character of a building. Such written consents shall be obtained and filed with the commissioner of buildings before a permit is issued for the construction of any building; provided, that in determining whether two-thirds of the buildings on both sides of such street are used exclusively for residence purposes, any building fronting upon another street and located upon a corner lot shall not be considered; and provided, further, that the word "block" as used in this section, shall not be held to mean a square but shall be held to embrace only that part of the street in question which lies between the two nearest intersecting streets.

1688 Undertaking Establishment Frontage Consents: It shall be unlawful for any person, firm or corporation to establish or maintain a morgue or to carry on the business of an undertaker, as defined in chapter 104 of this ordinance, that receives in connection with such business, at his, or their place of business, the body of any dead person for embalming or other purposes, on or along any street, without the written consent of a majority of the property owners according to the frontage on both sides of such street in the block in which such morgue or place of business is located; it shall also be unlawful for any person, firm or corporation to establish or maintain a morgue or to carry on the business of an undertaker, as defined in chapter LXVII of this ordinance, that receives, in connection with such business, at his, their or its place of business,the body of any dead person for embalming or other purposes, on or along any street in any block in which two-thirds of the buildings on both sides of the street are used exclusively for residence purposes, without the written consent of a majority of the property owners according to the frontage on both sides of such street in such block; provided that nothing herein contained shall apply to such location in the case of any person licensed as an undertaker and authorized to carry on such business at any such location at the time of the passage of this ordinance nor to any block in any street on which street cars are operated. Such frontage consents shall be obtained and filed with the department of health before a license shall issue for such business.

1689 Ice Plant Frontage Consents: It shall be unlawful for any person, firm or corporation to locate, establish, conduct or maintain any ice-making house or cooling plant, or any buildings used for the storage of ice, in any block in which two-thirds of the buildings fronting on both sides of the street on which the proposed plant shall be located are devoted exclusively to residence purposes, unless the owners of the majority of the frontage in said block on both sides of the street on which said plant is located shall consent in writing to the location, establishment, conducting or maintenance of such plant in such block. Such written consents of the majority of said property owners shall be filed with the Commissioner of Buildings before a permit shall be granted for the building or construction of any such ice-making house or cooling plant. Any person, firm or corporation violating any of the provisions of the section, or refusing, failing or neglecting to comply with any of the said provisions, shall be fined not less than five dollars nor more than one hundred dollars for each offense, and a separate offense shall be regarded as having been committed for each day during which such violation shall continue.

Billboards, Signboards.

1672 Frontage Consents Required: It shall be unlawful for any person, firm or corporation to erect or construct any billboard or signboard in any block on any public street in which one-half of the buildings on both sides of the street are used exclusively for residence purposes without first obtaining the consent in writing of the owners or duly authorized agents of said owners owning a majority of the frontage of the property on both sides of the street in the block in which such billboard or signboard is to be erected, constructed or located. Such written consents shall be filed with the Commissioner of Buildings before a permit shall be issued for the erection, construction or location of such billboard or signboard.

1687 Hospital or Home Frontage Consents: It shall be unlawful for any person, firm, association or corporation to build, construct, maintain, conduct or manage a hospital, or a home, as defined, in Part VIII of this ordinance, in any block in which two-thirds of the buildings fronting on both sides of the street or streets on or along which the proposed hospital or home may face are devoted exclusively to residence purposes, unless the owners of a majority of the frontage in such block and the owners of a majority of the frontage on the opposite side or sides of the street or streets on or along which said building faces consent in writing to the building, construction or maintaining, managing or conducting of any such hospital or home in such block; provided, however, that no new frontage consents shall be required if such hospital or home has heretofore been licensed by the city of Chicago as a hospital, home or nursery at the present location. Such written consents of the majority of said property owners shall be filed with the commissioner of health before a permit shall be granted for the building or construction of any such hospital or home, and before a license shall be issued for the maintaining, conducting or managing of any such hospital or home.

1232 Frontage Consents for Hospitals: It shall be unlawful for any person, firm or corporation to build, construct, maintain conduct or manage any hospital in any block in which two-thirds of the buildings fronting on both sides of the street or streets on which the proposed hospital may front are devoted to exclusive residence purposes, unless the owners of a majority of the frontage in such block and the owners of a majority of the frontage on the opposite side or sides of the street or streets on which said building fronts and faces consent in writing to the building, constructing or maintaining, managing or conducting of any such hospital in said block. Such written consents of the majority of said property owners shall be filed with the Commissioner of Health before a permit shall be granted for the building or constructing, or a license be issued for the maintaining, conducting or managing of any such hospital.

1297 Class IVc Defined—Moving Picture and Vaudeville Shows — Seating Capacity: Class IVc shall include every building hereafter erected used for moving picture or vaudeville shows and similar entertainments, where an admission fee is charged and regular performances are given, and where the seating capacity does not exceed three hundred, provided that every building of Class IVc existing at the time the passage of the ordinance known as The Chicago Code of 1911 shall comply with the provisions of Class IVb. All buildings hereafter erected for moving picture and vaudeville shows and similar entertainments, where an admission fee is charged and regular performances are given, with a seating capacity of over three hundred, and for the exhibition of moving pictures only, where the seating capacity is more than one thousand, shall be built to conform with the requirements for buildings of Class V hereafter erected as contained in this ordinance. Buildings for the exhibition of moving pictures only and with a seating capacity of over three hundred, but not to exceed one thousand, shall also be built to conform with the requirements for buildings of Class V hereafter erected, in all their structural requirements and equipment except in so far as such requirements and equipment are modified in Sections 1298 and 1299, hereof.

1320 Frontage Consents Required: No building of this class shall hereafter be constructed for, or converted to the use of said class, unless frontage consents are secured as required by the ordinances of the City of Chicago and filed with the Commissioner of Buildings.

1323 Grandstands — Frame Within Fire Limits—Grandstands Hereafter Constructed —Fireproof—Frontage—Consents:
(b) Every person, firm or corporation desiring a permit for the construction of a grandstand, except in connection with such as are now in existence, shall first obtain the consent in writing of the owners of a majority of the frontage on both sides of the street or streets on each side of the block or square in which it is desired to erect such grandstand.

Amusement Parks.

1327 Frontage Consents Required: It shall hereafter be unlawful for any person, firm or corporation, to build, construct, establish, produce or carry on, any amusement within any ground, garden or enclosure of the kind commonly known and described as amusement parks, wherein shows of different classes are offered or presented by one or more concessionnaires, without first securing written frontage consents as required by the ordinances of the City of Chicago. Such frontage consents shall be filed with the Commissioner of Buildings before a permit shall be issued for the construction of any building or structure connected in any way with such amusement or amusement park.

1637 Sheds—Coal, Brick, Stone, Cement and Salt Sheds and Sheds for Icing Cars Along Railroad Tracks and Navigable Stream: Open shelter sheds to be used for the storage or handling of coal, brick, stone, cement, salt or such commodities which are incombustible, or for the icing of cars, may be erected within or without the fire limits upon, along or adjacent to steam railroad tracks, or along or adjacent to navigable waters; provided, such sheds shall have incombustible roofing and shall not exceed 35 feet in height from the ground to the highest point of the roof, and provided, further, that said sheds shall be located at least 25 feet distant from any other structure and from any side lot line. If it is desired or intended to enclose any such sheds, the enclosing walls shall be of incombustible material. No such shed shall be built upon any lot or parcel of ground fronting upon any street within 200 feet of any building used exclusively for residence purposes, unless the consent of the owners of the majority of the frontage on both sides of such street between the two nearest intersecting cross streets shall first have been obtained by the person, firm or corporation desiring to erect and maintain such shed, and said written consents shall be filed with the Commissioner of Buildings before a permit shall be issued for such shed.

Automobile Repair Shop.

3260 Location and Frontage Consents: No person, firm or corporation shall locate, build, construct or maintain any automobile repair shop within two hundred feet of any building used as and for a hospital, church or public or parochial school or the grounds thereof, nor shall any person, firm or corporation locate, build, construct or maintain any automobile repair shop in the city on any lot in any block in which two-thirds of the buildings on both sides of the street are used exclusively for residence purposes, or within one hundred feet of any such street in any such block, without the written consent of a majority of the property owners according to frontage on both sides of the street; provided, that all lots which abut only on a public alley or court shall be considered as fronting on the street to which such alley or court leads. Such written consents shall be obtained and filed with the Commissioner of Buildings before a permit is issued for the construction of any such building or before a license is issued for the operation of any automobile repair shop in any existing building; provided, that in determining whether two-thirds of the buildings on both sides of such street are used exclusively for residence purposes, any building fronting upon another street and located upon a corner lot shall not be considered; and provided, further, that the word "block," as used in this section, shall not be held to mean a square, but shall be held to embrace only that part of the street in question which lies between the two nearest intersecting streets.

Automobile Salesroom.

3270 Location and Frontage Consents—Exception: No person, firm or corporation shall locate, build, construct or maintain any automobile salesroom within two hundred feet of any building used as and for a hospital, church or public or parochial school or the grounds thereof, nor shall any person, firm or corporation locate, build, construct or maintain any automobile salesroom in the city on any lot in any block in which two-thirds of the buildings on both sides of the street are used exclusively for residence purposes, or within one hundred feet of any such street in any such block, without the written consent of a majority of the property owners according to frontage on both sides of the street; provided, that all lots which abut only on a public alley or court shall be considered as fronting on the street to which such alley or court leads. Such written consents shall be obtained and filed with the Commissioner of Buildings before a permit is issued for the construction of any such building, or, in case a license is required for such business on account of the sale of second-hand cars, before a license is issued for the operation of any automobile salesroom in any existing building; provided, that in determining whether two-thirds of the buildings on both sides of such street are used exclusively for residence purposes, any building fronting upon another street and located upon a corner lot shall not be considered; and provided, further, that the word "block," as used in this section, shall not be held to mean a square, but shall be held to embrace only that part of the street in question which lies between the two nearest intersecting streets.

The provisions of this section shall only apply where the premises are of such a nature that automobiles, autocars or any similar self-propelled vehicles are or may be admitted thereto.

Filling Stations.

2670 Tanks—Pumps—Permits—Frontage Consents—Located Near School, etc.—Location Under Alleys, Inside Curb—Compensation: (a) It shall be unlawful for any person, firm or corporation to install a tank for the storage of any of the liquids mentioned in section 2667 without first obtaining a permit so to do from the division marshal in charge of fire prevention, as hereinafter provided.

(b) Any person, firm or corporation desiring to install a tank for the storage of any of the liquids mentioned in section 2667 shall first obtain a permit so to do from the division marshal in charge of fire prevention. The application for permit shall be made by the owner or his agent to the division marshal in charge of fire prevention. Before issuing such permit the division marshal in charge of fire prevention shall first inspect or cause to be inspected the location or site where such tank is to be installed, and if the site is satisfactory, the applicant shall pay to the city collector a fee of ten dollars for each tank of the capacity of five hundred gallons or less, and an additional fee of twenty-five cents for each additional one hundred gallons capacity or fraction thereof. Provided, however, that where one or more tanks are to be installed for the storage of fuel oil to be used for heating purposes only in the building or buildings for which such tank or tanks are installed, the permit fee shall be five dollars for each such tank of a capacity of five hundred gallons or less, and an additional fee of twenty-five cents for each additional one hundred gallons capacity or fraction thereof. Such permit fee shall be paid at the same time the written application for tank permit is filed and shall be forfeited to the city in case the site is not approved for permit. No such tank or equipment shall be covered or used until the installation, material and workmanship have been finally inspected, approved and certified by the division marshal in charge of fire prevention.

(c) Each application to install a tank or tanks where frontage consents shall be required as hereinafter provided in paragraph (d) shall be accompanied by a plat or survey showing the location and dimensions of all the property coming within the frontage area, the name and address of the owner or owners of each parcel of ground coming within such area, including the site, and the total frontage in feet and the required majority of such total frontage. In determining the area to be considered in computing frontage consents, the distance shall be taken from the boundaries of all the land used by such oil storage and incidental thereto, including the driveways and enclosing fences, if any

"In any location where a driveway or driveways across a public sidewalk are required permit shall not be issued until the applicant therefor has first obtained from the Commissioner of Public Works a permit for driveway or driveways as prescribed by Article II of Chapter 14 of this ordinance."

(d) It shall be unlawful to install any tank or tanks for the storage or sale of any of the liquids mentioned in section 2667 in any lot or plot of ground without first obtaining the written consents of property twners representing the majority of the total frontage in feet of any lot or plot of ground lying wholly or in part within lines one hundred and fifty feet distant from and parallel to the boundaries of the lot or plot of ground upon which said tank or tanks is or are to be installed, provided, however, that for the purpose of this section only the frontage of any such lot or plot of ground or that part of the frontage of any part of such lot or plot of ground as comes within the one hundred and fifty foot limit herein prescribed shall be considered.

Whenever the lot or plot of ground in which such tank or tanks is or are to be installed is in any shape other than a rectangle the one hundred and fifty foot limiting line aforementioned shall not exceed in distance one hundred and fifty feet from any point in the boundaries of such lot or plot of ground.

All petitions containing such consents of property owners shall be based on and contain the legal description of the property affected.

No such tank or tanks shall be installed in any lot or plot of ground where any of the boundaries of such lot or plot of ground are within two hundred feet of the nearest boundary of any lot or plot of ground used for a school, hospital, church or theatre. Provided, however, that the provisions of this paragraph shall not apply to the installation of a tank or tanks containing any of the liquids mentioned in section 2667 having a flash point above one hundred and sixty-five degrees fahrenheit when such liquids are to be used for heating purposes in the building or buildings for which said tank or tanks are installed.

(e) Tanks containing liquids of classes one or two shall not be located under any public street, alley or sidewalk or in the space between the building line and the curb. Permits may be issued to locate tanks to contain liquids of classes three or four, as mentioned above, under any public street, alley or sidewalk or in the public space between the building line and the curb upon obtaining a permit from the Commissioner of Compensation, upon the payment of the proper fee and the obtaining of a tank permit from the bureau of fire prevention, as provided by ordinance. Each application submitted for the location of a new filling station shall be accompanied by an additional fee of twenty-five dollars which is to be made payable to the City of Chicago and shall only cover the cost of verifying the frontage consents, as well as the signatures to the petition.

Frontage consents shall not be required for the installation of a tank or tanks containing any of the liquids mentioned in section 2667 when such liquids are to be used for heating purposes in the building or buildings for which such tank or tanks are installed or when such liquids are to be stored or sold within a garage building in connection with and incidental to the garage business therein conducted or where such liquids are used in a manufacturing plant and are essential to the manufacturing business therein conducted; provided, however, that where such liquids having a flash point below the hundred and sixty-five degrees fahrenheit are sold or offered for sale outside of a garage building or any other building such location shall be construed to be a filling station within the meaning of section 3273 of the Revised Chicago Code of 1931 and shall require frontage consents in accordance with section 2670 of said code.

For the purposes of this ordinance the lot or plot of ground used for filling station purposes shall not be less than twenty feet wide nor less than fifty feet in length.

(f) Every application for a permit for any such tank shall be in writing, stating specifically the location, the space desired to be used, the length, breadth and depth, together with the measurement in feet from the surface of the ground to the top of such tank, and shall contain the plans and specifications for the construction of said tank, its connections, fittings, openings and safety appliances.

CHAPTER 20.

Existing Buildings and Structures.

ARTICLE 2001.

General.

2001.01 Existing Buildings and Structures Defined:

(a) **General.** An existing building or structure is hereby defined as a building or structure in existence at the time of the passage of this ordinance.

(b) **Buildings and Structures Under Construction or Permit.** Any work of construction, alteration, repair or replacement under a permit issued by the Department of Buildings prior to the passage of this ordinance, may be completed in accordance with the terms of said permit and the ordinances relating thereto; provided however, that such completion shall be made prior to the expiration of said permit and any legal renewal thereof. Any building or structure so constructed, altered, repaired or replaced is hereby designated an existing building or structure.

2001.02 Alterations, Repairs and Replacements:

(a) **Cost Not to Exceed Fifty (50) Per Cent.** Any part of an existing building or structure, except as otherwise provided in the Occupancy Provisions of this ordinance, may be altered, repaired or replaced with the material and in the forms permitted in the original structure thereof, or with materials and in forms of construction of equal or greater strength and fire-resistance; provided however, that the combined cost of alterations, repairs and replacement during any period of thirty (30) months shall not exceed fifty (50) per cent of the cost of constructing a like building or structure new, entire, at the site of the original structure, assuming such site to be cleared; and provided further, that no building or structure shall be enlarged or increased in capacity or converted for any of the purposes of such units as are described in Chapter 11, Hazardous Use Units unless the entire building or structure shall be made to conform to the requirements of this ordinance for new buildings and structures. Such combined cost shall be the cost of all alterations, repairs and replacements made during the thirty (30) month period, whether such alterations, repairs and replacements are completed or not, and whether they are paid for or not. No other alteration, repair or replacement shall be made on or in such an existing building, unless the entire building or structure shall be made to conform to the requirements of this ordinance for new buildings and structures.

(b) **Affidavits Required.** Before the issuance of a permit for the alteration, repair or replacement of any part of an existing building or structure, the owner, the architect or engineer and the contractor shall file with the Commissioner of Buildings, a joint affidavit or separate affidavits stating the estimated cost of the proposed alteration, repair or replacement and the estimated cost of constructing the building or structure new, entire as provided in paragraph (a) of this section; and stating further that such estimated costs are true and correct to the best of affiant's knowledge and belief. Said affidavit of the architect or engineer shall be accompanied by plans and details adequately illustrating the construction of the existing building or structure. Upon completion of the alteration, repair or replacements covered by said permit, each of the aforesaid affiants shall file with the Commissioner of Buildings an affidavit or affidavits stating the actual cost of the alterations, repairs or replacements made under the permit; and stating further that no rebate, discount, gratuity or free service has been granted or received in connection with such alterations, repairs or replacements, unless it be such a rebate, discount, gratuity or free service as is regularly granted or received in connection with similar work in cases where no such affidavit is required.

2001.03 Adequate Exits Required: Whenever an inspector of buildings shall report to the Commissioner of Buildings that any existing building or structure has inadequate or insufficient means of exit, by which shall be understood means of exit not complying with the ordinance under which such building was constructed or exists which have since been reduced or rendered inadequate by disproportionately increased capacity or by conversion to uses not originally contemplated, the Commissioner of Buildings shall so notify the owner, agent or person in possession, charge or control of such building or structure, and shall direct him forthwith to make such alterations and changes in the construction or equipment of such building or structure as are necessary in order to make such building or structure comply with the requirements of this ordinance for exits. No exit or exit facilities in an existing building shall be reduced or decreased below the requirements in effect at the time when they were originally constructed.

ARTICLE 2002.

Buildings Moved to New Locations.

2002.01 Shall Conform to This Ordinance: It shall be unlawful for any person, firm or corporation to move any building from one (1) location to another, unless the same shall be altered or reconstructed so as to conform to the ordinances governing the construction of such building in its new location at the time of moving the same.

CHAPTER 21.
General Construction Provisions.

ARTICLE 2101.

General.

2101.01 Precedence of Chapters:

(a) **General and Special Provisions.** The general provisions of Chapters 21 to 33 inclusive, of this ordinance, shall govern the construction, alteration and conversion of all buildings and other structures, except as modified or enhanced by the special provisions for different occupancies contained in Chapters 7 to 18, inclusive, and the structural provisions of Chapters 34 to 43, inclusive.

(b) **Conflict Between Chapters.** In the case of any conflict between a general provision of Chapters 21 to 33, inclusive, and a special provision of any other chapter, the special provision shall govern.

2101.02 Scaffolds — Protection During Building Operations—Temporary Floors:

(a) **Scaffolds.** All scaffolds for use in the erection, repair, alteration or removal of buildings shall be so constructed as to insure the safety of persons working thereon or passing under or by the same and to prevent the falling thereof, and the falling therefrom of any material that may be used, placed or deposited thereon.

(b) **Temporary Floor.** It shall be the duty of every owner, person or corporation who shall have the supervision or control of the construction of or remodeling of any building having more than one (1) framed floor, to provide and lay upon the upper side of the joists or girders, or both, of the first floor below the riveters and structural steel setters, a plank floor, which shall be laid to form a good and substantial temporary floor for the protection of the employees and all persons engaged above or below or on such temporary floor in such building. A permanent or temporary floor shall be in place on the joists or girders of the next lower floor, while a temporary or permanent floor of the next floor or roof is being placed.

(c) **Roof or Planking.** In buildings more than one (1) story high, where scaffolding is used on the outside of such buildings in the course of erection, the workers thereon shall be protected by a roof of planking set not more than one (1) story's height above the floor of such scaffolding; provided however, that such roof shall not be required if no workmen are employed above such scaffolding.

(d) **Elevator and Flue Openings.** It shall be the duty of all owners, contractors, builders or persons having the control or supervision of all buildings in course of erection, to see that all stairways, elevator openings, flues and all other openings in the floors shall be covered or properly protected, and it shall be their further duty to comply with an act of the legislature of the State of Illinois, entitled "An Act Providing for the Protection and Safety of Persons in or about the Construction, Repairing, Alteration or Removal of Buildings, Bridges, Viaducts and Other Structures, and to Provide for the Enforcement Thereof", approved June 3, 1907, and in force July 1, 1907.

2101.03 Storage of Building Materials—Limitations: The occupation of the street for the storage of building material for any one (1) building or for temporary sidewalks, shall never exceed one-third (⅓) of the width of the roadway of the same, and in no event shall any material be stored or placed within four (4) feet of any steam or street railway track, and in all cases where such obstruction of the street is made, there shall be a clear space of not less than one (1) foot between such obstruction and the curb line; provided however, that the Commissioner of Buildings and the Commissioner of Public Works, or either of them, may limit or prohibit the storage of material on any street or alley where a tunnel, conduit, or any underground passageway or subway is located.

2101.04 Sidewalk and Street—Occupation of—Limitations: The extent of occupation of sidewalk and street, to be covered by the terms of a permit for street obstruction or building shall be as follows: Such permit shall not authorize the occupation of any sidewalk or street, or part thereof, other than that immediately in front of the lot or lots upon which any building is in process of erection and in relation to which such permit is issued. During the progress of building operation, a sidewalk not less than six (6) feet in width shall be at all times kept open and unobstructed for the purpose of passage in front of such lot or lots. Such sidewalk shall, if there are excavations on either side of the same, be protected by substantial railings which shall be built and maintained thereon so long as excavations continue to exist. It is not intended hereby to prohibit the maintenance of a driveway for the delivery of material across such sidewalk from the curb line to the building site.

2101.05 Temporary Sidewalks: It shall be permitted for the purposes of delivering material to the basements of buildings in process of erection to erect elevated temporary sidewalks to a height of not exceeding four (4) feet above the curb level of the street, and in case a sidewalk is so elevated it shall be provided with good substantial steps or easy inclines on both ends of the same and shall have railings on both sides thereof.

2101.06 Temporary Roof Over Sidewalks: When buildings are erected of a height greater than three (3) stories and such buildings are near the street line, there shall be built over the adjoining sidewalk a roof having a framework composed of supports and stringers of three (3) inch by twelve (12) inch timbers not more than four (4) feet from center to center, covered by two (2) layers of two (2) inch plank. Such framework and covering shall be of such construction and design as shall support safely a live load of two hundred fifty (250) pounds per square foot. Such roof shall be maintained as long as material is being used or handled on such street front above the level of the sidewalk. When additional stories are added to an existing building and such building is located near the street line, there shall be built over the sidewalk, at the point where the new stories commence, a scaffold not less than six (6) feet wide, which shall form a covering over the sidewalk composed of a framework of stringers and supports covered with two (2) layers of two (2) inch planks. Temporary sidewalks, their railings, approaches and roofs over same, shall be made with regard to ease of approach, strength and safety to the satisfaction of the Commissioner of Buildings.

2101.07 Frontage Adjacent: If the written consent of and a waiver of claims for damage against the City by the owners of properties adjoining the site of any proposed building is first obtained and filed with the Commissioner of Public Works, the permission to occupy the roadway and the sidewalk may be extended beyond the limits of such building in front of the property for which the consent of the owner or lessee thereof has been secured, upon the same terms and conditions as those herein fixed for the occupation of sidewalk and street in front of the building site.

2101.08 Use of Derricks: For all buildings more than four (4) stories in height, the use of derricks set upon the sidewalk or street is prohibited. In no case shall the guy lines be less than fifteen (15) feet over the roadbed.

2101.09 Use of Street—When Terminated—Red Lights:

(a) **Use of Streets.** The permission to occupy streets and sidewalks for the purpose of building is intended only for use in connection with the actual erection, repair, alteration or removal of buildings, and shall terminate with the completion of such operation. It shall be unlawful to occupy any sidewalk or street after the completion of the operation for which a permit has been issued by the Department of Buildings. It shall also be unlawful to occupy a street or sidewalk, under authority of such permit, for the storage of articles not intended for immediate use in connection with the operations for which such permit has been issued.

(b) **Red Lights.** Red lights shall be displayed and maintained during the whole of every night at each end of every pile of building material in any street or alley and at each end of every excavation.

2101.10 Building Operations at Night: It shall be unlawful for any person, firm or corporation, in conducting any building operations between the hours of ten (10) o'clock in the evening and six (6) o'clock in the morning, to operate or use any pile drivers, steam shovels, pneumatic hammers, derricks, steam or electric hoists in any block in which more than one-half (½) of the buildings on either side of the street are used for residence purposes.

2101.12 Display of Placard Indicating Floor Strength:

(a) **Where Required.** It shall be the duty of the owner, his agent, the occupant or person or corporation in possession, charge or control of every building, now in existence or hereafter erected, which is or shall be designed, erected, altered or used for the purposes of the storage for sale, storage or manufacture of merchandise or for the purposes of a Class 1 Garage or of a stable or hangar having a ground area more than eight hundred (800) square feet, to affix and display and maintain conspicuously on each floor of such building, a placard stating the uniformly distributed load per square foot of floor surface, which may be safely applied as provided by this ordinance and there shall be placards displayed to indicate the safe loading of each varying part of such floor. It shall be unlawful to load any floor, or any part thereof, to a greater extent than the loads indicated upon such placards.

(b) **Calculation of Strength.** The calculations and loads shall be made and determined in accordance with the Structural Provisions of this ordinance.

(c) **Approval Required.** It shall be the duty of the owner, his agent, the occupant or person or corporation in possession, charge or control of every building referred to in paragraph (a) of this section, to procure and submit evidence of the correctness of the figures on such placards to the Commissioner of Buildings. Whenever such evidence shall be satisfactory to the Commissioner of Buildings, he shall approve such placards. Such placards so approved shall then be affixed as hereinbefore required.

(d) **Unsafe Floors.** Whenever it shall be found by the Commissioner of Buildings that any floor is incapable of bearing, in addition to the weight of the floor construction, partitions, permanent fixtures and mechanisms that may be upon the same, a live load of forty (40) pounds for every square foot of surface, he shall condemn the same and order such floor to be repaired or reconstructed within a reasonable time by the owner or occupant thereof. In such case, it shall be unlawful for the owner or occupant to continue to use such building until the same floor shall be repaired or reconstructed in accordance with this section.

ARTICLE 2102.

Heights of Buildings.

2102.01 Designation of Stories: In every building the story, having its floor nearest the grade and not more than one (1) foot below the grade, shall be designated and known as the "first story", and as the "main exit story", the "main exit floor", or the "main exit level"; stories located above the first story shall be designated and known as "upper stories" and be distinguished from each other by number consecutively as the "second", "third", or "fourth" stories or floors and so on upward; and stories located below the first story shall be designated and known as "basements" and be distinguished from each other by letter alphabetically as "Basement A", "Basement B" or "Basement C" stories or floors, and so on downward; provided however, that in any building which is located on ground adjoining an existing or proposed and authorized two (2) level street, the story which is located between the upper and lower street levels shall be designated and known as th "ground" story or "ground floor"; the stories located above said ground story as the "first" or "second" stories or floors and so on upward; and the stories located below said ground story as "Basement A" or "Basement B" stories or floors and so on downward. Such a ground story shall not be considered a basement for the purposes of this ordinance.

2102.02 Height Limits:

(a) **General.** The height of any building shall not exceed the limit permitted by Chapters 7 to 18 inclusive, except as otherwise provided in this section; provided however, that this article shall not be construed as authorizing or permitting the erection of any building above the limits prescribed by the Chicago Zoning Ordinance.

(b) **Measurement of Height.** The height of a building shall be measured to the average finished ceiling line of the topmost habitable story. The height shall be measured from the established grade along the street line upon which a building abuts, or, if such grade varies, from the average elevation of such grade. In the case of an existing or proposed and authorized two (2) level street, the references of this section are to the upper level street.

(c) **Projection Above Height Limits.** Penthouses as provided in Section 2107.03 and spires, cupolas, domes and similar architectural features as provided in Section 2108.01 may project above the height limits prescribed in this section.

ARTICLE 2103.

Vertical Shafts.

2103.01 Inclosure for Wells:

(a) **Inclosures Required.** Every well in any building, except a building of wood frame construction, and except as otherwise provided in this section, shall be inclosed in a vertical shaft: provided however, that the requirements of this section shall not apply to a court or vent shaft open to the sky and meeting the requirements for a court or vent shaft as given elsewhere in this ordinance. Inclosing walls required by this paragraph shall be of not less than one (1) hour fire-resistive construction in buildings of fireproof and semi-fireproof construction and shall be of two (2) hour fire-resistive construction in buildings of heavy timber or ordinary construction. In all cases such inclosing walls shall be of reinforced concrete or shall be composed of units four (4) inches or more in thickness of burned clay, Portland cement concrete or solid partitions of Portland cement plaster and metal lath not less than two (2) inches in thickness. Every opening in such a wall shall be protected as provided in paragraph (d) of this section. Every inclosure required by this paragraph shall be complete and continuous for its entire height. Where a vertical shaft passes through a structural floor of combustible material, the required inclosure shall be carried through the well so as to completely isolate the shaft from the floor construction and shall be supported independent of all combustible construction. No stairways shall be installed in the same shaft with an elevator or in any vertical shaft shared with any similar or other device. Ash chutes, incinerator flues and vertical shafts serving as chutes for waste materials shall conform to the requirements for chimneys.

(b) **Inclosures Not Required.** No fire-resistive inclosure shall be required for a well in any of the following cases:

Case 1. A well for a sidewalk type elevator having a one (1) story travel, which is equipped with a trap door, or other closing member, of the strength and fire-resistive value equal to that required by this section for the floor or roof in which it occurs.

Case 2. A chase which is fire-stopped by required means at every floor or roof through which it penetrates.

Case 3. A roof scuttle or skylight, except between the roof soffit and any ceiling thereunder.

Case 4. A sleeve built solidly into floor, roof or ceiling construction and having not more than one-half (½) inch clearance around a pipe, conduit or cable passing through it.

Case 5. A stairwell or other well-hold occurring in a space which has an inclosure such as is required for a vertical shaft.

Case 6. A well for an escalator which is not used in lieu of a required stairway; provided however, that this exception shall apply only in buildings equipped throughout with a standard system of automatic sprinklers.

(c) **Reduction of Fire-Resistance Requirements for Inclosures.** The inclosure of a vertical shaft may be less fire-resistive than required by paragraph (a) of this section, but not less fire-resistive than metal studding with metal lath and plaster, if such shaft is located in a building of fire-proof or semi-fireproof construction and conforms to the provisions of Case 1 or Case 2 of this paragraph.

Case 1. A scuttle or skylight well hole extending from the ceiling immediately below the roof to a curb of non-combustible solid construction around the scuttle or skylight.

Case 2. Between a manhole, hatch, trap or other floor opening having a required closure and extending from the ceiling to the frame of the closure which shall be of non-combustible material.

(d) **Openings to Vertical Shafts.** Every opening into a vertical shaft, except in a building of wood frame construction, and except as otherwise provided in this ordinance, shall be equipped with a closing member not less fire-resistive than a Type A window or a forty-five (45) minute fire-resistive door.

2103.02 Hatchways for Elevators:

(a) **Fire-Resistive Inclosures.** Every hatchway for a passenger or freight elevator, except as otherwise permitted, shall be inclosed in a vertical shaft meeting the requirements of Section 2103.01, paragraph (a) and the further requirements of this section.

(b) **Non-Fire-Resistive Inclosures.** Hatchways for sidewalk type elevators, the travel of which does not exceed one (1) story, shall be inclosed for the full height of such story except on the sides used for loading and unloading. They shall be inclosed from the door lintels to the ceiling on the open sides and the clearance between the inclosure and the car platform at such sides shall not exceed five (5) inches. Inclosures shall be building walls, solid or latticed partitions, grillework, metal grating or expanded metal. Where wire grillework is used, the wire shall be not less than No. 13 steel wire gauge and the mesh shall be not more than two (2) inches. Where expanded metal is used, its thickness shall be not less than No. 13 U. S. Gauge. The spacing between vertical bars shall be not more than one (1) inch. Where the clearance between the inclosure and the car platform, counterweights or any sliding door, is less than one (1) inch, the openings in the inclosure shall be covered with a netting of one-half (½) inch square mesh of wire not smaller than No. 20 Steel Wire Gauge. Such netting shall extend from the floor to a height not less than six (6) feet above the floor.

(c) **Construction of Inclosures.** All projections inward from the general surface of the hatchway inclosure and which are opposite a car entrance shall be beveled on the underside or shall be guarded with metal plates or wood faced with metal of not less thickness than No. 11 U. S. gauge. The angle of such bevels or guard plates shall be not less than sixty (60) degrees.

(d) **Windows in Inclosures.** Windows shall be permitted only in exterior walls of hatchway inclosures. Windows in the inclosing walls of any elevator hatchway which are opposite any car entrance shall be provided with a grating or guard of vertical bars and the soffit of the recess formed by or between the vertical bars shall be leveled as hereinbefore required for other projections. Windows in hatchway inclosures less than seven (7) floors above grade or less than three (3) floors above an adjacent roof shall be fitted with metal guards having vertical bars not less than five-eighths (⅝) inch in diameter, spaced not more than ten (10) inches apart.

(e) **Pipes and Wiring in Hatchways.** No pipes conveying gases or liquids and no electric wires or cables shall be installed in any elevator or counterweight hatchway, except such cables as are required for the connections with the car.

(f) **Machine Room and Pits.** Safe and convenient means of access shall be provided to every elevator machine room and pit. Permanent provisions for adequate artificial lighting shall be made in every pit and machine room. Machine rooms shall provide a clear headroom of not less than six (6) feet. Substantial wire guards shall separate every counterweight hatchway from every pit and shall extend at least six (6) feet above the floor of such pit. All elevator machinery shall be separated from adjoining rooms or spaces by walls, partitions or substantial metal screens.

(g) **Platforms Under Machinery.** The supporting beams for overhead machines shall be of steel or reinforced concrete. The total load on overhead beams shall be assumed as equal to the weight of all superimposed loads plus twice the weight of all suspended loads. A non-combustible platform shall be provided under all sheaves and other parts of machines at the top of any elevator hatchway. Such platforms may be perforated to allow the minimum clearance required for necessary cables and shall be capable of sustaining a concentrated load of three hundred (300) pounds on any four (4) square inches of its area, a live load of one hundred (100) pounds per square foot or a total live load equal to one hundred fifty (150) per cent of the weight of the heaviest single unit or assembly of parts normally supported by the sheave or machine supports, if such total load exceeds a load of one hundred (100) pounds per square foot over the entire area of such platform.

(h) **Hatchway Doors.** No heat actuated automatic fire door shall be installed on any landing opening in a passenger elevator hatchway. The maximum distance between the edge of a landing threshold and the hatchway side of a door shall be four (4) inches. For automatic operation elevators the distance between the hoistway side of the landing door opposite the car opening and the hoistway edge of the landing threshold shall be not more than one (1) inch for swinging doors and two (2) inches for sliding doors. For existing installations of automatic operation elevators, where the clearance exceeds one and one-half (1½) inches for swinging doors or two and one-half (2½) inches for sliding doors, the space between the hoistway side of the landing door and the hoistway edge of the landing threshold door shall be filled in by suitable means. If the door slides in two (2) or more sections, the specified dimension applies to that section which closes against the lock jamb. No hardware, except that required for interlocking, indicator and signal devices, shall project into the hoistway beyond the line of the landing threshold. The lower edge of the interlocking devices shall be beveled. All hatchway doors shall be of not less than forty-five (45) minute fire-resistive value, except that vision panels of clear wired glass, not more than one hundred forty-four (144) square inches in area, may be provided in any hatchway door. Hatchway doors shall close the entire opening and shall be capable of withstanding a force of seventy-five (75) pounds, applied at any point without being sprung from their guides. Where a one (1) elevator hatchway extends through more than three (3) stories without a landing opening, emergency doors not smaller than two (2) feet, six (6) inches wide: nor less than six (6) feet, six (6) inches high, shall be provided at three (3) story intervals. Such doors shall be self-closing and shall be operative only by a key from the outside of the hatchway. Hatchway covers for sidewalk type elevators shall be self-closing.

(i) **Artificial Lighting.** Every elevator machine room, hatchway and pit and every landing threshold shall be adequately lighted from an electrical source.

ARTICLE 2104.

Guards to Prevent Falling.

2104.01 General Requirements: A guard to prevent persons from falling shall be provided at every point of danger, as required by the following sections of this article.

2104.02 Locations:

(a) **Balconies, Roofs and Miscellaneous Points of Danger.** Protection shall be afforded by guards not less than three (3) feet in height, at all edges of every balcony, mezzanine, space or other floor or roof surface which is at a height of more than two (2) feet above the floor, ground or pavement directly below, or which is accessible from any doorway, or from any window having a sill two (2) feet and six (6) inches or less above the floor inside of such opening. Protection shall be afforded by guards not less than one (1) foot, six (6) inches in height, at the edges of all other flat roofs except for one (1) story buildings and single dwellings; provided however, that buildings three (3) stories or less in height with flat roofs and having no public access by stairway, fire escape, fixed ladder, doors or windows shall not be required to have such guards.

(h) **Casement Windows and Doorways.** A casement window or doorway having a sill two (2) feet and six (6) inches or less above the floor, ground or pavement, inside or outside of such opening, shall be protected by a guard not less than three (3) feet in height unless such window or doorway opens directly upon such a guarded space as described in paragraph (a) of this section, the level of which is two (2) feet or less below the sill of such opening.

(c) **Areaways.** All sides of every open basement areaway except, one giving access to a stairway, shall have guards three (3) feet or more in height.

2104.03 Construction of Guards: The guards required under this article may be formed by walls, balustrades, grilles or railings. No portion of such a guard which is two (2) feet, six (6) inches or less from the floor or ground, shall have any opening therein more than six (6) inches wide. All required guards shall be designed to safely resist a force equivalent to a uniform horizontal pressure of two hundred (200) pounds against every lineal foot of such guards, applied at a height of three (3) feet above the ground level, pavement or floor of the guarded area.

ARTICLE 2105.

Exterior Openings.

2105.01 Facing of Openings: An opening in a wall shall be said to face a lot line, street or alley line, another wall or another opening when the plane of the wall containing such opeings makes an angle of sixty (60) degrees or less with such line or with the plane of such another wall or opening.

2105.02 Doors and Windows: In all exterior walls which are required to have a three (3) or four (4) hour fire-resistive value, except in single dwellings of any height and multiple dwellings of three (3) stories or less in height, a forty-five (45) minute fire-resistive door or a Type A window shall be installed in every opening through a wall, where said opening faces an alley at a distance of less than twenty-four (24) feet from the opposite side thereof; or where said opening faces a property dividing lot line at a distance of less than twelve (12) feet; or where window openings or doorways of two (2) or more areas of the same building, which are required by this ordinance to be separated by a fire division wall, are located on a court and face each other at a distance of less than twenty-four (24) feet; or where window openings or doorways of two (2) or more areas of the same building which are required by this ordinance to be separated by a fire-division wall, are located in planes forming an angle of one hundred fifty (150) degrees or less, measured outside the building, and are separated from one another by a distance of less than nine (9) feet; or where window openings or doorways of two (2) or more areas of the same building which are required by this ordinance to be separated by a fire-division wall, are located in the same plane or in planes forming an angle greater than one hundred fifty (150) degrees, and are separated from one another by a distance of less than four (4) feet, six (6) inches, or where the lowest point of said opening is less than forty (40) feet directly above a roof on the same premises sheathed with wood or other combustible material, except as provided in Chapter 25—MULTIPLE USE BUILDINGS.

ARTICLE 2106.

Roofs.

2106.01 Classification According to Slope:

(a) **Classification.** Roofs shall be classified according to the slopes of their surfaces as follows:

1. **Flat Roof.** A roof, the plane of which forms an angle of fifteen (15) degrees or less with the horizontal.

2. **Medium Pitched Roof.** A roof, the plane of which forms an angle of more than fifteen (15) degrees and not more than sixty (60) degrees with the horizontal.

3. **Steep Roof.** A roof, the plane of which forms an angle of more than sixty (60) degrees and not more than seventy-five (75) degrees with the horizontal.

(b) **Sloping Surfaces Not Classed As Roofs.** Any sloping inclosing surface of a building, the plane of which forms an angle of more than seventy-five (75) degrees with the horizontal, shall not be considered to be a roof, but shall be classed as a wall; provided however, that the exterior surface of a spire of any slope whatsoever shall be classed as a roof and not as a wall.

2106.02 Roof Coverings:

(a) **General.** Roof coverings shall be classified as either fire-retarding or combustible. Every building within the fire limits, and every building other than a wood frame building located outside the fire limits, shall have a roof of fire-retarding material, except as otherwise provided in paragraph (c) of this section, or elsewhere in this ordinance.

(b) **Fire-Retarding Roof Coverings.** The following roof covering materials are hereby designated as fire-retarding:

1. Compositions of rag or asbestos felt with asphalt or coal tar pitch, with or without a coating of gravel, crushed rock, slate or similar non-combustible material.
2. Compressed asbestos and cement sheets.
3. Compressed asbestos and cement shingles not less than one-eighth (⅛) inch thick, laid over felt, saturated with asphalt or coal tar pitch.
4. Slate.
5. Burned clay tile.
6. Concrete tile.
7. Any non-combustible composition tile.
8. Any ferrous metal covering not less than one-fortieth (1/40) inch thick.
9. Any non-ferrous metal covering not less than one-sixty-fourth (1/64) inch thick.

(c) **Combustible Roofs on Existing Buildings.** A wood shingle roof on a wood frame building not more than three (3) stories and basement high, existing at the date of passage of this ordinance, may be repaired with wood shingles or other materials, whether such building is located within or without the fire limits. No wood shingles or other combustible material shall be used on a part of a wood frame building located within the fire limits which shall be increased in height or area.

(d) **Roof Insulation.** Combustible or non-combustible insulating material shall be permitted between the required roof construction and the required roof covering in any type of construction.

2106.03 Wood Sleepers or Nailing Strips: All spaces between wooden sleepers or nailing strips on medium pitched or steep roofs of fireproof or semi-fireproof construction shall be filled solidly with non-combustible material.

2106.04 Eaves and Gutters: Eeaves and gutters on any building shall be not less fire-resistive than the roof surface required for such building, and it is further provided that no combustible material shall be used in the eaves or gutters of any building more than fifty (50) feet high.

2106.05 Monitors: The construction of every monitor shall be at least as fire-resistive as is required for the roof above which it is built. Every monitor which extends more than ten (10) feet above the roof at any point shall be constructed entirely of non-combustible materials.

2106.06 Scuttles and Hatches: A scuttle for the purpose of providing a means of access for persons to a roof, shall have an opening not less than twenty-four (24) inches by twenty-four (24) inches, nor more than forty-eight (48) inches by forty-eight (48) inches in clear dimension. The curb of any scuttle or hatch shall be of not less fire-resistive construction than the roof on which it is located and the cover shall be not less fire-resistive than a wood core, metal clad door. Every required scuttle for the access of persons to a roof shall be without any lock, screw or other fastening which will prevent any person from opening the cover from the inside at any time without a key or tool. Every cover of a roof scuttle or hatch shall be so fastened by hinges, chains or other means, that it cannot be blown from the roof when in either an open or a closed position.

2106.07 Dormers: Every dormer shall be of the same construction throughout, as required for the roof from which it projects, or of other not less fire-resistive construction; provided however, that in buildings of ordinary construction, or a superior type of construction, every dormer wall which in length exceeds fifty (50) per cent of the length of the parallel wall in the story below, shall be of masonry.

2106.08 Roof Construction for Certain Annexes: Any school or public assembly unit may have appended to it, a one (1) story, or one (1) story and basement structure used only for assembly, gymnasium, natatorium or similar purposes, having a roof and non-combustible roof-supporting members, including columns, unprotected by fire-resistive material; provided, **first**, that the capacity of such appended structure shall not exceed one thousand five hundred (1500) persons; provided, **second**, that the under side of its principal roof-supporting girders or trusses shall be not less than twenty (20) feet above the first floor; provided, **third**, that the roof sheathing, if of wood, shall be not less than two and five-eighths (2⅝) inches in one (1) thickness; and provided **fourth**, that such structure shall be separated from the remainder of the building by a solid masonry or reinforced concrete wall of not less than three (3) hour fire-resistive construction, any openings in which are protected by at least sixty (60) minute fire-resistive doors.

ARTICLE 2107.

Penthouses.

2107.01 Penthouse Defined: A penthouse is defined, for the purposes of this ordinance, as an inclosed space, located on a roof and used or designed for the housing of a stairway or of equipment used in the operation of the building, such as tanks, fans or elevator machinery, but not including spires, cupolas, domes or similar architectural features housing only clocks, bells, chimes or lights.

2107.03 Height and Area: A penthouse, or group of penthouses, constructed within the height and area limitations of this paragraph, and meeting the construction requirements of Section 2107.04, may be built upon the roof of any building. A penthouse so constructed shall not be considered in measuring the height of the building in accordance with Article 2102. The area occupied by such a penthouse or group of penthouses shall not exceed twenty-five (25) per cent of the area of the roof on which located. A plane forming an angle of forty-five (45) degrees with the horizontal shall be assumed to extend from each property dividing lot line of the premises at the allowable height for a main roof at such line and no part of any penthouse shall extend through or above such a plane; provided, however, that nothing in this paragraph shall be construed to prevent the construction of spires, cupolas, domes or similar architectural features to a greater height as provided in Article 2108.

2107.04 Construction:

(a) **Penthouse.** A penthouse, constructed within the limitations provided in Section 2107.03, shall be constructed entirely of non-combustible materials; provided however, that the roof of such a penthouse need not be more fire-resistive than is required for the main roof of the building on which it is located; and provided further, that where a penthouse is located on a roof surface less than fifty (50) feet above grade, every portion of such penthouse which is nearer than three (3) feet to a property dividing lot line shall have a wall not less fire-resistive than is required for the story immediately below the penthouse, and a roof not less fire-resistive than is required for the roof surface on which the penthouse is located.

ARTICLE 2108.

Spires and Similar Architectural Features.

2108.01 General: Spires, cupolas, domes and similar architectural features, designed solely for architectural effect, or for the support of or housing of clocks, bells, chimes or lights and their operating mechanisms, and which provide no habitable rooms or spaces, may be built above the roof of a building and above the height limits prescribed in Article

2102, except as provided in the Chicago Zoning Ordinance. Access through such a feature for purposes of cleaning or repairing its exterior, and for the care or operation of the devices or mechanisms located therein or thereon, shall be permitted and shall not be considered as habitable use within the meaning of this section.

2108.02 Construction:

(a) **Structural Frame.** The structural frame of any spire, dome or similar architectural feature, permitted under Section 2108.01, shall be of the same type of construction as required for the building supporting such feature.

(b) **Coverings.** The entire exterior surface of such features shall be of non-combustible material forming a structural part of such features or shall be of fire-retarding material if applied over a structural covering. Wood sleepers for the attachment of surface materials, if used in connection with a structural frame of fireproof or semi-fireproof construction, shall be as required by Section 2106.03.

ARTICLE 2109.

Acoustic Materials.

2109.01 General: Any material applied to a wall, ceiling or other surface for the purposes of sound-deadening, sound absorption, or the prevention of reverberations, shall be known as an acoustic material. Any acoustic material may be used in buildings for the aforesaid purposes, subject to the requirements of this article. Combustible acoustic material shall be applied directly to a non-combustible surface or to a ceiling of metal lath and plaster or to the members of heavy timber construction. Acoustic material and the support thereof when suspended, furred, or used as a non-required ceiling shall be of non-combustible material.

2109.02 Non-Combustible Acoustic Materials:

(a) **Qualification Tests.** Any acoustic material to be classed as non-combustible shall be capable of meeting all the requirements of the following tests:

Item 1. The material to be tested shall be not less than two hundred (200) square inches in area.

Item 2. The material to be tested shall be firmly mounted in the same manner as used in attaching the acoustic material to walls, ceilings or other surfaces in buildings.

Item 3. A thermo-couple shall be installed on the unexposed side of the material to be tested; said thermo-couple being located at the center of the area to be tested and in contact with the test material in a manner which will cause it to indicate the temperature of said test material.

Item 4. The material to be tested shall be exposed directly to the flame in a test furnace and said material shall be maintained at a temperature of not less than twelve hundred (1200) degrees Fahrenheit for a period of not less than ten (10) minutes.

Item 5. The material during the ten (10) minute test period shall not support combustion nor carry flame or fire.

Item 6. The test specimen shall be removed from the furnace immediately upon conclusion of the prescribed fire test, and shall not carry flame nor support combustion when so removed.

Item 7. Immediately after the test material is removed from the furnace, it shall be cooled in cold water, after which the test material shall not fall away from its support.

(b) **Qualification by Proof of Test.** A certificate of the responsible executive of a reputable testing laboratory, stating that a material has been tested and found to comply with the requirements of this section, shall be accepted by the Commissioner of Buildings, as proof that said material complies with the requirements of this section.

2109.03 Attachment of Acoustic Materials: All acoustic material shall be securely attached to the wall, ceiling or other surface by cement or by other non-combustible material, or may be nailed directly to heavy timber construction. Such cement, or other material, when heated, shall continue to hold the acoustic material in place. Non-combustible material may be supported by non-combustible furring strips.

ARTICLE 2110.

Walls and Partitions.

2110.01 Walls:

(a) **General.** All required and all non-required walls shall conform to the general provisions of this chapter, the special provisions of the chapter governing the particular class of occupancy of which such walls are a part and to the Structural Provisions of this ordinance.

(b) **Fire-Resistive Values.** Wherever a fire-resistive value of one (1), two (2), three (3) or four (4) hours is required, such wall shall conform to the provisions of Chapter 32—FIRE-RESISTIVE STANDARDS; provided however, that such compliance shall not exempt such walls from any further requirements of the structural provisions.

(c) **Standard Fire Division Walls.** Standard Fire-division walls shall be constructed of masonry or reinforced concrete in accordance with the following provisions:

1. Masonry shall be not less than twelve (12) inches thick and shall be constructed of solid units with no hollow spaces between units.
2. Reinforced concrete shall be not less than eight (8) inches thick and shall have no hollow spaces.
3. The wall shall extend from the foundation to the roof of the building and in a building of other than fireproof or semi-fireproof construction, shall be continued as a parapet wall to a height of not less than three (3) feet above the roof; provided however, that the wall may be offset at any point if the horizontal or inclined construction connecting the offset portions is of the same thickness and construction as required for the wall itself, and any structural member or members supporting any part of the wall or the connecting construction shall have a four (4) hour fire-resistive value. In buildings of wood frame construction, such walls shall project horizontally through the inclosing walls of the building and project therefrom for a distance of not less than one (1) foot.
4. Any opening through the wall shall be protected with double standard fire doors. The combined width of openings in any one (1) story in the wall shall not exceed twenty-five (25) per cent of the length of the wall; provided however, that where the building is equipped throughout with a standard system of automatic sprinklers, the combined width of openings in any one (1) story shall not exceed fifty (50) per cent of the length of the wall, except that one (1) opening not more than five (5) feet in width shall be allowed in any case. Any opening through offset portions of the wall shall have a closure which will furnish fire protection at least equal to that of double standard fire doors in a wall.
5. Mortar used in standard fire division walls shall be as required under Chapter 38—MASONRY CONSTRUCTION, for mortar 1-2-3 or 6.

(d) **Parapet Walls.** In every building of other than fireproof or semi-fireproof construction, every exterior wall, except a street front wall, which is nearer than fifteen (15) feet to a property dividing lot line, or nearer than twenty-four (24) feet to the opposite side of an alley, shall have a parapet carried above the roof as hereinafter provided. Every parapet wall shall be of the same thickness and of the same fire-resistive value as the wall of the topmost story immediately beneath it. Every parapet wall shall extend at least three (3) feet above the roof, measured at right angles to the roof surface; provided however, that a parapet three (3) feet or more from any dividing lot line shall extend at least two (2) feet above the roof; and provided further that a parapet wall at a gable end of a medium pitched or steep roof need not extend more than one (1) foot above the roof. Parapets shall not be required for any exterior wall which is covered by a roof surfaced with two (2) inches of slate, clay tile or not less fire-resistive material laid in mortar; provided however, that such roof extends at least fifteen (15) feet inside of or at right angles to such wall. (See Article 2104—Guards to Prevent Falling.)

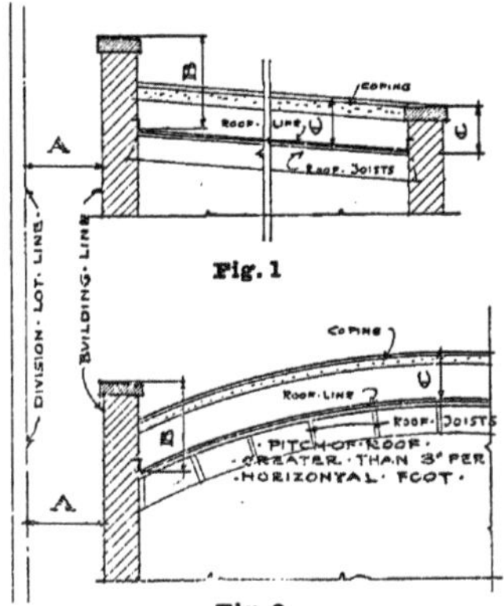

Fig. 1

Fig. 2

SECTION 2110.01 d.

Fig. No. 1.

A—distance from division lot line to building line.

B—height of parapet wall above roof on division lot line side.

C—parapet wall on other sides when required.

Explanation:

If A is 3'0" or more, B shall be 2'0".

If A is less than 3'0", B shall be 3'0".

C shall be not less than 2'0".

Fig. No. 2.

A—distance from division lot line to building line.

B—height of parapet wall above roof, with a greater pitch than 3" per horizontal foot, on division lot line side.

C—parapet wall on other sides when required.

If A is less than 3'0", B shall be 3'0".

C shall not be less than 18".

For exceptions where fireproof or semi-fireproof construction is used see ordinance Sec. 2110.01 d.

(e) **Sloping Walls.** Any exterior inclosing surface of a building, the plane of which forms an angle of more than seventy-five (75) degrees with the horizontal, shall be classed as a wall and shall be subject to the requirements of exterior walls; provided however, that an exterior inclosing surface, which is part of a street front of a building, or which is located more than fifteen (15) feet from a property dividing lot line; or more than twenty-four (24) feet from the opposite side of an alley; or which incloses an attic or similar space, not occupied as a place of human habitation, may be constructed as required for roofs.

2110.02 Standard Fire Separation: A standard fire separation shall consist of vertical or sloping roofs and horizontal or inclined floors or slabs of masonry or reinforced concrete, forming a complete separation between two (2) parts of a building and meeting the following provisions:

Item 1. Masonry shall be not less than twelve (12) inches thick of solid units with no hollow spaces between.

Item 2. Reinforced concrete shall be not less than eight (8) inches thick and shall have no hollow space.

Item 3. Any structural members supporting such construction shall have a four (4) hour fire-resistive value.

Item 4. Any opening through such separation shall have a closure equal in fire-resistive value to a double standard fire door.

2110.03 Partitions:

(a) **Required Separations.** Partitions forming required separations of a specified fire-resistive value shall conform to the requirements of Chapter 32 — FIRE-RESISTIVE STANDARDS.

(b) **Sub-Dividing Partitions.**

Item 1. Fireproof and Semi-Fireproof Construction. In buildings of fireproof or semi-fireproof construction, all partitions shall be non-combustible; except that in business units only, any area of six hundred (600) square feet or less inclosed by non-combustible walls or partitions, may be subdivided by partitions of wood or wood and glass.

Item 2. Heavy Timber Construction. In buildings of heavy timber construction every partition shall be of non-combustible material; provided that in business units only areas of four hundred (400) square feet or less inclosed by such non-combustible partitions, may be subdivided as permitted for fireproof and semi-fireproof buildings.

Item 3. Ordinary and Wood Frame Construction. In buildings of ordinary and wood frame construction solid or hollow partitions composed partially of combustible materials shall be permitted. Partitions of wood studs shall be lathed with wood or metal lath, gypsum, compressed fibre or similar plaster base which shall receive a coat or coats of lime, gypsum or cement plaster not less than one-half (½) inch in thickness, or, in lieu of said plaster base and plaster, other material of equal thickness and having equal fire-resistive and sanitation values may be used.

(c) **Fire Stops.** Wherever separations of a specified fire-resistive value are required, such value shall be maintained throughout the story height and throughout any hollow spaces in the floor or ceiling construction by fire stops of material at least equal in fire-resistive value to the value required for such separations. Hollow partitions of wood studding shall have top and bottom plates in each story.

(d) **Buck Frames and Adjacent Studding.** Buck frames for openings may be of wood. Wood studs may be used between bucks where the total distance across the adjacent bucks and the intervening space does not exceed one (1) foot, three (3) inches.

ARTICLE 2111.

Separation of Buildings from Public Spaces Below Grade.

2111.01 Separating Walls and Floors:

(a) **Where Required.** The interior of any building which adjoins a subway or other space below grade not open to the sky, under public control and open to the public, shall be separated from such space by walls or floors of solid masonry or reinforced concrete not less than twelve (12) inches thick.

(b) **Communicating Openings.** Every opening through a wall, required by this section, shall be provided with double standard fire doors. The combined area of all such openings in any wall shall not exceed twenty-five (25) per cent of the gross area of the wall. Every opening through a floor required by this section shall be provided with a bulkhead, vestibule or shaft of masonry or reinforced concrete not less than twelve (12) inches thick and equipped with fire doors, as required for an opening in a wall.

(c) **Display Space in Front of Walls.** No wall used in a separation required by this section shall be located more than twenty (20) feet back from the subway or other public space. Show window and display space may be located on the public side of such separating wall, but there shall be not more than one (1) doorway into each show window space through such wall.

(d) **Vehicular Passage.** Any public vehicular passage, not open to the sky, shall be separated from the interior of any adjoining building as required for subways in this section.

(e) **Railroad Right-of-Way Under Buildings.** Nothing in the preceding portions of this section shall prevent the entire space underneath a building, or any portion of such space, or any passage through a building, from being used as right-of-way or switching space for railroads or public carriers; provided however, that such space or passage shall not be designed nor used as a public thoroughfare, nor as a parking space for motor vehicles where such use is prohibited by other sections of this ordinance; and provided further, that such space is separated from the interior and adjacent parts of the building as required by this section for subways.

ARTICLE 2112.

Transformer Vaults.

2112.01 General: In every building, other than a manufacturing unit used exclusively for the production or distribution of electrical energy, all electrical transformers, capacitors and static condensers, which contain oil and are required by other ordinances of the City of Chicago, to have a fireproof inclosure, shall be housed in a room or vault meeting the provisions of this article.

2112.02 Construction:

(a) **Vaults Outside of Other Buildings.** Transformer vaults entirely outside of any other building, may be constructed of not less fire-resistive construction than semi-fireproof construction.

(b) **Vaults in Buildings.** Transformer vaults contained in any building, shall have inclosing walls, floors and ceilings of three (3) hour fire-resistive construction. Every door in every opening through such inclosure shall be a sixty (60) minute fire-resistive door and every window shall be a Type A window.

(c) **Curbs and Sills.** Every doorway in every transformer vault shall have a curb or sill of Portland cement concrete raised at least three (3) inches above the floor and forming a basin of sufficient capacity to contain all of the oil contained in the largest transformer.

(d) **Ceiling Height.** Every transformer vault shall have a clear ceiling height of not less than eight (8) feet.

(e) **Emergency Vents.** Every transformer vault shall have provisions for emergency ventilation through exhaust openings to the outside air. Exhaust openings shall be provided in area at least equivalent to three (3) square inches per k.v.a of maximum demand, as used in determining the size of the installation, but in no case less than one (1) square foot in area. If such opening is provided through a duct passing through another part of any building, said duct shall be inclosed in construction of a fire-resistive value at least equal to that required for interior chimneys, as provided in Chapter 41—CHIMNEYS. Every such opening may be provided with dampers of iron or steel not less than one-eighth (⅛) inch in thickness. Said dampers may be normally controlled manually, but shall be arranged to open automatically in case of fire, by means of a fusible link or by the actuation of a rate of rise thermostatic device. Supply openings in area, equivalent to that of the exhaust openings shall be provided. Nothing herein contained shall preclude the provision of a mechanical ventilating system in addition to the required emergency vents, nor to prohibit the installation of an exhaust fan in the required emergency exhaust opening. All vent openings through the vault inclosure, except the required emergency exhaust opening, shall be equipped with dampers at least equal in fire-resistive value to the type of fire door required for a doorway through such inclosure and such dampers shall be arranged to close automatically in case of fire.

ARTICLE 2113.

Rooms for Mechanical Equipment.

2113.01 Rooms for Furnaces and Boilers:

(a) **Low Pressure Plants.** Every room containing a furnace or boiler having a working pressure not to exceed ten (10) pounds, in all buildings, except single dwellings, shall be inclosed in walls, partitions and ceilings of not less than one (1) hour fire-resistive value; provided however, that where such rooms shall have no story over or above any portion of such rooms, the fire-resistive ceiling shall not be required unless otherwise provided for a specific occupancy. In any theatre and in any building devoted to the purposes of a place of public assembly, such room and every room containing any refuse burner or incinerator shall be inclosed in two (2) hour fire-resistive construction, with every doorway having a forty-five (45) minute fire-resistive door and every window being a Type A window.

(b) **High Pressure Plants.** No boiler room or other room in which there shall be a heating plant having a working pressure of more than ten (10) pounds shall be located under the auditorium of any theatre or within a distance horizontally of less than twenty (20) feet from such auditorium. Every such boiler room shall have an inclosure of not less than one (1) hour fire-resistive value and in theatres and other buildings devoted to the purposes of a place of public assembly shall have inclosures of not less than two (2) hour fire-resistive value with every doorway having a forty-five (45) minute fire-resistive door and every window being a Type A window.

2113.02 Elevator Machine Rooms: All rooms for elevator or dumbwaiter machines which are not a part of the hatchway for such elevator or dumbwaiter, shall have inclosures as provided for such rooms which are a part of such hatchway.

2113.03 Rooms for Cooling and Refrigerating Equipment: The plant of every direct mechanical system and the direct portion of every indirect mechanical system, which contains a highly flammable or irritant refrigerant in any quantity more than one hundred (100) pounds of Group "B" or fifty (50) pounds of Group "C" as classified under Section 4502.01 shall be inclosed in a room devoted to no other purpose, which room shall be separated from other parts of buildings by solid construction having a four (4) hour fire-resistive value. Except as otherwise provided, there shall be no direct means of communication between such room and any other portion of a building. The entrance or exit of such rooms shall be located in an outside wall not a part of the inclosure of an exit court.

ARTICLE 2114.

Construction of Porches—Verandas and Porticos—Inside Fire Lines.

2114.01 Combustible Construction:

(a) **Where Permitted.** Porches, verandas or porticos of combustible material, not more than three (3) stories in height, and extending not more than fifty (50) feet across the rear of buildings, may be of combustible materials. A storm door inclosure, not more than two (2) feet wider than the inner doorway, and projecting not more than four (4) feet from the face of the building and not more than twelve (12) feet high, may be of combustible materials.

(b) **Where Prohibited.** Except as permitted in paragraph (a) of this section, the inclosing walls and structural frame of every porch, veranda or portico within the fire limits, shall be of non-combustible material.

2114.02 Non-Combustible Construction: On buildings more than three (3) stories in height, porches, verandas or porticos of non-combustible construction, which are continuous and extend more than fifty (50) feet across the rear of buildings shall be divided into sections not more than fifty (50) feet in extent by partitions of non-combustible materials.

ARTICLE 2115.

Framing Around Chimneys.

(See Illustrations Paragraph 4102.)

2115.01 General:

(a) **Floors and Roofs.** No structural members of any building or structure shall rest or be supported on the walls of any chimney or smoke flue or of any shaft or chute, required to be constructed or inclosed as required for chimneys. No combustible framing shall be placed nearer than two (2) inches from the outside wall of such chimney inclosure nor closer than seven (7) inches from the inside of any such flue, shaft or chute.

(b) **Partitions.** No smoke pipe of an area greater than that of a six (6) inch diameter round pipe shall be permitted to pass through a combustible partition. Where a smoke pipe of not greater area than that of a six (6) inch diameter round pipe passes through a combustible partition, it shall be surrounded by a ventilated thimble of non-combustible material or by non-combustible material not less than four (4) inches thick.

ARTICLE 2116.

Cornices.

2116.01 General:

(a) **Materials.** No combustible material shall be used for any purpose in connection with cornices of buildings more than fifty (50) feet in height.

(b) **Support.** Cornices of non-combustible materials shall be supported on non-combustible structural members secured to the walls or structural frame of the building and the walls shall maintain the required fire-resistive value the full height behind every cornice throughout their entire length.

(c) No cornice, sign or other projected construction shall extend beyond any adjacent private property line.

ARTICLE 2117.

Fireplaces.

2117.01 Scope of This Article: The provisions of this article shall apply to open recesses or fireplaces in chimneys wherein fire may be made of combustible material of any kind, for domestic purposes and shall not be construed as applying to any fireplace or hearth used for any commercial or manufacturing purpose or process.

2117.02 Materials and Construction:

(a) **Hearths.** The floor or hearth of every fireplace shall be of brick, stone, burned clay tile or Portland cement concrete, or a combination of such materials not less than twenty (20) inches in width, measured from the face of the chimney breast and extending not less than twelve (12) inches beyond the jamb of the recess on each side thereof. Hearths shall be of minimum thickness of four (4) inches and shall be supported on a fireproof floor slab or on brick trimmer arches. Wood centering for trimmer arches shall be removed after the masonry has thoroughly set.

(b) **Walls and Lining.** The walls of fireplaces shall be not less than eight (8) inches thick, if of solid burned clay units and not less than twelve (12) inches thick if of stone or other hollow units. The faces of all such walls exposed to fire shall be lined with firebrick, soapstone, cast iron or other suitably fire-resistive material. When lined with four (4) inches of firebrick, such lining may be included in the required minimum thickness of walls. Mortar used shall be as required by the provisions of Chapter 38—MASONRY CONSTRUCTION.

(c) **Throats and Flues.** No flue from any fireplace shall incline more than forty-five (45) degrees from the vertical. Where the smoke flue ascends elsewhere than directly above the center line of the fireplace recess opening, the throat shall be gathered together with its outlet directly on such center line and any necessary offsets shall be affected only above the throat outlet. Smoke flues shall provide a minimum cross-sectional area equal to one-tenth (1/10) the area of the fireplace recess opening, and any damper or dampers used in the throat, shall not reduce such required flue area when in the opened position, nor provide less free opening than the required flue area.

(d) **Construction of Flues and Chimneys.** The construction of flues and chimneys for fireplaces, including ash chutes and pits, shall be as provided in Chapter 41—CHIMNEYS of the structural provisions.

ARTICLE 2118.

Protection Against Corrosion.

2118.01 Structural Members:

(a) **Structural Frame.** All structural steel and iron members, supporting exterior walls of any building, which are to be encased in a covering other than concrete, shall first receive a coat of mortar at least one-half (½) inch thick. Said mortar coat shall be of No. 1 Mortar, as described in Chapter 38—MASONRY CONSTRUCTION and shall be applied directly to such metal members. All spaces between any fire-resistive covering and such protected metal members shall be filled solidly with mortar.

(b) **Floor and Roof Systems.** All interior structural steel and iron members, which are not to be encased in concrete shall receive two (2) coats of a rust preventative paint, at least one of which shall be applied after erection; provided, however, that surfaces inaccessible for painting after erection, shall be given two (2) shop coats of paint. All trussed steel joist and steel floors and decks shall receive two (2) coats of rust preventative paint. All abrasions of the paint coating incurred during erection, shall be retouched after erection.

CHAPTER 24

Size and Location of Rooms

ARTICLE 2401

Capacity

2401.01 General: For particular classes of occupancy, the manner of computing capacity is prescribed in the chapters devoted to the occupancy provisions. Except as otherwise provided, the capacity of any room or space shall be determined in accordance with the provisions of this article.

2401.02 Computation by Capacity:

(a) **Rooms or Spaces Having Fixed Seats.** The capacity of every room or space equipped with fixed seats having arms or other positive divisions between sittings over the entire floor area, except for the required aisles and spaces between rows, shall be in number of persons equivalent to the number of such fixed seats. Where the fixed seats or benches are without arms or other positive divisions, the capacity of such room or space shall be in number of persons equivalent to the number of sittings, one (1) foot six (6) inches wide provided by such undivided seats.

(b) **Rooms or Spaces Without Fixed Seats.** The capacity of every room or space not equipped with fixed seats and of such portions of rooms or spaces not occupied by fixed seats or by the required aisles and spaces between rows of such fixed seats shall be in number of persons equivalent to the quotient obtained by dividing the area of such room or space in square feet by the lowest divisor given in Table 2401.02 for the respective purposes of use for which such room or space or portion thereof is designed, erected, altered, converted or used.

TABLE 2401.02

Purpose of Use	Remarks	Divisor
Armory or Rink	Drilling, Playing or Skating Space	10
Armory or Rink	Other Assembly Space	6
Art Rooms, Draughting Rooms and Libraries		25
Band or Orchestra Rooms		25
Billiards, Pool or Bowling Rooms	Including area of alleys and pits	25
Cafeterias	Seating space only	12
Circus	Exhibition space	10
Circus	Other assembly space	6
Class and Study Rooms		18
Cooking Rooms, Serving Rooms and Laboratories		30
Dance Floors		10
Eating Rooms		12
Gymnasiums	In public schools—Playing space	50
	Other Assembly Space	5
	In Other Buildings, entire space	10
Halls, Meeting Rooms		6
Lodge Rooms		10
Lecture Rooms	In public schools not more than 1500 sq. ft.	10
Locker Rooms		15
Manufacturing		100
Museum Rooms		10
Natatoriums	Including pool area	15
Office and Financial		100
Except Banking Rooms		50
Promenades		20
Reading Rooms		25
General Stores and Sales Units	Main Floor and Basement	30
	Upper floors	60
Stages and Platforms		6
Storage Rooms		100
Trading Rooms in Stock and Produce Exchanges		5
Waiting Rooms and Lounges		15
Except permitted standing spaces		4

2401.03 Net Floor Area: For the purposes of determining the capacity of a building, floor, room or space, the net area shall be assumed to be the floor area, as defined in Chapter 5—DEFINITIONS; provided however, that lobbies, in closed corridors and other inclosed means of exit and toilet rooms shall not be included.

2401.04 Ceiling Heights: Wherever a given ceiling height or a given volume is required for any room by this ordinance, beams or ducts shall be permitted to project not more than twelve (12) inches below the plane of such required ceiling height and the volume of such projections shall be included in the required volume of the room; provided however, that no such projection shall be located at less than seven (7) feet above the floor, nor occupy more than twenty-five (25) per cent of the area of the ceiling.

ARTICLE 2402
Fire Division Areas

2402.01 General: A fire division area, for the purposes of this ordinance, is hereby defined as an area within a building, separated from every other area in the same story by a standard fire division wall or walls meeting the requirements of Section 2110.01, paragraph (c).

2402.02 Fire Division Area Limits: No fire division area in a building not equipped with a standard system of automatic sprinklers shall exceed the area given in Table 2402.02 (1) or Table 2402.02 (2) for the occupancy use and construction type of such building:

TABLE 2402.02 FDA (1)
Maximum Allowable Floor Areas in One-Story Buildings Not Equipped with Automatic Sprinklers

Class of Occupancy	Fireproof Const. Sq. Ft.	Semi-Fire-proof Const. Sq. Ft.	Heavy Timber Const. Sq. Ft.	Ordinary Const. Sq. Ft.	Wood Frame Const. Sq. Ft.
Single Dwellings	U	U	U	U	U
Multiple Dwellings	U	U	*	*	*
Institutional Buildings	U	U	10,000	10,000	O
Office Units	U	U	UK*	UK*	400 L 1,250 N
Sales Units	U	U	UK*	UK*	400 L 1,250 N
Financial Units	U	U	UK*	UK*	400 L 1,250 N
Storage Units	*	*	*	*	400 L 5,000 N
Manufacturing Units	*			*	400 L 5,000 N
Hazardous Use Units	*	*	*	*	*
Garages—					
Class 1	30,000	20,000	20,000	20,000	O
Class 2	800	800	800	800	400
Theatres	U	U	O	O	O
Public Assembly Units	U	U	*	*	O
Churches	U	U	*	*	O
Schools	U	U	*	*	O

Note: The asterisk * denotes that area limits are given in the Chapter of requirements for this class of occupancy.

TABLE 2402.02 (2)

Maximum Allowable Floor Areas in Multi-Story Buildings Not Equipped with Automatic Sprinklers

Class of Occupancy	Fireproof Const. Sq. Ft.	Semi-Fire-proof Const. Sq. Ft.	Heavy Timber Const. Sq. Ft.	Ordinary Const. Sq. Ft.	Wood Frame Const. Sq. Ft.
Single Dwellings	U	U	U	U	*
Multiple Dwellings	U	U	12,000	9,000	*
Institutional Buildings	30,000	20,000	O	O	O
Office Units	U	15,000	12,000	9,000	O
Sales Units (2-story)	U	15,000	12,000	9,000	O
Sales Unit More Than (2-story)	30,000	15,000	12,000	9,000	O
Financial Units	U	15,000	12,000	9,000	O
Storage Units	30,000	15,000	12,000	9,000	O
Manufacturing Units	30,000	15,000	12,000	9,000	O
Hazardous Use Units	*	*	*	*	O
Garages—					
Class 1	30,000	18,000	9,000	9,000	O
Class 2	O	O	O	O	O
Theatres	U	*	O	O	O
Public Assembly Units	U	*	*	*	O
Churches	U	U	*	*	O
Schools	U	U	*	*	O

NOTES TO TABLE 2402.02 (1) and TABLE 2402.02 (2)

U Fire area is unlimited.
* Area limits are given in the chapter of requirements for this class of occupancy.
L For buildings inside fire limits.
N For buildings outside fire limits.
K This area shall be allowed for the business of one (1) person, firm or corporation only. Where an area is used for the business of more than one (1) person, firm, corporation or for more than one business enterprise conducted by the same person, firm or corporation the area shall be subdivided as provided in Occupancy Chapters.
O Buildings of this class of occupancy and type of construction are not permitted.

(b) **Sprinklered Areas.** Except as otherwise provided, the maximum allowable fire area in any building, which is equipped with a standard system of automatic sprinklers, shall be twice the area permitted under this section or under the provisions of Chapters 7 to 18, inclusive, for a like building not so equipped. Provided however, that the maximum allowable fire area in any garage of semi-fireproof, heavy timber or ordinary construction which is equipped with a standard system of automatic sprinklers may be increased not to exceed twenty-five (25) per cent of the area permitted under this section for a like garage not so equipped.

2402.03 Areas Below Sidewalks: In business units, any basement space located under a public sidewalk and entirely outside of the street line of the property shall be permitted in excess of the fire division areas allowed by this chapter.

CHAPTER 25
Multiple Use Buildings

ARTICLE 2501
Multiple Occupancy

2501.01 General: For any building which is occupied or used for different purposes in different parts, the provisions of Chapters 6 to 19, inclusive, applying to each class of occupancy shall apply to such parts of the building coming within that class; and where such provisions are dissimilar, the requirements securing the greater fire-resistive construction, strength and safety, shall apply.

2501.02 Multiple Occupancy Not Permitted:

(a) **General.** Certain classes of occupancy shall not be permitted within the same building or connected with other occupancies as follows:

(b) **Garage of Class 1.** A garage of Class 1 shall not be permitted in any building used for a single dwelling, institutional building, church, school, or open air assembly unit except as otherwise provided in Chapter 14, public assembly unit of other than fireproof construction.

(c) **Hazardous Use Unit.** A hazardous use unit shall not be permitted in any building or any other occupancy, except as otherwise provided in Chapter 11—HAZARDOUS USE UNITS, for the purposes auxiliary to the hazardous use; and except a room used for the storage or baling of waste paper, a standard drying room or a permissible standard fireproof vault, which may be located in a building of any occupancy.

(d) **Theatre.** A theatre shall not be permitted in any building used for a hazardous use unit or a Class One School.

2501.03 Separations Between Different Occupancies:

(a) **General.** The floors, walls and other separations between parts of a building or adjacent buildings, used for different purposes, shall have a minimum fire-resistive construction value as required by this section; provided however, that when the required type of construction requires walls and floors of a greater fire-resistive value, the higher required value shall govern.

(b) **Single Dwellings.** Four (4) hour fire-resistive separation from any theatre, open air assembly unit or school. One (1) hour fire-resistive walls and ceiling separation from any Class 2 garage. Two (2) hour fire-resistive separation from any church, multiple dwelling, business unit or public assembly unit.

(c) **Multiple Dwelling.** Four (4) hour fire-resistive separation from any Class 1 garage, theatre, open air assembly unit or school. Two (2) hour fire-resistive separation from any single dwelling, institutional building, business unit, church or public assembly unit.

(d) **Institutional Buildings.** Four (4) hour fire-resistive separation from any business unit, open air assembly unit, or public assembly unit. Two (2) hour fire-resistive separation from any single dwelling, multiple dwelling or church.

(e) **Business Units.** Four (4) hour fire-resistive separation from any institutional building, Class 1 garage, theatre or open air assembly unit. Two (2) hour fire-resistive separation from any single dwelling, multiple dwelling, public assembly unit, church or school.

(f) **Hazardous Use Units.** For separation between different hazardous use units, see Chapter 11. Three (3) hour fire-resistive separation from any standard fireproof vault in any occupancy. Two (2) hour fire-resistive separation from any room used for storage or baling of waste paper in any occupancy. One (1) hour fire-resistive separation from any standard drying room in any occupancy.

(g) **Class 1 Garage.** Four (4) hour fire-resistive separation from any heating plant, multiple dwelling, business unit, theatre where permitted by Chapter 13, or public assembly unit. Two (2) hour fire-resistive separation from any open air assembly unit.

(h) **Class 2 Garage.** One (1) hour fire-resistive separation from any single dwelling, business unit, open air assembly unit or public assembly unit. Three (3) hour fire-resistive separation from any multiple dwelling and from any basement or heating plant. Four (4) hour fire-resistive separation from any theatre.

(i) **Theatre.** Four (4) hour fire-resistive separation from any single dwelling, multiple dwelling, business unit, garage passage where permitted by Chapter 13—THEATRES, open air assembly unit, public assembly unit, church or Class Two school.

(j) **Open Air Assembly Unit.** Four (4) hour fire-resistive separation from any single dwelling, multiple dwelling, institutional building, business unit, theatre or school. Two (2) hour fire-resistive separation from any garage, public assembly unit or church.

(k) **Public Assembly Unit.** Four (4) hour fire-resistive separation from any institutional building, Class 1 Garage, theatre, or school. Two (2) hour fire-resistive separation from any single dwelling, multiple dwelling, business unit, open air assembly unit or church.

(l) **Church.** Four (4) hour fire-resistive separation from any theatre. Two (2) hour fire-resistive separation from any single dwelling, multiple dwelling, institutional building, business unit, open air assembly unit, public assembly unit or school.

(m) **School.** Four (4) hour fire-resistive separation from any single dwelling, multiple dwelling, theatre, open air assembly unit or public assembly unit. Two (2) hour fire-resistive separation from any church or business unit and between Class One and Class Two schools.

(n) **Other Buildings and Structures.** Four (4) hour fire-resistive separation between any building or structure included in Chapter 18 and any institutional building, Class 1 Garage, theatre, open air assembly unit, church or school.

2501.04. Ground Story Communication: Wherever under this ordinance, communication is permitted between different occupancies on the main exit level of any multiple use building, such communication shall be permitted at one (1) or both street levels of any such building which is located on ground adjoining an existing, proposed or authorized two (2) level street; provided however, that any such story shall have not less than two (2) outside exit doorways.

2501.05 Separate Wiring System: In any multiple use building, any part of the building which is used for an institution building, theatre, public assembly unit, church or school, shall have for each such occupancy an electric wiring system independent of the lighting system for any part of the building used for other occupancy; provided however, that current for the entire building may be taken through a common service connection, when the main service and each such independent service are equipped with automatic circuit breakers, or other automatic switching devices.

2501.06 Assembly Room Separation: Every assembly room in a building of multiple occupancy shall be separated from every other assembly room and other occupancy by not less than two (2) hour fire-resistive construction.

2501.07 Doors: All doorways or openings in fire-resistive walls between separate occupancies as required by this article shall have fire-resistive doors as provided for such walls in Chapter 33—ALL TYPES AND KINDS OF FIRE DOORS.

CHAPTER 26
Means of Exit

ARTICLE 2601
General

2601.01 Means of Exit Defined:

(a) **General.** A means of exit shall, for the purposes of this ordinance, be defined as any means providing a route of travel used by the occupants of a building for ingress thereto or egress therefrom, and may be one of the following:

(b) **Horizontal Exit Connection.** A horizontal exit connection shall include any aisle, doorway, passageway, corridor, foyer, lobby, vestibule, balcony or space other than a vertical means of exit, which leads to an exit from the building.

(c) **Vertical Means of Exit.** A vertical means of exit shall include any means of exit leading from one (1) level to a higher or lower level, and toward the main exit level of a building, and may include a stairway, ramp, fire escape, escalator or ladder as hereinafter provided.

(d) **Outside Exit Doorways and Exit Courts.** Outside exit doorways shall lead dirertly or by way of fireproof horizontal exit connections to an open space at grade not less than ten (10) feet wide which leads to a public alley not less than ten (10) feet wide or to a street. Exit courts shall lead to an open space at grade directly or by means of a fireproof passage having a width not less than the required width of the exit court, or to a public alley not less than ten (10) feet wide or to a street.

(e) **Normal and Emergency Exits.** A normal exit shall include only such means of exit which are also used as means of ingress by the occupants of a building; and an emergency exit shall include additional means of exit.

2601.02 Horizontal Exit Connections:

(a) **In Lieu of Stairways.** One (1) or more horizontal exit connections between adjoining or adjacent building units may be used in lieu of one (1) of every two (2) required stairways. Every such horizontal exit shall be so arranged that there will be continuously available paths of travel leading from each side of the exit to stairways or other means of exit which lead to outside the connected buildings. Such horizontal exit connections shall be by means of:

1. **Doorways.** A doorway on each floor, except the main exit floor, through the dividing walls of adjoining units may be employed for such purpose. Such a doorway in any department store shall be not less than five (5) feet in width, and not less than ninety (90) per cent of the width of the main aisle required by Section 1006.07.

2. **Bridges.** A bridge or passageway across an alley or other open space and connecting each habitable floor above the main exit level of one (1) such area with the corresponding floor of the other area may be employed for such purposes, if of non-combustible construction with every window therein a Type A window and with each end closed by walls having self-closing doors therein; provided, however, that such a bridge or passageway shall not be counted as a required stairway in a department store, unless the bridge or passageway be no longer than forty (40) feet measured between building lines and equipped with a standard system of automatic sprinklers. At the end or ends of every such bridge or passageway, connected with a building of fireproof or semi-fireproof construction, the end wall shall be of two (2) hour fire-resistive construction with forty-five (45) minute fire-resistive doors; and at the end or ends of every such bridge or passageway connected with a building of other than fireproof or semi-fireproof construction, the end wall shall be of four (4) hour fire-resistive construction, with sixty (60) minute fire-resistive doors.

3. **Tunnels.** A tunnel or passageway underground may be employed for such purpose,

if made to comply with the applicable provision of **Section 2501.03 Separations from Different Occupancies.**

4. **Ramped Floors.** The floor of any such bridge or tunnel may be ramped to meet differences in floor level between connected buildings or areas, subject to the provisions of Section 2603.02.

5. **Location of Exits.** Any such horizontal exit shall only count as a required exit when so located that no point on any floor shall be more distant therefrom than the permitted distance from a stairway.

(b) **Exceptions.** The provisions for horizontal exit connections between buildings of different occupancies shall not apply to hazardous use units, garages, theatres or schools.

(c) **Doorways and Doors.**

Item 1. Obstruction and Visibility of Exit Doorways. Every inside and outside doorway shall be so arranged as to be readily visible and no obstruction interfering with access or visibility shall be permitted. No drapery shall be permitted to conceal or obstruct the required width of any exit doorway. No mirror shall be placed in the door nor shall any miror be so arranged as to be mistaken for a means of exit.

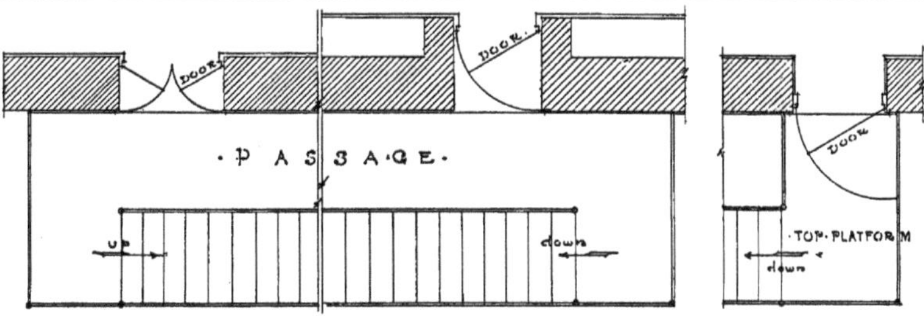

EMERGENCY EXITS.
Section 2601.02.
Suggestion how to swing doors, so as not to obstruct passageway.

Item 2. Swing of Interior Doors. Every swinging door at any doorway in an inclosure for a vertical means of exit shall be arranged to open only in the direction of the way out of the building. At every doorway which connects any habitable space with a corridor, lobby or other horizontal exit connection, it shall be permissible to arrange any swinging door to open into the habitable space, except in theatres, open air assembly units, public assembly units, churches and schools where such doors shall open in the direction of egress.

Item 3. Outside Exit Doorways and Doors. No door or other closure for any outside doorway shall, when opened, project over a public sidewalk, street, alley or other public space, except as otherwise provided in this ordinance. No door or other closure for any outside exit doorway, when opened, shall in any manner obstruct egress through any other outside exit doorway and, where necessary to avoid such obstruction, such a door or closure shall be recessed; provided however, that such recess or vestibule shall not encroach upon any required public space. Every swinging door which opens into an outside passage leading to a public sidewalk, street, alley or other public space, shall be arranged to swing only in such manner that, if it should be ajar when the occupants are passing through such passage on their way out of the building, it shall automatically be closed by the movement of the crowd, or else shall swing through an arc of at least one hundred eighty (180) degrees against the wall of the passage. Every outside exit door shall swing outward, except in single dwellings, mutiple dwellings, business units not more than one (1) story high with a floor area of five thousand (5,000) square feet or less and garages of Class II, unless otherwise provided in special occupancy chapters.

Item 4. Revolving Doors. Except in theatres and places of public assembly, a revolving door, meeting the requirements of this ordinance, may be used as a means of exit. The revolving wings of such revolving doors shall be so arranged that, by the application of a force slightly more than is necessary to revolve said doors and which one (1) person of ordinary strength is capable of exerting, all the wings of said door shall fold flat on each other in an outward direction, or unless the revolving wings of said revolving doors are so arranged that they may be readily collapsed or removed by pressure or simple mechanical means and leave sufficient opening for two (2) or more persons to pass through with a minimum width of not less than one (1) foot eight (8) inches on each side of said collapsed doors. Glass panels of revolving doors shall be of laminated shatter-proof glass or wired glass. Where revolving doors are used as exits they shall be so credited only to the extent of the clear space remaining when the doors are collapsed.

Item 5. Other Swinging or Sliding Doors. The requirements of this section for doors opening outward shall not be construed to prohibit double swing doors which swing both inward and outward, except to vertical means of exit; nor shall such requirements be construed to prohibit the use of sliding fire doors in openings not counted as exits.

Item 6. Width of Doorways, Allowance for Hardware. The clear width required for any exit doorway shall be the net unobstructed width of the passageway through the opening, measured across the doorway when the door is fully opened. No part of any door, hardware, or other object shall be permitted to encroach upon the required clear width of any exit doorway; provided however, that four (4) inches of the projection of any panic hardware or push bar may be disregarded in determining or measuring the required width of any such doorway.

(d) **Exit Locks.** Any lock, latch or other fastening device, used on any door, grille, gate, sash or other closure for a doorway, gateway or window, through which it is necessary to pass in order to reach a required exit or exit connection, in any building, except a single dwelling, correctional and penal institution, shall be arranged so that it may be easily opened by any person on the way out without a key or other detached opening device. Standard panic-proof hardware or devices shall be provided where required by the occupancy chapters.

(e) **Obstructions.** There shall be no radiators, doors, or other obstructions to exit travel in any horizontal exit connection which is seven (7) feet or less in width, in any hospital, theatre, open air assembly unit, public assembly unit, church or school, other than handrails having a projection of four (4) inches or less from any wall.

2601.03 Vertical Means of Exit:

(a) **General.** Every floor or occupied space more than one (1) foot above or below grade shall have vertical means of exit as hereinafter provided.

(b) **Multiple Occupancy.** Every required vertical means of exit in a building of multiple occupancy shall afford no communication except upon the main exit level, with any part of the building used for other occupancies, in the following classes of occupancy:

Hospitals.
Hazardous Use Units.
Garages.
Theatres.
Churches, except to a fireproof lobby in a business unit of fireproof construction.
Schools, except to a fireproof lobby or corridor in a business unit of fireproof construction.

2601.04 Each Fire Area Considered Separate Building. In every building each fire area shall be considered a separate and distinct building and, except as otherwise provided under this ordinance, shall have means of exit as required under this chapter which will be separate and distinct from those of any other fire area, except as follows:

One (1) stairway may be used in common for two (2), areas of the same occupancy in the same building or as provided by Section 2601.03 for multiple occupancy, when such areas are separated by a required fire division wall; provided however, that such stairway shall be not less than five (5) feet six (6) inches wide. Such a stairway shall be considered equivalent to one (1) required stairway from each of said areas; provided however, that the inclosing walls shall be of the same construction as required for fire division walls, and shall have one (1) door at each opening which shall be not less fire-resistive than a standard fire door.

2601.05 Vomitories: Any vomitory used as a means of exit or in lieu of a required stairway or ramp shall meet the requirements of Article 2601 for required stairways and ramps with respect to number, width, location, inclosure and construction, except as otherwise provided in Occupancy Chapters 7 to 18 inclusive.

2601.06 Exits from Penthouses: Every non-habitable penthouse or other inclosed structure on a roof of any building shall have exit by one (1) stairway not less than two (2) feet wide, and if on a flat roof shall have at least one (1) doorway to such roof.

2601.07 Exits from Roofs: Every building more than three (3) stories high and having a flat roof, shall be provided with one (1) means of access to such roof for each twenty thousand (20,000) square feet or fraction thereof of such roof area. Such access shall be either by a stairway from grade or from the main exit level of the building directly to said roof, or by a stairway from grade or from the main exit level of the building to a lower roof leading to a stairway to the higher roof. Any of said stairways may be located either inside or outside the building. Every such stairway shall be accessible from a street or public alley, or from an open space diectly connected with a street or public alley; and if located inside the building shall have the rise of its lowest step not more than fifty (50) feet by the most direct route of travel from an ouside doorway on such street, public alley or open space. Every building three (3) stories or less in height and having a flat roof, shall be provided with like access to such roof, except that a metal ladder may be used in lieu of the stairway from the topmost floor to the roof except in schools. Every penthouse with a flat roof shall be provided with either a stairway twenty-four (24) inches wide or a ladder fourteen (14) inches wide to such roof located either inside or outside the penthouse.

2601.08 Exits from Courts: Every court in any building shall have an exit at the bottom which shall meet the requirements for exits from flat roofs; provided however, that if a fire escape or stairway terminates at the bottom of an inner court or a closed lot line court, such exit shall meet the requirements for stairway exits. The floor of every court or open space upon which any exit doorways open shall be paved and shall be generally level or if ramped shall have a slope not greater than one (1) rise in six (6) horizontal, and shall meet the adjoining public alley, sidewalk or other open space without a step or curb or other obstruction. Every pipe, manhole, door or other object located in the floor or walls of an exit court shall be flush with the floor or walls to prevent obstructions.

2601.09 Areaways: Every areaway serving as a required means of exit from any basement shall have a clear width not less than that of any stairway, ramp or other vertical means of exit, or any corridor, passage, doorway or other horizontal exit connection with which it is connected; provided however, that such an areaway shall not in any case have a clear width of less than four (4) feet.

2601.10 Fire Escapes in Inner Courts: In any building of fireproof or semi-fireproof construction a fire escape stairway may be located in an inner court having a horizontal dimension of not less than fifty (50) feet, but such means of exit shall be connected at the bottom of the court with a stairway descending to an exit connection on the main exit floor, or to an outside exit doorway at ground level. Such a stairway shall be at least as wide as the fire escape served by it, but not less than three (3) feet wide if provided solely for the purpose of serving such a fire escape, and if otherwise required by this ordinance shall be increased in width by the width of the fire escape served. In a building of fireproof construction such junction may be at any level, but in a building of semi-fireproof construction such junction shall be no higher than thirty-five (35) feet above grade.

ARTICLE 2602

Stairways

2602.01 General:

(a) **Definitions.**

Item 1. Stairway. For the purpose of this ordinance, a stairway shall be defined as a system of three (3) or more steps and rises between two (2) or more different levels of any building either within or without its inclosing walls, inclined at an angle of forty-five (45) degrees or less from the horizontal and serving as a means of ingress or egress of the occupants of such building.

Item 2. Parts. The different parts of a stairway shall, for the purposes of this ordinance, be defined as follows:

Cut. The horizontal distance between the faces of successive rises of a flight of steps.

Flight. A series of steps between successive landings or between successive landings and platforms

Landing. The floor space adjoining the top or bottom of a flight, for a length equal to the width of stairs, or within the inclosure of an inclosed stairway.

Newel. An upright post at the end of a stair railing.

Nosing. A part of a tread projecting beyond the rise, but not a part of the required width of a cut.

Open Stairs. An interior stairway without a complete inclosure.

Platform. Any landing between flights other than at a floor level.

Rise. The vertical distance between two (2) successive treads or between a tread and a landing.

Soffit. The underside of a stairway.

Step. A tread and the rise beneath.

Stringer. A support for treads and rises, approximately at right angles thereto.

Tread. The horizontal surfaces of a step, including nosing.

Width of Stairs. The least clear unobstructed length of tread; provided however, that no handrail project more than four (4) inches into such stairway.

(b) **Required Stairways.**

Item 1. Type, Number and Width. The required type, number and width of stairways shall be as provided in the occupancy chapters 7 to 18, inclusive. Except as otherwise permitted for certain utility stairs for the use of employees only, the minimum width of required stairways shall be three (3) feet; and where handrails are provided on both sides of a stairway, the minimum width shall be three (3) feet and four (4) inches.

Item 2. Construction. Except as hereinafter provided for fire-shield stairways, the mateials used in the construction of stairways shall be those permitted under Chapter 31 — CONSTRUCTION TYPE PROVISIONS, for the type of building in which they occur.

Item 3. Inclosure of Stairways. The inclosure of fire-shield stairways shall be as required under Section 2602.04 of this Chapter. Other interior stairways, above the first story, and all basement stairways shall be inclosed as follows:

In buildings of fireproof or semi-fireproof construction, inclosing walls and ceilings shall be of non-combustible materials of at least one (1) hour fire-resistive value.

In buildings of other types of construction, inclosing walls and ceilings shall be of at least two (2) hour fire-resistive value. Such inclosing walls shall be of burned clay or Portland cement concrete units, four (4) inches or more in thickness.

Doors in stair inclosures shall be at least forty-five (45) minute fire-resistive doors; except in dwellings of four (4) stories or less where two and one-fourth (2¼) inch thick solid wood slab doors may be used in lieu of forty-five (45) minute doors.

Item 4. Location. Except as otherwise required or permitted under occupancy chapters 8 to 18, inclusive, stairways shall be so located that no part of any habitable floor shall be more than one hundred (100) feet distant from a stairway, measured upon the route of travel thereto.

(c) **Non-Required Stairways.** Non-required stairways shall be subject to the same requirements, except as to location, as required stairways.

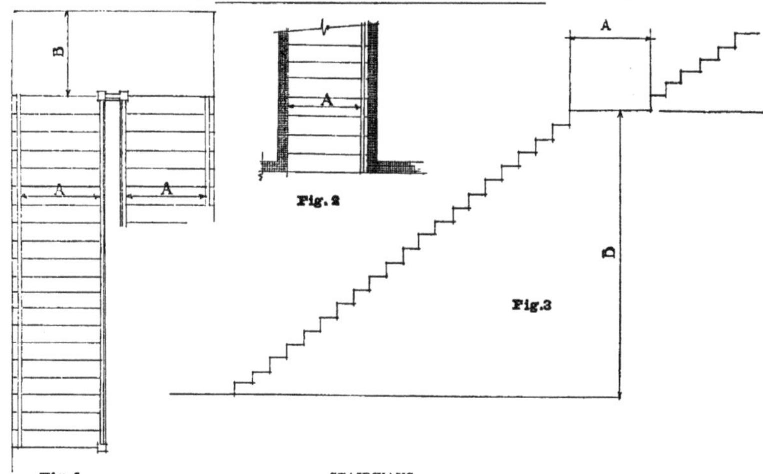

STAIRWAYS.
Sections 2602.01, 2602.02, 2602.03.

Fig. 1 (A) Shows measurement of stairways where hand rails are required on each side.

(B) Shows measurement of platform.

Fig. 2. Measurement of stairway where hand rail is required on one side only.

Fig. 3 (A) Platform.

(B) Stairways shall not ascend to an unlimited height (B) without a landing or platform (A), and (A) shall not be less in width and length than (A) Fig. 4 measurement of stairs.

Fig. 4

Fig. 5

Fig. 4. Over 7'0" (for exceptions see ordinance) wide stairways (C) shall have double intermediate hand rails. In plan (Fig. Sec. AA).

(B) Measurement of stairs where double intermediate hand rails occur.

(C) Measurement of stairs where double intermediate hand rails do not occur, as in general case, Fig. 4 (A).

Fig. 5. Newel post 5½'0" high (A) required for stairs as referred to in Fig. 8.

2602.02 Design of Stairways:

(a) **Rise, Tread and Cut.** No step in a required interior stairway shall have a rise of more than eight (8) inches and no step in a non-required stairway shall have a rise of more than eight and one-half (8½) inches. The cut of any stair shall be not less than nine (9) inches. The cut and rise of every stair shall be such that the product of the cut multiplied by the rise shall be not less than seventy (70) and not more than seventy-five (75). Rises and treads shall be of uniform dimensions throughout any flight. Any vertical steps not conforming to this paragraph shall be as required for ladders. In a hospital, theatre, public assembly unit, church or school, the rise shall be not more than seven and one-half (7½) inches and the cut not less than ten (10) inches in any required stairway.

(b) **Landings and Platforms.** Every floor landing and every intermediate platform shall be level and shall be at least as wide as the stairway; except that in a straight run of stairs, an intermediate platform shall be at least four (4) feet long. Every platform intermediate between flights where there is a change of direction of sixty (60) degrees or less shall be at least four (4) feet long for it sentire width and for a change of direction of more than sixty (60) degrees, such length shall be at least equal to the width. Every stairway shall have a level landing not less than four (4) feet long intervening between such stairway and any ramp or sloping floor.

(c) **Length of Flight.** No flight in any stairway shall have a total vertical rise of more than thirteen (13) feet six (6) inches without an intermediate level platform and no flight in any interior stairway shall contain less than three (3) rises. To overcome any lesser difference in level, an incline with a slope as required for ramps shall be used.

(d) **Handrails and Railings.** Each side of a stairway, except stairways in aisles between fixed seats, or between fixed seats and a wall, shall be protected with a railing. There shall be a handrail on at least one (1) side of every stairway, and on each side of every stairway which is more than three (3) feet six (6) inches in width. Every interior stairway which is more than seven (7) feet in width shall have double intermediate handrails to divide the stairway into clear widths of seven (7) feet or less. The height of a required handrail shall be not less than two (2) feet six (6) inches, nor more than three (3) feet, measured vertically from the front edge of any tread to the top of the handrail. Every required handrail and railing shall extend from the top of the stair to the front edge of the third tread from the bottom or to a lower point. Where a flight of stairs meets an intermediate platform, the ends of the side handrail or handrails shall be turned and joined to the wall, or else shall extend completely across or around said intermediate platform. Handrails may be of any material which is equal or superior in strength to wood. In a theatre, public assembly unit, church or school, there shall be a newel post at the upper and lower ends of every intermediate handrail, which shall have a height not less than five (5) feet above the landing, platform or tread. In a hospital, theatre, public assembly unit, church or school, every handrail mounted on a wall shall have its ends returned and joined to the wall.

(e) **Space Under Stairs.** Space under the bottom flight of any stair in any building of wood frame, unprotected metal, non-combustible frame, ordinary or heavy timber construction, shall be either entirely open or completely inclosed without any opening thereto.

(f) **Winders.** There shall be no winders in any required or non-required stairway; except as hereinafter provided. Any tread in any winder shall be not less than nine (9) inches wide at a distance of one (1) foot from the center of the newel post or from the wall or railing on the narrow side of the tread and the widths of successive treads at such distance shall not vary from one another by more than one (1) inch.

(g) **Stairways Combined.** Stairways in the main exit story may be combined into a single open flight; provided however, that the width of such single flight shall be not less than one hundred and twenty (120) per cent of the sum of the required width of all flights combined; and provided further, that the number of flights so combined shall not exceed fifty (50) per cent of the total number of required stairways.

(h) **Separation of Flights:** Two (2) flights of a required stairway may be separated by a distance not exceeding fifty (50) feet measured on the most direct route of travel from the center of the bottom riser of one (1) flight to the center of the top riser of the other flight, and such route of travel shall be inclosed in the same manner as required for the stairway; provided however, that fire-shield stairways may be separated any required distance. Any required stairway which is inclosed in the first story shall extend to the main exit floor of the building and shall be so arranged that it will be impossible to pass from any such stairway to a basement stairway without going outside the stairway inclosure. In any building abutting a two (2) level street, either street level may be used to determine the main exit level.

(i) **Elimination of Sharp Internal Wall Angles in Stairway Inclosures.** Wheer there is any change in the direction of a public stairway in a theatre, open air assembly unit, public assembly unit, church or school, the inclosing wall on the outer side of the turn shall be curved to a radius of not less than two (2) feet; or if such wall be not curved, it shall have no angle of less than one hundred thirty-five (135) degrees, measured between the interior faces of the walls and shall be without any straight run less than one (1) foot eight (8) inches horizontally.

2602.04 Fire-Shield Stairway:

(a) **Definition.** A fire-shield stairway is an interior stairway, constructed of non-combustible materials, in accordance with all of the provisions of this section; equipped with a vestibule or balcony vented to the outer air directly or by means of an interior smoke shaft; and with self-closing doors of at least a forty-five (45) minute fire-resistive value, at each entrance thereto.

(b) **Construction.** Every stair landing or platform in a fire-shield stairway shall be of fireproof construction and every other part shall be non-combustible except for handrails.

(c) **Extent.** Every fire-shield stairway providing exit from any story shall extend from such story to the main exit floor of the building, either directly or as hereinafter provided.

(d) **Width.** Every fire-shield stairway shall be not less than three (3) feet eight (8) inches wide at any point.

(e) **Inclosures.** Every fire-shield stairway shall be inclosed with solid walls of brick masonry built in accordance with the requirements of Chapter 38—MASONRY CONSTRUCTION of this ordinance, not less than eight (8) inches thick, or of reinforced concrete built in accordance with the requirements of Chapter 39—REINFORCED CONCRETE CONSTRUCTION of this ordinance, not less than six (6) inches thick. Every inclosing wall shall extend from the bottom of the stairway to the roof or a fireproof floor above the stairway; and in a building of other than fireproof or semi-fireproof construction, such walls shall extend at least three (3) feet above such roof. The thickness of such walls shall be maintained at all points, without any cut or chase to afford a bearing for any structural member or for any recesses. Any combustible structural member bearing in any way on such a wall, and any anchor of such a structural member, shall be of a self-releasing type. The floor beneath and the floor or roof above every fire-shield stairway and any soffit in an offset connection permitted by this section shall be of fireproof construction not less than six (6) inches thick. There shall be no openings for doors, windows, recesses or cabinets, or other openings in the inclosure other than the required exit doorways.

(f) **Vestibules or Balconies.** The entrance from each floor of the building to the fire-shield stairway shall be through a vestibule or balcony, not less than the required width of stairs in least dimension. The inclosing walls, roof and floor of any such vestibule or balcony shall conform to the requirements of this section for the inclosure of stairs, except that an outside wall shall conform to the requirements of Chapter 38—MASONRY CONSTRUCTION for masonry inclosing walls. There shall be no combustible material in any such vestibule or balcony. One (1) side of every vestibule or balcony required by this section shall face a street or an alley, or an open court or space leading directly to a street or to a public alley, or an interior smoke shaft conforming to the requirements of this section, or a fire-resistive tunnel leading directly to a street or public alley. The open side of every such entrance vestibule or balcony shall be closed for its full width up to a level four (4) feet above the floor by a solid wall. Extending from the top of such wall to the ceiling there shall be a space with a width not less than eighty (80) per cent of the width of the vestibule or balcony, or in any case not less than three (3) feet six (6) inches wide; which space shall be closed with a fire-shield as hereinafter provided, except that any such opening on a street or public space, not less than forty (40) feet wide, may be left open. Every balcony of a fire-shield stairway shall be constructed within the lot lines, and shall not encroach upon or overhang a public street, public alley or other public space.

(g) **Fire-Shields.** Every fire-shield required or permitted by this section shall be set in a solid frame of steel or iron, or in a hollow sheet steel or iron frame filled with concrete, with a horizontal cross member midway between the top and bottom of said frame, which frame and cross member shall have a strength as required by Chapter 34—STRUCTURAL PROVISIONS of this ordinance. Such frame shall be set with its inside surface flush with the inside face of the wall below the opening. Every fire-shield shall consist of two (2) or more metal sashes with wired glass, meeting the requirements of Chapter 27—WINDOWS, SKYLIGHTS AND VENTILATION for a Type A window. The upper half of every fire-shield sash shall be arranged to open automatically in case of fie to the full limit, and shall be held securely in such open position and shall close easily and quickly when desired. Every mechanism or device for the accomplishment of these purposes shall be positive in action, non-rusting and located inside the building and shall be as approved by the Commissioner of Buildings. Every sash shall be provided with a bar or lever for manual operation to release and open the sash; and shall have a sign placed near bar or lever which shall read, "IN CASE OF FIRE PUSH." Such mechanism for the opening of sash may be actuated by springs, electricity or other means which will set the mechanism in operation as a result of a difference in the rate of temperature rise, or other devices adjusted to operate when the surrounding air reaches a temperature of one hundred ten (110) degrees Fahrenheit. One (1) such temperature operated device shall be mounted in the ceiling outside every entrance vestibule or balcony of the fire-shield stairway at a distance of not more than five (5) feet from the entrance doorways, and the operation of any one (1) of these devices shall cause all the upper sashes of the fire-shield in such story to open immediately. Sashes may be closed manually. Where the sash would otherwise exceed five (5) feet in width, intermediate piers or mullions of the same construction as required for fireproof construction shall be installed.

(h) **Stair Landings and Floors.** In any fire-shield stairway there shall be no step between the floor of a vestibule or balcony and the floor landing of the fire-shield stair, nor between either of such floors and the adjoining floor of the building. Any difference in level between any such floors shall be overcome by a ramp with a slope not exceeding one (1) in twelve (12). Any exit landing shall be not less than the required width of the stairway.

(j) **Smoke Shafts.** Every interior smoke shaft used for an entrance vestibule or balcony of a fire-shield stairway, shall be at least five (5) feet wide at every point and shall be open and unobstructed over an area of not less than fifty (50) square feet from its bottom to the sky; provided however, that such shaft may be covered with a roof if the walls are open on all sides beneath such roof with a ttoal open area of not less than one hundred (100) square feet, either with or without fixed louvers. The bottom of such smoke shaft shall be not more than twelve (12) inches above the floor level of the lowest required vestibule or balcony. Any such smoke shaft in a building of semi-fireproof or fireproof construction may be offset at any point; provided however, that such lateral connection shall be not less in area than the required area of the shaft, and the soffit of said lateral connection shall be inclined at an angle of not less than forty-five (45) degrees to the horizontal. There shall be no other openings into any smoke shaft except as required by this section.

(k) **Exits.** Every fire-shield stairway shall terminate at the main exit level of the building and shall provide an exit to a street or alley, or to an open space leading to a street or public alley, either directly from the stairway without passing through any vestibule or by way of a tunnel of fireproof construction. The outside exit doorway from the fire-shield stairway shall have a clear width between jambs not less than the width of the stairway. Every exit doorway required by this section shall be equipped with a forty-five (45) minute fire-resistive door which shall open outward only. Every such door shall be self-closing, equipped with panic hardware.

(l) **Entrance.** In every building occupied, or arranged, designed or intended for more than one (1) tenant, every entrance required by this section shall be from a corridor, lobby or other space accessible to all occupants of the building and to the public. In every story occupied by only one (1) tenant, all or any of the entrances required by this section, may be directly from rentable area or other space, without passage through any corridor, lobby or other public space. Every doorway from the building into an entrance vestibule or balcony of a fire-shield stairway, and every doorway from such a vestibule or balcony into the stairway proper, shall have a clear width between jambs not less than ninety (90) per cent of the width of the stair, and shall be equipped with a forty-five (45) minute fire-resistive door.

(m) **Fire-Shield Stairways in Lieu of Required Stairways.** Two (2) fire-shield stairways not less than four (4) feet wide shall be considered equal to three (3) required interior stairways; provided however, that one (1) fire-shield stairway not less than four (4) feet wide shall be considered equal to two (2) required stairways in any building of fireproof or semi-fireproof construction except multiple dwellings, hazardous use units, garages and theatres. Any building of fireproof or semi-fireproof construction except hospitals, garages, theatres, public assembly units, churches, schools and multiple dwellings, in which the net area of the habitable part of any floor does not exceed three thousand (3,000) square feet, is required to have only one (1) stairway, if such stairway is a fire-shield stairway not less than four (4) feet wide. In every other building, in which a fire-shield stairway is used, there shall be not less than two (2) stairways. One (1) inch of width of a fire-shield stairway shall be considered equivalent to one and one-half (1½) inches of width of any other type of inclosed stairway, in determining the aggregate width of stairways required for any such building. A fire-shield stairway, serving an area of not more than three thousand (3,000) square feet may also be used in combination with two (2) interior stairways, when connected in the story served by it with both of the other two (2) stairways by means of a fireproof passage, which is no less in any horizontal dimension than is required for an entrance vestibule of such a stairway; and on either side of such junction, such passage shall be at least seventy-five (75) per cent of the required width of the fire-shield stairway, but in any case not less than three (3) feet six (6) inches wide at any point. Where a fire-shield stairway is used in lieu of interior stairways, the distance from such stairway doorway to the end of a corridor shall be not more than one hundred (100) feet in any business unit on any floor other than at the main exit level.

ARTICLE 2603

Ramps in Lieu of Stairways

2603.01 General: Ramps may be used in lieu of stairways when constructed in accordance with the provisions of this chapter.

2603.02 Design and Construction of Ramps:

(a) **Number, Width, Location and Inclosure.** The number, width, location and inclosure of ramps, used in lieu of stairways, shall be as required for the stairways displaced by such ramps; provided however, that ramps forming a required means of exit from a building shall be so inclosed as to prevent the accumulation or formulation of snow or ice thereon.

(b) **Construction.** Ramps shall be subject to all of the applicable provisions for the construction of stairways.

(c) **Slope.** Any ramp used in lieu of stairways shall have a slope not greater than one (1) unit rise in eight (8) units horizontal. Any ramp used as a horizontal exit connection shall have a slope not greater than one (1) unit rise in six (6) horizontal, except as otherwise provided in the occupancy chapters of this ordinance.

(d) **Non-Slip Surfaces.** Every ramp having a slope greater than one (1) rise in twelve (12) horizontal shall have a non-slip surface.

(e) **Handrails.** Where a ramp has a slope greater than one (1) rise in twelve (12) horizontal, handrails shall be required at both sides, but no intermediate handrails will be required.

ARTICLE 2604

Escalators in Lieu of Stairways

2604.01 Where Permitted in Lieu of Required Stairways: An escalator moving in the direction toward the main exit level and meeting the requirements of Chapter 44—ELEVATORS, DUMBWAITERS, ESCALATORS AND MECHANICAL AMUSEMENT DEVICES may be used in lieu of an equal width of required stairway, and may be substituted for one (1) out of four (4) required stairways. Such escalators shall be inclosed as required for the stairways displaced thereby. An escalator serving a basement or sub-basement area shall be at least one hundred (100) feet distant from any escalator serving floors above the main exit level.

ARTICLE 2605

Exterior Fire-Escape Stairways

2605.01 General: Every exterior non-combustible stairway and every exterior fire-escpae stairway shall be constructed in accordance with the provisions of this article.

2605.02 Where Permitted: Exterior fire-escape stairways shall not be placed on buildings hereafter erected or converted for department store use. Such exterior fire-escape staiways may be used on buildings classified as "Business Units" below a level of two hundred sixty-four (264) feet above grade, and on other buildings where permitted by occupancy provisions of this ordinance.

2605.03 Design and Construction:

(a) **In Lieu of Interior Stairways.** One (1) or more exterior fire-escape stairways, meeting the requirements of this article, may be used in lieu of a required interior stairway, as provided in the special requirements for exits in buildings of each occupancy group. Except for buildings designed, intended or used as department stores, one (1) or more exterior fire-escape stairways may be used in lieu of one (1) out of each three (3) required interior stairways for any building six (6) stories or less in height; provided however, that the total width of any such fire-escape stairway shall be not less than the required width of stairway which is displaced; and provided further, that a fire-escape stairway shall not be used in lieu of a fire-shield stairway or in lieu of any portion of the required width of a fire-shield stairway. A fire-escape stairway shall be not less than two (2) feet and not more than four (4) feet wide unless an intermediate handrail is provided.

(b) **Location.** Every fire-escape stairway shall be located outside the building on a wall facing a street or public alley, or on a wall of a private alley, court or other open space not less than ten (10) feet wide and having an unobstructed exit to a street or public alley at grade. A tunnel of two (2) hour fire-resistive construction shall be considered an unobstructed exit from an inner court or a closed lot line court if it is of the required exit width, and is not less than seven (7) feet high throughout its length and is without any door, gate or other means of closure. No part of any such stairway shall project beyond a street line, except as otherwise provided for theatres, but may project wholly or in part into a public alley. A fire-escape stairway, placed on an exterior wall adjoining a fire-division wall, shall be considered as a fire-escape stairway for each building area which it adjoins. In such cases there shall be at least one (1) door or window from each building area leading to the stairway platform.

(c) **Access to Stairway.** Wherever access is required from the interior of a building to a fire-escape stairway, such access shall be through a doorway or casement window not less than two (2) feet in clear width nor less than six (6) feet six (6) inches in clear height. The sill of any such doorway or window shall be not more than two (2) feet above the floor unless a stairway is provided leading to it. Such a stairway shall be at least as wide as the doorway or window. Any such required door or window shall be arranged to open from the inside without the use of a key. Successive flights of a fire-escape stairway may be located at a distance from one another and connected by a runway located outside the building and constructed as required by this section for a fire-escape runway.

(d) **Balconies, Landings, Platforms and Runways.** Any balcony, landing, platform or runway forming a part of a fire-escape stairway shall have a width or length not less than the required width of the stairway. No such balcony, landing or platform shall be less than three (3) feet long. No door, window or shutter shall open in such a manner as to reduce any required dimension of a balcony, landing, platform or runway, or obstruct the clear passage thereon, whether such clear passage is in excess of the required dimension or not. The frame of every balcony, landing, platform or runway shall be made of steel or wrought iron angles not less than two (2) inches by two (2) inches by one-fourth (¼) inch in size, or of other structural shapes of equivalent strength. Such angles or other shapes shall be securely riveted together with cross bars every two (2) feet. Such bars shall be punched one-half (½) inch square close to the top of the bar on two (2) inch centers and one-half (½) inch square bars shall be forced through the same. The cross bars shall be securely riveted to the angle iron frame. The cross bars for a balcony two (2) feet four (4) inches wide or less shall be two (2) inches by three-eights (⅜) inch. Balcony frames more than two (2) feet four (4) inches wide shall be made of not less than two (2) by three-eighths (⅜) inch iron, and shall be made to conform with the increased dimensions of iron in cross bars. For a balcony three (3) feet or more in width, balcony frames shall be two and one-half (2½) inches by three-eighths (⅜) inch. All balconies over this width shall have a two (2) inch "T" beam through the center of the balcony for the bars to rest upon; provided however, that such balconies and platforms of hospitals may have solid floors of non-combustible material. Such balconies shall have a substantial cast or wrought iron post every three (3) feet, bolted to the balcony. Every balcony shall have not less than three (3) guard rails which shall be of wrought iron or new iron pipe not less than three-fourths (¾) inch in diameter, not less than three (3) feet high, and securely anchored to the wall of the building.

(e) **Rise, Tread and Width of Stairs.** Rises and treads shall be uniform throughout any flight. The height of rises shall not exceed eight (8) inches and the treads shall be not less than ten (10) inches wide for any required stairway, and any tread may over hang the tread beneath not more than one (1) inch. The width of the stairway shall be at least two (2) feet except as otherwise provided in this ordinance.

(f) **Anchors and Braces.** Where the anchors and braces by which a fire-escape stairway is attached to or supported on a building, are of non-corroding metal, such members shall not be required to be larger than necessary to carry the computed stresses. Where such members are of metal, subject to rust or corrosion, they shall be one (1) inch square for a stairway, balcony, landing, platform or runway four (4) feet two (2) inches or less in width; one and one-fourth (1¼) inches square for a stairway, balcony, landing, platform or runway more than four (4) feet two (2) inches and not more than six (6) feet in width; and one and one-half 1½) inches square for a stairway, balcony, landing, platform or runway more than six (6) feet in width. Every such anchor shall be securely fastened to the building by being passed through a masonry wall not less than twelve (12) inches thick and fastened on the inside thereof; or by being inserted in masonry or reinforced concrete construction to a depth of not less than fourteen (14) inches at an angle of thirty-five (35) degrees with the horizontal and set with Portland cement mortar; or by being fastened to a channel set in a masonry or reinforced concrete wall, which channel shall be not less than four (4) inches wide and five (5) feet long; or by being attached to a reinforced concrete wall or structural member, when such wall or structural member and the attachment of the anchor thereto are designed in accordance with the Structural Provisions of this ordinance; or by being bolted or riveted to the structural metal frame of the building. All anchors shall be turned up at least eight (8) inches at their outer ends to provide for their attachment of posts. Braces shall be attached to the building either by being inserted in masonry or reinforced concrete to a depth of six (6) inches or more, and set with Portland cement mortar, or by bolting and riveting to the structural steel frame of the building. Single braces shall have a spread equal to the width of the platform. Braces for any platform more than five (5) feet wide shall be double, one (1) brace extending to the outside of the platform and one (1) to its center. Where an anchor or brace passes into or through any masonry or reinforced concrete construction, the portion of such anchor or brace immediately outside such masonry or reinforced concrete shall be bent downward, either vertically or at an incline, to a point at least two (2) inches lower than the bottom of the opening into the masonry or reinforced concrete.

(g) **Stringers.** Stringers shall be made of two (2) steel or wrought iron bars, three (3) inches by five-sixteenth (5/16) inch about one (1) inch apart, or four and one-half (4½) inches by three-eighths (⅜) inch flat iron, or of steel or wrought iron channels, angles or I-beams. Where over twelve (12) feet in length they shall have anchor and brace in the center. The tread shall be made of one-half (½) inch square steel or wrought iron bars, set corner upwards not to exceed one and five-eighths (1⅝) inches apart on centers, riveted to ends to two (2) by seven-sixteenths (7/16) inch flat iron or steel. Treads shall be supported by a truss made of steel or bar iron two (2) inches by three-eighths (⅜) inch in thickness welded to bars of treads in center supported by not less than two (2) seven-sixteenths (7/16) inch rods bolted at each end of treads. All stairs shall have three-bar railings made of one (1) inch bar iron for top rail, and three-fourths (¾) inch bar iron for lower rails and when such stairs are more than three (3) inches from the wall or pass any opening of the building, or are more than three (3) feet in width, there shall be one (1) or more handrails on the wall side of such stairs. The top rail shall be not less than three (3) feet six (6) inches high measured vertically from the nosing of the stair treads. The required railings shall be supported on posts made of one and one-half (1½) inch channels or angles or larger sections. Such posts shall be attached by bolts or rivets or by welding to the turned up ends of the braces of the fire-escape.

(h) **Counterbalances.** A counterbalanced stair, meeting the requirements of this section, where such requirements are not in conflict with other requirements of this ordinance, may be used for the bottom flight of any stairway fire-escape. The vertical height from top landing to bottom landing of such a counterbalanced section shall not exceed fourteen (14) feet. The stringer carrying the counterweight may be built of steel or wrought iron channels, angles or I-beams, or any combination thereof, not less than eight (8) inches deep and three-eighths (⅜) inch thick; and shall be so designed that the maximum fibre stress over the support shall not exceed eight thousand (8,000) pounds per square inch. The moment of inertia about the vertical axis parallel to the web of the stringer shall be not less than thirty-three (33) per cent of the moment of inertia about the horizontal axis perpendicular to the web passing through the center, such moment of inertia being obtained by riveting an angle or angles to the channel or I-beam stringer. The same section of stringer shall be continued for an equal distance on either side of the support and the reinforcement shall be extended as close to the counterweight as practicable. The truss rod from the counterweight to the opposite end of the stringer shall be used either as an independent brace or in connection with the railing to prevent any sag of the stringer. It shall be at least three-fourths (¾) inch in diameter, with connections sufficient to develop the strength of the rod, but in figuring stresses, the stringer must be assumed to carry the total dead and live load as required by this ordinance. The connection between the stringer and supporting rod must be designed to stiffen the stringer against horizontal or twisting motion by means of a steel casting or forging riveted to the stringer both through the web and the flange.

(i) **Clearance Above Public Spaces.** There shall be a vertical clearance of not less than twelve (12) feet, underneath any fire-escape, or any part thereof, which overhangs an alley or other public space, or an easement which is used as a public thoroughfare. This requirement shall apply to both fixed and counter-balanced section.

(j) **Windows and Doorways.** Wherever a stairway, balcony, landing, platform or runway of a fire-escape stairway passes, a window or doorway, or is located ten (10) feet or less from a window or doorway, such window or door shall be of wired glass and shall have metal frames and sash; and wherever such a stairway passes above a window, doorway or other opening not fitted with wired glass and metal frame, such stairway shall be protected on the underside by sheet metal of not less than one hundred twenty-five thousands (.125) inch thickness above such openings and for a distance of three (3) feet on each side thereof.

ARTICLE 2606

Chute Fire-Escapes

2606.01 General:

(a) **In Lieu of Stairways.** An inclosed chute, either straight or spiral, may be used in lieu of one (1) stairway out of three (3) required stairways from any story in any building four (4) stories or less in height, except hospitals, theatres, public assembly units or churches; provided however, that such chute fire-escape shall be credited as substitute for not more than the minimum width of a required stairway. An inclosed straight slide chute may be used in hospitals three (3) stories or less in height. An inclosed spiral chute within the building may be used in lieu of one (1) stairway out of three (3) required stairways within a building of any height, except in hospitals, sales units, theatres, public assembly units or churches; provided however, that in hazardous use units such spiral chute may be either within or without the building, and may be used in lieu of one (1) out of two (2) required stairways in hazardous use units when all such remaining stairways are inclosed. No open chute shall be used as a means of exit.

(b) **Location.** An inclosed chute may be located either inside or outside a building. Every inclosed chute shall provide an exit to a street or public alley or to an open space leading to a street or public alley, either directly from the bottom of the chute, or by a route of travel measuring not more than twenty-five (25) feet from the point of discharge of the chute to an outside exit doorway of the building, or by way of a tunnel of two (2) hour fire-resistive construction not less than four (4) feet wide.

(c) **Construction.** The side, the inclosure, all supporting members and all parts of auxiliary features of an inclosed chute shall be of non-combustible material. Sheet metal shall be no thinner than No. 16 U. S. standard gauge or .0625 inch for the slide or inclosure, except the inclosure of a stright slide which shall be no thinner than No. 20 U. S. Standard gauge. All sheet iron or steel in a slide and its inclosure located outside a building shall be galvanized. All sheet metal in the slide, whether inside or outside the building, shall be galvanized and unpainted, and shall be maintained bright and free from rust. The slide and any other part of the chute with which a user may come in contact shall be free of cracks, crevasses or any projection or roughness which may produce injury. The construction shall be without any projection which will afford a handhold whereby a user might stop his descent. The pitch and the design of the slide shall be such that a person using it will be discharged without injury. The height of the bottom of the discharge opening shall be not less than one (1) foot four (4) inches nor more than two (2) feet above the ground or the floor in front of it. Stresses in any part of a chute shall conform to the stresses required by the engineering sections of this ordinance. No galvanized metal shall be welded. An inclosed straight chute shall consist of a tube having an inside diameter not less than thirty-six (36) inches, with a flare to an inside height not less than forty-two (42) inches measured at the exit end, except that a straight chute for a school for children, or children's home may have an inside diameter of not less than two (2) feet six (6) inches. A straight tube shall be inclined at an angle not steeper than thirty (30) degrees to the horizontal for any building except a hospital and at an angle of forty (40) degrees to the horizontal for a hospital. Supports for the tube shall be spaced not more than twelve (12) feet apart on centers. The lower end of the tube shall be horizontal for a length of three (3) feet for any building except a hospital and for a length of six (6) feet for any hospital. Such horizontal section shall be connected to the main inclined section by an elbow to provide easy transition. A horizontal metal bar of three-fourths (¾) inch diameter shall be located near the top of the entrance for users to grasp with hands when entering chute. An inclosed spiral chute shall consist of a spiral slide within a tubular shell of not less than five (5) feet inside diameter. The pitch or rate of vertical descent of the slide shall be not more than seven (7) feet nor less than five (5) feet in one (1) full turn of thre hundered sixty (360) degrees. In a metal shell, the upper three (3) stories shall be not lighter than No. 16 U. S. standard gauge, and the next wo (2) stories not lighter than No. 14 U. S. standard gauge and so increased two (2) gauge numbers for each additional two (2) stories or part thereof, unless the chute is supported at successive floors; but the base section shall be in any case not lighter than No. 12 U. S. standard gauge; and the base section of any chute more than fifty (50) feet in height shall be not lighter than three-sixteenths (3/16) inch thick steel plate. Every exterior chute shall be anchored to the building, and shall be reinforced and braced to resist wind pressure as required by this ordinance. Every chute shall have angle rings at joints and every chute more than fifty (50) feet high shall have vertical angles not less than three (3) in number.

(d) **Inclosures.** Every chute and passage thereto located within a building shall be inclosed with fire-resistive construction of the value required for the stairway which it displaces. There shall be no opening into such inclosure except for entrance to, and exit from the chute. Access to an exterior spiral chute at any point shall be through a fully inclosed passage between the entrance doorway and the tubular shell. Such inclosure shall be of non-combustible material, and if of metal, it shall be not less than No. 16 U. S. standard gauge.

(e) **Live Load.** Every part of fire-escape chutes shall be designed to support a live load of not less than one hundred (100) pounds per square foot of slide surfaces projected horizontally.

(f) **Openings.** Every entrance doorway to a chute, either straight or spiral, shall have a clear width of at least thirty (30) inches and a clear height of at least six (6) feet six (6) inches, when such chute is entered by means of a platform between the entrance door and chute. When the sliding surface of the chute is extended to the threshold of the entrance door, the doorway shall have a clear width of not less than two (2) feet six (6) inches and a height of not less than three (3) feet six (6) inches for any building except a hospital and shall have a clear width of not less than three (3) feet four (4) inches for a hospital. Every such doorway shall be provided with a self-closing door or doors arranged to open inward toward the chute on simple pressure from the building side, and having a fire-resistance not less than a forty-five (45) minute fire-resistive door. Such door or doors may be locked so as to prevent entry from the chute to the building; provided however, that such lock cannot in any way interfere with entrance into the chute as required in this paragraph. Such a door, when opened, shall not project into the tubular shell of the spiral chute. Every exit doorway from a chute shall have a clear width of not less than the inside diameter of the chute, and a clear height of at least three (3) feet six (6) inches. Every exit door within the building shall be self-closing and arranged to open outward only, and shall not project over a sidewalk or other public space. Every such door shall open on a simple pressure from the inside, and shall not have any lock which can in any way interfere with exit from the chute.

ARTICLE 2607

Ladder Fire-Escapes

2607.01 General:

(a) **Definition.** Any stairway or vertical means of exit by steps inclined at an angle of more than forty-five (45) degrees from the horizontal shall be defined as a ladder.

(b) **Construction.** The clear width shall be not less than seventeen (17) inches. Side guards shall consists of flat steel or wrought iron not less than two (2) inches by three-eighths (⅜) inch in cross section. Rungs shall be spaced uniformly and not more than fourteen (14) inches on centers. Rungs shall be of square steel or wrought iron bars, of not less than one-half (½) inch size, set with corners upward and riveted to the side guards. The ladder shall be anchored to the wall of the building as required for stairway fire-escape, except that anchors may be inserted in masonry or reinforced concrete to a depth of eight (8) inches. The brace for the anchors shall have a spread of at least twenty (20) inches, and shall extend into the wall not less than four (4) inches. No pipe or other hollow section shall be used in the construction of a ladder fire-escape.

(c) **Extension and Counterbalances.** The lowest portion of a ladder fire-escape extending to street level or ground level shall consist of a counter-balanced stair. Such a counter-balanced stair shall conform to the requirements for a counter-balanced stair of a stairway fire-escape, except that the width need not be greater than is required for the ladder fire-escape.

CHAPTER 27
Windows, Skylights and Ventilation

ARTICLE 2701
General

2701.01 Natural Light and Ventilation: Every habitable room in any building hereafter designed, erected or converted, shall have at least one (1) window, or doors or transoms, opening directly upon a street, alley, yard or court, or other open space, or skylights opening through the ceiling and roof immediately above such a room. Except as otherwise required by Chapters 6 to 18 inclusive, for buildings of special occupancy, such openings shall have a combined glass area of not less than ten (10) per cent of the floor area of any habitable room, and in any case not less than ten (10) square feet; and shall have ventilating openings of not less than four (4) per cent of the floor area, unless otherwise provided in Chapter 47—VENTILATION, of this ordinance. If any habitable room is deficient in the requirements either as to glass or ventilating openings, a mechanical ventilating system shall be installed for the excess floor area, as provided by Chapter 47 — VENTILATION of this ordinance. Every room, other than a habitable room, which contains any water closet, urinal fixture, and every kitchen, shall have windows or skylights for natural ventilation, or a mechanical ventilating system as provided by this section; provided however, that any window or skylight so required shall have a glass area of not less than six (6) square feet and a glass width of not less than one (1) foot.

2701.02 Mechanical Ventilation Not Permitted In Lieu of Natural Lighting: Any windows or skylights required by this ordinance for light and ventilation shall not be reduced in number or area because of any mechanical ventilating system in the habitable rooms of any single dwelling or multiple dwelling or any hospital sleeping room, or any school class room or any bakery required by other ordinances to have such windows or skylights.

2701.03 Exposures Requiring Protected Openings:

(a) **General.** Windows and skylights shall not be required to be other than wood frames with ordinary glass in any building of wood frame construction, or in any other buildings, except as otherwise provided for certain locations, as follows:

(b) **Type A Windows Required.** Type A Windows, built as provided in Article 2702, shall be required for openings in all buildings more than one (1) story in height, excepting single dwellings of any height multiple dwellings three (3) stories or less in height and churches having double-glazed openings when located as follows:

Item 1. Alleys or Easements. Any window located less than twenty-four (24) feet from the opposite side of an alley or easement; provided however, that such windows facing on an alley or easement in the first or second story, if located not more than sixteen (16) feet from the street line, may be of any construction; except as otherwise provided for hazardous use units in Chapter 11 and garages in Chapter 12.

Item 2. Lot Lines. Any window located less than twelve (12) feet from a dividing lot line.

Item 3. Courts. Any window facing other openings in the opposite wall of a court less than twenty-four (24) feet in width.

Item 4. Fire Areas. Any windows in two (2) or more fire areas in walls forming an angle of less than one hundred fifty (150) degrees on the outside faces thereof, which are also separated from each other by a distance of less than nine (9) feet measured along the faces of walls; or in such walls forming an angle of from one hundred fifty (150) to one hundred eighty (180) degrees which are also separated from each other by a distance of less than four (4) feet, six (6) inches.

Item 5. Roofs. Any windows located less than forty (40) feet above a roof on the same premises having combustible sheathing.

Item 6. Stairs. Any window in a stairway, except where a fire shield is required and any window in a required stairway inclosure in a school or public assembly unit.

Item 7. Fire Escapes. Any window within ten (10) feet of a fire escape or exterior stairway.

(c) **Type A Skylights Required.** Type A skylights, built as provided in Article 2703, shall be required for any skylight in buildings more than one (1) story in height, or for any greenhouse or skylight more than thirty (30) feet above grade, or for any skylight twelve (12) feet or less from any dividing lot line, except as otherwise provided for hazardous use units in Chapter 11 and garages in Chapter 12.

ARTICLE 2702
Windows

2702.01 Size and Location: Any required window for a habitable room shall have a glass area measured between rebates of not less than ten (10) square feet, and for rooms other than habitable rooms, not less than six (6) square feet, unless such window is in excess of the area required by this ordinance for such room. The head of any required window shall be not less than six (6) feet, six (6) inches from the floor to the top of operating sash in any single or multiple dwelling, and not less than seven (7) feet in any other building, except as otherwise required by Chapters 6 to 18, inclusive; provided however, that any casement sash having a transom bar with glass above to the height of six (6) feet, six (6) inches, may be not less than six (6) feet from the floor to the top of the operating sash in any single or multiple dwelling. The width of glass in any required operating sash shall be not less than one (1) foot.

2702.02 Sills and Spandrels: Every window in a non-combustible wall shall have a non-combustible sill and spandrel wall equivalent in fire-resistive value to two (2) hour fire-resistive construction for a vertical distance not less than three (3) feet between such opening and any opening in the story next below such opening.

2702.03 Window Cleaning Devices: Every window above the first story of any building, which is not made reversible in construction, so that it may be cleaned on both sides, either from the inside or from an outside floor, roof or balcony, shall be equipped with devices for window washers' safety harness attachment. Such devices shall be of twin-head forged non-rusting metal secured by not less than two (2) bolts of three-eighths (⅜) inch diameter of the same metal through the wall or window frame at each side of the window. Any single dwelling and multiple dwelling three (3) stories or less in height, shall be exempt from the provisions of this section.

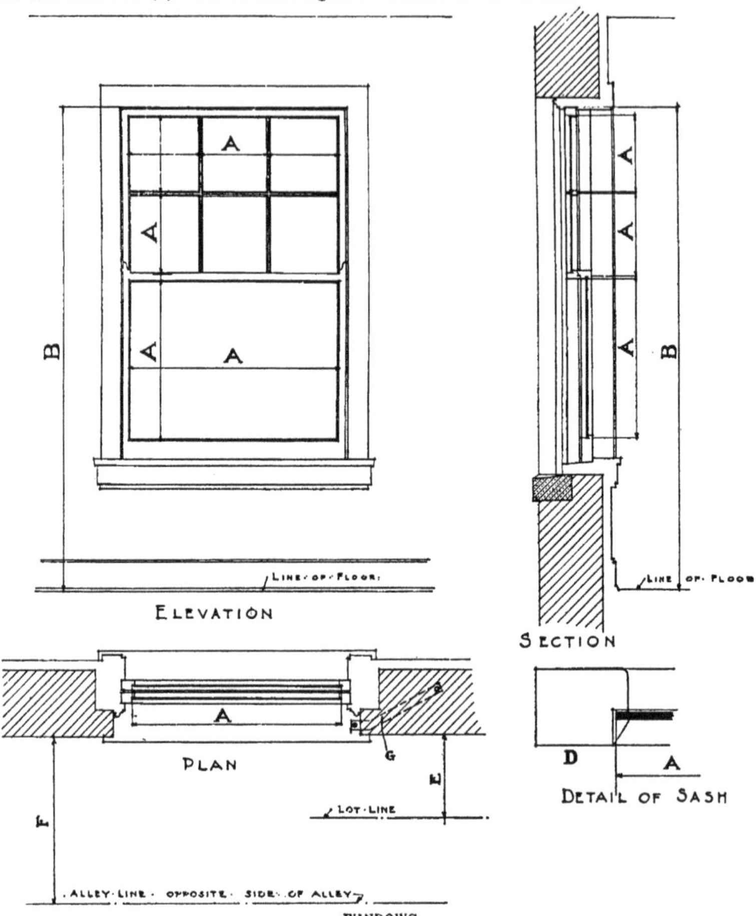

WINDOWS.

Sections 2105.01, 2105.02, 807.01, 1207.01, 1607.02, 2701.01, 2701.02, 2701.03, 2701.04, 2701.05, 4701.04.

(A) Where measurement of glass is taken.
(B) Top of window.
(D) Detail of sash showing where (A) is taken, under ordinary conditions.
Area of glass would be (A x A).
Total area would be summation of all (A x A).
Windows to be constructed so that 45% of the full area (C) can be opened.
(E) If E is less than 12 ft. wide, metal frames and wire glass to be used; and the glazed portion of frames to be set with fire resisting glass, as provided in ordinance. (Sec. 2105.02).
(F) If F is less than 24 ft., metal frames and wire glass to be used; and the glazed portion of frames to be set with fire resisting glass, as provided in ordinance. (Sec. 2105.02).
(G) Provision made for safety device in cleaning are used see ordinance. (Sec. 2702.03).

2702.04 Type A Windows:

(a) **General.** Wherever in this ordinance a Type A window is required, such window shall be constructed of non-combustible materials and glazed with wired glass as required by this section.

(b) **Size of Opening.** Any window opening in a masonry wall, which is more than eighty-four (84) square feet in superficial area, shall be divided by one (1) or more beams into spaces which are not more than eighty-four (84) square feet in area. Every beam required for this purpose shall have a moment of inertia of not less than twelve (12) about an axis parallel to the plane of the wall and shall have a two (2) hour fire-resistive covering as provided in Chapter 32—FIRE RESISTIVE STANDARDS. Such a beam shall extend from the sill to the head of the frame if serving as a mullion and from jamb to jamb if serving as a transom bar, and shall be securely fastened at each end.

(c) **Frames and Sashes.** Every frame shall be retained in the wall opening by a continuous flange or offsets at both jambs which shall have a bearing against the masonry of not less than seven-eighths (⅞) inch. Every movable sash shall be so constructed that it will engage properly with the frame at all points of contact and will maintain adequate protection from the weather. It shall not warp or bulge excessively under heat or rapid cooling.

(d) **Hollow Sheet Metal.** Hollow sheet metal frames and sashes shall be made of galvanized iron or non-rusting steel, not less than one-fortieth (1/40) inch thick and of a quality soft enough to permit of necessary bending without breaking, or shall be made of copper not less than one-thirty-seventh (1/37) inch thick or of other non-ferrous metal of equal thickness, of not less resistance to corrosion and fusibility. All joints shall be made with interlocking seams securely riveted together and in no case shall solder be used for other than weather-proofing purposes. The head of the frame shall be closed at the top and the piece forming this closure shall be securely fastened to each side at all points. Sills formed of sheet metal, having a thickness of less than seventy-five thousandths (.075) of an inch shall be filled solidly with non-combustible material. Every sliding or other movable sash shall have stiles, rails and muntins, if any, measuring not less than one and three-fourths (1¾) inches in either width or thickness. Every such sash shall be securely fastened at all corners and shall be so constructed that it will correspond with the frame at every place of contact. No single sash other than a casement sash shall be larger than five (5) feet by six (6) feet, and no single casement sash shall be larger than three and one-half (3½) feet by ten (10) feet.

(e) **Solid Rolled Steel.** Solid rolled steel frames and sashes shall be made of solid sections not less than seven-sixty-fourths (7/64) inch thick securely riveted, locked or welded together at all corners and junctions so as to have sufficient strength and rigidity to withstand handling and installation without distortion and also to withstand operation and wind pressure. Weathering strips and removable members for retaining the glass shall be not less than one-sixteenth (1/16) inch thick.

(f) **Heavy Gauge Steel.** Heavy gauge steel frames and sashes shall be made of steel not less than one-sixteenth (1/16) inch thick. Such frames and sashes shall conform in all other respects to the provisions for solid rolled steel frames and sashes, but shall not be used for any Type A windows opening exceeding thirty (30) square feet in area, nor for any single pane of glass exceeding five hundred sixty (560) square inches.

(g) **Sliding Sashes.** Lifting or sliding sashes shall be counterweighted so as to balance, and if double hung the sash weights shall be separated by parting strips in the weight boxes and the weights shall be accessible through the boxes. Such sashes shall be provided with metallic sash chain, cord or tape, and smooth running sash pulleys securely riveted or bolted into place. The sash chain, cord or tape shall be of sufficient strength to withstand severe heat without parting and shall be thoroughly protected against moisture or corrosion. Every sash shall be fitted into its frame with stops and parting beads of metal or their equivalent and shall be removable. Meeting rails of sashes shall be so constructed as to prevent the passage of heat and flame and shall be equipped with one (1) or more substantial sash locks securely riveted or bolted in place.

(h) **Pivoted Sashes.** Pivoted sashes shall be provided with metal pivots at least three-eighths (⅜) inch in diameter, securely attached. Pivots shall work in substantial metal eye plates bushed with non-ferrous metal and securely attached in place. Sheet metal frames shall be reinforced where the pivots enter by riveting on one-eighth (⅛) inch metal strips so drilled as to receive the pivots. Every such sash shall be provided with suitable stops and an effective attachment for holding it open or closed, and with a substantial gravity lock or ledge that will be positive in action and hold the sash tightly closed when exposed to heat. Where either sash is stationary or where two (2) pivoted sashes are used, the transom bar dividing such sashes shall be so constructed that it will not warp or bulge excessively under heat or rapid cooling. Rails or transom bars, where used, shall be made with washes and drips to shed water properly and to afford ample weatherproof qualities.

(i) **Hinged Sashes.** Hinged sashes shall be hinged with substantial metal hinges securely bolted or riveted in place, and provided with metal latches or locks securely fastened in place. Every such sash shall be constructed so as to fit the frame closely and shall be provided with stops and fastenings that will prevent excessive warping or bulging under heat or rapid cooling.

(j) **Painting.** Frames and sashes, except non-ferrous metals, shall have all their parts protected by a covering of enamel or mineral paint.

(k) **Glazing.** All glazing shall be with wired glass not less than one-fourth (¼) inch thick, with three-fourths (¾) inch or smaller mesh. No single pane shall have an exposed area of more than seven hundred twenty (720) square inches measured on the inner face of the window, nor a length or width of more than fifty-four (54) inches, except as otherwise provided in paragraph (d). Glass shall be held in position by cleats or other structural pieces or parts of the frame or sash, without dependence of any material used for waterproofing, extending completely around each pane on both faces. On the inner face of the glass, the width of such cleats or other structural pieces or parts shall be not less than three-fourths (¾) inch for a pane of seven hundred twenty (720) square inches; and for a pane of lesser area, said required width shall be reduced one-sixteenth (1/16) inch for each one hundred (100) square inches by which the area of the pane is reduced below seven hundred twenty (720) square inches; provided however, that the width of such cleats or other pieces or part of the inner face of the glass shall be not less than one-half (½) inch in any case. On the outer, or weather face of the glass, the width of such cleats or other structural pieces or parts shall be not less than one-half (½) inch for a pane of three hundred fifty (350) square inches or greater area; and for a pane of lesser area said width shall be not less than one-half (½) the width of the cleat or other structural piece or part of the frame or sash on the inner face of the glass; provided however, that the width of such cleats or other structural pieces or parts on the outer side of the glass shall be not less than one-fourth (¼) inch in any case. All glass shall be embedded in putty.

2702.05 Mullions and Transom Bars: In any building required to have exterior walls of non-combustible materials, any door or window mullion or transom bar over eight (8) inches wide shall be covered on the exterior side with non-combustible materials equivalent to two (2) hour fire-resistive covering as provided in Chapter 32—FIRE-RESISTIVE STANDARDS.

ARTICLE 2703

Skylights

2703.01 Size and Location: There shall be no limitations of size or location of any Type A skylight, built as provided by Section 2703.05, or for any greenhouse thirty (30) feet or less above grade, or for any skylight for a one (1) story building when such skylight is located more than twelve (12) feet from any dividing lot line.

2703.02 Construction: The curb of any skylight shall be of the same construction as the roof on which it is located, but if of wood shall be metal-clad. The frame and the sash shall be of non-combustible material, if any point of the skylight is more than fifty (50) feet above grade, and shall be of non-combustible material or of wood completely metal-clad if more than one (1) story and fifty (50) feet or less above grade; provided however, that wood frames and sashes may be used in lieu of non-combustible material in any wood frame building or any other building where metal corroding fumes are present as an incident to the occupancy of the building. All structural parts of any skylight shall meet the provisions of the structural provisions of this ordinance.

2703.03 Glazing: Every Type A skylight shall be glazed with wired glass not less than one-fourth (¼) inch thick. Every other skylight required to have non-combustible or metal-clad frames and sash shall be glazed with wired glass not less than one-fourth (¼) inch thick; provided however, that any part of a skylight more than thirty (30) feet above grade shall be glazed with wire glass. Any sidewalk type or pavement type skylight may be glazed with prism glass lights sixteen (16) square inches or less in area and not less than one-half (½) inch thick, with individual lights separated by non-combustible structural members where the roof or floor adjoining is required to be of heavy timber construction or a superior type of construction.

2703.04 Ventilation: Every skylight used in lieu of windows for ventilation shall be provided with a ventilating opening or openings having a combined area of not less than three (3) per cent of the base area of the skylight. The area of such opening or openings shall be counted as a part of the area of ventilating openings required by Chapter 47—VENTILATION of this ordinance.

2703.05 Type A Skylight:

(a) **General.** Whenever in this ordinance a Type A skylight is required, such skylight shall be constructed of non-combustible materials and glazed with wired glass as required by this section.

(b) **Sheet Metal.** Every Type A skylight of sheet metal shall be constructed of galvanized iron or non-rusting metal not lighter than No. 24 U. S. standard gauge or of copper weighing not less than sixteen (16) ounces per square foot.

(c) **Solid Rolled Metal.** Type A skylight members of solid rolled metal shall be not less than one-eighth (⅛) inch thick.

(d) **Painting.** Frames and sashes, except non-ferrous metals shall have all their parts protected by a covering of enamel or mineral paint.

(e) **Glazing.** All glazing shall be with wired glass not less than one-fourth (¼) inch thick with three-fourths (¾) inch or smaller mesh. No single pane shall have an exposed area of more than twelve hundred (1200) square inches nor a length of more than five (5) feet. Glass shall be held in position by cleats or other structural pieces or parts of the frame as required by Section 2702.04 for Type A Windows.

CHAPTER 30

Artificial Lighting and Exit Signs

3001.01 General: Every means of exit in buildings hereafter designed, erected, or altered shall be provided with electric lighting consisting of normal, additional or emergency illumination and signs, as provided by this article, except as otherwise required in Chapters 7 to 18, inclusive. Such means of exit in every existing building shall be provided with normal illumination, exit signs and fire escape signs as required by this article within one (1) year after the passage of this ordinance.

3001.02 Normal Illumination of Exits: Except as otherwise provided in occupancy chapters, every means of exit shall have a system of electric lighting for normal illumination of an intensity not less than five-tenths (.5) foot candle at every point of the floor therein, except vertical means of exit which shall have an intensity of one (1.0) foot candle in every part thereof. Every passanger elevator car floor and every doorway outside of a passenger elevator hatchway shall be illuminated as required for a vertical means of exit.

3001.03 Additional Illumination of Exits: Additional illumination of an intensity equal to twenty (20) per cent of that required by Section 3001.02 shall be provided in all means of exit which shall be supplied from the general lighting source, but controlled separately from the required lights and connected with such source ahead or outside of the main circuit breaker which controls the required general lighting circuits in the following buildings: Institutional buildings, Business Units, three (3) stories or more in height, Hazardous Use Units, three (3) stories or more in height; Class 1 Garages, Theatres, Public Assembly Units and every assembly room having a capacity of more than fifteen hundred (1500) persons in churches and schools. Such system of additional illumination shall be controlled only from a point within the vestibule, lobby or principal entrance hall of the building.

3001.04 Illuminated Signs: In every building exit signs, directional signs and fire escape signs, illuminated by electricity, shall be provided to mark all ways of egress as required by this section, except as otherwise provided for buildings of special occupancy in Chapters 7 to 18, inclusive.

3001.05 Exit Signs:

(a) **General.** Standard exit signs shall be required to mark the doors leading to all vertical means of exit in multiple dwellings of the corridor type and over every exit doorway from a lobby, dining room or assembly room having an area of more than five hundred (500) square feet in any multiple dwelling; and over every exit doorway or connection in institutional buildings, office, financial and sales units, two (2) stories or more in height; storage and manufacturing units and hazardous use units, three (3) stories or more in height; Class 1 garages, theatres, open air assembly units having a capacity of more than eight hundred persons, public assembly units having a capacity of more than two hundred fifty (250) persons, churches and schools.

(b) **Letters.** Every standard exit sign shall bear the word "EXIT" with letters not less than six (6) inches high if the sign is illuminated externally, and not less than four and one-half (4½) inches high if illuminated internally. Six (6) inch letters shall have the principal stroke three-fourths (¾) inch wide and four and one-half (4½) inch letters, one-half (½) inch wide.

(c) **Colors.** All externally illuminated signs shall have white letters on a red field, and internally illuminated signs shall have red letters of translucent material in an opaque field. The light for externally illuminated signs shall have a disc or lens of not less than twenty-five (25) square inches surface of tanslucent material to show red on the side of approach.

(d) **Illumination.** Every exit sign shall be illuminated by electricity, with an intensity of illumination of not less than five (5) foot candles on the surface of sign. Every such sign shall have lights supplied from circuits of the wiring system for additional illumination of exits where such illumination is required by Section 3001.03.

(e) **Exceptions for Fire Escapes.** Such exit signs shall not be required where fire escape signs are required by this article.

3001.06 Directional Signs:

(a) **General.** Standard directional signs shall be provided for such buildings as are required to have exit signs as provided in the following paragraphs.

(b) **Horizontal Exits.** At every intersection or turning of any bridge, passageway, or tunnel serving as a horizontal exit, and for every horizontal exit connection on the main exit level at every change in direction and every intersection of such exit connection, and at every doorway on the main exit level from an inclosure for any vertical means of exit other than an inclosure having an outside exit doorway therefrom.

(c) **Exterior Passages.** At every yard, court, passage or other exterior space leading from any public exit to a public space serving as a way of departure from a building, where there is any change in direction, intersection or other irregularity of arrangement.

(d) **Letters.** Every standard directional sign shall bear the words "TO EXIT" with a suitable arrow to point the way and such letters shall be as otherwise required by Section 3001.05 for exit signs.

(e) **Colors and Illumination.** The colors and illumination of all directional signs shall be as required by Section 3001.05 for exit signs.

3001.07 Fire Escape Signs:

(a) **General.** Standard fire escape signs shall be provided at every doorway or other opening leading to any fire escape, or to any exterior stairway and at every other doorway or passage leading to any such means of exit from any corridor, lobby or passage on the same floor.

(b) **Letters.** Every fire escape sign shall bear the words "FIRE ESCAPE" in letters as otherwise required by Section 3001.05 for exit signs.

(c) **Colors and Illumination.** The colors and illumination of all fire escape signs shall be as required by Section 3001.05 for exit signs.

3001.08 Floor Indicating Signs: In every building, required by this article, to have exit signs, there shall be a sign or number at every floor within any vertical means of exit to indicate the several respective floors: provided however, that such a sign shall not be required at the main exit level when the vertical means of exit terminates at such level. Letters or figures shall be of the size required for standard exit signs, and of any legible contrasting colors on walls, glass or other surface.

CHAPTER 31
Construction Type Provisions

ARTICLE 3101
General

3101.01 Classification:

(a) **Types.** All building construction shall be classified under the following five types:

1. Fireproof Construction.
2. Semi-Fireproof Construction.
3. Heavy Timber Construction.
4. Ordinary Construction.
5. Wood Frame Construction.

(b) **Relative Value.** The relative fire-resistive value of the types listed herein shall be deemed to decrease in the order given. Buildings and structures shall be classified as of the type of the least fire-resistive construction used therein.

3101.02 Conformity to Types: Every building and structure shall conform to the provisions of this chapter for the type of construction required by the Occupancy Provisions of this Ordinance for the respective occupancy uses of such buildings and structures.

3101.03 Fireproofing: The term "fireproofing" wherever used in this ordinance, or in other ordinances of the City of Chicago, where the term has reference to building or structures shall be understood to refer to the fire-resistive materials, coverings or values, required by this ordinance.

ARTICLE 3102
Fireproof Construction

3102.01 General: Every building or structure and every part thereof, required to be of fireproof construction shall meet the requirements of this article. Joists or studding of wood or other combustible material shall not be used in the construction of floors or partitions, except as door bucks, in buildings of fireproof construction.

3102.02 Structural Frame:

(a) **Materials and Coverings.** The structural frame shall be of metal, reinforced concrete, masonry bearing walls or piers, or a combination of such materials. All metal members of the structural frame shall be protected by coverings of the fire-resistive values required by this article and of materials and thickness required for such values by Chapter 32—FIRE-RESISTIVE STANDARDS. Except as hereinafter provided, all metal or reinforced concrete members of the structural frame shall be constructed with the fire-resistance indicated in Items 1 and 2 of this paragraph.

Item 1. All columns, all members supporting walls, all members supporting more than one (1) floor and all wind-bracing shall have four (4) hour fire-resistance.

Item 2. All members, other than columns and joists, supporting only one (1) floor or roof shall have three (3) hour fire-resistance.

(b) **Reduction or Omission of Fireproofing.** The fire-resistive covering of certain members of the structural frame, may be omitted or reduced in value as permitted under Items 1 to 4 of this paragraph.

Item 1. Where lintel or spandrel beams are protected, as required by paragraph (a), Item 1 of this section, the exterior wall facing of same, if not more than four and one-half (4½) inches thick, may be carried, for the height of the spandrel beam, on unprotected metal angles anchored to and supported by the protected member at not to exceed three (3) foot intervals.

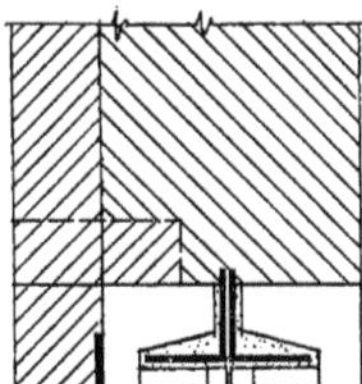

Item 2

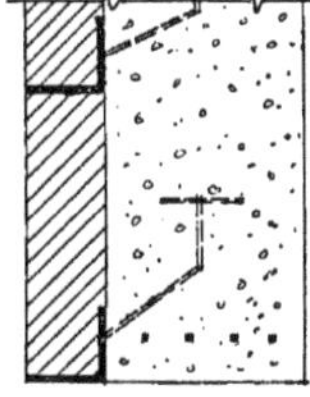

Item 1

Where lintels are fireproofed previously and independently, the Commissioner of Buildings has ruled that the application of the architectural facing may be supported as shown.

The ruling is only applied to openings not to exceed four feet in width, and the maximum depth of the beam is not to exceed eighteen inches.

Item 2. Loose lintels over exterior wall openings not exceeding three and one-half (3½) feet in clear width need not be fireproofed.

Item 3. Certain members forming part of the floor construction may have fireproofing reduced as hereinafter provided.

Item 4. Tie beams at floor levels between hatchways in elevator shafts, elevator sheave beams and beams forming elevator cable slots need not be fireproofed.

3102.03 Floor Construction:

(a) **General.** The floor construction shall include structural slabs, arches, joists and beams other than members of the structural frame named in Section 3102.02, paragraph (a).

(b) **Structural Parts.** Every floor shall have structural metal or reinforced concrete framing members with reinforced concrete or masonry slabs or arches forming a floor construction having a fire-resistive value of three (3) hours.

(c) **Fireproofing.** All structural metal shall be fireproofed with a covering of three (3) hour fire-resistive value; except that concrete fireproofing on the topsides of such members need not exceed one and one-fourth (1¼) inch in thickness. All reinforced concrete and masonry shall be constructed with a three (3) hour fire-resistive value.

(d) **Sleepers and Fill.** Sleepers of wood or not less fire-resistive material may be laid on the required slab or arch construction for the securing of a finish floor. Such sleepers shall not extend under any required partition wall or inclosure. The spaces between such sleepers shall be filled with a non-combustible material.

3102.04 Roof Construction:

(a) **General.** The roof construction shall include structural slabs, arches, joists, purlins and beams other than members of the structural frame named in Section 3102.02, paragraph (a).

(b) **Structural Parts.** Every roof shall have structural metal or reinforced concrete framing members, with metal plate roofs, reinforced concrete slabs or masonry arches, forming a roof construction having a fire-resistive value of three (3) hours.

(c) **Fireproofing.** All structural members shall be fireproofed with a covering of three (3) hour fire-resistive value.

3102.05 Exterior Inclosing Walls:

(a) **General.** Every exterior bearing and non-bearing wall, including exterior walls of courts, vent shafts and light shafts, shall be constructed of masonry or reinforced concrete, as provided in the General Provisions and the Structural Provisions of this ordinance and shall have a fire-resistive value of not less than four (4) hours.

(b) **Bay Windows and Similar Projections.** Every exterior inclosing wall of a bay window, oriel or other similar projection from the face of a main wall shall be constructed as required for the main wall construction.

(c) **Cornices.** Every cornice shall be of non-combustible material, including the backing, furring and supports thereof.

3102.06 Partitions: The provisions of Chapter 21—GENERAL CONSTRUCTION PROVISIONS, shall govern the construction of partitions in fireproof buildings.

3102.07 Grounds and Furring: Wood grounds and wood furring, projecting not more than three (3) inches from a non-combustible wall surface, shall be permitted for the attachment of finish wood trim and cabinet work permitted under Section 3102.08; provided however, that all hollow spaces so formed shall be subdivided by fire-stops spaced not more than six (6) feet apart in any direction. Such fire-stops shall be not less fire-resistive than one and five-eighths (1⅝) inch of solid wood. All other grounds and furring shall be non-combustible.

3102.08 Use of Combustible Materials: Wood or other not more combustible material may be used in buildings of fireproof construction, except as otherwise provided in this ordinance, for the following purposes:

1. Interior architectural trim.
2. Floor surfaces and sleepers.
3. Wood cabinet work for wall and ceiling paneling.
4. Backing, frames, platforms and aprons of exterior show windows at street level.
5. Handrails.
6. Doors, door frames and bucks.
7. Grounds, as permitted in Section 3102.07.

ARTICLE 3103

Semi-Fireproof Construction

3103.01 General: Every building or structure and every part thereof required to be of semi-fireproof construction, shall meet the requirements of this article. Joists or studding of wood or other combustible material shall not be used in the construction of floors or partitions, except as door bucks, in buildings of semi-fireproof construction.

3103.02 Structural Frame:

(a) **Materials and Coverings.** The structural frame shall be of metal, reinforced concrete, masonry bearing walls or piers, or a combination of such materials. All metal members of the structural frame shall be protected by coverings of the fire-resistive value required by this article and of materials and thickness required for such values by Chapter 32—FIRE-RESISTIVE STANDARDS. Except as hereinafter provided, all metal or reinforced members of the structural frame shall be constructed with the fire-resistance indicated in Items 1 and 2 of this paragraph.

Item 1. All exterior columns shall have a four (4) hour fire-resistive value. All interior columns, all members other than exterior columns, which support walls, all members supporting more than one (1) floor, and all wind-bracing, shall have three (3) hour fire-resistance.

Item 2. All members, other than columns and joists, supporting only one (1) floor or roof shall have two (2) hour fire-resistance.

(b) **Reduction or Omission of Fireproofing.** The fire-resistive covering of certain members of the structural frame may be omitted or reduced in value as permitted under Items 1 to 4 of this paragraph.

Item 1. Where lintel or spandrel beams are protected, as required by paragraph (a), Item 1 of this section, the exterior wall facing of same, if not more than four and one-half (4½) inches thick, may be carried, for the height of the spandrel beam, on unprotected metal anchored to and supported by the protected member at not to exceed three (3) foot intervals.

Item 2. Loose lintels over exterior wall openings not exceeding three and one-half (3½) feet in clear width need not be fireproofed.

Item 3. Certain members forming part of the floor construction may have fireproofing reduced as hereinafter provided.

Item 4. Tie beams at floor levels between hatchways in elevator shafts, elevator sheave beams and beams forming elevator cable slots need not be fireproofed.

3103.03 Floor Construction:

(a) **General.** The floor construction shall include structural slabs, arches, joists and beams other than members of the structural frame named in Section 3103.02, paragraph (a).

(b) **Structural Parts.** Every floor shall have structural metal or reinforced concrete framing members or trussed steel joists, with reinforced concrete or masonry slabs or arches, or metal plate floors, meeting the requirements of the structural provisions of this ordinance and forming a floor construction having a fire-resistive value of one (1) hour. All metal or reinforced concrete members shall be constructed with the fire-resistive values prescribed in paragraph (c) of this section.

(c) **Fireproofing.** All structural metal and trussed steel joists shall be fireproofed with a covering of one (1) hour fire-resistive value or by a ceiling of metal lath and plaster attached to or suspended below such members and shall be fire-stopped at points of support. Any side of a structural member not inclosed by the space between such fire-resistive floor and ceiling shall be protected with a covering of one (1) hour fire-resistive value.

(d) **Sleepers and Fill.** Sleepers of wood or not less fire-resistive material may be laid on top of the required slab or arch construction for the securing of a finish floor. Such sleepers shall not extend under any required partition wall or inclosure. The spaces between such sleepers shall be filled with a non-combustible material.

3103.04 Roof Construction:

(a) **General.** The roof construction shall include structural slabs, arches, joists, purlins and beams other than members of the structural frame named in Section 3103.02, paragraph (a).

(b) **Structural Parts.** Every roof shall have metal or reinforced concrete framing members or trussed steel joists, rafters or purlins, with reinforced concrete slabs or masonry arches, metal plate decks, meeting the requirements of the structural provisions of this ordinance and forming a roof construction having a fire-resistive value of one (1) hour.

(c) **Fireproofing.** All structural members shall be fireproofed with a covering of one (1) hour fire-resistive value in one (1) of the two (2) following ways:

1. By a complete covering of one (1) hour fire-resistive value.
2. By a ceiling of one (1) hour fire-resistive value attached to or suspended below the bottom of such members. Any side of a structural member, not inclosed by the space between such ceiling and the roof construction shall be protected by a covering of one (1) hour fire-resistive value.

3103.05 Exterior Inclosing Walls:

(a) **General.** Every exterior bearing and non-bearing wall, including exterior walls of courts, vent shafts and light shafts, shall be constructed of masonry or reinforced concrete as provided in the General Provisions and the Structural Provisions of this ordinance and shall have a fire-resistive value of not less than four (4) hours.

(b) **Bay Windows and Similar Projections.** Every exterior inclosing wall of a bay window, oriel or other similar projection from the face of a main wall shall be constructed as required for the main wall construction.

(c) **Cornices.** Every cornice shall be of non-combustible material, including the backing, furring and supports thereof.

3103.06 Partitions: The provisions of Chapter 21—GENERAL CONSTRUCTION PROVISIONS, govern the construction of partitions in semi-fireproof buildings.

3103.07 Grounds and Furring: Wood grounds and wood furring, projecting not more than three (3) inches from a non-combustible wall surface, shall be permitted for the attachment of finish wood trim and cabinet work permitted under section 3103.08; provided, however, that all hollow spaces so formed shall be subdivided by fire-stops spaced not more than six (6) feet apart in any direction. Such fire-stops shall be not less fire-resistive than one and five-eighths (1⅝) inch of solid wood. Other grounds and furring shall be non-combustible.

3103.08 Use of Combustible Materials: Wood or other not more combustible material may be used in buildings of semi-fireproof construction, except as otherwise provided in this ordinance, for the following purposes:

1. Interior architectural trim.
2. Floor surfaces and sleepers.
3. Wood cabinet work for wall and ceiling paneling.
4. Backing, frames, platforms and aprons of exterior show windows at street level.
5. Handrails.
6. Doors, door frames and bucks.
7. Grounds as permitted in Section 3103.07.

ARTICLE 3104

Heavy Timber Construction

3104.01 General: Every building or structure and every part thereof required to be of heavy timber construction shall meet the requirements of this article.

3104.02 Structural Frame:

(a) **Materials and Coverings.** All exterior walls shall be constructed of either a structural frame of metal or reinforced concrete with inclosing walls of masonry; or of masonry bearing walls or piers, or a combination of such construction. Other structural members shall be of metal, reinforced concrete, masonry, timber, or a combination of such materials. Except as otherwise provided in paragraph (b) of this section, all structural members shall be constructed with the fire-resistive values required by Items 1 to 6 of this paragraph.

Item 1. Exterior columns shall have four (4) hour fire-resistance.

Item 2. Members, other than exterior columns supporting walls, shall have three (3) hour fire-resistance.

Item 3. Interior columns, beams, girders and truss members, other than timber, shall have two (2) hour fire-resistance.

Item 4. Timber columns shall have minimum actual dimension of nine and one-half (9½) inches.

Item 5. Timber beams, girders and truss members supporting any load other than a roof shall have minimum actual dimensions of five and one-half (5½) inches with minimum actual area of cross-section of fifty-two (52) square inches.

Item 6. Timber beams, girders and truss members supporting only a roof shall have minimum actual dimension of three and five-eighths (3⅝) inches.

(b) **Members Not Requiring Fireproofing.** Except as otherwise provided in this ordinance, fireproofing shall not be required for the following members:

Item 1. Beams, girders and trusses supporting a roof only, where the clear height above the floor is twenty (20) feet or more.

Item 2. Loose lintels which support not more than one (1) story.

Item 3. Metal connections, fastenings and accessories for timber members such as bases, caps, stirrups, bolsters, tie rods, anchors and boxes.

3104.03 Floor Construction: In heavy timber construction, all members of the floor construction, except the flooring, are included in and shall be governed by the provisions for the structural frame. The floors shall be of wood not less than three and one-half (3½) inches in actual thickness, with the component pieces fastened together by spikes, tongues and grooves, splines or otherwise in a manner which will cause them to act in unison. Such floors may be of two (2) layers; provided however, that the lower layer shall be not less than two and five-eighths (2⅝) inches thick. If a ceiling is installed, it shall be placed directly against the floor above without an intervening air space of more than one-half (½) inch.

3104.04 Roof Construction: In heavy timber construction, all members of the roof construction, except the sheathing, are included in and shall be governed by the provisions for the structural frame. Every roof shall be sheathed with wood not less than two and five-eighths (2⅝) inches thick in one (1) layer, provided however, that in buildings equipped throughout with a standard system of automatic sprinklers, the roof sheathing may be not less than one and five-eighths (1⅝) inch in actual thickness.

3104.05 Exterior Inclosing and Bearing Walls:

(a) **General.** Every exterior bearing and non-bearing wall, including exterior walls of courts, vent shafts and light shafts shall be constructed of masonry or reinforced concrete as provided in the General Provisions and the Structural Provisions of this ordinance and shall have a fire-resistive value of not less than four (4) hours.

(b) **Bay Windows and Similar Projections.** Every exterior inclosing wall of a bay window, oriel or other similar projection from the face of a main wall shall be constructed as required for the main wall construction.

(c) **Cornices.** Every cornice shall be of non-combustible material, including the backing, furring and supports thereof.

3104.06 Partitions: The provisions of Chapter 21—GENERAL CONSTRUCTION PROVISIONS, shall govern the construction of partitions in buildings of heavy timber construction.

3104.07 Grounds and Furring: Grounds and furring in buildings of heavy timber construction shall be as required in buildings of fireproof construction.

3104.08 Use of Combustible Materials: Wood, or other not more combustible material may be used in buildings of heavy timber construction, except as otherwise provided in this ordinance, for the following purposes:

1. Interior architectural trim.
2. Flooring and roof sheathing.
3. Wood cabinet work for wall and ceiling paneling.
4. Backing, frames, platforms and aprons of exterior show windows at street level.
5. Handrails.
6. Doors and frames, except as otherwise required.

No combustible sleepers, combustible lath, nor combustible interior construction, forming hollow spaces, shall be used.

ARTICLE 3105

Ordinary Construction

3105.01 General: Every building or structure and every part thereof required to be of ordinary construction, shall meet the requirements of this article.

3105.02 Structural Frame: The structural members shall be of metal, reinforced concrete, masonry, wood or a combination of such materials.

3105.03 Floor and Roof Construction:

(a) **General.** The floor and roof construction shall include joists, flooring and sheathing. All floor and roof construction shall be of wood or other more fire-resistive material. Trussed steel joists, meeting the requirements of this ordinance, may be used in buildings of ordinary construction. Except as provided in Chapter 7—SINGLE DWELLINGS, no floor nor roof shall be supported on a wood stud partition.

(b) **Basement Ceilings.** The basement ceilings of buildings of ordinary construction more than two (2) stories in height shall be protected by metal lath and plaster; provided however, that this requirement shall not apply where the first floor construction is of at least one (1) hour fire-resistive value.

3105.04 Exterior Inclosing and Bearing Walls:

(a) **General.** Every exterior bearing and non-bearing wall, including exterior walls of courts, vent shafts and light shafts, shall be constructed of masonry or reinforced concrete, as provided in the General Provisions and the Structural Provisions of this ordinance, and shall have a fire-resistive value of not less than three (3) hours.

(b) **Bay Windows and Similar Projections.** Every exterior inclosing wall of a bay window, oriel or other similar projection from the face of a main wall shall be constructed as required for the main wall construction.

(c) **Cornices.** Every cornice projecting over a street or alley lot line, shall be of non-combustible material, including the backing, furring and supports thereof.

3105.05 Partitions: The provisions of Chapter 21—GENERAL CONSTRUCTION PROVISIONS, shall govern the construction of partitions in buildings of ordinary construction.

ARTICLE 3108

Wood Frame Construction

3108.01 General: Every building, or part thereof, designated as of wood frame construction shall meet the requirements of this article.

3108.02 Location and Area Provisions:

(a) **Within the Fire Limits.** No building, structure, shed or inclosure of wood frame construction shall be erected inside the fire limits, or provisional fire limits, except as permitted for a specified use under the Occupancy Provisions of this ordinance.

(b) **Outside the Fire Limits.** The area, height, use and other limits for buildings, structures, sheds or inclosures of wood frame construction, outside of the fire limits, or provisional fire limits, shall be as given in the Occupancy Provisions of this ordinance.

3108.03 Materials and Construction:

(a) **Exterior Walls and Roofs.** Exterior walls and roofs shall be framed with wood framing members. Exterior walls shall be surfaced with materials not less fire-resistive than wood. Roofs shall be sheathed with wood and shall be covered with a fire-retarding material. The outside of exterior wall framing shall be sheathed with wood. Sheathing of any material shall be covered with weather resisting surface material. Where exterior stucco finish is used, sheathing shall be covered with waterproof paper and the lath or other stucco base shall be furred out from the surface of such covering.

(b) **Partitions.** The provisions of Chapter 21—GENERAL CONSTRUCTION PROVISIONS, schall govern partitions in wood frame construction.

(c) **Floors.** Finish floors may be of wood or not less fire-resistive material. Structural floor joists, supported on wood construction, shall be of wood.

(d) **Veneered Construction.** Any veneer of masonry over wood frame construction shall be not less than three and three-fourths (3¾) inches thick. Such veneers shall be supported on a foundation of masonry or reinforced concrete and shall be secured to the frame by not less than two (2) anchors per square foot of wall area. Each anchor shall have a cross-sectional area of not less than two-tenths (.2) of one (1) inch and shall be secured to a stud. Masonry veneers shall be not more than two (2) stories or more than thirty (30) feet in height at any point measured from the top of their supporting foundations.

(e) **Fire-Stops.** All spaces between studs shall be fire-stopped top and bottom in each story. All spaces between joists and rafters shall be fire-stopped at the supports. Fire-stops shall consist of wood not less than one and five-eighths (1⅝) inch in thickness or of not less fire-resistive material.

CHAPTER 32

Fire-Resistive Standards

ARTICLE 3201

General

3201.01 General Requirements: Materials and methods of construction conforming to the provisions of this chapter shall be used in all buildings and structures for which fire-resistive materials are required and fire-resistive classifications are established by this ordinance.

3201.02 Fire-Resistive Classification: Building materials and methods of construction shall be classified for fire-protective and fire-resistive purposes, decreasing in relative fire-resistive value in the following order:

Four (4) hour fire-resistive construction.
Three (3) hour fire-resistive construction.
Two (2) hour fire-resistive construction.
One (1) hour fire-resistive construction.

3201.03 Fire-Resistive Materials:

(a) **General.** The materials which, for the purposes of this ordinance, shall be defined as fire-resistive, shall include materials which conform to the following requirements:

(b) **Brick.** Brick shall be at least equal to the requirements of one (1) of the following Standard Specifications of the American Society for Testing Materials: Specifications for Building Brick made from Clay or Shale (Serial Designation C62-30); Specifications for Concrete Building Brick (Serial Designation C55-34); or Specifications for Sand-Lime Building Brick (Serial Designation C73-30).

(c) **Burned Clay or Shale Tile.** Burned clay or shale tile shall be at least equal to the requirements of one (1) of the following Standard Specifications of the American Society for Testing Materials: Specifications for Structural Clay Load-Bearing Wall Tile (Serial Designation C34-36); Specifications for Structural Clay Non-Load Bearing Tile (Serial Designation C56-36); or Specifications for Structural Clay Floor Tile (Serial Designation C57-36).

(d) **Terra Cotta Units.** Terra Cotta units shall be at least equal to the requirements of Section 3802.04.

(e) **Concrete Units.** Hollow concrete units shall be at least equal to the requirements of one (1) of the following Specifications of the American Society for Testing Materials: Specifications and Tests for Load-Bearing Concrete Masonry Units (Serial Designation C90-36); or Specifications and Methods of Test for Concrete Units for Non-Load Bearing Masonry (Serial Designation C129-37T) Solid concrete units shall conform to the requirements of Section 3802.03.

(f) **Portland Cement Concrete.** Portland cement concrete for fire-resistive purposes shall conform to the provisions of Chapter 39—REINFORCED CONCRETE CONSTRUCTION, except that burned clay or shale, blast furnace slag, or other non-combustible material, approved by the Commissioner of Buildings may be used as an aggregate. Concrete for such purposes shall contain not more than seven (7) volumes of aggregate, fine and coarse measured separately, to each volume of Portland Cement.

(g) **Metal Lath.** Metal lath of either wire mesh or expanded metal, to be used as a fire-resistive plaster base, shall weigh not less than two and five-tenths (2.5) pounds per square yard when used on vertical surfaces and not less than three (3) pounds per square yard when used on horizontal or sloping surfaces. Metal lath shall not be considered as a fire-resistive material except when used as a plaster base in the same plane and integral with the plaster.

(h) **Plaster.** Wherever plaster is required or permitted as a fire-resistive material in this chapter, it shall consist of not less than two (2) coats having a total thickness of not less than one-half (½) inch applied directly to concrete or masonry and three-fourths (¾) inch applied to metal lath, measured from the face of the lath.

(i) **Cast-in-Place Gypsum.** Cast-in-Place Gypsum shall contain not more than five (5) per cent of wood chips, excelsior or fibre, measured in a dry condition as a percentage by weight of the dry mix.

(j) **Gypsum Tile or Block.** Gypsum tile or block shall conform to Standard Specifications of the American Society for Testing Materials (Serial Designation C52-33), except that it shall contain not more than five (5) per cent of wood chips, excelsior or fibre, measured in a dry condition as a percentage by weight of the dry mix.

ARTICLE 3202

Fire-Resistive Construction Provisions

3202.01 Walls and Partitions:

(a) **General.** The required thicknesses and systems for construction for walls, partitions and surfaces inclined fifteen (15) degrees or less from the vertical, for fire-resistive construction of any required value, shall conform to the provisions of this section, except as otherwise required in Chapter 21—GENERAL CONSTRUCTION PROVISIONS; Chapter 38—MASONRY CONSTRUCTION and Chapter 39—REINFORCED CONCRETE CONSTRUCTION. Gypsum shall not be used for exterior walls or as a fire-protective covering for that portion of any member which is in an exterior wall or below grade, except for ceilings which are part of a floor system.

(b) **Four (4) Hour Fire-Resistive Walls.** Four (4) hour fire-resistive walls or partitions shall include walls of the following construction:

- Brick, solid wall, eight (8) inches thick.
- Brick, hollow wall, twelve (12) inches thick with three (3) continuous vertical withes; or eight (8) inches thick, plastered both sides.
- Plain concrete, eight (8) inches thick.
- Reinforced concrete, six (6) inches thick.
- Solid concrete units, eight (8) inches thick, plastered one (1) side.
- Hollow concrete units, two (2) cells in wall thickness, twelve (12) inches thick; or eight (8) inches thick, plastered both sides.
- Hollow clay tile, two (2) units and three (3) cells in wall thickness, twelve (12) inches thick, or load bearing tile, two (2) cells, eight (8) inches thick, plastered both sides.
- Hollow units of clay, concrete or gypsum faced with brick, stone, terra cotta or cast stone, twelve (12) inches total thickness; or eight (8) inches thick, plastered one (1) side.

(c) **Three (3) Hour Fire-Resistive Walls.** Three (3) hour fire-resistive walls or partitions shall include any walls permitted under paragraph (b) or walls of the following construction:

- Plain concrete, seven (7) inches thick.
- Reinforced concrete, five (5) inches thick.
- Solid concrete units, eight (8) inches thick.
- Hollow concrete units, twelve (12) inches thick.
- Hollow concrete units, eight (8) inches thick, with one and one-half (1½) inch thick outer shell.
- Hollow clay tile, three (3) cells in wall thickness twelve (12) inches thick.
- Hollow gypsum single units, eight (8) inches thick.
- Hollow units of clay, concrete or gypsum faced with brick, stone, terra cotta or cast stone, eight (8) inches total thickness.

(d) **Two (2) Hour Fire-Resistive Walls.** Two (2) hour fire-resistive walls or partitions shall include any of the walls permitted under paragraphs (b) and (c or walls of the following construction:

- Brick, solid wall, three and three-fourths (3¾) inches thick, plastered both sides.
- Brick, hollow wall, eight (8) inches thick.
- Plain concrete, six (6) inches thick.
- Reinforced concrete, four (4) inches thick.
- Cast-in-Place Gypsum poured four (4) inches thick, or hollow units three (3) inches thick, plastered both sides.
- Solid concrete units, six (6) inches thick.
- Hollow concrete units, eight (8) inches thick; or four (4) inches thick, plastered both sides.
- Hollow clay tile, three (3) cells in wall thickness, eight (8) inches thick; or two (2) cells in wall thickness, eight (8) inches thick, plastered one (1) side; or two (2) cells in wall thickness, six (6) inches thick, plastered both sides.
- Solid plaster partition on metal studs and metal lath two (2) inches thick.

(e) **One (1) Hour Fire-Resistive Walls.** One (1) hour fire-resistive walls or partitions shall include any of the walls permitted under paragraphs (b), (c) or (d) or walls or partitions of the following construction:

- Brick, three and three-fourths (3¾ inches thick, or two and one-fourth (2¼) inches thick, plastered both sides.
- Concrete, three (3) inches thick.
- Gypsum, solid or hollow units, three (3) inches thick, or solid units two (2) inches thick, plastered both sides.
- Solid concrete units, three (3) inches thick.
- Hollow concrete units, four (4) inches thick; or three (3) inches thick, plastered both sides.

Hollow clay tile, three and three-fourths (3¾) inches thick; or load-bearing tile three (3) inches thick, plastered both sides.

Hollow partition of either non-combustible or combustible studding, with metal lath and three-fourths (¾) inch of plaster; or gypsum lath and one-half (½) inch of plaster, on both sides. Joints of gypsum lath shall be covered with strips of metal lath at least three (3) inches wide.

(f) **Cuts and Chases in Walls.** The minimum wall thicknesses, required by this section, shall not be reduced or cut into. Bearing walls shall comply in all respects with the requirements of Chapter 38—MASONRY CONSTRUCTION. A seperation of not less than four (4) inches of solid masonry shall be provided in all walls between combustible members which may enter such walls from opposite sides.

3202.02 Floors, Roofs and Ceilings:

(a) **General.** The required thicknesses and systems of construction for floors, roofs and ceilings or any separating construction between stories, other than walls, for fire-resistive construction of any required value, shall conform to the provisions of this section, except as otherwise required in the structural provisions.

(b) **Four (4) Hour Fire-Resistive Systems.** Four (4) hour fire-resistive floors or roof systems, shall include:

Reinforced concrete slabs, six (6) inches thick.

Solid masonry slabs or arches, six (6) inches thick.

Hollow tile arches, ten (10) inches thick, plastered on the underside and with a solid or loose floor fill of non-combustible material two (2) inches thick.

Reinforced concrete joists with filler units of hollow tile, concrete or gypsum, ten (10) inches total thickness.

(c) **Three (3) Hour Fire-Resistive Systems.** Three (3) hour fire-resistive floor or roof systems shall include any system permitted under paragraph (b) or the following:

Reinforced concrete slabs, four (4) inches thick.

Reinforced gypsum slabs, for roofs only, four (4) inches thick.

Solid masonry slabs or arches, four (4) inches thick.

Hollow tile arches, plastered, eight (8) inches thick for flat arches, six (6) inches thick for segmental arches, with a floor fill of solid or loose non-combustible material, two (2) inches thick.

Reinforced concrete joists with filler units of hollow tile, concrete or gypsum blocks, six (6) inches, total thickness.

Reinforced concrete joists in combination with three (3) inch thick reinforced concrete slab.

(d) **Two (2) Hour Fire-Resistive Systems.** Two (2) hour fire-resistive floor or roof systems shall include any system permitted under paragraphs (b) and (c) or the following:

Reinforced concrete slabs three (3) inches thick.

Reinforced gypsum slabs, for roofs only, three (3) inches thick.

Reinforced concrete joists in combination with two and one-half (2½) inch thick reinforced concrete slab.

Reinforced concrete joists with filler units of hollow tile, concrete or gypsum blocks, five (5) inches total thickness.

Steel joists with reinforced concrete slabs two and one-half (2½) inches thick protected on the underside by a ceiling of metal lath and plaster seven-eighths (⅞) inch thick.

Steel joists with reinforced gypsum slabs for roofs only, two and one-half (2½) inches thick, protected on the underside by a ceiling of metal lath and plaster seven-eighths (⅞) inch thick.

(e) **One (1) Hour Fire-Resistive Systems.** One (1) hour fire-resistive floor or roof systems shall include any systems permitted un-

Hollow units of concrete, three (3) inches thick, plastered.
Solid units of concrete, two (2) inches thick, plastered.
Hollow units of clay tile or concrete, four (4) inches and two (2) units thick, with staggered joints, not plastered.

For Other Metal Members:
Solid units of gypsum, three (3) inches thick.
Hollow units of clay tile, two (2) inches thick, plastered.
Solid units of concrete, two (2) inches thick, plastered.
Hollow units of clay tile or concrete, four (4) inches thick and two (2) units thick, with staggered joints not plastered.

(d) **Two (2) Hour Fire-Protective Covering.** Two (2) hour fire-protective coverings shall include any covering permitted under paragraphs (b) and (c) or the following:

For Metal Columns:
Brick, two and one-fourth (2¼) inches thick.
Hollow units of clay tile, two (2) inches thick.
Solid or hollow units of concrete, two (2) inches thick.

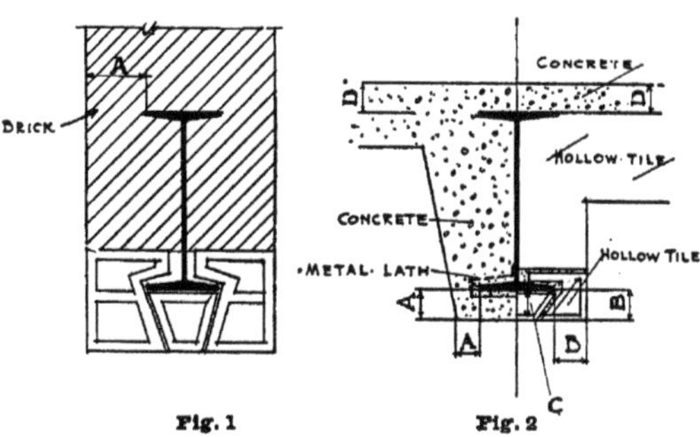

Fig. 1 Fig. 2

PROTECTION OF BEAMS.
Section 3202.03.

(A) Fire-proof covering for beams, girders, etc., for exterior structural parts, Sec. 3202.03.
Figs. 1 and 2. Necessary fire-proof covering for beams, girders, etc., for interior structural parts.
Fig. 1. (A) 4" for brick.
Fig. 2. (B) 2" for concrete or hollow tile if plastered.
(A) 2" for concrete.
(D) Concrete covering for tops of beams, girders, etc., to be 1¼" (Sec. 3102.03c).

Solid units of gypsum, two (2) inches thick, no fill.

For Other Metal Members:
Cast-in-place concrete one and one-half (1½) inches thick.
Cast-in-place gypsum, one and one-half (1½) inches thick.

For Reinforcement of Slabs, Joists and Walls:
Concrete three-fourths (¾) inch thick.

(e) **One (1) Hour Fire-Protective Covering.** One (1) hour fire-protective coverings shall include any covering permitted under paragraphs (b), (c) and (d) or the following:

For Metal Columns:
Hollow units of clay tile, two (2) inches thick, no fill.
Units of concrete, two (2) inches thick, no fill.
Portland cement, lime or gypsum plaster, one (1) inch thick on metal lath.

For Other Metal Members:
Portland cement, lime or gypsum plaster, seven-eighths (⅞) inch thick on metal lath.

For Reinforcement of Columns, Beams, Girders and Trusses:
Concrete, one (1) inch thick.

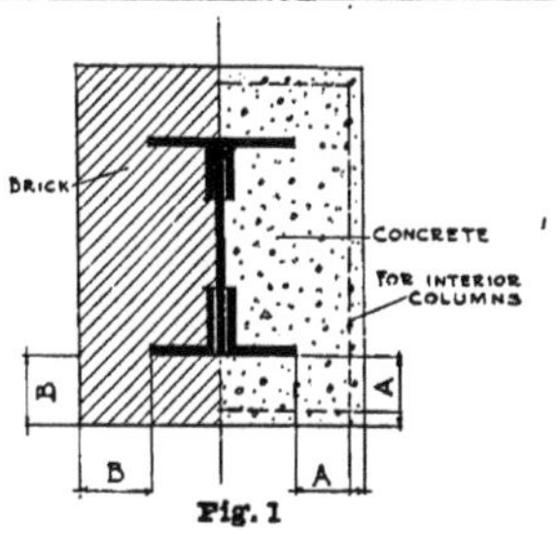

Fig. 1

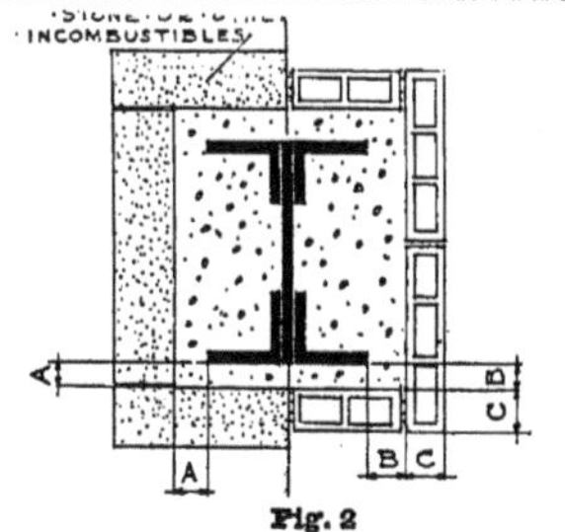

Fig. 2

Figs. 1 and 2. Requirement for protection of columns of building of fireproof construction from external change of temperature and fire.

Fig. 1. (A) 3" concrete required.

(B) 3¾" brick and ½" mortar.

Fig. 2. (A) Exterior masonry facing 3¾" thick is credited as one-half the required fire resistance. If stone or other incombustible material is used for exterior facing then (A) can equal 2".

(B) (C) Combination of materials in fireproofing, etc., is allowed as at (B plus C), Sec. 3202.04.

3202.04 Facing Materials as Fire-Protective Covering.

(a) **General.** Facing material may be considered as fire-protective covering when it meets the requirements of Section 3201.03.

(b) **Masonry Facings.** When materials specified in Chapter 38—MASONRY CONSTRUCTION, Section 3801.01, paragraph (a), of a minimum thickness of three and three-fourths (3¾) inches are used for the facing of a building, such facing, where covering or inclosing a structural member shall be assumed to be effective for only one-half (½) of the required fire-resistive value for such material and the remaining one-half (½) shall be provided by additional fire-resistive material applied in accordance with Section 3202.03.

3202.05 Construction of Fire-Protective Covering:

(a) **Application to Metal Members.** For beams, girders and trusses, the covering shall be bedded solidly against the webs and flanges of the member and shall completely surround or inclose the member. The fire-resistive covering for columns shall surround or inclose the column, and all interior spaces between the covering and the column shall be filled with concrete or the same material as the required fire-resistive covering; except as otherwise specifically permitted by the word "no fill" under Section 3202.03. Masonry units in all cases shall be well anchored or bonded by one (1) of the following methods:

1. Shape of unit giving positive anchorage to structural member or to other units.
2. Wall ties of corrugated metal, or strips of metal mesh, laid in all the horizontal joints, where such joints are ten (10) inches or more apart.
3. Metal clips providing a mechanical tie between units.
4. Outside tie wires not smaller than No. 10 W. & M. gauge, with at least one (1) tie around each course, and protected by plaster.

Gypsum units shall be anchored by Method No. 3 or No. 4. All joints in unit coverings shall be solidly filled with mortar. Before cast-in-place concrete or gypsum is placed, all members to be protected with these materials shall be wrapped with woven or welded wire mesh weighing not less than one and one-half (1½) pounds per square yard, or with wire not smaller than No. 12 W. & M. gauge spaced six (6) inches on centers, or provided with fabricated flange reinforcement of equivalent weight.

(b) **Mortar for Fire-Protective Covering Units.** Units of burned clay or shale, or concrete when used as a fire-protective material, shall be bedded with full joints in mortar equivalent in strength to No. 3 or No. 4 mortar, as provided in Chapter 38—MASONRY CONSTRUCTION. Units of gypsum shall be bedded wtih full joints in gypsum mortar composed of one (1) part of neat gypsum plaster and not more than three (3) parts of clean, sharp, well graded sand by weight.

(c) **Beam Flanges.** The fire-protective covering beyond the edge of any beam flange, less than three-fourths (¾) inch thick at the thin edge may be reduced to one and one-half (1½) inches when such covering is solid concrete.

(d) **Exceptions.** The minimum thickness of fire-protective covering required by this section shall be measured exclusive of any plaster, and shall apply to the thickness of covering over metal members under stress, not including rivet heads.

(e) **Pipes and Ducts.** Pipes or ducts, excepting electric conduits shall not be embedded in the required fire-protective covering of any members; provided however, that where steel beams or girders are embedded in concrte, pipes or conduits may rest on the top flanges of such beams or girders but shall not encroach upon the required covering for the sides and bottom.

ARTICLE 3203

Standard Fire Tests

3203.01 General: Wherever in this ordinance reference is made to a standard fire test, such test shall be construed to mean and is required to mean a test conducted in accordance with all of the provisions of the "Standard Specifications for Fire Tests of Building Construction and Materials (A. S. T. M. Designation C19-33) of the American Society for Testing Materials."

CHAPTER 33

All Types and Kinds of Fire Doors

ARTICLE 3301

General

3301.01 Definition: Any movable barrier which may be opened or closed to allow or to prevent passage through a doorway in a wall or inclosure, which wall or inclosure is required by this ordinance to be of fire-resistive construction, shall be classed as a fire door. Wherever the words "Fire Door" or "Fire-Resistive Door" are used in this ordinance, they shall be construed to mean and are required to mean acceptable protection assemblies consisting of the following:

1. Door.
2. Frame, where required.
3. Sill, where adjoining floors are combustible.
4. Hardware.
5. Automatic Closing Device, where required.
6. Self-Closing Device, where required.
7. Thermostatic Releasing Device, where required.
8. Guides and all necessary operating parts of rolling or sliding doors.
9. Safety Governors, on vertically descending underbalanced doors.
10. Protection, for sliding doors and around counterweights.
11. Automatic Releases, for power operated doors.

3301.02 Classification of Fire Doors:

(a) **Types.** All fire doors shall be classified as of the following four (4) types:

1. Double Standard Fire Doors.
2. Standard Fire Doors.
3. Sixty (60) Minute Fire-Resistive Doors.
4. Forty-five (45) Minute Fire-Resistive Doors.

(b) **Relative Value.** The relative fire-resistive value of the four (4) types of fire doors shall be deemed to decrease in the order given in paragraph (a) of this Section.

3301.03 Qualification:

(a) **General.** Every required fire door, including the frame, hardware and operating mechanisms, shall be an exact duplicate, except in size, of a model assembly which has been subjected to the Standard Fire Test described in Chapter 32 for the period of time and with the performance under test as hereinafter required. That the door assembly is a true duplicate, except in size, of a duly tested model, shall be evidenced by an attached label issued by the testing laboratory, which label shall attest that such door assembly meets the requirements of this ordinance for the specified type.

(b) **Double Standard Fire Doors.** A double standard fire door shall be an assembly providing a complete standard fire door on each face of a wall.

(c) **Standard Fire Doors.** The test models for all standard fire doors shall have been subjected to the standard fire test for a period of not less than sixty (60) minutes, followed within two (2) minutes by the fire stream test and shall have at least equalled the standards of performance which are hereinafter required.

(d) **Sixty (60) Minute Fire-Resistive Doors.** The test models for all sixty (60) minute fire-resistive doors shall have been subjected to the standard fire test for a period of not less than sixty (60) minutes, followed within two (2) minutes by the fire stream test and shall have at least equalled the standards of performance which are hereinafter required.

(e) **Forty-five (45) Minute Fire-Resistive Doors.** The test models for all forty-five (45) minute fire-resistive doors shall have been subjected to the standard fire test for a period of not less than forty-five (45) minutes, followed within two (2) minutes by the fire stream test and shall have at least equalled the standards of performance which are hereinafter required.

(f) **Standards of Performance Under Test.** All types and kinds of fire doors, qualified or hereafter qualifying as of the several types required by this ordinance, shall be considered to have successfully withstood the required tests only in case of performance under said tests, as follows:

1. The automatic closing device operated promptly as intended, the door closed properly and was tightly held in the closed position. Automatic closing devices may be tested and approved without subjecting the door assembly to the standard fire test.

2. The door and its assembly shall have complied with at least one of the following two requirements:

First. The said fire door and its arrangement has not permitted the passage of fire and flame through or around said door.

Second. The said fire door and its arrangement has not permitted the transmission of heat through the door, so as to cause the temperature on the unexposed side of the fire door to exceed five hundred (500) degrees Fahrenheit in the case of standard fire doors, eight hundred (800) degrees Fahrenheit in the case of sixty (60) minute and forty-five (45) minute fire doors.

3. In order that the fire door and its arrangement comply with the standard fire test and the herein prescribed fire stream test, said fire door must not be materially weakened either in the door structure or in its arrangement on the fire wall after its exposure to the standard fire test and the fire stream test.

(g) **Test Laboratory.** All required tests shall be made by a laboratory or engineering experimental station qualified as required by Chapter 32—FIRE-RESISTIVE STANDARDS.

ARTICLE 3302

Type Requirements

3302.01 General: Except as otherwise provided by this ordinance, the required type of fire door for any doorway shall be determined from the required fire-resistive value of the wall, separation or inclosure in which such doorway occurs in accordance with the provisions of this article.

3302.02 In Standard Fire Division Walls: Every doorway in a required standard fire-division wall shall be protected by double standard fire doors; provided however, that where a vestibule or stair inclosure intervenes between adjacent fire-division areas, each doorway in such vestibule or inclosure shall be protected by a standard fire door.

3302.03 In Four (4) Hour Fire-Resistive Walls:

(a) **Separations Between Occupancies.** Every doorway in a required four (4) hour fire-resistive separation between occupancies, shall be protected by double standard fire doors, except as otherwise provided in Section 3302.02 for doorways in vestibules and stair inclosures.

(b) **Other Required Separations.** Every doorway in a required four (4) hour fire-resistive separating wall, other than a separation between occupancies, shall be protected by standard fire doors.

3302.04 In Three (3) Hour Fire-Resistive Walls: Every doorway in a required three (3) hour fire-resistive separating wall shall be protected by sixty (60) minute fire-resistive doors; provided however, that forty-five (45) minute fire-resistive doors shall be permitted in doorways in thre (3) hour fire-resistive inclosures of vertical shafts.

3302.05 In Two (2) Hour Fire-Resistive Walls: Every doorway in a required two (2) hour fire-resistive wall or inclosure shall be protected by forty-five (45) minute fire-resistive doors.

3302.06 In One (1) Hour Fire-Resistive Walls: Every doorway in a required one (1) hour fire-resistive wall or inclosure shall be protected by forty-five (45) minute fire-resistive doors.

3302.07 In Exterior Walls and Fire Escape Doorways: Doorways affording egress to an exterior emergency exit stairway or fire escape shall be protected by forty-five (45) minute fire-resistive doors.

ARTICLE 3303

Construction of Doors

3303.01 General:

(a) **Materials.** Except as hereinafter provided, all fire doors, except tin-clad and metal-clad doors shall be composed entirely of metal or of metal and a core of non-combustible insulating material; provided however, that combustible or destructible washers may be employed to prevent distortion by expansion and that fusible link releasing devices shall be permitted. Any fire door, except a steel plate door, otherwise meeting all of the requirements of this chapter for construction and performance under test, may have a decorative veneer applied to either or both sides of such approved door and the application of such veener shall not be deemed to affect the approval or classification of such door, for the purposes of this ordinance.

(b) **Construction.** Where frames are used in masonry or concrete walls, eight (8) inches or more in thickness, they shall be secured by anchors built into the walls or bolted through the walls. Frames shall be required in other walls and shall be secured at the floor and through extension members to the ceiling. Sills shall be bedded in cement mortar and shall be anchored to the walls or frames. Where no frames or sills are used, all tracks and binding or supporting hardware shall be secured to the walls by bolts, one 1) of which, at each point of support, shall extend through the wall.

3303.02 Standard Fire Doors:

(a) **Kinds of Doors.** Standard fire doors shall be tin-clad three-ply wood core doors, sheet metal and hollow metal doors, steel plate doors and rolling steel shutters. All such doors, except rolling steel shutters, shall be hinged or arranged to slide or roll, either singly or in pairs.

(b) **Construction.** Where frames or sills are used they shall be secured as required by Section 3303.01, paragraph (b). Where standard fire doors are permitted to be installed without frames, such doors shall lap four (4) inches over the opening at both sides and at the top and means shall be provided to bind the doors firmly to the wall. No glass panels shall be permitted in standard fire doors. Adequate provision, by means of slotted holes or combustible or destructible washers or inserts, shall be made to prevent distortion from unequal expansion. Where slots in sills are used as guides, such slots shall be closed when doors are open and such closing member shall operate with the automatic door closing mechanism. All such mechanism shall be subject to the approval of the Division Fire Marshal in charge of Fire Prevention.

(c) **Hardware and Operating Devices.** Doors held normally in the open position shall be provided with an automatic closing device containing a fusible link and tripping arrangement which at temperature of one hundred eighty (180) degrees Fahrenheit or less, shall actuate the closing mechanism or containing a thermostatic device which shall actuate said mechanism as a result of change in the rate of rise of temperature. Such link shall be exposed within the area of the opening when the door is in the open position. Doors held normally in the closed position shall be provided with door checks which may be set to hold the doors open and containing a one hundred eighty (180) degree Fahrenheit or lower temperature fusible link or rate of rise thermostatic releasing device. Single swinging doors shall be provided with three (3) point latches having not less than a three-fourth (¾) inch engagement with the frames and sills. Doors swinging in pairs shall be provided with astragals. The active door of the pair shall have three (3) point latches and the other door shall have two (2) point latches engaging at top and bottom. An interference device, to prevent the wrong door from closing first, shall be secured to the wall or frame. Vertically descending doors and rolling shutters shall be provided with safety devices for controlling the speed of descent.

3303.03 Sixty (60) Minute Fire-Resistive Doors:

(a) **Kinds of Doors.** Sixty (60) minute fire-resistive doors shall be tin-clad three (3) ply wood core doors, sheet metal, hollow metal and steel plate doors. All such doors shall be hinged either singly or in pairs; provided however, that sliding doors opening either horizontally or vertically shall be permitted in elevator hatchway openings.

(b) **Constrution.** No glass panels shall be permitted in sixty (60) minute fire-resistive doors. Frames and sills, as required under Section 3303.01 shall be provided for all sixty (60) minute fire-resistive doors. Adequate provision shall be made to prevent distortion from unequal expansion, as required for standard fire doors.

(c) **Hardware and Operating Devices.** All sixty (60) minute fire-resistive doors shall be held normally in the closed position and except in elevator hatchway openings, shall be provided with a device which may be set to hold the doors open and containing a one hundred eighty (180) degree Fehrenheit or lower temperature fusible link or rate of rise thermostatic releasing device. Latches for single swinging doors and latches and interference devices for doors swinging in pairs shall be the same as required for standard fire doors.

3303.04 Forty-five (45) Minute Fire-Resistive Doors:

(a) **Kinds of Doors.** Forty-five (45) minute fire-resistive doors shall be metal clad two ply wood core doors, sheet metal, hollow metal and steel plate doors. All such doors shall be hinged, either singly or in pairs; provided however, that sliding doors opening horizontally or vertically shall be permitted in elevator hatchway openings.

(b) **Wired Glass Panels.** Panels of wired glass, not exceeding one hundred forty-four (144) square inches in individual and total area shall be permitted in any forty-five (45) minute fire-resistive door used in a doorway in a vertical shaft inclosure.

(c) **Hardware and Operating Devices.** All forty-five (45) minute fire-resistive doors shall be held normally in the closed position, and except in elevator hatchway openings, shall be provided with a device or devices which may be set to hold the doors open and containing a one hundred eighty (180) degree Fahrenheit or lower temperature fusible link or rate of rise thermostatic releasing device. Doors swinging in pairs shall be equipped with astragals and interference device. Single point latches shall be permitted for single swinging doors and for the active door of a pair of doors and two (2) point latches shall be required for the inactive door of a pair.

3303.05 Steel Plate Doors: Steel plate doors, qualifying as standard fire doors or sixty (60) minute fire-resistive doors shall be constructed of plates not thinner than No. 12 U. S. Gauge. Forty-five (45) minute fire-resistive steel plate doors shall be constructed of plates of not less thickness than No. 14 U. S. Gauge.

3303.06 Protection of Other Openings: Openings in required fire-division walls and required four (4) hour fire-resistive walls or construction, for the passage of ventilating ducts or inclosed conveyors, shall be protected by closing members at least equal in fire-resistive value to the type of fire door required for any doorway in such location; provided however, that in lieu of such a closing member, a sprinkler head connected with a standard system of automatic sprinklers, may be installed inside such duct or inclosure at such wall or separating construction.

3303.07 Thermostatic Releasing Device: Wherever in this ordinance, the words "rate of rise thermostatic releasing device" or the words "automatic heat actuated controlling device" or the words "heat actuated device" or the words "thermostatic releasing device" are used, said words are hereby defined, construed to mean and require a device which is constructed of such material and so arranged that said device is placed in operation as a result of a difference in the rate of rise of the temperature rather than as a result of a change in the temperature.

ARTICLE 3304

Size Limitations

3304.01 General: All types and kinds of fire doors, required by this ordinance shall be limited in size to the maximum dimensions and areas permitted by Table 3304-SOFD.

TABLE 3304—SOFD

Kind of Door	Area (Sq. Ft.)	Maximum Height of Opening (Feet)	Width of Opening (Feet) Maximum
Tin Clad, three ply Wood Core, Counterbalanced..........		12	8
Tin Clad, three ply Wood Core, Sliding..................	120	12	12
Tin Clad, three ply Wood Core, Pairs Swinging.........		12	10
Tin Clad, three ply Wood Core, Single Swinging.........		12	6
Metal Clad, two ply Wood Core, Sliding.................		10	10
Metal Clad, two ply Wood Core, Counterbalanced.........		10	8
Metal Clad, two ply Wood Core, Pairs Swinging.........		10	10
Metal Clad, two ply Wood Core, Single Swinging.........		10	6
Sheet Metal, Pairs Sliding...........................	120	12	12
Sheet Metal, Single Sliding..........................	80	10	10
Sheet Metal, Pairs Swinging..........................	80	10	10
Sheet Metal, Single Swinging.........................		10	6
Hollow Metal, Sliding................................		8	8
Hollow Metal, Pairs Swinging.........................		8	8
Hollow Metal, Single Swinging........................		8	4
Steel Plate	40	10	4
Rolling Steel Shutters...............................	80	12	12

ARTICLE 3305

Fire Doors as Means of Exit

3305.01 General: Sliding steel plate doors, ertically descending doors of any type, and olling steel shutters are hereby prohibited or use in any opening, which is a required eans of exit. All doors used in connection rith required means of exit shall swing in he direction of egress except as horizontally liding doors may be permitted under other ections of this ordinance. Doors from ividual rooms to hallways or corridors shall wing in the direction of egress where such ooms are used for the purposes of public ssembly or where such rooms are occupied by fifty (50) or more persons or where such rooms contain any hazardous occupancy. All doors used in connection with exits shall be so arranged as to be readily opened at all times from the side from which egress is made without the use of a key or any separate operating device.

ARTICLE 3306

Protection of Doors

3306.01 General: All required fire doors shall be protected against mechanical injury or obstruction which would interfere with the operation thereof in any emergency.

3306.02 Guards:

(a) **Sliding Doors.** Sliding doors shall be protected by a substantial screen, railing or other guard which will prevent the contact of furniture, stock or other material with such doors. All counterweights, guides and tracks shall be provided with inclosures or guards to prevent interference with or obstruction thereof. Slots in sills or floors shall have automatic closing members which shall form substantial trucking thresholds when doors are open. All such mechanism shall be subiect to the approval of the Division Fire Marshal in Charge of Fire Prevention.

(b) **Swinging Doors.** Swinging doors shall have floor stops or other devices to prevent such doors from swinging through an arc of more than ninety (90) degrees.

ARTICLE 3307

Other Types and Kinds of Doors

3307.01 General: Fire doors of types and kinds other than those required by any described in this chapter, shall be permitted as acceptable substitutes for the types and kinds required by this chapter only if such substitutes shall have been proven or shall be proved by the standard fire test to be at least equal in fire-resistive value and performance to the required types and kinds.

3307.02 Non-Combustible Doors: Non-combustible doors, where required or permitted by this ordinance, shall be composed or constructed entirely of non-combustible materials or with combustible cores inclosed in metal or asbestos or in metal and asbestos. A finish or veneer of combustible material not more than one-fourth (¼) inch thick shall be permitted, if applied to a non-combustible core or if applied to a metal or asbestos inclosure surrounding a combustible core. Frames and sills for such doors shall be entirely of non-combustible materials or shall be of the same construction as such doors.

CHAPTER 34
Structural Provisions

ARTICLE 3401

Loads

3401.01 Live Load for Each Occupancy of Building—Roof Live Load:

(a) **Minimum Live Load.** All buildings and structures shall be designed for the safe support of live loads in addition to all dead loads as specified in Section 3401.05 and wind loads as specified in Sections 3401.06 to 3401.08 inclusive. Minimum live loads for which buildings and structures may be designed for each type of occupancy are given in Table 3401.01 (a) with permissible reductions as specified in Sections 3401.02 to 3401.04 inclusive. These requirements for live loads are to be applied to every portion of each floor according to occupancy.

Table 3401.01 (a)

Minimum Live Loads to Be Used in Design

Building Classification	Occupancy	Minimum Live Loads (Pounds per Sq. Ft.)
Single Dwellings	All Purpose	40
Multiple Dwellings	Habitable Rooms and Private Utility Rooms.	40
	Public Assembly Rooms, Public Corridors and Stairways	100
Hospitals	Patients' Rooms or Wards	40
	Public Assembly Rooms, Corridors and Stairways	100
Business Units	Office Space, Class Rooms, Bowling, Billiard and Pool Rooms, except Spaces for Spectators	50
	Exhibition Rooms, Public Assembly Rooms with Fixed Seats	75
	Auto Truck Storage Space, Bleachers and Grandstands. Bus Terminal Space, Gymnasiums. Loft Space, Manufacturing Space, Power Plant Space, Public Assembly Space without Fixed Seats, Public Corridors and Stairways, Stores, Retail and Wholesale	100
Hazardous Use Units	All Purpose	100
Garages	First and Second Floors	100
	Floors above the Second	50*
Theatres	Private Offices	50
	Spaces with Fixed Seats	75
	Other Public Rooms and Public Spaces	100
Open Air Assembly Units	All Purpose	100
Public Assembly Units	Spaces with Fixed Seats	75
	Other Public Rooms and Public Spaces	100
Churches	Class Rooms, Choir and Vestry Rooms	50
	Spaces with Fixed Seats	75
	Other Public Rooms and Public Spaces	100
Other Buildings and Structures	All Purpose	100

*These floors are to be placarded as follows: "This floor to be used for the storage of passenger automobiles only."

(b) **Special Loads.** Any space to be occupied by tanks, special equipment, or by driveways, private sidewalks, loaded truck storage space, tracks, shall be designed for the actual weight of the superimposed loads and for impact, where this exceeds the minimum requirements, but the space occupied by such equipment needs not be figured for any other live load.

(c) **Safes.** Floors of offices, or other space, in which safes are used shall be designed to accommodate a concentrated load of two thousand (2000) pounds on any selected area three (3) by three 3) feet, but in this calculation no other live load shall be considered on the same members at the same time.

(cc) **Concentrated Loads for Passenger Auto Storing Space.** Floors of passenger auto storage space shall be designed for a concentrated load of three thousand (3000) pounds on any area four (4) feet by four (4) feet square, but in this calculation, no other live load shall be considered on the same members at the same time.

(d) **Roof Loads.** Roofs shall be designed for a live load, acting normal to the roof surface, of twenty-five (25) pounds per square foot, including wind load. No reduction shall be made in roof loads, as specified in Section 3402.03 or on account of the type of c , or of ratio of dead to live load.onstruction

3401.02 Reduction of Live Load on Slabs and Joists:

(a) **General.** Every slab, or joist shall be designed for the full live load specified in Section 3401.01 except as provided in Paragraph (b) of this Section.

(b) **Dead Load Greater Than Live Load.** When the dead load exceeds the live load, the live loads specified in Section 3401.01 may be reduced to a load obtained by multiplying by the ratio of the specified live load to the dead load, but the live load so computed shall in no case be less than sixty-six and two-thirds (66⅔) per cent of live load as specified in Section 3401.01.

(c) **Floor Placards.** Where floor load placards are required by this ordinance, floors designed in accordance with this Chapter shall be placarded for the full live load specified and not for the reduced load.

3401.03 Reduction of Live Loads on Beams, Girders and Trusses: Beams, girders and trusses shall be designed for a live load not less than as provided in Section 3401.01 and 3401.02, except that a further reduction of the live load on floors shall be permitted on the basis of tributary floor area, but no reduction shall be permitted on the live load on tributary roof area, in accordance with the following schedule:

Tributary Floor Area	Allowable Reduction (Per Cent)
Not more than 100 square feet..	4
More than 100 square feet and not more than 200 square feet	5
More than 200 square feet and not more than 300 square feet	10
More than 300 square feet.......	15

3401.04 Reduction of Live Load on Columns, Piers, Walls and Foundations: The total live load carried by columns, piers, walls and foundations, may be reduced according to the following schedule, the reductions being based on the live loads either as specified in Section 3401.01 or as computed in accordance with Section 3401.02 applied to the entire tributary floor and roof areas:

	Allowable Reduction	
Members Carrying	For Storage and Manufacturing Units (Per Cent)	For All Other Space (Per Cent)
The roof	0	0
1 floor and roof.................................	15.0	15
2 floors and roof................................	17.5	20
3 floors and roof................................	20.0	25
4 floors and roof................................	22.5	30
5 floors and roof................................	25.0	35
6 floors and roof................................	25.0	40
7 floors and roof................................	25.0	45
8 or more floors and roof........................	25.0	50

3401.05 Dead Loads: In addition to live loads and to wind loads to be provided for, each portion of every structure shall be designed for the entire dead weight of the structure superimposed upon it.

3401.06 Wind Load on Buildings and Structures:

(a) **General.** The inclosing walls and surfaces of every building or structure shall be designed to resist a horizontal wind pressure acting in any direction, and on either side of the inclosing wall, or surface in the amounts specified in Paragraphs (b) and (c) of this Section in addition to all other loads coming upon them.

(b) **Walls, Joists and Studs.** The wind load on the exterior structural walls, slabs, joists and studs shall be taken as not less than twenty-five (25) pounds per square foot, up to an elevation of two hundred seventy-five (275) feet above the sidewalk, and as not less than thirty-five (35) pounds per square foot above that elevation. In buildings of skeleton construction, panel or other walls supported on the structural frame, shall be designed to carry the wind loads specified in this paragraph and shall be securely attached to the frame so as to insure complete structural stability.

(c) **Beams, Columns and Other Members.** The wind load on beams, girders, trusses, posts, columns and piers supporting the exterior walls, slabs, joists and studs, the wind load acting on the structural framework and the wind load considered as an overturning force, shall be taken as not less than the values given in the following table:

WIND LOADS

For Structural Members Supporting a Tributary Surface	On a surface less than 275 feet above grade (Pounds per Sq. Ft.)	On a surface more than 275 feet above grade (Pounds per Sq. Ft.)
Not more than 100 square feet....................	25	35
More than 100 square feet and not more than 200 square feet	25	32½
More than 200 square feet..........................	25	30
For structural framework..........................	25	30
For overturning forces............................	25	30

(d) **Walls and Partitions as Resisting Members.** Walls and partitions shall not be taken into consideration as resisting wind loads, except that walls, or equivalent bracing continuous from the foundations to the highest level where they are considered effective, may be used in the design; provided however, that the effect of openings in such walls is taken into account; and that such walls are located so as to be effective for the purpose of resisting wind.

(e) **Sloping Roofs.** For the design of sloping roofs, the provisions of Section 3401.01, paragraph (d) shall govern.

(f) **Overturning Moment.** In no case shall the overturning moment, due to wind pressure, exceed seventy-five (75) per cent of the moment of stability of the building due to the dead loads only. Dead loads may include properly anchored foundations.

3401.07 Wind Loads on Special Structures:

(a) **Signs, Tanks, Towers and Similar Structures.** All structures, such as signs, tanks, tank towers, radio towers and similar structures, shall be designed so as to safely resist a wind pressure of thirty (30) pounds per square foot, on the vertical projection of every surface exposed to the action of the wind. This does not apply to ordinary masonry inclosed towers for which wind load is given by Section 3401.06.

(b) **Flag Poles.** All flagpoles shall be designed to resist a wind pressure of one and one-half (1½) pounds per square foot of flag area, applied at the top of the pole and an additional pressure of fifty (50) pounds per square foot, on the vertical projection of the pole and shall be so braced, supported or stayed as to withstand this wind pressure.

3401.08 Wind Loads on Chimneys and Smoke Stacks: All isolated chimneys and smoke stacks and all portions of smoke stacks projecting above the roofs of buildings shall be designed to safely resist wind loads calculated as follows:

On a square, or rectangular chimney, thirty (30) pounds per square foot on one (1) side of the square, or on the greater side of the rectangle;

On an octagonal chimney, twenty-five (25) pounds per square foot, on the projected short diameter;

On circular chimneys, or metal smoke stacks, twenty (20) pounds per square foot on the projected diameter.

Recommended for repeal. Superseded by 4102.02.

ARTICLE 3402
Materials and Stresses

3402.01 Standard Specifications for Materials: Except as hereinafter specified for particular materials, every material permitted to be used in buildings or structures in the City of Chicago, shall meet the standard specifications for that material as prepared by the American Society for Testing Materials and as adopted by that Society as a Standard, or as a Tentative Standard, on the date of the enactment of this ordinance. Where such standards require acceptance tests for the determination of the quality and properties of the material proper evidence of the making of such acceptance tests shall be submitted to the Commissioner of Buildings upon request from him. Where in this ordinance some other specification, or standard of quality, is set up for any particular material, the same shall take precedence over the said standard of the American Society for Testing Materials, but any standard requirements of the American Society for Testing Materials not in conflict with the same shall remain in full force and effect.

3402.02 Stresses Due to Live, Dead and Wind Loads:

(a) **Combined Loads.** All structural members in buildings shall be of the proper dimensions, and with the proper reinforcement, if of reinforced concrete, to carry the superimposed loads at stresses that do not exceed the values permitted in this ordinance. Stresses due to live and dead loads combined shall not exceed the allowable values as stated in the various sections. Stresses produced by live, dead and wind loads combined, may exceed the allowable values by fifty (50) per cent, except for rivets or welds, in wind connections, which may exceed the allowable value by thirty-three and one-third (33⅓) per cent; provided however, that the resulting sections shall be not less than those required to carry the combined live and dead loads without considering the wind load.

3402.03 Moments and Shears, General and Special Cases:

(a) **General.** Stresses in structural members shall be computed in general, from the usual mechanics of static structures, taking into account the actual conditions of restraint at supports and connections. For members carrying walls without openings, the load causing flexure may be considered as that within a triangle whose base is the span of the member and with base angles of sixty (60) degrees.

(b) **Complex Structures.** For the more complex structures, moments shall be taken as given elsewhere in this ordinance for particular conditions. The moments specified for reinforced concrete structures shall apply to any structure with rigidly constructed joints.

ARTICLE 3403
Special Design and Construction Provisions

3403.01 Working Drawings: All working drawings which are used in the construction of buildings or structures shall be either made or checked by the architect, or engineer responsible for the design, who shall certify thereon that such drawings conform to the approved design drawings on which the permit is based. This certification shall cover the size of wood or concrete members; the size and weight of structural steel members and the size, length and bending of concrete reinforcement, together with the adequacy of all connections.

3403.02 Design of Buildings Resting on Bedrock: Where structures rest on foundations carried to rock, so that settlement is precluded, the structural framework may be designed for smaller moments than those specified herein; provided however, that a complete analysis of forces and stresses, including those in the foundations, shows that the moments used in design of each member are the greatest moments that may exist in that member under the required design loads, distributed in the most unfavorable manner that may reasonably occur in the use of the structure and that the maximum stresses allowed in this chapter will not be exceeded. The analysis required above shall take account of wind loads as well as of static loads and of all combinations of loads that may exist.

3403.03 Design of Skylights and Monitors: Skylights, monitors and similar roof constructions shall be designed to meet the same structural provisions as required for the roof of the building of which they form a part.

3403.04 Sheeting, Shoring, Underpinning and Retaining Walls: Where the construction or alteration of a building requires that an existing construction or the street or earth adjoining the new building, be temporarily or permanently supported, the sheeting, shoring, underpinning or retaining wall, shall be designed for the loads they have to support using materials of the quality specified in this ordinance for new buildings. For temporary construction, the stresses specified in this ordinance may be increased twenty-five (25) per cent greater than the stresses permitted in new buildings. If such sheeting, shoring or underpinning is to remain in place over one (1) year, no increase in design stresses shall be taken. Sheeting, shoring or underpinning designed for increased stresses and held in place over one (1) year shall be strengthened at the end of the year to reduce all stresses to the values permitted for new buildings. Plans prepared by the architect or engineer of the project together with the computations for and details of such sheeting, shoring, underpinning or retaining walls shall, upon demand by the Commissioner of Buildings, be presented for his approval.

CHAPTER 35
Foundations

ARTICLE 3501
General

3501.01 Requirements: Every building, or structure shall rest upon one (1) or more of the types of foundations included in Items 1 to 7, inclusive, of this section.

Item 1. Footings on Soil.
Item 2. Wood Piles.
Item 3. Concrete or Composite Piles.
Item 4. Metal Pipe Piles.
Item 5. Foundation Piers or Caissons.
Item 6. Foundation Columns.
Item 7. Steel Beams Used as Piling.

3501.02 Encroaching Foundations: Foundations may project into, or encroach upon public streets, alleys or other public thoroughfares, except as otherwise specifically provided.

3501.03 Protection of Footings:

(a) **Exposure.** All footings exposed to frost shall be carried down at least four (4) feet below the adjoining ground surface unless bearing upon solid rock; except that a reinforced concrete slab foundation extending over the entire area below a one (1) story building, which building does not exceed four hundred (400) square feet in area, shall be permitted at a lesser depth below the adjoining ground surface.

(b) **Freezing Weather.** Footings shall not be placed in freezing weather unless adequate precautions are taken against frost action. Footings shall not be placed upon frozen soil.

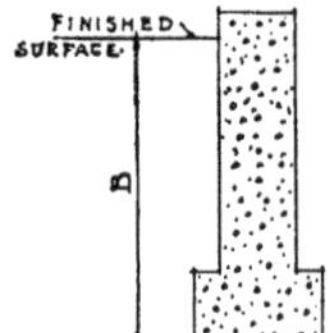

FOOTINGS.
Section 3501.03.

(B) Shall in all cases extend 4'0" below finished grade at building, unless footings rest on bed rock.

ARTICLE 3502
Footings and Foundations

3502.01 Materials and Design: Footings and foundations shall be built of masonry or concrete, designed in accordance with Chapter 38, or of reinforced concrete designed in accordance with Chapter 39, or of structural steel or other metal designed in accordance with Chapter 40 and encased in concrete with at least three (3) inches of covering over all steel. All materials used shall comply with the requirements for such materials given in Chapters 38, 39 and 40 of this ordinance.

3502.02 Loads Permitted on Various Soils:

(a) **Unit Pressure Allowable.** All foundation plans shall show the maximum pressure in pounds per square foot transmitted to the soil by footings and foundations. The pressure on various soils shall be taken in accordance with the following table:

Table 3502.02(a)
Loads Permitted on Various Soils

1. On filled ground or loam, not over five hundred (500) pounds per square foot.

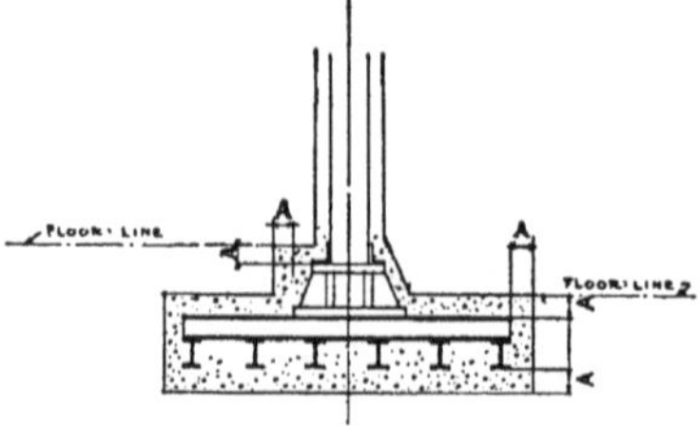

Section 3502.01.

(A) Steel and iron rails and beams to be imbedded inconcrete, extending not less than 4 inches beyond metal.

2. On sand or clay having an admixture of loam, and containing not over fourteen (14) per cent of moisture by weight, not over fifteen hundred (1500) pounds per square foot.
3. On dry yellow clay not over three thousand five hundred (3,500) pounds per square foot.
4. On soft blue clay containing not over fourteen (14) per cent of moisture by weight, not over three thousand (3,000) pounds per square foot.
5. On stiff blue clay containing not over fourteen (14) per cent of moisture by weight, not over forty-five hundred (4500) pounds per square foot.
6. On stiff clay with an admixture of gravel and stone, not over four thousand (4,000) pounds per square foot.
7. On sand and clay mixed, not over three thousand (3,000) pounds per square foot.
8. On coarse sand, or fine gravel, not over six thousand (6,000) pounds per square foot.
9. On fine wet sand, not over three thousand (3,000) pounds per square foot, unless confined by solid walls of masonry or concrete, or by sheathing extending the full depth of the layer and designed to resist the lateral pressure, in which case six thousand (6,000) pounds per square foot may be used.
10. On coarse gravel, not over ten thousand (10,000) pounds per square foot.
11. On hardpan, not ofer twelve thousand (12,000) pounds per square foot; except where hardpan extends under the entire area of the structure in which case not over fifteen thousand (15,000) pounds per square foot shall be allowed.
12. On solid rock, not over one hundred (100) tons per square foot.

When the classification of soft soil or stiff soil is in question it shall be determined from the average results of three (3) tests in the air at seventy (70) degrees Fahrenheit, in a Hubbard Stability Testing Machine, as described in the technical papers included in the proceedings of the A.S.T.M., volume 25. Elastic soil showing a load of less than forty (40) pounds shall be considered soft soil, and soil showing a load of forty (40) pounds or more shall be considered stiff soil.

(b) **Varying Soils.** When soils decrease in carrying capacity below the underside of the footing, the bearing value selected shall be that of the weakest soil encountered within ten (10) feet from the bottom of the footing, adjusted for the depth below the footing, using an angle of spread of not more than thirty (30) degrees from the vertical.

(c) **Hardpan Defined.** Hardpan is defined as a very dry mixture of sand, pebbles and clay, naturally cemented, with or without an admixture of boulders, difficult to remove by picking, and containing no pliable material.

(d) **Tests.** The values given in Table 3502.02(a) are for average conditions of soil. In case there is uncertainty as to the soil being of average bearing value, or as to the classification of any given soil, the Commissioner of Buildings shall permit or may require a test to be made, described as follows:

Soil bearing test shall be made by loading no less than two (2) square feet of soil at the elevation of the bottom of the proposed footing. The initial test load shall be not more than fifty (50) per cent of the desired working load on the soil, and each further increment shall not exceed five hundred (500) pounds per square foot of soil under test. A continuous record of the settlement shall be made to determine the point at which the rate of settlement increases in greater proportion than the increment of loading. The point where this rapid increase of settlement takes place shall be called the "yield point." The working load shall be taken as one-third (⅓) of the load at the above described yield point. The ultimate settlement shall be taken as twice the settlement recorded for the load taken as the working load. A drill hole shall be made in the bottom of the test pit, no less than fifteen (15) feet in depth below the elevation of the test. Proper samples shall be taken at frequent intervals to determine the character of the strata for this depth, and the permissible working load under any and all footings shall be determined by a registered engineer or architect and shall be approved by the Commissioner of Buildings.

3502.03 Footings at Different Levels: Except in the case of bearing on rock, the difference in elevation of the bottoms of any two (2) footings shall be such that a line drawn between the lower adjacent edges shall not incline at an angle more than the angle of repose of the soil, or greater than forty-five (45) degrees with the horizontal, unless provisions are made by means of retaining walls, or otherwise, adequately to restrain the soil.

ARTICLE 3503

Wood, Concrete and Composite Piles

3503.01 Jetting: It shall be permitted to jet piles through sand strata. Where jetted piles are seated in a sand stratum, at least one (1) drill hole shall be sunk in the area, at least ten (10) feet below the bottom of the piles. If the sand stratum continues to the bottom of the drill hole or is replaced by gravel, stiff clay or its equivalent, no impact driving shall be required. A load test of piles which are jetted to a bearing shall be made at least every seventy (70) feet in each direction and not less than two (2) such tests shall be made in any case.

3503.02 Materials, Design, Loading:

(a) **Concrete Caps.**

Item 1. Concrete. Concrete having a minimum strength of two thousand (2,000) pounds per square inch shall be used for all pile caps and shall contain not less than six (6) bags of cement per cubic yard of concrete.

Item 2. Reinforcement. Reinforcing or steel in caps shall be supported at least three (3) inches above the tops of piles and reinforcing shall have a covering of at least three (3) inches of concrete.

Item 3. Design. Concrete caps shall be designed in accordance with the provisions of Chapter 39—REINFORCED CONCRETE CONSTRUCTION.

(b) **Piles.**

Item 1. Wood Piles. Wood piles shall be single sticks cut from sound, live trees and shall be free from defects, such as injurious shakes, unsound or loose knots, or decay, which materially impair their strength or durability. Piles must be butt-cut above the ground swell. A straight line drawn from the center of the butt to the center of the top shall be within the body of the pile at all points. Piles shall have an approximately uniform taper from butt to tip, and shall be not less than six (6) inches in diameter at the small end, and twelve (12) inches in diameter at the cut-off.

Item 2. Concrete Piles. Concrete and reinforcing used in concrete, or composite piles, shall comply with the requirements of Chapter 39—REINFORCED CONCRETE CONSTRUCTION. It is further required that such concrete shall be of plastic concrete consistency and shall have a minimum compressive strength of two thousand (2,000) pounds per square inch and shall contain not less than six (6) bags of cement per cubic yard of concrete. The diameter or least lateral dimension of any concrete pile shall be not less than eight (8) inches at the point and not less than twelve (12) inches at the finished top. The length shall not exceed fifty (50) times the average diameter. Precast concrete piles shall be designed and reinforced to permit handling and driving without injury. The amount of longitudinal reinforcing employed shall be not less than two (2) per cent, nor more than four (4) per cent. Hoops or ties shall be not less than one-fourth (¼) inch in least dimension and spaced not farther apart than twelve (12) inches, except that the spacing of hoops or ties shall not exceed three (3) inches within a distance of three (3) feet from either end of the pile.

Item 3. Composite Piles. Composite piles of wood and concrete shall meet the following requirements: The wood portion shall meet all the requirements of this ordinance for wood piles. The diameter of the concrete portion shall be not less than the top diameter of the wood pile. The concrete pile and the splices between the wood section and the concrete section shall be designed and constructed so that the concrete section and the joint will be at least forty (40) per cent as strong in cross bending as a wood pile of twelve (12) inches diameter. A look shall be provided capable of resisting a direct tensile force of fifteen (15) tons between the wood and the concrete sections, so that the upper section shall not lift and the two (2) sections shall be maintained in alignment. The metal shell shall be left in place and shall be of sufficient strength to resist distortion and collapse.

(c) **Maximum Unit Stress on Concrete and Composite Piles.** The maximum unit stress on any concrete or composite pile at a section taken six (6) feet below the surface of the ground in immediate contact with the pile, shall not exceed one-fifth (1/5) of the ultimate strength of the concrete.

(d) **Allowable Design Load.** The allowable design load on a wood, concrete or composite pile laterally supported for its entire length shall be determined by the driving formula given in paragraph (e) of this section, or by load test, but shall not exceed fifty thousand (50,000) pounds on any wood or composite pile nor a value based on the maximum unit stress or a load of sixty thousand (60,000) pounds on any concrete pile.

(e) **Driving Formulae.**

Item 1. Wood Piles.

For Drop Hammer, $$P = \frac{2\ W\ h}{S + 1.0\ \text{inch}}$$

For Single-Acting Hammer $$P = \frac{2\ Wh}{S + 0.1\ \text{inch}}$$

For Double-Acting Steam Hammer $$P = \frac{2h\ (Am+W)}{S + 0.1\ \text{inch}}$$

which formulae the notation is as follows:

P = allowable design load in pounds.
S = average set in inches for last six (6) inches of driving.
h = fall or stroke in feet.
W = weight of hammer, or moving part in pounds.
A = area of piston in square inches.
m = mean effective steam pressure in pounds per square inch and assumed at eight-five (85).

A follower shall not be used in determining allowable loads. Where followers are used in driving, load tests shall be made in accordance with Paragraph (f) of this section, or a longer pile driven to the same depth without follower, shall determine the load. Such longer pile shall be driven at least every thirty (30) feet in each direction. In determining allowable load from driving formula, the length of the pile in contact with the earth above the final cut-off elevation shall not exceed five (5) feet.

Item 2. Concrete and Composite Piles.

For Single-Acting or Double-Acting Steam Hammer,

$$P = \frac{2E}{S\left\{1 + \frac{0.3\ w}{W}\right\}}$$

in which formula the notation is as follows:

P = allowable design load in pounds.

E = energy in foot-pounds = Wh for a single-acting hammer.
= h (Am + W) for a double-acting hammer.

h = stroke in feet.

A = area of piston in square inches.

m = mean effective steam pressure in pounds per square inch and assumed at eighty-five (85).

S = average set in inches for last six (6) inches of driving, for a pile driven without a follower or for a pile driven with an all-steel follower, so designed that the strength in cross-bending of the joint between the follower and the pile shall be not less than the strength of the pile itself.

w = weight of pile, or weight of pile and follower, or weight of steel core for piles moulded in place in pounds.

W = weight of moving parts in pounds.

In no case shall any precast pile be driven with the value for E less than three thousand five hundred (3,500) foot-pounds, per blow, for each cubic yard of concrete in the precast pile and in no case shall E be less than six thousand (6,000) foot-pounds.

(f) **Load Test.** When a load greater than that permitted by the driving formula is desired and wherever a load test is required by this ordinance, the allowable design load on the pile shall be determined as follows:

1. A test shall be made at least every seventy (70) feet in each direction, but not less than two (2) tests shall be made in any case.
2. The pile to be tested shall be loaded with at least twice the proposed allowable design load.
3. The settlement shall be measured daily until a period of twenty-four (24) hours shows no additional settlement.
4. One-half (½) of the test load shall be the allowable design load, if the test shows no settlement for twenty-four (24) hours and the total settlement has not exceeded one one-hundredth (1/100) inch multiplied by the test load in tons; otherwise the test shall be considered to have failed and new tests shall be made to determine the allowable load on the pile.

Tests on piles moulded in place shall not be started until ten (10) days or more after the piles are cast unless a high early strength cement, which meets the approval of the Department of Buildings, has been used, in which case the Department of Buildings will determine the length of setting time before test. Wood piles and precast concrete piles that have been cast and set before driving may be tested as soon as practicable after driving.

(g) **Piles in Unstable Ground.** The allowable load on any pile extending through a layer of quicksand, or other unstable soil, and which is not supported laterally throughout its length, shall not exceed the value determined as for a fully supported pile, nor a value determined in accordance with the provisions for strength of columns, using the unsupported height of the pile as the length in the formula prescribed in this ordinance.

(h) **Spacing.** The spacing of piles in any direction shall be not less on centers than two (2) times the diameter at the cut-off, nor less than two (2) feet, six (6) inches.

(i) **Column Footings.** There shall be not less than three (3) piles in a group under a single column.

(j) **Wall Footings.** Piles for wall foundations must be driven either in two (2) or more continuous lines, not less than one (1) foot three (3) inches apart on pile centers, or in groups of not less than two (2) piles to a group with spacing as required in paragraph (h). Piles driven in line shall be spaced not more than ten (10) feet on centers in each line.

(k) **Cut-Off and Caps.** All wood piles shall be cut off to a level not less than one (1) foot below city datum; provided however, that when it appears that the ground water is different from city datum, it shall be permissible to cut off piles one (1) foot below the ground water level, as determined by any architect or engineer responsible for the design of the building, subject to the approval of the Commissioner of Buildings. Pile groups shall be covered with concrete capping, which shall extend six (6) inches below top of pile, and then shall project at least six (6) inches beyond the outer face of the outside piles. All wood pile heads shall be cut off below broomed sections.

3503.03 Construction Provisions:

(a) **Piles Moulded in Place.** Concrete piles moulded in place shall be so made and placed as to insure the exclusion of foreign matter and secure a perfect, full sized shape. The shell used in forming the shaft shall remain in place and shall be of adequate strength to resist collapse, under the conditions of its use.

(b) **Precast Piles.** All precast concrete piles shall be protected against damage in driving by the use of a cap of approved design and when driven to rock or hardpan shall be provided with a metal shoe having ample bearing surface. If a cushion is used in driving, the average set for the last six (6) inches of driving shall be determined without the use of the cushion.

ARTICLE 3504

Metal Pipe Piles

3504.01 Materials, Design, Loading:

(a) **Types.** Metal pipe piles may be of either the closed end, or the open end type, driven or jacked into place and shall be filled with concrete.

(b) **Metal Thickness.** The metal pipe shall have a wall thickness of not less than five-sixteenths ($\frac{5}{16}$) inch and an outside diameter of not less than ten (10) inches. The outer one-eighth (⅛) inch of metal shall not be considered in the design.

(c) **Concrete Strength.** The ultimate strength of the concrete, by test at twenty-eight (28) days, or at the age at which the concrete is to be subject to load, if less than twenty-eight (28) days, shall show a minimum value of two thousand (2,000) pounds per square inch, and shall contain not less than six (6) bags of cement per cubic yard of concrete.

(d) **Tests and Loads.** For metal pipe piles the allowable load per pile shall not exceed that computed in accordance with the requirements of this ordinance for reinforced concrete columns, and shall not exceed the values determined by test, conducted as follows:

1. Two (2) tests shall be made on each job of two hundred (200) piles, or less, and one (1) additional test for each additional one hundred (100) piles, or fraction thereof, in the foundation.
2. The test load on a driven steel pipe pile shall be maintained until there is no settlement for a period of twenty-four (24) hours. One-half (½) of the test load shall be allowed for the carrying load, if the test shows no settlement for twenty-four (24) hours and the total settlement does not exceed one one-hundredth (1/100) inch, multiplied by the test load in tons.

3. The test load on a jacked-down steel pipe shall be measured by a pressure gauge on the jack and shall be maintained until there is no settlement for a period of twenty-four (24) hours. One-half (½) of the test load shall be allowed for the carrying load, if the test shows no settlement for twenty-four (24) hours and the total settlement has not exceeded one one-hundredth (1/100) inch multiplied by the test load in tons.

(e) **Length and Diameter.** The length of a steel pipe pile shall not exceed sixty (60) times the outside diameter of the pipe.

(f) **Piles in Unstable Ground.** The allowable design load on any steel pipe pile extending through a layer of quicksand, or other unstable soil and which is not supported laterally throughout its length, shall not exceed the value determined as for a fully supported pile according to paragraph (d) of this section, nor a value determined in accordance with the requirements of this ordinance for reinforced concrete columns, using the unsupported height of the pile as the length of the column.

(g) **Caps.** The design load on the cap shall be transferred to the pipe pile by one of the following methods:

1. By bond on the exterior surface of the pipe inclosed within the cap, at a bond unit stress not exceeding four one-hundredths (0.04) of the ultimate compressive strength of the concrete.
2. By compression on the concrete and bond on the reinforcing bars embedded in the pile and in the cap, computed in accordance with the provisions of Chapter 39—REINFORCED CONCRETE CONSTRUCTION.
3. By a bearing plate bearing on the milled end of the pipe and grouted solidly to the concrete filling of the pile.
4. By a combination of (1) and (2) above.

(h) **Splices.** Splices may be used for pipe piles; provided however, that the splices used between sections of the pipe are adequate to insure alignment and to resist both the direct stress and the bending stress due to loading.

(i) **Alignment.** All pipe piles shall be driven so that a vertical line from the center of the top shall not come closer to the outside shell at any point than one-third (⅓) the diameter of the shell.

(j) **Spacing.** The spacing of pipe piles of the closed end type on centers in any direction shall be not less than two and one-half (2½) times the diameter of the pile, nor less than two (2) feet six (6) inches.

(k) **Cleaning.** After the steel pipe shell is driven, all water and foreign matter shall be removed from the shell and the space within the shell shall be kept free from water and foreign matter.

ARTICLE 3505

Foundation Piers or Caissons

3505.01 Materials and Stresses: Materials for concrete provisions for mixing and placing, except as hereinafter provided, and stresses permitted in design, shall be as given in Chapter 39—REINFORCED CONCRETE CONSTRUCTION; provided however, that the minimum compressive unit strength shall be not less than three thousand (3,000) pounds per square inch.

3505.02 Design Provisions:

(a) **Design Loads.** The load used in determining the areas of the pier and of the belled bottom shall be the load supported at the top of the pier determined in accordance with the provisions of Chapter 34—GENERAL STRUCTURAL PROVISIONS.

(b) **Belled Bottoms.** Piers on hardpan may be belled out; provided however, that the inclination of bell is not less than sixty (60) degrees with the horizontal and that there is a non-belled base with a minimum depth of nine (9) inches.

(c) **Piers in Unstable Ground.** In unstable ground the load determining the area of the pier at its base shall include the weight of the pier.

3505.03 Construction Provisions:

(a) **Digging.** Digging shall be done in a manner which will leave the finished pier so that the center shall be in a vertical line from top to bottom.

(b) **Lagging and Rings.** Excavations for piers shall be lagged and braced with rings or otherwise. Lagging and rings shall be of sufficient strength to prevent collapse or distortion and tight enough to prevent loss of ground through the lagging. Lagging and rings shall be left in place.

(c) **Piers to Hardpan.** Where piers are to be supported on hardpan, sample drillings shall be made to a depth of six (6) feet below the bottom of each excavation. Samples of the hardpan must be submitted to the Commissioner of Buildings for approval as to soil classification.

(d) **Piers to Rock.** Where piers extend to bedrock, the rock bottom of ten (10) per cent of the total number of piers evenly distributed over the site shall be drilled to a depth of eight (8) feet and if fissures are encountered, all rock shall be removed from all piers on the site to the fissure and new borings shall be made until eight (8) feet of solid rock is found.

(e) **Concrete Filling.** In filling with concrete, top dumping will not be permitted unless the excavation has been dewatered, and is kept dewatered. Whenever filling is discontinued for more than one (1) hour all laitance shall be removed and a firm surface of concrete exposed. If the excavation cannot be kept dewatered it shall be filled with concrete by means of a water-tight tremie, or by a bottom dump bucket when the length of the tremie would exceed twenty (20) feet, or by means of half-filled bags of concrete. All concrete deposited by any of these three (3) latter described methods shall contain not less than six (6) bags of cement per cubic yard of concrete.

(f) **Coffer Dams.** Piers shall be protected against the seepage of surface water and storm water with coffer dams, the installation of which must precede the excavation for the piers and must reach blue clay or other impervious soil.

ARTICLE 3506

Foundation Columns

3506.01 Materials and Stresses: Qualities of materials and stresses permitted in design shall be as given in Chapter 39—REINFORCED CONCRETE CONSTRUCTION and Chapter 40—STEEL AND METAL CONSTRUCTION. The metal pipe shell shall have a wall thickness of not less than five-sixteenths (5/16) of an inch. The outer one-eighth (⅛) inch of metal shall be neglected in the design computations.

3506.02 Design Provisions:

(a) **Columns to Rock.** Foundation columus shall extend to solid rock. The rock bottom of ten (10) per cent of the foundation columns, evenly distributed over the site, shall be drilled to a depth of eight (8) feet below the base of the columns and if fissures are encountered all rock shall be removed to the fissure and new borings shall be made until eight (8) feet of solid rock is found. If a structural steel core is used it shall extend to the bottom of the rock excavation and shall be provided with a steel base plate grouted onto the rock.

(b) **Bearing Loads.** The bearing load on solid rock shall not exceed one hundred (100) tons per square foot or the value determined by test as provided in Section 3502.02, if the base is placed less than one (1) foot below the surface of the solid rock. If the base is placed one (1) foot or more below the surface of the solid rock, the bearing load shall be based on the following formula:

$$\frac{R}{a} = \frac{P}{a} \quad \frac{(3 + d)}{2}$$

d = depth in feet of the base below the surface of solid rock as defined in paragraph (a) of this section.

$\frac{P}{a}$ = bearing pressure according to paragraph (b) of this section.

$\frac{R}{a}$ = allowable bearing pressure.

No increase shall be permitted for values of "d," less than one (1) foot or greater than three (3) feet.

(c) **Columns in Unstable Ground.** In foundation columns the full permissible load, as computed in accordance with Section 3506.01 and Chapter 39—REINFORCED CONCRETE CONSTRUCTION, shall be used except that when the columns extend through a layer of quicksand, or other unstable soil of a depth equal to, or greater than, seven (7) times the least lateral dimension of the column section, the permissible load shall be taken as that for a long column using, in the computation a length equal to the depth of the unstable layer of soil, plus four (4) times the least lateral dimension of the column section.

(d) **Diameter of Foundation Columns.** The diameter of foundation columns, including the shell, shall not be less than two (2) feet.

(e) **Protection of Metal.** The metal inclosed within the steel pipe shell shall be not closer than three (3) inches to the outside of the steel pipe shell.

(f) **Bedding.** Wherever base plates are used they shall be bedded onto the rock before placing the encasing concrete.

3506.03 Construction Provisions:

(a) **Alignment.** The steel pipe shell shall be so driven that a vertical line from the center of the top shall not come closer to the outside of the shell, at any point, than one-third (⅓) the diameter of the shell.

(b) **Cleaning.** After the steel pipe shell is driven, all water and foreign matter shall be removed from the shell and the space within the shell shall be kept free from water and foreign matter.

(c) **Concrete Filling.** All space within the steel pipe shell shall be completely filled in one (1) continuous operation with concrete having a slump of not more than six (6) inches.

ARTICLE 3507

Steel Beams Used as Piling

3507.01 General: Steel beams used for piling shall not be less in dimension than six (6) inches by six (6) inches. The minimum thickness of metal in steel beam piles shall be three-eighths (⅜) of one (1) inch. The moment of inertia about the gravity axis parallel with the web shall not be less than one-fourth (¼) of the moment of inertia about a gravity axis perpendicular to the web. The unit stress in piles shall not exceed fifteen thousand (15,000) pounds per square inch. Piles in unstable ground not supported for the full length shall be figured as steel columns. Permissible bearing load shall be determined as given in Section 3503.02, paragraph (e), for wood piles. Test loads shall be as given in Section 3503.02, paragraph (f). The allowable load per pile shall not exceed sixty thousand (60,000) pounds for piles not driven to rock.

CHAPTER 36

Wood Construction

ARTICLE 3601

Materials and Stresses

3601.01 Quality of Timber:

(a) **General.** All timber used for building purposes shall be manufactured in accordance with the Standard Specifications for Grades of Longleaf and Shortleaf Southern Pine Lumber and Timber Conforming to American Lumber Standards of the Southern Pine Association effective September 1, 1932, and with the Standard Grading and Dressing Rules for Douglas Fir and West Coast Hemlock of the West Coast Lumbermen's Association effective July 1, 1934. For other species the stresses therefor shall be established in accordance with the "Guide to the Grading of Structural Timbers and Determination of Working Stresses," Miscellaneous Publication No. 185 of the U. S. Department of Agriculture.

(b) **Density.** Dense timber shall show, at least on one (1) end of the piece, not less than six (6) annual rings per inch, and at least one-third (⅓) summer wood.

3601.02 Working Stresses in Timber:

(a) **Actual Dimensions Shall Govern.** The strength of wooden members shall be determined from actual dimensions of the pieces, and not from nominal dimensions.

(b) **Bending Stress.** The maximum allowable stresses in pounds per square inch on actual sections for timber in flexure shall be as in Table 3601.02 (b).

Table 3601.02 (b)

MAXIMUM ALLOWABLE STRESSES ON TIMBER IN FLEXURE.

Species and Grades	Extreme Fibre Stress and Tension with grain lb. per sq. in.	Compression across grain lb. per sq. in.	Sheer with grain lb. per sq. in.	Modulus of Elasticity lb. per sq. in.
Prime Structural Long Leaf Yellow Pine	1800	375	125	1,600,000
Selected Douglas Fir having knots limited to one-half the size permitted in Select Structural Douglas Fir and slope of grain limited to 1 to 18	1800	345	100	1,600,000
Merchantable Structural Long Leaf Yellow Pine	1600	375	125	1,600,000
Dense Structural Square Edge and Sound Southern Yellow Pine	1600	375	125	1,600,000
Select Structural Douglis Fir	1600	345	100	1,600,000
No. 1 Structural Long Leaf Yellow Pine	1400	250	100	1,600,000
No. 1 Dimension Southern Yellow Pine	1000	250	100	1,200,000
No. 1 Dimension Douglas Fir	1000	250	100	1,200,000
Oak (White or Red)	1000	500	125	1,500,000
Western Hemlock	1000	250	80	1,000,000
Norway Pine	800	250	80	1,000,000
White Pine	800	200	80	1,000,000
Eastern Hemlock	700	200	60	800,000

(c) **Compressive Stress.** Working stresses in compression parallel to grain for column shall not exceed those given in Table 3601.02 (c) for the respective species and ratios of unsupported length to least dimension. The ratio of unsupported length of column to least dimension shall not exceed fifty (50).

Table 3601.02 (c)

MAXIMUM ALLOWABLE COMPRESSIVE STRESSES ON TIMBER COLUMNS.

Species and Grades	Ratio of Length to Least Dimension (L/D) 10 lb.per sq. in.	12 lb.per sq. in.	14 lb.per sq. in.	16 lb.per sq. in.	18 lb.per sq. in.	20 lb.per sq. in.	22 lb.per sy. in.	24 lb.per sq. in.
Prime Structural Long Leaf Yellow Pine	1300	1265	1235	1189	1123	1029	904	762
Selected Douglas Fir having knots limited to one-half the size permitted in Select Structural Douglas Fir and slope of grain limited to 1 to 18	1300	1265	1235	1189	1123	1029	904	762
Merchantable Structural Long Leaf Yellow Pine	1200	1172	1149	1113	1060	987	889	762
Dense Structural Square Edge and Sound Southern Yellow Pine	1200	1172	1149	1113	1060	987	889	762
Select Structural Douglas Fir	1200	1172	1149	1113	1060	987	889	762
No. 1 Structural Long Leaf Yellow Pine	1000	984	970	950	919	877	820	745
No. 1 Dimension Southern Yellow Pine	900	879	862	835	795	740	666	571
No. 1 Dimension Douglas Fir.	900	879	862	835	795	740	666	571
Oak (White or Red)	900	887	875	858	833	798	750	688
Western Hemlock	800	779	761	734	694	638	563	476
Norway Pine	700	686	674	656	629	592	542	476
White Pine	700	686	674	656	629	592	542	476

(d) **Ratios Intermediate in Table.** For columns with ratio of length to least dimension intermediate between those which are given in Table 3601.02 (c), the safe loads in pounds per square inch shall be determined by interpolation.

(e) **Ratios Greater than Given in Table.** For columns with ratio of length to least dimension greater than is given in Table 3601.02 (c), but in no case exceeding fifty (50), the safe load shall be determined by the following formula:

$$\frac{P}{A} = \frac{0.274E}{\left(\frac{L}{D}\right)^2}$$

in which

P = Total Load in pounds.

A = Cross Sectional area in square inches.

E = Modulus of Elasticity in pounds per square inch (Table 3601.2 (b)).

L = Unsupported length in inches.

D = Least dimension in inches.

(f) **Grade Certification.** Wherever a structural grade of Southern Yellow Pine or of Douglas Fir is used, with working stresses of fourteen hundred (1,400) pounds or more per square inch, as given in Table 3601.02 (b) or of one thousand (1,000) pounds or more per square inch as given in Table 3601.02 (c), the Commissioner of Buildings shall be furnished with a certificate from the designing architect or engineer of the building or structure that he has superintended its construction to the extent of being able to state without reservation that the timber specified has been used and is in place. The foregoing certificate shall be filed with the data submitted when application is made for the approval of floor load placards, and if floors are not required to be placarded such certificate shall be filed permanently in the file where the record of floor load placards is kept. The Commissioner of Buildings may establish such rules to enforce the provisions of this paragraph, not inconsistent herewith, as will prevent substitution of an inferior grade of timber for that grade which it is represented will be used at the time plans are presented and approved.

3601.03 Design Provisions:

(a) **Bearings.** Girders and beams where entering or resting on masonry or concrete walls, shall have a bearing of at least three and three-fourths (3¾) inches.

(b) **End Requirements.** A wood member entering a party or division masonry or concrete wall shall be separated from the opposite side of the wall, and from any beam entering the opposite side of the wall by at least six (6) inches of masonry or concrete. The ends of the joists shall be splayed or fire-cut, to a bevel of **not less** than three (3) inches in their depth.

(c) **Span.** Span of timber shall be taken as the distance between supports, plus one-half (½) of the required bearing at each end.

(d) **Horizontal Shear.** In computing horizontal shear the load at each end of the span may be neglected for a distance from the end of the required span equal to the depth of the beam.

(e) **Bridging.** Joists having a height of more than three (3) times their width shall be provided with cross-bridging of wood or metal if the joist span is more than eight (8) feet. The distance between successive bridgings, or between a bridging and a bearing shall not exceed eight (8) feet. The ratio of height to width shall not exceed eight and one-half (8½) to one (1), (actual sizes). Where the ratio of height to width exceeds seven and one-half (7½) to one (1), the distance between successive bridgings shall not exceed 6'-0".

(f) **Bases of Columns and Posts.** Bottoms of wood columns and posts shall be at least two (2) inches above finished concrete floor line, and shall be provided with base plates or other water stops.

(g) **Ventilation.** Wood construction shall not be inclosed without sufficient ventilation provisions to prevent rot. There shall be not less than one (1) inch of air space at the sides of truss members and girders entering masonry.

3601.04 Wood Trusses:

(a) **Design.** Tension members of all wood trusses shall be of lumber dried to fifteen (15) per cent or less in moisture content and shall be of a Structural Grade of Southern Yellow Pine or Douglas Fir. Splices shall develop full load at point of splice. All wood trusses in manufacturing units and garages shall be designed to support a concentrated load of not less than two thousand (2,000) pounds suspended from any point on the lower chord. All trusses shall be provided with lateral bracing of the top chords to prevent distortion.

(b) **Bolted Connections.** Allowable working loads per bolt for Southern Yellow Pine and Douglas Fir in double shear with steel or wood splice plates are given in Table 3601.04 (b). Allowable loads given are for dry timber, and are for use only where continuously dry conditions will obtain. Allowable loads per bolt for timber with a moisture content of more than eighteen (18) per cent when placed, or where there will be wet or alternately wet and dry conditions, are to be reduced by one-third (⅓). One-half (½) of values given in Table 3601.04 (b) may be used for connections in which bolts act in single shear. Twice the values given in Table 3601.04 (b) may be used for connections in which bolts act in quadruple shear. Recommendations of U. S. Department of Agriculture Technical Bulletin No. 332 are to be followed for allowable loads per bolt for other species of timber, for other conditions of use, for bolts acting at an angle to the grain, and for allowable bolt spacing, edge margins and other details of construction not covered or modified herein.

Table 3601.04 (b)

MAXIMUM ALLOWABLE LOADS IN BEARING PER BOLT IN DOUBLE SHEAR IN BOLTED TIMBER SPLICES AND CONNECTIONS.

For dry Southern Yellow Pine or Douglas Fir used in continuously dry locations.

Thickness of main member	Parallel to Grain — Bolt Size						
	½"	⅝"	¾"	⅞"	1"	1⅛"	1¼"
1⅝"	1000	1300	1600	1900	2100	2400	2600
2⅝"	1300	1900	2500	2900	3400	3800	4300
3⅝"	1300	2100	2900	3700	4500	5200	5800
5½"	1300	3100	3000	4000	5200	6500	7900
7½"	1300	3100	3000	4000	5200	6600	8200
9½"	1300	2100	3000	4000	5200	6600	8200
11½"	1300	2100	3000	4000	5200	6600	8200

Thickness of main member	Perpendicular to Grain — Bolt Size						
	½"	⅝"	¾"	⅞"	1"	1⅛"	1¼"
1⅝"	400	400	500	500	600	600	700
2⅝"	600	700	800	800	900	1000	1100
3⅝"	800	900	1000	1200	1300	1400	1500
5½"	800	1100	1500	1800	1900	2100	2200
7½"	800	1100	1500	1900	2400	2800	3100
9½"	800	1100	1500	1900	2400	3000	3600
11½"	800	1100	1500	1900	2400	3000	3600

(c) **Truss Ends.** Wood trusses shall be securely anchored to the wall and shall be provided with metal bearing plates.

3601.05 Wooden Flag Pole:

(a) **Design Load.** Wooden flag poles shall be designed in accordance with Section 3401.07 (b).

(b) **Diameter and Quality.** Wooden flag poles shall have a diameter at the base of not less than one-fiftieth (1/50) of their unsupported length and a diameter at top of not less than five-twelfth (5/12) of the diameter at the base. They shall be straight grained and free from loose knots.

CHAPTER 38

Masonry Construction

ARTICLE 3801

Definitions

3801.01 Definitions: For the purposes of this ordinance the terms herein defined shall be construed to mean and are required to mean as follows:

(a) **Masonry.** Units of stone, brick, concrete, terra cotta, medium or hard burned hollow clay tile not less than three and three-fourths (3¾) inches in thickness on the bed, laid in mortar with all joints filled and so bonded together as to exert common action under load, or monolithic concrete, or a combination of these materials.

(b) **Solid Wall.** A wall built of stone, brick, terra cotta, concrete or other solid units, or a combination of these materials, laid in mortar with all joints filled and so bonded together as to exert common action under load; or a wall of monolithic concrete.

(c) **Hollow Wall.** A wall built of solid units, laid in mortar with all joints filled, and bonded together in such a manner as to exert common action under load, and so constructed as to provide one (1) or more vertical air spaces within the wall; or a wall of monolithic concrete so constructed as to provide similar air spaces.

(d) **Wall of Hollow Units.** A wall consisting wholly or in part of hollow units of structural clay tile, terra cotta or concrete block, or a combination of these materials, laid in mortar with all joints filled, and bonded in such a manner as to exert common action under load.

(e) **Bearing Wall.** A wall which supports any load in addition to its own weight, or a wall which is more than one (1) story in height above its support.

(f) **Non-Bearing Wall.** A wall, other than a partition, which is supported at each structural floor and roof system.

(g) **Faced Wall.** A wall in which the masonry facing and backing are so bonded as to exert common action under load.

(h) **Veneered Wall.** A wall having a facing of non-combustible material which is not attached and bonded so as to form an integral part of the wall for the purposes of load bearing and stability.

(i) **Party Wall.** A wall used or adapted for joint service between two (2) buildings.

ARTICLE 3802

Quality of Materials

3802.01 Brick:

(a) **Solid Brick.** All brick of clay, shale, concrete or sand-lime shall have a minimum compressive strength of eighteen hundred (1800) pounds per square inch of gross area when tested flat, and a minimum modulus of rupture of four hundred (400) pounds per square inch and shall otherwise conform to the Standard Specifications of the American Society for Testing Materials for Building Brick Made from Clay or Shale (Serial Designation C62-30), for Concrete Block (Serial Designation C55-34) or for Sand-Lime Brick (Serial Designation C73-30).

(b) **Hollow and Perforated Brick.** Brick having two (2) longitudinal corings, each not more than three-fourths (¾) inch square or having not over twenty-five (25) per cent of their sectional area removed by vertical corings, where such corings are at least one-half (½) inch from the face of the brick, at least one-half (½) inch apart, may be used as permitted by this ordinance for solid brick; provided however, that such brick conforms to all the other requirements of this ordinance for solid brick.

3802.02 Structural Clay Tile:

(a) **Standard Sizes.** Structural clay tile for bearing walls, or for walls wholly or in part exposed to the weather shall at least comply with the standard specifications of the A. S. T. M. Serial Designation C34-36. The standard weight of load bearing structural tile per square foot of wall surface as laid up, shall be taken as twenty (20) pounds for four (4) inch thickness, thirty (30) pounds for six (6) inch thickness, thirty-six (36) pounds for eight (8) inch thickness, forty-two (42) pounds for ten (10) inch thickness and fifty-two (52) pounds for twelve (12) inch thickness. A maximum variation of two (2) pounds under this standard weight may be allowed. The minimum thickness of a single shell or the minimum aggregate thickness of a double shell shall be one (1) inch for tile laid in wall with cells vertical, and seven-eighths (⅞) inch for tile laid in wall with cells horizontal. With double shell tiles the maximum width of void between shells shall be five-eighths (⅝) inch. The minimum thickness of webs shall be five-eighths (⅝) inch.

(b) **Special Sizes.** Structural Clay Tile of sizes 4x5x12 and 8x5x12 when used in exterior walls shall conform to the standard specification for Structural Clay Load-Bearing Wall Tile, Grade LBX of the American Society for Testing Materials, serial designation C34-36, and when used in interior walls or in exterior walls where protected with a facing of three and three-quarters (3¾) inches or more of stone, brick, terra cotta, or other masonry shall conform to the standard specifications for Structural Clay Load-Bearing Wall Tile Grade LB of the American Society for Testing Materials, Serial Designation C-34-36. 4x5x12 size shall weigh nine (9) pounds per unit and 8x5x12 size shall weigh sixteen (16) pounds per unit.

(c) **Structural Clay Non-Load Bearing Tile.** All structural clay tile used for partitions or non-load bearing interior walls shall be at least equal to the Specifications for Structural Clay Non-Load Bearing Tile of the American Society for Testing Materials (Serial Designation C56-36).

3802.03 Concrete Units:

(a) **General.** Concrete units may be either solid or hollow and shall be made of concrete materials meeting the requirements of Chapter 39—REINFORCED CONCRETE CONSTRUCTION, except that burned clay or shale, blast furnace slag, cinders or other approved aggregate may be used as an aggregate. When cinders are used as an aggregate, the combustible content shall not exceed twenty (20) per cent by weight of the dry mixed cinders nor shall any unit contain in excess of one and one-fourth (1¼) per cent by weight of sulphur.

(b) **Solid Units.** Concrete units having voids of not more than twenty-five (25) per cent of core area shall be considered as solid units and shall have a minimum crushing strength of eighteen hundred (1800) pounds per square inch of gross area at an age of twenty-eight (28) days or when delivered on the job.

(c) **Hollow Units.** Hollow concrete units used in bearing walls or in exterior walls shall be at least equal to the requirements of the specifications and tests for Load Bearing Concrete Masonry Units of the American Society for Testing Materials (Serial Designation C90-36).

(d) **Marking.** All concrete units shall bear a distinctive mark of the manufacturer or shall otherwise be readily identified as to origin. Manufacturer's markings shall be registered with the Department of Buildings.

3802.04 Architectural Terra Cotta Units: All units of architectural terra cotta shall have a minimum ultimate compressive strength of fifteen hundred (1500) pounds per square inch of net area and seven hundred (700) pounds per square inch of gross area and a maximum absorption after forty (40) hours' immersion in cold water, of ten (10) per cent of the weight of the dry unit.

3802.05 Stone Units: All units of stone, when counted as a part of the required wall thickness shall have a minimum compressive strength of five thousand (5000) pounds per square inch and when showing natural cleavage planes or seams shall be laid on their natural beds.

3802.06 Architectural Cast Stone: All units of architectural cast stone shall be at least equal to the requirements of the Federal Specifications (SS-S721) for Architectural Cast Stone adopted November 10, 1931.

3802.07 Glass Masonry Windows: Glass block construction consisting of glass block units not less than three and three-fourths (3¾) inches thick, laid in approved masonry mortar and properly bonded or adequately reinforced with metal, may be inserted wherever window openings are permitted, if properly constructed as a substitute for glazed sash and window frames. Glass block construction shall have no performance in carrying weight or transmitting strains induced by the loads of the building or its contents. Glass block construction in any form of panels, may be continuous in any one (1) story but shall be supported laterally by masonry piers, steel or other pier or column construction complying with the requirements for the building at intervals not exceeding twenty-five (25) feet and no panel shall exceed one hundred fifty (150) square feet in area. Panels shall be protected from contraction and expansion of the structure by proper expansion joints.

3802.08 Materials for Masonry Mortars:

(a) **Portland Cement.** Portland Cement shall be at least equal to the requirements of the Standard Specifications for Portland Cement (Serial Designation C9-37) or the Specifications for High Early Strength Portland Cement (Serial Designation C74-36) of the American Society for Testing Materials.

(b) **Masonry Cement.** Masonry cement shall be at least equal to the requirements of the Specifications and Tests for Masonry Cement of the American Society for Testing Materials (Serial Designation C91-32T).

(c) **Lime.** Lime shall be at least equal to the requirements of the Standard Specifications for Hydrated Lime for Structural Purposes (Serial Designation C6-31) or Standard Specifications for Quicklime for Structural Purposes (Serial Designation C5-26) of the American Society for Testing Materials. Quicklime shall be slaked for at least forty-eight (48) hours before using, unless report of approved testing laboratory is submitted to the Commissioner of Buildings showing that complete hydration is accomplished in a shorter period of time.

(d) **Sand.** Sand shall be hard, clean and free from foreign materials which will reduce the strength of the mortar. Sand for mortar shall have a minimum fineness modulus of two (2.0).

3802.09 Plain Monolithic Concrete: Plain concrete used to form any part of a wall, pier, buttress or other structural member shall be made and placed in accordance with the provisions of Chapter 39—REINFORCED CONCRETE CONSTRUCTION of this ordinance.

ARTICLE 3803

Allowable Unit Stresses

3803.01 General Requirements: The thickness of masonry walls shall be sufficient at all points to keep the combined stresses due to live, dead and wind loads for which the building is designed, within the limits specified in this article. The area of all recesses, chases and other reductions of horizontal area shall be deducted from the effective area in making stress computations. In every case where the center of gravity of the loads does not coincide with the center line of the wall or the center of the pier or buttress, the stresses due to eccentric loading shall be included in the computations.

3803.02 Masonry:

(a) **Compression.** The maximum compressive stresses in masonry shall not exceed the values given in Table 3803.02 (a).

Table 3803.02 (a)

Maximum Allowable Compressive Stresses on Load-Bearing Masonry

Mortar Mixes	Mortar Number					
By Volume—	1	2	3	4	5	6
Portland Cement	5	1	1	1		
Lime	1	1		2	1	
Masonry Cement			10			1
Sand	15	6	30	9	3	3
Material	Allowable Masonry Stress (Lbs. per Sq. In.)					
Coursed Limestone Rubble	200	200	150	150	100	100
Ashlar Granite	400	200	150	150	100	100
Ashlar Limestone, Cast Stone, Brick Testing 5000 pounds per square inch	350	200	150	150	100	100
Brick or other solid units testing 3500 pounds per square inch	250	200	150	150	100	100
Brick or other solid units testing 1800 pounds per square inch	200	200	150	150	100	100
Hollow Units	80	80	60	Not Allowed		

(b) **Bending.** The tension on the extreme fibre in bending shall not exceed one-tenth (0.1) the maximum allowable compression stress as given in Table 3803.02 (a) nor twenty (20) pounds per square inch.

(c) **Composite Walls.** In walls composed of more than one (1) type of masonry the maximum unit stress shall not exceed the allowable unit stress for the weaker material.

3803.03 Plain Concrete:

(a) **Compression.** The maximum compressive stress for plain concrete shall not exceed two-tenths (0.2) of the ultimate compressive strength of the concrete at twenty-eight (28) days.

(b) **Bending.** The tension on the extreme fibre in bending shall not exceed two (2) per cent of the ultimate compressive strength of the concrete at twenty-eight (28) days.

ARTICLE 3804

Minimum Wall Thicknesses

3804.01 Distance between Lateral Supports:

(a) **General.** Masonry walls shall be supported at right angles to the face of the wall at intervals not exceeding eighteen (18) times the wall thickness. Such lateral support may be obtained by cross walls, piers, buttresses or columns when the limiting distance is measured horizontally or by floors when the limiting distance is measured vertically. Sufficient bonding or anchorage shall be provided between the wall and the supports to resist the assumed wind force acting in an outward direction. The length shall be considered as the clear distance between cross walls, piers, buttresses or columns. The height shall be considered as the clear distance between structural floor or roof systems.

(b) **Floors.** When walls are dependent upon floors for their later support, provision shall be made in the building to transfer the lateral forces resisted by the floors to the ground.

(c) **Piers or Buttresses.** Piers, pilasters or buttresses shall be designed to resist all lateral forces acting thereon.

3804.02 Bearing Walls:

(a) **General.** The minimum thickness of masonry bearing walls, except as modified in this article, shall be as given in Table 3804.02.

Table 3804.02

Minimum Thickness of Bearing Walls in Inches

Single and Multiple Dwellings

Height	Basement	1st Story	2nd Story	3rd Story	4th Story	5th Story	6th Story	7th Story
1 Story	8	8						
2 Story	12	8*	8					
3 Story	16	12	12	8				
4 Story	16	12	12	12	8			

*Solid Units Only.

Other Occupancies

Height	Basement	1st Story	2nd Story	3rd Story	4th Story	5th Story	6th Story	7th Story
1 Story	12	8						
2 Story	12	12	8					
3 Story	16	12	12	8				
4 Story	16	16	12	12	12			
5 Story	16	16	16	12	12	12		
6 Story	20	16	16	16	12	12	12	
7 Story	20	20	16	16	16	12	12	12

(b) **Height Limit.** Bearing walls of hollow units or hollow walls of solid units shall not exceed fifty (50) feet in height above the top of the foundation wall.

(c) **Concrete Basement Walls.** Where solid monolithic plain concrete basement walls extend up to the bottom of the structural system of the first floor, such basement walls may be of the same thickness as the first story walls.

(d) **Walls Below Grade for Single and Multiple Dwellings.** For single and multiple dwellings, two (2) stories or less in height, the walls below grade may be of solid masoury not less than eight (8) inches thick.

(e) **Walls Below Grade for Buildings of Wood Frame Construction.** The minimum thickness of walls below grade for buildings of wood frame construction shall be as required by paragraph (d) of this Section for walls below grade for single and multiple dwellings.

3804.03 Non-Bearing Walls: The minimum thickness of non-bearing walls shall be four (4) inches less than specified in Table 3804.02 but not less than eight (8) inches. Exterior non-bearing walls in skeleton construction shall be not less than eight (8) inches thick and shall, in each story, have a bearing of not less than seven and one-half (7½) inches upon a structural member which shall be fireproofed as required for members of the structural frame.

3804.04 Party Walls: All party walls shall be of solid masonry not less than twelve (12) inches thick.

3804.05 Parapet Walls:

(a) **Thickness and Construction.** Parapet walls shall be solid walls not less in thickness than required for the walls immediately below and shall be laid in No. 1 Mortar as specified in Table 3803.02. Except where common brick is used, throughout, as facing, parapet walls shall be constructed for their entire thickness and height, with building brick units, Grade "A," A. S. T. M. Designation C62-30 for ultimate compressive strength and with a maximum absorption of not more than eight (8) per cent.

(b) **Coping.** All parapet walls shall be capped with an impervious coping material so laid as to prevent water penetration.

(c) **Facing.** Parapet walls may be faced in accordance with Section 3805.01 (c).

3804.06 Retaining Walls: All basement and retaining walls of masonry shall be of sufficient strength to resist the lateral pressure transmitted to them without exceeding the stresses allowed by Article 3803.

3804.07 Existing Walls:

(a) **General.** An existing masonry wall may be used in the renewal or extension of a building if it is structurally sound or can be made so by reasonable repairs. Where it is not of sufficient thickness to meet the requirements of this ordinance it shall be reinforced by a lining or facing of masonry or by pilasters, buttresses or columns.

(b) **Lining or Facing.** Where existing walls are to be lined or faced, the lining or facing shall be not less than eight (8) inches thick and the total thickness shall be not less than four (4) inches greater than required by this ordinance for new walls. All linings and facings shall be thoroughly bonded to existing walls by toothings of not less than fifteen (15) per cent of the surface area uniformly distributed over the entire area. All such linings and facings shall be laid in No. 1 Mortar.

(c) **Pilasters, Buttresses and Columns.** Where existing walls are reinforced by new pilasters, buttresses or columns, the entire additional load shall be carried by the new construction. Pilasters or buttresses shall be thoroughly bonded into the old wall by building them into chases cut at least eight (8) inches into the existing wall throughout its height. The new masonry shall be laid in No. 1 Mortar as specified in Table 3803.02.

(d) **Footings.** Footings for the wall lining or facing, pilasters, or columns shall be combined with the existing footings so as to carry the total load in accordance with Chapter 35.

3804.08 Reinforced Concrete Walls: Reinforced concrete walls shall be designed in accordance with Chapter 39—REINFORCED CONCRETE CONSTRUCTION.

ARTICLE 3805

Details of Construction

3805.01 Bond:

(a) **Walls of Brick.** In every brick wall, pier, pilaster or buttress at least every sixth course shall be a full header course or there shall be at least one (1) full length header in every seventy-two (72) square inches of each surface of the wall.

(b) **Walls of Hollow Units.** Where two (2) or more hollow units are used in the thickness of a wall the inner and outer courses shall be bonded at vertical intervals not exceeding three (3) courses or sixteen (16) inches by lapping not less than four (4) inches.

(c) **Faced Walls.** Where walls are faced with brick, terra cotta, stone, architectural cast stone, concrete units or burned clay or shale units, bonded as provided in this paragraph, the facing may be considered as part of the required thickness. Brick facing shall be bonded with full headers equivalent in number and area to one (1) course composed exclusively of headers to each five (5) stretcher courses. For facings of terra cotta, stone, architectural cast stone or similar units, at least twenty (20) per cent of the surface area shall be composed of uniformly distributed bond units extending into the backing at least four (4) inches. All units more than one-half (½) square foot in face area, except bond units, shall be anchored into the backing with metal anchors at least three-sixteenths (3/16) inch thick. Such anchors shall be of solid non-rusting metal or shall be asphalt coated after bending. There shall be no less than two (2) anchors for each unit more than two (2) feet in length or three (3) square feet in face area. Units more than twelve (12) square feet in area shall have at least one (1) anchor in each four (4) square feet of face area; provided however, that a masonry wall consisting of a three and three-fourths (3¾) inch thickness of open back terra cotta and eight (8) inches or more of brick, shall be considered the equivalent of a brick wall of the same thickness as the combined thickness of brick and terra cotta; provided, that brick shall be bonded into all open spaces of terra cotta, and that each piece of terra cotta shall be additionally bonded to brickwork with two (2) or more metal anchors as required by this paragraph. The height of any facing unit shall not exceed eight (8) times its thickness, and no unit shall be less than three and three-fourths (3¾) inches thick. Isolated piers not more than twenty-four (24) inches in width faced with stone, terra cotta or similar units shall have bond units in every alternate course but the area of bond units need not exceed twenty (20) per cent of the face area of the pier.

(d) **Veneered Walls.** Veneers shall not be considered to be part of the required wall thickness. Veneers shall be subject to the requirements of the general provisions of this ordinance.

(e) **Veneered Frame Buildings.** Masonry veneers for buildings of wood frame construction shall be not less than three and three-fourths (3¾) inches thick, shall be supported on solid masonry basement or foundation walls below grade, shall be securely tied to the studding and shall extend not more than thirty (30) feet in height above grade.

3805.02 Joist Supports and Anchorage:

(a) **Minimum Bearing.** The minimum bearing of joists on masonry walls shall be three and three-fourths (3¾) inches. The section of any wall or pier shall not be reduced more than twenty (20) per cent to provide bearing for wood joists or other wood floor construction. Properly designed metal wall boxes or stirrups shall not be considered as reducing wall or pier section. There shall be at least six (6) inches of solid masonry separating the ends of wood joist carried on a fire division wall and between the ends of such joist and the opposite side of the wall.

(b) **Corbelling.** No load bearing corbel shall project more than four (4) inches from the face of the wall and no one (1) course shall project more than three-fourths (¾) inch beyond the course immediately below it. Corbels shall be bonded to the wall using alternate courses of headers and stretchers with the top course of headers. There shall be at least three (3) courses, the upper one (1) of which shall be a header course, laid flush with the top projecting course of every corbel which supports any load.

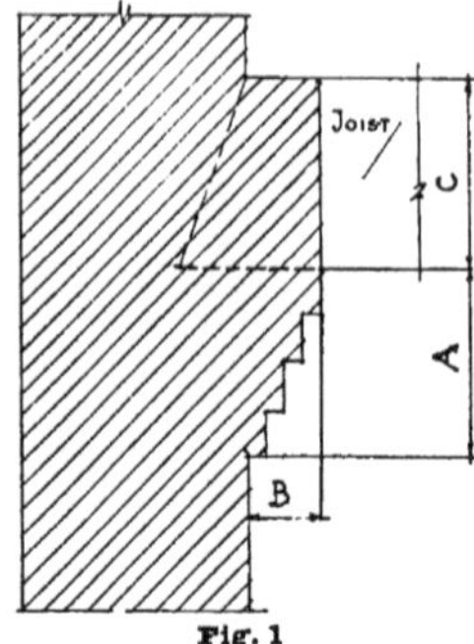

Fig. 1

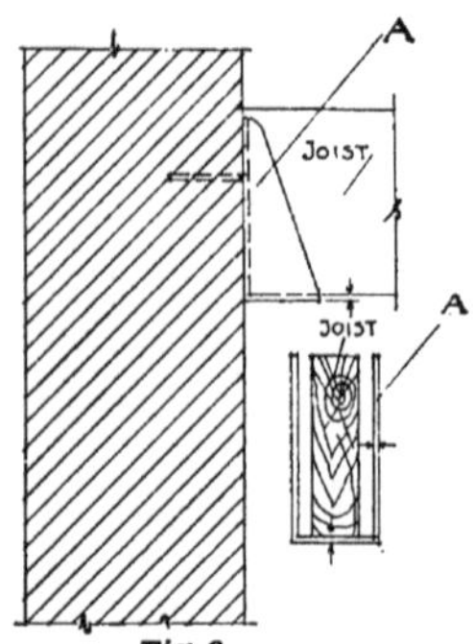

Fig. 2

LEDGES—JOIST SUPPORTS.
Section 3805.02.

Fig. 1 (A) Corbelling to be not less than three courses of brick.
(B) Upper course shall project not more than four inches.
(C) The joists shall be protected from top to bottom by brick.
Fig. 2. (A) Metal joist hanger allowable.
(B) ¼-inch metal required.

(c) **Anchorage.** All floor and roof systems shall be anchored to walls at intervals of not more than seven (7) feet with metal anchors so placed as to provide a continuous tie across the building. Where such a tie is composed of more than one (1) joist, girder or truss, the abutting members shall be joined by tie or splice plates equal in strength to the anchors to form a tie across the building. Anchors for the ends of joists or girders shall be fastened below the center. Anchors in joists parallel to the wall shall be fastened to the top of the joists and shall engage sufficient joists to have a combined thickness of at least five and one-half (5½) inches. The anchors shall extend to within four (4) inches of the opposite face of the wall. Every anchor and its fastenings shall be adequate to resist an outward pressure upon that portion of the wall anchored by it equal to the wind pressure specified in Article 3401 of this ordinance. Anchors shall have a minimum cross-sectional area of four-tenths (0.4) square inch and a minimum thickness of three-sixteenths ($\frac{3}{16}$) inch.

3805.03 Isolated Piers: The height of isolated masonry piers shall not be more than twelve (12) times their least dimension. Wall sections twenty-four (24) inches or less in width between openings shall be considered as isolated piers. All isolated load bearing piers shall be of solid masonry.

3805.04 Chases: There shall be at least eight (8) inches of masonry on three (3) sides of any chase. Horizontal or diagonal chases more than twenty-four (24) inches long shall be considered as openings entirely through the wall.

3805.05 Openings in Walls. Openings in walls shall have wetll buttressed arches of masonry, or lintels of reinforced concrete or metal with hearings at each end of not less than four (4) inches. In buildings of ordinary construction, wood lintels may be provided back of masonry arches.

3805.06 Arches: Masonry arches shall be of solid masonry and shall have a minimum rise of one-twelfth (1/12) of clear span. The units shall be laid on a radius of the arch curve. The minimum total depth of all the radially laid courses shall be one-twelfth (1/12) the span. Provision shall be made in the wall for resisting the arch thrust.

3805.07 Masonry in Contact with Earth: Masonry in contact with the earth shall be laid in Mortar No. 1 as given in Table 3803.02. Hollow units of clay tile shall not be used for exterior walls below grade.

3805.08 Progress of Work: All masonry shall be protected against freezing for at least forty-eight (48) hours after being laid. No frozen material shall be built upon.

CHAPTER 39

Reinforced Concrete Construction

ARTICLE 3901

General

3901.01 General Requirements:

(a) **Design and Construction.** With the specific exceptions enumerated in Section 3902.01, and except as otherwise provided in this ordinance the design and construction of reinforced concrete shall be in accordance with the provisions of the "Building Regulations for Reinforced Concrete" (A.C.I. 501-36-T) adopted by the American Concrete Institute on February 25, 1936, and published in the proceedings of that Institute, Volume 32.

(b) **Standards.** Wherever in said Building Regulations for Reinforced Concrete reference is made to standards or tentative standards of the American Society for Testing Materials, such standards or tentative standards are hereby made a part of this ordinance and shall remain as of February 25, 1936, except as other or later designated standards are specified elsewhere in this ordinance.

ARTICLE 3902

Exceptions and Additions

3902.01 Nullification: The following provisions, contained in the Building Regulations for Reinforced Concrete referred to in Section 3902.01 (a), are hereby nullified within the jurisdiction of the City of Chicago:

101: Scope:
103: Special Systems of Reinforced Concrete:
205: Concrete Aggregates:
406: Cold Weather Requirements: par (b)
603: Resistance to Wind Forces: par (c)
705: Requirements for T-Beams: par (c)
902: Ordinary Anchorage Requirements: par (b)
1103: Spirally Reinforced Columns:
1104: Tied Columns:
1105: Composite Columns:
1106: Combination Columns:

1107: Long Columns:
1108: Bending Moments in Columns:
1109: Combined Axial and Bending Stress:
1110: Permissible Combined Compressive and Tensile Stress:
1111: Wind Stresses:
1112 Monolithic Walls:

3902.02 Standards: The following designated standards or tentative standards of the American Society for Testing Materials, are hereby made a part of this ordinance and shall take precedence over other designations in the cases of the materials hereinafter enumerated:

High Early Strength Portland Cement: "Standard Specifications for High Early Strength Portland Cement" (Serial Designation C74-36).

3902.03 Concrete Aggregates:

(a) **Sand.** The sand to be used for concrete shall be clean, hard, coarse sand, of the grade known as torpedo sand, and free from loam or dirt, and not less than forty-five per cent shall be returned on a screen of four hundred mesh to the square inch.

(b) **Stone.** The stone to be used in concrete shall be clean crushed hard stone or clean crushed blast furnace slag or gravel of a size to pass through a one-inch square mesh. If limestone or slag is used, it shall be screened to remove all dust; if gravel is used, it shall be thoroughly washed. Stone shall be drenched immediately before using. If slag is used, it shall be of such character that when made into concrete the concrete will develop a crushing strength equal to that specified for stone or gravel concrete.

3902.04 Allowable Unit Stresses in Reinforcement: The following unit stresses in reinforcing steel shall not be exceeded:

Tension: f_s or f_v
Cold Drawn Wire, f_s................25,000 lb. per sq. in.

3902.05 Cold Weather Requirements: All concrete materials and all reinforcement, forms, fillers and ground with which the concrete is to come in contact, shall be free from frost. Whenever the temperature of the surrounding air is below 40 degrees Fahrenheit all concrete placed in the forms shall have a temperature of between 70 degrees Fahrenheit and 100 degrees Fahrenheit, and shall be maintained at a temperature of not less than 70 degrees Fahrenheit for 3 days or 50 degrees Fahrenheit for 5 days for standard concrete and not less than 70 degrees Fahrenheit for 2 days or 50 degrees Fahrenheit for 3 days for high early strength concrete or for as much more time as is necessary to insure proper curing of the concrete.

3902.06 Requirements for Tee Beams: Where the principal reinforcement in a slab which is considered as the flange of a T-beam (not a rib in ribbed floors) is parallel to the beam, transverse reinforcement shall be provided in the top of the slab. This reinforcement shall be designed to carry the load on the portion of the slab assumed as the flange of the T-beam. The spacing of the bars shall not exceed three (3) times the thickness of the flange, or in any case, twelve (12) inches.

3902.07 Composite Beams:

(a) **General.** Where structural steel members are encased in concrete beams, and protected by not less than the minimum thickness of concrete specified elsewhere in this Code for the fireproofing of structural steel, the resulting composite beam may be computed as a reinforced concrete beam in accordance with the assumptions stated in Section 601 (a) of the Building Regulations for Reinforced Concrete referred to in Section 3901.01 of this ordinance. The properties of sections of composite beams shall be computed on the basis of the transformed area of steel.

(b) **Reinforcing Bars.** In composite beams in which reinforcing bars are used in conjunction with structural steel members the properties of the cross section of the beam shall be computed on the basis of transformed area of steel and in accordance with the assumptions stated in paragraph (a) of this Section.

(c) **Design Provisions.** All provisions of this Chapter shall apply to composite beams. Standard beam connections shall be considered as affording slight restraint only, and the span of composite beams supported by such connections shall be taken as the overall length of the encased steel member measured out-to-out of the connection angles. Where end connections properly designed to take the end moment of a continuous beam or the moment at the joint of a rigid frame are provided, or where proper supplementary reinforcement in the form of bars or plates is supplied to take this moment at the support, the span and moments of composite beams may be taken as specified for continuous or restrained beams.

(d) **Built-Up Steel Sections.** Built-up steel sections used in composite beams shall be designed in accordance with Chapter 40—STEEL AND METAL CONSTRUCTION, for the forces to which they are subjected in the composite beam.

(e) **Construction Load Design.** In all cases of composite beams, where forms are suspended from the steel beam, the steel beam shall be designed in accordance with Chapter 40 to carry all loads coming upon it during construction; while the entire composite beam shall be designed to carry all loads after construction.

3902.08 Ordinary Anchorage Requirements: In continuous beams every positive reinforcing bar shall extend 12 diameters beyond the point at which it is no longer needed to resist stress. In cases where the length from the point of maximum tensile stress in the bar to the end of the bar is not sufficient to develop this maximum stress by bond, the bar shall extend into a region of compression and be anchored by means of a standard hook or it shall be bent across the web and made continuous with the negative steel or anchored in a region of compression. At least one-quarter (¼) the area of positive reinforcement shall extend along the same face of the beam into the support at least four (4) inches.

3902.09 Reinforced Concrete Columns:

(a) **Spirally Reinforced Columns.** The maximum permissible axial load (P) on columns reinforced with longitudinal bars, structural steel or cast iron cores and closely spaced spirals inclosing a circular core shall be as follows:

$P = A_c\,(.225f'_c + 2f_r p')$. The ratio, p, of the effective cross-sectional area of the vertical reinforcement to the core area, A_c, of the column, shall not be less than 0.01. The minimum numbers of bars shall be six (6) and the smallest size shall be three-fourths (¾) inch round. The cross-sectional area of the longitudinal reinforcement, where the stresses at splices are transferred by the bond with the concrete, shall not exceed six (6) per cent of the core area of the column. The center to center spacing of bars within the periphery of the column core shall not be less than two (2) times the diameter for round bars, nor two and one-half (2½) times the side dimension for square bars. The clear spacing between bars shall not be less than one (1) inch, nor one and one-third (1⅓) times the maximum size of the coarse aggregate. These spacing rules apply to the bars at a lapped splice. The clear opening between two (2) adjacent rings or rows of bars, or between a ring or row of bars and any metal core or structural shape, or between any two (2) types of longitudinal reinforcement, or between metal core or structural shape and spiral shall not be less than three (3) inches. No longitudinal bars shall be required in columns reinforced by a structural steel or cast iron core. The spiral ratio, p', shall not be less than the value given by the formula

$$P' = \frac{0.125f'_c\,(R-1)}{f_r}$$ nor shall it be less in any case than one-half (½) of one (1) per cent, nor

shall the value used in the formula for allowable avial load exceed 2/3p, nor two (2) per cent. R is the ratio between the gross area and the core area. For square or rectangular columns the gross area shall be considered as the area of the circle having a diameter three (3) inches larger than the diameter of the core. The spiral reinforcement shall consist of evenly spaced continuous spirals held firmly in place and true to line by at least three (3) vertical spacer bars. All splices in lateral reinforcement at points where the column is not laterally supported shall be made by welding in accordance with the provisions of Chapter 40. Column spirals may be omitted within that part of the column which lies within the floor slab provided that the length of column over which the spiral is omitted shall not exceed one-third (⅓) of the least lateral dimension of the column section. The center to center spacing of the spirals shall not exceed three (3) inches, nor one-sixth (⅙) of the core diameter, nor shall it be less than three and one-half (3½) times the diameter of the spiral rod. In no case shall the clear distance between spirals be less than one and one-third (1⅓) times the maximum size of aggregate nor less than one (1) inch. All openings within the section and all recesses in the exterior of metal cores or structural shapes shall be so designed as to be readily and completely filled with concrete when the column is concreted. Cores and shapes shall be disposed symmetrically within the column section in the case of axially loaded columns. When the eccentricity of loading is fixed in direction, cores and shapes shall be arranged accordingly.

(b) **Tied Columns.** The maximum permissible axial load (P) on columns reinforced with longitudinal bars and separate lateral ties shall be as follows:

$$P = A_g (0.2f'_c + 0.9f_r p_g)$$

The ratio, p_g, of the effective cross-sectional area of the vertical reinforcement to the gross area, A_g of the column, shall not be less than 0.01 nor more than 0.04. The reinforcement shall consist of not less than four (4) bars, not less than three-fourths (¾) inch in diameter, placed with a not less than one and one-half (1½) inches clear distance from the face of the column of All longitudinal reinforcement shall be tied in two (2) directions by lateral reinforcement at least one-quarter (¼) inch in diameter at not more than twelve (12) inch centers.

(c) **Combination Columns.** Structural steel columns of any rolled or built-up section wrapped with the equivalent of No. 8 W. & M. Standard gauge wire spaced four (4) inches on centers and encased in concrete not less than two and one-half (2½) inches thick over all of the metal, except rivet heads and connections, will be permitted to carry a load equal to $1 + \frac{A_c}{100A_s}$ times the permissible load for unencased steel columns as given in Chapter 40.

(d) **Pipe Columns.** The maximum permissible axial load (P) on columns composed of a metal shell completely enclosing a concrete core shall be as follows:

$$P = 0.25f'_c A_c + f_r A_s$$

The minimum thickness of the steel pipe shell shall be two-tenths (0.2) of one (1) inch.

(e) **Allowable Steel Stress.** The allowable stress, f_r, used in the design of spirally reinforced, tied and pipe columns shall not exceed the following stresses unless the material is identified by marking rolled or cast into the surface during manufacture in which case a stress of 0.36 of the yield point may be used.

Intermediate grade billet steel	15,000
Rail Steel	15,000
Structural Shapes	12,000
Silicon Steel	18,000
Cast Iron	9,000
Steel or wrought iron pipe	10,000
Cold Drawn Wire	20,000

(f) **Transfer of Load on Reinforcement.** Columns in any story that form a part of tiers of columns extending through two (2) or more stories shall be designed for continuity of the longitudinal reinforcement from story to story. Longitudinal bars shall be spliced by welding or by lapping into the columns above. When spliced by lapping, the length of the lap shall be not less than twenty-four (24) inches and not less than the length computed to transfer the entire load on the bars of the column above to bars extended from the column below.

In this computation, the unit bond stress on bars not inclosed within a column spiral shall be taken in accordance with Section 305 of the building regulations referred to in Section 3901.01 of this Chapter, and at forty (40) per cent more than those values for bars enclosed within a column spiral. Longitudinal metal may be spliced by means of welding or by bolted or riveted connections or otherwise in any manner that will fully provide for all stresses to which the particular reinforcement is subjected as a result of its position in the column section. Details of connections and splices shall be such as to permit of easy and complete encasement in concrete and of placing both columns and floor steel in the designed location where the latter passes through the column section. Metal column cores, pipe shells, or structural steel shapes used as longitudinal reinforcement, shall be provided with mechanical means for receiving load wherever the increment of load received in any story exceeds the amount of load that can be transmitted from the concrete to the metal by bond at unit values permitted by this paragraph for lapped splices.

(g) **Long Columns.**

Item 1. Permissible Axial Load on Concrete Columns. The permissible axial load on reinforced concrete or composite columns having an unsupported length, greater than ten (10) times the least lateral dimension of the column section, shall not exceed the value computed by the following formula:

$$P' = P\,(1.3 - .03\frac{h}{d})$$

Item 2. Permissible Load on Combination Columns. The permissible axial load on columns composed of structural metal sections solidly incased in concrete (of the thickness required for fireproofing) or of metal pipe shells filled with concrete, and in which one-half (½) or more of the total column load is carried by the metal sections, and having an unsupported length, greater than ten (10) times the least lateral dimension of the concrete section, shall not exceed the value computed by the following formula:

$$P' = P\,(1.16 - \frac{h}{70d})$$

(h) **Combined Axial Load and Bending.** The recognized methods of analysis shall be followed in calculating stresses due to combined axial load and bending. The column section shall not be less than that required by the design for axial loads alone. The maximum fiber stresses in compression in the concrete shall not exceed the permissible values under axial loads only by more than sixty (60) per cent. The tensile stresses in the longitudinal reinforcement shall not exceed the allowable unit stresses permitted in tension.

3902.10 Concrete Studs in Bearing Walls:

(a) **Where Permitted.** In single or multiple dwellings, two (2) stories or less in height, bearing walls constructed with reinforced concrete studs may be used, if designed and constructed in accordance with the provisions of paragraph (b) of this section.

(b) **Design and Construction.** Where concrete studs in bearing walls are designed in accordance with the provisions of this Chapter for long columns, and are made an integral part of the wall and are tied together by concrete members at each floor level, the restrictions as to minimum diameter or thickness of columns, shall not apply.

3902.11 Reinforced Concrete Walls:

(a) **Permissible Load.** The permissible axial load on reinforced concrete walls, in which the vertical reinforcement is tied through the wall by means of one-fourth (¼) inch diameter ties spaced not more than twelve (12) inches apart vertically and eighteen (18) inches apart horizontally, shall be computed as for tied columns having the same h/d ratio. Where the amount of vertical reinforcement in the two (2) faces is not equal the amount considered for vertical load shall be twice the amount in the side having the lesser reinforcement. The permissible axial load on reinforced concrete walls in which the reinforcement is placed in a single layer, or in which ties are not provided between the two (2) layers to meet the above requirements shall be taken as $0.2f'_cA_c$

(b) **Minimum Thickness of Bearing Walls.** Bearing walls of reinforced concrete shall be not less than six (6) inches in thickness for the uppermost fifteen (15) feet of their height, and for each successive twenty-five (25) feet downward, or fraction thereof, the minimum thickness shall be increased one (1) inch.

(c) **Non-Bearing Spandrel Walls.** Reinforced concrete spandrel walls carried on the structural frame of buildings of skeleton construction shall be designed to resist the wind load as required by Chapter 34. Such spandrel walls shall have a thickness of not less than four (4) inches, except where greater thickness is required to meet the requirements of this ordinance for fire-resistive value.

(d) **Reinforcement of Concrete Walls.** Exterior walls of reinforced concrete above the basement story shall be provided with the minimum percentage of reinforcement as required for roof slabs. Other walls shall be provided with the minimum percentage of reinforcement as required for floor slabs. The spacing of bars in each layer and each direction shall not exceed eighteen (18) inches. Walls more than eight (8) inches in thickness shall have the reinforcement in each direction placed in two (2) layers. One (1) layer consisting of not less than one-half (½) nor more than two-thirds (⅔) of the total required shall be placed not less than one and one-half (1½) inches, nor more than one-third (⅓) of the thickness of the wall, from the exterior surface. The other layer comprising the balance of the reinforcement shall be placed not less than three-fourths (¾) inches and not more than one-third (⅓) the thickness of the wall, from the interior surface.

CHAPTER 40

Steel and Metal Construction

ARTICLE 4001

General

4001.02 General Requirements:

(a) **Structural Steel.** The design, fabrication and erection of steel and metal construction shall be in accordance with the "Specification for the Design, Fabrication and Erection of Structural Steel for Buildings" as adopted by the American Institute of Steel Construction and revised June 24, 1936, for steel made by the open-hearth process, except as otherwise provided in this ordinance.

(b) **Welding.** The design and execution of welding shall be in accordance with "Code I, Part A, Code for Fusion Welding and Gas Cutting in Building Construction" of the American Welding Society, Edition of 1934, including "Appendices I, II, III and IV." Welding, or any combination of riveting, bolting and welding, may be used for the connecting or assembling of component parts of metal beams, columns, trusses or other members, and for connecting such members to one another to make up the frame of a building or a structure; except that the strength of any one (1) joint shall be considered as the strength of the welded joints alone or as the strength of the rivets or bolts. Surfaces to be welded shall be free from loose mill scale, rust, paint or other foreign matter. No welding shall be performed in a strong wind or in rain until adequate protection has been provided.

ARTICLE 4002

Materials and Stresses

4002.01 Quality of Materials: Steel and iron used in building and structures shall be at least equal to the requirements of the Standard and Tentative Standard Specifications of the American Society for Testing Materials as follows:

Structural Steel: Serial Designation A9-36 for Structural Steel for Buildings. Manufactured only by the open hearth process.

Rivet Steel: Serial Designation A141-36 for Structural Rivet Steel.

Silicon Steel: Serial Designation A94-36 for Silicon Steel for Buildings.

Cast Steel: Serial Designation A27-36T for Steel Casting.

Wrought Iron: Serial Designation A85-36 for Wrought Iron Bars.

Cast Iron: Serial Designation A48-36 for Iron Casting.

Welding Materials: Electrodes, welding wire, and welding rods used for welding shall be at least equal to the requirements of the Code of the American Welding Society, edition of 1934, for such materials.

4002.02 Stresses:

(a) **Maximum Allowable Tension.** The tensile stress in pounds per square inch of net section of any member shall not exceed the value given in the following table for the material of such member:

Structural Grade Steel	18,000
Silicon Steel	24,000
Cast Steel	15,000
Wrought Iron	12,000
Cast Iron	None
Rivet Steel (heads not countersunk or flattened to less than three-eighths (⅜) inch)	13,500
Metal in welds, bare wire	13,000
Metal in welds, shielded wire	16,000
Bolts in tension of structural grade steel	16,000
Bolts in tension of special high strength alloy steel	½ yield point as of A. S. T. M. specification.

(b) **Maximum Allowable Compression on Short Members.** The compressive stress in pounds per square inch on the gross section of any compression member with a ratio l/r not greater than twelve (12), shall not exceed the value given in the following table for the material of such member:

Structural Grade Steel	18,000
Silicon Steel	24,000
Cast Steel	15,000
Wrought Iron	10,000
Cast Iron	10,000
Metal in welds, compression on section through throat of butt weld	15,000

(c) **Maximum Allowable Compression on Long Members.** In formulae (1) to 5) inclusive of this section, the symbols are defined as follows:

P = Axial load in pounds on compression member on basis of equivalent static load where impact of vibration is involved.

A = Gross section of metal in square inches at minimum cross section of member.

l = Unsupported length of member in inches.

r = Least radius of gyration of gross section in inches.

The compressive stress in pounds per square inch on the gross section of any compression member with a ratio l/r greater than twelve (12), shall not exceed the value given by the formulae, nor the maximum value stated below each formula. The ratio l/r for any compression member shall not exceed the value given in the following table:

Structural Grade Steel or Silicon Steel—
- For columns and other compression members 120
- For bracing 200

Cast Steel 120
Wrought Iron 120
Cast Steel 70

Structural Grade Steel

$$(1)\qquad \frac{P}{A} = \frac{18{,}000}{1 + \dfrac{l^2}{18{,}000\ r^2}}$$

with a maximum of...15,000

Silicon Steel

$$(2)\qquad \frac{P}{A} = \frac{24{,}000}{1 + \dfrac{l^2}{18{,}000\ r^2}}$$

For l/r not more than 120, with a maximum of 20,000

Cast Iron

$$(3)\qquad \frac{P}{A} = 10{,}000 - \frac{60l}{r}$$

Cast Steel

$$(4)\qquad \frac{P}{A} = \frac{15{,}000}{1 + \dfrac{l^2}{15{,}000\ r^2}}$$

with a maximum of....12,000

Wrought Iron

$$(5)\qquad \frac{P}{A} = 12{,}000 - \frac{60l}{r}$$

with a maximum of....10,000

Cast iron, cast steel, and wrought iron compression members shall not be used in any building of height greater than twice the least width, or greater than one hundred (100) feet; provided however, that this provision shall not apply to specially reinforced combination, pipe columns, and steel or iron columns filled with concrete which are designed in accordance with the provisions of Chapter 39—Reinforced Concrete Construction.

(d) **Maximum Allowable Bending Stresses.** The total stress in pounds per square inch in the extreme fiber of any rolled shape or built up member in bending shall not exceed the values given in the following table, when lateral deflection is prevented. Stresses in the steel section of a composite beam, as specified in Chapter 39—Reinforced Concrete Construction, shall be subject to the above limitation.

Structural Grade Steel...........	18,000
Silicon Steel	24,000
Cast Steel	15,000
Wrought Iron	12,000
Cast Iron, tension fibers.........	3,000
Cast Iron, compression fibers....	10,000
Metal in welds, bare wire, tension	13,000
Metal in welds, shielded arc, tension	16,000
Metal in welds—compression.....	15,000

(e) **Maximum Allowable Ratio l/b.** The laterally unsupported length of any beam or girder shall not exceed forty (40) times the width of the compression flange in the middle half of the span.

(f) **Stress on Beams with Unsupported Length Exceeding 15b.** The stress in pounds per square inch on the extreme fiber of the gross section of compression flange of any rolled sections, plate girders, and built up members of structural grade steel, when the unsupported length of the compression flange exceeds fifteen (15) times the least width of the compression flange, shall not exceed the value given by formula (6), in which S is the stress in pounds per square inch, l is the unsupported length in inches, and b is the least width of the compression flange in inches.

$$(6)\qquad S = \frac{20{,}000}{1 + \dfrac{l^2}{2{,}000\ b^2}}$$

(g) **Shear Values.** The maximum allowable shearing stress in poupnds per square inch shall not exceed the value given therefor in the following table:

Pins	13,500
Power-driven rivets	13,500
Turned bolts in reamed holes, with a clearance of not more than one-fiftieth (1/50) inch...	13,500
Hand-driven rivets	10,000
Unfinished bolts	10,000
Metal in welds.................	11,3000
Shear on gross section of girder webs—	
Structural grade steel...........	13,000
Silicon steel	16,000

ARTICLE 4003

Design Provisions

4003.01 Thickness of Metal: For members carrying the load of more than one hundred (100) square feet of floor or roof area, no steel less than five-sixteenths ($\frac{5}{16}$) inch thick shall be used in exposed exterior construction, and no steel less than 0.24 inch thick in the interior construction, except for linings or fillers, and in the webs of rolled structural shapes in interior construction.

4003.02 Connections:

(a) **Beam Connections.** The use of continuous beams and girders, designed in accordance with accepted engineering principles, shall be permitted provided that their connections be designed to carry all stresses to which they may be subjected. The connections at the ends of non-continuous beams shall be designed so as to avoid excessive secondary stresses due to bending.

(b) **In Buildings More Than Fifty (50) Feet in Height.** In buildings and structures over fifty (50) feet high, or of lesser height where the height exceeds one and one-half (1½) times the least width, all connections of columns to columns, all connections of beams, girders, and trusses to columns, all wind bracing connections, and all members of trusses, shall be power riveted, welded or bolted with turned bolts in drilled or reamed holes with not more than one-fiftieth (1/50) inch clearance. Other connections, except as provided in Paragraph (d) of this section, may be bolted with unfinished bolts in punched holes.

(c) **In Buildings Less Than Fifty (50) Feet in Height.** In buildings and structures less than fifty (50) feet in height, connections may be made with unfinished bolts in punched holes; except as provided in paragraphs (b) and (d) of this section.

(d) **Connections Subject to Impact of Reversal of Stress.** Connections of members carrying live loads other than wind loads which produce impact or reversal of stresses, shall be power riveted, welded, or bolted with turned bolts in drilled or reamed holes, with not more than one-fiftieth (1/50) inch clearance.

(e) **Bolts.** If value of bolts is figured on the nominal diameter, no threaded parts shall be in bearing, and lock washers or cupped nuts must be used to give full grip when turned tight. Bolts used in tension, or with threads in bearing, shall be figured on the net diameter of the smallest cross section. Finished bolts in reamed or drilled holes with greater clearance than one-fiftieth (1/50) inch shall be classified as unfinished bolts.

4003.03 Lateral Support of Compression Members: In order to be considered effective in reducing the unsupported length of a compression member, bracing must be adequate to resist a lateral force equal to two (2) per cent of the axial force.

4003.04 Steel and Composite Beams: Steel members encased in concrete may be figured as composite beams in accordance with the provisions of Chapter 39—Reinforced Concrete Construction.

ARTICLE 4004

Construction Provisions

4004.01 Test of Welded Structures: Welds shall be subjected to test in such manner as to be stressed in shear to one and one-half (1½) times the computed shear reaction. At least three (3) per cent of the welds in primary members, and not less than four (4) tests of primary members, and at least one (1) per cent of the welds in secondary members in each story shall be so tested as the work progresses and before it is encased in concrete or otherwise concealed. Should any of the primary joints fail under these tests, additional tests of five (5) per cent of the total number of primary joints shall be made for each failure. Should any of the secondary joints fail under these tests, additional tests of one (1) per cent of the total number of secondary joints shall be made for each failure. Any joints that have failed or show any signs of material damage as a result of such tests, shall be cut ont, rewelded and retested. In the case of shop welding such tests may be required before the member is erected at the building. All welding shall be done under supervision of a laboratory of recognized standing.

The Commissioner of Buildings shall be furnished, at the expense of the owner, with a certificate, from a laboratory of recognized standing, certifying that the tests required above have been made and giving the results of said tests.

Upon completion of all structural welding operations and required tests thereof, the architect or engineer responsible for the design of the structure, shall furnish to the Commissioner of Buildings a certificate, showing that the fabrication and erection of such welded structure has fulfiled the requirements of his design in every particular.

CHAPTER 41
Chimneys

ARTICLE 4101

General

4101.01 Definitions:

(a) **Interior Chimney.** A chimney built wholly within the walls of a building and having lateral support from the building.

(b) **Exterior Chimney.** A chimney so built that at least one (1) of its sides is coincident with the exterior walls of a building, by which it has lateral support.

(c) **Isolated Chimney.** Any chimney, other than an interior or exterior chimney. Also that part of a chimney above the building or walls which give it lateral support. Guys or struts shall not be considered as lateral supports for the purpose of this definition.

4101.02 Materials and Design:

(a) **Materials.** Every chimney shall be built of masonry, reinforced concrete or metal, with lining, where required, as hereinafter provided. Combustible materials shall not be included in the structure of any chimney nor in any chimney lining.

(b) **Wind Loads.** Every chimney shall be designed, constructed and maintained to resist a wind pressure of twenty (20) pounds per square foot on the portion less than one hundred seventy-five (175) feet above the ground, and thirty (30) pounds per square foot on the higher portion. The area for wind pressure may be considered as acting on two-thirds (⅔) of the projected area for round chimneys, and on five-sixths (⅚) of the projected area for octagonal chimneys. The stresses, due to combined wind and other loads, shall be provided for without increasing the allowable unit stresses.

(c) **Foundations and Supports.** Every chimney shall rest directly on the ground or shall be supported on structural members having four (4) hour fire-resistive rating.

(d) **Height Above Roof.**

Case 1. Where There Are No Other Roof Structures.

Every exterior and interior chimney in a building of other than fireproof or semi-fireproof construction shall extend not less than three (3) feet above any portion within five (5) feet of said chimney; and if said chimney projects through or adjoins a gable or hip roof, it shall project at least two (2) feet above the ridge of the roof.

Case 2. In Relation to Other Roof Structures.

Where a wooden tank, penthouse or other combustible structure is located above the roof of the same building with a chimney or upon an adjoining building the required minimum height of such chimney shall be five (5) feet above the top of such other structure; provided, however, that for each one (1) foot of horizontal distance intervening between such chimney and such other structure, said required minimum height shall be reduced by four (4) inches; provided, further, that the minimum required height above the roof shall be five (5) feet.

ARTICLE 4102

Special Provisions

4102.01 Isolated Masonry Chimneys:

(a) **Structural Materials and Stresses.** Isolated masonry chimneys shall be built of solid or hollow units of burnt clay or shale, or a combination thereof. Hollow units for such chimneys shall be of special design adapted to chimney construction and subject to approval by the Commissioner of Buildings. The bond for such units shall be at least equivalent to the bond required for bearing walls. Mortar used in masonry chimneys shall be Mortar No. 1 or Mortar No. 2 as given in Table 3803.02 (a) of Chapter 38—MASONRY CONSTRUCTION. Lime, sand and Portland Cement shall conform to the requirements given in Section 3802.08. The maximum compressive stress in any isolated masonry chimney shall not exceed two hundred (200) pounds per square inch, computed on the cross section of the wall of the chimney. The maximum compressive stress in any unit in such a chimney shall not exceed one-tenth (1/10) the "Minimum Ultimate Compressive Strength of Individual Units" as set forth in Table 3803.02 (a). When chimneys are built of hollow masonry units, the weight of the masonry wall shall not be assumed to exceed the weight of the individual units by more than five (5) per cent of the individual units. Unit stress due to wind pressure shall not exceed seventy-five (75) per cent of the unit compressive stress due to dead load.

(b) **Chimneys Joined to Building Walls.** Any masonry chimney outside or partly outside a building, but having one (1) or more walls joined to a wall of the building, shall be built as required for an interior masonry chimney.

4102.02 Interior and Exterior Masonry Chimneys:

(a) **Structural Materials and Stresses.** All the provisions of Section 4102.01, Paragraph (a), for isolated masonry chimneys, shall apply to interior and exterior masonry chimneys, with the added provisions that the walls of an interior masonry chimney shall be not less than eight (8) inches thick in any case, and not less than twelve (12) inches thick where they form a portion of a bearing wall of the building; except that small interior chimneys, lined as required by Section

4102.06, Paragraph (i), may have walls four (4) inches less in thickness than required by this paragraph and except that solid concrete units may be used for chimneys built in connection with exterior walls of hollow units in buildings not exceeding three (3) stories in height.

4102.03 Reinforced Concrete Chimneys:

(a) **General.** Chimneys of reinforced concrete shall be constructed in accordance with the provisions of Chapter 39, Reinforced Concrete Construction, and shall have a minimum wall thickness of four (4) inches exclusive of any lining.

4102.04 Isolated Metal Chimneys:

(a) **Metal Thickness.** The thickness of metal in isolated metal chimneys shall be not less than the following:

For chimneys of not more than fourteen (14) inches inside diameter. $\frac{1}{16}$ inch

For chimneys more than fourteen (14) inches and not more than twenty-four (24) inches inside diameter .... $\frac{3}{16}$ inch

For chimneys of more than twenty-four (24) inches inside diameter .... ¼ inch

(b) **Stresses.** No stress in any isolated metal chimney shall exceed the allowable stress given for such metal in Chapter 40, Steel and Metal Construction.

4102.05 Interior Metal Chimneys:

(a) **Metal Thickness.** The thickness of metal in interior metal chimneys shall be not less than one-fourth (¼) inch.

(b) **Stresses.** No stress in any interior metal chimney shall exceed the allowable stress given for such metal in Chapter 40—STEEL AND METAL CONSTRUCTION.

(c) **Bracing.** Every interior metal chimney shall be braced to the floor and roof construction of the building at intervals not exceeding ten (10) times the diameter of such chimneys nor more than forty (40) feet.

(d) **Surrounding Air Space.** Every interior metal chimney shall be surrounded by continuous air space from the lowest story through the roof, not less than four (4) inches across at any point, and said air space shall be surrounded by brick, hollow tile, or reinforced concrete. No structural metal in such air space shall be without fireproof cover ng.

4102.06 Linings:

(a) **General.** Chimney linings shall meet the provisions for one (1) or more of the following six (6) types:

Type 1. Fire clay brick, fire clay blocks, or other fire clay products, meeting the requirements of Paragraph (b) of this Section.

Type 2. Uncalcined diatomaceous earth blocks or bricks meeting the requirements of Paragraph (c) of this Section.

Type 3. Calcined diatomaceous brick meeting the requirements of Paragraph (d) of this Section.

Type 4. Fused asbestos blocks or boards meeting the requirements of Paragraph (e) of this Section.

Type 5. Shale brick meeting the requirements of Paragraph (ee) of this Section.

Type 6. Paving brick meeting the requirements of Paragraph (f) of this Section.

All bricks, blocks, boards or other products of fire clay, diatomaceous earth, asbestos, or other material, and all pointing cement required or permitted by this Section shall resist the disintegrating action of moist steam and the acid and gaseous fumes present in the chimneys.

(b) **Fire Clay.** Fire clay bricks, fire clay blocks, or other fire clay products shall be not less than three and three-fourths (3¾) inches thick, unless supported on structural steel ledges at intervals of not more than twenty (20) feet, in which case the thickness shall be not less than two (2) inches. All such fire clay bricks, blocks, or other products shall be laid up in No. 1 mortar or acid resisting, plastic, refractory cement.

(c) **Uncalcined Diatomaceous Earth.** Uncalcined diatomaceous earth blocks or bricks shall contain not less than seventy-five (75) per cent by weight of diatomaceous earth and fifteen (15) per cent by weight of asbestos fiber, and shall be not less than one and one-fourth (1¼) inches thick. After such diatomaceous earth blocks or bricks have been set, they, and all metal bands and ties exposed within the chimney, shall be plastered with semi-refractory cement, not less than one-half (½) inch thick, thus making the total thickness of insulation not less than one and three-fourths (1¾) inches. Uncalcined diatomaceous earth chimney lining shall not be used in any chimney in which the temperature exceeds sixteen hundred (1,600) degrees Fahrenheit.

(d) **Calcined Diatomaceous Brick.** Calcined diatomaceous brick shall be semi-refractory, not less than four and one-half (4½) inches thick, free from shrinkage at temperatures below twenty-five hundred (2,500) degrees Fahrenheit, and shall have a crushing strength of not less than three hundred and fifty (350) pounds per square inch. All such calcined diatomaceous brick shall be laid up in acid resisting, plastic, refractory cement, and the surface shall be given a wash coating of such cement. Calcined diatomaceous chimney lining shall not be used in any chimney in which the temperature exceeds twenty-five hundred (2,500) degrees Fahrenheit.

(e) **Fused Asbestos or 85% Magnesia.** Chimney linings of fused asbestos or eighty-five (85) per cent magnesia boards, sheets or blocks shall meet the requirements for one (1) of the following two (2) types, but shall not be used in any chimney in which the temperature exceeds seven hundred (700) degrees Fahrenheit.

Type a. Solid blocks not less than one and one-half (1½) inches thick. After such solid blocks have been set, they, and any metal bands and ties exposed within the chimney shall be plastered with refractory cement not less than one-half (½) inch thick, thus making the total thickness not less than two (2) inches.

Type b. Alternate flat and corrugated sheets of asbestos paper cemented together and fused to a minimum thickness of two (2) inches. After such fused asbestos boards have been set into the chimney, they and all exposed metal bands or ties shall be pointed with refractory cement.

(ee) **Shale Brick.** Shale brick shall be not less than three and three-fourths (3¾) inches thick. Shale brick shall not be used in chimneys in which the temperature exceeds seven hundred (700) degrees Fahrenheit.

(f) **Paving Brick.** Paving brick shall be not less than three and three-fourths (3¾) inches thick. Paving brick shall not be used in chimneys in which the temperature exceeds seven hundred (700) degrees Fahrenheit.

(g) **In Isolated Masonry Chimneys.** Every isolated masonry chimney shall be provided with an independent lining of fire brick, fire tile or other approved refractoy material, not less than three and three-fourths (3¾) inches thick. Such lining shall extend at least two (2) feet below the bottom of the smoke inlet and at least thirty (30) feet above the top of the smoke inlet. There shall be an air space of at least two (2) inches between such lining and the chimney wall. The thickness of the independent lining shall not be considered as a part of the required thickness of the wall.

(h) **In Larger Interior Masonry Chimneys.** Every circular or octagonal interior masonry chimney having an inside diameter or least dimension of more than thirty (30) inches, and every non-circular or non-octagonal interior masonry chimney having any interior side dimension more than twenty-four (24) inches, shall be provided with an independent lining and air space as required by Paragraph (g) of this section for an isolated masoury chimney. The dimensions referred to in this Paragraph shall be measured in the clear space inside the lining.

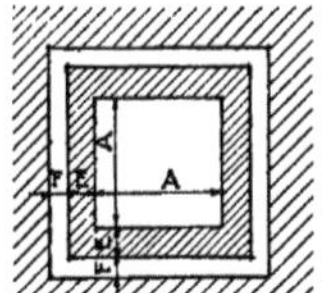

Fig. 1

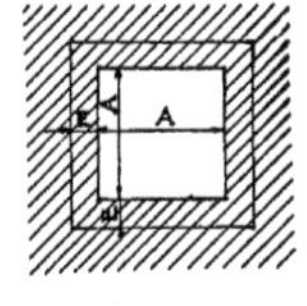

Fig. 2

Section 4102.06h.

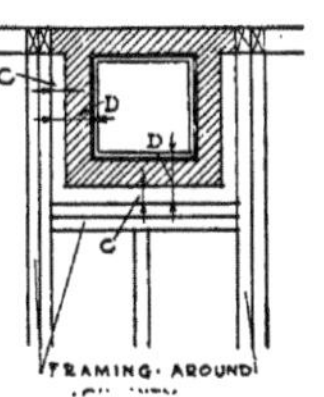

Fig. 3

Section 2115.00.

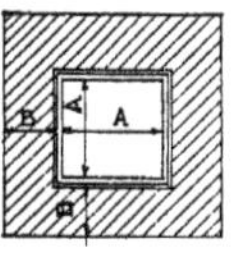

Section 4102.06i.

Fig. 4

CHIMNEYS—INSULATING CAVITIES WHERE REQUIRED.

Figs. 1, 2 (A) Maximum inside dimension.
(E) Insulating lining.
(F) Insulating cavity.

Explanation: If A is more than 24 in. an insulating lining (E) is required. (See Section 4102.06h) If A is more than 24 in. the walls surrounding shall have an insulating cavity F not less than 2" wide.

If E in Fig. 2 is of fire brick of 4" or more in thickness it may be considered as a portion of thickness required for walls surrounding.

See Chapter 2115. Framing Around Chimneys.

Fig. 3. (C) Is distance joists or timbers are to be kept away from walls of chimneys = 2".

(D) Is distance to be kept away from inside of flue lining = 7".

Fig. 4. Shows Chimney with flue lining, not less than ¾" thickness.

If A is 24 in. or less, walls surrounding may be (B) or 4".

(i) **In Small Interior Masonry Chimneys.** Every circular or octagonal interior masonry chimney having an inside diameter or least dimension not exceeding thirty (30) inches and every non-circular or non-octagonal interior masonry chimney having no interior side dimension exceeding twenty-four (24) inches, shall be provided with a lining of burnt fire clay not less than three-fourths (¾) inch thick, extending from a point at least two (2) feet below the bottom of the smoke inlet to the top of the chimney. No chimney lining required by this Paragraph shall be installed by being dropped or lowered into a previously constructed section of the structural part of the chimney; but said structural part shall be paid up with mortar around each successive length of lining as said lining is set. All joints and spaces between the masonry and the chimney lining shall be thoroughly slushed and grouted full as each course of masonry is laid.

(j) **In Concrete Chimneys.** Every reinforced concrete chimney shall be lined as provided in Paragraph (g) of this Section, for isolated masonry chimneys.

(k) **In Interior Metal Chimneys.** Every interior metal chimney shall be provided with a lining for its entire height.

CHAPTER 42.

Floor and Roof Construction.

ARTICLE 4201.

Flats and Segmental Clay Tile Arches.

4201.01 Hollow Units for Arches: All structural clay tile used in arch construction shall be at least equal to the requirements of the Standard Specifications of the American Society for Testing Materials for "Structural Clay Floor Tile (Serial Designation C57-37T). All structural clay tile shall be bedded in Mortars 1, 2 or 4, as indicated in Chapter 38—MASONRY CONSTRUCTION. All joints between tile shall be fully and evenly bedded with mortar. Tile slab fillers, the full depth of the tile, shall be used where necessary to close and key the arch. Arch centers shall not be removed until the mortar has set.

4201.02 Limiting Proportions:

(a) **Flat Arches.** Flat arches of structural clay tile shall have a thickness above the bottom of the supporting beams of at least one and one-fourth (1¼) inch per foot of span, but not less than six (6) inches.

(b) **Segmental Arches.** Segmental arches of structural clay tile shall have a rise of at least one (1) inch per foot of span, and shall have a least thickness of one-half (½) inch per foot of span, and not less than six (6) inches.

4201.03 Skewbacks and Tie Rods:

(a) **General.** Both flat and segmental arches shall have skewbacks carefully fitted to the supporting beams and designed to resist both the horizontal thrust of the arch and vertical loads. Both flat and segmental arches shall have the tie rods designed to resist the thrust. These tie rods shall be of not less than three-fourths (¾) inch diameter and shall be protected the same as is required for other metal members of the floor system and shall be located below the center of arch at support and as much below as practicable.

(b) **Steel for Tie Rods.** All tie rods shall be of steel conforming to the requirements of the Standard Specifications of the American Society for Testing Materials for Structural S ee for Buildings (Serial Designation A9-36).t l

(c) **Stresses Permitted.** The tensile stress at the root of threads or at minimum section shall not exceed that allowed under Chapter 40—STEEL AND METAL CONSTRUCTION.

ARTICLE 4203.

Precast Concrete Joist Floors.

4203.01—General: Precast concrete joist floor construction includes floor systems in which precast reinforced concrete joists are used in conjunction with precast or cast-in-place floor slabs.

4203.02 Materials: Materials shall conform to the provisions of Chapter 39—REINFORCED CONCRETE CONSTRUCTION.

4203.03 Design: The design and construction of these floors shall conform to the provisions of Chapter 39—REINFORCED CONCRETE CONSTRUCTION, except as provided in this Article. When the top slab is cast in place and adequately reinforced and bonded to the joist by steel stirrups designed to resist the horizontal shear, the construction may be considered as a T-beam. The allowable designed stress for bond between slab and joist shall not exceed sixty (60) pounds per square inch.

4203.04 Minimum Requirements: The depth of joists, exclusive of the floor slab, shall be not more than four (4) times the width of the top or bottom flanges nor less than one twenty-fourth (1/24) of the span length. The thickness of the top slab shall be not less than one-twelfth (1/12) clear span between joists nor less than two (2) inches and not less than one and one-half (1½) inches over the joists. Stirrups shall be not less than one-fourth (¼) inch round steel bars spaced not farther apart than six (6) inches at the end one-fourth (¼) length of the joist and eighteen (18) inches in the middle half of the joist. All stirrups shall project into the slabs and looped stirrups shall be set parallel with the length of the joists. The requirements of Chapter 39—REINFORCED CONCRETE CONSTRUCTION for shear shall be met in all precast concrete joists.

CHAPTER 43.

Trussed Steel Joist and Other Construction.

ARTICLE 4301.

Trussed Steel Joists.

4301.01 General Requirements:
(a) **General.** Subject to the limitations of use prescribed by this ordinance, trussed steel joists made of rolled shapes, round bars or specially rolled bars riveted or welded together, or made by expanding rolled shapes, may be used for supporting floors or roofs between walls or between main girders or beams, where the clear span of the joist does not exceed 32 feet. Exposed trussed steel joists shall not be used for first floor panels except over accessible basement spaces, nor in any floor or roof above any room or space where a condition of high relative humidity occurs.

(b) **Specifications.** All steel used shall conform to the requirements of the Standard Specifications of the American Society for Testing Materials for Structural Steel for Buildings (Serial Designation A9-36).

(c) **Minimum Thickness.** The thickness of any material in trussed steel joists shall be not less than one-eighth (⅛) inch and in bridging systems not less than one-sixteenth (1/16) inch.

4301.02 Loads and Stresses:
(a) **Concentrated Loads.** Trussed steel joists, when erected and bridged, shall be designed to sustain safely eight hundred (800) pounds at any panel point on any one (1) joist.

(b) **Stresses.** All stresses used in the design of trussed steel joists shall be limited by the requirements for structural grade steel as given in Chapter 40—STEEL AND METAL CONSTRUCTION.

4301.03 Limiting Provisions: The maximum horizontal projection of any part of the compression flange from the plane of the web shall not exceed twelve (12) times the thickness of the flange metal. Top chords may be considered as fixed laterally by a concrete slab above, or as supported laterally at points of attachment to a wood floor above. The ratio of 1 to r shall not exceed one hundred twenty (120). In computing the resistance of joists to construction loads, the top chord shall satisfy the additional requirement that it shall safely carry the required compression, using 1 as the distance between lines of bridging and r as the least radius of gyration around the vertical axis; but the ratio of 1 to r in this case shall not exceed two hundred (200). No bending stresses shall be assumed in top chords of trusses supporting concrete slabs which have a thickness of more than one-fourteenth (1/14) of the distance between supports of the top chord; but for joists supporting concrete slabs thinner than this, and joists supporting wood floors, the theoretical bending stress for a uniform load shall be computed and the combined stress at any point in the top chord shall not exceed eighteen thousand (18,000) pounds per square inch.

4301.04 Spacing: The spacing of the trussed steel joists is dependent on the design of the top slab computed in accordance with the provisions of this ordinance for the materials used and in no case shall it be greater than twenty-four (24) inches in floors or thirty (30) inches in roofs; except that trussed steel joists may be used to support wood or sheet metal roofs, when spaced not more than seven (7) feet apart.

4301.05 Connections: All joints and connections in steel joists shall be made by connecting the members directly to one another by electric arc, spot or pressure welds, or by rivets. In the case of expanded joists, a portion of the metal may be left intact to form a connection. All joints and connections shall be capable of withstanding a stress at least three (3) times the design stress, and shall be sufficiently rugged to resist the stressed incident to handling and erection. All intermediate joints shall be designed so as to be approximately symmetrical. The end portions may be designed as projecting beams. Stresses arising from eccentricity shall be included with other stresses in designing all members. The Commissioner of Buildings shall be furnished with a certificate from the designing architect or engineer, that the design and fabrication of welded members has been done in accordance with the provisions of this ordinance.

4301.06 Bridging:
(a) **Number of Lines Required.** The number of lines of bridging provided shall be not less than required by the following table:

Clear Span	Number of Lines of Bridging
7 feet or less	None.
Over 7 feet and not over 14 feet	One row, near center.
Over 14 feet and not over 21 feet	Two rows.
Over 21 feet and not over 32 feet	Three rows.

(b) **Stress Transfer and Design.** Each line of bridging shall be capable of transferring not less than five hundred (500) pounds from each joist to the adjoining joists. All bridging shall be designed in accordance with the structural provisions of this ordinance.

(c) **Deck as Top Member.** In the case of joists provided with nailer strips carrying a wood deck, or of joists clipped to a concrete deck, the deck may be used as the top member of the bridging system.

(d) **Nailing Strip Fastenings.** Nailing strips, when used, shall be attached continuously throughout their entire length to the top chords of steel joists, or shall be bolted or otherwise attached at intervals not exceeding eighteen (18) inches on centers.

4301.07 Erection:
(a) **Bearing Pressures.** Joists shall have a bearing of at least four (4) inches on masonry or concrete, and at least two and one-half (2½) inches on steel; but the unit bearing pressure shall in no case exceed that allowed for the materials used as stated elsewhere in this ordinance.

(b) **Anchors.** The size and arrangement of anchors shall be equivalent to that specified for wood and steel joists as called for in Chapter 38—MASONRY CONSTRUCTION.

CHAPTER 44.

Elevators, Dumbwaiters, Escalators and Mechanical Amusement Devices.

ARTICLE 4401.

General.

4401.01 Scope: The following provisions of construction shall apply to all elevators, dumbwaiters and escalators; to all mechanical equipment used for, or in connection with, the raising or lowering of any stage, or orchestra floor, or any platform and to all mechanical amusement devices and apparatus in parks, carnivals and the like, which may hereafter be installed. They shall also apply to all existing aforementioned devices, whenever such existing devices are altered in any way which may change the capacity or speed of, the purpose for which the device is used, or its operating, controlling or safety equipment, but they shall not apply to other existing installations, except as hereinafter prescribed, nor shall they apply to repairs in such existing devices which are necessary to keep same in safe operating condition. When the alteration consists solely of a change in the character of electric power supply, only such parts, the operation of which is thereby affected, need be changed, except that in all cases electrically released brakes shall be installed. These provisions shall not apply to other raising, lowering or conveying devices; nor shall they apply to elevators used only for handling building materials, or tools in any building in course of construction, but the Commissioner of Buildings shall make such reasonable requirements as he may deem necessary for public safety in the operation of such elevators.

4401.02 Definitions: For the purposes of this ordinance, terms used in this chapter are hereby defined as follows:

Buffer. A device designed to absorb the impact of an elevator, or dumbwaiter car, or counterweight, at the extreme limits of travel.

Cab, Elevator. An inclosure consisting of walls and top built upon a car platform.

Car. The load carrying unit.

Car Door or Gate. A door, or gate in or on an elevator or dumbwaiter car.

Car Leveling Device. Any mechanism, or control, which will move a car within a limited zone toward, and stop the car at, the landing. For an elevator, the device may also be used for emergency operation of the car throughout its entire travel and for safe lifting purposes.

Car Frame. A supporting frame to which the platform upper and lower sets of guide shoes and the hoisting cables are usually attached.

Car Platform. The structure which forms the floor of the car and directly supports the load.

Clearance, Bottom. The vertical distance between any obstruction in the pit, exclusive of the compensating device, buffer and buffer supports, and the lowest point of the understructure of the car exclusive of the safeties, car frame channels and guide shoes and other necessary equipment attached to the underside of the platform, when the car floor is level with the bottom terminal landing.

Clearance, Top.

Item 1. Car. The top clearance of a car is the distance of the car floor can travel above the level of the top terminal landing without any part of the car, or devices attached thereto, coming in contact with the overhead structure.

Item 2. Counterweight. The top clearance of a counterweight is the shortest vertical distance between any part of the counterweight structure and the nearest part of the overhead structure, or any other obstruction when the car is level with the bottom terminal landing.

Door Closer. A device, operated by gravity or other means, which will automatically close a door when released by the operator, or by suitable automatic means.

Dumbwaiter. A raising and lowering mechanism equipped with a car, the floor area of which does not exceed nine (9) square feet, whose compartment height does not exceed four (4) feet, the capacity of which does not exceed five hundred (500) pounds and which is used exclusively for carrying freight.

Electric Contact, Car Door or Gate. A device to open the control circuit, or an auxiliary circuit, when the car door or gate is open more than two (2) inches from full closure and thus prevent operation to move the car away from the landing.

Electric Contact, Hatchway Door or Gate. A device to open the control circuit, or an auxiliary circuit, when the hatchway door or gate, at which the car is standing, is open more than two (2) inches from full closure and thus prevent operation to move the car away from the landing.

Door Unit Contact System. A contact system which requires that the hatchway door or gate, at which the elevator is standing, must be closed before the elevator can leave the landing, but which does not prevent the operation of the car if other doors in the hatchway are not closed.

Hatchway Unit Contact System. A contact system which will prevent the operation of the car unless all hatchway doors are closed.

Elevator. A raising and lowering mechanism equipped with a car or platform which moves in guides in a substantially vertical direction, further defined as of the following types:

(a) **Auxiliary Power Elevator.** An elevator having a source of mechanical power in common with other machinery.

(b) **Chain Driven Elevator.** An elevator having its machine connected to a reversible motor, engine, or turbine by a chain.

(c) **Double-Belted Elevator.** An auxiliary power elevator in which the direction of travel is changed without reversal of the prime mover.

(d) **Electric Elevator.** An elevator operated by an electric motor directly applied to the elevator machinery.

(e) **Freight Elevator.** An elevator designed for carrying freight and the operator and persons necessary for loading and unloading.

(f) **Gravity Elevator.** An elevator in which gravity is the source of power.

(g) **Hand Elevator.** An elevator driven by manual power.

(h) **Hydraulic Elevator.** An elevator operated by liquid under pressure.

Item 1. Plunger Elevator. A hydraulic elevator having a ram or plunger directly attached to the under side of the car platform.

Item 2. Rope Geared Hydraulic Elevator. A hydraulic elevator in which the movement of the car is obtained by multiplying the travel of a piston or ram by a system of sheaves over which the hoisting ropes operate.

(i) **Passenger Elevator.** An elevator which is designed to carry persons.

(j) **Power Elevator.** An elevator operated otherwise than by gravity or manually.

(k) **Private Residence Elevator.** A power passenger elevator serving a single family, installed in a dwelling and having a rated capacity of not more than seven hundred (700) pounds and a rated speed of not more than fifty (50) feet per minute.

(l) **Sidewalk Type Elevator.** A freight elevator having a speed of not more than fifty (50) feet per minute and having the top landing not more than four (4) feet above grade level at the point where elevator is located, the platform of which elevator is suspended or supported at or below the platform level and in such a manner as will not permit tipping of the platform.

(m) **Single-Belted Elevator.** An elevator machine connected to a reversible motor, engine or turbine by a belt.

(n) **Steam Elevator.** An elevator operated by a steam engine directly applied to the elevator machinery.

Elevator Control. A system of regulation by which the starting, stopping, direction of motion, acceleration, speed and retardation of an elevator are governed and further defined as of the following types:

(a) **Generator Field Control.** A system in which control is accomplished primarily by the use of an individual generator for each elevator, in which the voltage applied to the hoisting motor is adjusted by varying the strength and direction of the generator field.

(b) **Multi-Voltage Control.** A system in which control is accomplished primarily by impressing successively on the armature of the hoisting motor a number of substantially fixed voltages.

(c) **Rheostatic Control.** A system in which control is accomplished primarily by varying resistance or reactance in the armature or field circuit of the hoisting motor, or by any combination thereof.

Emergency Release. A device to make the door or gate electric contacts or door interlocks inoperative in case of emergency.

Emergency Stop Switch. A device in a car used to cut off the power from the machine independently of the operating devices.

Escalator. A moving inclined continuous stairway or runway used for raising or lowering passengers.

Hatchway. The space in which an elevator or dumbwaiter is designed to operate.

Hatchway Door or Gate. The hinged or sliding portion of the hatchway inclosure for access to the car at any landing.

Hatchway Door Interlock. A device, the purpose of which is:

Item 1. To prevent the operation of the machine to move the car away from a landing unless the hatchway door at that landing is locked in the closed position; and

Item 2. To prevent the opening of the hatchway door from the landing side, except by special key, unless the car is at rest within the landing zone, or is coasting through the landing zone, with its operating device in the stop position.

(a) **Door Unit Interlock System.** An interlock system which requires that the hatchway door at which the elevator is standing must be locked in the closed position before the elevator can leave the landing, but which does not prevent the operation of the car if other doors in the hatchway are not locked.

(b) **Hatchway Unit Interlock System.** An interlock system which will prevent the operation of the car unless all hatchway doors are locked in the closed position.

Hatchway Inclosure. Any structure which separates the hatchway, either wholly or in part, from the floors or landings through which the hatchway extends.

Landing. That portion of a floor or platform used to receive and discharge passengers or freight.

Landing Zone. The space from a point not more than eighteen (18) inches below the landing to a point not more than eighteen (18) inches above the landing.

Load, Rated. The load which the elevator or dumbwaiter is designed to carry at rated speed.

Machine. The machinery and its equipment used in raising or lowering the car or platform, of the following types:

(a) **Direct-Drive Machine.** A machine in which the power is transmitted directly to the driving sheave or sheaves without intermediate mechanism or gears.

(b) **Spur-Geared Machine.** A machine in which the power is transmitted to the driving sheaves or drum through spur gearing.

(c) **Traction Machine.** A machine in which the movement of the car and counterweight is obtained by means of traction between the driving drum, sheave or sheaves and the hoisting cables.

(d) **Winding Drum Machine.** A machine in which the cables are fastened to and wind on a drum.

(e) **Worm-Geared Machine.** A machine in which the power is transmitted to the driving sheaves and drum through worm gearing.

Mechanical Amusement Device. Any device designed or used to move a person, or persons, or to permit the movement of a person, or persons, by mechanical means, in any direction for amusement and operated within a space or over a route devoted exclusively to such use.

Operating Device. The device used to actuate the control.

Operation. The method of actuating the control, of the following types:

(a) **Automatic Operation.** Operation by means of buttons or switches, both in the car and at the landings, the momentary pressing of which will cause the car to start and automatically stop at the floor corresponding to the button pressed.

(a-1) **Single Automatic Operation.** Operation by means of one (1) button in the car for each landing level served and one (1) button at each landing so arranged that if any car or landing button has been pressed, the pressure of any other car or landing operating button will have no effect on the operation of the car until the response to the first button has been completed.

(a-2) **Non-Selective Collective Automatic Operation.** Operation by means of one (1) button in the car for each landing level and one (1) button at each landing, wherein all stops registered by the momentary pressure of landing or car buttons are made, irrespective of the number of buttons pressed, or the sequence in which the buttons are pressed. With this type of control the car stops at all landings for which buttons have been pressed, making the stops in the order in which the landings are reached after the buttons have been pressed, but irrespective of its direction of travel.

(a-3) **Selective Collective Automatic Operation.** Operation by means of one (1) button in the car for each landing level served and by "up" and "down" buttons at the landings, wherein all stops registered by the momentary pressure of the car buttons are made as defined under non-selective collective automatic operation, but wherein the stops registered by the momentary pressure of the landing buttons are made in the order in which the landings are reached in each direction of travel after the buttons have been pressed. With this type of control, all "up" landing calls are answered when the car is traveling in the "up" direction and all "down" landing calls are answered when the car is traveling in the "down" direction, except the uppermost or lowermost calls which are answered as soon as they are reached, irrespective of the direction of travel of the car.

(b) **Car-Switch Automatic Floor Stop Operation.** Operation in which the stop is initiated by the operator from within the car with a definite reference to the landing at which it is desired to stop, after which the slowing down and stopping of the elevator is automatically effected.

(c) **Car-Switch Operation.** Operation wherein the movement of the car is directly and solely under the control of the operator by means of a switch or buttons in the car.

(d) **Continuous-Pressure Operation.** Operation with "up" and "down" buttons or an "up" and "down" switch in the car and at each landing, except the terminal landings, at each of which there may be one (1) button or switch, any one of which may be used to control the movement of the car in the direction for which the button is pushed or the switch actuated, but only so long as the button or switch is manually held in the operation position. Landing stops are not is automatic.

(e) **Dual-Operation.** Operation whereby the elevator is arranged to be operated as an automatic operation elevator through landing and car buttons or switches, or as a manual operation elevator by an operator in the car, who may either use a car switch or the buttons provided in the car. When operated by an operator, upon the throwing of a suitable switch, or switches, the car can no longer be started by the landing button. These buttons may, however, be used to signal the operator that the car is desired at a certain landing.

(f) **Pre-Register Operation.** Operation in which signals to stop are registered in advance by buttons in the car and at landings. At the proper point in the car travel, the operator in the car is notified by a signal to initiate the stop, after which the landing stop is automatic.

(g) **Signal Operation.** Operation by means of signal buttons or switches in the car and "up" and "down" direction buttons or switches at the landings, by which predetermined landing stops may be set up or registered for an elevator or for a group of elevators. The stops set up by the momentary pressure of the car buttons are made automatically in succession as the car reaches those landings, irrespective of its direction of travel or the sequence in which the buttons are pressed. The stops set up by the momentary pressure of the "up" and "down" buttons at the landing are made automatically by the next car in the group approaching the landing in the corresponding direction, irrespective of the sequence in which the buttons are pressed. With this type of operation the car can be started only by means of a starting switch or button in the car.

Overhead Structure. All of the equipment supporting structure and platforms at the top of the hatchway.

Overtravel, Bottom.

Item 1. Car. The bottom overtravel of the car is the distance the car floor can travel below the level of the bottom terminal landing until the fully loaded car rests on the buffers, and includes the resulting buffer compression.

Item 2. Counterweight. The bottom overtravel of the counterweight is the distance the counterweight can travel below its position when the car platform is level with the top terminal landing and until the counterweight rests on the buffers, and includes the resulting buffer compression.

Overtravel, Top. The distance provided for the car floor to travel above the level of the upper terminal landing until the car is stopped by the normal terminal stopping device.

Pit. That portion of a hatchway extending below the level of the bottom landing to provide for bottom overtravel and clearance and for parts which require space below the bottom limit of car travel.

Power Operated Door or Gate. A door or gate opened and closed by power as hereinafter described, of the following types:

(a) **Power Closed Door or Gate.** A door or gate which is manually opened and is closed by power other than by hand, gravity, springs, or the movement of the car.

(b) **Power Opened, Self-Closing Door or Gate.** A door or gate which is opened by power other than by hand, gravity, springs, or the movement of the car, and when released by the operator is closed by energy stored during the opening operation.

(c) **Power Operated Door or Gate, Automatically Opened.** A door or gate which is opened other than by hand, gravity, springs or the movement of the car, the opening of the door being initiated by the arrival of the car at or near the landing. The closing of such door or gate may be under the control of the operator or may be automatic.

(d) **Power Operated Door or Gate, Manually Controlled.** A door or gate which is opened and closed by power other than by hand, gravity, springs or the movement of the car, the door movement in each direction being controlled by the operator.

Power Operated Door or Gate Device. A device to operate the hatchway door and car door or gate, by power other than by hand, gravity, springs or the movement of the car.

Safety, Car or Counterweight. A mechanical device attached to the car or counterweight frame to stop and hold the car or counterweight in case of predetermined overspeed free fall, or through slackening of the cables.

Speed, Rated. The speed which the car is designed to attain when carrying its rated load in the "up" direction.

Terminal Stopping Device, Final. An automatic device for stopping the car and counterweight from rated speed, within the top clearance and bottom overtravel, independently of the operation of the normal terminal stopping device, and the operating device.

Terminal Stopping Device, Normal. An automatic device for stopping the car within the overtravel independently of the operating device.

Travel or Rise. The vertical distance between the bottom terminal landing and the top terminal landing.

ARTICLE 4402.

Hatchway Requirements for Power Elevators and Mechanical Equipment for Raising or Lowering any Stage or Orchestra Floor or Platform.

4402.01 Clearance Between Cars, Counterweights and Hatchway Inclosures: The minimum clearance between cars, counterweights and hatchway inclosures shall be as follows:

(a) **Car and Hatchway Inclosure.** Between the sides of the car and the hatchway inclosure three-fourths (¾) inch.

(b) **Car and Counterweight.** Between the car and its counterweight one (1) inch.

(c) **Car and Thresholds.** Between the car platform and the landing thresholds for an elevator using side post construction one-half (½) inch and for an elevator using corner post construction three-fourths (¾) inch.

(d) **Car Landing and Inclosure.** The maximum clearance between the landing side of the car platform and the hatchway inclosure shall be as follows:

Item 1. Between the loading side of the car platform at the car entrance and the inclosure, five (5) inches, except where doors are installed wholly within the hatchway where this distance may be increased to seven and one-half (7½) inches.

Item 2. Between the car platform and the landing threshold, one and one-half (1½) inches.

4402.02 Pits: The depths of pits measured from the thresholds of the bottom landing shall be not less than the following:

(a) Elevators having a total travel of not more than fifteen (15) feet, a speed not more than fifty (50) feet per minute and not equipped with buffers, and private residence elevators, two (2) feet.

(b) Elevators equipped with spring buffers or their equivalent, except private residence elevators, three (3) feet, six (6) inches.

(c) Elevators equipped with oil buffers or their equivalent, the distance between the upper surface of the car platform and the bottom of the buffer strike plate plus the fully extended buffer, plus three (3) inches.

(d) Additional pit depth shall be provided if necessary to provide clearance for compensating rope sheaves and any vertical movement thereof, and for any obstruction in the pit when the buffers are fully compressed. When an elevator car, except of the sidewalk type and private residence elevator, rests on the fully compressed buffer, there shall be a clear space vertically of not less than two (2) feet between the lowest projection of the underside of the car platform, except guide shoes and aprons attached to the car sills, safeties, car frame channels, junction boxes and other necessary equipment attached to the underside of the platform and any obstruction in the pit, exclusive of the compensating device, buffers and buffer supports.

(e) The depth of pits, measured from the threshold of the bottom landing, may be reduced where structural conditions make this necessary, if the construction used provides a two (2) foot clearance, as hereinbefore prescribed. Where precompression is used, the depth of the pit may be reduced to correspond with CAR CLEARANCES AND

COUNTERWEIGHT CLEARANCES hereinafter prescribed.

4402.03 Car Clearance: When an elevator car is at its top landing, the clear distance between the top of the crosshead of the car and any obstruction in the hatchway above it shall be not less than:

(a) If spring buffers, or their equivalent, are used the clearance between the bottom of the counterweight buffer and its striking block, in any case not less than six (6) inches, plus the compression of the buffer, plus two (2) feet; provided however, that if the car is a private residence elevator and is equipped with an emergency stop switch installed in an accessible point on top of the car, the clearance may be reduced one (1) foot.

(b) If oil buffers or their equivalent, are used the clearance between the bottom of the counterweight buffer and its striking block, in any case not less than six (6) inches, plus one and one-half (1½) times the stroke of the buffer corresponding to tripping speed of the elevator at full speed, plus two (2) feet. If retarders are installed which reduce the speed of the car at terminal landings, and the stroke of the buffer is reduced correspondingly, not less than the clearance between the bottom of the counterweight buffer and its striking block, in any case not less than six (6) inches, plus the stroke of the buffer corresponding to tripping speed of the elevator at full speed, plus two (2) feet, plus one-half (½) of the stroke of the buffer used. If provision is made to eliminate the jump of the car at counterweight buffer engagement, not less than the clearance between the bottom of the counterweight buffer and its striking block, in any case not less than six (6) inches, plus the stroke of the buffer used, plus two (2) feet. If precompression is used, the clearance may be reduced by the clearance between the bottom of the counterweight buffer and its striking block, plus the amount of precompression provided a spring return type of buffer is used.

(c) When the car crosshead is two (2) feet from the nearest obstruction above it, no projection on the car shall strike any part of the overhead structure.

(d) Where the depth of an elevator pit is reduced, as hereinbefore described the top clearance may be reduced accordingly.

4402.04 Counterweight Clearance: When an elevator car is level with the bottom landing, the clear distance between the top of the counterweight and any obstruction in the hatchway above it shall be not less than:

(a) If spring buffers or their equivalent are used, the clearance between the top of the car buffer and its striking block, in any case not less than three (3) inches, plus the compression of the buffer, plus six (6) inches.

(b) If oil buffers or their equivalent are used, the clearance between the top of the car buffer and its striking block, in any case not less than three (3) inches, plus one and one-half (1½) times the stroke of the buffer corresponding to the governor tripping speed of the car at full speed, plus six (6) inches. If retarders are installed which reduce the speed of the car at terminal landings and the stroke of the buffer is reduced correspondingly, not less than the clearance between the top of the car buffer and its striking block, in any case not less than three (3) inches, plus the stroke of the buffer corresponding to the governor tripping speed of the car at full speed, plus six (6) inches, plus one-half (½) the stroke of the buffer used. If provision is made to eliminate the jump of the counterweight at car buffer engagement not less than the clearance between the top of the car buffer and its striking block, in any case not less than three (3) inches; plus the stroke of the buffer, plus six (6) inches. If precompression is used, clearance may be reduced by the clearance between the top of the car buffer and its striking block, plus the amount of precompression.

(c) When the depth of an elevator pit is reduced, as hereinbefore described, the top clearance may be reduced accordingly.

4402.05 Overtravel for Sidewalk Type Elevators: For sidewalk type elevators there shall be provided an overtravel of not less than six (6) inches at the top and of not less than three (3) inches at the bottom.

4402.06 Installation of Machinery: Machinery and sheaves shall be so supported and held as to prevent any part from becoming displaced. No elevator machinery, except overhead sheaves for private residence elevators, and idler and deflecting sheaves with their guards or frames and devices for limiting or retarding the car speed and their accessories, shall be hung underneath the supporting beams at the top of the hatchway. Supporting members for elevator machinery hung underneath beams shall not depend solely on cast iron in tension.

4402.07 Protection Around Equipment: Exposed gears, sprockets, tape sheaves and ropes, and tapes passing through secondary levels shall be protected with substantial metal guards securely fastened in place and arranged to provide adequate protection for attendants and the public against injury. Deflecting sheaves extending below the machine level shall be provided with adequate guards.

4402.08 Stops for Couterweights: Where winding drum machines are used, a permanent beam or bar shall be provided at the top of the counterweight guides and beneath the counterweight sheaves to prevent the counterwights from being drawn into the sheaves. It shall be so arranged, and of such strength, that the cables will be pulled out of the sockets before there is any undue deflection of the beam.

4402.09 Wiring: All wiring in connection with elevators shall be done to conform with the requirements of the Chicago Electrical Code.

4402.10 Hatchway Door Equipment for Passenger Elevators:

(a) **Interlocks.** Hatchway door interlocks shall be provided on the hatchway doors of passenger elevators.

(b) **Door Closers.** The hatchway doors of hydraulic passenger elevators shall also be provided with an automatic door closer which will close the doors should the car creep more than six (6) inches away from the landing.

(c) **Emergency Release.** An emergency release shall be installed in each car, which is arranged for operation by an operator in the car, by which such operator can operate the car from within the car only, independent of the position of the hatchway doors. The emergency release in dual operated elevators shall be arranged so as to be inoperative when the car is operated as an automatic elevator.

(d) **Opening Devices.** Hatchway doors shall be arranged to be opened by hand from the hatchway side, except when locked out of service. Neither the main exit doors, nor the doors at the bottom terminal landing shall be locked out of service while the elevator is in operation. Hatchway doors for passenger elevators shall be arranged so that it is unnecessary to reach back of any panel, jamb or sash to operate them. If the entire control of a passenger elevator is located in the car, the hatchway doors shall be so arranged that they cannot be opened from the landing side, except by service or emergency keys as hereinafter prescribed. If the control is not located entirely in the car, the hatchway doors shall be so arranged that unless the car is in the landing zone, the doors cannot be opened from the landing side, except by service or emergency keys.

(e) **Door Hangars.** Hangars for power operated hatchway doors shall be designed to withstand a downward thrust of five (5) times and an upward thrust of four (4) times the weight of the door. Means shall be provided to prevent hangars for all sliding hatchway doors from jumping the tracks. Stops shall also be provided to prevent the hanger carriage from leaving the ends of the track, or suitable stops may be provided on the door only.

4402.11 Hatchway Door Equipment for Freight Elevators:

(a) **Interlocks or Electric Contacts.** Hatchway doors for freight elevators, where the elevator can be operated only from inside the car, shall be provided with interlocks or electric contacts and locks. Hatchway doors for automatic and double button elevators, having a speed not to exceed one hundred fifty (150) feet per minute and equipped with vertical type doors or gates, shall be provided with interlocks or electric contacts and locks. Hatchway doors for other freight elevators, except hatch covers for sidewalk type elevators, shall be provided with interlocks. Where electric contacts are provided on a hatchway door and are not a part of an interlocked system, the lock or latch and contact shall be so arranged as to insure the door being in position to be locked or latched when or before the contact is closed. Door electric contacts and door locks or latches shall be so located as normally to be inaccessible from the landing side.

(b) **Emergency Release.** An emergency release shall be installed in each car, except a sidewalk type elevator, arranged for operation by an operator in the car, by which such operator can operate the elevator from within the car only, independent of the position of the hatchway doors. The emergency release in dual operated elevators shall be arranged so as to be inoperative when the car is operated as an automatic elevator.

(c) **Opening Devices.** If the entire control of a freight elevator is located in the car the hatchway doors shall be so arranged that they cannot be opened from the landing side except by service or emergency keys hereinafter prescribed. If the control is not located entirely in the car, the hatchway doors shall be so arranged that unless the car is in the landing zone, the doors cannot be opened from the landing side, except by service or emergency keys.

4402.12 Service and Emergency Keys:

(a) **Service Key.** A service key shall be provided for every elevator, except of the automatic or the continuous control type, to open the hatchway door from the landing side at the landing where the car is normally parked out of service. This key shall open this door only when the car is at the landing and shall open no other door in the hatchway. Where two (2) or more cars are normally parked out of service at the same landing, the service key may be arranged to open all the hatchway doors of such cars at that landing.

(b) **Emergency Key.** An emergency key shall be provided for every elevator which, irrespective of the position of the car, will from the landing side open the hatchway door only at the landing where the car is normally parked and at the lowest landing; or if for an elevator operating in a blind hatchway, it shall also open the first hatchway door above the blind portion and no other door in the hatchway. For an elevator operating in a single hatchway it shall open all hatchway doors.

(c) **Key Receptacle.** The emergency key shall be placed in a break glass receptacle clearly marked "ELEVATOR DOOR KEY, FOR FIRE DEPARTMENT AND EMERGENCY USE ONLY", at the landing that is nearest the main street entrance to the building. If there is more than one street entrance, an emergency key in a receptacle marked as described shall be located in a position conveniently reached from each entrance. The same service and emergency keys shall fit the hatchway doors of all passenger and freight elevators in the building.

4402.13 Hatchway Door Interlocks: The door interlock system shall prevent the opening of any hatchway door from the landing side except by a special key as hereinbefore described, unless the elevator is at rest within the landing zone, or is coasting through the landing zone with its operating device in the "STOP" position. The interlock shall permit the operation of the car when the emergency release is in temporary use, or when the car is being moved by a car leveling device. An interlock system of the door unit type is prohibited.

(a) **Closed Position.**

Item 1. For elevators employing a type of operation that does not require the presence of an operator in the car, and for all elevators where the hatchway door is not equipped with a door closer, the door shall be considered in the closed position only when the door is within three-eighths (⅜) inch of contact with the door jamb, or, if the doors are of the bi-parting type, only when the doors are within three-eighths (⅜) inch of contact with each other.

Item 2. Where the hatchway door of an elevator, requiring the presence of an operator in the car, is equipped with a door closer, the door shall be considered to be in the closed position and the car may be started when the door is within four (4) inches of being fully closed against the jamb, or, if the doors are of the bi-parting type, when the sections are within four (4) inches of contact with each other, if at this position and any other up to full closure, the door cannot be opened from the landing side more than four (4) inches from the jamb, or bi-parting sections more than four (4) inches from each other, provided the door closer is of a type which will eventually close the door to the fully closed position and lock it in this position.

(b) **Operation.** The interlocks for all hatchway doors shall be so designed that the doors are locked in the closed position before the car can be operated, except that for private residence elevators the interlocks may be so arranged that after the doors are closed the car may move away from a landing a distance of not more than twelve (12) inches before the locking operation takes place, providing that should the locking operation fail to take place, further movement of the car will open the circuit and apply the brake. If a private residence elevator is arranged to be operated in this manner, an apron shall be installed on the front of the car platform extending below the car floor a distance equal to that which the car can travel before the door is locked.

(c) **Springs and Electric Circuits.** The functioning of door interlocks to prevent the movement of a car shall be independent of the action of a spring in tension, or of the closing of an electric circuit. If springs are used they shall be in compression. If an electric circuit is used, its interruption shall prevent the movement of the car.

(d) **Marks of Approval.** The type and make of door interlock shall be approved by the Commissioner of Buildings on the basis of the "Tests of Interlocks" hereinafter described, made by or under the supervision of a competent recognized laboratory. Approved interlocks shall be suitably and plainly marked for identification. The marking shall be permanent and so placed as to be readily visible when the interlocks are mounted in position. Auxiliary appliances forming a part of, or used in conjunction with, an interlock shall be similarly marked. Marking shall include the manufacturer's name or trade-mark, type or style letter or number and rated voltage.

4402.14 Hatchway Door Electric Contacts: The contact shall permit the operation of the car when the emergency release is in temporary use, or when the car is being moved by a car-leveling device. An electric contact of the door unit type is prohibited. Hatchway door contacts shall be designed so that they are positively opened by a lever or other device attached to and operated by the door.

(a) **Springs and Electric Circuits.** The functioning of a hatchway door electric contact to prevent the movement of the car shall be independent of the action of a spring or springs in tension and of the closing of an electric circuit. If springs are used, they shall be in compression. If an electric circuit is used, its interruption shall prevent the movement of the car.

(b) **Marks of Approval.** Each type and make of hatchway door electric contact shall be approved by the Commissioner of Buildings on the basis of "Endurance Test", "Current Interruption Test", "Test in Moist Atmosphere", "Mis-Alignment Test" and "Insulation Test" hereinafter described, made by or under the supervision of a competent recognized laboratory. Approved contacts shall be suitably marked for identification as required for interlocks.

4402.15 Emergency Releases: The emergency release shall be in the car, plainly visible to the occupants of the car; shall be easily accessible to the operator and shall be provided with a break glass cover and with means for breaking the glass. It shall be of such a design that the operator must hold the emergency release in the operative position to operate the car. Emergency releases shall be constructed so that they cannot be readily tampered with, or plugged in the release position. Rods, connections and wiring used in the operation of the emergency release, which are accessible from the car shall be inclosed and protected from injury. Each make and type of emergency release shall be tested and approved by the Commissioner of Buildings if found to show compliance with the foregoing requirements and the "Insulation Test" hereinafter described.

4402.16 Door Counterweight Inclosures: Door counterweights shall run in metal guides from which they cannot become dislodged or shall be inclosed or boxed in. The bottoms of the guides or inclosures shall be so constructed as to retain the counterweight if the counterweight rope breaks.

4402.17 Use of Space Under Elevator Hatchways and Counterweights: If the space under elevator hatchways and counterweights is used for any purpose, buffers equipped with safety devices conforming to the requirements prescribed for "Car and Counterweight Buffers" and "Car and Counterweight Safeties and Speed Governors" shall be provided for both cars and counterweights.

ARTICLE 4403.

Power Elevators, Except of Special Character.

4403.01 General: The provisions of this Article apply to all types of power elevators except stage and orchestra floors and other elevators of a special character, and dumbwaiters, all of which are treated in separate articles.

4403.02 Guide Rails:

(a) **Material and Weight.** Car and counterweight guide rails shall be of rolled steel, except where steel would constitute an accident hazard, where wood guide rails may be used. The weights of steel guide rails, except for sidewalk elevators having a travel of not more than fifteen (15) feet, shall be not less than the following:

Maximum Permissible Total Weight of Car and Load or Total Weights of Counterweights Per Pair of Rails	Minimum Weight of Each Car Guide	With Guide Rail Safeties	Minimum Weight of Each Counterweight Guide Rail	
			1 to 1 Roping	Without Guide Rail Safeties 2 to 1 Roping
(Pounds)	(Pounds Per Foot)		(Pounds Per Foot)	
4,000	7½	7½	6½	6½
15,000	14	14	7½	7½
27,500	22½	22½	7½	14
40,000	30	30	7½	14

Where seven and one-half (7½) pound rails are effectively bracketed or tied at intervals of not more than six (6) feet, the load permitted under the preceding table may be doubled for counterweights with guide rail safeties. Where seven and one-half (7½) pound rails are effectively bracketed or tied at intervals of not more than seven (7) feet six (6) inches, the load permitted under the preceding table may be increased to five thousand (5,000) pounds for cars with guide rail safeties.

(b) **Installation.** Guide rails shall be securely fastened with iron or steel brackets, or their equivalent, of such strength, design and spacing that the deflection of the rails and their fastenings will not be more than one-fourth (¼) inch under normal operation. Where the distance between rail supports is more than fourteen (14) feet, the rails shall be suitably backed or bracketed to secure rigidity. Joints of guide rails shall be accurately machined with tongue and groove, or other substantial construction which will maintain true alignment and shall be fitted with fishplates each secured with at least four (4) substantial bolts through each rail. Guide rails and their fastenings shall withstand the application of the safeties when stopping a fully loaded car or the counterweight. Rails shall be extended at the top and bottom to prevent guide shoes running off within the limits of the bottom overtravel and the top clearance. Guide rails for sidewalk type elevators, the top landing of which is not more than four (4) feet above grade shall be arranged to be extended to permit the elevator to rise to the top limit of its travel.

(c) **Double Safety Devices.** The maximum weights of car and load, as given in the preceding table for each pair of guide rails shall apply when only one (1) safety device gripping both rails in a horizontal plane is used. When two (2) such safety devices are used on the same guide rail and arranged so that both will be applied at practically the same time and with substantially equal retarding force, the total weight may exceed that shown in the table, but shall in no case, exceed the maximum weight given in the table multiplied by the following factors, based on distance between safeties:

CAR SAFETIES.

Distance between Safeties	Factor
18 (or over)	2.0
15	1.83
12	1.67
9	1.50
6	1.33

COUNTERWEIGHT SAFETIES.

Distance between Safeties	Factor
15 (or over)	2.0
10	1.67
3	1.33

4403.03 Car and Counterweight Buffers:

(a) **Car Buffers.** Buffers of the spring, oil or equivalent type shall be installed under the cars of elevators, except sidewalk elevators having a travel of not more than fifteen (15) feet.

Item 1. Spring Type. Spring buffers, or their equivalent, may be used with elevators having a rated speed of two hundred (200) feet per minute or less. The maximum retardation shall be not more than eighty and five-tenths (80.5) feet per second per second, based on governor tripping speed.

Item 2. Oil Type. Oil buffers, or their equivalent shall be used with elevators having a rated speed greater than two hundred (200) feet per minute. The minimum total stroke of oil buffer shall be based on an average retardation of thirty-two and two-tenths (32.2) feet per second per second based on governor tripping speed. Where precompression is used, buffer shall be so installed that when the car is level with the terminal landing, the remaining buffer stroke shall not be less than fifty (50) per cent of the gravity stopping distance corresponding to the governor tripping speed used. Where the car or counterweight oil buffer is compressed more than three (3) inches when the car is level with the lower or upper terminal landing, respectively, buffers shall be provided with a switch which will prevent movement of the car in a direction to compress the buffers at a speed greater than one-half (½) the rated speed until the buffers are restored to their normal position. Oil buffers shall be provided wi means for gauging the amount of oil in them.

(b) **Counterweight Buffers.** Buffers similar to those required for cars shall be installed under the counterweights.

(c) **Speed Retarding Devices.** Where speed retarding devices, independent of normal and final stop switches are provided for retarding car or counterweight, or both, to a definite limiting speed before the buffer is engaged, the required buffer stroke may be based only on retardation from the governor tripping speed corresponding to such speed. Such speed retarding devices shall be so designed that the retarding force is quickly but gradually applied, substantially constant and the retarding distance is not less than the sliding distance hereinafter prescribed for Undercar Safeties. For rated speeds in excess of five hundred (500) feet per minute, the corresponding reduced buffer stroke shall be not less than eighteen (18) inches.

(d) **Location and Access.** Except where precompression is used, buffers shall be located so that under ordinary operating conditions, the car or counterweight does not strike them. Where the buffer cylinder is over five (5) feet high a fixed metal ladder shall be provided.

(e) **Marks of Approval.** Each type and size of oil buffer used shall be approved by the Commissioner of Buildings after tests hereinafter described "Tests of Buffers" made by or under the supervision of a competent recognized testing laboratory. Approved buffers shall be marked by the manufacturer with the range of speed and load for which they have been approved.

4403.04 Counterweights: Counterweights shall run in guides and shall be guided at top and bottom of counterweight. If two (2) counterweights run in the same guides the car counterweights shall be above the machine counterweight and there shall be a clearance of not less than eight (8) inches between the counterweights. The cables of the machine counterweight shall be covered or protected by suitable sleeves not less than six (6) inches longer than the car counterweight and firmly attached to the cables. The ends of the sleeves shall be carefully reamed before being placed on the cables. If an independent car counterweight is used it shall not be of sufficient weight to cause undue slackening in any of the cables during acceleration or retardation of the car. Counterweight sections shall be secured by at least two (2) tie rods passing through holes in all the sections, or by other approved means. The tie rods and suspension rods shall have lock nuts and cotter pins at each end. Suspension rods shall be free from welds.

4403.05 Car Construction:

(a) **Frames.** Elevator cars shall have metal car frames and metal outside frames of platforms and shall be provided with top and bottom guide shoes. Non-metallic shoes may be used. Where wood platform flooring is used, the underside of car platforms shall be covered with metal not less than No. 27 U. S. Gauge, except for sidewalk type elevators the travel of which does not exceed one (1) story.

() **Steel Stresses.** The allowable working stresses of rolled steel sections or annealed cast steel in the construction of car frames and platforms, except for elevators of the plunger type, which have no counterweights, based on the static load imposed on them, shall not be more than the values given in the following table for steels meeting the standard specifications of the American Society for Testing Materials, Designation A 7-34, for steel having an ultimate strength of from fifty-five thousand (55,000) to sixty-five thousand (65,000) pounds, per square inch for rolled sections or cast steel; and forty-six thousand (46,000) to fifty-six thousand (56,000) pounds per square inch, for rivets. For steels of greater strength the allowable working stresses may be increased proportionately, based on ultimate strength.

TABLE 4403.05 (b)

Passenger Elevators.

Loading	Maximum Allowable Stress (Pounds per Sq. In.)	Basis
Tension	10,000	Net Area
Bending	10,000	Gross Section
Shear on Shop Rivets	8,000	Net Area
Bearing on Shop Rivets	16,000	Net Area
Shear on Bolts in Clearance Holes	7,000	Gross Section
Bearing on Bolts in Clearance Holes	14,000	Gross Section
Bolts or threaded portions of rods in tension	6,000	Gross Section
Compression	11,700-49 L/R	Gross Area
Freight Elevators.		
Tension	12,000	Net Area
Bending of car frame member and platform framing at entrance	12,000	Gross Section
Bending of platform stringers	15,000	Gross Section
Shear on Shop Rivets	9,500	Net Area
Bearing on Shop Rivets	19,000	Net Area
Shear on Bolts in Clearance Holes	8,000	Gross Section
Bearing on Bolts in Clearance Holes	16,000	Gross Section
Bolts or threaded portions of rods in tension	8,000	Gross Section
Compression	14,000-59 L/R	Gross Area
L = effective free length of member in inches. R = least radius of gyration in inches.		

(c) **Plunger Heads.** Plunger heads in tension shall be cast steel.

(d) **Cast Iron.** No cast iron shall be used in the construction of any member of car frame or platform subject to tension, torsion or bending, except for compensating cable anchorages, releasing carriers and guide shoe stands.

(e) **Other Metals.** When material other than steel is used in the construction of car frames or platforms, the maximum allowable working stresses shall be thirteen (13) per cent of the ultimate strength of the material for passenger elevators, and fifteen and six-tenths (15.6) per cent for freight elevators.

(f) **Platform Aprons.** When a car is equipped with a car leveling device, the car platform shall be provided with a substantial vertical apron flush with its outer edge extending a sufficient distance below the car floor so that there shall be no horizontal opening into the hatchway while the car is within the landing zone and the hatchway door is fully or partially open.

(g) **Sidewalk Type Elevators.** A sidewalk type elevator, equipped with a hinged hatch cover, shall be provided with a bow iron not less than seven (7) feet six (6) inches high. A sidewalk type elevator, equipped with a vertical lifting cover shall be provided with stanchions framed together at the upper ends to be of sufficient strength to lift and support the hatch cover. Stanchions shall be provided with suitable buffer springs and shall be of such height as to permit the cover to be completely closed when the car platform is level with the first landing below the grade. Guide shoes for sidewalk type elevators shall be not less than twenty-four (24) inches long unless two (2) sets of shoes are used, and spaced eighteen (18) inches on centers. Where a vertical lift cover is used for a sidewalk type elevator, the vertical distance between the centers of the guide shoes remaining on the guide rails when the car platform is level with the top landing shall be not less than one-third (⅓) of the height of the hatch cover stanchions. Where single guide shoes, not less than twenty-four (24) inches long are used, six (6) inches of the shoe may be off the rails when the platform is level with the top landing.

(h) **Tracks on Cars.** If there is a railroad track on the elevator car, the tops of the rails shall be flush with the car floor.

(i) **Lighting Fixtures.** Elevator cars, except sidewalk type elevators, shall be equipped with electric lamps which will provide adequate illumination at the landing edge of each car platform. A light socket or receptacle shall be provided under the car platform and on top of the car for inspection purposes, except sidewalk type elevators. For passenger elevators each lighting device provided with a glass or metal shade or reflector, shall have an integral metal base, husk and spring-clamp holder. If suspended glass bowls or glass plates are used, each such bowl or plate shall rest in, and be fastened to, a metal supporting frame provided with at least three (3) point suspension; glass bowl or plate shall not be drilled for attachment to frame suspensions. Glass bowls, larger than ten (10) inches in diameter, shall be of shatter-proof glass or surrounded by a guard made of wire not less than No. 22 Steel Wire Gauge and of a mesh which will reject a one-half (½) inch diameter ball. Guards shall be securely fastened to the holder or suspension.

(j) **Glass.** No glass shall be used in any elevator car except to cover certificates, directories, lighting fixtures and appliances necessary for the operation of the car and as a vision panel in the car door. No piece of glass exceeding one (1) square foot in area shall be used unless it is shatter-proof; no piece of such glass used in the car in connection with lighting fixtures shall exceed four (4) square feet.

4403.06 Car Compartments: No elevator car shall be constructed with more than one (1) compartment on the same landing level. If an elevator car has upper and lower compartments, each compartment shall be equipped with an operating device. There shall be an operator in each compartment which is in use with exclusive control of that car door or gate and hatchway door and it shall be impossible to start the car unless both operating devices are in the starting position. When any compartment is not of use, its car door or gate shall be locked in the closed position and when so locked the car may be started with the operating device in that compartment in the "off" position. Each compartment shall be equipped with an emergency stop switch and with an emergency release effective for the door or gate of that compartment and the corresponding hatchway door. Each compartment shall be provided with emergency exits as hereinafter prescribed for car inclosures. If impossible to provide the required exits in the sides of the inclosure the upper compartment shall have a trap door in the floor connecting with the top exit of the lower compartment.

4403.07 Cars Counterbalancing One Another: Elevator cars shall not be arranged in such a manner as to counterbalance one another.

4403.08 Two (2) Elevators in a Single Hatchway: Two (2) single elevators may be used in a single hatchway provided both elevators are equipped with all the safeguards prescribed herein for a single elevator both with relation to each other and to top and bottom terminals.

4403.09 Car Inclosures:

(a) **Passenger Elevators.** The car for every passenger elevator shall be inclosed over the top and at the sides, except the opening necessary for entrance or exit in normal operation.

(b) **Freight Elevators.** The car for every freight elevator shall be inclosed at the sides, except at the opening necessary for loading and unloading. The inclosure, except of a sidewalk type elevator, shall extend to a height or not less than six (6) feet above the platform or to the crosshead if the crosshead is lower. The section of such a car inclosure opposite the counterweight shall extend to the crosshead or car top. The inclosure of a sidewalk type elevator shall extend to a height of not less than six (6) feet above the platform.

(c) **Construction.** No part of a power elevator car inclosure shall deflect so as to reduce the actual running clearance between cars, counterweights and hatchway inclosures hereinbefore prescribed. The car inclosure shall be secured to the car floor and sling or frame in such a manner that it cannot work loose or become dislodged in ordinary service. Cast iron shall not be used for a car top. The inclosure, including the sides, ceiling and car doors for a passenger and freight elevator, except a sidewalk type elevator, the travel of which does not exceed one (1) story, may be of solid or open work. If of combustible material it shall be covered on the exterior, including the top, with sheet metal not less than No. 27 U. S. Standard Gauge. Ventilating openings less than seven (7) feet above the car platform shall reject a ball two (2) inches in diameter.

(d) **Open Grill Work.** If an inclosure, except of a sidewalk type elevator, is of open work, it shall reject a ball two (2) inches in diameter at sides of inclosure and one-half (½) inch in ceiling. Where the clearance from any part of the hatchway structure or the counterweight is less than five (5) inches, openings which will pass a ball one-half (½) inch or larger in diameter shall be covered to a height of not less than six (6) feet above the car platform with wire netting with a mesh not larger than one-half (½) inch square made of wire not less than No. 20 Steel Wire Gauge. The inclosure of a sidewalk type elevator, if of open work, shall reject a ball four (4) inches in diameter. **If the car in-**

closure on a freight elevator is cut away to provide access to the hand-rope, the inclosure shall be cut large enough to prevent injury to the operator's hand.

(e) **Inclosure Top.** Every freight elevator car twelve (12) feet or more long and having hatchway openings at only one end shall be provided with a top on the end where the landings occur to a line six (6) feet from the opposite side of the crosshead and the full width of the platform; and every other freight elevator car, except sidewalk type elevators, shall be provided with a top over the entire area of the platform. Top shall be solid or wire grille work having a mesh which will reject a ball one and one-half (1½) inches in diameter and made of wire not less than No. 10 Steel Wire Gauge or its equivalent. The top shall be sufficiently strong to sustain a concentrated load of one hundred fifty (150) pounds applied on any four (4) square inches and a distributed load of seventy-five (75) pounds per square foot over the entire area. Where no car gate is provided, at least the front section of the elevator car top shall be hinged along a line approximately eighteen (18) inches from the edge of the car. A top may be provided on a sidewalk type elevator inclosure only if the clearance between the top and any obstruction above it is at least two (2) feet when the car is at the limit of its top overtravel.

(f) **Emergency Exits.** Every passenger elevator car shall be provided with an emergency exit. Where there are elevators in adjoining hatchways without an intervening inclosure, the emergency exits may be located in the sides of the adjacent cars directly opposite each other. Where there is no such elevator in an adjoining hatchway, or where it is not practicable to provide the emergency exit in the side of the car, it shall be located in the ceiling of the car. Where the size of the car permits, the emergency exit, if in the side of the car, shall have a clear width of not less than sixteen (16) inches and shall extend from the floor or base to the soffit moulding frame; and in no case shall the clear height be less than three (3) feet. It shall be located so that it is not obstructed by car frame members; and traveling cables and other hatchway equipment shall not be located in front of an exit if such obstruction can be avoided. A side emergency exit panel shall be held securely in place and arranged so that it can be opened from the inside of the car by a key kept in the car and from the outside by a non-removable key. If the car is of the automatic operation type, the side exit panel shall be provided with an electric contact to prevent the operation of the car when the exit panel is open. Where the size of the car platform permits, the emergenoy exit, if in the ceiling of the car, shall be not less than sixteen (16) inches wide and four hundred (400) square inches in area. A ceiling exit panel shall be held in place in such a manner that the exit cover can be readily opened from both the inside and outside of the car. All equipment and working platform mounted above the top of a car shall be so located as not to obstruct access to or from the emergency exit. If a working platform is placed so as to cover any of the required area of an emergency exit it shall be provided with a trap door, without a catch, opening upward.

4403.10 Car Doors and Gates:

(a) **General.** A car door or gate shall be provided at each entrance on both passenger and freight elevator cars, except freight elevators having a type of operation requiring the presence of an operator in the car and handling motor vehicles or hand trucks of over two thousand (2000) pounds capacity. Each sidewalk type elevator car, the top hatchway opening of which is located in an area accessible to the public, shall have a gate on the sides used for loading or unloading at the grade level, which gate shall extend from the car platform to the top of the inclosure. Car doors or gates for passenger elevators, when closed, shall guard the full height and width of the opening; for freight elevators, when closed, they shall guard the full opening to a height of at least four (4) feet. If a weight is used to close a car door or gate automatically it shall run in metal guides from which it cannot become dislodged, or it shall be inclosed. The bottom of the guides or inclosures shall be so constructed as to retain the weight if the rope breaks.

(b) **Power Doors or Gates.** For passenger and freight elevators employing a type of operation not requiring the presence of an operator in the car, power car doors or gates other than those closed by hand shall be driven by a mechanism so designed and set that the force necessary to prevent the closing of the gate on any car shall not be more than thirty (30) pounds, and the kinetic energy of the gate plus all parts connected rigidly thereto, computed for the average closing speed shall not be more than five (5) foot-pounds, except that if the same mechanism also closes the hatchway door, the total kinetic energy may be increased to not more than seven (7) foot-pounds. For automatic operation passenger elevators having power closed, power operated or automatically released self-closing car doors or gates, and manually closed or self-closing hatchway doors, arrangement shall be made so that the door or gate on any car cannot be closed unless the hatchway door is closed.

(c) **Operation.** Car gates of the scissors or pantograph type, used for passenger elevators shall be of such design that when fully expanded they shall reject a ball of three (3) inches in diameter. For freight elevators, except sidewalk type elevators, such gates when fully expanded shall reject a ball four and one-half (4½) inches in diameter. Every car gate shall be of such design and so constructed that it will not deflect beyond the line of the car platform threshold when a force of approximately fifty (50) pounds is applied across two (2) adjacent bars approximately at the center of the bars when the gate is fully extended. Sliding car doors shall be guided top and bottom.

(d) **Hangers.** Hangers for power operated car doors shall be designed to withstand a downward thrust of five (5) times and upward thrust of four (4) times the weight of the door.

(e) **Contacts.** An electric contact shall be provided on every car door or gate which will prevent the operation of the car unless the door or gate is closed within two (2) inches, except that when the car is fitted with a door closer and attended by an operator and the hatchway door is fitted with a door closer, this distance can be increased to four (4) inches. The car door or gate electric contact shall permit the operation of the car when the emergency release is in use or when the car is being moved by a leveling device. An automatic operation elevator may be operated with an open gate if there is no person in the car. Freight elevator cars operating in hatchways outside the structure, which are inclosed only at the ground landing shall be protected on the exposed side or sides by independently operated gates equipped with electric contacts. The car door or gate contact shall be designed so that it is positively opened by a lever or other device attached to and operated by the door or gate. The functioning of a car door or gate contact to prevent the movement of the car shall be independent of the action of a spring or springs in tension, and of the closing of an electric circuit. If springs are used they shall be in compression. If an electric circuit is used, its interruption shall prevent the movement of the car.

(f) **Marks of Approval.** Each type and make of car door and gate contact shall be tested and approved by the Commissioner of Buildings on the basis of "Endurance Test", "Current Interruption Test", "Test in Moist

Atmosphere", "Misalignment Test", and "Insulation Test" hereinafter described, made by or under the supervision of a competent recognized testing laboratory. Approved contacts shall be suitably marked for identification.

(g) **Emergency Release.** An emergency release shall be installed in each car equipped with a gate or door, except sidewalk type elevators, by which an operator can operate the elevator from within the car independent of the position of the car door or gate. Such an emergency release shall conform to the requirements for emergency releases for hatchway doors. This emergency release shall be independent from the emergency release installed on the hatchway doors. The emergency release in dual operated elevators shall be arranged so as to be inoperative when the car is operated as an automatic elevator.

4403.11 Car and Counterweight Safeties and Governors: Every elevator car suspended by cables shall be provided with a car safety or safeties capable of stopping and sustaining the car with rated load, attached to the car frame, except sidewalk type elevators not inside of a building and having a travel of not over twenty (20) feet. When one (1) safety is used, it shall be located beneath the car frame.

(a) **Car and Counterweight Safeties.** The application of the safety or safeties shall not cause the car platform to become out of level more than one-half (½) inch per foot measured in any direction. When the car safety or safeties are applied, no decrease in the tension of the governor cable or motion of the car in the descending direction shall release the car safety or safeties. No car safeties which depend for application upon the completion or maintenance of an electric circuit shall be used. Car safeties shall be applied mechanically. The gripping surfaces of car or counterweight safeties shall not be used to guide the car or counterweights. Pawls and ratchets alone shall not be used as safety devices. Counterweight safeties shall be capable of stopping and sustaining the weight of the counterweight. The application of counterweight safeties shall not cause the counterweight frame to become out of level more than one-half (½) inch per foot in any direction. When a counterweight safety is applied, no decrease in the tension of the governor cable or motion of the counterweight in the descending direction shall release the counterweight safety. Jaws and other parts of car and counterweight safeties of the sliding type shall be made of forged steel of an ultimate strength of not less than fifty-five thousand (55,000) pounds per square inch, and cast steel of an ultimate strength of not less than sixty-five thousand (65,000) pounds per square inch, in which case they may in action be stressed to seventeen thousand (17,000) pounds per square inch. If steels of greater strength are used, the allowable stress may be increased proportionately, based on ultimate strength.

(b) **Governors.** Car safeties shall be operated by speed governors, except that instantaneous safeties of the broken rope type may be used, (1) for private residence passenger elevators; (2) for sidewalk type elevators not inside of a building and having a travel of not over twenty (20) feet; and (3) freight elevators inside of a building having a travel of not more than fifteen (15) feet, a maximum platform area of fifty (50) square feet and a maximum speed of fifty (50) feet per minute. Instantaneous safeties of the governor controlled type may be used on elevators having a rated speed of not more than one hundred (100) feet per minute provided that the elevator speed is not more than one hundred ten (110) feet per minute on the up travel with rated load in the car. On overspeed such safeties shall be applied by the governor. On the parting of the hoisting cables such safeties shall apply instantly and independently of the speed action of the governors. Car safeties shall not be installed for stopping ascending cars. If an ascending car is to be stopped on account of overspeed, a safety shall be applied to the counterweights for this purpose. The car safety may be permitted to stop the ascending car above the top terminal landing, provided the retardation of the ascending car under such conditions is within thirty-two and two-tenths (32.2) feet per second. The governor may open the motor circuit and apply the brake in case of overspeed in the up direction. The governor shall be located where it cannot be struck by the car in case of overtravel and where there is sufficient space for the full movement of governor parts. For elevators having a rated speed of four hundred seventy-five (475) feet per minute, or more, the pull-out of the governor cable from the normal running position until the safety jaws begin to apply pressure to the guide rails shall not be more than thirty (30) inches. The car speed governor shall be set to cause the application of the safety at a speed not less than fifteen (15) per cent nor more than forty (40) per cent above the rated speed, except that no governor shall be required to trip at a car speed less than one hundred seventy-five (175) feet per minute. For rated speed exceeding five hundred (500) feet and not exceeding seven hundred (700) feet per minute the maximum governor tripping speed shall be thirty-three and one-third (33⅓) per cent above rated speed and for rated speed exceeding seven hundred (700) feet per minute, twenty-five (25) per cent above rated speed. A governor for the operation of a counterweight safety shall comply with the requirements for a governor used with a car safety of the same type, except that it shall be adjusted to trip at a speed not over ten (10) per cent in excess of the tripping speed of the car governor. A single governor may operate car and counterweight safety, but the counterweight safety must be tripped at a speed ten (10) per cent in excess of the tripping speed of the car governor. Broken rope safeties of the instantaneous type may be used on counterweights within the limits of the following:

Rated Speed (Feet per Minute)	Total Weight of Counterweight (Pounds)
250	2,000
200	3,000
160	4,000
125	5,000

(c) **Required Types of Safeties.** Safeties shall be of the following types:

1. Type 1. Instantaneous.
2. Type W.C. Wedge Clamp, with constant retarding force.
3. Type G.W.C. Gradual Wedge Clamp. with gradually increasing retarding force.
4. Type F.G.C. Flexible Guide Clamp with Constant retarding force.

Each safety shall be marked for identification with letters "I", "W.C.", "G.W.C.", or "F.G.C.", depending on whether it is type 1, 2, 3, or 4. The distance between the safety jaws shall be not less than the thickness of the guide rail plus three thirty-seconds (3/32) inch and the jaws shall not drag against the rail.

(d) **Governor Construction.** Where governor controlled safeties are used, the motor control circuit and the brake control circuit shall be opened before or at the time the governor trips by a switch located on the governor or car safety device. Governor cables shall be of iron, steel, monel metal or phosphor bronze. The cable shall be at least three-eighths (⅜) inch in diameter. Tiller rope construction shall not be used for governor cables. The portion of the cable wound on the safety drum shall be of corrosion resisting metal. Governor ropes shall run clear of governor jaws during the normal operation of the elevator. The size, material and construction of the governor rope, together with the proper tripping speed of the governor, shall be stamped on the governor stand or stated on a brass name plate attached to it in letters at least one-fourth (¼) inch in height. The arc of contact between the governor rope and its driving sheave shall, in conjunction with a tension device, provide sufficient traction to cause proper operation of the governor. The design and length of governor jaws shall be such that no serious cutting, tearing or deformation of the rope shall result from the operation of the safety. The governor shall so function that the safety rope will pull through the governor jaws on the application of a stress exceeding that required to operate the safety so as to stop the car, except in the case of instantaneous type safeties. Governors shall have a protective covering over sheave and exposed gears. Winding drum machines shall be provided with a slack-cable device which will cut off the power and stop the elevator machine if the car is obstructed in its descent. Slack-cable switches shall be so constructed that they will not automatically reset when the slack in the cable is removed.

(e) **Marks of Capacity.** Safeties shall be marked by the manufacturers with the range of weight and speed for which they are designed; such weight shall include the complete car structure, the safety, the rated load in the car, and all moving equipment, the weight of which is borne by the safety.

(f) **Replacements.** Replacements of governor cables on elevators installed after the date of the adoption of this Code shal be of the size, material and construction stamped on the governor stand or name plate. Replacements of governor cables on elevators installed prior to the adoption of this Code, shall be of the material and design required by this Code and of the size and construction required by the governor, except that in cases where the present governors, if in satisfactory condition, will not permit the use of iron, steel, monel metal or phosphor bronze cables, the existing type of cables may be used.

4403.12 Capacity and Loading:

(a) **Rated Load.** The rated load of a passenger elevator in pounds shall be not less than the following:

Maximum Horizontal Free Area Inside of Car (Square Feet)			Rated Load (Pounds)
Up to		5.0	310
5.0	to	6.0	380
6.0	to	7.0	450
7.0	to	8.0	520
8.0	to	9.0	590
9.0	to	10.0	650
10.0	to	11.0	740
11.0	to	12.0	820
12.0	to	13.0	900
13.0	to	14.0	980
14.0	to	15.0	1050
15.0	to	17.0	1250
17.0	to	20.0	1500
20.0	to	23.5	1750
23.5	to	26.5	2000
26.5	to	29.0	2250
29.0	to	32.0	2500
32.0	to	34.0	2700
34.0	to	37.0	3000
37.0	to	39.0	3250
39.0	to	42.0	3500
42.0	to	46.5	4000
46.5	to	51.0	4500
51.0	to	56.0	5000
56.0	to	60.5	5500
60.5	to	65.0	6000
65.0	to	70.0	6500
70.0	to	74.5	7000
74.5	to	79.0	7500
79.0	to	84.0	8000
84.0	to	93.0	9000
93.0	to	102.0	10000
102.0	to	111.0	11000
111.0	to	120.0	12000

The rated load of a freight elevator is the load to be carried at rated speed. The elevator may carry a heavier load at lower speed if all parts are designed for the heavier load.

(b) **Capacity Displayed.** A metal plate bearing (1) the weight of the complete car including the safeties, (2) the rated capacity in pounds and the rated speed in feet per minute at which the car is designed to travel, and (3) the cable data prescribed hereinafter, shall be placed on the crosshead of passenger and freight elevators and on the bow iron or other conspicuous place of sidewalk type elevators. The capacity of freight elevators shall be indicated in a conspicuous place in the car in letters and figures at least one (1) inch high by the word "CAPACITY", followed by figures giving the rated load in pounds.

(c) **Special Loads.** Passenger and freight elevators, if designed for carrying safes or other one piece loads greater than the rated load of the elevator, shall be provided with the following features:

Item 1. A locking device so designed that it will hold the car at any landing independently of the hoisting cables while the safe or other object is being loaded or unloaded, and that it cannot be unlocked unless the entire weight of the car and load is suspended on the cables. The wrench or other device for operating the locking device shall be removable. The locking device shall be designed to withdraw the bars should it come in contact with the landing locks if the car is operated on the "up" motion.

Item 2. A metal plate in the elevator car bearing the words "Capacity Lifting Safes" in letters followed by figures giving the capacity in pounds for lifting safes for which the machine is designed. The letters and figures shall be at least one-fourth (¼) inch high, stamped, etched, or raised on the surface of the plate.

Item 3. A car platform, car frame, sheaves, shafts, cables, guide rails, and locking device designed for the specified "Capacity Lifting Safes" with a factor of safety of at least five (5).

Item 4. Car safeties designed to stop and hold the specified "Capacity Lifting Safes" with the aid of the cables and counterweights.

Item 5. A machine designed to operate with the "Capacity Lifting Safes" at low speed and the car safety designed to stop and hold the specified "Capacity Lifting Safes" independently of the cables where the space under the hatchway is used for any purpose.

Item 6. Additional counterweights added for traction machines so that the total overbalance is at least equal to forty-five (45) per cent of the "Capacity Lifting Safes".

Item 7. The locking device hereinbefore prescribed shall be provided for any passenger elevator installed for carrying safes or other one (1) piece loads where the weight of such safes or one (1) piece load equals or exceeds seventy-five (75) per cent of the rated load.

Item 8. The maximum "Capacity Lifting Safes" of any traction elevator shall be one and one-third (1⅓) times the rated load of the elevator.

Item 9. Elevator machines equipped for carrying safes or other concentrated loads greater than the rated load of the elevators shall be provided with special switches for operating the cars under such conditions.

(d) **Use of Freight Elevators for Carrying Passengers.** Passengers may be carried on freight elevators installed after the adoption of this Code, provided the elevators comply in all respects with the requirements herein prescribed for passenger elevators. Passengers may be carried in freight elevators installed prior to the adoption of this Code, subject to the restrictions for new freight elevators, provided the elevators comply in all respects with the rules for passenger elevators installed prior to the adoption of this Code.

4403.13 Machines, Stopping Devices, Control and Operation:

(a) **Drums and Sheaves.** Drums and loading sheaves shall be of cast iron or steel and shall have finished grooves which may be faced with materials other than iron and steel having sufficient traction. U Grooves shall be not more than one-sixteenth (1/16) inch larger than the cables. Hoisting rope sheaves for traction machines shall have sheave grooves designed so that the traction will not materially be decreased by the wear of the grooves. The diameter of sheaves or drums for hoisting or counterweight cables for private residence elevators shall be at least thirty (30) times the diameter of the cables. For other elevators the diameters of sheaves or drums for hoisting or counterweight cables shall be at least forty (40) times the diameter of the cables, except for sidewalk type elevators; provided however, that where structural conditions make this impracticable, the diameter of sheaves or drum may be reduced to not less than thirty (30) times the diameter of the cables. Openings in drums shall be drilled at an angle of less than forty-five (45) degrees with the run of the ropes and be provided with a rounded corner with a radius at least equal to that of the ropes.

(b) **Factor of Safety.** The factors of safety based on the static loads including the rated loads plus the weight of the car, cables, counterweights and other appurtenances, to be used in the design of any elevator hoisting machine shall be not less than eight (8) for wrought iron or wrought steel and ten (10) for cast iron, cast steel, or other materials.

(c) **Construction.** Keys shall be used for fastenings except that set screws may be used where the connection is not subject to torque. No friction gearing or clutch mechanism shall be used for connecting the drums or sheaves to the main driving gear. No elevator machines shall be belt or chain-driven nor shall any worm gearing with cast iron teeth be used.

(d) **Brakes.** Winding drum and traction machines shall be equipped with brakes applied automatically by springs or gravity when the operative device is at the "stop" position. Electric elevator machines shall be equipped with electrically released brakes. Brakes shall be of such design that they cannot be released before power has been applied to the motor. The brake construction shall be such that no single ground short-circuit or counter-voltage will prevent the setting of the brake during normal operation, and no motor field discharge, counter-voltage, single ground or accidental short-circuit will retard its setting during emergency stops.

(e) **Special Construction Applying to Hydraulic Elevators.** Hydraulic elevator machines shall be so constructed that the piston will be stopped before the car can be drawn into the overhead work. Stops of ample strength shall be provided to bring the piston to rest when under full pressure without causing damage to the cylinder or cylinder head. Traveling sheaves for vertical hydraulic elevators shall be guided. Guide rails and guide shoes shall be of metal. Side frames of traveling sheaves for vertical hydraulic elevators shall be either structural or forged steel. The construction commonly known as the "U-strap connection" between the piston rods and the traveling sheaves shall not be used. Where more than one piston rod is used on the vertical pulling type hydraulic elevators, an equalizing crosshead shall be provided for attaching the rods to the traveling sheave frame to insure an equal distribution of load on each rod. Equalizing or cup washers shall be used under the piston rod nuts to insure a true bearing. Cylinders of elevator machines shall be provided with means for releasing air or other gas. Piston rods of tension type hydraulic elevators shall have a factor of safety of at least eight (8) based on the cross-sectional area at the root of the thread. A true bearing shall be maintained under the nuts at both ends of the piston rods to prevent eccentric loading on the rods. Automatic stop valves shall be packed with cup leathers, or other means shall be used to prevent sticking of the valve stems. In the design of a plunger elevator, provision shall be made to stop both the plunger and the car.

Item 1. Pumps. Each pump connected to the pressure tank of a hydraulic elevator shall be equipped with one (1) or more relief valves piped to discharge into the discharge tank or the pump suction and so installed that they cannot be shut off. The relief valve, or valves, shall be of sufficient size and so set as to pass the full capacity of the pump at full speed without exceeding the safe working pressure of the pump or tank. Elevator pumps, unless equipped with pressure regulators which control the motive power, shall be equipped with automatic by-passes.

Item 2. Pressure Tanks. Pressure tanks shall be made and tested in accordance with the requirements prescribed in Chapter 52 for Steam Boilers and Unfired Pressure Vessels. Each pressure tank shall be provided with a water gauge glass having brass fittings and valves attached directly to the tank and so located as to show the level of the water when the tank is more than half filled. Each pressure tank shall have a pressure gauge which correctly indicates pressure to not less than one hundred fifty (150) per cent of the normal working pressure allowed in the tank; this gauge shall be connected by a corrosion resisting pipe equipped with an indicating shut-off cock. Each pressure tank shall be provided with a one-fourth (¼) inch pipe size valved connection for attaching an inspector's gauge while the tank is in service. Any pressure tank that may be subject to vacuum shall be provided with one or more vacuum relief valves having openings of sufficient size to prevent the collapse of the tank if a vacuum occurs. Outlets of pressure tanks shall be so located as to prevent the probability of the entrance of air or other gas into the elevator cylinder. Pressure tanks shall be so located and supported that inspection may be made of the entire exterior.

Item 3. Discharge Tanks. Discharge tanks of hydraulic elevators, open to the atmosphere, shall be so designed that when completely filled the factor of safety shall be at least four (4) based on the ultimate strength of the material. Each discharge tank shall be provided with a cover and with a suitable vent to the atmosphere.

4403.14 Terminal Stopping and Safety Devices: Each elevator shall be provided with upper and lower terminal normal stopping devices arranged to stop the car automatically from any speed attained in normal operation within the top and bottom overtravel independent of the operating device inside the car, the final terminal stopping device and the buffers; except that in the case of a hand-rope or rod operating device the normal terminal stopping device may operate in conjunction with the operating device.

(a) **Normal Terminal Stopping Devices.** Normal terminal stopping devices shall be installed as follows:

Item 1. Winding Drum Machines. Each electric elevator having a winding drum machine, except a sidewalk type elevator, and except an elevator operated by a hand-rope, wheel or lever device, shall have stopping switches on the car or in the hatchway operated by the movement of the car. Each electric elevator having a winding drum machine, with a lever or wheel operating device shall have a device to center the operating device automatically.

Item 2. Traction Machines. Each electric elevator having a traction machine, except an elevator operated by a hand-rope device, shall have stopping switches on the car or in the machine room, or in the hatchway, operated by the movement of the car. When located in the machine room, the stopping contacts shall be mounted on and operated by a stopping device mechanically connected to the car and designed so that it is not dependent on friction as a driving means. An automatic safety switch shall be provided which will stop the car if the means for mechanically connecting the stopping device to the car should fail.

Item 3. Hand Ropes or Rods. Each electric elevator with a hand-rope or rod operating device shall have stop balls securely fastened to the rope or rod arranged to center the operating device, and, except for a sidewalk type elevator, if a winding drum machine is used it shall also have an additional device to center the operating device automatically.

Item 4. Hydraulic Elevators. Each hydraulic elevator having a rated speed of more than one hundred (100) feet per minute shall have an automatic stop valve independent of the normal control valve or valves operated either by the car or the machine. Each hydraulic elevator, having a rated speed of not more than one hundred (100) feet per minute, with a hand-rope or rod operating device, shall have stop balls on the operating device.

Item 5. Sidewalk Type Elevators. Each electric sidewalk type elevator having a winding drum machine and hand-rope or pull chain operating device shall have a stopping device on the machine and on the operating device. Each electric sidewalk type elevator having a winding drum machine and either automatic or continuous pressure operation shall have a stopping device on the machine and in the hatchway. These stopping devices shall not control the same switches unless two (2) or more separate and independent switches are provided, two (2) of which shall be closed to complete the motor and brake circuit in each direction of travel.

(b) **Final Terminal Stopping Devices.** Each electric elevator, except a sidewalk type elevator, shall be provided with upper and lower final terminal stopping devices arranged to stop the car and counterweight automatically from rated speed within the top clearance and bottom overtravel independently of the operation of the normal terminal stopping devices and the operating device in the car, but with buffers operative. Final limit switches and oil buffers shall be located so that the engagement of the buffer and the opening of the limit switch will occur as nearly simultaneously as possible. When spring buffers are provided, the final limit switches shall be opened before the buffer is engaged. Where means are provided to prevent jumping of the car or counterweight it shall only be necessary that the limit switch open before the buffer is fully compressed. Final terminal stopping devices shall be installed in connection with electric elevators, except sidewalk type elevators, as follows:

Item 1. Winding Drum Machines. Each electric elevator, having a winding drum machine, shall have stopping switches on the machine and also in the hatchway operated by the movement of the car.

Item 2. Traction Machines. Each electric elevator having a traction machine shall have stopping switches in the hatchway operated by the movement of the car.

(c) **Operation.** Final terminal stopping devices shall act to prevent movement of the car in both directions. The normal and final terminal stopping devices shall not control the same switches on the controller unless two (2) or more separate and independent switches are provided, two (2) of which shall be closed to complete the motor and brake circuit in each direction of travel. When 2-phase or 3-phase alternating current is used to operate the elevator the above switches shall be of the multiple type. In the case of hand-rope, rod, wheel or lever operating devices the normal and final terminal stopping devices may control the same switch on the controller. When the final terminal stopping device controls the same controller switch, or switches, as the operating device or the normal terminal stopping device, it shall be connected into the control circuit on the opposite side of the line. No chain, rope or belt-driven machine terminal stopping device shall be used on elevators having winding drum machines. Each electric elevator having a winding drum machine, driven by 2-phase or 3-phase alternating current, shall have an automatic terminal stopping device mounted on the machine or in the hatchway operated by a cam attached to the elevator car except where a direct current brake is used and a direct current main line or potential switch. This stopping device shall be arranged to open the main line circuit to the motor and brake directly. This device shall prevent the movement of the machine in either direction before or coincident with the operation of the final terminal stopping device.

(d) **Contacts and Switches.** The contacts of all terminal stopping devices shall be directly opened mechanically without the use of springs or gravity, or both. Normal and final terminal stopping switches unless located in the machine room, shall be of the inclosed type. Normal and final terminal stopping devices, where on the car or in the hatchway, shall be securely mounted in such a manner that the movement of the switch lever or roller to open the contacts shall be as nearly as possible at right angles to a line drawn between the car guide rails. The cams for operating the terminal stopping switches shall be of metal and shall be so located and of sufficient length to maintain the switch in the open position when the car is in contact with the overhead structure or resting on the fully compressed buffer with the overhead structure and the buffer in their normal position. Each electric elevator, equipped with a floor controller, or other similar device for automatic stops at landings, need be provided with only one set of stopping contacts for the terminal floors, provided these contacts and the means of operating them comply with the following:

Item 1. Metal Contacts. If metal to metal contacts, either gravity or spring opened, or a combination of these are used, there shall be at least two (2) independent breaks.

Item 2. Mechanically Opened. If contacts are directly opened mechanically.

Item 3. Circuit Interrupted. If breaking of the circuit to stop the car is independent of the operation of springs in tension or the completion of another electric current.

4403.15 Operation and Control: No elevator having a rated speed of more than one hundred (100) feet per minute shall be operated by direct hand operated ropes, cables or rods. No elevator having a rated speed of more than one hundred fifty (150) feet per minute shall be operated by wheel or lever mechanism except hydraulic elevators. Each hydraulic elevator, operated by a wheel operating device, shall be provided with an indicating device in the car to show the position of the control valve. Such device shall be marked to indicate "up", "down" and "off".

(a) **Hand-Ropes.** No elevator shall be operated by a rope or cable which is accessible from the outside of the hatchway. Overhead tension weights for hand-ropes shall be secured by chains or cables attached to the weights and to a suitable anchorage. When hand-ropes are used, guards shall be provided which will keep the hand-ropes on the sheaves. Each freight elevator, except sidewalk type elevator, operated by means of a direct operated hand-rope, shall be provided with a centering device which will insure the operating mechanism being placed in the stop position when it is desired to stop the car.

(b) **Car Switch Devices.** The handle of car switch operating devices and the switch operating devices used with continuous pressure operation, except push button, shall be arranged to return to and lock automatically in the stop position when the hand of the operator is removed. Where more than one operating device is used in a car, except in automatic operation elevators, the operating devices shall be so interlocked that only one can be used at a time. If a single operating device is used, it shall be located near the car opening, or if for more than one (1) opening, near the car opening serving the greatest number of landing openings. An emergency stop-switch, which will cut off the source of power, shall be provided in the car adjacent to the operating device for each electric elevator. If the stop button of an automatic operation or continuous pressure operation elevator is a rod button marked "stop", it may be used as an emergency stop-switch. One lead to the emergency stop-switch shall run to the car through a separate and independent traveling cable where electric elevators have winding drum machines.

(c) **Disconnecting Switch.** A manually operated multi-pole disconnecting service switch shall be installed in the main line of electric elevator machines and motor generator sets. This switch shall be located adjacent to and visible from the elevator machine or motor generator set. It shall be so arranged that the disconnecting switch cannot be closed from any other part of the building.

(d) **Construction.** Where metal to metal contacts, gravity or spring opened or a combination of the two (2), are used on controller switches for stopping elevator machines there shall be at least two (2) independent brakes. Breaking the circuit to stop an automatic control elevator at the terminals shall be independent of the operation of springs in tension or the completion of any other electric circuit. If springs are used they shall be in compression. If an electric circuit is used, its interruption shall prevent the movement of the car. The frame of an electric elevator machine, the frame of the controller, the operating rope if used, and the frames of electric appliances in or on the elevator car shall be effectively grounded. Electric slack cable switches shall be inclosed. No control system shall be used which depends on the completion or maintenance of an electric circuit for the interruption of the power and for the application of electro-mechanical brakes at the terminals, for the operation of safeties or the closing of a contact or by an emergency stop button, except for dynamic braking and speed control devices.

(e) **Operating Levers and Switches.** Car switching and hand-operating levers shall be so arranged that the movement of the lever toward the opening which the operator usually faces will cause the car to descend and the movement of the lever away from the opening will cause the car to ascend. A freight elevator operated by a hand-rope, if not equipped with an emergency switch, shall be provided with a centering device which will insure the operating mechanism being placed in the stop position when it is desired to stop the car; if it is not equipped with interlocks or electric contacts or is not a sidewalk type elevator, it shall also be provided with rope locks for holding the car at any landing. On a mechanically operated passenger elevator the operation of directional switches or operating valves shall not depend solely upon bolts or upon cast or malleable iron chains. If a hand-rope is used the cable shall be securely anchored to the operating sheave or drum. No circuit breaker operated automatically by a fire alarm system shall be used to cut off the power, or to interrupt the operating circuit of a power elevator.

(f) **Special Construction Applying to Automatic Operation Elevators.** An automatic operation elevator, except a sidewalk type elevator the travel of which does not exceed one (1) story, shall comply with the following requirements:

Item 1. Non-Reversing. If the car has started for a given landing it shall be impossible for an impulse to be given from any landing to send the car in the reverse direction until it has reached the destination corresponding to the first impulse.

Item 2. Continuity of Direction. If the car has been stopped at an intermediate landing and is to continue in the direction determined by the first impulse, the car may be started by closing the door or car gate.

Item 3. Hatchway Door Locks. It shall be impossible to start the car under normal operation unless every hatchway door is closed and locked in the closed position. No devices employing locks and contacts of a type where the contact is made when the door is closed and the locking of the door takes place subsequently shall be used.

Item 4. No Sending Buttons. Where the elevator is used by the general public no buttons shall be provided at any landing to send the car to any other landing.

Item 5. Floating Platform. The floating platform construction may be used to permit operation of the car from the landing buttons with the car gate open when there is no passenger in the car if: (1), each landing floor is flush with the hatchway edge of its landing threshold and lintel within a tolerance of one-fourth (¼) inch, plus or minus; (2), the platform will prevent the operation of the car from the car push buttons unless the platform is depressed; (3), it is impossible to operate the elevator from the car push buttons unless the car gate is closed; (4), the platform will operate when a weight of thirty (30) pounds is placed on it at any point; (5), the platform is so constructed that there is no pocket or recess capable of holding refuse or dirt beneath the platform; and (6), the entire platform within the inclosure and the platform threshold sill will float. No handrails shall be provided in the elevator car.

Item 6. Inspector's Switch. An inspector's switch shall be installed on the machine controller to render all landing buttons inoperative and when the opening of the emergency stop switch or button does not cancel all registered car and hall calls. the inspector's switch on the controller shall also render all car buttons inoperative and there shall be furnished into the car an "Up" and "Down" inspection switch or button which will enable the inspector to operate the elevator in either direction as long as the switch or button is held intact.

(g) **Continuous Pressure.** Continuous pressure operation shall not be used for a passenger elevator except when it is provided with all of the safety devices required for an automatic operation elevator.

(h) **Phase Rotation or Failure.** Each electric elevator driven by a polyphase alternating current motor shall be provided with a device which will, except in the case of an alternating current motor used in a motor generator set, prevent starting the motor if the phase rotation is in the wrong direction, or if there is a failure in any phase.

(i) **Hand Operating Devices.** Each electric elevator operated by a hand cable, lever or wheel operating device, shall be so arranged that in case of failure of power, or the opening of a car gate, landing door or limit switch, it will be necessary to return the operating device to the "Off" position before the elevator can again be started.

(j) **Circuit Breakers.** If an overload circuit-breaker is used for a direct current rheostatic control electric elevator, the wiring shall be arranged so that the circuit of the brake magnet coil is opened at the same time that the line circuit is opened.

(k) **Sidewalk Elevators.** Each sidewalk type elevator, the hatch opening of which is located in area accessible to the public shall be provided with a device which normally will prevent the car from opening or closing the hatch cover. A switch which can be operated only by a special key and which when closed will permit the car to open and close the hatch cover shall be installed at an easily accessible point above grade level and adjacent to the elevator. It shall be so arranged that it is necessary to hold the key or button manually in position to keep the circuit closed. The installation may be so made that the elevator will be operated by the key switch or button.

(l) **Condensers.** No condensers, the operation or failure of which will hold in any magnet, or keep alive any circuit so as to interfere with the proper operation of any elevator apparatus, shall be installed in connection with any elevators after the adoption of this Code. All such condensers installed in connection with any elevators prior to the adoption of this Code shall be removed.

4403.16 Limits of Speed: The maximum rated speed of passenger freight elevators, except as otherwise prescribed herein, shall be limited only by the top and bottom clearances in the hatchway. The maximum rated speed of freight elevators without a regular operator, except sidewalk type elevators, unless provided with automatic operation or continuous pressure operation, shall be one hundred (100) feet per minute. The maximum rated speed of electric freight elevators with continuous pressure operation, except sidewalk type elevators, shall be one hundred fifty (150) feet per minute.

4403.17 Cables: Car and counterweight cables shall be of iron or steel without covering except that where liability to excessive corrosion or other hazard exists, marline covered cables may be used for freight elevators only. Chains shall not be used for hoisting, except for sidewalk type elevators the travel of which does not exceed twenty (20) feet.

(a) **Plates and Tags.** The capacity plates hereinbefore prescribed shall bear the following legend with blanks filled in:

For Winding Drum Machines.

Cable Specifications.

Cable	Number	Diameter in Inches	Rated Ultimate Strength in Pounds
Hoisting	....	....	....
Car Counter-weights	....	....	....
Machine Counter-weights	....	....	....

For Traction Machines and Drum Machines Without Counterweights.

Cable Specifications.

	Number	Diameter in Inches	Rated Ultimate Strength in Pounds
Hoisting Cables	....	....	....

For Hydraulic Machines.

Cable Specifications.

Cable	Number	Diameter in Inches	Rated Ultimate Strength in Pounds
Hoisting	....	....	....
Car Counter-weights	....	....	....
Machine Counter-weights	....	....	....

In addition, a metal tag stating the diameter, rated ultimate strength, material of the cable, and the date of the cable installation shall be attached to the cable fastenings.

(b) **Factor of Safety.** The factor of safety based on static loads for car and counterweight cables for elevators shall be not less than the following, corresponding to the rated speed of the car:

Car Speed (Feet per Minute) Passenger Elevators	Freight Elevators	Factor of Safety
	50	6.6
	100	7.0
	150	7.4
50	200	7.7
100	250	8.0
150	300	8.2
	350	8.4
200	400	8.6
250	450	8.8
	500	9.0
300	550	9.25
350	600	9.5
	650	9.6
400	700	9.75
	750	9.9
450	800	10.0
500	900	10.2
	1000	10.3
550	1100	10.4
	1200	10.5
600		10.7
650		10.9
700		11.0
750		11.1
800		11.2
900		11.4
1000		11.5
1100		11.6
1200		11.7

The factor of safety based on static loads used for hoisting chains and cables of sidewalk type elevators shall be not less than five (5) for chains and seven (7) for cables.

(c) **Number and Size.** The number and diameter of the cables shall be determined by using the required factor of safety and the rated ultimate strength of the cable. The computed load on the cables shall be the weight of the car, plus its rated load, plus the weight of hoisting cables and the compensation. The minimum number of cables used with traction elevators shall be three (3). The minimum number of cables used with winding drum elevators and for private residence elevators shall be two (2). No hoisting ropes for elevators shall be less than one-half (½) inch in diameter, except for private residence elevators where ropes may be reduced to not less than three-eighths (⅜) inch.

(d) **Fastenings.** Cables anchored to winding drums shall have not less than one (1) complete turn of each cable on the winding drum when the car or counterweight has reached the extreme limit of its overtravel. No car or counterweight cables shall be repaired or lengthened by splicing. The winding drum ends of car or counterweight cables shall be secured by clamps on the inside of the drums or by one (1) of the methods hereinafter described for fastening cables to cars or counterweights. The car and counterweight ends of cables shall be fastened by individual tapered babbitted sockets. Other fastenings may be used for compensating counterweight cables and for plunger elevators. Adjustable shackle rods shall be used to attach cables to cars and counterweights in such a manner that all portions of each cable, except the portion in the socket, shall be readily visible.

Item 1. Sockets. The length of a babbitted socket shall be at least four and seventy-five hundredths (4.75) times the diameter of the cable. The hole at the small end shall be not more than one-sixteenth ($\frac{1}{16}$) inch larger in diameter than the actual cable diameter for cables having a nominal diameter from one-fourth (¼) to seven-sixteenths ($\frac{7}{16}$) inch; three thirty-seconds ($\frac{3}{32}$) inch larger for cables having a nominal diameter of one-half (½) inch to three-fourths (¾) inch; one-eighth (⅛) inch larger for cables having a nominal diameter of seven-eighths (⅞) inch to one and one-eighth (1⅛) inch and three-sixteenths ($\frac{3}{16}$) inch larger for cables having a nominal diameter of one and one-fourth (1¼) inches to one and one-half (1½) inches. The small end of the socket shall be free from cutting edges. The hole at the large end of the socket shall be at least two and one-fourth (2¼) times the diameter of the cable. A socket shall be drop forged steel, steel casting, or formed in a substantial block of cast iron. The socket shall be of such strength that the cable will break before the socket is perceptibly deformed.

Item 2. Method of Attachment to Sockets. The ends of wire rope shall be served with three (3) seizings at each side of any point at which the rope is to be cut. Only annealed iron wire shall be used as seizing wire. The wires shall be wound tight and even. The twisted ends of the seizing shall be so placed that they fall into the valley between strands and away from the center of the rope. For five-eighths (⅝) inch and smaller cables, the first two (2) seizings shall be at least one-half (½) inch long and the third seizing at least three-fourths (¾) inch long; for larger cables, seizings shall be increased accordingly. The first seizing shall be close to the cut and the second seizing shall be spaced within two and one-fourth (2¼) inches from the first seizing. The third seizing shall be at a distance from the second seizing equal to the length of the socket. Tape shall not be used for annealed iron wire seizing. The ropes thus served shall be slipped into the socket a sufficient distance for manipulating and after removing the first two seizings the strands shall be opened up and the hemp center cut out as close as possible to the remaining seizing. All grease shall be wiped off the extended strands and the lubricant carefully removed by washing with gasoline. The ends of the strands shall then be bent in and bunched close together, and the rope pulled back as far as possible so that the strands rest in the basket with the third seizing slightly projecting outside the mouth of the socket. The socket shall be warmed and shall be held vertical and truly axial with the rope to be socketed. Tape or waste may be wound around the rope at the base of the socket to prevent the metal from seeping through, but must be removed after the babbitt has cooled off. Pure babbitt only, free of dross, shall be used, heated to a temperature just sufficient to produce fluidity. The seizing and socketing shall be done so that there will be no loss of rope lay.

4403.18 Emergency Signals: Each automatic operation passenger or freight elevator, except a sidewalk type elevator the travel of which is not more than twenty (20) feet, shall be provided with an audible emergency signal operative from the car and located outside of the hatchway or shall be provided with a telephone. The emergency alarm shall be clearly audible in a room in which an employe is ordinarily located. Each automatically operated elevator installed in a private residence shall be provided with a telephone permanently connected to a central exchange. The hatchway of each freight elevator, except automatic operation, continuous pressure operation and sidewalk type elevators, shall be provided with a signal system by means of which signals can be given from any landing whenever the elevator is desired at that landing.

ARTICLE 4404.

Stage and Orchestra Elevators and Other Elevators of Special Character.

4404.01 General: Stage and orchestra elevators and other elevators of special character, shall comply with all of the foregoing requirements for elevators, which are applicable to the type of equipment used and for the purpose for which the elevator is installed; also all additional parts and accessories necessary for their full operation.

ARTICLE 4405.

Hand Elevators.

4405.01 Clearances and General Construction: The clearance between a car platform and the landing threshold shall be not more than two (2) inches for passenger elevators and three and three-fourths (3¾) inches for freight elevators, except that where the operating rope is located at the side of the platform, this clearance shall not exceed one (1) inch. An overtravel of not less than eighteen (18) inches shall be provided at the top for hand elevators, except for sidewalk type elevators. No overtravel is required at the bottom. Machinery and sheaves shall be so supported and held as to prevent any part from becoming displaced. The supporting beams shall be of steel or reinforced concrete. No elevator machinery, except idler or deflecting sheaves with their guards or frames and devices for limiting and retarding the car speed and their accessories shall be hung underneath the supporting beams at the top of the hatchway.

4405.02 Hatchway Door Hangers: Means shall be provided to prevent hangers for all sliding hatchway doors from jumping the tracks. Stops shall also be provided to prevent the hanger carriage from leaving the ends of the track, or suitable stops may be provided on the door only. Door counterweights shall run in metal guides from which they cannot become dislodged, or shall be boxed in. The bottoms of the guides or boxes shall be so constructed as to retain the counterweight if the counterweight rope breaks.

4405.03 Car Construction and Safeties:

(a) **Inclosures.** Cars, except for sidewalk type elevators, shall be inclosed on the top and sides not used for entrance. Inclosures shall be of solid or openwork rigidly braced with steel. Where slats, bars or wire mesh are used, the openings shall reject a ball two (2) inches in diameter. Where sheet metal is used, it shall be not less than No. 16 U. S. Standard Gauge. Where wire mesh is used, the wire shall be not less than No. 10 Steel Wire Gauge. A car inclosure shall not deflect more than one-fourth (¼) inch when a force of seventy-five (75) pounds is applied perpendicularly to the inclosure at any point. The car inclosure shall be securely fastened to the car platform or frame so that it cannot work loose or become displaced in ordinary service. Cars for sidewalk type elevators shall be inclosed on the sides not used for entrance to a height of not less than four (4) feet or to the spring of the bow iron, if higher. If the inclosure is openwork of bars, slats or wire mesh the openings shall reject a ball four (4) inches in diameter. Wire mesh inclosures shall be of wire not less than No. 13 Steel Wire Gauge. Hand elevator cars upon which persons are permitted to ride shall have only one (1) compartment.

(b) **Frames and Platforms.** Car frames and platforms shall be of metal or sound seasoned wood designed with a factor of safety of at least four (4) for metal or six (6) for wood on the rated load uniformly distributed. The frame members shall be securely bolted or braced.

(c) **Glass.** No glass shall be used in an elevator car except to cover the certificate, directory, light fixtures and appliances necessary for the operation of the car. No piece of glass over one (1) square foot in area shall be used unless it is shatterproof; the total area of such glass used in the car in connection with lighting fixtures, whether in one (1) or more pieces, shall not exceed four (4) square feet.

(d) **Gates.** Elevator cars, operating in hatchways outside of structures, which are inclosed only at the grade landing, shall be protected on the exposed sides by independently operated gates or by self-closing gates.

(e) **Safety Devices.** If the rise of an elevator exceeds fifteen (15) feet it shall be equipped with an approved safety device attached to the underside of the car which will immediately stop and hold the car and rated load if the suspension means breaks.

(f) **Capacity.** The rated load of a passenger elevator shall be not less than fifty (50) pounds per square foot of maximum clear horizontal area inside of the car inclosure. A metal plate bearing (1) the rated capacity of the elevator in pounds and (2), if a passenger elevator, the maximum number of passengers to be carried, based on one hundred fifty (150) pounds per person, in letters or figures not less than one-fourth (¼) inch high, etched or raised on the surface of the plate shall be fastened in a conspicuous place in the elevator car.

4405.04 Guides, Buffers and Counterweights:

(a) **Guides.** Car and counterweight guide rails shall be of rolled steel except where steel would constitute an accident hazard, where wood guide rails may be used, and except further, that wood guide rails may be used where the car travel is not more than thirty-five (35) feet. Joints in steel rails shall be either tongued and grooved or doweled and fitted with splice plates. Joints in wood rails shall be tongued and grooved or doweled and screwed to backing pieces or brackets. Guide rails shall be securely fastened with through bolts, wood screws, or clips of such strength, design and spacing that the maximum deflection of the guide rails and their fastenings will not be more than one-fourth (¼) inch under normal operation. Guide rails, and their fastenings, shall withstand the application of the safety when stopping a fully loaded car or the counterweight. The guiding surfaces of the guide rails for elevators requiring safeties shall be finished smooth. Guide rails shall be bottomed on suitable supports and extended at the top to prevent guide shoes running off in case the overtravel is exceeded.

(b) **Buffers.** Car buffers of the spring type or their equivalent shall be installed in the pits of passenger elevators.

(c) **Counterweights.** Counterweights shall run in guides; they shall not be boxed unless incombustible material is used. Sections of counterweights for passenger elevators, whether carried in frames or otherwise, shall be secured by at least two (2) tie rods passing through holes in the sections. The tie rods shall have lock nuts at each end, secured by cotter pins.

4405.05 Machines:

(a) **Brakes.** Automatic brakes shall be provided on all elevators having a travel of thirty-five (35) feet or more. Hand brakes operating in both directions of motion or combined automatic brakes and speed retarders operating in both directions of motion, except when motive power is derived through use of a self-locking or non-overhauling worm gear drive, shall be provided on all other elevators.

(b) **Construction.** The factors of safety, based on the static loads to be used in designing parts of hoisting machines shall be not less than eight (8) for wrought iron or wrought steel and ten (10) for cast iron or other materials. Keys or pins shall be used for fastenings except that set screws may be used where the connection is not subject to torque. No sheaves or idlers in cast iron stirrups shall be suspended from the underside of the supporting beams. No hand elevator machine shall be equipped with any means or attachment for supplying any other power, unless such elevator is permanently and completely converted into a power elevator complying with the requirements for power elevators. No rope gripping attachments or clutch mechanisms shall be used as a means of applying power to hand elevators.

4405.06 Suspension Members: A metal tag shall be attached to the suspension fastenings stating the size, rated ultimate strength and material of the suspension and the date of its installation. The number of suspension members for both car and counterweight shall be at least two (2). Suspension members shall be of iron, steel or marline covered, and shall be installed in the manner hereinbefore prescribed for power elevators. The factor of safety used in determining the size of the suspension member shall be at least five (5), based on the weight of the car and its rated load. Suspension members shall be so adjusted that either the car or the counterweight shall be bottomed before the counterweight or the car strikes any part of the overhead structure. Suspension members secured to a winding drum shall have not less than one complete turn of the suspension member on the winding drum when the car or counterweight has reached the extreme limit of its overtravel. The drum end of cables shall be secured by clamps or sockets inside the drum.

ARTICLE 4406.

Dumbwaiters.

4406.01 General: Machinery and sheaves shall be so supported and held as to prevent any part from becoming displaced.

4406.02 Electric Contacts and Locks: Hatchway doors for a power dumbwaiter shall be equipped with electric contacts and locks which will prevent the operation of the machine while any hatchway door is open and prevent the opening of a door unless the car is at a landing.

4406.03 Dumbwaiter Construction: Dumbwaiter cars shall be of such strength and stiffness that they will not deform appreciably if the load leans or falls against the sides of the car. Cars shall be made of wood or metal, reinforced at the point of suspension. Metal cars, if sectional, shall be rigidly riveted, welded or bolted together. Dumbwaiter cars, machines and suspension means shall at least be capable of sustaining the rated load.

(a) **Capacity.** Cars having a total clear platform area, including shelves, if any, of four (4) square feet or more, shall be capable of sustaining the loads given in the following table, but the motive power need not be sufficient to raise the structural capacity load:

Area (Square Feet)	Structural Capacity Loads (Pounds).
4.0	100
5.0	150
6.25	300
9.0	500

A metal plate bearing the name of the manufacturer and the rated load shall be placed in a conspicuous place in each dumbwaiter car and on its machine.

(b) **Factor of Safety.** Dumbwaiter machines shall be securely fastened to their supports. The factors of safety, based upon the ultimate strength of the material, and the rated load plus the weight of the car, suspension means, counterweights, and similar apparatus used in the design of dumbwaiter machines shall be not less than six (6) for steel and nine (9) for cast iron or other materials. Keys or pins shall be used for fastenings except that set screws may be used where the connection is not subject to torque. Sheaves or idlers shall not be suspended in cast iron stirrups from the underside of supporting beams.

(c) **Guides.** Guides shall be of wood or metal, except that metal guides shall be used where the rated speed is more than one hundred (100) feet per minute unless the use of steel would constitute an accident hazard, in which case wood may be used. Guides shall be rigidly secured to the hatchway. Joints in metal guides shall be either tongued or grooved or doweled and fitted with splice plates. Joints in wood guides shall be either tongued and grooved or doweled and screwed to backing pieces or brackets. One set of guides may be used for both the car and counterweight.

(d) **Counterweights.** Counterweights for hand dumbwaiters and for power dumbwaiters with a rated load of not more than one hundred (100) pounds and a rated speed of not more than one hundred (100) feet per minute, if sectional, shall be carried in suitable frames. Counterweight sections for power dumbwaiters, having a capacity of more than one hundred (100) pounds or a speed of more than one hundred (100) feet per minute, shall be secured by not less than two (2) tie rods passing through holes in all sections, unless suitable counterweight frames or boxes are provided. The tie rods shall have lock nuts at each end; the lock nuts shall be secured by cotter pins.

(e) **Means of Suspension.** Suspension means for hand dumbwaiters may be of hemp. Suspension means for power dumbwaiters shall be of metal and where exposed to corrosion shall be provided with suitable protective covering. Suspension means for a dumbwaiter may consist of a single number. The minimum factor of safety of suspension means for hand dumbwaiters shall be five (5). The minimum factor af safety of the suspension means for power dumbwaiters shall be not less than the following:

Factor of Safety for Suspension Means for Power Dumbwaiters.

Rated Speed (Feet per Minute)	Factor of Safety (Except for Tapes)
Up to 50	5.0
Over 50 to 75	5.2
75 to 100	5.3
100 to 150	5.4
150 to 200	5.6
200 to 250	5.8
250 to 300	6.05
300 to 350	6.3
350 to 400	6.55
400 to 450	6.8
450 to 500	7.0
500 to 550	7.25
550	7.50

Add 25% to the above minimum factors of safety for tapes. The number and size of the suspension means shall be determined by using the required factor of safety and the rated ultimate strength of the suspension means. The computed load on the suspension means shall be the sum of all suspended weights plus the rated load. No suspension means shall be repaired or lengthened by splicing. The winding drum ends of the car and counterweight suspension means shall be secured by clamps or sockets inside the winding drum. Suspension means secured to a winding drum shall have not less than one (1) turn on the winding drum when the car or counterweight has reached the extreme limit of its over travel.

4406.04 Speed, Control and Safeties for Power Dumbwaiters: The maximum speed for power dumbwaiters controlled by hand ropes shall be fifty (50) feet per minute. Guards which will keep the ropes on the sheaves shall be installed unless means are used to maintain the hand ropes in proper tension automatically. Power dumbwaiters, except hydraulic dumbwaiters, shall be equipped with brakes which are automatically applied when the power is cut off the motor. A power dumbwaiter having a travel of more than thirty (30) feet, a capacity of more than one hundred (100) pounds and operated by a winding drum machine, except a hydraulic dumbwaiter, shall be provided with a slack cable device which will cut off the power and stop the machine if the car is obstructed in its descent. Each power dumbwaiter shall be provided with means independent of manual operation, to stop the car automatically at each terminal within the limits of overtravel.

ARTICLE 4407.

Escalators.

4407.01 General: The maximum angle of inclination of an escalator shall be thirty (30) degrees from the horizontal in new buildings and thirty-three (33) degrees in existing buildings. The width of an escalator shall be not less than twenty-four (24) inches nor more than forty-eight (48) inches measured between the balustrading at a vertical height of twenty-four (24) inches above the nose line of the treads. All escalators shall have a horizontal tread formation. The maximum speed of an escalator, measured along the angle of inclination, shall be one hundred twenty-five (125) feet per minute, except that if the line of entrance and exit is not in the vertical plane of travel, the maximum speed shall be one hundred (100) feet per minute.

4407.02 Construction:

(a) **Balustrades.** Every escalator shall be provided on each side with solid balustrading. On the escalator side the balustrading shall be smooth without depressed or raised paneling or molding, and without glass panels. There shall be no abrupt changes in the width of the balustrading; should any change be necessary it shall be not more than eight (8) per cent of the greatest width and shall be made at an angle of not more than fifteen (15) degrees from the line of escalator travel. Each balustrading shall be equipped with a handrail moving at the same speed and in the same direction as the travel of the escalator.

(b) **Treads and Landings.** Escalator treads and landings shall be of material affording secure foothold. If the landing is of concrete it shall have edge insertions of metal, wood or other material to prevent slipping. The track arrangement shall be designed to prevent the displacement of the treads and running gear if a tread chain breaks.

(c) **Construction.** The rated load in pounds on an escalator shall be four and six-tenths (4.6) times the width of the escalator in inches times the horizontal projected length of the exposed treads in feet. The factor of safety to be used in the design of an escalator truss or girder shall be not less than five (5) based on the static loads. The escalator truss or girder shall be so designed that it will safely retain the steps and running gear in case of failure of the track system to retain the running gear in its guides. Chains shall have a factor of safety of not less than ten (10) except where the chain is composed of cast steel links thoroughly annealed, when the factor of safety shall be not less than twenty (20). Access to the interior of the escalator shall be provided for inspection and maintenance.

(d) **Drive.** Escalators shall be driven by individual electric motors.

(e) **Marks of Approval.** Each escalator shall be marked by the manufacturer with the rated load and speed for which that size and type has been tested and approved in accordance with "Escalator Tests" hereinafter described.

4407.03 Safety Requirements for Escalators:

(a) **Stop Buttons.** An emergency stop button or other type of switch accessible to the public shall be conspicuously located at the top and bottom of each escalator landing. The operation of either of these buttons or switches shall cause the interruption of power to the escalator. It shall be impossible to start an escalator by means of these buttons or switches. The buttons or switches shall be marked "ESCALATOR STOP BUTTON" or "ESCALATOR STOP SWITCH". Where starting buttons or switches are accessible to the public they shall be either of the key operated type or enclosed in a box provided with a lock and key. Each escalator shall be equipped with means to cause the interruption of power to the escalator in case of accidental reversal of travel of an escalator operating in the ascending direction.

(b) **Speed Governors and Other Devices.** Each escalator shall be provided with a speed governor which will cause the interruption of power to the escalator in case the speed exceeds a pre-determined value which shall be not more than forty (40) per cent in excess of the normal running speed. Each escalator shall be provided with a broken chain device which will cause the interruption of power to the escalator in case a tread chain breaks. Where an escalator is equipped with a tightening device, operating by means of tension weights, provision shall be made to retain these weights in the escalator truss in case the weights should fall.

(c) **Brakes.** Each escalator shall be provided with an electrically released and mechanically applied safety brake of sufficient power to stop the fully loaded escalator mounted on the main drive shaft of the escalator. This brake shall automatically stop the escalator when operating, or tending to operate, in the descending direction in case any of the safety devices function, except that if the escalator drive machine is equipped with an electrically released mechanically applied brake of sufficient power to stop the fully loaded escalator, the above safety devices may apply this brake in lieu of the safety brake, if a device is furnished which will apply the safety brake in case the connection between the escalator drive machine and the main drive shaft parts.

(d) **Electrical Phase Protection.** Each escalator, operated by a polyphase alternating current motor, shall be provided with a device which will prevent starting the motor (1) while the phase rotation is in the wrong direction, or (2) if there is a failure of any phase.

ARTICLE 4408.

Tests of Equipment.

4408.01 Tests of Interlocks Before Approval of Type and Makes:

(a) **General:** Hatchway door interlock devices shall be examined with respect to their proper performance of the prescribed functions at temperatures from twenty-five (25) to one hundred forty (140) degrees Fahrenheit. Where the functioning of any such device might be affected by change of temperature, and coefficients of thermal expansion of the affected parts are known or measured, the effect of temperature may be computed and tests need not be carried out at more than one (1) temperature. For interlocks employing a single switch operated by wire or tape to protect several hatchway doors, the switch shall be marked with the physical properties of the wire or tape used, and the certificate of approval shall cover the allowable maximum length of tape or wire of a given material which may be used with the device. During tests for (c) Endurance, (d) Current Interruption, (e) In Moist Atmosphere and (f) Without Lubrication, interlocks shall have electrical parts connected in a non-inductive electric circuit having a constant resistance, in which a current of two (2) amperes from a source of two hundred twenty (220) volts direct current is flowing. During tests c, e and f, the electrical circuit shall be closed but shall not be broken at the contact within the device on each cycle of operation.

(b) **Separate Devices.** A separate device shall be used in each test described as: (c) "Endurance Test", (e) "Tests in Moist Atmosphere", and (f) "Tests Without Lubrication".

(c) **Endurance Test.** The interlock device with initial lubrication and adjustment only, shall complete one hundred thousand (100,000) cycles of operation without failure of any kind and without evident indications of approaching failure. If an interlocking device is not a complete and separate unit for each hatchway door, but includes any part which is common to the interlock operation of more than one (1) hatchway door, that portion of the device shall complete four hundred thousand (400,000) additional cycles of operation without failure of any kind and without evident indication of approaching failure.

(d) **Current Interruption Test.** One thousand (1,000) cycles of operation shall be performed by the device used in the "Endurance Test", while making and breaking the circuit within the device.

(e) **Tests in Moist Atmosphere.** Preliminary to this test the device shall be given a wearing-in run of ten thousand (10,000) cycles of operation fully lubricated. The interlocking device, except self-lubricating bearings, and bearings of a type not requiring frequent replenishment of lubricant, shall be taken apart and freed of lubricant by washing in gasoline. The device, after reassembling, shall be subjected continuously in a closed hatchway to an atmosphere saturated with a three and one-half (3½) per cent solution of sodium chloride for seventy-two (72) consecutive hours being operated for only ten (10) consecutive cycles at the end of each of the first two (2) twenty-four (24) hour periods and allowed to stand exposed to the air for an additional twenty-four (24) hours, and shall not fail in a manner to create an unsafe condition. After having been lubricated it shall then, without adjustment and without further attention, complete fifteen thousand (15,000) cycles of operation without failure of any kind.

(f) **Tests Without Lubrication.** Preliminary to this test the device shall be given a wearing-in run of ten thousand (10,000) cycles of operation fully lubricated. The interlocking device shall then be taken apart and all bearings, except self-lubricating bearings and bearings of a type not requiring frequent replenishment of lubricant shall be freed of lubricant by washing in gasoline, and after reassembling without other attention than the usual initial adjustment; that is, without adjustment especially made to meet the conditions of this particular test, and without further attention, shall complete twenty-five thousand (25,000) cycles of operation without failure of any kind or without evident indication of approaching failure.

(g) **Misalignment Test.** The device shall operate successfully when the car, cam or other equivalent operating device, used in making the test, has been displaced horizontally from its normal position, successively as follows:

1. In a direction perpendicular to the edge of the landing—backward one-fourth (¼) inch; forward one-fourth (¼) inch.
2. In a direction parallel with the edge of the landing—to the right one-fourth (¼) inch; to the left one-fourth (¼) inch. For horizontal sliding doors the device shall operate successfully:
3. When the bottom of the hatchway door has been displaced horizontally from its normal position in a direction perpendicular to the edge of the landing—backward one-fourth (¼) inch; forward one-fourth (¼) inch.
4. When the top of the hatchway door has been displaced horizontally from its normal position in a direction perpendicular to the edge of the landing — backward one-eighth (⅛) inch, forward one-eighth (⅛) inch.

(h) **Insulation Test.** Insulation of electrical parts shall be tested with a sixty (60) cycle effective voltage twice the rated voltage plus one thousand (1,000) volts, applied for one (1) minute.

(i) **Force and Movement Test.** If the interlock is of the type which is released by a car cam, before and after Endurance Test, the force required to release the interlock and the movement of the element engaged by the cam, shall be measured. The force recorded in each case shall be the maximum, acting in a horizontal plane, which must be applied to that member of the interlock which is directly actuated by the cam, to release the door locking member of the interlock from locking engagement. The movement recorded shall in each case be the distance vertically projected on a horizontal plane, which the member of the interlock directly actuated by the cam travels from its position when the cam is retired and the door locked to its position when the door locking member is first released from locking engagement. The force and movement readings shall be determined with the interlock mounted in its normal position, as determined by the manufacturer. The test certificate shall state the average of the recorded forces and movements.

4408.02 Tests of **Electric Contacts Before Approval** of **Type and Makes:** Electric door and gate contacts shall be subjected to the "Endurance Test", "Current Interruption Test", "Tests in Moist Atmosphere", "Misalignment Test" and "Insulation Test' hereinbefore described for interlocks.

4408.03 Test of **Emergency Releases Before Approval** of **Type and Makes:** Emergency releases shall be subjected to the "Insulation Test" hereinbefore described for interlocks.

4408.04 Tests of **Buffers Before Approval** of **Type and Makes:** Each type and size of oil buffer shall be subjected to the tests herein described.

(a) **Retardation Test.** The buffer shall be installed upon a suitable foundation so that the axis of the cylinder is vertical. It shall be filled with oil provided by the manufacturer. An elevator car of suitable size shall be dropped from two (2) different heights, as specified below, freely in its guides, upon the buffer, and readings of the travel of the car after it comes in contact with the plunger, and of the plunger for its entire stroke shall be recorded on a drum chronograph or by photographing a calibrated tape. The error in the time readings shall not exceed 0.0005 seconds. The peripheral speed of the drum shall be approximately that of the car at the instant of impact. Pressure travel or pressure time records shall be taken simultaneously. From the time travel curve the velocity and the retardation of the car shall be computed, and plats made of the car travel, car velocity and car retardations, together with the pressure time curve.

Schedule of Drops.

Test Drop in Inches (bottom of car to striker or top of buffer).	Total load in pounds (weight of car plus loading).
(A) S (such a distance that the maximum velocity attained by the car shall be equal to the governor tripping speed for which the buffer is rated).	1. Manufacturer's rated minimum. 2. Manufacturer's rated maximum. 3. One hundred ten (110) per cent manufacturer's rated maximum.
(B) Buffer stroke under twenty four (24) inches.. .51S twenty-four (24) to thirty (30) inches .... .56S exceeding thirty (30) inches...... .64S	1. Manufacturer's rated minimum. 2. Manufacturer's rated maximum. 3. One hundred ten (110) per cent manufacturer's rated maximum.

No acceleration peak having a duration greater than one-twenty-fifth (1/25) second shall exceed two and one-half (2½) times gravity, eighty and five-tenths (80.5) per second per second for tests A-1 and A-2 and B-1 and B-2. Results of A-3 and B-3 shall be recorded for the purpose of examination to detect any abnormal performance. Upon completion of tests no part of the buffer shall show any deformation or injury.

(b) **Oil Leakage Test.** The oil leakage test shall be made simultaneously with the retardation tests. The oil level in the buffer, when filled prior to test, shall be carefully marked. At the completion of the six (6) drops, three (3) different loads at each of two (2) speeds, the buffer shall be allowed to stand one-half (½) hour to permit the return of the oil to the reservoir and to permit the escape of any entrained air, after which the oil level shall again be measured. The oil level at the completion of these tests should be unchanged, and shall in any case be less than one-sixteenth (1/16) inch lower than the level at the start of the test for each foot of buffer stroke.

(c) **Churning Test.** In the churning test the time of the buffer stroke after the car has dropped a distance equivalent to the stroke of the buffer shall be determined either from the chronograph or photographic record or by means of an automatic timer. Any automatic timing device shall be accurate, having the minimum possible lag, and be capable of being read to 0.01 seconds. The car shall then be run on to the buffer with cables attached at approximately one-half (½) the velocity used in test B at intervals of one (1) minute until ten (10) such strokes have been made. The oil shall then be examined for foam. No oil foam shall appear on the outside of the buffer following this test. Upon completion of the ten (10) strokes at one-half (½) speed another free-fall equivalent to the buffer stroke shall be made immediately and the time of the stroke taken. The time of this second free-fall test shall be at least seventy-five (75) per cent of that of the drop test made prior to the churning.

(d) **Plunger Return Test.** In the buffer test the buffer shall not stick on the return stroke after removal of the load. In case of sticking the manufacturer shall submit either a duplicate buffer or a new pressure cylinder and piston upon which equipment a second test shall be run. If sticking again results, the buffer shall be rejected.

(e) **Test for Lateral Movement** of **the Plunger.** With the buffer casing clamped or otherwise securely fastened to a firm base, the lateral movement of the top of the plunger shall be accurately measured, the plunger head being moved from its extreme right to the extreme left in a vertical plane. This total movement shall be divided by two (2) to determine the movement from the vertical position. The maximum permissible movement from the vertical shall be one-sixteenth (1/16) inch per foot of buffer stroke.

4408.05 Field Tests of **Elevators:** A test shall be made of every new elevator with rated load in the car and the brakes, limit switches, buffers, safeties and speed governor shall be caused to function. Field tests of buffers and car safeties shall be made as follows:

(a) **Buffers.** Run on to buffers with rated load at rated speed with final limit switches operative, except that if buffer stroke has been reduced due to the use of a speed retarding device the car or counterweight shall be run on the buffer at the speed corresponding to the buffer stroke used.

(b) **Car Safeties.** An overspeed test with rated load in the car shall be made of the safeties, except that governor controlled instantaneous type safeties shall be tested at rated speed, the governor being tripped by hand and broken rope instantaneous type safeties shall be tested by obtaining the necessary slack rope to cause them to function. For wedge clamp, gradual wedge clamp and flexible guide clamp safeties, this test shall be made to determine whether the safety will operate within the allowable limits of the maximum and minimum stopping distances. Overspeed tests shall be made with all electric apparatus intact, except for the overspeed contact on the governor. For alternating current elevators, where the rated load is unable to bring about overspeed, the safety governor shall be tripped by hand at maximum obtainable speed. No test of the safeties with safe lifting load in the car is required.

Item 1. Wedge Clamp Type. The maximum stopping distances of car and counterweight safeties of the wedge clamp type (W.C.) shall be not more than those given in the following table:

Maximum and Minimum Stopping Distances at Various Governor Tripping Speeds at Runaway Test—Wedge Clamp (W.C.) Type Safeties.

Governor Tripping Speed (Feet per Minute)	Maximum Stopping Distance (feet) for car with rated Load and for Counterweight	Minimum Stopping Distance (feet) for car with rated 150-lb. Load	Minimum Stopping Distance (feet) for car with rated Load and for Counterweight
300	2.0	1.0	1.3
400	2.8	1.2	1.6
500	4.0	1.4	2.0
600	5.2	1.6	2.4
700	6.8	1.9	3.0
800	8.6	2.3	3.6
900	10.7	2.7	4.4
1000	13.0	3.0	5.2
1200	18.4	4.0	7.0
1500	28.2	5.7	10.4

Item 2. Gradual Wedge Clamp Type. The maximum stopping distances of car and counterweight safeties of the Gradual Wedge Clamp Type (G.W.C.) shall be not more than those given in the following table:

Maximum and Minimum Stopping Distances at Various Governor Tripping Speeds at Runaway Test—Gradual Wedge Clamp (G.W.C.) Type Safeties.

Governor Tripping Speed (Feet per Minute)	Maximum Stopping Distance (feet) for car with rated Load and for Counterweight	Minimum Stopping Distance (feet) for car with 150-lb. Load	Minimum Stopping Distance (feet) for car with rated Load and for Counterweight
300	7.0	1.5	2.2
400	7.8	1.6	2.5
500	8.6	1.8	2.8
600	9.9	2.1	3.3
700	11.0	2.4	3.8
800	12.2	2.7	4.5
900	13.5	3.0	5.2
1000	14.6	3.5	6.1
1200	17.3	4.5	8.0
1500	21.2	6.2	11.2

Item 3. Flexible Guide Clamp Types. The maximum stopping distances of the car and counterweight safeties of the Flexible-Guide Clamp Type (F.G.C.) shall be not more than those given in the following table:

Maximum and Minimum Stopping Distances at Various Governor Tripping Speeds at Runaway Tests—Flexible Guide Clamp (F.G.C.) Type Safeties.

Governor Tripping Speed (Feet per Minute)	Maximum Stopping Distance (feet) for car with rated Load and for Counterweight	Minimum Stopping Distance (feet) for car with 150-lb. Load	Minimum Stopping Distance (feet) for car with rated Load and for Counterweight
300	1.6	.6	.8
400	2.5	.8	1.2
500	3.6	1.0	1.5
600	4.8	1.2	2.0
700	6.4	1.5	2.6
800	8.2	1.8	3.2
900	10.4	2.2	4.0
1000	12.8	2.6	4.8
1200	18.0	3.5	6.7
1500	28.0	5.2	10.0

(c) **Stopping Distance.** Stopping distance is actual slide as indicated by the marks on the rails. For elevators having a rated speed of four hundred and seventy-five (475) feet per minute, or more, the pull-out of the governor cable from its normal running position until the safety jaws begin to apply pressure to the guide rails shall be not more than thirty (30) inches.

4408.06 Field Tests of Dumbwaiters: A test shall be made of every new dumbwaiter with rated load in the car and the brake and all other safety devices shall be caused to function.

4408.07 Field Tests of Escalators: Each escalator shall be subjected to the following tests, without load:

(a) **Speed.** The application of the overspeed safety device shall be obtained by causing the escalator to travel at the governor tripping speed. With escalators driven by alternating current motors, the governor may be tripped by hand with the escalator traveling at its normal rate of speed.

(b) **Reversal.** The accidental reversal device prescribed shall be made to function by manually operating, or attempting to operate, the escalator in the reverse direction.

(c) **Broken Chain.** The application of the broken chain device shall be obtained by operating the device by hand.

(d) **Miscellaneous Safety Devices.** Tests of escalator emergency stop buttons or switches shall be made to determine whether they function properly. Where the device which applies the "safety brake" in case the connection between the escalator drive machine and the main drive shaft fails, is required, it shall be tested by operating the device by hand.

Mechanical Amusement Devices.

ARTICLE 4409.

Roller Coasters, Scenic Railways and Other Devices.

4409.01 General: All mechanical amusement devices shall be built of the material hereinafter enumerated, or of other materials approved by the Commissioner of Buildings, substantially constructed and designed to withstand shocks and to afford adequate protection for passengers riding thereon; structural features shall meet the requirements prescribed elsewhere in this Code. Handrails, handles, safety straps, or other protective devices of suitable design shall be provided in all cars of roller coasters, scenic railways, ferris wheels, whips and other riding, sliding, rotating, and rolling devices of similar type. Each horse on a merry-go-round shall be equipped with a stirrup and a bridle, also a strap on the horse rod to snap or buckle under the arms of the rider.

(a) **Construction.** No device shall extend more than three (3) feet below the ground level unless the sides and bottoms of all pits are built of concrete; all pits shall be provided in the bottom with drains connected to the sewers. If pits are too deep to drain to the sewer by gravity, a syphon, automatic electric pump or other device shall be installed in the drain connection. The structure shall be of wood, steel or other serviceable material substantially fabricated and braced; no permanent structure more than thirty-five (35) feet in height shall be of wood. In an amusement device of the dip type, the up grade in each dip shall be so constructed that the cars will run up the structure at a speed such that the cars will run over the top of the next dip without having a tendency to raise the passengers out of their seats or throw them out of the cars. The cars shall be of substantial construction; they shall be equipped with dogs to drop into a sprocket chain or other approved device to pull the car or train to the starting point of its travel. Ferris wheels, except of the portable type, shall have steel frames and steel tripods supported upon, and anchored to concrete piers. Cars shall be of all steel construction or of wood reinforced with steel. Ferris wheels of portable type used in carnivals and under similar conditions, shall be of steel construction set on suitable bases under the towers and the side tripods.

(b) **Safeties.** Every device shall be provided with a terminal brake; if designed for more than two (2) car trains, it shall also be provided with an emergency brake, release of which will immediately stop the train, which shall be placed in some level spot on the structure; or, if approved by the Commissioner of Buildings, on one of the curves. The emergency brake shall be under the control of the brakeman or other attendant at the loading platform. Each car shall also be equipped with a safety device arranged to catch and hold the train at any point on the road should the chain break or any other accident occur to the machinery while a car or train is in transit.

(c) **Lighting.** All mechanical amusement devices shall be provided with electric lighting if to be in use after sunset.

4409.02 Test: A test shall be made of every new mechanical amusement device and all safety devices shall be caused to function.

ARTICLE 4410.

Requirements Governing Operation of Elevators, Dumbwaiters, Escalators and Mechanical Amusement Devices.

4410.01 Operation:

(a) **General.** It shall be unlawful for any operator of any elevator in the city, wherein passengers are conveyed, to start such elevator until all doors of such elevator and leading into such elevator are closed and locked, or to open the doors of such elevator until said elevator has come to a full and complete stop, unless the elevator is equipped with interlocks and a slow speed automatic levelling or landing device, which will stop the car at the floor.

(b) **Reporting Accidents.** Whenever any accident shall occur, causing injury to life or limb to any person in or about an elevator, dumbwaiter, escalator or mechanical amusement device, or while getting on or off of same, or which shall in any way impair the safety of the equipment, such accident shall be reported at once by the operator of the equipment, owner, superintendent, lessee or manager of the building to the Commissioner of Buildings. No broken or damaged parts of such elevator, dumbwaiter, escalator or mechanical amusement device shall be moved or displaced, nor shall repairs be made thereon, nor shall said elevator, dumbwaiter, escalator or mechanical amusement device be operated until an investigation into such accident has been made by the Commissioner of Buildings or his duly authorized agent. A full report in writing of the result of such investigation shall be filed in the Department of Buildings and the Commissioner of Buildings shall keep a complete record of all such accidents and reports thereon.

(c) **Penalty.** Any person, firm or corporation violating any of the provisions of this Chapter of the Building Code pertaining to Elevators, Dumbwaiters, Escalators and Mechanical Amusement Devices, or failing to or neglecting to comply therewith, shall be fined not less than Twenty-five $25.00) Dollars nor more than Two Hundred ($200.00) Dollars for each offense.

CHAPTER 46.

Heating Provisions.

ARTICLE 4601.

General.

4601.01 Requirements: All heat producing appliances used within buildings or structures shall conform to the general and structural provisions of this Chapter, except as special provisions of this Chapter, except as otherwise provided in Occupancy Chapters 7 to 18, inclusive. Plans shall be required to show the general location of all heat producing appliances and equipment conforming with these provisions before the issuance of building permits.

4601.02 Heat Producing Appliances, Definition: For the purposes of this ordinance, a heat producing appliance is defined as any device used for the production of heat by the combustion of fuel.

4601.03 Minimum Temperature: All multiple dwellings, hospitals, office and financial units, theatres, public assembly units, churches, schools, jails, police stations, prisons and reformatories shall be provided with heating appliances and equipment capable of maintaining within all habitable rooms thereof, a minimum temperature of sixty-five (65) degrees F., when the outside temperature is ten (10) degrees below zero (0) F.

ARTICLE 4602.

Construction.

4602.02 Foundations:

(a) **Up to Three Hundred (300) Degrees F.** Every heat producing appliance, the exposed bottom surface of which cannot attain a temperature of more than three hundred (300) degrees Fahrenheit shall be placed on:

1. Foundation on soil;
2. A standard furnace foundation as provided in Section 1108.15;
3. A non-combustible floor; or
4. Non-combustible supports providing a clearance of not less than four (4) inches between the floor and the bottom of the appliance open on at least three (3) sides.

(b) **Three Hundred One (301) to Eight Hundred (800) Degrees F.** Where the exposed bottom surface temperature of the appliance is more than three (300) degrees and not more than eight hundred (800) degrees Fahrenheit, the appliance shall be placed on:

1. A foundation on soil;
2. A standard furnace foundation as provided in Section 1108.15;
3. A non-combustible floor; or
4. Non-combustible supports providing a clearance of not less than twenty-two (22) inches between the floor and the bottom of the appliance, open on all sides.

(c) **Over Eight Hundred (800) Degrees F.** Where the exposed bottom surface temperature of the appliance is more than eight hundred (800) degrees Fahrenheit, the appliance shall be placed on:

1. A foundation on soil;
2. A non-combustible floor and a standard furnace foundation as provided in Section 1108.15; or
3. A non-combustible floor and non-combustible supports providing a clearance of not less than twenty-two (22) inches between the floor and the bottom of the appliance, open on all sides.

4602.03 Ash Pits: All ash pits or receptacles for ashes from heat producing appliauces, including floors, walls and covers, if any shall be of non-combustible materials.

ARTICLE 4603.

Smoke and Gas Disposal.

4603.01 Smoke Pipes, Breechings, Flues and Vents: Every heat producing appliance using solid fuel and every heat producing appliance using liquid fuel with a supply tank of more than two (2) gallons capacity shall be provided with a smoke pipe, breeching, flue or vent which will carry the products of combustion to the outside air. Every heat producing appliance using gas, except as follows, shall be provided with a similar smoke pipe, breeching, flue or vent, which, if for a cooking appliance, may be connected to the suction of a mechanical ventilating exhaust system provided for the space in which the appliance is located.

1. A water heater having an input at maximum rating of not more than 10,000 British Thermal Units per hour.
2. A space heating appliance located in the space to be heated and having an input at maximum rating of not more than 10 British Thermal Units per hour per cubic foot of space to be heated.
3. A cooking appliance having an input, exclusive of the top burner section, of more than 50,000 British Thermal Units per hour.
4. Any other household appliance not located in a habitable room and having an input at maximum rating of not more than 30 British Thermal Units per hour per cubic foot of space in the room in which the appliance is located.

The chimney to which any smoke pipe, breeching, or other flue or vent from a heat producing appliance is connected shall be constructed in accordance with the requirements of Chapter 41. No such smoke pipe, breeching, flue or vent shall pass through any floor or roof unless such floor or roof is of non-combustible construction, and then only through one (1) such floor; not through any combustible partition unless the clearances hereinafter prescribed are provided; except that a smoke pipe for a domestic stove or range may pass through such partition if it is guarded by a double metal ventilated thimble not less than six (6) inches larger than the pipe or by a steel tube built in brick work or other equivalent fireproofing material extending not less than eight (8) inches beyond all sides of the tube. Every connection of a smoke pipe or breeching to a chimney shall be made tight with non-combustible material. Not more than eighty (80) per cent of the internal cross sectional area of any smoke pipe or flue shall be closed off by any damper or automatic draft regulating device except where two (2) or more heat producing devices are connected to a common smoke pipe or flue. There shall be no construction which will impede the free circulation of air around the entire surface of any smoke pipe, breeching, flue or vent.

4603.02 Hoods Over Ranges and Other Fixtures: Every range or other heat producing appliance where food is cooked in any restaurant or hotel kitchen or food establishment, or where from any cause, grease or other flammable substances are produced, shall be provided with a non-combustible hood or curtain wall inclosure. Every such hood or cur-

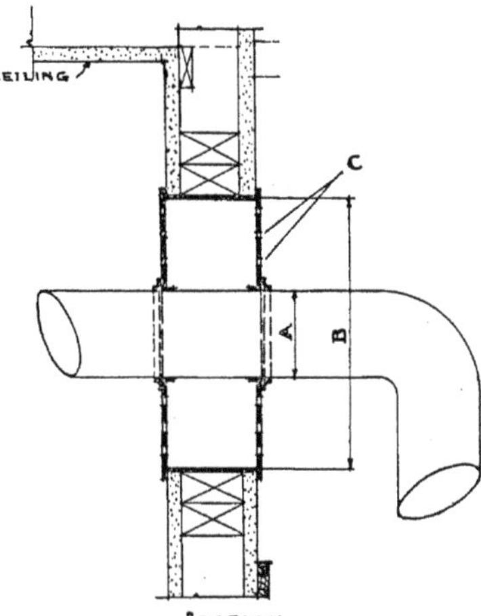

SMOKE PIPES PASSING THROUGH PARTITIONS AND WOOD WORK AROUND.

Section 4603.01.

(A) Diameter of smoke pipe, 6″ or less.
(B) Diameter of thimble required 6″ greater than diameter of smoke pipe.
(C) Ventilation holes required.

tain wall inclosure shall be connected by an independent duct or other suitable means to a mechanical ventilating exhaust system which shall discharge independently above the roof or by a flue connection to a chimney which will take off all smoke, gases and vapors. Hoods, curtain walls, ceiling inside of curtain walls, and all ducts or flue connections shall be of non-combustible material and if sheet metal they shall be of not less weight than No. 14 U. S. Gauge securely riveted or welded. If a mechanical ventilating exhaust system is installed, ducts exhausting from other portions of the room may be connected to the hood or curtain wall inclosure exhaust duct through a section of No. 14 U. S. Gauge sheet metal not less than four (4) feet long, equipped with a damper with fusible link. Every fan used to exhaust from hoods or spaces inclosed by curtain walls shall be provided with a by-pass around the fan, which shall be equipped with a fire damper held in place by a fusible link arranged in such manner that in normal operation all exhaust gases will pass through the fan, but in case of fire in the duct the damper will close against the inlet of the fan and open the by-pass so that products of combustion will pass directly to the atmosphere. Means shall be provided for replacing fusible links and for cleaning the inside of ducts.

4603.03 Vents for Gas Fired Heat Producing Appliances: Flues and vents to which gas fired heat producing appliances are connected shall have cross-sectional areas not less than the aggregate areas of the vent outlets of the appliances connected thereto. Every flue connected gas fired heat producing appliance, except an incinerator, unless its construction serves the same purpose, shall be provided with a draft hood or equivalent device, designed:

1. To insure the free escape of the products of combustion if there is no draft, if there is any back draft, or if any stoppage occurs between the draft head and the outlet to the outside air.

2. To prevent a back draft from entering the appliance, and
3. To neutralize the effect of stack action of the flue upon the operation of the appliance.

4603.04 Covering or Lining: Every breeching, smoke pipe, flue, vent or other duct carrying gases or other products of combustion, the temperature of which may exceed eight hundred (800) degrees Fahrenheit, shall be lined with a refractory material not less than two (2) inches thick or covered with an insulating material equivalent to not less than two (2) inches of eighty-five (85) per cent magnesia; provided however, that any such ducts which are used in industrial processes within any building of non-combustible construction shall be exempt from the requirements of this section.

ARTICLE 4604.

Clearances.

4604.01 Clearances: The exposed surface of any heat producing appliance or any part thereof, shall be not less than the following distances from any wall or ceiling, depending on the temperature of the exposed surface:

	Combustible		Non-Combustible	
	Walls	Ceiling	Walls	Ceiling
Heat producing appliances except domestic stoves, ranges, water heaters and space heaters located in the space to be heated, the maximum temperature of exposed surface of which does not exceed 300° Fahrenheit..........	15″	18″	4″	6″
Heat producing appliances except domestic stoves, ranges, water heaters and space heaters located in the space to be heated, the maximum temperature of exposed surface of which is from 300° to 800° Fahrenheit..........	22″	22″	8″	8″
Heat producing appliances the maximum temperature of exposed surface of which exceeds 800° Fahrenheit..........	Not permitted		12″	12″
Domestic stoves, ranges, and water heaters using solid fuel, without shields..........	12″	..	No Requirements	
Domestic stoves, ranges and water heaters using solid fuel with shields of metal or non-combustible insulating material so attached as to form an open space behind and to extend from the bottom of such an appliance to one (1) foot above the top..........	6″	..	No Requirements	
Domestic stoves, rages and water heaters using gas or liquid fuel..........	6″	..	No Requirements	
Smoke pipes and breechings, except domestic stoves and ranges, unlined and uncovered or insulation less than prescribed..........	24″	24″	8″	8″
Smoke pipes and breechings lined with a refractory material not less than two (2) inches thick or the covering of smoke pipes and breechings covered with insulating material equivalent to not less than two (2) inches of 85 per cent magnesia...	12″	12″	No Requirements	
Ducts and flues from heat producing appliances where food is cooked or where from any other cause, grease or other flammable substances are produced..........	12″	12″	No Requirements	

No heat producing appliance or any part thereof shall be placed close enough to a combustible wall, ceiling or other part of any building to cause the temperature of the surface of such combustible wall, ceiling or other part to rise more than ninety (90) degrees Fahrenheit above the temperature of the room; minimum distance between a heat producing appliance and any combustible construction shall be three (3) inches.

ARTICLE 4605.

Warm Air Heating.

4605.01 Furnace: Every warm air heating furnace shall be set on a foundation as hereinbefore prescribed in Section 4602.02. Warm air furnaces shall be inclosed in metal casings or walls of masonry.

(a) **Metal Casings.** Sheet metal casings including casing tops, shall be made of metal not lighter than No. 26 Gauge.

(b) **Masonry Casings.** Brick or masonry casings shall have walls not less than eight (8) inches thick. A metal casing bonnet may be used on a masonry set furnace. Any warm air heating furnace, the casing top of which is one (1) foot six (6) inches or less from a combustible ceiling or joist, shall be protected by a metal shield extending not less than eighteen (18) inches beyond the casing of said furnace. This shield shall be suspended at least two (2) inches below woodwork allowing free air space between the shield and woodwork. No furnace casing or top shall be less than twelve (12) inches from the ceiling.

4605.02 Warm Air Pipes and Stacks: All warm air pipes and stacks shall be made of bright tin not lighter than IC, or galvanized iron; any pipe twelve (12) inches or greater in diameter shall not be lighter than IX tin or No. 26 U. S. Standard Gauge galvanized iron. Where warm air pipes pass through combustible walls there shall be a clearance of not less than one (1) inch between the pipe and the wall. All openings around first floor, wall and floor boxes and stacks to upper floors shall be sealed dust tight and no metal surface of warm air pipes, stacks, heads, boots or fittings containing heated air shall come in contact with any wood or other combustible construction. All wall stacks or wall pipes, heads, boots, elbows, tees, angles and other connections shall be covered with not less than one (1) thickness of twelve (12) pounds per one hundred (100) square feet of asbestos paper, and shall have a clearance of not less than five-sixteenths (5/16) inch on all sides from any combustible material; or shall be double with a uniform air space or not less than five-sixteenths (5/16) inch between the outer and inner metal walls. No warm air pipe or stack shall be located in a combustible floor or partition unless it is at least six (6) feet distant in a horizontal direction from the furnace.

4605.03 Registers: All registers located in woodwork or combustible floors shall be surrounded with a border of non-combustible material not less than two (2) inches wide. Register boxes, if single, shall be covered with asbestos paper, as required for pipes and stacks, and shall be not less than two (2) inches from any other wood surface unless such wood is covered with sheet metal; if double they shall have an air space of not less than five-sixteenths (5/16) inch between inner and outer boxes. In any furnace system having not more than two (2) warm air registers, at least one (1) of the registers shall be without valve or louvres and the pipe thereto shall be without damper.

4605.04 Cold Air Ducts: All cold air ducts shall be non-combustible in all buildings except single dwellings; in single dwellings they shall be non-combustible for a distance of at least six (6) feet from the furnace.

ARTICLE 4606.

Oil Burners.

4606.01 Oil Burners Defined: For the purposes of this ordinance oil burners shall be defined as any devices designed to burn fuel oil, having a flash point of one hundred (100) degrees, Fahrenheit, or higher, and having a connected fuel tank or container with a capacity of more than ten (10) gallons.

4606.02 Storage Tanks and Piping: All fuel oil storage tanks and piping connected to oil burners shall be as required by Chapter 11 for flammable liquids.

4606.03 Controls: Oil burners shall be provided with means for manually stopping the flow of oil to the burners, from a point at a safe distance from the burner. Automatically operated oil burners used in steam or water boilers or fan type forced warm air furnaces shall be equipped with automatic devices which will shut down the burners in the event of undue steam pressure, excessive water temperature or excessive air temperature within the heat producing appliance. Such systems shall also be equipped with automatic devices to shut off the oil supply to the burners in case of failure of ignition, or interruption of the supply or atomization of oil. All electric controls and electric wiring shall conform to the requirements of the electrical code of the City of Chicago.

4606.04 Installation: Where oil burners are hereafter installed in heat producing appliances designed for solid fuel, bottom ventilation shall be provided to prevent the accumulation of vapors in the ash pit, unless such pit is automatically purged by the burner.

ARTICLE 4607.

Gas Burners.

4607.01 Control: Gas burners provided with automatic control to turn on and off the gas automatically, shall be equipped with a device to shut off the gas supply to the main burner or burners in case of failure of ignition or interruption of the gas supply. Automatically operated gas burners used in steam or water boilers or fan type forced warm air furnaces shall be equipped with automatic devices which will shut down the burners in the event of undue steam pressure, excessive water temperature or excessive air temperature within the heat producing appliance.

ARTICLE 4608.

Boilers.

4608.01 Material and Construction: The material and construction of boilers shall be in accordance with the Rules for the Construction of Power Boilers and Other Pressure Vessels and for their Care in Service including Sections 1 to 7, inclusive, of the American Society of Mechanical Engineers' Boiler Construction Code, dated 1937, where same apply to the type of boilers which is used.

ARTICLE 4609.

Pipe and Duct Coverings.

4609.01 Pipes and Ducts: All pipes or ducts carrying steam, hot gases or any other vapor or medium having a temperature of more than two hundred fifty (250) degrees Fahrenheit, shall have a clearance of not less than one (1) inch from all combustible materials. All such pipes shall be surrounded by metal collars where they pass through combustible floors, ceilings or partitions.

4609.02 Coverings: All covering or insulation used on pipes or ducts extending vertically or approximately vertically within a pipe shaft or other open space to a height of fifty (50) feet or more shall be of non-combustible material, except for a surface covering of single ply fabric. The covering or insulation for pipes or ducts run horizontally or approximately horizontally in a story of any height shall not be required to be of non-combustible material, except as otherwise required for furnace piping.

CHAPTER 47-A.

Administration of Ventilating and Plumbing Provisions.

ARTICLE A-4701.

Officers, Powers and Duties.

A-4701.01 Administration by Board of Health: The provisions of this ordinance for the ventilation and industrial sanitation, for the plumbing and drainage, and for the sanitation of all buildings, both public and private, affecting the community in the City of Chicago, shall be administered and enforced by the Board of Health, an executive department of the Municipal Government established by ordinance, except such matters as are hereinafter required to be administered and enforced by the Commissioner of Buildings, or the Commissioner of Public Works. The Board of Health and its officers and employees, or anyone authorized to act for it, shall be permitted and empowered to enter any building or structure or premises, whether completed or in the process of erection, for the purpose of determining whether the same has been or is being constructed in accordance with the provisions of this ordinance and the provisions contained in other general ordinances of the City regulating matters pertaining to sanitary conditions of the City.

A-4701.02 Duties of the President of the Board of Health: The President of the Board of health shall institute such measures and prescribe such rules and regulations for the control and guidance of his subordinate officers and employees as shall secure the careful inspection of all buildings while in process of construction, alteration, repair, or removal, and the strict enforcement of the several provisions of this ordinance. It shall be the duty of the President of the Board of Health to administer and enforce the provisions of this ordinance, and to administer and enforce the provisions of all general ordinances now in force or hereafter to be passed by the City Council, which relate to the erection, construction, alteration, repair and safety of the ventilation system and industrial and community sanitation, or plumbing and drainage of all buildings and premises, except as hereinafter provided in Section A-4701.17.

A-4701.03 Power to Pass On Ordinances: The Board of Health shall have full power to pass upon any question under the provisions of this ordinance or any of the provisions relating to the ventilation and industrial sanitation and sanitation affecting the community, or plumbing and drainage of buildings, as contained in any general ordinances of the City, subject to the conditions, modifications and limitations contained therein.

A-4701.04 Power to Make Rules: The Board of Health and the Commissioner of Public Works may adopt reasonable rules not inconsistent with this Chapter, Chapters 47 and 48, to determine the quality of materials and workmanship in plumbing construction and equipment and in the repair thereof; provided however, that any standards adopted by such rules shall be according to the practice, custom and usage prevailing in the plumbing industry. The Commissioner of Buildings may adopt reasonable rules not inconsistent with this Chapter to determine the quality of materials and workmanship in the construction of ventilating systems for buildings and premises; provided however, that any standards adopted by such rules shall be according to the practice, custom and usage prevailing in the ventilating industry. Copies of said rules shall be published and shall be kept always on file in the offices of the Board of Health, the Commissioner of Public Works and Commissioner of Buildings. The Board of Health, the Commissioner of Public Works and Department of Buildings shall publish in a convenient form and as a unit the ordinance coming within their jurisdiction and the rules pertaining to ventilating, plumbing and drainage which such departments may have adopted and approved. Copies of all such rules so adopted shall be transmitted to the City Council at the first regular meeting held after the adoption of same.

A-4701.05 Inspection Where Complaint Is Made: It shall be the duty of the Board of health to make an examination of any ventilating system or the industrial and community sanitation, or the plumbing and drainage of any building or premises where any citizen represents that either the ventilating system, or the plumbing and drainage, does not comply with the requirements of this ordinance or the requirements contained in other ordinances of the City, or that such building or premises, or part thereof, is in an insanitary or dangerous condition. If such representation is found to be true, the Board of Health shall give notice, in writing, to the owner, occupant, lessee, or person in possession, charge or control of such building or structure to make such changes, alterations or repairs as safety or the the ordinances of the City may require. It shall be unlawful to continue the use of such building until the changes, alterations, or repairs found necessary by the Board of Health to make such building, or part thereof, safe or to bring it into compliance with this ordinance and with the provisions of all other ordinances of the City, shall have been made. Upon the failure of parties so notified to comply with the requirements of said notice, legal action shall be recommended against such parties.

A-4701.06 Records: The Board of Health shall keep a record of all its transactions and operations relating to this ordinance, which shall be at all times open to the inspection of the Mayor, Comptroller, Superintendent of Police, Commissioner of Buildings, Fire Commissioner and members of the City Council. Said Board of Health shall keep in proper books for that purpose, an accurate account of all fees charged, giving the name of the person to whom same is charged, the date on which said charge is made, and the amount of each such fee. The Board of Health shall cause a complete record to be kept showing the location and character of every building or other structure, for which a permit is issued, and shall cause to be filed, every report of inspection made on such building, which report shall bear the signature of the inspectors making such inspections.

A-4701.07 Classification of Buildings—Definitions: Whenever reference is made in this ordinance to any class of building or building occupancy, such reference and classification shall conform to the classification of buildings and building occupancy in the Building Ordinance of the City of Chicago.

(a) **Definitions.** Any definitions of terms not specially defined in this ordinance, shall be as defined in the Building Ordinance when included therein.

A-4701.08 Other Ordinances: The administration of the provisions of this ordinance shall be governed by the special provisions of the Building Ordinance relating to ventilation, industrial and community sanitation, or plumbing and drainage, and the general ordinances of the City of Chicago subject to the conditions, modifications and limitations contained therein.

A-4701.09 Bureau of Public Health Engineering: There is hereby established the Bureau of Public Health Engineering, as an administrative unit of the Board of Health, an executive department of the municipal government established by ordinance. The Bureau of Public Health Engineering shall embrace the Division of Plumbing, and New Buildings, the Division of Heating, Ventilation and Industrial Sanitation, and the Division of Community Sanitation, including such officers, assistants, and employees as may from time to time be provided for in the annual appropriation ordinance. The Chief of the Bureau of Public Health Engineering shall be an engineer trained in the theory and practice of public health engineering, including water safety, housing, ventilation, plumbing, industrial and community sanitation and nuisance control. The chief of the Bureau of Public Health Engineering shall, under the direction of the Board of Health, have general control of all matters and things pertaining to the administration of the Bureau of Public Health Engineering, and shall perform such other duties as may be required of him by the President of the Board of Health.

A-4701.10 Duties of the Division of Heating, Ventilation and Industrial Sanitation: The Division of Heating, Ventilation and Industrial Sanitation shall enforce the provisions of Chapter 47, of this ordinance, pertaining to the ventilation of rooms or spaces to provide air conditions which will protect the health and comforts of the occupants thereof, and shall enforce the general ordinances of the City of Chicago pertaining to ventilation, industrial sanitation and air conditions which are noxious, dangerous or detrimental to health and those pertaining to heating conditions insofar as they affect ventilating and air conditions. This division shall also enforce all rules and regulations of the Board of Health based on the requirements contained in the ordinances of the City of Chicago and the statutes of the State of Illinois pertaining to industrial sanitation and air conditions. This division shall examine and approve drawings and plans and issue permits for the installation of all equipment required for ventilation, air conditioning and systems for the removal of dust, smoke or gas in accordance with the provisions of this ordinance. This division shall inspect and test all new and remodeled installations, of mechanical ventilating or air conditioning systems, and systems for the removal of dust, smoke or gas, and if such are found to fulfill the requirements contained in this ordinance, shall issue a certificate of approval to the owner or the contractor who made the installation. This inspection shall be made as soon as possible after the person or persons, firm or corporation responsible for the installation, adjustment and test of the system has notified the Board of Health in writing that the system has been adjusted and tested and is in good operating condition and that it fulfills all the requirements prescribed herein. This division shall inspect and test all ventilating equipment annually to see that it is in proper condition to comply with the provisions applying thereto, and shall also make such additional inspections and tests as the Board of Health may direct. This division shall investigate and enforce all license regulations of the general ordinances prescribing ventilation and industrial sanitation standards, and heating as it affects ventilation and industrial sanitation.

A-4701.11 Employees of Division of Heating, Ventilation and Industrial Sanitation: The head of the Division of Heating, Ventilation and Industrial Sanitation shall be an engineer, qualified and technically trained in the theory and practice of ventilating, industrial sanitation and air conditioning, and of heating as it affects ventilating and air conditioning. Under the direction of the Chief of the Bureau of Public Health Engineering, he shall have charge of this division and direct its operation, and shall perform such other duties as may be required of him by the Chief of the Bureau of Public Health Engineering. There shall also be employed in the Division of Heating, Ventilation and Industrial Sanitation such engineers, inspectors, plan examiners, and other employees as shall be provided for by the appropriation ordinances.

A-4701.12 Duties of the Division of Plumbing and New Buildings: Except where similar duties are by this ordinance imposed upon other departments of the City, the Division of Plumbing and New Buildings shall enforce the provisions of Chapter 48 of this ordinance, pertaining to the plumbing and drainage of all buildings and premises, both public and private, in the City of Chicago, and shall enforce the general ordinances of the City of Chicago pertaining to the plumbing and drainage of all buildings. This division shall also enforce all rules and regulations of the Board of Health based on the requirements contained in the ordinances of the City of Chicago and the statutes of the State of Illinois pertaining to plumbing and drainage of buildings. This division shall examine and approve drawings and plans and issue permits for plumbing and drainage in accordance with the provisions of this ordinance, and shall perform all inspections and tests and issue certificates of approval as required by ordinances. This division shall investigate and enforce all license regulations of the general ordinances prescribing plumbing and drainage.

A-4701.13 Employees of Division of Plumbing and New Buildings: The head of the Division of Plumbing and New Buildings shall be a licensed plumber. Under the direction of the Chief of the Bureau of Public Health Engineering, he shall have charge of this division and direct its operation, and shall perform such other duties as may be required of him by the Chief of the Bureau of Public Health Engineering. There shall also be employed in the Division of Plumbing and New Buildings such engineers, plumbing inspectors, plan examiners and other employees as shall be provided for by the appropriation ordinances.

A-4701.14 Duties of Division of Community Sanitation: The Division of Community Sanitation shall enforce the provisions of all building ordinances regulating the sanitary conditions in buildings and not otherwise required to be enforced by the Division of Plumbing and New Buildings, or the Division of Heating, Ventilation and Industrial Sanitation. This division shall examine and approve drawings and plans for buildings which involve community sanitation problems, such as garbage reduction plants, incinerators, night soil disposal plants, asphalt plants, lodging houses, morgues, and vaults used for fumigating purposes using poisonous gases. This division shall investigate and enforce all license regulations and general ordinances prescribing sanitary provisions for buildings and premises not otherwise assigned to the Division of Plumbing and New Buildings or the Division of Heating, Ventilation and Industrial Sanitation. This division shall enforce the ordinances preventing nuisances, dealing with rat proofing of buildings, prevention of waters flooding habitable premises, the proper construction of roofs to prevent leaks, vermin control and sound-proofing of buildings. This division shall enforce all ordinances pertaining to the condemnation or demolition of buildings because of insanitary conditions, pertaining to housing, and pertaining to the abatement of nuisances and regulation of sanitation, and which are not placed under the jurisdiction of the Division of Plumbing and New Buildings, or the Division of Heating, Ventilation and Industrial Sanitation.

A-4701.15 Employees of Division of Community Sanitation: The head of the Division of Community Sanitation shall be a trained sanitarian. Under the direction of the Chief of the Bureau of Public Health Engineering he shall have charge of this division and direct its operation and shall perform such duties as may be required by the Chief of the Bureau of Public Health Engineering. There shall be employed in the division of community sanitation such engineers, sanitary inspectors and other employees as shall be provided for by the appropriation ordinances.

A-4701.16 Approval of Plumbing: The Board of Health or the Commissioner of Public Works shall not approve any plumbing installed in violation of this ordinance. Except as otherwise provided in Section 4801.02—Minor Repairs, the Board of Health or the Commissioner of Public Works shall not approve any plumbing work installed unless the master plumber installing such work has in effect a certificate as master plumber, or unless such journeyman plumber employed in such installation has in effect a certificate as a journeyman plumber, or unless each plumber's apprentice employed in such installation is registered as a plumber's apprentice, all as required by "An Ordinance Providing for a Board of Plumbing Examiners to Conduct Examinations for Journeymen Plumbers, Master Plumbers, to Register Plumbers' Apprentices and to Issue and Revoke Plumbers' Licenses."

A-4701.17 Approval by the Department of Public Works: The installation of all house sewers and tile house drains, and the connections of all house sewers and private sewers to the public sewers of the City or to the Chicago River or to any of its branches or to any slip, channel, canal, ditch or other waterway, the connections of all house sewers to private sewers, the extensions of and connections to private sewers and house sewers, and the extensions of authorized connections to any sewer controlled by the Sanitary District of Chicago and all water supply systems shall conform to the provisions of the general ordinances, and the rules and regulations of the Commissioner of Public Works, and when so installed shall be approved by authorized officers of the Department of Public Works.

CHAPTER 47.
Ventilation.

ARTICLE 4701.
Definitions.

4701.01 Ventilation Defined: Ventilation, for the purposes of this ordinance, is hereby defined as the providing and maintaining in rooms or spaces, by natural or mechanical means, air conditions which will protect the health and comfort of the occupants thereof.

The ventilating requirements, as herein stated, shall apply to every room hereafter designed, erected, altered or converted for the purposes enumerated.

Openings required by hazardous room purposes; to remove heat or gases which may accumulate; for fire prevention, and for any other cause except health and comfort of the human occupants of the rooms shall be provided as required by other Chapters of this Ordinance.

4701.02 Methods of Producing Ventilation: Ventilation may be produced by

(a) **A Natural Ventilating System,** which for the purposes of this ordinance, is hereby defined as a ventilating system, the effectiveness of which depends upon natural atmospheric conditions and upon the operation of windows, doors, transoms and other openings, the operation of which is in control of the person or persons in the room or space which is ventilated.

(b) **A Mechanical Ventilating Supply System,** which for the purposes of this ordinance, is hereby defined as a system for forcing air into a room or space by artificial means combined with the removal of air through windows, skylights, transoms, doors, grilles, shafts, ducts or other openings.

(c) **A Mechanical Ventilating Exhaust System,** which for the purposes of this ordinance is hereby defined as a system for removing air from a room or space by artificial means combined with a supply of air through windows, skylights, transoms, doors, grilles, shafts, ducts or other openings.

4701.03 Ventilating Openings: Ventilating openings in any room or space for the purposes of this ordinance, are hereby defined as apertures opening directly upon a street, alley, yard, court, public park, public waterway, or onto a roof of a building or structure in which the room or space is situated. They shall be windows, skylights or transoms, or auxiliary openings which are provided for ventilating purposes and which are equipped with adjustable louvres, dampers or other devices to deflect or diffuse the air currents.

4701.04 Area of Ventilating Openings: The area of ventilating openings shall be computed as follows (see Illustration 2702.04):

Item 1. Windows. The maximum area that can be opened.

Item 2. Skylights. The area of the maximum opening to the outer air, provided that it does not exceed the area of the sashed openings to the outer air, or the area of the skylight well. If this area exceeds either the area of the sashed openings or the skylight well, the smaller area is the ventilating area.

Item 3. Transoms. The free area through the sashed opening if the transom swings through an arc of not less than sixty (60) degrees. It is the same percentage of the free area as the maximum angle of the transom when open is to sixty (60) degrees if the transom swings through an arc of less than sixty (60) degrees.

Item 4. Auxiliary Openings. The free area when louvres, dampers, or other devices are in position to deflect or diffuse the air currents in such a manner that there will be no objectionable drafts.

ARTICLE 4702.

Ventilation Requirements.

4702.01 Basis of Requirements: The ventilating requirements shall be based on the purposes for which the rooms are used, regardless of the class of building or structure in which it is located.

The method of producing ventilation and the quantities of air to be supplied and exhausted by mechanical ventilating systems stated in the following table, are the minimums permitted. Where it is stated that natural ventilation is required, this may be supplemented but not replaced by a mechanical ventilating supply system or a mechanical ventilating exhaust system, or both.

4702.02 TABLE OF VENTILATING REQUIREMENTS

Room	Conditions			Requirements
	Areas of Ventilating Openings in Percentage of Floor Area		Other Conditions	Cubic Feet of Air per Minute supplied or exhausted per square foot of floor area of rooms. Except as otherwise noted.
Purpose	Less Than	Not Less Than		S indicates Mechanical Supply E indicates Mechanical Exhaust
Anaesthetizing Rooms				S 1.2 and E 1.2
Aquariums				No Requirements
Armories (Drilling		5		No Requirements
Spaces)	5			S .8
Art Rooms				Same as for "Reading Rooms"
Auditoriums, Except			Capacity 601 or more persons	S 3.0 and E 1.5 in open spaces having no fixed seats plus S 30. and E 15. for each fixed seat.
Those Used for Worship Only		5	Capacity 600 or less persons	S 3.0 in open spaces having no fixed seats plus S 30. for each fixed seat. E through shafts or other openings for natural exhaust having a total ventilating opening of not less than .5 of 1% of the floor area as uniformly distributed as practicable and open when the room is in use without causing objectionable drafts.
	5			S 3.0 and E 1.5 in open spaces having no fixed seats plus S 30. and E 15. for each fixed seat.
Auditoriums		3		No Requirements
Used for Worship Only	3			S 1.5
Auditoriums in		3		No Requirements
Connection with Private Schools	3			S 1.5
Autopsy		5		No Requirements
Rooms	5			S 1.0 or E 1.0
Bakeries				Same as for "Food Baking Rooms"
Ball Rooms				Same as for "Dance Halls"
Band Rooms				Same as for "Reading Rooms"
Banking Rooms (Public and Teller			Stories below that nearest to grade	S 1.2 and E.1.2
Spaces)		5	Other	No Requirements
	5		Stories	S 1.0 or E 1.0

Room Purpose	Conditions: Areas of Ventilating Openings in Percentage of Floor Area — Less Than	Not Less Than	Other Conditions	Requirements: Cubic Feet of Air per Minute supplied or exhausted per square foot of floor area of rooms. Except as otherwise noted. S indicates Mechanical Supply E indicates Mechanical Exhaust
Bank Vaults Attended				S 1.6 or E 1.6
Banquet Halls				S 2.0 and E 2.0
Bath		5		No Requirements
Rooms	5			E 1.0
Beauty		5		No Requirements
Parlors	5			S 1.2 and E 1.2
Billiard Rooms			Having more than four tables	S 1.6 and E 2.0
		5	Having four or less tables	No Requirements
	5	3		S 1.6 or E 1.6
	3			E 1.6 or E 2.0
Board of		5		No Requirements
Trade	5	3		S 2.0 or E 2.0
Trading Rooms	3			S 2.0 and E 1.0
Boiler Rooms				No Requirements
		5		No Requirements
Bowling	5	3	Disregard floor area from foul line to pit	S 3.0 in open spaces having no fixed seats plus S 30. for each fixed seat or E 3.0 in open spaces having no fixed seats plus E 30. for each fixed seat.
Alleys	3			S 3.0 and E 3.0 in open spaces having no fixed seats plus S 30. and E 30. for each fixed seat.
Brokers Board Rooms				Same as for "Board of Trade (Trading Rooms)"
Cab Stands				See Note No. 1 at end of Table
Cabarets				Same as for "Dance Halls"
Cafeterias				Same as for "Dining Rooms in which there is Cooking Equipment, Public"
Chapels				Same as for "Auditoriums used for Worship Only"
Chart Rooms				Same as for "Offices"
Circuses (Arena)				No Requirements
Circuses (Spectators' Spaces)				Same as for "Auditoriums, except those used for Worship Only"
Class		8		No Requirements
Rooms in	8	5		S 1.5 or E 1.5
Private Schools	5			S 1.5 and E .75
Class Rooms in Public Schools				S 1.5 and E .75
Closets				No Requirements
Coat		5		No Requirements
Rooms	5			E .8
Cold Storage Rooms				No Requirements
Community Halls			1500 square feet Floor area more than	S 2.0 and E 1.0
		5	Floor area not more than 1500 square feet	S 1.5
	5			S 1.5 and E .75
Concert Halls				Same as for "Auditoriums Except those used for Worship Only"
Conservatories				No Requirements
Convention Halls				Same as for "Auditoriums Except those used for Worship only"
Cooking Rooms		5		E 1.2
for instruction purposes only	5			S 1.0 and E 1.2
Cooling and Refrigerating Machinery Rooms				No Requirements
Corridors				No Requirements

4702.02 TABLE OF VENTILATING REQUIREMENTS—Continued

Room Purpose	Conditions: Areas of Ventilating Openings in Percentage of Floor Area — Less Than	Not Less Than	Other Conditions	Requirements: Cubic Feet of Air per Minute supplied or exhausted per square foot of floor area of rooms. Except as otherwise noted. S indicates Mechanical Supply E indicates Mechanical Exhaust
Council Chambers				Same as for "Auditoriums Except those used for Worship only"
Court Rooms				Same as for "Auditoriums Except those used for Worship only"
Dance Halls		4		S 3.0 and E 1.5
	4	3		S 3.0 and E 2.0
	3	2		S 3.0 and E 2.5
	2			S 3.0 and E 3.0
Delivery Rooms				Same as for "Operating Rooms"
Dining Rooms in which there is no Cooking Equipment, Public		5	Floor area not more than 400 square feet	No Requirements
			Floor area more than 400 square feet	S 1.6 or E 1.6
	5	3		S 1.6 or E 1.6
	3			S 1.6 and E 1.6
Dining Rooms in which there is Cooking Equipment, Public		3		E 2.0
	3			S 1.5 and E 2.0
Doctors' and Dentists' Examination Rooms				Same as for "Offices"
Dormitory Rooms				Natural Ventilation
Drawing Rooms				Same as for "Reading Rooms"
Dressing Rooms		5		No Requirements
	5	3		S 1.2 or E 1.2
	3			S 1.2 and E 1.2
Dry Cleaning Rooms				E 4.0
Electric Sub-Stations Attended				No Requirements
Electric Sub-Stations Unattended				No Requirements
Electric Transformer Vaults				No Requirements
Enameling Rooms				Same as for "Work Shops"
Engine Rooms				No Requirements
Excelsior Storage Rooms				No Requirements
Exercise Rooms				Same as for "Gymnasiums"
Exhibition Rooms Except Picture Galleries (for Permanent Exhibits)				No Requirements
Exhibition Rooms Except Picture Galleries (and Except those for Permanent Exhibits)		5		S 2.0
	5			S 2.0 and E 2.0
Explosive Gas Storage Rooms				No Requirements
Exposition Rooms				Same as for Exhibition Rooms
Fan Rooms				No Requirements
File Rooms				Same as for "Offices"

Room Purpose	Conditions: Areas of Ventilating Openings in Percentage of Floor Area — Less Than	Not Less Than	Other Conditions	Requirements: Cubic Feet of Air per Minute supplied or exhausted per square foot of floor area of rooms. Except as otherwise noted. S indicates Mechanical Supply E indicates Mechanical Exhaust
Food Baking Rooms			Stories below that nearest to grade	S 1.2 and E 1.2
		5	Other Stories	No Requirements
	5			E .6 See Note No. 2 at end of Table
Foyers in Theatres having a Single Performance				S 1.5 or E 1.5
Foyers in Theatres having a Continuous Performance				S 3.0 or E 3.0
Foyers in Club Houses and Hotels		5		No Requirements
	5			S .8
Foyers, except in Theatres, Club Houses and Hotels				No Requirements
Fraternal Halls				Same as for "Community Halls"
Freight Handling Rooms				No Requirements
Game Rooms				Same as for "Billiard Rooms"
Garage Spaces for Automobiles operated under own power, Single Floor or Elevator Type, Capacity 5 or more cars.			Entrance Story	E 3.0 in Main Entrance drive plus E .5 in car storage space
			Any Story Except Entrance Story	E .5 in car storage space
Garage Spaces for Automobiles operated under own power, Ramp Type, Capacity 5 or more cars			Any Story below Entrance Story	E 2.0 in ramps and drives between ramps in first story below entrance story, which may be reduced .2 for each story below (Minimum E .5) plus E .5 in car storage space.
			Entrance Story	E 3.0 in main entrance drive to ramp plus E .5 in car storage space.
			Any Story Above Entrance Story	E 2.0 in ramps and drives between ramps in second story, which may be reduced .2 for each story above (Minimum E .5) plus E .5 in car storage space.
Garage Communication Vestibule, Pedestrian				S .6
Passages (Enclosed) for vehicles using Internal Combustion Engines				E 3.0 See Note No. 3 at end of Table
Green Rooms				Same as for "Dressing Rooms"
Grills				Same as for "Dining Rooms in which there is Cooking Equipment, Public".
Gymnasiums		5		No Requirements
	5			S .8 to which the following shall be added if seats are provided for spectators. S 2.0 in open spaces with no fixed seats and S 20. for each fixed seat.
Halls (Corridors)				No Requirements
Hangars				No Requirements
Inflammable Gas Storage Rooms				No Requirements

Room Purpose	Conditions: Areas of Ventilating Openings in Percentage of Floor Area — Less Than	Not Less Than	Other Conditions	Requirements: Cubic Feet of Air per Minute supplied or exhausted per square foot of floor area of rooms. Except as otherwise noted. S indicates Mechanical Supply E indicates Mechanical Exhaust
Instruction Rooms Not Otherwise Specifically Noted				Same as for "Class Rooms"
Janitors' Closets				No Requirements
Japanning Rooms				Same as for "Work Shops"
Kitchens,		3		E 4.0
Public	3			S 1.2 and E 4.0
Kitchens, Non-Public				Same as for "Living Quarters"
Laboratories		5		No Requirements
Chemical	5			E .6
				See Note 2 at end of Table
Laboratories, Except Chemical				Same as for "Offices" See Note 2 at end of Table
Lacquering Rooms				Same as for "Work Shops"
Laundries containing equipment		2½		No Requirement
which can be used by not more than one family at one time	2½		Minimum area of ventilating openings 1½% of floor area	E 1.0
Laundries containing equpiment		5		No Requirements
which can be used by more than one family at one time but not serving the general public	5		Minimum area of ventilating openings 1½% of floor area	E 1.0
Laundries, Serving the		10		No Requirements
General Public	10			E 1.5
Lecture Rooms				Same as for "Community Halls"
Libraries				Same as for "Reading Rooms"
Linen Rooms				Same as for "Storage Rooms, Active Storage"
Living			Sitting Rooms, Living Rooms, Parlors and other rooms of similar use; Dining Rooms, Bed Rooms and Kitchen (except Closet Kitchens) having a floor area of not more than sixty square feet	Natural Ventilation
Quarters			Closet Kitchens having a floor area of not more than sixty (60) square feet	Natural Ventilation or, in lieu of this, Natural Exhaust thirty-two (32) sq. in. for each kitchen served connected to a vertical duct leading directly to atmosphere, or E 1.0 and either S .9 or Natural Supply eighty (80) sq. in. per one hundred (100) sq. ft. of floor area.
Lobbies in Theatres having a Single Performance				S 1.2 and E 1.5
Lobbies in Theatres having Continuous Performance				S 3.0
Lobbies in Club Houses		5		No Requirements
and Hotels	5			S .6

Room Purpose	Conditions: Areas of Ventilating Openings in Percentage of Floor Area — Less Than	Not Less Than	Other Conditions	Requirements: Cubic Feet of Air per Minute supplied or exhausted per square foot of floor area of rooms. Except as otherwise noted. S indicates Mechanical Supply E indicates Mechanical Exhaust
Lobbies, except in Theatres, Club Houses and Hotels				No Requirements
Locker Rooms		5		No Requirements
	5			E 1.2
Lodge Halls				Same as for "Community Halls"
Lounges				Same as for "Lobbies in Club Houses and Hotels".
Lumber Storage Rooms				No Requirements
Lunch Counters				Same as for "Dining Rooms in which there is Cooking Equipment, Public".
Lunch Rooms				Same as for "Dining Rooms in which there is Cooking Equipment, Public".
Machinery Rooms for Building Equipment				No Requirements
Medicine Rooms				Same as for "Laboratories, Chemical"
Morgues				Natural Ventilation
Motion Picture Studios				S 1.5 and E 1.5 See Note 2 at end of Table
Museums				No Requirements
Music Rooms				Same as for "Reading Rooms"
Musicians' Rooms				Same as for "Dressing Rooms"
Natatoriums			Disregard area of swimming pool	Same as for "Gymnasiums"
Non-Habitable Rooms, except as Noted Otherwise				No Requirements
Offices, except		5		No Requirements
Intermediate	5	2½		S .6
Rooms	2½			S .6 and E .3
Offices, Intermediate			If ventilating openings having 5% of floor area of intermediate room are located in the partitions separating said room from another room having ventilating openings not less than 5 per cent of the floor area of both rooms and if ventilating openings having 2½ per cent of floor area of intermediate room are located in wall or corridor	No Requirements
Rooms			If ventilating openings of less area than noted above are located in either wall above mentioned	S 1.0 or E 1.0
			If no ventilating openings in either wall above mentioned	S 1.2 and E .9
Operating Rooms				S 1.2 and E 1.2
Packing		5		No Requirements
Rooms	5			S .6 or E .6
				See Note 2 at end of Table
Paint Mixing Rooms				Same as for "Work Shops"
Paint Shops				Same as for "Work Shops"
Paint Spraying Rooms				Same as for "Work Shops"

Room Purpose	Conditions: Areas of Ventilating Openings in Percentage of Floor Area — Less Than	Not Less Than	Other Conditions	Requirements: Cubic Feet of Air per Minute supplied or exhausted per square foot of floor area of rooms. Except as otherwise noted. S indicates Mechanical Supply E indicates Mechanical Exhaust
Pantries, Butlers'				No Requirements
Pantries, Serving		5		No Requirements
	5	2		E 2.0
	2			S 1.2 and E 2.0
Pantries, other than Butlers' and Serving				No Requirements
Paraffin, Pitch, Resin, Tar and Similar Heating Rooms				No Requirements See Note 2 at end of Table
Patients' Private Rooms				Natural Ventilation
Picture Galleries				S .6 or E .6
Picture Projection Rooms			One Machine	Natural exhaust one hundred sixty sq. in. area of duct and an independent mechanical exhaust from machine housing one hundred fifty.
			Two or More Machines	E 1.0 and independent mechanical exhaust from machine housing total based on one hundred fifty (150) for each housing. See Note 4 at end of Table
Pool Rooms				Same as for "Billiard Rooms"
Pressing Rooms				Same as for "Work Shops"
Printing Shops				Same as for "Work Shops"
Police Station		5		No Requirements
Cell Rooms	5			S 1.2
Property Rooms				Same as for "Storage Rooms
Promenades, Inclosed				Same as for "Foyers in Club Houses and Hotels"
Reading		5		No Requirements
Rooms	5			S 1.2
Receiving Rooms				Same as for "Packing Rooms"
Reception Rooms in connection with Offices				Same as for "Offices, Intermediate Rooms"
Reception Rooms except in connection with Offices, Public				Same as for "Lobbies in Club Houses and Hotels"
Recreation Rooms				Same as for "Gymnasiums"
Repair Shops, Automobile				Same as for "Work Shops"
Rest Rooms				Same as for "Offices"
Retiring Rooms				Same as for "Offices"
Safe Depositories (Coupon Spaces)				Same as for "Banking Rooms, (Public and Teller Spaces)"
Sales Rooms,			Stories below that nearest to grade	S 1.5 and E 1.5
Department			Story nearest to grade	S 1.0
		5	Stories above that	No Requirements
Stores	5		nearest to grade	S .6
Sales Rooms			Stories below that nearest to grade	S 1.2 and E 1.2
(except Depart-		5	Stories nearest to	No Requirements
ment Stores)	5		grade and above	S .5
Serving Rooms				Same as for "Pantries, Serving"
Shipping Rooms				Same as for "Packing Rooms"
		5		No Requirements
Skating Rinks,	5	4		S 1.8 and E .9
Ice	4	3		S 1.8 and E 1.2
	3	2		S 1.8 and E 1.5
	2			S 1.8 and E 1.8

Room Purpose	Conditions: Areas of Ventilating Openings in Percentage of Floor Area — Less Than	Not Less Than	Other Conditions	Requirements: Cubic Feet of Air per Minute supplied or exhausted per square foot of floor area of rooms. Except as otherwise noted. S indicates Mechanical Supply E indicates Mechanical Exhaust
Skating Rinks, Roller				Same as for "Dance Halls"
Sleeping Rooms (except sleeping stall rooms and rooms otherwise specifically noted)				Natural Ventilation
Sleeping Stall		5		No Requirements
Rooms	5			S .5 and E .5
Slop Sink Rooms				No Requirements
Smoking		5		No Requirements
Rooms	5			S 1.0 and E 1.5
Stables				No Requirements
Stage Hands Rooms				Same as for "Dressing Rooms".
Storage Battery Rooms				E 1.0
Storage Rooms		2		No Requirements
Active Storage	2			E .4
Storage Rooms, Inactive Storage				No Requirements
Study Rooms				Same as for "Class Rooms"
Sunday School Rooms				Same as for "Auditoriums used for Worship Only".
Teaching Amphitheatres, (Instructors' Spaces)				S 1.2 and E 1.2
Teaching Amphitheatres, (Spectator's Spaces)				S 3.0 and E 3.0 in open spaces with no fixed seats and S 30. and E 30. for each fixed seat.
Telephone Apparatus		5		No Requirements
Rooms, Attended	5			S .4
Telephone Apparatus Rooms Unattended				No Requirements
Telephone Exchange Board Rooms		5		No Requirements
Manually Operated	5			S 1.2 and E .9
Toilet		5		No Requirements
Rooms	5			E 2.0
Transformer Rooms				Same as for "Electric Transformer Rooms".
Treatment Rooms				Same as for "Offices"
Urinal Rooms				Same as for "Toilet Rooms"
Ushers' Rooms				Same as for "Dressing Rooms"
Waiting Rooms in Railway Stations and similar occupancies				Same as for "Lobbies in Club Houses and Hotels".
Waiting Rooms,		5		No Requirements
except in Railway Stations and similar occupancies	5			S 1.0 and E 1.0
Wards				Same as for "Dormitory Rooms"
Waste Paper Baling Rooms				Same as for "Packing Rooms"
Work		5		No Requirements
Shops	5			S 1.0 or E 1.0
				See Note 2 and Note 5 at end of Table.
X-Ray Operators Rooms				Same as for "Offices"

Where, in the foregoing table, requirements for room purposes are stated as "No Requirements," rooms or spaces used for the designated purposes shall be subject to all general provisions applicable thereto.

The minimum quantity of air supplied to or exhausted from a room shall be fifty (50) cubic feet per minute.

Note 1. Cab Stands for Cabs Using Internal Combustion Engines.

If driveways and other apertures open directly to atmosphere said apertures having an area not less than twenty (20) per cent of the floor area of the cab stand—No Requirements.

If apertures have an area of less than twenty (20) per cent of the floor area of the cab stand—same as for "Passages (Enclosed), for vehicles using internal combustion engines."

Note 2. In rooms or apartments of any factory, mercantile establishment, mill or workshop, the requirements given in the foregoing table shall be increased, if necessary, to conform to the following:

Rooms in which the total area of outside windows and doors is not less than twelve and one-half (12½) per cent of the floor area and in which the air space per person employed therein is not less than two thousand (2,000) cubic feet—No Requirements.

Rooms in which the total area of outside windows and doors is not less than twelve and one-half (12½) per cent of the floor area and in which the air space per person employed therein is less than two thousand (2,000) cubic feet and more than five hundred (500) cubic feet—S .25 per person employed.

Rooms in which there are no outside windows or doors:

Rooms in which there are no outside windows or doors; rooms having less than two thousand (2,000) cubic feet of air space per person employed therein and in which the area of outside windows and doors is less than twelve and one-half (12½) per cent of the floor area, and rooms having less than five hundred (500) cubic feet of air space per person employed therein—S .30, per person employed.

Note 3. Passages (Inclosed) for Vehicles Using Internal Combustion Engines.

If openings, each having an area not less than twenty-five (25) per cent of the cross-sectional area of the passage, are provided to atmosphere in both end walls of the passage —Natural Ventilation.

If openings having a combined area not less than fifty (50) per cent of the area of a side wall of the passage, uniformly distributed are provided to atmosphere—Natural Ventilation.

Note 4. Picture Projection Rooms.

The ventilating system exhausting from a picture projection room shall be independent of any other system, except that if the projection room is one of the rooms in a projection block, the system may serve the entire block.

The system which exhausts from the picture machine housings shall be equipped with an independent fan which discharges directly out-of-doors or to the fan discharge of any other ventilating exhaust system, if such system is so arranged that the air cannot be recirculated.

The natural exhaust outlet from a picture projection room containing one machine shall be in the ceiling of the room; the natural exhaust duct shall be carried to the atmosphere in an upward direction only and as nearly vertical as possible.

Dampers in the ducts, if any, shall be arranged to open automatically in case of fire within the room.

Note 5. Work Shops.

When the work in the room is of such a character that dangerous or noxious dust or fumes are given off, the requirements, as stated, shall be supplemented by local exhaust sufficient to remove such dust or fumes.

The ventilating requirements for rooms used for purposes similar to those enumerated in the foregoing table but not specifically named therein, shall be the same as those for room purposes of similar character.

4702.03 Additional Ventilation Required if Air Conditions Become Objectionable Due to Causes Other Than Occupancy by Human Beings: If the air conditions in any habitable room become objectionable due to causes other than occupancy by human beings and the Board of Health finds that the health and comfort of the human occupants is endangered thereby, additional ventilation by natural or mechanical means, approved by the Board of Health, shall be provided.

4702.04 Source of Air Supply: The air supply for every ventilating system, either natural or mechanical, shall be taken directly from out-of-doors, except that

(a) When air is supplied by a mechanical ventilating supply system, a portion of the required air supply may be recirculated provided the system is equipped with such devices for the control of temperature, humidity and dust content in the spaces to be ventilated that the physical properties of the air so supplied are substantially the same as though all of the supply were taken from out-of-doors. The quantity so recirculated may be considered as exhaust from the rooms from which it is withdrawn. The quantity of air withdrawn from any room and recirculated shall not exceed sixty-six and two-thirds (66⅔) per cent of the air supplied to the room by the ventilating system. All equipment shall be so arranged that at least fifty (50) per cent of the air supplied by the system can be taken from out-of-doors. The fresh air intake shall be so arranged that at least one hundred (100) per cent of the air supplied by the system can be taken from out-of-doors.

(b) When air is supplied by a mechanical ventilating supply system which is not equipped with devices prescribed in paragraph (a), a portion of the air supplied may be recirculated during the time that the rooms are not occupied. The quantity of air so recirculated shall not exceed sixty-six and two-thirds (66⅔) per cent of the air normally supplied by the ventilating system. The intake and all equipment and ducts shall be so arranged that all of the air supplied by the system can be taken from out-of-doors, and that the air permitted to be recirculated as herein described, can be discharged to the atmosphere when the rooms are occupied.

(c) When air is supplied by a mechanical ventilating supply system, either with or without recirculation, a secondary intake may be provided, taking air indirectly from out-of-doors through an underground tunnel or other construction which has a sufficient number of openings to the outer air to provide an adequate supply of fresh, uncontaminated air. The main intake shall be so arranged that it will supply all the fresh air required for the systems; the quantity of air that can be taken through the secondary intake shall not exceed that which can be taken through the main intake. Provision shall be made so that the secondary intake can be closed tight and the full quantity of air required can be taken through the main intake.

(d) The intake drawing air from out-of-doors shall be at such a point that the air supply will be uncontaminated and that the openings will be unobstructed at all times. The intake opening shall be not less than fifteen (15) feet from the discharge outlet of any exhaust fan, and, unless adequate means is provided for the removal of dust from the air, the bottom of the opening shall be not less than ten (10) feet above the surface of any abutting sidewalk, gangway, street, alley, driveway or grade or of any abutting roof. No intake opening shall be placed in a horizontal position in any sidewalk, or in the pavement of any street, alley or driveway or level with any other surrounding grade nor so as to take air from the lower level of any two level street or similar construction.

(e) No air exhausted from bath, toilet, urinal or similar room, lavatory, locker or coat room, kitchen, boiler room or other room in which such air might be contaminated by smoke, gases, or dust which might be noxious, dangerous or detrimental to health shall be recirculated at any time.

4702.05 Air Inlets and Outlets: The air inlets and outlets in every system of ventilation, either natural or mechanical, shall be so located and constructed as to insure, when said systems are properly operated, circulation of air throughout each room; reasonable precaution shall be taken in the design and installation of the system to prevent the air from striking the occupants of the room under such conditions of direction, temperature or velocity as to cause discomfort.

If a mechanical ventilating supply system only is installed for a room, or if a greater quantity of air is supplied by a mechanical ventilating supply system than is removed by a mechanical ventilating exhaust system for a room, adequate means shall be provided for the natural exit of the excess air supplied. If a mechanical ventilating exhaust system only is installed for a room or if a greater quantity of air is removed by a mechanical ventilating exhaust system than is supplied by a mechanical ventilating supply system for a room, adequate means shall be provided for the natural supply of the deficiency in the air supplied.

Openings for natural ventilation shall be adjustable and located in both upper and lower portions of the room. A window having adjustable openings in the upper and lower portions shall be considered as complying with this requirement.

4702.06 Point of Exhaust Discharge: The air removed by every mechanical ventilating exhaust system shall be discharged out-of-doors at a point where it will not cause a nuisance, and from which it cannot again be readily drawn in by a ventilating system, excepting that:

(a) Air which is to be used for recirculation may be discharged to a supply system.

(b) Air which will not cause a nuisance may be discharged into a boiler room in such quantity as is required to supply the needs of combustion.

4702.07 Toilet Room Systems: Mechanical ventilating exhaust systems for bath, toilet, urinal and similar rooms shall be independent from those for rooms of other character, except that:

(a) Exhaust ducts from janitors' closets containing slop sinks or similar fixtures may be connected to and made a part of toilet room systems.

(b) Exhaust ducts from private bath, toilet and urinal rooms and from isolated public rooms of the same nature may be connected to and made a part of the exhaust system for rooms of other character or exhaust ducts from rooms other than bath, toilet and urinal rooms may be connected to and made part of the exhaust system for toilet rooms, provided:

1. That the exhaust fan for the system is in operation all of the time that the building is occupied;

2. That the branch duct from each bath, toilet or urinal room or group of such rooms shall be run parallel and adjacent to the duct from other rooms for a distance of not less than five (5) feet within which distance there are no exhaust openings in either duct and the connection between the ducts made with an easy curve having its outlet toward the exhaust fan, and

3. That the total quantity of air exhausted from private bath, toilet and urinal rooms and from isolated public rooms of same nature shall not exceed ten (10) per cent of the capacity of the fan.

4702.08 Heating Requirements of Mechanical Ventilating Systems: If an installation is a mechanical ventilating exhaust system without a mechanical ventilating supply system, or if it is a mechanical ventilating supply system, either with or without a mechanical ventilating exhaust system, the equipment shall be installed so that the supply shall be heated to such a temperature as will provide proper conditions in the room.

The heating elements and all equipment and connections required therefor shall be based on maintaining the required conditions when the out-of-doors temperature is ten (10) degrees below zero (0) Fahrenheit.

4702.09 Structural Requirements of Mechanical Ventilating Systems: The materials used in every mechanical ventilating supply system and every mechanical ventilating exhaust system shall be incombustible and of a moisture resisting character. The design and construction of all equipment and the weight and bracing of all duct work shall be such as will operate under all conditions without causing vibration. Ducts shall be substantially air tight.

ARTICLE 4703.

Interpretation of Requirements.

4703.01 Source of Fresh Air Supply for Rooms In Which a Preponderance of Exhaust, or Exhaust Only Is Required: In rooms which are required to be provided with mechanical ventilating exhaust systems, either with or without mechanical ventilating supply systems, the fresh air to replace the air exhausted from each room shall be obtained from ventilating openings in that room, or from a mechanical ventilating supply system installed for that room, except

(a) In Foyers, Living Quarters, Lobbies, Locker Rooms, Offices, Medicine Rooms and Picture Galleries, it may be obtained (1st) from ventilating openings in each of said rooms; (2nd) from ventilating openings in uncontaminated rooms adjacent to the designated rooms through unobstructed openings, having a total area not less than two (2) per cent of the designated rooms, provided the area of the ventilating openings is not less than five (5) per cent of the combined floor areas of the adjacent room and the designated room which it serves; (3rd) from a mechanical ventilating supply system serving rooms adjacent to the designated room, either alone or in combination with the designated room, provided that there are unobstructed openings having a total area not less than one (1) per cent of the floor area of the designated room between such adjacent rooms and the designated room, and that the quantity of air supplied to the adjacent rooms is not less than that required for both the designated room and the adjacent rooms.

If an adjacent room requires a preponderance of mechanical exhaust when the ventilating openings are not adequate for natural ventilation, or if the air in such adjacent rooms might contain dangerous or noxious dust or fumes, such rooms shall not be used as a source of supply to the designated room.

(b) In Public Kitchens and Serving Pantries it may be obtained (1st) from ventilating openings in each of said rooms; (2nd) from dining rooms adjacent to the designated rooms, provided the area of the ventilating openings in the dining room is not less than five (5) per cent of the combined floor area of the dining room and the designated room served, or (3rd) from dining rooms supplied by a mechanical ventilating supply system either alone or in combination with the designated room, provided that there are doors or other openings having a total area of not less than two (2) per cent of the floor area of the designated room between such dining room and the designated room, and that the quantity of air exhausted from the dining room through the designated room, plus the quantity of air supplied directly to the designated room is not less than the quantity of air required to be exhausted from that room.

(c) In Banking Vaults in connection with Banking Rooms and Safety Depositories it may be obtained (1st) from natural ventilating openings in adjacent uncontaminated banking spaces, provided the area of the natural ventilating openings is not less than five (5) per cent of the combined floor area of the banking spaces and the vault, or (2nd) from a mechanical ventilating supply system serving the adjacent banking spaces, provided that the quantity of air supplied to the banking spaces is not less than that required for both the vault and the banking spaces.

(d) In Picture Projection Rooms, it may be obtained from openings to uncontaminated rooms adjacent to the picture projection room, which have ventilating openings or which are provided with a mechanical ventilating supply system of the capacity required for such adjacent room.

(e) In Bath, Toilet and Urinal Rooms, Check Rooms, Machinery Rooms, Storage Rooms, Storage Battery Rooms and other rooms of similar character, it may be obtained (1st) from ventilating openings in the rooms, or (2nd) from openings to uncontaminated rooms adjacent to the designated rooms which have ventilating openings or which are provided with a mechanical ventilating supply system of the capacity required for the adjacent rooms, provided that the quantity of air supplied to the adjacent rooms is not less than that exhausted through the designated rooms.

4703.02 Requirements If Ventilating Openings Are Kept Closed: If the installation incindes an air cooling system, or if, for any other reason it is impracticable to open the ventilating openings, the ventilation requirements shall be the same as though no such openings were provided.

4703.03 Rooms Used for More Than One Room Purpose: If a room is used for two (2) or more room purposes, having different ventilating requirements, and it is possible to determine the space which will be used for each purpose, each such space may be considered as a separate room and ventilated in accordance with the Table of Ventilating Requirements; if it is not possible to determine the space which will be used for each purpose the entire room shall be ventilated as required for the most severe room purpose in the room.

4703.04 Partial Story Height Partitions: A partition which is not more than one-half (½) the story height of a room and which is open at both the floor and ceiling shall not be considered a partition forming an independent room, but the space on both sides of the partition shall be considered as one room. Grilles having openings of not less than seventy-five (75) per cent free area may be placed in the openings both below and above such partitions.

ARTICLE 4704.

Method of Determining Compliance With Mechanical Ventilating Requirements.

4704.01 Adjustment of Air Supply and Exhaust: The air inlets of every mechanical ventilating supply system and the air outlets of every mechanical ventilating exhaust system shall be adjusted to supply or exhaust the quantities of air prescribed with approximately uniform distribution over the entire area of such inlet or outlet.

4704.02 Instruments to Be Used and Points at Which Readings Are to Be Taken: Compliance with mechanical ventilating requirements shall be determined by readings indicating the velocity of the air in feet per minute; they shall be taken with a calibrated anemometer unless the conditions surrounding the installation are not within the range of accuracy of this instrument, in which case a Pitot tube, or other device approved by the Board of Health, shall be used. Readings shall be taken at the openings, except (1st) where these are inaccessible or if accessible only at great hazard, and (2nd) where the design or construction of the openings is such as to make it impossible to determine the free area.

4704.03 Method of Making Readings: Readings shall be taken by holding the anemometer close to the face of the opening and moving it at a uniform rate of speed over the entire surface of the opening. If the area of the opening is more than one hundred forty-four (144) square inches and the velocity of air is not approximately uniform over the entire area, the opening shall be subdivided into equal areas, the longest side of which shall not exceed twelve (12) inches and each subdivision shall be considered as an independent opening, and the average velocity of all of the subdivisions shall be considered as the velocity through the opening.

If the openings are inaccessible or if accessible only at a great hazard or if the design or construction of the openings is such as to make it impossible to determine the free area, readings shall be taken over areas as hereinbefore described through handholes placed in the branch duct behind each opening or as close as practical to the opening.

4704.04 Correction of Readings: The velocity readings, as indicated by the anemometer shall be corrected to conform to the calibration curve for the instrument used.

4704.05 Method of Determining Area of Openings: The area of openings shall be taken as (1st) the full opening in the duct if there is no grille, register face or screen over same, or (2nd) the average betwen the gross area and the total free area of the grille, register face or screen if such is installed in the opening.

4704.06 Determination of Quantities of Air Supplied and Exhausted: The quantities of air supplied and exhausted through openings shall be determined by multiplying the area of each opening by the corrected reading of the anemometer or other instrument used.

The sum of the quantities of air delivered at all the openings of a system shall be considered the total capacity of the fan at the speed at which readings are taken.

ARTICLE 4705.

Operation of Ventilating Systems.

4705.01 Times When Ventilating Systems Shall be Operated: Every ventilating system, either natural or mechanical, required by the ordinances of the City of Chicago, shall be kept in good repair and in operation so as to insure the required ventilation of all rooms and spaces to be ventilated thereby during all hours of human occupancy; no ventilating system shall be operated so as to cause objectionable drafts upon occupants.

The ventilating system in every garage and every passage for vehicles, using internal combustion engines, shall be operated at such times and in such a manner that the amount of carbon monoxide in the air in the garage or passage shall not exceed a maximum concentration of two (2) parts in ten thousand (10,000) parts of air.

CHAPTER 48.

Plumbing Provisions.

ARTICLE 4801.

General.

4801.01 Plumbing and Drainage to Comply With Ordinance: The plumbing and drainage of all buildings and premises, both public and private, in the City of Chicago, shall be constructed in accordance with the provisions of this ordinance. All repairs and alterations in the plumbing and drainage of existing buildings shall also comply with this ordinance.

4801.02 Definitions: For the purposes of this ordinance, terms and words shall be construed as defined in this section.

Approved. The word "approved" as applied to a material, device or mode of construction in this ordinance, shall mean that the material or method has received approval and authorization by the respective municipal department or board having jurisdiction.

Back Pressure. A force exerted causing or tending to cause water or air to flow in a pipe opposite to the normal direction of flow.

Bilge Pump. See "Ejector".

Board of Health. A board of five (5) members appointed by the Mayor with the approval of the City Council by ordinances of the City of Chicago, with powers as established by ordinance.

Catch Basin. A receptacle which separates and retains greases, oil, dirt, gravel and all other substances lighter or heavier than the liquid waste which bears them in order to prevent their entrance into the house sewer. A catch basin may perform the functions of a gravel or grease basin, or both, except that the liquid waste which it receives shall not contain focal matter.

Cesspool. A receptacle in the ground which receives crude sewage and is so constructed that the organic portion of such sewage is retained while the liquid portion seeps through its walls or bottom.

Circuit or Loop Vent. A system of vent pipes arranged as a substitute for individual revents when two (2) or more fixtures are located on a branch between a soil or waste stack and a vent riser.

Combined Sewer. A sewer or drain which receives storm water, other liquid wastes and sewage.

Common Trap. A trap having a water seal of not less than two (2) inches or not more than four (4) inches.

Continuous Vent. The continuation of a vertical soil or waste pipe above the point of entrance of the pipe from a fixture trap.

Cross Connection. A physical arrangement whereby one (1) system of piping is connected to another system of piping in such a way that the contents of the two (2) systems may become mixed.

Dead End. A pipe leading from a soil waste vent pipe, house drain or house sewer which is terminated at a developed distance of two (2) feet or more by means of a cap, plug or other fitting not used for admitting sewage or water to the pipe.

Deep Seal. A term applied to a trap having a water seal of more than four (4) inches.

Downspout. A leader or conductor pipe which carries water from the roof or gutter to the ground or to any part of the drainage system.

Drainage System. A system of piping or conduits through which are conveyed liquids or liquid borne solids from all plumbing fixtures and appurtenances in buildings and structures and discharges such liquid or solids into the house sewer. The drainage system of buildings or structures includes the house drain and its branches and the sewer and its branches.

Ejector. A device operated to elevate water, sewage or liquid wastes from a lower level to a point of discharge into a sewer or drain.

Floor Drain. A receptacle fitted with a strainer or grate and a trap or seal and connected to the plumbing or drainage system.

Gravel Basin. A receptacle through which roof water flows and which is designed to retain sediment.

Grease Basin or Intercepter. A receptacle designed to cause separation and retention of oil or grease from liquid wastes.

Horizontal. Level or with a slope of not more than one (1) inch per foot.

House Drain. That part of the horizontal piping of a plumbing system which receives the discharge from soil, waste, downspout and other drainage pipes inside the walls of any building and conveys it to the house sewer terminating three (3) feet outside of the building walls.

House Sewer. That part of the horizontal piping of a plumbing or drainage system extending from the house drain to its connection with the main sewer or other place of sewage disposal.

Indirect Connection. A connection in which there is a break in a line of pipe through which the water, sewage or other liquid may be discharged from one (1) pipe to another by gravity and open to the atmosphere for a sufficient altitude to permit visibility of such discharge and to prevent a back flow into the pipe above the connection.

Main. That part of any system of piping which receives the soil or waste pipes, vent pipes or revent pipes from fixture outlets or traps, direct or through branch pipes as tributaries.

Minor Repairs. The mending of leaks in drains, soil, waste, vent, water supply piping, faucets and valves; the replacing of a fixture in the old location, forcing out an obstruction and the replacement of a faucet or valve or not more than ten (10) feet of either soil, waste or vent piping.

New and Existing. The word "new" used in connection with a building, structure or plumbing system is equivalent to the phrase "constructed since the passage of this ordinance" unless a new provision is made otherwise and the word "existing" is equivalent to the phrase "in existence at the time of the passage of this ordinance". If neither of these qualifying words is used, the ordinance shall be construed as applying both to "new" and "existing" structures or conditions.

Open Plumbing. A plumbing system in which no plumbing fixtures, except a built-in bathtub, is so inclosed as to form a space in which air does not circulate.

Owner. The term "owner" for the purposes of this ordinance shall mean the owner or owners of the freehold of the premises or of a lesser estate therein, a vendee in possession or the lessee or joint lessees of the whole thereof.

Person. The word "person" shall apply to any individual, association, corporation (municipal or private) or firm acting either directly or through a duly authorized agent.

Place of Employment. Any dwelling, shop, store, factory or other building or place in which one (1) or more persons are employed for hire or wage.

Plumbing. The word or term "plumbing", as used in this ordinance, means and shall include all piping, fixtures, appurtenances and appliances for a supply of water for all personal and domestic purposes in and about buildings, structures, and public places where a person or persons, live, work or assemble and shall also include all piping, fixtures, appurtenances and appliances for a sanitary drainage and related ventilating system within a building, and all piping, fixtures, appurtenances and appliances outside a building connecting the building with the source of water supply on the premises or the main in the street, alley or at the curb. Also all piping, fixtures, appurtenances, appliances, drains or waste pipes carrying sewage from the foundation walls of a building to the sewer service lateral at the curb or in the street or alley or other disposal terminal, holding private or domestic sewage, excepting any underground house drain or other underground sewer of vitrified tile or masonry construction or any catch basin or cesspool of masonry. Plumbing shall also include the installation, repair and maintenance work upon and in connection with such piping, fixtures, appurtenances, appliances, drain or waste pipes, except minor repairs by a person upon his own premises.

Plumbing Fixture. A water supplied receptacle intended to receive and discharge water, liquid or water carried wastes into a plumbing or drainage system with which it is connected.

Private Sewer. A sewer built in a street, alley or granted easement not dedicated for public use and serving or intended for the service of two (2) or more pieces of property.

Public Sewer. A sewer built by or constructed under the authority of the municipality in a public place or places, such as streets or alleys or in and through land or lands for which an easement or easements have been granted for the purpose, for the common use of the property abutting on such public place, places, easement or easements.

Revent Pipe. A pipe which connects directly at or near the junction of an individual trap outlet with a waste or soil pipe underneath or back of a fixture and extends to a connection with the main or branch vent above the top of the fixture.

Roof Gutter. A receptacle either suspended from the edges of a roof or constructed in a roof to convey roof water to the downspout rain leader or conductor pipe.

Sanitary Sewer. A house drain or house sewer designed and used to convey only sewage.

Septic Tank. A water tight reservoir or tank which receives sewage and by sedimentation and bacterial action effects a process of clarification and partial purification.

Size and Length. The given caliber or size of pipe is for a nominal internal diameter except that other than iron pipe size brass pipe is measured by outside diameter. The developed length of a pipe is its longitudinal length along the center line of pipe and fittings.

Soil Pipe. Any pipe which conveys the discharges of one (1) or more water closets or bedpan sterilizers with or without the discharges from other fixtures to the house drain.

Soil or Waste Vent. That part of the main, soil or waste pipe which extends above the highest installed branch or fixture connection.

Stack. Any vertical line of soil, waste or vent pipe.

Sub-Soil Drain. That part of a drainage system which conveys sub-soil ground or seepage water to the house drain or house sewer.

Terminal. The upper portion of a soil, waste or vent pipe which projects above or through the roof of the building.

Tile Sewer or Drain. A sewer or drain of tile, terra cotta or cement pipe.

Trap. A fitting or device so constructed as to prevent the passage of air or gas through a pipe or fixture by means of a water seal.

Trap Seal. The vertical distance between the overflow and the dip separating the inlet and outlet arms of the trap.

Vent Pipe or Vent. Any pipe provided to ventilate a plumbing system, to prevent trap siphonage and back pressure and to equalize the air pressure within and without the piping system.

Waste Pipe and Special Waste Pipe. Any pipe which receives the discharge from any fixture or device except water closets and bedpan sterilizers and conveys the same to the house drain, sewer or waste pipe. When such pipe does not connect directly with the house drain or soil pipe, it is termed special waste pipe.

Water Mains. The pipes through which Lake Michigan water is distributed from the Chicago Water Works System Pumping Stations to all water service connections.

4801.03 Plans Required: No person shall construct, add to, alter or use any part of a plumbing system within buildings, public or private, until plans have been examined and approved by the Board of Health, Department of Public Works, Department of Buildings and other authorized departments and permit is issued and the fees paid. No plans shall be required for such portions of a plumbing system as are classed by this ordinance as minor repairs, which do not include the extension or reconstruction of soil, waste and vent pipes or supply pipes or the addition of fixtures. Plans of each floor, or typical floors, shall be presented in duplicate and shall fully show the plumbing system, including drains, soil, waste, vent, revent pipes, traps and all plumbing fixtures and all main water supply pipes within the building. Plans shall be legibly drawn and reproduction shall be clear at a scale of not less than one-eighth (⅛) inch to the foot.

4801.04 Vertical Elevations: One (1) or more vertical elevations or diagrams shall be provided which shall show the complete system of soil, waste, vent and main water supply pipes through the several stories of the building. Special construction, such as filters, condensers, surge and storage tanks, bathing pools and sterilizers, so far as they are connected to the plumbing, shall be shown in detail. Such detail plans shall be drawn to a scale.

4801.05 Conformity to Plans—Deviations: All parts of the plumbing system shall be constructed in conformity with the approved plan. If any deviations affecting plumbing plans and details are desired a supplementary plan covering that portion of the work involved shall be filed for approval and shall conform to this ordinance.

4801.06 Approval and Permits:

(a) **General.** The approval by the Board of Health, Department of Public Works and other authorized departments shall be stamped and dated upon the required plans of every plumbing system, and this approval, together with the receipt for payment of fees, shall constitute the permit required for such work. No permit shall be issued for the installation of plumbing except to a licensed master plumber, licensed architect or a licensed structural engineer.

(b) **Permits by Other Departments.** The approval and permit of the Board of Health may be withheld until all permits have been issued by the Department of Public Works or other authorized departments, as required by the general ordinances of the City; provided however, that such permit shall be issued before the building permit is issued by the Commissioner of Buildings.

(c) **Permit Required for Changes in Water System.** Whenever in any building, structure or premises receiving its water supply directly from the city system, a pump or any other device for increasing the water pressure is to be installed, plans of such installation shall first be submitted to the Commissioner of Public Works for approval.

ARTICLE 4802.

General Regulations.

4802.01 Water Closet and Connection to Sewer Required: Every new and existing dwelling, or place of employment, located on a lot abutting on a street, alley, granted easement or thoroughfare in which a public sewer is available, shall be provided with a plumbing system including water closet facilities connected to such sewer.

4802.02 Plumbing Pipes Within Buildings: Every soil, waste, vent and revent pipe shall be placed within the building and shall remain exposed to view until inspected and approved by the Board of Health.

4802.03 Extensions and Repairs:

(a) **General.** When old or defective plumbing is to be remodeled or replaced, or additional fixtures installed in any building, or house drains or house sewers extended, the modified system shall conform to the requirements of this chapter.

(b) **Extensions and Remodeling.** Any remodeling or extension of existing soil, waste and vent pipes shall be subject to the requirements of this ordinance.

(c) **Repairs, Emergency.** Repairs are to be made without a permit only in case of leaks and emergencies, in which cases, the extent need not be reported in advance to the Board of Health, Department of Public Works and other authorized departments. If the repair in any such emergency exceeds the limit of the definition of minor repairs, it shall be reported to the Board of Health, Department of Public Works and other authorized department within twenty-four (24) hours thereafter and the requisite permit applied for.

4802.04 Rarely Used Fixtures: Every plumbing fixture which is so rarely used that there is danger that the seal of the trap will be lost, shall be removed and the outlets securely closed when so ordered by the Board of Health.

4802.05 Plumbing by Licensed Plumber: All work of plumbing within buildings in the City of Chicago shall be performed by a master plumber, or a journeyman plumber, under the direction of a licensed master plumber, in accordance with an Ordinance Creating a Board of Plumbing Examiners, Enacted August 2, 1935, except as otherwise provided in Section 4802.06.

4802.06 Drain Laying: All work of tile drain laying underground, and the building of masonry catch basins or cesspools, shall be performed by a drain layer licensed to conduct, carry on, or engage in the business of drain laying as provided by the general ordinances of the city.

4802.07 Property Lines: All appurtenances to house drains and house sewers, including catch basins, manholes, cleanouts, backwater valves and fittings shall be located within the property lines of the premises served by such drains and sewers.

4802.08 Penalty: Any person violating, failing or refusing to comply with any of the sections of this chapter shall be fined not less than Twenty-five Dollars ($25.00) and not more than Two Hundred Dollars ($200.00) for each offense.

ARTICLE 4803.
Materials.

4803.01 Quality of Materials: All materials used in any drainage or plumbing system, or part thereof, shall be free from defects and shall conform to the requirements of this Article. The abbreviation "A.S.T.M." hereinafter employed, refers to the American Society for Testing Materials.

4803.02 Label, Cast or Stamped: Each length of pipe, fitting, trap, fixture or device used in any plumbing or drainage system shall be stamped or indelibly marked with the weight or quality thereof, and the maker's name or mark.

4803.03 Pipe:

(a) **Vitrified Clay Pipe.** All vitrified clay pipe shall conform to the A.S.T.M. "Standard Specifications for Clay Sewer Pipe" (serial designation C-13-33T).

(b) **Cast-Iron Pipe.** All cast-iron soil pipe and fittings shall conform to the A.S.T.M. "Standard Specifications for Cast Iron Soil Pipe and Fittings" (serial designation A74-29). All cast iron pipe and fittings for underground use shall be coated with asphaltum or coal-tar pitch.

(c) **Cast Iron Threaded Pipe.** All cast iron threaded pipe shall conform to "Cast Iron Threaded Pipe Manufacturers' Specifications" (Serial Designation CITP-1).

(d) **Cast Iron Water Service Pipe.** All cast iron water pipe and special castings shall conform to the American Water Works Association Standard Specifications for Cast Iron Water Pipe and Special Castings, adopted May 12, 1908. All cast iron pipe and special castings shall be coated inside and out with coal tar pitch varnish.

(e) **Wrought Iron Pipe.** All wrought iron pipe shall conform to the A.S.T.M. "Standard Specifications for Welded Wrought Iron Pipe" (serial designation A72-33) and shall be galvanized.

(f) **Mild Steel Pipe.** All steel pipe shall conform to the A.S.T.M. "Standard Specifications for Welded and Seamless Steel Pipe" (serial designation A120-34T-) and shall be galvanized.

(g) **Brass Pipe.** All brass pipe shall conform to the A.S.T.M. "Standard Specifications for Brass Pipe, Standard Sizes" (serial designation D43-33).

(h) **Hard Drawn Copper Tubing.** All copper tubing shall conform to the A.S.T.M. "Standard Specifications for Copper Water Tube" (serial designation B88-33). All sizes two (2) inch and smaller shall be Class L; sizes over two (2) inch shall be of Classes, K, L or M.

(i) **Concrete Sewer Pipe.** All concrete sewer pipe shall conform to the A.S.T.M. "Standard Specifications for Concrete Sewer Pipe" (serial designation C14-35).

(j) **Reinforced Concrete Sewer Pipe.** All reinforced concrete sewer pipe shall conform to the A.S.T.M. "Standard Specifications for Reinforced Concrete Sewer Pipe" (serial designation C75-35).

(k) **Lead Pipe, Diameter Weights.** All lead pipe shall be of the best quality of drawn pipe, of not less weight per linear foot than shown below.

Item 1. Lead soil, waste, vent or flush pipes, including bends and traps, extra light:

Internal Diameter Inches	Weights Per Foot Lbs.	Ozs.
1	2	
1¼	2	8
1½	3	8
2	4	12
3	6	
4	7	14

Item 2. Lead Water Supply Pipe.

Internal Diameter Inches	Weights Per Foot Lbs.	Ozs.
½	1	8
⅝	2	8
¾	3	
1	4	
1½	8	
2	13	12

4803.04 Sheet Lead: Sheet lead shall weigh not less than four (4) pounds per square foot.

4803.05 Sheet Copper or Brass: Sheet copper or brass shall be not lighter than No. 18 B. & S. Gauge.

4803.06 Threaded Fittings:

(a) **Plain Screwed Fittings.** Plain screwed fittings shall be of cast iron, malleable iron or brass of standard weight and dimensions.

(b) **Drainage Fittings.** Drainage fittings shall be of cast iron, malleable iron or brass with smooth interior water way, with threads tapped out of solid metal. Cast Iron fittings shall be galvanized or coated with asphaltum. Malleable iron fittings shall be galvanized.

(c) **Malleable Iron Fittings.** Malleable fittings shall be used on steel or wrought iron pipe and shall be galvanized.

4803.07 Calking Ferrules: Brass calking ferrules shall be of the best quality red cast brass with weights and dimensions in accordance with the following:

Pipe Size (Inches)	Actual Inside Diameter (Inches)	Length (Inches)	Weight Pounds	Ounces
2	2¼	4½	1	
3	3¼	4½	1	12
4	4¼	4½	2	8

4803.08 Soldering Nipples and Bushings:

(a) **Nipples.** Soldering nipples shall be of bronze of the following composition:

	Per Cent
Copper	.84 to .86
Tin	.04 to .06
Zinc	.04 to .06
Lead	Not more than .06

All threads shall be cut full and perfect to the American Standard (Briggs) pipe threads. Only nipples which are finished by workmanship of high standard shall be used. Nipples shall conform to the following requirements:

Diameter (Inches)	Minimum Thickness of Metal (Inches)	Minimum Length Soldering Surface (Inches)
½	3/32	1 11/16
¾	3/32	1 11/16
1	3/32	1 11/16
1¼	7/64	1¾
1½	7/64	2
2	⅛	2 7/16

(b) **Bushings.** Soldering bushings shall be of the material required under paragraph (a) of this Section for nipples.

4803.09 Backwater Valves: Backwater valves shall have all bearing parts or balls of non-ferrous metal and shall be so constructed as to insure a positive mechanical seal and remain closed except when discharging wastes.

4803.10 Soldered Fittings: Soldered fittings for copper tubing shall be of cast brass or of wrought copper and shall be capable of withstanding a pressure test of not less than o hundred fifty (250) pounds per square inch.

ARTICLE 4804.

House Drains and Sewers.

4804.01 Independent System:

(a) **General.** The drainage and plumbing system of each new building and of any work installed in any existing building shall be separate from and independent of that of any other building, and every building shall have an independent connection with a public or private sewer when available, except as hereinafter provided.

(b) **Buildings on Interior Lot.** Where any building stands in the rear of another building on an interior lot, a single house drain may be used for both buildings; provided however, that the size of drain is sufficient for the requirements herein provided and except when the lot abuts on a public alley in which there is a public sewer suitable for the drainage of the said rear building.

(c) **Group of Buildings Used as a Unit.** Where two (2) or more buildings are on land in one (1) ownership and such buildings are used as a unit and where division of property is impossible, a single house sewer may be used for the drainage of two (2) or more of such buildings; provided however, that such installation shall first receive the approval of the Commissioner of Public Works.

4804.02 Connections With Sewage Disposal Systems: When a public sewer is not available, drain pipes from buildings shall be connected to an approved system of sewage disposal.

4804.03 Excavations: Each system of piping shall be laid in a separate trench. Where a double system of drainage is installed, the sanitary sewers and storm water drains may be laid side by side in one (1) trench. Tunneling for distances not greater than six (6) feet is permissible in yards, courts or driveways of any building site. All excavations required to be made for the installation of a house drainage system, or any part thereof, within the walls of a building, shall be open trench work. All such trenches and tunnels shall be kept open until the piping has been inspected, tested and approved.

4804.04 Material:

(a) **House Sewer.** The house sewer shall be of extra heavy cast iron soil pipe or vitrified clay pipe, or concrete sewer pipe, or reinforced concrete sewer pipe or brick. Whenever the excavation for a house sewer is made in unstable ground, the material for such house sewer shall be extra heavy cast iron pipe.

(b) **House Drain.** The house drain, or any part of it, when exposed within a building, shall be of extra heavy cast iron soil pipe, galvanized wrought iron, steel or brass pipe. House drains, when underground, shall be of the materials required by paragraph (a), except that house drains in institutional buildings, office and financial units, hotels, theatres, public assembly units churches and schools and any part of house drains in any other building, having less than six (6) inches of cover between the top of the pipes and the underside of the floor shall be of brass, or of cast iron soil pipe, or of acid-resisting metal pipe of equal strength.

4804.05 Fixture Units: The following table shall be employed in determining the fixture unit values assigned to fixtures, in computing the required sizes of sewers, drains, soils and waste branches and stacks:

TABLE 4804.05.

Kind of Fixture	Equivalent Fixture Units
Bath Tub	2
Bath Tub with Shower over Tub	3
Bidet	2
Combination Sink and Laundry Tray, each compartment....	2
Dishwasher, Dwellings	2
Dishwater, Restaurant	3
Drinking Fountain, Cuspidor or Aspirator	½
Floor Drain	2
Floor Drain, Car Wash.......	6
Fountain Cuspidors, group of two	1
Laundry Tray, 1 or 2 on one trap	1½
Laundry Trays, 3 on one trap	2
Lavatory, 1 only............	1
Lavatory, group of two.......	1½
Refrigerator	½
Shower Stall	3
Showers, Gang	3 ea. head.
Sink, Dwelling or Surgeon's, Pantry, Bar, Glass or Silver	2
Sink, Public, Kitchen or Scullery	3
Sink, Pantry (Large for Hotel or Institutions)	3
Sink, Slop (with trap combined)	3
Sink, Slop, flush rim, siphon jet or bedpan.............	6
Sink, Slop, Ordinary.........	2
Sitz Bath	1
Sterilizer, instrument, utensil, water	1
Sterilizer, Bedpan	1
Urinal, Pedestal	2
Urinal, Wall Type	2
Urinal, Floor	3
Water Closet	6

4804.06 Size of Drains and Sewers:

(a) **Sanitary Sewers.** The required size of sanitary house drains and sanitary house sewers shall be determined on the basis of the total number of fixture units drained by them in accordance with the following:

TABLE 4804.06 (a).
Sanitary System Only.

Diameter of Pipe, Inches	Maximum Number of Fixture Units Allowed ½" Slope	¼" Slope	⅛" Slope	1/16" Slope	1/32" Slope	1/64" Slope
3*	25	19				
4	120	50				
5	650	325	90			
6		900	475			
8		2,400	1,650	1,100		
9			2,500	1,750		
10			3,500	2,500		
12			6,500	4,100	3,000	
14				7,000	5,000	
15				8,750	6,000	
16				10,000	7,000	
18				14,000	10,000	7,000
20				19,000	13,000	9,500
21				22,000	15,500	10,500
24					23,000	16,000

*A minimum of four (4) inches in diameter is required for soil pipe and water closet branches. See Section 4806.03.

(b) **Storm Water Sewers.** The required sizes of storm water house drains and house sewers and other lateral storm water drains shall be determined on the basis of the total drained area in horizontal projection in accordance with the following:

TABLE 4804.06 (b).
Storm Drains Only.

Size of Drain (Inches)	Minimum Slope Inches per Ft.	Maximum Area Drained	Double Standard Slope	Maximum Area Drained
3	¼	1,600	½	2,400
4	¼	3,000	½	4,000
5	¼	6,000	½	8,000
6	⅛	7,000	¼	9,500
8	⅛	16,000	¼	20,500
9	1/16	16,000	⅛	22,000
10	1/16	20,500	⅛	29,000
12	1/16	33,000	⅛	49,000
14	1/32	37,000	1/16	53,000
15	1/32	45,000	1/16	64,000
16	1/32	54,000	1/16	77,000
18	1/32	75,000	1/16	100,000
20	1/32	97,000	1/16	140,000
21	1/64	79,000	1/32	113,000
24	1/64	114,000	1/32	164,000

(c) **Combined Sewers.** Whenever a combined sewer system is employed the required size of the combined house sewer shall be determined by adding to the total drained area in square feet in Table 4804.06 (b) an equivalent area for the number of fixture units in accordance with the following:

TABLE 4804.06 (c).

Number of Fixture Units	Equivalent Area Sq. Ft.	Number of Fixture Units	Equivalent Area Sq. Ft.	Number of Fixture Units	Equivalent Area Sq. Ft.
1	165	21	2,250	50	3,530
2	325	22	2,310	55	3,665
3	475	23	2,360	60	3,790
4	615	24	2,440	65	3,900
5	750	25	2,500	70	4,000
6	875	26	2,550	75	4,090
7	1,000	27	2,600	80	4,175
8	1,115	28	2,660	85	4,250
9	1,225	29	2,710	90	4,320
10	1,330	30	2,770	95	4,390
11	1,435	31	2,820	100	4,450
12	1,530	32	2,870	105	4,500
13	1,620	34	2,955	110	4,550
14	1,710	36	3,040	115	4,600
15	1,800	38	3,125	120	4,645
16	1,880	40	3,200	125	4,690
17	1,960	42	3,270	130	4,725
18	2,040	44	3,340	140	4,800
19	2,110	46	3,400	145	4,830
20	2,180	48	3,465	150	4,850

For more than 150 Fixture Units the Equivalent Area shall be determined by adding seven and two-tenths (7.2) square feet for **each fixture unit over 150 to 4850 square feet.**

(d) **Minimum Sizes.** The minimum required sizes for house drains and branches underground shall be as required by Section 4806.02, paragraph (b) for soil or waste underground.

4804.07 Sumps and Receiving Tanks:

(a) **General.** Drains below the level of house sewer shall discharge into a tightly covered sump or receiving tank, so located as to receive the discharge of such drains by gravity. Pumps, ejectors or other mechanical devices shall be provided to lift and discharge the contents of sumps or receiving tanks into the house sewer. Such sumps, ejectors or other mechanical devices shall operate automatically or the sumps or receiving tanks shall be of sufficient capacity to contain the wastes accumulating during a period of twenty-four (24) hours. A backwater valve shall be provided in the outlet line of every sump or receiving tank.

(b) **Construction.** Every sump or receiving tank within a building, receiving the discharge from sanitary drains, storm water or combined drains, shall be of cast iron, equally durable metal or reinforced concrete. Every sump or receiving tank receiving the discharge from open jointed subsoil drainage systems only, shall be of cast iron, equally durable metal, reinforced concrete, vitrified clay tile or masonry.

(c) **Venting.** Every sump or receiving tank shall have a vent pipe equal in size to the largest entering waste pipe and not less than four (4) inches in diameter where soil waste is received; provided however, that sumps or receiving tanks for subsoil drainage systems only, or floor drains only, or combination of the two (2) only need not be vented.

4804.08 Motors, Compressors and Pressure Tanks: All motors, air compressors and air pressure tanks shall be so located as to be accessible for inspection and repair at all times.

4804.09 Hot Water or Steam Discharge Prohibited: Liquid wastes of a temperature exceeding one hundred sixty (160) degrees Fahrenheit shall not discharge directly into any house drain or house sewer. Wastes of a higher temperature shall be intercepted and cooled to one hundred sixty (160) degrees Fahrenheit, or less. The material and construction of blow-off basins and systems shall be in accordance with the provisions of the American Society of Mechanical Engineers "Rules for the Construction of Unfired Pressure Vessels, Section VIII dated 1934 and addendum dated July 23, 1935."

4804.10 Water Pressure Ejectors: Water pressure ejectors or siphons shall not be installed for the discharging of any sewage or waste.

ARTICLE 4805.

Roof, Storm Water and Seepage Drains.

4805.01 Drainage of Roofs, Areas and Yards:

(a) **Roofs and Downspouts.** All roofs exceeding five hundred (500) square feet in area shall be drained to a sewer, where such sewer is available in any adjoining street, alley or public place. Every connecting roof downspout having the open roof connection, located nearer than twelve (12) feet to any inside lot line or any door or window on the same premises, shall be trapped; provided however, that this provision shall not apply to any single dwelling, nor to any roof which is at the maximum height limit for such roof as regulated by the volume provisions of the Zoning Ordinance. One (1) trap may serve more than one (1) downspout, and any such trap shall be on the downspout side of the connection to any sanitary sewer or any combined sewer or drain, and shall be set where not subject to frost.

(b) **Areas and Yards.** Outside areas other than roof areas may be drained to a sewer and when paved shall be so drained where necessary to avoid the discharge of water onto adjoining premises. Paved areas of three hundred (300) square feet or less where connected to the sewer, shall be provided with trapped connections before connecting to any sanitary sewer or combined sewer, with traps placed where not subject to frost. Outside areas exceeding three hundred (300) square feet, and not more than four thousand (4000) square feet, where connected to sewers, shall be connected through a catch basin, not less than three (3) feet in diameter and not less than three (3) feet deep below the bottom of the trap, except as otherwise provided in Section 4808.14. Areas of more than four thousand (4000) square feet and not more than eight thousand (8000) square feet shall be provided with a catch basin not less than four (4) feet in diameter and not less than three (3) feet six (6) inches deep below the bottom of the trap.

4805.02 Downspouts:

(a) **Without Increasers.** No downspout shall be of smaller size than shown in the following table, except as hereinafter provided for downspouts with increasers at the roof:

TABLE 4805.02 (a).
Without Increasers.

Area of Roof in Horizontal Projection (Square Feet)	Diameter of Downspout (Inches)
Up to 250	2
Up to 400	2½
Up to 650	3
Up to 1,350	4
Up to 2,400	5
Up to 3,800	6
Up to 5,600	7
Up to 7,800	8

(b) **Downspouts With Increasers at the Roof.** When the diameters of downspouts are increased at the roof for a length of at least twice the diameter of the downspout, the following areas in horizontal projection may be drained to them:

TABLE 4805.02 (b).
With Increasers.

Area of Roof in Horizontal Projection (Sq. Ft.)	Diameter of Downspout (Inches)	Diameter of Increaser (Inches)
650	2	3
800	2	4
950	2½	3½
1,350	2½	4
1,350	3	3½
2,250	3	4
2,400	4	5
3,800	4	6
3,800	5	6
7,750	5	8
7,750	6	8
10,500	6	9
12,700	6	10

The above sizes of downspouts are based on the diameter of circular downspouts and other shapes shall have equivalent cross-sectional area. All downspouts from gravel roofs shall be fitted with gravel basins or equally serviceable devices to screen out loose gravel.

(c) **Inside Downspouts.** When placed within the walls of any building, all downspouts shall be constructed of cast iron or galvanized wrought iron pipe or galvanized steel pipe. A roof connection to a downspout shall be made gas and watertight by the use of aluminum cast iron or copper downspout head, or by means of heavy lead or copper drawn tubing, wiped and soldered to a brass ferrule, calked or screwed into the pipe. A downspout ferrule head of aluminum, brass or cast iron shall be provided with a dome strainer, having clear openings at least equal to the area of the downspout head.

4805.03 Waste or Vent Connections With Downspouts Prohibited: Downspout pipes shall not be used as soil, waste or vent pipes, nor shall any soil, waste or vent pipes be used as downspouts.

4805.04 Overflows. Overflow pipes from cisterns, condensing or supply tanks, expansion tank, condenser coils, filter wash pipes and drip pans, water-jackets, or other fixtures or equipment, shall connect only indirectly with any house sewer, house drain, soil or waste pipe.

4805.05 Sub-Soil Drains: Where subsoil drains are placed under the cellar floor, same shall be made of drain tile or earthenware pipe, not less than four (4) inches in diameter; shall be properly trapped, and before entering the house sewer or drain, protected against back pressure by an automatic back pressure valve accessibly located. Access for the purpose of cleaning and removing obstructions in subsoil drains shall be provided at every change of direction.

ARTICLE 4806.

Soil, Waste and Vent Pipes.

4806.01 Material: Every new main or branch soil waste and vent pipe within a building shall be of cast iron, galvanized steel, galvanized wrought iron, brass or copper.

4806.02 Fixture Traps and Branches: The following table shall be employed in determining the minimum diameters of fixture traps and branch waste pipes from single fixtures:

TABLE 4806.02.

Kind of Fixture	Minimum Size of Trap (Inches)	Branch Inlet Size (Inches)
Bath Tub	2	2
Bath Tub, with Shower over Tub	2	2
Bidet	1½	2
Combination Sink and Laundry Tray—each compartment	1½	2
Dishwasher, Dwellings	1½	2
Dishwasher, Restaurant	2	2½
Drinkingi Fountain, Cuspidor or Asp rator	1	1¼
Floor Drain	2	2
Floor Drain, Public Car Wash	4	4
Fountain Cuspidors, Group of Two	1	1½
Laundry Tray, 1 or 2 on one trap	1½	2
Laundry Tray, 2 on one trap	2	2
Lavatory, 1 only	1¼	1½
Lavatory, Group of Two	1½	1½
Refrigerator	1¼	1½
Shower Stall	2	2
Showers, Gang	3	3
Sink, Dwelling or Surgeon's Pantry, Bar glass or silver	1½	1½
Sink, Public Kitchen or Scullery	2	2
Sink, Pantry (Large for hotel or institutions)	2	2
Sink, Slop (with trap combined)	2¼	3
Sink, Slop, flush rim, siphon jet or bedpan		4
Sink, Slop, ordinary	2	2
Sitz Bath	1½	1½
Sterilizer, instrument, utensil water	1¼	1½
Urinal, Pedestal	2	2
Urinal, Wall Type	2	2
Urinal, Floor	2	2
Water Closet	2¼	4

4806.03 Soil and Waste Branches:

(a) **General.** The minimum required sizes of soil and waste branches, except as provided in paragraph (b), and except that the size in no case shall be less than for a single fixture as given in Table 4806.02 shall be determined on the basis of the total number of fixture units drained by them in accordance with the following:

TABLE 4806.03 (a).

Minimum Sizes of Soil and Waste Branches.

Diameter (Inches)	Slope in Inches Per Foot	Maximum Fixture Units Allowed
1½	For sizes more than	2
2	four inches the	3
2	number of fixture	6
2½	units shall vary	12
3	with the slope of	19
4	the pipe	120
5	¼	325
5	½	650
6	⅛	475
6	¼	900
8	⅛	1,650
8	¼	2,400

(b) **Exceptions.** The minimum required sizes of underground soil and waste branches, or of underground branches of storm water, sanitary or combined house drains, shall be four (4) inches for soil branches, and three (3) inches for branches from downspouts, floor drains, laundry trays and kitchen stacks. The minimum required size of soil pipe and branches for water closets, above ground, shall be four (4) inch diameter.

4806.04 Soil and Waste Stacks:

(a) **General.** Every building in which a plumbing fixture is installed shall have one (1) or more soil or waste stacks, each of which shall extend full size through the roof. Every soil or waste stack shall be carried up as direct as possible and free from sharp angles and turns. No stack shall be of lesser diameter than the largest branch connected to it. Not more than two (2) branches shall connect to any three (3) inch stack at the same level. No kitchen waste stack shall be less than two (2) inch diameter.

(b) **Minimum Sizes.** The required size of each soil or waste stack shall be independently determined by the total number of fixture units of all fixtures connected to the stack in accordance with the following table:

TABLE 4806.04 (b).

Soil and Waste Stacks.

Diameter of Stack (Inches)	Waste Stack. Allowed Fixture Units
1½	8
2	16
2½	34
Soil and Waste.	
4	500
5	1,000
6	1,800
8	3,500
10	6,000

4806.05 Soil and Waste Stages, Angle of Connections: All soil and waste stacks and branches shall be provided with fixture connections at an angle of forty-five (45) degrees or by combination Y and one-eighth (⅛) bend inlets.

4806.06 Prohibited Connections: No other fixture connection shall be made to a lead closet bend. No soil or waste vent, circuit or loop vent above the highest installed fixture on the branch or main shall thereafter be used as a soil or waste pipe.

4806.07 Plumbing Protected From Frost: No soil or waste or vent stack shall be installed outside of a building. All plumbing pipes or fixtures shall be adequately protected from freezing.

4806.08 Roof Terminals:

(a) **General.** Vertical soil, waste and vent pipes shall extend through and above the roof at least six (6) inches and shall have a diameter at least one (1) inch greater than that of the pipe proper; but in no case shall it be less than four (4) inches in diameter through and above the roof. When roof is used as a floor, such extension shall be not less than seven (7) feet above the roof. The increaser shall extend at least one (1) foot below the roof and shall have a slope of not less than forty-five (45) degrees from the horizontal. Where it is desirable to avoid many openings in the roof, vertical soil, waste and main vent pipes may be connected to a horizontal vent pipe which shall be connected to a main soil or waste vent at least one (1) foot below the roof. The diameter of the horizontal vent and its extension through the roof shall equal the combined areas of the vertical vent pipes connected therewith.

(b) **Location.** No roof terminal for any soil, waste or vent stack shall be located nearer than twelve (12) feet to any inside lot line or any door or window on the same premises; provided however, that this provision shall not apply to any single dwelling, nor to any roof which is at the maximum height limit for such roof as regulated by the volume provisions of the Zoning Ordinance.

4806.09 Traps Protected by Vents Every new fixture trap shall be fully protected against siphonage and back pressure and air circulation assured by means of a soil or waste stack vent, a continuous waste or soil vent. No crown vent shall be installed.

4806.10 Distance of Vent From Trap Seal: No trap shall be placed more than five (5) feet, horizontal developed length from its vent. The distance shall be measured along the central line of the waste or soil pipe from the vertical inlet of the trap to the vent opening. The vent opening from the soil or waste pipe, except for water closets and similar fixtures, shall not be below the dip of the trap. No fixture shall be installed in any building more than forty (40) feet distant, measured along the central line of the soil waste or vent pipe.

4806.11 Main Vents to Connect at Base—Cross Connection of Vents: Every main vent or vent stack shall connect full size at its base to the main soil or waste pipe at or below the lowest fixture branch, and shall be reconnected with the main soil or waste vent not less than three (3) feet above the highest fixture branch or shall extend undiminished in size above the roof. In buildings more than eighty (80) feet in height a cross-connection between the soil or waste stack and the main vent stack shall be made at intervals of not more than fifty (50) feet. Such a cross-connection shall be equal in diameter to the main vent stack and shall be at an angle of forty-five (45) degrees upward from the soil or waste stack at least three (3) feet above the floor level.

4806.12 Vents, Required Sizes: The required sizes of main vents or vent stacks shall be determined on the basis of the size of the soil or waste stack vented, the number of fixtures or fixture units drained thereby, and the developed length of the main vent or vent stack in accordance with the following tables:

TABLE 4806.12 (a).
Waste Stack.

Diameter of Stack (Inches)	Fixture Units on Stack	Dimensions of Vent Maximum Diameter (Inches)	Maximum Length (Feet)
1½	2- 8	1½	50
2	9-16	1½	60
2½	17-34	1½	45
2½	17-34	2	60
2½	17-34	2½	105
3	6-18	1½	20
3	19-42	2	45
3	19-42	2½	150
3	43-76	2	30
3	43-76	2½	90
3	43-76	3	150

TABLE 4806.12 (b).
Soil or Waste Stack.

Diameter of Stack (Inches)	Fixture Units on Stack	Water Closets Only	Dimensions of Vent Maximum Diameter (Inches)	Maximum Length (Feet)
4	24- 42	4- 7	2	20
4	24- 42	4- 7	2½	45
4	24- 42	4- 7	3	100
4	43- 72	8- 12	2½	30
4	43- 72	8- 12	3	75
4	43- 72	8- 12	3½	150
4	43- 72	8- 12	4	300
4	73- 150	13- 25	3	60
4	73- 150	13- 25	3½	120
4	73- 150	13- 25	4	*225
4	151- 500	26- 50	3	20
4	151- 500	26- 50	3½	50
4	151- 500	26- 50	4	100
4	151- 500	26- 50	5	225
5	301- 480	51- 80	2½	20
5	301- 480	51- 80	3	50
5	301- 480	51- 80	3½	100
5	301- 480	51- 80	4	175
5	301- 480	51- 80	5	*300
5	481-1000	81-166	3½	25
5	481-1000	81-166	4	50
5	481-1000	81-166	5	125
5	481-1000	81-166	6	300
6	721- 840	121-140	3	20
6	721- 840	121-140	3½	40
6	721- 840	121-140	4	75
6	721- 840	121-140	5	225
6	721- 840	121-140	6	*400
6	841-1800	141-300	4	50
6	841-1800	141-300	5	125
6	841-1800	141-300	6	300
6	841-1800	141-300	8	400
8	1080-3500	181-583	4	20
8	1081-3500	181-583	5	60
8	1081-3500	181-583	6	150
8	1081-3500	181-583	8	*600

*This is limit in height of soil stack, but the vent may exceed this by fifty (50) per cent. See Section 4806.10.

4806.13 Branch and Individual Vents: No vent shall be less than one and one-half (1½) inches in diameter. In no case shall a branch or main event have a diameter less than one-half (½) that of the soil or waste pipe served, and in no case shall the length of a branch vent of given diameter exceed the maximum length permitted for the main vent serving the same size soil or vent stack.

4806.14 Vent Pipe Grades and Connections: Every vent pipe shall be free from drops or sags and shall be so graded and connected as to drip back to the soil or waste pipe. Where a vent pipe connects to a horizontal soil or waste pipe, the vent branch must rise vertically or at an angle of forty-five (45) degrees to the vertical to a point three (3) feet, six (6) inches above the floor before offsetting horizontally or connecting to the branch, main, waste or soil vent.

4806.15 Vents Not Required: No vent will be required on a downspout trap, a backwater trap, a sub-soil catch basin trap draining by gravity to the house drain or sewer or on a cellar floor drain which branches into the house drain on the sewer side at a distance of not less than five (5) feet from any stack or waste. Where bathrooms or water closets or other fixtures are located on opposite sides of a wall or partition or directly adjacent to each other within the prescribed distance, such fixtures may have a common soil or waste pipe and common vent. Where a water closet or other plumbing fixture is located above any other fixture branch on its soil or waste stack, and the trap of such fixture is not more than five (5) feet from its soil or waste vent, a revent pipe is not required.

4806.16 Changing Soil Vent or Waste Vent Pipe: In any existing building where the soil vent or waste vent pipe is not extended undiminished through or above the roof, or where there is a sheet metal soil or waste vent pipe and the fixture is changed in style or location, or is replaced, a soil vent or waste vent pipe of the size and material prescribed for new work shall be installed.

4806.17 Special Wastes—Indirect Connections Required: No new or existing waste pipe from a drip pan serving an air washer, condenser, dehumidifier or any heat transfer apparatus, or from a refrigerator or ice box drain or cooler or any other receptacle where food is stored, shall connect directly with any house drain, soil or waste pipe. Such waste pipe shall discharge over an open floor drain, or with an open end above a sink that is properly supplied with water, connected, trapped and vented or over a trapped drain. Such a waste connection shall be accessibly located and protected against back pressure. When located above any plumbing fixture, there shall be an open interval of not less than two (2) inches between the discharge end of waste and the overflow level of the fixture.

4806.18 Refrigerator Waste Sizes: Refrigerator waste pipes shall not be smaller than shown in the following:

ARTICLE 4807.

Catch Basins, Traps and Cleanouts.

4807.01 Traps, General: Every new trap shall be self-cleaning, shall be set true with respect to its water seal and shall be protected against freezing. Every trap for any plumbing fixture shall be of lead, brass, copper, cast or malleable iron, except for vitreous earthenware fixtures with integral traps and visible water seal. If of iron, every trap shall be extra heavy, shall have a full bore smooth interior waterway with threads tapped out of the solid body of the trap and may be galvanized or lined with vitrified enamel.

4807.02 Protection from Rats: Every concealed new lead pipe or lead trap shall be protected from rats by a covering of concrete not less than two (2) inches thick, or heavy galvanized wire netting of one-half (½) inch mesh.

4807.03 Traps Prohibited: No new bell, bottle or D trap, or any trap which depends for its seal upon the action of movable parts or mechanism shall be installed or used. No form of trap with any concealed interior partition shall be installed or used unless constructed of vitrified earthenware having a visible seal. No trap made of pipe fittings or in which the inlet and outlet legs on a trap for a plumbing fixture are a greater distance apart than twelve (12) inches, shall be installed or used. No metal trap having a cleanout secured otherwise than by bolts or screw thread shall be installed or used. No trap shall be installed at the foot of a soil or waste stack in a house drain or house sewer. No fixture shall be double-trapped.

4807.04 Traps, Where Required: Each fixture shall be separately trapped by a water sealed trap placed as near to the fixture as possible, except that a set of not more than three (3) laundry trays, two (2) lavatories, or a combination sink and dishwasher may connect with a single trap not more than two (2) feet from the most distant fixture waste outlet.

4807.05 Water Seal of Trap: Each new fixture trap, except a refrigerator trap, shall shall have a water seal of not less than two (2) inches and not more than four (4) inches.

4807.06 Open Plumbing: All sinks, lavatories and laundry trays shall be so installed as to provide for access to traps and waste connections. All access panels below such fixtures shall contain openings to permit circulation of air, in area equivalent to at least twenty-five (25) per cent of the area of such panels.

4807.07 Trap Cleanouts: Each new trap except those in combination with fixtures in which the trap seal is plainly visible shall be provided with an accessible brass cleanout not less than one (1) inch diameter, with American Standard screw thread protected by a water seal.

4807.08 Pipe Cleanouts: The body of every cleanout ferrule shall be of size and thickness for the standard pipe sizes of the metal used and extend not less than one-fourth (¼) inch above the hub. The cleanout cap or plug shall be of heavy red brass not less than one-eighth (⅛) inch thick, provided with a heavy raised nut or recessed socket.

4807.09 Pipe Cleanouts, Where Required: Every horizontal soil or waste pipe shall be provided with a readily accessible cleanout at every ninety (90) degree turn at each junction and at the end of each branch. All floor and wall connections of fixture traps shall be considered cleanouts. An accessible cleanout shall be provided at the foot of each waste and soil stack. House drains of a diameter of eight (8) inches or less shall be provided with cleanouts at least four (4) inches in diameter and not more than fifty (50) feet apart and one (1) such cleanout with full size wye branch shall be located near to the connection between the house sewer and the house drain. House drains and house drain branches more than eight (8) inches in diameter and all house sewers and house sewer branches shall be provided with cleanouts or manholes not more than one hundred fifty (150) feet apart.

4807.10 Manholes: All underground traps and cleanouts, except those which are set flush with the floor surface, and all outdoor underground traps shall be made accessible through manholes with proper covers.

4807.11 Barn Drainage: All wastes from barns, stables or yards for animals shall be intercepted before entering the sewer by a catch basin, which shall be trapped, and covered with a tight iron cover; and such catch basins, if within buildings, shall be provided with a vent not less than three (3) inches in diameter carried through the roof. All floor drains and wash racks shall be provided with deep seal traps and heavy strainers.

4807.12 Basement Floor Drains: Basement floor drains shall connect into a trap so located and constructed that it can be readily cleaned and a size to serve the purpose for which it is intended but in no case less than three (3) inches. The drain inlet and cleanout openings shall be accessibly located. Floor drains located where they may be subject to loss of water seal, because of unusual conditions of atmosphere, temperature or infrequency of use, shall be provided with suitable devices to maintain a constant water seal.

4807.13 Back-Water Valves: Where the plumbing system of a building, the basement floor drains, or subsoil drainage system is subject to back-flow or back-pressure, a suitable back-water valve shall be installed in the house drain on the sewer side. Back-water valves shall have all bearing parts and balls of heavy non-corrodible metal and shall be so constructed as to insure positive action. If of the flap type, the flap shall be so attached to the thimble as to insure a permanent connection and both flap and seat shall be accurately machine faced. All back-water valves, except those operated by power other than that of the sewage or waste, shall normally remain closed and shall open only for the purpose of discharging wastes.

4807.14 Grease Interceptor or Catch Basin Required: Every single dwelling and every multiple dwelling building not containing a restaurant or hotel kitchen, shall be provided with an approved grease interceptor or an outside catch basin for all kitchen wastes. Sinks, dishwashers, or other fixtures in restaurants, hotels, club-houses, public institutions, butcher shops or other establishments from which greasy wastes are discharged, shall be connected to an approved grease interceptor located as close as possible to the fixtures served, and within the limits of the same room. All grease interceptors located within buildings shall be constructed of heavy cast iron or equally durable metal and shall be provided with a tight metal cover, securely fastened. All grease interceptors shall have a capacity sufficient to intercept and retain not less than ninety-five (95) per cent of the grease received. All grease interceptors shall be provided with suitable vents to prevent loss of seal, as required for plumbing fixtures. Traps shall have not less than a four (4) inch seal.

4807.15 Grease Catch Basins Located Outside: A grease catch basin located outside a building shall be of concrete, precast in blocks or monolithic, or of brick construction with the block or brick laid up in Portland cement mortar with walls not less than seven (7) inches thick. The basin shall be water tight. The basin shall be not less than thirty-six (36) inches in diameter below the top of the highest inlet pipe. It shall be partitioned by a vertical baffle plate of cast iron. One (1) face of the baffle shall be perpendicular to the line of the outlet pipe and eight (8) inches from it. The top of the baffle shall be twelve (12) inches above the top of the outlet pipe and the bottom of the baffle shall be twelve (12) inches below the invert of the pipe. The bottom of the basin shall be twelve (12) inches below the bottom of the baffle. The outlet pipe shall not be trapped. The basin shall be covered with an all cast iron cover or cast iron lid in a stone or concrete ring and with a minimum opening of sixteen (16) inches. The lid shall not be less than one-fourth (¼) inch thick. The inverts of the inlet pipes carrying grease bearing wastes shall be not less than six (6) inches above the outlet pipe. Downspouts may be connected to basins and shall connect on the ouelet side of the baffle. The connections shall be trapped at the basin, where traps are required as provided in Section 4804.01. Yard drains, draining not more than five hundred (500) square feet of surface may be connected to a grease catch basin without any other intercepting basin and the same shall connect to the basin on the outlet side of the baffle and shall be trapped at the basin. Garage drains may be connected to a grease catch basin without any other intervening basin when the capacity of the garage does not exceed two (2) cars and the roof area is not more than five hundred (500) square feet. Such drains shall connect on the inlet side of the grease basin and shall be trapped at the basin unless traps are provided elsewhere. Traps at the grease basin for downspouts, yard drains and garage drains shall be cast iron elbows, cast iron offsets, or cast iron, tile or concrete half traps, all of them to be built into the wall of the basin and so placed as to give at least a two (2) inch seal for downspout connections and yard drains and a four (4) inch seal for garage drains.

4807.16 Car Wash Mud Basins: Wherever car washing facilities are furnished in any Class 1 garage, the wash water, if drained to a sewer, shall be drained through a catch basin provided to intercept and retain all earth, sand and similar material. Such catch basin shall have a grated cover, and a trapped outlet, and a minimum diameter of thirty (30) inches and shall have a capacity for dirt below the bottom of the trap of not less than five (5) cubic feet. Every such catch basin shall be independent of and shall not discharge into any basin for volatile wastes required by Section 4806.22.

ARTICLE 4808.

Plumbing Fixtures.

4808.01 Materials: All new fixtures used as water closets, urinals, bedpan sinks or otherwise for the disposal of excreta, shall be of vitrified earthenware, hard natural stone, or cast iron white enameled on the inside. Bedpan washers may be of non-corrodible alloys. Where used fixtures are to be installed in any new or existing building, such fixtures shall be subject to the inspection and approval of the plumbing inspector. No used plumbing fixture shall be installed in any building in this city unless such fixture is structurally sound and free from cracks or other defects. No sink, toilet, urinal, bath tub, laundry tray, slop sink or wash bowl which has been used in any building or elsewhere, shall be installed in any building in this cityy, unless such fixture has before installation been thoroughly washed and disinfected in a solution approved by the Board of Health. The plumbing inspector shall see to it that this provision is strictly enforced and that no used fixture which may cause the spread of infection or disease is installed in any building in this city. No used fixture, even if structurally sound and in sanitary condition, shall be installed in any building in this city, unless such fixture is of the design required by this ordinance for a similar new fixture.

4808.02 How Installed: All plumbing fixtures installed after the passage of this ordinance shall be installed in a manner to afford access for cleaning and air circulation. All traps not integral with the fixtures shall be exposed for access.

4808.03 Number of Fixtures Required: Plumbing fixtures and equipment shall be provided in every building for human habitation as required by the Building Ordinance of the City of Chicago and as otherwise required by this ordinance.

4808.04 Fixtures Prohibited: A fixed wooden wash tray or wooden sink shall not be installed or maintained in any building designed or used for human habitation. No metal lined wooden bathtub shall be installed or reconnected. No water closet of the pan and valve plunger type, offset washout trough and range type or other water closet having an invisible seal or unventilated space, or wall not thoroughly washed at each flush shall be installed or reconnected. Old fixtures of this kind shall not be reset or reconnected.

4808.05 Water Closets: In any building where the location is protected against freezing, every new water closet bowl and trap shall be made in one (1) piece and of such form as to hold a sufficient quantity of water when filled to the trap overflow to prevent fouling of surfaces and shall be provided with integral flushing rims.

4808.06 Closet Seats: Seats for water closets shall be constructed of or surfaced with non-absorbent material. Seats for water closets for the use of employees or the public, shall be open-front seats.

4808.07 Flushing Devices and Connections: Every water closet or other plumbing fixture or appliance designed to be cleansed by flushing with water shall, at each flush be supplied with a sufficient quantity of water to remove quickly all waste matter and properly cleanse the interior surfaces exposed to atmosphere. No water closet, or other plumbing fixture or appliance shall be supplied directly from a drinking and domestic water supply system through a flush valve or other valve, except as otherwise provided elsewhere in this ordinance. No water closet shall have a submerged inlet.

4808.08 Long Hopper Closets: No long hopper water closet shall hereafter be installed except for the service of existing buildings not used for dwelling purposes and then only in compartments which have no direct entrance from the building. No long hopper closet shall be installed to be flushed from a stop and waste cock but every such closet shall be provided with flush tank, the flush pipe of which shall be brought to the seat level inside of the building. No such closet shall be installed except at or below the gound level where it shall be provided with an impervious floor of cement or like material. The inclosure for such hopper closet shall be provided with door and window openings and shall be fly-tight. The door and window openings shall be screened and the door and seat cover shall be screened and the door and seat cover shall be self-closing.

4808.09 Workmen's Temporary Closets: It shall be unlawful for the owner of any building, or any person, firm or corporation, employing or in charge of any persons, to begin the construction, alteration or repair of any building, or the construction of any public or private works without having provided proper and sufficient toilet facilities consisting of water closets, chemical closets, privies or incinerators of a type to be approved by the Board of Health for the use of all employees engaged in the construction, alteration or repairs of such building, or the construction of any public or private works. There shall be at least one (1) such water closet, chemical closet, privy or incinerator for every thirty (30) employees or fraction thereof. Such toilet facilities in multiple story buildings shall be so located that no floor is more than four (4) stories from a story containing toilet facilities.

4808.10 Chemical Closets: No chemical closet or toilet shall be installed within any building for human habitation, nor on any premises abutting upon a street, alley or thoroughfare in which a public sewer is available. Wherever any chemical closet or toilet is installed it shall conform to the provisions of Section 4811.01 for Privy Vaults.

4808.11 Temporary Toilet Facilities: Every tent or structure provided for the temporary accommodations of the public, shall be provided, if access can be made to sewer and water, with not less than one (1) water closet for males, and one (1) water closet for females for each three hundred (300) persons or fraction thereof of the total capacity of the tent or structure. If access cannot be had to sewer and water, chemical toilets and incinerators shall be provided in the ratio above required.

4808.12 Fixture Strainer: Every fixture other than a water closet, pedestal urinal, clinic service sink or hospital fixture and every floor drain, shall be provided with a metallic strainer.

4808.13 Fixture Overflow: A new or existing overflow pipe if provided for a fixture, shall be connected on the inlet side of the trap and shall be accessible for cleaning.

4808.14 Water Closet and Urinal Compartments: The floor of every public water closet or urinal compartment shall be of concrete or other non-absorbent material and may be arranged to drain into any floor type urinal waste opening.

4808.15 Exposed Lavatory in Food Establishments: Wherever a water closet or urinal is installed to afford toilet facilities for a bakery or other food dispenser or establishment there shall also be provided a lavatory so located as to be in plain sight of the proprietor or person in charge of the establishment.

4808.16 Urinals: All new urinal fixtures, troughs and gutters shall be constructed of materials impervious to moisture and that will not corrode under action of urine, such as vitreous earthenware, vitreous enameled cast iron, hard slate, glass or nonferrous metal other than copper. Metal lined wooden urinals shall not be installed or maintained except in premises where sewer connection and water supply are not accessible.

4808.17 Bath Tubs, Sinks and Laundry Tubs: All bath tubs, sinks and laundry trays shall be made of vitreous chinaware, earthenware, metal or other impervious material. Wooden trays or sinks with metallic lining shall be permitted only for bar sinks and soda fountain sinks. Wooden sinks or trays may be used for special purposes and then only by special written permission by the President of the Board of Health.

4808.18 Bubblers and Drinking Devices: All drinking fountains shall be made of earthenware, vitreous chinaware, enameled iron or other impervious material. The nozzle shall be of non-oxidizing impervious material of the angle stream type. The jet or orifice shall be higher than the rim of the waste paper receiving bowl, and shall be protected with an approved mouth guard.

ARTICLE 4809.

Joints, Connections and Fittings.

4809.01 Water and Gas Tight Joints: Every joint and connection mentioned under this Article shall be made permanently gas tight and water tight.

4809.02 Prohibited Joints and Connections: Any fitting or connection which forms an enlargment chamber or recess with a ledge, shoulder or reduction of the pipe area in the direction of the flow on the outlet or drain side of any fitting or trap is prohibited.

4809.03 Vitrified Pipe:

(a) **General.** Every joint between vitrified clay pipes and between vitrified clay pipes and metal pipes shall be either a cement joint or hot poured or bituminous joint made as follows:

(b) **Cement Joints.** Plastic mortar shall be spread in the bottom one-third (⅓) of the bell of the first pipe laid. A closely twisted gasket of hemp or jute of proper thickness and of sufficient length to pass around the pipe and lap at the top shall be neatly squeezed into the aforesaid mortar. The spigot of the next pipe shall then be entered and pressed home. After the pipe has been bedded to line and grade, the gasket shall be calked into the annular space and the space shall be filled with plastic mortar bevelled against the outside of the pipe. Mortar for pipe joints shall be made of one (1) part Portland Cement and one (1) part of fine sand by volume.

(c) **Hot-Poured or Bituminous Joints.** Joints shall be carefully centered and calked with a gasket of jute or untreated calking yarn. The depth of the calking shall be sufficient to leave a space of at least one (1) inch for pipes of twelve (12) inch and less diameters. The depth shall be measured from the end of the bell. A suitable runner shall then be placed and the bituminous compound, heated to flowing consistency, shall be poured in such a manner that the remaining annular space shall be filled. Where bituminous joints are used, it shall be permissible to make alternate joints before lowering the pipe into the trench.

4809.04 Calked Joints for Metal Pipe: All calked joints shall be firmly packed with oakum or hemp, and shall be secured only with hot poured pure lead, not less than one (1) inch deep, well calked. No paint, varnish or putty or other material shall be applied until after the joint is tested.

4809.05 Screwed Joints: Every screwed joint for iron and steel, brass or copper shall be American Standard tapered screw thread and all burrs or cuttings shall be removed.

4809.06 Cast Iron Pipe Joints in Buildings: Joints in cast iron pipe shall be either calked, screwed or flanged welded or brazed joints.

4809.07 Wrought Iron, Steel or Brass to Cast Iron: The joints between a wrought iron, steel, copper or brass pipe and a cast iron pipe shall be either welded, brazed, screwed, flanged or calked joints. A calking ferrule shall be used where necessary.

4809.08 Lead Waste Pipe Connections: Every joint in a lead pipe or between a lead pipe and a brass or copper pipe, ferrule, soldering nipple, bushing or trap, shall be a burned or full-wiped joint with an exposed surface of the solder extending to each side from the end of the pipe not less than three-fourths (¾) of an inch and with a minimum thickness at the center of not less than three-eighths (⅜) of an inch.

4809.09 Lead to Cast Iron, Steel or Wrought Iron: The joint between lead pipe and iron, steel or wrought iron pipe or fittings shall be made by means of a calking ferrule, soldering nipple or soldering bushing.

4809.10 Slip Joints and Unions: Slip joints shall be permitted only in trap seals or on the inlet side of the trap. Unions on the sewer side of the trap shall be ground faced.

4809.11 Vertical Expansion: In buildings more than one hundred fifty (150) feet in height, all vertical lines of piping shall be provided with swing sections or traverse joints or other devices which shall absorb the strains or stresses due to the expansion and contraction or vibration of the vertical pipe lines.

4809.12 Welded Joints: Joints and connections of waste pipes and water supply pipes of brass, copper, black steel, black wrought iron lead or combinations thereof may be made by the welding, brazing or solder sweating process. Welding may be by the electric arc or oxy-acetylene process conforming to the requirements of the American Bureau of Welding Specifications.

4809.13 Earthenware Trap Connections. Every water closet trap or other earthenware trap shall be connected to a waste or soil pipe by inserting into the base of the fixture a brass or iron flange not less than three-sixteenths (3/16) of an inch thick which shall be screwed, welded or calked to an iron or brass bend or fitting, or shall be wiped or soldered to a lead bend. The flange shall be securely attached to the water closet or other earthenware fixture. The joint between the fixture trap and the mental flange shall be made tight by seating the trap flange on a gasket of soft metal or other durable material. All such flanges and gaskets shall be installed and tested with the soil, waste and vent pipes. No rubber, leather, putty, plaster, or other plastic compound shall be applied to make the joints tight.

4809.14 Closet, Pedestal Urinal and Trap Standard Slop Sink Floor Connections: A brass floor connection shall be wiped or soldered to lead pipe, an iron floor connection shall be calked to cast iron pipe or an iron floor connection calked or screwed to wrought iron or steel pipe and the floor connection bolted to an earthenware trap flange. A metal to earthenware, a metal to metal union, or a gasket of soft metal or other imperishable material shall be used to make the joint tight.

4809.15 Piping and Fittings:

(a) **Pipe Supports.** Vertical lines of soil, waste, vent or downspout pipes shall be adequately supported.

(b) **Change of Direction.** All changes in direction shall be made by the appropriate use of forty-five (45) degree Ys, half Ys, long sweep quarter bends, sixth, eighth, or sixteenth bends, except that single sanitary tees may be used on vertical stacks and short quarter bends may be used in soil and waste lines where the change in direction of flow is from the horizontal to the vertical. Straight tees and crosses may be used only in vent pipes.

(c) **Prohibited Fittings.** No double hub, double T or double sanitary T branch shall be used on soil or waste lines. The drilling and tapping of house drains, soil, waste or vent pipes and the use of saddle hubs and bands are prohibited.

(d) **Protection of Material.** All pipes passing under or through walls shall be protected from breakage. All metal pipes passing through or in contact with cinder concrete or other corrosive material shall be protected against external corrosion.

(e) **Workmanship.** Workmanship shall be of such character as to secure the results sought to be obtained in all of the sections of this ordinance.

ARTICLE 4810.

Water Supply and Distribution.

4810.01 Water Systems. Water piping taking supply from the mains of the Chicago Water Works System and all equipment in connection therewith shall be provided in accordance with the requirements of the separate chapter of this ordinance pertaining to that supply. All water piping taking supply from any other source and all equipment in connection therewith shall be provided with independent systems and in such a manner that there will be no cross-connection with any system supplied from the city mains. Such independent systems shall comply with all requirements of a system supplied from the city mains insofar as such requirements are applicable thereto.

ARTICLE 4811.

Inspection-Test of Plumbing and Drainage Systems.

4811.01 General:

(a) **General.** The entire plumbing system, when roughed in, in any building, shall be tested by the plumber in the presence of the plumbing inspector and as directed by him, under either a water pressure or air pressure.

(b) **Method of Inspection-Test.** The water pressure test for plumbing shall be applied by closing the lower end of the vertical pipes and filling the pipes to the highest opening above the roof with water. The air pressure test for plumbing shall be applied with a force pump and mercury column equal to ten (10) inches of mercury. The use of spring gauges is prohibited. Special provision shall be made to include all joints and connections to the finished line or face of floors or side walls, so that all vents or revents, including lead work, may be tested with the main stacks. All pipes shall remain uncovered in every part until they have successfully passed the test. After the completion of the work, and when fixtures are installed, either a smoke test under a pressure of one (1) inch water column shall be made of the system, including all vent and revent pipes, in the presence of the plumbing inspector and as directed by him, or a peppermint test made by using five (5) fluid ounces of oil of peppermint for each line up to five (5) stories and basement in height, and for each additional five (5) stories, or fraction thereof, one (1) additional ounce of peppermint shall be provided for each line.

(c) **Correction of Defects.** All defective pipes and fittings or fixtures shall be removed and all defective work shall be made good so as to conform to the provisions of this chapter.

(d) **Test of Drainage System.** The tile drainage system inside any building shall be tested by the drainage layer or sewer builder in the presence of the house drain inspector, by closing up the end of the drains two (2) feet outside the building and filling the pipes inside the building with water to a height of at least two (2) feet above the highest point of the tile drainage system.

(e) **Test After Alterations.** In the case of an extension or alteration of any existing plumbing system, if new stacks are run, it shall be tested when roughed in and when completed, as hereinbefore provided. In other alteration work, a peppermint test, and only this test, shall be applied, by using five (5) fluid ounces of oil of peppermint for each line up to five (5) stories and basement in height and for each additional five (5) stories or fraction thereof one (1) additional ounce of peppermint shall be provided for each line.

ARTICLE 4812.

Protection Against Freezing.

4812.01 General: Protection for plumbing pipes and fixtures. In all new plumbing installation concealed water pipes, storage tanks, flushing devices or systems and all pipes or tanks exposed to low temperatures shall be protected from freezing.

CHAPTER 49.

Water Supply and Distribution Systems.

4901.01 Administration by the Department of Public Works: The provisions of this ordinance for the installation in any building or structure in the City of Chicago of any water pipe or pipes or systems of water piping which receives or is intended to receive, its service from the Chicago Water Works system shall be administered and enforced by the Department of Public Works an executive department of the City Government established by ordinance.

4901.02 Inspection of Water Supply by the Department of Public Works:

(a) **General.** It shall be the duty of the Commissioner of Public Works to inspect the installation of, extension to, or any alterations in all water service, water supply and/or water distribution piping system in all new buildings, structures and premises having service from the Chicago Water Works system.

(b) **Notice to Make Alterations or Repairs.** The officers and employees of the Department of Public Works or anyone authorized to act for it, shall have free entry and access to any new building, structure or premises or part thereof, whether completed or in the process of erection, for the purpose of determining whether the provisions of this ordinance are being observed, and wherever it is found that such installation, extension or alteration does not conform, it shall be his duty to serve written notice on owner, occupant, agent, or person in possession of such building, structure or premises of such non-compliance to make such alterations or repairs as are necessary to eliminate the cause or causes of non-conformity, and in case of failure to do so within ten (10) days from the date of such notice, the Commissioner of Public Works may cause the water supply from the Chicago Water Works System to be shut off until the requirements of this ordinance are complied with. Nothing herein contained shall be deemed to apply to any building, structure or premises existing as of the date of the adoption of this ordinance, unless the plumbing or water distribution systems in said existing buildings are changed in a degree exceeding "minor repairs" as defined in Chapter 48.

4901.03 Authority to Enter Premises: The officers of the Department of Public Works and any and every person delegated or authorized by the Commissioner of Public Works shall have free entry and access to every part of any building, structure or premises whenever such entry or access is deemed necessary or advisable. Wherever any person, in possession, charge or control of any such building, structure or premises, into which any such officer or person shall desire entry or access shall refuse to permit such entry or access, or shall do or cause to be done, any act or thing for the purpose of preventing such entry or access, the Commissioner of Public Works may turn off the water service from said building, structure or premises, until notice shall have been given the Commissioner in writing, that entry or access will be permitted or provided and until such entry or access shall have been accomplished.

4901.04 Premises: Definition of the word "premises" wherever used in this article shall be held to include a lot, or part of a lot, a building, or part of a building or any parcel or tract of land whatever.

4901.05 Water Service From Chicago Water Works System; Commissioner of Public Works to Enforce: Every person who shall construct, enter, alter or use any part of the Chicago Water Works System and every consumer of water and owner, occupant or person in possession, charge or control of any building, structure or premises having service therefrom, shall be governed and be subject to the provisions of this article and also such other rules and regulations governing the use of water as may from time to time be adopted and approved by the City Council. It is hereby made the duty of the Commissioner of Public Works to enforce the provisions of this article and also any rules and regulations that may be adopted as aforesaid.

4901.06 Permits to Install Water Supply Systems in Buildings:

(a) **Application.** No person, firm or corporation shall install in any building or structure in the City of Chicago, any pipe or pipes or system of piping which receives its service from the Chicago Water Works System, nor shall make any alterations in, or additions or extensions, to, any existing pipe or system of piping in any building or structure which was erected previous to the passage of this ordinance, until such person, firm or corporation, shall have made application to the Department of Public Works for permission for such installation, alteration, addition or extension; provided however, that wherever such installation or construction work is done, wherever an emergency exists for the purpose of preventing the loss or damage to property, such application may be dispensed with.

(b) **Application in Writing.** All applications for permits for the installation in any building or structure, of water supply or water distribution pipes, or system of piping, shall be in writing upon printed forms furnished by the Department of Public Works.

(c) **Service Shut Off Until Permit Has Been Issued.** The Commissioner of Public Works may withhold or shut off service from any building, structure or premises, or to any portion thereof, in which shall be found any plumbing work, fixture or any apparatus which has not been installed in accordance with the provisions of this article, until such plumbing work, fixture or apparatus has been disconnected or until a proper permit has been issued.

4901.07 Permits Issued to Licensed Plumbers Only: No permit shall be issued by the Department of Public Works to any person, firm or corporation, for the installation of any service or supply pipe in the streets or alleys, or other public places in the City of Chicago; or for the alteration extension, installation or repair of any service pipe or any system of water supply piping in connection with any plumbing system in any building, structure or premises where such pipe or system of piping is connected, or intended for connection to the pipes of the Chicago Water Works system unless such person, firm or corporation be duly licensed and bonded by the City of Chicago as a qualified Master Plumber.

4901.08 Plans and Drawings to Accompany Application for Permit: No permit for the installation of any pipe or pipes, or system of piping, taking water from the Chicago Water Works System for distribution in any building or structure shall be granted until plan of such water supply or distribution system has been examined and approved by the Department of Public Works. Such plan or plans shall be presented with the application for permit to the Department of Public Works in duplicate, and shall clearly show the complete water supply piping system from the service to the plumbing fixtures and other appliances to which such water supply piping system is connected, together with detail drawings of connections to surge tanks, storage tanks, pressure tanks, filter, swimming pools, bathing and display pools, sterilizers, condensers, compressors, reservoirs and washers. Said plans shall be drawn to a scale of not less than one-eighth (⅛) inch to the foot; provided however, that detail plans shall be drawn to a scale and may be presented in either horizontal or vertical plans or isometric form.

4901.09 Permits—Master Plumber: No Master Plumber shall install any plumbing work in any building, structure or premises with the intention of connecting such plumbing work with pipes, mains or conduits of the Chicago Water Works System or make any alteration of, or addition to any pipe, valve, fixture or apparatus already installed, unless such Master Plumber has first obtained permit for the doing of such work from the Commissioner of Public Works.

4901.10 Permits for Opening Paved Streets: No permit shall be granted for the opening of any paved street for the tapping of mains, or laying of service pipe when the ground is frozen to a depth of twelve (12) inches or more; except when in the opinion of the Commissioner of Public Works there exists sufficient emergency to justify it. All of the provisions and restrictions of this ordinance relating to the opening of streets, alleys and public ways and doing of underground work therein as set forth, shall be complied with by every plumber performing any such work.

4901.11 High Pressure Steam Boilers—Direct Connection Prohibited: All persons are prohibited from connecting pipes whereby high pressure steam boilers may be supplied with water direct from City water mains.

4901.11½ Reserve Water Supply—Interruption of Distributing System: Wherever a continuous supply of water is deemed indispensable by owner or occupant in any building, structure or premises the owner or occupant shall provide a tank or other receptacle of capacity which he deems sufficient to supply the needs of such buildings, structures or premises during the period that the pipe section to which the service pipe is connected is shut off for repairs, connections, extensions or testing purposes; or else provide for the water supply through an auxiliary or emergency connection and service pipe with separate meter control taken from distribution system so that each service pipe shall have an independent source of supply so arranged as to insure a continuous supply of water in case of such contingency.

4901.12 Notification of Wrecking of Buildings So Water May Be Shut Off: No building, structure or premises shall be permanently abandoned, wrecked or destroyed without first giving notification, in writing, to the Commissioner of Public Works of such abandonment, wrecking or destroying, in order that the water service may be shut off and leaking or wasting water shall be eliminated or prevented. Such notification shall be given by the person or persons in charge of the wrecking or destroying of the building or by owner of the building, structure or premises.

4901.13 Notification of Construction Work Being Started: No foundations, caissons, walls, ramps or abutments for buildings, structures or premises of any kind, shall be constructed without notification first being given by the person or persons in charge of such construction to the Commissioner of Public Works, in writing, so that the proper inspection may be made to guard against damage of any kind to the Chicago Water Works system, mains, services and appurtenances which might result from settlement, vibration or other causes before, during or after such construction.

4901.14 Metered Service from Chicago Water Works System: No building, structure or premises shall be allowed to have service from the Chicago Water Works system until an application, in writing, shall first have been made to the Commissioner of Public Works by the person desiring such service, and until said Commissioner shall have given permission. Where application shall have been made for permission for connection at the main for water service for any projected building or structure, or for any alteration or addition in any building or structure, which from the information given in such application, would appear to come under any of the provisions of any ordinance which requires such building, structure or premises to have the water service controlled by meter, said permit shall be authority to install metered service only.

4901.15 Connection of Non-Metered Service to Metered Service Forbidden: Where the ordinances of the City of Chicago require the water supply for any building, structure or premises, or any part thereof, to be under meter control, it shall be unlawful for any person to connect any building, structure or premises, or any part thereof, or addition thereto, with any service or supply pipe other than a service or supply controlled by meter.

4901.16 Location of Water Meter: Whenever the law or ordinances of the City of Chicago or the rules or regulations of the Department of Public Works provide for the installation of a water meter on any service pipe supplying water from the Chicago Water Works system to any building, structure or premises, such meter shall be considered a part of such service pipe. The location of such meter, either inside or outside of any building or structure, shall be determined by the Commissioner of Public Works. The installation of any water meter at any location other than that determined by the Commissioner of Public Works is prohibited.

4901.17 Meter Vaults — Construction Thereof; Wherever a water meter is installed in the ground, either inside or outside of any building or structure, on public or privately owned property, it shall be inclosed in a meter vault which shall be built of hard pressed common brick or Portland cement concrete blocks, laid up in Portland cement mortar, poured Portland cement concrete, extra heavy salt-glazed vitrified clay tile pipe, Portland cement concrete pipe, or other equally durable material. Such meter vault shall be provided with a cast iron cover frame and removable cast iron cover. Each meter vault shall be built to conform to the specifications and dimensions for meter vaults on file at the offices of the Department of Public Works.

4901.18 Service Pipes of Adequate Size: The service pipe is the pipe which conveys the water from the mains of the Chicago Water Works system to the building, structure or premises served. Each service pipe shall be of sufficient size to permit the continuous and ample flow of water to supply adequately all floors at any given time.

4901.19 Water Mains—Definition: Water mains are the pipes through which Lake Michigan water from the Chicago Water Works System pumping stations is distributed to any and all service pipes, fire hydrants, fire cisterns and sprinkling systems.

4901.20 Permit to Install Service Pipes:
(a) **Application.** No water mains shall be tapped nor shall any service be installed, nor shall any connection be made with any service or supply pipe which is a part of the Chicago Water Works System, until an application, in writing, shall have been made to the Commissioner of Public Works, and until said Commissioner has given his permission.

(b) **Application to Contain Description.** Such application shall set forth a true legal description of the premises which it is proposed to serve, the name and address of the person, firm or corporation about to perform the work and the size of the tap desired. It shall set forth fully, the kind of building for which such service is intended and the applicant shall be required to answer all questions regarding such application which may be put to him by any officer or authorized employee of the Department of Public Works.

4901.21 Taps Installed by Department of Public Works: No water main shall be tapped other than by a tapper employed by the Department of Public Works and all tapping shall be performed only under the authority of the Commissioner of Public Works. All service cocks or ferrules must be inserted at or near the top of the street main, and not nearer than six (6) inches from the bell of the pipe. The size of the cock shall be that specified in the permit. Each service pipe shall have its own independent tap at the main and said tap shall be of the type in use by the Department of Public Works.

4901.22 Minimum Size of Service Pipe; No service pipe of an internal diameter less than one (1) inch shall be installed in any street, alley or other public place of the City of Chicago, nor connected to the mains of the Chicago Water Works System.

4901.23 Minimum Weights of Lead Service Pipes: Service pipes of one (1) inch internal diameter, one and one-half (1½) and two (2) inch internal diameter shall be lead pipe of the following minimum weights:

TABLE 4901.23 LEAD PIPE SIZE.

Internal Diameter	Minimum Weight Per Lineal Foot	
(Inches)	Pounds	Ounces
1	4	0
1½	8	0
2	13	12

4901.24 Joints in Lead Service Pipes: All joints in lead service pipes shall be of the wiped solder type with an exposed surface of the solder to each side of the joint of not less than three-fourths (¾) inch and a minimum thickness at the thickest part of the joint of three-eighths (⅜) inch.

4901.25 Cast Iron Service Pipes: Service pipes larger than two (2) inch internal diameter shall be American Water Works Association, Class B, cast iron pipe of the bell and spigot type and shall be coated internally and externally with asphaltum or coal tar pitch and shall conform to the City of Chicago specifications for cast iron water pipe and for special castings.

4901.26 Joints in Cast Iron Service Pipes: All joints in cast iron service pipes shall be of the lead and gasket calked type. Each joint shall be firmly packed not more than one-fourth (¼) of its depth with hand picked oakum or hemp, flushed filled with poured molten pure lead and firmly calked to withstand hydrostatic pressure of one hundred (100) pounds per square inch.

4901.27 Increased Service Pipes — Cross Connections: In any building, structure or premises where the consumption of water requires a larger service than can be supplied by the existing tap or connection, a new tap or connection shall be made with the City water main to correspond to the size of the service desired. The joining of two (2) or more smaller service pipes for the purpose of connection to one (1) larger service is prohibited. The size of such service pipe shall be determined by the Commissioner of Public Works and it shall be supplied from the mains of the Chicago Water Works system through a single tap of the same internal diameter as the service pipe, and shall extend from the water main to the lot line without inlet or outlet connections.

4901.28 Lead Service Pipes to be Provided With Swing Sections: No lead service pipe shall be connected to the mains of the Chicago Water Works system unless such service pipe is of sufficient length to permit the construction of a non-rigid swing section at its connection to such main. Such swing section shall be constructed in such manner as will completely absorb all strain to service pipe and main which may be caused by any shock, strain or vibration to which said service pipe or main may be subjected, and each such swing shall move on an axis composed of the tap coupling and the threaded and screwed joint at the water main and shall conform to the rules and regulations of the Department of Public Works for such construction.

4901.29 Service Pipes Required With Stop Cocks: Each service pipe shall have its own independent stop cock or valve, the location of which in relation to any pavement, curbing, sidewalk, parkway or other improvement in any street, alley or other public place shall be determined by the Department of Public Works and each stop cock shall be set exactly as so determined and shall conform to the rules and regulations for the location of stop cocks and valves on file at the office of said department. Each stop cock or valve shall have a clear waterway equal in area to that of the service pipe which it controls. It shall be of the type in use by the Department of Public Works of the City of Chicago and shall be obtained from said department.

4901.30 Stop Cocks Shall be Equipped With Shut Off Boxes: Each stop cock shall be provided with a shut-off rod box of a type which shall conform to the drawings and specifications on file at the office of the Department of Public Works and shall be obtained from said department.

4901.31 Service in Unimproved Streets—Extension Thereof: Wherever any service pipe is installed in any street or alley or other public place previous to the permanent improvement thereof, such service pipe shall be extended not less than one (1) foot beyond the location of the stop cock or valve which controls such service pipe and in the direction of the lot line or building line of such street, alley or other public place. Wherever an extension into any building or structure is made to any existing service pipe which terminates in any street, alley or other public place, such extension shall be constructed of the same materials, shall have the same internal diameter and shall be laid at the same distance below the surface of the ground or pavement as is provided in this article for the installation of service pipes.

4901.32 Service Pipes, Individual Trenches and Depth and Back-Filling Thereof: All service pipes in streets or alleys or other places, exposed to the weather, shall be installed at a depth of not less than five (5) feet below the surface of the ground or pavement and at right angles to the main to which they are connected. Each service pipe shall be installed in a separate individual trench, such trench shall be not less than six (6) feet distant from any other trench or excavation of greater depth than that in which service pipe is laid, except in streets or alleys where permanent pavement exists. Where permanent pavement exists service may be laid on solid earthen shelf not less than one (1) foot from trench in which house sewer or other parallel conduit is installed; provided however, that service pipe may cross at right angles any trench which may be excavated on a line parallel with the center line of the street or alley in which service pipe is installed. No ashes, cinders or refuse shall be used in back-filling any trench or excavation in which service pipes are installed. Each service pipe shall be covered with not less than eighteen (18) inches of bank sand or coated with coal tar or asphaltum paint.

4901.33 Service Shall Be Exposed for Inspection: Each service pipe shall be left exposed to view for its entire length until after inspection of such installation has been made by the Department of Public Works. No concealment of such service pipe by the filling of the trench in which it has been laid or the application of any other type of covering shall be permitted until permission for such concealment of covering is granted by the Department of Public Works.

4901.34 Connection of Chicago Water Works System to Another Water System Prohibited: Wherever a system of water supply piping is installed either inside or outside of any building or structure, or in any street or alley, or other public or private property, which receives its supply from any well, cistern, river, pond, lake or any other source except the Chicago Water Works System, such system shall be kept entirely separated from and no connections of any kind, either direct or indirect, shall be made with any pipe or system of piping which receives its supply from the Chicago Water Works system.

4901.35 Direct Connection of Chicago Water Works System to Various Fixtures and Appliances Prohibited: No pipe, or system of piping, which receives its supply from the Chicago Water Works System, shall be directly connected to any processing tank, vat, mixer, heater, cooker, washer, pump appliance or equipment used for storing, holding or conveying fluids or materials, or for manufacturing or food processing, or washing purposes. Such appliances and equipment shall be supplied from the Chicago Water Works System through an open funnel connection or an open tank which shall be located not less than six (6) inches above the overflow rim of such container, appliance or equipment. No pipe or system of piping in any building, structure or premises, which receives its supply from the Chicago Water Works System shall be directly connected to any device, appliance or apparatus in which such water supply is used to provide power through a water jet or other device to create vacuum or partial vacuum with which to operate any aspirator, syphon, cellar drainer, ejector, cleaner, sweeper, conveyor or washer of any kind or description.

4901.36 Secondary Water—Its Definition: Secondary water is any water from a private water system of pipes or piping which receives its water supply from the Chicago or Calumet Rivers or their tributaries, shore water from Lake Michigan or Calumet Lake or from any well or cistern or any ground water or rain water reservoir. Secondary water is also water from the mains of the Chicago Water Works System which has been used for any purpose within any building, structure or premises, or which has been discharged from any type of condenser coils, or cooling systems, drinking fountains, hydraulic lifts, boilers, linotype machines, dye casting machines, metal rolling or pressing machines or other crushing or rolling machines or apparatus, or which has been stored in such a manner as to expose it to possible contamination.

4901.37 Purposes for Which Use of Secondary Water is Prohibited: No secondary water shall overflow into or be discharged into any surge tank, storage tank or reservoir, or shall in any way be piped or conveyed into the water supply system of any building, structure or premises to become a part of or be mixed with the fresh water supply from the mains of the Chicago Water Works System either inside of the premises or in the water service pipe. Secondary water shall not be piped to or used for cooling crushers, rollers, or mixers where foods, candies, liquids or materials are manufactured for human or animal consumption. No connection, tap or opening shall be made in a water distribution system other than an approved water distribution system which will permit such water being used for drinking. Wherever the fire protective equipment in any building, structure or premises has service from the Chicago Water Works System, no pipe or other conduit which conveys secondary water shall be cross-connected to the fire protective equipment. All fire protective equipment connected to the Chicago Water Works system shall be constructed in such manner that all tanks, pipes, pumps, surge tanks and fire hydrants can be thoroughly drained, flushed and cleaned by the owners of such equipment and premises and there shall be no direct connections from the tanks, pipes and other equipment to any drainage pipes or sewer.

4901.38 Fire Extinguishing Equipment Shall be Cleaned: All fire protective equipment that is supplied with water from the Chicago Water Works system shall be drained and flushed at least every six (6) months and kept free from accumulations of sand and silt and stagnant water which would nullify the action of the chlorine content of city water.

4901.39 Fire Extinguishing Equipment Cross Connection Prohibited: No city water pipe lines in any building, premises, material or storage yard on railroad property which receives service from the Chicago Water Works system shall have for fire extinguishing purposes a siamese or other connection that is or has been installed near any river or lake waterway where a city fire boat or city fire engine or pump may pump river or lake shore water into the city water pipes through a cross-connection of any kind. A check valve or any other kind or type of valve on any cross-connection is prohibited. No stationary pump or privately owned fire equipment shall be maintained for emergency use as described above in any premises adjacent to the rivers or lakes, nor shall they use cisterns or wells adjacent to the rivers and lakes for cross-connections to city water supplied pipes for emergency fire-extinguishing purposes.

4901.40 Surge Tanks for Fire Protective Equipment: Wherever in any buildings or structures or premises, water is drawn through a service pipe three (3) inches or larger in diameter from city water mains to supply any pump or other appliance which may be used to increase water pressure for fire protection purposes an open surge tank or reservoir shall be provided.

4901.41 Surge Tank for Pumps Where Installed—Construction and Connections:

(a) **For Supply from the Chicago Water Works System.** Wherever in any building, structure or premises, water is drawn from the city mains to supply any pump or other appliance which may be used to increase the water pressure for any purpose, a surge tank shall be provided for the purpose of preventing the water from flowing back into the city mains. Such tank shall receive the supply of water direct from the city mains and shall serve as a reservoir from which the pump or other appliance shall draw the supply of water. The tank shall have a dust-tight cover and shall be open to the atmosphere through an overflow pipe or vent pipe not less than one (1) pipe size larger than the water supply pipe and shall be connected to the city water main through one (1) or more slow-acting automatic controlling cock, valve or other device. The inlet pipe to the tank shall be so installed that the discharge opening cannot become submerged. No automatic device shall be used which has not been approved by the Commissioner of Public Works.

(b) **For Secondary Water Supply.** Wherever in any building, structure or premises a secondary water system is maintained a similar surge tank shall be provided to receive and store the supply which shall serve as a reservoir for the system. There shall be no connection to any system supplied with water from the mains of the Chicago Water Works system which might cause contamination.

4901.42 Surge Tanks Shall be Provided With Sludge Drain Pipe: Each surge tank shall be provided with a sludge drain pipe which shall be connected thereto at the lowest level of the bottom section thereof and shall not extend above such level. Such drain shall be provided with a control valve, with a clear waterway equal in area to that of the sludge drain pipe and shall discharge through an indirect connection into the drainage system of the building.

4901.43 Water Supply Pipe Connections to Gravity Storage Tank: Wherever any pipe which is a part of the water supply system of any building or structure, receives its service from a gravity storage tank, it shall be connected thereto in such a manner as will prevent any water stored within four (4) inches of the bottom of such tank from entering such water supply pipe. Every gravity storage tank shall be provided with an overflow pipe of a cross area of not less than twice that of the pipe which supplies water to such tank. Said overflow pipe shall discharge through an open connection into the drainage system. Every gravity storage tank shall be supplied with water through an approved slow-acting automatic control valve or control cock, the discharge outlet of which shall be installed at a distance not less than six (6) inches above the top of the overflow pipe connection to such gravity storage tank.

4901.44 Gravity Storage Tank Shall Have Sludge Drain Pipe: Each gravity storage tank shall be provided with a sludge drain pipe which shall be connected thereto at the lowest level of the bottom section thereof, and shall not extend above such level. Such drain pipe shall be provided with a control valve, with a clear waterway equal in area to that of the sludge drain pipe and shall discharge through an indirect connection into the drainage system of the building.

4901.45 Gravity Storage Tank Location and Protection Thereof: Every house storage tank shall be so located and shall be housed in a manner which shall prevent contamination of the water therein. Such tank shall be protected with a cover sufficiently tight to exclude dust. Approval of the proposed support for a new tank shall be obtained from the Department of Buildings before the permit for its installation shall be issued.

4901.46 Compression Devices for Increasing Water Pressure:

(a) **Where Required.** Wherever in any building, the service pipe supplying water thereto is two (2) inches or less in internal diameter and the pressure of the water supplied through such service is insufficient to furnish an adequate supply, an automatic compression pressure increasing device or high level gravity storage tank may be installed. Each such compression device shall consist of a water pump and compression tank, and shall be provided with all the regulating and control appurtenances and devices which are necessary to insure complete automatic operation.

(b) **Construction and Connections.** Each compression tank shall be so constructed and connected that the air content thereof shall be not less than twenty-five (25) per cent of its total capacity. Each compression tank shall be provided with a sludge drain pipe which shall be connected to such tank at the lowest level of the bottom section thereof and shall discharge through an open connection into the drainage system of the building. Each sludge drain pipe shall be provided with a control valve with a clear waterway equal in area to that of the sludge drain pipe. No water shall enter into any water supply pipe at a lower level than four (4) inches above the bottom of such compression tank.

4901.47 Plumbing Fixtures Shall Be Flushed: All plumbing fixtures shall be provided with a sufficient supply of water for flushing to maintain them in a sanitary condition. Every water closet or other plumbing fixture or appliance designed to be cleansed by flushing with water shall, at each flush, be supplied with a sufficient quantity of water to remove quickly all waste matter and properly cleanse the interior surfaces exposed to atmosphere. Every water closet or urinal shall be flushed by means of an approved flushing tank, except as hereinafter provided. Each water closet flushing tank shall have a flushing capacity of not less than four (4) gallons and each urinal flushing tank a capacity of not less than two (2) gallons and shall be adjusted to prevent waste of water by excess flushing. No water closet flushing pipe shall be less than one and one-fourth (1¼) inches internal diameter and no water from flushing tanks shall be used for any other purpose than that herein described. Water closets and urinals may be flushed by means of an approved flushing valve.

4901.49 Submerged Water Supply Inlets Prohibited: No plumbing fixture shall be installed unless the water supply enters said fixture at least two (2) inches above any overflow connections. All submerged water supply inlet connections are hereby prohibited, and no tanks, vats, utensils or other water supply devices used for other than drinking purposes and having submerged water supply inlets shall be directly connected to city water supply; except any water heating device or heated water storage tank or low pressure boiler.

4901.50 Definitions of the Names Used in this Ordinance for Various Water Supply Pipes:

(a) **Main Supply Pipe.** A main supply pipe is one which is connected to the service pipe of any building, structure or premises and conveys the water therefrom to the principal supply pipe.

(b) **Principal Supply Pipes.** The water supply arteries in buildings and structures. They are connected to the main water supply pipe and convey the water therefrom to pumps, tanks, filters, heaters and other equipment together with all their appurtenances and to the branch supply pipes.

(c) **Branch Supply Pipe.** A pipe which is connected to a principal supply pipe and conveys the water therefrom to the riser pipe or distributing pipe.

(d) **Distributing Pipe.** A pipe which is connected to a riser pipe or branch supply pipe and conveys the water therefrom to the branch distributing pipe.

(e) **Riser Pipe.** That pipe which is installed perpendicular to the horizontal through the floors, stories and other open spaces of buildings and structures and conveys the water from the main or branch supply pipes to the distributing pipes or branch distributing pipes.

(f) **Branch Distributing Pipe.** A pipe which is connected to a distributing pipe or riser pipe and conveys the water therefrom to the plumbing fixture.

4901.51 Size of Various Water Supply Pipes: Each main supply pipe, principal supply pipe, branch supply pipe, riser pipe, distributing pipe or branch distributing pipe shall be of a size which under normal pressure shall deliver a full volume of water to each and all of its outlets.

4901.52 Minimum Sizes of Branch Distributing Pipes for Plumbinr Fixtures and Appliances: The minimum internal diameter of any branch distributing pipe which supplies water and is directly connected to any valve, cock, bibb, spigot, faucet or other device which is attached to and becomes a part of any plumbing fixture or appliance shall be as shown in Table 4901.52.

TABLE 4901.52.

Device or Appliance	Size in Inches
Lavatory or Basin Cocks	⅜
Water closet and urinal flushing tank ball cocks	⅜
Drinking Fountain Jet Valves	⅜
Domestic kitchen sink faucets	½
Bath tub cocks or valves	½
Shower bath cocks or valves	½
Domestic laundry tub faucets	½
Bar sink faucets	½
Scullery sink faucets	¾
Slop sink faucets	¾
Water closet flushing or flushometer valve	1
Urinal Flushing or flushometer valves	1
Sill cocks or sprinkling valves	¾

The maximum length of such branch pipes shall be:

⅜-inch pipe.......... 2 feet
½-inch pipe..........10 feet
¾-inch pipe..........30 feet

All water supply pipes and control valves shall be of sufficient size and capacity to supply water to all fixtures. All water supply pipes, riser pipes and distributing pipes shall be graduated as to size and shall be interconnected in such manner that a full volume of water may be discharged into twenty (20) per cent of the plumbing fixtures of any building when operated at any given time without causing loss of more than ten (10) pounds pressure at the plumbing fixtures which are located on upper floor of such building for a length of time not less than sixty (60) minutes. In the graduation of the sizes of such pipes and their connections the following table shall be observed; provided however, that no riser pipe of an internal diameter less than three-fourths (¾) of an inch and no branch distributing pipe of three-eighths (⅜) inches internal diameter of a length greater than two (2) feet shall be installed:

Mult.	Size of Pipe Maximum number of discharge openings therefrom.	⅜"	½"	¾"	1"	1¼"	1½"	2"
1.	No. of ⅜" openings	1	3	12	28	55	80	142
2.	No. of ½" openings		2	7	16	31	45	80
3.	No. of ¾" openings			2	7	13	20	35
4.	No. of 1 " openings				2	6	11	20
5.	No. of 1¼" openings					2	7	13
	Mult.	1	2	3	4	5	5	5

4901.53 Control Valves for Various Water Supply Pipes: Valves which shall control the water supply to riser pipes shall be installed in each branch supply as near as practicable to its connection with the riser pipe or branch supply pipe; provided however, that where a plumbing fixture or water supplied appliance or group of plumbing fixtures or water supplied appliances are equipped with individual control valves, the installation of valves in said distributing pipe or branch distributing pipe shall not be required. Such control valves may be located outside of the apartment, suite, store or loft which they serve.

4901.54 Control Valves: An approved control shut-off valve shall be installed in each main water supply pipe. Such valve shall be located inside of building or structure as near as practicable to the wall thereof under which or through which it enters; provided however, that where the service is a meter service, such shut-off valve shall be installed in the outlet pipe of such meter and as near as practicable thereto. Each such shut-off valve shall be accessible for operation at all times.

4901.55 Materials for Water Supply Pipes and Fittings: All water supply pipes shall be of lead, galvanized wrought iron, galvanized steel, brass copper or cast iron, with brass, copper or galvanized malleable iron fittings. No pipe or fittings that have been used for other purposes shall be used for distributing water. No steel or wrought iron pipe or fittings shall be installed in the ground. Lead water pipes, supply pipes or distributing pipes whether installed in the ground or in the open spaces of buildings or structures either publicly or privately owned shall conform to the following minimum weights.

TABLE 4901.55.

Internal Diameter (Inches)	Weight Per Lineal Foot Pounds	Ounces
½	1	8
⅝	2	8
¾	3	
1	4	
1¼	6	
1½	8	
2	13	12

Cast iron pipe and fittings shall conform in every respect to the City of Chicago specifications for cast iron water pipe and for special castings.

4901.56 Protection for Pipes and Fixtures: All new concealed water pipes, storage tanks, flushing services or systems and all pipes or tanks exposed to low temperatures shall be protected from freezing.

4901.57 Hydrant and Pump Protection: Every pump or hydrant for providing drinking water supply shall be protected from surface water and contamination.

4901.58 Water Supply Pipe Protection: All water supply pipes or distributing pipes that are installed in the ground in any street, alley or other public place or in buildings, structures or privately owned premises shall be covered to a depth of eighteen (18) inches with bank sand or coated with a coal tar paint. No ashes, cinders or refuse shall be used in back-filling any trenches or excavations in which water pipes are installed. All such pipes shall be coated with a coal tar paint and insulated where they pass through concrete floors, walls or new mason work.

4901.59 Definitions of Pipes in Heated Plumbing Water Supply Systems in Buildings:

(a) **Circulation Pipe.** One which is connected to a heater water riser pipe, distributing pipe or branch distributing pipe and conveys the water therefrom to the principal heated water return pipe or branch heated water return pipe.

(b) **Branch Heated Water Return Pipe.** A pipe which is connected to a circulation pipe and conveys the water therefrom to the principal heated water return pipe.

(c) **Principal Heated Water Return Pipe.** A pipe which receives the discharge from one (1) or more circulation pipes, or branch heated water return pipes and conveys the water therefrom to the water heater or heated water storage tank.

(d) **Relief Valve Required.** Wherever a check valve is installed on the cold water supply pipe between the street main and the hot water tank in any building, there shall be installed on the hot water tank system a pressure relief valve not smaller than one (1) pipe size less than that of the cold water supply to the tank.

4901.60 Heated Plumbing Water Pipes Shall Have Straight Lines and Uniform Grade: All heated plumbing water supply pipes, riser pipes, circulation pipes and return pipes shall be installed in a manner which shall maintain a straight line in their several directions. Any installation which permits the forming of air pockets or traps, or other conditions within the heated water supply system which would retard or prevent the free unrestricted flow of water therein is prohibited. No pipe or pipes which are a part of the heated water supply system shall be installed on an exact horizontal line. A grade of not less than one (1) inch to each ten (10) feet of such installation shall be maintained.

4901.61 Heated Plumbing Water Supply Circulating System: In every hot water plumbing supply piping system hereafter installed in any building or structure, which is also a circulating system, through which heated water which has not been discharged from the system into plumbing fixtures or other heated water supplied appliances, the water shall be returned to water heater or heated water tank with a minimum loss of heat. Where heated water is conveyed to fixtures or appliances through an upfeed riser pipe, the circulation return pipe shall be connected to such riser pipe at a point therein not greater than six (6) inches below the highest connection of such riser pipe to any distributing pipe, or branch distributing pipe and shall be extended therefrom to its connection to the branch heated water return pipe. Where heated water is conveyed to fixtures or appliances through a down-feed riser pipe, the circulation return pipe shall be connected to such riser pipe at a point below the lowest connection of such riser pipe to any distributing pipe, or branch distributing pipe and shall be extended therefrom to its connection with the branch heated water return pipe or principal water return pipe.

4901.62 Valves Shall Be Installed:

(a) **Control and Check Valve for Cold Water Supply Pipe:** An approved control valve and an approved check valve shall be installed in every cold water supply pipe which is connected to any water heating device or heated water storage tank or boiler in such manner as will prevent the heated water from entering any cold water supply pipe or system of cold water supply piping.

(b) **Pressure Relief Valve.** An approved pressure relief valve, the discharge from which shall drain through an indirect connection into a plumbing fixture or floor drain shall be installed in the principal domestic hot water supply pipe as near to its connection with such water heating device or storage tank as is practicable. Such pressure relief valve shall not be smaller than one (1) pipe size less than that of the cold water supply pipe to the tank.

(c) **Shut-off Valve.** Shut-off valves shall be installed in each branch heated plumbing water return pipe as near as is practicable to its connection with the circulation return pipe.

(d) **Control Valve for Principal Heated Plumbing Water Return Pipe.** A control valve shall be installed in each principal heated plumbing water return pipe as near as is practicable to the connection thereof, to the water heater or heated water storage tank.

4901.63 Swing Shall Be Installed: Each hot water riser pipe shall be provided with a swing section for every fifty (50) feet, or fraction thereof, of its length. Such swing section shall be located midway between the clamps which anchor such riser pipe to the structural members of the building. Each swing section shall be installed at right angle to the riser pipe, and shall be so constructed that it shall absorb the strains and stresses of expansion and contraction thereof. Each hot water distributing pipe shall be provided with a swing section which shall be located as near as practicable to its connection with the hot water riser pipe, or branch hot water supply pipe. Such swing section shall move with the contraction and expansion of the hot water riser pipe, or one (1) or more axes of threaded and screwed joints in such swing section.

4901.64 Safety of Heated Water Supply System Shall Be Assured: Every device for heating water or for storing heated water in tanks or boilers, shall be so designed and constructed and shall be installed in such manner in any building or structure, that any condition which would cause or contribute to the cause of any excess pressure, strain, stress or explosions therein, or leakage in any pipe or system of piping connected thereto, or shall permit any heated water to flow into or through the meter shall be eliminated. No water in any plumbing system shall be heated to a higher temperature than two hundred (200) degrees Fahrenheit in any water heating device, or shall be stored in any tank, boiler or reservoir, or shall be discharged into or conveyed in any heated water supply

pipe or system of heated water supply piping to any plumbing fixture.

4901.65 Drain Pipes for Heated Water Storage Tanks: Each heated water storage tank shall be provided with a sludge drain pipe which shall be connected thereto at the lowest level of the bottom section thereof and shall not extend above such level. Such drain shall be provided with a control valve with a clear waterway, equal in area to that of the sludge drain pipe and shall discharge through an indirect connection into the drainage system of the building.

4901.66 Connections Prohibited: Any water pipe connection to any pump or other appliance through which any vibration, shock, strain, stress or pulsation which may originate or develop a water hammer, or any excess noises or sounds in the water supply system of any building or structure, and which may be communicated through such connection or service pipe of the Chicago Water Works system is prohibited.

4901.67 Check Valves Shall Be Installed: Check valves which are noiseless in their operation, and air chambers of sufficient air cushion capacity, shall be installed in the suction pipe and the discharge pipe of each house pump or fire pump, or other water pressure increasing device; provided however, that where such check valves are of the slow acting type, and the operation thereof entirely eliminates the shocks, strains, stresses and excess noises caused by the operation of such pump or device, the installation of air chambers may be dispensed with. No check valve shall be installed as herein provided unless the design and construction thereof shall first have been approved by the Commissioner of Public Works.

4901.68 Air Chambers Shall Be Installed: Air chambers shall be installed at the upper terminals of all up feed riser pipes and in branch distributing pipes contiguous to, and directly above, the connection of such branch distributing pipes to the plumbing fixture or other water supplied appliance. Such air chambers shall be installed in a direct line with the flow of water through such pipes and shall be of sufficient capacity to provide an air cushion which wil absorb shock, stress or strain and eliminate all excess noises which may be caused by the operation of any valves, faucets, bibbs or cocks in the water supply system. No air chamber which is constructed of pipe shall be of a size less than the pipe which it serves and to which it is connected and shall be not less than two (2) feet in length.

4901.69 Water Supply Pipes—Distinctive Colors: Wherever in any building or structure there is a secondary water supply system, all water pipes shall be painted with distinctive colors on every fitting and for at least six (6) inches from the intersection of every pipe with any wall, floor or ceiling, as follows:

(a) City Water Supply pipes.. Blue
(b) Fire protection Lines.... Red
(c) Secondary W a t e r supply pipes and fittings...... Yellow.

4901.70 Test of Water Distribution Pipes: The entire water distribution system within buildings shall be tested, in the presence of an inspector of the Department of Public Works and proven tight under a water or air pressure of fifty (50) per cent more than the maximum pressure of such system and not less than one hundred (100) pounds.

CHAPTER 51.

Administration of Ordinances Regulating Steam Boilers, Unfired Pressure Vessels and Mechanical Refrigerating Systems.

ARTICLE 5101.

Administration.

5101.01 General: The provisions of this ordinance for the regulation of steam boilers, unfired pressure vessels, refrigerating systems and cooling plants in the City of Chicago shall be administered and enforced by the Department for the Inspection of Steam Boilers, Unfired Pressure Vessels and Cooling Plants, an executive department of the Municipal Government established by Sections 538 to 546 of the Revised Chicago Code of 1931 as amended October 21, 1931. The Department for the Inspection of Steam Boilers, Unfired Pressure Vessels and Cooling Plants, its officers and employees, or anyone authorized to act for it, shall be permitted and empowered to enter any building or structure or premises whether completed or in process of erection, for the purpose of determining whether any boilers, unfired pressure vessels or refrigerating systems are being constructed and installed in accordance with the provisions of the City Ordinance.

(a) **Exceptions.** The provisions of this chapter shall not apply to single dwellings, nor to any multiple dwelling having not to exceed three (3) apartments.

5101.02 Power to Pass on Ordinances: The Department for the Inspection of Steam Boilers, Unfired Pressure Vessels and Cooling Plants shall have full power to pass upon any questions arising under the provisions of this ordinance, or any of the general ordinances of the city, relating to the construction, installation and operation of steam boilers, unfired pressure vessels and mechanical refrigerating systems.

5101.03 Power to Make Rules: The Department for the Inspection of Steam Boilers, Unfired Pressure Vessels and Cooling Plants shall have the power to make reasonable regulations or rules interpreting or clarifying the requirements which are definitely prescribed, and shall have the further power to adopt reasonable rules to prescribe the quality and regulate the application of the materials and combination of materials not inconsistent with the provisions of this ordinance; provided however, that the standard adopted in such rules shall be consistent with the practice, custom, and usage prevailing in the various branches of industry affected. A copy of these said rules shall be published and shall be kept always on file in the office for the Department for the Inspection of Steam Boilers, Unfired Pressure Vessels and Cooling Plants. The department shall publish in a convenient form, and as a unit, the ordinance coming within its jurisdiction and the rules and regulations pertaining to steam boilers, unfired pressure vessels and refrigerating systems which have been formulated and adopted.

5101.04 Certificates and Records: When an inspection of a boiler, unfired pressure vessel or mechanical refrigerating system has been made, and the same has been approved by the Chief Inspector or Supervising Mechanical Engineer and Chief Deputy Inspector of Steam Boilers and Cooling Plants, the said inspector shall make and deliver to the person for whom the inspection was made, upon the payment of the fees hereinafter provided, a certificate of inspection together with a general description of such apparatus stating for what purpose the appliances are to be used and the pressure in pounds at which they may be safely used. This certificate shall be framed and hung in a conspicuous place in the machinery or boiler room, and a record of the same shall be made and kept by said department, indexed alphabetically or by locality.

ARTICLE 5102.

Permits, Plans, Inspections and Tests.

5102.01 Permits and Plans Required: Before any owner or agent shall proceed with the installation or alteration of any boiler or unfired pressure vessel, he shall place on file in the Department for the Inspection of Steam Boilers, Unfired Pressure Vessels and Cooling Plants, plans and specifications of the same.

Upon approval of such plans and specifications, a duplicate set of which shall be left on file in said office, and the payment of fees as hereinafter provided, said department shall issue a permit for the installation or alteration of such apparatus.

(a) **Boilers and Pressure Vessels.** It shall be unlawful for any person to use any steam boiler, or tank, or tanks, subject to pressure other than the pressure in the City water mains, until he shall first have procured a certificate from the Chief Inspector of Steam Boilers and Cooling Plants that such apparatus may be safely used, and that the boiler or boilers, boiler setting, means of producing draft, smoke connections and furnace or firebox are of such size and capacity that they will do the work required.

If such owner, agent or person using a steam boiler or tank shall fail to notify said Chief Inspector of his intention to make any alteration or enlargement of such steam plant or tank, and shall fail to file plans and specifications for the enlargement or alteration of the same, and shall proceed to make such alterations or enlargement without a permit therefor, he shall be liable to a fine of Twenty-five Dollars ($25.00) for each day on which he shall have prosecuted such alteration or enlargement without said permit, and each day's violation shall constitute a separate offense.

(b) **Refrigerating Systems.** Before any owner or agent shall proceed with the installation or alteration of any mechanical refrigerating system, he shall secure the approval of the Department for the Inspection of Steam Boilers, Unfired Pressure Vessels and Cooling Plants, and obtain a permit for the work. No refrigerating system shall be installed, erected or maintained, nor shall any reconstruction of old apparatus or old system for mechanical refrigerating or cooling purposes be undertaken, unless plans and specifications for the same shall be filed in the office of the Department for the Inspection of Steam Boilers, Unfired Pressure Vessels and Cooling Plants and approved by the Chief Inspector of said department. Such plans and specifications shall show the kind and amount of refrigerant used. Upon approval of said plans, a duplicate set of which shall be left on file in the office of the Chief Inspector, and the payment of the respective fees, as hereinafter provided, said Chief Inspector shall issue a permit for the installation of said apparatus. No permit shall be required for emergency repairs. Where water is to be used for cooling the refrigerant or for other cooling purposes no permit for the installation of a refrigerating plant shall be granted unless the plans and specifications have been submitted to the Department of Public Works and Board of Health and said departments have issued a permit for the use of water from the City water mains.

5102.02 Boiler Tests:

(a) **Hydrostatic Test.** It shall be the duty of the Chief Inspector of Steam Boilers and Cooling Plants and his deputies to inspect all boilers, tanks, jacketed kettles, generators or other apparatus used for generating or transmitting steam for power, or using steam under pressure for power, or using steam under pressure for heating or steaming purposes, and all other tanks, jacketed kettles and reservoirs under pressure of whatsoever kind, except as hereinafter provided, as often as once in each and every year, by making a hydrostatic pressure test where such test shall be deemed necessary; provided however, that the hydrostatic pressure used in such test shall not exceed the maximum working pressure of such apparatus by more than fifty (50) per cent; and by making a careful external and internal examination. In all cases where hydrostatic pressure test is used, an internal examination of such apparatus shall afterwards be made.

(b) **Drilling Test.** Any boiler, tank, jacketed kettle, generator or reservoir having been in use eight years or more and in such condition that in the opinion of the inspector the same should be drilled in order that the exact thickness and condition may be ascertained, shall be reported to the Chief Inspector of Steam Boilers and Cooling Plants, who shall serve the owner or agent with a written notice to show cause to the Chief Inspector within five (5) days why such boiler, tank, jacketed kettle, generator or reservoir should not be drilled. If, after the owner or agent has been heard, or at the end of five (5) days, the Chief Inspector deems it necessary, then such boiler, tank, jacketed kettle, generator or reservoir may be drilled at points near the water line, and at the bottom of the shell of the boiler, or at such other points in the boiler, tank, jacketed kettle, generator or reservoir as the inspecting officer may direct, and the thickness of said material shall be determined thereafter at such annual inspection as the inspecting officer may deem necessary, and the steam pressure or other pressure allowed shall be governed by such ascertained thickness and general condition of boiler, tank, jacketed kettle, generator or reservoir. The drilling and plugging of said holes shall be done at the expense of the owner.

5102.03 Manufacturers and Dealers to Notify Department—Second Hand Boilers, Unfired Pressure Vessels and Refrigerating Systems: Any person, firm or corporation manufacturing, dealing in selling or erecting boilers, unfired pressure vessels, refrigerating systems or apparatus as defined by this ordinance, shall, on the sale or delivery of any such system or apparatus, at any point or locality within the City, notify the Department for the Inspection of Steam Boilers, Unfired Pressure Vessels and Cooling Plants, giving the name of the purchaser, his street address, and the street addrss to which such system is to be delivered. Any person, firm or corporation selling a second hand or used refrigerating system, boiler or unfired pressure vessel shall before painting same, have it inspected by the Department for the Inspection of Steam Boilers, Unfired Pressure Vessels and Cooling Plants, and before offering for sale any such system, boiler or unfired pressure vessels shall have in his, their or its possession a certificate issued by said department to the effect that said system is in such condition that it can be safely used. The fee for such inspection shall be the same as provided for each classification as hereinafter provided.

5102.04 Inspection of Mechanical Refrigerating Systems: It shall be the duty of the Chief Inspector of the Department for the Inspection of Steam Boilers, Unfired Pressure Vessels and Cooling Plants, and his assistants, in addition to the duties otherwise required in this Chapter, to inspect all parts of all mechanical refrigerating systems employing any refrigerant which is expanded, vaporized, liquefied or compressed in its refrigerating cycle, including piping machinery, boilers, tanks, jacketed kettles, generators, shell brine coolers, shell condensers, shell absorbers, purifiers, pipe condensers, compressors and pipes used therein, and the apparatus connected therewith and the extensions thereunto. Said Chief Inspector shall make such inspection once in each year. Said Chief Inspector is further authorized and required to inspect any refrigerating system or apparatus whenever, in his judgment, inspection is necessary for the protection of life and property. Whenever such inspection discloses that on account of age, obsolescence, wear and tear or for any other cause such refrigerating system has become or is likely to become dangerous to life and health, the said Chief Inspector shall give notice in writing to the person owning, leasing or controlling such refrigerating system directing him to make such changes, alterations or repairs as in the judgment of the Chief Inspector are necessary to make the said refrigerating system safe for the occupants of the premises; such notice

shall state briefly the nature of the work required to be done and shall specify the time in which the said work shall be completed, which shall be fixed by the Chief Inspector upon consideration of the condition of such refrigerating system, or parts thereof and the danger to life or property which may result from its unsafe condition. Upon failure of such person to comply with the request set forth in such notice within the time fixed in such notice, the Chief Inspector is hereby authorized to order the system shut down and the refrigerant gas pumped from such system and to prohibit its further use until the aforesaid directions are complied with. Any expense or outlay incurred by the Chief Inspector in shutting down such refrigerating system shall be a charge upon and be collected from the owner, lessee or person controlling such system by legal proceedings prosecuted by the Department of Law.

(a) **Certificate of Inspection.** It shall be unlawful for any person, firm or corporation to use or operate any refrigerating system required to be inspected under this ordinance until a certificate of inspection has been issued by the Department for the Inspection of Steam Boilers, Unfired Pressure Vessels and Cooling Plants. It shall be unlawful for any person, firm or corporation to use or operate any refrigerating system that has been repaired, reconditioned or reconstructed, until a certificate shall have been procured from the Department for the Inspection of Steam Boilers, Unfired Pressure Vessels and Cooling Plants, to the effect that said system is in such condition that it can be safely used.

Inspection Waived. No inspection shall be required on a single unit system containing six (6) pounds or less of refrigerant with power supplied by a motor not exceeding one-third (⅓) H. P. or on a remote system using not more than a one-third (⅓) H. P. motor and containing six (6) pounds or less of refrigerant. Such remote system shall supply not more than one (1) evaporator in a separately refrigerated space; provided however, that nothing herein contained shall prevent the department from inspecting any such single unit or remote system whenever, in its judgment, inspection is necessary for the protection of life and property.

5102.05 Inspection of Repairs: It shall be the duty of the Chief Inspector of Steam Boilers and Cooling Plants, upon an application in writing, made by any person or corporation owning, leasing or controlling the use of any boiler, unfired pressure vessel or mechanical refrigerating system, stating that the same is out of repair or has been repaired, to examine the same when so repaired, and determine if such repairing has been properly done; and it shall be unlawful for any person or corporation to use any boiler, unfired pressure vessel or mechanical refrigerating system, after the same has been repaired, until a certificate shall have been procured from the inspector to the effect that such repairing has been properly done and such boiler, unfired pressure vessel or mechanical refrigerating system may be safely used.

ARTICLE 5103.

Licenses.

5103.01 Erecting or Repairing — License Required: Any person, firm or corporation engaged in or desiring to engage in the work of repairing or erecting steam boilers, steam kettles, pressure tanks, superheaters or generators in the City of Chicago shall submit to an examination and shall obtain a license from the Chief Inspector of Steam Boilers and Cooling Plants in the manner hereinafter provided; except that whenever a firm or corporation consists of more than one member, it shall not be necessary for more than one member of said firm, or one officer of said corporation, to undergo such examination in order to obtain a license for said firm or corporation. The word "erect" as herein used shall refer only to such steam boilers, steam kettles, pressure tanks, superheaters or generators as are assembled at the place of installation and shall not include such steam boilers, steam kettles, pressure tanks, superheaters or generators as are constructed at the place of manufacture and delivered assembled to the place of installation. It shall not include sectional heating boilers constructed of cast iron.

5103.02 Application for License—Examination: Any person, firm or corporation desiring to procure a license to repair or erect steam boilers, steam kettles, pressure tanks, superheaters or generators shall make application to the Chief Inspector of Steam Boilers and Cooling Plants, and shall at such time and place as said inspector may designate, undergo an examination as to qualifications and competency to properly repair or erect steam boilers, steam kettles, pressure tanks, superheaters or generators. Said examination shall be made in whole or in part in writing, and shall be of practical and elementary character, sufficiently strict to test the qualifications of the applicant. Where the applicant is a firm or corporation, such applicant shall state in writing the name or names of the person or persons connected therewith who will submit to such examination as to qualifications, and in case such firm or corporation receives a license and thereafter severs its connection with such person or persons, so that no member of said firm or officer of said corporation has qualified as required under this Article, the license granted to such firm or corporation shall be void and such firm or corporation shall be required to make a new application for license in the same manner as before.

5103.03 License Fees: The Chief Inspector of Steam Boilers and Cooling Plants shall examine such applicants as to their practical knowledge of the construction and repair of steam boilers, steam kettles, pressure tanks, superheaters or generators, and if satisfied of the competency of such applicant, shall issue a license to such applicant authorizing him to repair or erect steam boilers, steam kettles, pressure tanks, superheaters or generators. The fee for such examination, including the first year's license fee, shall be Fifty Dollars ($50.00) and thereafter the annual license fee shall be Twenty-five Dollars ($25.00). Said license shall be valid for a period of one (1) year from the date of issuance, except as herein otherwise provided, and may be renewed upon its expiration by paying in advance the annual renewal fee. All fees provided for in this Article shall be paid to the City Collector.

5103.04 Penalty: Any person, firm or corporation repairing or erecting any steam boiler, steam kettle, pressure tank, superheater or generator, that shall fail to procure a license as herein provided, or any person, firm or corporation that shall violate any of the provisions of this Article shall be fined not less than Twenty-five Dollars ($25.00) nor more than One Hundred Dollars ($100.00) for each offense and such license may be revoked at the discretion of the Mayor.

ARTICLE 5104.

Fees.

5104.01 Permit and Inspection Fees: Upon approval of the plans and specifications by the Department for the Inspection of Steam Boilers, Unfired Pressure Vessels and Cooling Plants and the issuance of a permit for the installation of any boiler, unfired pressure vessel or mechanical refrigerating system, the applicant for such permit shall pay a permit fee as follows:

(a) **Boilers and Unfired Pressure Vessels.**

Item 1. Permit Fees, Including First Inspection:

For each low pressure boiler	$ 8.00
" " miniature boiler	8.00
" " high pressure boiler containing not more than 250 square feet of heating surface	10.00
" " high pressure boiler containing more than 250 square feet and not more than 1500 square feet of heating surface	11.00
" " high pressure boiler containing more than 1500 square feet and not more than 5000 square feet of heating surface	12.00
" " high pressure boiler containing more than 5000 square feet of heating surface	13.00
" " unfired pressure vessel designed for 15 pounds pressure or less	8.00
" " unfired pressure vessel designed for more than 15 pounds pressure	10.00

Item 2. Annual Inspection Fees shall be as follows:

For each low pressure boiler	$ 3.00
" " miniature boiler	3.00
" " high pressure boiler containing not more than 250 square feet of heating surface	5.00
" " high pressure boiler containing more than 250 square feet and not more than 1500 square feet of heating surface	6.00
" " high pressure boiler containing more than 1500 square feet and not more than 5000 square feet of heating surface	7.00
" " high pressure boiler containing more than 5000 square feet of heating surface	8.00
" " unfired pressure vessel carrying 15 pounds pressure or less	3.00
" " unfired pressure vessel carrying more than 15 pounds pressure	5.00

The fee for inspection of boilers and other apparatus above provided for shall be double the respective amounts above specified when an inspection is made on Sunday or Legal Holiday at the request of the person or corporation owning or operating said boiler or other apparatus.

(b) **Mechanical Refrigerating Systems.** The fees provided for the issuance of a permit for the installation of a refrigerating system and the inspection of a refrigerating system shall be as follows:

Permit Fees—Including First Inspection.
Multiple Dwelling and Remote Systems.

Class B and C Ten Dollars ($10.00) for each compressor or generator unit and Twenty-five cents ($.25) for each evaporator.

Class D—Six Dollars ($6.00) for each compressor or generator unit and Twenty-five cents ($.25) for each evaporator.

Commercial and Industrial Systems.

Class A—Twenty Dollars ($20.00) for each compressor or generator unit.

Class B—Seventeen Dollars ($17.00) for each compressor or generator unit.

Class C—Ten Dollars ($10.00) for each compressor or generator unit.

Class D—Six Dollars ($6.00) for each compressor or generator unit.

Class E—Six Dollars ($6.00) for each compressor or generator unit with power supplied by a motor larger than one-third (⅓) H. P.

The permit fee for a single unit system containing six pounds (6 lbs.) or less of refrigerant with power supplied by a motor not exceeding ⅓ H. P. shall be $1.00 and the permit fee for a remote system serving not more than one evaporator containing less than six pounds (6 lbs.) of refrigerant with power supplied by motor not exceeding ⅓ H. P. shall be $3.00.

Annual Inspection Fees.

The fee for annual inspection for **multiple dwelling** and **remote** systems shall be as follows:

Class B—$10 for each compressor or generator unit.

Class C—$5 for each compressor or generator unit.

Class D—$3 for each compressor or generator unit.

The fee for annual inspection for **commercial** and **industrial** systems shall be as follows:

$3.00 for each compressor or generator unit of 5 tons or less capacity.

$5.00 for each compressor or generator unit over 5 tons and not over 35 tons capacity.

$10.00 for each compressor or generator unit over 35 tons and not over 100 tons capacity.

$12.00 for each compressor or generator unit over 100 tons capacity.

Compressor capacity shall be based on five degrees evaporator temperature and eighty-six degrees condenser temperature, and compressor displacement shall conform to the following table for refrigerants named:

Carbon Dioxide CO_2—1,625 cu. in. per Min. per Ton.

Ammonia NH_3—6,912 cu. in. per Min. Per Ton.

Dichlorodifluoromethane CCL_2F_2 — 12,528 cu. in. per Min. per Ton.

Methyl Chloride CH_3CL—13,824 cu. in. per Min. per Ton.

Sulphur Dioxide So_2—20,736 cu. in. per Min. per Ton.

All fees required hereunder shall be paid to the City Collector.

5104.02 Exemptions—Charitable, Religious and Educational Institutions: The Chief Inspector may and he is hereby directed and instructed to remit all inspection fees charged, or that may hereafter be charged against any and all charitable, religious and educational institutions, when the boiler, refrigerating system or other apparatus inspected is located in or upon premises used or occupied exclusively by such charitable, religious or educational institutions; provided however, that such charitable, religious or educational institution is not conducted or carried on for private gain or profit.

5104.03 Charges in Excess of Fees—Penalty: If any person acting on behalf of the City under the provisions of this Article shall take or receive any money or any valuable thing from any person for the purpose of deceiving or defrauding any person or persons, or for the purpose of favoring any person or persons, or if any inspector shall recommend the issuance of any certificate of inspection without having, at the time stated, thoroughly examined and tested the apparatus so certified, he shall be fined One Hundred Dollars ($100.00) for each offense.

5104.04 Penalty: Any person, firm or corporation violating any of the provisions of the foregoing sections of this Ordinance shall be fined not less than Twenty-five Dollars ($25.00) nor more than Two Hundred Dollars ($200.00) and each day's violation shall constitute a separate offense.

CHAPTER 52.

Steam Boilers and Unfired Pressure Vessels.

ARTICLE 5201.

General.

5201.01 Scope: The following provisions shall apply to every steam boiler and every unfired pressure vessel except tanks containing only water under the pressure in the city mains. It does not apply to boilers on railroad locomotives or to boilers which are subject to Federal Inspection and Control.

5201.02 Definitions: For the purposes of this ordinance, terms used in this Chapter are hereby defined as follows:

Steam Boiler. A device used for the purpose of evaporating water into steam under pressure, further defined as follows:

Brinks Express Co. J. L. McConnell, Engineer.

Ciné Theatre. C. W. & Geo. L. Rapp, Inc., Architects.

(a) **High Pressure Boiler.** A boiler in which steam pressure is more than fifteen (15) pounds per square inch above atmospheric pressure. This classification is sub-divided as follows:

Item 1. Miniature Boiler. A high pressure boiler, the diameter of the shell of which does not exceed sixteen (16) inches; the length between heads of which does not exceed forty-two (42) inches; the water heating surface of which is not more than twenty (20) square feet and

Item 2. Other High Pressure Boilers.

(b) **Low Pressure Boiler.** A boiler in which steam pressure is not more than fifteen (15) pounds per square inch above atmospheric pressure.

Unfired Pressure Vessel. Any tank or pressure vessel used to contain air, water or other substance under pressure, except tanks containing only water under pressure in the city mains unless otherwise prescribed in other Sections of this Ordinance.

ARTICLE 5202.

Material and Construction.

5202.01 Material and Construction: The material and construction of boilers shall be in accordance with the American Society of Mechanical Engineers "Rules for the Construction of Stationary Boilers", as follows:

Section I. Power boilers, dated 1933, including addenda dated 1937.

Section II. Material Specifications dated 1933, including addenda dated 1937.

Section III. Locomotive Boilers dated 1937.

Section IV. Low Pressure Heating Boilers dated 1937.

Section V. Miniature Boilers dated 1937.

Section VI. Rules for Inspection dated 1937.

ARTICLE 5203.

Unfired Pressure Vessels.

5203.01 Unfired Pressure Vessels: The material and construction of tanks and other vessels under pressure shall be in accordance with the American Society of Mechanical Engineers "Rules for The Construction of Unfired Pressure Vessels," Section VIII dated 1934 and addendum dated 1937.

CHAPTER 53.

Mechanical Refrigeration.

ARTICLE 5301.

Definitions.

5301.01 General: For the purpose of this ordinance the following definitions shall apply:

Approved. Official approval by the Department for the Inspection of Steam Boilers, Unfired Pressure Vessels and Cooling Plants.

Brine Cooler. An evaporator for cooling brine in an indirect system.

Check Valve. A valve allowing refrigerant flow in one (1) direction only.

Commercial System. A refrigerating system assembled or installed in a building used for business or commercial purposes.

Compressor. A compressor is any device having one (1) or more pressure imposing elements, used in a refrigerating system to increase the pressure of the refrigerant in its gas or vapor state for the purpose of liquefying the refrigerant.

Compressor Relief Device. A valve or rupture member located between the compressor and the stop valve on the discharge side arranged to relieve the pressure at a predetermined point.

Condenser. A vessel or arrangement of pipe or tubing in which the vaporized refrigerant is liquefied by the removal of heat.

Direct System of Refrigeration. A system in which the evaporator is located in the material or space refrigerated or in air circulating passages communicating with such space.

Emergency Relief Valve. A manually operated valve for the discharge of refrigerant in case of fire or other emergency.

Evaporator. That part of a system in which refrigerant is expanded or vaporized to produce refrigeration.

Flammable Refrigerant. Any refrigerant which will burn when mixed with air.

Fusible Plug. A device for the relief of pressure having a fusible metal that will melt at a maximum temperature of two hundred (200) degrees Fahrenheit.

Generator. A device equipped with a heating element used in a refrigerating system to increase the pressure of the refrigerant in its gas or vapor state for the purpose of liquefying the refrigerant.

Indirect System of Refrigeration. A system in which a liquid, as brine or water, cooled by the refrigerant, is circulated to the material or space refrigerated or is used to cool air applied to such space.

Indirect Open Spray System. A system in which a liquid, such as brine or water, cooled by an evaporator located in an inclosure external to a cooling chamber, is circulated to such cooling chamber and is sprayed therein.

Industrial System. A refrigerating system used in the manufacture, processing, or storage of materials located in a building used exclusively for industrial purposes.

Irritant Refrigerant. Any refrigerant which has an irritating effect on the eyes, nose, throat or lungs.

Liquid Receiver. A vessel permanently connected to the high pressure side of a system for the storage of refrigerant.

Mixer. A vessel or device for mixing the refrigerant with another substance.

Multiple System. A system employing the direct system of refrigeration in which the refrigerant is delivered to two (2) or more evaporators in separately refrigerated spaces.

Multiple Dwelling System. A refrigerating system employing the direct system in which the refrigerant is delivered by a pressure imposing element to two (2) or more evaporators in separate refrigerators or refrigerated spaces located in rooms of separate tenants in multiple dwellings.

Pressure Limiting Device. A pressure—or temperature-responsive mechanism for automatically stopping the operation of the pressure imposing element at a predetermined pressure.

Pressure Relief Device. A pressure relief valve, a rupture member, or other approved device for relieving pressure.

Pressure Relief Valve. A valve held closed by a spring or other means, which automatically relieves pressure in excess of its setting.

Pressure Vessel. Any refrigerant containing receptacle of a refrigerating system other than expansion coils, headers and pipe connections.

Refrigerant. A substance used to produce refrigeration by absorbing heat in its expansion or vaporization.

Refrigerating System. A combination of parts in which a refrigerant is circulated for the purpose of extracting heat.

Remote System. A refrigeration system in which the compressor or generator is located in a space other than the cabinet or fixture containing the evaporator.

Rupture Member. A pressure relief device having a diaphragm or member which will rupture or blow out at a predetermined pressure.

Sealed Unit. A pressure imposing element which operates without stuffing box, or which does not depend upon contact between moving and stationary surfaces for refrigerant retention.

Shell Type Apparatus. A refrigerant-containing pressure vessel having tubes for the passage of a heating, cooling or refrigerating fluid.

Ton of Refrigeration. A ton of refrigeration is heat removal at the rate of twelve thousand (12,000) B.T.U. per hour. Compres-

sor capacity shall be based on five (5) degrees evaporator temperature and eighty-six (86) degrees condenser temperature except the capacity of a compressor when used for comfort cooling or air conditioning purposes shall be based on forty (40) degrees Fahrenheit evaporator temperature.

For the purpose of fees charged, see Fee Section 5104.01.

Unit System. A system which can be removed from the user's premises without disconnecting any refrigerant-containing parts, water connections, or fixed electrical connections.

ARTICLE 5302.

Classification.

5302.01 Classification of Refrigerating Systems: Refrigerating systems shall be classified according to the total weight of refrigerant contained in or required for their proper operation.

Class "A" system is one containing one thousand (1000) pounds or more of refrigerant.

Class "B" system is one containing more than one hundred (100) pounds but less than one thousand (1000) pounds of refrigerant.

Class "C" system is one containing more than twenty (20) pounds but not more than one hundred (100) pounds of refrigerant.

Class "D" system is one containing more than six (6) pounds but not more than twenty (20) pounds of refrigerant.

Class "E" system is one containing six (6) pounds or less of refrigerant.

5302.02 Classification of Refrigerants: For the purpose of this ordinance refrigerant shall be classified as follows:

Non-Irritant and Class 1	Non-Flammable Class 2	Flammable Class 1	Class 2	Irritant
Carbon Dioxide CO_2	Dichlorodifluoro methane CCL_2F_2 Monofluorotri chloromethane $CFCL_3$ Dichlorotetra fluoroethane $C_2CL_2F_4$ Dichloromenthane CH_2CL_2	Methyl Chloride CH_3CL Dichloroethylene $C_2H_2CL_2$	Ethane C_2H_6 Propane C_3H_8 Isobutane C_4H_{10} Butane C_4H_{10} Ethyl Chloride C_2H_5CL Methyl Formate $C_2H_4O_2$	Sulphur Dioxide SO_2 Ammonia NH_3

Flammable refrigerants, or refrigerants of a poisonous nature, when used in systems containing more than six (6) pounds of refrigerant, shall have added a substance which will impart a pungent odor or an irritating quality to the refrigerant, unless vapors are readily apparent to human sense.

ARTICLE 5303.

Limitations As to Use and Special Requirements.

5303.01 Institutional Occupancies:

(a) **General.** Direct refrigerating systems shall not be placed in wards, private and operating rooms of hospitals or asylums, cell blocks of prisons, homes for infants, aged or infirm, or any building where people are confined or are helpless, except Class E Unit Systems containing a non-irritant and non-flammable refrigerant. Refrigerating systems installed in any such rooms or buildings shall be subject to the following provisions:

(b) **Direct Systems.** Direct systems wholly confined to a kitchen or laboratory may be installed; provided however, that the system does not contain more than twenty (20) pounds of a non-irritant and non-flammable refrigerant. Such kitchen or laboratory shall be equipped with tight-fitting self-closing doors and shall have natural or mechanical means of ventilation to the outside.

(c) **Indirect Systems with Not More than 2000 Lbs. of Class I.** Indirect systems containing not more than two thousand (2000) pounds of a Class I non-irritant and non-flammable refrigerant may be installed; provided however, all refrigerating containing equipment is placed in a machinery room located on the first floor or in a basement. Such machinery room shall have tight-fitting self-closing doors and shall be provided with natural or mechanical means of ventilation.

(d) **Indirect Systems Operating Above Atmospheric Pressure.** Indirect systems operating above atmospheric pressure containing not more than three hundred fifty (350) pounds of a Class II non-irritant and non-flammable refrigerant may be installed; provided however, that all refrigerant containing parts are placed in a machinery room located on the first floor or in a basement as provided above.

(e) **Indirect Systems Operating Below Atmospheric Pressure.** Indirect systems operating below atmospheric pressure containing not more than two thousand (2000) pounds of a Class II non-irritant and non-flammable refrigerant may be installed; provided however, that all refrigerant containing equipment is placed in a machinery room located on the first floor or in a basement as provided above.

(f) **Indirect Systems Containing 100 Pounds or Less.** Indirect systems containing one hundred (100) pounds or less of Class I or Class II non-irritant and non-flammable refrigerant may be installed on the roof or in a penthouse; provided however, that all refrigerant containing parts are placed in a machinery room having tight-fitting self-closing doors and provided with natural or mechanical means of ventilation.

5303.02 Industrial Systems:

(a) **General.** A refrigerating system used in the manufacture, processing or storage of materials located in a building used exclusively for industrial purposes shall be classified as an industrial system. Industrial systems may be located without restriction in separate buildings, groups of buildings or separate sections of buildings, provided:

(b) The pressure imposing element, receiver and shell type apparatus of Class A systems except brine or water cooler, connecting pipes and evaporators, are placed in a ventilated machinery room.

(c) No Class A system using a flammable refrigerant shall be permitted unless the entire building is of fireproof construction; no Class B system using such a refrigerant shall be permitted unless the machinery room is of fireproof construction.

(d) Machinery rooms of Class A systems using an irritant or flammable refrigerant shall have two (2) exits, one (1) leading directly to the outside.

(e) If an irritant or flammable refrigerant is used in a direct system, the number of workmen shall not exceed one (1) for each one hundred fifty (150) square feet of floor

cal Education Building,
rn Illinois State Teachers College,
eston, Illinois.

C. Herrick Hammond, Supervising Architect,
State of Illinois, Springfield.
Hewitt, Emerson and Gregg, Associate Architects,
Peoria, Illinois.

area on the floors above the first, or ground level.

(f) Systems containing flammable Class 2 refrigerants shall be limited to ten (10) pounds of refrigerant except in industries engaged in the manufacture and liquefication of flammable gases.

In all other details such systems must conform to the provisions contained in other sections of this ordinance.

5303.03 Commercial Systems: A refrigerating system assembled or installed in a building used for business or commercial purposes shall be classified as a Commercial system.

Class 2 flammable refrigerants shall not be permitted in any installation.

Commercial systems shall be limited on the basis of occupancy and location as follows:

Places of Public Assembly: Refrigerating systems installed in rooms or buildings for public assembly, for example, auditoriums, theatres, dance halls, exhibition halls, schools, except laboratories used for teaching refrigeration, court houses, churches, entrances and exits of business units and all similar occupancies except eating places, shall be subject to the following limitations:

(a) Unit systems containing a non-irritant or non-flammable refrigerant may be used in such locations.

(b) Direct systems wholly confined to a kitchen may be installed; provided however, the system does not contain more than twenty (20) pounds of non-irritant and non-flammable refrigerant. Such kitchen shall be equipped with tight-fitting, self-closing doors and shall have natural or mechanical means of ventilation.

All other refrigeration in the above locations must be supplied by the indirect system of refrigeration, except as permitted in Section 5303.06 with the entire refrigerant containing equipment placed in a machinery room. These systems shall be subject to the following provisions:

(a) If an irritant or Class 1 flammable refrigerant is used, system shall not contain over five hundred (500) pounds of refrigerant and entire refrigerant-containing equipment shall be placed in a machinery room devoted to no other purpose, located on the first floor or basement. This room shall be separated from other parts of building by solid construction having a four (4) hour fire-resistive value and there shall be no direct means of communication between such room and any other portion of building. The entrance or exit of such room shall be located in an outside wall but not in a wall forming a part of the enclosure of an exit court.

(b) If a Class 1 or Class 2 non-irritant and non-flammable refrigerant is used, the machinery room shall be provided with natural or mechanical means of ventilation directly to the outside and shall be equipped with self-closing tight-fitting doors. Such systems when installed above the first floor shall be limited in quantity to three hundred (300) pounds of refrigerant.

Other Occupancies: Refrigerating systems installed in retail stores, restaurants, wholesale stores, beauty parlors, barber shops, offices, sales rooms, manufacturing spaces, and all similar occupancies shall be subject to the following limitations. Direct systems of refrigeration may be installed subject to the following provisions:

(a) Systems containing a Class 1 non-irritant and non-flammable refrigerant shall be limited in refrigerant content to not more than one (1) pound for each one hundred (100) cubic feet by volume in the total space, and the total refrigerant content in such systems shall not exceed two thousand (2000) pounds.

(b) Systems containing a Class 2 non-irritant and non-flammable refrigerant shall be limited in refrigerant content to not more than one (1) pound for each fifty (50) cubic feet by volume of total space and the total refrigerant content in such systems shall not exceed three hundred and fifty (350) pounds.

(c) The compressor, condenser and shell type apparatus of systems, containing more than one hundred (100) pounds of refrigerant shall be placed in a machinery room, with natural or mechanical means of ventilation directly to the outside, and shall be equipped with self-closing tight-fitting doors.

(d) Systems containing Class 2 non-irritant and non-flammable refrigerant shall not be located in a space in which there is apparatus to produce an open flame, unless such space is provided with means of adequate mechanical ventilation or the apparatus for producing the open flame is vented by means of a hood, so as to carry the products of decompositiou to the outside atmosphere. Flames made by matches, cigarette lighters, small alcohol lamps and similar devices shall not be considered as open flames.

(e) Systems containing not more than two hundred (200) pounds of an irritant or Class 1 flammable refrigerant may be installed in manufacturing portions of a commercial building normally frequented by employes only; provided however, that where system contains over fifty (50) pounds of refrigerant, compressor, condenser and shell type apparatus shall be placed in a machinery room with natural or mechanical ventilation and openings to other rooms be fitted with close-fitting self-closing doors.

(f) Systems containing not more than one hundred (100) pounds of an irritant or Class 1 flammable refrigerant may be installed on first floor or in basement of buildings where the public is permitted, except commercial residential buildings; provided however, that where systems contain over fifty (50) pounds, the compressor, condenser and shell type apparatus shall be placed in machinery room, with natural or mechanical ventilation and openings to other rooms fitted with tight-fitting, self-closing doors and the first floor of buildings shall have required exits, opening directly out of doors or through corridors or other horizontal means of exit directly to an outside exit doorway.

(g) Systems containing fifty (50) pounds or less of an irritant or Class 1 flammable refrigerant may be installed on any floor above the first floor of commercial buildings where the public is permitted, except that unit systems only may be used in retail stores, restaurants, and other locations where the number of employees serving the public exceed ten (10).

(h) Systems containing fifty (50) pounds or less of an irritant or flammable refrigerant may be installed in the basement or on the first floor only of multiple use buildings containing dwellings; provided however, that the compressor, condenser and shell type apparatus of such systems containing more than twenty-five (25) pounds shall be located in the machinery room. Such machinery room shall have natural or mechanical means of ventilation and shall be fitted with tight-fitting, self-closing doors.

Indirect systems may be installed in the above locations subject to the following limitations:

(a) Systems containing a non-irritant and non-flammable refrigerant may be installed in a basement or on the first floor; provided however, all refrigerant-containing parts of a system containing more than one hundred (100) pounds of refrigerant are installed in a machinery room having natural or mechanical means of ventilation. Such machinery room shall be equipped with tight-fitting, self-closing doors.

(b) Systems containing five hundred (500) pounds or less of a non-irritant and non-flammable refrigerant may be installed between the first floor and the roof; provided however, all refrigerant-containing parts of such sys-

tem are installed in a machinery room having natural or mechanical ventilation. Such machinery room shall be equipped with tight-fitting, self-closing doors.

(c) Systems containing more than three hundred (300) pounds of non-flammable and non-irritant refrigerant may be installed above the roof; provided however, all refrigerant-containing parts are located in the machinery room. All openings from such machinery room shall be made to the outside and shall not be located adjacent to the entrance or exit of any stairway, elevator shaft or fresh air intake.

(d) Systems containing an irritant or a Class 1 flammable refrigerant may be installed subject to the limitations as to quantity of refrigerant and location as required for direct system.

(e) Indirect systems containing a flammable or irritant refrigerant in quantities greater than that permitted for direct systems may be installed; provided however, that they do not contain more than five hundred (500) pounds of refrigerant; and provided further, that all refrigerant-containing parts of the system are placed in a machinery room located on the first floor or basement. Such machinery rooms shall be separated from other parts of the building by solid construction having four (4) hour fire-resistive value and there shall be no direct means of communication between such room and other portions of the building.

In all other details direct and indirect systems of refrigeration must conform to the provisions contained in other sections of this ordinance.

5303.04 Multiple Dwelling Systems: No multiple dwelling system shall contain over fifty (50) pounds of sulphur dioxide, fifty (50) pounds of methyl chloride or one hundred (100) pounds of a Class 1 or Class 2 non-flammable refrigerant. All other refrigerants listed in refrigerant classification section are prohibited in multiple dwelling systems.

No evaporator shall be placed in a sleeping room.

No evaporator shall be placed in a room having adjoining rooms, the combined cubical contents of which has less than four thousand (4000) cubic feet and any part of which is used for sleeping purposes, unless such rooms are provided with a window or windows the total window area of such being not less than forty (40) per cent of the floor area of such rooms. And unless such window, or windows, shall open to the outside air and be so constructed that at least one-half (½) of such window, or windows, can be readily opened.

Shut-off valves for multiple dwelling systems shall be installed at the following locations:

In each branch liquid and suction line at or near compressor.

At the bottom of each riser or manifold connection of any riser or any branch connecting manifold extending over forty (40) feet in length.

At each service outlet in liquid and suction lines.

These valves shall be fitted with a hand wheel as a means of ready operation, and shall be located outside of refrigerating unit and at such distance above the floor as will provide ready accessibility.

Such valves, where located in living quarters shall be placed in a metal box, or other suitable inclosure and shall be rigidly attached thereto or to the supports thereof. Every such outlet box shall have an accessible door or removable cover.

Not more than one (1) refrigerator shall be connected to one (1) outlet box.

Every such outlet box shall be located within the premises containing the refrigerator and not within the refrigerated space.

No outlet box shall be located in any hallway, stairway or vertical shaft.

Service valves shall be installed in both connections to every multiple dwelling system evaporator of the flooded type that can be removed from the refrigerator as a unit, in such manner as to permit the removal of the evaporator with valves attached.

Every evaporator, unless constructed of material of sufficient strength to prevent injury in the ordinary and customary use thereof, shall be protected by a suitable shield to assure protection against such injury.

Every evaporator shall be firmly anchored or secured in such a manner as to make it rigid.

Every refrigerator cabinet, box or refrigerated space, containing any evaporator, using the direct system of refrigeration shall be firmly and securely anchored and fixed to a wall, floor or other fixed object in such manner as to hold such refrigerator box or casing securely in place.

Compressors, condensers and shell type apparatus shall be located in a machinery room provided with natural or mechanical ventilation or in an accessible part of the basement not used for habitation, workshop or laundry purposes; provided however, that where compressors are not located in a machinery room they shall be protected by the use of heavy wire netting secured to metal posts or two by four (2x4) wooden studding and shall not be located under stairways, near dumbwaiters or elevator shafts.

In all other details such systems must conform to the provisions contained in other sections of this ordinance.

5303.05 Unit Systems:

(a) **General.** A system which can be removed from the users' premises without disconnecting any refrigerant-containing parts, water connections or fixed electrical connections. Unit systems may be installed subject to the following limitations:

(b) Unit systems containing an irritant or flammable refrigerant shall not be placed in:

Wards or private rooms of hospitals, sleeping quarters of asylums, cell blocks of prisons or any building where persons are confined or helpless.

Room or buildings for public assembly, for example, auditoriums, theatres, dance halls, exhibition halls, schools (except laboratories for teaching refrigeration), court houses, churches, entrances and exits of business units.

Sleeping Rooms.

(c) Unit systems containing not more than six (6) pounds of a non-flammable or non-irritant refrigerant may be installed in any location.

(d) Unit systems containing more than six (6) pounds of non-flammable and non-irritant refrigerant may be installed in any location except wards or private rooms of hospitals, sleeping quarters of asylums or any room or building where persons are confined or helpless.

(e) No unit system may be installed in any location unless the room is adequately ventilated to the outside air and unless sufficient clearance is provided to make them accessible.

(f) Unit systems unless so constructed that they will not burst due to the expansion of the refrigerant when subjected to an abnormal outside temperature, such as that created by fire, shall be protected by a pressure relief device as follows:

Item 1. Safety valves, if used, shall be one-fourth (¼) inch in size.

Item 2. Rupture members, if used, shall have an opening at least one-sixteenth ($\frac{1}{16}$) inch in diameter.

Item 3. Fusible plugs, if used, shall have a maximum fusing point of two hundred (200) degrees Fahrenheit and a free opening of one-sixteenth ($\frac{1}{16}$) inch in diameter.

(g) Unit systems employing an air cooled condenser and containing not more than six

(6) pounds of refrigerant, may have the discharge from the pressure relief device piped into the low pressure side of system; provided however, that the volume of the low pressure side of system, excluding the compressor crank case, is capable of holding the entire refrigerant charge of system.

(h) Every evaporator, unless constructed of sufficient strength to prevent injury in the ordinary and customary use thereof, shall be protected by a suitable shield to assure protection against such injury. In all other details such systems must conform to the provisions contained in other sections of this ordinance.

5303.06 Refrigerating Systems Used in Connection with Air Conditioning Installations:

(a) **General.** Refrigerating systems used in connection with the control of temperature or humidity of air in or circulated to space, or spaces, occupied by people, or where used for comfort cooling, shall be subject to the following requirements:

(b) **Limitations.** Direct systems in which the refrigerant is circulated through coils or evaporators located in the path of the air used for air conditioning may be installed subject to the following limitations:

Item 1. Where a Class 1 non-irritant and non-flammable refrigerant is used the following provisions shall apply:

1. The refrigerant of the system shall not exceed one (1) pound for each one hundred (100) cubic feet of volume in total space to be cooled and total refrigerant content shall not exceed two thousand (2000) pounds.
2. Non-corroding material shall be used in coils unless permanent protection against corrosion is provided.
3. Coils or coolers shall be tested to the test pressure specified for the refrigerant; provided however, the department for the inspection of steam boilers, unfired pressure vessels and cooling plants may accept from the manufacturer a certificate copy of test report showing that tests of said apparatus, as required by this ordinance, were made at the place of manufacture.

(c) **Class 2 Refrigerant.** Where a Class 2 non-irritant and non-flammable refrigerant is used the following provisions shall apply.

1. The refrigerant content of the system shall not exceed one (1) pound for each fifty (50) cubic feet of volume of total space to be cooled and the total refrigerant content shall not exceed three hundred fifty (350) pounds.
2. Any joint or connection in the refrigerant-containing part of the system shall be threaded, welded, brazed sweated or of such other type of construction as approved by the department.
3. No expansion valve or connection thereto shall be located in a ventilating system.
4. No joint or connection shall be located in air stream except connections to coil headers may be located in air stream; provided however, such connections are welded, brazed or sweated and one and one-half (1½) times the pressure specified for this refrigerant is applied in the presence of the inspector.
5. Coils or coolers shall be tested to one and one-half (1½) times the test pressure specified for the refrigerant; provided however, the Department for the Inspection of Steam Boilers, Unfired Pressure Vessels and Cooling Plants may accept from the manufacturer a certified copy of test report showing that tests of said apparatus, as required by this ordinance, were made at the place of manufacture.
6. Refrigerating system shall not be located in a space in which there is apparatus to produce an open flame unless such space is provided with means of adequate mechanical ventilation, or the apparatus for producing the open flame is vented by means of a hood so as to carry the products of decomposition to the outside atmosphere. Flames made by matches, cigarette lighters, small alcohol lamps, and similar devices shall not be considered as open flames.
8. In every refrigerating system containing twenty-five (25) pounds of Class 2 non-irritant and non-flammable refrigerant there shall be provided a single level signal device or such other means as will produce a signal sufficient to attract the attention of the person or persons responsible for the operation of such equipment, in case of refrigerant loss from the refrigerating system during operation of a predetermined amount according to the following schedule:

Less than eight (8) ton capacity (air conditioning rating) ten (10) pounds.

Eight (8) to fifteen (15) ton capacity (air conditioning rating) fifteen (15) pounds.

Above fifteen (15) ton capacity (air conditioning rating). Permissible loss shall not exceed fifteen (15) pounds plus one (1) pound for each ton capacity of the system in excess of fifteen (15) tons.

Where evaporators are located in apartment or hotel rooms used for sleeping purposes or in ducts leading thereto, systems shall be limited to fifty (50) pounds of a Class 1 or 2 non-irritant and non-flammable refrigerant; provided however, that the total refrigerant content of the system shall not exceed one (1) pound for each one hundred (100) cubic feet of volume in total space to be cooled. The evaporators of such systems shall be subjected to the provisions specified above.

(d) Where an irritant or Class 1 flammable refrigerant is used, the system shall be limited to industrial locations and shall be subject to the following provisions:

1. The number of employees shall not exceed one (1) for every two hundred (200) square feet of floor area of the space to which the air is circulated.
2. The coils or evaporators shall be constructed of non-corroding materials or permanent protection against corrosion provided.
3. All joints in the evaporators or connections thereto located in the air stream shall be welded and subjected to a pressure of one and one-half (1½) times the test pressure specified for the refrigerant.
4. Unit coolers used in connection with air conditioning systems serving one (1) room may be used; provided however, that where an irritant or flammable refrigerant is used the number of workmen shall not exceed one (1) for each one hundred fifty (150) square feet of floor area on the floors above the first or ground level.

(e) Indirect systems may be installed for air conditioning subject to the following limitations:

1. All such systems shall be subject to the limitations for institutional occupancies or industrial systems or commercial systems as they may apply.
2. Systems in multiple dwellings shall not contain more than three hundred fifty (350) pounds of a Class 2 non-flammable and non-irritant refrigerant provided however, that where system contains over one hundred (100) pounds, refrigerant-containing parts shall be placed in a machinery room with tight walls and natural or mechanical ventilation direction to the outside and with tight-fitting self-closing doors.
3. Liquids cooled by an irritant or flammable refrigerant shall not be used in a spray system to cool the air except in manufacturing units.

In all other details such systems must conform to the provisions contained in other sections of this ordinance.

5303.07 General Limitations: Spaces subject to limitation as to the type of equipment and the kind and amount of refrigerant permitted, which adjoin horizontally or vertically a space having greater restrictions and are not separated therefrom by a floor or wall having no openings, shall be subject to the greater restrictions.

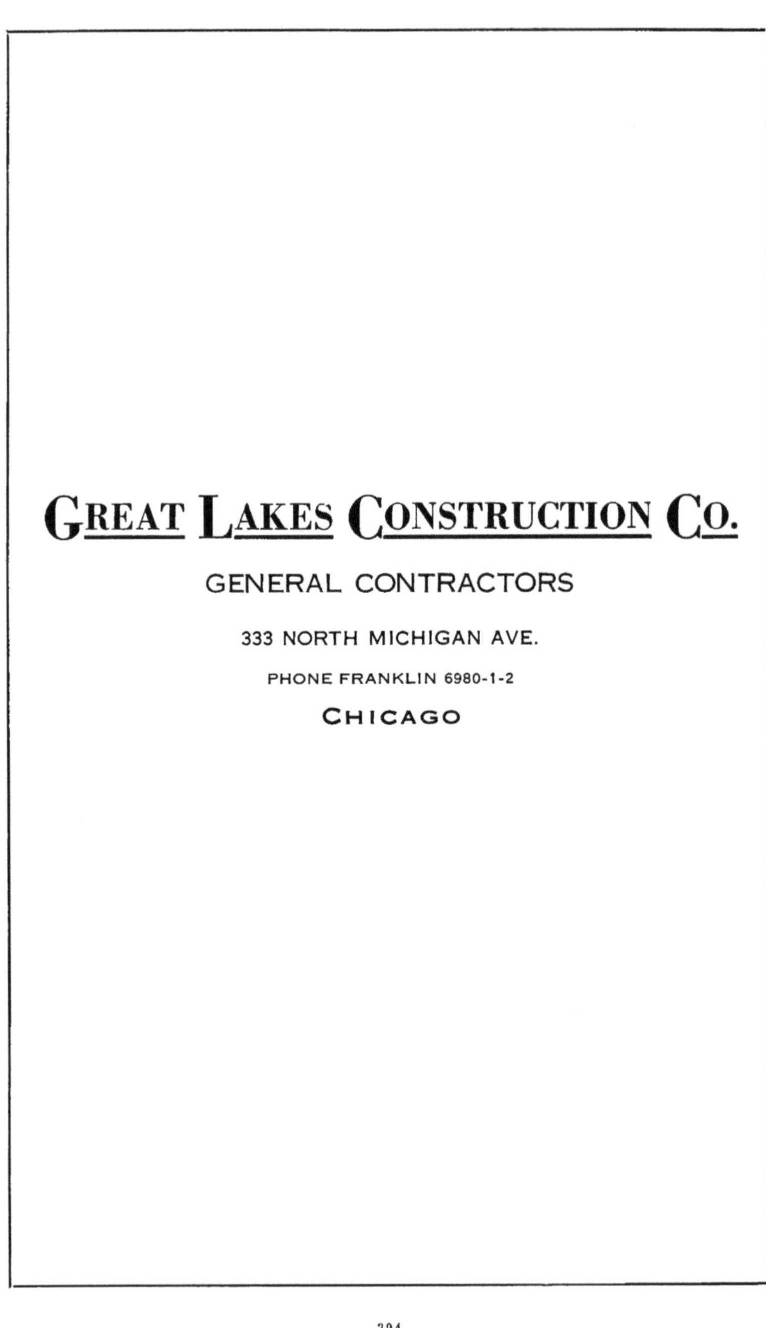
GREAT LAKES CONSTRUCTION CO.
GENERAL CONTRACTORS
333 NORTH MICHIGAN AVE.
PHONE FRANKLIN 6980-1-2
CHICAGO

ARTICLE 5304.

Construction and Installation.

5304.01 General Requirements: All materials used in the construction and installation of refrigerating systems shall be suitable for the refrigerant used and no material shall be used that will deteriorate due to the chemical action of the refrigerant or the oil, or the combination of both.

Every part of a refrigerating system shall be designed, constructed and assembled to withstand safely and without injury the following required test pressures:

Test Pressures in Pounds Per Square Inch.

Refrigerant	Symbol	Test Pressure
Carbon Dioxide	CO_2	1500
Ethane	C_2H_6	1100
Propane	C_3H_8	325
Ammonia	NH_3	300
Dichlorodifluoromethane	CCL_2F_2	235
Methyl Chloride	CH_3CL	215
Sulphur Dioxide	SO_2	170
Isobutane	C_4H_{10}	130
Butane	C_4H_{10}	90
Dichlorotetrafluoroethane	$C_2CL_2F_4$	80
Ethyl Chloride	C_2H_5CL	60
Methyl Formate	$C_2H_4O_2$	50
Dichloroethylene	$C_2H_2CL_2$	50
Trichloromonofluoromethane	CCL_3F	50
Dichloromethane	CH_2CL_2	30

Where the low pressure side of Class A systems are protected by a safety valve, the test pressure shall be twice the pressure at which the safety valve is set.

Castings used in the manufacture of compressors and other parts of refrigerating systems shall be of sufficient thickness and close-grain structure to prevent porosity.

5304.02 Pressure Vessels: The material and construction of liquid receivers, shell condensers, shell coolers, shell evaporators, absorbers and all other pressure vessels used in refrigerating systems shall be in accordance with the American Society of Mechanical Engineers Code for Unfired Pressure Vessels dated 1937.

Liquid receivers of direct systems used for air conditioning and for multiple dwelling systems, except those using Class 1 non-irritant and non-flammable refrigerant, shall be of sufficient capacity to contain the total charge of refrigerant. Eighty (80) per cent of the net content of shell and tube condenser can be considered as liquid receiver capacity.

Every liquid receiver, shell and tube condenser must be equipped with means of readily determining the refrigerant content therein.

5304.03 Coils: Refrigerant-containing parts of coils shall have not less than thirty-four one-thousandths (.034) of an inch wall thickness for tubing sizes three-fourths (¾) outside diameter or smaller. For outside diameters over three-fourths (¾) inch the wall thickness shall be increased in the ratio of the new diameter to three-fourths (¾) inch.

Unless coils are constructed of a non-ferrous metal they shall be protected against corrosion by an approved process. All connections shall be welded or brazed. Solder used for brazing shall have a melting point of at least one thousand (1000) degrees.

5304.04 Piping and Fittings: No refrigerant lines shall be located in any elevator dumbwaiter or other shaft containing moving objects.

Steel and Iron Pipe.

(a) Standard weight steel or wrought iron pipe may be used for working pressures not exceeding one hundred fifty (150) pounds for pipe sizes two (2) inches and over.

(b) Extra heavy steel or wrought iron pipe shall be used for all pipe one and one-half (1½) inch and under.

(c) Extra heavy steel or wrought iron pipe shall be used for all pipe sizes two (2) inches and over for working pressures exceeding one hundred fifty (150) pounds.

(d) Pipe joints may be screwed, flanged or welded. Screw joints shall conform to the American Standard Association pipe thread Standard No. B-2-1919. Exposed threads shall be tinned or otherwise coated to efficiently inhibit corrosion.

(e) Flanges may be screwed or welded to pipes. All flange fittings shall be of the recess gasket type. Flange bolts shall project through nuts.

(f) Welds may be made by the fusion process, both acetylene and electric arc. Pipe may be butt end welded by the electric resistance process. All welding shall be in accordance with the standards issued by the Department for the Inspection of Steam Boilers, Unfired Pressure Vessels and Cooling Plants.

(g) No cast metal fittings, except steel, semi-steel malleable iron or bronze shall be used.

(h) Reducing fittings shall be used for one (1) or more pipe size reductions.

(i) No bends exceeding ten (10) degrees, made at place of installation, shall be made in standard butt welded pipe. All bends exceeding ten (10) degrees which shall have a center radius of not less than six (6) pipe size diameters shall be not less than extra heavy pipe. Bends shall be substantially circular and free from wrinkles and kinks.

(j) Piping shall be supported to prevent excessive vibration and strain at joints or connections by hangers or braces of solid wrought iron or steel, not less than one by one-eighth (1 x ⅛) inches cross-section. And for horizontal lines, shall be spaced not farther than eight (8) feet apart.

(k) Seamless steel tubing, if used, shall conform to the thicknesses for standard and extra heavy pipe.

Copper Tubing and Piping.

Seamless soft copper tubing may be used provided:

(a) Systems contain not over one hundred (100) pounds of refrigerant.

(b) Tubing does not exceed five-eighths (⅝) inch in diameter nor is less than thirty-four one-thousandths (.034) of an inch in wall thickness.

(c) Copper tubing is inclosed in iron pipe or tubing or other rigid metal inclosures, except flexible metal inclosure may be used at bends or terminals if not exceeding six (6) feet in length.

(d) No inclosure shall be required for connections between compressor, condenser or box; provided however, that such connections other shell type apparatus and nearest riser do not exceed six (6) feet in length and are located within the refrigerating machinery room.

(e) Every opening from or into an enclosed conduit for refrigerant lines shall be free from sharp edges which might injure tubing.

(f) Flared, flanged or soldered joints may be used in connections.

(g) All valves and fittings, except those on the evaporator and at compresser and all connections of tubing shall be arranged in or on a metal box or other suitable inclosure and shall be rigidly attached thereto or to the supports thereof. Every such box or inclosure shall have an accessible door or removable cover.

(h) Piping shall be supported by hangers or other supports so as to sustain a weight of two hundred (200) pounds applied at the point of support.

(i) Fittings shall be of forged brass or wrought copper.

(j) No tubing joint shall be placed in conduit.

(k) Flexible metal inclosures, where used, shall be constructed of a non-ferrous material.

Godair Memorial Old Peoples Home — N. Max Dunning, Architect

American Medical Association Bldg. — Holabird & Root, Architects

Hard copper pipe may be used without being inclosed on conduit, provided:

(a) Pipe shall conform to the following thicknesses:

Nominal Pipe Size (Inches)	Decimal Equivalent of Wall Thickness
¼	.065
⅜	.065
½	.065
¾	.065
1	.065
1¼	.065
1½	.072
2	.083
2½	.095
3	.109
3½	.120

Note: The outside diameter of nominal size pipe is one-eighth (⅛) inch larger than the pipe size shown above.

(b) Pipe shall be made from copper having a purity of at least 99.90 per cent, as determined by electrolytic assay, silver being counted as copper. It shall be free from cuprous oxide, as determined by a microscopic examination at a magnification of seventy-five (75) diameters.

(c) Pipe shall have a tensile strength ranging from thirty-six thousand (36,000) pounds per square inch to not more than forty-three thousand (43,000) pounds per square inch and shall be legibly marked with manufacturer's trade mark and symbol on each length of tubing for the purpose of identification.

(d) Connections shall be made by the sweated type fittings and solder used shall be made of clean, unused metals having a melting point of not less than four hundred (400) degrees Fahrenheit and not more than five hundred (500) degrees Fehrenheit.

(e) Fittings shall be of forged brass, wrought copper or bronze.

(f) Pipe shall be supported by solid hangars spaced not more than eight (8) feet apart and hangars so placed that no soldered joints shall be under tension. Vertical risers shall be supported at each floor and at base of riser.

5304.05 Valves: Stop or service valves shall be provided on systems in accordance with the following schedule:

Location of Valves	Class of System A	B	C	D	E	
Compressor, Generator or other pressure imposing device	X	X	X	X	X	
Condenser	X	X	Outlet	Outlet		
Liquid Receiver	X	X	Outlet	Outlet	Outlet	
Dry Evaporator	X	X	Inlet	Inlet		(See Note)
Flooded Evaporator	X	X	X	X	X	
Branch Headers	X	X				(See Note)

Where "X" is shown in the above schedule valves shall be provided on inlet and outlet refrigerant connections.

In a Class C and Class D system having two (2) or more evaporators in separately refrigerated spaces valves will be required on both inlet and outlet of evaporator.

Where branch headers are used on a Class C or D system with two (2) or more evaporators located in separately refrigerated spaces valves shall be required on the inlet and outlet of the branch headers.

5304.06 Safety Devices:

(a) **General.** Refrigerating systems normally operated above atmospheric pressure shall be equipped with safety devices according to the following table:

Safety Device Location and Venting	Non-Flammable and Non-Irritant: Class 1 System Class A	B	C	D	E	Class 2 System Class A	B	C	D	E	Irritant or Flammable Refrigerants: System Class A	B	C	D	E
Pressure Limiting Device	X	X	X	X		X	X	X	X		X	X	X	X	
Pressure Vessel Relief Device	X	X	X	X	X	X	X	X	X	X	X	X	X	X	X
Vent to Atmosphere	X	X				X	X	X	X		X	X	X	X	X
Compressor or other pressure imposing element Relief Device	X	X	X			X	X	X			X	X	X		
Emergency Relief Valve	X					X	X				X	X			
Discharge Check Valve						X					X	X			
Quick Closing Suction Valve						X					X	X			

No connection shall be made with the public water supply which will impair the purity thereof. Water used for removing heat from a refrigerating system shall not thereafter be used for drinking purposes. When a regulating valve is used it shall be connected between compressor or condenser and city water supply.

(b) **Pressure Vessel Relief Devices.** Shell type apparatus such as condensers, liquid receivers, evaporators, liquid separators and absorbers, shall be equipped with an approved pressure relief device set at a pressure not to exceed the pressure for which vessel is designed. The inlet size of these relief devices shall be based on the refrigerating capacity for which the pressure vessel is designed and shall be sized in accordance with the following table:

Refrigerating Capacity	Relief Device Pipe Size (Inches)
Up to 15 tons	⅜
16 tons to 30 tons	½
31 tons to 60 tons	¾
61 tons to 100 tons	1
101 tons to 175 tons	1¼
176 tons to 250 tons	1½
251 tons to 450 tons	2
451 tons to 900 tons	Two 2

The discharge connection from these relief devices shall be equal to or greater than the inlet connection to the relief device.

Fusible plug shall not be permitted as of relief device.

Goodyear Tire & Rubber Co., 25th and South Parkway, Chicago
L. G. Hallberg & Co., Architects

(c) **Compressor or Other Pressure Imposing Device, Relief Devices.** Compressors on Class A, B and C refrigerating systems, except those containing fifty (50) pounds or less of non-irritant and non-flammable refrigerant shall be equipped with a pressure relief device located between the pressure imposing element and the stop valve on the discharge side. Such valves shall be set at not more than ninety (90) per cent of the test pressure specified for the refrigerant used. Where a flammable or irritant refrigerant or a Class 2 non-flammable or non-irritant refrigerant is used, this device shall be vented into the low pressure side of the system. Where a Class 1 non-flammable and non-irritant refrigerant is used this device may discharge to the atmosphere but shall be piped to discharge toward the ceiling, not less than seven (7) feet above the floor. Systems containing one hundred (100) pounds or more of Class 1 non-flammable and non-irritant refrigerant shall be vented to outside atmosphere, or to suction side. In systems where the pressure imposing element is a compressor, the inlet pipe size of these relief devices shall be based on the gross displacement of the compressor, and shall conform to the following table:

Cubic Feet Per Minute	Relief Device Size (Inches)
Up to 25	⅜
26 to 100	½
101 to 200	¾
201 to 350	1
351 to 580	1¼
581 to 850	1½
851 to 1500	2

In systems where the pressure imposing element is other than a compressor, the Department for the Inspection of Steam Boilers, Unfired Pressure Vessels and Cooling Plants, shall determine the size of the relief devices to be used.

The discharge connection from these relief devices shall be equal to or greater than the inlet connection to the relief device.

(d) **Pressure Limiting Devices.** Pressure limiting devices, where required, shall stop the action of the pressure imposing element at a pressure less than eighty (80) per cent of the test pressure for the refrigerant used.

(e) **Stop Valve.** No stop valve shall be located between a pressure relief device or pressure limiting device and the part of the system protected thereby, unless two (2) devices of required size are used, and so arranged that only one (1) can be shut off at any one (1) time.

(f) **Emergency Relief Valves.** A hand operated relief valve, or other approved means for transferring refrigerant from building or buildings where an emergency condition may exist, must be provided in accordance with table of safety devices.

Where a hand operated relief valve is used it shall be of two (2) inch diameter for Class B system and three (3) inch diameter for Class A systems. Such valves shall be connected to the low pressure side of the system and arranged or located so that they may be operated outside of the entrance to machinery room.

These valves shall be vented to the atmosphere as further provided.

Valves shall be identified with a permanent tag containing instructions for operation.

(g) **Check Valves.** Absorption systems shall be provided with check valves located between the rectifier and condenser and in the discharge line of the aqua pump as close to the pump as is practical.

(h) **Quick Closing Suction Valves.** A quick closing suction valve shall be installed in the suction pipes of each pressure imposing element in accordance with the above schedule. These valves shall be arranged for manual closing from a point near the entrance to the room in which the element is located, or to automatically close in case of sudden failure of pressure in the suction pipe at the element.

(i) **Pump Out Connection.** Compressors of Class A and B refrigerating systems, except those operating below atmospheric pressure or those containing a Class 1 non-irritant and non-flammable refrigerant, shall be equipped with pump out connections. These connections shall be designed to reverse the normal flow of refrigerant so that the refrigerant contained in the high pressure side may be pumped into the low pressure side of the system.

(j) **Pressure Gauges.** Class A, B, C and D refrigerating systems shall be equipped with pressure gauges on the high and low pressure side. Every liquid level gauge glass shall be of sufficient strength to withstand the high side test pressure and, except those of the bull's eye type, shall have automatic closing shut-off valves and shall be adequately protected against injury.

(k) **Discharge of Refrigerant.** Refrigerant shall be discharged as follows:

1. Where the discharge of a refrigerant from a pressure relief valve to the outside atmosphere is required, the discharge pipe outlet shall be not less than twelve (12) feet above the ground and not closer than ten (10) feet to any opening in building, or closer than twenty (20) feet to any fire escape, and shall be turned downward. The discharge pipe shall be not less than the size of the relief device outlet.

2. Where ammonia is used in a Class B or C system, refrigerant may be discharged into a tank of water which shall be used for no purpose except ammonia absorption. At least one (1) gallon of fresh water shall be provided for every one (1) pound of ammonia in the system. The water used shall be prevented from freezing without the use of salt or chemicals. Tank shall be substantially constructed of not less than one-eighth (⅛) inch steel. No horizontal dimension shall be greater than one-half (½) the height. The tank shall have a hinged cover or, if closed, shall have a vent of ample capacity at the top of the tank only. The discharge pipe for the pressure relief valve shall be equipped with a diffuser and shall discharge the ammonia near the bottom of the tank.

3. An emergency relief valve on a system using ammonia shall discharge into such tank of water or into an ammonia mixer consisting of a closed cylinder designed for fifty (50) pounds per square inch working pressure with ammonia and water inlet connections and a valveless connection to sewer, so arranged that the ammonia will be completely absorbed before leaving the mixer. The connection shall be so arranged that it is impossible to operate the hand operated ammonia relief valve without first opening the valve supplying the water to the ammonia mixer.

4. Relief devices located on the pressure vessels may be vented to the low pressure side of a system; provided however, that the low pressure side is equipped with relief device vented to the atmosphere.

5. Discharge from more than one (1) relief valve may be run into a common header, the area of which shall be equal to the area required for the total refrigerating capacity of all the vessels.

ARTICLE 5305.

Tests and Operation.

5305.01 Tests:

(a) **General.** It shall be the duty of any person, firm or corporation, installing refrigerating systems to apply tests to said systems as required by the chief inspector of the Department for the Inspection of Steam Boilers, Unfired Pressure Vessels and Cooling Plants.

Tests shall be made in the presence of an inspector of the Department for the Inspec-

tion of Steam Boilers, Unfired Pressure Vessels and Cooling Plants and shall be applied to every refrigerating system, or parts thereof, installed before being put into use; provided however, that the department for the inspection of steam boilers, unfired pressure vessels and cooling plants may accept from the manufacture of a single unit system a certified copy of test report showing that tests of said system, as required by this ordinance were made at the place of manufacture.

It shall be the duty of every person, firm or corporation installing refrigerating systems, as aforesaid, to notify the chief inspector of the department for the inspection of steam boilers, unfired pressure vessels and cooling plants whenever any system has reached the stage of construction where it is ready for inspection tests.

The testing of every refrigerating system shall include a pressure or vacuum test of the complete piping system preferably with the evaporator installed but valves thereon may be closed to prevent withdrawal of the refrigerant.

Under these tests a partial vacuum of twenty (20) inches of mercury shall be produced within the system and shall be held for a period of at least one (1) hour with no detectable drop.

Every part of every refrigerating system shall be designed, constructed and assembled to withstand safely, and without injury, the required minimum test pressures, which shall be applied for the pressure test. These test pressures shall be held by every refrigerating system under test for a period of one (1) hour or over without any appreciable fall.

(b) **Gases to Be Used for Testing.** Pressure tests shall be made with air, carbon dioxide, nitrogen or other inert gas approved by the department. When practical, water may be used. When testing with air, care shall be taken to prevent the temperature at any point from rising above one hundred thirty (130) degrees Fahrenheit.

5305.02 Operating Precautions: It shall be the duty of the person in charge of the premises wherein any refrigerating system is located to exercise due diligence to see that the refrigerating system is properly maintained and operated at all times.

Two (2) gas helmets or masks suited to the refrigerant used shall be provided with every Class A system and with every Class B system that operates above atmospheric pressure except when carbon dioxide is used.

Every gas helmet or mask shall be of a type approved by the Chief Inspector of the Department for the Inspection of Steam Boilers, Unfired Pressure Vessels and Cooling Plants as being suitable for the refrigerant used, shall be inspected annually by said Chief Inspector and shall be kept in operative condition in an easily accessible case or cabinet outside of entrance to machinery room.

Whenever the losses of refrigerant from a system are such as to endanger the health or lives of the human occupants of any room or structure in which such refrigeration system, or any part thereof, is located, it shall be the duty of the owner of such refrigeration system to apply suitable pressure or other tests to prove the system tight.

Every system which may be charged after installation shall have the charging connection located on its low pressure side. No container shall be left connected to a system except while charging or withdrawing refrigerant.

Refrigerants withdrawn from refrigerating systems shall be transferred to containers as prescribed by the Regulations of the Interstate Commerce Commission for the transportation of such refrigerant.

5305.03 Instructions and Refrigerant Charges to be Posted: It shall be the duty of the installer or operator of any refrigerating system, except a Class D and E system, to post and keep conspicuously posted as near as practicable to the pressure imposing element of such system a card giving operating directions for such system, including precautions to be observed in case of a breakdown or leak. The following shall be included in the instructions and information listed on such cards:

1. Instructions for shutting down system in case of emergency.
2. The name, address and telephone number of every engineer or operator in charge.
3. The name, address and day and night telephone number for service.
4. The location of the nearest fire alarm box.
5. The telephone number of the Department for the Inspection of Steam Boilers, Unfired Pressure Vessels and Cooling Plants, and instructions to notify said department immediately in case of serious leakage or other emergency.
6. The date, amount and kind of refrigerant placed in the system at the time of initial charging and every subsequent recharging and the name and address of the person, firm or corporation charging or recharging the system.

Every single unit refrigerating system shall have attached in a permanent manner and in a conspicuous place within or on the refrigerated space, a metal tag or other emblem containing the following information, in legible words and figures: maker's name, kind and amount of refrigerant used, test pressures applied.

Every Class A, B, C and multiple Class D system shall have the name of the refrigerant painted or affixed in a permanent manner either to the pressure imposing element or to the piping in proximity thereto. The kind of refrigerant and instructions for shutting off refrigerant shall be prominently posted at the branch valves used for shutting off each evaporator or set of evaporators contained within each refrigerated space.

ARTICLE 5306.

Existing Refrigerating Systems.

5306.01 General: Existing refrigerating systems installed in accordance with the Revised Chicago Code of 1931, amendments added thereafter and ordinances enacted prior thereto and in operation previous to the date of passage of this ordinance, shall be maintained in a safe condition; and alterations, replacement and changes shall be made in such refrigerating systems, when, due to deterioration, wear and tear, such refrigerating systems have become or are likely to become dangerous to public health or public safety.

Section 2. Any person, firm or corporation that violates, neglects or refuses to comply with, or who resists or opposes the enforcement of any of the provisions of this ordinance, where no other penalty is provided, shall be fined not less than Twenty-five Dollars ($25.00) nor more than Two Hundred Dollars ($200.00) for each offense and every such person, firm or corporation shall be deemed guilty of a separate offense for every day on which such violation, neglect or refusal shall continue; and any builder or contractor who shall construct any building in violation of the provisions of this ordinance, and any architect designing, drawing plans for, or having charge of such building, or who shall permit it to be constructed, shall be liable to the penalties imposed and provided by this section.

Section 3. All provisions of the building ordinances of the city in effect prior to the passage of this ordinance shall remain in full force and effect except insofar as they conflict with any of the foregoing provisions.

Section 4. This ordinance shall be in full force and effect from and after its passage and due publication.

CHICAGO ZONING ORDINANCE

AN ORDINANCE.

An Ordinance establishing a plan for dividing the City of Chicago into districts for the purpose of regulating the location of trades and industries and of buildings and structures designed for dwellings, apartment houses, trades, industries, and other specified uses, for regulating the height, volume, and size of buildings and structures, and intensity of use of lot areas, for determining building lines, and for creating a board of appeals.

Be it ordained by the City Council of the City of Chicago:

1859. Section 1. **Interpretation; Purpose.** In interpreting and applying the provisions of this ordinance such provisions shall in every instance be held to be the minimum requirements adopted for the promotion of the public health, safety, comfort, morals or welfare.

1860. Section 2. **Definitions.** Certain words in this ordinance are defined for the purposes thereof (unless there is express provision excluding such construction or the subject matter or context is repugnant thereto) as follows:

(a) Words used in the present tense include the future; the singular number includes the plural and the plural the singular; the word "building" includes the word "structure".

(b) **Alley**—A narrow thoroughfare upon which abut generally the rear of premises, or upon which service entrances of buildings abut, and is not generally used as a thoroughfare by both pedestrians and vehicles, or which is not used for general traffic circulation, or which is not in excess of 30 feet wide at its intersection with a street.

(c) **Apartment House**—A building which is used or intended to be used as a home or residence for two or more families living in separate apartments.

(d) **Auxiliary Use**—A use customarily incidental to and accessory to the principal use of a building or premises located on the same premises with such principal use.

(e) **Block**—A block shall be deemed to be that property abutting on a street on one side of such street and lying between the two nearest intersecting or intercepting streets, or nearest intersecting or intercepting street and railroad right of way or waterway.

(f) **Building**—A building is a structure entirely separated from any other structure by space or by walls in which there are no communicating doors or windows or similar openings.

(g) **Depth of Lot**—The depth of a lot is the mean distance from the front street line of the lot to its rear line measured in the general direction of the side lines of the lot.

(h) **Dwelling House**—A building used or intended to be used as a home or residence in which all living rooms are accessible to each other from within the building and in which such living rooms are accessible without using an entrance vestibule, stairway or hallway that is designed as a common entrance vestibule or common stairway or common hallway for more than one family, and in which the use and management of all sleeping quarters, all appliances for cooking, ventilating, heating, or lighting, other than a public or community service, are under one control.

(i) **Family**—One or more individuals living, sleeping, cooking and eating on the premises as a single housekeeping unit.

(j) **Grade**—The finished grade of premises improved by a building is the elevation of the surface of the ground adjoining the building The established grade of premises whether vacant or improved is the elevation of the sidewalk at the property line as fixed by the City. Where the finished grade is below the level of the established grade, the established grade shall be used for all purposes of this ordinance.

(k) **Garage**—A public garage, except as otherwise provided by this paragraph, is a building or premises arranged, designed, and intended to be used for the storage of motor vehicles for hire or reward, or which does not come within the definition of a private or community garage as herein set forth. A private garage is a building with ground area not in excess of 80 square feet arranged, designed, and intended to be used for the storage on the ground floor of not more than 4 individually owned passenger automobiles devoted to the private use of the owner, when such garage is located on the same premises, as an auxiliary use, with the residence or apartment or business of the owner of such automobiles so stored, and where no fuel is sold. A use as a private stable shall be subject to the same ground area regulations for the purposes of this ordinance as the regulations controlling the ground area of a private garage. Where two or more separate private garages, each having a ground area not in excess of 200 square feet, are located on the rear half of the premises, not more than one of such garages having a vehicle entrance on a public street, such garages collectively shall be deemed a community garage, but a group of two or more private garages on a single lot not so located or arranged or any one of which is in excess of 200 square feet in area shall be deemed a public garage.

(l) **Height of Building**—The height of a building shall be the vertical distance measured in the case of flat roofs from the mean level of the established grade to the level of the highest point of the under side of the ceiling beams adjacent to the street, and in the case of a pitched roof from such grade to the mean height level of the under side of

the rafters of the gable. Where a block has a frontage on a two-level street the upper street level may be used to determine the height of buildings for a distance back from such frontage not in excess of one-half the depth of the block at right angles to such frontage, but not farther back than the alley most nearly parallel to such street in any case. Where a structure is set back from the street line, the mean level of the finished grade of the premises along the line of that part of the structure nearest the street line may be substituted for the established grade for the purpose of determining the height of a building. Where no roof beams exist or there are structures wholly or partly above the roof, the height shall be measured from the established grade or finished grade to the level of the highest point of the building.

(m) **Lot**—A parcel of land or premises occupied, or which it is contemplated shall be occupied, by one building with its usual auxiliary buildings or uses customarily incident to it, including such open spaces as are required by this ordinance and such open spaces as are arranged and designed to be used in connection with such building, shall be deemed a lot for the purposes of this ordinance. A corner lot shall be deemed to be that property which has an area not in excess of 8,000 square feet, and which abuts on two streets making an angle on the lot side of not greater than 120 degrees.

(n) **Non-conforming Use**—A non-conforming use is a use which does not comply with the regulations of the use district in which it is situated.

(o) **Public Space**—A park, public square, or submerged land under the jurisdiction of a park district shall be deemed a public space.

(p) **Street**—A thoroughfare used for public foot and vehicle traffic other than an alley as herein defined, shall be deemed a street.

(q) **Street Line**—The street line is the dividing line between a street and the lot. The front street line shall be deemed to be the shortest street line.

(r) **Street Wall**—The street wall, for the purposes of this ordinance, shall be deemed that wall or part of a wall of a building, or that part of the wall of a porch or other structure, nearest to and most nearly parallel with the street, extending more than 4 feet 6 inches above the finished grade.

(s) **Volume of Building**—The volume of a building shall be the contents in cubic feet of that space between the grade used in determining the height of buildings and the mean level of the roof (except as otherwise specifically provided by Section 16, Paragraph (a),) including scenery lofts and other storage spaces, cooling towers, elevator bulkheads, towers, penthouses, water tanks or water towers, dormers, bays, covered ways, covered porches or other spaces not open to the sky, and courts, provided that certain courts or certain parts thereof opening on thoroughfare or public spaces, cornices projecting beyond the exterior walls, piers or columns, or the space under the projection of a cornice, chimneys, parapet walls, structures extending into thoroughfares or public spaces, architectural finials or open framework wireless towers shall not be included as a part of the volume of a building. No court except an open court unobstructed from the street or alley or other public place by walls for its full width shall be excluded from the volume of a building. An offset court opening on an open court but having a wall between the offset court and the thoroughfare or public place, or that part of a court not open to the sky, shall not be within the definition of an open court or of a part of an open court. The distance between the mean level of the top of the enclosing walls of the court and the mean level of the bottom of the court shall be used to determine the volume of such cour .

DISTRICTS AND USES.

1861. Section 3. **Use of Districts.** For the purpose of classifying, regulating and restricting the location of trades and industries and the location of buildings designed for specified industrial, business, residential, and other uses, the City of Chicago is hereby divided into four classes of districts: (1) Residence districts, (2) Apartment districts, (3) Commercial districts, and (4) Manufacturing districts; as shown on the use district map which accompanies this ordinance. The said use district map, consisting of forty-nine separate parts all of which are sections of the same map covering the entire territory of the City, the volume district map, also containing forty-nine parts each of which relates to the corresponding part of the use district map, as amended and the index map and chart containing the explanation of symbols and indications which appear on said use district and said volume district maps, are hereby made a part of this ordinance. The use districts designated on said map are hereby established. No building shall be erected nor shall buildings or premises be used for any purpose other than a purpose permitted by this ordinance in the use district in which such buildings or premises is or are located.

1862. Section 4. **Residence Districts.** (a) In a Residence district no building or premises shall be used nor shall a building be erected, altered, or enlarged which is arranged, intended or designed to be used for an A, C, or M use as defined hereinafter. In a Residence district no building or premises shall be used nor shall any building be erected, altered, or enlarged which is arranged, intended, or designed to be used except for R uses or special uses exclusively as hereinafter provided.

(b) For the purposes of this ordinance, R uses are hereby defined as uses designed for and permitted in Residence districts and conforming to the provisions relating to such districts; and all R uses are classified as R1, R2, R3, or R4 uses as follows:

R1 Use—An R1 use shall include every use as a dwelling house.

R2 Use—An R2 use shall include every use as golf or tennis grounds or similar use, church, convent, parish house, public recreation building, community center building, music school, university, public school, juvenile dancing school, or a private or boarding school or college unless such private or boarding school or college is operated so as to bring it within the definition of a C use

R3 Use—An R3 use shall include every use as a public park, public playground, or railway passenger station.

R4 Use—An R4 use shall include every use as a tree or plant nursery, farm, truck garden, greenhouse (unless such greenhouse is operated as a retail business), and a railway right of way not including yard tracks or industrial tracks.

1863. Section 5. **R Use Limitations.** In a Residence district no building shall be erected or used and no building shall be erected which is arranged, intended, or designed for an R2 use unless such building or use is located—

On premises adjoining a street under the jurisdiction of a park district;

On premises adjoining or across a street or alley from a railway right of way;

On premises on the same street and adjoining premises or directly across a street from premises where there exists a building devoted to an R2 or R3 or special use as hereinafter defined;

On corner premises diagonally or directly across a street from premises upon which is maintained an R2 or R3 or special use;

On premises entirely surrounded by streets or alleys;

On premises three sides of which adjoin streets;

On premises adjoining or immediately across a street from an Apartment, Commercial or Manufacturing district;

On premises adjoining on the same street premises where there exists a building devoted to a non-conforming use;

On premises already devoted to an R2 or R3 or special use; or

On premises located in a block in which there are no premises devoted to dwelling house purposes.

1863a. Section 6. **Apartment Districts.** (a) In an Apartment district no building or premises shall be used nor shall a building be erected, altered, or enlarged which is arranged, intended, or designed to be used for a C or M use as defined hereinafter. In an Apartment district no building or premises shall be used nor shall any building be erected, altered, or enlarged which is arranged, intended, or designed to be used except for R or A uses or special uses exclusively as hereinafter provided.

(b) For the purposes of this ordinance, A uses are hereby defined as uses other than R uses, designed for and permitted in Apartment districts and conforming to the provisions relating to such districts; and all A uses are classified as A1, A2, or A3 uses as follows:

A1 Use—An A1 use shall include every use as an apartment house.

A2 Use—An A2 use shall include every use as a boarding house, lodging house, or a hotel which is maintained within the limitations in Apartment districts imposed thereon by this ordinance.

A3 Use—An A3 use shall include every use as a public library, public museum, public art gallery, hospital or sanitarium, an eleemosynary institution except as otherwise classified, or a private club excepting a club the chief activity of which is a service customarily carried on as a business.

1864. Section 7. **Auxiliary Uses in Residence or Apartment Districts.** (a) Auxiliary uses which do not alter the character of the premises in respect to their use for residential purposes shall be permitted in Residence and Apartment districts. Auxiliary uses shall include the following, but the enumeration of such cases shall not be deemed to prevent proper auxiliary uses that are not referred to:

Signs not over 12 square feet in area advertising the premises for sale or for rent which are located (if space occupied by buildings does not prevent) not nearer to adjoining premises than 8 feet or nearer to a street line than the building line established by this ordinance;

The office of a surgeon, physician or dentist, clergyman, lawyer, artist or other professional person located in the dwelling or apartment used as the private residence of such persons.

Customary home occupation located in a dwelling, studio, or apartment and carried on only by the members of the husehold of the person occupying such dwelling, studio, or apartment as his private residence, provided no window or other display or sign is used to advertise such occupation other than a window card 1 square foot in size;

The renting of one or more rooms or the providing of table board in a dwelling or apartment occupied as a private residence, provided no window or other display or sign is used to advertise such use;

A public dining room or restaurant located in a hotel provided that the public entrance to such dining room or restaurant is from the lobby of the hotel, and further provided that no window or other display or sign is used to advertise such use;

Such facilities or retail shops as are required for the operation of a hotel or apartment house, or for the use or entertainment of guests or tenants of the hotel or apartment house, when conducted and entered only from within the building; provided no street window or other exterior sign is used to advertise such use; and further provided that in an apartment district which is also in a 4th or 5th volume district, at any time after May 15, 1933, but not previously (and no construction shall be given to the following language which would permit the uses therein named or any of them before the expiration of said period), an auxiliary use shall be deemed to include a retail shop on the ground floor of an apartment house or hotel (which apartment house or hotel is not less in height than 120 feet), and such shop having a store front with show windows on and an entrance from a street, with such signs only as are on the glass of said window or entrance door; provided, however, that no such retail shop, such store front or entrance, or such sign shall be used for any purpose or business (1) which is not suitable to the neighborhood and to the main occupancy of said apartment house or hotel, (2) which involves the trucking of material through the abutting or adjacent streets or alleys in sufficient quantities to produce undue congestion in such streets or alleys or to interfere with the usual functioning of those streets or alleys, or (3) which is of such character as an automobile or automobile tire or accessory business, or heavy machinery display or sales room, garage, meat market, bakery, grocery store, hardware store, ice cream parlor, soda water fountain, gasoline filling station, street front lunch room or cafeteria, undertaking establishment, laundry, amusement place, or any other use of an objectionable character; and the specific enumeration above of certain uses shall not be held to exclude other uses which are unsuited to the neighborhood although not specifically enumerated.

Private dining halls, dormitories, printing presses, students' laboratories or workshops, playgrounds, athletic fields, or other customary facilities in connection with an R2 use;

A news or refreshment stand or restaurant in connection with a passenger station;

Recreation and service buildings in a public park or public playground;

A private garage or private stable in connection with an R use, limited in ground area to 10 per cent of the area of the lot, but not in excess of the ground area prescribed for or in excess of the capacity limits of a private garage; provided, however, that a private garage or private stable in connection with an R use shall not be located on the same lot with another private garage or private stable or community garage;

A private garage or private stable or community garage in connection with an A use in an Apartment district, limited in ground area to 15 per cent of the area of the lot, provided that a community garage auxiliary to an A1 use shall not be composed of a greater number of private garages than the number of separate dwelling apartments located on the same lot.

In a Residence district on a lot occupied by a dwelling one sign only, which sign shall not exceed one square foot in area, bearing the name or occupation of the occupant of such dwelling, placed not nearer to the street than the building line established by this ordinance.

A sign in an Apartment district not over two square feet in area, placed not nearer to the street than the building line established by this ordinance, announcing the existence of an enterprise permitted on the premises.

Public charitable or religious institutions in Residence or Apartment districts may have a sign or bulletin board not over twelve square feet in area displaying the name or services therein provided.

(b) Auxiliary uses shall not include:

A garage or stable in connection with a non-conforming use except a private garage or private stable whose ground area does not exceed 10 per cent of the area of the lot:

A driveway or walk used for access to a C or M use:

A billboard, signboard or advertising sign, store, trade, business, garage or stable, except such as are hereinbefore specifically permitted.

1865. Section 8. **Commercial Districts.** (a) In a Commercial district no building or prem-

ises shall be used nor shall a building be erected, altered, or enlarged which is arranged, intended, or designed to be used for M uses as defined hereinafter. In a Commercial district no building or premises shall be used nor shall any building be erected, altered, or enlarged which is arranged, intended, or designed to be used except for R, A, or C uses or special uses exclusively as hereinafter provided.

(b) For the purposes of this ordinance, C uses are hereby defined as uses other than R and A uses, designed for and permitted in Commercial districts, and conforming to the provisions relating to such districts; and all C uses are further defined and classified as C1, C2, or C3 uses as follows:

C1 Use—A C1 use shall include every use as

Airplane hangar or airplane repair shop;

Amusement park or pier, skating rink, baseball park, or race track, if such park or pier, rink, baseball park, or track is operated as a business for purposes of private profit;

Armory or arsenal, except where ammunition is manufactured;

Automobile repair shop, automobile parts or tire repair or vulcanizing shop, public garage, automobile fuel or service station;

Advertising sign;

Convention hall;

Driveway or walk used for access to any C or M use;

Financial institution;

Greenhouse operated as a retail business;

Internal combustion engine operated in connection with any use permitted in a Commercial district, provided such engine is equipped and operated only with a competent muffling device;

Office;

Public or private institution, except an institution otherwise classified;

Railroad or water freight station, or storage, team, loading or unloading track or private track, or wharf; provided that the handling of materials, products, or articles at such station, track, or wharf shall be subject to the same limitation and restrictions as apply to the district in which the station, track, or wharf is located;

Restaurant, laundry, theatre, dance hall, billiard room or bowling alley, if such restanrant, laundry, theatre, dance hall, billiard room or bowling alley is operated as a business for purposes of private profit.

Retail store, retail trade, vocation, profession, or shop for custom work or the making of articles to be sold at retail on the premises to the ultimate consumer, storage in warehouse of materials or products permitted as a C2 use; provided the operation of such store, trade, vocation, profession, shop, or storage does not involve the handling of materials, products, or articles across the public sidewalks in sufficient or considerable amounts so as to interfere with the free, safe, and continuous passage of pedestrians along such walks; and provided such store, trade, vocation, profession, shop, or storage does not involve the handling or trucking of materials, products, or articles, through the abutting or adjacent streets or alleys in sufficient quantities as to produce undue congestion in such streets and alleys or interfere with the usual functioning of those streets or alleys;

School for dancing except as hereinbefore classified, trade or vocational school other than an M use, horseback riding school;

Wholesale sales office or sample room;

Provided the operation of any such specified use is not offensive or noxious by reason of the emission of odors, fumes or gases, dust, smoke, noise or vibrations.

C2 Use—A C2 use shall include all uses not otherwise classified, provided all materials and products are stored and all manufacturing operations are carried on entirely within substantial buildings completely enclosed with walls and roof, and provided no operations are of such a nature as to become offensive or noxious to the occupants of adjoining residence or apartment uses by reason of the emission of odors, fumes or gases, dust, smoke, noise, or vibrations; and C2 uses shall include such uses as

Carpet cleaning, provided no dust is permitted to escape from the building;

Cigars, cigarettes, or smoking tobacco manufacturing;

Clay or glass products manufacturing, decorating, or assembling, provided no individual kiln capacity exceeds 200 cubic feet and no kiln is fired except by oil, gas, or electricity;

Cotton, wool, flax, hair, hemp, leather, felt, paper, cardboard, cork, rubber, fur, feathers, horn, bone, shell, celluloid, fiber articles or products manufacturing, or the manufacturing of articles or products from similar materials, but not including uses otherwise classified;

Felt manufacturing, provided no dust is permitted to escape from the building;

Ink manufacturing, not including the preparation of linseed or resin oils;

Lumber sawing, planing, dressing, shaping, pressing, turning, bending, carving, assembling, including carpenter shop for any kind of repairing or manufacturing except as otherwise classified whether or not the product is sold at retail on the premises;

Metal planing, shaping, bending, grinding, milling, drilling, die sinking, forging (except an M use), coring, punching, stamping, pressing, soldering, welding, riveting (other than snap riveting), buffing, polishing, or finishing, plating, galvanizing, sherardizing, tempering, annealing, hardening, other than by processes or operations which emit odor or noise of a disagreeable or annoying nature for the manufacturing of metal products; casting of aluminum, babbitt, brass, bronze, iron, lead, white metal for the manufacture of metal products, provided no metals are melted except in melting pot the capacity of which does not exceed 500 pounds or in electric furnace the capacity of which does not exceed 500 pounds, and further provided no pneumatic chippers are employed; assembling, not including an M2 or M3 use, of metal products or parts, or of metal assembled with other materials, except by processes or operations which emit noise of a disagreeable or annoying nature; sheet metal, tin, copper, brass workers' shop, plumbing shop, wagon shop, or machine shop, whether or not the product or service is sold at retail on the premises;

Painting, enameling, japanning, lacquering, oiling, staining, or varnishing shop, whether or not the product or service is sold at retail on the premises;

Pharmaceutical products, toilet preparations, patent or proprietary medicines, or baking powder manufacturing, provided no toxic or corrosive fumes, offensive odors or dust are permitted to escape from the building;

Rubber products manufacturing from Para, plantation or non-ill-smelling African rubbers in which sulphur chloride is not used;

Shoddy or shoddy felt manufacturing, provided no dust is permitted to escape from the building;

Storage of such materials or products as acids, bark, broom corn, cotton, chemicals, clothing, drugs, dry goods, eggs, farm products, feed, food products, fruits, furniture, glass, groceries, hardware, hemp, hops, household goods, ice, junk, jute, liquors machinery metals, millinery, naval or ship stores, paint, paper, pipes, plaster, produce, rags roofing materials, rice, rope, rubber, scenery, shop or mill supplies sugar, tobacco, textiles, vegetable fibre such as hemp, jute or others not specifically mentioned, waste paper, wines, wood; storage in underground tanks of oils, petroleum or inflammable fluids in quantities and under conditions permitted by other ordinances;

Wholesale produce salesroom or market;

Wholesale, packing, repacking, labeling, consigning or storage warehouse; also

Every use of manufacturing, assembling, repairing, packing, finishing, or storage, or any legal use not otherwise classified, if conducted wholly within a building generally occupied by more than one manufacturing use and customarily called a loft building, without serious annoyance or injury to other usual occupants of the same building and without affecting by reason of noxious odors, fumes or gases, or excessive dust, noise, vibration, or danger, a business or other use or activity which is customarily carried on or may be carried on wholly within the same loft building with the C2 use or which may be conducted on adjacent premises.

C3 Use—A C3 use shall include, provideo all materials and products are stored and a manufacturing operations are carried on entirely within substantial buildings completely enclosed with walls and roof, and provided no operations are of such a nature as to become offensive or noxious to the occupants of adjoining premises devoted to or adapted for other uses, by reason of the emission of odors, fumes or gases, dust, smoke, noise, or vibrations, the following uses:

Brewery;

Cement products such as concrete blocks, pipe, garden furniture manufacturing;

Custom dyeing or cleaning, clothes cleaning, steam cleaning;

Distilled liquors or spirits manufacturing except an M use;

Feed manufacturing, except from refuse, offal or tankage;

Food products, beverages, confections manufacturing, preparation, compounding, baking, canning, packing, or bottling, including the grinding, cooking, roasting, preserving, drying, smoking, or curing of meats, fruits, or vegetables, except a C1 use or a use otherwise classified;

Fuel distributing station (except a C1 use) from which fuel is sold at retail and where all fuel is unloaded from carriers and loaded upon carriers and stored entirely within substantial enclosed buildings, provided the operation of said station is carried on without the emission of dust or noise;

Ice manufacturing for purposes of sale;

Milk or ice distributing station from which truck or wagon deliveries are customarily made;

Poultry killing, packing, or storage for purposes of sale at wholesale;

Paint or enamel blending, including all operations except operations which are M uses or other processes from which offensive or noxious odors, gases or fumes escape from the building;

Soap manufacturing from refined oils or fats, provided competent condensers or other appliances shall be operated where necessary to comply with the definition or the intended definition of a C3 use, and excepting the use of low grade greases, oils or tallow or other ingredients which emit noxious odors;

Stable for the housing of more than 8 horses or cows, livery or boarding or sales stable.

1866. Section 9. **Auxiliary Uses in Commercial Districts.** (a) Auxiliary uses shall be permitted in a Commercial district. An auxiliary use to a C1 or C2 use shall not include a stable for the housing of more than 8 horses or cows or a livery or boarding stable.

(b) An auxiliary use in a Commercial district shall include an M1 storage use as hereinafter defined, provided such M1 storage use shall not occupy in excess of 50 per cent of that part of any premises wholly within a Commercial district, nor shall such M1 storage use be located nearer to a Residence or Apartment district than 50 feet, and further provided that such M1 use shall not be located nearer to a street upon which the C use abuts than 50 feet where a Manufacturing district does not adjoin the same street in the same block or in a block directly across the street from the C use, but an auxiliary M1 use shall in any case be permitted in that part of a Commercial district within 50 feet of a railroad right of way other than a street railway.

(c) An auxiliary use shall not include an M1 use other than storage, nor an M2 or M3 use as hereinafter defined.

1867. Section 10. **C Use Limitation.** (a) A C1 use shall not include a C2 or C3 use. A C2 use shall not include a C3 use.

(b) No C2 use which is not auxiliary to and incidental to a C1 use, if such C2 use is located in that part of a Commercial district which is nearer at any point to a Residence or Apartment district than 125 feet, shall be operated between the hours of 8 P. M. and 6 A. M., if such operation involves the trucking or hauling of materials or products during such hours or if such operation involves processes of a nature such as to disturb the occupants of said Residence or Apartment districts between the hours of 8 P. M. and 6 A. M.

(c) No C2 use or part thereof, except a storage warehouse or more than one of such uses collectively or individually, together with auxiliary uses thereto, shall be established on more than one-half of the total floor space of a building located in that part of a Commercial district which is nearer than 125 feet at any point to a Residence or Apartment district, but floor space equal to the ground area of any premises in such part of a Commercial district may be occupied by C2 uses in any case although in excess of the said one-half; and such part of a Commercial district located within 125 feet of a railroad right of way other than a street railway, or located adjoining or across a street or across an alley from a Commercial district which is not restricted by the provisions of this paragraph or from a Manufacturing district, shall be exempt from the floor space restrictions of this paragraph. That portion of a building or premises wholly within such part of a Commercial district shall be deemed a separate building or separate premises for the purpose of determining the areas limited by the provisions of this paragraph.

(d) No C3 use or part thereof, together with auxiliary uses thereto, shall be established in that part of a Commercial district which is nearer at any point to a Residence district or Apartment district than 125 feet.

(e) No opening in the side or rear wall or roof of a public garage shall be nearer to the boundary line of a Residence or Apartment district than 16 feet.

1868. Section 11. **Manufacturing Districts.** (a) In a Manufacturing district no building or premises shall be used nor shall a building be erected, altered, or enlarged which is arranged, intended, or designed to be devoted to a use prohibited in the City of Chicago by any other ordinance. In a Manufacturing district no building or premises shall be used nor shall any building be erected, altered, or enlarged which is arranged, intended, or designed to be used except for R, A, C, or M uses or special uses exclusively as hereinafter provided.

(b) For the purpose of this ordinance, an M use is hereby defined as any use for an occupation, business or activity other than an R, A, or C use, that may lawfully be carried on within the city and shall include every lawful use except an R, A, or C, or special use. All M uses are further defined and classified as M1, M2, or M3 uses as follows:

M1 Use—An M1 use shall include such storage, manufacturing or other uses of property coming within the definition of an M use as do not injuriously affect the occupants of adjacent uses and are so operated that they do not emit dust, gas, smoke, noise, fumes, odors, or vibrations of a disagreeable or annoying nature.

An M1 storage use shall include such uses as

Above ground tanks for the storage of oils, petroleum or other inflammable fluids in quantities not greater than 3,000 cubic feet, except as prohibited or otherwise regulated by other ordinances;

Wholesale lumber yard, retail or mill lumber yard; wood yard; the storage in bulk or in yard or in shed of such products or materials as articles manufactured or in the process of manufacture (except as otherwise classified), asphalt, bark, barrels, boxes, brick, cement, cord wood, cotton, contractor's equipment, crates, creosoted products, gravel, iron, junk, lime, machinery, pipe, plaster, rags, roofing, sand, scrap iron, scrap paper, stone, tar, terra cotta, timber, vehicles; or the storage of any other products or materials which do not emit dust, gas or odors of a disagreeable or annoying nature.

An M1 use, provided such use does not customarily emit dust, gas, smoke, noise, fumes, odors, or vibrations which may be offensive or noxious to the adjacent R or A or C uses and does not injure the operation of adjacent C or M uses, shall include also every such use as

Bleaching and dying of yarns, textiles, or felt in case sulphur colors or materials which create offensive odors are not used.

Chalk, graphite, emery, corundum, carborundum, whiting, mercury salts, white lead, red lead, zinc salts, lithopone, plaster, pumice, or talc products manufacturing from the dry materials, or the manufacturing of products from other dust producing materials; provided no operation is contrary to the general definition of an M1 use;

Clay, glass, or shale products manufacturing except a C2 use, including the refining or blending of the raw materials;

Crematory except a crematory located in a cemetery;

Fabricating, other than snap riveting or processes used in bending and shaping of metal which emit noises of a disagreeable or annoying nature, for assembling metal products; forging of metals, melting, casting of metals or manufacturing of steel or alloys of steel from iron, provided no cupola is employed; and further provided no operation is contrary to the general definition of an M1 use;

Paper or strawboard manufacturing from waste paper stock or pulp board;

Railroad freight, storage or classification yard; railroad shop or roundhouse;

Stone, marble or granite grinding, dressing, or cutting; provided no operation is contrary to the general definition of an M1 use;

Varnish or enamel manufacturing from balsam gums, copal, or spar and turpentine, alcohol or benzine and other ingredients which do not emit disagreeable or noxious fumes or gases;

Or any use not otherwise classified which is not contrary to the general definition of an M1 use and not contrary to the classification of such uses herein made.

M2 Use—An M2 use shall include the uses set forth hereunder, provided such use does not customarily emit dust which is not controlled by competent dust collecting appliances, or such use is one which from the nature of the materials handled or processes customarily employed, emits dust, gas, smoke, noise, fumes, or odors, to such an extent as to affect the health, safety, comfort, morals or welfare of occupants of R or A or C uses located not farther than 400 feet from the M2 use, and which use does not customarily emit corrosive or tarnishing gases or fumes which injure C or M uses distant 100 feet or more from the M2 use, or which does not create vibrations to an extent that would damage buildings or affect the position or alignment of machinery erected with usual permanency on premises distant 100 feet or more from the M2 use; in which classification, subject to the conditions named, is every such use as

Bone grinding from soft bone;

Carpet beating or cleaning;

Chalk, graphite, emery, corundum, carborundum, whiting, mercury salts, white lead, red lead, zinc salts, lithopone, plaster, pumice or talc products manufacturing from the dry materials, or the manufacturing of products from other dust producing materials;

Chewing tobacco or snuff manufacturing;

Coffee roasting or manufacturing of coffee substitutes where roasting of cereals is done;

Dyes manufacturing from coal tar derivatives;

Emery, corundum or carborundum, graphite products manufacturing by the employment of grinding processes;

Foundry compound or parting sand manufacturing;

Fuel gas or illuminating gas manufacture or purification;

Fuel gas or illuminating gas storage or the storage above ground of other inflammable fluids except as otherwise classified;

Fuel pocket, tipple, trestle, dump or yard, wholesale or retail, other than a C3 use;

Grain elevator;

Grease, lard, fat, or tallow rendering or refining, except from refuse or rancid fats;

Linseed oil, or similar oils, manufacturing, boiling, or refining;

Lithopone manufacturing;

Live stock corrals or pens, stock yards;

Metal fabricating processes or the assembling of materials where snap riveting is done, or where processes creating noises permitted in the general definition of an M2 use are carried on, for the manufacturing of such products as locomotive or power plant boilers or similar boilers; cranes, dredges, derricks, excavating buckets, locomotives, railroad and electric cars, ships, steel and wood cars, steel truck bodies; structural and reinforcing steel for buildings, bridges, ships and other structures; wire fence, wire lath and reinforcing wire; forging, melting, heating or casting of metals or their alloys, employing all processes, except a use otherwise classified, for the manufacturing of such prodnets as armor plate, automobile or wagon springs, brake shoes, cast iron pipe, cast iron safes, drop forgings, furnaces, ingot-molds, iron or steel billets, plates, sheets, structural shapes, rails, tubes, molding machinery, railroad car wheels, axles, or springs;

Nail, tack or rivet manufacturing where heading or cutting machines are employed;

Operation of internal combustion engines without competent muffling devices;

Paper manufacturing, except as otherwise classified:

Planing mill;

Plaster or plaster of Paris manufacturing;

Pumice stone grinding or refining;

Rubber products manufacturing from Para, plantation or non-ill-smelling African rubbers, in which sulphur chloride is used;

Sausage casings, gut strings or similar products manufacturing;

Sewage purification by Imhoff, activated sludge or similar processes;

Shellac refining;

Slaughtering;

Shoddy or shoddy felt manufacturing;

Soap manufacturing, except a use otherwise classified;

Soya bean oil, or china wood oil manufacturing or refining;

Stone crushing and screening; stone grinding, cutting or buffing not otherwise classified; stone quarry;

Varnish or enamel manufacturing if animal glues or shellac are used as ingredients of the varnish or enamel;

Vinegar or yeast manufacturing;

White lead or red lead manufacturing; whiting manufacturing;

Or any use not otherwise classified or which is not contrary to the general character of M2 uses as indicated by the classification herein contained and the conditions imposed.

M3 Use—An M3 use shall include all M uses which are excluded from the M1 and M2 classification; including every such use as:

Animal black, bone black or lamp black manufacturing;

Asphalt manufacturing or refining; asphalt or similar preservative coating or impregnation of fibre materials or wood where heat is applied;

Cattle or sheep dip manufacturing;

Chlorine or bleaching powder manufacturing; electrolysis of brine;

Coal distillation, including derivation of such products as gas, ammonia, or coal tar;

Coal tar, refuse grain, fermented refuse grain, bones or wood distillation;

Cottonseed oil, or similar oils, manufacturing, boiling or refining;

Creosote manufacturing or refining;

Dyeing of yarn, textiles, or felt, except a use otherwise classified;

Fertilizer manufacturing from organic matter or minerals;

Fish curing, cooking, smoking or canning; fish oil manufacturing or refining;

Glue, size or gelatine manufacturing, where the processes include the refining or recovery of products from fish or animal refuse or offal;

Grain drying or poultry feed manufacturing from refuse mash from breweries or from refuse grain;

Gypsum refining;

Hydrochloric, nitric, sulphuric, or sulphurous acid manufacturing;

Incineration, drying, or reduction or storage, of garbage, offal, refuse, dead animals or other refuse;

Lime kiln;

Ore or slag pile or dock;

Petroleum or kerosene refining or distillation or derivation of by-products;

Portland, slag, or natural cement manufacturing;

Rubber products manufacturing from or the refining of ill-smelling African or similar rubbers;

Slaughter house refuse, or other refuse, or rancid fats, or refuse dead animals, cooking, boiling, or rendering;

Smelting or refining of such metals or their alloys as aluminum, iron, lead, steel, tin, zinc, from the ores;

Starch, dextrine, or glucose manufacturing; sugar refining;

Tanning of hides or pelts, also storage, curing or cleaning of raw hides or pelts;

Wool scouring, washing of hair from tanneries, or from slaughter houses; washing of feathers or similar operations;

Or any other use that is lawful within the city, which would be harmful by reason of dust, gas, smoke, noise, fumes, odors, vibrations, soot, sudden fire or explosion or any other causes to a use otherwise classified at a distance of 2,000 feet or more from the M3 use; provided the uses set forth hereunder are not contrary to the provisions of any other ordinance of the City of Chicago.

1869. Section 12. **M Use Limitations.** (a) An M1 use shall not include an M2 or M3 use and shall not be classified as an M1 use if such M2 or M3 use is present; an M2 use likewise shall not include an M3 use.

(b) No M2 use shall be established nearer to a Residence or Apartment district than 400 feet nor nearer to a Commercial district than 125 feet.

(c) An M3 use shall not be established nearer to a Residence, Apartment or Commercial district than a distance at which the M3 use would not from any cause be offensive or noxious to the occupants of such Residence, Apartment or Commercial district, but the distance of an M3 use from a Commercial district shall not be less than 500 feet nor shall the distance from a Residence or Apartr n district be in any case less than 2,000 feet.t

1870. Section 13. **Special Uses.** (a) For the purposes of this ordinance all special uses are classified as follows:

Airdrome;

Street car barn;

Cemetery;

Circus, carnival, carousal, open air or tent show or similar use, operated for purposes of private profit;

Hospital or sanitarium for the care of contagious diseases or incurable patients;

Institution for the care of the insane or feeble-minded;

Penal or correctional institution;

Police or fire station;

Public service water reservoir, filtration plant, or pumping station;

Public service or institutional light, heat or power plant except auxiliary use;

Public utility gas plant, electric station or substation:

Telephone exchange.

(b) A special use or the extension of an existing special use may be located in any district without restriction as to the distance from any other district, provided such location or such extension will not seriously injure the appropriate use of neighboring property.

1871. Section 14. **Non-conforming Uses.** (a) A non-conforming use existing at the time of the passage of this ordinance may be continued.

No exterior sign aggregating more than 12 square feet in area shall hereafter be erected to advertise a non-conforming use.

(b) A non-conforming use shall not be extended, but the extension of a use to any portion of a building which was arranged or designed for such non-conforming use at the time of the passage of this ordinance shall not be deemed the extension of a non-conforming use.

(c) A building other than an A3 use arranged, designed or devoted to a non-conforming use at the time of the passage of this ordinance may not be reconstructed or structurally altered to an extent exceeding in aggregate cost, during any ten-year period, 50 per cent of the value of the building unless the use of such building is changed to a conforming use.

(d) A non-conforming A3 use may be enlarged or extended within the limitations of the volume district in which it is located.

(e) A non-conforming yard storage use shall not be expanded in area of storage space so used.

(f) A non-conforming advertising sign use if removed from the premises may not be replaced.

(g) A non-conforming use shall not be changed unless changed to a more restricted use; provided, however, that in a Residence district an M use shall not be changed unless changed to a conforming use.

(h) A non-conforming use if changed to conforming use shall not thereafter be changed back to any non-conforming use.

(i) A non-conforming use if changed to a more restricted non-conforming use shall not thereafter be changed unless to a still more restricted use.

(j) In a Residence district an A1 use shall not be changed to an A2 use.

(k) In a Manufacturing district no existing M use shall be deemed to be non-conforming except where such use is nearer at the time of the passage of this ordinance to a Residence or Apartment or Commercial district, as the case may be, than the minimum distance as prescribed by this ordinance.

(l) For the purposes of this ordinance a use shall be deemed to be changed if changed from a use included in a use class to a use not included in such class.

(m) A non-conforming use except as hereinbefore provided shall be deemed to be changed to a more restricted use if the use to which such non-conforming use is changed is a use included in a use class that in the arrangement of classes precedes the class in which such non-conforming use is included. The classes shall be deemed to be arranged

in order of precedence as R, A, C1, C2, C3, M1, M2 and M3, as hereinbefore defined.

1872. Section 14A. **Junk Yards—Locations.** It shall be unlawful for any person, firm or corporation to carry on or engage in the business of keeping a junk store or a junk yard upon any street in the city upon which is located a street railway line.

1873. Section 15. **Size of Building.** For the purpose of regulating and limiting the height and bulk of buildings hereafter to be erected, of regulating and limiting the intensity of the use of lot areas, and of regulating and determining the area of open spaces within and surrounding such buildings, the City of Chicago is hereby divided into five classes of districts: 1st Volume district, 2nd Volume district, 3rd Volume district, 4th Volume district and 5th Volume district, as shown on the volume district map which accompanies this ordinance, such volume district map being referred to in Section 3, and by said Section 3 made a part of this ordinance. The volume districts designated on said map are hereby established. No building or part of a building shall be erected except in conformity with the regulations herein prescribed for the volume district in which said building is located. No lot area shall be so reduced or diminished nor shall a building be so enlarged that the volume of the building shall be greater or the open spaces shall be smaller than hereinafter prescribed. The open spaces required for a particular building shall not be included as a part of the required lot or yard areas of any other building.

1874. Section 16. **1st Volume District.** In a 1st Volume district, (except as provided by Section 21 of this ordinance):

(a) No building, except a building in a Commercial or Manufacturing district, shall occupy more than 50 per cent of the area of a lot if an interior lot or 65 per cent if a corner lot, exclusive of the area hereinbefore provided for a garage, and the aggregate volume in cubic feet of all buildings on a lot exclusive of the volume of certain attic spaces or spaces above the ceiling level of the story next below the roof and exclusive of the ground story of a garage shall not exceed the area of the lot in square feet multiplied by 10 feet where the lot is not a corner lot, or by 13 feet in the case of a corner lot, or by 36 feet in a Commercial or Manufacturing district; provided that 2/10 feet but not more than a total of 2 feet in any case shall be added to the 10 feet or to the 13 feet for each 100 square feet that the lot of record prior to the date of the passage of this ordinance in a Residence or Apartment district is less in area than 3,600 square feet. Attic space, space above the ceiling level of the story next below the roof of a building or any part of a building, space above the enclosing walls of a church or auditorium, or room or that part thereof contained wholly within the roof space above the level of the enclosing wall or walls, may be enclosed in addition to the volume of a building, provided the cubic content of such space or room or such part thereof is not in excess of the cubic content of the space which would be enclosed by a hip roof making angles of 60 degrees with the horizontal springing from a horizontal plane on the enclosing walls or part thereof of such building, church, or auditorium;

(b) At any street line no building or part thereof shall exceed a height of 33 feet. For each 1 foot that a building or portion of it sets back from any street line, such building or such portion thereof may be erected 2 feet in height in excess of 33 feet. No part of a building shall be erected to a height at any point in excess of 66 feet;

(c) For each 1 foot that a building or portion of it is distant from the center line of an alley, such building or such portion thereof may be erected 3 feet in height. No building or portion thereof shall be erected nearer the center line of an alley than 8 feet;

(d) For each 1 foot that a building or portion of it sets back from all lines of adjacent premises, such building or such portion thereof may be erected 3 feet in height in excess of 30 feet, provided that along lines of adjacent premises in a 2nd Volume district the setback regulation required along lines of adjacent premises in a 2nd Volume district shall apply, and further provided that along the lines of adjacent premises in a 3rd or 4th or 5th Volume district no setback shall be required. For the purpose of this paragraph the height of a building shall be the mean level of the top of a parapet wall or the mean level of the top of the structure. Chimneys are exempt from the provisions of this paragraph.

1875. Section 17. **2nd Volume District.** In a 2nd volume district (except as provided by Section 21 of this ordinance):

(a) Located within a Residence or Apartment district no building shall occupy more than 60 per cent of the area of a lot if an interior lot or 75 per cent if a corner lot, exclusive of the area hereinbefore provided for a garage, and the aggregate volume in cubic feet of all buildings on a lot exclusive of the ground story of a garage shall not exceed the area of the lot in square feet multiplied by 40 feet, or by 50 feet in the case of a corner lot, except that 1 per cent but not more than a total of 5 per cent shall be added to the 60 per cent or 75 per cent respectively for each 100 square feet that the lot of record prior to the date of the passage of this ordinance is less in area than 3,600 square feet;

(b) Located within a Commercial or Manufacturing district the aggregate volume in cubic feet of all buildings on a lot shall not exceed the area of the lot in square feet multiplied by 72 feet.

(c) At any street line no building or any part thereof shall exceed a height of 66 feet. For each 1 foot that a building or portion of it sets back from any street line, such building or portion thereof may be erected 2 feet in height in excess of 66 feet. No part of a building shall be erected to a height at any point in excess of 132 feet;

(d) For each 1 foot that a building or portion of it is distant from the center line of any alley, such building or such portion thereof may be erected 5 feet in height and no building or portion thereof shall be erected nearer to the center line of an alley than 8 feet; provided these regulations shall not be applied along that part of an alley for the 55 feet of its length nearest the street which the alley intersects;

(e) Located within a Residence or Apartment district for each 1 foot that a building or portion of it sets back from any line of adjacent premises, such building or such portion thereof may be erected 3 feet in height in excess of 44 feet, provided that along lines of adjacent premises in a 3rd or 4th or 5th Volume district, this setback regulation shall not be required. Chimneys are exempt from the provisions of this paragraph.

1876. Section 18. **3rd Volume District.** In a 3rd Volume district (except as provided by Section 21 of this ordinance):

(a) Located within a Residence or Apartment district no building shall occupy more than 75 per cent of the area of a lot if an interior lot or 90 per cent if a corner lot, exclusive of the area hereinbefore provided for a garage, and the aggregate volume in cubic feet of all buildings on a lot exclusive of the ground story of a garage shall not exceed the area of the lot in square feet multiplied by 100 feet, or by 120 feet in the case of a corner lot;

(b) Located within a Commercial or Manufacturing district the aggregate volume in cubic feet of all buildings on a lot shall not exceed the area of the lot in square feet multiplied by 144 feet;

(c) At any street line no building or part thereof shall exceed a height of 132 feet. For each 1 foot that a building or portion of it sets back from any street line, such building or such portion thereof may be erected 2 feet in height in excess of 132 feet. No part of a building shall be erected to a height at any point in excess of 198 feet;

(d) For each 1 foot that a building or portion of it is distant from the center line of any alley, such building or such portion thereof may be erected 7 feet in height and no building or portion thereof shall be erected nearer to the center line of an alley than 8 feet, provided these regulations shall not be applied along that part of an alley for the 55 feet of its length nearest the established building line on the street which the alley intersects.

1877. Section 19. **4th Volume District.** In a 4th Volume district (except as provided by Section 21 of this ordinance):

(a) The aggregate volume in cubic feet of all buildings on a lot shall not exceed the area of the lot in square feet multiplied by 216 feet, or by 240 feet in the case of a corner lot which is located also in a Residence or Apartment district; in a Residence or Apartment district the area provisions of Section 18, paragraph (a) shall apply;

(b) At any street line no building or part thereof shall exceed a height of 198 feet. For each 1 foot that a building or portion of it sets back from any street line, such building or such portion thereof may be erected 3 feet in height in excess of 198 feet. No part of a building shall be erected to a height in excess of 264 feet;

(c) For each 1 foot that a building or portion of it is distant from the center line of any alley, such building or such portion thereof may be erected 9 feet in height, provided this regulation shall not be applied along that part of an alley for the 55 feet of its length nearest the street which the alley intersects;

1878. Section 20. **5th Volume District.** In a 5th Volume district (except as provided by Section 21 of this ordinance):

(a) No building or part thereof shall be erected to a height at any street line or alley line in excess of 264 feet, provided, however, that back from the street line or alley line such building or part thereof may be erected so as not to protrude above a plane sloping up at an angle of 30 degrees with the horizontal from such street line or alley line at the height limit a distance from such street line or alley line of 32 feet measured on the slope. The height of such sloping plane shall be the ultimate height of the structure. In a Residence district or Apartment district the area provisions of Section 18, paragraph (a) shall apply;

(b) For each 1 foot that a building or portion of it is distant from the center line of any alley, such building or such portion thereof may be erected 10 feet in height, provided this regulation shall not be applied along that part of an alley for the 55 feet of its length nearest the street which the alley intersects.

1879. Section 21. **General Volume District Provisions.** (a) Where all parts of a cornice of any building or structure are more than 12 feet above the grade as defined in Section 2, paragraph (1) and below a height of 120 feet in a 3rd Volume district or below 186 feet in a 4th Volume district or below 252 feet in a 5th Volume district, and where such cornice extends in whole or in part along the street frontage of a building and where the return of such cornice, if any, along an alley wall is not longer than a distance equal to the width of the alley, such cornice may project into the street a distance of 5 feet and into the alley a distance of 3 feet, but for each 1 foot above the height of 120 feet or 186 feet or 252 feet in the 3rd or 4th or 5th Volume districts respectively, the projection of the cornice shall be reduced 3 per cent of the prescribed 5 feet or 3 feet until a projection of 2 feet shall have been reached. Above the height of a parapet as provided for by paragraph (b) of this section, no part of a structure shall project into a street or alley a greater distance than 2 feet.

(b) Nothing in this ordinance shall prevent the erection above the street line height limit of such structural members as are required to support the roof, or a parapet wall or cornice solely for ornament and without windows, extending above such height limit not more than 5 per cent of such height, but such parapet wall or cornice may in any case be at least 5½ feet high but shall not be higher than 8 feet above such height limit.

(c) Nothing in this ordinance shall prevent the erection in a Manufacturing or Commercial district, above the height and in excess of the volume as provided by this ordinance of grain elevators, conveyors, derricks, gas holders, or other necessary appurtenances to manufacturing or storage operations in connection therewith.

(d) In a 1st or 2nd Volume district which is also in a Commercial or Manufacturing district, or in a 3rd, 4th or 5th Volume district; if the area of a building is reduced so that above the street line height limit it covers in the aggregate not more than 25 per cent of the area of the premises, the building above such height shall be excepted from the volume and street line height limit regulations. The aggregate volume in cubic feet of all such portions of the building shall not exceed one-sixth of the volume of the building as permitted by this ordinance on the premises upon which such portions are erected; provided that for each 1 per cent of the width of the lot on the street line that the street wall above the street line height limit is greater in length than 50 per cent of the width of the lot, such wall shall be erected not nearer to such street line than 1 foot; and further provided that for each 10 feet in height that any such portion of the building is erected above the street line height limit, such portion of the building shall be set back 1 foot from all lines of adjacent premises. For purposes of this paragraph, the permitted volume of a building in the 5th Volume district shall be the cubic contents of the space which may be occupied under the provisions of Section 20 of this ordinance.

(e) The street line height limit in a 2nd, 3rd, or 4th Volume district shall be increased 33-1/3 per cent of such height limit on that frontage of premises which abuts on a street greater in width than 120 feet, or on that frontage of premises directly across the street from a public park, public playground, public waterway, or cemetery, or railroad right of way other than a street railway. The same increase in the street line height limit shall apply to the frontage on a street which intersects or intercepts such street or park or playground or waterway or cemetery or railroad right of way for a distance from such street or park or playground or waterway or cemetery or railroad right of way equal to the depth of the lot under one ownership at the time of the passage of this ordinance but not beyond the boundary of the volume district which that part of the frontage of the lot is in (1) which abuts on a street greater in width than 120 feet, or (2) which is directly across the street from such park, playground, waterway, cemetery, or right of way, in any case. But the provisions of this paragraph shall not be so construed as to increase the ultimate height limit or the volume limit as provided by this ordinance.

(f) In a 3rd, 4th or 5th Volume district which is also in an Apartment district, the entire ground area of the lot up to an ultimate height of 30 feet may be occupied, provided such space shall be used only as a waiting room, lobby, or lounging room or auditorium or service rooms, auxiliary to an R2 or A use, and further provided that the volume as permitted by this ordinance shall

not be increased and further provided that the provisions of Section 22 shall take precedence over all provisions of this paragraph.

(g) Nothing in this ordinance shall prevent the erection above the street line height limit, of spires in connection with an R2 use.

(h) Where premises in one volume district are directly across an alley from a less restricted volume district, all the regulations prescribed by this ordinance pertaining to the distance of a building or part thereof from the center line of an alley for that less restricted district shall be applied to such premises.

(i) Where premises or any portion thereof abut on an alley which also adjoins a railroad right of way, public park, playground or cemetery, or which abut on the end of what is commonly known as a blind alley, the provisions pertaining to distance of a building or part thereof from the center line of an alley shall not apply for such premises or such portion thereof.

(j) A fire escape as required by other ordinances of the City of Chicago, fire-proof outside stairway or solid floor balcony to a fire tower if projected not more than 4 feet into a court or yard, the ordinary projections of window sills, belt courses, if such projections do not exceed 6 inches, shall not be deemed to reduce the area or volume of open spaces. Cornices or similar ornamental features projecting not over 4 feet into courts, which open on a street or alley shall not be deemed to reduce the area or volume of open spaces for the purpose of determining the volume of a building.

(k) Where a lot greater in area than 8,000 square feet, located in a Residence or Apartment district abuts on two intersecting streets at their intersection, the area and volume of the building as permitted by this ordinance may be distributed over the lot.

(l) The provisions of Section 22 shall take precedence over the area provisions of all Volume district sections of this ordinance.

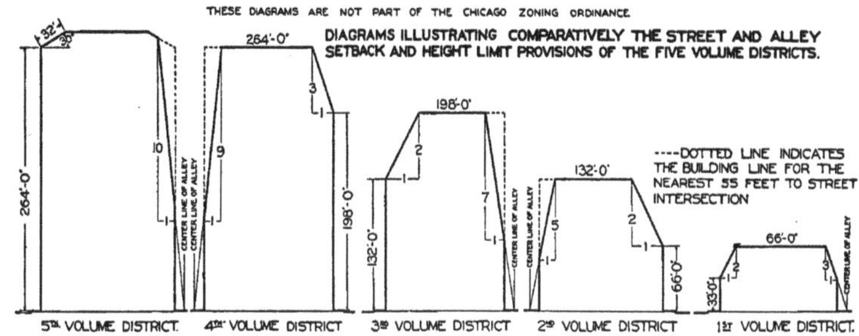

Volume Districts	Use Districts	Lot	Occupancy of lot in per cent of lot area	Volume of building, area of lot times	1 ft. setback from side lot lines for each 3 ft. above	Height limit at street line in feet	1 ft. setback from street line above height limit for added height of	1 ft. setback from center line of alley at grade for height of	Ultimate height of building in feet	No building nearer the center line of alley than
1st	Res or Apt. Res. or Apt. Com. or Mfg.	Interior Corner	50 65 100	10 (D) 13 (D) 36 (D)	30 ft. (G)	33 (F)	2 ft.	3 ft	66 (K)	8 ft
2nd	Res. or Apt. Res or Apt Com. or Mfg.	Interior Corner	60 (B) 75 (B) 100	40 50 72	44 ft. (G) None	66 (F)	2 ft	5 ft. (L)	132 (K)	8 ft. (M)
3rd	Res. or Apt. Res or Apt. Com. or Mfg.	Interior Corner	75 90 100	100 120 144	None	132 (F)	2 ft	7 ft. (L)	198 (K)	8 ft. (M)
4th			100	216	None	198 (F)	3 ft.	9 ft. (L)	264 (K)	
5th			100	No Volume Provision	None	264	(H)	10 ft. (L)	(H) (K)	

NOTES.

(Not a part of the ordinance.)

A—Corner lot maximum area 8,000 square feet.

B—1% (maximum 5%) may be added to the 60% or 75% of the area in a 2nd Volume Residence or Apartment district for each 100 square feet that the lot is less in area than 3,600 square feet.

C—Private or community garage 1 story not included in area or volume limits in 1st, 2nd or 3rd Volume Residence or Apartment districts.

D—Volume of a building includes courts not open to a street or alley; in a 1st Volume district space under a pitched roof (equal in volume to a 60 degree hip roof) may be erected in addition to Volume; 2/10 foot (maximum 2 feet) may be added to Volume factors of 10 feet and 13 feet in a 1st Volume Residence or Apartment district for each 100 square feet that the lot is in area than 3,600 square feet.

E—Height limit at street line is to under side of ceiling beams; parapet (maximum height 8 feet) may be added.

F—Street line height limit may be relaxed where frontage is on a public space.

G—Side lot line set back height limit is to the mean level of the top of fire wall; at a district boundary the least restrictive rule applies.

H—In a 5th Volume district the slope up from the street and alley lines above 264 feet is 30 degrees, for a distance of 32 feet up the slope.

I—Cornices with 5 feet projection are permitted (3 feet projection in an alley back from the street a distance equal to the width of the alley) to a height 20 feet below the height limit of the parapet; for each 1 foot above that height the cornices are reduced in projection 3%.

J—Grain elevators, derricks, gas tanks, etc., are allowed above height limit in a 1st, 2nd or 3rd Volume manufacturing use.

K—Towers in 1st and 2nd Volume commercial or manufacturing districts and in 3rd, 4th or 5th Volume districts. (See paragraph (d) Section 21.)

L—Alley set back in a 2nd, 3rd, 4th or 5th Volume district does not apply for nearest 55 feet to the street which the alley intersects.

M—Distance of buildings from center line of alleys in a 2nd or 3rd Volume district does not apply for nearest 55 feet to the street which the alley intersects.

OTHER PROVISIONS.

1880. Section 22. **Building Lines.** (a) For the purpose of preventing the obstruction to light and air for adjoining premises in Residence and Apartment districts by establishing building lines along the street frontage, no building shall be erected or altered in a Residence or Apartment district which is also in a 1st, 2nd or 3rd volume district or as provided by paragraph (e) of this section in a Commercial district, except in such a manner as to conform to the provisions of this section.

(b) In a Residence district no building shall be erected whose street wall is nearer the front street line than a distance equal to 15 per cent of the average depth of the lots in a block except as hereinafter provided. In an Apartment district no building shall be erected whose street wall is nearer to the front street line than a distance equal to 10 per cent of the average depth of the lots in a block except as hereinafter provided.

(c) Where a block is occupied or partially occupied by buildings which existed in the block at the time of the passage of this ordinance, the average of the distances from the street line of the front street walls of buildings shall be the established building line; but where this average distance does not exceed 10 feet in a block in which the street wall of any existing building is nearer along the front line to the street than 5 feet the street wall may be erected at the street line. Lots occupied by buildings designed for residence uses permitted in a Residence district, unless the aggregate frontage of such lots exceeds 50 per cent of the total frontage in the block, shall be considered as though vacant where located in an Apartment district for the purpose of establishing the building line.

(d) For the purpose of computing the average of the distances of street walls of buildings from the street line, the street wall nearest the street shall be considered as though it were continuous across the entire lot frontage and such average shall be based upon units of lot frontage, but buildings whose street walls are distant from the street line in excess of the provisions of paragraph (b) of this section shall be deemed to exactly conform to the provisions of paragraph (b), and existing auxiliary buildings, temporary buildings, fences, advertising signs, retaining walls, steps, balustrades, or similar existing structures shall not be considered in computing such average.

(e) Along the side of a corner lot in a Residence district or Apartment district which is not known as the front line and which generally is the side having the greatest dimension along a street line and which side line is in the same block with a lot or lots whose street line is the front line, no building shall be erected whose street wall is nearer the street at the rear end of such line than the established building line in the block and for each 1 foot that the building or part thereof is distant from the adjoining lot line exclusive of the width of an intervening alley, if any, such building or such part thereof may be erected 1 foot nearer to the street line. The provisions of this paragraph shall apply to a Commercial district which is also in a 1st or 2nd Volume district and which is in the same block with a Residence district or Apartment district.

(f) Where a lot adjoins premises, the street wall line of which is unrestricted or less restricted by this section, the street wall line of such lot for that 75 per cent of the lot frontage nearest to such unrestricted or less restricted street wall line but not in excess of 30 feet in any case, may conform to the provisions of this section as they apply to such unrestricted or less restricted street wall which it adjoins.

(g) Where any existing building erected prior to the time of the passage of this ordinance has its street wall nearer to the street line than the building line as established by this section, then the street wall of any building erected or altered on that 75 per cent of the frontage not in excess of 30 feet of the adjacent lot which immediately adjoins the lot occupied by such existing building may approach not nearer the street line than the street wall of such existing building.

(h) Cornices, belt courses, an entrance canopy or similar roofed space having not more than 20 square feet of horizontal area covered by roof for each 25 feet of lot frontage, porches or bays projecting not more than 3 feet 6 inches exclusive of cornice and having an aggregate volume at any story not in excess of 35 per cent of the area of that part of the street wall of a building at such story multiplied by 3½ feet, and steps and landings below the level of the first floor, and their balustrades and open fences or railings or similar structures hereafter erected, provided such fences or railings or structures do not obstruct vision to an extent in excess of 40 per cent above a height of 4 feet 6 inches above the established grade, shall be exempt from the restrictions provided by this section.

(i) The premises of each building, with its usual auxiliary buildings, existing at the time of the passage of this ordinance, or premises or part thereof which may hereafter be occupied by buildings, or additions to existing buildings, shall be deemed a lot for the purposes of this section. Lots separated by an alley shall be deemed to be adjoining. All measured distances shall be to the nearest integral foot. If the fraction is ½ foot or less the integral foot next below shall be taken.

1888. Section 23. **District Boundaries.** (a) Whenever a portion of any district is indicated upon the use or volume district map as a strip paralleling an opened or unopened street, the width of this strip, unless delimited on the map by dimensions, lot lines, alleys, railroad or elevated railway rights of way, or otherwise, shall be assumed to be 125 feet measured at right angles from the nearest street line of the street to which it is parallel and adjacent.

(b) The district boundaries are, unless otherwise indicated, either street lines or lines drawn parallel to and 125 feet back from one or more of the street lines bounding a block. Where two or more district designations are shown within a block 250 feet or less in width the boundary of the less restricted district shall be deemed 125 feet

C. A. KLOOSTER

BUILDING CONSTRUCTION

CHICAGO

back from its street line. Where two or more district designations are shown within a block more than 250 feet in width the boundary of the more restricted district shall be deemed 125 feet back from its street line.

(c) Where the street layout actually on the ground varies from the street layout as shown on the use or volume district map, the designation shown on the mapped street shall be applied to the unmapped streets in such a way as to carry out the manifest intent and purpose of the plan for the particular section in question.

(d) Where a district boundary line as defined in this section or as shown on the use or volume district map divides a lot in single ownership at the time of the passage of this ordinance, the use or volume authorized on the least restricted portion of such lot shall be construed as extending to the entire lot, provided this does not extend more than 25 feet beyond the said boundary line of the district in which such use is authorized. The use or volume so extended shall be deemed to be conforming.

(e) The space above the surface of streets, alleys or waterways are to be regarded merely as explanatory of the maps and shall not be deemed to be a part of the use district to which it is adjacent.

(f) Submerged lands which may hereafter be reclaimed, unless otherwise indicated on the use or volume district maps, shall be deemed to be in the same use and volume district as premises not now submerged to which such submerged lands are contiguous.

(g) Areas on the use and volume district maps along the margin of such maps outside of the border line streets are to be regarded merely as explanatory of the maps and shall not be considered as indicating the use or volume indicated thereon.

1882. Section 24. **Completion and Restoration of Existing Buildings.** Nothing herein contained shall require any change in the plans, construction or intended use of a building for which a building permit has been heretofore issued and the construction of which shall have been diligently prosecuted within one year of the date of such permit, and the ground story of which, including the second tier of beams shall have been completed within such year, and which entire building shall be completed according to such plans as filed within three years from the date of the passage of this ordinance; provided the time shall be extended for not to exceed one year, or in cases where one such extension may have been granted the time shall be further extended for one year within which such ground story framework, including the second tier of beams shall be completed in any case where actual construction or fabrication was begun early enough to allow, under the then existing conditions, adequate time for completion as above specified and where such construction or fabrication was diligently prosecuted and where such completion has been prevented by conditions impossible to foresee and beyond the control of the owner or builder. Nothing in this ordinance shall prevent the restoration of a building or an advertising sign destroyed by fire, explosion, act of God or act of the public enemy, not in excess of 50 per cent of the value of the building, or prevent the continuance of the use of such building or part thereof as such use existed at the time of such destruction of such building or part thereof or prevent a change of such existing use under the limitations as hereinbefore provided.

1883. Section 25. **Administration.** This ordinance shall be enforced by the Commissioner of Buildings. The Commissioner of Buildings is hereby empowered and it shall be his duty to administer this ordinance in conjunction with the administration of such portions of the general ordinances of the City of Chicago as are commonly designated as the building code of the City of Chicago in such a manner as to facilitate their joint administration. For the purpose of enforcing this ordinance the authority vested in him under the said building code is hereby declared to be vested in him under this ordinance.

1884. Section 26. **Certificates of Occupancy.** (a) It shall be unlawful to use or permit the use of any building or premises or part thereof, hereafter created, erected, changed or converted wholly or partly in its use or structure, until a certificate of occupancy, to the effect that the building or premises or the part thereof so created, erected, changed or converted, and the proposed use thereof, conform to the provisions of this ordinance, shall have been issued by the Commissioner of Buildings. No change or extension of use and no alterations shall be made in a non-conforming use or premises without a certificate of occupancy having first been issued by the Commissioner of Buildings that such change, extension or alteration is in conformity with the provisions of this ordinance.

(b) Certificates of occupancy shall be applied for at the same time that the building permit is applied for and shall be issued within 10 days after the erection or alteration of the building shall have been completed. A record of all certificates shall be kept on file in the office of the Commissioner of Buildings and copies shall be furnished upon request to any persons having a proprietary or tenancy interest in the building affected.

(c) Pending the issuance of a regular certificate, a temporary certificate of occupancy may be issued for a period not exceeding six months, during the completion of alterations or during partial occupancy of a building pending its completion. Such temporary certificates shall not be construed as in any way altering the respective rights, duties or obligations of the owners or of the city relating to the use or occupation of the premises or any other matter covered by this ordinance, and such temporary certificate shall not be issued except under such restrictions and provisions as will adequately insure the safety of the occupants. No temporary certificate shall be issued if prior to its completion the building fails to conform to the provisions of the building code or of this ordinance to such a degree as to render it unsafe for the occupancy proposed.

1885. Section 27. **Plats.** Each application for a build permit shall be accompanied by a plat in duplicate, drawn to scale and in such form as may be prescribed by the Commissioner of Buildings, showing the actual dimensions of the lot to be built upon, the size of the building to be erected, and such other information as may be necessary to provide for the enforcement of the regulations contained in this ordinance. A careful record of such applications and plats shall be kept in the office of the Commissioner of Buildings.

AN ORDINANCE

Amending the Chicago Zoning Ordinance

Passed January 3, 1934 (Proc. p. 1362-65), amended February 13, 1935 (p. 3725-6); March 27, 1935, (p. 3944-5).

Be it ordained by the City Council of the City of Chicago:

Section 1. That Section 28 of an ordinance entitled, "An ordinance establishing a plan for dividing the City of Chicago into districts for the purpose of regulating the location of trades and industries and of buildings and structures designed for dwellings, apartment houses, trades, industries, and other specified uses, for regulating the height, volume, and size of buildings and structures, and intensity of use of lot areas, for determining building lines, and for creating a board of appeals", passed by the City Council on April 5, 1923, approved April 16, 1923, and published on pages 2396 to 2515, both inclusive, of the printed Journal of the Proceedings of the City Council, as amended by an ordinance passed by the City Council on July 30, 1931, and published on pages 925 to 927 inclusive, of the printed Journal of the proceedings of the City Council, be and the same is

hereby further amended so that said Section 28 shall hereafter read as follows:

"Section 28. **Board of Appeals.)** That there be and there is hereby created a Board of Appeals consisting of five members, in conformity with and under the provisions of the Act of the General Assembly of the State of Illinois, entitled as amended: "An Act to confer certain additional powers upon city councils in cities and presidents and boards of trustees in villages and incorporated towns concerning buildings and structures, the intensity of use of lot areas, the classification of trades, industries, buildings, and structures, with respect to location and regulation, the creation of districts of different classes, the establishment of regulations and restrictions applicable thereto, the establishment of boards of appeals and the review of the decisions of such boards by the court.

The Board of Appeals shall be appointed by the Mayor with the approval of the City Council, provided, however, that a majority of said members at the time of appointment shall be members of one or more of the following organizations: Illinois Society of Architects, Western Society of Engineers, Chicago Real Estate Board, Cook County Real Estate Board, Building Manager's Association of Chicago, Building Construction Employer's Association, Chicago Building Trades Council; or shall be the incumbent of one of the following positions in the City of Chicago: Commissioner of Public Works, City Architect, Commissioner of Police, Corporation Counsel; or shall be a citizen who has had outstanding experience in zoning administration. The members of said Board shall be appointed to serve respectively for the following terms: One for 1 year, one for 2 years, one for 3 years, one for 4 years and one for 5 years, the successor to each member so appointed to serve for a term of five years.

The members of said Board of Appeals shall receive no compensation whatsoever from the City for their services as members of the Board of Appeals. One of the members so appointed to such Board of Appeals shall be designated by the Mayor as Chairman at the time of his appointment. The Mayor shall have the power to remove any member of the board for cause and after a public hearing. Vacancies shall be filled for the unexpired term of the member whose place has become vacant. All meetings of the Board of Appeals shall be held at the call of the chairman and at such other time as such board may determine. Such chairman or in his absence the acting chairman may administer oaths and compel the attendance of witnesses. All meetings of such board shall be open to the public. Such board shall keep minutes of its proceedings, showing the vote of each member upon every question, or if absent or failing to vote, indicating such fact, and shall also keep records of its examinations and other official actions. The Board of Appeals shall designate one of its employes to act as its secretary who has had experience in zoning matters and who is qualified to make stenographic reports of the record of all proceedings of said board and whose duty it shall be to keep a full and detailed record on file in the office of the board of all its proceedings. Every rule, regulation, every amendment or repeal thereof and every order, requirement, decision or determination of the board shall immediately be filed in the office of the board and shall be a public record. Such Board of Appeals shall hear and decide appeals from and review any order, requirement, decision or determination made by the Commissioner of Buildings with reference to this ordinance. It shall also hear and decide all matters referred to it or upon which it is required to pass under this ordinance. The concurring vote of four members of the board shall be necessary to reverse any order, requirement, decision or determination of the Commissioner of Buildings, or to decide in favor of the applicant any matter upon which it is required to pass under this ordinance. Such an appeal may be taken by any person aggrieved or by an officer, department, board or bureau of the municipality. Such appeal shall be taken within such time as shall be prescribed by the Board of Appeals by general rule by filing with the Commissioner of Buildings from whom the appeal is taken and with the Board of Appeals a notice of appeal specifying the grounds thereof. The Commissioner of Buildings from whom the appeal is taken shall forthwith transmit to the board all the papers constituting the record upon which the action appealed from was taken.

An appeal shall operate to stay all proceedings in furtherance of the action appealed from, unless the Commissioner of Buildings from whom the appeal is taken certifies to the Board of Appeals after the notice of appeal has been filed with him that by reason of facts stated in the certificate a stay would in his opinion, cause imminent peril to life or property, in which case proceedings shall not be stayed otherwise than by a restraining order which may be granted by the Board of Appeals or by a court of rècord on application, on notice to the Commissioner of Buildings and on due cause shown.

The Board of Appeals shall fix a reasonable time for the hearing of the appeal and give due notice thereof to the parties and decide the same within a reasonable time. Upon the hearing, any party may appear in person or by agent or by attorney. The Board of Appeals may reverse or affirm, wholly or partly, or may modify the order, requirement, decision or determination as in its opinion ought to be made in the premises and to that end shall have all the powers of the Commissioner of Buildings from whom the appeal is taken.

After a public hearing before the Board of Appeals the City Council may by ordinance determine and vary the application, in harmony with their general purpose and intent, of any of the regulations of this ordinance relating to the use, construction or alteration of building or the use of land in cases where there are practical difficulties or particular hardship in the way of carrying out the strict letter of any of such regulations. Variations in specific cases of practical difficulties or particular hardship may include the following:

(1) Granting of permission to devote premises in a Residence or Apartment District to a non-conforming A use or C use, except a billboard, in a block, or in a block directly across a street from a block in which there exists a non-conforming A use or C use respectively of a similar nature, provided that such permission shall not be so exercised as to permit either such use of premises in blocks where no such use existed at the time of the passage of this ordinance on either side of the street, and further provided that a non-conforming use herein permitted shall not exceed in area of premises or cubical contents of structures the similar non-conforming use then existing. In granting such permission the building line regulation provided by this ordinance for the block shall be maintained, and the use permitted shall be deemed to be non-conforming in the same sense as though it were erected prior to the time of the passage of this ordinance.

(2) The extension of a non-conforming use or building upon the lot occupied by such use or building at the time of the passage of this ordinance. The erection of an additional building upon a lot occupied at the time of the passage of the ordinance by a business or industrial establishment in case such additional building is a part of such establishment, when carrying out the strict letter of the other provisions of this ordinance would result in practical difficulties or extreme and unnecessary hardship.

(3) In undeveloped sections of the city the issuance of temporary and conditional permits

for structures and uses in contravention of the use regulations controlling residence or apartment districts; provided that such uses are important to the development of such undeveloped sections and also provided such uses are not prejudicial to the adjoining and neighboring sections already developed.

(4) In a residence district the location of an R2 use contrary to the provisions of Section 5 provided that such R2 use will not injure neighboring property for dwelling house purposes.

(5) In a 1st, 2nd or 3rd Volume District, the moderate relaxation of the area or volume provisions of Sections 16, 17 or 18 for the erection of a building for an R2 use or A2 use.

(6) Variation in the application of Section 10 or Section 14 in a Commercial District or part thereof in blocks adjoining the city limits of Chicago or on premises across the street from blocks in which the limitations provided by Section 10 do not apply or in neighborhoods where uses existing at the time of the passage of this ordinance are contrary to the provisions of such section or where by reason of amendments to the ordinance the said district or part thereof comes to be within 125 feet of a residence or apartment district, taking into consideration the conditions then existing in the blocks affected by the amendment.

(7) Variation in the application of Section 12 or Section 14 in a Manufacturing District or part thereof in locations where uses existing at the time of the passage of this ordinance are contrary to the provisions of such sections or where by reason of amendments to this ordinance any part of the then existing M2 or M3 uses in such district or part thereof come to be nearer to a Residence or Apartment or Commercial District than permitted by Section 12.

(8) Permission to maintain a C2 or C3 use anywhere in a Commercial District or an M2 or M3 use anywhere in a Manufacturing District which otherwise would not be permitted by this ordinance, where clearly the appropriate use of neighboring property is not injured thereby.

(9) Granting of permission to devote premises in a Commercial District to an M use where clearly the appropriate use of neighboring property is not injured thereby.

(10) Permit the extension of an existing or proposed use or building into a more restricted district under such conditions as will safeguard the character of the more restricted district.

(11) Variation of the area or volume provisions of this ordinance in a block where there exists a structure which exceeds the area or volume requirements respectively of this ordinance, provided, however, that such variation shall not be construed to permit the erection of a structure in excess of the area or volume of any such existing structures.

(12) In a Residence or Apartment District where lots are irregular in shape or where obviously no building line is required by reason of the peculiar conditions or where all light is obtained from public spaces, variation of the area requirements of this ordinance and in such cases a proportionate variation in volume.

(13) Variation in the definition in the height of building where a building is erected with a frontage on a public waterway or on a natural hillside, but such variation shall be made only for the purpose of adjusting the height limits so as to conform with that of neighboring structures.

(14) Variation in the height of building regulations for the purpose of permitting the erection of additional stories to an existing building where it can be shown that the erection of such additional stories are contemplated, and where the original foundations were designed to carry such additional stories.

(15) Granting of permission in an Apartment District which is also in a Second, Third, Fourth or Fifth Volume District to occupy space on the lot in addition to the area or volume permitted by Sections 17, 18, 19, 20, of this ordinance, provided such additional space shall occupy lower floor only and further provided that such additional space shall be used only as a waiting room, lobby or lounging room or auditorium or service rooms auxiliary to an R2 or A use. In granting such permission the building line regulations provided by this ordinance for the block shall be maintained, but the volume permitted by this ordinance may be correspondingly increased.

(16) Alterations or relaxation of the provisions of Section 22, to the extent necessary to prevent undue or peculiar hardship where in any block or portion of a block there are lots not of uniform depths, or irregular shapes or peculiar proportions, forms or topography, or fronting on more than one street, or where any frontage less in length than 100 feet has adjoining it on each side permanently less restricted frontage, or when clearly the general purposes and intent thereof will be better served thereby.

(17) Where a district boundary line divides a lot in single ownership at the time of passage of this ordinance, the extension of the use or volume authorized on the least restricted portion of such lot over the entire lot, provided this does not extend more than 100 feet beyond the boundary line of the district in which the use is authorized.

(18) Granting of permission, for auxiliary uses in Residence and Apartment Districts which do not alter the character of the premises in respect to their use for Residential purposes, for such facilities or retail shops as are required for the operation of a hotel or apartment house or for the use or entertainment of guests or tenants of a hotel or apartment house when conducted and entered only from within the building, provided no street window or other exterior display or other exterior sign is used to advertise such use.

No variation, however, shall be made except by ordinance in a specific case and after a public hearing before the Board of Appeals of which there shall be at least fifteen days' notice of the time and place of such hearing published in an official paper or a paper of general circulation in the City of Chicago, said notice to contain the particular location for which the variation is requested as well as a brief statement of what the proposed variation consists.

Upon the report of the Board of Appeals the City Council may by ordinance without further public hearing adopt any proposed variation or may refer it back to the board for further consideration. Every such variation shall be accompanied by a finding of fact specifying the reason for making such variation. Any proposed variation which fails to receive the approval of the Board of Appeals shall not be passed except by the favorable vote of two-thirds of all the elected members of the City Council.

Section 2. This ordinance shall take effect and be in force from and after its passage and due publication.

1887. Section 29. **Amendments.** The regulations imposed and the districts created by this ordinance may be amended from time to time by ordinance but no such amendment shall be made without a hearing before the committee on buildings and zoning of the city council, which committee is hereby designated by that purpose. At least fifteen days' notice of the time and place of such hearing shall be published in an official paper or a paper of general circulation in the city of Chicago.

In case of written protest against any proposed amendment, signed and acknowledged by the owners of twenty per cent of the

frontage proposed to be altered, or by the owners of twenty per cent of the frontage immediately adjoining or across an alley therefrom, or by the owners of twenty per cent of the frontage directly opposite the frontage proposed to be altered as to such regulations or district, filed with the city clerk, such amendment shall not be passed except by the favorable vote of two-thirds of all of the members of the city council.

1888. **Change of Districts.** If any area is hereafter transferred to another district by a change in district boundaries, by an amendment as above provided, the provisions of this ordinance in regard to buildings or premises existing at the time of the passage of this ordinance shall apply to buildings or premises existing at the time of passage of such amendment in such transferred area.

1889. Section 30. **Violations and Penalties.** For any and every violation of the provisions of this ordinance, the owner, general agent or contractor of a building or premises where such violation has been committed or shall exist, and the lessee or tenant of an entire building or entire premises where such violation has been committed or shall exist, and the owner, general agent, contractor, lessee or tenant of any part of a building or premises in which part such violation has been committed or shall exist, and the general agent, architect, builder, contractor or any person who commits, takes part in or assists in such violation or who maintains any building or premises in which any such violations shall exist, shall for each and every violation and for each and every day or part thereof that such violation continues, be subject to a fine of not more than $200.00. Any person violating the provisions of this ordinance by pursuing a C or M1 use which without operation of approved nuisance prevention equipment or without certain nuisance eliminating processes or methods of operation would be classified as an M2 or M3 use, or an M2 use which without such equipment, processes or methods would be classified as an M3 use, shall be deemed to have committed a separate violation of this ordinance for each day or part thereof that such C or M1 or M2 use is operated in such a manner as to violate the manifest purpose and intent of the definition of a C or M1 or of an M2 use respectively, and each complete unit of equipment shall be deemed a separate use for the purposes of this paragraph and shall be subject to the same penalty as provided herein. Legal remedies for violations shall be had and violations shall be prosecuted in the same manner as is prescribed by law or ordinance for the prosecution of violations of other ordinances, effective in the City of Chicago.

1890. Section 31. **Remedies.** In case any building or structure is erected, constructed, reconstructed, altered, repaired, converted, or maintained, or any building, structure or land is used, in violation of this ordinance or any other ordinance or lawful regulation, the proper authorities of the City of Chicago, in addition to the remedies herein provided for may institute any appropriate action or proceeding to prevent such unlawful erection construction, reconstruction, alteration, repair, conversion, maintenance or use, or to impose a penalty for such violation, or to restrain, correct or abate such violation, in order to prevent the occupancy of said building, structure or land contrary to the provisions hereof, or to prevent any illegal act, conduct, business or use in or about such premises.

1891. Section 32. **Validity of Ordinance.** If any section, paragraph subdivision, clause, sentence or provision of this ordinance shall be adjudged by any court of competent jurisdiction to be invalid, such judgment shall not affect, impair, invalidate or nullify the remainder of this ordinance but the effect thereof shall be confined to the section, paragraph, subdivision, clause, sentence or provision immediately involved in the controversy in which such judgment or decree shall be rendered.

1892. Section 33. **Effect on Present Ordinances.** This ordinance shall not be construed as repealing or modifying any valid ordinances of the City of Chicago now in effect which restrict the location of industries, entertainments, occupations, establishments or enterprises of any kind, either by requiring frontage consents from property owners or residents affected by such location, or by prohibiting or restricting the location of same within a fixed distance from a hospital, church, public school or parochial school, or the grounds thereof, or on or near any class of streets or boulevards or any parks, playgrounds or bathing beaches. As to all other ordinances or parts of ordinances in conflict with any of the provisions of this ordinance, the same are hereby repealed.

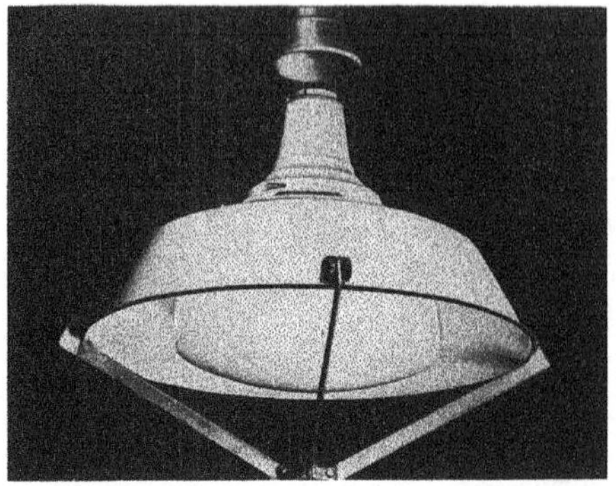

"FLOOR-SERVICE" TURNLOX

The answer to the

Maintenance Problem in Lighting Equipment

Now to Benjamin Turnlox equipment's outstanding superiorities in quick and convenient removal of reflector, lamp and globe for easy cleaning on the floor is added the extra feature of safe and easy removal from the floor by means of a dependable, quick-acting reflector changer and bell-mouth hoods of Benjamin Floor-Service Turnlox construction.

All of the familiar Turnlox advantages are retained. Standing safely on the floor—with a Floor-Service reflector changer clamped to the reflector rim—a slight upward pressure on the changer shaft and less than a quarter turn to the left, releases the lighting fixture from its hood. It is just as simple to put up; just raise the fixture to the hood—the bell mouth makes it easy to locate—and a quarter turn right secures it in place.

OTHER RECENT DEVELOPMENTS

400-Watt Mercury-Mazda Diffuser

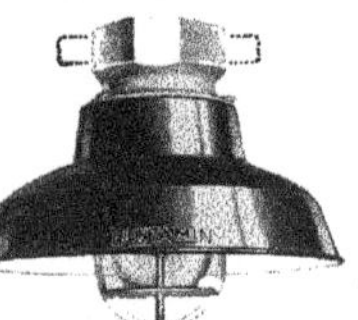

Industrial Vapolet Dome

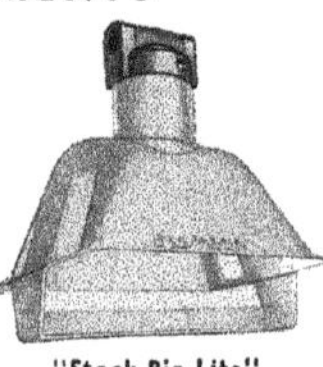

"Stock-Bin-Lite" Reflector

"Silver Lamp" Diffuser

Hinged Dust-Tight Cover

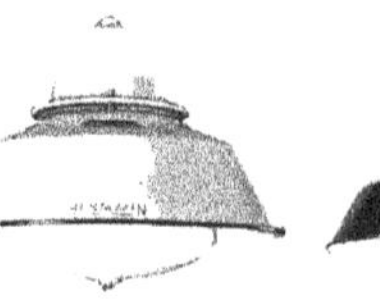

250-Watt Mercury-Mazda Diffuser

Explosion-Proof Dome Unit

WRITE FOR COMPLETE CATALOG

BENJAMIN ELECTRIC MFG. COMPANY

DES PLAINES, ILLINOIS

ELECTRICAL ORDINANCE OF THE CITY OF CHICAGO

Passed August 3, 1938—Extracts From.

ELECTRICAL INSTALLATION AND EQUIPMENT.

ARTICLE I.

Administrative Provisions.

1810. **Title of Ordinance:** This chapter shall be known as the "Electrical Code" of the City of Chicago.

1811. **Administration of Electrical Code:** The Chief electrical inspector of the electrical inspection division shall administer the provisions of this electrical code.

1812. **Non-Liability of City for Damages:** This electrical code shall not be construed to relieve from or lessen the responsibility of any person owning, operating or installing any electrical wires, appliances, apparatus, construction or equipment, for damages to anyone injured by any defect therein by reason of the inspection authorized herein or the certificate of inspection issued by the electrical inspection division; nor shall the City of Chicago be held liable for any damages resulting from the enforcement of the provisions of this electrical code.

1813. **Personal Liability:** In all cases where any action is taken by the chief electrical inspector of the department of streets and electricty to enforce the provisions of any of the sections contained in this electrical code, such acts shall be done in the name of and on behalf of the City of Chicago, and the said chief electrical inspector in so acting for the City shall not render himself liable for any damage that may accrue to persons or property as a result of any such act committed in good faith in the discharge of his duties, and any suit brought against said chief electrical inspector of the department of streets and electricity, by reason thereof, shall be defended by the department of law of said city until final termination of the proceedings contained therein.

1814. **Special Permission to Waive Code Requirements:** The requirements of this code may be modified or waived by special permission in particular cases where such modification or waiver is permitted by "Special Permission" in this code. Such permission shall in all cases be obtained from the Chief Electrical Inspector in writing prior to the commencement of the work.

1815. **Penalty for Violation of This Code:** Any person, firm or corporation that violates any of the provisions of this code, or who maintains any electrical wiring and apparatus found to be dangerous to life and property, shall, upon conviction, be punished by a fine of not less than five dollars and not more than fifty dollars, and every person, firm or corporation shall be deemed guilty of a separate offense for every day such violation shall contimue, and shall be subject to the penalty of this section for each and every separate offense, and so much of any electrical installaion as may be erected or altered and maintained in violation of this ordinance shall be condemned and the electrical inspecion division is hereby empowered to cut off and discontinue current to such electrical wires and apparatus in violation of this ordinance.

ARTICLE II.

Electrical Inspection Department—Electrical Commission—Installation, Alteration, Use and Inspection and Reinspection of Electrical Equipment.

1818. **Duties of the Electrical Commission:** It shall be the duty of the said commission to formulate and recommend safe and practica lstandards and specifications for the installation, alteration and use of electrical equipment designed to meet the necessities and conditions that prevail in the City of Chicago, to recommend reasonable rules and regulations governing the issuance of permits by the electrical inspection division, and to recommend reasonable fees to be paid for inspections made by the electrical inspection division. The standards and specifications, rules and regulations governing the issuance of permits and the fees so recommended, shall become effective upon the passage of an ordinance adopting some by the City Council. All such fees shall be paid to the City Collector.

1819. **Electrical Equipment Defined:** The term "electrical equipment" as used herein is hereby defined as meaning conductors and equipment installed for the utilization of electricity supplied for light, heat or power, and signal installations, but does not include radio apparatus or equipment for wireless reception of sounds and signals and does not include apparatus, conductors and other equipment installed for or by public utilities, including common carriers, which are under the jurisdiction of the Illinois Commerce Commission, for use in their operation as public utilities.

1820. **Electrical Contractor Defined:** The term "electrical contractor" as used in this article shall be understood to many any person, firm or corporation engaged in the business of installing or altering by contract electrical equipment for the utilization of electricity supplied for light, heat or power, not including radio apparatus or equipment for wireless reception of sounds and signals, conductors and other equipment installed for or by public utilities including common carriers which are under the jurisdiction of the Illinois Commerce Commission, for use in their operation as public utilities; but the term "electrical contractor" does not include employees employed by such contractors to do or superwise such work.

1821. **Permits:** No electrical equipment shall be installed or altered except upon a permit first issued by the electrical inspection division authorizing the installation, alteration or repair of electrical equipment.

1822. **Applications:** The electrical inspection division shall issue permits for such installation and alteration of electrical equipment in all cases where application for such permit shall be made in accordance with the rules and regulations applicable thereto; provided, however, that no permit shall be issued for installing or altering by contract, electrical equipment, unless the person, firm or corporation applying for such permit is registered as an electrical contractor as required in Section 1839 of Article IV of this electrical code, and further provided, that the fee as provide for in Article IX shall have been paid in advance upon filing the application.

1823. **Inspection:** The chief electrical inspector or the electrical inspectors of the electrical inspection division shall inspect all electrical equipment installed or altered, except such electrical equipment as may be lawfully exempt, and shhll require that it conform to the provisions of this electrical code.

1824. **Certificate:** Upon completion of such installation or alteration in compliance with this code, the chief electrical inspector shall, on request may by a registered electrical contractor, issue a certificate of inspection covering such installation or alteration; provided, however, that no such certificate shall be issued until all inspection fees for such installation have been paid.

1825. **Reinspection:** The chief electrical inspector or the electrical inspectors of the electrical division are hereby empowered to reinspect any electrical equipment within the scope of this electrical code, and when said equipment is found to be unsafe to life or

property, shall notify in writing the person, firm or corporation owning, using or operating same to place such electrical equipment in a safe and secure condition in compliance with the provisions of this electrical code within such time as the chief electrical inspector shall consider just and reasonable. Refusal or failure to comply with the requirements of such notification shall subject the person, firm or corporation owning, using or operating such electrical equipment to the penalties provided for in Section 1815 of Article I of this electrical code, and in addition thereto the chief electrical inspector or the electrical inspectors of the electrical inspection division are hereby empowered to cut off and stop current to any such electrical equipment found unsafe to life or property.

1826. **Records of Permits:** The electrical inspection division shall keep complete records of all permits issued and inspections made and other official work performed under the provisions of this ordinance.

ARTICLE III.

General Provisions.

1829. **Power to Stop Work:** No registered electrical contractor shall install any electrical conduits, electrical wires, equipment or apparatus in any building or structure, for which a permit is required, until such permit shall have been secured. In case any work is begun on the installation of electrical conduits, raceways or the installation, alteration or repair of electrical wires or apparatus in any building or structure without a permit authorizing the same, being first issued therefor, or the aforesaid installations are being made in violation of the provisions of this code, the chief electrical inspector and the electrical inspectors shall have the power to stop such work at once and to order any and all persons engaged therein, to stop and desist therefrom until the proper permit is secured.

1831. **Use of Equipment:** Whenever any electrical equipment has been installed or altered, no electrical current shall be supplied to or used on such equipment previous to the inspection of such equipment by the chief electrical inspector or the electrical inspectors and the issuance of temporary current permit covering such installation or alteration; provided, however, that the chief electrical inspector may issue a temporary current permit for the use of electrical current during the course of construction or alteration of buildings, which temporary permit shall expire when the construction or aleration of such building is complete.

1832. **Unlawful to Install Meters Previous to the Issuance of Certificate or Temporary Current Permit:** It shall be unlawful for any person, firm or corporation to install electrical energy recording meters on any electrical equipment that has been installed, previous to the issuance, by the chief electrical inspector, of a temporary current permit or a certificate of inspection authorizing the use of current on such installation.

1833. **Breaking Seals:** The chief electrical inspector of the division of electrical inspection, or the electrical inspectors, are hereby empowered to attach to electrical cabinets and equipment, any official notice or seal to prevent use of electricity, and it shall be unlawful for any person to put or attach such seal, or to break, change, destroy, tear, mutilate, cover or otherwise deface or injure any such official notice or seal posted by an inspector of the division of electrical inspection.

1835. **Unlawful to Disturb Existing Wiring:** It shall be unlawful for any person in any way to cut, disturb, alter, or change any electrical wiring or to permit such electrical wiring to be cut, distrubed, altered, or changed, unless done in conformity with the provisions of this code.

1836. **Unlawful to Overfuse Conductors or Apparatus:** It shall be unlawful for any person to overfuse any conductor, motor or apparatus in excess of the maximum allowed by this code for such conductor, motor or apparatus, or to install any substitute in lieu of an approved fuse or device so as to remove or reduce the factor of safety of the same.

1837. **Subcontracts:** When contracts to install electrical work have been obtained by persons, firms, or corporations, that are not registered as electrical contractors, as provided for in Article IV of this code, and the contract is assigned or sublet to a registered electrician on a sub-contract or percentage basis, the name of such registered electrical contractor shall immediately be disclosed by the registered electrical contractor to the other party to the contract in writing.

1838. **Penalties:** Any person, firm or corporation violating any of the sections of this article, shall be subject to the penalties provided for in Article I—Section 1815.

ARTICLE IX.

Inspection Fees.

Inspection Fees as Prescribed for by the Electrical Commission, Adopted as Ordinances by the City Council: That the fees prescribed by the Electrical Commission of the City of Chicago to be paid for the inspection made by the electrical inspection division of all electrical equipment installed or altered within the city of Chicago, be and the same are hereby adopted as reasonable fees for such inspection in the city of Chicago, said fees being as follows:

1847. **Wiring, Not Including Fixtures, Sockets or Receptacles, for Lighting Circuits:** For the inspection of each branch lighting circuit of 1,000 watts or less; one dollar and fifty cents for one circuit, one dollar and twenty cents for each of the next four circuits, one dollar for each of the next five circuits, eighty-five cents for each of the next five circuits, seventy-five cents for each of the next five circuits, sixty-five cents for each of the next five circuits and sixty cents for each succeeding circuit.

1847-A. **Lighting Circuit Defined:** The term circuit as used in sections 1847, 1847-B of this article, shall mean any set of branch lighting conductors which have been extended from a distribution center, and which may be utilized for the transmission of electrical energy.

1847-B. **Branch Lighting Circuits:** For the inspection of each two wire branch lighting circuit the fee shall be as provided for in section 1847. For each two-wire branch circuit of larger capacity than 1000 watts and not more than 2000 watts, the fees shall be as for two 1000 watt circuits; for each three wire circuit of 1000 watts on each side of neutral, the fee shall be as for two 1000 watt circuits; for each three wire circuit of over 1000 watts and not more than 2000 watts on each side of a neutral, the fee shall be as for four 1000 watt circuits; for each three phase four wire circuit of a capacity of 1000 watts or less from phase to neutral, the fee shall be as for three 1000 watt circuits; for each three phase four wire circuit of more than 1000 watts and less than 2000 watts from phase to neutral, the fee shall be as for six 1000 watt circuits.

1847-C. **Outlets:** For the inspection of outlets on existing circuits; ten cents for each outlet on which a socket, receptacle or fixture may be attached.

1847-D. **Electrical Fixtures, Sockets and Receptacles, not Including the Circuit Feeding Same:** For the inspection of fixtures and sockets for lamps of 50 watts, or less, rating, and for inspection of receptacles for current-consuming devices of one-quarter horse power or less, inspection fees shall be: one to fifteen sockets or receptacles, fifty cents; sixteen to twenty sockets or receptacles, seventy-five cents; twenty-one to twenty-five sockets or receptacles, one dollar; twenty-six to thirty sockets or receptacles, one dollar and twenty-five cents; thirty-one to forty sockets or receptacles, one dollar and fifty cents; forty-one

to fifty sockets or receptacles, one dollar and seventy-five cents; fifty-one to sixty sockets or receptacles, two dollars; sixty-one to seventy sockets or receptacles, two dollars and twenty-five cents; seventy-one to eighty sockets or receptacles, two dollars and fifty cents; eighty-one to ninety sockets or receptaeles, two dallars and seventy-five cents; ninety-one to one hundred sockets or receptaeles, three dollars; one hundred and one to one hundred and ten sockets or receptacles, three dollars and twenty cents; one hundred and eleven to one hundred and twenty sockets or receptacles, three dollars and forty cents; one hundred and twenty-one to one hundred and thirty sockets or receptacles, three dollars and sixty cents; one hundred and thirty-one to one hundred and forty sockets or receptacles, three dollars and eighty cents; one hundred and forty-one to one hundred and fifty sockets or receptacles, four dollars; one hundred and fifty-one to one hundred and sixty sockets or receptacles, four dollars and twenty cents; one hundred an sixty-one to one hundred and seventy sockets or receptacles, four dollars and forty cents; one hundred and seventy-one to one hundred and eighty sockets or receptacles, four dollars and sixty cents; one hundred and eighty-one to one hundred and ninety sockets or receptacles, four dollars and eighty cents; one hundred and ninety-one to two hundred sockets or receptacles; five dollars; above two hundred sockets or receptacles, five dollars for the first two hundred sockets or receptacles and twenty-five cents for each group of twenty-five sockets or receptacles or less. For sockets for lamps of rating greater than fifty watts, the fee shall be in proportion to the wattage of the lamps. For receptacles for current-consuming devices of rating greater than fifty watts and not in excess of ¼ horse-power, the fee shall be in proportion to the wattage of the current-consuming device. For the purpose of this section one horse-power shall be equivalent to 746 watts.

1847-E. **Wiring and Fixtures:** For the inspection of both circuit wiring and fixtures, sockets or receptacles: The aggregate sum of the fees as shown above for wiring and for electrical fixtures.

1847-F. **Motors and Other Forms of Power:** For the inspection of electric motors and current-consuming devices of rating in excess of one-quarter horse-power, the inspection fee shall be: First motor or current-consuming device, $2.00, plus ten cents per horse-power; additional motors or current-consuming devices on same permit, fifty cents each plus ten cents per horse-power; for the inspection of receptacles for motors and current-consuming devices of rating greater than one-quarter horse-power, inspection fee shall be based on the total rating in watts of receptacles, transformed to horse-power and computed as stated in this section for motors. The charge for current conversion equipment shall be based on the rating of the driving motor or on the input rating of the conversion equipment. For the purpose of computing inspection fees one horse-power shall be equivalent to 746 watts. Inspection fees as stated in this section shall not apply to illuminated signs outside of buildings.

1847-G. **Temporary Work, Outside Work, Etc.:** Inspections of temporary installations, underground or overhead wires and apparatus, and all other inspections not specifically provided for herein shall be at the rate of $2.00 per hour.

1847-H. **Reinspections:** Inspection fees for reinspection of any electrical apparatus altered, changed or repaired shall be at the rate of $2.00 per hour.

1847-I. **Extra Inspections:** Where extra inspections are made because of inaccurate or incorrect information, failure to make necessary repairs, or faulty construction, a charge of one dollar and fifty cents shall be made for each such inspection.

1847-J. **Minimum Fee:** No inspection shall be made for a less amount than one dollar and fifty cents.

1847-K. **Fees for Projecting Signs Over Public Property — Attachments — Additions:** The feet for all signs projecting at right angles or obliquely from the building or structure against which same are placed, as defined in section 1848, Article X of this electrical code, whether such signs are vertical or horizontal and not being flat signs as hereinafter described, shall be computed at the rate of twenty cents per annum per square foot of sign area, this area to include all area within the perimeter design on each illuminated side of such signs, provided, however, that no fee shall be less than three dollars. Minimum fee for attachments or additions to illuminated signs shall be one dollar and fifty cents.

1847-L. **Fees for Flat Signs:** The fee for all signs placed against a building or strueture, with the face running parallel to the lot line and not projecting obliquely or at right angles therefrom containing twenty-five nominal fifty watts, or fifty volt-ampere lamps or less, shall be two dollars and fifty cents per annum, to which shall be added nine cents for each of the next twenty-five lamps, eight cents for each of the next twenty-five lamps, seven cents for each of the next twenty-five lamps, six cents for each of the next one hundred lamps, five cents for each of the following one hundred lamps, and four cents for each additional lamp above three hundred. Fee if lamps are other than fifty watts or fifty-volt ampere rating shall be based on the total connected load reduced to fifty-volt ampere units and the above schedule applied.

1847-M. **Fees for Temporary Signs:** The fee for illuminated signs installed for temporary use for special occasions, not to exceed thirty days, shall be computed at one-fourth of the annual rate fixed for the particular type or style of sign whether projecting or flat, provided, however, that no fee shll be less than two dollars and fifty cents.

1847-N. **Fee for Re-Erection — Alteration Without Removal:** A minimum fee of two dollars and fifty cents shall be charged for a permit to re-erect any sign removed for repairs, or altered without removal, as defined in Section 1848-I of Article X of this electrical code.

1847-O. **Fees for Illuminated Signs—Not Over Public Property:** The fee for illuminated signs, billboards, signboards, and roof signs not over public property, illuminated by twenty-five fifty-watt, or fifty-volt ampere lamp, or less, shall be two dollars and fifty cents, to which shall be added nine cents for each of the next twenty-five lamps, eight cents for each of the next twenty-five lamps, seven cents for each of the next twenty-five lamps, six cents for each of the next one hundred lamps, five cents for each of the following one hundred lamps, and four cents for each additional lamp above three hundred. Fees, if lamps are of other than fifty watts, or fifty volt-ampere rating, shall be based on the total connected load reduced to fifty-volt ampere unis and the above schedule applied.

1847-P. **Fees for Original Inspection—Annual Reinspection—Electric Lamp Posts—Festoons:** Inspection fees for the original installation of commercial street-lighting equipment shall be at the rate of $2.00 for each lamp post or festoon. Annual compensation and reinspection fee shall be at the rate of $1.50 for each lamp post or festoon.

1847-Q. **Fee for Registration—Term:** The fee for registration as an electrical contractor shall be twenty-five dollars per annum, which sum shall be paid by the applicant to the city collector in advance upon filing the application; provided, that when such application is made by an applicant, not previously registered, in this city, on or after July 1st of any year, the fee for registration shall be $12.50 for the remainer of such calendar year.

1847-R. **Fee for License to Act as Moving Picture Projecting Machine Operator:** The fee to be paid to the city collector before taking the examination for moving picture projecting machine operator shall be fifty (50) dollars. The annual license fee, after the first year, shall be fifteen (15) dollars.

ARTICLE X.

Illuminated Signs Over Public Property.

1848. **General Requirements—Definition:** It shall be unlawful for any person, firm or corporation to erect, attempt to erect or cause to be erected or maintained, any illuminated sign over any sidewalk, street, avenue, alley or public way in the city except in accordance with the ordinances of the city of Chicago.

For the purpose of this article, illuminated signs shall be deemed to be signs constructed as follows: signs, all or any part of the letters or characters of which are made in an outline of electric lamps; signs with painted, flushed or raised letters, or characters lighted by an electric lamp or lamps attached thereto; signs having a border of electric lamps attached thereto and reflecting light thereon; glass signs whether lighted by electricity or other illuminant; signs with painted, flush or raised letters or characters illuminated by electric lamps or lamps placed for the purpose of projecting or reflecting light thereon; and signs having any electrical equipment attached thereto.

1848-A. **Compensation and Inspection Fees —Illuminated Signs Over Public Property:** The owner or person having charge of any illuminated sign authorized by this article which projects in whole or in part over any sidewalk, street, avenue, alley or public way in the city, shall pay for the use of the city, as compensation for the maintenance of same in such place, and to cover the expense of inspection, an annual fee to be computed according to the classification and schedule as provided for in sections 1847L to P inclusive of Article IX of this code.

1848-B. **Permit.** No illuminated sign shall be erected, altered or maintained over any sidewalk, street, avenue, alley or public way in the city except upon a permit first issued by the electrical inspection division.

1848-C. **Application:** The electrical inspection division shall issue permits for such erection, alteration or maintenance of illuminated signs in all cases where application for such permit shall be made in accordance with the rules and regulations applicable thereto; provided, however, that no permit shall be issued for the erection, alteration or maintenance by contract unless the person, firm or corporation applying for such permit is registered as an electrical contractor as required by sections 1839 and 1840 of Article IV of this electrical code. Application shall be made to the electrical inspection division for that purpose on a printed form to be furnished therefor by the electrical inspection division. When such application is in accordance with the rules and regulations applicable thereto, the electrical inspection division shall issue to such applicant, upon the payment by such applicant to the city collector of the compensation and inspection fees as heretofore fixed, a permit in writing, authorizing such applicant to erect the sign at the location designated in such application and of the style or design described therein.

1848-D. **Notice of Completion:** Upon the completion of the work of erecting such sign under sign permit the applicant shall notify the electrical inspection division.

1848-E. **Inspection:** Upon receipt of notice of completion of the work of erecting and connecting an illuminated sign, the electrical inspection division shall inspect said sign with respect to safety, construction, supporting and electrical equipment.

1848-F. **Certificate:** When the electrical inspection division shall find that such sign has been constructed, erected and connected electrically in accordance with the ordinances of the city of Chicago, the chief electrical inspector shall upon request issue to such applicant a certificate in writing authorizing the operation and maintenance of said sign for the period of one year from the date of the filing of application therefor; such certificate to be issued without further cost or expense other than the fees hereinbefore provided for in Section IX of this electrical code.

1848-G. **Unlawful to Use Current, Etc.:** The use of electrical current in connection with such sign, previous to the issuance of the certificate last described, is prohibited, except by order of the electrical inspection division, for the purpose of testing the same to determine whether it is connected electrically in accordance with and pursuant to the provisions of the electrical code of the city of Chicago.

1848-H. **Attachment or Addition to Signs:** No attachment or addition thereto shall be made to any sign erected or maintained under the authority of this article unless all the provisions herein are fully complied with, and unless a permit be first issued by the electrical inspection division for the express purpose of allowing such attachment or addition. The fee shall be as provided for in Section 1847-L Article IX of this electrical code.

1848-I. **Re-Hang Permit:** No sign erected or maintained under the authority of this article which has been removed temporarily for repairs or alteration, shall be re-erected in the location from which said sign was removed, nor shall any sign be altered while in place, unless a permit be first issued for this express purpose by the electrical inspection division. The fee for this permit shall be as provided for in Section 1847-0 Article IX of this electrical code.

1848-J. **Location of Sign:** The supporting structure for any sign erected under and pursuant to the provisions of this article, shall be wholly inside of the lot line. The lowest part of the sign shall be at least nine feet above the surface of that part of the public way which any such sign overhangs, and the part of any projecting sign nearest to the building or structure to which it is attached, shall be not more than two feet from lot line. Flat signs shall be installed with the face parallel to the lot line and as close to the building as the contour of the building will permit. No sign erected under the authority of this article shall be permitted to project beyond the curb line.

1848-K. **Time of Illumination:** All sides of every such sign which are designed to be illuminated shall be illuminated each and every night for a period of not less than from dusk until the hour of 9:30 P. M.

1848-L. **Revocation of Permit:** The authority granted for the erection and maintenance of any such sign may be revoked at any time by order of the Mayor or of the City Council, and any compensation or permit fees paid to the city for such sign shall not be refunded in case of any such revocation.

1848-M. **Compliance with Ordinances:** Every sign erected and maintained under and pursuant to the provisions of this article shall comply with the provisions of this article and with all other ordinances of the city relating to the erection, maintenance and installation of such sign and the installation and use of electrical equipment, and shall be installed and maintained in all other respects in a safe and secure manner.

1848-N. **Annual Inspection of Signs:** The electrical inspection division shall make an annual inspection of each sign erected under provisions of this article with regard to its mechanical and electrical safety, and when said sign is found to be in compliance with this electrical code, and upon payment to the city collector of the compensation and inspection fees, as provided for in Section 1847 Article IX of this electrical code, the electrical inspection division shall issue to the person, firm or corporation, owning or oper-

ating said sign, a receipted bill, which shall authorize the maintenance of such sign for the period of time stated in said receipt, provided such sign complies in all other respects with all city ordinances.

1848-O. **Abandoned Illuminated Sign:** Any illuminated projecting sign erected under authority of this article, that is not maintained in strict conformity with all the provisions of this electrical code, shall be declared to be abandoned or a hazard over the public way, and the division of electrical inspection is hereby empowered to remove or cause to be removed any such abandoned or hazardous illuminated projecting sign.

1848-P. **Removal of Signs:** It shall be the duty of the electrical inspection division to remove or cause the removal of any sign not in compliance with any of the provisions of this article, and any compensation or inspection fees paid to the city of Chicago for such sign shall not be refunded.

1848-Q. **Penalty:** Any person, firm or corporation that shall violate any of the provisions of this article, shall be subject to the penalty provided for in Article I Section 1815 of this electrical code; in addition to such penalties the electrical inspection division shall, for violation of any of the provisions of this article, compel the cutting off and stopping of electrical current supplied to any such sign, and if deemed necessary or advisable shall remove or cause such sign to be removed.

ARTICLE XI.

Illuminated Signs Outside of Buildings and Not Projecting Over Public Property.

1849. **Permit Required:** No person, firm or corporation shall erect or cause to be erected or maintain any illuminated sign, which is attached to the front, sides or rear wall of any building or to any structure, and which does not extend over any sidewalk, street, alley or public way; or shall install or alter or cause to be installed or altered any electrical equipment for the purpose of illuminating any billboard, sign board or roof sign erected under the provisions of Chapter 28, Sections 1660 to 1673 inclusive as amended, and Section 1676 of the Chicago Municipal Code of 1931 as amended; except upon a permit first issued by the Electrical Inspection division.

1849-A. **Permits:** The electrical inspection division shall issue permits for the installation, or alteration, of the electrical equipment of such signs in each case where application for such permit shall be made in accordance with the rules and regulations applicable thereto; provided, however, that no permit shall be issued for the erection or alteration by contract, unless the person, firm or corporation applying for such permit is registered as provided for in Section 1839, Article IV of this electrical code.

1849-B. **Illuminated Sign Defined:** For the purpose of this article, illuminated signs shall be deemed to be signs constructed as follows: signs, all or any part of the letters or characters of which are made in an outline of electric lamps; signs with painted, flush or raised letters, or characters lighted by an electric lamp or lamps attached thereto; signs having a border of electric lamps attached thereto and reflecting light thereon; glass signs whether lighted by electricity or other illuminant; signs with painted, flush or raised letters or characters illuminated by electric lamp or lamps placed for the purpose of projecting or reflecting light thereon; and signs having any electrical equipment attached thereto.

1849-C. **Inspection—Certificate:** The electrical inspection division shall inspect all such illuminated signs, and electrical equipment installed or altered for the purpose of illuminating billboards, sign boards, and roof signs, and shall require conformity with the standards and specifications of the Chicago Electrical Code applicable thereto, and upon completion of such installation or alteration in compliance with such standards and specifications, shall issue upon receipt of fee hereinafter provided a certificate of inspection covering each installation or alteration.

1849-D. **Annual Inspection — Stopping of Electrical Current:** It shall be the duty of the electrical inspection division to make an inspection annually of all such signs and electrical equipment of all billboards, sign boards, and roof signs, and whenever said electrical inspection division shall find that any electrical equipment on any illuminated sign, billboard, sign board, or roof sign has been installed or altered in violation of this electrical code, it shall notify in writing the person, firm or corporation owning, using or operating the same, to place such electrical equipment or sign in compliance with the standards and specifications applicable thereto, within such time as the electrical inspection division shall consider just and reasonable. Upon refusal, neglect, or failure to comply with the requirements of such notification, the electrical inspection division may order, and compel, the cutting off and stopping of electrical current on such installation until electrical equipment has been placed in compliance with this electrical code.

1849-E. **Fees for Inspection:** The person, firm or corporation having charge, or control, of the electrical equipment on any illuminated sign, billboard, sign board, or roof sign, maintained under the provisions of this article, shall pay an annual fee for inspection of such electrical equipment to be computed according to Section 1847, Article IX of this electrical code.

1849-F. **Penalty:** Any person, firm or corporation that shall violate any of the provisions of this article, shall be subject to the penalty provided for in Article I, Section 1815 of this electrical code, and in addition to such penalty, the electrical inspection division, shall, for the violation of any of the provisions of this article, compel the cutting off and stopping of electrical current, supplied to any electrical equipment on such sign.

ARTICLE XV.

Poles, Wires and Conductors.

1853. **Permit to Erect Required—Penalty:** No person, firm or corporation shall erect, construct, maintain, use, alter or repair any pole, line or wire, underground conductors or electric conductors of any description whatever on, over or under any street, sidewalk, avenue, alley or public place within the city, without first having obtained a permit therefor from the department of streets and electricity, which permit shall be countersigned by the commissioner of streets and electricity.

1853-A. **Requirement Before Permits Can Be Issued:** All applications for permits to erect poles in the streets and alleys of the city shall provide that the city may use the poles to be so erected and may attach thereto such necessary cross arms, wires or other electrical appliances as may be deemed necessary for the electrical service of the city, and no permit shall be issued by the Commissioner of streets and electricity for the erection of such poles in which the application and permit does not provide for the privileges required by the city as herein contained.

Applications for permits shall be made in duplicate on a form to be approved by the commissioner of streets and electricity.

Applications shall be submitted to the commissioner of streets and electricity who shall cause the necessary inspection to be made.

1853-B. **Division of Electrical Inspection to Inspect—Fees:** All apparatus installed under authority of permit issued in accordance with this article, shall be inspected by the division of electrical inspection. The fee for such inspection shall be in conformity with Section 1847, Article IX of this electrical code.

Said fees shall be demanded by the commissioner of streets and electricity before he

countersigns any such permit, and shall be paid to the city collector.

1853-C. **Removal of Poles — Wires Underground:** No permission or authority shall be given to any person or corporation to erect any pole or poles for telegraph, telephone or electric light or power purposes or for the purpose of stringing thereon wires, cables or conveyors for the transmission of sounds or signals, or of heat, light or power, upon or along any street, alley or public way, within the city, except upon the express provision that such poles and conductors are to be and will be removed forthwith whenever the City Council shall order such removal; provided, however, that nothing in this article shall apply to any pole or poles used solely for the carrying and support of its overhead contact trolley wires by any street railway company where it is claimed by such company that it is operating its cars under the authority of any ordinance of the city.

1853-D. **Location:** Such wires or conductors shall in no case be placed at a greater distance from the curbstone separating sidewalk from carriageway than four feet, except in crossing streets running transversely to the direction of the said lines when such crossing shall be made in the shortest straight line, or in making necessary connections with buildings and stations.

1854. **Traffic Not to Be Impeded:** The method employed of laying said conductors shall be such that it will at no time be necessary to remove so much of the pavement, or to make such excavation, as to materially impede traffic or passage upon sidewalk or street during the operation of laying or repairing said conductors, except when crossing streets transversely, where authority may be granted to remove the pavement for a width not exceeding two feet in the nearest straight line from corner to corner. In no case during the general hours of passage and traffic shall passage be interrupted thereby for a longer period than one hour.

1855. **Supervision — Space:** The work of removal and replacement of the pavements in any and all of the streets, avenues, highways and public places in and through which the wire of any company shall be laid, shall be subject to the control and supervision of the commissioner of streets and electricity; excavations in any and all of the unpaved streets, avenues, highways or public places shall also be subject to like control and supervision. The space selected for placing said wires, the same in every case being limited as to direction and general positoin by the foregoing provisions, shall be of sufficient size to permit the installation of the necessary conductors and equipment.

1856. **Penalty:** Any person, firm or corporation violating any of the provisions of this article or failing to comply with the same, shall be subject to the penalties provided for in Article I, Section 1815 of this electrical code.

SUB-ARTICLE 20.

Wiring Installation Design.

2001. **Lighting and Appliance Branch Circuits:**

(a) For the purposes of this section, the terms "Outlet" and "Appliances" are defined as follows:

"Outlet"—An outlet is that fixed point on a branch circuit at which current is taken to supply fixtures and appliances. An outlet having a fixture and more than one socket attached shall be considered as one outlet. An outlet having a multiple receptacle installed therein shall be considered as one outlet.

"Appliances"—Appliances are current-consuming devices for domestic or general commercial use, such as heating, cooking and small motor-operated devices, etc., suitable for use on branch circuits.

2002. **Polarity Identification of Systems and Circuits:**

(a) All interior wiring systems, unles otherwise excepted, shall have a grounded conductor which is continuously identified throughout the system.

(b) All ungrounded wires of a branch circuit shall be protected by fuses or circuit-breakers. When the grounded conductor is identified and properly connected, branch circuits shall be so protected in the ungrounded wires only. In locations where the conditions of grounding or the liability of the reversal of connections warrant, the electrical inspection division may require, on systems having a grounded neutral or having one side grounded, that both wires of two-wire branch circuits shall be so protected even though the grounded conductor is identified and properly connected.

(c) Unidentified, ungrounded systems and circuits may be permitted by special permission of the division of electrical inspection.

(d) Wires having white or natural gray covering shall not be used in identified systems or circuits other than as conductors for which identification is required by this section, except under the following conditions:

1. Identified wires rendered permanently unidentified by painting or other effective means at such outlet where the wires are visible and accessible may be used as unidentified conductors.

2. Cable containing an identified wire may be used for single-pole switch loops if the connections are so made that the unidentified wire is the return conductor from the switch to the outlet.

(e) No interior wiring system shall be electrically connected to a supplying system unless the latter contains, for any identified conductor of the interior system, a corresponding wire which is grounded.

(f) No switch or automatic over-current protective device shall be placed in a permanently grounded conductor unless the device simultaneously opens all conductors of the circuit, except as permitted in paragraph (e) of Section 805.

(g) Screw shells of sockets in fixtures or pendants shall be connected to the identified grounded conductor of the circuit.

(h) Identification for conductors shall be secured as follows:

1. Rubber-covered conductors of No. 6 or smaller shall have identifying outer covering as specified in paragraph (f) of Section 602.

2. Conductors larger than No. 6 and conductors of other than rubber covering shall have identifying outer covering as specified in paragraph (f) of Section 602 or shall be identified by a distinctive marking at terminals during the process of installation.

3. Flexible cords shall be identified as provided in paragraph (g) of Section 609.

4. Terminals of devices shall be identified as provided in Section 207.

(i) Wiring systems and circuits derived from transformers in which a part of the turns are common to both primary and secondary alternating current circuits, ordinarily known as auto-transformers, shall not be permitted for any interior wiring system except as follows:

1. Unless the system supplied contains an identified grounded wire which is solidly connected to a similar identified grounded wire of the system supplying the auto-transformers.

2. Auto-transformers used for starting and controlling induction motors whether included in starter cases or installed as separate units.

3. For supplying circuits wholly within apparatus which also contains the auto-transformer.

4. Auto-transformers as fixed voltage adjustment on existing unidentified power circuits.

2003. **Method of Calculation:**

(a) The total current in amperes for all loads shall determine the size of feeders in accordance with table of Section 612.

1. For a lighting and appliance load, the current in amperes shall be not less than that

required for the demand load as determined in accordance with paragraph b. In the case of any unusual type of load for stores or offices, such load shall be added to the feeder capacity with demand of 100% for such unusual load after demand factor has been established.

2. For a combined lighting and appliance load and a power load, the current in amperes for the lighting and appliance load shall be that required for the demand load determined as specified in sub-paragraph 1 of this section and the current in amperes for the power load shall be as specified in sub-paragraph 2 of paragraph (a) of Section 807.

(b) **Minimum Demand Load:**

The minimum requirements for watts or fraction of a watt per unit of area and the demand factor applying thereto for each kind of building and occupancy served are as specified in the following sub-paragraphs. The required minimum demand load shall be the computed load as determined herein multipled by the indicated demand.

(c) **Feeders and Sub-Feeders:**

1. **Buildings constructed and used for single family dwellings:** One watt per square foot, plus 1000 watts for small appliances. For area of 2000 or less square feet, demand 100 per cent; for all excess over 2000 square feet, 75 per cent. No demand shall be applied in connection with small appliance loads.

(d) **Apartment Buildings:**

To estimate the size of feeders and sub-feeders for apartment buildings, the sum of the following three factors shall be considered the total watt load:

Factor 1—Apartment Lighting Load—the total floor space, including partitions, shall be estimated as one watt per square foot. Should the layout or design of the apartment lighting load in watts, exceed that estimated on the basis of one watt per square foot, then the lighting load shall be estimated on the actual watt load to be connected.

Factor 2—The appliance load shall be estimated as 1000 watts per apartment, to which shall be added the watt load of any appliance over 660 watts rating intended to be connected and used in the apartment.

Factor 3—The public halls and basements watt load shall be estimated as 60 watts per outlet, except that for outlets to be used for appliances, the watt load of such appliances shall be added.

The estimated maximum demand shall be the sum of the three above factors multiplied by a demand factor in the following table corresponding to the number of apartments wired.

Number of Apartments	D.F.
1 to 5	75.00%
6 to 10	71.25%
11 to 15	67.50%
16 to 25	63.75%
26 to 35	60.00%
36 or more	56.25%

Where feeders are to serve loads other than apartments, such loads shall be computed by the use of the proper demand factor and shall be added to the estimated maximum demand of the apartment feeders.

(e) **Hotels (having no provisions for individual cooking):**

One watt per square foot, except for the ballrooms. For areas 10,000 square feet or less per feeder, demand 100 per cent.

For that part of the area in excess of 10,000 square feet and not more than 50,000 square feet per feeder, a demand of 80 per cent.

For the excess above 50,000 square feet per feeder, a demand of 70 per cent.

(f) **Stores and Department Stores:**

In the wiring of stores, feeders and sub-feeders shall be provided on a basis of two watts per square foot of sales space; 100 watts capacity per lineal foot of show windows, measured horizontally along the base of the show window and a further allowance of 1500 watts capacity for store for sign. Demand factor for store feeders 100 per cent.

(g) **Office Buildings:**

In the wiring of office buildings, feeders and sub-feeders shall be provided on a basis of 2 watts per square foot. For areas 10,000 square feet or less per feeder—demand 100 per cent.

For all excess above 10,000 square feet per feeder—demand 70 per cent minimum.

(h) In the case of any unusual type of load, for stores or offices, such load shall be added to the feeder capacity with demand of 100 per cent for such unusual load, after demand factor has been established.

(i) **Industrial Commercial (Loft) Buildings:** One watt per square foot—demand 100 per cent.

(j) **Garages:** ½ watt per square foot, exclusive of the machine shop or display rooms, if any—demand 100 per cent.

(k) **Hospitals** (except in the operating suites and X-Ray department): ¾ watt per square foot.

For areas of 25,000 square feet or less per feeder—demand 100 per cent.

For the excess area above 25,000 square feet per feeder—demand 60 per cent.

1. **Schools:** 1½ watts per square foot. For areas of 10,000 square feet or less per feeder—demand 100 per cent.

For the excess area above 10,000 square feet per feeder—demand 50 per cent.

(m) **Storage Warehouses:** ¼ watt per square foot.

For areas of 50,000 square feet or less per feeder—demand 100 per cent.

For the excess area above 50,000 square feet per feeder—demand 50 per cent.

(n) **Factory Buildings:** Feeder sizes shall be based on the specific load which they are to serve.

(o) **Other Kinds of Building and Occupancies:** Theaters, churches, and other places of public assemblage, ballrooms, dance halls, restaurants, club and lodge rooms, community centers, armories, libraries, operating suites and X-ray departments in hospitals, etc., and buildings for special purposes, such as banks, motion picture studios, etc., vary so widely due to individual requirements, architectural and ornamental treatment, that no standard has been established upon which the watts per square foot may be determined with accuracy. Therefore, the feeders for these and other buildings or occupancies not listed above, shall be determined by specific load which they are to serve and as ordinarily computed. This applies also to special uses, such as flood and outline lighting, signs, etc.

(p) Every portion of the above building devoted to the use or accommodation of the public shall be well and thoroughly lighted.

(q) **Electrically Heated Ranges:** The sizes of feeders supplying electrically heated ranges, each rated at more than 5½ K.W., may be determined on the basis of the demand values shown in the following table:

Number Ranges	Demand Per Cent	Number Ranges	Demand Per Cent
1	70	10	34
2	60	11 to 15	32
3	50	16 to 20	28
4	45	21 to 25	26
5	42	26 to 30	24
6	40	31 to 40	22
7	38	41 to 50	20
8	36	51 to 60	18
9	35	61 and over	16

(r) **Neutral Conductor—Special Provisions:**

The size of the neutral and the demand factor applying thereto shall be determined, as specified in the previous paragraphs, for the

tacles are of No. 14 wire, and where the location of sockets or receptacles is such as to render unlikely the attachment of flexible cords thereto, the wattage shall not exceed 2000 and the number of mogul sockets or receptacles shall not exceed 6 on two-wire branch circuits, or on either side of a three-wire or four-wire branch circuit. Such circuits shall be protected by fuses of no greater rated capacity than 20 amperes at 125 volts or less and 10 amperes at 126 to 250 volts.

BRANCH LIGHTING CIRCUITS.

Capacity of Branch Circuit	Smallest Wire Allowed	Number of Mogul Sockets	Number of Med. Base Sockets	Number of Int'med. Candelabra Sockets	Size of Fuse
125 Volts or Less.					
1000 watts	No. 14	3	16	25	15
2000 watts	No. 12	6	32	Not Allowed	20
126-250 Volts.					
1000 watts	No. 14	3	16	Not Allowed	10
2000 watts	No. 14	6		Not Allowed	10

(h) In the wiring of apartments and residences, all rooms shall be wired and a sufficient number of branch circuits shall be provided to allow for at least one circuit to every 1000 square feet or fraction thereof of floor area (1 watt per square foot). Each apartment shall be supplied independently of any other apartment. The floor area is to be determined by the total floor area, including partitions. The socket limitation for branch circuits as stated above shall not apply in the case of apartments and residences. There shall also be provided at least one additional ordinary appliance branch circuit, not smaller than No. 12 wire, for each apartment or residence to supply only the convenience outlets in the kitchen, pantries, dining rooms and laundries.

(i) Convenience outlets shall be installed in residences and apartment buildings as follows:

Dining Rooms Laundries Kitchens Pantries	At least one outlet.
Living Rooms Sun Parlos Living Porches Bedrooms Bathrooms Parlors	One outlet to every twenty (20) lineal feet of wall space, or fraction thereof, such wall space to include all openings. Receptacles shall be so spaced that no point on the perimeter of the room will be more than ten feet from a receptacle.

Duplex receptacles shall be provided in all cases where convenience outlets are required in apartment buildings or residences.

(j) Fireplace outlets shall be wired for the capacity of the appliance to be used.

(k) In the wiring of stores and offices a sufficient number of branch circuits shall be provided to allow for at least one circuit to every 500 square feet or fraction thereof. The floor area is to be determined by the total floor area, including partitions. The socket limitation shall apply in the cases of offices and stores. In fixture display rooms the number of outlets per circuit shall not exceed ten (10).

(l) In all buildings wired for electric light, lights shall be so placed as to illuminate the front of every furnace and every heating boiler, and, where boilers are cleaned from the rear, the rear of such boiler. Lights shall also be provided to illuminate coal bins. Lights shall be provided in all accessible coal bins. Lights shall be provided in all accessible attics and in all clothes closets.

(b) A garage shall be deemed to be a building or portion of a building in which one or more self-propelled vehicles carrying volatile, inflammable liquid for fuel or power are kept for use, sale, storage, rental, repair, exhibition or demonstration purposes, and all that portion of a building which is on or below the floor or floors on which such vehicles are kept and which is not separated therefrom by tight, unpierced fire walls and fire-resistive floors.

3302. **Wiring:**

(a) Approved conduit, armored cable, metal surface wiring raceways, electrical metallic tubing or wireways and busways as specified in Section 511, shall be employed as the wiring method.

(b) Cutouts, switches, attachment plug receptacles and fixed sockets, unless of a type approved for use in Class I locations as defined in sub-article 32, shall be located at least four feet above the floor, except in a showroom separated by a partition from the garage proper.

(c) Approved reinforced cord shall be used for pendant lamps.

(d) In private garages, an outlet shall be provided at the ceiling for general illumination and a side wall receptacle installed for the attachment of a portable lamp.

3303. **Portables:**

(a) Approved portable cord designed for rough usage, such as hard-service cord, Type S, shall be used to connect portable lamps, motors or other appliances. The portable cord shall carry the male end of an approved polarity-type. The receptacle shall be kept at least four feet above the floor.

(b) Flexible cord leads for portable lamps shall be equipped with handle, socket hook, and substantial guard, the guard being securely attached to the socket or the handle. Approved keyless sockets of moulded compositiou or metal-sheathed porcelain type or other keyless sockets approved for the purpose shall be used. Brass shell paper-lined sockets, either key or keyless, shall not be used.

3304. **Charging Cables:**

(a) Approved Type S cord shall be used for charging purposes.

(b) Connectors shall be of approved type and of at least 50 amperes capacity, and shall be so designed or so hung that at least one will break apart readily at any position of the charging cable. Live parts shall be guarded from accidental contact. The fixed, or wall connector shall be kept at least four feet above the floor, and, if not located on a switchboard or charging panel, shall be guarded from accidental contact. Where plugs for direct connection to vehicles are suspended from overhead wiring they shall hang at least six inches above the floor and no connector need be placed in the cable or at the outlet.

3305. **Switchboards and Charging Panels:**

(a) Switchboards and charging panels carrying spark-producing devices not located at least four feet above the floor and which are not of types approved for use in Class I Locations as defined in sub-article 32, shall be located in a room or enclosure provided for the purpose.

3306. **Generators, Motors and Control Apparatus:**

(a) Generators, motors and control apparatus, that embody the use of commutators, collector rings or other make-and-break or sliding contacts shall either be of the totally-enclosed type or be located at least four feet above floor.

(b) Generators, motors, control apparatus and the like, having commutators, collector rings, or other make-and-break or sliding contacts located more than four feet above the floor shall, unless of the totally-enclosed type, have wire screens or perforated metal placed at commutator or brush ends to prevent the falling of particles. No dimension of any opening in the wire screen or perforated metal shall exceed .05 inch, regardless of the shape of the opening and of the material used.

3307. **Special Precautions:**

(a) Cutouts and switches shall be enclosed in approved boxes or cabinets unless placed on switchboards or charging panels in the manner prescribed in Section 3305 of this code.

(b) Hatch limit switches of elevators shall be located at least four feet above the lowest floor level.

(c) Batteries shall not be tested by short-circuiting their terminals in such a manner as to produce sparks.

(d) Garages containing electrical machines only, shall conform to the above rules as far as they apply except that vapor-proof receptacles will not be required and electrical apparatus may be located any desired distance from the floor. The switchboard or charging panel will be enclosed, as provided in paragraphs (e) and (f).

(e) In public or private garages containing both electrical and gas cars the enclosure shall be a screen of not less than No. 6 gauge steel wire with a mesh not smaller than 2 inches. This screen shall be rigidly fastened at the floor and sides and if necessary also at the top. The screen shall extend to a distance of not less than 6 feet from the floor.

(f) In public garages containing electric cars only the switchboard or charging panel shall be enclosed or otherwise arranged to protect it from mechanical injury. The enclosure or other protection shall be of such design and construction as to protect the board from injury by being run into by cars.

3308. **Grounding:**

(a) Conduit, metal sheaths, raceways and exposed metal frames and enclosures of equipments and the like shall be grounded when and in the manner prescribed in sub-article 9 of this code. This shall apply to all devices except pendent and portable lamps operating on grounded circuits of not more than 150 volts to ground.

3309. Flexible Cords:

(a) Flexible cords used to supply pendent or portable lamps or portables which include sockets on polarized wiring systems, shall have one conductor identified, and such identified conductor shall be connected to the screw shell of the socket. Receptacles, attachment plugs, connectors, and similar devices used with such cord shall be of the polarity type.

SUB-ARTICLE 35.

Motion Picture Projectors and Equipment.

3501. **General:**

(a) The requirements of this sub-article shall be deemed to be additional to, or amendatory of, those prescribed in sub-articles 1 to 20 inclusive, of this code.

(b) All conduit, armored cable, metal raceways, exposed metal frames and enclosures of equipment shall be grounded by the method as prescribed in sub-article 9 of this code. This shall apply to all devices except pendant and portable lamps operating at not more than 150 volts to ground.

3502. **Projectors of Professional Type:**

(a) The professional types of projectors, such as are commonly used in theaters and motion picture houses, shall be located in fireproof enclosures as required by the code of the department of buildings.

(b) The arc lamp house shall be composed entirely of metal having a thickness not less than No. 24 U. S. sheet metal gauge (.025 inch), except where the use of approved insulating material is necessary. An automatic overload protective device and a manually operable switch shall be provided for each ungrounded conductor supplying the lamp. Incandescent lamp enclosures shall conform to the above requirements as far as may be practicable. Arc or incandescent lamp enclosures shall be marked with the name of the maker and with the current and voltage rating for which they are designed.

(c) The enclosure for the arc switch or the projector shall be of metal of a thickness not less than No. 16 U. S. sheet metal gauge (.0625 inch).

(d) Wires not smaller than No. 4 shall be employed to supply the projector outlet in a theater. Wiring shall, in all cases, be installed for the full rated current of the lamp to be used.

(e) Rheostats, transforming devices and any substitute therefor shall be of types expressly designed and approved for the purpose.

(f) Top and bottom magazines shall be so designed with a rabbet on the inside or a flange on outside of the door in such a manner as to prevent the entrance of flame and for the purpose of reinforcing the door. No solder shall be used in their construction. The front side of each magazine shall consist of a door swinging horizontally and equipped with a substantial latch. The top and bottom magazines shall be constructed of metal, reinforced in an approved manner and having a thickness not less than No. 22 U. S. sheet metal gauge (.032 inch). Top and bottom magazines shall be equipped with fire trap and fire rollers approved by the division of electrical inspection.

(g) An automatic shutter shall be provided and permanently attached to the gate frame. The construction of the shutter shall be such as to shield the film from the beam of light whenever the film is not running at operating speed.

(h) Motion picture projectors shall be so constructed that the film shall be entirely enclosed within the magazines, the machine head and the sound head, except at the aperture during operation.

(i) The optical shutter on the projector head shall be so mounted as to intercept the beam of light between the source of light and the aperture.

(j) The projector shall be equipped with a film take-up device approved by the division of electrical inspection.

(k) In the top and bottom magazines there shall be a clearance of not less than one inch between the outer edge of the reel and the inner wall of the magazine.

(l) Reels containing more than 2000 feet of film capacity shall not be used.

3503. **Projection Room for Projectors of Professional Type:**

(a) Plans for the projection room shall be submitted to the division of electrical inspection for approval before any construction is started. These plans shall show the dimensions of the projection room, the layout of all wiring and location of all electrical equipment; also shall show the location and size of all openings in the walls and ceilings. Rigid conduit or electrical metallic tubing shall be employed as the wiring method. Conduits shall be concealed and all boxes shall be of the flush mounting type. One outlet connected to the emergency lighting system shall be installed in the projection room. Motors, generators, rotary converters, rectifiers, rheostats and sound amplifying equipment shall be marked with the name of the maker and with the current and voltage rating for which they are designed. Facing the picture screen, there shall be a spacing of not less than forty-two (42) inches between the lens center of the right hand projector and the right end wall, sixty (60) inches between lens centers of projectors and forty-eight (48) inches between lens center of left hand projector and left end wall. There shall be a spacing of not less than forty-eight (48) inches between other light projectors. The projection room inside dimensions shall be not less than ten (10) feet from front wall to rear wall, twelve (12) feet six (6) inches between end walls and eight (8) feet six (6) inches between the floor and the ceiling.

(b) In addition to the gravity vent required by the code of the building department of the city of Chicago, there shall be provided, when the port holes are closed, an out draft of air in each projection room by means of an air duct and an exhaust fan having a capacity of at least 200 cubic feet of air per minute for each 80 square feet of floor area of the projection room. The exhaust fan motor shall be so installed that fumes passing through the exhaust air duct cannot come in contact with the motor. Current for the motor shall be supplied from the general lighting panel. There shall be a dual switch control for this motor—one switch shall be so connected to the master shutter control that the exhaust fan motor will be put in operation automatically when the port hole master shutter control is released—the second switch shall be connected in parallel and be of a manually operable type and shall be located outside of the projection room near the exit door.

(c) The doors on main openings to the projection room shall be metal clad and swing outwards and arranged to be held normally closed by spring hinges or door checks. If a door check is used it shall be located outside of the projection room. All other openings in the projection room except the ventilator opening, shall be protected by sliding shutters constructed of sheet metal of a thickness not less than No. 14 U. S. sheet metal gauge (.0781 inch). The slides shall be arranged to close by gravity and shall operate freely in guides which are continuous along both sides and bottoms of openings. These guides shall be built up of strap iron two (2) inches wide and one-eighth (⅛) inch thick with spacers one (1) inch wide and twice as thick as the shutter. These shutters shall be held up, by a master shutter control cord which passes over the center line of the aperture of each projector and seventy-eight (78) inches above the projection room floor. This cord shall terminate in a metal ring placed over a steel pin at the exit door and so arranged that all shutters may be readily manually closed at either or both exit doors. An approved fusible link with a maximum rating of 165 degrees F. shall be inserted in the master cord above each projector. The metal frame for each observation or light projection opening shall be provided with a sill having a downward pitch from inside to outside. This pitch shall be not less than 30 degrees from the horizontal.

(d) Rewinding of films shall be performed in the projection room. Reels carrying films in process of rewinding shall be enclosed in magazines of approved design. The rewind film shall not be mounted within reach of the projector head.

(e) Extra films shall be kept in approved fireproof metal cabinets designed for the storage of each reel in a separate fireproof compartment. Each compartment shall have a separate self-closing cover.

(f) Rectifiers, motor generators, rheostats, storage batteries, switchboards, dimmers or other similar devices shall not be located in the projection room.

(g) Each incandescent lamp not otherwise protected by non-combustible shades or enclosures shall be provided with an approved lamp guard.

(h) No switches or controls of any kind except for the operation or control of projection, sound, or accessory projection equipment shall be located in the projection room or the generator room, provided, however, a switch or switches for the control of a motor operating the curtain at the picture screen may be located in the projection room.

(i) Smoking is prohibited at all times within the projection room.

(j) The use of any fire or open light is prohibited in the projection room during the time the audience is in the building.

(k) It shall be unlawful for anyone other than a person licensed by the city of Chicago as a moving picture operator or holding a permit as an assistant or apprentice or officer or employe of the city, while acting in the discharge of his duty to enter any projection room where a moving picture machine or device is in operation, or to operate or in any way handle or manage such machine or device while the same is being operated during an exhibition for the public; provided, that this shall not apply to the proprietor, owner or manager in charge of the premises, who may enter same for the purpose of giving necessary orders and directions. In no case shall more than four persons be within such projection room at one time while such exhibition is going on.

(l) The projection room shall contain nothing but the moving picture machines, the sound reproducing equipment and the necessary accessories, and the room shall be kept clean at all times.

(m) The operator in charge of the moving picture machines shall, before every presentation, carefully examine the machine and its devices, including the films, and ascertain if the same comply with the rules and ordinances, and that the said machine is in a safe condition to operate. Also, he shall, before every presentation, examine the projection room shutters and shutter control apparatus and ascertain if the same are in proper condition to be operated manually or automatically.

3504. **Non-Professional Motion Picture Projectors and Equipment:**

(a) **Definitions:** The following definitions shall apply for the purpose of this sub-article:

(b) **Non-Professional Motion Picture Projector:** A motion picture projector intended for use with slow-burning (acetate cellulose or equivalent) film only.

(c) **Miniature Non-Professional Motion Picture Projector:** A non-professional motion picture projector whose construction provides for the use of films of a width less than one and three-eighths (1⅜) inches which film is regularly supplied only as slow-burning (cellulose acetate or equivalent) film.

(d) **Slow-Burning Film:** Motion picture projection film printed on film stock of cellulose acetate composition or on other stock approved as slow-burning.

(e) **Identification of Non-Professional Motion Picture Projectors:** Such projectors shall be marked with the name or trade mark of the maker, and with the voltage and current rating for which they are designed, and shall also be plainly marked, "for use with slow-burning films only."

(f) **Identification of Slow-Burning Film:** Slow-burning (cellulose acetate or equivalent) film shall have a permanent distinctive marker for its entire length identifying the manufacturer and the slow-burning character of the film stock.

(g) **Approval Required:** Non-professional motion picture projectors and equipment including all auxiliary mechanical and electrical projection appliances intended for installation and use within the jurisdiction of the city of Chicago shall be approved.

(h) **Connection to Supply Power:** The wiring, and the means and manner of connection thereto for any non-professional motion picture projector shall be as provided in sub-articles 1 to 20 inclusive of this code.

(i) **Highly Flammable Film Prohibited:** No film other than approved slow-burning (cellulose acetate or equivalent) film shall be used with any non-professional motion picture projector.

(j) **Fireproof Booth Unnecessary:** The location of a non-professional motion picture projector in a fireproof booth shall not be required.

(k) **Location in Assembly Halls and Rooms:** The installation and operation of non-professional motion picture projectors located in assembly halls and assembly rooms shall be as provided in section 3504 and Article XIV of this electrical code.

(l) **Permit Required:** A permit shall be first obtained for the operation of a non-pro-

(Continued on page 649)

Information relative to services may be secured at the local offices or the Architects and Builders Service Division, Public Service Co. of Northern Illinois, 72 West Adams Street, Chicago, Illinois, 'phone Randolph 2500.

E—Electricity. G—Gas. W—Water. H—Heat.

Acacia Park ......EG
Addison ....E
Alden ....E
Algonquin ....E
Alsip ....E
Ancona ....E
Andres ....E
Antioch ....EG
Aptakisic ....E
Arlington Heights.EG
Aroma Park ....E
Bannockburn ....E
Barrington ....EG
Bartlett ....E
Beach ....E
Bedford Park .....EG
Beecher ....E
Bellwood ....EG
Bensenville ....E
Benson ....E
Berkeley ....EG
Berwyn ....EG
Big Foot Prairie...E
Blackstone ....E
Bloomingdale ......E
Blue Island ......EG
Bluff Lake ....E
Bonfield ....E
Bourbonnais ....E
Braceville ....E
Bradford ....E
Bradley ....EG
Braidwood ....E
Bristol ....E
Bristol Station ....E
Broadmoor ....E
Broadview ....EG
Brookfield ....E
Buckingham ....E
Buffalo Grove .....E
Burnham ....E
Cahery ....E
Calumet City......EG
Calumet Park .....EG
Camp Grove ....E
Campus ....E
Carbon Hill ....E
Carpenterville ....E
Cary ....E
Castleton ....E
Cayuga ....E
Cazenovia ....E
Channahon ....E
Channel Lake ......E
Chicago Heights ..EG
Chicago Highlands .E
Chicago Ridge ....EG
Chillicothe ......EW
Cicero ....EG
Clearing Industrial
District ....EG
Cloverdale ....E
Coal City ....E
Cornell ....E
Crestwood ....E
Crate ....EG
Crooked Lake ......E
Crystal Lake ......E
Custer Park ....E
Dana ....E
Deerfield ....E
Deselm ....E
Des Plaines ......EG
Diamond ....E
Diamond Lake.......E
Dixmoor ....EG
Dolton ....EG
Druces Lake ....E
Duncan ....E
Dwight ....E
East Brooklyn .....E
East Chicago
Heights ....E
East Dundee ....E
East Hazelcrest ..EG
East Wenona ....E
Edelstein ....E
Eileen ....E

Elk Grove ....E
Elmhurst ....E
Elmwood Park .....EG
Elwood ....E
Emington ....E
Essex ....E
Evanston ....EGH
Evergreen Park ...EG
Fairview ....EG
Faithorn ....E
Fannette ....E
Flossmoor ....EG
Forest Park ......EG
Forestview ....E
Fort Sheridan .....E
Fox Lake ....EG
Fox River Grove....E
Fox River Springs..E
Frankfort ....EG
Franklin Park ....EG
Fremont Center ....E
Gages Lake ....E
Gardner ....E
Garfield ....E
Gilberts ....E
Gilmer ....E
Glencoe ....E
Glenview ....EG
Glenwood ....EG
Godley ....E
Golf ....EG
Goodenow ....E
Goodrich ....E
Grand Ridge ....E
Grant Park ....E
Grass Lake ....E
Grays Lake ....E
Great Lakes ....E
Greenwood ....E
Gurnee ....E
Hainesville ....E
Halfday ....E
Harvey ....EG
Hazelcrest ....EG
Hebron ....E
Helmar ....E
Henry ....E
Herscher ....E
Hickory Corners ...E
High Lake ....E
Highland Lake....EG
Highland Park .....E
Highwood ....E
Hillside ....EG
Hodgkins ....E
Holbrook ....EG
Homewood ....EG
Indian Oaks ....E
Ingalton ....E
Irwin ....E
Itasca ....E
Ivanhoe ....EG
Johnsburg ....E
Joliet ....E
Justice ....EG
Kangley ....E
Kankakee ....EG
Kempton ....E
Kenilworth ....EG
Kernan ....E
Kinsman ....E
Koster ....E
Lacon ....E
LaGrange ....EW
LaGrange Park .....E
Lake Bluff ....E
Lake Catherine ....E
Lake Forest ....E
Lake Marie ....E
Lake Villa ....EG
Lakewood ....E
Lake Zurich ......EG
Lansing ....EG
La Rose ....E
Lawn Ridge ....E
Lehigh ....E
Lemont ....E
Leonore ....E

Libertyville ......E
Lincolnwood ......EG
Lisbon ....E
Lockport ....E
Lombard ....E
Lombardville ......E
Long Grove ....E
Long Lake ....EG
Long Point ....E
Loon Lake ....E
Lorenzo ....E
Loretta ....E
Lostant ....E
Lowell ....E
Low Point ....E
Lyons ....EG
Manhattan ....E
Manteno ....E
Manville ....E
Markham ....EG
Marley ....E
Marseilles ....G
Matteson ....EG
Maywood ....EG
Mazon ....E
McCook ....EG
McDowell ....E
McHenry ....E
Medinah ....E
Melrose Park .....EG
Midlothian ....EG
Milburn ....E
Minonk ....E
Minooka ....E
Mokena ....EG
Momence ....EG
Monaville ....E
Monee ....E
Morris ....EG
Morton Grove .....EG
Mount Prospect ...EG
Mundelein ....E
Munger ....E
Munster ....E
New Lenox ....E
Niles ....EG
Niles Center .....EG
Normantown ....E
Northbrook ....EG
North Chicago .....E
North Chillicothe.EW
Northfield ....EG
North Riverside ..EG
Oak Forest ....EG
Oak Lawn ....EG
Oak Park ....EGH
Ocoya ....E
Odell ....E
Olympia Fields ...EG
Ontarioville ......E
Orchard Place .....E
Orland Park ......EG
Ottawa ....G
Otto ....E
Palatine ....EG
Palos Park ....EG
Park Ridge ....EG
Peotone ....E
Petite Lake ....E
Phoenix ....EG
Pingree Grove .....E
Pistakee Bay .....EG
Pistakee Lake ....EG
Plainfield ....E
Plato Center ......E
Plattville ....E
Pontiac ....EGW
Posen ....EG
Prairie View ......E
Putnam ....E
Ransom ....E
Reddick ....E
Richmond ....E
Richton Park ......E
Ridgefield ....E
Ringwood ....E
Ritchie ....E
Riverdale ....EG

River Forest .....EG
River Grove ......EG
Riverside ....EG
Robbins ....E
Rockdale ....E
Rockville ....E
Rome ....E
Romeoville ....E
Rondout ....E
Rosecrans ....E
Roselle ....E
Round Lake ....EG
Round Lake Beach...E
Rowe ....E
Russell ....E
Rutland ....E
St. Anne ....EG
St. George ....E
Sand Lake ....E
Saunemin ....E
Schaumberg ....E
Schiller Park ....EG
Seneca ....EG
Sherburnville .....E
Sollitt ....E
Solon Mills ....E
South Chicago
Heights ....EG
South Holland ....EG
South Wilmington ..E
Sparland ....E
Speer ....E
Spring Grove ......E
Stark ....E
Steger ....EG
Stickney ....EG
Stockdale ....E
Streator ....EG
Summit ....EG
Symerton ....E
Techny ....EG
Terra Cotta ....E
Third Lake ....E
Thornton ....EG
Tiedtville ....E
Tinley Park ......EG
Toluca ....E
Tonica ....E
Toulon ....E
Troy ....E
Union Hill.........E
Varna ....E
Verona ....E
Villa Park ....E
Volo ....EG
Wadsworth ....E
Washburn ....E
Wauconda ....EG
Waukegan ....E
Wayne ....E
Wayne Center ......E
Wenona ....E
Westchester ......EG
West Chicago ......E
West Dundee ....E
West Jersey ....E
Western Springs ...E
Wheeling ....EG
Whitaker ....E
Wichert ....E
Wilburn ....E
Wildwood ....E
Willow Springs...EG
Wilmette ....EG
Wilmington ....E
Wilson ....E
Wing ....E
Wilton Center .....E
Winfield ....E
Winthrop Harbor ...E
Wood Dale ....E
Wooster Lake .....EG
Worth ....EG
Wyoming ....E
York Center ....E
Yorkville ....E
Zion ....E

RULES AND INFORMATION PERTAINING TO ELECTRIC SERVICE, METERS AND WIRING OF COMMONWEALTH EDISON CO.

(Corrected to December, 1938)

FOREWORD

INFORMATION

The Company is desirous of serving its customers promptly and satisfactorily. It will endeavor to coperate with customers, contractors, engineers, and architects to the fullest extent in completing service connections with as little delay and inconvenience as possible, and will gladly give special attention to any particularly difficult situation confronting a customer.

With regard to matters concerning customers' service or installations, information will be gladly given if a telephone call is made on Randolph 1200 with a request for connection to the following bureaus or divisions:

Customers' Order Department.
Service Bureau.
Contract Service Division.
Wiring Inspection Division.
Secondary Line Design Office, Engineering Department.
Line Installation Section, Engineering Department.
Service Investigation Section.

INSPECTION.

1. All wiring which is to be connected to the Company's service shall be inspected and approved by the Department of Streets and Electricity of the City of Chicago, and shall conform to the rules and regulations established by the Company from time to time.

2. A temporary current-permit, or final certificate of inspection issued by the Department of Streets and Electricity of the City of Chicago shall be forwarded to the Wiring Inspection Division before the electricity can be turned on. This applies to additional wiring which may be connected at any time, wiring of a temporary nature, and wiring for use during construction work as well as to original installations. The Company reserves the right to make final connections of all wiring to its mains and, in case any damage results from unauthorized connections, the customer will be held responsible.

3. The Company furthermore reserves the right to discontinue service, after due notice to the customer (or without notice in the event of emergency), in case the Company deems the condition of the customer's wiring, apparatus or motors either hazardous or not in compliance with the Company's rules and regulations. The Company reserves the right to discontinue the customer's service on orders from the Department of Streets and Electricity or the Department of Public Service of the City of Chicago when such a step is deemed necessary.

Section 1.

FORMS OF SERVICE.

General. 1. The electricity furnished by the Company is generated in the form of alternating current. For the most part it is distributed as alternating current, but it is also in certain localities converted into and distributed as direct current. In accordance with the form of service predominating therein, the city is divided by the Company into two zones, an Outer Zone where the form of service is almost entirely alternating current—the few remaining installations of direct current being in the areas in the vicinity of or bordering upon the Inner Zone—and an Inner Zone where the form of service is still largely direct current. The dividing line between the Outer Zone and the Inner Zone is shown on the map.

Effective—June 8, 1938.

CHANGE-OVER TO ALTERNATING CURRENT.

At the Option of the Company. 2. When in the Company's judgment it is economically desirable to do so, the Company shall have the right to replace its direct current service to any customer with alternating current service. The Company will give written notice in such cases. Details and terms for changing can be secured from the Company.

Forms of Service Available. Subject to the provisions of the foregoing paragraphs of this Section, the forms of service available in the Outer and Inner Zones are:

Outer Zone—Alternating Current 60 Cycles

Single-phase, 3-wire at approximately 120-240 volts for light and power of 5 hp or less;

3-phase, 3-wire at approximately 240 volts for connected loads of 5 hp and larger;

3-phase, 3-wire at approximately 480 volts upon request for power installations having an aggregate rated motor capacity of 250 hp or more;

3-phase service at higher voltages may at times be furnished for large installations under certain conditions.

Direct Current in Limited Areas

At approximately 120-240 volts, for light and power.

Inner Zone—Alternating Current 60 Cycles

3-phase, 4-wire, at approximately 120-208 volts, for light and power.

Direct Current

At approximately 120-240 volts, for light and power.

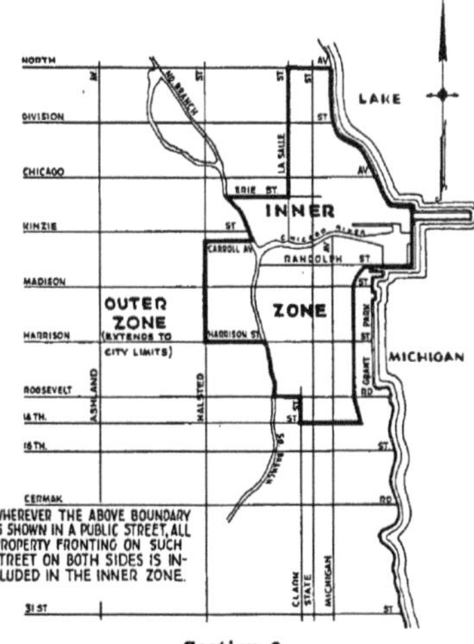

Section 2.

SERVICE CONNECTIONS.

General. 1. A service connection, or a service drop, as it is commonly designated, is that portion of the supply conductors between the Company's pole or underground mains and the customer's service outlet. If taken from an underground main, it is known as an underground service connection and if taken from an overhead main, it is known as an overhead service connection or drop. In the

case of an overhead service drop, if the distance between the Company's pole and the customer's service outlet is in excess of 115 feet, that portion of the supply conductors between the Company's pole and the nearest customer's pole will be considered the service drop. Service connections from an underground main are sometimes run overhead from a rear lot line to the building, and service connections or drops from an overhead main may be carried down a pole and taken into a building underground.

2. It is essential, in order to avoid error, that the customer secure written information from the Company as to the location at which the Company's service lines are to be brought to his building.

3. Where it is necessary, on account of size, voltage, or other conditions, that the transformers be placed on the customer's premises, the customer shall provide, at his expense, the necessary space and enclosures on his property to enable the Company to install the transformer equipment. Such equipment shall be installed either on poles or in an enclosure as may be determined by the Company after investigation of the conditions on the customer's premises.

4. One set of service equipment will be provided by the Company for a customer in the Outer Zone and for small loads in the Inner Zone. For service to large loads, the Engineering Department will determine the number of service installations, depending upon the total load and its distribution, which the Company will provide at its own expense. Should the customer request additional service equipment, the Company will provide such additional equipment together with the necessary cables required to connect such equipment to the Company's system, provided that the customer reimburses the Company for the extra cost occasioned by the additional equipment. The service equipment at all points of supply shall remain the property of the Company.

Overhead. 1. In case a pole line from which service is to be given is not in position at the time the interior wiring is being done, inquiry should be made of the Line Installation Section for information as to the location of the service outlet.

2. The Company will provide at its own expense, for any building, one overhead service drop for light and, where required by these rules, one overhead service drop for power. The length of this service drop shall in no case exceed 115 feet. Where the distance to be spanned between the Company's pole and the customer's service outlet exceeds 115 feet, a pole shall be provided by and at the expense of the customer, for each 110 feet or fraction thereof, provided, however, that where, due to special conditions, 110 foot spans are impossible, single spans between poles may be increased to 125 feet in length.

3. The pole shall be of cedar, at least 25 feet in length, with a minimum diameter of 6 inches at the top and set in the ground at least 4½ feet. A square timber will not be approved as an intermediate support in place of a pole. If a steel pole is desired, this should be a two-section tubular steel pole, made up of two lengths of standard steel tubing, having nominal diameters of 4 inches and 5 inches. The tubes should be joined by a swedge joint. Where the length of the service drop does not exceed 100 feet and the service conductors are 3 No. 6 American gauge wires or less, a single piece of 4-inch standard pipe 25 feet long may be used. This pole shall be set 4½ feet in the ground in a concrete collar which is at least 12 inches in diameter and extends from the bottom of the pole to 3 inches above the ground line. At the top of the pole, drillings are to be made and a standard spool-type bracket shall be mounted in the proper position. This bracket is to be 2 or 3-wire, depending on the character of the customer's installation.

4. Where not more than one pole is required, the necessary service wires between the building outlet and this pole and from this pole to the Company's lines will be installed by the Company at its expense. Where more than one pole is required, weatherproof wire of the proper size, depending on the capacity of the customer's installation, but in no case smaller than No. 6 American gauge, and all necessary poles, complying with the above specifications, spaced at intervals, as provided under this section and subject in paragraph 2, page 10, shall be installed by the customer at his expense, from the building outlet to a point accessible to and ordinarily not more than 15 feet from the Company's lines. In such case, the service wires between the Company's lines and the first pole will be installed by the Company. The foregoing applies only to the construction of the service drop and is not to be construed as having any application to the extension of the Company's distribution lines to serve any customer.

5. The requirements in the above paragraphs refer to the installation of service connections which do not require a service in excess of No. 6 American gauge wires. On installations requiring a service in excess of the above, the Line Installation Section shall be consulted for specifications covering the length of spans between poles and the size of poles.

6. If possible, the position of the service outlet should be such that service wires can be brought from the Company's nearest pole without crossing adjacent property. Crossing adjacent property with service wires is, of course, contingent upon there being no objection to such crossing by the owner of such property. Whenever such owner objects to the installing or to the maintaining of service wire over his property, the customer, and not the company, shall, if possible, satisfy such objections.

7. Where it is necessary, for any reason except the Company's convenience, to move a service drop or service outlet, the Company will reroute the service drop at its own expense and the customer shall move the service outlet at his expense. The new location of the service outlet shall comply with the Company's rules for new outlets.

8. Service outlets should never be more than 30 feet nor less than 10 feet from the ground. For buildings of two or more stories in height, the outlet should ordinarily be brought out at the ceiling of the second floor, provided that this point is not higher than the allowable maximum height of 30 feet from the ground.

9. Where the position of a building is such that the service outlet cannot be located at a point which may be reached from the pole line, or where the outlets of a low building must be brought out less than 10 feet above the ground, some form of support of suitable strength and height shall be provided by the customer. When such support is a pole, it shall comply with the specifications noted in paragraph 3.

10. **Risers.** (a) Risers used for the support of service wires are objectionable and should be used only in cases where their use cannot be avoided. Methods of supporting other than those specified above, shall be specially approved by the Wiring Inspection Division. The service conduit shall run up the riser, terminating at a point 24 inches below the top of the riser or other form of support.

11. The use of a building or a support attached to a building as an intermediate support for the attachment of service wires is not permitted by the Department of Streets and Electricity of the City of Chicago.

12. Where a one-story building is located on the alley lot line, usually the service, if brought out on the alley side of the building, will not clear the telephone wires when the pole is located on the same side of the alley as the building and will not give the required 18-foot clearance over the alley (see para-

THE CLARK ELECTRIC WATER HEATER

The Clark Electric Water Heater is entirely automatic, requiring no attention or maintenance, and because of the exclusive "Black Heat" element—so called because it does not glow even faintly red—it will give remarkably long and trouble-free service. The low temperatures used, spread over a wide area, heat the water quickly, yet are not high enough to cause lime and scale precipitation in hard water, nor oxidation with pitting and corrosion in soft water.

Features of the Clark

The Clark "Black Heat" element consists of a number of porcelain unit elements, each containing over 6 ft. of nichrome wire, connected together in series to form a broad belt of heat 6 in. wide and several feet long. It is clamped to the outside of the tank in a special steel channel which allows it to be removed or replaced in a few minutes' time without draining the tank or disrupting service. Clark Heaters are regularly equipped with extra heavy steel tanks (two and a half times as thick as ordinary tanks), electrically welded and heavily galvanized. They are equipped with built-in heat traps and cold water baffles. A fully automatic and remarkably reliable thermostatic heat control is used on all Clark Heaters. A knob on the outside of the heater allows setting of thermostat to maintain any desired temperature of water between 125° F. and 165° F. A minimum of over 4 inches of mineral wool closely packed on all sides of heater prevents heat losses and enables water to be kept hot for days.

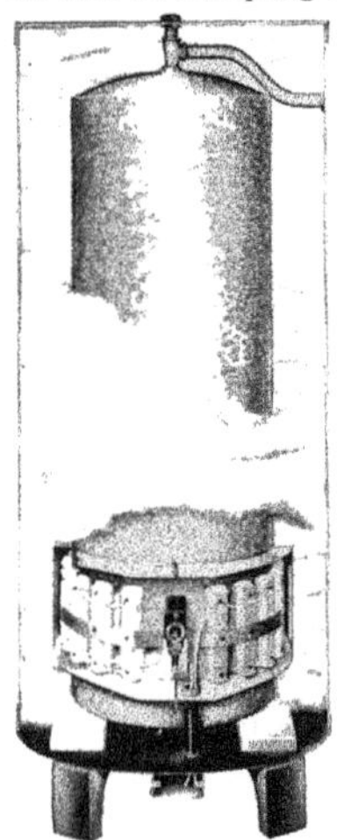

Cutaway View Showing Single Unit "Black Heat" Element

Outer shell is heavy gauge steel electrically welded. Top and base are of extra heavy spun steel. Finished in multiple coats of high gloss baked enamel. Standard color is white with black top, base, and legs.

Water and Electrical Connections

Water inlet is at bottom of heater. Outlet is at back of heater near top. Both are standard thread ¾ in. pipe. Electrical connections are made through convenient knock-out box located on bottom of heater at front, except on the square heater, where the outlet box is at the top and back of the heater.

50-Gallon Clark Square Type Heater

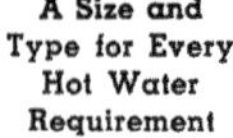

A Size and Type for Every Hot Water Requirement

As shown in table below, Clark Heaters are available in a wide range of sizes, either round or square and in either single or double element.

Capacity gals.	Model No. with galvanized tanks		Dimensions, in.		
	Single Unit	Double Unit	Diameter		Height
Clark Round Type Heaters					
30	30RS	30RD	22		59
50	50RS	50RD	26		60
67	67RS	67RD	28		63
82	82RS	82RD	30		65½
82	82KS	82KD	30		65½
100	100KS	100KD	32		65½
125	125KS	125KD	34		69½
Clark Square Type Heaters					
			Width	Depth	Height
5	5W	....	16	16	37
10	10W	....	20	20	34½
15	15W	....	20	20	43½
30	30RSW	30RDW	21	22¼	57
40	40RSW	40RDW	23	24¾	57
50	50RSW	50RDW	25	26¼	58

Suggested Wattages: 500, 600, 750, 900, 1000, 1250, 1340, 1500, 1640, 1750, 2000, 2250, 2460, 2500, 3000.

Any of the above wattages may be specified on either unit. Wattage and voltage must be specified on each order. Heaters will be furnished for 230-volt A. C. operation, unless otherwise specified.

Size 82, 100 and 125-gal. will be shipped "knocked-down" for assembly in the customer's home, unless otherwise specified on order.

MANUFACTURED BY

McGRAW ELECTRIC COMPANY • CLARK WATER HEATER DIVISION

5201 WEST 65TH STREET | PHONE PORTSMOUTH 6735

graph 18 (d)), when the pole is located on the opposite side of the alley. For this reason, the Department of Streets and Electricity of the City of Chicago will permit a variation from the standard rule requiring services brought to the alley side of the building. On such one-story buildings, the service should be brought out on the side of the building, but in no case shall the service outlet be more than 15 feet from the alley line. The service outlet should be placed on the same side of the building as the pole so that the service drop will extend away from the building and not over it. Where a pole is located directly back of a building of the above type, the service outlet, if located away from the alley side of the building, will in some cases, eliminate the necessity of conduit being carried down the pole. The above modification of the rule of the Department of Streets and Electricity of the City of Chicago applies only to one-story buildings and will, in many cases, avoid the use of objectionable risers. In many cases, risers and the installing of wires down a pole may be avoided by connecting a rear building to the service drop installed for a front building. In such cases, the customer shall install either weatherproof or rubber-covered wire between the rear building and the front building, but the final connection on the front building service will be made by the Company.

13. Service outlets shall not be located on chimneys or on fire or parapet walls extending above a roof.

14. Service outlets shall never be terminated within 1 foot of a downspout.

15. Where a service outlet is terminated on a post supporting a porch, galvanized steel straps or braces shall be fastened in such a manner that the post will be firmly held to the joist. Each strap shall be at least 1 inch by 12 inches and ⅛-inch in thickness, and shall be fastened to the post and joist by lag screws, 2 into the post and 2 into the joist, such lag screws to be ¼-inch by 2½ inches. Two straps shall be used, one on each side of the post.

16. Where wiring is being installed in buildings under construction which are to have a stucco, stone-coat, brick veneer finish or equivalent, a substantial form of support for the service bracket shall be provided; this support to consist of a ¼-inch by 6-inch by 30-inch steel plate securely fastened either on the inside of the wall or studding of the building with ⅝-inch hot-galvanized bolts projecting through and 2 inches beyond the wall surface, and provided with two ⅝-inch by 1$\frac{1}{16}$-inch square nuts. When the service outlet wires are larger than 4/0, or where the outlets are so arranged that more than one outlet can be connected to one service drop, and where the total carrying capacity of these several outlets is more than 225 amperes, the customer shall install a steel plate and bolts. If the Company, in order to provide adequate capacity, installs more than 3 service wires, the length of the plate shall be increased by 9 inches for each additional service wire provided in excess of 3. An additional bolt shall also be provided for each such additional service wire.

17. Where, in order to obtain capacity, paralleling of wires is employed, all conduits for a single service outlet shall be terminated in the same box service-head. The different phase wires shall be grouped so that, in any one conduit, there are all 3 phases, 3 phases and a neutral, or on single-phase, a neutral and conductors for each side of the circuit.

18. **Clearances.** (a) The service wires shall, in no case, be within easy reach from porches, windows, or any other part of the building ordinarily accessible to the occupants.

(b) The service wires, when passing over a rear or adjacent building to the one being served, shall clear a pitched roof 2 feet, and a flat roof 8 feet. Any roof on which a person can walk with ease shall be considered a flat roof.

(c) The service outlet shall be so located that there will be at least 24 inches clearance between it and any telephone or signal wires where attached to the building, and at least 3 feet clearance shall be provided between the service drops of both systems in the open span.

(d) When the Company's pole line is on the opposite side of the street or alley from that of the building to which service is to be given, the service outlet for such a building shall be of sufficient height to give at least an 18 foot clearance between any point of the street, driveway or alley and the service drop.

Underground. 1. If the Company has an existing service within 50 feet of the customer's premises, the customer's service mains shall be brought to this service. If no service is available, an application should be made to the Company to have a service installed.

2. The Line Installation Section of the Engineering Department, upon request, will furnish a service location sketch designating the point at which the service will be available. The junction cabinet and the service switch cabinet should not be installed until after the Company has completed the installation of its service conduit as obstructions in the street may necessitate some deviation from the point designated.

3. **Termination of Service Conduit.** (a) If the sidewalk space is excavated, the Company will extend its service conduit through the curb wall.

(b) If the sidewalk space is unexcavated and the basement extends to the property line, the Company will extend its service conduit to the property line.

(c) If the customer desires the service extended beyond the points designated in (a) and (b), this service extension shall be at the customer's expense.

(d) A service will not be terminated in any place which shall not be readily accessible at all times.

(e) When an underground service conduit enters a building at a level below the sewer and water elevation in a street or alley, the customer shall provide and install conduit end seals. These conduit end seals consist of brass caps threaded for attachment on the end of the service conduits and so designed that they can be wiped to the cable sheath.

4. **Termination of Service Cables.** (a) If a single set of service cables is to be installed, the Company will extend the service cables 3 feet beyond the end of the service conduit for the customer's connections. A set of service cables shall be defined as either one set of lighting cables, one set of power cables or one set of each.

(b) If more than one set of service cables is installed, the Company will extend the service cables an average of 10 feet beyond the end of the service conduit for the customer's connections.

(c) If the customer desires the service cables extended beyond the points designated in (4a) and (4b) this service cable extension less quantities in either (a) or (b) shall be at the customer's expense.

5. **Junction Cabinets.** (a) All underground service cables shall, except for switchboard installations, be terminated in a service junction cabinet. Where the service cables supply not more than one building, the junction cabinet shall be installed by the customer. Where the service cables supply more than one building owned by different people, the junction cabinet will be installed by the Company.

(b) The junction cabinet shall be installed at the terminus of the underground service duct, and arranged to provide a complete enclosure for the lead-covered service cables which shall enter the cabinet through the rear wall at one end in such a manner that they can be readily removed for repairs or

Specify Jeff

Jefferson-Union Renewable Fuses

For over 40 years Union Fuses have established their worth—simple, rugged, easily renewable, Jefferson-Union Renewable Fuses are made in all standard capacities.

Jefferson Fustats

Safer than conventional plug fuses. Thermal cutout provides a long time-lag and prevents needless blowing on harmless motor starting loads. Use of Jefferson Fustats stops over-fusing because, for example, a 20 or 30-ampere Fustat will not fit into a 15-ampere adapter. Similar limitations apply to all sizes. Made in sizes to 30 amperes.

replacement. The junction cabinet shall be securely clamped to the incoming service duct with a locknut and bushing. The Company's service cables shall not be directly connected to the customer's service-switch except by special permission of the Wiring Inspection Division. The customer shall install from the service-switch into the junction cabinet, the proper size rubber-covered wire of sufficient length to permit the Company to make the connection to its lead-covered service cables. By this method, cables can be more safely handled, and future customers in the vicinity can be readily connected without disturbing the service-switch connections.

(c) The junction cabinet shall be of sufficient size to permit connections to be readily made to the underground service cables and also to permit the cable to be bent properly. Such cabinets should have minimum dimensions as shown in the following table for services consisting of only three cables. For large size service mains the contractor should secure the approval of the Wiring Inspection Division as to the type and size of junction cabinet to be used.

(d) Dimensions

Size of Service Cable	Cabinet Dimensions Lgth.	Width	Deth	Cover
3—No. 6	24 in.	12 in.	6 in.	Screw
4—No. 6	24 in.	12 in.	6 in.	Screw
3—1/0	24 in.	14 in.	8 in.	Screw
4—1/0	30 in.	14 in.	10 in.	Screw
3—4/0	30 in.	18 in.	12 in.	Screw
3—4/0 and 1-1/0	30 in.	18 in.	12 in.	Screw
4—4/0	30 in.	18 in.	12 in.	Screw
3— 500,000 cir mils	36 in.	18 in.	12 in.	Screw
3— 500,000 cir mils and 1-4/0	36 in.	24 in.	14 in.	Hinged
4—1,000,000 cir mils and 1,500,000 cir mils	48 in.	36 in.	20 in.	Hinged
4—1,500,000 cir mils and 1-1,000,000 cir mils	48 in.	36 in.	24 in.	Hinged

6. In an installation in which the service conduit enters from the bottom of the cabinet, such as in an unexcavated basement where the service conduit is brought vertically through the floor, the junction cabinet shall be wide enough to permit the installation of two service conduits of the same size.

7. In cases where required, the contractor shall furnish with the junction cabinet, a locknut and bushing of the proper size for clamping the service conduit into the cabinet.

8. For underground direct current services, a junction cabinet shall be made so as to insulate the incoming service conduit and cables from the customer's conduit. This can be done by providing a fibre plate in the side or bottom of the junction cabinet into which the outgoing conduits can be clamped. This is to prevent electrolysis.

9. If the service-switch and junction cabinet are combined, the cabinet shall be of ample size to permit the safe handling of the service cables, and the switch-panel arranged in such a manner as to permit the removal of the service cables for repairs or replacement without removing the panel. All connections for service cables shall be made on the front of the panel.

10: **Conduit for Service Cables.** (a) The standard size of conduits to be used for extending the service cables used by the Company from a junction cabinet at curb or building wall into a customer's premises is given in the following table:

Number and Size of Service Cable	Conduit Size
3—# 6	3 in.
2—#1/0 & 1—#6	3 in.
3—#1/0	3 in.
4—#1/0	3 in.
3—#4/0	3 in.
4—#4/0 or 3—#4/0 & 1—#1/0	4 in.
1—500,000 cir mils	2 in.
2—500,000 cir mils	3 in.
2—500,000 cir mils and 1—#4/0	3½ in.
3—500,000 cir mils	3½ in.
3—500,000 cir mils and 1—4/0	4½ in.
4—500,000 cir mils	4½ in.
1—1,000,000 cir mils	3½ in.
1—1,500,000 cir mils	3½ in.
1—2,000,000 cir mils	4 in
1—3 Cond. #12 (Pressure cable)	1½ in.

(b) When cables are extended into a building from the end of the underground service conduits, such extension shall be in conduits of the same size, number and the same formation as those of the Company's service conduits. The Company's service cables, when extended into a building, shall not be installed under or over boilers or in any other location where they would be subjected to excessive heat. The conduits containing the extension of the service cables, shall be covered with 3 inches of concrete over and around them. The use of stone or fiber conduits on the inside of the building is prohibited except with special approval from the Wiring Inspection Division. Plans of a conduit installation for the Company's service cables on private property shall be submitted to and approved by the Wiring Inspection Division, before installation.

11. **Cables and Disconnects.** (a) For alternating current installations, cables are to be grouped as follows: In any one conduit, all 3-phases, or 3-phases and a neutral, and on single-phase a neutral and conductors for each side of the circuit.

(b) Where the size of the installation requires more than one set of underground direct current service cables into the customer's premises, the customer shall install, at his own expense, a fuse-extension switch on each cable except the neutral cable, which shall be provided with a bolted disconnecting link. To facilitate operation in an emergency, a nameplate holder shall be provided on the service panel so that each of the cables may be readily identified. This nameplate holder should have dimensions to contain a card 3¾ inches by 1¼ inches.

12. **Underground Service from Overhead Lines.** Where the customer desires underground service from an overhead line, he shall furnish and install the lead-covered cable and conduit from the service-switch to a location on the pole to be designated by the Company.

Section 3.

TRANSFORMER INSTALLATIONS.

Scope. 1. The following rules cover the general requirements which apply to either 4,000-volt or 12,000-volt transformer installations in fireproof enclosures on a customer's premises. Specific requirements for each installation will be furnished by the Company after a study of conditions existing at that location.

2. The Company reserves the right to use these transformer installations to serve customers on the same or other premises.

Enclosure. 1. The fireproof enclosure complete with all necessary facilities, including concrete bus and switch structure when required, shall be furnished by and at the expense of the customer and be constructed in accordance with these rules and specific instructions furnished by the Company. These enclosures shall be accessible at all times to properly authorized persons. The enclosure shall be located accessible to a driveway and shall be constructed of brick or concrete, with a fireproof ceiling and with a concrete floor cable of sustaining the weight of all necessary equipment. The enclosure shall have adequate ventilation, cooling water facilities, drainage, lighting, suitable entrances equipped with fireproof doors and be of sufficient size to properly accommodate all the necessary equipment.

2. Adequate passageways shall be provided between a driveway and the enclosure, together with facilities for raising or lowering the equipment if necessary. Clearances and minimum headroom of the passageways, entrances and shafts, will be specified in each case by the Company.

3. All plans of enclosures prepared by the

American Blower
VENTILATING APPARATUS

Ventura Home Conditioner for Homes

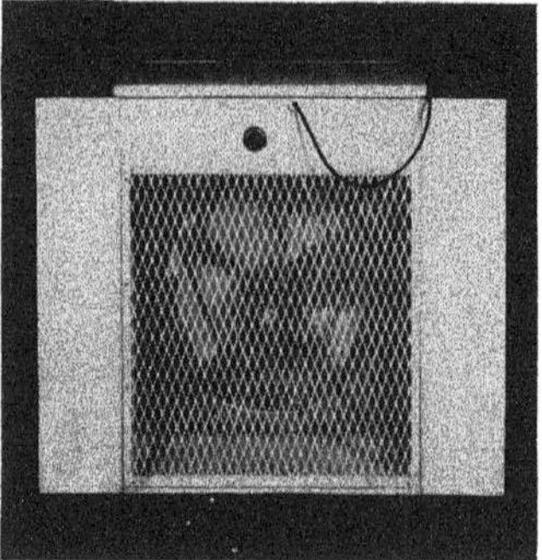

Ventura Portable Conditioner for Apartments

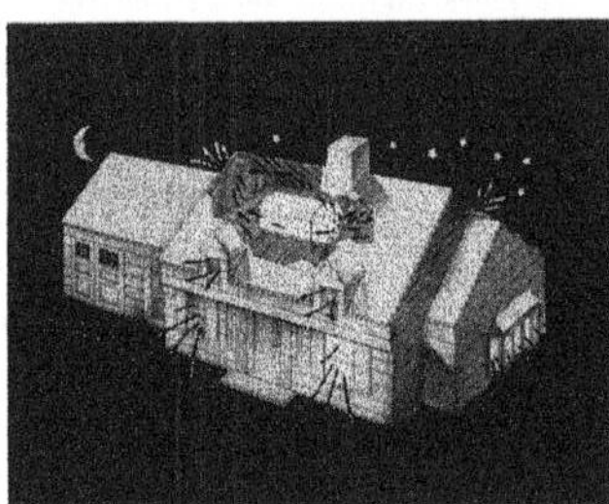

A Ventura Home Conditioner quickly exhausts the hot, stifling air

American Blower Ventura Fans (above) and Sirocco Utility Blowers

Aeropel Electric Ventilators

Manufacturer of

UNIT HEATERS · UNIT VENTILATORS

FANS and BLOWERS

AIR WASHERS

AIR CONDITIONING EQUIPMENT

For Over 50 Years

American Blower Corp.

6000 Russell Street, Detroit, Michigan

Division of the American Radiator and Standard Sanitary Corporation • Canadian Sirocco Co., Ltd., Windsor, Ontario

customer shall be submitted in duplicate to the Company for approval.

4. The enclosure for each installation shall comply with the Electrical Code of the Department of Streets and Electricity and with the Building Code of the Department of Buildings of the City of Chicago.

Equipment. 1. The Company will, at its own expense, furnish, install, connect, and maintain, within the enclosure, the necessary service equipment, such as transformers, high-voltage switches, primary fuses, protective relays, control equipment, and related apparatus.

2. The customer shall provide a suitable panel or space on his switchboard outside of the enclosure for the installation of the Company's meters, and provide thereon the wiring, test-links and terminals required for such meters.

3. The customer's secondary service-switch shall be placed outside the enclosure. Whenever a customer's bus is supplied from more than one transformer installation, each secondary supply shall be provided with a breaker, so equipped as to automatically disconnect such transformer installation from the customer's bus. Such service-switches or breakers shall be provided by and at the expense of the customer.

Cable and Conduit. 1. The necessary conduits for high and low-voltage cables from the enclosure to the service point selected by the Company at the property line, with the necessary splicing chambers and pull boxes, shall be provided by and at the expense of the customer and installed under the supervision of the Company in accordance with plans approved by the Company.

2. The primary cable required on the customer's premises to connect the transformer installation to the Company's system will be installed and maintained by the Company at the expense of the customer. Where a customer is supplied by a loop from a 12,000-volt ring feeder, the Company will furnish, install and maintain at its own expense, for one transformer installation, the 12,000-volt cable forming a part of such ring feeder including any portion thereof located on the customer's premises.

3. The customer shall extend all secondary service connections equipped with the necessary lugs, to points designated by the Company, inside the enclosure ready for connection. The final connection of the service to the supply buses will be made by the Company.

Section 4.

SERVICE WIRING AND EQUIPMENT FOR SPECIAL VOLTAGES ABOVE 480.

General. 1. These requirements refer to all installations where a service voltage above 480 volts is used on customer's equipment.

2. Arrangements shall be made witn the Company both as to form of contract and method of obtaining electricity before equipment is purchased or wiring is planned. As the equipment furnished by the customer may affect the operation of the Company's system, tests shall be made by the Company, at the expense of the customer, on all such equipment, before it is accepted for connection on the system.

Section 5.

WIRING.

General. The following rules cover the general requirements for wiring. Special rules covering the requirements for wiring of installations connected on 4-wire, 3-phase network service are given on page 469.

The following synopses of provisions effecting wiring requirements were taken from the Company's schedules now in force. These schedules are subject to amendment from time to time. While at present no change in such provisions is contemplated, the Company's schedules should be consulted before installing or changing wiring.

1. **Rating.** Wiring shall be so arranged that a separate meter may be installed for each class of service supplied under the Company's schedule of rates, which follows. Apparatus not rated in horsepower, shall be computed on the basis of 1 kv-a as equivalent to 1 hp. One hp is computed as equivalent to 1,000 watts input.

2. **General Residential Service.** (a) Available for any customer using the Company's standard service for lighting purposes or for both lighting and power purposes in his residence, provided that electricity will not be furnished hereunder for commercial use nor for wireless telegraph apparatus, or other power apparatus in which the use of electricity is intermittent or subject to violent fluctuation and the operation of which may interfere with lighting service.

(b) Service hereunder will be furnished only to a single occupancy. Where an owner or a tenant is operating a building containing more than one apartment, or is operating several apartments in a building and desires service under this rate, a separate meter must be arranged for and installed and the Company will render separate bills for each apartment.

(c) Where a residence and business are combined in one premises, service will not be furnished under General Residential Service.

(d) The lighting for hall or halls, stairways or basements of an apartment building and service for equipment operated by such building may be served under this rate as a separate customer on a separate meter or meters. In case an owner or lessee occupies an apartment in the building, the hall lighting for this building may be connected to his apartment meter when the building contains not more than six apartments.

3. **Small Commercial Light and Power Service.** (a) Available for any commercial customer using the Company's standard service for lighting purposes or for both lighting and power purposes, except for residential service. "Power" is defined as electric service used for any purpose other than lighting. Service for photographic printing, bath cabinets and other kinds of equipment which are not used for general illumination, shall be considered as power service.

(b) Where apartments and stores, offices or shops are in the same building, the store, office or shop lighting shall be considered as commercial lighting, and the wiring should accordingly be arranged for separate meters, provided, however, that if the customer desires, both installations may be combined on one meter, and the combined installation shall be considered as commercial lighting.

(c) Where a portion of a store or shop is used as living quarters, and the wiring is arranged for a single meter for both the store or shop and the living quarters, the installations shall be considered as commercial lighting.

(d) Where the rated capacity in connected load of a commercial installation is more than 6½ kilowatts the maximum demand is measured, and provisions shall be made for setting a demand-meter.

(e) In the case of welders, hoists and similar intermittently operated apparatus, the Wiring Inspection Division shall be consulted in connection with metering equipment.

4. **Large Light and Power Service.** On service connections and meters provided for the installations served under these rates, information shall, in all cases, be obtained from the Wiring Inspection Division prior to the installation of the wiring.

5. **Lighting Service.** On the alternating current system a separate service and meter may, at the option of the Company, be provided for the lighting service, mostly for purposes of lighting regulation, and in such cases, the customer shall arrange the wiring accordingly for such meters.

6. **Air Conditioning Rider 49-A.** Where the

customer contracts with the Company for using air-conditioning motors on Rider 49-A the motors driving refrigerating units used solely for air-conditioning (not cold storage or ventilating) equipment and such auxiliaries used in connection therewith as are only operated simultaneously with, and as a part of such air-conditioning equipment, shall be wired for a separate meter independent of a meter supplying other power.

Residence and Apartment Loads. Every residence and apartment installation which does not exceed 3,000 watts, or 50 sockets shall have a 2-wire service main and 2-wire meter loops, as such an installation will be connected to the Company's system by two service wires at 120 volts. Where the installation exceeds 3,000 watts, or 50 sockets, it shall be wired for a 3-wire service and a 3-wire meter. All single or duplex wall receptacles, other than brackets, shall be figured at not less than 60 watts each, or the equivalent of 1 socket.

Commercial Loads. 1. On every new commercial installation, a 3-wire meter-connection block and cabinet shall be provided and when the load is over 2000 watts, a 3-wire service main shall be provided. Motors and heating appliances operating on the lighting service, where the rated capacity of such equipment exceeds 25% of the lighting load, shall be considered in determining the size of wire to be installed and also in determining if the installation should be 2 or 3-wire.

2. All signs exceeding 1 circuit shall be connected 3-wire and shall be properly balanced.

Commercial and Residence Loads in Same Building. On installations where there are stores or offices and apartments in the same building, the rule governing the service and meter installation is as follows: Where the total connected load is 3000 watts or 50 sockets or less, add to the total connected load of the stores or offices ¼ of the connected load of the apartments, and if this load is in excess of 2000 watts, service mains shall be 3-wire but the meter-connection blocks and cabinets will be governed according to the rules above on residence and commercial loads. Where the total connected load is in excess of 3000 watts or 50 sockets, the service shall be 3-wire.

Motor Loads. 1. Direct current motors should usually be connected on the lighting meter. The Company should be consulted when large direct current motor loads are contemplated.

2. In alternating current territory, the fluctuation in voltage caused by the starting currents prevents the connecting of motors larger than 1 hp to the lighting service, except in special cases approved by the Wiring Inspection Division. An example of such a case would be a small motor somewhat exceeding 1 hp installed in connection with a large capacity lighting service. Several motors, none of which is larger than 1 hp, may be connected to the lighting service where the aggregate does not exceed 3½ hp.

Miscellaneous Loads. 1. Single direct current stereopticons, outlets for battery charging, and other devices which are operated most economically at 120 volts will be approved for this voltage, provided that such devices or outlets do not require or supply a load exceeding 2 kilowatts. If such devices or outlets require or supply a load exceeding 2 kilowatts, such devices and outlets shall be connected for operation at 240 volts. Where there is an installation of more than one such device in the same premises, they shall, if the total wattage of the installation exceeds 2000, be connected to a 3-wire main and be balanced as nearly as possible.

2. All rectifiers requiring an input of more than 2 kv-a shall be operated at 240 volts.

3. In theatres, all 220-volt motion-picture and spot arcs shall be connected to the power meter.

4. Alternating current arcs requiring more than 2 kv-a shall be operated at 240 volts.

5. Where transformers are used in connection with motion-picture arcs or spot arcs they shall be operated at 240 volts.

6. In the case of welding machines, X-ray machines, hoists, elevator motors, compressor motors, furnaces, flashing signs, and other installations of similar character, where the use of electricity is intermittent or subject to violent fluctuation, the Company reserves the right to require the customer to install, at his own expense, suitable wiring or equipment to, in a reasonable degree, limit such intermittence or fluctuation, where good engineering practice would require such wiring or equipment to prevent undue interference with the Company's service. The Wiring Inspection Division should be consulted prior to wiring for such equipment in order that the customer may obtain information regarding the necessary provision for properly connecting such apparatus.

Transformers. 1. Any new customer or any existing customer who desires to secure a simultaneous recording of the light and power demands when the light and power loads are serviced and metered separately, shall provide one or more ratio transformers as required by the Company and connect his light to the power service and meter. If this is done, the customer shall agree to accept the voltage and voltage variations due to the operation of motors and other equipment on the power transformers. The Company shall be consulted and shall approve all plans for changing any lighting service to a power service.

2. The use of the so-called auto-transformers will not be permitted. A type of transformer shall be used having primary and secondary windings. Such a transformer, unless installed in a specially constructed fireproof enclosure, shall be air-cooled.

3. In general when the connected lighting load to be connected on the power meter is less than 20 kilowatts, this load may be connected on a single phase of the power using one transformer. The transformer in this case shall be connected between the two mains having the least potential to ground. When the load is 20 kilowatts and less than 40 kilowatts, it shall be balanced on two phases of the power using two transformers. When the connected lighting load is 40 kilowatts or more the load shall be balanced on the three phases of the power using three transformers.

Voltage Regulation. The wiring installed in the customer's premises shall be of such capacity that the entire connected lighting load (as given to the contractor by the owner or architect) can be carried with a loss in voltage of not more than 2% between the service entrance and the most remote lamp on the premises. The loss in voltage for a power installation shall not exceed 5%.

Switchboards. 1. The Wiring Inspection Division will upon request, supply a special set of requirements with prints containing wiring diagrams pertaining to switchboards and panel-boards.

2. Specifications and prints for service and meter-switchboard installations shall be submitted to the Wiring Inspection Division for approval before construction of the switchboard is begun.

3. The attention of electrical contractors and switchboard manufacturers is called to the following requirements pertaining to the Company's approval of plans for switchboards and panel-boards. The plans shall furnish the following information in all cases:

(a) Name of owner, address (with correct street number) of buildings.

(b) Name of architect or electrical engineer and the electrical contractor.

(c) Ratings in amperes of all main switches, cutouts, and circuit-breakers.

(d) Size in inches of bus-bars.

(e) In addition to the usual front, rear, and elevation views of switchboards, there will be

required a floor plan showing the location in which the switchboard is to be installed, with clearances, distances to walls, and to other objects.

(f) The plans shall show the manner of bringing the service cables to the panel or switchboards, with the size, location of conduits, size of elbows, pull-boxes, and junction cabinets.

(g) If the lead-covered service cables are to be furnished by the Company, the size of conduits required shall be obtained from the Company.

4. In connection with the construction of switchboards, cables shall be brought to the switchboard so that the lugs clamping the cables to the bus work or switches, are accessible. This will not permit installation of cables among bus-bars or between the front slate and the rear fuse panels. Where single-pole, fuse-extension, disconnecting switches are required with direct-current switchboards, when more than one set of underground cables is required, an approved copper link fuse may be installed on the service side of the switch. This will simplify the bus construction.

5. Fuses shall be so arranged that they will be readily accessible for the purpose of replacement, and to this end, it is recommended that no more than three rows of switches be placed on a switchboard.

6. To prevent overheating of switches, fuses, and cables, it is recommended that all the lugs have a conductivity of not less than 60% of that of pure copper and that their cross-sectional area be such that they will not be required to carry continuously more than 600 amperes per square inch. They should have a bolting contact surface of not less than 1 square inch for each 100 amperes of current.

7. The general arrangement of the connections on the back of the board shall be such as to render it possible to make repairs or alterations with a reasonable degree of facility and safety while the board is in service.

8. The bus-bars shall be rigidly supported and anchored. The arrangement of the feeder cables between the terminal of the conduit system and the back of the switchboard shall be made in a systematic and orderly manner and the cables shall be segregated as far as possible, with a view to minimizing the possibility of serious interruption to the service. No supporting framework for electrical busses or conductors shall be installed in such a manner that there will be a completely closed magnetic circuit of steel or iron about any single conductor or portion of an electrical circuit.

9. For details concerning the installation of meter test links and other matter pertaining to switchboard meters, see paragraph on "Switchboard Meters."

Service Switches and Cutouts. 1. The customer shall provide and install an approved service-switch or circuit-breaker for disconnecting each service connection from the Company's mains or transformers, and locate the same as close as possible to a point where the service cables or bus enter a premises. Fuse blocks and service-switches shall be equipped with fuses of approved type and capacity at the time of their installation.

2. The neutral wire of a 3-wire single-phase or direct-current service-switch and fuse-block for branch mains, shall not be fused. For installations where the service wire is number 4/0 or smaller, the neutral service wire may be connected to the building wiring neutral by lugs anchored on an insulated solid base. When the service wire is larger than number 4/0 a bolted disconnect link is required.

3. The neutral wire shall be connected to the center blade of all 3-pole switches except for 3-phase. On 3-phase installations the two phases having the least difference of potential to ground shall be connected to the two outer blades of the service-switch.

4. For 4-wire, 3-phase services, a 3-pole fused switch or an approved circuit-breaker shall be provided at the point of service entrance, to control the 3 ungrounded wires. The neutral shall be solidly connected as explained above. The service-switch or circuit-breaker installed for use on the 4-wire, 3-phase service cables from the network system shall be provided with approved 2-hole lugs. These lugs shall be attached to the switch or circuit-breaker terminals by means of ½-inch bolts.

5. Switches and fuse-blocks shall not be installed above or in close proximity to laundry tubs, sinks, or other plumbing fixtures.

Service Neutral and Conduit Ground. 1. The customer shall ground the neutral wire of his installation by installing a grounding wire separate from the grounding conductor provided for the conduit system. This neutral ground shall be installed according to the rules of the Department of Streets and Electricity of the City of Chicago. A 3-phase and a 2-wire, 230-volt single-phase system having no wires within the building at ground potential, are not to be grounded at the building service. A direct current system is not to be grounded at the building service.

2. The 4-wire, 3-phase network services are to have the neutral grounded as specified under grounding of neutral and conduit.

Auxiliary or Breakdown Service. 1. Where a customer contracts to use the Company's service as an auxiliary or breakdown service in connection with his source of supply other than that from the Company's lines, he shall, in case the number of kilowatts which the Company is obliged to stand ready to supply under the contract, be less than the estimated maximum of the customer's plant, as estimated by the Company, furnish and install a circuit-breaker of a type approved by the Company. This circuit-breaker shall be set to break the connection with the Company's service in case the maximum demand shall at any time materially exceed the number of kilowatts which the Company has agreed to supply.

2. The circuit-breaker shall be installed by the customer at a suitable location between the Company's meter and the customer's load, and shall be in a steel cabinet so constructed that it can be sealed by the Company.

3. Customers receiving part of their electric service from any other source than the Company shall not, in general, be allowed to operate such other source in multiple with the Company's service due to possible hazard to both the customer's and Company's equipment.

In special cases, however, the Company may give written permission under the following conditions:

(a) When both rated capacity of the customer's generators and his guarantee under Rider 27 are 200 kilowatts or more.

(b) When, in the Company's opinion, such source of supply is of the proper type and so operated as not to jeopardize the Company's system or service.

(c) When the customer assumes full responsibility for all and any damage from such multiple operation, and agrees to hold the Company harmless.

For all such installations the Company will equip its meters with ratchets, so as to measure only energy input and lagging reactive component.

Additions and Alterations. When any change in the size of a customer's installation is made, the Company shall be informed, so that it may inspect the installation and provide service and meter of the proper capacity. If alterations are to be made in a building which may disturb the electric wiring and require the re-location or removal of the Company's meter, the Company shall be notified in advance, in order that the changes may be given proper attention. If it is necessary to move the meter to a new

location, this change will be made if meter-connection cabinets are provided. A temporary location and meter-connection cabinet shall be provided by the customer, if electricity is desired during such alterations, but under no circumstances will the use of electricity be allowed without a meter.

Primary Service. Where primary cables are brought into a customer's premises, the Company should be consulted as to the manner of installing cables before the installation is planned and all plans for this type of service installation should be submitted to the Wiring Inspection Division in duplicate for approval before the installation is made.

Section 6.

ALTERNATING CURRENT 4-WIRE, 3-PHASE NETWORK SERVICE.

General. The following wiring and equipment requirements will apply only on installations which are to be connected on the alternating current 4-wire, 3-phase network system. In the territory where this system is available, inquiry should be made of the Company as to the kind of service which will be furnished for the installation before equipment is purchased or wiring installed. Installations for customers to be served from the network system are to be wired according to the following regulations and limitations:

2-Wire Service. On new wiring installations, connected loads of less than 5 kilowatts of light and power with no motor larger than 1 hp and no other equipment larger than 2 kv-a shall be wired with a 2-wire service main and wired for a 2-wire, 120-volt meter.

3-Wire Service. Connected loads of 5 kilowatts, but less than 15 kilowatts of light and single-phase power, with no motor of 5 hp or larger, shall be wired with a 3-wire service main and wired for a 3-wire, 120-208-volt meter. Motors of 1 hp or smaller and other equipment of 2 kv-a or less, may be connected for operation on either 120 or 208-volts. Motors larger than 1 hp and equipment of over 2 kv-a, shall be connected for operation on 208-volts. All 120-volt load shall be properly balanced on the 3-wire service mains. On old wiring installations cut-over to the network service, loads of 3 kilowatts and larger may be connected for 3-wire service mains and wired for a 3-wire, 120-208-volt meter.

4-Wire Service. Connected loads of 15 kilowatts or more of light and power shall be wired with a 4-wire service main and wired for a 4-wire, 3-phase, 120-208-volt meter. Motors of 5 hp and larger shall be wired for operation on 3-phase, 208-volts. On 4-wire installations, motors smaller than 5 hp may be wired for single-phase or 3-phase operation. When the combined light and power load includes an aggregate motor load of 5 hp or more in 3-phase motors, the entire load shall be wired on 4-wire, 3-phase service. All single-phase load shall be properly balanced. When the connected load is less than 15 kilowatts, 3-phase service will not be furnished for motors aggregating less than 5 hp in 3-phase motors unless the customer deposits with the Company, the excess cost of a 4-wire meter over that of a 3-wire meter. An installation of 3-phase motors shall be wired with a 4-wire meter-connection cabinet and the neutral wire shall be extended from the service cables or 4-wire service mains into the meter-connection cabinet.

Meter, Service, Fuse Sequence Wiring. On all new wiring installations connected on the network system, no fuses other than the main service fuses shall be installed ahead of the meters. When requested to do so the Wiring Inspection Division will check and approve wiring plans before installation.

Total Load, X-Ray and Special Equipment. 1. In calculating total connected load, all motors, appliances and lamps are to be included.

2. Special equipment, such as X-rays welders, etc., larger than 2-kv-a, shall be wired for 208-volt operation.

Wiring. 1. In buildings where there are two or more tenants and where there is a likelihood of an increase in the total connected load to 15 kilowatts or more, it is recommended that 4-wire mains and submains be provided.

2. All phase wires of the 4-wire system shall be protected with an approved 3-pole fused service-switch or an approved circuit-breaker.

3. The general rules governing the installation of meter-connection cabinets, shall apply to all network installations.

4. Wiring connected on the 3-phase, 4-wire network system, shall be identified, using a white wire for the neutral or grounded conductor and a different colored wire for each of the respective phases.

5. Service-switches or switchboards should be located as close as possible to the point where the Company's cables enter the premises.

6. When service cables are extended into a building from the end of the Company's ducts or from a junction cabinet over the end of the Company's ducts, the conduits are to be installed with 3 inches of concrete over and around them.

Grounding of Neutral and Conduit. 1. The neutral of a 4-wire network service shall be grounded on a cold water pipe. The ground connection shall be made to the neutral cable inside the junction cabinet which encloses the service cables.

2. The Company will arrange to install the neutral ground when they install the junction cabinet supplying more than one building.

3. The size of the grounding conductor shall be the same as that of the neutral service wire to which it is connected but need not be larger than number 4/0. This wire shall be installed in conduit of size in accordance with the Chicago Electrical Code and it shall be securely bonded to each end of the conduit enclosing the grounding conductor. The bonding of the grounding conductor and the grounding conduit shall be accomplished by the use of a lug and bolt at the cabinet end and a ground clamp at the water pipe end.

4. The grounding conductor shall be installed and connected to a water pipe having requirements in accordance with the Chicago Electrical Code. The size of the water pipe to which the grounding conductor and conduit is connected shall not be smaller than the conduit carrying the grounding conductor.

Motors. 1. Starting current limitations shall be as specified in Section 8 under Starting Current, with corrections made for the rated voltage of the motor.

2. Motors of ¾ hp and 1 hp in size, shall be limited to a locked-rotor current of 57 amperes at 120-volts.

3. Special starting-current limitations for motors of 50 hp and larger will be furnished by the Wiring Inspection Division upon request.

Section 7.

APPARATUS.

General. The Company reserves the right, at any time, to inspect, in order to obtain nameplate data, and to test, all motors and other devices and apparatus which are owned by the customer and are or may be connected to the Company's lines. Such tests are for the purpose of determining starting current, efficiency, power factor or other characteristics which may affect the service of other customers or cause undue disturbance to the Company's system.

Nameplates. All electrical equipment such as welders, furnaces, heating appliances, X-ray and radio apparatus, and the like, shall be provided with nameplates showing the rating of the apparatus in kv-a. The nameplate shall show the capacity of the apparatus or device in full-load amperes. The volts, phase

Actual photo of Model Jr. 3-B Sink installed in Hotel Alexander Hamilton, Patterson, N. J.

and cycles shall be given and, in the case of motors, in addition, there shall be given the horsepower, speed, type, model, frame, style, form or class, whichever may be the particular method of a manufacturer, in designating a motor. When a nameplate on electrical equipment is located so that the data on the plate can be obtained only with considerable difficulty or by removing housing around the motor, the nameplate shall be moved or a duplicate plate installed, by the customer in a location where it can be readily inspected.

Section 8.

MOTORS.

General. 1. The following motor regulations are necessary for the purpose of supplying uniform service to all customers, as the successful operation of lighting equipment on the same mains with motors requires that the normal voltage of the supply circuit be closely maintained.

2. In the case of hoists, elevator motors, compressor motors or other installations having similar load characteristics, where the use of electricity is intermittent or subject to violent fluctuation, the Company reserves the right to require the customer to install, at his own expense, suitable wiring or equipment to, in a reasonable degree, limit intermittence or fluctuation, where good engineering practice would require such wiring or equipment to prevent undue interference with the Company's service.

3. The Company reserves the right to inspect, in order to obtain nameplate data, and to test, all motors and other devices and apparatus which are owned by the customer and are or shall be connected to the Company's lines. Such tests are for the purpose of determining starting current, efficiency, power factor or other characteristics which may affect the service of other customers or cause undue disturbance to the Company's system.

4. The Company reserves, in the case of any customer, the right to place special limitations, other than those specified in the section entitled "Starting Current" on starting and pulsating currents of motors and other power equipment. Such limitations are necessary to prevent objectionable disturbance on its lines. In any event, the difference between the maximum and minimum effective value of the pulsating current of any large motor shall not exceed 66% of the rated full-load current of the motor.

Direct Current. 1. Direct current motors of 1½ hp and smaller may be operated on either 120 or 240 volts. Motors larger than 1½ hp shall be operated on 240 volts.

2. A starting resistance is recommended in connection with all direct current motors, but motors not larger than ½ hp of the shunt type, ¾ hp compound, and 2 hp series wound, not requiring a starting-current in excess of the values given in the starting-current tables, may be installed without starting resistance.

3. Direct current motors should usually be wired, except under Rate B, so that they may be connected to the lighting meter.

ALTERNATING CURRENT.

1. **General.** (a) Motors of less than ½ hp may be operated at either 120 or 240 volts on the lighting service and meter but the locked-rotor current under any operating condition shall not exceed 20 amperes at 110 volts when measured with a well-damped ammeter.

(b) Motors of ½ hp may be operated on either 120 or 240 volts, on the lighting service and meter and the locked-rotor current under any operating condition shall not exceed 26.6 amperes at 110 volts and 20 amperes at 220 volts when measured with a well-damped ammeter.

(c) Motors of ¾ and 1 hp shall be operated on 240 volts on the lighting service and meter, and the locked-rotor current under any operating condition shall not exceed 20 amperes at 220 volts for ¾ hp and 27 amperes at 220 volts for 1 hp when measured with a well-damped ammeter.

(d) Motors of over 1 hp shall be operated on 240 volts and wired for a separate service and meter. If there is a power service on the building, the motors are to be wired to this service.

(e) When the aggregate rating of several motors, of which the largest motor does not exceed 1 hp, amounts to 3½ hp or less, they shall be wired on the lighting service and meter.

(f) When the aggregate rating of several motors exceeds a total of 3½ hp, they shall be wired on a separate service and meter.

2. **Single Phase.** When single-phase motors or other apparatus are connected to only one phase of a 3-phase installation, they shall be connected between the two wires having the least difference of potential to the ground. When the service is not alive for test, the single-phase motor shall be connected on the wires feeding from the outside blades of the 3-pole service-switch. The Wiring Inspection Division shall be notified when any single-phase motor is to be connected on a 3-phase installation.

3. **Three Phase.** (a) Motors of 5 hp or more are supplied from the 3-phase system in a large part of the alternating current territory, but inquiry should be made of the Power Sales Division as to the proximity of 3-phase lines to any particular location at which such power may be desired.

(b) 3-phase service will not be furnished for installations aggregating less than 5 hp of 3-phase motors, unless the customer deposits with the Company, a sum equivalent to the excess cost to the Company of installing a 3-phase service and meter, above the cost of installing a single-phase service and meter, plus the excess cost of installing a 3-phase line extension over that of a single-phase extension, provided such an extension is required, and provided further that such service will in no case be furnished for any installation aggregating less than 2 hp of 3-phase motors. In case the customer's power installation shall subsequently be increased to a total rated capacity of 5 hp or more of 3-phase motors, which have been in use for a period of not less than 6 months, the Company will return the amount of money advanced on account of the excess cost to the Company of installing a 3-phase service and meter. If the 3-phase power load of 5 hp or more is subsequently reduced to less than 5 hp, a deposit is again required if the customer desires to continue the use of the 3-phase service.

(c) On all new installations of motors or other power equipment aggregating 100 hp or over and on all existing installations to which power load is added, where the sum of the original and the new loads equals or exceeds 100 hp, a power factor of at least 85% at normal full load shall be maintained by the customer for the combined light and power installation. Welders and other intermittently operated apparatus on separate meters and fire-pump motors are not included in this requirement. Installations which include auxiliary equipment which must be in service to bring the power factor of the total power installation at normal full load to 85%, will be considered as having a power factor of 85% at normal full load, only if the corrective equipment is interconnected with the power equipment in such a way that the corrective equipment must be connected to the line when the normal full-load condition is reached.

(d) On any installation requiring motors of 40 hp and larger, a synchronous motor is recommended, when practical, because this type of motor has power-factor corrective characteristics, which will materially aid in raising the customer's overall power factor.

(e) Reverse-phase relays or series-wired, hatch-limit switches shall be installed on all 3-phase elevator, crane, and similar instal-

lations, as required by the Department of Streets and Electricity of the City of Chicago.

Fractional Horsepower Motors. 1. The use of the fractional horsepower motor for many applications in the home, store and factory, is increasing very rapidly. As many such motors are connected to the Company's lighting mains, it is important from both the customer's and the Company standpoint for such motors to conform to the following requirements, in order that the customer may not only have his lighting equipment free, from annoying fluctuations but that he may also be assured of a motor which will give him the highest economy in operation.

2. The locked-rotor current of fractional horsepower motors of ⅓ hp and less, connected to the Company's lighting service, shall not exceed, under any operating conditions, a maximum of 20 amperes at 110 volts, as measured with a well-damped ammeter.

3. The full-load power factor, efficiency and apparent efficiency (product of true efficiency + power factor) shall conform to the values in the following tables. These tables are divided into long-hour and short-hour usage.

(a) Long-hour is considered as any usage of over 1,000 hours annually. Short-hour is 1,000 hours and less annually.

(b) For the guidance of motor and appliance manufacturers, there is listed not only the full-load efficiency and power factor required by both long-hour and short-hour usage for the various sizes of fractional horsepower motors, but there is also included a table of appliances operated by such motors listed under both long-hour appliances and short-hour appliances. This list is necessary tentative and is subject to changes and additions as may be demanded by experience and advances of the art.

(c) Minimum specifications for motors operating on 110-220 volts, 60-cycle circuits, at 1800 rpm.

LONG-HOUR USAGE						220 Volts
Horsepower	⅛	⅙	¼	⅓	½	¾
Power Factor.......	52	56	60	61	63	65
Efficiency	53	58	62	63	65	67
Apparent Efficiency.	30	36	42	44	47	49
Locked-Rotor Amperes	20	20	20	20	26.6	20

Minimum specifications for motors operating on 110-220 volts, 60-cycle circuits at 1800 rpm.

SHORT-HOUR USAGE						220 Volts
Horsepower	⅛	⅙	¼	⅓	½	¾
Power Factor	50	52	56	58	60	62
Efficiency	41	46	51	54	58	61
Apparent Efficiency.	24	27	32	35	39	42
Locked-Rotor Amperes	20	20	20	20	26.6	20

Note: The values of power factor and efficiency are minimum. Where the product of the minimum values of power factor and efficieney as shown in the table does not equal the apparent efficiency, it is required that either the power factor or the efficiency values shall be increased to make the product of efficiency and power factor equal the stated apparent efficiency.

(d) Appliances operated by fractional horsepower motors.

LONG-HOUR USAGE

Adding machines
Addressograph machines and similar office equipment
Advertising signs
Air compressors (small)
Automobile washers
Binding and trimming machines
Bottling machinery
Candy machines
Carbonating machines
Conveyors and loaders (small)
Cutting machines
Dairy machinery
Garage equipment
Heating equipment
Ice cream machinery
Knitting machinery
Labeling machines
Laundry machinery
Oil burners and oil pumps
Packing machines
Printing machines
Refrigerating machines
Sewing machines—factory and domestic
Shoe machinery—manufacturing and repair shop
Stokers—domestic and commercial
Traffic signals and similar applications
Ventilating and air conditioning equipment
Wrapping machines

SHORT-HOUR USAGE

Coffee mills
Dish washers—commercial and domestic
Exercisers
Floor polishers
Meat slicers and grinders
Motion picture machines
Organ blowers
Pianos—electric
Pumps—sump, booster, house, condensation and gasoline dispensing
Surfacing machines
Vacuum cleaner (fixed type)
Washing machines

Starting Equipment. Every motor of 7½-hp rating and above shall be provided with no-voltage-release starting equipment which will cause the motor to be disconnected from the line in case of an interruption to the power supply. For motors of large capacity which the slow starting, the no-voltage-release should have a time-element relay, which will prevent the opening of the circuit in the event of momentary voltage fluctuation.

Starting Current. The instantaneous current (determined by test or based on the value guaranteed by the manufacturers) drawn from the lines by any motor, during the starting cycle, shall not exceed the locked-rotor current value for the rated horsepower of such motor, as shown in the following tables:

(a) Single-Phase—60-Cycle

Horsepower	R Volts	Locked-Rotor Amperes†
⅓ and below....................	110	20
½	110	26.6
½	220	20
¾	220	20
1	220	27
1½	220	30
2	220	40
3	220	60
5	220	100

(b) Three-Phase—60-Cycle

Horsepower	R Locked Rotor Amperes at 220 Volts	N Locked-Rotor Amperes† at 208 Volts
1	26.6	28.1
1½	36.6	38.7
2	46.6	49.3
3	60	63.5
5	86.6	91.6
7½	115	121.6
10	141	149
15	197	208.2
20	251	265.4
25	304	321.5
30	360	381
35	370	391.4
40	380	402

50 to 75 inc.... (220 v) 8 amps. per hp.
50 to 75 inc.... (208 v.) consult wiring Inspection Division for motor limitations.

On motors larger than 75 hp, the customer shall obtain the starting current limitations from the Wiring Inspection Division for each installation.

† Locked-rotor amperes are those indicated by an ammeter equipped with a special device to eliminate damping and inertia effects. In general, these values will be equivalent to the locked-rotor current, with the starter in the first position.

Locked rotor current values for 440-volt

fire-pump installation. A booster-pump installation may be connected to the service side of a main switch, provided the service mains are of sufficient capacity to carry both the house load and the fire-pump without overloading. In industrial plants it is always best to have separate service mains for a booster pump, because a future load may overload the service mains.

3. The service mains shall be extended underground from the fire-pump control cabinet and meter to the outside of the building and to the same location as the general power service mains for an overhead service installation. Where the general power service mains are underground, the fire-pump service shall be extended to the manhole or transformer vault designated by the Company as that form which the service is supplied to the building.

4. All service cables for fire-pumps shall be lead-covered and shall be installed in conduits imbedded in concrete of not less than 3 inches over and around them.

Service Cables. 1. The service-cable sizes as given in the following table shall be used for fire-pumps serviced by the Company's underground, direct current or alternating current systems, or from a "Street Service Vault."

Horsepower	Number and Size of Cable For D-C Motor 230 Volts	Number and Size of Cable For A-C Motor 230 Volts
25	2-1/0	3-1/0
40	2-4/0	3-4/0
50	2-500,000 cir mils	3-4/0
60	2-500,000 cir mils	3-4/0
75	2-500,000 cir mils	3-500,000 cir mils
100	2-1,000,000 cir mils	3-500,000 cir mils
125	2-1,000,000 cir mils	3-500,000 cir mils
150	2-1,500,000 cir mils	6-500,000 cir mils
175	2-1,500,000 cir mils	6-500,000 cir mils
200	2-2,000,000 cir mils	6-500,000 cir mils
225	2-2,000,000 cir mils	6-500,000 cir mils
250	2-2,000,000 cir mils	6-500,000 cir mils
275	2-2,000,000 cir mils	6-500,000 cir mils

2. The cables used by the Company on underground services are lead-paper insulated. This type of cable shall be used by the customer for the fire-pump service mains, when connecting to the Company's underground system, in preference to lead-rubber insulation, because the rubber insulation deteriorates when spliced to a paper-insulated cable due to oil which runs from the paper-insulated cable along the copper of the rubber-insulated cable.

3. The use of smaller cable sizes than those in the table above shall be approved by a letter from the insurance authorities having jurisdiction. Copies of such letter of approval shall be sent to the Company and to the Department of Streets and Electricity of the City of Chicago.

4. When a fire-pump installation is supplied from an overhead pole, the underground wires installed by the customer between the Company's pole and the fire-pump shall be lead-rubber covered and shall be of the size required by the rules of the Department of Streets and Electricity of the City of Chicago for a single motor installation.

5. Where a separate service connection, independent of the general power service connection, is required for a fire-pump, the customer shall pay the cost of such service installation.

Meter Installations. 1. Meters for fire-pump installations are to be of the following types: On the direct current system, shunt-type meters are used. On the alternating current system, self-contained-type meters are used for 30 hp or smaller 220-volt motors, and current transformers are used for motors larger than 30 hp in size and for all 440-volt motors. Meter cabinets with the dimensions, mounting of equipment, and wiring as specified by the Company shall be furnished and installed by the customer.

2. The standard dimensions in inches of re-pump meter cabinets are are as follows:

ype of nstallation	Cabinet			Slate Panel		
	Width	Height	Depth	Width	Height	Thickness
	30 x	48 x	12	20 x	38 x	1
	30 x	36 x	15	28 x	28 x	1
	36 x	36 x	15	32 x	32 x	1

3. The cabinet shall be arranged with two ongue-sealing devices, one at the top and one t the bottom of the door so that the doors an be sealed by padlock seals.

4. The cabinet may be located at any point 1 the service run provided it is readily accesible and not subject to moisture or vibration. clearance of 36 inches shall be provided in ront of the fire-pump meter cabinet and an nobstructed passageway shall be maintained o the cabinet. The conduits entering the abinet shall have cables long enough to conect with lugs on the shunts or current transormers. The lugs for the shunts or current ransformers will be furnished by the Comany, but shall be installed by the customer t his expense.

5. A slate panel, 1 inch in thickness, shall e mounted in the cabinet and the Company's ieters shall be mounted on this panel and ired at the expense of the customer.

6. There shall be a 6-inch clearance beween the cabinet walls and the meter. For elf-contained meters used for motors of 30 p and smaller operated at 240 or 208 volts, 4-terminal test block of 200-ampere size hall be provided and installed by the electrical contractor on a slate panel.

7. It will be necessary that fire-pump meer panels be completely wired by the manufacturer in the shop. This fact should be taken into consideration when giving cost of fire-pump meter cabinets and panels.

8. There shall be 10-ampere cartridge fuses with a cutout-block installed in the pressurewire circuits of the meters as shown in the standard wiring diagrams for fire-pump meter installations.

9. On all fire-pump meter installations, ¼-inch fibre washers shall be provided with the meter panel and shall be installed by the contractor so that in mounting the panel there is provided a ¼-inch air space between the panel and the back of the cabinet to ventilate and prevent sweating and grounding of the slate panel.

10. All 480-volt fire-pump service panels shall have standard 480-volt pressure fuses with 480-volt spacing and have the following words painted in white on the face of the cabinet and the slate of the panel: "DANGER —480 VOLTS."

11. Where a meter has been furnished by the Company for regular power load, the usual rental will be charged for the separate fire-pump meter, as provided under the Company's schedule of rates. In case a printometer or other type of demand indicator is required the usual rental will be charged and the customer shall pay for the installation of such meter.

12. All service cables to any fire-pump shall pass through the meter cabinet before reaching the fire-pump control panel.

13. On alternating current fire-pump installations, the meter and current coils of the current transformers shall be connected in the service wires that terminate on the outside blades of the double-throw switch on the fire-pump control panel.

Section 10.

METERS.

General. 1. The Company will install one meter or one unified set of meters for one class of service.

2. A monthly rental charge for each additional watt-hour or demand-meter is made by the Company when, at the request of the customer, and for his convenience, the Company approves an installation of more than one meter on his premises for one class of service. The rental varies with the type and capacity of the meter installed.

3. The Company shall be consulted whenever it is necessary to know in advance the type and capacity of meter which a given installation will require.

Meter Specifications. 1. Various details, such as the method of metering, the type and capacity of watthour meters and maximum-demand meters, and the testing facilities required, will be determined by the Company for each installation. These details shall, on the larger installations, be taken up with the Wiring Inspection Division by the customer or his representative before the installation is designed and sufficiently in advance of its construction to give the Company sufficient time to obtain the metering equipment. Blueprints or sketches showing the proposed location and connections of meters and equipment on switchboards and panels shall also be submitted in duplicate to the Wiring Inspection Division for approval, before the switchboard is constructed. For "Switchboards" see Section 5 above. For fire-pump meter installations, see Section 9 above.

2. Dimensions of the meters and equipment to be used on panels and switchboards may be secured from the Wiring Inspection Division.

Types. 1. On direct current, standard front-connected types of meters are used up to a capacity of 400 amperes, 240 volts, 2 and 3-wire. Larger capacity meters of the back-connected switchboard type, will be furnished, and will require mounting on slate panels.

2. On alternating current, standard front-connected meters are provided for all installations. Self-sustained meters are used on alternating current installations for capacities up to and including 100 amperes. Meters with current transformers are used for capacities above 100 amperes and for all 480-volt installations.

3. Current and potential transformers are used with meters for all voltages above 480.

Demand-Meters. 1. On all commercial installations of over 6½ kilowatts, provisions must be made for a demand-meter.

2. On alternating current installations of 60 amperes and less, where the demand is to be measured, a demand-register is furnished in all watthour meters. On installations of over 60 amperes, when the Company's rate schedule requires it, separate demand-meters will be installed. Provisions shall be made for the installation of such demand-meters.

3. If a customer elects to use a printometer or type G demand-meter, when his load does not warrant it, he shall arrange for this instrument on a rental contract. The rental charge for meters installed at the customer's request to replace meters installed by the Company as its standard practice, will be equivalent to the difference of the rental charges of the equipment furnished and the equipment replaced.

4. Where the customer's installation consists of both power and lighting loads, and where the maximum demand of each such load is 100 kilowatts or more, provision shall be made for the Company to install, on limited-hour alternating current installations, demand meters suitable for the determination of the simultaneous demand. Where the maximum demand of the lighting load is less than 100 kilowatts, such demand will be assumed as occurring during the peak period unless the customer pays to the Company a monthly meter-rental charge sufficient to cover the additional cost of providing such recording maximum demand-meters as are necessary for the determination of the simultaneous maximum demand.

mounted on separate panels located where they are not affected by the magnetic field of buses or conductors carrying heavy currents, and where they are not subjected to extreme changes in temperature.

9. When current transformers or other metering equipment are mounted at the rear of the switchboard, there shall be left a clear space of not less than 30 inches between current-carrying parts and the wall, so as to permit free access to this equipment.

M. S. F. Sequence Metering. 1. "Meter, Service Switch, Fuse sequence" metering requires the installation of the meter ahead of the service-switch and fuses, and provides maximum convenience to the customer for reading and testing by company employes.

Outdoor meter-connection cabinets will be furnished without charge by the Commonwealth Edison Company and shall be installed outside the building for alternating-current service from overhead pole connections on the following types of installations:

New Buildings.

1. New residences.
2. New apartment buildings not over three stories high.
3. Small commercial and industrial buildings requiring not over 100-ampere capacity in wiring for either single or three phase.
4. Private garages.

Old Buildings.

1. Reconnections.
2. Remodeled buildings
3. Additions or alterations to present buildings requiring service conduit and/or major wiring changes.
4. Electric range installations in single residences and apartment buildings not over three stories high.

Outdoor meter-connection cabinets will be furnished without charge by the Commonwealth Edison Company for the following special types of installations:

New sequence; (no fuses ahead of meters) inside the building by special permission of the Wiring Inspection Division of the Meter Department.

Modified sequence; (service-switch inside building and meter outside or inside) by special permission of the Wiring Inspection Division of the Meter Department.

It is suggested that contractors call with plans of new buildings and discuss the meter wiring with the Wiring Inspection Division before the work is started.

2. On new wiring installations of 60 amperes or less connected on single-phase or 3-phase, alternating current service, the meter or meters shall be installed in outdoor meter-connection cabinets. At the request of the Company, installations over 60 amperes and not exceeding 100 amperes shall be wired with meters connected ahead of the service-switch and fuses.

3. The meter-connection cabinets in sizes of 30 to 100 amperes inclusive will be furnished by the Company but it is to be installed by the customer at his expense, in a location satisfactory to the owner of the building and to the Company.

4. The outdoor meter-connection cabinet shall be installed on the outside of the building and connected ahead of the service-switches and fuses, but shall not be installed on any building wall that is located on the alley or property line.

5. Meter-connection cabinets are to be installed with the top supports of the cabinet not over 7 feet nor the bottom supports less than 3 feet, 6 inches from the ground. When cabinets are ganged beside each other, a minimum clearance of 1 inch shall be provided between them.

6. An approved service-switch shall be installed inside the buildings on each such meter installation.

7. Arrangements for obtaining meter-connection cabinets are to be made with the Wiring Inspection Division.

8. Outdoor meter-connection cabinets,

by the Wiring Inspection Division, depending on the character of the installation. Meters placed behind heating tanks which are not insulated to prevent radiation of heat, or behind machines in motion, shall be at least 3 feet from such equipment and in such cases the amount of space is to be determined by consultation with the Wiring Inspection Division.

(h) Meters shall not be located on platforms which are not accessible by stairs. Ladders can not be accepted in place of stairs. When located on platforms, there shall be a space in front of the meter at least 3 feet wide, protected by a suitable railing.

(i) Large capacity and switchboard direct current meters requiring the use of heavy testing equipment, such as storage batteries, shall be placed in a location where proper facilities can be provided for moving the testing equipment to and from the meters.

3. When meters of over 60 ampere capacity are to be installed for construction work or when the location is to be out-of-doors or in non-weatherproof buildings, a meter housing or substantial cabinet of weatherproof construction shall be provided by the customer to protect them from injury. For temporary and construction work when it is possible, a meter location should be selected at the outset which can be used throughout the construction period.

4. Cabinets to contain meters shall be of ample size to permit the safe handling of wires for connecting, disconnecting, or testing the meters. If a metal cabinet is used, the inside shall be lined with suitable insulating material, such as wood or transite boards.

Meter Connections. 1. For alternating current installations requiring a capacity of more than 100 amperes, current transformers are generally used with the meter. The transformers shall be installed in a switchboard bus or in a current-transformer cabinet on the house side of the service-switch. For switchboards, the secondary wires of the current transformers shall be extended to the test-links and continued to the meter. When meter-connection cabinets are installed, the secondary wires of the current transformers shall be terminated in the meter-connection cabinet.

2. For all direct current meters and those alternating current meters not wired under the MSF sequence, adequate meter protection shall be provided for each meter.

3. Service and house leads, for front-connected direct current meters of 200 amperes capacity, and up to and including 400 amperes capacity, shall be carried in a metal trough not higher than 42 inches from the floor and to a point directly beneath the meter lugs. They shall be brought outside the trough through bushings spaced far enough apart so that the loops may be run in a direct vertical line to the meter terminals. The loops shall be so anchored that the weight of the cables will not rest on the meter terminals. The length of the meter loop required outside the trough is determined by the type of meter to be installed.

Inside Meter-Connection Cabinets. 1. Meter-connection cabinets of the following two types have been approved by the Company for inside use on direct current installations, and are to be provided and installed by the customer:

(a) A cabinet containing a combined fused service-switch and meter-connection block.

(b) A cabinet containing a meter-connection block.

2. These cabinets are of a type which permit the mounting of the Company's meter in combination with the cabinet so that by means of suitable adapters or end walls, all connecting wires are completely enclosed. They also permit disconnecting the meter for exchange or test without interruption to the customer's service. All necessary trims will be furnished by the Company and installed with the meter.

3. Combined fused service-switch and meter-connection-block-type cabinets are approved in the 30-ampere and 60-ampere sizes. This type of equipment will be approved by special permission of the Wiring Inspection Division on certain alternating current installations.

4. For direct current installations, meter-connection-block-type cabinets are approved in the 30 and 60 ampere size. For 100 and 200 ampere sizes, this type also shall be installed, but they are to be used with a separate fused service switch or separate fuses. Where more than one meter is connected on a service-switch, separate meter-protection fuses shall be installed with the meter-connection block for each meter.

5. All meter-connection cabinets for inside a building furnished by a contractor are to be equipped with shutter-type end walls and to be closed with a blank shutter having standard twist-outs.

6. A complete list of all meter-connection cabinets approved for use on the Company's lines, giving the manufacturer's name and the catalogue number, will be kept on file and issued on request to anyone having occasion to use such a list.

7. On the network system, meter-connection cabinets shall be used. See Section 6 above.

8. The sizes of meter-boards required for mounting meters and meter-connection cabinets, and general wiring diagrams will be furnished on request.

9. For the mutual protection of the customer and the Company, all meter-connection cabinets will be kept sealed in order to aecomplish the full safety features of the equipment. Circuit fuses, connected on the load side of the meter, shall be in a separate compartment or cabinet and shall be accessible to the customer. When meter-connection cabinets of the fused type are used, the service fuses in the cabinets shall be accessible to the customer without the necessity of opening the main door of the cabinet and should be of a size determined by the capacity of the connection-block and shall in all cases be larger than the circuit fuses.

10. A card-holder shall be provided on the front of every meter-connection cabinet. The contractor shall insert in this holder a card showing the complete address and the location in the building of the premises connected to each meter-connection cabinet.

Mounting of Meter-Connection Cabinets. 1. The method of mounting outdoor meter-connection cabinets shall be such as to insure permanent attachment to the suporting wall. The use of wood plugs driven into brick walls for anchoring will not be approved.

2. Meter-connection cabinets inside a building of 30 and 60-ampere capacity shall be installed so that the top of the cabinets will not be more than 6 feet above the floor. For meter-connection cabinets of 100 amperes or above, this distance shall not be more than 4 feet 6 inches.

3. The minimum distance between the floor and the bottom of a meter-connection cabinet shall not be less than 2 feet.

4. When a meter-connection cabinet is used with a current-transformer meter or on 4-wire, 3 phase service it shall be installed so that the top of the cabinet is not more than 4 feet 6 inches above the floor.

5. Meter-connection cabinets installed inside of a building, shall be so arranged that the meters can be placed at least 6 inches away from metal cabinets, cutout boxes, conduits, or walls, so as to permit the safe handling and accessibility of wires in making conthe top of the cabinet 6 inches for alternating nections and tests. On installations requiring 200-ampere meter-connection cabinets and on direct current installations where meters of 200-ampere capacity or larger are installed with troughs, this clearance shall be 12 inches.

6. Where 30-ampere meter-connection cabinets are used inside of a building on alternating current installations, there shall be 14 inches clearance and on direct current installations 28 inches clearance, from the top of the meter-connection cabinet to the bottom of the next one above or to the bottom of a cutout cabinet or any grounded surface. The minimum clearance between the top of a meter-connection cabinet and a ceiling shall be 18 inches for alternating current and 32 inches for direct current meters. Where larger than 30-ampere meter-connection cabinets are used, the clearance should be approved.

7. The distance between centers of meter-connection cabinets of 30-ampere capacity used inside of a building, shall not be less than 10 inches for alternating current and 15 inches for direct current.

8. The distance between centers of meter-connection cabinets of 60-ampere capacity and larger used inside of a building, either for alternating or direct current, shall not be less than 24 inches and the leads of one meter shall not be run within 12 inches of another meter.

Meter Wiring. 1. All meter-connection cabinets used inside of a building, should be connected according to the wiring diagrams which are inside of the cabinet covers, according to the standard wiring diagrams furnished by the Company for a particular installation. For outdoor meter-connection cabinets and when meters are installed ahead of the service switch and the fuses, wiring diagrams will be furnished by the Wiring Inspection Division upon request.

2. For interior use, a 3-wire meter-connection cabinet shall be provided for all commercial and office installations.

3. Where a 60-ampere switched-type meter-connection cabinet is used inside of a building for single-phase or 3-phase service, the service wires shall be brought into and the load wires shall leave the lower portion of the cabinet below the connection block, using the knockouts provided. This is because, in the limited space provided at the top of the cabinet above the block, there is room only for the meter connections and the space necessary for the opening of the meter-terminal chamber cover when connections or tests are being made on the meter.

4. When external resistances, current or potential transformers are used in connection with meters, they shall be located where they are accessible for inspection and can be removed without danger of making a short circuit.

5. No. 12 wire with colored outer braid shall be used for the secondary current and potential wires for meters installed with instrument transformers, and for the wires connecting the demand-meter equipment on such an installation. Wiring diagrams and colors to be used for such meter connections may be secured from the Wiring Inspection Division.

6. Circuits for operating demand-meters shall be connected ahead of the service-switches and protected by fuses of suitable capacity.

7. When the demand on a 2-wire, 240-volt alternating current installation exceeds 100 amperes, a current transformer will be used in one side of the line and a potential tap shall be provided from the other side When the demand on a 2-wire, 240-volt direct current installation exceeds 50 amperes, one side of the line shall be connected through the meter with a potential tap provided for the demand-meter.

8. The neutral load-wire shall be terminated in the meter-connection cabinet with a free end of sufficient length to extend above current meters and 24 inches for direct current meters. This rule applies to all 2-wire installations of 30 and 60 ampere meter-connection cabinets. It will apply for inside use on the replacement of an old-type cabinet or on additional-load installations requiring meter-connection-cabinet changes or changes in the wiring of the cabinet. Where outdoor

meter-cabinets are ganged, the neutral wire shall be continuous and looped around terminals.

9. For 3-wire, 240-volt alternating current and direct current meters, both sides of the supply line and the two load wires shall be terminated in the meter-connection cabinet or meter trough and a pressure tap shall be taken from the neutral wire in the meter-connection cabinet or meter trough.

10. For 6-terminal, 3-wire, single-phase, alternating current meters, three potential wires shall be provided, one from each outside wire and one from the neutral.

11. For 3-phase meters, the two wires from the outside blade terminals on the load side of the service-switch shall be brought to the service terminals in the meter-connection cabinets. The wire from the third phase or middle blade terminal of the service-switch shall be brought to the meter-connection cabinet, connected to the terminal connection provided for the wire and carried on to the distribution center.

12. Where current transformers are used they shall be installed in a separate cabinet in the bus from the outside blade terminals of the service-switch and potential wires shall be provided. For single-phase meters, one potential wire shall be connected to each outside bus and the neutral and for 3-phase meters, one potential wire shall be connected to each phase.

All meter potential wires shall be connected in the same cabinet in which the current transformers are mounted.

13. For 3-element, 4-wire, 3-phase meters, the 3-phase wires from the service-switch together with the neutral wire shall be extended to the meter-connection cabinet. Where current transformers are used with 4-wire, 3-phase meters, they shall be installed in the load bus from the 3-phase blade terminals of the service-switch and 4 potential wires shall be provided.

14. Three-phase meters for 480-volt potential are installed only with current transformers, and shall preferably be mounted on slate or asbestos-board panels. In addition to the potential wires, one for each phase, a special one-half voltage tap shall be brought to the meter to operate a standard 240-volt demand-meter. The service connection which is required for the one-half voltage tap shall be arranged for with the Wiring Inspection Division before such service connection is installed.

15. Potential wires for all meters shall be so installed that they cannot become disconnected. The connecting wire should be as short as possible, and shall be installed without a break and without any fuses other than those of the service-switch and the meter-connection cabinet. All potential wires are to be connected to wires or terminals inside the meter-connection cabinets and cabinets containing current transformers. Such cabinets shall be equipped with sealing facilities. If the continuity of the potential wire is broken in any place, the wire shall be soldered at the point where the break in connection is made.

Line and Load Wires. 1. Service or line wires and metered load wires shall not be installed in the same conduit, panel box, junction box, header box, or pull box. No condulet or junction boxes are to be used in the service run to the service switch or meter fitting except those of design approved by the Company. All devices must be capable of being sealed when installed ahead of any meter.

2. Conduits containing service or line wires shall be installed directly from the service mains and enter meter-connection cabinets in the space provided below the terminal block. All load wires shall leave the meter-connection cabinets in conduit or troughs from the space provided below the terminal block, and thence to the distribution or circuit fuse center.

3. On a meter installation requiring current transformers, the service-switch and current transformers shall be installed in separate cabinets or in separate compartments of the same cabinet, provided each compartment is equipped with a separate door. Current transformers shall not be installed in the same cabinet with distribution fuses, load fuses, or load wires.

4. A service-switch shall have the service cables connected on the front of the panel.

Meter Fuse Protection. 1. 30-ampere fuses shall be installed in switched type meter-connection and service-switch cabinets of 30-ampere capacity. All fuses shall be installed by the contractor before leaving the job, or, fuses may be left with someone on the premises and a notice placed in the meter service-switch showing with whom they were left.

2. When required by the Wiring Inspection Division, meter-protection fuses shall be omitted on alternating current installations, when the meter is installed in an approved outdoor meter-connection cabinet or on the inside of the building.

3. The fuses provided for demand-meters shall be mounted in an accessible place so they may be replaced without danger to the Company's representative.

Meter-Boards. 1. On installations where inside meter-connection cabinets of 200 amperes capacity and smaller are installed, a suitable meter-board of white pine, or other soft, well-seasoned wood, not less than ⅞ inch in thickness, or transite board (or equivalent), not less than ½ inch in thickness, shall be provided by the customer and fastened rigidly and in a vertical plane to the wall or other support. If transite board or equivalent is used it shall be so mounted that it will be accessible from the back of the board in order to permit a nut to be fastened to the machine bolt which is used for support of the meter. Where the meter-board is mounted on metal lath or other metal structure, all supporting screws or bolts shall be countersunk.

2. Meter-boards in general should be made up of strips of matched pine or other soft wood, 5¼ inches by ⅞ inch. This lumber shall not be green or wet. The furring strip or 2 by 4, to which the board shall be fastened to the wall shall be secured by expansion bolts. The meter-boards shall then be nailed to these supports by at least two nails, one on each edge of the board. The meter-boards shall be level and plumb.

3. When the strips of a meter-board are mounted horizontally, meter-connection cabinets are to be mounted so that any crack between the boards is located one inch below the top of the cabinet. With vertical mounted boards, the cracks shall be located so that mounting screws for the different types of meters which might be used, will not come at a crack.

4. On direct current installations of 200 to 400-ampere meters requiring troughs, the meter-board, if of wood, shall be not less than 1½ inches in thickness. If transite board (or equivalent) is used, the board shall be not less than 1 inch in thickness.

5. Metal meter-boards will not be approved as a mounting for the Company's meters.

Tampering with Meters or Closing Meter-Loops. All meters and metering equipment are sealed by the Company. The breaking of meter seals by unauthorized persons, tampering with the meters, metering equipment or cutouts, or tampering with any wires or switches in connection with the meter wiring or the unauthorized closing of meter-loops will not be permitted by the Company. Attention is called to Revised Statutes of Illinois, Chapter 38, Section 117, in force July 1, 1895. The penalty for the breaking of this law is a fine or imprisonment, or both.

MECHANICAL REFRIGERATION

DEANE E. PERHAM, Mech. Engr.
Pres. Chicago Section American Society of Refrigerating Engineers.

MECHANICAL REFRIGERATION.

Food Preservation.

During the past ten years refrigeration has made rapid strides. Formerly it was considered of so little importance to residential and many commercial buildings that architects and builders gave it little consideration in the preparation of building plans. During the past ten years comparatively few buildings have been erected but the rapidly growing need for refrigeration of residential and small commercial types and more recently for comfort cooling has created a necessity for definite allotment of adequate space for mechanical refrigeration. Existing buildings are limited to the use of space available, which in most cases is not adequate, nor suitably located to obtain the best results from the refrigeration equipment.

Obviously refrigeration is now a definite and necessary part of every residential building and most commercial buildings. The value of a building is enhanced by up-to-date equipment correctly installed to provide good performance with low depreciation and maintenance.

Domestic refrigeration, for homes and apartments, can be of the self contained type with all apparatus enclosed in the over all dimension of the refrigerator; the remote type having a condensing unit located apart from the refrigerator and the central plant type cooling two or more refrigerators from one condensing unit. Certain fundamental requirements are common to all three types. Their purpose is to store and preserve foods in the form received from the source of supply. Their equipment consists of heat handling apparatus.

Purpose: The average refrigerator should have adequate space for storage and preservation of more than one day's food supply. Practice indicates that a family of two or three need at least six cubic feet of food storage space. For larger families one additional cubic foot of food space per person over three seems to be a practical minimum. Experience indicates that the average six cubic foot refrigerator costs its owner no more per year for refrigeration than a smaller one.

Heat Handling: Refrigeration is a general term for extraction of heat. The operating range of mechanical refrigeration may be from comfort cooling temperatures of around 80 F. to industrial process temperatures of minus 100 F. or lower. The heat extracted from the substance cooled plus the heat equivalent of the work done to operate the system must all be expelled or thrown away by means of air, water or both.

The self contained unit is usually air cooled and must liberate its heat in the space immediately surrounding the refrigerator. A small size unit working under the heat load of a hot summer day must expel heat at approximately the hourly rate of three to five square feet of radiation. If the apparatus is not permitted to expel the heat freely, its capacity to cool is reduced. Therefore, the

COMPRESSOR RATING.

Table No. 1 (italics) is the displacement for rating a compressor for a given refrigerant at 5 F. suction temperature. For other suction temperatures multiply the rating (at 5 F.) by the factor (italics) shown in Table No. 2 opposite the suction temperature desired.

Example: What is the capacity of a two cylinder, single acting compressor 2½" bore, 3" stroke, 350 R.P.M., using Freon as the refrigerant?

The displacement per cylinder from Table 11 is 14.72 cubic inches. Multiply 14.72 by 2 cylinders and 350 R.P.M. Divide the result (10304) by 12500 from column 2 of Table 1. The answer is .82 tons or 9840 Btuh capacity at 5 F. For 40 F. suction multiply by the factor 1.90 from Table 2, column 2, which is 1.56 tons or 18720 Btuh.

TABLE NO. 1 Displacement for Rating Compressors at 5 F. Suction		1 Ammonia NH_3	2 Freon CCL_2F_2	3 Methyl Chloride CH_3CL	4 Sulphur Dioxide SO_2	5 Carbon Dioxide CO_2
Suction Pressure	Lbs. Gage	20	12	6	—3	317
Suction Temp.	Deg. F.	5	5	5	5	5
Condensing Pres.	Lbs. Gage	153	93	80	51	1025
Condensing Temp.	Deg. F.	86	86	86	86	86
Maximum R.P.M. Recommended		375	375	375	375	375
Displacement:	Cubic Inches Per Min. Per Ton	*6900*	*12500*	*13800*	*20700*	*1620*
One Ton=12000 Btuh						

TABLE NO. 2	Suction Temp. Deg. F.	1 Lbs. Gage	1 Factor	2 Lbs. Gage	2 Factor	3 Lbs. Gage	3 Factor	4 Lbs. Gage	4 Factor	5 Lbs. Gage	5 Factor
*Inches Mercury	— 50	*14.3	*.25*	...	...	...	...	...	...	103	*.33*
Vacuum	— 40	* 8.7	*.30*	*10.9	*.35*	*15.5	...	*23.1	*.25*	131	*.40*
	— 30	* 1.6	*.40*	* 5.4	*.45*	*11.2	...	*20.7	*.35*	157	*.50*
	— 20	3.6	*.55*	.5	*.60*	* 6.9	*.55*	*17.6	*.50*	205	*.65*
	— 10	9.0	*.70*	4.5	*.75*	.3	*.70*	*13.6	*.60*	247	*.75*
	0	15.7	*.90*	9.2	*.90*	4.1	*.90*	* 8.7	*.85*	293	*.90*
†Rating at 5 F.	† 5	19.58	*1.00*	11.8	*1.00*	6.2	*1.00*	* 5.6	*1.00*	319	*1.00*
obtained from	10	23.8	*1.15*	14.6	*1.10*	8.6	*1.10*	2.3	*1.10*	347	*1.10*
Table No. 1	15	28.4	*1.25*	17.7	*1.20*	11.2	*1.20*	.4	*1.25*	376	*1.20*
	20	33.5	*1.40*	21.0	*1.30*	14.1	*1.30*	2.4	*1.40*	407	*1.30*
	25	39.0	*1.50*	24.6	*1.45*	17.1	*1.50*	4.6	*1.60*	440	*1.40*
	30	45.0	*1.70*	28.5	*1.60*	20.5	*1.60*	7.0	*1.80*	474	*1.55*
	35	51.5	*1.90*	32.5	*1.70*	24.0	*1.80*	9.5	*2.00*	511	*1.70*
	40	58.6	*2.00*	37.0	*1.90*	27.9	*2.00*	12.4	*2.20*	550	*1.85*
	45	66.3	*2.25*	41.6	*2.00*	32.2	*2.10*	15.5	*2.40*	591	*2.00*
	50	74.5	*2.40*	46.6	*2.20*	36.7	*2.35*	18.7	*2.70*	635	*2.20*
	55	84.0	*2.70*	52.0	*2.35*	42.0	*2.50*	23.0	*2.90*	681	*2.40*

importance of adequate, well ventilated space for the refrigerator is obvious.

The remote and central plant systems liberate their heat where the condensing unit is located making it unnecessary to place the refrigerator in ventilated space.

Frozen Foods: Scientific research of international scope has made available frozen foods in packages of convenient size for household use. The neighborhood store has available a practical unit for handling frozen foods in a form that appears to have public acceptance to the extent that every domestic refrigerator should provide one or more cubic feet of frozen food storage that can be maintained at zero degrees F. when desired. This space is in addition to the present higher temperature food space and ice cube space, resulting in a need for larger refrigerators if a building is to keep its facilities up-to-date.

Each refrigeration installation of any type is an individual engineering problem of fitting the proper type and capacity to the requirements. The recommendations of an engineer experienced with installation and operation of refrigerating equipment should be followed for each project. Some of the engineering data involved is contained on the following pages:

COMPRESSOR DISPLACEMENT.

Table shows (for given bore stroke) the displacement in cubic inches. Multiply by number of cylinders and R.P.M. to obtain displacement per minute. The table will furnish the displacement of cylinders with bore not listed by using a bore one-half of the given bore and multiply by four. Example: The displacement of a 6 inch bore can be obtained by taking the displacement of a 3 inch bore and multiply by four. For strokes longer than listed simply add two or more that total the given stroke. Example: A 9 inch stroke has the same displacement as a 5 inch plus a 4 inch.

TABLE NO. 11.

Stroke Inches	Cylinder Bore Inches.																
	1"	1⅛"	1¼"	1⅜"	1½"	1⅝"	1¾"	1⅞"	2"	2⅛"	2¼"	2⅜"	2½"	2⅝"	2¾"	2⅞"	3"
1"	.78	.99	1.22	1.48	1.76	2.07	2.40	2.76	3.14	3.54	3.97	4.43	4.90	5.41	5.93	6.49	7.06
1⅛"	.88	1.17	1.38	1.66	1.98	2.33	2.70	3.10	3.53	3.99	4.47	4.98	5.52	6.08	6.68	7.30	7.95
1¼"	.98	1.24	1.53	1.85	2.21	2.59	3.00	3.45	3.92	4.43	4.97	5.53	6.14	6.76	7.42	8.11	8.83
1⅜"	1.07	1.36	1.68	2.04	2.43	2.84	3.30	3.80	4.32	4.87	5.47	6.09	6.75	7.48	8.16	8.92	9.71
1½"	1.17	1.49	1.84	2.22	2.65	3.10	3.60	4.14	4.71	5.31	5.96	6.65	7.37	8.12	8.90	9.73	10.60
1⅝"	1.27	1.61	1.99	2.41	2.87	3.36	3.90	4.48	5.11	5.76	6.45	7.20	7.98	8.79	9.64	10.54	11.48
1¾"	1.37	1.73	2.14	2.59	3.09	3.62	4.21	4.83	5.50	6.21	6.95	7.76	8.60	9.46	10.39	11.35	12.36
1⅞"	1.47	1.86	2.27	2.78	3.31	3.88	4.51	5.17	5.89	6.65	7.45	8.32	9.21	10.14	11.13	12.16	13.25
2"	1.57	1.98	2.45	2.96	3.53	4.14	4.81	5.52	6.28	7.09	7.95	8.86	9.82	10.82	11.87	12.98	14.13
2⅛"	1.66	2.11	2.60	3.15	3.75	4.40	5.11	5.87	6.67	7.53	8.44	9.42	10.43	11.50	12.61	13.79	15.01
2¼"	1.76	2.23	2.76	3.33	3.97	4.66	5.41	6.22	7.07	7.97	8.94	9.97	11.05	12.17	13.35	14.60	15.90
2⅜"	1.86	2.36	2.91	3.52	4.20	4.92	5.71	6.57	7.46	8.43	9.44	10.52	11.67	12.85	14.10	15.41	16.78
2½"	1.96	2.48	3.07	3.71	4.42	5.18	6.01	6.91	7.86	8.87	9.94	11.06	12.28	13.52	14.84	16.22	17.66
2⅝"	2.06	2.60	3.22	3.89	4.64	5.44	6.31	7.25	8.25	9.31	10.43	11.62	12.90	14.20	15.58	17.03	18.55
2¾"	2.15	2.73	3.47	4.08	4.86	5.70	6.61	7.60	8.64	9.75	10.93	12.17	13.50	14.87	16.32	17.84	19.43
2⅞"	2.25	2.85	3.52	4.26	5.08	5.96	6.91	7.95	9.04	10.20	11.42	12.73	14.12	15.55	17.06	18.65	20.31
3"	2.35	2.98	3.68	4.45	5.30	6.21	7.21	8.28	9.42	10.63	11.92	13.29	14.72	16.23	17.81	19.47	21.20
3⅛"	2.46	3.10	3.84	4.63	5.52	6.47	7.51	8.63	9.82	11.08	12.42	13.84	15.36	16.90	18.55	20.28	22.08
3¼"	2.55	3.22	3.99	4.82	5.75	6.74	7.81	8.97	10.21	11.52	12.92	14.40	15.96	17.58	19.29	21.09	22.97
3⅜"	2.64	3.35	4.14	5.00	5.96	7.00	8.11	9.32	10.60	11.96	13.42	14.95	16.57	18.26	20.03	21.90	23.85
3½"	2.74	3.47	4.29	5.18	6.18	7.25	8.41	9.67	11.00	12.41	13.91	15.50	17.20	18.94	20.77	22.71	24.73
3⅝"	2.84	3.60	4.45	5.37	6.41	7.51	8.71	10.02	11.39	12.85	14.40	16.05	17.80	19.60	21.52	23.52	25.61
3¾"	2.94	3.72	4.60	5.56	6.62	7.77	9.01	10.35	11.78	13.30	14.90	16.60	18.41	20.28	22.26	24.33	26.50
3⅞"	3.04	3.85	4.75	5.75	6.85	8.03	9.31	10.70	12.17	13.74	15.40	17.15	19.03	20.95	23.00	25.15	27.38
4"	3.14	3.97	4.90	5.94	7.06	8.29	9.62	11.04	12.56	14.18	15.90	17.72	19.63	21.64	23.74	25.96	28.25
4⅛"	3.23	4.10	5.06	6.12	7.29	8.55	9.92	11.39	12.95	14.62	16.39	18.27	20.26	22.30	24.48	26.77	29.13
4¼"	3.33	4.22	5.21	6.30	7.51	8.81	10.22	11.74	13.35	15.06	16.89	18.83	20.87	23.00	25.23	27.58	30.01
4⅜"	3.43	4.34	5.37	6.49	7.73	9.07	10.52	12.07	13.75	15.50	17.39	19.37	21.48	23.67	25.97	28.39	30.90
4½"	3.52	4.47	5.52	6.67	7.95	9.32	10.82	12.42	14.14	15.94	17.89	19.94	22.10	24.35	26.71	29.20	31.78
4⅝"	3.64	4.60	5.68	6.87	8.17	9.60	11.14	12.78	14.54	16.42	18.41	20.51	22.72	25.05	27.39	30.05	32.73
4¾"	3.73	4.72	5.83	7.05	8.39	9.85	11.42	13.11	14.92	16.84	18.89	21.04	23.31	25.70	28.20	30.93	33.57
4⅞"	3.83	4.85	6.09	7.24	8.62	10.12	11.73	13.47	15.32	17.31	19.40	21.62	23.95	26.41	28.98	31.67	34.49
5"	3.92	4.97	6.14	7.42	8.84	10.36	12.02	13.82	15.72	17.75	19.88	22.15	24.54	27.05	29.69	32.45	35.34

CONDENSING UNITS.

This section headed Condensing Units ©C.M.S.F. Assn. is general information pertinent to practically every type of a refrigeration installation.

Tables Nos. 1, 2, 11 and 17 are preceded by an example showing how to use the data. Table No. 3 shows compressor capacities recommended for economical operation of certain sizes of domestic refrigerators insulated as specified.

Compressor ratings for a given bore, stroke, and R.P.M. at pressures corresponding to 5 F. evaporator temperature and 86 F. condensing temperature may be calculated from Tables Nos. 1 and 11. For other suction pressures multiply by the proper factor from Table No. 2.

Displacement was selected as the method of rating a compressor, because, for a ton of cooling performed, a definite weight of the refrigerant used having a known volume at a given suction pressure must be handled by the compressor cylinders. It is a practical method to apply. It is a uniform and fair method for all concerned. It should encourage improved design, good workmanship and better installations because of the fact that the compressor performance is dependent upon design and proportion of other apparatus of the condensing unit, such as compressor valves, refrigerant containing parts, condenser cooling surface, refrigerant receiver volume, and power to drive. Since practical experience has quite definitely established the proportions of the foregoing apparatus necessary for a given compressor displacement and known operating conditions, there seems to be no valid reason for doing otherwise than to make each installation a most satisfactory one.

Compressor Speed should not exceed the limits of its design for quiet and efficient operation. Increasing the speed of a compressor requires that valves, ports, refrigerant lines, condenser, and motor power be increased. Therefore, the possible saving in cost of condensing units due to increased speed is small while unlimited trouble and expense to the contractor can occur from noise and excessive wear that may result from speeds that are too high. Experience leads your association to recommend compressor speeds under 400 R.P.M. and piston speed not exceeding 500 feet per minute.

Condensers water cooled should have adequate cooling surface and sufficient water tube area to permit efficient cooling at the normal condensing pressure corresponding to 90 F. for the refrigerant used when supplied with 76 F. entering condensing water.

For each 12000 Btub heat transfer the effective water cooling surface should be not less than the following:

Double pipe type, copper....6 square feet
Shell and tube type.........9 square feet

If a cooling tower is used, condenser should be designed for a minimum of 80 F. entering water.

Condenser should be located in space not exceeding outdoor temperatures.

Receiver for refrigerant should be of sufficient internal volume to permit pumping out the evaporator and liquid lines without increasing the condensing pressure more than 50 per cent above the normal pressure—corresponding to 86 F.—for the refrigerant used. (80% of the net internal volume of a

DOMESTIC REFRIGERATION.

TABLE NO. 3.

The following table is average data for selecting equipment for domestic central systems, domestic remote individual installations and domestic self contained installations.

Gross Cubic Feet Inside Refrigerator	4	5	6	7	8	9	10	12	14	16	20	25	30
1. Square Feet Exterior Surface—Average	24	28	30	33	36	40	44	49	54	58	67	75	84
2. Wall Heat Load—Pounds Refrigeration ..	30	34	36	40	44	49	54	60	66	71	82	91	102
3. Food and Service Load—Pounds Refrigeration ..	5	6	7	7	9	9	11	12	13	14	16	18	22
4. Ice Load 2 Freezes per 12 Hours—Pounds Refrigeration	15	15	20	20	30	30	35	45	45	52	76	76	76
5. Average Pounds Ice per Freeze	3	3	4	4	6	6	7	9	9	10	15	15	15
6. Average Square Feet Evaporator Surface Cooling Cabinet	3.5	4	4.3	4.6	5.3	5.6	6.5	7.2	8	8.5	9.7	11	12.5
7. **Compressor Capacity,** Total of Lines 2-3-4. Pounds Refrigeration ..	50	55	63	67	83	88	100	117	124	137	174	185	200
Btuh	**300**	**330**	**380**	**400**	**500**	**530**	**600**	**700**	**750**	**820**	**1050**	**1100**	**1200**

LINE 2: Based upon 3 inches of insulation with a heat transfer not to exceed .30 Btuh per square foot, per degree, per inch thickness.

LINE 7: For water cooled condensing units. For air cooled, add 25 per cent.

NOTE: Pounds of refrigeration as used above may be converted to Btuh by multiplying by 6. In line 7 the Btuh is made even figures for convenience.

REFRIGERANT PIPE SIZE.

TABLE NO. 17.

TABLE No. 17 shows the STANDARD pipe size required for refrigerant lines. The LARGE TYPE FIGURES are BTU per hour. Gas line capacities are in LARGE BLACK TYPE. *Liquid line capacities are in italics.*

The small figures at the right in each column are the pounds of refrigerant to fill 100 feet of pipe.

Maximum velocities are shown in feet per minute BLACK TYPE for gas—*italics for liquid.*

Use column headed 40° F. for suction gas temperature OVER 25 degrees F.

Use column headed 5° F. for suction gas temperature UNDER 25 degrees F.

EXAMPLE: Assume an AIR CONDITIONING load of 34 000 BTU per hour with FREON. What size pipe lines are required and how much refrigerant is necessary to fill the liquid line?

ANSWER: Under Freon read down the 40° F. column to the first black number that equals or exceeds 34 000, which is 36 000. Read right to the pipe size column showing 1" pipe for the suction line.

For the liquid line read down the Freon column to the first italic number that equals or exceeds 34 000, then right to the pipe size column which shows ⅜" pipe required, and it will hold six pounds of Freon per 100 linear feet.

NOTE: Refrigerant lines ¾" and under, exceeding 25 linear feet should increase one size, 26 to 75 linear feet; two sizes, 76 to 150 linear feet.

Refrigerant lines 1" to 2" inclusive, exceeding 50 linear feet should increase one size, 51 to 100 linear feet; two sizes, 101 to 200 linear feet.

Refrigerant lines 2½" and over, exceeding 100 linear feet should increase one size, 101 to 200 linear feet; two sizes, 201 to 300 linear feet.

A TON OF REFRIGERATION IS HEAT REMOVAL AT THE RATE OF 12 000 BTU'S PER HOUR. TO CONVERT THE BTU FIGURES TO TONS OF REFRIGERATION, DIVIDE BY 12 000.

AMMONIA NH_3			SULPHUR DIOXIDE SO_2			METHYL CHLORIDE CH_3CL			FREON CCL_2F_2			CARBON DIOXIDE CO_2			Read Note Above
Specific Gravity 0.68			*Specific Gravity 1.46*			*Specific Gravity 0.99*			*Specific Gravity 1.44*			*Specific Gravity 0.99*			
5000	*225*	3600	4600	*150*	3000	3800	*180*	2700	2200	*150*	1600	900	*225*	700	← Maximum Velocities
5° F.		40° F.	5° F.		40° F.	5° F.		40° F.	5° F.		40° F.	5° F.		40° F.	PIPE SIZE NOMINAL Tube O.D. ⅛ Larger

receiver may be considered as liquid refrigerant capacity.) Refrigerant shall be pumped out of evaporators and stored in the receiver during the period when the apparatus functions for heating or is exposed to room temperature of 70 F. or higher.

Motor should be of a size that the power required does not exceed the manufacturer's normal rating.

Condensing Unit Complete: The same condensing unit used for various commercial purposes without objection to noise may, when used for comfort cooling, meet with serious objection because of mechanical noise and vibration, the reason being that condensing units are placed in or near the space occupied by people desiring comparative quiet. Assuming that a compressor is reasonably well fitted and balanced mechanically, noise and vibration, if existing, can uusually be traced to one or both of two sources:

1. Speed too high for the moving parts and valve design.

2. Condensing pressures in excess of normal due to insufficient condenser surface, lack of refrigerant receiver volume or restricted refrigerant passages.

In the interest of economical and quiet operation, we emphasize the importance of providing each condensing unit with adequate cooling surface and refrigerant containing parts to permit operation at a normal pressure corresponding to 90 F. for the refrigerant used.

Gauges with shut off valves should be installed on high and low sides in such a manner that the compressor pulsations will not cause excessive vibration of the gauge hand.

In General the complete installation must conform to laws and ordinances governing design, materials, and methods of installation for the area in which the equipment is to operate.

AIR CONDITIONING.

Refrigeration for comfort cooling is a growing part of building equipment, both residential and commercial. Comfort cooling compared with domestic refrigeration requires greater capacity and much larger equipment. Adequate machine room space is important with provision for power service, water lines and drain lines. To conserve water, an evaporative type of condenser may be used. This type requires space conveniently located for air supply and exhaust. Modern equipment includes types for year around air conditioning, which can be made a part of the radiator heating system.

Careful consideration of year around air conditioning during the plan stage of the building permits the necessary provisions for installation of the radiator heating system in a manner for winter and summer use. If summer cooling is not desired at the time of building erection provision can be made for addition at a later date. This flexibility permits the building to be modernized by stages, in a practical and economical manner.

Illinois Society of Architects

1906, 134 N. La Salle Street, Chicago

The following is a list of the publications of the Society; further information regarding same may be obtained from the Financial Secretary.

FORM NO. 21, "INVITATION TO BID"—Letter size, 8½x11 in., two-page document, in packages of fifty at **75c**, broken packages, two for **5c**.

FORM NO. 22, "PROPOSAL"—Letter size, 8½x11 in., two-page documents, in packages of fifty, at **75c**, broken packages, two for **5c**.

FORM NO. 23, "ARTICLES OF AGREEMENT"—Letter size, 8½x11 in., two-page document, in packages of fifty, at **75c**, broken packages, two for **5c**.

FORM NO. 24, "BOND"—Legal size, 8x13 in., one-page document, put up in packages of twenty-five, at **25c** per package, broken packages, three for **5c**.

FORM NO. 25, "GENERAL CONDITIONS OF THE CONTRACT"—Intended to be bound at the side with the specifications, letter size, 8½x11 in., ten-page document, put up in packages of fifty at **$2.50**, broken packages, three for **25c**.

FORM 26, CONTRACT BETWEEN ARCHITECT AND OWNER. Price, **5c** each, in packages of fifty, **$1.25**.

FORM 1, BLANK CERTIFICATE BOOKS—Carbon copy, from 3¾x8½ in., price, **50c**. Two for **5c**.

FORM 4, CONTRACT BETWEEN THE OWNER AND CONTRACTOR—(Old Form.) Price, two for **5c**. Put up in packages of 50 for **$1.00**.

FORM E, CONTRACTOR'S LONG FORM STATEMENT—As required by lien law. Price, two for **5c**.

FORM 13, CONTRACTOR'S SHORT FORM STATEMENT—Price, **1c** each.

CODES OF PRACTICE AND SCHEDULE OF CHARGES—8½x11 in. Price, five for **10c**.

These documents may be secured at the Financial Secretary's office, suite 1906, 160 N. La Salle St., telephone Cent. 4214. We have no delivery service. The prices quoted above are about the cost of production. An extra charge will be made for mailing or expressing same. Terms strictly cash, in advance, with the order; except that members of the Society may have same charged to their account.

GAS FITTERS' RULES

Of the Peoples Gas, Light and Coke Company

Rules and Regulations for the Installation of Gas Piping—December 21, 1938.

Office Buildings, Dwelling Houses and Flats

Manufactured Gas for Light

GENERAL.

1. Importance of Rules.

All gas piping and the installation thereof shall conform to the ordinances and regulations of the City of Chicago and to the rules and regulations of The Peoples Gas Light and Coke Company. No approval or certificate of inspection by the Company will be given, no meter will be set, and no gas will be turned on unless and until the Company's rules relating to the installation of piping are complied with in every respect. The Company reserves the right to change or amend these rules at any time and from time to time.

2. Initial Inspection of Gas Piping.

On other than single or 2 family dwellings an inspection must be made before the interior of the building is lathed or covered with sealing board.

Any underground piping, which the contractor may install in accordance with these rules, must be inspected and approved by the Company before the trench is back-filled.

Twenty-four hours' notice will be required for each inspection. The work shall be completed and the piping tested for tightness before notifying Gas Company to make the inspection.

3. Final Inspection and Test of Gas Piping.

Before piping is finally approved, and where meter bars and cocks have been installed, an inspection and test for tightness will be made by the District Shop.

Before appliances are installed, the piping is required to stand a pressure of six inches on a column of mercury without showing any drop in the column for a period of ten minutes. Appliances shall not be installed until piping has been tested and sealed by a Gas Company Inspector.

After appliances are installed piping is required to stand a pressure of one inch on a column of mercury without showing any drop for the same period of time.

A mercury gauge shall be used in testing house piping, using an inert gas, such as air, as a pressure medium. The use of water in the pipe for testing is strictly prohibited.

The above methods of testing piping by air apply only to pipe sizes of 6 inch and under and to gas working pressures of 15 inches water column or less.

For all pipe sizes above 6 inch, or for pressures over 15 inches of water column, the following will apply:

The installation shall withstand an air test of twice the gas working pressure, but in no case less than 10 lbs. per square inch test pressure. The test pressure, as measured by a column of mercury, shall show no drop during a period of 10 minutes after the air in the pipe has been given time to arrive at the same temperature as the room.

For all sizes of pipes and all pressures, the approval of the District Shop will be based on a pressure test showing that at the time of the test the installation was air tight under the conditions of the test.

Testing for gas leaks should be done by applying a solution of soap and water or by using sense of smell. No flame or fire in any form shall be used in testing for gas leaks.

4. Notification to Piping Contractor.

Upon receipt of any telephone or other notification stating the corrections required in piping, the piping contractor shall make such corrections as soon as possible and notify the District Shop by mail that such corrections have been completed.

5. Understanding Rules.

If, in any instance, the rules governing the sizes of pipe to be installed are not clearly understood, or if unusual conditions not covered by the rules arise, the District Shop should be consulted.

6. Inspection Notification.

The District Shop shall be notified upon completion of house piping.

District Shop Boundaries:

North District Territory—This district comprises all territory North of Central District. Both sides of North Avenue are included in this district.

Telephone—Avenue 1300. Address—4643 W. Irving Park Blvd.

Central District—This territory is bounded on the north by North Avenue, on the east by Lake Michigan, on the south by 22nd Street, from the lake to the point of intersection with the south branch of the Chicago River, and from that point on by the south branch of the Chicago River and the Sanitary District Drainage Canal to the western city limits. Both sides of 22nd street are included in this district. Any frontage on North Avenue is not included in this district.

Telephone—Haymarket 3802. Address—164 N. Loomis Street.

South District—This district comprises all territory south of the Central District.

Telephone—Normal 5860. Address—38 West 64th Street.

Sales Division Boundaries:

North Division—North of Madison Street.

Telephone—Diversey 6464. Address—1608 N. Larrabee Street.

South Division—South of Madison Street.

Telephone—Boulevard 2831. Address—45 E. Pershing Road.

7. Work Reserved by Gas Company.

The Gas Company reserves the right to install all gas piping which will convey unmetered gas to the meter.

Only authorized employes of the Gas Company are permitted to install any piping or connections on any part of either the outlet or inlet meter connections, or on any piping conveying unmetered gas, break seals or remove lock plugs, turn on gas to premises, disconnect, move, or interfere in any way with its meters, or connections. The Gas Company will promptly, upon receipt of notice, attend to any work required of it under these rules.

The owner or occupant of the building shall pay the Gas Company for the installation of all such piping installed by the Gas Company from the lot line to the meter and also all piping within any premises outside the lot line which may be occupied or controlled by the owner or occupant.

GENERAL PIPING REQUIREMENTS.

8. Piping Material.

All piping for building services, risers and fuel lines, should be of standard full weight wrought iron or steel pipe with malleable fittings, free from defects. Cast iron fittings are prohibited in either exposed or concealed work, except as cited in Rule No. 9.

9. Fitting Material.

Only approved malleable iron or steel fittings shall be used on pipe sizes up to 6 inch. For sizes 6 inch or over, fittings of cast iron, malleable cast iron, or cast steel designed for 125 lbs. per sq. in. Working Pressure must be used.

trapping so as to avoid impeding the flow of gas.

Pipe supports should be located approximately every 8 feet for ½ inch, ¾ inch and 1 inch pipe, and at least every 10 feet for 1¼ inch, 1½ inch and 2 inch pipe. Larger size pipe should be supported at sufficient intervals to prevent sagging.

A building service, a riser, or a fuel run must not be hung or strapped to a water pipe, a steam pipe, a sewer pipe, waste pipe or any other pipe or electric conduit.

23. Weakening Building Structure.

The piping shall be installed so that it will not weaken the building structure or any part thereof.

24. Piping on Outside Walls.

When it is necessary to run pipe on or along an outside wall a furring strip shall be placed between the pipe and the wall, and if subjected to extremely cold temperature the pipe should be covered with a suitable covering.

25. Piping on Masonry Walls.

All piping run on masonry walls shall be furred out and shall be securely fastened thereto with proper supports.

26. Piping in Unsatisfactory Locations.

Gas piping shall not be exposed to varied or sudden changes of temperatures, as will be obtained by running through hot air furnace pipes, cold air ducts, refrigerator plants, ice boxes, or to any outside exposures.

27. Trapping Piping.

To avoid trapping pipe, the pipe shall be graded to risers or drops, to the meter or to the appliance.

28. Meter Inlet Connections.

All meter inlet connections shall be taken from the sides or tops of running lines, never from below.

29. Determination of Pipe Sizes.

One hundred cubic feet shall be considered as a "unit" of gas measurement in all pipe sizing calculations. The normal full load on a modern gas range, having four top burners, an oven and a broiler, is about 100 cubic feet per hour gas consumption. Therefore an average range is equivalent to 1 "unit".

Appliances other than ranges will have their gas load expressed in tenths of a "unit" depending on their normal consumption. For instance, a laundry stove, which normally burns about 30 cubic feet per hour, will be considered as equal to 0.3 of a "unit".

Table No. 1 on appliances shall be consulted to obtain the total number of "units" to be supplied to a given pipe and this sum total of "units" will be used to determine the sizes and length of pipe to be run as shown in tables No. 2, 3 and 4 of rules 53, 69 and 70.

Table No. 1.

Appliances Commonly Used in Residences and Apartments.

Appliance Domestic Range:—	Equivalent "Units"
4 top burners, oven and broiler...	1.0
6 top burners, oven and broiler...	1.4
3 top burners and oven...........	0.6
Water heater (sidearm)...........	0.3
Laundry Stove	0.3
Gas-Fired Steam Radiator..........	0.5
Radiant heater	0.3
Space heater	0.6
Washing machine	0.3
Ironing machine	0.2
Clothes dryer	0.3
Incinerator	0.5
Single light opening	0.2
Electrolux refrigerator	0.03

NOTE: For information on other appliances consult the Architects and the Builders Service Section of the Domestic Sales Department.

STREET SERVICES.

30. Apartment Buildings.

In apartment buildings, only one street service will be installed. The building services supplying the various groups of risers should be connected regardless of fire walls, and extended to the point where the street service will enter the building.

If an unusual condition arises where more than one street service is required to supply a building, special instructions should be obtained from the District Shop.

A building service shall not be terminated in a finished room.

31. Corner Buildings.

To avoid complications when working on a corner building, information shall be obtained from the District Shop as to the exact location where the street service will enter the building.

32. Rear Buildings on Corner Lot.

A building located on the rear of a corner lot shall be supplied from the side street if a gas main is on that street. If there is no main on the side street, the rear building may be supplied either from the front building or by a long street service run in the parkway from the main supplying the front building. The long street service shall have a cover of not less than two (2) feet at each point and, if possible, should be extended in such a manner as to avoid installing a drip.

33. Rear Buildings on Inside Lot.

When a rear building on an inside lot (not a corner lot) is to be supplied, a separate street service should be used wherever possible, and in no case shall be less than 1½ inch in size.

In all cases where a supply to a rear building is desired, the District Shop shall be consulted.

34. Opening in Foundation or Floor for Street Service.

When a street and building service connection is to be made above the floor level, a temporary opening shall be left in the floor for connecting such services. If it is necessary to make an opening in the floor or foundation to install street service the expense of repair and resultant responsibility shall be borne by the owner or occupant.

35. Service Entering Building.

For definite location where the street service will enter the building consult the District Shop.

36. Right of Way.

Any right of way or easement required by the Gas Company for the installation of a street service shall be acquired by and at the expense of the owner or occupant.

BUILDING SERVICES.

37. Connecting to Street Service.

When the house piping has been approved by the Gas Company Inspector, the building service will be connected to the street service by the Gas Company. See rule 3 and 7.

38. Building Service in Sub-Basement.

In buildings where there is a sub-basement and provisions have been made to locate the gas meters in the sub-basement, the building service shall extend to the ceiling of the first basement and a loop provided in the manner indicated in Figure 1.

39. Cocks on Building Service Branches.

In buildings where there are several branches off a building service and provisions have been made for the installation of shut-offs on the various lines, it is permissible only to use gas cocks without lever handles. The use of valves in cases of this kind is prohibited. See Rule No. 11.

40. Size of Openings.

All openings in a building service shall be of the same size as the riser, which in no case shall be less than ¾ inch in size. The openings in the building service shall in all cases be located on the left hand side of the riser. See table No. 5, Rule No. 98 for dimensions for distance to be spaced. No building service for stores shall be less than 1¼ inch in size. Full 1¼ inch opening shall be left at the meter end.

41. Location of Building Service.

When risers are located in an open basement, in a room provided for that purpose, or on the various floors, the building service shall be brought to within 18 inches of the wall at the point where the street service will enter the building.

A building service must not terminate in a finished room, or be run in such a manner that the street service or the building service will be opposite to or under a coal chute, or any opening in the sidewalk or any other location where it is liable to be broken or damaged by falling material.

42. Building Service Size for Large Buildings.

For size of building service for large buildings where consumption is not known and cannot be readily estimated, consult the Architects and Builders Service Section of the Domestic Sales Department — Wabash 6000, Local 208.

43. Building Service Location for Buildings Without Basement.

In commercial or factory buildings where there is no basement, the building service shall be run overhead as closely as possible to the front building wall and dropped down to within 2 feet of the floor and clearing doorways by at least 2 feet.

An opening in the floor, not less than 2 feet square, shall be left at the point where the street service enters the building, so that the street service can be connected to the building service.

Where a solid wooden floor or a bulk-head exists or is to be installed, the street end of the building service shall terminate outside the building wall at a depth to be determined by the District Shop. The portion under the floor and bulkhead shall be in glazed tile pipe with cemented joints and the pipe shall be extended to one foot outside of the front building wall.

A building service shall not be run through notched joists.

A building service shall not be run under a basement floor or under the first floor of a residential building without a basement, unless there is at least a 30-inch clearance under the floor or the service is encased in glazed tile pipe with cemented joints to a point one foot outside of the front building wall at a depth to be determined by the District Shop.

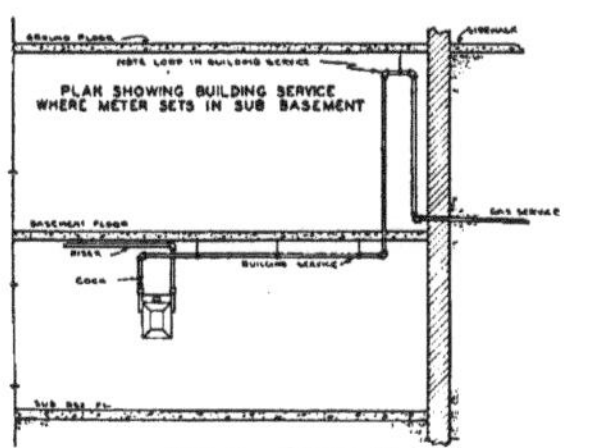

Fig. 1.—Rule 38.

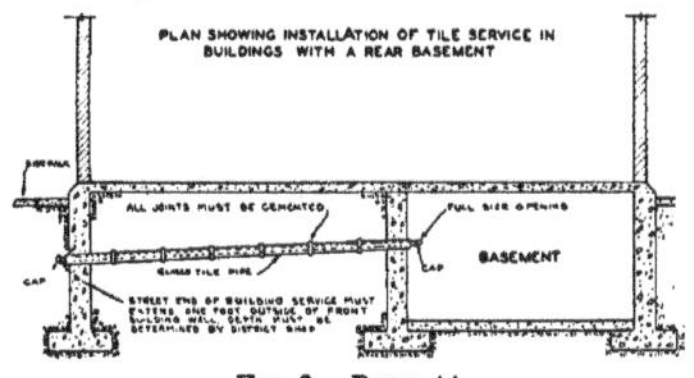

Fig. 2.—Rule 44.

No Cooking Task is too LARGE nor too SMALL for MAGIC CHEF

Magic Chef Heavy Duty Equipment offers the busy hotel, restaurant, hospital, and other institutions, savings . . . of time, labor, and food . . . that in many cases is sufficient to pay for the equipment. Features include perfect control of gas flow provided by the famous Red Wheel Oven Regulator; non-clog burners; special alloy quick-heating cooking tops; streamline design; automatic lighting; speed; safety; dependability and durability.

A typical Magic Chef Battery installed in a large western hospital. The equipment includes Ranges, Broilers, and Bakers.

The domestic line of Magic Chef gas ranges embraces every price class from the lowest to the highest. In addition to the regular exclusive Magic Chef features, some models have two features never before seen on a gas range . . . SWING-OUT BROILER attached to the door which, when opened, swings the contents out and away from the heat zone. No stooping to broil. The HIGH-SPEED OVEN bakes biscuits in 12 minutes from a cold start . . . less time than it takes to preheat an ordinary oven for baking. Remember, all Magic Chefs are backed up with national advertising and a company that has been in the stove business for over 50 years.

Series 3600 was designed for the small domestic kitchen and apartments . . . it is priced to suit the small budget.

AMERICAN STOVE COMPANY

NORTHERN SALES DIVISION

179 NORTH MICHIGAN AVENUE CENtral 2007

44. Underground Building Service.

If the type of building is such that a cement or tile flooring is laid on the ground to the front of the building then the building service shall be run in glazed tile pipe with cemented joints under this portion of cement flooring and shall be extended to 1 foot outside of the front building wall clearing doorways by at least 2 feet, at a depth to be determined by the District Shop. Building service should be graded downward toward the street service. See Figures 2 and 3.

The District Shop shall be notified before a building service laid in glazed tile pipe with cemented joints is covered.

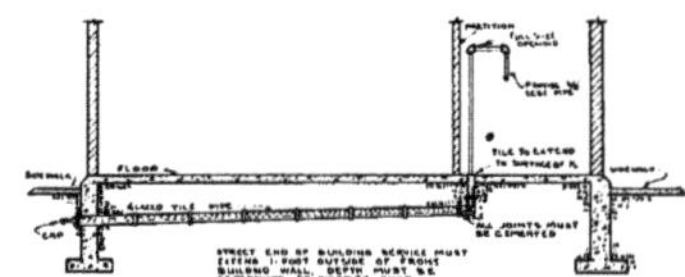

FIG. 3.—RULE 44.

45. Building Service in Pipe Ducts.

A building service may be run in pipe ducts provided it is installed to terminate at least one foot outside the front building wall, and provided the ducts are ventilated at both ends and do not contain light and power cables. Special instructions should be obtained from the District Shop in regard to the cover of the building service terminating at the outside wall.

46. Solid Wall Porch.

In a building with a solid wall porch, the building service shall be run to the side wall near the front of the building where the street service will enter.

47. Building Service in a 2, 3 or 4 Apartment Building.

Buildings containing 2, 3 or 4 apartments shall have a building service not less than 1¼ inch in size. A 1¼ inch building service shall not be connected to a load in excess of 5.3 "units."

If a larger number of "units" is used, then the correct larger size building service shall be installed (See Rules Nos. 29, 53 and 69).

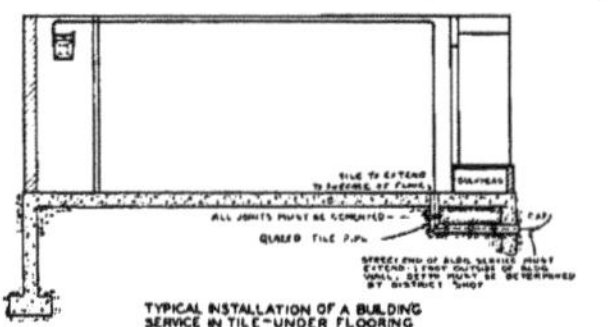

FIG. 4.—RULE 48.

48. Stores Under One Roof With Common Party Wall.

(a) Street Services.

1. Only one street service, which shall be of sufficient size to supply all stores, will be installed.

2. Where the building service piping inside of the store is run in glazed tile pipe with cemented joints, it shall be extended to 1 foot outside of the front building wall, at a depth to be determined by the District Shop, by the contractor for connection to the street service by Gas Company.

(b) Building Services.

1. The building service may be installed as follows: exposed within the store, exposed within the basement, overhead above ceiling joists, under the floor in glazed tile with cemented joints.

In a partial basement the building service may be run under the floor suspended from the floor joists if a clearance of 30" is available, or in glazed tile underground if no clearance exists.

2. Where bulkheads are installed in a store with no basement, the building service shall be run in glazed tile pipe with cemented joints under bulkhead and shall be extended to 1 foot outside of the front building wall, at a depth to be determined by the District Shop. See Fig. 4.

3. Under the above conditions, the building service may be branched and extended to the various stores exposed through the common party walls, overhead above ceiling joists, in open basements through common party walls or under floor joists where sufficient clearance is available. Pipe above ceiling joists shall be protected.

4. Where the building service is installed in glazed tile, extend the service exposed or in party wall, up for branching in the stores or above the ceiling joists. However, where such service terminates in a partial basement, it may also be branched in the partial basement.

(c) General.

In permitting the installation of piping and meters under the above conditions, the Gas Company will arrange to tag building service at the meter inlet through the use of a special fitting with a metal tag securely attached so as to indicate what the building service supplies.

49. High Pressure Territory.

In high pressure territory, where governors are used, building services m stores without basements shall be run overhead and space provided for the governor where the street service enters the building.

50. Sleeve on Building Service Through Masonry Walls.

A building service hereinafter laid through a masonry wall shall be encased within a sleeve and centered with a constant clearance of ½ inch.

51. Structures Having Insulated Ceiling.

In structures having insulated ceilings, building services, risers or fuel runs shall not be run in the space above the insulation unless piping is properly insulated.

52. Sleeve on Building Service Connecting Separate Building Sections.

A building service run under an open porch connecting separate buildings or sections shall be continuously sleeved through an iron pipe by means of spacers to obtain symmetrical spacing, and shall have at least two inches of clearance inside of sleeve to allow a circulation of air to pass through the sleeve. Where basements are not heated the ends of sleeve should be sealed.

The sleeve shall be left open at both ends where basements are heated.

53. Building Service Sizing.

In Schools, Hospitals, Office Buildings, Stores, Apartment Buildings, etc., size the building service according to the following table:

Table No. 2.

Number of ¾" Openings	Size of Pipe	Feet of Pipe Allowed	Number of "Units" Allowed
1 or 2	1	70	2
3 to 5	1¼	80	5
6 to 10	1½	100	10
11 to 20	2	150	20
21 to 35	2½	175	35
36 to 65	3	225	65
66 to 130	4	325	130

NOTE: See Rule No. 29 for determination of number of "Units" and Rules Nos. 69 and 70 for riser sizes.

67. Riser Sizing for Master Meters in Apartment Buildings.

In all apartment buildings, regardless of the number of apartments, 0.5 "unit" shall be used for each ¾ inch opening and 0.3 "unit" for each laundry header, when master meters are used. See Rule No. 29 for explanation of "Units" and Rule No. 70 for riser sizes.

68. Riser Sizing for Apartment Buildings With Heavy Duty Appliances.

In apartment buildings where the manifold connection on the appliance is larger than ¾ inch, indicating a heavy duty appliance, the piping shall be sized to supply the demand of the larger manifold connection. See Rules Nos. 69 and 70 for riser sizes.

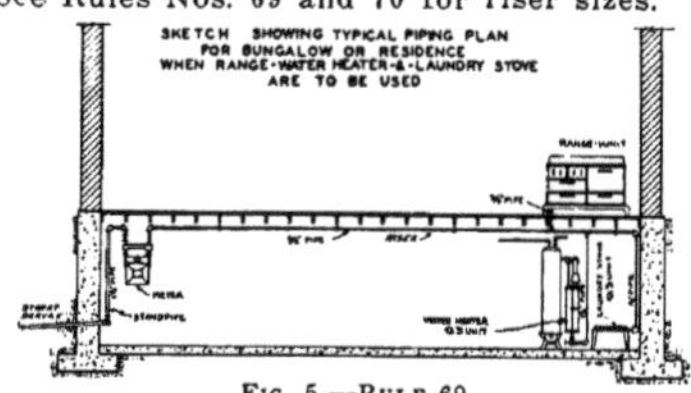

FIG. 5.—RULE 69.

69. Riser Sizing for Individual Meters.

In office Buildings, Schools, Hospitals, Stores, Residences and Apartments, size the risers according to the following table.

Table No. 3.

Number of ¾" Openings	Size of Pipe Inches	Feet of Pipe Allowed	Number of "Units" Allowed
1	¾	60	1.6
2	1	70	2.6
3 to 5	1¼	80	5.3
6 to 10	1½	100	10
11 to 20	2	150	20
21 to 35	2½	175	35
36 to 65	3	225	65
66 to 130	4	325	130

NOTE: See Rules Nos. 29 and 66 for determination of number of "units" and Rule No. 53 for size of building service. A ¾ inch riser is the smallest size allowed.

Equivalent Pipe Capacities for lengths not over 30 ft.:

4—⅜" openings are equivalent to one ¾" opening.

2—½" openings are equivalent to one ¾" opening.

Branches of ⅜ inch and ½ inch may be taken from a running fuel line providing they are not over 30 feet long. See Figures 5 and 6.

70. Riser Sizing for Apartments With Master Meter.

In apartments with master meter, size the risers according to the following table.

Table No. 4.

Number of ¾" Openings	Size of Pipe Inches	Feet of Pipe Allowed	Number of "Units" Allowed
2	1	70	2
3 to 7	1¼	80	7
8 to 15	1½	100	15
16 to 35	2	150	35
36 to 60	2½	175	60
61 to 120	3	225	120
121 to 225	4	325	225

NOTE: A 1-inch riser is the smallest size permitted. See Rules Nos. 67 and 29 for determination of number of "units". See Rule No. 53 for size of building service.

71. Test Pipe.

A ⅜ inch test pipe shall be installed on a riser or a fuel run in all cases where the size of the pipe is larger than ¾ inch in size. See Rule No. 56 for test pipe on building service.

72. Riser Sizing for Restaurants, Hotels, Commercial Establishments.

Because of the many different types and sizes of appliances used in restaurants, hotels and commercial establishments, and their wide variation in capacity, a schedule of gas consumption is not shown. If gas consumptious of these appliances are known, one "unit" shall be considered for every 100 cubic feet consumption per hour. For explanation of "unit" see Rule No. 29. If consumption of these appliances are not known, the number of "units" equal to or allowable for the size of the gas connection on the appliances shall be taken as the load on any one appliance, and a total made of all the involved appliances in terms of "units", which total will determine the starting size of the riser which will supply all these appliances.

If any appliance has a ⅜ inch connection it shall be considered equal to a ¼ "unit". If the connection is ½ inch, it shall be considered equal to a ½ "unit". See Figure 7.

Consult Riser Sizing Table in Rule 69.

73. Capping Riser Outlets.

All outlets shall be securely closed with iron caps or plugs until appliances or fixtures have been installed.

74. Underground Fuel Lines.

Fuel lines are preferably not to be installed underground. However, where an underground installation is unavoidable, the District Shop should be consulted. See Rules Nos. 2 and 7.

75. Compressed Air.

When compressed air is used to support combustion at a pressure higher than the normal gas pressure of ¼ lb., an approved check valve shall be installed in the piping close to the outlet of the gas meter or at the appliance using compressed air. If such a valve is not used a gas meter will not be set.

76. Oxygen or Other Gases.

When oxygen or other gases are used on an appliance to support combustion an approved device must be installed to prevent the oxygen or other gases from entering gas piping.

77. Location of Fuel Outlet in Kitchen.

When piping enters kitchen through the floor, the fuel line shall be extended at least 3 inches above the finished floor and 2 inches clear of the baseboard.

When piping enters kitchen through the wall, it shall terminate not over 3 feet above the finished floor and 2 inches clear of the finished wall.

78. Laundry Header.

In a flat building where appliances such as laundry stoves, clothes driers, etc., are installed for the joint use of tenants, a ½ inch pipe from each tenant's meter shall be run to the laundry room and a header provided on the wall adjacent to the appliance. Each riser shall be equipped with a lever handle lock cock. A metal tag with the flat number plainly stamped thereon shall be securely fastened to each cock. See Figure 7.

One outlet for a light fixture in the laundry may be taken from the end of the laundry header.

79. Breaking Pipe Sizes.

In every case, when remodeling or extending old house piping, the pipe shall be broken at a point where the full size can be maintained. No extension shall be made from a pipe of a smaller size. Where sizes cannot be maintained from old piping, the piping extensions shall be supplied by a separate run of pipe from the meter.

80. Light Drops From Branch Lines.

Drops on branch lines for lights should have a 90 degrees bend four inches long and they shall be dropped square, extending 2 inches below the finished ceiling. Outlets for light side brackets may be either square bends or long drop ells. The use of nipples is prohibited.

81. Exit Lights.

When running pipe for exit lights in theatres, schools, amusement or assembly halls, it shall be done in accordance with the ordinances and regulations of the City of Chicago. In all cases drop ells or bends shall used. The use of nipples is strictly prohibited.

82. Typesetting Machines.

A linotype or monotype machine shall be supplied by a separate fuel run from the meter to the appliance, unless the machine is equipped with a gas pressure regulator.

83. Piping for Larger Appliances.

Any appliance that will consume a considerable amount of gas and operates intermittently and is controlled by a thermostat or any other controlling device should preferably be supplied by a separate fuel run from the meter which will insure a uniform supply of gas at the appliance at all times.

84. Piping Requirements for Appliance Connections.

All stationary gas burning appliances shall be connected with iron pipe, except in those cases where the approval of the District Shop has been obtained to use approved A.G.A. Semi-Rigid Copper tubing.

In such cases where copper tubing is used for appliance connections, a riser of iron pipe shall be run from the meter location to the apartment or place where the appliance is to be installed. It may then be connected with an approved copper tubing.

Copper tubing shall not be concealed under a finished floor or in the walls, except in the case of Electrolux refrigerator installations in old buildings where it is extremely difficult to install iron pipe. The use of copper tubing is permitted in all types of buildings for the immediate connection at the refrigerator.

Portable appliances may be connected with approved A.G.A. flexible tubing preferably with union connectors. A shutoff cock shall be installed in the iron pipe line at the inlet connection to the flexible tubing and there shall be no complete shutoff cock at the appliance to shut off the gas supply.

85. Location of Shut-Offs on Fuel Runs.

Stop cocks shall not be placed in any gas line beyond the meter, except those provided at appliances for turning gas off or on in which case they should be installed on the meter side of the union, and those provided on separate risers from a master meter installation. See Rules Nos. 11, 39, 94 and 97.

METERS.

86. Location.

The Company reserves the right to determine in all cases the location of meters.

In installing pipes to be connected with the meter, it should be borne in mind that no meter should, without the consent or direction of the Gas Company, be installed in any of the following locations:

(a) Under a bulkhead or show window, in an attic, bedroom, closet, stairway closet, under a work bench, under plumbing fixtures or over a laundry tub.

(b) A gas meter should never be set over a gas fixture, range or laundry stove, over a radiator or any heating appliance, or where it would be subjected to excessive temperatures.

(c) In a passageway or a hallway where the clearance is less than 4 feet from any wall to an opposite wall.

(d) In a basement where the clearance is less than 6 feet from the floor to the ceiling.

(e) In a hallway or passageway which is outside of the building proper.

(f) Closer than 3 feet to any electrical appliance, switch, coutout box or electric meter.

(g) Under a hanging sewer pipe.

(h) In a fuel bin or on the partitions forming such bins.

(i) In a room which is not properly ventilated either by a window to the outside or a ventilator installed in the top and bottom of the door of the meter room.

(j) In a room where acids or other corrosive chemicals are used or stored.

(k) In any damp or unventilated location.

(l) In any location where it may be bumped or damaged.

(m) Or in any other location which may be hazardous or detrimental to the meter or metering.

(n) In a living room, bathroom, over a door or window, over an electric light fixture, or over a work bench, or in any location where the visits of the Meter Reader will cause annoyance to the customer.

87. Accessibility of Meter Location.

A meter shall be located in a suitable accessible location. No meter supplying one tenant shall be set in a flat or a store in any location that is occupied by another tenant.

88. Meters Not to Be Exposed to Heat and Cold.

Meters shall not be set so close to any source of artificial heat that the accuracy of the meter will be affected.

Meters shall not be set in any location where a temperature below 40 degrees might occur during the winter.

89. Covering Cold Water Pipes.

Where cold water pipes run directly over the top of a meter location, cold water pipes shall be covered with a suitable pipe covering to prevent water from dripping on meters.

90. Covering Steam Pipes.

Where a steam line runs directly over, under or along side of a meter location the steam pipe shall be covered with a suitable covering to protect meters from heat.

91. Working Space Between Gas and Electric Meters.

Where provisions have been made to locate all the gas meters in the same room with the electric meters, a space of no less than three feet shall be provided between the face of the gas meters and the face of the electric meters to provide ample working conditions.

92. Apartment Building Service Halls.

Where provisions have been made to locate the gas meters in the service halls on the various floors they shall be located on a wall in a suitable location not closer than 3 feet from an electric equipment, and in no case shall be set over the doors of the service elevators or over a radiator.

93. Meters in Boiler Room.

Where it is necessary to locate a meter in a boiler room the meter shall not be set in front of the boiler nor over space used for storage of ashes. The meter must not be set where it will be subjected to abnormal room temperatures. Approval for meter location in a boiler room must be obtained from the District Shop.

94. Location of Shut-Offs on Fuel Run for Apartments on Master Meter.

On every fuel run where a master meter is to be used, a shut-off cock shall be installed in the apartment to be supplied at each of the appliances. If one riser supplies all the gas to any one apartment only one shut-off cock may be used, but it shall belocated so as to be readily accessible. This permits control of gas in each apartment without shutting off the master meter. See Rules Nos. 11, 39, 85, 95 and 97.

95. Location of Shut-Offs on Fuel Run for Offices on Master Meter.

Where a master meter supplies a building containing offices, a shut-off cock shall be installed on each fuel run in the office. See Rules Nos. 11, 39, 85, 94 and 97.

96. Meter for Theatre.

A meter to supply a theatre may be set in a public meter room with other meters and may be supplied by the same service supplying these meters.

97. Location of Shut-Offs on Risers in Basement, Supplied From Master Meter.

In the event that shut-off cocks are to be installed on the risers in the basement, instead of on the fuel run at the appliance where a master meter is used, each riser shall be tagged with a metal tag, plainly stamped or marked in such a manner to designate the correct apartment which is being supplied. See Rules Nos. 11, 39, 85, 94 and 95.

98. Master Meter Header.

In all cases where provisions have been made to supply an apartment building from a master meter, a header shall be provided both on the building service and the riser. See following table No. 5 for proper space between building service and riser openings. For information on size of meters and spacing contact the District Shop.

Table No. 5.

Building Service and Riser Headers for Master Meters.

Size Opening	Space Between Building Service Opening and Riser in Inches
1	14
1¼	16
1½	24
2	24
2½	30
3	34
4	42

99. Building Service Header for Multiple Meters.

When it is necessary to set two or more meters together, a building service header must be supplied with an opening for each meter. See Rule No. 98.

100. Meter Bar Installation.

An approved type of meter bar and cock shall be installed in all buildings requiring three or more meters of the 5 or 10 light size, which are located in a public meter room.

These bar connections can be obtained from the District Shop, but shall remain the property of the Gas Company.

The meter bars shall be in alignment, level, and with a space of not less than 7 inches (7") between the center of the bar and the wall or partition back of the location provided for gas meters. Meters bars shall not be more than 9 feet from the floor but of sufficient height so that the bottom of meter will not be closer than 12 inches from the floor.

There shall be proper clearance for the meter, with no interference with steam or water pipes, electric conduits, switch boxes, etc., before installing the meter bar.

The meter cocks shall be installed on the left hand side of the meter bar in a vertical position with the stationary lug of the cock to the left, and pointing away from the wall or partition back of the location where the meters are to be set. The inlet and outlet of the bar shall be plugged with ½ inch iron plugs.

Meter bars shall be installed and marked with metal tags to designate the apartments they supply. It is understood that the Gas Company representative and the fitter shall cooperate and mutually agree on the tagging of risers designating the apartments they supply.

See Figure 8, showing the correct manner of meter bar and cock installation.

101. Conversion to Master Meter.

In the event the piping in a building is to be rearranged from single meters to a master meter, and there are meter bars installed on the piping, the District Shop should be notified so that gas can be shut off and meter bars and meters removed prior to alterations.

Exterior view of our modern plant

Interior view of north half of plant

SPECIFICATION FOR THE DESIGN, FABRICATION AND ERECTION OF STRUCTURAL STEEL FOR BUILDINGS

AMERICAN INSTITUTE OF STEEL CONSTRUCTION.

PART I. GENERAL.

Section 1. Scope.

(a) **Scope.**
This Specification defines the practice adopted by the American Institute of Steel Construction in the design, fabrication, and erection of structural steel for buildings.

(b) **Code.**
In the execution of contracts entered into under this Specification, the Code of Standard Practice for Buildings of the American Institute of Steel Construction shall apply unless otherwise specified or required.

Section 2. Plans and Drawings.

(a) **Plans.**
The plans shall show a complete design with sizes, sections, and the relative location of the various members. Floor levels, column centers, and off-sets shall be dimensioned. Plans shall be drawn to a scale large enough to convey the information adequately.

(b) **Shop Drawings.**
Shop drawings shall be made in conformity with the best modern practice and with due regard to speed and economy in fabrication and erection.

PART II. MATERIAL.

Section 3. Material.

(a) **Structural Steel.**
Structural steel shall conform to the Standard Specifications of the American Society for Testing Materials for Steel for Buildings, Serial Designation A 9 (or, if so specified by the Buyer, for Steel for Bridges, Serial Designation A 7), as amended to date.

(b) **Rivet Steel.**
Rivet steel shall conform to the Standard Specifications of the American Society for Testing Materials for structural Rivet Steel, Serial Designation A 141, as amended to date.

(c) **Other Metals.**
Alloy steels, cast steel, cast iron and other metals shall conform to the applicable Specifications of the American Society for Testing Materials, as amended to date.

(d) **Stock Material.**
Stock material shall be of a quality equal to that called for by the specifications of the American Society for Testing Materials for the classifications covering its intended use; and mill test reports shall constitute sufficient record as to the quality of material carried in stock.

Unidentified stock material, if free from surface imperfections, may be used for short sections of minor importance, or for small unimportant details, where the precise physical properties of the material would not affect the strength of the structure.

PART III. LOADS AND STRESSES.

Section 4. Loads and Forces.

(a) **Loads and Forces.**
Steel structures shall be designed to sustain the following loads and forces:

1. Dead Load.
2. Live Load.
3. Impact.
4. Wind and other Lateral Forces.
5. Erection Loads.
6. Other Forces.

(b) **Dead Load.**
The dead load shall consist of the weight of the steelwork and all material fastened thereto or supported thereby.

(c) **Live Load.**
The live load shall be that stipulated by the Code under which the structure is being designed or that required by the conditions involved. In general, the live loads should not be less than those recommended by the Building Code Committee of the National Bureau of Standards, November, 1924, under the caption "Minimum Live Loads for use in the Design of Buildings."

(d) **Impact.**
For structures, carrying live loads inducing impact or vibrational forces, the design stresses shall be increased by a percentage of the live load stresses sufficient to suitably provide for such forces.

(e) **Wind.**
Proper provision shall be made for stresses caused by wind both during erection and after completion of the building. The wind pressure is dependent upon the conditions of exposure and geographical location of the structure. The allowable stresses specified in Sections 6 (c) and 7 are based upon the steel frame being designed to carry a wind pressure of not less than twenty (20) pounds per square foot on the vertical projection of exposed surfaces during erection, and fifteen (15) pounds per square foot on the vertical projection of the finished structure.

(f) **Erection.**
Proper provision shall be made for temporary stresses caused by erection.

(g) **Other Forces.**
Structures in localities subject to earthquakes, hurricanes, and other extraordinary conditions shall be designed with due regard for such conditions.

Section 5. Reversal of Stress.

(a) **Reversal of Stress.**
Members subject to live loads producing alternating tensile and compressive stresses shall be proportioned as follows:

To the net total compressive and tensile stresses add 50 per cent of the smaller of the two and proportion the member to resist either of the increased stresses resulting therefrom.

Connections shall be proportioned to resist the larger of the two increased stresses.

Section 6. Combined Stresses.

(a) **Axial and Bending.**
Members subject to both axial and bending stresses shall be so proportioned that the quantity

$$\frac{f_a}{F_a} + \frac{f_b}{F_b}$$ shall not exceed unity, in which

F_b = bending unit stress that would be permitted by this Specification if axial stress only existed.

F_b = bending unitstress that would be permitted by this Specification if bending stress only existed.

f_a = axial unit stress (actual) = axial stress divided by area of member.

f_b = bending unit stress (actual) = bending moment divided by section modulus of member.

(b) **Rivets.**
Rivets subject to shearing and tensil forces shall be so proportioned that the combined unit stress will not exceed the allowable unit stress for rivets in tension only.

(c) **Wind and Other Forces.**
Members subject to stresses produced by a combination of wind and other loads may be proportioned for unit stresses 33⅓ per cent greater than those specified in Section 10, provided the section thus required is not less than that required for the combination of dead load, live load, and impact (if any).

inches (l) divided by the least-radius-of-gyration-in-inches (r) of cross section does not exceed 120 shall be loaded no more on the cross-sectional-area than determined by the equation which follows

$$\frac{P}{a} = \left[1700 - 0.485 \frac{l^2}{r^2}\right] \text{ lbs. per sq. in.}$$

Axiliary Loaded Columns when (l) divided (r) is greater than 120 shall be loaded no more in pounds per square inch of cross section area than determined by the equation which follows:

$$\frac{P}{a} = \left[\frac{18,000}{1 + \left(\frac{l^2}{18,000\ r^2}\right)}\right] \text{ lbs. per sq. in.}$$

Plate Girder Stiffeners shall be loaded on gross cross sectional area no more than 20,000 lbs. per sq. in.

Webs of Rolled Sections at toe of fillet shall be loaded on gross-cross-sectional area no more than 24,000 lbs. per sq. in. (See Sec. 19h, Web crippling of Beams).

For Bending:

Tension on Extreme Fibers of rolled sections, plate girders, and built-up members (See Sec. 19a 20,000 lbs. per sq. in.)

Compression on Extreme Fibers of rolled sections, plate girders, and built-up members for values of the ratio of the laterally-unsupported length-in-inches (1) divided by the width-of-the-compression flange (b) not greater than 40 shall be loaded no more on the cross-sectional-area than determined by the equation which follows:

$$\frac{P}{a} = \left[\frac{22,500}{1 + \frac{l^2}{1800\ b^2}}\right]$$

Stress on Extreme Fibers of pins. 30,000

SHEARING:

Rivets	15,000
Pins, and turned bolts in reamed or drilled holes	15,000
Unfiinished bolts	10,000
Webs of beams and plate girders, gross section	13,000

BEARING:	Double Shear	Single Shear
Rivets	40,000	32,000
Turned bolts in reamed or drilled holes	40,000	32,000
Unfiinshed bolts	25,000	20,000
Pins	32,000	
Contact Area:		
Milled Stiffeners and other Milled Surfaces	30,000	
Fitted Stiffeners	27,000	
Expansion rollers and rockers (pounds per linear inch)...	600d	
in which d is diameter of roller or rocker in inches.		

(b) Cast Steel.

Compression and Bearing same as for Structural Steel. Other Unit Stresses, 75 per cent of those for Structural Steel.

(c) Masonry [Bearing].

Granite	800
Sandstone and Limestone	400
Concrete, unless otherwise specified	600
Hard Brick in Cement Mortar	250

PART V. DESIGN.

Section 11. Slenderness Ratio.

(a) The ratio of unbraced length to least radius of gyration $\frac{l}{r}$ for compression members shall not exceed:

For main compression members	120
For bracing and other secondary members in compression	200

Stair Well, Campana Building — Frank D. Chase, Engineer — Childs & Smith, Architects

ALLOWABLE WORKING STRESS FOR AXIALLY LOADED COLUMNS.

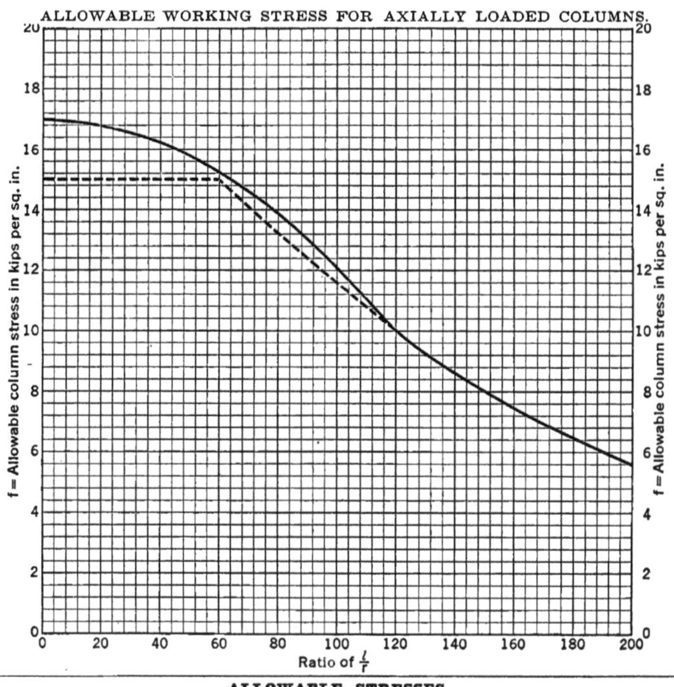

ALLOWABLE STRESSES

Per Square Inch for Beams and Girders Laterally Unsupported

$$f = \frac{22500}{1 + \frac{l^2}{1800\,b^2}}$$

l = Unsupported length in inches.
b = Width of flange in inches.

$\frac{l}{b}$	Unit Stress f (Kips)	Ratio	$\frac{l}{b}$	Unit Stress f (Kips)	Ratio	$\frac{l}{b}$	Unit Stress f (Kips)	Ratio
15.0	20.00	1.000	23.5	17.22	.861	32.0	14.34	.717
15.5	19.85	.993	24.0	17.05	.852	32.5	14.18	.709
16.0	19.70	.985	24.5	16.87	.844	33.0	14.02	.701
16.5	19.54	.977	25.0	16.70	.835	33.5	13.86	.693
17.0	19.39	.969	25.5	16.53	.826	34.0	13.70	.685
17.5	19.23	.961	26.0	16.36	.818	34.5	13.54	.677
18.0	19.07	.953	26.5	16.19	.809	35.0	13.39	.669
18.5	18.91	.945	27.0	16.01	.801	35.5	13.23	.662
19.0	18.74	.937	27.5	15.84	.792	36.0	13.08	.654
19.5	18.58	.929	28.0	15.67	.784	36.5	12.93	.647
20.0	18.41	[illegible]	28.5	15.50	.775	37.0	12.78	.639
20.5	18.24	.912	29.0	15.34	.767	37.5	12.63	.632
21.0	18.07	.904	29.5	15.17	.758	38.0	12.49	.624
21.5	17.90	.895	30.0	15.00	.750	38.5	12.34	.617
22.0	17.73	.887	30.5	14.83	.742	39.0	12.20	.610
22.5	17.56	.878	31.0	14.67	.733	39.5	12.05	.603
23.0	17.39	.869	31.5	14.50	.725	40.0	11.91	.596

(d) **Angles.**

For angles, the gross width shall be the sum of the widths of the legs less the thickness. The gage for holes in opposite legs shall be the sum of the gages from back of angle less the thickness.

(e) **Splice Members.**

For splice members, the thickness considered shall be only that part of the thickness of the member which has been developed by rivets beyond the section considered.

(f) **Size of Holes.**

In computing net area the diameter of a rivet hole shall be taken as ⅛ inch greater than the nominal diameter of the rivet.

(g) **Pin Holes.**

In pin connected tension members, the net section across the pin hole, transverse to the axis of the member, shall be not less than 140 per cent, and the net section beyond the pin hole, parallel with the axis of the member, not less than 100 per cent, of the net section of the body of the member.

In all pin-connected riveted members the net width across the pin hole, transverse to the axis of the member, shall preferably not exceed 12 times the thickness of the member at the pin.

Section 15. Expansion.

(a) **Expansion.**

Proper provision shall be made for expansion and contraction.

Section 16. Connections.

(a) **Minimum Connections.**

Connections carrying calculated stresses, except for lacing, sag bars, and girts, shall have not fewer than 2 rivets.

(b) Eccentric Connections.

Members meeting at a point shall have their gravity axes meet at a point if practicable; if not, provision shall be made for their eccentricity.

(c) **Rivets.**

The rivets at the ends of any member transmitting stresses into that member should preferably have their centers of gravity on the gravity axis of the member; otherwise, provision shall be made for the effect of the resulting eccentricity. Pins may be so placed as to counteract the effect of bending due to dead load.

(d) **Unrestrained Members.**

When beams, girders or trusses are designed on the basis of simple spans in accordance with Section 9 (a), their end connections may ordinarily be designed for the reaction shears only. If, however, the eccentricity of the connection is excessive, provision shall be made for the resulting moment.

(e) **Restrained Members.**

When beams, girders or trusses are subject both to reaction shear and end moment, due to the restraint specified in Section 9 (b), their connections shall be specially designed to carry both shear and moment without exceeding at any point the unit stresses prescribed in Section 10. Ordinary end connections comprising only a pair of web angles, with not more than a nominal seat and top angle, shall not be assumed to provide for this kind of end moment.

(f) **Fillers.**

In truss construction when rivets carrying computed stress pass through fillers, the fillers shall be extended beyond the connected member and the extension secured by sufficient rivets to develop the stress in the filler.

Fillers under plate girder stiffeners at end bearings or points of concentrated loads shall be secured by sufficient rivets to prevent excessive bending and bearing stresses.

(g) **Splices.**

Compression members when faced for bearing shall be spliced sufficiently to hold the connecting members accurately in place. Other joints in riveted work, whether in tension or compression, shall be spliced so as to transfer the stress to which the member is subject.

compression members composed of plates and shapes shall not exceed 16 times the thickness of the thinnest outside plate or shape, nor 20 times the thinnest enclosed plate or shape with a maximum of 12 inches, and at right angles to the direction of stress the distance between lines of rivets shall not exceed 30 times the thickness of the thinnest plate or shape. For angles in built sections with two gage lines, with rivets staggered, the maximum pitch in the line of stress in each gage line shall not exceed 24 times the thickness of the thinnest plate with a maximum of 18 inches.

(c) **End Pitch Compression Members.**

The pitch of rivets at the end of built compression members shall not exceed four diameters of the rivets for a length equal to 1½ times the maximum width of the member.

(d) **Two-Angle Members.**

In tension members composed of two angles, a pitch of 3'-6" will be allowed, and in compression members, 2'-0", but the ratio *l*-r for each angle between rivets shall be not more than ¾ of that for the whole member.

(e) **Minimum Edge Distance.**

The minimum distance from the center of any punched rivet hole to any edge shall be that given in Table I.

TABLE I

Rivet Diameter Inches	Minimum Edge Distance (Inches) for Punched Holes		
	In Sheared Edge	In Rolled Edge of Plates and Sections with Parallel Flanges	In Rolled Edge of Sections with Sloping Flanges
½	1	⅞	¾ *
⅝	1⅛	1	⅞ *
¾	1¼	1⅛	1 *
⅞	1½	1¼	1⅛ *
1	1¾	1½	1¼ *
1¼	2	1¾	1½ *
1⅛	2¼	2	1¾ *

*May be decreased ⅛ inch when holes are near end of beam.

(f) **Minimum Edge Distance in Line of Stress.**

The distance from the center of any rivet under computed stress, and that end or other boundary of the connected member toward which the pressure of the rivet is directed, shall be not less than the shearing area of the rivet shank (single or double shear respectively) divided by the plate thickness.

This end distance may however be decreased in such proportion as the stress per rivet is less than that permitted under Section 10 (a); and the requirement may be disregarded in case the rivet in question is one or three or more in a line parallel to the direction of stress.

(g) **Maximum Edge Distance.**

The maximum distance from the center of any rivet to the near edge shall be 12 times the thickness of the plate, but shall not exceed 6 inches.

Section 19. Plate Girders and Rolled Beams.

(a) **Proportioning.**

Riveted plate girders, cover-plated beams, and rolled beams shall in general be proportioned by the moment of inertia of the gross section. No deduction shall be made for standard shop or field rivet holes in either flange; except that in special cases where the reduction of the area of either flange by such rivet holes, calculated in accordance with the provisions of Section 14, exceeds 15 per cent of the gross flange area, the excess shall be deducted. If such members contain other holes, as for bolt, pins, or countersunk rivets, the full deduction for such holes shall be made. The deductions thus applicable to either

nge shall be made also for the opposite nge if the corresponding holes are there esent.

(b) **Web.**
Plate girder webs shall have a thickness of t less than 1/170 of the unsupported dis- nce between flanges.

(c) **Flanges.**
Cover plates, when required, shall be equal thickness or shall diminish in thickness ·om the flange angles outward. No plate ıall be thicker than the flange angles.
Unstiffened cover plates shall not extend ore than 6 inches nor more than 12 times ıe thickness of the thinnest plate beyond ıe outer row of rivets connecting them to ıen angles.
The total cross-sectional area of cover ates shall not exceed 70 per cent of the tal flange area.

(d) **Rivets.**
Rivets connecting the flanges to the web ıall be proportioned to resist the horizontal ıear due to bending as well as any loads plied directly to the flange.

(e) **Stiffeners.**
Stiffeners shall be placed on the webs of ate girders at the ends and at points of ncentrated loads. Such stiffeners shall have close bearing against the flanges, shall ex- nd as closely as possible to the edge of the ange angles, and shall not be crimped. They ıall be connected to the web by enough vets to transmit the stress. Only that por- on of the outstanding legs outside of the llets of the flange angles shall be considered ffective in bearing.
If h/t is equal to or greater than 70, inter- ıediate stiffeners shall be required at all oints where h/t exceeds $\frac{8,000}{\sqrt{v}}$, in which

h = the clear depth between flanges, in inches
t = the thickness of the web, in inches
v = greatest unit shear in panel, in pounds per square inch under any condition of complete or partial loading.

The clear distance between intermediate tiffeners, when stiffeners are required by the oregoing, shall not exceed 84 inches or that ·iven by the formula $d = \frac{270,000t}{v}\sqrt[3]{\frac{vt}{h}}$

d = the clear distance between stiffeners, in inches.

Intermediate stiffeners may be crimped ver the flange angles.
Plate girder stiffeners shall be in pairs, one n each side of the web, and shall be con- ected to the web by rivets spaced not more han 8 times their normal diameter.

(f) **Splices.**
Web splices in plate girders shall be pro- ·ortioned to transmit the full shearing and ·ending stresses in the web at the point of plice.
Flange splices shall be proportioned to de- ·elop the full stress of the members cut.

(g) **Lateral Forces.**
The flanges of plate girders supporting ranes or other moving loads shall be pro- ·ortioned to resist any lateral forces pro- ·uced by such loads.

(h) **Web Crippling of Beams.**
Rolled beams shall be so proportioned that he compression stress at the web toe of the illets, resulting from concentrated loads, hall not exceed the value of 24,000 pounds ·er square inch allowed in Section 10 (a). The governing formulas shall be:

For interior loads $\frac{R}{t(N + 2k)}$ = not over 24,000

For end reactions $\frac{R}{t(N + k)}$ = not over 24,000

R = concentrated interior load or end re- actions, in pounds.
t = thickness of web, in inches.
N = length of bearing, in inches.
k = distance from outer face of flange to web toe of fillet, in inches.

Section 20. Tie Plates.

(a) **Compression Members.**
The open sides of compression members shall be provided with lacing having tie plates at each end, and at intermediate points if the lacing is interrupted. The plates shall be as near the ends as practicable. In main mem- bers carrying calculated stresses the end tie plates shall have a length of not less than the distance between the lines of rivets con- necting them to the segments of the member, and intermediate ones of not less than one- half of this distance. The thickness of tie plates shall be not less than one-fiftieth of the distance between the lines of rivets con- necting them to the segments of the members, and the rivet pitch shall be not more than four diameters. Tie plates shall be connected to each segment by at least three rivets.

(b) **Tension Members.**
Tie plates shall be used to secure the parts of tension members composed of shapes. They shall have a length not less than two-thirds of the length specified for tie plates in com- pression members. The thickness shall be not less than one-fiftieth of the distance between the lines of rivets connecting them to the segments of the member and they shall be connected to each segment by at least three rivets.

Section 21. Lacing.

(a) **Spacing.**
Lacing bars of compression members shall be so spaced that the ratio l/r of the flange included between their connections shall be not over ¾ of that of the member as a whole.

(b) **Proportioning.**
Lacing bars shall be proportioned to resist a shearing stress normal to the axis of the member equal to two per cent of the total compressive stress in the member. In deter- mining the section required the compression formula shall be used, l being taken as the length of the bar between the outside rivets connecting it to the segment for single lacing and 70 per cent of that distance for double lacing. The ratio l/r shall not exceed 140 for single lacing nor 200 for double lacing.

(c) **Minimum Proportions.**
The thickness of lacing bars shall be not less than one-fortieth for single lacing, and one-sixtieth for double lacing, of the distance between end rivets; their minimum width shall be three times the diameter of the rivets connecting them to the segments.

(d) **Inclination.**
The inclination of lacing bars to the axis of the members shall preferably be not less than 45 degrees for double lacing and 60 de- grees for single lacing. When the distance between the rivet lines in the flanges is more than 15 inches, the lacing shall be double and riveted at the intersection if bars are used, or else shall be made of angles.

Section 22. Adjustable Members.

(a) **Initial Stress.**
The total initial stress in any adjustable member shall be assumed as not less than 5,000 pounds.

Section 23. Column Bases.

(a) **Loads.**
Proper provision shall be made to transfer the column loads, and moments if any, to the footings and foundations.

(b) **Alignment.**
Column bases shall be set level and to cor- rect elevation with full bearing on the masonry.

(c) **Finishing.**
Column bases shall be finished to accord with the following requirements:

1. Rolled steel bearing plates, 2" or less in thickness, may be used without plan- ing, provided a satisfactory contact bear-

ing is obtained; rolled steel bearing plates, over 2", but 4" or less in thickness, may be straightened by pressing (planed on all bearing surfaces if presses are not available) to obtain a satisfactory contact bearing; rolled steel bearing plates, over 4" in thickness, shall be planed on all bearing surfaces (except as noted under 3).

2. Column bases other than rolled steel bearing plates shall be planed on all bearing surfaces (except as noted under 3).
3. The bottom surfaces of bearing plates and column bases which rest on masonry foundations and are grouted to insure full bearing contact need not be planed.

Section 24. Anchor Bolts.

(a) **Anchor Bolts.**

Anchor bolts shall be of sufficient size and number to develop the computed stress.

PART VI. FABRICATION.

Section 25. Workmanship.

(a) **General.**

All workmanship shall be equal to the best practice in modern structural shops.

(b) **Straightening.**

All material shall be clean and straight. If straightening or flattening is necessary, it shall be done by a process and in a manner that will not injure the material. Sharp kinks or bends shall be cause for rejection.

(c) **Heating.**

Rolled sections, except for minor details, shall preferably not be heated or, if heated, shall be annealed but this restriction does not apply to gas cutting, Section 25 (1).

(d) **Holes.**

Holes for rivets or unfinished bolts shall be $\frac{1}{16}$ inch larger than the nominal diameter of the rivet or bolt. If the thickness of the material is not greater than the nominal diameter of the rivet or bolt plus ⅛ inch, the holes may be punched. If the thickness of the material is greater than the nominal diameter of the rivet or bolt plus ⅛ inch, the holes shall be either drilled from the solid, or sub-punched and reamed. The die for all sub-punched holes, and the drill for all sub-drilled holes, shall be $\frac{1}{16}$ inch smaller than the nominal diameter of the rivet or bolt.

Holes for turned bolts shall be 1/50 inch larger than the external diameter of the bolt. If the bolts are to be inserted in the shop, the holes may be either drilled from the solid, or sub-punched and reamed. If the bolts are to be inserted in the field, the holes shall be sub-punched in the shop and reamed in the field. All drilling or reaming for turned bolts shall be done after the parts to be connected are assembled.

Drifting to enlarge unfair holes shall not be permitted. Holes that must be enlarged to admit the rivets shall be reamed. Poor matching of holes shall be cause for rejection.

(e) **Planing.**

Planing or finishing of sheared plates or shapes will not be required unless specifically called for on the drawings.

(f) **Assembling.**

All parts of riveted members shall be well pinned or bolted and rigidly held together while riveting. Drifting done during assembling shall not distort the metal or enlarge the holes.

(g) **Riveting.**

All rivets are to be power-driven hot. Rivets driven by pneumatically or electrically operated hammers are considered power-driven. Standard rivet heads shall be of approximately hemispherical shape and of uniform size throughout the work for the same size rivet, full, neatly finished, and concentric with the holes. Rivets after driving, shall be tight, completely filling the holes, and with heads in full contact with the surface. Rivets shall be heated uniformly to a temperature not exceeding 1950° F.; they shall not be driven after their temperature is below 1000° F.

Loose, burned, or otherwise defective rivets shall be replaced.

(h) **Finishing.**

Compression joints depending upon contact bearing shall have the bearing surfaces truly machined to a common plane after the members are riveted. All other joints shall be cut straight.

(i) **Lacing Bars.**

The ends of lacing bars shall be neat and free from burrs.

(j) **Tolerances.**

Finished members shall be true to line and free from twists, bends, and open joints.

Compression members may have a lateral variation not greater than 1/1000 of the axial length between points which are to be laterally supported.

A variation of $\frac{1}{32}$ inch is permissible in the overall length of members with both ends milled.

Members without milled ends which are to be framed to other steel parts of the structure may have a variation from the detailed length not greater than $\frac{1}{16}$ inch for members 30 feet or less in length, and not greater than ⅛ inch for members over 30 feet in length.

(k) **Castings.**

All steel castings shall be annealed.

(l) **Gas Cutting.**

The use of a cutting torch is permissible if the metal being cut is not carrying stresses during the operation. To determine the effective width of members so cut, ⅛ inch shall be deducted from each gas cut edge. The radius of re-entrant gas cut fillets shall be as large as practicable, but never less than 1 inch.

Section 26. Shop Painting.

(a) **Shop Coat.**

Before leaving the shop all steel work shall be thoroughly cleaned, by effective means, of all loose mill scale, rust and foreign matter. Except where encased in concrete, all steel work shall be given one coat of approved metal protection, applied thoroughly and evenly and well worked into the joints and other open spaces. All paint shall be applied to dry surfaces.

(b) **Inaccessible Parts.**

Parts inaccessible after assembly shall be given two coats of shop paint, preferably of different colors.

(c) **Contact Surfaces.**

Contact surfaces shall be cleaned by effective means, before assembly, but not painted.

(d) **Finished Surfaces.**

Machine finished surfaces shall be protected against corrosion by a suitable coating.

PART VII. ERECTION.

Section 27. Erection.

(a) **Bracing.**

The frame of all steel skeleton buildings shall be carried up true and plumb, and temporary bracing shall be introduced wherever necessary to take care of all loads to which the structure may be subjected, including erection equipment, and the operation of same. Such bracing shall be left in place as long as may be required for safety.

(b) **Bolting Up.**

As erection progresses the work shall be securely bolted up to take care of all dead load, wind and erection stresses.

(c) **Erection Stresses.**

Wherever piles of material, erection equipment or other loads are carried during erection, proper provision shall be made to take care of stresses resulting from the same.

(d) **Alignment.**

No riveting shall be done until the structure has been properly aligned.

(e) **Riveting.**

Rivets driven in the field shall be heated and driven with the same care as those driven in the shop.

(f) **Turned Bolts.**

Holes for turned bolts to be inserted in the field shall be reamed in the field as specified in Section 25 (d).

(Continued on Page 649)

CODE OF STANDARD PRACTICE FOR STEEL STRUCTURES OTHER THAN BRIDGES

As Adopted June, 1937.

AMERICAN INSTITUTE OF STEEL CONSTRUCTION

This Document Includes the A. I. S. C. Standard Riders for Delivered and for Erected Work and Standard Release and Indemnity Agreement.

Section 1. General.

(a) **Scope.**

The rules and practices hereinafter defined are adopted by the American Institute of Steel Construction as standard for the industry and shall govern all conditions where the contract between the Buyer and Seller does not specify otherwise and where they do not conflict with local or state requirements.

(b) **Design, Fabrication and Erection.**

Unless otherwise specified or required, the Standard Specification for the Design, Fabrication and Erection of Structural Steel for Buildings of the American Institute of Steel Construction, as amended to date, shall apply.

(c) **Plans and Specifications for Bidding.**

The plans shall show a complete design with sizes, sections and the relative location of various members with floor levels column centers and offsets figured, and shall show the character of the work to be performed with sufficient dimensions to permit the making of an accurate estimate of cost. Plans shall be made to scale not less than ⅛ inch to the foot, and large enough to convey the information adequately.

Wind bracing and special details when required shall be shown in sufficient detail regarding rivets and construction to permit an accurate estimate of cost.

(d) **Responsibility for Design and Erection.**

If the design, plans and specifications are prepared by the Buyer, the Seller shall not be responsible for the suitability, strength and rigidity of design or the practicability or safety of erection.

Section 2. Classification.

The Steel and Iron items entering into the construction of a structure are divided into the following classes:

CLASS "A"—Structural Steel and Iron.
CLASS "B"—Ornamental Steel and Iron.
CLASS "C"—Steel Floor Joists.
CLASS "D"—Miscellaneous Steel and Iron.

In contracting to furnish the material for a structure where the material to be furnished is designated as structural steel and iron, ornmental steel and iron, steel floor joists, or miscellaneous steel and iron, the Seller will furnish only such items under each classification as are listed below, and no other items will be included unless by special agreement. In cases where materials in excess of minimum requirements are furnished to provide for waste or loss, all unused material remaining after completion of work shall be the property of the Seller and returned to him.

Unless specifically agreed to in the contract, the Seller of the structural steel "Class A" will not provide field connections or field holes for the ornamental steel and iron "Class B", the miscellaneous steel and iron "Class D", nor the materials for any other trades.

(a) **Class "A" Structural Steel and Iron.**

Contracts taken to furnish the structural steel and iron for a building are based on furnishing the following items only:

Anchors for structural steel only
Bases of steel or iron only
Beams of rolled structural steel
Bearing plates for structural steel
Brackets made of structural steel sections
Channels of rolled structural steel
Channels and angle supports only for suspended ceilings where they attach to structural steel, but not including small channel or angle furring
Columns, structural steel, cast iron and pipe
Girders of structural steel
Grillage beams and girders — structural steel
Hangers of structural steel
Lintels as shown or enumerated
Marquees (structural frame only)
Rivets and bolts for field connections, as follows:

1. The Seller shall furnish sufficient rivets of suitable size, plus at least 10 per cent to cover waste for all field connections of steel to steel which are designated as riveted field connections.
2. The Seller shall furnish sufficient bolts of suitable size, plus 5 per cent to cover waste for all field connections of steel to steel which are designated to be bolted.
3. No fitting-up bolts or washers will be included unless specifically called for.

Separators, angles, tees, clips, bracing and detail fittings in connection with structural steel frame
Tie rods
Trusses of structural steel.

(b) **Class "B" Ornamental Steel and Iron.**

Contracts taken to furnish the ornamental steel and iron for a building are based on furnishing the following items only:

All bronze and brass work, except hardware fittings
Balconies
Cast iron cornices
Curtain guides
Elevator fronts and enclosures
Grilles and gratings
Iron store fronts
Lamp standards and brackets
Marquees (steel or iron, except frame) see Class "A"
Ornamental brackets, steel or iron
Ornamental inside stairs, steel or iron
Ornamental outside steel or iron stairs, including fire escapes
Safety treads
Railings (gas pipe, ornamental or brass)
Sills and thresholds (brass, steel or iron)
Spiral stairs, steel or iron
Window sills and frames, steel or iron
Wire work, ornamental steel or iron

(c) **Class "C" Steel Floor Joists.**

Contracts taken to furnish the steel floor joists for a building are based on furnishing the following items only:

Steel joists which are not a part of the structural steel frame for the building, and which are devised to carry the floor or roof panels
Bracing and bridging for floor joists, clips for fastening floor joists
Stirrup and hanger for floor joists
Ties for floor joists

(d) **Class "D" Miscellaneous Steel and Iron.**

The nature and character of the material of this classification makes it impossible to cover all items and it is recommended that the Seller taking the contract to furnish the miscellaneous steel and iron work for a building specify all items in detail which it is intended to furnish. The general list of items under this classification is as follows:

Area gratings
Cast iron cover and frames
Cast iron rainwater receivers
Cast iron downspout shoes
Cleanouts
Coal chutes
Column guards
Door frames and bucks
Foot scrapers
Furnace or fireplace dampers
Flag pole
Ladders
Pin rails
Sidewalk doors

Sills and curb angles, and anchors for same	Steel and cast iron platforms
Special bolts or anchors where distinctly shown on the plans	Steel or iron chimney caps
Stairs made of plain structural steel — not including treads of other materials	Thimbles
Stacks	Wall plate anchors
	Wheel guards
	Window guards
	Wire screens for partions, door and window guards (this does not include fly screens)

(e) **Materials not classed under above headings.**

The following items are not coverd by classifications A-B-C and D and will in no case be furnished by the Seller unless specifically agreed to and mentioned in the contract. It is not possible to designate every detail and the list is typical of material not included in classifications A-B-C and D. It is shown here to assist the Architect and Engineer in avoiding confusion:

Ash hoists	Rolling doors
Awning boxes	Sheet metal work or corrugated sidings and roofing
Boilers	Sidewalk lights
Elevators or accessories	Steel sash and steel sash partitions
Elevator guides or sheave beams	Spiral slides
Expanded metal	Suspended ceiling, except as noted under Class "A"
Furring	Tanks and pans
Glass for any purpose whatever	Toilet partitions
Hollow metal doors or frames	Treads, except steel or iron
Hoppers	Vault doors
Mail chute	Ventilating brick
Metal lockers	Wall, ceiling and floor registers
Miscellaneous carpenter or masonry bolts for connecting wood to wood, steel to wood, or wood to stone, etc.	Wood handrails
Name plates	Wood handrail brackets
Patented devices	And all other material not mentioned.
Reinforcing steel	

Section 3. Invoicing.

When conditions make it possible to award contracts on a lump sum basis the confusion of determining weights will be avoided. Scale weights involve a variation which frequently lead to a compromise based on calculated weights.

The rules hereinafter established, while not giving exact weights, are the basis upon which the Seller must make a lump sum or a pound price bid and they eliminate the necessity of increased cost of shop drawings and other refinements of manufacture which would very materially increase costs if exact weights were required.

(a) **Weights.**

Structural steel and iron sold at a unit price per pound, hundred weight (100 pounds) or ton (2000 pounds) shall be invoiced on the calculated weights of shapes, plates, bars, castings, rivets and bolts, based on the detailed shop drawings and shop bills of material which show actual dimensions of materials used as follows:

Dimensions:—

The weight will be figured on the basis of rectangular dimensions for all plates, and ordered overall lengths for all structural shapes and with no deductions for copes, clips, sheared edges, punchings, borings, milling or planing. When parts can be economically cut in multiples from material of larger dimension the calculated weight shall be taken as that of the material from which the parts are cut.

Over-run, as follows:—

1. To the nominal theoretical weight of all universal mill and sheared plates or slabs will be added one-half the allowance for variation or over-weight in accordance with the specifications of the American Society for Testing Materials. For the determination of over-run all plates less than 5 feet in length shall be classed as sheared plates. (See table in A. S. T. M. Specification.)
2. Reinforcing bars when not sold on a basis of scale weights shall be invoiced by the Seller at the theoretical weights plus 1½ per cent to allow for over-run weight of deformations, etc.
3. The calculated weights of castings shall be the weights determined from the drawings of the pieces including standard fillets for such pieces. To this an average over-run of 10 per cent shall be added.

Rivets, as follows:—

1. The weight of shop rivets will be based on the weights shown in the following table:
   - Rivets ½" in diameter 20 pounds per 100 rivets
   - Rivets ⅝" in diameter 30 pounds per 100 rivets
   - Rivets ¾" in diameter 50 pounds per 100 rivets
   - Rivets ⅞" in diameter 100 pounds per 100 rivets
   - Rivets 1" in diameter 150 pounds per 100 rivets
   - Rivets 1⅛" in diameter 250 pounds per 100 rivets
   - Rivets 1¼" in diameter 325 pounds per 100 rivets
2. Field rivets and bolts shall be invoiced at their actual weight.

Painting and Galvanizing:—

A percentage of the theoretical weight of the material protected shall be added as follows:

One-half of 1 per cent for each coat of paint.

One-fourth of 1 per cent for each coat of oil.

3½ per cent for galvanizing by hot dipping.

Welds, as follows:—

The following weights include allowance for over-run in size of welds, but no allowance for waste.

WEIGHTS OF STANDARD FILLET WELDS

Specified Size, W, Inches	Recommended Weight, Lbs. per Lin. Ft.
¼	.20
5/16	.25
⅜	.35
½	.55
⅝	.80
¾	1.10
⅞	1.50
1	2.00

Section 4. Drawings and Specifications.

(a) To enable the Seller to enter properly upon the execution of the work the Buyer shall furnish the Seller within a time agreed to in the contract, a survey of the lot lines, together with a complete and full design of the structural steel frame definitely locating all openings, levels, etc.; and showing all material to be furnished by the Seller with such information as may be necessary for the completion of the shop drawings by the Seller. All such information and drawings shall be consistent with the original drawings and specifications. Any expense caused by changes or omissions in such information and drawings, referred to above shall be at the Buyer's expense.

(b) In case of discrepancies between the drawings and the specifications prepared by either the Seller or the Buyer, the specification shall govern; and in case of discrepancies between the scaled dimensions of the drawings and the figures written on them, the figures shall govern.

Should the Seller in the regular progress of his work find discrepancies in the information furnished by the Buyer, he shall refer such discrepancies to the Buyer before proceeding further with work which would be affected.

The Seller shall be reimbursed by the Buyer for damage resulting from changed manufacturing operation caused by delays.

(c) Shop drawings shall be made and submitted to the Buyer's representative who shall examine the same and in five days return them approved with such corrections as he may find necessary. They shall be corrected by the Seller if necessary and returned for the Buyer's file as finally approved. The Seller may proceed with shop work, but in so doing he shall assume responsibility for having properly made the corrections indicated by the Buyer.

In addition to the set of blue prints of approved shop drawings for the Buyer's file as above referred to, the Buyer may require the Seller to furnish without cost to the Buyer, one additional set of shop drawing blue prints, but any further additional sets shall be paid for by the Buyer. All drawings or tracings made by the Seller for the execution of his work shall remain his property unless otherwise specifically agreed to.

(d) Shop Drawings prepared by the Seller and approved by a representative of the Buyer shall be deemed the correct interpretation of the work to be done, but does not relieve the Seller of responsibility for the accuracy of details.

(e) After the shop drawings have been "approved" or "approved as noted" by the authority designated in the contract, any changes required by said authority shall be made at the expense of the Buyer.

(f) When detailed shop drawings are furnished by the Buyer no responsibility for misfits due to errors in the drawings will be assumed by the Seller.

Section 5. Stock Material.

Stock material shall be of a quality substantially equal to that called for by the specifications of the American Society for Testing Materials for the classifications covering its intended use and mill test reports shall constitute sufficient record as to the quality of material carried in stock. It is obviously impossible for the Seller to maintain records of heat or blow numbers of every piece of material in his stock, and the same shall not be required if all his stock purchases are made under an established specification as to grade and quality.

Whenever a shop maintains such a practice in carrying a stock of material, it is deemed good practice to permit the use of such stock material in its fabricating operations whenever the shop desires to do so, instead of ordering items from the mill for a specific operation. Stock materials bought under no particular specifications, or under specifications materially less rigid than those mentioned above, or stock material which has not been subject to mill or other recognized test reports, shall not be used, except as noted below, without the approval of the Buyer and under rigid inspection.

It is permitted to use unidentified stock material free from surface imperfections for short section of minor importance or for small unimportant details, where the quality of the material could not affect the strength of the structure.

Section 6. Inspection and Delivery.

(a) **Inspection.**

The Seller's shop service includes inspection by his own inspectors, and shop or mill inspection other than this shall be paid for by the Buyer.

(b) **Acceptance of Materials.**

When material is inspected by a representative of the Buyer at the Shop, the acceptance of such material by the Buyer's representative shall be considered the Buyer's final approval; but the Seller shall be responsible for the accuracy of the work and for defective materials or workmanship which may be discovered and condemned within a reasonable [illegible] after its incorporation into the structure.

(c) **Order of Delivery.**

Unless the order or sequence of delivery is specifically arranged for before the work is undertaken, it will be at the convenience of the Seller.

(d) **Materials sold delivered.**

When material is sold delivered on cars or trucks at the site of the structure, all unloading shall be done by the Buyer, and all responsibility to persons or property during such unloading shall be at the Buyer's risk.

(e) **Loss in shipment where material is sold fabricated only.**

The quantities and weights of material shown by the shipping statement will in all cases govern settlements unless notice of shortage is immediately reported to the agent of the delivering carrier, and his signed verification obtained, and like notice sent to the Seller within 48 hours after receipt of the shipment, in order that the alleged shortage may be investigated by the Seller.

(f) **Storage of Material.**

Where conditions make it necessary that material be stored for any length of time, and the contract does not provide for such storage, payments are to come due and be payable the same as if the material had been delivered at the building site; and the Seller shall be compensated for handling, storage, and other increased expenses that may result from such conditions.

Section 7. Erection.

(a) **Method of Erection.**

The Seller shall base his price on the most economical method of erection, consistent with the plans and specifications and such information as may have been furnished prior to the execution of the contract, and the interposition of erection conditions other than those specified or agreed in the contract, shall entitle the Seller to an addition to the contract price to compensate for such changes.

(b) **Foundations.**

The Buyer shall be responsible for the location, strength and suitability of the foundations.

(c) **Building Lines and Bench Marks.**

Building lines and bench marks at the site of the structure shall be accurately located by the Buyer, who shall furnish the steel erector a plot plan containing all such information.

(d) **Steel and Cast Iron Bases.**

All steel grillage, rolled steel bearing plates, cast iron, or steel bases, shall be set and wedged or shimmed by the steel erector to grade or level lines, which shall be determined and fixed by the Buyer, who shall grout all such parts in place. Before grouting the Buyer shall check the grades and levels of the parts to be grouted, and shall be responsible for the accuracy of the same. For steel columns or girders with bases fabricated as an integral part of the member the foundation shall be finished to exact grade and level, ready to receive the steel work so that shims or wedges shall not be required for plumbing or leveling of steel work.

(e) **Anchor Bolts.**

All anchor or foundation bolts, expansion bolts, cinch bolts or other connections between the work to be performed by the Seller and the work of other trades shall be located and set by the Buyer.

(f) **Working Room.**

The erection contractor shall be entitled to sufficient space at the site of the structure at a place convenient to him to place his derricks and other equipment necessary for erection. When conditions at the site provide working space not occupied by the structure, the erection contractor shall be entitled to storage space for sufficient material to keep his working force in continuous operation.

Before the date set for starting erection the Buyer shall complete and have ready all foundations, accessible and free from obstructions.

(g) **Plumbing Up.**

The temporary guys and braces shall be the property of the Seller, and if after the steel has been plumbed and leveled, the work of completing the structure by other contractors is suspended or delayed, the owner of the temporary guys and braces shall receive reasonable compensation for their use. The guys shall be removed by the Buyer at his expense, and returned to the Seller in as good condition as when placed in the building with a reasonable depreciation.

Immediately upon completion by the steel erector, the Buyer shall assure himself by whatever agencies he may elect, that the steel erector's work is plumb and level, and properly guyed. If it is not, he should immediately notify the erector and direct him to perfect his work. After the steel erector has guyed and plumbed the work once to the satisfaction of the Buyer, his responsibility ceases. Any further work in guying or plumbing shall be performed entirely at the Buyer's expense.

In the setting or erecting of structural steel work, the individual pieces shall be considered plumb or level where the error does not exceed 1 to 500.

For exterior columns and columns adjacent to elevator shafts of multiple story building, the error from plumb shall not exceed 1 to 1000 for the total height of the column.

(h) **Opportunity to Investigate Errors.**

Correction of minor misfits and a reasonable amount of reaming and cutting of excess stock from rivets will be considered as a legitimate part of erection. Any error in shop work which prevents the proper assembling and fitting up of parts by the moderate use of drift pins, or a moderate amount of reaming and slight chipping or cutting, shall be immediately reported to the Seller and his approval of the method of correction obtained.

(i) **Wall Plates.**

All loose masonry bearing plates for beams, lintels, trusses or columns shall be set and grouted to grade and line by the Buyer ready for the steel erector to set his work, without unnecessary interruption or delay.

(j) **Loose Lintels.**

Loose lintels or pieces of all kinds and descriptions required by the design of a building to carry brick work over openings, and which can be set without a motor hoisting apparatus, and which lintels or pieces are not attached in any way to the rest of the steel structure, and cannot be placed except as the masonry work advances, will not be erected by the steel erector unless by special agreement.

(k) **Ornamental Iron and Bronze.**

Fine ornamental iron and bronze work is considered as finishing material, and shall not be set in a building until after the marble, plaster, and other work, except decorating, is in place.

(l) **Elevator Framing.**

The setting or erection of guides, cars, machinery, cables, sheaves, pans, etc., for elevators, is not to be required of the steel erector.

(m) **Field Assembling.**

The size of assembled pieces of structural steel is fixed by the permissible weight and clearance dimensions of transportation. Unless such conditions are provided for by the Buyer or his engineer, the Seller shall provide for such field connections as will require the least field work; and such field connections shall be a part of the erection work.

(n) **Cutting. Drilling and Patching.**

The Seller shall not be required to cut, drill or patch the work of others or his own work for the accommodation of other trades except at the Buyer's expense, unless such cutting, drilling or patching of the steel for the accommodation of other trades as shall be specified as part of the Seller's Work shall be shown on the plans so that it may be done at the time of fabrication or paid for as extra work. The number, size and location of any tapped holes shall be given at the time of the signing of the contract or be paid for as extra work.

(o) **Insurance.**

Until completion and acceptance of the work the Seller shall maintain workmen's compensation insurance as may be required by law, as well as the standard form of public liability insurance, in such limits as may be agreed upon between the Buyer and Seller, protecting the Seller against claims for damages for personal injuries or death arising out of his negligent acts or omissions. All other forms of insurance coverage shall be provided by the Buyer, who shall protect the Seller against loss of or damage to the work performed and material delivered at the site. All insurance covering such loss or damage shall be payable to the parties as their interests may appear. In no event shall the Seller indemnify the Buyer against loss or expense other than by reason of and to the extent of, the liability imposed by law upon the Seller for damages resulting from death or injury to persons or destruction of property occasioned by the prosecution of the work.

(p) **Temporary Floors.**

The Buyer shall provide plank, and cover all floors required by municipal or state laws, excepting the floor upon which the erecting derricks are located. This floor will be covered by the steel erector for working purposes.

(q) **Field Paint.**

Unless specifically agreed to in the contract, field paint shall be considered a phase of maintenance, and shall be provided for by the Buyer.

Section 8. Delays in Prosecution of Work.

(a) **Causes not controlled by Seller of Buyer.**

Neither the Buyer nor the Seller shall be responsible for delays in performance caused by delays in transportation by common carriers or at rolling mills. In case of delay to work due to any of the above causes, a reasonable extension of time shall be given for the completion of the work.

(b) **Delays caused by the Seller.**

Should the Seller at any time, except as provided in the preceding paragraphs, refuse or neglect to supply enough workmen of proper skill or material of proper quality, or to carry on the work with promptness and diligence, the Buyer, if not in default, may give the Seller ten days' written notice, and at the end of that time if the Seller continues to neglect the work the Buyer may provide such labor or materials and deduct a reasonable cost from any money due or to become due the Seller under the contract, or may terminate the employment of the Seller under the agreement and take possession of the premises and of all materials, tools, and appliances thereon and employ any other person to finish the work. In the latter case, the Seller shall receive no further payment until the work be finished; then if the unpaid balance that would be due under the contract exceeds the cost to the Buyer of finishing the work, such excess shall be paid to the Seller, but if such cost exceeds unpaid balance, the Seller shall pay the excess to the Buyer.

(c) **Delays caused by the Buyer.**

Should the Buyer fail to furnish the plans and other data described in Section 4 (a) at the time agreed, or should he delay or obstruct the work in any way so as to cause loss or damage to the Seller, he shall reimburse the Seller for such loss or damage.

Should the Buyer cancel the contract or delay the fabrication or erection of the work for more than thirty (30) days at any time, the Seller may, upon five (5) days' written notice to the Buyer, terminate the contract and tender all undelivered material, whether worked or unworked, to the Buyer, who shall, upon transfer of title to such material, pay all cost or expense which the Seller shall have paid or be obligated to pay, together with all loss or damage to the Seller, including but not limited to, the value of all drawings prepared, materials purchased or fabricated, or shipped, stored, delivered, erected or in process of fabrication or erection, together with the value of the labor performed, and the loss of profits suffered by the Seller. Should the Buyer fail to accept such material, the Seller shall, at the expiration of such five (5) day period, in order to save the continuing storage and handling charges, proeeed to sell all materials in his possession and control at not less than the then market value for scrap material and shall credit the sums received therefor against the amounts due from the Buyer as above set forth.

The contract price and contract dates of delivery and erection shall not be binding upon the Seller unless the Buyer shall provide the information and work to be performed by him at the time and in the manner agreed.

Section 9. Extra Work.

(a) **General.**

Charges for extra work, or work not covered by the contract, shall be made on a basis that is definitely and mutually understood between the Buyer and the Seller at the time the occasion for such extra expense arises.

In the absence of such an understanding between the Buyer and Seller, the following are listed as proper items to be included.

(b) **Material.**

All extra material required shall be invoiced out at current warehouse prices, plus cost of fabrication, including regular overhead, insurance and administration costs, plus transportation costs, and ten per cent for profit.

(c) **Drafting Labor.**

All extra labor in the drafting room shall be invoiced out at cost plus overhead, insurance and administration costs plus ten per cent for profit.

(d) **Shop Work.**

All extra shop labor shall be charged at actual cost as shown by the time cards; to this shall be added the overhead, insurance and administration expense. The sum of these charges shall be considered the actual shop cost, to which shall be added ten per cent for profit.

(e) **Field Work.**

All extra labor required in the erection of structural steel shall be invoiced as follows:

The actual labor cost shall be that shown by the time cards, to which shall be added the cost of insurance at established rates, the cost of labor transportation when necessary, and an additional allowance for overhead, and administration expense. The sum of these shall be considered the actual cost, to which shall be added ten per cent for profit.

Use of the Seller's equipment by the Buyer for work not included in the Seller's contract shall be billed at proper rental rates. Hoists for other trades employed on the structure shall be billed at a rate per hoist to be agreed upon.

(f) **Miscellaneous.**

Any additional cost, such as hauling, painting, crating, freight, etc. shall be charged at actual cost plus overhead, plus administration, plus insurance, plus ten per cent for profit.

(g) **Overtime.**

On contract work where the Seller has not agreed to work overtime he shall not be required to do so without being paid for his extra expense and a profit.

(h) **Extra Cleaning.**

If because of continued storage, or for any other reason not the fault of the Seller, it becomes necessary to clean and repaint the steel work, the cost of this additional cleaning and painting shall be paid for as an extra, including regular overhead charges as specified for extra work elsewhere in this section.

Section 10. Proposals and Contracts.

(a) **Conflicts.**

In the event of a conflict between the terms and conditions of the proposal, or contract, and the terms and conditions stated in the plans and specifications, the terms of the proposal or contract shall govern.

(b) **Price for Additions or Deductions.**

The Seller is not to be required nor expected to make the same unit price for additions to as for deductions from the work included in the contract. Such unit prices when made are intended to apply to incidental changes made in the design drawings prior to the ordering of the material.

The price of all changes in the work which may be required after the material has been ordered should be agreed upon at the time such changes are made, or paid for as extra work as defined in this code.

(c) **Material not shown or called for.**

Clauses in the specification to the effect that all steel and iron items necessary to complete the structure shall be furnished by the Seller, whether or not they are shown on the plans or called for in the specifications, being obviously unfair, will not be recognized or subscribed to. The Seller shall, however, furnish all material and labor for details that may be required for such steel and iron work as is shown on the drawings or called for in the specification, although such details may themselves not be shown or called for.

(d) **Items not to be furnished.**

Unless specifically mentioned in the request for bids, or specifically agreed to, the bidders do not estimate or include the following items in their proposals:

Any charges for surety bonds or insurance not required by law, or any other general charge such as building pemits, license fees or taxes for permission to work in city or state, engineering fees, removal of rubbish, cutting, patching or repairing of plaster or masonry work, office or telephone service, light, heat, fire insurance, or the erection of temporary structures enclosures or stairs, and the like.

(e) **Terms.**

The following terms of payment are adopted as standard and shall govern in all cases, except when otherwise agreed to in the contract.

1. All payments shall be made in funds current at part in the Federal Reserve Bank District in which the principal office of the Seller is located.

2. All materials for export, net cast in exchange for shipping documents shall be required.

3. For all materials to be erected by the Seller, the Buyer shall on the 10th day of each month pay an amount equal to not less than 90 per cent of the contract value of all materials shipped, stored or ready for shipment; and not less than 90 per cent of the contract value of the erection performed during the preceding month; and shall pay the remainder within 30 days after the completion of the steel contract; but the amount reserved by the Buyer shall at no time exceed double the contract value of work remaining yet to be done.

4. When the material which is not to be erected by the Seller is sold to a Buyer whose credit has been established with the Seller, terms shall be net cash for contract value of each shipment; payments to be made on the 10th day of the month following shipment.

5. Unless otherwise agreed to, when material is sold delivered at, or freight is allowed to destination, the Buyer shall pay freight charges and the Seller shall accept receipted freight bills as cash to apply on matured payments due on or after arrivals at destination of materials covered by such freight expense bills.

6. Payments shall all be considered to be due and shall be paid at the time specified, regardless of the final settlement for the building as a whole, or for the work of any other trade; and when the contract is with a general contractor the payment for steel shall not be delayed by such general contractor pending his receiving estimates of payments from the owner.

7. Amounts past due shall bear interest at the maximum lawful rate.

STANDARD SPECIFICATIONS FOR RAIL-STEEL CONCRETE REINFORCEMENT BARS

American Society for Testing Materials.

A.S.T.M. Designation: A16-35

Scope.

1. These specifications cover three classes of rail-steel concrete reinforcement bars: namely, plain, deformed, and hot-twisted.

MANUFACTURE.

Manufacture.

2. The bars shall be rolled from standard section Tee rails. No other materials such as those known by the terms "rerolled," "rail-steel equivalent," and "rail-steel quality" shall be substituted.

Hot-twisted Bars.

3. Hot-twisted bars shall have one complete twist in a length not over 12 times the thickness of the bar.

PHYSICAL PROPERTIES AND TESTS.

Tension Tests.

4. (a) The bars shall conform to the following minimum requirements as to tensile properties:

	Plain Bars	Deformed and Hot-twisted Bars
Tensile strength, lb. per sq. in.	80,000	80,000
Yield point, lb. per sq. in.	50,000	50,000
Elongation in 8 in., per cent	1,200,000* / tens. str.	1,000,000* / tens. str.

* See Section 5.

(b) The yield point shall be determined by the drop of the beam or halt in the gage of the testing machine.

Modifications in Elongation.

5. (a) For bars over ¾ in. in thickness or diameter, a deduction from the percentages of elongation specified in Section 4 (a) of 0.25 per cent shall be made for each increase of 1/32 in. of the specified thickness or diameter above ¾ in.

(b) For bars under 7/16 in. in thickness or diameter, a deduction from the percentages of elongation specified in Section 4 (a) of 0.5 per cent shall be made for each decrease of 1/32 in. of the specified thickness or diameter below 7/16 in.

Bend Tests.

6. (a) The test specimen shall stand being bent cold around a pin without cracking. The following requirements for degree of bending and sizes of pins shall be observed:

Thickness or Diameter of Bar	Plain Bars.	Deformed and Hot-twisted Bars
Under ¾ in.	180 deg. d=3 t	180 deg. d=4 t
¾ in. or over	90 deg. d=3 t	90 deg. d=4 t

Explanatory Note: d = the diameter of pin about which the specimen is bent; t = the thickness or diameter of the specimen.

(b) Bend tests shall be made on specimens of sufficient length to insure free bending and with apparatus which provides:

(1) Continuous and uniform application of force throughout the duration of the bending operation;

(2) Unrestricted movement of the specimen at points of contact with the apparatus:

(3) Close wrapping of the specimen about the pin or mandrel during the bending operation.

(c) Other methods of bend testing may be used but failures due to such methods shall not constitute a basis for rejection.

Test Specimens.

7. (a) Tension and bend test specimens from plain or deformed bars shall be of the full section of bars as rolled. For tension tests of deformed bars the sectional area used for unit stress determination shall be calculated from the length and weight of the test piece.

Note: The area in square inches may be calculated by dividing the weight per linear inch of specimen in pounds by 0.2833 (weight of 1 cu. in. of steel); or by dividing the weight per linear foot of specimen in pounds by 3.4 (weight of steel 1 in. square, 1 ft. long).

(b) Tension and bend test specimens of hot-twisted bars shall be taken from the finished bars, without further treatment.

Number of Tests.

8. (a) One tension and one bend test shall be made from each lot of ten tons or less of each size of bar rolled from rails varying not more than 10 lb. per yd. in nominal weight.

(b) If any test specimen develops flaws, it may be discarded and another specimen substituted.

(c) If the percentage of elongation of any tension test specimen is less that that specified in Section 4 (a) and any part of the fracture is outside the middle third of the gage length, as indicated by scribe scratches marked on the specimen before testing, a retest shall be allowed.

PERMISSIBLE VARIATIONS IN WEIGHT.

Permissible Variations in Weight.

9. The weight of any lot (Note) of bars shall not vary more than 3½ per cent over or under the theoretical weight for bars ⅜ in. and over in diameter; nor more than 5 per cent over or under for bars under ⅜ in. in diameter. The weight of any individual bar shall not vary more than 6 per cent under the theoretical weight for bars ⅜ in. and over in diameter; nor more than 10 per cent under the theoretical weight for bars under ⅜ in. in diameter. The theoretical weight of deformed bars shall be the theoretical weight of plain round or square bars of the same nominal size.

Note: The term "lot" used in this paragraph means all the bars of the same nominal weight per linear foot in a carload.

FINISH.

Finish.

10. The finished bars shall be free from injurious defects and shall have a workmanlike finish.

INSPECTION AND REJECTION.

Inspection.

11. The inspector representing the purchaser shall have free entry, at all times while work on the contract of the purchaser is being performed, to all parts of the manufacturer's works which concern the manufacture of the bars ordered. The manufacturer shall afford the inspector, without charge, all reasonable facilities to satisfy him that the bars are being furnished in accordance with these specifications. All tests and inspection shall be made at the place of manufacture prior to shipment, unless otherwise specified, and shall be so conducted as not to interfere unnecesarily with the operation of the works.

Rejection.

12. Bars which show injurious defects subsequent to their acceptance at the manufacturer's works will be rejected, and the manufacturer shall be notified.

[1] Under the standardization procedure of the Society, these specifications are under the jurisdiction of the A.S.T.M. Committee A-1 on Steel.

CODE FOR FUSION WELDING AND GAS CUTTING IN BUILDING CONSTRUCTION

AMERICAN WELDING SOCIETY

Extracts from Part A, Structural Steel, Edition of 1934.

FOREWORD

(A. I. S. C. Manual Committee, 1937).

The use of the fusion welding process (sometimes electric arc, sometimes oxy-acetylene) plays an increasing part in the fabrication of structural steel,

(a) as an aid to shop assembling processes ("tack-welding"),

(b) as the final method of joining parts for transfer of stresses, in shop or field ("strength-welding").

The art of designing and detailing for the safe and economical employment of strength-welding is a relatively new and changing one; the economy of a welded structure as compared with a riveted or riveted-and-bolted one, is as yet a subject not for general statements but for estimate of the individual case.

The A. I. S. C. Manual Committee reserves for future editions of this Manual the possibility of publishing tabular data and recommendations for practice in the field of strength-welding.

At present, it suggests that strength-welding of structures be permitted only on condition that the designing, detailing and erection are in the hands of personnel well trained in the best practice of the art. It considers the best exposition of that art is to be found in the reports and recommendations of the American Welding Society, 33 West 39th St., New York City. The "Code for Fusion Welding and Gas Cutting in Building Construction" of that Society, with Appendices, is authoritative with respect to materials, technique and specifications for design.

For convenience, a digest of permissible unit stresses for welded connections, as established by the American Welding Society Code, is appended.

EXTRACTS AND REFERENCES, A. W. S. CODE.

Shear on section through throat of weld. . . . . 11300 pounds per sq. in.

Note: This gives for fillet welds, working values per linear inch of weld as follows:

Leg in Inches	Pounds per Lin. In.
¼	2000
[illegible]	2500
⅜	3000
7/16	3500
½	4000

Tension on section through throat of weld. . . . . 13000 pounds per sq. in.

Compression (crushing) on section through throat of butt weld. . 18000 pounds per sq. in.

"Fiber stresses due to bending shall not exceed the values prescribed above for tension and compression respectively.

"Stress in a fillet weld shall be considered as shear, for any direction of the applied stress.

"Adequate provision shall be made for bending stresses due to eccentricity, if any, in the disposition or section of base metal parts."

The A. W. S. has compiled a code covering electric resistance welding, as applied to shop fabrication of building parts, such as bar joists.

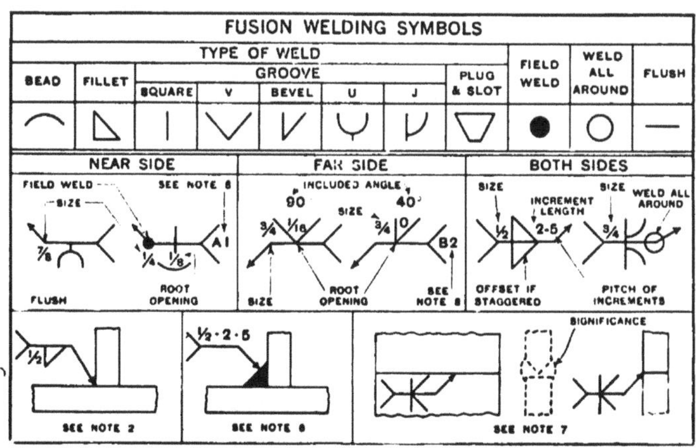

1. In plan or elevation, near, far and both sides locations refer to nearest member parallel to plane of drawing and not to others farther behind.
2. In section or end views only, when weld is not drawn, the side to which arrow points is considered near side.
3. Welds on both sides are of same size unless otherwise shown.
4. Symbols govern to break in continuity of structure or to extent of hatching or dimension lines.
5. All welds are continuous and of user's standard proportions and all except V- and bevel grooved welds are closed unless otherwise shown.
6. When welds are drawn in section or end views obvious information is not given by symbol.
7. In joints in which one member only is to be grooved, arrows point to that member.
8. Tail of arrow used for specification reference.

Legend for Use on Drawing Specifying Fusion Welding.

The above symbols are recommended by American Welding Society for incorporation on drawings specifying fusion welding. For more detailed instruction in the use of these symbols refer to the publications of American Welding Society.

STRUCTURAL CLAY BUILDING TILE SPECIFICATIONS

Application and Installation

1. Work and Materials required to be furnished under this division of the specifications comprehends and includes everything in the way of material, transportation and labor required for furnishing and erecting in place in the building complete ready for use of all structural "Hollow Burned Clay Building Tile" of every sort including all mortar, scaffolding, forms, centers, hauling, hoisting, placement and cleaning up after completion, for clay tile foundations, walls including party division and fire, partitions, furring, fireproofing, floor and roof construction, combination hollow clay tile and concrete floor construction, etc., placed under "General Conditions of the Contract" as defined in documents known as the "Illinois Building Contract Documents" and special general conditions enumerated as follows:

A. Skilled Masons, Mechanics and Laborers, as required for the proper execution of same shall be employed on this construction work who shall be fully and carefully supervised by responsible, competent authority.

B. Apparatus including hoisting devices, machinery, tools, equipment, forms, centering, necessary to carry on this work shall be furnished so as not to unnecessarily delay the progress of the building or other contractors and without retarding the rate of progress stipulated in the contract.

C. Drawings shall be furnished as necessary to accurately locate setting and detail adaptation to structural shapes consisting of large scale details or full sized drawings for all special shapes required, including column coverings, girder covers, lintel covers and general type of arch. Drawings shall be submitted to the architect for approval before tile is burned.

2. Common Physical Characteristics of Structural Clay Tile.—See paragraph 3802.02 Chicago Building Code in this volume.

Color shall not be taken as indicative of classification as different types of clay produces different colors of the same relative hardness and density and different types of clay are used in the manufacture of tile of same physical qualities.

A. HOLLOW BURNED-CLAY LOAD-BEARING BUILDING TILE

Class	Absorption, per cent		Compressive Strength, Based on Gross Area, lb. per sq. in.			
			End Construction		Side Construction	
	Mean of 5 tests	Individual Maximum	Mean of 5 tests	Individual Minimum	Mean of 5 tests	Individual Minimum
Hard	12 or less	15	2000 or more	1400	1000 or more	700
Medium	16 or less	19	1400 or more	1000	700 or more	500
Soft	25 or less	28	1000 or more	700	500 or more	350

B. End Construction Tile when used on the side shall meet the requirements of that construction and vice versa.

C. Masonry Strength for Clay Tile Walls of varying thickness shall vary uniformly in proportion to wall area. Tile for different wall thicknesses shall be designed to meet this requirement.

D. Dry Weight for tiles of various sizes and cell numbers shall not vary more than 5% from the enumeration which follows:

Size of Unit in.	Number of Cells	Standard Weight Lb.
4 by 12 by 12..............	3	20
6 by 12 by 12..............	6	30
8 by 12 by 12..............	6	36
10 by 12 by 12..............	6	42
12 by 12 by 12..............	6	48
12 by 12 by 12..............	9	52
3¾ by 5 by 12..............	1	9
8 by 5 by 12..............	2	16
8 by 5 by 12..............	3	16
8 by 5 by 12 ("L" Shaped).		16
8 by 6¼ by 12 ("T" Shaped).	4	16
8 by 7¾ by 12 (Square).....	6	24
8 by 10¼ by 12 ("H" Shaped)	7	32
8 by 8 by 8 (Cube).......	9	18

E. Dimensions in any particular shall not vary more than 3% for any form of tile.

F. Weather Resistance of Tile for exterior work shall be such that it will be able to withstand 100 alternate freezings and thawings. Tile classed as "hard" or "medium" by these specifications may be classified as meeting weathering requirements provided they are burned to the normal maturity for the given clay. Tile classed as "soft" shall be classified according to results of freezing tests.

G. Fire Resistance rating shall be in strict accord with the Bureau of Standards requirements, serial 617-BS and the standards established by special report of Underwriter's Laboratories. Special shapes where asking for ratings higher than indicated on this report shall carry certificate of the manufacturer guaranteeing such higher ratings.

3. Clay for Manufacture of Hollow Clay Tile shall be fire clay, shale or admixtures of same. These clays may be used in any kind of hollow clay tile. Surface clay of approved quality may be used only for the manufacture of interior non-load bearing tile, floors, girder and column covers, and for fireproofing. Exposed or exterior wall tile facing or veneering regardless of kind, shall be of low absorption ratios and manufactured of fire clay shale or admixtures of same.

4. Classification and Grade Marking of Clay Tile shall be legibly imprinted or indented in the exterior wall of each tile, together with the trade mark or name of the manufacturer so as to identify the character of the tile and its appropriate use. Manufacturers of tile shall guarantee the replacement of same without cost to users if the tile does not comply with the grade mark stamped on same by them. All tile used in construction shall conform to the standards established by the American Society for Testing Ma-

reinforced concrete material before applying cinder fill.

2nd. **Fill** on top of same with a 6" layer of bituminous cinders. Level off to a true even level and tamp or roll free from voids which might cause settlement.

3rd. **Tiles** shall be laid over same in true uniform manner with joints perfectly aligned and grouted with a Portland Cement grout consisting of one part Portland Cement to one part torpedo sand to 1½ parts clean roofing gravel, well grouted into the joints and built up on surface to 1¼" in thickness above the tile and troweled off to a true even surface in strict accord with best cement finisher's practice. Tile for paving purposes shall be not less than 4" in thickness and of the kind known as "hard." It shall be dovetailed scored on the top surface so as to form a mechanical bond with cement top finish.

7. Exterior and Load Bearing Clay Tile Walls, where required to be constructed shall comply with standard specifications as to absorption, compressive strength area and weight for "Load-Bearing Building Tile" (C-34-24-T) and also shall meet special conditions as per enumeration which follows:

A. Placement shall be such as will develop the full strength of the tile when laid in the wall.

B. Exposed Surfaces of Walls either inside or outside shall be faced with tile of color and texture to match in spirit and intent the sample approved by the Architect.

C. Mortar Joints shall be of color, type and character as per samples laid up and approved by the Architect. It shall be the duty of the contractor to lay up different samples of tile for the Architect's approval.

D. Allowable Working Stress on hollow clay tile laid with the cells vertical shall in no case exceed 120 pounds and when laid with the cells horizontal 80 pounds, in each case per square inch gross horizontal cross sectional area of walls.

E. Allowable Thickness of walls shall in no case be less than the thickness prescribed by the law and good practice under similar placement for brick masonry.

F. Corners shall be carefully formed plumb on both sides with perfect tile free from defects, fully bedded in mortar and perfectly bonded to secure straight and true corners which will develop full strength.

G. Bonding of Face Brick Stone or Other Facing to hollow clay tile backing shall be with at least one row of full length headers in every seventh course of brick facing, or there shall be at least one full length header in every 90 square inches of wall surface area.

H. Facing Material Thickness shall be considered a part of the wall thickness but the maximum stress in such facing material shall not be assumed to exceed that allowable for tile backing (see Sec. 7-D).

I. Changes in Thickness of Walls where required to be made shall be so made as to transmit loads from vertical webs directly on top of webs below; where this is not practical or possible, tile slabs or solid brick masonry shall be introduced to distribute the load and not put strain on horizontal shells.

J. Concentrated Loads, wherever they occur on walls, shall be so distributed by means of spread metal bearing plates, beams or lintels or by brick concrete or other solid masonry as not to impose a unit stress on wall of more than that allowable for the type of tile used (see Sec. 7-D).

K. Excess Loads on Pilasters where more than proper allowable tile bearing load (see

Sec. 7-D) shall be provided for by strengthening the pilaster in the following manner:

1st. **Insert Metal Rods** in the four corners of the pilaster of the size specified and if no size is mentioned, not less than ⅜".

2nd. **Fill** all the cells of the pilaster and the wall back of same with concrete composed of one part Portland Cement to two parts clean sharp sand to three parts of gravel or crushed stone which will pass a ¾" mesh screen and be retained on a ¼" mesh.

L. Joist Bearings wherever joists are required to rest on walls shall be not less than 4" and shall be composed of bearing slabs of tile not less than 1" in thickness or solid brick masonry placed for bearing purposes. The sides, top and back of all floor joists where they rest on walls shall be surrounded with solid tile construction, solid brick masonry or closed tile, so that combustible construction does not come in contact with any openings in wall.

M. Joist Anchors where required to be used shall be placed at the bottom of joists so that if the joists fall for any cause whatsoever, they can drop down without putting leverage strain on the wall.

N. Sills of Clay Tile shall be formed with special shapes designed for that purpose of "dense" grade tile. These shapes shall be of such a nature as to form a shoulder or interlock underneath window frames so as to dam off and prevent water from running underneath same and they shall also be so designed as to drip free from wall. Joints shall be filled with Portland Cement mortar and wood sub-sills or frames shall be set in a thick heavy bed of mastic or rock putty.

O. Lintels not exceeding 5' in clear span shall be formed with load bearing tile reinforced with rods through cells and filled solid with concrete same as specified for filling pilaster (see Sec. 7-K). Such lintels shall be cast and allowed to season at least ten days before placing. They shall be reinforced in the top as well as in the bottom so as to facilitate placement without fracture. For size and amount of reinforcement see notes on drawings or in schedules or reinforcing.

P. Openings in clay tile walls shall have straight true tile jambs laid in workmanlike manner to provide a solid exterior surface using special shapes where required. When special shapes or jamb blocks are used, the installation shall be according to manufacturer's details and instructions. Wood Frames for such openings shall be carefully caulked around with oakum and mastic cement so as to make a perfectly water and wind tight bond between frames and walls. These frames will be designed with some form of wooden interlock or metal bond strip to form an interlocking bond between the wood and tile.

Q. Arch Openings shall be built to radius and camber as indicated by drawings using tile units small enough to attain proper curvature in order that the top of mortar joints may not be too heavy. Arch framing forms shall be placed which shall not be removed until work has thoroughly set up. The width of abutment must be sufficient to resist the thrust of arch and mortar used for arches shall be composed of Portland Cement and sand mixed in the proportions of one to three. Walls immediately above arches shall have one ⅜" bar laid horizontally in the joint of same for each 4" in thickness, requiring three bars for 12" wall and two bars for 8" wall.

R. Special Shapes where used to provide special bond features or for built in insulation shall be furnished and laid strictly in accord with manufacturer's directions.

S. Furring Tile wherever indicated on plans for exterior walls of building shall be 2" furring tile. These tile shall be anchored to the walls with metal ties built into the masonry every two courses in height and at intervals not greater than 36" in horizontal direction.

8. Interior Non-Load Bearing Clay Tile Walls where required to be constructed shall comply with standard specifications as to absorption, compressive strength area and weight for non-load bearing tile as prepared by the American Society for Testing Materials, Serial designation C-56-26T, with all amendments thereto to date, and shall also meet special conditions as per enumeration which follows:

A. Tile for Walls and Partitions shall be sound, uniform in shape, and free from imperfections which will impair its fire resisting quality, permanence, or durability for the purposes intended.

B. Support for non-load bearing tile partitions shall be arranged so that all partitions shall start on solid bearings and continue in straight and level courses, being firmly restrained at sides and top where same come in contact with other materials. In no sense shall non bearing clay tile partitions or walls carry any distributed and/or concentrated loading in excess of their own weight.

C. Nailing Strips shall be provided for the attachment of grounds at not to exceed 24" o.c. and so as to carry out the spirit and intent of Architect's detail drawings. Exception: if spot grounds, toggle bolts or clinch nails are specified or detailed, then nailing strips may be omitted.

9. Chases shall be built into load bearing walls using, if necessary, special shapes to be provided. Where the strength of walls will not be impaired, tile with vertical cells may be cut for the installation of vertical chases. Tile damaged in this process must be replaced satisfactorily before proceeding with other work. Under no circumstances shall tile walls or partitions be cut into for horizontal lines of pipe or conduit and any material so treated shall be replaced at the cost of the mechanic installing such pipe or conduit work to make a perfect complete and workmanlike job.

10. Decorative Wall Treatment for sanitary walls, special glazed tile, together with coves, special corners, caps or mouldings, shall be provided in accordance with architect's specifications. Texture surfaces, colors, kind, quality and manner of making mortar joints shall be in accordance with samples furnished and approved by the Architect.

11. Floor and Roof Arch Clay Tile Construction where required shall comply with standard specifications as to absorption, compressive strength, area, weight and design for floor and roof arch construction as prepared by the American Society for Testing Materials, serial designation C-57-26T with all amendments thereto to date, and shall also meet special conditions as per enumeration which follows:

A. Spans of hollow clay tile flat arch construction between steel floor or roof beams shall not exceed 8' 6". Span of segmental arch construction shall not exceed 12' except where specially designed in accordance with accepted engineering formula. The supporting beams for arch floor construction shall be tied together with steel tie rods.

B. Tie Rods in no case shall be less than ¾" in diameter and spaced as required to resist the thrust of arch at not to exceed eight times the depth of the beam between tie rods or tie rod and girder. Tie rods shall be placed as near the point of thrust of the arch as practicable, and shall be completely incased within the construction to a depth of at least two inches unless the tie rods are otherwise protected with fireproof covering below the soffit of arch.

C. Quality and Manufacture. Hollow clay tile for this form of construction shall be sound, hard, uniformly burned and shall be free from imperfections which would impair the structural properties, permanence, or fire-resistive quality of the construction, and shall have an (average) crushing strength of at least 2,000 lbs. per sq. inch of net sectional area when tested on end with the cells in alignment with the direction of pressure. The shells of hollow tile for arch construction shall not be less than ⅝" thick and the webs not less than ⅝" thick.

D. Flat Arch Construction shall not be less than 6" and all arches shall have at least two cellular spaces in the depth and unless reinforced by steel shall have a depth of not less than 1½" per foot of span. All arches shall be set on properly designed skewbacks that are cut to fit and protect the various beam sections. Keys of the sizes required to fit the different spans shall be provided, all arches to be keyed up in the middle third span, and any additional wedging up required, to be done with tile slabs or slate. End construction arches shall be set with the tile in a direct line between beams.

E. Segmental Arches shall have a total depth of not less than 6" having two cellular spaces in the depth and no segmental arch shall have a rise of less than ¾" per ft. of span. Segmental arches shall be set with broken joints and be securely wedged up with tile or slate.

F. Skewbacks for all arches shall be designed to receive the thrust of arch and properly transmit same to flange of beam, the point at which thrust is supported in all cases to be above the top of beam flange supporting the arch.

12. Fireproofing where required to be furnished shall comply with the standard specifications as to absorption, compressive strength, area, weight and design for fireproofing tile as prepared by the American Society for Testing Materials serial designation C-56-26T, and all amendments thereto to date, and shall also meet special conditions as per enumerations which follow:

A. Supporting Beams, Girders, Lintels and other steel members shall be encased in hollow tile fireproofing throughout their entire length. All important beams and girders shall be incased independently of the floor arch construction, and wherever possible any openings that are required shall be left in the arch construction to avoid cutting. The minimum thickness of hollow tile fireproofing shall be as follows:

(a) **Soffit Covering** on beams carrying flat arches 1½".

(b) **Covering** on beams, girders, etc., 2".

(c) **Covering** on inside projections of wall beams, lintels, etc., 2".

B. Wall Columns and Girders wherever hollow clay tile is used for fireproofing of wall columns, wall girders, or spandrel beams, etc. in exterior walls, the minimum thickness of hollow tile covering on these steel members shall be the same as that required for brick masonry covering and the thickness of solid material be at least equal to that specified for hollow tile fireproofing of interior columns. Hollow tile covering shall be well bonded and tied around columns and into the enclosing walls, and all channel spaces in wall columns shall be filled solid, same as required for interior columns. In no case shall any projection of the steel beyond the edge or face of columns, or the extreme outer edge of flanges of wall beams, or the plates and angles attached to same come within less than 2" of the exterior face of fireproof covering.

C. Interior Columns, Struts, or Other Vertical Supports of structural steel, wrought iron or cast iron, including the connection plates, shall be entirely incased and protected with hollow tile not less than 3" in thickness.

D. Beam Connections, Lugs, Brackets, Etc., attached to columns shall be covered with fireproofing of not less than 1" thickness of solid material over the extreme outer edge of metal.

E. Channel Spaces in Columns if not specified or shown to be filled with other fireproof material shall be filled with hollow tile, and the fireproof covering applied around the filled column. In all such cases all exposed steel shall be plastered with a full ½" coating of Portland Cement mortar as the channel filling and column covering is applied so that every bit of metal is first covered with a ½" coat of Portland Cement plaster.

F. Chases in Fireproof Covering shall not be cut under any circumstances and all pipe and conduit shall be kept outside fireproof covering excepting that outlets on the face of columns may have one electric conduit not exceeding ¾" in size built into the channel filling when placed with at least 3" thickness of fireproofing material between same and the steel columns and securely built in before the fireproof covering is erected. The outside elbow shall extend out to face of fireproof covering and be solidly built in with not less than 3" thickness of tile brick or cement mortar between back and sides of outlet box and any metal.

G. Hollow Clay Tile Column Fireproofing shall be set in cement mortar as hereinafter specified and shall be securely bound with not less than No. 12 gauge wires once in every course or be tied together in every course with "U" shaped clips of No. 18 gauge band iron and shall be plastered at least ⅛" in thickness with mortar or heat retarding composition.

Note. **Column Fireproofing in Warehouses** where subject to damage from trucking will be protected to a height of at least 3' by metal plates or cast iron covering, but this metal covering will be furnished by the contractor for miscellaneous iron and this contractor shall co-operate with him in setting same.

13. Combination Hollow Clay Tile and Concrete Floor Construction where required shall comply with standard specifications as to material, density, absorption, compressive strength, area, weight and arrangement for combination hollow clay tile and concrete construction as prepared by the American Society for Testing Materials, serial designation C-57-26T, with all amendments thereto to date, and shall also meet special conditions as per enumeration which follows:

A. In General hollow clay tile and concrete floor construction is understood to consist of reinforced concrete joists between rows of hollow tile. When concrete topping over the tile is figured as part of the structural slab, same shall be cast monolithic with the concrete joist system. All floor slabs and beams shall have at least 4" bearing on the wall and if the bearing is on a tile wall, then a section of concrete the full length of tile wall and not less than 4" wide shall be cast over the tile wall to spread the bearing between beams and this section shall be reinforced in the direction of the length of the wall.

B. Forms shall be of such a size, lumber, and so placed as to prevent deflection and shall be provided in such quantity as not to delay the progress of the work. Care shall be taken not to remove the forms before the concrete is set. Under long spans the center row of supports shall be maintained for at least three weeks after the concrete has been poured. In cold weather, the contractor shall

(Continued on page 719)

LUMBER STANDARDS

By L. KRAEMER, Engineer, In Forest Products CHICAGO LUMBER INSTITUTE

For the Specification of Lumber Meeting the Requirements of the Chicago Building Code and American Lumber Standards

1. The following information and comparative tables on lumber and lumber grading are recommended for use in the Chicago Metropolitan area and the Middle West.

DESCRIBE SPECIES AND GRADES ACCURATELY.

2. The grading of lumber has not been considered an exact science, because it is based on a visual inspection of each piece and on the judgment of the grader. Each grade in the Manufacturers' rule book covered a wide range in quality because they were formulated to supply material for a large variety of uses.

3. Lumber to be suitable for the purpose intended must be of an accurately described species and should be selected from the grade that most nearly lends itself to the desired results. Selection is necessary because seldom is a grade wholly suitable for a given use. In addition, the condition of use should be definitely known so that the lumber may be properly conditioned to avoid unwarranted change in form after installation.

SHRINKAGE IN LUMBER (Seasoning).

4. Unanticipated changes in dimension have caused distortion and rupture in finished structures and have proved detrimental to the successful use of lumber. Such changes may be avoided by requiring lumber to be seasoned, either naturally or artificially, to a condition that does not vary appreciably from the normal moisture condition in the finished structure.

MILL WORK.

5. The selection and conditioning of lumber in planing mill products is of equal importance and requires a thorough understanding of the performance of different species and conditions of use. Recommendations should not be made without definite knowledge of all requirements. Particular consideration should be given to joinery methods as most of the difficulties with planing mill products occur at joints.

Accurate quality specifications and species identification are essential to an intelligent agreement and contract.

SPECIFICATIONS FOR LUMBER IN RESIDENTIAL, APARTMENT, COMMERCIAL AND INDUSTRIAL BUILDINGS.

It is recommended that the following "General Lumber Requirements" be included in all specifications controlling building construction within the Chicago area.

"General Lumber Requirements—All lumber used in the construction of this building shall be selected from the manufacturers' grades indicated in these specifications and shall meet the approval of the architect and the Chicago Lumber Institute. A certificate of inspection issued by the Chicago Lumber Institute shall be delivered to the architect before final payment is made."

There is no charge to the owner for this inspection and lumber of equal quality and condition of seasoning costs the same with or without this inspection. It provides positive assurance that the quality of a material that is difficult to evaluate, is exactly as specified.

RESIDENCE SPECIFICATIONS.

For Rigid, Stabilized Construction Free of Shrinkage and Distortion Where Long Life and Low Maintenance Costs Are Required.

Floor Joists—Selected close grain lumber of not more than 12 per cent moisture content, in Douglas fir or southern yellow pine valued at 1600 pounds in extreme fiber stress.

Studding—Selected close grain, No. 1 Common lumber of not more than 15 per cent moisture content, in West Coast hemlock, Douglas fir or southern yellow pine.

Ceiling Joists and Roof Rafters—Medium grain southern yellow pine or close grain Douglas fir in selected No. 1 Common lumber of not more than 15 per cent moisture content.

Sub-flooring, Sheathing, Roof Boards—Standard 25/32" No. 2 Common lumber of not more than 15 per cent moisture content in yellow pine with knot holes cut out in laying.

Other Framing Lumber — No. 1 Common lumber of not more than 15 per cent moisture content in southern yellow pine or Douglas fir.

Flooring Under Linoleum — Vertical grain. 2⅜" face flooring of not more than 10 per cent moisture content in Douglas fir or equal (grade optional).

Finish Flooring Oak or Maple—Appalachian Oak or Hard Maple. 2¼" Strip Flooring between 5 and 8 per cent in moisture content (grade optional).

Siding and Exterior Trim—All heartwood of not more than 15 per cent moisture content in red cypress, redwood, or western red cedar.

Porch Railings and Shutters—All heartwood in red cypress or redwood.

Exterior Posts or Timbers—All heartwood in red cypress, redwood or side cut Douglas fir.

Exterior Flooring—Vertical grain all heartwood in Douglas fir or red cypress.

Exterior Stairs—All heartwood in Douglas fir or red cypress with vertical grain stepping, if of fir, and bark face up stepping, if of red cypress.

SECONDARY RESIDENCE SPECIFICATIONS.

For Less Exacting Work Where Some Sacrifice to Initial Economy Is Necessary.

Floor Joists—Medium grain southern yellow pine or close grain Douglas fir in No. 1 Common lumber of not more than 15 per cent moisture content.

Studding—Close grain West Coast hemlock or Douglas fir or medium grain southern yellow pine in No. 1 Common lumber of not more than 15 per cent moisture content.

Ceiling Joists and Roof Rafters — No. 1 Common kiln-dried Douglas fir or southern yellow pine of 15 per cent moisture content.

Sub-flooring, Sheathing, Roof Boards—No. 2 Common kiln-dried yellow pine of 15 per cent moisture content.

Other Framing Lumber—No. 1 Common kiln-dried Douglas fir or southern yellow pine of 15 per cent moisture content.

Flooring Under Linoleum — Vertical grain Douglas fir flooring of 10 per cent moisture content.

Finish Flooring Oak or **Maple**—(grade optional) Oak or maple 2¼" face strip flooring between 5 and 8 per cent moisture content.

Siding and Exterior Trim—Red cypress, redwood, red cedar or white pine of 15 per cent moisture content.

Porch Railings and Shutters—All heartwood in red cypress or redwood.

Exterior Posts or Timbers—All heartwood in red cypress, redwood or side cut Douglas fir.

Exterior Flooring — All heartwood in red cypress or vertical grain Douglas fir.

Exterior Stairs — All heartwood in red cypress or vertical grain Douglas fir.

APARTMENT BUILDING SPECIFICATIONS.

For Rigid Substantial Construction Free of Excessive Shrinkage Where Low Maintenance Costs are Required.

Floor Joists—Selected medium grain southern yellow pine or close grain Douglas fir of not more than 15 per cent moisture content value at 1600 pounds in extreme fiber stress.

Studding—Close grain West Coast hemlock or Douglas fir or medium grain southern yellow pine in selected No. 1 Common lumber, of not more than 15 per cent moisture content.

Ceiling and Roof Joists—No. 1 Common lumber of not more than 15 per cent moisture content in medium grain southern yellow pine or close grain Douglas fir.

Sub-flooring and Roof Boards—Standard 25/32" No. 2 Common yellow pine of not more than 15 per cent moisture content.

Flooring Under Linoleum—2¼" face hard maple between 5 and 8 per cent moisture content or vertical grain Douglas fir of not more than 10 per cent moisture content.

Finish Flooring—(grade optional) Oak 2¼" face strip flooring between 5 and 8 per cent in moisture content.

Open Porch Posts—Stair stringers, girders and joists of medium grain No. 1 Common Douglas fir or southern yellow pine of 15 per cent moisture content. Rails of all heartwood in red cypress or redwood. Treads of all heartwood in red cypress with bark face up or vertical grain Douglas fir.

Open Porch Flooring—All heartwood in vertical grain Douglas fir or red cypress.

COMMERCIAL BUILDINGS.

For Substantial Construction Where Low Maintenance Costs are Required.

Floor Joists in Dimension Sizes—Selected medium grain southern yellow pine or close grain Douglas fir of not more than 15 per cent moisture content valued at 1600 pounds in extreme fiber.

Floor Joists in Timber Sizes—Selected close grain Douglas fir of approximately 22 per cent moisture content valued at 1600 pounds in extreme fiber.

Ceiling and Roof Joists—No. 1 Common lumber of not more than 15 per cent moisture content in medium grain southern yellow pine or close grain Douglas fir.

Sub-flooring and Roof Boards — Standard 25/32" No. 2 Common southern yellow pine of not more than 15 per cent moisture content.

Floor and Roof Plank—Close grain No. 1 Common southern yellow pine of not more than 15 per cent moisture content.

Floor Sleepers in Concrete—Southern yellow pine impregnated under pressure with creosote, zinc chloride, or Wolman Salts.

Finish Flooring—Hard maple select strip flooring between 5 and 8 per cent in moisture content laid on sleepers spaced not more than 12 inches apart. (Thickness of flooring according to nature of loads.)

INDUSTRIAL BUILDINGS.

For Substantial Construction Where Low Maintenance Costs are Required.

Timber Girders and Joists—Selected close grain side cut structural Douglas fir of approximately 24 moisture content valued at 1600 pounds in extreme fiber.

Posts—No. 1 Common structural longleaf yellow pine.

Floor and Roof Plank—Close grain No. 1 Common southern yellow pine of not more than 15 per cent moisture content.

Trucking Platforms — Close grain No. 1 Common longleaf yellow pine laid with the bark face for the wearing surface.

Finish Flooring — Hard maple select strip flooring between 5 and 8 per cent in moisture content laid on floor strips not more than 12 inches apart. (Thickness of flooring according to loading.)

WINDOW FRAMES AND SASH.

For all Buildings.

The following specification is suggested as providing sound frame and sash construction: "All window frames shall have southern yellow pine pulley stiles and all heartwood sills in red cypress, redwood or western red cedar. All sash shall be true white pine or may be of western pine if bottom rails and muntin bars are of all heartwood in red cypress, redwood or red cedar."

Note: In industrial construction, where occupancy conditions develop high moisture conditions, the entire frame and sash shall be all heartwood in red cypress, redwood, or red cedar.

AMERICAN COMMERCIAL WOODS AND ASSOCIATIONS RESPONSIBLE FOR GRADING RULES.

Hardwoods — National Hardwood Lumber Association.

Maple Flooring—Maple Flooring Manufacturers' Association.

Oak Flooring—National Oak Flooring Manufacturers' Association.

Douglas Fir — West Coast Lumbermen's Association (Coast region) Western Pine Association ("Inland Empire" and California region).

Eastern Hemlock—Northern Hemlock and Hardwood Manufacturers' Association.

Eastern Red Cedar (Aromatic Red Cedar)[8] —National Hardwood Lumber Association.

Eastern Spruce (Red Spruce, Black Spruce, White Spruce)[9]—Northern Pine Manufacturers.

Englemann Spruce—Western Pine Association.

Incense Cedar—Western Pine Association.

Northern White Pine[9]—Northern Pine Manufacturers.

Norway Pine (Red Pine)[9]—Northern Pine Manufacturers.

Ponderosa Pine—Western Pine Association.

Port Orford Cedar—West Coast Lumbermen's Association.

Redwood—California Redwood Association.

Sitka Spruce — West Coast Lumbermen's Association.

Southern Cypress (only 1 type—Tidewater Red Cypress)—Southern Cypress Manufacturers' Association.

Southern Cypress (no grading distinction as to types—rules cover White Cypress, Yellow Cypress, and Red Cypress—(National Hardwood Lumber Association.

Southern Pine (includes the various commercial pines of the Southeast, such as Longleaf, Shortleaf, Slash, Loblolly, etc.)—Southern Pine Association.

Sugar Pine—Western Pine Association.

Tamarack (Eastern Larch)—Northern Hemlock and Hardwood Manufacturers' Association.

Western Hemlock (West Coast Hemlock)—West Coast Lumbermen's Association.

Western Larch—Western Pine Association.

Western Red Cedar—West Coast Lumbermen's Association.

Western White Pine (Idaho White)—Western Pine Association.

White Fir—Western Pine Association.

[8] Although a softwood, Eastern Red Cedar is graded under Hardwood Association Rules.

[9] The rules of the former Northern White Pine Manufacturers' Association are still in use by manufacturers of Northern White and Norway Pine and Red, Black and White Spruce.

TABLE OF STRENGTH
YELLOW PINE & DOUGLAS FIR BEAMS

Prepared by

CHICAGO LUMBER INSTITUTE November 1938

CHICAGO BUILDING ORDINANCE (1938)

LOAD IN POUNDS (UNIFORMLY DISTRIBUTED)
BASED ON ACTUAL DRESSED SIZES
Weight of Beam Included

STRENGTH
UNPLASTERED CONSTRUCTION

Fiber Stress 1600 lbs. In^2 Shear 125 lbs. In^2
For 1400 lbs. stress—Multiply by 0.875
For 1000 lbs. stress—Multiply by 0.625
*=s—Limited by Shear (125 pounds)

DEFLECTION
PLASTERED CONSTRUCTION

E equals 1,600,000
For E = 1,200,000—Multiply by 0.75
†=b—Limited by Bending (1600 lbs.)

Width in Inches							Span in Feet	Width in Inches						
2 in.	3 in.	4 in.	6 in.	8 in.	10 in.	12 in.		2 in.	3 in.	4 in.	6 in.	8 in.	10 in.	12 in.
6 in. Beam = 5⅝ in. Load in Pounds Except 6 x 6 = 5½ x 5½ in.								6 in. Beam = 5⅝ in. Load in Pounds Except 6 x 6 = 5½ x 5½ in.						
1525*	2461*	3400*	5040*	......	......	......	5	1525*	2461*	3400*	5040*	......	......	......
1524	2460	3399	4930	......	......	......	6	1525*	2461*	3400*	4930†	......	......	......
1306	2109	2913	4226	......	......	......	7	1165	1883	2601	3688	......	......	......
1143	1841	2549	3697	......	......	......	8	892	1442	1991	2824	......	......	......
1016	1640	2266	3287	......	......	......	9	705	1139	1573	2232	......	......	......
914	1476	2039	2958	......	......	......	10	571	922	1274	1808	......	......	......
831	1342	1854	2689	......	......	......	11	472	762	1053	1494	......	......	......
761	1230	1699	2465	......	......	......	12	396	640	885	1255	......	......	......
8 in. Beam = 7½ in.								8 in. Beam = 7½ in.						
2031*	3281*	4531*	6875*	9375*	......	......	7	2031*	3281*	4531*	6875*	9375*	......	......
2031	3281	4530	6875	9375	......	......	8	2031*	3281*	4531*	6875*	9375*	......	......
1806	2916	4028	6111	8333	......	......	9	1672	2701	3729	5658	7716	......	......
1625	2625	3624	5500	7500	......	......	10	1354	2188	3021	4583	6250	......	......
1477	2386	3295	5000	6818	......	......	11	1119	1808	2497	3788	5165	......	......
1354	2187	3020	4583	6250	......	......	12	940	1519	2098	3183	4340	......	......
1250	2019	2788	4231	5769	......	......	13	801	1294	1787	2712	3698	......	......
1160	1875	2589	3929	5259	......	......	14	690	1116	1541	2338	3189	......	......
1083	1750	2416	3667	5000	......	......	15	601	972	1343	2037	2778	......	......
1016	1640	2265	3437	4687	......	......	16	528	854	1180	1790	2441	......	......
10 in. Beam = 9½ in.								10 in. Beam = 9½ in.						
2572*	4156*	5740*	8707*	11875*	15040*	......	8	2572*	4156*	5740*	8707*	11875*	15040*	......
2572*	4156*	5740*	8707*	11875*	15040*	......	9	2572*	4156*	5740*	8707*	11875*	15040*	......
2572*	4156*	5740*	8707*	11875*	15040*	......	10	2572*	4156*	5740*	8707*	11875*	15040*	......
2370	3829	5287	8022	10939	13856	......	11	2274	3674	5074	7698	10498	13297	......
2173	3509	4847	7354	10030	12701	......	12	1911	3087	4263	6468	8821	11173	......
2005	3239	4474	6788	9256	11724	......	13	1628	2631	3633	5512	7516	9520	......
1863	3008	4154	6303	8595	10887	......	14	1404	2268	3132	4752	6481	8209	......
1738	2807	3877	5883	8022	10161	......	15	1223	1976	2729	4140	5645	7150	......
1629	2631	3635	5515	7512	9526	......	16	1075	1737	2398	3639	4962	6285	......
1533	2477	3421	5191	7078	8966	......	17	952	1538	2124	3223	4395	5567	......
1448	2339	3231	4903	6685	8468	......	18	849	1372	1895	2875	3920	4966	......
1372	2216	3061	4645	6333	8022	......	19	762	1231	1700	2580	3518	4457	......
1304	2105	2908	4411	6017	7621	......	20	688	1111	1534	2329	3175	4022	......

Continued on next page.

2 in.	3 in.	4 in.	6 in.	8 in.	10 in.	12 in.	Span in Feet	2 in.	3 in.	4 in.	6 in.	8 in.	10 in.	12 in.
12 in. Beam = 11½ in.								12 in. Beam = 11½ in.						
3115*	5031*	6948*	10540*	14375*	18207*	22040*	9	3115*	5031*	6948*	10540*	14375*	18207*	22040*
3115*	5031*	6948*	10540*	14375*	18207*	22040*	10	3115*	5031*	6948*	10540*	14375*	18207*	22040*
3115*	5031*	6948*	10540*	14375*	18207*	22040*	11	3115*	5031*	6948*	10540*	14375*	18207*	22040*
3115*	5031*	6948*	10540*	14375*	18207*	22040*	12	3115*	5031*	6948*	10540*	14375*	18207*	22040*
2939	4747	6555	9947	13564	17182	20799	13	2889	4666	6444	9777	13333	16888	20443
2729	4408	6087	9237	12595	15954	19313	14	2491	4023	5556	8430	11496	14562	17627
2547	4114	5681	8621	11755	14890	18025	15	2170	3505	4840	7344	10014	12684	15355
2388	3857	5326	8022	11021	13960	16899	16	1907	3080	4254	6454	8801	11149	13495
2248	3630	5013	7607	10372	13138	15905	17	1689	2729	3768	5717	7796	9875	11954
2123	3428	4734	7184	9796	12408	15021	18	1507	2434	3361	5100	6954	8808	10663
2011	3248	4485	6806	9281	11755	14231	19	1353	2184	3017	4577	6241	7906	9570
1910	3085	4261	6466	8817	11168	13519	20	1220	1971	2722	4131	5633	7135	8637
1820	2938	4058	6158	8397	10636	12875	21	1107	1788	2469	3747	5109	6472	7834
1737	2805	3873	5878	8015	10152	12290	22	1009	1629	2250	3414	4655	5897	7138
1661	2683	3705	5622	7667	9711	11756	23	923	1491	2059	3124	4259	5395	6531
1592	2571	3551	5388	7347	9306	11266	24	848	1369	1891	2869	3912	4955	5998
14 in. Beam = 13½ in.								14 in. Beam = 13½ in.						
3656*	5906*	8156*	12375*	16875*	21375*	25875*	11	3656*	5906*	8156*	12375*	16875*	21375*	25875*
3656*	5906*	8156*	12375*	16875*	21375*	25875*	12	3656*	5906*	8156*	12375*	16875*	21375*	25875*
3656*	5906*	8156*	12375*	16875*	21375*	25875*	13	3656*	5906*	8156*	12375*	16875*	21375*	25875*
3656*	5906*	8156*	12375*	16875*	21375*	25875*	14	3656*	5906*	8156*	12375*	16875*	21375*	25875*
3510	5669	7830	11880	16200	20520	24840	15	3510	5670	7830	11880	16200	20520	24840
3291	5315	7340	11138	15187	19237	23287	16	3085	4983	6882	10441	14239	18036	21832
3097	5002	6908	10482	14294	18106	21917	17	2733	4414	6096	9249	12613	15976	19339
2925	4724	6525	9900	13500	17100	20700	18	2438	2937	5437	8250	11250	14250	17250
2771	4476	6181	9379	12790	16200	19611	19	2188	3534	4880	7405	10097	12790	15482
2633	4252	5872	8910	12150	15390	18630	20	1974	3189	4404	6682	9112	11543	13973
2507	4049	5592	8486	11572	14657	17743	21	1791	2893	3995	6061	8265	10470	12674
2393	3865	5338	8100	11045	13991	16936	22	1632	2636	3640	5523	7531	9539	11547
2289	3697	5106	7748	10565	13383	16200	23	1493	2412	3330	5053	6890	8728	10565
2194	3543	4893	7425	10125	12825	15525	24	1371	2215	3059	4641	6328	8016	9703
16 in. Beam = 15½ in.								16 in. Beam = 15½ in.						
4198*	6781*	9365*	14207*	19375*	24542*	29707*	13	4198*	6781*	9365*	14207*	19375*	24542*	29707*
4198*	6781*	9365*	14207*	19375*	24542*	29707*	14	4198*	6781*	9365*	14207*	19375*	24542*	29707*
4198*	6781*	9365*	14207*	19375*	24542*	29707*	15	4198*	6781*	9365*	14207*	19375*	24542*	29707*
4198*	6781*	9365*	14207*	19375*	24542*	29707*	16	4198*	6781*	9365*	14207*	19375*	24542*	29707*
4082	6595	9107	13818	18843	23868	28893	17	4082†	6595†	9107†	13818†	18843†	23868†	28893†
3856	6228	8601	13051	17796	22542	27288	18	3689	5959	8229	12487	17027	21568	26108
3653	5900	8148	12364	16860	21355	25852	19	3311	5349	7386	11207	15282	19357	23433
3470	5605	7741	11745	16016	20287	24559	20	2988	4827	6666	10114	13793	17470	21148
3305	5338	7372	11186	15254	19322	23390	21	2711	4378	6046	9174	12510	15846	19181
3155	5096	7037	10678	14561	18444	22326	22	2470	3989	5509	8359	11399	14438	17478
3018	4874	6731	10214	13927	17642	21356	23	2260	3650	5041	7648	10429	13209	15991
2892	4671	6451	9788	13347	16906	20466	24	2075	3352	4629	7024	9578	12132	14686
2776	4484	6193	9397	12814	16231	19647	25	1913	3089	4266	6473	8827	11181	13534
2670	4312	5954	9035	12321	15606	18892	26	1768	2856	3944	5985	8161	10337	12514
18 in. Beam = 17½ in.								18 in. Beam = 17½ in.						
....	7656*	10573*	16042*	21875*	27707*	33542*	15	....	7656*	10573*	16042*	21875*	27707*	33542*
....	7656*	10573*	16042*	21875*	27707*	33542*	16	....	7656*	10573*	16042*	21875*	27707*	33542*
....	7656*	10573*	16042*	21875*	27707*	33542*	17	....	7656*	10573*	16042*	21875*	27707*	33542*
....	7656*	10573*	16042*	21875*	27707*	33542*	18	....	7656*	10573*	16042*	21875*	27707*	33542*
....	7521	10387	15760	21492	27223	32954	19	....	7521†	10387†	15760†	21492†	27223†	32954†
....	7145	9868	14973	20416	25862	31305	20	....	6947	9593	14556	19850	25143	30436
....	6805	9398	14260	19444	24630	29814	21	....	6301	8701	13204	18004	22805	27606
....	6496	8971	13612	18561	23511	28460	22	....	5741	7928	12030	16404	20780	25154
....	6213	8581	13020	17754	22488	27223	23	....	5253	7254	11007	15010	19012	23014
....	5954	8223	12477	17014	21551	26088	24	....	4824	6662	10108	13784	17461	21136
....	5716	7894	11977	16334	20689	25044	25	....	4446	6140	9316	12704	12092	19479
....	5496	7591	11517	15705	19894	24082	26	....	4111	5676	8613	11746	14878	18010

LUMBER STANDARDIZATION.

Simplified Practice Recommendations on Lumber Classifications.

Use Classification.

For the purpose of simplification of sizes and grades and of equalizing, among species used for similar general purposes, the grades of a similar name, lumber shall be classified by principal use into three classes as shown below. Different grading rules may apply to each class.

Yard Lumber. Lumber of all sizes and patterns which is intended for general building purposes. The grading of yard lumber is based on the intended use of the particular grade and is applied to each piece with reference to its size and length when graded without consideration to further manufacture.

Structural Lumber. Lumber that is 2 or more inches thick and 4 or more inches wide, intended for use where working stresses are required. The grading of structural lumber is based on the strength of the piece and the use of the entire piece.

Factory and Shop Lumber. Lumber intended to be cut up for use in further manufacture. It is graded on the basis of the percentage of the area which will produce a limited number of cuttings of a specified, or of a given minimum, size and quality.

STRUCTURAL LUMBER.

Basic Provisions for the Selection and Inspection of Softwood Lumber Where Working Stresses are Required.

56. The following provisions are the basis for the preparation of grading rules for structural lumber. They are not grading rules, being merely a standardized working basis for the coordination of the grades of the various species.

BASIC CLASSIFICATION.

57. The basic classification of structural lumber is as follows:

Joist and Plank — Lumber of rectangular cross section, 2 inches to but not including 5 inches thick and 4 or more inches wide, graded with respect to its strength when loaded either on the narrow face as joist or on the wide face as plank.

Beams and Stringers—Lumber of rectangular cross section, 5 or more inches thick and 8 or more inches wide, graded with respect to its strength in bending when loaded on the narrow face.

Posts and Timbers — Lumber of square or approximately square cross section, 5 by 5 inches and larger, graded primarily for use as posts or columns carrying longitudinal load but adapted to miscellaneous uses in which strength in bending is not especially important.

OPTIONAL PROVISIONS.

58. No heartwood requirements are provided in these basic provisions. When desired, heartwood shall be specified in terms of heartwood required on the girth, or on each face, side or edge. Girth shall be measured at the point where the greatest amount of sapwood occurs.

59. For structural lumber to be treated, a large amount of sapwood is preferable. It is not practicable to specify a minimum sapwood requirement, but it may be provided that there shall be no restriction on sapwood.

60. Wane is permissible in all grades, as far as strength properties are concerned, but square edges may be specified when required by appearance, bearing, or other factors of use.

DENSITY.

61. The following methods for determining density of Douglas fir and southern pine are for use with the grades where density is required.

62. Dense Douglas fir and southern pine shall average either on one end or the other, not less than 6 annual rings per inch, and in addition, one-third or more summerwood (the darker, harder portion of the annual ring), over a 3-inch portion of a radial line located as described below. The contrast in color between summerwood and springwood shall be distinct.

63. Coarse grained material excluded by this rule shall be accepted as dense if averaging one-half or more summerwood.

64. In inspection for density, reasonable variation of opinion between inspectors must be recognized. In re-inspection of a particular lot of lumber for density, for every three pieces accepted as having one-third or more summerwood, one of the remaining pieces shall be accepted if agreed upon as having between 30 and 33⅓ per cent summerwood.

CLOSE GRAIN

65. The following methods for determining rate of growth of Douglas fir and redwood are for use with grades where close grain is required.

66. Close grained Douglas fir shall average on either one end or the other not less than 6 nor more than 20 annual rings per inch, over a 3-inch portion of a radial line located as described below.

67. Pieces averaging 5 rings or more than 20 shall be accepted as equivalent of close grain if having one-third or more summerwood.

68. Close grained redwood shall average on either one end or the other not less than 10 nor more than 35 annual rings per inch, over a 3-inch portion of a radial line located as described below.

RADIAL LINE.

69. The radial line shall be representative of the average growth on the cross section.

70. When the radial line specified is not representative, it shall be shifted sufficiently to present a fair average, but the distance from the pith to the beginning of the 3-inch portion of the line in boxed-pith pieces shall not be changed.

71. In case of disagreement, two radial lines shall be chosen, and the number of rings per inch and percentage of summerwood shall be the average determined on these lines.

Location of Radial Line in Douglas Fir and Redwood.

72. In boxed-pith pieces the radial line shall run from the pith to the corner farthest from the pith. When the least dimension is 6 inches or less, the 3-inch portion of the line shall begin at a distance of 1 inch from the pith. When the least dimension is more than 6 inches, the 3-inch portion of the line shall begin at a distance from the pith equal to one-fourth the least dimension of the piece.

73. In pithless (side cut) pieces the center of the 3-inch portion of the radial line shall be at the center of the end of the piece.

74. If a 3-inch portion of a radial line cannot be obtained, the measurement shall be made over as much of a 3-inch portion as is available.

Location of Radial Line in Southern Pine.

75. In boxed-pith pieces the measurement shall be made over the third, fourth, and fifth inches from the pith along the radial line.

76. In pieces containing the pith, but not a 5-inch radial line, which are less than 2 by 8 inches in section or less than 8 inches in width, that do not show over 16 square inches on the cross section, the inspection shall apply to the second inch from the pith. In larger pieces that do not show a 5-inch radial line, the inspection shall apply to the 3 inches farthest from the pith.

77. In cases where pieces do not contain the pith and it is impossible to locate it with any degree of accuracy, the same in-

spection shall be made over 3 inches on an approximate radial line beginning at the edge nearest the pith in pieces over 3 inches in thickness and on the second inch nearest the pith in pieces 3 inches or less in thickness.

GENERAL PROVISIONS.

Quality of Wood.

78. No pieces of exceptionally light weight shall be permitted in any structural grade.

Sound Wood.

79. All structural grades shall contain only sound wood, free from any form of decay, incipient or advanced, except that pick is permitted to the same extent as holes.

Nominal Dimensions.

80. Rough lumber shall be sawn full to nominal dimensions except that occasional slight variations in sawing is permissible. No shipment shall contain more than 20 per cent of pieces of minimum dimension due to such variation in sawing.

81. Surfaced lumber shall be dressed to standard dimensions as hereinafter provided. Such standard dimensions shall be minimum dimensions when measured green.

Definition of Faces and Edges.

82. The faces of a piece of lumber are the four longitudinal surfaces of the piece, sometimes further designated as "wide" faces or "narrow" faces.

83. In a piece of lumber graded for use in bending, wide faces shall be taken as vertical faces, and narrow faces as horizontal faces, unless otherwise noted.

84. When the faces of a piece of lumber are of equal or approximately equal width, post and timber shall be used unless otherwise noted. When such a piece of lumber is graded for use in bending, the better pair of opposite faces shall be taken as the horizontal faces.

85. The edges of a piece of lumber are understood to be the narrower faces, and the sides the wider faces. In describing the locations of knots and other grade limitations in structural lumber, however, the edges of a given face are understood to be the intersection of two adjacent faces, commonly called corners.

Knots.

86. Cluster knots are not permitted in structural grades.

87. Knot holes and holes from causes other than knots are measured and limited as are knots.

STRUCTURAL GRADE LIMITATIONS AND WORKING STRESSES.

The characteristics permitted and limitations prescribed in any structural grade are based on the working stress requirements of the grade and its strength ratio, and shall be determined as indicated in the **Guide to the Grading of Structural Timbers,** Miscellaneous Publication No. 185 of the U. S. Department of Agriculture, pertinent provisions of which are published in the following pages.

STRENGTH RATIOS AND WORKING STRESSES FOR PUBLISHED GRADES.

In table 18 are listed grades produced by a number of associations of lumber manufacturers whose rules describe structural grades in accordance with the principles of strength grading as presented herein. The working stresses recommended by the respective associations for material that will be continuously dry are listed, and strength ratios as found by comparing the grade descriptions with the system presented on following pages are shown for each grade.

Consideration should be given to reducing some of the values in table 18 if the material is to be used where there is a decay hazard; that is, where it will not be continuously dry or wet.

TIMBERS USED UNDER CONDITIONS FAVORABLE TO DECAY.

Timber that remains dry, such as that in most covered structures, or that is constantly wet, such as parts that are below permanent water level, is not subject to decay, and stresses derived as outlined in the preceding paragraph apply. Under other circumstances timber is, in varying degrees, subject to decay, and allowance for the resulting deterioration is advisable. This may be accomplished by increasing the sizes of members either arbitrarily or through the lowering of design stresses.

Decay progresses most rapidly in places that are warm, humid, damp, or poorly ventilated and varies to moderate or very slight in places where lower temperatures prevail or where occasional dampness or wetting is offset by good ventilation that leads to quick drying.

Table 21 presents suggested average ratios among stresses for no decay hazard, moderate decay hazard, and severe decay hazard. The magnitude of the allowance to be made for deterioration from decay is a moot question, and the ratios of table 21 are therefore to be taken only as a general guide and as applicable to species of average durability. They should be modified in accordance with such factors as the natural resistance of the heartwood to decay, the proportion of sapwood permitted in the timber, the expected life of the structure or part, the frequency and thoroughness of inspection, and the cost of making replacements. In any instance in which the decay hazard is obviously high, treated material or the heartwood of a highly decay-resistant species should be employed unless the hazard can be reduced by attention to such features as drainage and ventilation. Nondurable woods should never be used where there is a decay hazard. In all instances design details should be such as to minimize the danger of decay.

Table 21.—Average ratios among stresses for 3 decay hazards.

Kind of stress	Decay hazard		
	None	Moderate	Severe
	Percent	*Percent*	*Percent*
Stress in extreme fiber in bending	100	85	71
Stress in compression perpendicular to grain	[1] 100	70	58
Stress in compression parallel to grain	100	92	78
Stress in horizontal shear in beams	100	100	100
Modulus of elasticity	100	100	100

[1] Value applicable only to material continuously dry; for material continuously wet the value is 70 percent.

FACTORS AFFECTING THE USE OF WORKING STRESSES.

The working stresses given in table 18 are for long-continued load and are considered safe under circumstances such that failure would cause personal injury or large prop-

NEW LONDON
GREEN BAY
AMERICAN PLYWOOD CORPORATION
NEW LONDON, WISCONSIN
LA CROSSE
MILWAUKEE
Many of America's finest homes and office buildings have that "extra rich" appearance because the architect wrote "Ply-rite panels and doors" in his specifications. You will discover the real meaning of dependability when you select this twenty-year old organization as your source for fine plywood panels and doors.
The "New Londoner", a Ply-rite product, is the newest development in flush doors. Write for full details.
PLYWOOD of RECOGNIZED QUALITY

Table 18.—Strength ratios for various lumber-association grades, with working stresses recommended by the producers for material used where it will be continuously dry.[1]

Name of association and effective date of grading rules	Species	Grade	Beams and stringers: Stress in extreme fiber: Strength ratio	Beams and stringers: Stress in extreme fiber: Stress	Beams and stringers: Stress in horizontal shear: Strength ratio	Beams and stringers: Stress in horizontal shear: Stress	Joist and plank: Stress in extreme fiber: Strength ratio	Joist and plank: Stress in extreme fiber: Stress	Joist and plank: Stress in horizontal shear: Strength ratio	Joist and plank: Stress in horizontal shear: Stress	Posts and timbers: Compression parallel to grain: Strength ratio	Posts and timbers: Compression parallel to grain: Stress	Stress in compression perpendicular to grain	Modulus of elasticity
			Percent	*Lb. per sq. in.*	*Percent*	*Lb. per sq. in.*	*Percent*	*Lb. per sq. in.*	*Percent*	*Lb. per sq. in.*	*Percent*	*Lb. per sq. in.*	*Lb. per sq. in.*	*Lb. per sq. in.*
California Redwood Association, San Francisco, Calif., Jan. 15, 1930.	Redwood	Prime structural [2]	88	1,494	88	82	83	1,707	88	82	88	1,245		
		Select structural [2]	76	1,322	75	70	67	1,280	75	70	78	1,100	267	1,200,000
		Heart structural [2]	68	1,150	60	56	57	1,024	60	56	70	1,000		
Northern Hemlock and Hardwood Manufacturers' Association, Oshkosh, Wis., Nov. 28, 1932.	Eastern Hemlock	Select structural	75	1,100	75	70	67	1,100	75	70	75	700	300	1,100,000
Southern Pine Association, New Oreleans, La., Sept. 1, 1932.	Longleaf pine [3]	Select structural	86	2,000	67	125	(4)		(4)		86	1,450		
		Prime structural	76	1,800	67	125	67	1,800	67	125	77	1,300		
		Merchantable structural	69	1,600	67	125	65	1,600	67	125	70	1,200		
		Structural square edge and sound.	69	1,600	67	125	65	1,600	56	125	70	1,200	380	1,600,000
		No. 1 structural	60	1,400	56	105	(4)		(4)		60	1,000		
	Shortleaf pine [3]	Dense select structural	86	2,000	67	125	(4)		(4)		86	1,450		
		Dense structural	76	1,800	67	125	69	1,800	67	125	77	1,300		
		Dense structural square edge and sound.	69	1,600	67	125	65	1,600	67	125	70	1,200	380	1,600,000
		Dense No. 1 structural [5]		1,200		125	57	1,200	56	125		850		
Southern Cypress Manufacturers' Association, Jacksonville, Fla., May 11, 1933.	Southern cypress	Select structural	75	1,300	75	100	67	1,300	75	100	75	1,100		
		Select structural heart	75	1,300	75	100	67	1,300	75	100	75	1,100		
		Common structural	60	1,040	60	80	57	1,040	60	80	60	880	300	1,200,000
		Common structural heart	60	1,040	60	80	57	1,040	60	80	60	880		
West Coast Lumbermen's Association, Seattle, Wash., Aug. 30, 1932.	Douglas fir (coast region).	Select structural [2] [6]	76	1,600	75	100	69	1,600	75	120	80	1,200	345	1,600,000
		Dense select structural [6] [7]	76	1,800	75	120	69	1,800	75	140	80	1,300	380	

[1] These stresses agree closely with the values computed from the strength ratios given in this table and the basic stresses of table 20 by the procedure given on p. 105 except that the horizontal shear stresses for joist and plank of select structural and dense select structural Douglas fir, structural square edge and sound longleaf pine, and dense no. 1 structural shortleaf pine are about 30 percent higher; other horizontal shear stresses for longleaf and shortleaf pines and Douglas fir are about 10 percent higher.

[2] Close-grain material required.

[3] Dense material required in all grades. Cluster knots and knots in groups not prohibited. Material up to 5 inches thick graded as joist and plank.

[4] Joist and plank are not described under these grade names.

[5] Admits in beams and stringers and in posts and timbers unsound knots up to 1½ inches in diameter and pin wormholes.

[6] In beams and stringers and in joists and planks knots are restricted throughout the length as required for the middle third of the length. No shakes permitted.

[7] Requires dense material.

erty damage. No actual failure would be expected with loads 50 per cent in excess of those computed from the tabulated stress values or with decrease in strength of the timber through decay or other deterioration to two-thirds its original value, but numerous failures are to be expected if such loads are doubled or if the strength of the timber decreases to one-half its original strength.

Timber Constantly Yields Under Long Continued Loading, acquiring a permanent set. This set with a fully loaded beam is about equal to the deflection using the molulus of elasticity as given in the tables. In order to minimize the results of sag, it is advisable to use values one-half those given in the tables.

Quality of Wood—The strength of the clear wood of any species varies over a considerable range; wood of the lowest strength is undesirable in a strength grade. On the other hand, recognition of the higher strength of the better wood is desirable. Strength is closely related to the weight or density of the wood. Higher strength may be obtained by excluding pieces that are obviously of exceptionally light weight and by using rate of growth and percentage of summerwood in selecting pieces of superior strength from those species in which these characteristics are acceptable criteria. Selection for rate of growth requires the number of annual rings per inch to be within a specified range. Selection for density imposes in addition to limitations of rate of growth the requirement of a minimum percentage of summerwood. It is applied only to those species in which summerwood is well differentiated and is known to be an efficient criterion of strength.

DETERMINATION OF WORKING STRESS.

The values listed in table 20 are basic stresses for material which is free of defects that affect the strength and which is used under such conditions that no deterioration will occur. They are termed "basic" because they are the working stress values for pieces of timber having a strength ratio of 100 per cent. In deriving them, allowances have been made for the variability of the strength of clear wood and for the effect of long-continued stress, and a factor of safety has been introduced. For example, in deriving the basic stresses in extreme fiber in bending the aver-

Table 20.—Basic stresses for clear material.[1]

Species (1)	Extreme fiber in bending (2)	Compression perpendicular to grain [2] (3)	Compression parallel to grain L/d=10 (4)	Maximum horizontal shear (5)	Modulus of elasticity (6)
Ash, black	1,333	300	866	120	1,100,000
Ash, commercial white	1,866	500	1,466	167	1,500,000
Beech	2,000	500	1,600	167	1,600,000
Birch, sweet and yellow	2,000	500	1,600	167	1,600,000
Cedar, Alaska	1,466	250	1,066	120	1,200,000
Cedar, northern and southern white	1,000	175	733	93	800,000
Cedar, Port Orford	1,466	250	1,200	120	1,200,000
Cedar, western red	1,200	200	933	106	1,000,000
Chestnut	1,266	300	1,066	120	1,000,000
Cypress, southern	1,733	300	1,466	133	1,200,000
Douglas fir, coast region	2,000	325	1,466	120	1,600,000
Douglas fir, coast region, close-grained	2,133	345	1,565	120	1,600,000
Douglas fir, Rocky Mountain region	1,466	275	1,066	113	1,200,000
Douglas fir, dense, all regions	2,333	380	1,711	140	1,600,000
Elm, American and slippery [3]	1,466	250	1,066	133	1,200,000
Elm, rock	2,000	500	1,600	167	1,300,000
Fir, balsam	1,200	150	933	93	1,000,000
Fir, commercial white	1,466	300	933	93	1,100,000
Gum, black and red	1,466	300	1,066	133	1,200,000
Hemlock, eastern	1,466	300	933	93	1,100,000
Hemlock, western [4]	1,733	300	1,200	100	1,400,000
Hickory, true and pecan	2,533	600	2,000	187	1,800,000
Larch, western	1,600	325	1,466	133	1,300,000
Maple, sugar and black [6]	2,000	500	1,600	167	1,600,000
Oak, commercial red and white	1,866	500	1,333	167	1,500,000
Pine, western white,[7] northern white, ponderosa, and sugar	1,200	250	1,000	113	1,000,000
Pine, Norway	1,466	300	1,066	113	1,200,000
Pine, southern yellow [8]	2,000	325	1,466	146	1,600,000
Pine, southern yellow, dense	2,333	380	1,711	171	1,600,000
Redwood	1,600	250	1,333	93	1,200,000
Redwood, close-grained	1,707	267	1,422	93	1,200,000
Spruce, Engelmann	1,000	175	800	93	800,000
Spruce, red, white, and Sitka	1,466	250	1,066	113	1,200,000
Tamarack	1,600	300	1,333	126	1,300,000
Tupelo	1,466	300	1,066	133	1,200,000

(All values are in pounds per square inch and are for material that is continuously dry or continuously wet.)

[1] Basic stresses are for determining design or working stresses according to the grade of timber and conditions of exposure.

[2] For material that is continuously wet take 70 percent or these values.

[3] Sold as white elm or soft elm.

[4] Also sold as west coast hemlock.

[5] In setting up basic working stresses consideration has been given to results of tests on small clear specimens and to tests on full-sized timbers of the species when the latter are available. That the value for stress in extreme fiber of western larch ought to be higher than that listed here is indicated by tests of small specimens but is not confirmed by available tests on structural sizes.

[6] Sold as hard maple.

[7] Also sold as Idaho white pine.

[8] Also sold as longleaf or shortleaf southern pine.

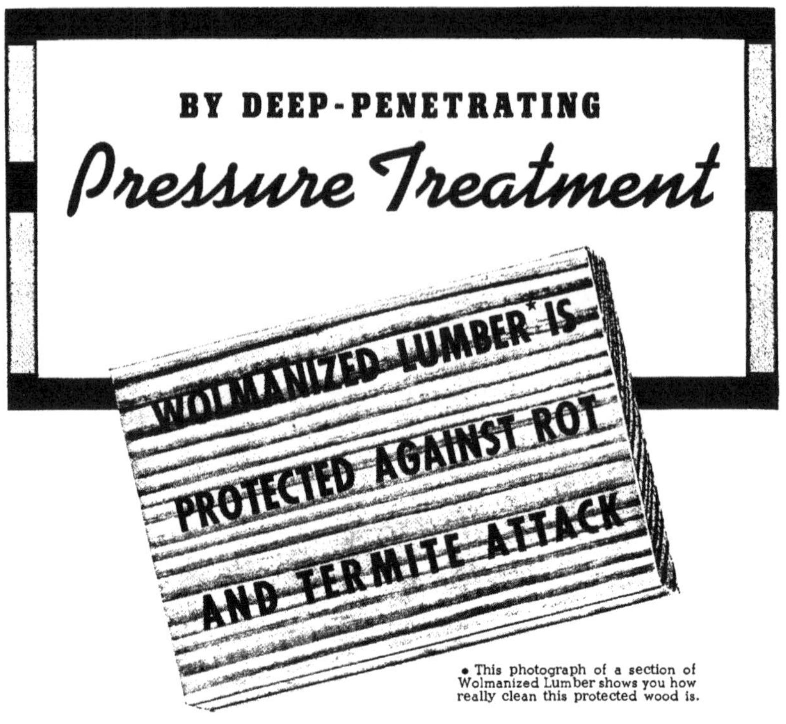

BY DEEP-PENETRATING
Pressure Treatment
WOLMANIZED LUMBER* IS
PROTECTED AGAINST ROT
AND TERMITE ATTACK

• This photograph of a section of Wolmanized Lumber shows you how really clean this protected wood is.

IT'S CLEAN, ODORLESS, PAINTABLE

For the jobs where chemically treated wood naturally belongs, architects are specifying Wolmanized Lumber, pressure-treated with a preservative which does not change the physical characteristics of the wood.

For substructure framing, sub-floors, industrial roof decks, platforms, sleepers and nailers in concrete and the number of other types of service where untreated lumber is subject to rot and termite attack, Wolmanized Lumber can be depended upon for protection.

"Wolman Salts"* preservative is "dyed-in-the-wood," and can't be leached out by rain or ground water. It leaves the wood clean and odorless, and allows painting with the assurance that the finish will hold as well as on untreated wood. It does not corrode metal.

Write for data on the technique of use of Wolmanized Lumber. Address American Lumber & Treating Company, Old Colony Building, Chicago

* Registered Trademark

age ultimate strength values as found from tests of clear wood in the green condition have been reduced by one-fourth to allow for the effect of variability and multiplied by nine-sixteenths to allow for the effect of long-continued stress and then divided by one and two-thirds as a factor of safety. Mutually consistent and equitable stress values cannot be derived solely by systematic computation from recorded data on strength and the basic stresses of table 20 therefore vary from systematically computed values according to favorable or unfavorable behavior of structural timbers of the species as observed in tests and under actual service conditions.

SPECIFICATION REQUIREMENTS FOR STRUCTURAL GRADES.

Information is presented here for use in drafting descriptions of timber grades or for determining the strength ratios of a grade or of a timber from its description.

Classification of Timbers.

The effects of knots, deviations of grain, shakes, and checks on the strength of a timber vary with the loading to which the piece is subjected. Also the effect of seasoning varies with the size of the timber. Consequently, efficiency in grading necessitates classifying timbers according to their size and use. Such a classification is as follows:

Beams and Stringers—Large pieces (nominal dimensions 5 by 8 inches and up) ot rectangular cross section graded with respect to their strength in bending when loaded on the narrow face.

Joist and Plank—Pieces (nominal dimensions 2 to 4 inches in thickness by 4 inches and wider) of rectangular cross section graded with respect to their strength in bending when loaded either on the narrow face as joist or on the wide face as plank.

Posts and Timbers—Pieces of square or approximately square cross section, 4 by 4 inches and larger, in nominal dimension, graded primarily for use as posts or columns but adapted to miscellaneous uses in which strength in bending is not especially important.

Actual Dimensions.

In accordance with American lumber standards, rough (unsurfaced) pieces shall be sawn full to nominal dimension except that occasional slight variation in swing is permissible. At no part of the length shall any piece because of such variation be more than three-sixteenths inch under the nominal dimension when this dimension is 3 to 7 inches, inclusive, nor more than one-fourth inch under the nominal dimension when this dimension is 8 inches or greater. The actual thickness of nominal 2-inch material shall not be less than 1⅞ inches at any part of the length. Further, no shipment shall contain more than 20 per cent of pieces of minimum dimension.

Surfacing of beams and stringers, whether on one or both of a pair of opposite faces, shall leave the finished size not more than one-half inch under the nominal dimension.

Surfacing of joist and plank, whether on one or both of a pair of opposite faces, shall leave the finished size not more than three-eighths inch under the nominal dimension when this dimension is 7 inches or less and not more than one-half inch under the nominal dimension when this dimension is 8 inches or greater.

Surfacing of posts and timbers, whether on one or both of a pair of opposite faces, shall leave the finished size not more than three-eighths inch under the nominal dimensions when this dimension is 4 inches and not more than one-half inch under the nominal dimension when this dimension is 5 inches or greater.

Quality of Wood.

No piece of exceptionally light weight foi the species is permitted.

Decay.

Only pieces consisting of sound wood, free from any form of decay, including firm red heart and dote, are acceptable.

Slope of Grain.

Table 22 gives the strength ratios corresponding to various slopes of grain. Slope of grain is to be measured over a distance sufficiently great to define the general slope disregarding short local deviations.

Table 22—Strength ratios corresponding to various slopes of grain (Beams and stringers or joist and plank—strength ratio for stress in extreme fiber in bending. Posts and timbers—strength ratio for stress in compression parallel to grain).

Slope of grain	Strength ratio		Slope of grain	Strength ratio	
	Beams and stringers or joist and plank	Posts and timbers		Beams and stringers or joist and plank	Posts and timbers
	Per cent	Per cent		Per cent	Per cent
1 in 6		50-56	1 in 15	74- 76	87-100
1 in 8	50-53	56-66	1 in 16	76- 80	....
1 in 10	53-61	66-74	1 in 18	80- 85	....
1 in 12	61-69	74-82	1 in 20	85-100	....
1 in 14	69-74	82-87			

The limitations given for beams and stringers and joist and plank in table 22 apply only within the middle half[24] of the length of the piece; the slope in other parts is disregarded. The limitations for posts and timbers apply throughout the length of the piece.

The limitations herein stated for the sizes of knots in the middle third of the length and for slope of grain in the middle one-half of the length of beams and stringers or joists and planks assume that such pieces of timber will be used on single spans. They should be applied to the middle two-thirds of the length of pieces to be used over double spans and to the entire length of pieces to be used over three or more spans.

Knots.

Sizes of knots permissible in grades having various strength ratios are determined according to tables 23, 24 and 25 supplemented by the following definitions, explanations, and limitations.

Cluster knots and knots in groups are not permitted.

Knot holes and holes from causes other than knots are measured and limited as provided for knots.

Measurement of Knots

Beams and Stringers.

On a narrow face of the piece the size of a knot is taken as its width between lines enclosing the knot and parallel to the edges of the piece; except that when a knot on a narrow face extends into the adjacent one-fourth of the width of a wide face its least dimension is taken as its size.

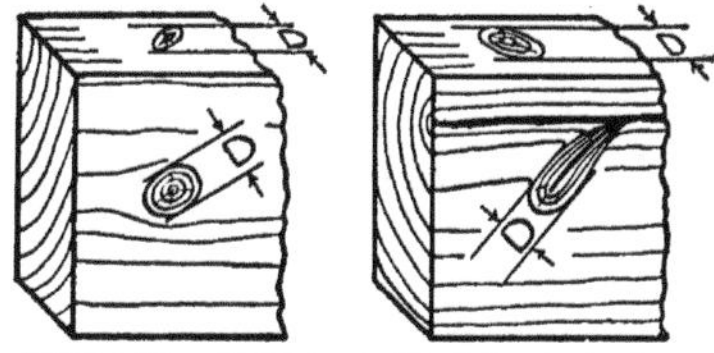

Figure 10.—Measurement of knots in beams and stringers.

taken as half the sum of its length and greatest width.

The sizes of knots on narrow faces and at edges of wide faces may increase proportionately from the size permitted in the middle third of the length to twice that size at the ends of the piece.

The size of knots on wide faces may increase proportionately from the size permitted at the edge to the size permitted along the center line.

The sum of the sizes of all knots within the middle half of the length of any face measured as specified for the face under consideration shall not exceed 4½ times the size of the largest knots allowed on that face.

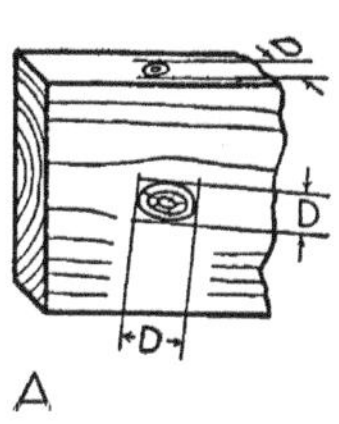

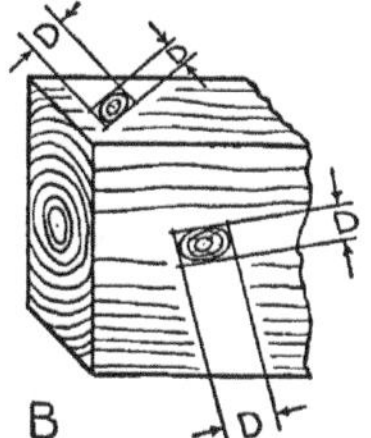

Figure 12.—Measurement of knots: A, On joist and plank; B, on posts and timbers.

On a wide face the size of a knot is taken at its smallest diameter.

At the edges of wide faces knots are limited to the same sizes as on the narrow faces of the same piece.

The sizes of knots on narrow faces and at edges of wide faces may increase proportionately from the size permitted in the middle third of the length to twice that size at the ends of the piece except that the size of no knot shall exceed the size permitted along the center line of the wide face.

The size of knots on wide faces may increase proportionately from the size permitted at the edge to the size permitted along the center line.

The sum of the sizes of all knots within the middle half of the length of any face measured as specified for the face under consideration shall not exceed four times the size of the largest knot allowed on that face.

Joist and Plank.

On a narrow face of the piece, the size of a knot is taken as its width between lines parallel to the edges of the piece. The only knots measured on the narrow faces of a piece are those that do not show on the wide faces.

On a wide face the size of knot is taken as half the sum of its largest and smallest diameters and the size of a spike knot is

Posts and Timbers.

On any face the size of a knot is taken as half the sum of its largest and smallest diameter, and the size of a spike knot is taken as half the sum of its length and greatest width.

The sum of the sizes of all knots in any 6 inches of length of the piece is not permitted to exceed twice the maximum permissible size of knots. Two knots of maximum permissible size are not allowed in the same 6 inches of length on any face.

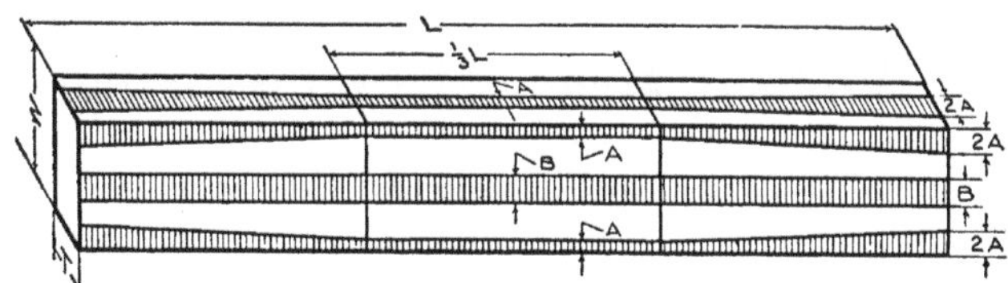

Figure 11.—Maximum size of knots permitted in various parts of beams and stringers: A, Maximum size on narrow face or edge of wide face, middle third of length with a gradual increase to 2A at ends of piece; B, maximum size at center line of wide face; L, length; W, width; T, thickness.

Tables of Knot Sizes and Their Use.

The use of tables 23, 24 and 25, which give sizes of knots, is illustrated by the following example. The sizes of knots permissible in a nominal 8 by 16 inch piece in a grade having a strength ratio of 70 per cent are desired. The smallest ratio in the column for 8-inch face in table 23 (narrow face) that equals or exceeds 70 per cent is opposite 2⅛ inches in the size-of-knot column and a similar rate in the column for 16-inch face in table 24 (wide face) is opposite 4¼ inches. Hence, the permissible sizes are 2⅛ inches on the 8-inch face and 4¼ inches along the center line of the 16-inch face.

SHAKES AND CHECKS.

Beams and Stringers or Joist and Plank.

Limitations for shakes in beams and stringers and in joist and plank are indicated by table 26.

Shakes are measured at the ends of the piece. In beams and stringers and joist and plank only those within the middle half of the height of the piece are considered. (Height equals width of wide face.) The size of a shake is the distance between lines enclosing the shake and parallel to the wide faces of the piece. The permissible size is determined by the width of the narrow face of the piece.

Checks and splits are limited in the same way as shakes. The following limitations apply to both ends but only within the middle half of the height of the piece and within three times the height from the end. (Height equals width of wide face.) The size of checks within this portion of the piece shall be taken as their estimated area, along the horizontal section showing maximum area, divided by three times the height of the piece. When checks on two parallel faces are opposite or approximately so, the sum of their sizes is taken. The sum of the sizes of shakes, checks and/or splits shall not exceed the permissible size of shake.

Checks extending entirely across the end within the middle half of the height shall not extend into the piece at the center of the width of the end a distance greater than the size of the allowable shake.

Table 23.—Strength ratios for stress in extreme fiber in bending corresponding to various combinations of size of knot and width of face.

(Beams and stringers—narrow face or edge of wide face within middle third[1] of length of piece. Joist and plank—narrow face within middle third[1] of length of piece.)

Size of knot (inches)	Percentage strength ratios for nominal face width (inches) indicated									
	2	3	4	5	6	8	10	12	14	16
¼	90	93	95	96	96	97	97	97	98	98
⅜	83	89	92	93	94	95	96	96	96	97
½	77	85	88	91	92	93	94	95	95	95
⅝	71	81	85	88	90	92	92	93	94	94
¾	65	76	82	86	88	90	91	92	92	93
⅞	58	72	79	83	86	88	89	90	91	91
1	52	68	76	81	84	86	88	89	89	90
1⅛	.....	64	73	78	82	84	86	87	88	89
1¼	.....	60	70	76	80	83	84	86	87	88
1⅜	.....	56	67	73	78	81	83	84	85	86
1½	.....	51	63	71	76	79	81	83	84	85
1⅝	.....	.....	60	68	74	77	80	81	83	84
1¾	.....	.....	57	66	71	75	78	80	81	83
1⅞	.....	.....	54	63	69	73	76	78	80	81
2	.....	.....	51	61	67	72	75	77	79	80
2⅛	.....	.....	.....	58	65	70	73	75	77	79
2¼	.....	.....	.....	56	63	68	71	74	76	77
2⅜	.....	.....	.....	53	61	66	70	72	74	76
2½	.....	.....	.....	51	59	64	68	71	73	75
2⅝	.....	.....	.....	.....	57	63	67	70	72	74
2¾	.....	.....	.....	.....	55	61	65	68	70	72
2⅞	.....	.....	.....	.....	53	59	63	67	69	71
3	.....	.....	.....	.....	51	57	62	65	68	70
3⅛	.....	.....	.....	.....	.....	55	60	64	66	68
3¼	.....	.....	.....	.....	.....	54	59	62	65	67
3⅜	.....	.....	.....	.....	.....	52	57	61	64	66
3½	.....	.....	.....	.....	.....	50	55	59	62	65
3⅝	.....	.....	.....	.....	.....	.....	54	58	61	63
3¾	.....	.....	.....	.....	.....	.....	52	56	59	62
3⅞	.....	.....	.....	.....	.....	.....	50	55	58	61
4	.....	.....	.....	.....	.....	.....	.....	53	57	60
4⅛	.....	.....	.....	.....	.....	.....	.....	52	55	58
4¼	.....	.....	.....	.....	.....	.....	.....	50	54	57
4⅜	.....	.....	.....	.....	.....	.....	.....	.....	53	56
4½	.....	.....	.....	.....	.....	.....	.....	.....	51	54
4⅝	.....	.....	.....	.....	.....	.....	.....	.....	50	53
4¾	.....	.....	.....	.....	.....	.....	.....	.....	.....	52
4⅞	.....	.....	.....	.....	.....	.....	.....	.....	.....	51

[1] The limitations assume that such pieces of timber will be used on single spans. They should be applied to the middle ⅔ of the length of pieces to be used over double spans and to the entire length of pieces to be used over 3 or more spans.

Table 24.—Strength ratios corresponding to various combinations of size of knot and width of face.

(Beams and stringers or joist and plank—along center line of wide face at any point in the length of the piece. Strength ratios for stress in extereme fiber in bending. Posts and timbers—at any point on any face. Strength ratios for stress in compression parallel to grain.)

Size of knot (inches)	Percentage strength ratios for nominal face width (inches) indicated											
	4	5	6	8	10	12	14	16	18	20	22	24
¼	95	96	96	97	98	98	98	98	99	99	99	99
½	88	91	92	94	95	96	96	97	97	97	97	97
¾	82	86	88	91	93	94	94	95	95	95	96	96
1	76	81	84	88	90	92	93	93	93	94	94	94
1¼	70	76	80	85	88	90	91	91	92	92	93	93
1½	63	71	76	82	85	88	89	89	90	91	91	91
1¾	57	66	71	79	83	86	87	88	88	89	89	90
2	51	61	67	75	80	84	85	86	87	87	88	88
2¼	.....	56	63	72	78	82	83	84	85	86	86	87
2½	.....	51	59	69	75	79	81	82	83	84	85	85
2¾	.....	.....	55	66	73	77	79	80	82	82	83	84
3	.....	.....	51	63	70	75	77	79	80	81	82	83
3¼	.....	.....	.....	60	68	73	75	77	78	79	80	81
3½	.....	.....	.....	57	65	71	73	75	76	78	79	80
3¾	.....	.....	.....	54	63	69	71	73	75	76	77	78
4	.....	.....	.....	50	60	67	69	71	73	74	76	77
4¼	.....	.....	.....	.....	58	65	67	70	71	73	74	75
4½	.....	.....	.....	.....	55	63	66	68	70	71	73	74
4¾	.....	.....	.....	.....	53	61	64	66	68	70	71	72
5	.....	.....	.....	.....	50	59	62	64	66	68	70	71
5¼	.....	.....	.....	.....	.....	57	60	62	65	66	68	69
5½	.....	.....	.....	.....	.....	54	58	61	63	65	66	68
5¾	.....	.....	.....	.....	.....	52	56	59	61	63	65	66
6	.....	.....	.....	.....	.....	50	54	57	59	61	63	65
6¼	.....	.....	.....	.....	.....	.....	52	55	58	60	62	63
6½	.....	.....	.....	.....	.....	.....	50	53	56	58	60	62
6¾	.....	.....	.....	.....	.....	.....	.....	52	54	57	59	60
7	.....	.....	.....	.....	.....	.....	.....	50	53	55	57	59
7¼	.....	.....	.....	.....	.....	.....	.....	.....	51	53	56	57
7½	.....	.....	.....	.....	.....	.....	.....	.....	.....	52	54	56
7¾	.....	.....	.....	.....	.....	.....	.....	.....	.....	50	53	55
8	.....	.....	.....	.....	.....	.....	.....	.....	.....	.....	51	53
8¼	.....	.....	.....	.....	.....	.....	.....	.....	.....	.....	.....	52
8½	.....	.....	.....	.....	.....	.....	.....	.....	.....	.....	.....	50

Posts and Timbers.

Limitations for shakes in posts and timbers are indicated in table 27. Shakes do not greatly affect the strength of members subjected to longitudinal compression. Their effect on appearance and the opportunity they afford for the start of decay form the principal basis for the limitations given in table 27.

Shakes are measured at the ends of the piece. The size of a shake is the distance between the maximum spaced pair of lines exactly enclosing the shake and parallel to two opposite faces.

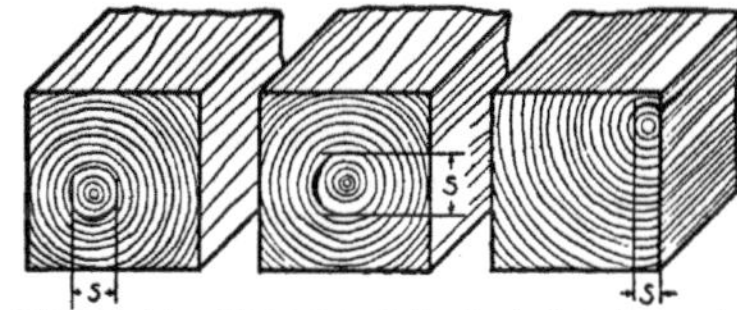

Figure 14.—Measurement of shakes in posts and timbers.

Fraction of width of face occupied by width of wane:	Strength ratio of grade
¼	50 per cent up to and including 60 per cent.
⅕	Above 60 per cent up to and including 66 per cent.
⅙	Above 66 per cent up to and including 75 per cent.
⅛	Above 75 per cent up to and including 87 per cent.
1/10	Above 87 per cent.

The schedule of permissible sizes of wane has been made to conform with usual practice in grading although in some instances more strict limitations are imposed because of appearance or for other reasons. Undue importance is often attached to wane, but its actual effects are much less than is implied. For example, the loss in strength due to wane one-fourth the width of two adjacent faces of a beam is only about 9 per cent; and if such wane extended the full length of a beam, the deflection of the beam under load would be only about 7 per cent greater than if no wane existed. It is safe to assume that the percentage reduction in bending strength due to wane does not exceed three times the percentage reduction in area of cross section of the piece. Wane at the ends of beams is often undesirable because of the eccentric and reduced area available for bearing.

YARD LUMBER.

Definition.

9. The term "yard lumber" as here used means lumber that is manufactured and classified into those sizes, shapes, and qualities required for ordinary construction and general purpose uses. (Structural lumber, softwood factory and shop lumber, hardwood factory lumber, and other special-use lumber are not considerable yard lumber.)

Manufacturing Classification.

5. Lumber is classified, according to the extent to which it is manufactured, into 3 classes as shown below, the third class being further subdivided.

Rough Lumber—Lumber undressed as it comes from the saw.

Surfaced Lumber—Lumber that is dressed by running it through a planer. It may be surfaced on one side (S1S), two sides (S2S), one edge (S1E), two edges (S2E), or on a combination of sides and edges: (S1S1E), (S2S1E), (S1S2E), or (S4S).

Worked Lumber—Lumber which has been run through a matching machine, sticker, or molder. Worked lumber may be:

Matched Lumber — Lumber that is edge dressed and shaped to make a close tongued and grooved joint at the edges or ends when laid edge to edge or end to end.

Ship-lapped Lumber—Lumber that is edge dressed to make a close rabbeted or lapped joint.

Patterned Lumber—Worked lumber that is shaped to a patterned or molded form.

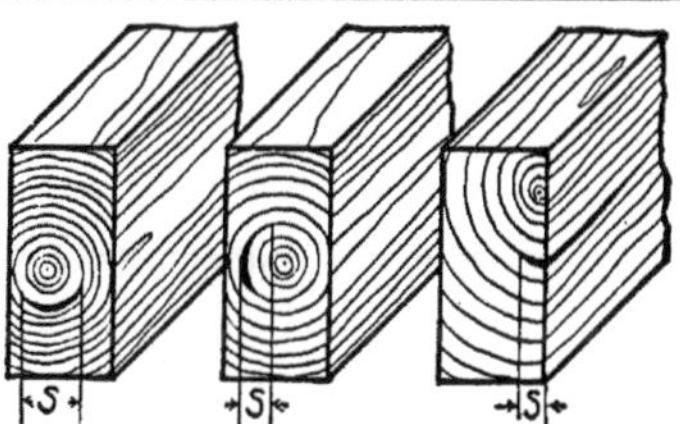

Figure 13.—Measurement of shakes in beams and stringers or joists and plank.

Table 26.—Strength ratios for stress in horizontal shear corresponding to various combinations of size of shake and width of end of piece.
(Beams and stringers or joist and plank.)

Percentage strength ratios for nominal end width of piece (inches) indicated

Green material

Size of shake (inches)	2	3	4	5	6	8	10	12	14	16
1/4	90	93	95	96	96	97	98	98	98	99
3/8	83	89	92	93	94	96	97	97	98	98
1/2	77	85	88	91	92	94	95	96	97	97
5/8	71	81	85	88	90	93	94	95	96	96
3/4	65	76	82	86	88	91	93	94	95	96
7/8	58	72	79	83	86	90	92	93	94	95
1	52	68	76	81	84	88	90	92	93	94
1 1/8		64	73	78	82	86	89	91	92	93
1 1/4		60	70	76	80	85	88	90	91	92
1 3/8		56	67	73	78	83	87	89	90	92
1 1/2		51	63	71	76	82	85	88	90	91
1 5/8			60	68	74	80	84	87	89	90
1 3/4			57	66	71	79	83	86	88	89
1 7/8			54	63	69	77	82	85	87	88
2			51	61	67	75	80	84	86	88
2 1/8				58	65	74	79	83	85	87
2 1/4				56	63	72	78	82	84	86
2 3/8				53	61	71	77	81	83	85
2 1/2				51	59	69	75	79	82	85
2 5/8					57	68	74	78	81	84
2 3/4					55	66	73	77	81	83
2 7/8					53	65	72	76	80	82
3					51	63	70	75	79	81
3 1/4						60	68	73	77	80
3 1/2						57	65	71	75	78
3 3/4						54	63	69	73	77
4						50	60	67	72	75
4 1/4							58	65	70	74
4 1/2							55	63	68	72
4 3/4							53	61	66	71
5							50	59	65	69
5 1/4								57	63	67
5 1/2								54	61	66
5 3/4								52	59	64
6								50	57	63
6 1/4									56	61
6 1/2									54	60
6 3/4									52	58
7									50	56
7 1/4										55
7 1/2										53
7 3/4										52
8										50

Seasoned material

Size of shake (inches)	2	3	4	5	6	8	10	12	14	16
1/4	100	100	100	100	100	100	100	100	100	100
3/8	94	100	100	100	100	100	100	100	100	100
1/2	87	95	100	100	100	100	100	100	100	100
5/8	80	91	96	99	100	100	100	100	100	100
3/4	73	86	93	97	99	100	100	100	100	100
7/8	66	81	89	94	97	100	100	100	100	100
1	59	77	85	91	94	99	100	100	100	100
1 1/8	52	72	82	88	92	97	100	100	100	100
1 1/4		67	78	85	90	95	99	100	100	100
1 3/8		62	75	82	87	94	97	100	100	100
1 1/2		58	71	80	85	92	96	99	100	100
1 5/8		53	68	77	83	90	95	98	100	100
1 3/4			64	74	80	88	93	96	99	100
1 7/8			61	71	78	87	92	95	98	100
2			57	68	76	85	90	94	97	99
2 1/8			54	66	73	83	89	93	96	98
2 1/4			50	63	71	81	88	92	95	97
2 3/8				60	69	80	86	91	94	96
2 1/2				57	66	78	85	89	93	95
2 5/8				54	64	76	83	88	92	94
2 3/4				52	62	74	82	87	91	93
2 7/8					59	73	81	86	90	93
3					57	71	79	85	89	92
3 1/4					52	67	76	82	87	90
3 1/2					48	64	74	80	85	88
3 3/4						60	71	78	83	86
4						57	68	75	81	85
4 1/4						53	65	73	79	83
4 1/2						50	62	71	77	81
4 3/4							59	68	75	79
5							57	66	73	78
5 1/4							54	64	71	76
5 1/2							51	61	69	74
5 3/4								59	67	72
6								57	65	71
6 1/4								54	63	69
6 1/2								52	61	67
6 3/4								50	59	65
7									57	64
7 1/4									55	62
7 1/2									53	60
7 3/4									51	58
8										56
8 1/4										55
8 1/2										53
8 3/4										51
9										49

GRADE STANDARDS.

Basic Grade Classification.

10. On the basis of quality, yard lumber is divided into 2 main divisions: (a) Select lumber, and (b) Common lumber. These are again divided into 2 classes: Select lumber, into (1) that suitable for natural finishes, and (2) that suitable for paint finishes; Common lumber, into (1) that which can be used without waste, and (2) that which permits some waste. Each of these 4 classes is further divided into quality classes or grades. The grade names of Select lumber are A, B, C, and D, and of Common lumber, No. 1, No. 2, No. 3, No. 4, and No. 5. These divisions, classes, and grades are differentiated as shown below.

Total products of a typical log arranged in series according to quality as determined by appearance and use.

- SELECT. Lumber of good appearance and finishing qualities.
  - Suitable for natural finishes.
    - Grade A. Practically clear.
    - Grade B. Of high quality—generally clear.
  - Suitable for paint finishes.
    - Grade C. Adapted to high quality paint finishes.
    - Grade D. Intermediate between higher finishing grades and common grades, and partaking somewhat of the nature of both.
- COMMON. Lumber not of finishing quality, but which is suitable for general utility and construction purposes.
  - Suitable for use without waste.
    - No. 1 Sound and tight knotted. May be considered water-tight.
    - No. 2 Less restricted in quality than No. 1, but of the same general character.
  - Permitting some waste.
    - No. 3 Prevailing grade characteristics larger than in No. 2.
    - No. 4 Low quality.
    - No. 5 Lowest recognized grade, but must be usable.

Table 27.—Strength ratios for stress in compression parallel to grain corresponding to various combinations of size of shake and width of end of piece.
(Posts and timbers.)

Percentage strength ratios for nominal end width of piece (inches) indicated

Green material							Seasoned material						
Size of shake (inches)	4	5	6	8	10	12	Size of shake (inches)	4	5	6	8	10	12
¾	100	100	100	100	100	100	¾	100	100	100	100	100	100
1	95	100	100	100	100	100	1	100	100	100	100	100	100
1¼	87	95	100	100	100	100	1¼	100	100	100	100	100	100
1½	79	89	95	100	100	100	1½	92	100	100	100	100	100
1¾	72	82	89	98	100	100	1¾	84	95	100	100	100	100
2	64	76	84	94	100	100	2	76	88	97	100	100	100
2¼	56	70	79	90	97	100	2¼	68	82	91	100	100	100
2½	48	63	74	87	94	99	2½	61	76	86	99	100	100
2¾		57	69	83	91	97	2¾	53	70	81	95	100	100
3		51	63	79	88	94	3	45	63	76	91	100	100
3¼			58	75	85	92	3¼		57	71	87	97	100
3½			53	71	82	89	3½		51	65	83	94	100
3¾			48	67	79	86	3¾			60	80	91	99
4				63	75	84	4			55	76	88	96
4¼				59	72	81	4¼			50	72	85	94
4½				55	69	79	4½				68	82	91
4¾				51	66	76	4¾				64	79	88
5					63	73	5				60	75	86
5¼					60	71	5¼				56	72	83
5½					57	68	5½				52	69	81
5¾					54	65	5¾				48	66	78
6					50	63	6					63	75
6¼						60	6¼					60	73
6½						58	6½					57	70
6¾						55	6¾					54	68
7						52	7					50	65
7¼						50	7¼						62
7½							7½						60
7¾							7¾						57
8							8						55
8¼							8¼						52
8½							8½						49

BASIC SIZES OF LUMBER—AMERICAN LUMBER STANDARDS.

Structural Lumber.

The thickness and widths adopted as standard for each of the three classes of structural material are given in Table 16.

Table 16.—American standard thicknesses and widths for structural material.
(The thicknesses apply to all widths and the width to all thicknesses.)

Structural item	Thicknesses			Widths		
	Nominal thickness	Permissible minimum rough thickness[1]	Dressed thickness (S1S or S2S)	Nominal width	Permissible minimum rough width[1]	Dressed width (S1E or S2E)
	Inches	*Inches*	*Inches*	*Inches*	*Inches*	*Inches*
Joist and plank	2	⅛ off	⅜ off	4 to 7	$\frac{3}{16}$ off	⅜ off.
	3 and 4	$\frac{3}{16}$ off		8 and wider	¼ off	½ off.
Beams and stringers	5 and 6	do	½ off	do	do	Do.
	7 and thicker	¼ off				
Posts and timbers	5 by 5, 6 by 6, 8 by 8, and larger.	$\frac{3}{16}$ off ¼ off	do	5 by 5, 6 by 6, 8 by 8, and larger.	$\frac{3}{16}$ off ¼ off	Do.

[1] No shipment shall contain more than 20 percent of pieces of the permissible minimum rough thicknesses or widths specified here.

YARD LUMBER.

Basis of Measurement of Sizes.

22. Dressed dimensions, shall apply to lumber in the condition of seasoning as sold and shipped.

Dressed Sizes.

23. The terms "standard yard board" and "standard yard dimension" shall be the designations for 1-inch boards and 2-inch dimension, respectively.

24. $\frac{25}{32}$ inch, S1S or S2S, shall be the thickness for the standard yard board.

25. 1⅝ inches, S1S or S2S, shall be the thickness for standard yard dimension not more than 12 inches wide.

26. The finished widths of finish S1E or S2E shall be ⅜ inch off on lumber of standard width of 3 inches; the finished widths of finish S1E or S2E (based on kiln-dried lumber) shall be ½ inch off on lumber of standard widths of 4 to 7 inches, inclusive, and ¾ inch off on lumber of standard widths of 8 to 12 inches, inclusive; and the finished widths of strips, boards and dimension S1E or S2E shall be ⅜ inch off on lumber of standard widths of 8 to 12 inches, inclusive; and the finished widths of strips, boards and dimension S1E or S2E shall be ⅜ inch off on lumber of standard widths less than 8 inches and ½ inch off on lumber of standard widths of 8 to 12 inches.

27. The minimum thickness and widths of finished lumber, S1S or S2S and/or S1E or S2E, shall be as follows:

FINISH, COMMON STRIPS, BOARDS, AND DIMENSION.
(The thicknesses apply to all widths and the widths to all thicknesses.)

Product	Size, Board Measure		Dressed Dimensions	
	Thickness	Width	Standard thickness, yard	Standard width
	Inches	Inches	Inches	Inches
Finish	....	3	5/16	2⅝
	....	4	7/16	*3½
	....	5	9/16	*4½
	....	6	11/16	*5½
	1	7	25/32	*6½
	1¼	8	1 1/16	*7¼
	1½	9	1 5/16	*8¼
	1¾	10	1 7/16	*9¼
	2	11	1⅝	*10¼
	2½	12	2⅛	*11¼
	3	..	2⅝	.....
Common:				
Strips and boards	1	3	25/32	2⅝
	1¼	4	1 1/16	3⅝
	1½	5	1 5/16	4⅝
	....	6	....	5⅝
	....	7	....	6⅝
	....	8	....	7½
	....	9	....	8½
	....	10	....	9½
	....	11	....	10½
	....	12	....	11½
Dimension	2	2	1⅝	1⅝
	2½	4	2⅛	3⅝
	3	6	2⅝	5⅝
	4	8	3⅝	7½
	....	10	....	9½
	....	12	....	11½

*Based on kiln-dried lumber.

DOUGLAS FIR PLYWOOD STANDARDS.
Effective for New Production Beginning Nov. 10, 1938.

General Requirements.

4. All Douglas fir plywood sold as of commercial standard quality shall meet the following general requirements:

5. **Workmanship** — It shall be smoothly sanded on two sides unless otherwise specified. It shall be well manufactured and free from blisters, laps, and defects, except as permitted in the specific rules for the various grades.

6. **Construction**—Veneers 1/12 inch or more shall be used in the construction of panels ¼ inch and upward in the thickness. The veneer thickness shall be measured before the panel is sanded.

7. **Bonding**—Plywood shall be bonded in an approved manner with material best adapted to each use classification.

8. **Loading or packing**—It shall be securely loaded or packed to insure delivery in a clean and serviceable condition.

Detail Requirements.

9. Douglas fir plywood shall be graded according to both sides of the piece into the following standard grades. The grade descriptions set forth the minimum requirements, and therefore, the majority of panels in any shipment will exceed the specification given.

10. **Good 2 Sides (G2S)**—Each face shall be of a single piece of smoothly cut veneer of 100-per cent heartwood, free from knots, splits, checks, pitch pockets, and other open defects. The faces shall be a yellow or pinkish color without stain. Shims that occur only at the ends of panels and inconspicuous well-matched small patches not to exceed ⅜ inch wide by 2½ inches long shall be admitted. This grade is recommended for uses where a light stain or natural finish is desired.

11. **Good 1 Side (G1S)**—One face shall be equal to that described under "Good 2 Sides" grade, while the opposite face shall be equal to the "Sound 2 Sides" grade described below.

12. **Sound 2 Sides (SO2S)**—Each face shall be of one or more pieces of firm, smoothly cut veneer. When of more than one piece, it shall be well joined and reasonably matched for grain and color at the joints. It shall be free from knots, splits, checks, pitch pockets, and other open defects. Streaks, discolorations, sapwood, shims, and neatly made patches shall be admitted. This grade shall present a smooth surface suitable for painting.

13. **Wallboard (WB)** — This is a 3-ply board of ¼ inch or ⅜ inch sanded, or 5-ply ½ inch sanded thickness, made only in standard wallboard sizes, the face of which shall be of one or more pieces of firm, smoothly cut veneer. When of more than one piece it shall be well joined and reasonably matched for grain and color at the joints. It shall be free from knots, splits, pitch pockets, and other open defects. Streaks, discolorations, sapwood, shims, and neatly made patches shall be admitted. The face on this grade shall present a smooth surface suitable for painting. The backs shall contain knot holes or pitch pockets, splits, and other defects in number and size that will not seriously affect the strength or serviceability of the panel and which cannot reasonably and economically be repaired to make a sound face. All wallboard panels shall be so designated by grade marking each panel.

14. **Sheathing (SH)**—This is an unsanded plywood made only in the following sizes: Thicknesses 5/16 inch and ⅜ inch 3 ply, ⅝ inch 3 or 5 ply; widths 32 and 48 inches; length 96 inches. One face shall present a solid surface except that the following will be permitted: (a) Not more than 6 knotholes ⅜ inch or less in greatest dimension, (b) splits 1/16 inch or less in width, (c) one or two strips of paper tape. There may be any number of patches and plugs in the face but the face may not be of such quality that, if sanded, it will pass for a wallboard face. No belt sanding is permissible. The back shall contain solid knots, knotholes or pitch pockets, splits and/or other defects in number and size

IDING, FLOORING, CEILING, PARTITION, SHIP-LAP, AND DRESSED AND MATCHED.
The thicknesses apply to all widths and the widths to all thicknesses except as modified.[1])

Product	Size, Board Measure		Dressed Dimensions	
	Thickness	Width	Standard thickness	Standard face width
	Inches	Inches	Inches	Inches
evel siding	...	4	7/16 by 3/16	3½
	...	5	⅝ by 3/16	4½
	...	6	....	5½
'ide bevel siding	...	8	7/16 by 3/16	7¼
	...	10	9/16 by 3/16	9¼
:ustic and drop siding (ship-lapped)	...	12	11/16 by 3/16	11¼
	...	4	9/16	3⅛
	...	5	¾	4⅛
	...	6	....	5 1/16
:ustic and drop siding (dressed and matched)	...	8	....	6⅞
	...	4	9/16	3¼
	...	5	¾	4¼
	...	6	....	5 3/16
	...	8	....	7
'looring	...	2	5/16	1½
	...	3	7/16	2⅜
	...	4	9/16	3¼
	1	5	25/32	4¼
	1¼	6	1 1/16	5 3/16
	1½	..	1 5/16	....
'eiling	...	3	5/16	2⅜
	...	4	7/16	3¼
	...	5	9/16	4¼
	...	6	11/16	5 3/16
'artition	...	3	¾	2⅜
	...	4	....	3¼
	...	5	....	4¼
	...	6	....	5 3/16
'hip-lap	1	4	25/32	3⅛
	...	6	....	5⅛
	...	8	....	7⅛
	...	10	....	9⅛
	...	12	....	11⅛
)ressed and matched	1	4	25/32	3¼
	1¼	6	1 1/16	5¼
	1½	8	1 5/16	7¼
	...	10	....	9¼
	...	12	....	11¼

[1] In tongued-and-grooved flooring and in tongued-and-grooved and ship-lapped ceiling 5/16, 7/16, and 9/16 inch thick, board measure, the tongue or lap shall be 3/16 inch wide, with the lumber, 11/16, ¾, 1, 1¼, and 1½ inches thick, board measure, the tongue shall be ¼ inch wide in tongued-and-grooved lumber, and the lap ⅜ inch wide in ship-lapped lumber, with the over-all widths ¼ inch and ⅜ inch wider, respectively, than the face widths shown above.

that will not seriously affect the strength or serviceability of the panel. No tape shall be permitted in the glue line. All sheathing panels shall be scored or marked for nailing to conform to standard spacing of lumber studding.

15. **Automobile and industrial stock (rough).** Faces of panels shall be free from knot holes. Tight knots, straight and tight checks shall be admitted. Pieced faces constitute no defect. Core and cross bands shall be of firm stock. Knot holes in cores and cross bands, up to 1¼ inches in diameter are permitted.

16. **Concrete form plywood**—Concrete form plywood shall be built up of three or five thicknesses of veneer, of which the two outside plies are at least ⅛ inch thick before sanding. An occasional knot hole is permissible in the center or core of 5-ply panels only but no knot holes are permitted in cross banding. Faces shall be free from knots or open defects. The bonding agent used shall be especially prepared for this purpose and be very highly water-resistant. All concrete form plywood shall be so designated by grade marking each panel. (When so ordered, concrete form plywood will be treated with a satisfactory form oil or other preparation.)

FACTORY FLOORING, HEAVY ROOFING, DECKING AND SHEET PILING.
(The thicknesses apply to all widths and the widths to all thicknesses.)

Size, Board Measure		Dressed dimensions			
Thickness	Width	Standard thickness	Standard face width		
			D & M	Ship-lapped	Grooved for splines
Inches	Inches	Inches	Inches	Inches	Inches
2	4	1⅝	3⅛	3	3½
2½	6	2⅛	5⅛	5	5½
3	8	2⅝	7⅛	7	7½
4	10	3⅝	9⅛	9	9½
....	12	...	11⅛	11	11½

Note: In patterned lumber 2 or more inches thick, board measure, the tongue shall be ⅜ inch wide in tongued-and-grooved lumber and the lap ½ inch wide in ship-lapped lumber, with the over-all widths ⅜ inch and ½ inch wider, respectively, than the face widths shown above.

Door Panels.

17. **Number 1 door panel (No. 1 D. P.)**—The grade of No. 1 door panels shall be the same as for Good 2 Sides panels.

18. **Number 2 door panel (No. 2 D. P.)**—Each face shall be of a single piece of veneer that is free of knots and other open defects, but may admit medium stain and discoloration. Patches not exceeding ⅝ by 2½ inches and shims of any size, when reasonably selected for color and grain, are admissible.

Moisture Resistance Requirements.

19. Douglas fir plywood is made in either of two classes, namely, Moisture-Resistant (M. Res.) and Exterior (Ext.), the test requirements of which are set forth below.

20. **Moisture-Resistant (M. Res.)** — This class represents the majority of production and consists of plywood with a high degree of moisture resistance where its application requires that it shall retain its original form and practically all its strength when occasionally subjected to a thorough wetting and where subjected to occasional deposits of moisture by condensation through walls or leakage or from other sources.

21. **Tests**—Five samples 6 by 6 inches shall be taken from each test panel. They shall be submerged in water at room temperature for a period of 4 hours, followed by drying at a temperature not to exceed 100° F for a period of 20 hours. This cycle shall be repeated a second time, after which the samples must show not more than 2 inches of delamination along the edge.

22. **Exterior (Ext.)**—This class represents the ultimate in moisture resistance, a plywood that will retain its original form and strength when repeatedly wet and dried and otherwise subjected to the elements, and suitable for permanent exterior use.

23. **Tests**—Five samples shall be cut from each test piece. They shall be submerged in water at room temperature for a period of 48 hours and dried for 8 hours at a temperature of 145° F (+ 5° F) and then followed by two cycles of soaking for 16 hours and drying for 8 hours under the conditions described above. The samples shall again be soaked for a period of 8 hours and tested while wet in a shear testing machine by placing them in the jaws of the device to which a load shall be applied at the rate of 600 to 1,000 lbs. per minute until failure. The test specimens must show no less than 30 per cent minimum and 60 per cent average wood failure, and no delamination. If the number of plies exceeds 3, the cuts shall be made so as to test any two of the joints, but the additional plies need not be stripped except as demanded by the limitations of the width of the retaining jaws on the testing machine. When desired, special jaws may be constructed to accommodate the thicker plywood. If number of plies exceeds 3, the choice of joints to be tested shall be left to the discretion of the inspector, but at least one-half the tests shall include the innermost joints.

24. **Alternate Test**—An alternate test applicable at the manufacturer's option to the one above mentioned consists of taking the samples as described above and boiling them in water for 4 hours, followed by a drying of 20 hours at the above mentioned temperature. They shall be boiled again for a period of 4 hours and the samples tested while wet, as above described. The test specimens must show no less than 30 per cent minimum and 60 per cent average wood failure, and no delamination.

25. **Sampling**—Samples for testing shall be taken from one per cent of the panels in any shipment, but not less than 5 and not more than 10 panels shall be selected. Test specimens shall be cut from the ends and the top and bottom of the panel at or near the middle and the edge; a fifth sample shall be taken from somewhere near the middle of the panel.

26. **Interpretation of tests**—If there is failure of more than one test specimen from any panel, that specific panel shall be rejected. If there is a failure in any of the panels tested, 5 additional panels shall be selected and tested under the conditions described, and all of these 5 panels must pass the required test in order that the shipment be accepted.

STANDARD SIZES.

27. Douglas fir plywood is made in the following standard sizes.

Table 1—Standard Douglas Fir Plywood Sizes.

Item	Width (Inches)	Length (Inches)	Thickness (Inches)
Standard Panels (G2S) (G1S) (SO2S)	12 26 14 28 16 30 18 36 20 42 22 48 24	48 60 72 84 96	(after sanding) 3/16 (3 ply) ¾ (5 ply) ¼ (3 ply) 13/16 (5 or 7 ply) 5/16 (3 ply) ⅞ (7 ply) ⅜ (3 ply) 15/16 (7 ply) 7/16 (5 ply) 1 (7 ply) ½ (5 ply) 1 1/16 (7 ply) 9/16 (5 ply) 1⅛ (7 ply) ⅝ (5 ply) 1 3/16 (7 ply) 11/16 (5 ply)
Wallboard	48	60 72 84 96	¼ (3 ply sanded 2 sides) ⅜ (3 ply sanded 2 sides) ½ (5 ply sanded 2 sides)
Sheathing	32 48	96	5/16 (3 ply unsanded) ⅜ (3 ply unsanded) ⅝ (3 ply unsanded) ⅝ (5 ply unsanded)
Automobile and Industrial	As ordered	As ordered	½ (5 ply unsanded) 9/16 (5 ply unsanded) ⅝ (5 ply unsanded) 11/16 (5 ply unsanded) ¾ (5 ply unsanded) ⅞ (5 or 7 ply unsanded)
Concrete form panels	Same as standard panels	Same as standard panels	½ (3 or 5 ply sanded 2 sides) 9/16 (5 ply sanded 2 sides) ⅝ (5 ply sanded 2 sides) 11/16 (5 ply sanded 2 sides) ¾ (5 ply sanded 2 sides)

Size Tolerances.

28. A tolerance of 1/64 (0.0156) inch over or under the specified thickness shall be allowed on sanded panels and a tolerance of 1/32 (0.0312) inch on unsanded panels.

29. A tolerance of 1/32 (0.0312) inch over or under the specified length and/or width shall be allowed but all panels shall be square within ⅛ (0.1250) inch.

Haskelite Phemaloid Compound Lumber Flooring

Beautiful in appearance—costs less than other high grade flooring—low installation cost—vermin- and rat-proof—highly resistant to moisture.

Practical for laying on concrete or wood subfloor. Fully tested and approved by major flooring contractors. Protected by Meyercord Patents and Patents Pending.

PARQUET, BASKET WEAVE PATTERN AND PLANK FLOORING

Haskelite Phemaloid Compound Lumber Flooring is available in the following species: Hard Elm, Sycamore, Santa Maria.

SIZES AVAILABLE: Wood Block: 12"x12" squares. Parquet and Basket Weave: 6" and 12" width units and lengths in multiples thereof, up to eight units. Planks: 6" and 12" widths with six and eight foot lengths.

THICKNESS: All sizes are factory finished to full ½".

FINISHES: Accurately machine sanded. Filler applied on face, back and all edges.

GRADES: Available in light, medium and dark brown.

Haskelite Phemaloid Compound Lumber Flooring successfully limits the effect on floors of varying humidity in the air. Flat cut lumber 12" wide, with an increase in moisture content of 6%, will expand across the grain one-quarter of an inch. Under similar conditions, a Haskelite Phemaloid Compound Lumber plank of the same width will expand only one-fiftieth of an inch, assuring close fitting joints. Warping and swelling are a negligible factor. In this 3-ply laminated flooring with the grain of the core running at right angles to the grain of the two face plies, a bond of water-proof phenolic resin glue prevents a separation of wood plies regardless of temperature or moisture conditions.

THE FINEST FLOOR AT MODERATE COST

The large areas of handsome graining permit highly modern effects. Special hand-pegging can be simulated in 12" widths as desired, or narrow planking effects can be produced on face plies.

Convenience of installation keeps cost down. There is practically no waste. No doty or defective wood can be used in this flooring since all veneers are rotary cut.

LONG LIFE

Although well maintained floors seldom require sanding, the thickness of the face veneers enable a sander to remove .02 of an inch eight times and still have a perfect surface. Haskelite Phemaloid Flooring under severe wear, lasts from 30 to 50 years in perfect condition. Rodents, vermin and other wood pests are destroyed by the phenolic resin in the glue before they cause damage.

Over wood subflooring, the material lies flat and the squeaks caused by warped or buckled floors are satisfactorily eliminated. When laid over concrete, it does not creep or warp once the mastic bond is set.

CONSTRUCTION

As shown in the diagram, Phemaloid Compound Lumber is laminated of three 3/16" veneers. All three are produced under a patented Meyercord drying system at 130° Fahrenheit, conserving the natural mineral oils of the wood, thus prolonging its life.

For flooring laid over wood or concrete, a tongue and groove joint is provided as shown and the pieces are put down in accordance with standard flooring practice. Since the nails are driven through two plies, there is small chance of the wood of the tongue splitting. For laying in mastic, the tongue and groove forms a secure close fitting joint under all humidity and temperature conditions.

HASKELITE MANUFACTURING CORP.

FLOORING DIVISION

208 WEST WASHINGTON STREET — CHICAGO, ILL., U. S. A.

Product	Size, Board Measure		Dressed Dimensions	
	Thickness	Width	Standard Thickness	Standard Face Width
	In.	In.	In.	In.
Bevel siding	......	4	* 7/16 (mir.) by 3/16	3 1/2
	......	5	10/16 by 3/16	4 1/2
	......	6	......	5 1/2
Wide beveled siding	......	8	7/16 (mir.) by 3/16	7 1/4
	......	10	9/16 by 3/16	9 1/4
	......	12	11/16 by 3/16	11 1/4
Rustic and drop siding (shiplapped)	......	4	9/16	3 1/8
	......	5	3/4	4 1/8
	......	6	......	5 1/16
	......	8	......	6 7/8
Rustic and drop siding (dressed and matched)	......	4	9/16	3 1/4
	......	5	3/4	4 1/4
	......	6	......	5 3/16
	......	8	......	7
Flooring	......	2	5/16	1 1/2
	......	3	7/16	2 3/8
	......	4	9/16	3 1/4
	1	5	25/32	4 1/4
	1 1/4	6	1 1/16	5 3/16
	1 1/2	......	1 5/16	......
Ceiling	......	3	5/16	2 3/8
	......	4	7/16	3 1/4
	......	5	9/16	4 1/4
	......	6	11/16	5 3/16
Partition	......	3	3/4	2 3/8
	......	4	......	3 1/4
	......	5	......	4 1/4
	......	6	......	5 3/16
Shiplap	1	4	25/32	3 1/8
	......	6	......	5 1/8
	......	8	......	7 1/8
	......	10	......	9 1/8
	......	12	......	11 1/8
Dressed and matched..	1	4	25/32	3 1/4
	1 1/4	6	1 1/16	5 1/4
	1 1/2	8	1 5/16	7 1/4
	......	10	......	9 1/4
	......	12	......	11 1/4

*Minimum 7/16.

In tongued and grooved Flooring and in tongued and grooved and Shiplapped Ceiling 5/16", 7/16", and 9/16" thick, board measure, the tongue or lap shall be 3/16" wide, with the over-all widths 3/16" wider than the face widths shown above.

In all other patterned lumber, 11/16", 3/4", 1", 1 1/4", and 1 1/2" thick, board measure, the tongue shall be 1/4" wide in tongued and grooved lumber, and the lap 3/8" wide in shiplapped lumber, with the over-all widths 1/4" and 3/8" wider, respectively, than the face widths shown above.

Factory Flooring, Heavy Roofing, Decking and Sheet Piling.

(The thicknesses apply to all widths and the widths to all thicknesses)

Size, Board Measure		Dressed Dimensions			
			Standard Face Width		
Thickness	Width	Standard Thickness	D & M	Ship-lapped	Grooved for Splines
Inches	Inches	Inches	Inches	Inches	Inches
2	4	1 5/8	3 1/8	3	3 1/2
2 1/2	6	2 1/8	5 1/8	5	5 1/2
3	8	2 5/8	7 1/8	7	7 1/2
4	10	3 5/8	9 1/8	9	9 1/2
......	12	......	11 1/8	11	11 1/2

In patterned lumber 2 inches and thicker, the tongue shall be 3/8 inch wide in tongued-and-grooved lumber and the lap 1/2 inch wide in shiplapped lumber, with the over-all widths 3/8 inch and 1/2 inch wider, respectively, than the face widths shown above.

Weil-McLain Company

Manufacturing Division: **Michigan City, Ind. and Erie, Pa.**

General Offices: **641 W. Lake Street, Chicago**

NEW YORK OFFICES: **501 Fifth Avenue**

Prompt Weil-McLain Boiler and Radiator service is made conveniently available through local stocks carried by Weil-McLain Distributors in most of the important distributing centers.

No. 77 All-Fuel Boiler
Conversion type boiler with insulated enameled jacket. For hand or automatic firing. Connected Load Ratings: Steam 410 to 950 sq ft, Water 655 to 1,520 sq ft.

No. 78 Boiler for Automatic Firing
Boiler has insulated enameled de luxe jacket. Front or rear jacket extension available. Connected Load Ratings: Steam 400 to 1,030 sq ft, Water 640 to 1,650 sq ft.

"RO Series" Boiler for Automatic Firing
Jacketed and insulated round boiler for small homes. Connected Load Ratings: Steam 420 and 520 sq ft, Water 630 and 790 sq ft.

Concealed Raydiant
Raydiant is a convector type all cast iron radiator made in Concealed, Partially Exposed and Cabinet Types. Raydiant Radiators, however, differ from conventional convectors in

Partially Exposed Raydiant

that they supply not only convected heat but also sun-like radiant warmth from their heat radiating "live" panel front. A second important advantage is their ability to hold heat longer.

Cabinet Raydiant
This helps increase the comfort of (on and off) automatic heating and makes mixed installation of Raydiant and standard radiation practicable.

Square-Type Boilers
Sectional boilers for larger installations. Complete range of sizes. Connected Load Ratings: Steam 650 to 9,300 sq ft, Water 1,050 to 14,900 sq ft.

Self-Feed Boiler
Magazine type boiler for small inexpensive sizes hard coal or coke. Connected Load Ratings: Steam 240 to 785 sq ft, Water 385 to 1,260 sq ft.

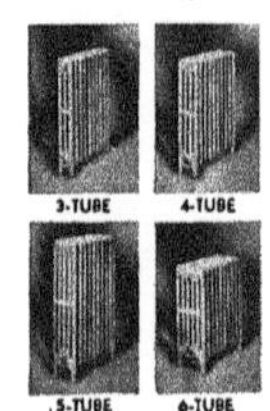

Junior Radiators
Occupy 40 per cent less space than conventional radiators of same rating. 3-tube, $3\frac{3}{8}$ in. wide; 4-tube, $4\frac{9}{16}$, in. wide; 5-tube, $5\frac{3}{4}$ in. wide; 6-tube, $6\frac{15}{16}$ in. wide.

CHICAGO MASTER STEAM FITTERS' ASSOCIATION STANDARDS

For Computing Boiler Sizes and Radiation Quantities for Buildings of Average Construction.

RULE FOR COMPUTING RADIATOR QUANTITIES FOR HEATING PLANTS.

The following are rules compiled and recommended by the Chicago Master Steam Fitters' Association. However, they should not control against the best judgment of the competent designing engineer.

Factors for Multiplying Square Feet of Surface or Lineal Feet of Crack, to Figure Square Feet of Cast Iron Steam Radiation Required for Heating to 70°.

	Glass	Infiltration	Outside Walls	Roofs	Ceilings	Basement Floors	Interm. Floors	Cold Partitions
Glass—								
Single	.293							
Double or Storm	.147							
Sky-Light	.346							
Infiltration, Stationary—								
Sash	.293	.12						
Double Hung Wood or Steel Sash	.293	.24						
Casement-Winds	.293	.48						
Outside or French Doors	.293	.48						
Transoms	.293	.48						
Outside Door With Inner Vest. Door	.293	.48						
Store Doors	.293	.96						
Outside Walls—								
8" Plain Brick			.111					
12" Plain Brick			.085					
16" Plain Brick			.069					
8" Brick and Plaster			.10					
12" Brick and Plaster			.077					
16" Brick and Plaster			.067					
8" Brick Fur Lath and Plaster			.072					
12" Brick Fur Lath and Plaster			.062					
16" Brick Fur Lath and Plaster			.056					
4" Brick, 4" Tile and Plaster			.08					
8" Brick, 4" Tile and Plaster			.069					
12" Brick, 4" Tile and Plaster			.059					
8" Plain Concrete			.16					
12" Plain Concrete			.14					
16" Plain Concrete			.109					
8" Concrete Fur Lath and Plaster			.133					
12" Concrete Fur Lath and Plaster			.107					
16" Concrete Fur Lath and Plaster			.091					
Frame Studding			.064					
Frame No Sheating			.082					
Frame On Lath and Plaster			.093					
Roofs—								
Tar and Gravel on 1" Boards				.08				
Tar and Gravel on 4" Concrete				.16				
Shingle on Sheating				.106				
Shingle Sheating, Lath and Plaster				.08				
Basement Floors—								
Concrete on Earth						.041		
Wood on Sleepers						.017		
Intermediate Floors—								
4" Concrete							.026	
4" Concrete, 3" Fill., 1" Fin.							.02	
Double Wood							.026	
Ceilings—								
Lath and Plaster					.065			
Lath and Plaster Wood Floor Over					.037			
Cold Partitions Stud. L. and P. 1 Side								.08
Cold Partitions Stud. L. & P. 2 Sides								.044

Gas Fired Boiler Ratings—See Page 625.

New AMERICAN RADIATOR CONDITIONING SYSTEM

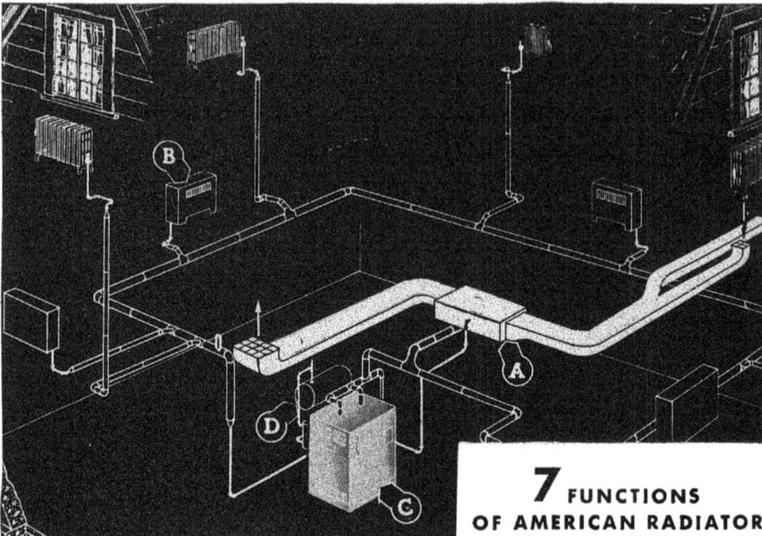

A—conditioned air—humidified, filtered, circulated.
B—modern, sun-like radiant heat.
C—a source of controlled warmth.
D—domestic hot water supplied by installation of a Taco-Abbott System for Summer and Winter hot water supply.

● Here is the home conditioning equipment that marks a home modern—today and for years to come. It is backed by national advertising. It has national acceptanco. It is equipment your client knows and will accept. It provides a better method of home conditioning because the heating is a modern radiator system, and the air conditioning unit operates independently. The results are best summed up in the phrase "New Home Comfort you never Dreamed Possible."

AMERICAN RADIATOR COMPANY
DIVISION OF AMERICAN RADIATOR & STANDARD SANITARY CORPORATION
40 West 40th Street, New York, N. Y.

7 FUNCTIONS OF AMERICAN RADIATOR CONDITIONING SYSTEMS*

1. **HUMIDIFICATION** . . . Restores proper moisture content to indoor air.
2. **AIR CIRCULATION** . . . maintains stimulating, refreshing air motion.
3. **AIR CLEANING** . . . Filters dust, soot, pollen, etc., out of air.
4. **VENTILATION** . . . Distributes fresh outdoor air without drafts.
5. **RADIATOR HEATING** . . . An independent heat source in each room.
6. **CONTROLLED HEAT DISTRIBUTION** . . . Assures warmth in every part of the house, in any weather.
7. **YEAR-'ROUND DOMESTIC HOT WATER**

*Mechanical summer cooling and dehumidification may be added if desired

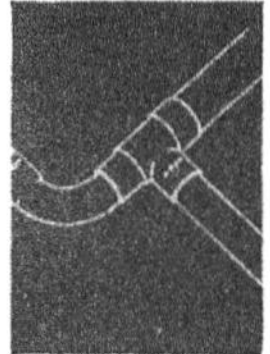
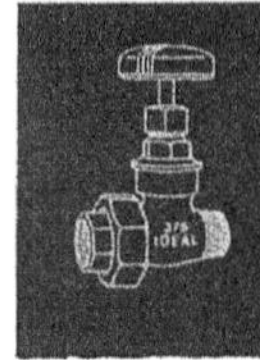
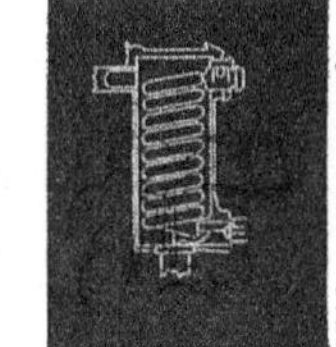

FULL AREA OF TWO-PANE WINDOWS

GIVING THE TOTAL AREA OF TWO-PANE WINDOWS, BRICK OPENING.

Height of Glass / Width of Glass	16″	18″	20″	22″	24″	26″	28″	30″	32″	34″	36″	38″	40′
	TOTAL AREA IN SQUARE FEET												
12″	5.9	6.4	7.	7.5	8.	8.6	9.2	9.8	10.2	10.9	11.4	12.	12.5
14″	6.5	7.1	7.7	8.3	9.	9.6	10.2	10.8	11.4	12.	12.6	13.2	13.8
16″	7.	7.7	.4	9.1	9.8	10.5	11.2	11.9	12.6	13.3	14.	14.4	15.
18″	7.6	8.4	9.1	9.8	10.5	11.2	12.	12.7	13.	14.1	14.8	15.6	16.3
20″	8.2	9.	9.8	10.5	11.3	12.1	12.9	13.7	14.5	15.2	16.	16.8	17.5
22″	8.8	9.6	10.4	11.2	12.1	13.	13.8	14.6	15.4	16.2	17.	17.8	18.8
24″	9.4	10.3	11.1	12.	12.9	13.8	14.7	15.6	16.5	17.4	18.3	19.2	20.
26″	10.	10.9	11.8	12.7	13.7	14.7	15.6	16.6	17.5	18.5	19.4	20.4	21.3
27½″	10.4	11.4	12.5	13.4	14.3	15.3	16.3	17.3	18.3	19.3	20.3	21.3	22.2
28″	10.5	11.5	12.5	13.5	14.5	15.5	16.5	17.5	18.5	19.5	20.5	21.5	22.5
30″	11.1	12.2	13.2	14.3	15.4	16.4	17.5	18.5	19.6	20.6	21.7	22.7	23.8
32″	11.7	12.8	13.9	15.	16.1	17.2	18.4	19.5	20.9	21.7	22.8	23.9	25.
34″	12.3	13.5	14.6	15.8	17.	18.1	19.3	20.5	21.6	22.9	24.	25.2	26.3
36″	12.9	14.1	15.3	16.5	17.8	19.	20.9	21.4	22.7	23.9	25.1	26.3	27.5
38″	13.5	14.7	16.	17.3	18.6	19.9	21.1	22.4	23.7	25.	26.2	27.5	28.9
40″	14.	15.4	16.7	18.	19.4	20.7	21.6	23.4	24.7	26.	27.4	28.7	30.
44″	15.2	16.7	18.1	19.5	21.	22.4	23.9	25.3	26.8	28.2	29.7	31.1	32.5
48″	16.4	17.9	19.5	21.	22.6	24.2	25.7	27.3	28.8	30.4	31.9	33.5	35.

Sizes not shown, figure brick opening.

SIZES OF LOW PRESSURE STEAM MAINS ONE PIPE CIRCUIT SYSTEM DRIPPED AT END

1″	up to 60 sq. ft.
1¼″	60 sq. ft. to 100 sq. ft.
1½″	100 sq. ft. to 200 sq. ft.
2″	200 sq. ft. to 400 sq. ft.
2½″	400 sq. ft. to 600 sq. ft.
3″	600 sq. ft. to 900 sq. ft.
3½″	900 sq. ft. to 1,400 sq. ft.
4″	1,400 sq. ft. to 2,000 sq. ft.
4½″	2,000 sq. ft. to 2,600 sq. ft.
5″	2,600 sq. ft. to 3,300 sq. ft.
6″	3,300 sq. ft. to 4,500 sq. ft.
7″	4,500 sq. ft. to 7,000 sq. ft.
8″	7,000 sq. ft. to 9,000 sq. ft.
9″	9,000 sq. ft. to 11,000 sq. ft.
10″	11,000 sq ft. to 15,000 sq. ft.
12″	15,000 sq. ft. to 24,000 sq. ft.

On all piping, proper provision shall be made for expansion and contraction.

All piping shall be properly pitched.

Supply mains shall not be reduced more than one size larger than one-half the diameter of the largest main.

Dry returns shall be not less than one-half the diameter of the supply.

Wet returns may be one size smaller than one-half the diameter of the supply pipe. By supply pipe is meant the size of main at the point of leaving boiler.

All horizontal branches more than 16 feet in length shall be properly dripped.

Supply mains shall not be reduced more than one-half the diameter of the largest main.

Dry returns shall be not less than one-half the diameter of the supply.

Wet returns may be one size smaller thar one-half the diameter of the supply pipe. By supply pipe is meant the size of the main at the point of leaving boiler.

PIPE SIZES FOR UP-FEED RISERS

1″	30 square feet or under.
1¼″	30 to 60 square feet
1½″	60 to 100 square feet
2″	100 to 200 square feet
2½″	200 to 350 square feet
3″	350 to 900 square feet
3½″	900 to 1,200 square feet
4″	1,200 to 2,000 square feet

RADIATOR CONNECTIONS

Up to and including 30 square feet......1″
Above 30 and including 60 square feet..1¼″
Above 60 and including 100 square feet.1½″
Above 100 square feet................2″

PIPE SIZES FOR ARMS TO RADIATORS AND BRANCHES TO UPFEED RISERS

1″	up to and including 20 square feet.
1¼″	— 21 and including 40 square feet.
1½″	— 41 and including 80 square feet.
2″	— 81 and including 150 square feet.
2½″	— 151 and including 275 square feet.
3″	— 276 and including 625 square feet.
3½″	— 626 and including 1,050 square feet.
4″	—1,051 and including 1,600 square feet.

All horizontal branches or arms more than 8 feet in length and not over 12 feet in length shall be increased one size larger than given above.

All horizontal branches or arms more than 12 feet in length and not over 16 feet in length shall be increased two sizes larger than given above.

A New Factor in Building Valuations

It is felt that there is need for a new viewpoint in building valuation. In the past it would seem that too much stress has been placed on age and immediate revenue in building valuation. It would seem that the architect, owing to his peculiar training, is in a position to point out other very important factors entering into value make-up, which have not heretofore been considered in this locality; such as, balance of structural design, adaptability to purposes intended, character of construction as influencing cost of up-keep.

Feeling that the stability of building investment can be better assured by taking the before mentioned factors into consideration, the Illinois Society of Architects has appointed a Building Valuation Committee to furnish the public with competent architectural valuation service. This Committee should form a real asset to the community.

Rule for Computing Net C. I. Column Radiation Equivalent Load for Boilers Selected from Net Load Chart

EXAMPLE—

(1) 500 sq. ft. of direct cast iron column radiation in room to be heated to 70° F.

(2) 500 sq. ft. of direct cast iron column radiation in room to be heated to 50° F.

(3) 500 sq. ft. of cast iron wall radiation or wall pipe coils in room to be heated to 50° F.

(4) 500 sq. ft. of gravity indirect radiation.

(5) 500 sq. ft. of direct-indirect radiation.

(6) 250-gal. hot water tank. Water to be heated with steam coil.

(7) 500 sq. ft. of cast iron hot blast radiation, having a condensation rate of 1.92 lbs. of steam per hour per sq. ft. with incoming air at —10° F.

SOLUTION—

(1)	500 sq. ft. x 1.0	500 sq. ft.
(2)	500 sq. ft. x 1.13	565 " "
(3)	500 sq. ft. x 1.25x1.13	707 " "
(4)	500 sq. ft. x 1.5	750 " "
(5)	500 sq. ft. x 1.25	625 " "
(6)	250 gal. x 2	500 " "
(7)	(500x1.92) divided by .375	2560 " "

C. I. column radiation equivalent load ....6207 sq. ft.

CHIMNEYS

Due to the wide variation in boiler design, the length and nature of the gas passage, the nature of the fuel burned and the rate of combustion all of which affects directly the draft pressure required, it is recommended that the chimney sizes given by the various manufacturers for their boilers be used for both round and square sectional cast iron boilers. It is advisable that chimney have approximately 25 per cent excess area of smoke collar on the boiler.

A poor draft means imperfect combustion, therefore it is highly important that all boilers be attached to chimneys providing sufficiest draft to consume with proper combustion the required amount of fuel per hour.

It is also important that the chimney be so located with reference to adjacent buildings or objects nearby that draft will not be interfered with.

Round flues will give a better draft than a square or other rectangular shape, having the same cross-sectional area. Round flues are recommended where it is practical to obtain them.

To secure the most satisfactory draft conditions, the area and the height of a chimney must be proportioned to the size and character of heating appliance attached to it and all flue chimney connections made perfectly tight.

RECOMMENDATIONS

It is recommended that no boiler be installed having a grate longer than 72 inches.

Also that in all installations of steam boiler that drain valves be placed on the returns and that the condensation from such returns be discharged into the sewer for a period of from three days to one week after starting fire, thereby clearing system of grease and dirt. At the end of this period boiler should be thoroughly washed and blown out.

For net loads for boilers communicate with the Chicago Master Steam Fitters' Association. Phone Franklin 6280—228 N. La Salle Street.

KEWANEE TYPE RESIDENCE "R" BOILERS Steel

with SMARTLINE JACKETS

REG. U. S. PAT. OFF

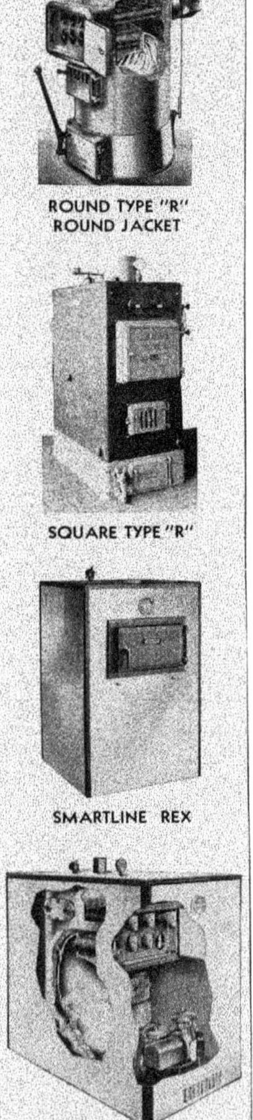

ROUND TYPE "R" ROUND JACKET

SQUARE TYPE "R"

SMARTLINE REX

SMARTLINE REGAL

18 SIZES KEWANEE ROUND & SQUARE "R" BOILERS RATED 320-2924 FT.

ROUND TYPE "R"

	734	735	736	..
Number of Boiler.........Coal	734	735	736	..
Oil-Gas	1734	1735	1736	1737
Stoker	0734	0735	0736	
Cap'ty, Steam.......Coal sq.ft.	320	410	550	..
O, G, & S sq. ft.	400	540	680	900
Height Over-all..............in.	54¾	55¼	59½	61
Water Line..................in.	44	45	47	49
Diameter of Boiler...........in.	23¾	27½	30½	30½
Shipping Wt................lbs.	880	1040	1280	1400

ALL FUELS:— HAND-FIRING, MECHANICALLY-FIRED. STEAM, VAPOR, HOT WATER

SQUARE TYPE "R" / 83R SQUARE

	742	743	745	746	747	748	..	..	..	..	..	..	..	..
Number of Boiler..Coal	742	743	745	746	747	748	..	..	..	..	..	..	..	..
Oil-Gas	1742	1743	1745	1746	1747	1748	83R1	83R2	83R3	83R4	83R6	83R7	83R8	83R9
Stoker	0742	0743	0745	0746	0747	0748	..	..	..	..	..	..	..	..
Cap., Steam. Coal sq.ft.	790	1000	1350	1600	1780	1960	..	..	..	..	..	..	..	..
O, G, & S sq. ft.	840	1120	1470	1900	2160	2380	901	1105	1326	1513	2091	2363	2652	2924
Height Over-all.....in.	59½	59½	70¼	70¼	70¼	70¼	60¾	60¾	60¾	60¾	71½	71½	71½	71½
Width of Boiler Over-all..........in.	32½	32½	32	32	32	32	32½	32½	32½	32½	32	32	32	32
Length of Boiler....in.	39½	45½	45½	51½	57½	63½	27½	33½	39½	45½	45½	51½	57½	63½
Shipping Wt.......lbs.	2150	2360	2800	3050	3300	3550	1600	1800	2000	2200	2800	2900	3100	3300

For Anthracite Coal, Boilers 2734-2748.

Domestic Hot Water with KEWANEE Coils in Type "R" Boilers 65-720 gals. Storage Tank or Instantaneous Flow.

For largest Residence jobs and buildings demanding greater capacities up to 42,500 sq. ft. Type "C" welded Boilers— also Kewanee Riveted Firebox Boilers 300, 400, 500 and Portable type "K". Codes: ASME-SHBI.

STANDARD of the TRADE for 45 years . . . KEWANEE BRICKSET RIVETED BOILER, 16 sizes 1240-20,000 sq. ft. for Coal . . . or converts to Oil, Gas or Stoker.

The enduring reputation of Kewanee Brickset Boilers is solidly established in the city Apartment size and class of heating job through years and years of reliable service.

PHONE: CANAL 8860, OR WRITE OUR CHICAGO BRANCH. 1858 SO. WESTERN AVENUE, FOR CATALOGS ON KEWANEE BOILERS, TABASCO HEATER, GARBAGE BURNERS AND TANKS.

KEWANEE BOILER CORPORATION KEWANEE ILLINOIS

BOILER CAPACITIES

OW PRESSURE STEEL AND CAST IRON BOILER RATINGS AND METHODS OF SELECTING CAPACITY REQUIREMENTS.

n order to provide Architects with a simple method of selecting Steel and Cast Iron Boiler Capacities the editors have compiled the following article based on the Table of the Steel Heating Boiler Institute. The section on Steel Boilers was prepared by **W. E. Posket, M. E.** and the section on Cast Iron Boilers by **P. W. Vandenberg, M. E.**

OW PRESSURE STEEL BOILER RATINGS AND METHODS OF SELECTING CAPACITY REQUIRED.

Steam ratings of low pressure steel fire-;ox heating boilers and the method of de-ermining the actual connected standing oads which are allowable in accordance with he Heating, Piping, Air Conditioning Con-ractors National Association working in ·onjunction with the Steel Heating Boiler nstitute and the Bureau of Mines at Car-ıegie Tech at Pittsburgh, Pennsylvania.

The methods of establishing the catalog ating of low pressure steel heating boilers is s follows:

Hand Fired Coal Burning Boiler Ratings.

For the hand fired coal burning rating, the oilers are rated on the basis of 14 sq. ft. f rating for each sq. foot of heating sur-ace, provided of course that the grate area s large enough to burn the coal required at reasonable rate to produce the heat release equired. For boilers with ratings of 4,000 q. ft. of steam radiation and larger the grate rea equals the sq. root of the catalog rating n sq. ft. of steam radiation minus 1,500, di-'ided by 16.8. And in the case of boilers ated from 300 to 4,000 sq. ft. grate area quals the sq. root of catalog rating in sq. t. of steam radiation, minus 200, divided by !5.5.

Mechanically Fired Boiler Ratings.

The rating of a low pressure steel boiler vhen mechanically fired by a stoker or an oil ›urner, the rating is based on 17 sq. ft. of 'ating for each sq. foot of heating surface, ›rovided the furnace contains enough volume n cubic feet so that in the case of stoker iring the rating would not be more than 125 ;q. ft. for each cubic foot of furnace volume ınd in the case of oil fired boilers, enough 'olume so that the rating would not exceed .40 sq. ft. of rating for each cubic foot of fur-ıace volume. In connection with the above t is understood that the boiler will be con-ıected by a breeching to the stack in accord-ınce with the Manufacturers Catalog re-ıuirements and that a sufficient supply of ›utside fresh air is available in the boiler 'oom to support combustion.

Selection of Boiler Capacity.

With relation to the amount of actual con-ıected load allowable on boilers rated in ıccordance with the Steel Heating Boiler 'nstitute the method is as follows:

By actual connected load is meant the act-ıal sq. ft. in standing column radiation and ts equivalent load in terms of sq. ft. of radia-ion of unit heaters; blast coils, and hot vater load.

The amount of load allowable for loads of 000 ft. or under would be the actual sq. ft. ›f radiation plus a safety factor of 30% of .his amount to be added to the actual load vhich would determine the catalog rating ›f the boiler required in accordance with the ;. H. B. I. rating.

Example: 1000 sq. ft. of actual radiation ›lus 30%, which would be 300 sq. ft., would make the boiler requirement for 1000 sq. ft. of radiation a 1300 sq. foot rated boiler.

From this point up to loads of 10,000 sq. ft. the amount of the safety factor would decrease so that at a load of 2500 sq. ft., a safety factor of 27.5% would be required. At 5000 sq. ft. a 25% safety factor would be required. At 7500 sq. ft. a 22.5% safety factor would be required. And at 10,000 sq. ft. a safety factor of 20% would be required. And from 10,000 sq. ft. up a straight 20% safety factor is to be added.

Examples: 2500 sq. ft. of radiation plus a 27.5% safety factor of 687.5 sq. ft. the S. H. B. I. boiler rating required would be a 3187 sq. ft. boiler. At 5000 sq. ft. a 25% safety factor amounting to 1250 sq. ft. would make an S. H. B. I. boiler rating requirement of 6250 sq. ft. And at 7500 sq. ft. load, a safety factor of 22.5% in the amount of 1687.5 sq. ft. would make the S. H. B. I. boiler rating requirement 9187.5 sq. ft. In the case of 10,000 sq. ft. load with a 20% safety factor in the amount of 2000 sq. ft. would make the S. H. B. I. rating of 12,000 sq. ft.

In the case of loads larger than 10,000 sq. ft., say a 15,000 sq. ft. load, your 20% safety factor in this case would amount to 3000 sq. ft., and an S. H. B. I. boiler rated at 18,000 sq. ft. would be required.

As an explanation of the above for amounts of radiation between the points above mentioned, your attention is called to the chart called Method of Boiler Selection adopted by the S. H. B. I. This chart will simplify the method of selecting the proper size boiler for a certain load. You will note that there are columns headed with the letters "D", "E", and "F", representing the actual load in terms of sq. ft. of radiation, "E" the safety factor required, and "F" the S. H. B. I. boiler rating required.

Boiler Ratings as defined in the Steel Heating Boiler Institute's Boiler Rating Code are intended to correspond to the estimated design load, which is the sum of items A, B and C.

A. The estimated normal heat emission of the connected radiation required to heat the buildings as determined by accepted practice, expressed in square feet of radiation or B.T.U. per hour.

B. The estimated heat required by water heaters or other apparatus connected to the boiler, expressed in square feet of radiation or in B.T.U. per hour. To determine the Maximum Gals. per hour × temperature rise. connected to the boiler expressed in square feet of radiation, use the following formula.

$$\text{Maximum heat required} = \frac{\text{Maximum Gals. per hour} \times \text{temperature rise.}}{25}$$

C. The estimated heat emission of piping, connecting radiation and other apparatus to the boiler, expressed in square feet of radiation or in B.T.U. per hour;

Or when the estimated heat emission of the piping (connecting radiation and other apparatus to the boiler) is not known, the net load to be placed upon the boiler shall be determined from the following table:

			Nett† Load Steam Sq. Ft. F÷E=D	Factor‡	Boiler* Rating Steam Sq. Ft. D×E=F	Nett† Load Steam Sq. Ft. F÷E=D	Factor‡	Boiler* Rating Steam Sq. Ft. D×E=F
		F	D	E	F	D	E	F
1000	1.3000	1300	3200	1.2756	4082	5800	1.2467	7231
1050	1.2994	1364	3250	1.2750	4144	5900	1.2456	7349
1100	1.2988	1429	3300	1.2745	4206	6000	1.2445	7467
1150	1.2983	1493	3350	1.2738	4267	6100	1.2433	7584
1200	1.2978	1557	3400	1.2733	4329	6200	1.2422	7702
1250	1.2972	1622	3450	1.2728	4391	6300	1.2411	7819
1300	1.2967	1686	3500	1.2722	4453	6400	1.2400	7936
1350	1.2961	1750	3550	1.2717	4515	6500	1.2389	8053
1400	1.2956	1814	3600	1.2711	4576	6600	1.2378	8169
1450	1.2950	1878	3650	1.2706	4638	6700	1.2367	8286
1500	1.2944	1942	3700	1.2700	4699	6800	1.2356	8402
1550	1.2939	2006	3750	1.2695	4761	6900	1.2345	8518
1600	1.2933	2069	3800	1.2689	4822	7000	1.2333	8633
1650	1.2928	2133	3850	1.2683	4883	7100	1.2322	8749
1700	1.2922	2197	3900	1.2678	4944	7200	1.2311	8864
1750	1.2917	2260	3950	1.2672	5005	7300	1.2300	8979
1800	1.2911	2324	4000	1.2667	5067	7400	1.2289	9094
1850	1.2905	2387	4050	1.2661	5128	7500	1.2278	9208
1900	1.2900	2451	4100	1.2656	5189	7600	1.2267	9323
1950	1.2895	2515	4150	1.2650	5250	7700	1.2256	9437
2000	1.2889	2578	4200	1.2644	5310	7800	1.2245	9551
2050	1.2883	2641	4250	1.2639	5372	7900	1.2233	9656
2100	1.2878	2704	4300	1.2633	5432	8000	1.2222	9778
2150	1.2872	2767	4350	1.2628	5493	8100	1.2211	9891
2200	1.2867	2831	4400	1.2622	5554	8200	1.2200	10004
2250	1.2861	2894	4450	1.2617	5615	8300	1.2189	10117
2300	1.2856	2957	4500	1.2611	5675	8400	1.2178	10230
2350	1.2850	3020	4550	1.2606	5736	8500	1.2167	10342
2400	1.2845	3083	4600	1.2600	5796	8600	1.2156	10454
2450	1.2839	3146	4650	1.2595	5857	8700	1.2145	10566
2500	1.2833	3208	4700	1.2589	5917	8800	1.2133	10677
2550	1.2828	3271	4750	1.2583	5977	8900	1.2122	10789
2600	1.2822	3334	4800	1.2578	6037	9000	1.2111	10900
2650	1.2817	3397	4850	1.2572	6097	9100	1.2100	11011
2700	1.2811	3459	4900	1.2567	6158	9200	1.2089	11122
2750	1.2806	3522	4950	1.2561	6218	9300	1.2078	11233
2800	1.2800	3584	5000	1.2556	6278	9400	1.2067	11343
2850	1.2795	3647	5100	1.2545	6398	9500	1.2056	11453
2900	1.2789	3709	5200	1.2533	6517	9600	1.2045	11563
2950	1.2783	3771	5300	1.2522	6637	9700	1.2033	11672
3000	1.2778	3833	5400	1.2511	6756	9800	1.2022	11782
3050	1.2772	3895	5500	1.2500	6875	9900	1.2011	11891
3100	1.2767	3958	5600	1.2489	6994	10000	1.2000	12000
3150	1.2761	4020	5700	1.2478	7112			

† The "Net Load" is made up by the sum of items A and B. All "Net Loads" are expressed in 70° F.

‡ These "Factors" apply not only to the net loads opposite them, but also to those 10 sq. ft. less and 40 sq. ft. greater. Thus, 1.2889 is applied not only to 2,000 sq. ft. to determine the necessary boiler rating for this net load, but to all loads from 1990 sq. ft. to 2040 sq. ft. steam.

* For water boilers use the same factor as for steam boilers of like physical size.

To net loads of less than 1,000 sq. ft. steam (1,600 sq. ft. water) apply 1.3 to determine the size boiler necessary.

To net loads of more than 10,0000 sq. ft. steam (16,000 sq. ft. water) apply 1.2 to determine the size boiler necessary.

These factors have been established through tests made by the S. H. B. I. in collaboration with the Bureau of Standards at Pittsburgh, Pa.

LOW PRESSURE CAST IRON BOILER RATINGS AND METHODS OF SELECTING CAPACITY REQUIRED.

The methods of establishing ratings of loss for low pressure cast iron steam heating boilers, as outlined for steel boilers, is the same for cast iron boilers with a few changes, which are as follows:

Hand-Fired Coal Burning Boiler Ratings.

Method of Choosing Boiler Size for Fuels Other Than 12,500 B.t.u. per Pound Calorific Value.—The performance data given for boilers are based on fuel having a calorific value of 12,500 B.t.u. per pound. If the fuel value varies from this figure an allowance of 1% per 100 B.t.u. should be made; for example, if the fuel to be used has a calorific value of 12,000 B.t.u instead of 12,500 B.t.u. the output required of the boiler should be increased 5%. On the other hand, if the fuel should have a calorific value of 14,000 B.t.u. per pound the required output should be decreased 15%.

To explain further, let us assume the following case:

Radiator Load ....500 sq. ft.
Piping Load (in equivalent direct radiation) ....100 sq. ft.
Fuel to be used.......14,000 B.t.u. per pound
Radiator Load ....500 sq. ft.
Piping Load ....100 sq. ft.
Total equivalent radiator load.....600 sq. ft.
Less 15% fuel correction factor.... 90 sq. ft.
Equivalent boiler output for 12,500 B.t.u. fuel ....510 sq. ft.

This is the proper figure for "output in equivalent direct radiation" to be used for selecting a boiler.

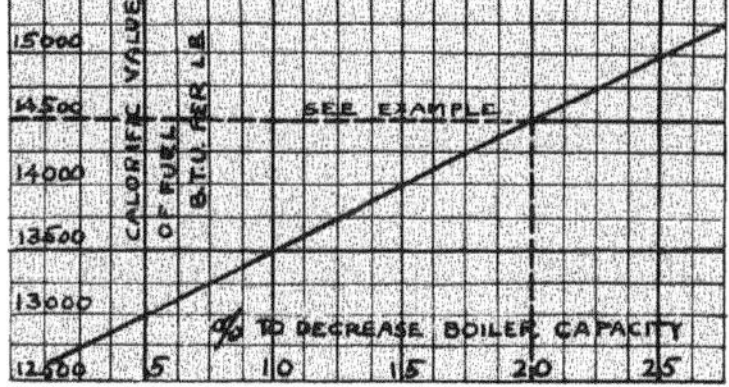

To simplify and quicken the calculations given above, a condensed method is shown in the accompanying chart. Determine the B.t.u. value of the fuel to be used. Where the horizontal line, denoting the fuel value, intersects the diagonal, it also intersects a vertical line, which represents the percentage by which the boiler size may be decreased.

Example (see chart):

14,500 B.t.u. per lb. coal to be used. Steam Radiation and Piping—550 sq. ft.

Note: According to chart, on 14,500 B.t.u. coal, boiler size may be reduced 20%.

Therefore: 550 sq. ft. of steam radiation, less 20% = 440 sq. ft., total load.

Heat Transfer Rates.—The practical rate of heat transfer in heating boilers will average about 3300 B.t.u.'s per sq. ft. per hour for hand-fired boilers.

Automatic-Fired Boiler Ratings. — For a given boiler the average difference in temperature between the flue gas and the heating surface is the factor which governs the transmission rate. As the average flue gas temperature is increased by increasing the oil burning rate, the heat transmission rate will be increased. **In general, it may be said that the lower the heat transmission rate the lower will be the average flue gas temperature and the more efficiently the boiler will perform.**

Many of the manufacturers are now rating their boilers on heat transmission rates higher than were heretofore employed. This is done to secure a high rating for their boilers. However, for economical operation and consumer satisfaction, we recommend that boilers always be selected as follows:

To supply a total boiler output up to 2,000 sq. ft. of E.D.R. a heat transmission rate of 4,000 B.t.u. per hour per sq. ft. of surface should not be exceeded. For greater boiler output heat transmission rates up to 5,000 B.t.u. per hour may be used.

Example:

(1) Assume there is a total of 500 sq. ft. of steam radiation installed in a building. If 20% be allowed for piping and and 25% for pick-up, the total boiler output figure will be 750 sq. ft. The boiler selected for this job should have a total boiler output of 750 sq. ft. and a heat transmission rate of 4,000 B.t.u. or below.

Selection of Boilers.

Estimated Design Load.—The load stated in B.t.u. per hour on equivalent direct radiation, as estimated by the purchaser for the conditions of the inside and outside temperature for which the amount of installed radiation is determined, is the sum of the heat emission of the radiation to be actually installed, plus the allowance for the heat loss of the connecting piping, plus the heat requirement for any apparatus requiring heat connected with the system. The estimated design load is the sum of the following three items:

(1) The estimated heat emission in B.t.u. per hour of the connected radiation (direct, indirect, or central fan) to be installed.

(2) The estimated maximum heat in B.t.u. per hour required to supply water heaters or other apparatus to be connected to the boiler.

(3) The estimated heat emission in B.t.u. per hour of the piping connecting the radiation and other apparatus to the boiler.

Radiation Load.—The connected radiation (Item No. 1) is determined by calculating the heat losses of the building and dividing by 240 to change to sq. ft. of equivalent radiation. For hot water the emission commonly used is 150 B.t.u. per hour per sq. ft., but the actual emission depends on the temperature of the medium in the heating units and of surrounding air.

The radiator ratings of the various manufacturers are expressed in terms of equivalent direct radiation—240 B.t.u.'s per hour per sq. ft. of rating. This output is developed when the radiators are supplied with dry steam at one pound pressure 215 degrees Fahr. When radiators are to be used as hot water heating systems, allowances must be made in selecting the radiators to get that size which will produce the equivalent total heating capacity and equivalent total radiation as is required to supply the estimated radiator load at the average water temperature in the radiator.

Therefore in selecting the size of the radiator or convector to be used, it is necessary to correct for this difference. The table below contains the factors by which radiation requirements, as determined by dividing heat load by 240, shall be multiplied to obtain proper radiator or convector sizes from published rating tables for room temperatures ranging between 50 and 80 F as well as for water temperatures from 150 to 300 F.

Therefore, to determine the radiator sizes for a given space divide the heat loss in BTU's per hour by 240 and then multiply the result by the proper factor from the table below.

Domestic Hot Water.—An allowance of 120 B.T.U.'s per hour per gallon of rated heater capacity should be made in determining the total boiler load. Rated heater capacity is based on 100 degree temperature rise in three hours. Therefore, with a 400-gallon rated capacity heater the domestic hot water load on the boiler is 400 times 120, which equals 48,000 B.T.U.'s per hour.

Piping Tax.—To the design load, it is customary to add an allowance for piping. The most common value for piping is 25%.

Pick-up Load.—To select a boiler of suitable size for an automatic fired installation, the radiation and piping loads should be calculated in the same manner as for a coal-burning boiler. To this must be added the pick-up load. The most common value used for the pick-up load is 25%.

WATER TEMP. F.	FACTORS FOR DIRECT CAST-IRON RADIATORS							FACTORS FOR CONVECTORS						
	ROOM TEMPERATURE F.							INLET AIR TEMPERATURE F.						
	80	75	70	65	60	55	50	80	75	70	65	60	55	50
150	2.58	2.36	2.17	2.00	1.86	1.73	1.62	3.14	2.83	2.57	2.35	2.15	1.98	1.84
160	2.17	2.00	1.86	1.73	1.62	1.52	1.44	2.57	2.35	2.15	1.98	1.84	1.71	1.59
170	1.86	1.73	1.62	1.52	1.44	1.35	1.28	2.15	1.98	1.84	1.71	1.59	1.49	1.40
180	1.62	1.52	1.44	1.35	1.28	1.21	1.15	1.84	1.71	1.59	1.49	1.40	1.32	1.24
190	1.44	1.35	1.28	1.21	1.15	1.10	1.05	1.59	1.49	1.40	1.32	1.24	1.17	1.11
200	1.28	1.21	1.15	1.10	1.05	1.00	0.96	1.40	1.32	1.24	1.17	1.11	1.05	1.00
215	1.10	1.05	1.00	0.96	0.92	0.88	0.85	1.17	1.11	1.05	1.00	0.95	0.91	0.87
230	0.96	0.92	0.88	0.85	0.81	0.78	0.76	1.00	0.95	0.91	0.87	0.83	0.79	0.76

MINIMUM HEAD ROOM REQUIREMENTS FOR SMOKELESS SETTINGS.

	BOILERS								
Furnaces	Horizontal Return Tubular				Water Tube				Scotch Marine
	54"	60"	66"	72"	Hor. Baff 1"-1½" Pitch	Vert Baff 1"-1½" Pitch	Hor Baff 3¼" Pitch	Vert Baff 3¼" Pitch	
	Shell to Dead Plate	Shell to Dead Plate	Shell to Dead Plate	Shell to Dead Plate	Front Header to Floor				
No. 6	32"	34"	34"	36"	**6'0"	*	**6'6"	*	##
No. 7 (Modified)	32"	34"	34"	36"	=	=	=	=	##
No. 8	32"	34"	34"	36"	6'0"	*	6'6"	*	##
Hand Stoker	26"	28"	28"	30"	5'6"	*	6'0"	*	Full Extension
	Shell to Floor	Shell to Floor	Shell to Floor	Shell to Floor					
Down Draft	60"	60"	60"	60"	6'0"	*	6'6"	*	Full Extension
Twin Fire	58"	60"	62"	64"	6'0"	*	6'6"	*	*
Semi Ext. Refuse Burning	44"	46"	46"	48"	7'0"	*	7'6"	*	*
Feed: Burke	48"	48"	50"	54"	5'0"	*	5'6"	*	Full Extension
McMillan	48"	48"	50"	54"	5'0"	*	5'6"	*	Full Extension
Twin Fire (Gravity)	48"	48"	50"	54"	5'0"	*	5'6"	*	Full Extension
ain Grates	72"	72"	78"	78"	7'0"	9'0"	8'0"	10'0"	##
Feed: Moore	48"	54"	60"	60"	6'0"	8'6"	6'6"	9'0"	##
Roney	60"	60"	60"	72"	7'0"	9'0"	7'6"	10'0"	##
Wetzel	60"	60"	60"	72"	7'0"	9'0"	7'6"	10'0"	##
Detroit	66"	72"	78"	84"	7'6"	*	8'0"	*	Full Extension
Model	66"	72"	78"	84"	7'6"	*	8'0"	*	Full Extension
McKenzie	66"	70"	70"	70"	7'6"	*	8'0"	*	Full Extension
Murphy	66"	72"	78"	84"	7'6"	*	8'0"	*	Full Extension
Type "E"	##	##	##	##	6'6"	8'6"	7'6"	9'0"	##
Jones	36"	38"	40"	42"	6'0"	8'0"	7'0"	8'6"	Min Diam.
Detroit	42"	44"	46"	48"	6'6"	8'6"	7'6"	9'0"	Furnace 36"
Taylor	##	##	##	##	6'6"	8'6"	7'6"	9'0"	##
Sanford-Riley	##	##	##	##	6'6"	8'6"	7'0"	9'0"	##
Westinghouse	##	##	##	##	6'6"	8'6"	7'6"	9'0"	##

NOTES

* Combinations not recommended as smokeless settings
= Not adapted to water tube boilers.
\## Combinations not ordinarily met with in practice
** Omit double arches—using only deflection arch
Setting heights for Jones stoker refer to standard stoker

The accompanying table is intended to how the minimum setting heights for the arious combinations of boilers and furnaces ound in use in Chicago.

These settings are not intended for high apacities, but have proven satisfactory for ormal loads where draft is sufficient and roper methods of operation used.

The setting heights shown for side feed tokers are for furnace widths of 7' 0" or less.

For wider furnaces the heights must be increased to allow for increased arch spring necessitated by the wider span.

Combinations of vertically baffled water tube boilers noted as not being recommended as smokless settings have been found in actual operation to produce too much smoke to comply with the smoke ordinance in its strictest interpretation, and have proven unsatisfactory from the Smoke Inspector's viewpoint.

FOR

Proved Economy

IN AIR CONDITIONING

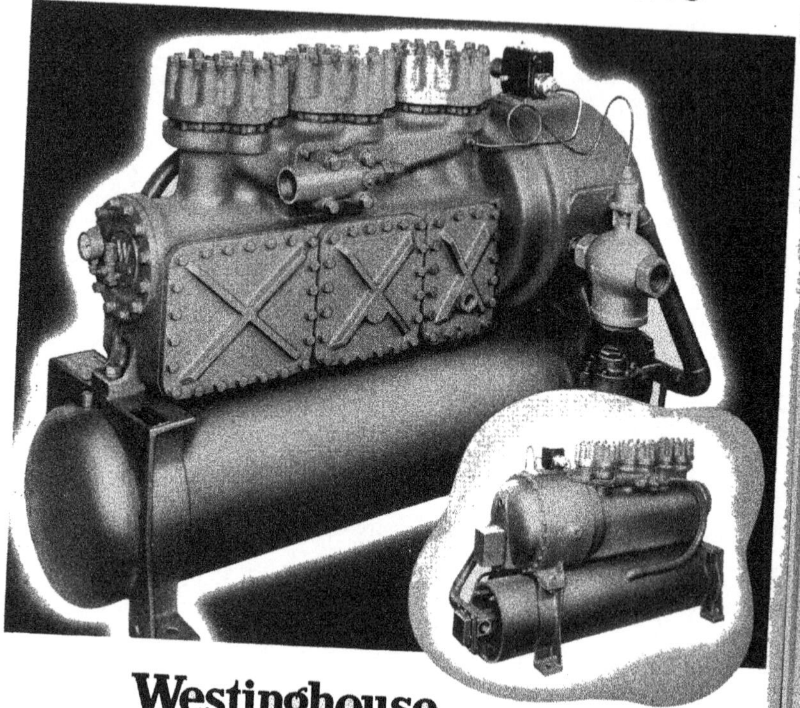

Westinghouse

Seal-less CONDENSING UNITS

★ Lighter—more compact—no foundations required—can be mounted in upper floors without vibration. No flywheel—no visible moving parts—many exclusive features, such as water-cooled motor which eliminates necessity of ventilated space. Wide range of sizes.

WESTINGHOUSE ELECTRIC & MANUFACTURING COMPANY

20 NORTH WACKER DRIVE CHICAGO, ILLINOIS

RESIDENTIAL AIR CONDITIONING

ITS SCOPE & FUNCTIONS

by **JOHN J. DAVEY, Architect**

I. THE TERM DEFINED.

To the average layman, "Air Conditioning" and "Summer Cooling" are synonymous. He has not yet been educated to appreciate the true meaning of the term and is generally unaware of its limitations although recognizing that it should provide greater bodily comfort.

To the Architect and Engineer, "Air Conditioning" is the process by which the temperature, moisture content, movement and cleanliness of the air in enclosed spaces are simultaneously controlled and maintained within definite specified limits regardless of outside conditions.

For bodily comfort, the equipment controls the temperature by heating in winter and cooling in summer; humidity by the addition or elimination of moisture, and movement by the uniform distribution of clean, filtered air. It also provides sufficient ventilation to produce a mild circulation and enough fresh or revitalized air to eliminate objectional odors.

Efforts have been made to introduce a term which would more clearly define the functions performed but the designation "Air Conditioning" will probably continue in general use indefinitely.

II. STATUS OF THE INDUSTRY.

Only within recent years has the general public become acquainted with the subject, but "Air Conditioning" is not a recent development as the industry would have us believe.

Air Conditioning has been used extensively in Manufacturing and Industrial process work for more than half a century. The term as applied to bodily comfort is of later origin, first appearing about 1903-4.

In the Chicago area, probably the first mechanically controlled system for this purpose was supplied by the Wittenmeier-Kroschell Company for the Congress Hotel in 1906.

The first "year-round" residential system was designed by Neiler, Rich & Co., Consulting Engineers, Chicago; and installed in the Mrs. C. H. McCormack home at Lake Forest in 1915.

This firm also designed the system installed in the Northern Trust Company of Chicago, the first year-round completely air conditioned commercial building in the loop district.

The "Old" Randolph Street Suburban Station of the Illinois Central Railroad, long advertised as Chicago's first completely "Air Conditioned" building, was a so-called "natural system" utilizing chilled air from the Chicago Tunnel Company subway.

There is no mystery about "Air Conditioning." For almost a century Ventilation has provided tempered, humidified air circulation and cooling is but another step forward.

To date installations are few. More persons are becoming interested and "Summer-Cooling" has caught the public fancy, but the great majority still have to be sold on the idea.

The present wave of publicity can be largely attributed to efforts on the part of the Industry to create a market for the apparently unlimited range of equipment offered for the purpose.

A recent survey indicating that, at this time, out of "22 million electrically wired homes in the United States" less than ¼ of 1% can boast of even one conditioned room.

Advertisements fill the pages of every printed medium. Exaggerated claims as to performance initial and maintenance costs, etc., prevail. To quote the Modern Brickbuilder of Sept. 1937: "A great deal of faking is being done. So much, in fact, that the Federal Trade Commission has begun issuing cease and desist orders to makers of some so-called Air Conditioning devices that are anything but what they are represented to be."

Free engineering service, free layouts, the installment plan with three or more years in which to pay, and "cut-rate" prices are but a few of the inducements offered.

The results have been the direct opposite of those anticipated. 1st—Its scope has been extended but the volume of business has not materially increased. 2nd—The majority of systems installed under such conditions are faulty in design, generally unsatisfactory and, as might be expected, the re-action most unfavorable. 3rd—It has clearly demonstrated that continuation of such a policy can result only in the complete disruption of an Industry which, properly conducted, has unlimited possibilities.

The Industry has now entered the second stage of its development, that of "standardization" and has recognized the necessity of a basic "Code of Minimum Requirements" acceptable to all concerned, viz. the Manufacturer, Architect, Engineer and Owner. The result is The Code of "Minimum Requirements for Comfort Air Conditioning" as sponsored by the Joint Committee of the American Society of Heating and Ventilating Engineers and the American Society of Refrigerating Engineers. Modifications as required by Ordinance to meet local conditions are fully covered in the supplement published by the Chicago Committee on Air Conditioning Standards. Various Technical Societies and Engineering Organizations have made available much authentic information; but the so-called "Standards" published solely in the interest of a single Manufacturing or Trade group are of little, if any, practical value.

III. GENERAL DIVISIONS AND FUNCTIONS.

Installations may be divided into three general groups: I.—Residential; II.—Commercial; III.—Manufacturing and Industrial process work. Groups I. and II. will be considered only as applied to bodily comfort. Group III. most important in the manufacture of candy, weaving and spinning of textiles, color-printing and many other products are outside the scope of this article.

Air Conditioning for comfort may be installed for "year-round" service or in part for either winter-conditioning or summer-cooling; also one or more rooms or the entire building may be conditioned.

Each installation presents its own individual problem and must be so treated. The type of equipment and control naturally varies with the specific functions performed.

Except in minor details, the major functions of Air Conditioning are heating and humidifying in winter, and cooling and de-humidifying in summer; plus the reduction or elimination of odors and the circulation of clean, vitalized air at all times. Of these "Heating" is by far the most important.

IV. THE HEATING SYSTEM.

According to the "National Bureau of Heating and Air Conditioning," proper heating is 80% of winter conditioning. The ac-

curacy of this may be questioned, but it is certain that no part of the home is more important than a dependable, efficient and economically operating heating system.

There are three general types of residential installations. The duct system for warm air, the pipe system for steam, vapor and hot-water heat; and the split-system which, as its name implies, is a combination of both.

A national survey of residential dwellings in towns of 2500 persons or more, made by the United States Department of Commerce in 1914, is authority for the statement that these are distributed approximately as follows: Warm air systems—73%; Steam and vapor—15%; and hot water—12%. It is further stated that 60% of all installations were over 16 years old, and of these approximately 50% practically obsolete. This is also true at the present time.

Under the caption "Radiator Heat for Air Conditioning" there is now offered a so-called improved system. This consists of the standard steam or hot-water installation plus a return duct system for air circulation. Additional equipment may be installed to provide summer cooling, winter conditioning, or both. It however, is but a modified "Split-system" and presents nothing new.

In initial cost, gravity warm-air heat is cheapest, the one-pipe steam system next, and hot-water most expensive. This ratio also holds good for all variations of the respective types.

V. THE DUCT SYSTEM.

The process of evolution rather than applied engineering principles is largely responsible for the development of residential warm-air heating. The earlier installations were strictly one-pipe, gravity systems. The brick enclosed, or sheet-metal jacketed, coal-burning furnace centrally located to permit so far as possible duct runs of uniform length.

The return duct or two-pipe system was the first major improvement. This provided more positive heat although one or more registers usually failed to function properly.

Fresh air was supplied from outside the building or the chilled inside air recirculated, a combination of both being generally used. Hand operated dampers in the horizontal pipe runs and registers were the only means of heat regulation.

The oil and gas burning units introduced the fan or blower and forced circulation replaced the gravity system. The round supply pipe with a maximum efficient length of approximately fifteen feet and a pitch of one inch per lineal foot was discarded for the smaller rectangular duct. Less head room was required, longer duct runs possible, and the heater located at any convenient point. The new burners were indiscriminately installed in heating units entirely unsuited for the purpose, and only in recent years has an efficient furnace been perfected which permits the economical use of gas and oil.

The latest development is the "oil-fired conditioner." The majority are combination units equipped with humidifier, blower and filters for winter conditionnig. Cool air entering either at the back or top, is filtered at point of entrance, heated, and the moist tempered air forced through the system from the top or either side of the casing. The humidifier is located above, below or at the side of the combustion chamber.

The new equipment is compact and well designed, but few so constructed as to permit the addition of refrigeration for summer cooling. Close inspection further indicates that internal repairs and replacements will be both difficult and expensive.

At present, complete with automatic control, the initial cost of such installations range from $175.00 to $200.00 per room. Operating costs are from 50% to 100% greater than for the simple heating system.

For homes of moderate cost, compact in plan and the Heating Plant centrally located, the duct system of direct warm-air heat is undoubtedly the best.

For larger homes and rambling structures the so-termed indirect or "split-system" has many advantages. A steam or hot-water boiler supplies a centrally located unit-heater, the heated air is forced over or through the heating coil, through the duct system, re-circulated, and the process repeated, or several independent units, supplied from the one boiler, may be installed at different points, long duct runs eliminated and better circulation assured.

For "Year-round" residential heating and conditioning a correctly designed and properly installed duct system, of the same capacity and an initial cost equal to that of a corresponding pipe installation, will give much better results at a minimum of operating expense.

The greatest objection to direct warm-air heat is its inability to successfully provide an adequate all-year-round supply of hot-water for domestic purposes. This must be obtained from an independent source, usually an instantaneous gas fired heater.

Ducts in general should be of bright tin, either single or double wall construction, pre-fabricated, with "stream-line" fittings. They should be air and moisture tight, properly insulated to prevent heat loss, and noise caused by the operation of mechanical equipment used in connection with the heating system.

Metal cold air ducts should be well insulated to prevent corrosion, and inlets, outlets and duct lines large enough to permit the introduction of 100% of the design air quantity from outside the building if desired, and the escape of an equal amount. Dampers should be provided in each branch duct and trunk line, and at air inlets and outlets to permit proper regulation. The required area in square inches of a warm-air supply duct of nominal length equals the B.t.u. heat-loss in the occupied area divided by 250. For long runs the duct area must be proportionately increased.

VI. PIPE SYSTEMS.

The majority of residential installations for steam and hot-water heating are of the one or two-pipe, up-feed, gravity return type although the down-feed system is occasionally found.

Vapor heat is a latter development, using steam at pressures slightly higher than atmosphere. The supply is controlled at the inlet valve and only the amount of steam actually required admitted to each radiator.

A combination of warm-air and steam or hot-water heat or the "split-system" is rapidly gaining in popularity, and forced hot-water circulation by means of air pressure or by the introduction of an electric pump in the returu line has generally replaced the old gravity system.

There are round and square boilers and also especially designed units for the use of fuel-oil, gas combustion, or coal consumption; the latter for hand-firing, magazine feet or stoker firing. In all types the heating surfaces are water-jacketed, a feature not usually found in the warm-air furnace.

The latest models have a fire travel several times the boiler length. Those burning oil are constructed with narrower passways, for the retarding or trapping of the combustion gases. Many are provided with build-in-units which furnish a year-round supply of hot water for domestic purposes by the indirect heating of same from the hot water in the boiler itself.

The principle of sustained heat has also been developed in the oil-fired unit, the boiler transmitting heat to the water after the flame is cut off.

The most objectional feature of "radiator-heat" has been the radiator itself.

The vertical pipe-radiator, pressed-steel and the various types of cast-iron "floor radiation" were unsightly, dirty and, generally inefficeint.

Of improved design, smaller sections and greater efficiency, the new radiator is far superior to those of former years but still retains many of its objectional features.

There are three types of modern radiation, the cast-iron radiator, and all cast-iron and the "tube-and-fin type" convectors.

The so-called free standing direct radiator, which strictly speaking is not a radiator at all, supplies both radiated and convected heat in approximately equal amounts but air circulation in the room itself is negligible.

The convector or indirect type is fully enclosed, provides a definite circulation of air over the heated surface, and radiated heat is practically eliminated.

Convectors may be free-standing, fully or partially recessed, with grilles or perforated one or more piece face panel screening the heating element, and are of more pleasing design and less obnoxious than the free-standing radiator.

The tube-and-fin type have no hold-over heat capacity, cooling rapidly when the heat supply is cut-off, wich is not true of the cast-iron units.

The initial cost of a pipe system is higher than that of a corresponding warm-air installation, but radiator heat has many advantages. Distant units are heated as readily as those adjacent to the heating plant. It also has a hold-over heat capacity and provides an adequate, year-round supply of domestic hot water.

VII. GENERAL REQUIREMENTS:

In all central installations both heating unit and pipe or duct system should be of ample size and the boiler or furnace especially designed for consumption of the fuel selected. Experience has clearly demonstrated that to operate a gas or oil burner in a unit built solely for coal-consumption is neither satisfactory or economical, and an all-purpose boiler which operates efficiently with either gas, oil or solid fuels has not yet been perfected.

Do not spend money uselessly in reconditioning an out-of-date heater, replace it with a modern unit—it is less expensive and more satisfactory; neither rig up an obsolete boiler or furnace with a lot of so-called air conditioning gadgets and expect efficiency. It is an impossibility.

Pipe and duct lines should be properly graded, and so sized that only the proportionate amount of heat required to maintain a specific temperature is provided by the individual radiator or warm-air register.

Finally; purchase quality, it is less expensive in the end and the relatively small additional cost required to secure a high grade heating installation is the home-owner's best possible investment. The results obtained, however, will depend almost entirely upon the ability of the designer.

VIII. THE SYSTEMS COMPARED.

The modern duct-system provides positive circulation of filtered, tempered, and humidified warm air heat. Without additional equipment, the straight radiator system heats only and the initial cost is greater.

With the introduction of residential "Air-Conditioning" this became an important factor, warm-air heat apparently occupying a most enviable position. But we must not overlook the fact that each type has its own particular field, its apparent advantages and its undesirable features.

Claims advanced by various Associations and "Trade-Groups" as to the absolute superiority of any one type over all others are absolutely unwarranted and reflect no credit upon the sponsor. The alleged desirability of "radiator-heat" and of certain fuels has received much publicity but, there is no material difference in the quality of heat thrown off by a radiator and that taken from the bonnet of a warm-air furnace. Likewise the quality of the steam passing through a radiator is the same regardless of the fuel used.

The warm-air system is particularly adapted to the addition of conditioning equipment, although same may be readily installed and successfully operated in connection with practically any type of haeting system. The modern warm-air heater requires no additional equipment for "winter-conditioning." For pipe systems a blower, filter, and humidifier or vaporizer, are required. For summer cooling, refrigerating equipment must be provided for all systems.

A forced air heating system, with each room connected to the cold-air return, requires only the addition of cooling coil, pump, and equipment for refrigerating cold-water; to meet all demands of "Summer Conditioning."

Warm air systems filter the air before same reaches the register, and the forced draft type provide ample ventilation and air circulation. The straight radiator system does neither without auxiliary equipment, at additional cost.

Auxiliary equipment for either or both winter conditioning or summer cooling usually consist of one or more combination units. These function successfully with either new or existing installations operating preferably in tandem with, but not forming an integral part of the heating system.

IX. FUEL ECONOMY.

The amount of fuel required for adequate heating, and the capacity of the structure to retain the heat generated, largely determine the relative efficiency and operating cost of the Heating System.

In the Chicago area the heating system functions some 6000 degree days based on temperatures below 65° F.; or approximately nine months of the year. The average winter temperature is approximately 35°F.

The fuel required to properly heat the average seven-room residence of ordinary construction will approximate 1⅓ tons per room, a total of ten tons of coal or its equivalent in heat units per season.

Coal is still the most economical. Under the same operating conditions the cost of gas being approximately three times, and fuel-oil twice that of "stoker-fired" coal at $6.50 per ton.

The increasing cost of suitable coal and the more general use of higher priced gas and oil demonstrates the need of fuel conservation. With coal this can readily be done.

Hand-fired installations consume up to 100% more fuel than actually required for adequate heating.

Using the same grade of coal, the underfeed forced-draft type, mechanical stoker will show a saving of from 35-50% in actual tonnage per season over that required by the hand-fired unit. It also permits the use of cheaper coal, produces a hotter fire, uniform meat, less ash, soot and clinker. Stokers for residential installation are now available for practically all types of heaters are most efficient, comparatively inexpensive and their initial cost soon repaid by the economies effected.

Gas and oil provide clean, uniform heat. Dirt, ashes and the drudgery associated with hand-fired coal are practically eliminated. Fuel costs are greater but largely offset by the convenience of operation.

Forced circulation permits the use of smaller equipment, pipe or duct lines and less fuel is required; also water at a lower temperature can be used in the hot-water heating system.

their particular uses. Of these insulating lumber is probably best adapted for all round purposes and bulk-fillers least desirable. The latter should always be of a mineral substance. Most insulating materials are effective only when dry and all insulation should be dust, water and vermin proof, odorless and readily installed in a manner to provide for movement in the structure due to shrinkage and swelling of other materials.

Air-conditioned homes require additional protection to prevent excessive condensation on inside wall and roof surfaces.

Unlined air-spaces in the house of ordinary construction offer little resistance to the passage of heat.

Assuming that 60% of the heat transfer is due to radiation rather than convection, and these spaces lined with bright, metallic material would reflect back to its source most of the radiated heat striking it, "Reflective Metallic Insulation" has been developed. This in general consists of a thin polished metal foil, backed up or used in combination with other building material, and was introduced from Germany some seven years ago.

It is equally efficient in summer or winter; in air spaces ¾" or more in width the percentage of radiant heat being always in excess of 70% when either the inside or outside temperature is 70° F. or more.

The presence of dirt, dust or other foreign substances have little or no effect as the material works on infra-red reflectivity rather than on visible light; thus its efficiency when used in wall or ceiling spaces which visible light cannot contact.

Several metals have relatively the same insulating value but the high tensile strength, light weight and low cost of aluminum foil-faced Kraft paper makes its use more general. The purity of the metal reduces the possibility of its tarnishing to a minimum, an aluminum oxide forming on the surface of the foil as soon as the material is rolled.

The "Bureau of Standards' rates each air space faced with this material the same as ¾"-thick cork, thus a single air space divided by a double-faced foil sheet would have an increased resistance valve equal to that of 1½" of cork. Its resistance value does not depend upon the thickness of the foil, which is not true of insulating materials in general. Metallic insulation is of value only when used in conjunction with air spaces.

When placed in contact with solid materials on both sides it is far superior to the ordinary felt or sheathing paper, providing permanent protection against the passage of dust, air, moisture, vermin, or water-vapor at but little additional cost.

Another type of metal insulation consists of lead-and-tin oxide coated, 4-ounce, ribbed steel sheets shaped to fit standard stud and rafter spacing. In frame an veneered construction the interlocking sheets are set ½-inch back of the stud or rafter face; and the same distance inside the face of wood furring strips applied to the inside surface of masonry walls. Wood cross furring at the horizontal metal joints are required in all cases. Its thermal conductivity is low, the material substantial and easily installed but, locally at least, has not been used to any great extent for general insulating purposes.

The insulation of buildings during construction is not expensive; insulating lumber (¾"-thick) costing approximately 1½ to 2 cents, and lath (½"-thick) 2½ to 3 cents per square foot more than ordinary sheathing and wood-lath, a total of $150.00 to $200.00 for the average dwelling.

To properly insulate an old building involves considerable expense, and where the saving effected does not show a minimum re-

most important factor. From the standpoint of comfort and health this consists largely in providing air in the proper quantity rather than any given quality, although it is highly desirable that the air supply conforms to an acceptable standard of purity and freshness.

A condition where all are equally comfortable cannot be provided, but "comfort-zones" acceptable to the majority of persons have been definitely established. Exhaustive tests by various technical research groups show that in winter the greater number are comfortable at 70 degrees and 50%, 68 degrees and 70%, and 72 degrees F. dry-bulb inside temperature and 30% relative humidity. Other combinations of approximately 10% increase in relative humidity per 1° F. drop in temperature or vice-versa, were found equally acceptable. Complete information may be found in the copyrighted "Comfort-Zone and Effective-Temperature Chart" of the Am. Soc. Heating & Ventilation Engrs. and Psychrometric Charts published by The Carrier Engineering Corporation, and others.

Humidification:

Humidity is the moisture content of the atmosphere. Absolute humidity the point of saturation, or maximum weight of water-vapor retained per unit volume of space occupied, as expressed in grains per cubic foot of air. Relative humidity is the percentage indicating the degree of saturation due to the moisture present in the unit of volume.

Air at any given temperature will absorb only a specific amount of water-vapor. The cooler the air the lower the moisture content and vice-versa. At sea level, absolute humidity at various dry-bulb temperatures are as follows.

Temperature Degree F.	Moisture Gr. per C. F.	Temperature Degree F.	Moisture Gr. per C. F.
—10°	0.285	50°	4.106
0°	0.475	60°	5.795
10°	0.776	70°	8.05
20°	1.242	80°	11.04
30°	1.946	90°	14.94
40°	2.863	100°	19.95

The dew-point, or point of visible condensation naturally varies with the temperature and humidity of the air volume.

When introduced by infiltration into a heated area, cold outside air with its low moisture content mixes with warmer air of greater humidity and as it becomes heated absorbs all possible moisture from persons-or-objects which it surrounds creating a dry condition of relative low humidity. This imparts a feeling of chilliness to the occupants even though the thermometer shows no change in temperature. To overcome this condition it is necessary to increase the moisture content of the air within the enclosed area.

Healthful comfort is possible with a relative humidity of from 30-70%, but 40% is generally considered the practical maximum.

With outside temperatures at zero, or below, the relative humidity should not exceed 25% to eliminate excess condensation.

At 35-40° F. outside temperature 40% relative humidity may be maintained and up to 50% under certain conditions.

The maximum humidity possible with single glazed windows is 15% in zero weather. This condition is also frequently found at other temperatures in homes where humidifiers are not provided.

Moisture from cooking, washing, etc. will raise the humidity approximately 5-15%; and from persons, by evaporation, 700-grains or more per hour per person. These factors however may be usually ignored in residential installations.

The general accepted standards of minimum lesign requirements for Winter Conditioning re:

(a) Dry-bulb temperature for Heating, —10° F. outside, 70° F. inside.

(b) Humidity for Heating, 35% at 70° F. dry-bulb, inside temperature, or an effective temperature of 66° F. and the dew-point at 40% relative humidity.

This prevents excessive condensation with n outside temperature of —10° F. providing he structure is reasonably tight, windows .ouble glazed or protected by storm sash.

The weight of water at 62° F., sea level, s 8⅓ pounds or approximately 58,330 grains f moisture per gallon. To provide 35% relaive humidity with an inside temperature of 70° F. a moisture content of 2.817 grains per cubic foot of air must be maintained.

Theoretically, a building with 20,000 cubic foot content and an infiltration loss equivalent to one complete air change per hour should require the evaporation per twenty-four hour day of approximately 21 gals. at —10° F., 19 gals. at zero, and 3½ gals. of water at 35° F. outside temperature. Actually more will be required as outside air entering the building at temperatures of 30° F. and below is seldom more than 50% saturated.

Research by the "University of Illinois" would indicate that for the average house of 15,000 cubic foot content, from 8 to 72 gallons of water per twenty-four hour day may have to be evaporated at times to provide a relative humidity of 40% in zero weather due to the heat loss by infiltration. These amounts may be excessive but show that adequate evaporating equipment is necessary to insure proper humidity. Due to the widely fluctuating demands caused by varying weather conditions, full automatic regulation is also essential to secure same at all times. The automatic control should be as simple as possible that its functions and operation may be readily understood by the average, non-technical home owner.

The majority of humidifiers provided for residential installations are too small for the purpose. The ordinary warm-air furnace pan will evaporate only from 2 to 5 gallons of water per twenty-four hours, a pan placed in the furnace dome a maximum of 14 gallons and open pans used in connection with Steam Radiators ¾ of a gallon per square foot of water surface.

To provide additional evaporating capacity auxiliary equipment must be installed.

Many efficient humidifying devices are available but those using applied heat to cause evaporation are much more efficient than units using unheated water.

In warm air installations—evaporator pans placed in the path of the outgoing-air; or heated directly from the fire pot, and the water level maintained by a suitable float-valve, are in general use. They are not particularly efficient. Lack of sufficient humidity in cold, and over humidification in moderate weather causing excessive condensation on walls and glass surfaces being their general tendency.

For direct radiation and air circulating systems a vertical stack of shallow, steam-heated pans or troughs with a down-feed top supply is commonly used; the water overflowing from the upper pan to the one immediately below. These may also be used with a fan system, the unit being placed in the inlet path on the inlet side of the fan. A "Spray-type" humidifier will be equally satisfactory.

The "atomizing-type" has met with little favor in the residential field but excellent results can be had with a unit of the proper capacity. Water pans above or behind the radiators are still fairly effective within their limits if given proper attention.

Air-Cleaning:

The air is cleaned by blowing it through air washers or filters, the latter being the most efficient.

The air-washer is primarily a device for regulating temperature and moisture content and its efficiency as a cleaning unit is limited. It will not remove soot or the fine dust content in the air.

Filters are usually of the dry, or of the oil-coated or viscous type. The filters are of fibre, glass or mineral-wool, fabric; removable cloth bags that can be beaten or washed and the "throw-away" type of various kinds.

Equipment should be large enough to clean air volumes up to 100% of the fan capacity. This to include both the fresh and re-circulated air and also the removal of a minimum of 75% of the dust and other impurities from both. Filters should be of ample size, for the larger the filter area the lower the resistance which varies with the square of the air velocity. It also reduces operating costs and prolongs the life of the filter screen. The total filter area naturally depends upon the volume of air handled. To eliminate odors from 10 to 20 cubic feet per minute per person will be required.

Air-Circulation:

Proper air-circulation insures practically uniform heat distribution. The temperature difference between the inside air currents and the average temperature of the room should not exceed 2 or 3 degrees with an air current velocity of 50 lineal feet per minute, nor more than 3° F. between any two points located the same distance above the floor and 24" or more away from the heat transfer surface. Without proper circulation, the inside air becomes stratified and an 18° F. temperature difference between floor and ceiling levels of the same room are not uncommon.

In the average home from 6-8 complete changes per hour and an air-velocity of 15-25 lineal feet per minute will be found generally satisfactory during the heating season. Recirculated air is not desirable under all conditions but in the home, particularly during severe weather, re-circulation of a large percentage of the air-volume is not detrimental to the general health of the occupants. Kitchen, Laundry, Bath and Toilets should be ventilated independent of the house system, exhaust direct to atmosphere, and fresh outside air introduced into these rooms.

The volume of air varies with the size and type of heating equipment, the requirements of warm-air installations being four or more times greater than those of a split-system.

For summer conditioning, warm-air and split systems require additional ducts for introduction of the cooled air into the room at a height of 7 or 7½ feet above the floor wherever possible.

Ducts of a central forced air heating system, suitably located and controlled, may also be successfully used as a distributing system for cold-air during the warm season without material changes.

Ducts must also be provided with direct radiator heat but these may likewise function as part of the heating system.

For economy, where summer cooling is contemplated later, all necessary duct work should be installed during building construction.

With proper controls, the split system of steam or hot-water heat will also provide a continuous supply of clean humidified air during the spring and fall seasons independent of the heating system.

Both fan or blower and motor should form a complete, fully enclosed, self-oiling, well-insulated unit; the blower of sufficient capacity to easily handle the design load at a nor-

The effects of solar-heat are responsible for 50% or more of the cooling load. On a clear, bright summer day, temperatures of 180° to 200° F. are common on dark-colored roof areas, and unprotected windows transmit from 70 to 150 B.t.u. per hour, per square foot of glass area directly exposed to the sun rays.

As previously stated, proper insulation and protection against loss by infiltration materially reduce "heat-gain." On windows and other glass areas, suitable awnings will eliminaté the effects of solar radiation and further reduce the cooling-load by approximately 20%, and in single rooms up to 30% depending upon its size, number of windows, orientation, etc. although it is not always advisable to install smaller cooling equipment.

De-Humidification:

This is accomplished either by cooling or by adsorption methods. With the former the excess moisture is precipitated by cooling the air to temperatures below the dew-point. This may be done with a surface type unit where the moist air is forced over "finned-coils" through which cold-water, brine, refrigerant, or a combination of both is circulated or the refrigerant expanded. Initial and operating costs are low but the equipment does not permit accurate humidity control. Where the cooling coils function only during the warm season, and at certain times evaporative cooling will suffice, a "spray-type" unit is often preferable. Cold spray water is usually provided by mechanical refrigeration, although under favorable conditions deep-well or city water at temperatures of 50° F. or lower may be used to remove the excess heat and moisture from the air.

Adsorption processes de-humidity without materially changing the temperature of the air; all or part of the moisture content being extracted by the appication of Silica Gel, Activated Alumina, or similar substances of a mineral nature which inherently absorb moisture and also retain same in liquid form within the material itself.

The essential equipment viz.:—A series of fully enclosed adsorption beds or trays, a gas or oil heater for drying purposes, motor driven fans for conveying the untreated air and automatic controls required for continuous operation, are all self-contained and fully enclosed. Two or more adsorption chambers are required to permit the operation of one unit while the others are being reactivated. This is accomplished by driving out the accumulated moisture by the application of direct heat and then cooling the adsorption beds or trays to atmospheric temperature. The units may then be put back into service. The adsorption beds can also be re-used indefinitely, deterioration is negligible and their efficiency in no way lessened.

Lithium-chloride solutions may also be used for either dehumidifying purposes or as a humidifying agent, the principle utilized being the difference in vapor pressure between the non-conditioned air and the chloride solution over which the air is passed. The vapor pressure of the solution is controlled by regulating its temperature. When the vapor pressure of the solution is lower than that of the air the process is one of dehumidification; at a pressure higher than that of the air, the process is reversed. Replacement or replenishment of the solution is also seldom required.

Cooling:

The moisture content of the inside air largely determines the amount of refrigeration required.

The elimination of one pound of water necessitates the extraction of 1,000 B.t.u. or 1/12 ton of refrigeration per hour, or the total "heat-gain" divided by 12,000 will give the required tonnage. The rate of refrigeration

of the basement air-conditioning plant or the warm-air furnace.

Approximately nine complete air changes per hour are necessary before night-cooling becomes reasonably effective and from 17 to 20 changes for the best results. Oversized ducts and greater blower capacity than required for "winter-conditioning" must also be provided to handle the increased volume of air circulated.

The well insulated house cools approximately at the rate of 1.5° F. and the non-insulated structure at 2.2° F. per hour; both however have sufficient mass and heat resisting capacity so that once cooled to the minimum night temperature they will absorb the solar heat of the following day without a corresponding rise in the inside temperature which will remain at from 8 to 15° F. lower than the outside daylight temperatures.

Night-air cooling may also be supplemented by mechanical refrigeration. Installations are not numerous to date but this type of cooling apparently has many advantages, particularly in the residential field. The method advocated by Samuel R. Lewis is a representative one, the cooling being accomplished by operating the refrigerating plant only during those hours warmer than 85° F. outside after three preceding hours of higher outside temperatures. The inside house-circulating fan is operated not only while the refrigerating plant is in service but also during all hours where outside daylight temperatures are 80° F. or higher. With a maximum outside temperature of 75° F. for one or more hours the operation of the night-fan alone for a minimum of six hours per night is usually all that is required.

In existing buildings heated by steam hot-water or other pipe systems the cost of a central refrigerating plant is usually prohibitive but night-air cooling may be supplemented by individual unit-type coolers or refrigerating units.

The adaptability of mechanical refrigeration, night-air, ice, cold-water and other methods of residential cooling has been investigated for a number of years by the Engineering Experimental Station of the University of Illinois in co-operation with the American Society of Heating and Ventilating Engineers and others. These cover experiments, operating costs, and investigations of actual installations of various types and provide much valuable detailed information which is available through the various bulletins issued by the University from time to time.

FRICTION AND DUCT SIZING CHART
FOR ROUND AND RECTANGULAR DUCTS

CHART No. 1

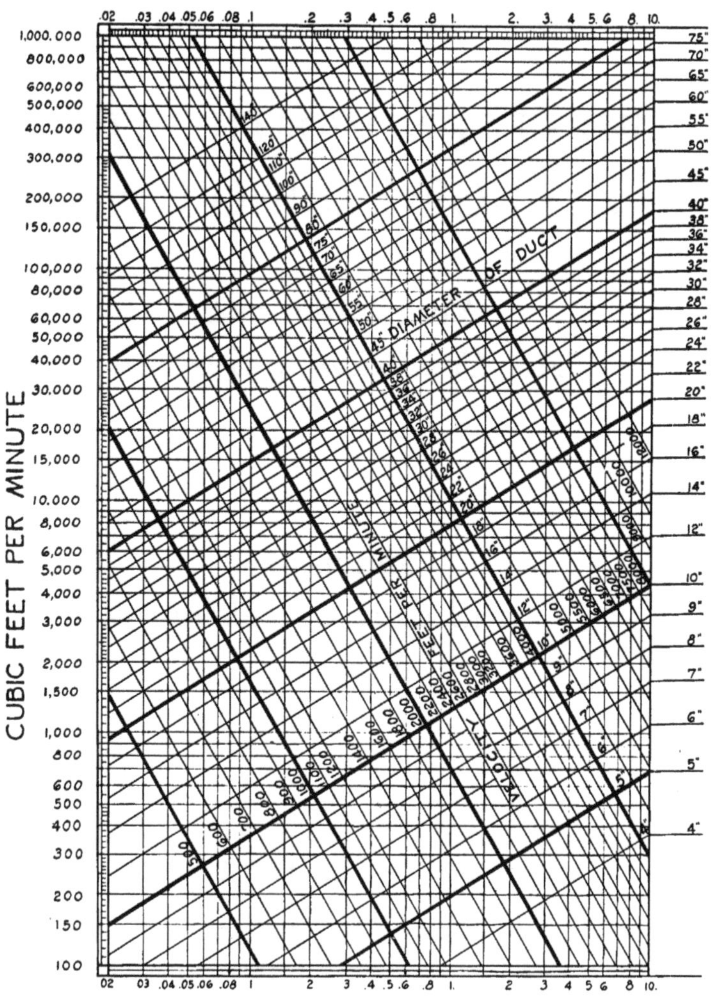

FRICTION IN INCHES WATER GAUGE PER 100 FEET

Courtesy American Blower Corp.

CONTINUED NEXT PAGE

CHART No. 2

5 6 7 8 9 10 12 15 20 25 30 35 40 45 50 55 60 65 70 75 80

EQUIVALENT DIAMETER OF ROUND DUCT

4" 5" 6" 7" 8" 9" 10" 12" 15" 20" 25" 30" 35" 40" 45" 50" 55" 60" 65" 70" 75" 80"

WIDTH OF RECTANGULAR DUCT IN INCHES

Courtesy American Blower Corp.

OIL HEATING

By HARRY F. TAPP

Oil is, each year, being more widely ac-pted as the fuel best suited to modern eds. Many reliable oil burners are being anufactured but certainly in choosing the e you will purchase it is important to know at the design, workmanship and materials e of the best. And because you are buying installation, not merely an oil burner, our choice should be influenced by the echanical skill of the man who will make ie installation and by his financial standing your community.

Burner Types and Characteristics.

There are two distinct types of oil burners ed for heating—the natural draft burner id the mechanical draft burner. Their names dicate the manner in which the air for com-stion is obtained.

The natural draft burner requires no motor d usually has no moving parts. The air r combustion is supplied by the pull of the aft and is therefore dependent upon the de-gn of the chimney. This burner is often correctly referred to as a gravity type. ravity indicates the manner in which the iel is fed to the burner and a gravity system applicable to either the natural draft or the echanical draft burner.

The mechanical draft burner is motor :iven and the air for combustion is supplied y a fan or blower. Generally where a fan used it is of sufficient capacity to supply e entire amount of air for combustion when e burner is operating at the maximum rate. here a blower, either centrifugal or posi-ve pressure, is used, only a portion of the r required is supplied under pressure, the mainder being induced by the injector ac-on of the air from the blower, plus the pull f the natural draft from the chimney. The ir from the blower is generally used to aid n the atomization of the fuel.

The natural draft burner is more sensitive han the mechanical draft burner to changes n the weather, wind currents about the chim-ey and other factors which cause a varia-ion in the draft intensity. A well designed himney and the use of a special draft reg-ilator will, to some extent, overcome this ondition.

The mechanical draft burner, equipped with blower, produces a more constant supply of ir under varying draft conditions and, there-ore, maintains a uniform, efficient combus-ion condition.

The second classification of burners is isually made with reference to the means mployed to prepare the fuel for combustion. The distinguishing terms are "atomizing" nd "vaporizing."

In the vaporizing burner the fuel is pre-iared for combustion by the addition of heat. The heat serves to convert the liquid fuel nto a vapor which is mixed with the air for ombustion either just before or during the ombustion process. Vaporizing burners re-iuire a light fuel and there is always the iossibility of carbon trouble due to cracking ir decomposition of the oil in the vaporizing hamber.

In the atomizing burner the fuel is broken nto a fine mist of small particles which are nixed with the air for combustion either just iefore or at the time they are forced into the ombustion chamber. The particles are so ine that they are quickly vaporized by the ieat of combustion and, if properly mixed vith the air for combustion, burn with a clean iot flame.

There are many ways to atomize oil—under ressure through a small orifice, by com-ressed air or steam, by centrifugal force rom the edge of a rapidly rotating cup or disc and numerous other methods equally efficient. All the methods will adequately atomize the oil, and when applied with intel-ligence will give satisfactory results.

Atomizing burners successfully utilize slightly cheaper oils. They will start readily and can be economically used in large in-stallations.

Ignition Systems.

In most natural draft burners the oil is ignited manually with a torch through the firing door, although some of them are pro-vided with a gas pilot. Full automatic burn-ers are ignited with either an electric spark from a spark transformer or from a gas pilot light. This gas pilot in some designs burns constantly, while in others a combination of the electric and the gas system is used, the spark igniting the gas and the gas flame igniting the oil.

The type of ignition system that is used is dependent, to some extent, upon the design of the burner and upon the personal opinion of the designer. Several burners are being designed to make the means for ignition op-tional with the purchase. Where gas is used, the application of the burner may be limited to a territory having a gas supply, although it can be used with gas supplied in containers. Under average conditions the cost of ignition with gas is slightly greater than with electricity.

Automatic Burner Controls.

A room thermostat is used to indicate and control the operation of the burner and main-tain the desired temperature between a limit-ing plus or minus one degree.

The boiler or furnace is provided with a control to prevent overheating and, in the case of steam boilers, to prevent the develop-ment of abnormal pressures.

A safety control is provided to establish a time limit within which the oil must be ig-nited every time the burner is started and shut the burner down in the event ignition does not take place.

There are two systems of control instru-ments—the low voltage (15-20 volts) and the high voltage which is the voltage at which the motor operates, usually 110 volts. Both systems have been used with complete satisfaction and opinion as to which system is the best is evenly divided.

In order to obtain the best results from the thermostat it should be located with care. Most people prefer to have it in the living room. It should be located on an inside wall, about four to five feet from the floor at the breathing level, protected from abnormal drafts such as stairways or entrances. It should not be near the chimney, radiators or registers, hot water pipes or steam pipes, or other sources of heat. Avoid concealed hot water or steam pipes, and warm air ducts.

In designing new homes, the location of the thermostat can be provided and a panel, in keeping with the general scheme of decora-tion, designed for its mounting. The mount-ing panel can be provided with an electric conduit to carry the control wires from the basement.

To make the operation of the thermostat effective, it is necessary that the radiation installed in each room be carefully propor-tioned. This establishes the desired temper-ature in each room when the room in which the thermostat is located is at the correct temperature.

The majority of the burners operate on the intermittent principle, but a few operate on a graduated control system and the flame in-tensity is varied to meet the temperature variation as indicated by the thermostat.

ıere is considerable discussion among de-
gning engineers as to the relative merits of e two systems, but while the continuously erating burner may have a slight advantage ıring colder weather, the intermittent would more economical during the mild weather, that over the entire heating season the tal amount of fuel used will be very nearly e same.

Any good boiler can be used satisfactorily ith oil burning equipment but, naturally, a iler that has been specially designed for l will have the advantage. In selecting a iler for use with an oil burner, make sure at it has long flue passes and that the com-
stion gases are not short circuited from e combustion chamber to the stack outlet.

In applying oil burning equipment to exist-
g coal burning boilers it is sometimes ad-
sable to baffle the passes to increase the avel of the hot flue gases and keep them closer contact with the heating surface. fire-tube boilers retarders are used in place baffles. These retarders inserted in the bes give the flue gases a spiral motion and ep them in contact with the surface of the bes, increasing the heat transfer in this ıy.

An automatic feedwater regulator should included for hot water or steam installa-
)ns as the boiler is often neglected. The tomatic features eliminate the necessity daily inspections of the boiler room.

Welded steel warm air furnaces are gen-
ally more suitable, although satisfactory stallations are made in cast iron furnaces hen all joints have been sealed to prevent akage of combustion gases. It is recom-
ended that an inspection of the furnace be ade at the beginning of each heating sea-
n.

Domestic Hot Water Supply.

Any automatic burner may be used in con-
ection with steam, vapor, or hot water eating system for furnishing hot water for ousehold use both summer and winter by ıeans of an indirect heater on the boiler. A emperature control set at 180 degrees Fahrenheit maintains the temperature of the water in the boiler at all times and insures an ample heat transfer to the hot water tank through the indirect heater coil.

In hot water boilers an automatically operated valve is placed in each one of the risers. The actuating mechanism of the valve is electrically connected to the room thermostat and only opens the valve when the thermostat calls for heat. There are also indirect heaters having an automatically operated valve which controls both the water flow tc the radiators and the supply of domestic hot water.

Some of the distinct advantages to an arrangement of this kind are:

1. Year round operation keeps the burner and boiler in good condition.
2. Only one central heating plant is necessary, conserving space and increasing the efficiency of the boiler.
3. Uniformly hot water is available at all hours.

For warm air systems—or when desired—separate automatic oil burning hot water heaters with attractive insulated jackets are used.

Actual results have shown that the total operating cost for the service rendered with either indirect heaters or separate oil burning units is lower than that of any other type of automatic water heating equipment.

INSTALLATION

Installations should always be made by men trained under the direction of the burner manufacturer, as every burner has peculiarities that should be given consideration when the installation is made. It is also important that the burner be correctly adjusted for each installation as greater losses in efficiency can result from poor adjustment than from poor design. A burner should have sufficient capacity to develop full rating of the boiler and it is preferable that it have some excess; it should also be adjustable over a range from 50 to 100 per cent rating. See Figure 1. The flame adjustment is best determined

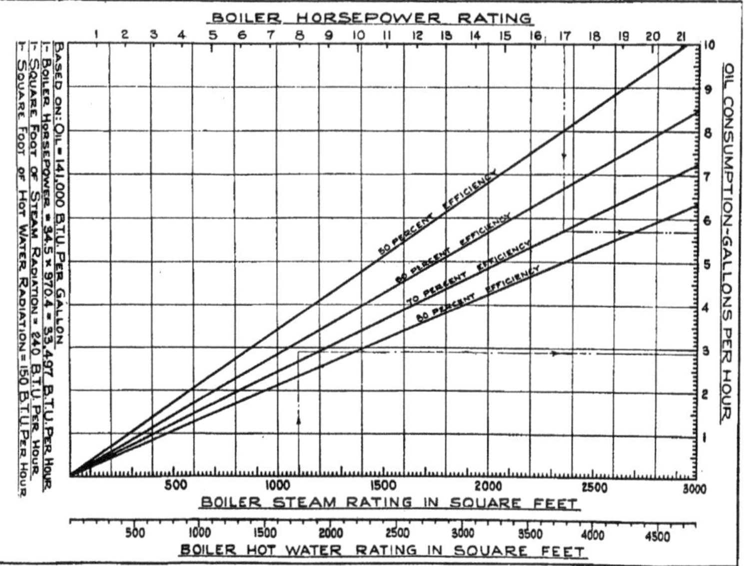

Fig. 1.

y a flue gas analysis, but a good check can e made by noting the color of the flame. A hite flame indicates excess air or insuffiient oil, a red smoky flame indicates insuffiient air or excess oil, and an orange flame ust tipped with red indicates an efficient nd clean combustion.

The boiler room should be well ventilated o that the burner can obtain an adequate upply of fresh air at all times.

Chimney Design.

Although slightly less draft is required for il heating equipment and a smaller flue area vill suffice, it is recommended that the chimey be designed to meet the boiler manufacurers' specifications. The oil burner should ave a separate flue to provide a uniform raft and to prevent any mechanical sound rom being amplified and transmitted to the ooms of the house. A well designed chimey is as important for an oil burner as it is or a coal fired heating plant.

Fuel Oil Specifications.

In April, 1928, the American Oil Burner Asociation adopted uniform fuel oil specifications as recommended by the Association Committee on Fuel Oil Specifications. On January 9, 1929, a joint conference of representative refiners, distributors, and consumers of fuel oil, manufacturers of oil burners, and general interests, adopted these specifications as commercial standard for domestic and industrial fuel oils, and they have since been accepted by the Department of Commerce and the entire industry.

These specifications make no reference to the grayity of an oil, a quality that has been widely used in the past as an index of volatility. Gravity is not a reliable indicator of the quality of a fuel oil and so has no place in the specifications.

All listings of burners published by the Underwriters' Laboratories give the minimum grade number of oil suitable for each burner.

Detailed Commercial Standard Specifications, publication CS12-38, may be secured from the United States Government Printing Office, Washington, D. C. Below is a summary of the specifications.

Grade of oil	Flash point		Water and sediment maximum	Pour[1] point, maximum	Distillation test			Viscosity, maximum
	Minimum	Maximum			10 per cent point, maximum	End point, maximum	90 per cent point, maximum	
			Per cent					
No. 1 Fuel Oil	100° F. or legal	165° F.	*Trace*	15° F.	410° F.	560° F.	....	
No. 2 Fuel Oil	110° F. or legal	190° F	.05	15° F.	440° F.	....	600° F.	Saybolt universal at 100° F. 55 seconds
No. 3 Fuel Oil	110° F. or legal	230° F.	.1	20° F.	....	....	675° F.	

[1] Lower or higher pour points may be specified whenever required by conditions of storage and use. However, these specifications shall not require a pour point less than 0° F. under any conditions.

Grade of Oil	Flash point		Water and Sediment, maximum		Pour point, maximum	Viscosity, maximum
	Minimum	Maximum				
			Per cent			
No. 5 Fuel Oil	150° F.	....	1.0		....	Saybolt Furol at 122° F. 40 seconds.
			Water	Sediment		
No. 6 Fuel Oil	150° F.	....	(3)	0.50	....	Saybolt Furol at 122° F. 300 seconds.

[2] Pour point may be specified whenever required by conditions of storage and use. However, these specifications shall not require a pour point less than 15° F. under any conditions.
[3] The total water plus sediment shall not exceed 2.0 per cent.

The following table gives the approximate gravity range for the various grades, together with the approximate heat content:

Oil No.	Approximate Gravity Range—A. P. I.	Approximate B.T.U. per gallon
1	36°-40°	136,000
2	32°-36°	138,000
3	28°-32° and 25°	141,000
5	18° plus	146,000
6	12° plus	150,000

Comparative Costs

Comparative cost figures for various fuels depend entirely upon the heat content of each fuel and the efficiency with which each is utilized. With oil it is reasonable to assume an increase in efficiency of 10 to 15 per cent over coal. Charts shown in Figures 3 and 4 give comparative consumption of oil against seasonal coal or gas requirements. These figures are based on an oil containing 141,000 B.t.u. per gallon. By referring to Figure 2, these figures can be corrected for other grades of oil. When making corrections for other grades of oil the possibility of varying

operating efficiencies should be taken into consideration.

When larger installations are made, such as in apartment houses and office buildings, the saving in labor, effected by the use of oil, will offset considerable increase in the cost of the oil over the cost of the coal. This saving in labor permits oil heating equipment to be installed, with its advantages of storage, handling and cleanliness, and operated at an equal cost with the coal fired installation, even though the actual cost of the oil may be more than the cost of the coal.

Tank Installations.

Tank installations should always be made in accordance with local regulations, or in the absence of these, the regulations of the National Board of Fire Underwriters should be followed. Copies of these regulations may be obtained from the Underwriters' Laboratories, 207 East Ohio Street, Chicago, Illinois. amount of fuel on hand. All tanks must be vented and they should be located so that the fill line terminal is near the drive, or curb, to facilitate delivery from the tank truck.

The vent line from the tank is usually located as inconspicuously as possible on the side of the building. In constructing a new home, it is possible to provide a channel or duct in the wall of the building so that the only exposed part is the vent cap where the line terminates.

The tank location should be near the drive or curb so that the oil can be delivered through a hose from the delivery truck tank.

Important Considerations in Selecting Oil Burning Equipment.

In specifying an oil burner the highest degree of satisfaction is experienced when the following points are carefully considered:

1. Consider only equipment listed as standard by the Underwriters' Laboratories

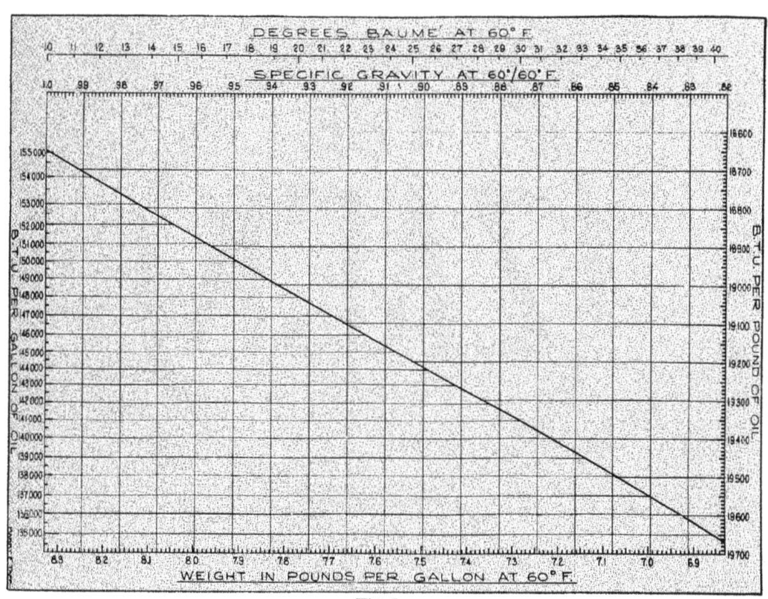

Fig. 2.

These latter regulations permit the installation of two exposed 275 gallon tanks with a three-way valve. This three-way valve permits gravity feed from either tank and insures an adequate fuel supply without a buried tank. This type of installation has many advantages and is rapidly coming into favor with local authorities having jurisdiction over the installation of oil heating equipment.

A very desirable installation for larger homes is an outside buried tank of at least 550 gallons capacity. The use of a large buried tank eliminates the necessity of constantly watching the fuel supply and in some cases, will permit the purchase of fuel at a price, enough lower, to pay for the difference in installation cost. All tanks should be provided with a direct reading gauge to give a constant check on the

and manufactured by nationally known responsible concerns.

2. Consider only financially sound and mechanically competent dealers located within a reasonable distance from the installation, who will install the equipment in accordance with all local and State regulations, as well as in accordance with manufacturer's specifications.
3. Insist that your heating system be carefully checked by the dealer before the installation is made.

Architects' Counsel To Clients.

A proper appreciation by the home and building owner of the nature of his oil heating equipment, its capabilities and limitations, and the care it will require, will do more than anything else to assure satisfying results. The essential points for the client are summarized here:

2311 VAN BUREN ST.
SEELEY 4254-5-6
VENTILATION
THE
ZACK
CO.
CHICAGO
AIR CONDITION

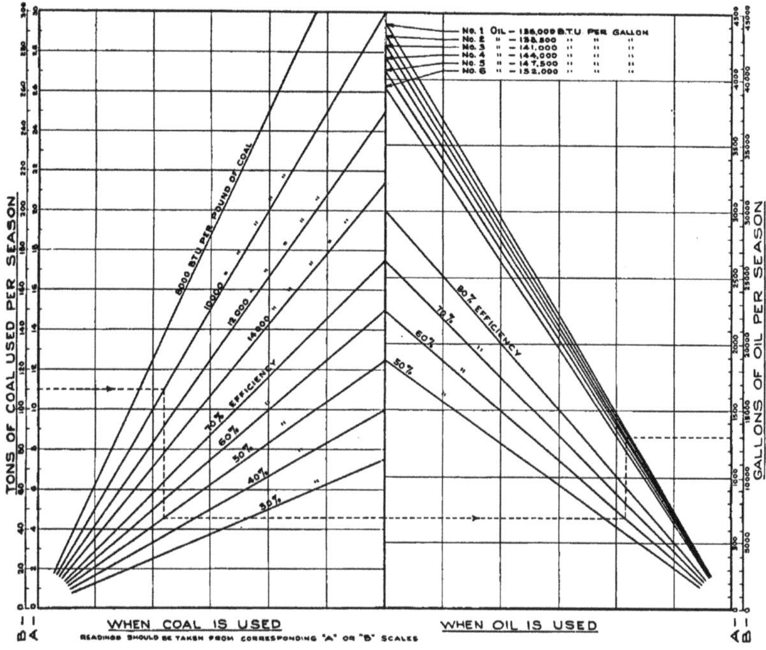

Fig. 3.

1. Oil heating equipment functions only to provide heat. It will not take care of the water level in the boiler. It will not operate when the current is shut off, nor when the fuel is exhausted. The entire plant should have regular inspection to see that these conditions are correct.
2. Oil burners require reasonable care—oiling of moving parts and occasional cleaning. In this respect they are like clocks, automobiles, fans and electric refrigerators.
3. Oil burners like all other machines, are subject to adjustment, and they operate best when perfectly adjusted. Expert service men can make the occasional adjustments far better than a layman or ordinary mechanic.
4. Once the correct adjustments are made, they can be disturbed only by the following methods:
   - (a) Manually, as when someone attempts to interfere with the automatic operation of the burner or to adjust the parts. Do not disturb the apparatus except for regular inspection, oiling and cleaning.
   - (b) By changes in fuel. Stick to the same grade and quality of fuel, or have a service man readjust the burner when changes are necessary.
   - (c) By pressure of foreign matter—particularly in the fuel.
   - (d) By natural wear. Periodic inspections will take care of these changes.
5. Automatic operation does not permit neglect. Give to your heating apparatus the reasonable care and attention any mechanical equipment requires.

These reasonable suggestions, if followed, will assure the maximum benefits and lowest operating costs to the owner.

These suggestions are nearly all given from the view-point of domestic and small commercial building installations because this is the most active field at the present time.

Outline of Architect's Specifications. Covering the Installation of Oil Heating Equipment.

Scope of Contract.

These specifications cover the complete installation of an oil heating apparatus and fuel oil storage system, for the boiler installed in .... The apparatus shall consist of a .... oil burner, fuel oil storage tank (tanks); room thermostat, boiler control, burner safety device, necessary and adequate installation of refractory lining in the combustion chamber of the boiler; all necessary piping, valves, electric wiring, switches, etc., all tested and ready for operation.

The oil heating installation shall comply with all local ordinances (or rules of National Board of Fire Underwriters), and must meet available electric current facilities.

Liability.

The contractor shall assume, etc., etc.

Completion of Work and Payment.

The work is to be completed, etc.

Materials.

All materials, etc. (recommended furnished by contractor.)

Cutting and Patching.

The contractor shall do all, etc.

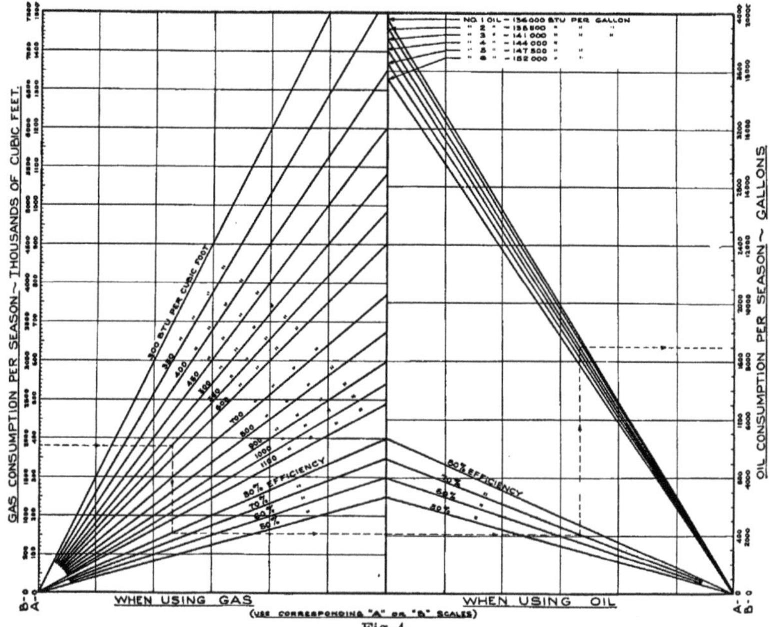

Fig. 4.

Cleaning up.

The contractor shall promptly, etc.

Additional Data.

All visible piping and scratched places will be painted to match other new similar adjacent material.

Oil Burning Apparatus.

The contractor shall furnish, make all necessary changes in the boiler (or furnace) and install, one completely equipped............ oil burner, etc.

Fuel Oil Storage.

The contractor shall furnish and install (one .... gallon inside) fuel oil storage tank. Tank shall be manufactured, tested and installed in accordance with local regulations (or rules of National Board of Fire Underwriters).

Piping.

Piping shall be installed in accordance with local regulations (or rules of National Board of Fire Underwriters).

Wiring.

All wiring shall be done in accordance with the National Electric Code, and local regulations.

Thermostat Control.

The thermostat shall be installed in .... room, five feet from floor, removed as far as is practical from any and all warming influences such as radiators, hot water pipes, etc., or possible cooling drafts.

Warm Air Furnace Control.

Furnaces shall be equipped with thermostatic warm air jacket control, wired in connection with room thermostat.

Boiler Control.

A maximum pressure or temperature control shall be installed in the boiler according to manufacturer's printed instructions. This control shall be wired so as to automatically prevent creation of excessive pressure or temperature in the boiler.

Burner Safety Device.

A burner safety device shall be installed in connection with the burner, so designed as to make the burner inoperative if for any reason the burner does not function properly.

In General.

The omission from these specifications of any minor detail of construction, installation, material, specialties, etc., shall not relieve the contractor from furnishing same in place complete, and such omissions shall not entitle contractor to make claims or demands for extra materials or labor. However, in the event that unusual water is struck or if quicksand, rock or other unusual obstructions are encountered, the contractor shall proceed with the necessary special construetion that is involved for which the contractor will receive sum equal to the actual cost of such special work plus .... per cent. The word "cost" as hereinabove used shall be understood to consist of actual field cost and overhead.

Adjustment.

The contractor shall agree to provide free inspection and adjustment of the oil burner installation for the first ninety days of the heating season during which the installation is made. The heating season shall be considered as beginning September first for installation made during the summer.

Guarantee.

The contractor shall guarantee to make good by replacement or repair, any original defects in parts, material or workmanship previously specified or described; provided that this obligation is assumed only in the event that written notification of such alleged defect be given the contractor within a period of one year after said equipment has been installed.

HEATING BY GAS

By C. W. BERGHORN, Executive Secretary
Association of Gas Appliance and Equipment Manufacturers

During the past few years there has been a great increase in the number of central plant gas heating installations in operation in this country. This rapid growth in the use of gas for house heating is, in part, due to the development of the distribution of Natural Gas to our larger cities. Probably the most important factor in this growth has been the increased public acceptance of the use of gas for heating purposes, as both Natural and Manufactured Gas, tend to make this use more desirable from the customer's standpoint.

Gas Designed Equipment.

The combustion characteristics of gas are such that generally it can be utilized most economically and satisfactorily in a heating device, be it a steam or hot water boiler, or a warm air furnace, when that device has been designed solely for the use of gas. Flame characteristics and freedom from fuel bed draft losses dictate design features that are not found in the device originally designed to burn solid fuel. These peculiar design features are incorporated in a number of well-known gas burning heating appliances, which differ in details of refinement, to a somewhat greater extent than in fundamental principles.

Many Sizes Available.

Gas burning heating devices designed solely for the utilization of gas as a fuel, are available for steam, hot water and warm air heating systems, in a size range extending from the small bungalow to the large office building. Small graduations between successive sizes in any given line of appliances, enable heating requirements to be met with a minimum amount of excess capacity being necessitated. Absolute certainty as to the performance and heat generating capacity of a gas burning appliance enables the choice of the proper type and size to be made with the highest degree of accuracy. Freedom from the effect of variable drafts, high winds, low barometric pressure make the selection of the proper size of gas burning appliance a problem that can be solved with scientific accuracy.

Construction of Boilers.

The usual heating boiler designed for gas fuel is sectional in construction, made of a number of parallel and substantially vertical cast iron sections. Different makers resort to different expedients to render the heating surfaces as effective as possible. Some use a tubular design, other use extended surface in the form of lugs or ribs, some may incorporate both of these features or other features peculiar to their individual design. Horizontal burners of the atmospheric Bunsen type, usually with drilled ports, are the general means of burning the gas. The burners are arranged directly beneath the sections. The entire assembly is almost invariably enclosed in a well-insulated sheet metal jacket, generally finished in a more or less attractive surface and color.

High Efficiency.

Due to the absence of any draft loss, due to fuel bed, the gas passage can be designed to provide the high gas velocity essential to heat transmission, while at the same time, keeping the actual frictional effect to such a low figure that pressures within the combustion and flue spaces are practically atmospheric. The gas is delivered to the burners by the gas company at a definite and positive pressure, no mechanical devices such as motors or blowers are necessary for the delivery of the fuel and for its mixture with the air for combustion. The usual building heating appliance designed originally for the use of gas is free of moving parts or of the necessity of electric current to insure steady operation. Due to the fact that combustion and flue gas travel take place at practically atmospheric pressure, such an appliance will operate successfully on practically no draft. Devices to be placed in the flue connection for the purpose of reducing the pull of the chimney and to minimize the effect of variable atmospheric conditions, are standard parts of gas designed heating appliances. These are commonly called Down Draft Diverters. Due to the freedom of the effect from variable chimney draft, constant efficiencies of 80% or better can be maintained throughout the heating season.

Gas Boiler Controlling Devices.

Gas designed heating appliances are provided with controlling devices that insure continued automatic operation at a constant efficiency, free from possibility of accidental shut-down. They should always be operated under the primary control of a room thermostat. The quick response of the gas burning heating appliance to a demand for heat, together with its small heat storage capacity, enables the temperature of the premises served by a well designed system to be held within a two degree range. Standard controlling devices on gas designed heating appliances include a gas pressure regulator that maintains a constant rate of fuel supply, steam pressure or water temperature limit control, low water cut-off devices on steam boilers and thermostatic pilots that prevent waste of unburned gas in the event of pilot outage.

All of these controlling devices may be absolutely independent of electrical current, so that in the event of current failure, the heating of the premises is uninterrupted.

Construction of Gas Warm Air Furnaces.

A number of gas warm air furnaces, designed exclusively for gas fuel are available. There is a greater variation in types of warm air furnaces than in steam and hot water boilers. Furnaces are made with cast iron heating surfaces, sheet metal surfaces or combinations of the two constructions. Straight upward flue travel is found in some, while others are constructed with revertible flues. In recent years, furnaces have appeared on the market, which are designed solely for forced air operation, secured with a fan.

Cleanliness, automatic control, quick response and possibility of artificial humidification render a warm air system, served by a gas burning furnace, the type of heating plant suitable for installation in the highest class of residence.

Warm Air Furnaces are capable of control by room thermostats and in addition are provided with gas pressure regulators, air temperature limiting devices and thermostatic pilots.

BRYANT
CONDITIONED AIR HEATING

for
Homes, Apartments,
Public Buildings

BRYANT DUALATOR

A complete air conditioning unit adaptable either to air conditioning exclusively or to any desired combination of air conditioning and radiator heating.

MODEL 78
AIR CONDITIONING UNIT

For warm air heating systems, in a complete range of sizes. Heats, cleans, humidifies and circulates the air. Available with air washer if desired.

MODELS 23 AND 25
GAS BOILERS

For steam and hot water heating. Especially designed for small and medium size homes. Tubular construction. Complete with automatic controls.

MODELS 443 AND 445
GAS BOILERS

For use with either natural or manufactured gas. Sectional construction, water tube type. For steam, hot water or vapor heating systems.

The recognized superiority of Conditioned Air Heating has brought about a rapid change in heating thinking and heating practice. Fully alive to this modern trend, the Bryant line of Gas Fired Air Conditioning Equipment measures up to every requirement which science and good Air Conditioning practice demand.

Bryant Air Conditioning Equipment is not a result of the present trend. Rather, it is the outgrowth of years of intensive research, development and manufacturing experience combined with an intimate knowledge of heating requirements.

For homes of any type or size—for apartments, stores or public buildings—there is Bryant Air Conditioning Equipment to do the job economically and dependably. A Bryant installation assures complete satisfaction for years to come.

We invite Architects to consult with our Air Conditioning Staff on any heating problems they may have.

For complete data on sizes, ratings, etc., write

THE BRYANT AIR CONDITIONING CORP.

122 South Michigan Avenue — **Chicago, Illinois**

Phone: Wabash 8291

(a) **Gas Supply.**

Obviously, the pipe carrying the gas from the meter to the appliance must be large enough to deliver the required amount of fuel at the prevailing pressures. The valve and connections that are sold as part of the appliance are proportioned to carry the required gas at the lowest supply pressure that can be reasonably expected to exist. That is a responsibility of the appliance manufacturer. If the line running from the meter to the appliance is one pipe-size larger than the gas control valve, it will have adequate capacity.

(b) **Air Supply for Combustion.**

Gas, like any other fuel, requires air for its combustion. The air necessary to sustain combustion, and to maintain adequate boiler room ventilation amounts to at least 7.5 cu. ft. for each cubic foot of 500 to 550 B.T.U. Gas. Failure to provide for the entrance of sufficient air into the boiler room, to meet this requirement, may result in symptoms that may be mistaken for those of insufficient draft. The effect of insufficient air is less evident to the eye, with gas as fuel, than with solid or liquid fuel, but it is no less destructive to efficiency of combustions.

(c) **Venting Flues.**

Although a gas designed appliance operates with practically atmospheric pressure, within its combustion and flue spaces, it nevertheless needs adequate provisions for venting the products of combustion. Poor chimney conditions are prolific causes of condensation of water vapor within chimneys and chimney connections as well as of poor combustion. Although the products of combustion issuing under normal conditions from an appliance designed for gas burning, such as a standard make of gas boiler or gas furnace, are totally harmless to life. Every effort shold be made to insure their adequate venting from the premises through a suitable and effective chimney.

(d) **Small Fire Hazard.**

When a heating appliance designed for gas burning is installed, no particular precautions in the direction of fire prevention are necessary. Standard makes of appliances are thoroughly insulated and give off very little heat to surrounding objects. In most appliances, all of the electrical wiring involved can be low voltage. There is no fuel storage required nor are chimney temperatures high.

Operating Costs.

In evaluating the cost of heating a given space, the following are among the items that should be given consideration:

(a) Initial investment for apparatus.

(b) Value of the space occupied by the apparatus.

(c) Rate of depreciation of the apparatus.

(d) Cost of attendance, including ash-handling.

(e) Reliability in operation, possibility of service being required and cost of that service.

(f) Space required for fuel and ash storage and its value.

(g) Inconvenience and damage from dust and soot.

Without going into the calculations of a concrete example, let it be said that proper calculation of comparative heating cost, including the factors mentioned above, will often reveal gas to be actually cheaper than other fuels selling at a much smaller price per million B.T.U.

(Kathene), or Calcium Chloride. Silica Gel, a powerful adsorbent, is a synthetic quartz which always retains its same physical structure and has a high affinity for moisture due to capillary attraction. The Lithium and Calcium Chlorides are made up into a spray solution through which the air is passed thus absorbing the moisture in the air. Activation or regeneration in all cases is accomplished by means of a gas flame. In the Silica Gel method heat is merely passed through the trays containing the saturated material driving off the moisture in the form of hot wet air. The Lithium and Calcium Chlorides are heated in a boiler to concentrate the solutions and the moisture collected from the air is dissipated in the form of mild steam.

The conventional method which is fast being supplanted is that of super-cooling the air below its dew point, thus condensing the moisture from the air. With this process, the dehumidifying is always "a by-product" of cooling while in the methods described above it is always a prime function.

The cooling (sensible heat removal) can then be accomplished by whatever means are available. Whether it be cold (65° F. or less) tap or well water or refrigeration. Then too, if refrigeration is used an evaporator temperature of 55° F. can be used for the sensible heat removal, thus allowing smaller compressor equipment to be utilized. The water can be pumped either with a gas operated engine or an electric motor. The same holds true for the compressor.

With the gas method of air conditioning independent control of humidity and temperature can be enjoyed at all times regardless of the fluctuating loads, as well as providing a more economical operating cost.

Conditions to Which Best Adapted.

This type of system is particularly adaptable to spaces that have a large latent heat load where the relative humidity might be high, such places as restaurants, theatres, night clubs, stores, showrooms, etc., where a large number of people congregate. It is also especially valuable in industrial applications where a positive control or humidity is essential to production or the product. It has been successfully applied to industrial uses such as printing and lithographing, effervescent salt products, textile factories, and others.

Gas Engine Application.

A comparatively recent development in the Gas field of air conditioning is the Gas Engine. This engine utilizes gas much in the same manner that an internal combustion engine utilizes gasoline as a fuel, and this engine can drive compressors, blowers, and so forth, in the place of using an electric motor. This engine is particularly applicable where electric costs are high and gas costs are low. This machine has a further advantage in that it can be used in localities where there is no electric service available and can even be used where no gas service is available where it would utilize bottled gas for a fuel.

Reasons for Air Conditioning Development.

The rapid development and progress made in air conditioning can be more readily appreciated when these facts are considered. Air is more important to sustain life than either food or water. Sixty per cent of a man's energy comes from the air he breathes. He breathes on the average of 24,000 times a

Compact, handsome, the Janitrol Conditioner enhances basement surroundings.

AIR CONDITIONING IS *Simplified* BY SPECIFYING

Janitrol WINTER AIR CONDITIONERS

● For small homes, large homes and remodeling, you can standardize on the gas-fired Janitrol Winter Air Conditioner confident in the reputation, responsibility and wide experience of the manufacturer. Handsome, compact, moderate in price, the Janitrol Conditioner is easily installed. Mechanically circulating filtered, humidified air at desired even temperature in every room, under full automatic control, Janitrol assures your client of complete, winter time comfort.

One of 14 models for any heating requirement.

Janitrol

UNIT HEATERS FOR COMMERCIAL AND INDUSTRIAL HEATING

● Positive, trouble-free, highly efficient, the gas-fired Janitrol Unit Heater is the modern method for heating commercial and industrial establishments. Suspended overhead, valuable floor space is saved. A completely self-contained heating unit, it requires no boiler, no steam lines.

COMPLETE, YEAR 'ROUND AIR CONDITIONING WITH KATHABAR

● For comfort in residences and commercial buildings and for atmosphere control in industry Kathabar accurately controls humidity of air using a liquid, Kathene. Temperature regulation, air filtration and circulation can also be provided.

12,000 homes in greater Chicago have installed Janitrol gas heating equipment. Janitrol is approved by the Peoples Gas Light and Coke Company. Competent heating and air conditioning engineering service is available to architects and engineers, without obligation. . . . Write or phone.

SURFACE COMBUSTION CORPORATION

122 South Michigan Avenue **Chicago, Illinois**

Phone HArrison 8225

ay, and with each breath he inhales 20 to 5 cubic inches of air, a total of 480,000 cubic nches, or 278 cubic feet a day. A man can ive forty days without food, three to four ays without water, but only a few minutes rithout air, and add to this fact that 60% iore people die from respiratory ills caused y contaminated air than from any other ailment, plus the additional facts that 90% of the air breathed by the average person is indoor air. It can be easily understood that Summer and Winter or Year-round Air Conditioning has at last found its place under the sun and is here to stay. It will mean a healthier and more comfortable world for the human race.

GAS FIRED BOILER RATINGS

Tested under the American Gas Association Code
1930

Radiation Load	Safety Factor	Radiation Load	Safety Factor	Radiation Load	Safety Factor
100	.56	**1500**	.486	**2900**	.427
200	.56	**1600**	.478	**3000**	.424
300	.56	**1700**	.472	**3100**	.422
400	.56	**1800**	.464	**3200**	.42
500	.56	**1900**	.457	**3300**	.417
600	.552	**2000**	.45	**3400**	.414
700	.545	**2100**	.447	**3500**	.412
800	.537	**2200**	.445	**3600**	.409
900	.53	**2300**	.443	**3700**	.407
1000	.523	**2400**	.439	**3800**	.405
1100	.515	**2500**	.437	**3900**	.402
1200	.508	**2600**	.435	**4000**	.40
1300	.501	**2700**	.432	**and Over**	.40
1400	.494	**2800**	.43		

CHIMNEY SPECIFICATION

The walls of the chimney shall be of brick nd shall be lined with approved fire clay flue lining. The joints of the flue lining shall be made air tight. The cleanout space at the bottom of the chimney shall be air tight when the cleanout door is closed. Flue lining shall start at least 4" below the bottom of the smokepipe intake and shall be continuous the entire height of the flue and project at least 6" above the chimney top to allow for a 3" wash and a 3" projection of the lining. The wash shall be formed of a rich cement mortar.

Chimneys shall not rest upon or be carried by wooden floors, beams, nor be hung from wooden rafters, but shall be built upon concrete or masonry foundations properly designed to carry the weight imposed without danger of settling or cracking.

Flues shall be .. in. x .. in. x .. ft. high (in no case shall the area of the flue be less than 12" x 12"), built vertical, without offsets and full size from the smokepipe inlet to the top of chimney.

The top of the chimney shall be at least 3 feet higher than the highest point of the building and in no case shall it be less than 30 feet above the boiler or furnace grates.

There shall be but one connection to the flue to which the boiler or furnace smokepipe is connected. The boiler or furnace smokepipe shall be thoroughly grouted into the chimney and shall not project beyond the inner surface of the flue lining.

The chimney flue to which the heating boiler is connected shall be subjected to a smoke test by the mason contractor in the presence of the architect, or his representative, after the mortar has thoroughly hardened, and must be SMOKE TIGHT.

The method of conducting this test shall be as follows: With a good fire in the boiler or furnace, or in the base of the chimney, put about a square yard of tar paper on the fire. As soon as smoke appears at the top of the chimney, close the top of the flue with a piece of old carpet or wet newspapers held down by a weighted board. Keep the tar paper burning in the firepot for five minutes. The architect or his representative shall sign an acceptance in triplicate, stating that the chimney was tight under the above test, and shall give one copy to the mason contractor; one copy to the heating contractor and one copy to the owner.

All work done under this specification must be in accordance with the requirements of the National Board of Fire Underwriters.

The above data is for all chimneys in many cases but as to size is reserved to coal burning boilers or furnaces. (For areas of chimney consult the boiler or furnace manufacturers' catalog.)

Gas Fuel Boiler or Furnace Chimneys.

In order to be sure that joints in the burned fire clay lining are properly made, it is desirable to have a bell joint tile in preference to the straight butt tile with the bell end up as an assurance against defective joints.

As additional precautionary measure is that after the chimney is constructed a metal liner of smaller size than the actual chimney to prevent action of flue gases. These linings may be removed without impairing the chimney. Metal liners should be a stainless steel from at least a 26 gauge sheet. Other liners are of a vitreous enamel over a base of 20 gauge sheets.

Recommended sizes for chimneys to handle flue products of gas fuel and with a stainless steel or vitreous enamel lining are as follows:

Gas Input per Hour	Size of Liner
Up to 200 cu. ft.	5"
200 to 280 cu. ft.	6"
280 to 375 cu. ft.	7"
375 to 500 cu. ft.	8"

(Rules for Brick Stacks and Fireplace flue areas see page 764.)

ILLINOIS

SELECTIVE PRESSURE CONTROL SYSTEMS

Represent an entirely new and unique development ... Heating Systems that set new standards in luxurious comfort, surprising economy, simplicity and convenience of operation.

Illinois Selective Pressure Control Systems automatically control steam circulation in vacuum heating systems by regulating the pressure maintained and the combustion rate. Suitable for buildings of every type, these control systems can be installed either in new buildings or adapted to existing vapor and vacuum systems. They can be applied to any form of controlled combustion and any type of fuel. Each system is individually engineered to meet exact requirements.

Illinois Flow Control Valves are used where operating conditions make it advisable to control the flow of steam after generation, rather than by control of combustion. They are made in a full range of sizes for manual, pneumatic or electric operation.

Of the full floating type, these valves furnish exactly the quantity of steam required for every varying demand and insure true "Heat Conditioning for Comfort."

Write for bulletins 16 and 517 which illustrate and describe these systems and valves, together with Illinois Thermostats, Infiltration Mounted, in both the pneumatic and integrating types, and show their application to these systems.

ILLINOIS ENGINEERING COMPANY

MANUFACTURERS OF A COMPLETE LINE OF STEAM SPECIALTIES

2043 SOUTH RACINE AVENUE CHICAGO

TWO-PIPE NON-SHORT CIRCUIT HOT WATER HEATING SYSTEM

HOMER R. LINN, Consulting Engineer, Member American Society Heating and Ventilating Engineers

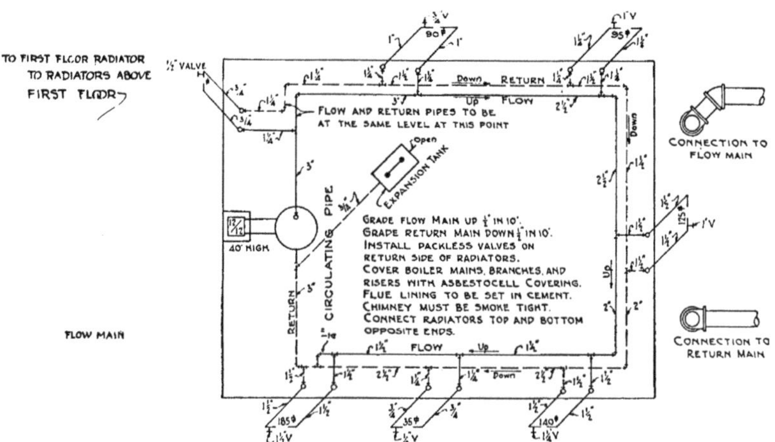

Gravity hot water heating systems may be divided into two general heads, viz.: Short circuit systems and non-short circuit systems. These may be subdivided into up feed, and down feed, etc.

In the short circuit system the flow and return mains run parallel, grade up away from the boiler and are of corresponding sizes where any radiator is taken off. The first radiator taken off of the flow main is also the first radiator on the return main. We, therefore, have the greatest push or pressure on the flow main and the greatest pull on the return main at this point. The result is that the tendency of all of the hot water is to go through this radiator, while the one on the farther end of the main has less pressure and therefore is sluggish. Various means are resorted to in overcoming this error, such as putting lead washers in the valve unions, taking flow connection off of the side of the main, etc. Any of these are uncertain and often cause trouble which is hard to locate.

In the non-short circuit system the flow main grades up away from the boiler. Where the connection to the first radiator is taken off of the flow main, the connection from the return of this radiator is brought into the end of the return main. In other words, this is the smallest diameter of the return main. It will be seen that we have here the greatest push on the flow main and the least pull on the return main. At this point both mains should be on the same level, but from here on the return main should grade down one-half inch in ten feet, while the flow main should continue to grade up one-half inch in ten feet. The last radiator taken off of the flow main will be the nearest radiator to the boiler on the return main: Therefore, at this point we have the least push in the flow main but the greatest pull in the return main. Thus it will be seen we have no short circuits, but a balanced condition throughout.

The accompanying sketch illustrates how the proper sizes of valves and pipes may be selected from the table. It also shows the best method of making connections to the flow main and also to the return main.

Unless boilers are furnished with integral metal insulating jackets, they should be well covered with a plastic covering having an air space between the boiler and the covering. All mains, branches and risers should be covered with a good grade of moulded covering. The expansion tank pipe may be taken off from either the flow or the return main, whichever is most convenient.

GRAVITY HOT WATER HEATING.

Sizes of mains for basement **two-pipe non-short circuit system** where mains are not over 100 feet long.

1¼" pipe,	0 sq. ft. to	100 sq. ft.
1½" pipe,	101 sq. ft. to	250 sq. ft.
2" pipe,	251 sq. ft. to	400 sq. ft.
2½" pipe,	401 sq. ft. to	650 sq. ft.
3" pipe,	651 sq. ft. to	1000 sq. ft.
3½" pipe,	1001 sq. ft. to	1900 sq. ft.
4" pipe,	1901 sq. ft. to	2500 sq. ft.
4½" pipe,	2501 sq. ft. to	3100 sq. ft.
5" pipe,	3101 sq. ft. to	4000 sq. ft.
6" pipe,	4001 sq. ft. to	5600 sq. ft.

Sizes of Risers.

¾" pipe,	0 sq. ft. to	70 sq. ft.
1" pipe,	71 sq. ft. to	120 sq. ft.
1¼" pipe,	121 sq. ft. to	180 sq. ft.
1½" pipe,	181 sq. ft. to	250 sq. ft.

Sizes of Valves.

½" valve,	0 sq. ft. to	60 sq. ft.
¾" valve,	61 sq. ft. to	90 sq. ft.
1" valve,	91 sq. ft. to	130 sq. ft.
1¼" valve,	131 sq. ft. to	180 sq. ft.
1½" valve,	181 sq. ft. to	250 sq. ft.

Grade Flow main **up** and Return main **down** ½" in 10 ft.

MECHANICALLY CIRCULATED HIGH TEMPERATURE HOT WATER HEATING SYSTEMS

The A. S. H. V. E. Guide for 1932, page 98, published a table of conversion factors from which could be determined the BTU output per hour per square foot of equivalent radiation for steam and water of various degrees of temperature ranging from 150° F. to 240° F. This table, and the explanatory information which accompanies it, demonstrates that there is an equal heat emission from any given radiator surface where either water or steam is the heating medium, providing that the temperature of the water or steam is equal. In other words, if an average water temperature of 215° is maintained in the radiator, the heat emission will be equal to that of the same radiator using steam with one pound gauge pressure as the heating medium, or an hourly emission of 240 BTU per equivalent square foot in air at 70° F.

For various other water (or steam) temperatures, various emissions per square foot are shown. Using the factors in the A. S. H. V. E. table, and converting them in terms of BTU per hour, the following table results:

TABLE No. 1

Temperature of Water in Radiator	Heat Emission Per Square Foot
150°	110 BTU per hour
160°	130 BTU per hour
170°	150 BTU per hour
180°	170 BTU per hour
190°	190 BTU per hour
200°	210 BTU per hour
215°	240 BTU per hour
225°	260 BTU per hour
240°	295 BTU per hour

Note—The above emissions are for a radiator in air at 70° F.

From this table it is readily seen that the hourly delivery of heat of a water radiator can be varied at will from an extremely low output, suitable for mild weather, to a delivery considerably higher than that ordinarily credited to steam.

The higher water temperatures can be obtained only under a pressure higher than the atmospheric pressure. If a boiling point for water under atmospheric pressure is assumed to be 210° F., the pressure necessary to obtain water temperature above 210° F., would be as follows:

TABLE No. 2

Water Temp.	BTU Hour	Pressure
215°	240	1. lbs. per sq. in.
225°	260	4.3 lbs. per sq. in.
240°	295	10. lbs. per sq. in.

The aforementioned pressures are those necessary in the top radiator, or the highest point in the system. The pressure in the boiler, or lowest point in the system, will be increased one pound (1 lb.) per square inch for every 2.3 feet of altitude. Hence, if the top radiator were 23 feet above the boiler, an over-all pressure of $\frac{23}{2.3}$ plus 4.3 lbs. or 14.3 lbs. at the boiler would permit a temperature of 225° in the top radiator.

All of the aforementioned tables and figures are based upon the assumption that the temperature at all points of the radiator surface is constant. Obviously, since the water will be cooled while passing through the radiator, this condition would never be present. However, the **average** between the temperature of the water entering the radiator, and the temperature when leaving the radiator will give the mean effective temperature of the radiator. In other words, if the water enters the radiator at 190° and is cooled in passing to 170° the mean average temperature of the water, and of the radiator, would be 180°. Hence, with a temperature of 190° at the boiler, and with a 20° drop through the system the mean radiator temperature would be 180° and could be so considered in referring to the aforementioned tables.

The amount of water to be pumped through any given system depends first upon the total amount of heat to be delivered by the system, and second, upon the amount of heat to be extracted from each pound of water as it passes through the system. Assume that the total heat requirement is 160,000 BTU per hour, and that the proper radiation and boiler has been provided to supply this demand. The water will give up approximately 1 BTU per pound per degree of heat loss. Hence, 160,000 pounds of water would have to be circulated through the system if each pound was cooled one degree in passing; 16,000 pounds would be required if 10 degrees were extracted, 8,000 pounds if 20 degrees were extracted, 4,000 pounds if 40 degrees were extracted, and so forth. While there is no rule which can be applied to govern the temperature drop of the water in every system, it is considered good practice in most instances to use 20°. Any less requires excessive pumping, larger pipes and hence a more costly installation. A greater temperature drop would permit smaller pipes and less pumping, but would require a higher boiler temperature to maintain the desired mean average radiator temperature, and hence a more costly operation.

For the balance of this discussion, therefore, a temperature drop of 20° will be assumed. Tables No. 1 and No. 2 can then be combined and revised on this basis as follows:

TABLE No. 3

Boiler Temp.	Radiator Temp.	BTU Hour Per Sq. Ft.	Pressure Required
160°	150°	110 BTU	
170°	160°	130 BTU	
180°	170°	150 BTU	
190°	180°	170 BTU	
200°	190°	190 BTU	
210°	200°	210 BTU	1.0 lb.
225°	215°	240 BTU	4.3 lbs.
235°	225°	260 BTU	8.5 lbs.
250°	240°	295 BTU	16.0 lbs.

It will be noted that the pressure shown for 215° mean average in the radiator is 4.3 lbs., which was indicated in Table No. 2 as being sufficient to permit a temperature of 225°. In this case, the boiler temperature will be 225°, and the entering temperature of the water in the top radiator will be approximately the same. Consequently, the boiler water temperature would dictate the pressure required to prevent boiling.

Unlike gravity water systems, the water in a mechanically circulated system can be made to supply each radiator equally regardless of the temperatures of the water in the boiler and the balance of the system. Consequently, in mild weather, water at a very low temperature can be supplied to each radiator, and as colder weather sets in, and a greater demand is imposed upon the system, water of higher temperature can be supplied each radiator until it is delivering the maximum total for which the radiator was selected. Assuming this total to have been 240 BTU per hour, and that the system was designed to deliver its maximum when the outside temperature is —10° F., the following Table No. 4 illustrates the manner in which the total heat delivered by each radiator could be varied to meet various outside temperatures. This control of the system could be accomplished automatically, or by manual adjustments to the boiler temperature control equipment.

TABLE No. 4

Outside Temperature	Required Delivery Per Sq. Ft.	Average Radiator Temperature	Boiler Temperature	Pressure in Top Radiator
—10°	240 BTU	215°	225°	4.3 lbs.
0°	210 BTU	200°	210°	1.0 lbs.
10°	180 BTU	185°	195°	
20°	150 BTU	170°	180°	
30°	120 BTU	155°	165°	
40°	90 BTU	135°	145°	
50°	60 BTU	115°	125°	
60°	30 BTU	95°	105°	

It will be noted that some interpolation was necessary to derive Table No. 4 from the preceding tables, but the results are essentially correct and based upon the original table extracted from the A. S. H. V. E. Guide for 1932.

This extreme flexibility of the mechanically circulated hot water heating system eliminates many of the objectionable features of other systems. Overheating in mild weather is a constant and necessary fault in gravity water systems and all steam systems not equipped with a vacuum pump and thermostatic traps. Even the vacuum system has no such range of operating temperature as can be found in the mechanically circulated water system. On the other hand, almost all heating systems have a definite maximum limitation in heat delivery. As shown in the tables, this is not true of mechanically circulated water. If the design demands on a system are exceeded either because of faulty engineering or because of unusual weather conditions, the excessive demand upon the system can be easily met by increasing the water temperature. Lowering or raising the total heat delivery imposes no strain upon the system nor materially affects its efficiency, requires no expensive nor critical equipment, and adds nothing to the depreciation and maintenance cost. Hence, the owner, contractor and engineer are equally protected against unsatisfactory results due to minor errors in design or workmanship.

In any forced circulation hot water heating system, the pipe sizes for mains and risers bear an important relation to operating results. If the pipes are too small, pumping cost will be excessive, if too large first cost will be needlessly high, and if not properly proportioned, the system will be unbalanced. The A. S. H. V. E. guides of recent years have included charts and tables governing main and riser sizes for any system. These charts and tables were devised to assist the engineer in selecting the proper pipe sizes when various temperature drops were selected for various systems. To simplify this discussion, the 20° temperature drop previously referred to will be maintained.

The Guide recommends that the length of pipe from the boiler to the farthest radiator in the longest circuit, and the return, shall be measured for its equivalent length; that is for the length of straight pipe plus an amount to express the friction in all fittings in terms of straight pipe. These friction equivalents are reduced to elbow equivalents (see Table No. 5), the resistance in one elbow being equal to the resistance found in 25 diameters of the same size pipe. For instance—

Resistance of 1″ Elbow = 25 x 1″ = 25″ of 1″ pipe.

In order to calculate the equivalent length, it is necessary to assume an average pipe size for the entire circuit. This of course must be done with some consideration for the size of the system. Next, the required gallons per minute are calculated; and finally the **desired velocity in feet per second** for the water in the mains is determined.

These four factors then—equivalent length of main—average pipe size—gallons per minute—and velocity in feet per second—can be applied to the various charts and tables in the Guide to determine the actual pipe sizes, the over-all friction head, and the capacity of the pump. (See A. S. H. V. E. Guide 1934, chapter 33.)

Since the information in the Guide is principally designed for Engineers, it is the opinion of the Committee that some simplification is desirable. Consequently, the following procedure, based entirely upon information taken from the Guide, has been devised to govern pipe sizing for both one and two pipe systems.

The selection of an average pipe size is entirely arbitrary, but should be made in such manner as to afford the most economical installation both from the standpoint of first cost and operation. The following sizes are recommended for trial calculations based on a 20° temperature drop:

120,000 BTU......... ¾″ average pipe size
250,000 BTU......... 1″ average pipe size
400,000 BTU.........1¼″ average pipe size
600,000 BTU.........1½″ average pipe size
1,000,000 BTU......... 2″ average pipe size

The length of pipe in the longest circuit is measured, and the number of fittings checked. The total equivalent length of these fittings in terms of the selected average pipe size is calculated and added to the actual pipe length to give the total equivalent length. For this purpose, the following table of elbow equivalents has been taken from the Guide:

TABLE No. 5
Elbow Equivalents

	Elbow Equivalent
1—90° Elbow	1
1—45° Elbow	0.7
1—90° Long Turn Elbow	0.5
1—Open Return Ben	1
1—Open Gate Valve	0.5
1—Open Globe Valve	12.0
1—Angle Radiator Valve	2.0
1—Stop Cock, Open	1
1—Radiator	3
1—Boiler	3
1—Tee: 25% Water to Branch	16
33% Water to Branch	9
50% Water to Branch	4
100% Water to Branch	1.8

Note—Equivalents for fittings and units of various types can be obtained from the respective manufacturers.

The gallons per minute required for any given system may be calculated from the following formula:

$$\frac{\text{Total BTU per hour}}{20 \times 60 \times 8} = \text{G.P.M.}$$

20 = 20° temperature drop.
60 = minutes per hour.
8 = ponnds per gallon—water at 215°.

With the equivalent length of pipe in the longest circuit, and the total gallons per minute for the entire system as known factors, all that remains is to determine the actual pipe size, the over-all friction head, and the required pump capacity. For this purpose, the following Table No. 6 has been devised from the complicated charts and tables in the Guide. Its use is simple.

TABLE NO. 6

G.P.M.	EQUIVALENT LENGTH OF PIPE (AVERAGE SIZE)											
2.5	100	120	150	200	225	250	300	375	400	450	510	600
5.	84	100	125	167	188	208	250	312	333	375	428	500
7.5	66	80	100	133	150	167	200	250	270	300	340	400
10.	116	140	175	223	263	291	350	437	463	525	595	700
12.5	100	120	150	200	225	250	300	375	400	450	510	600
15.	150	180	225	300	337	375	450	562	593	675	758	900
17.5	140	150	210	280	318	353	425	530	563	632	720	850
20.	133	160	200	266	300	333	400	500	533	600	685	800
22.5	116	140	175	223	263	291	350	437	463	525	595	700
25.	283	340	425	566	638	706	850	1062	1103	1275	1417	1700
30.	274	330	412	550	620	695	825	1030	1085	1235	1390	1650
35.	266	320	400	533	600	666	800	1000	1070	1200	1370	1600
40.	249	300	375	500	563	623	750	937	973	1125	1252	1500
45.	217	260	325	433	488	540	650	812	843	975	1088	1300
50.	183	220	275	366	413	457	550	687	713	825	923	1100
55.	390	470	587	782	882	972	1175	1468	1566	1757	1903	2350
60.	383	460	575	766	863	955	1150	1437	1533	1725	1897	2300
65.	366	440	550	733	825	916	1100	1375	1466	1650	1885	2200
70.	349	420	525	700	788	872	1050	1312	1363	1575	1737	2100
75.	333	400	500	666	750	833	1000	1250	1333	1500	1715	2000
80.	317	380	475	633	713	789	950	1187	1233	1425	1577	1900
85.	300	360	450	600	675	750	900	1125	1200	1350	1540	1800
90.	266	320	400	533	600	666	800	1000	1070	1200	1370	1600
95.	233	280	350	465	525	580	700	875	933	1050	1200	1400
100.	200	240	300	400	450	500	600	750	800	900	1030	1200
110.	400	480	600	800	900	1000	1200	1500	1600	1800	2030	2400
120.	333	400	500	666	750	833	1000	1250	1333	1500	1715	2000
130.	266	320	400	533	600	666	800	1000	1070	1200	1370	1600
140.	200	240	300	400	450	500	600	750	800	900	1030	1200
150.	517	620	775	1023	1163	1290	1550	1935	2060	2325	2620	3100
160.	466	560	700	933	1050	1167	1400	1750	1870	2100	2370	2800
170.	400	480	600	800	900	1000	1200	1500	1600	1800	2030	2400
180.	333	400	500	666	750	833	1000	1250	1333	1500	1715	2000
190.	266	320	400	533	600	666	800	1000	1070	1200	1370	1600
200.	200	240	300	400	450	500	600	750	800	900	1030	1200
	FRICTION HEAD IN MILINCHES PER FOOT OF PIPE											
	360	300	240	180	160	144	120	96	90	80	70	60
Pipe Sizes	BTU CAPACITY OF VARIOUS PIPE SIZE IN M.B.H.											
½″	15.5	13	12.4	10	9.6	9	8.2	6.8	6.7	6.3	6.0	4.8
¾″	31	28	26.4	22	20.6	19.6	17.2	14.6	14.	13.	12.5	11
1″	59	53	47	40	38	36.	32	30	28	27	24	23.5
1¼″	124	118	101	85	80	77	69	62	58	55.6	51	48
1½″	193	175	155	132	124	120	106	93	90	83	78	71
2″	360	322	287	248	235	223	198	180	170	160	148	137
2½″	620	551	500	440	395	380	340	300	290	270	250	230
3″	1100	1000	900	760	710	660	600	540	510	480	440	410
3½″	1700	1500	1320	1120	1050	1000	900	800	770	730	670	610
4″	2350	2100	1900	1610	1500	1420	1260	1100	1080	1000	940	850
4½″	3300	3000	2700	2300	2000	1800	1650	1450	1400	1340	1250	1200
5″	4600	4100	3600	3000	2750	2600	2300	2200	2100	1950	1800	1700

$$\text{TOTAL FRICTION HEAD IN FEET} = \frac{\text{EQUIVALENT LENGTH} \times \text{MILINCHES}}{12 \times 1000}$$

Taking as an example a system of 235,000 BTU total radiation and a calculated equivalent length of 285 feet, the procedure would be as follows:

$$\frac{235{,}000}{20 \times 60 \times 8} = 24.5 \text{ Gallons per minute.}$$

Reading down the left hand column of Table No. 6, 25 gallons per minute is found to be the closest figure. Reading across to the right, 283 feet appears to be the equivalent length most nearly that of the one under consideration. Reading down from this figure to the column of friction resistances, it is found that the resistance in milinches per foot is 360. Using this figure, all of the various pipe sizes can be selected for their actual capacity in the lower part of the table. The final calculation for total resistance in feet of head is made by multiplying milinch resistance by the equivalent length of the longest circuit and converting to feet. In this instance—

$$\frac{283 \times 360}{12 \times 1000} = 8.5 \text{ feet}$$

12 = inches per foot.
1000 = milinches per inch.

Consequently, the pump in this instance should be one capable of circulating 25 G.P.M. against an 8.5 foot head. It is particularly important that the pump installed shall have actual capacity equal to that required. Otherwise, regardless of boiler capacity, pipe sizes and radiation amounts, the system will not deliver the required heat.

In conclusion, this committee desires to submit definite recommendations on the basis of the foregoing information.

1. **Forced Circulation High Temperature Hot Water Heating:** Because of the very real advantages to the Association, the Trade, and the public which this heating medium offers, we believe that the Association should accept these findings, and should make certain additions to the Code of Standards which will permit this Association and its members to take advantage of the opportunities involved. We believe that such action will result in a greater volume of new business, and will encourage the reconditioning of many inefficient water and steam heating systems.

2. **Radiation Sizes:** While in accordance with these findings it would be possible to select a maximum heat emission per square foot of equivalent radiation which would be higher than that ordinarily accredited to steam, we believe that the standard should be 240 BTU per square foot. Such a standard would permit a most economical installation, and would provide a system which would operate well within the limits of the various units involved.

3. **Control Equipment:** Control equipment should be provided to permit the operation of the system at various temperatures depending upon the demand. This equipment can be for automatic or manual adjustment; and in the event that the latter is contemplated it should be accompanied by full operating instructions to guide the owner. These operating instructions should include a chart of boiler temperatures which must be maintained for various outside temperatures.

4. **Pressure:** All high temperature water systems, designed in accordance with these standards, must be provided with suitable equipment for maintaining necessary pressure and providing necessary expansion. This equipment should invariably include an expansion tank. The following formulae should be used for computing the proper tank size:

Systems up to 2000 square feet—

$$\frac{R \times 2 \times .125}{23} \times 2 = \text{gallon tank capacity.}$$

Systems up to 5000 square feet—

$$\frac{R \times 2 \times .125}{23} \times 1.75 = \text{gallon tank capacity.}$$

Systems over 5000 square feet—

$$\frac{R \times 2 \times .125}{23} \times 1.5 = \text{gallon tank capacity.}$$

NOTE—R = square feet of radiation.

5. **Boiler Size:** The boiler for high temperature water systems must be selected on the basis of total net load in BTU per hour. In the case of systems designed with 240 BTU output per square foot of radiation, the boiler would be selected on its steam rating. For systems designed with a higher or lower output per square foot, the total net load should be calculated in terms of BTU per hour and converted to a standard steam or water net load on the boiler. For conversion to steam ratings the formulae would be—

$$\frac{\text{Total Radiation x BTU per square foot}}{240} = \text{Total in terms of standard steam (240).}$$

For conversion to water rating the formulae would be—

$$\frac{\text{Total Radiation x BTU per square foot}}{150} = \text{Total in terms of standard water (150).}$$

6. **Pipe Sizes:** Pipe sizes should be selected in accordance with the tables based upon charts taken from the A.S.H.V.E. Guide. In presenting plans for the approval of the Association, the contractor should give full information concerning the derivation of pipe sizes, friction head per foot of pipe, equivalent length of pipe, and over-all friction head.

7. **Pumps:** Pumps must be of a guaranteed capacity sufficient to deliver the required gallons per minute against the friction head present. In this regard, the Association should require of each pump manufacturer full information concerning its guaranteed ratings for pumps of various sizes, and should arrange some method whereby these ratings can be accurately checked under operating conditions.

It is not the intention of this committee to limit the engineer or contractor in designing mechanically circulated systems. When such systems are designed on some other basis than that suggested by the committee, it will be necessary for the engineer or contractor to advise the Association of such deviation, that each case may be decided upon its merits.

SMOKE INSPECTION AND ABATEMENT ORDINANCE

(Passed August 5, 1937.)

Be it Ordained by the City Council of the City of Chicago:

546A. **Department of Smoke Inspection and Abatement Established.)** There is hereby established a department of the municipal government of the City of Chicago which shall be known as the Department of Smoke Inspection and Abatement. The said department shall embrace a smoke inspection and abatement commission composed of the Commissioner of Health as Chairman, the Commissioner of Buildings, the Corporation Counsel, the Commissioner of Police, the Chief Inspector of steam boilers and steam plants. There shall also be employed according to law, a Deputy Smoke Inspector in Charge and such other employees as the City Council may provide in the annual appropriation ordinance.

546B. **Duties.)** It shall be the duty of the Smoke Inspection and Abatement Commission to establish standards, rules and regulations for the inspection and control of the installation, reconstruction, alteration, repair and maintenance of heating, power and fuel burning equipment; the prevention and abatement of smoke and noxious gases, and nuisances arising therefrom, the examination and approval of plans of all heating, power and fuel burning installations and of all smoke prevention and abatement installations installed or reconstructed in any building, location or on any premises within the jurisdiction of the City of Chicago.

546C. **Deputy Smoke Inspector in Charge—Qualifications—Duties.)** The Deputy Smoke Inspector in Charge shall be an engineer qualified by technical training and experience in the theory and practice of the construction and operation of steam boilers and fuel burning equipment and also in the theory and practice of smoke abatement and prevention. It shall be the duty of the Deputy Smoke Inspector in Charge to supervise the work of all employees of the Department herein created, and to carry into execution the laws and ordinances pertaining to smoke prevention and abatement and the installation, reconstruction, alteration and repair of fuel burning equipment, and the standards, rules and regulations adopted by the Smoke Inspection and Abatement Commission.

546D. **Bond.)** The Deputy Smoke Inspector in Charge, before entering upon the duties of his office, shall execute a bond to the City in the sum of five thousand dollars ($5,000.00) with sureties to be approved by the Comptroller, conditioned for the faithful performance of the duties of his office.

546E. **Construction and Reconstruction of Plans and Specifications—Permit.)** No new fuel burning plants nor reconstruction of any existing fuel burning plants for producing power and heat, nor either of them, nor refuse burning, nor any new chimney connected with such fuel burning plant, shall be installed, erected, reconstructed or maintained in the City until plans and specifications of the same have been filed in the office of and approved by the Deputy Smoke Inspector in Charge and a permit issued by him for such installation, erection, reconstruction, or maintenance. Plans and specifications shall be filed with the Deputy Smoke Inspector in Charge, which said plans shall show the type of installation, the amount of work and the amount of heating to be done by such fuel burning plant and all appurtenances thereto, including all provisions made for the purpose of securing complete combustion of the fuel to be used and for the purpose of preventing smoke. Said plans and specifications shall also contain a statement of the amount and kind of fuel proposed to be used and said plans and specifications shall also show that the room or premises, in which fuel burning plant shall be located, is provided with doors, windows, air-shafts, fans and other means of ventilation sufficient to prevent the temperature of such room, basement or other portion of such building wherein such fuel burning apparatus is to be used, from rising to a point higher than 120 degrees Fahrenheit, and sufficient also to provide that the atmosphere of any such room, basement or other part of the building wherein such fuel burning apparatus may be located, may be changed every ten minutes. Such plans shall further show the dimensions of the room in which such fuel burning apparatus is to be located, the location and dimensions of all chimneys and smokestacks used in connection with or as a part of said fuel burning plant. Upon the inspection and approval of such plans and specifications by the Deputy Smoke Inspector in Charge, a duplicate set of said plans shall be left on file in the office of said Deputy Smoke Inspector in Charge and upon the payment of the fees as hereinafter provided, and if such plans and specifications shall show that proper provisions for the purpose of securing complete combustion of the fuel to be used and for the purpose of preventing and abating smoke have been made, he shall issue a permit for the installation or for the construction, erection, reconstruction or maintenance of such fuel burning plant. As soon as the Deputy Smoke Inspector in Charge has issued the permit as above provided, it shall be the duty of the various departments having charge of the inspection of the premises wherein said fuel burning apparatus is located to cooperate with the said Deputy Smoke Inspector in Charge, to see that the execution of the work so permitted by said permit shall be done in conformity with the plans and specifications submitted and approved, and the standards, rules and regulations fixed by said Smoke Inspection and Abatement Commission.

546F. **Use of Plant—Certificate Required.)** It shall be unlawful for any person to use any new or reconstructed plant for the production and generation of heat and power, or either of them, until he shall have first procured a certificate from the Deputy Smoke Inspector in Charge certifying that the plant is so constructed that it will do the work required, and that it can be so managed that no dense smoke shall be emitted from the chimney connected with the furnace or fire box in violation of the provisions of this ordinance.

546G. **Chimneys and Furnaces—Repairs—Permit.)** No owner shall install, alter or repair any chimney or any furnace or device, which alteration, change or installation would affect the method or efficiency of preventing and abating smoke, without first submitting plans and specifications to the Deputy Smoke Inspector in Charge and securing a permit therefor, nor shall any owner alter or repair any brick-work on or about a high pressure boiler without first submitting plans and specifications to the Deputy Smoke Inspector in Charge and securing a permit therefor. Any person, firm or corporation who shall violate this Section, shall be fined twenty-five dollars ($25.00) for each day upon which he or they shall prosecute such alteration, change or installation without a permit and each day's violation shall constitute a separate offense.

546H. **Fees—When Remitted.)** The fees for the inspection of plans and issuing or permits, and for the inspection of furnaces or other fuel burning apparatus or devices, and issuing of certificates, shall be as follows:

For inspecting plans of new plants and of plants to be reconstructed, two dollars.

For inspecting plans for repairs and alteration, one dollar.

For permits for the erection, installation, reconstruction, repair or alteration of any furnace or other fuel-burning apparatus, smoke prevention device or chimney, five dollars for each unit or single apparatus.

For examining or inspecting any new or reconstructed furnace connected to a high pressure boiler after its erection or reconstruction and before its operation and maintenance, five dollars for the first unit or single apparatus and three dollars for each additional unit or single apparatus.

For examining or inspecting any new or reconstructed furnace connected to a low pressure boiler or any other fuel-burning equipment, or any smoke prevention device, after its erection or reconstruction and before its operation and maintenance, three dollars for each unit or single apparatus. Provided, however, that this Section shall not apply to furnaces or other fuel burning apparatus or device installed or used to heat private residences, tenements or buildings consisting of two apartments or less.

The aforesaid fees shall be paid to the City Collector prior to the approval of plans for such installations by the Deputy Smoke Inspector in Charge. The fee for the examination or inspection shall include the issuing of a certificate for operation in case such certificate for operation is granted and shall be paid at the time the permit is secured.

The Deputy Smoke Inspector in Charge may and he is hereby directed and instructed to remit all inspection or examination fees charged against any and all charitable, religious and educational institutions, when the furnace or other device or apparatus inspected is located in or upon premises used and occupied exclusively by such charitable, religious or educational institution; provided that such charitable, religious or educational institution is not connected or carried on for private gain or profit; and provided further that the Deputy Smoke Inspector in Charge may require every application for the remission of such fees to be verified by the affidavit of one or more taxpayers of the City.

546I. **Emission of Dense Smoke.)** The emission of dense smoke within the City from the smokestack of any locomotive, steam boat, steam tug, steam roller, steam derrick, steam pile driver, tar kettle or other similar machine or contrivance or from any open fire or from the smokestack or chimney of any building or premises except for a period of or periods aggregating six minutes in any one hour at the time when the fire-box is being cleaned out or a new fire being built therein, is prohibited and is hereby declared to be a nuisance and may be summarily abated by the Deputy Smoke Inspector in Charge or by any one whom he may duly authorize for such purpose. Such abatement may be in addition to the fine hereinafter provided. Any person, firm or corporation owining, operating or in charge or control of any locomotive, steamboat, steam tug, steam roller, steam derrick, steam pile driver, tar kettle or other similar machine or contrivance or any open fire or of any building or premises who shall cause or permit the emission of dense smoke within the city from the smokestack or chimney of any such locomotive, steam boat, steam tug, steam roller, steam derrick, steam pile driver, tar kettle or other similar machine or contrivance or any open fire or from the smokestack or chimney of any building controlled or in charge of him, her or them except for a period of or periods aggregating six minutes in any one hour at the time the fire-box is being cleaned out or a new fire being built therein, shall be deemed guilty of a violation of this Section and upon conviction thereof shall be fined not less than ten dollars nor more than one hundred dollars for each offense; and each emission of dense smoke in violation of the provisions of this Section shall constitute a separate offense for each and every day on which such violation shall continue.

546J. **Dust, Soot, Noxious Gases.)** No person, or persons, firm or corporation shall cause, permit or allow the escape from any smokestack or chimney into the open air of such quantities of ash dust, soot, cinders, acid or other fumes, dirt, or other material or noxious gases in such place or manner as to cause injury, detriment, nuisance or annoyance to any person or persons or to the public or to endanger the comfort and repose, health or safety of any such person or persons or the public or in such manner as to cause or have a natural tendency to cause injury or damage to business or property.

Any person or persons, firm or corporation causing, permitting or allowing the escape from any smokestack or chimney into the open air of such quantities of ash dust, soot cinders, acid or other fumes, dirt or other material or noxious gases, in such place or manner as to cause injury, detriment, nuisance or annoyance to any person or persons or to the public or to endanger the comfort and repose, health or safety of any such person or persons or the public or in such a manner as to cause or have a natural tendency to cause injury or damage to business or property, shall be deemed guilty of a violation of this Section and upon conviction thereof shall be fined not less than ten dollars nor more than one hundred dollars for each offense and each violation of the provisions of this Section shall constitute a separate offense for each and every day upon which such violation shall continue. All persons participating in any violation of this provision either as owners, proprietors, lessees, agents, tenants, managers, superintendents, captains, engineers, firemen or janitors or otherwise shall be severally liable therefor, and to the penalties fixed in this article.

546K. **Violations—Prosecutions.)** Prosecutions for all violations of this article shall be instituted by the Deputy Smoke Inspector in Charge and shall be prosecuted in the name of the City of Chicago.

The issuance and delivery by the Deputy Smoke Inspector in Charge of any permit or certificate for the construction or reconstruction, or any permit for the alteration or repair of any plant or chimney, connected with a plant, shall not be held to exempt any person or corporation to whom any such permit has been issued or delivered, or who is in possession of such permit, from prosecution on account of the emission or issuance of dense smoke caused or permitted by any such person or corporation.

546L. **Penalty.)** Any person who shall violate any of the provisions of this article (except as herein otherwise provided) shall be fined not less than twenty-five dollars nor more than one hundred dollars for each offense.

546M. **Fraud—Favors—Penalty.)** If any person acting on behalf of the City under the provisions of this article shall take or receive any money or any valuable thing for the purpose of deceiving or defrauding any person or persons, or for the purpose of favoring any person or persons, or if any employe shall recommend the issuance of any certificate of inspection without having at the time stated, thoroughly examined and tested the furnace, device or apparatus so certified, he shall be fined one hundred dollars for each offense.

Amendments to Section 546F and 546H of the Smoke Ordinance.

AN ORDINANCE

Amending Article XVI of Chapter 4 of the Code by providing for the annual inspection of fuel burning equipment.

Be it Ordained by the City Council of the City of Chicago:

Section 1. That section 546F of the Revised Chicago Code of 1931 be and the same is hereby amended by adding thereto the following:

An annual inspection of fuel burning equipment for which certificates allowing use of plant have been issued by the Department of Smoke Inspection and Abatement shall be made to see that such equipment will do the work required and can be so managed that no dense smoke shall be emitted from the chimney connected with the furnace or firebox in violation of the provisions of this chapter. If at the time of the annual inspection or of any inspection subsequent to the issuance by the department of the certificate allowing use of the plant, it is found that the fuel burning equipment is in such condition that it cannot do the work required or cannot be so managed that no dense smoke will be emitted from the chimney connected to the furnace or fire-box in violation of the provisions of this chapter, the deputy smoke inspector in charge shall give notice in writing to the person, firm or corporation owning, operating or in charge of such fuel burning equipment of the defects found and an order to correct, repair or replace the defective equipment, and if within ten days from the date of said notice said order be not complied with then the deputy smoke inspector in charge may at his discretion revoke said certificate allowing use of the plant.

Section 2. That section 546H of said Code be and the same is hereby amended by inserting therein, between the seventh and eighth paragraphs of said section, as printed in the Council Journal of October 21, 1931 at page 1023, the following:

The schedule of fees for annual inspection shall be as follows:

Boilers (Heating) Low Pressure.

Of a capacity to supply a net load of more than 1200 sq. ft. and less than 10,000 sq. ft of direct steam radiation or its equivalent: $3.00 for each unit or single apparatus.

Of a capacity to supply a net load of 10,000 sq. ft. and less than 25,000 sq. ft. of direct steam radiation or its equivalent: $4.00 for each unit or single apparatus.

Of a capacity to supply a net load of 25,000 sq. ft. or more of direct steam radiation or its equivalent: $5.00 for each unit or single apparatus.

Boilers (Power) High Pressure.

Of a capacity of more than 12 horse power and less than 100 horse power: $3.00 for each unit or single apparatus.

Of a capacity of 100 horse power and less than 250 horse power: $4.00 for each unit or single apparatus.

Of a capacity of 250 horse power or more: $5.00 for each unit or single apparatus.

Furnaces (other than boiler furnaces).

Warm air furnaces; incinerators; bake-ovens; metallurgical furnaces: Of a capacity having more than 7 sq. ft. and less than 20 sq. ft. of grate area: $3.00 for each unit or single apparatus.

Of a capacity having 20 sq. ft. and less than 50 sq. ft. of grate area: $4.00 for each unit or single apparatus.

Of a capacity having 50 sq. ft. or more of grate area: $5.00 for each unit or single apparatus.

Powdered Coal Burning Furnaces.

Where effective furnace volume exceeds 36 cu. ft. and is less than 300 cu. ft.: $3.00 for each unit or single apparatus.

Where effective furnace volume is 300 cu. ft. or less than 750 cu. ft.: $4.00 for each unit or single apparatus.

Where effective furnace volume is 750 cu. ft. or more: $5.00 for each unit or single apparatus.

Oil Burners or Gas Fired Furnaces:

Where effective furnace volume exceeds 18 cu. ft. and is less than 150 cu. ft.: $3.00 for each unit or single apparatus.

Where effective furnace volume is 150 cu. ft. and less than 375 cu. ft.: $4.00 for each unit or single apparatus.

Where effective furnace volume is 375 cu. ft. or more: $5.00 for each unit or single apparatus.

The term "heating boilers" (low pressure) shall be construed to mean all boilers carrying not in excess of ten pounds per square inch steam or water pressure.

The capacity of heating boilers (low pressure) shall be calculated in accordance with the rules of the Heating, Piping and Air Conditioning Contractors' National Association Code.

The term "power boilers" (high pressure) shall be construed to mean all boilers carrying in excess of ten pounds per square inch steam or water pressure.

The capacity of power boilers (high pressure) shall be calculated on the basis of ten square feet of heating surface per boiler horse power.

Section 3. This ordinance shall take effect and be in force from and after its passage and due publication, but not before January 1, 1938.

Page 4210—Journal, City Council, Chicago. August 5, 1937.

DEPARTMENT OF SMOKE INSPECTION AND ABATEMENT

Standards, Rules and Regulations

Introduction by
FRANK A. CHAMBERS, Deputy Smoke Inspector in Charge.

In the enforcement of smoke ordinances dealing with the problem of abating smoke, the policy of some municipalities in the past has been to permit and encourage the installation of fuel burning equipment in which the use of smokeless fuels was necessary if smoke were to be prevented. It is evident from the results obtained from this practice that efforts to abate smoke by following this program have been unsuccessful so far as permanency is concerned and any benefits accruing have been short-lived and ineffective.

The development of a large number of small automatic stokers has made available to the coal industry a weapon that is rapidly coming into public favor. Such weapon offers a satisfactory solution for the problem of abating smoke from the smaller size heating plant. It essentially reduced the price of sized coals and encouraged the use of small coal, and makes possible the use of local in place of imported coals.

In a ward the small stoker properly applied and operated can make two important contributions to social progress: first, it provides a means by which coal can be burned more economically and conservatively; second, it places in the field new educational forces to teach the public the proper handling of it.

INDEX

Pursuant to provisions of Section 2 of an Ordinance passed by the City Council of the City of Chicago, December 31, 1928, creating the Department of Smoke Inspection and Abatement the following standards, rules and regulations are hereby adopted by the Smoke Inspection and Abatement Commission.

DEFINITIONS.

Stack.
Stack means any chimney or smoke stack or other structure whether of brick, tile, concrete, metal or other material or a combination of these materials intended for emission of smoke or products of combustion. Smoke stacks commonly known as smoke jacks attached to locomotive round houses shall be deemed stacks and a part of the locomotive beneath them for the time being.

Dense Smoke.
Dense smoke is smoke, the density or shade of which is equal to or greater than number three of the Ringleman Chart published and used by the U. S. Geological Survey and the U. S. Bureau of Mines. This shall be considered to mean smoke of 60% or greater density. Smoke of such density that cannot be seen through clearly as it leaves the top of the chimney is considered dense smoke.

Cleaning Fires.
The term in Section 9 "when the fire box is being cleaned out" shall mean the period during which the fuel bed, including ash and clinker, is being completely removed from the grate surface. This operation may be done by cleaning portions of the grate at different times. This does not mean that the act of shaking the grates to remove ash or removal of individual clinkers from the fuel bed constitutes an act of cleaning the fires as interpreted in this section of the ordinance.

Building Fires.
The term, a new fire being built shall be held to mean the period during which a fresh fire is being started and does not mean the process of replenishing an existing fuel bed with additional fuel.

Heating Surfaces.
Heating surfaces shall be construed to mean all boiler surfaces with water on one side and hot gases on the other side not excepting such surfaces as are covered by arches or tile used to cover tubes.

REGULATIONS FOR THE ISSUANCE OF PERMITS.

Plan Examination Division.
A Junior Mechanical Engineer shall be assigned to the plan examination division and all plans and specifications shall be submitted to him for inspection and examination. Copies of plans and specifications approved shall be kept on file in the office of the Department of Smoke Inspection and Abatement until work has been completed and a certificate of operation has been issued authorizing the use of the plant. Permit forms shall be available and a convenient means provided for the payment of fees for permits and inspection in accordance with the regulations of the City Comptrollers Office. Any employee handling and responsible for the collection of moneys shall be bonded in accordance with the regulations of the office of the City Comptroller.

Requirements For the Issuance of Permits.
Proper provision for the purpose of securing complete combustion of the fuel to be used and for the purpose of preventing and abating smoke shall be shown in the plans and specifications before a permit is issued for any fuel burning equipment. Such plans and specifications shall show that provisions are made to secure an intimate mixture of the combustible products with a sufficient amount of air in a furnace or fire box of ample volume to allow sufficient time for the smokeless combustion of the fuel or refuse being burnt.

Approval of Plans.
The following general rules shall govern the issuance of permits on approved plans for the installation or reconstruction of fuel and refuse burning equipment and plants.

Building Plans—Requirements.

1. The chimney or smoke stack must be of sufficient height and cross sectional area and so located as to permit of a well designed breeching of adequate area and containing a minimum number of bends or turns.

2. Sufficient head room must be provided to allow for ample combustion space and to allow for the installation of boilers or furnaces of sufficient capacity to do the work required. Sufficient head room must be provided for installation of a breeching of proper design and shape and to allow for accessibility to man holes and valves connected to boiler.

3. Sufficient floor space must be provided to allow for proper operation of furnace, removal of boiler tubes and making necessary repairs, cleaning boiler tubes, gas passages, furnace fire box, combustion chambers and to allow for easy access to furnace doors and other openings in and about walls of boiler setting or furnaces, and for the cleaning and removal of soot and ash from the breeching and stack. Sufficient space must be provided for the storage of fuel so as not to interfere with the operation of the furnace.

4. Provision must be made to permit the admission to boiler or furnace room of a sufficient amount of air to secure smokeless combustion of the fuel and to properly ventilate the room or premises in which the fuel burning equipment is located. The means of ventilation must be sufficient to prevent the temperature of any boiler or furnace room from rising to a point higher than 120° Fahrenheit and must be sufficient also to provide that the atmosphere of any such room may be changed every ten minutes.

Boiler Requirements.

Tube areas and gas passages through heating surfaces shall be of sufficient area to permit the ordinary accumulation of soot and fly ash without such restriction to draft which would interfere with smokeless operation of the fuel burning equipment. Sufficient means must be provided for the removal of soot and fly ash from the heating surfaces, tube region and gas passages so as to avoid an accumulation which would interfere with the draft. Baffling of such design must be installed and maintained so as not to interfere with smokeless combustion due to impinging the flame on heating surfaces where such impingement would result in reduction in temperature below ignition point of the combustible gases. Boilers should be rated on the basis of ten square feet of heating surface per boiler horse power.

Hand fired high pressure boilers should not be permitted to project out over the firing doors (when closed) except where a minimum head room of six feet is provided below such projection, to allow ample space for firing of furnace and cleaning of fires without obstruction.

Chimney Requirements.

Chimneys shall be of such height as to produce sufficient available draft to supply a proper amount of air to burn smokelessly the maximum amount of fuel that is to be consumed and to overcome draft losses or restrictions in furnace, boiler setting, gas passages, damper openings, breechings and stack. The chimney must be of sufficient cross section area to allow the free unrestricted discharge of the products of combustion. The following method can be used in determining the height of stack necessary to produce the required amount of available draft. The draft over the fire plus the losses through the furnace setting, boiler, gas passages, and breeching should equal the available draft required at the base of the stack.

The following method to obtain heights and areas of chimneys for boilers operating at a given rating is approved. Determine the rate of combustion of coal per square foot of grate surface per hour.

For rates of fuel consumption up to 24 # coal per square foot of grate surface per hour the draft over the fire measured by standard draft gauge in inches of water shall be .01" per pound of coal burned per square foot of grate surface per hour.

For rates of combustion above 25 # coal per square foot of grate surface per hour the following over fire draft shall be provided:

30 lbs. of coal per sq. ft. of grate surface per hr. .33 in. of water.
35 lbs. of coal per sq. ft. of grate surface per hr. .40 in. of water.
37 lbs. of coal per sq. ft. of grate surface per hr. .45 in. of water.
40 lbs. of coal per sq. ft. of grate surface per hr. .50 in. of water.
45 lbs. of coal per sq. ft. of grate surface per hr. .60 in. of water.

The draft loss through the boiler shall be estimated from the following table, the draft required at the boiler side of the damper to produce the necessary over fire draft to be calculated from the percentage loss shown.

Type of Boiler		Percentage of Draft at Damper Lost Through Boiler
Hawkes	Horizontal Baffle.	60%
Stirling	(5 pass)	81%
Stirling	(3 pass)	65%
Stirling	(4 pass)	86%
Erie City	(Vertical Boiler).	60%
Erie City	Horizontal Baffle.	60%
Keeler	Vertical Baffle	51%
Keeler	Horizontal Baffle.	60%
H R T		47%
Cahall	Vertical Boiler	65%
Atlas	Horizontal Baffle.	75%
Atlas	Vertical Baffle	60%
Lyons		60%
Oil City	Vertical Baffle	50%
Oil City	Horizontal Baffle.	60%
Wicks	Vertical Boiler	58%
B. & W.	Vertical Baffle	50%
B. & W.	Sewell Baffle	65%
B. & W.	Horizontal Baffle.	60%
Kroeschell Comb.		60%
Edgemoor	(4 pass) Vertical Baffle	64%
Edgemoor	Horizontal Baffle.	60%
Edgemoor	Vertical Baffle	50%
Heine	Horizontal Baffle.	65%
Scotch Marine		65%

To the draft required at the boiler side of the damper add the losses encountered between boiler and chimney. Allow a loss of .05 for each right angle turn in breeching and for each 90° turn in direction of gases. Allow a draft loss of .003 per foot of horizontal breeching. The total is the draft required at the base of the chimney. The formula for calculating the necessary height of stack to produce the required draft is as follows:

$$D = .52\ H.P. \left(\frac{1}{T} - \frac{1}{T_1}\right)$$

in which D = Theoretical draft in inches of water
H = Height of stack above grate in feet
P = Atmospheric pressure in pounds per square inch
T = Atmospheric temperature absolute
T_1 = Absolute temperature of stack gases (average)

The value of $.52 \times P \left(\frac{1}{T} - \frac{1}{T_1}\right) = K$.

Therefore $D = H \times K$.

At 60° Fahrenheit boiler room temperature and at sea level for different average stack gas temperatures the value of K is shown as follows:

Average Temperature of Stack Gas.	K.
750° Fahrenheit	.0084
700° Fahrenheit	.0081
650° Fahrenheit	.0078
600° Fahrenheit	.0075
550° Fahrenheit	.0071
500° Fahrenheit	.0067
450° Fahrenheit	.0063
400° Fahrenheit	.0058
350° Fahrenheit	.0053

Allow .001 loss per foot of stack for frictional resistance to be subtracted from K.

$K - .001 = U$ Avoidable draft $D_1 = H \times U$.

To find the height of stack above grate to produce an available given draft at a given stack temperature $H = \frac{D}{U}$

Example: To produce an available draft of 1" of water at 500° Fahrenheit stack temperature a stack height $H=\frac{1}{.0057}=175$ feet, is necessary height. The free area of the stack at the smallest point shall be not less than ⅙ the area of the total connected grate surface when the stack is less than 150' high. Where the stack height exceeds 150' the area shall be not less than ⅛ of the total connected grate surface. Horizontal return tubular boilers—stacks shall have an area in free opening at smallest point 25% in excess of combined area of tubes served.

Breeching Requirements.

Breechings or smoke pipes must be as nearly square or round as possible. Where the breeching is rectangular or oval in cross section the greatest dimension shall not exceed twice the smallest. The draft loss resulting from breechings deflecting downward below the horizontal shall be compensated for by requiring an additional height of stack sufficient to produce an amount of draft equal to that lost. The minimum radius of all bonds shall not be less than the width of breeching at point where bend occurs, except where provision is made to compensate for loss in draft.

Furnace Requirements.

The furnace shall be of sufficient volume and provided with the necessary mixing arrangements and equipped with means to permit the entry of sufficient amount of air, to consume the fuel without the producton of dense smoke.

A furnace volume of one cubic foot of combustion space for each developed horse power for coal or refuse burning furnaces shall be provided.

A furnace volume of one and one-half cubic feet for each developed horse power for oil burning furnaces shall be provided.

A furnace volume of three cubic feet of combustion space for each developed horse power for pulverized coal burning furnaces shall be provided.

The grate surface of any hand fired boiler using coal or refuse as fuel should not be less than 1/50 of the heating surface. On hand fired boilers the minimum space provded in front of boiler for firing room must be equal to the distance from the boiler front to rear end of grate plus two feet.

Scotch Marine Boilers containing more than 400 square feet of heating surface must be equipped with mechanical stokers, approved type of extension furnaces or fired with gas or oil. In Scotch Marine Boilers equipped with extension furnaces no boiler tubes should be located within four and one-half inches of the main flue of boiler and no more than three rows of tubes should be located below the lowest full row.

Table of Setting Heights and Space Requirements at the Rear of Horizontal Return Tubular Boilers for Hand Fired Furnaces.

Boiler Diameter—Length	Setting Height Distance From Dead Plate to Shell	Space Between Rear of Boiler and Back Wall
42"x12"	28"	22"
42"x14"	28"	22"
48"x12"	30"	24"
48"x14"	30"	24"
48"x16"	30"	24"
54"x16"	32"	24"
54"x18"	32"	24"
60"x16"	34"	26"
60"x18"	34"	26"
66"x16"	34"	28"
66"x18"	34"	28"
72"x16"	36"	30"
72"x18"	36"	30"
78"x18"	38"	32"
78"x20"	38"	32"
84"x18"	38"	32"
84"x20"	38"	32"

The following table of minimum requirements for smokeless settings shall be established except where owing to extraordinary conditions the setting heights shown cannot be obtained, but where adequate provision for purpose of preventing smoke is provided by other means.

TYPE OF BOILER.

Types of Furnaces and Stokers	Horizontal Return Tubular Boilers				Water Tube Boilers				Scotch Marine Boilers
	54"	60"	66"	72"	Hor. Baffl. Tube Pitch 1"-1½"	Vert. Baffl. Tube Pitch 1"-1½"	Hor. Baffl. Tube Pitch 3¼"	Vert. Baffl. Tube Pitch 3¼"	
Hand Fired Furnaces	Shell to Dead Plate				Front Header to Floor				
No. 6	32"	34"	34"	36"	**6'0"	*	**6'6"	*	‡‡
Nos. 8 and 9	32"	34"	34"	36"	6'0"	*	6'6"	*	‡‡
Hand Stoker	26"	28"	28"	30"	5'6"	*	6'0"		Full ext.
	Shell to Floor								
Downdraft	60"	60"	60"	60"	6'0"	*	6'6"	*	Full ext.
Twin Fire	58"	60"	62"	64"	6'0"	*	6'6"	*	*
Semi Ext. Refuse Burning	44"	46"	46"	48"	7'0"	*	7'6"	*	*
Gravity Feed	48"	48"	50"	54"	5'0"	*	5 6"	*	Full ext.
Stokers									
Chain Grates	72"	72"	78"	78"	7'0"	9'6"	8'0"	10'0"	‡‡
Front Feed	60"	60"	60"	72"	7'0"	9'0"	7'0"	10'0"	‡‡
Side Feed	66"	72"	78"	84"	7'6"	*	8'0"	*	Full ext.
	Shell to Stoker (Highest Point)								
Underfeed	36"	38"	40"	42"	"A"6'0"	"A"8'0"	"A"7'0"	"A"8'6"	Min. Dia. of Furnace 36"
	Shell to Dead Plate								
Underfeed	42"	44"	46"	48"	6'6"	8'6"	7'6"	9'0"	

* Combinations not recommended as smokeless settings.
‡‡ Combinations not ordinarily met with in practice.
** Omit double arches—using only defleetion arch.
"A" Setting height for Jones Standard Stoker or similar type.

FUEL AGREEMENT.

The following regulations shall be enforced permitting boilers of the types and sizes listed below to be installed under a fuel agreement. By fuel agreement is meant an agreement to use smokeless fuels such as hard coal, coke or semi-bituminous coal containing less than 20% volatile matter analyzed on a dry basis.

Fuel Agreement.

Portable Boilers that are to be used for locomotive cranes, construction work, or street work. All sizes for temporary use in any location.

High Pressure Boilers.

Only one to be allowed in any plant. This includes:

Fire Box Boilers as large as 42x11½" with straight grate.

Scotch Marine Boilers containing no more than 400 square feet of heating surface.

Vertical Tubular Boilers containing no more than 250 square feet of heating surface.

Economic Boilers as large as No. 7 (36x9) containing no more than 300 square feet of heating surface.

H. R. T. Boilers containing no more than 350 square feet of heating surface.

Low Pressure Boilers.

The Department of Smoke Inspection and Abatement will not approve or issue permits for the installation of low pressure heating boilers or hot water heaters of the surface burning type for burning bituminous, semi-bituminous or semi-anthracite coal, in sizes larger than those having a capacity to supply steam to 1,200 square feet of direct steam radiation or its equivalent or to supply hot water to 2,000 square feet of hot water radiation. All sizes below these are permitted under fuel agreement. The surface burning type of coal burning furnace is defined as a hand fired furnace in which the fresh fuel is thrown directly on the hot fuel bed.

SPECIAL RULING

Statistics of the Department of Smoke Inspection and Abatement show that smoke from heating plants is a major factor in contributing to air pollution of the City. Believing that this source of smoke can be greatly reduced by eliminating the use of surface burning type of heating boilers for burning both bituminous and semi-bituminous coal the department is promulgating the following ruling. This is adopted after a careful study of the situation and has the approval of the Advisory Smoke Abatement Board of Engineers.

(Continued from page 451)

fessional motion picture projecting machine for each location whether permanent or temporary. A miniature non-professional motion picture projector does not require a permit for its operation. Permit to operate shall be posted, at all times, in a conspicuous place on the premises appertaining where such non-professional motion picture projector is in use.

(m) **Licensed Operator Required:** Non-professional motion picture projectors, except miniature non-professional motion picture projectors, shall be operated by licensed operators as provided for in Article XIV of this electrical code.

SUB-ARTICLE 36.

Organs.

3601. **General:**

(a) The requirements of this sub-article shall be deemed to be additional to, or amendatory of those prescribed in sub-articles 1 to 20, inclusive, of this code. They shall be deemed to apply to those electrical circuits and parts of electrically operated organs which are employed for the control of the sounding apparatus and keyboards.

3602. **Source of Energy:**

(a) The source of energy shall be either a self-excited generator rated at not over 15 volts, or a primary battery.

(b) The generator shall either be permanently and effectively insulated both from ground and from the motor driving it, or both generator and motor frames shall be grounded as prescribed in sub-article 9 of this code.

3603. **Cables:**

(a) All conductors, except common return conductors and conductors inside the organ proper, the organ sections and the organ console, shall be cabled.

(b) The separate conductors of the cable shall be not smaller than No. 26, and shall have either rubber, cotton or silk insulation. The cotton or silk may be saturated with paraffin, if desired.

(c) The separate conductors shall be either bunched or cabled. In either event they shall be enclosed in one or more braided outer coverings. A tape may be substituted for an inner braid. The outside covering of a cable not run in conduit shall either be flameproof, or covered with a closely wound fireproof tape.

(d) The common return conductor shall be not smaller than No. 14, shall be of either the rubber-covered type, Type R, the asbestos covered type, Type A, or the slow-burning type, Type SB, and shall not be contained in the cable. It may be run in contact with the cable or placed under an additional covering enclosing both cable and return conductor.

3604. **Workmanship and Materials:**

(a) All wiring and devices within the organ or any of its parts shall be neatly disposed and securely fastened.

(b) Cables between parts of the organ and between the console and the organ shall be installed in a workmanlike manner, shall be securely fastened in position and shall be kept from contact with other wires. Conduit may be used, but shall not be required.

3605. **Fuses:**

(a) Feed conductors shall be protected at the source by a fuse of suitable capacity and, except common return conductors, shall be so subdivided and protected at the organ sections or distribution points by approved fuses of not over 15 amperes rating that every conductor will be protected by one or more such fuses.

(Continued from page 519)

(g) **Field Painting.**

All field rivets and bolts, also all serious abrasions to the shop coat, shall be spot painted with the material used for the shop coat, or an equivalent, and all mud and other firmly attached and objectionable foreign materials shall be removed, before general field painting.

Responsibility for this touch-up and cleaning, as well as for general field painting, shall be allocated in accordance with accepted local practices and this allocation shall be set forth explicitly in the contract.

PART VIII. INSPECTION.

Section 28. Inspection.

(a) **General.**

Material and workmanship at all times shall be subject to the inspection of experienced engineers representing the purchaser.

(b) **Cooperation.**

All inspection as far as possible shall be made at the place of manufacture, and the Contractor or Manufacturer shall cooperate with the Inspector, permitting access for inspection to all places where work is being done.

(c) **Rejections.**

Material or workmanship not conforming to the provisions of this Specification may be rejected at any time defects are found during the progress of the work.

NON-CONDUCTING INSULATING MATERIALS PIPE, DUCT AND BOILER COVERING

By FRED MORGAN, Secretary, Insulation Contractors' Association of Chicago

In an effort to assist the Architect and others who buy and specify Non-Conducting Insulating Materials and their Application in Connection with Heating, Ventilating, Air Conditioning, Plumbing and Refrigeration the following is offered:

One of the crying needs of all specification work stands out in the use of "Non-Conducting Insulating materials" and that is first the writing of a separate and distinct specification and second the completeness of that specification for the sake of clarity and the removal of all ambiguities. In few trades are so many various interpretations possible as with that of Insulation and Pipe Covering.

While this work has frequently, in fact in most cases been let with the pipe or duct work and then sublet by that contractor it is felt that with specifications written for a separate contract for the furnishing and installation of non-conducting insulating materials applied by Asbestos Workers that whether or not the contract is let direct to an Insulation Contractor or through other contractors the Architect will secure a better grade of work and this phase of his job will not be used as a price cushion by other contractors seeking full compensation for their work but the lowest figures on pipe covering.

The surfaces to be insulated should be specifically mentioned, their location given, and a correct statement made of the kind and thickness of insulation.

Ambiguous statements should be avoided, as far as possible, as for instance "all steam piping shall be covered." This does not specify whether or not steam return piping shall be covered. If it is desired to insulate only the steam supply system, it should be so stated as in the folowing manner: "Cover all steam supply mains and branches of the heating system in the boiler room, basement (or wherever they are located—mention location) including steam supply risers and steam supply radiator branches, with (state what kind or type and thickness of insulation shall be used and how it shall be finished). If it is desired to insulate the steam return piping in the heating system the specifications should state: All return mains and branches (state location), including return risers and radiator branches shall be insulated with (state what kind or type and thickness of insulation shall be used and how it shall be finished).

If certain portions of the steam supply and return heating piping are to be left uncovered, this should be mentioned by specifying "all steam and return or all steam and return mains and branches except the following (state location) shall be insulated".

If it is desired to finish pipe covering with a special jacket such as, an 8 oz. sewed canvas jacket over rosin sized paper, it should be so stated, mentioning the location where this work is to be done. If it is desired to finish only exposed pipe covering, the specification and drawings should state the exact location of such exposed covering. The insulation contractor has no way of determining, in general, from the mechanical drawings or specifications where the exposed pipe covering is located unless it is specifically given.

The specifications should state specifically that all fittings and valve bodies in the piping systems shall be covered (state type of material and thickness) and should also mention that the insulation on fittings and valve bodies shall be finished with a canvas jacket, of the same weight as the canvas on the adjacent covering, neatly pasted on.

If it is desired to have flanges insulated it should be so stated giving type of material, thickness, and how the insulation shall be finished. The insulating contractor does not consider a flange as a fitting, and fitting insulation does not include the flanges on flanged fittings. When flanges are to be insulated the specifications should mention the size of the piping and fittings on which flanges will occur.

If it is desired to have the insulation painted it should be specifically mentioned, stating kind of paint and number of coats. The insulation contractor does not include painting with insulating work unless it is definitely specified.

When piping or other surfaces are placed in locations subject to freezing the specifications should state the type and thickness of insulation and should also definitely mention the locations. It is not sufficient to state "all piping and surfaces exposed to low temperatures shall be properly insulated to prevent freezing". The architect or engineer should assume the responsibility for interpreting such a specification and should definitely state the locations on the drawings.

Where it is desired to insulate cold water piping to prevent sweating the specifications should state whether all or a part of the piping shall be insulated. If only a part of the system is to be covered the locations should be definitely mentioned. It is not sufficient to state 'all cold water piping shall be covered where liable to sweat".

For further clarification the drawings and specifications should specifically state how steam and return radiator branches, hot and cold water branches, and fixtures shall be run; whether in floor-fill, above the floor, on the ceiling of the floor below, in partitions, behind fixtures, etc. If a run is concealed in the floor and a different type of covering is desired from that on exposed lines it should be so stated, giving type, thickness, and finish of the covering. This is important as work on ceilings is more difficult to do than work located in or near the floor.

Where piping is run concealed in chases or trenches sufficient space should be provided around the pipes for applying full thickness of covering in order that the designed efficiency of the insulation may be maintained. The location of such chases and trenches should be specifically shown on the drawings.

When specifying insulation for ventilating ducts, air washers, or other flat surfaces, the architect or engineer should provide for punched angles to support the insulation. These angles should be provided by the sheet metal contractor who is equipped to do the work. The provision of such supports will eliminate the necessity of drilling or punching holes in the sheet metal and will reduce the cost of applying the insulation.

A large variety of insulating materials have been developed to meet the requirements of conditions and tempratures encountered in modern architectural and engineering practice. The following are most commonly used to meet the service in Heating, Ventilating, Air-Conditioning, Plumbing and Refrigerating Systems. Other materials are available for special conditions and data and descriptive literature of such products will be furnished by the various manufacturers, if desired, to help determine the proper insulation to specify for special and unusual conditions.

STANDARD ASBESTOS MANUFACTURING CO.

822 W. LAKE ST. CHICAGO, ILL.

MONroe 6475

SUPER-DISTRIBUTORS FOR

KEASBEY & MATTISON CO. INSULATIONS

K & M "FEATHERWEIGHT" 85% MAGNESIA . . . K & M HY-TEMP COMBINATION

A FEW REPRESENTATIVE INSTALLATIONS RECENTLY COMPLETED:

ROSENWALD MUSEUM OF SCIENCE AND INDUSTRY
JACKSON PARK—CHICAGO, ILL.

JANE ADDAMS HOUSING PROJECT—U. S. GOVERNMENT
CHICAGO, ILL.

MOODY BIBLE INSTITUTE—OFFICE BUILDING AND POWER HOUSE
CHICAGO, ILL.

INTERNATIONAL HARVESTER CO. PLANT AND POWER HOUSE
EAST MOLINE, ILL.

WIEBOLDT DEPARTMENT STORE
RIVER FOREST, ILL.

LINCOLN PARK POWER HOUSE
CHICAGO, ILL.

INLAND STEEL CO.—BLAST FURNACE NO. 5
INDIANA HARBOR, IND.

ACME STEEL CO.—OFFICE AND WAREHOUSE
RIVERDALE, ILL.

ST. GILES CHURCH AND CONVENT
OAK PARK, ILL.

DURKEE FAMOUS FOODS
CHICAGO, ILL.

STATE OF ILLINOIS INSTITUTIONS
EAST MOLINE
ELGIN
MACOMB
ST. CHARLES
JACKSONVILLE
DUNNING
NORMAL

BOARD OF EDUCATION—CITY OF CHICAGO
BETSY ROSS SCHOOL
NEWBERRY ELEMENTARY SCHOOL
WALLER HIGH SCHOOL
HAMMOND ELEMENTARY SCHOOL
FROEBEL ELEMENTARY SCHOOL
CALHOUN ELEMENTARY SCHOOL
BRADWELL ELEMENTARY SCHOOL

UNITED AIR LINES OFFICE BUILDING
CHICAGO, ILL.

CONSOLIDATED PAPER CO.
AURORA, ILL.

VILLAGE OF HINSDALE—MUNICIPAL POWER PLANT
HINSDALE, ILL.

VILLAGE OF WINNETKA—MUNICIPAL POWER PLANT
WINNETKA, ILL.

Asbestos Air Cell Type — Temperature limitation 350 degrees F.

Manufactured and furnished in standard 3 ft. sections to fit pipe size, finished with a canvas jacket.

Made in thicknesses from ¾″ to 3″ of alternate layers of plain and corrugated asbestos paper. Corrugations run four to the inch; six to the inch, and eight to the inch.

Also furnished in sheet and block form.

85% Magnesia Type—Temperature limitation 600 degrees F.

Manufactured and furnished in standard 3 ft. sections to fit pipe size, finished with a canvas jacket weighing approximately four (4) ounces to the square yard.

Made in thicknesses from Standard to 3″. Standard thick is as follows:

½″ to 1½″ pipe	⅞″ thick
2″ to 3½″ pipe	1 1/32″ thick
4″ to 6″ pipe	1⅛″ thick
7″ to 10″ pipe	1¼″ thick
12″ to 30″ pipe	1½″ thick

Also furnished in block form and cement.

Laminated Asbestos Type — Temperature limitation 600 degrees F.

Manufactured and furnished in standard 3 ft. sections to fit pipe size, finished with a canvas jacket weighing approximately four (4) ounces to the square yard.

Made in thicknesses from 1″ to 3″.

Also furnished in sheet and block form.

High Temperature Type—For temperatures from 600 degrees to 1900 degrees F.

Manufactured and furnished in standard 3 ft. sections to fit pipe size, finished with a canvas jacket weighing approximately four (4) ounces to the square yard.

Made in thicknesses from 1″ to 3″.

Also furnished in block form.

High temperature insulation can be used singly, but more often it is used in combination with 85% Magnesia or Sponge Felt. The high temperature insulation is always used as the inner layer and the other material as the outer layer.

The thickness of this combined insulation conforms with temperature requirements.

Rock Wool—Mineral Wool—Glass Wool Type.

These types of insulations are made in various thicknesses and shapes—in sectional pipe covering, block and sheet forms. Recommendations as to their uses and thicknesses should be obtained from the manufacturers.

Wool Felt Type—For temperatures from 35 degrees to 80 degrees F.

Manufactured and furnished in standard 3 ft. sections to fit pipe size, finished with a canvas jacket.

Regular or ordinary type of Wool Felt is made in thicknesses from ½″ to 1½″ of solid construction with a waterproof paper lining.

Anti-sweat coverings are made with an inner liner of water repellent felt and an outer layer of the same material having a two-inch lap running horizontal along the length of the section, which lap is sealed with cement after the section is applied to the pipe. This material is made of solid construction in ½″ and ¾″ thicknesses, and in double shell form in 1″ thicknesses and over.

Hair Felt Type—For temperature 50 degrees F. and lower.

This insulation is recommended for piping and apparatus conveying liquids or gases at low temperatures, and for the resiliency it offers to withstand expansion, contraction and vibration.

It is also used as an insulation to insure against freezing.

Various methods of application and construction are in use, and the Insulation Contractor should be consulted so he may assist in determining what methods to use for any particular application.

Cork Type—Vegetable and Rock—For temperatures 45 degrees and lower.

Manufactured and furnished in moulded sectional covering, fitting and flange covers, with necessary accessories for sealing joints, wiring and painting surfaces.

The purchase of insulating materials for mechanical equipment is a very good investment for the owner as they definitely increase the operating efficiency of the systems, improve working conditions for the operating staff and effect very great economies in fuel costs due to reduction in heat loss and heat absorption. The real standard by which original costs should be judged is in terms of the return on the investment.

For low temperature heating systems the insulation of piping and ducts will effect an annual return on the investment, in heat losses alone, equal to approximately fifty per cent. For boilers, tanks and high temperature systems the savings will effect a return of 50 to 200 per cent and for refrigeration piping and apparatus from 100 to 200 per cent per annum.

The economies effected depend upon the quality of materials used. In general the better grades of materials will result in the greatest economies.

The following are recommended specifications which are based on standard insulation practice.

1. For low temperature heating systems the supply and wet return mains, risers, and branches inside of buildings shall be covered with 4-ply asbestos air cell covering 1″ thick applied with canvas laps pasted. Fittings shall be insulated to the same thickness as the covering on adjacent lines with asbestos cement applied in two coats and finished with canvas jacket pasted on. Flanges shall not be insulated.

Where additional economies are desired, 6 or 8-ply asbestos air cell, or Standard Thick 85% Magnesia coverings may be used.

2. Low pressure supply and return heating mains in trenches or tunnels between buildings shall be insulated with standard thick Laminated Asbestos, 85% Magnesia, or other molded coverings as approved, applied with canvas laps pasted. Fittings shall be insulated to the same thickness as the covering on adjacent lines as follows: For pipe sizes 4½″ and larger the insulation shall consist of block material the same as the covering, wired on and finished smooth with cement and canvas jacket pasted on. For pipe sizes 4″ and smaller the insulation shall consist of two coats of cement finished with canvas jacket pasted on. Flanges shall be insulated with block material the same as the covering, wired on and finished smooth with cement and canvas jacket pasted on.

3. High pressure saturated and superheated steam piping shall be insulated with molded sectional covering or Laminated Asbestos covering of the following thicknesses applied with the laps pasted:

Size Pipe	85% Magnesia or Laminated Asbestos
Total Steam Temperature 250 degrees to 350 degrees F.	
½″ to 1½″	Standard
2″ to 4½″	1½″
5″ to 8″	2″
9″ and over	2″

Total Steam Temperature
351 degrees to 450 degrees F.

½" to 1½"	1½"
2" to 4½"	2"
5" to 8"	2"
9" and over	Double Standard

Total Steam Temperature
451 degrees to 550 degrees F.

½" to 1½"	1½"
2" to 4½"	2"
5" to 8"	Double Standard
9" and over	Double 1½"

Size Pipe High Temperature Insulation	Inner Layer	Outer Layer 85% Magnesia or Laminated Asbestos
Total Temperature 551 degrees to 650 degrees F.		
½" to 1½"	1½" Single layer	
2" to 4½"	1"	1½"
5" to 8"	1½"	1½"
9" and over	1½"	2"
Total Temperature 651 degrees to 750 degrees F.		
½" to 1½"	2" Single layer	
2" to 4½"	1"	2"
5" to 8"	1½"	2"
9" and over	1½"	2½"

All double layer coverings shall be applied with circumferential and longitudinal joints staggered. Canvas laps shall be pasted.

Fittings shall be insulated to the same thickness as the covering on adjacent lines with insulating block, ½" cement finish and canvas jacket for sizes 4½" and larger and with cement and canvas jacket on sizes 4" and smaller. Block materials shall be of the same type as specified for the covering. Fittings 4" and smaller shall be insulated with Magnesia or Asbestos Cement for temperatures up to 550 degrees, above this with 1" of high temperature insulating cement and Magnesia or Asbestos cement finish.

Flanges shall be insulated to the same thickness as the covering on adjacent lines with block materials of the same type as specified for the covering finished with ½" Magnesia or Asbestos cement and canvas jacket pasted on.

4. Domestic hot water mains, risers, and branches shall be insulated with 4-ply asbestos air cell covering 1" thick in accordance with specifications under Item 1 above or with 85% Magnesia or other moulded coverings in accordance with Item 2.

5. Cold water lines not subjected to freezing temperatures shall be insulated with wool felt type covering 1" thick on sizes up to and including 5" and with double ¾" thick on sizes 6" and larger. Double layer coverings shall be applied with joints staggered. Canvas laps shall be pasted.

Fittings and flanges shall be insulated to the same thickness as the covering on adjacent lines with hair felt tied on with jute twine and finished with asbestos cement and canvas jacket pasted on.

6. Cold water lines subjected to freezing temperatures in unheated portions of buildings shall be insulated as specified under Item 5 and over this shall be wrapped one layer of 1" hair felt securely tied on with jute twine and finished with heavy felt paper and canvas jacket pasted on.

For cold water lines running outdoors subjected to freezing temperatures the insulating contractor should be consulted for proper specifications.

7. Refrigerated ice water lines shall be insulated with ice water thick moulded cork covering on piping and fittings. All joints of the covering shall be tightly fitted together and cemented with waterproof cement and the covering securely wired on with copper clad steel wire. Seams shall be filled with asphalt seam filler and the surface of the cork painted one coat of black asphalt paint.

As an optional specification, refrigerated ice water lines shall be insulated with double shell anti-sweat felt covering double ¾" thick on pipe sizes ½" to 4½" and double 1" thick on sizes 5" and larger. Joints of the covering shall be staggered both circumferentially and logitudinally and the canvas laps pasted. Fittings and flanges shall be insulated to the same thickness as the covering on piping of corresponding size with hair felt tied on with jute twine and finished with asbestos cement and canvas jacket pasted on.

8. Refrigerated brine and ammonia suction lines shall be insulated with moulded cork covering on piping and fittings, standard brine thick for temperatures 0 degrees F. to 35 degrees F. and special brine thick for temperatures 0 degrees F. to minus 25 degrees F. All joints of the covering shall be tightly fitted together and cemented with waterproof cement and the covering securely wired on with copperclad steel wire. Open spaces between cork covers and fittings shall be filled with molten paraffin and cork dust poured into place through holes in the covers. Seams shall be filled with asphalt seam filler and the surface of the cork painted one coat of black asphalt paint.

As an optional specification, refrigerated brine and ammonia suction lines shall be covered with hair felt insulation 2" thick for temperatures 20 degrees F. to 35 degrees F., 3" thick for temperatures 0 degrees F. to 20 degrees F. and 4" thick for temperatures 0 degrees F. to minus 25 degrees F. The covering on piping shall be built up of 1" layers of hair felt alternated with asphalt saturated paper and each layer of felt and paper shall be applied in hot asphalt. Over the outer layer of felt shall be applied asphalt saturated paper followed by dry wool felt paper and canvas jacket pasted on. Fittings and flanges shall be insulated the same as the piping except for the outer finish which shall consist of asbestos cement troweled smooth and canvas jacket pasted on.

For extremely low temperatures below minus 25 degrees F. the insulating contractor should be consulted for suitable specifications.

9. Canvas over insulation on fittings and flanges shall be of the same weight and texture as that on the sectional covering in the adjacent lines. If special weight canvas is desired on covering, fittings and flanges, it shall be so stated in the specifications.

10. On lines running outdoors exposed to the weather the covering shall be protected with 3-ply asphalted rag felt roofing jacket applied with 2½" circumferential and longitudinal laps so placed as to properly shed water. This jacket shall be securely wired on with copper or copperclad steel wire on 6" centers and stapled along the laps with flat steel staples on 4" centers. Insulation on fittings and flanges shall be finished with ¼" of asphalt mastic neatly troweled on and tightly sealed against the edges of the roofing jacket.

11. Hot water heaters and storage tanks shall be insulated with 1½" thick 85% Magnesia block or other moulded insulation as approved, securely wired on. The surface shall be finished smooth with ½" of Magnesia (or asbestos) cement applied in two coats and over this a 6 ounce canvas jacket pasted on.

As an optional specification hot water heaters and storage tanks shall be insulated with four ply (1" thick) asbestos air cell blocks, securely wired on. The surface shall be finished smooth with ½" asbestos cement ap-

plied in two coats and over this a 6 ounce canvas jacket pasted on.

12. Low pressure steam and hot water heating boilers shall be insulated with 1½" thickness of asbestos air cell blocks, securely wired in place and finished with ½" of asbestos cement applied in two coats and 6 ounce canvas jacket pasted on.

For high pressure brick set type boilers the insulation contractor should be consulted for proper specifications.

13. Air washers or air coolers for air conditioning systems shall be insulated with 1½" or 2" thick cork board, as conditions dictate, applied in hot asphalt or suitable adhesive cement and wired to the stiffener angles. The outside shall be finished smooth with ½" of asbestos cement applied in two coats and over this 6 ounce canvas jacket pasted on.

For conditions of high humidity the cork board should be finished with ¼" of asphalt mastic in lieu of asbestos cement and canvas.

14. Cold air ducts shall be insulated with 1" or 1½" pure cork board, as conditions dictate, wired on to the stiffener angles. The surface shall be finished smooth with ½" of asbestos cement applied in two coats and over this 6 ounce canvas pasted on. In unfinished portions of buildings the cement finish may be eliminated, the joints stripped with asbestos paper pasted on and no canvas jacket applied. If desired the cork board may be applied in hot asphalt or suitable adhesive cement.

As an optional specification, cold air ducts shall be insulated with hair felt 1" or 1½" thick as conditions dictate. The felt shall be securely wired on and finished with heavy felt paper and canvas jacket pasted or sewed on.

15. Hot air ducts shall be insulated with 1" or 1½" asbestos air cell board, as conditions dictate. The board shall be securely wired on to the stiffener angles and finished smooth with ½" of asbestos cement applied in two coats and over this 6 ounce canvas jacket pasted on. In unfinished portions of buildings the cement finish may be eliminated, the joints stripped with asbestos paper pasted in place and over this canvas jacket pasted on.

The above specifications are for ordinary conditions as commonly met in architectural and engineering practice and are offered as a guide in the selection of proper insulating materials.

It is hoped that they will be of assistance to the architects and engineers in the preparation of comprehensible plans and specifications.

Illinois Society of Architects

1906, 134 N. La Salle Street, Chicago

The following is a list of the publications of the Society; further information regarding same may be obtained from the Financial Secretary.

FORM NO. 21, "INVITATION TO BID"—Letter size, 8½x11 in., two-page document, in packages of fifty at **75c**, broken packages, two for **5c.**

FORM NO. 22, "PROPOSAL"—Letter size, 8½x11 in., two-page documents, in packages of fifty, at **75c**, broken packages, two for **5c.**

FORM NO. 23, "ARTICLES OF AGREEMENT"—Letter size, 8½x11 in., two-page document, in packages of fifty, at **75c**, broken packages, two for **5c.**

FORM NO. 24, "BOND"—Legal size, 8x13 in., one-page document, put up in packages of twenty-five, at **25c** per package, broken packages, three for **5c.**

FORM NO. 25, "GENERAL CONDITIONS OF THE CONTRACT"—Intended to be bound at the side with the specifications, letter size, 8½x11 in., ten-page document, put up in packages of fifty at **$2.50**, broken packages, three for **25c.**

FORM 26, CONTRACT BETWEEN ARCHITECT AND OWNER. Price, **5c** each, in packages of fifty, **$1.25.**

FORM 1, BLANK CERTIFICATE BOOKS—Carbon copy, from 3¾x8½ in., price, **50c.** Two for **5c.**

FORM 4, CONTRACT BETWEEN THE OWNER AND CONTRACTOR—(Old Form.) Price, two for **5c.** Put up in packages of 50 for **$1.00.**

FORM E, CONTRACTOR'S LONG FORM STATEMENT—As required by lien law. Price, two for **5c.**

FORM 13, CONTRACTOR'S SHORT FORM STATEMENT—Price, **1c** each.

CODES OF PRACTICE AND SCHEDULE OF CHARGES—8½x11 in. Price, five for **10c.**

These documents may be secured at the Financial Secretary's office, suite 1906, 160 N. La Salle St., telephone Cent. 4214. We have no delivery service. The prices quoted above are about the cost of production. An extra charge will be made for mailing or expressing same. Terms strictly cash, in advance, with the order; except that members of the Society may have same charged to their account.

MODERN SANITATION OF BUILDINGS

By LEO H. PLEINS, Architect and Sanitary Engineer

The primary object of this article is to present to Architects in as brief a form as possible, data, which the writer trusts may be of service in their office practice in the preparation of plans and specifications covering plumbing work.

The great importance of sanitary plumbing work is daily becoming more and more recognized and hence if the Architect is to give his client full service, plumbing must be given the same careful consideration as the other structural parts of the building.

For convenience of reference the article is arranged under four headings—"Drainage of Building";—"The Water Supply";—"Arrangement of Toilet and Bath Rooms"; and "Plumbing Fixtures". Space does not permit of covering all that may be said under each heading, but endeavor has been made to mention characteristic features of importance, that should be given consideration in the proper analysis of each particular problem.

DRAINAGE OF BUILDINGS.

I. **Proper Fall to Main Sewer.** When a survey is made the location and size of main sewer should be indicated thereon. If stubs to curb are in place their location, size and grade should be shown. The basement floor grade should always be given and also grade of main sewer at curb or street. The desirable grade for house sewer connections is ¼" to one foot. If this cannot be obtained, the grade may be reduced but in this case the size of the tile pipe must be increased according to the length of the connection from building to main sewer.

See Table I for carrying capacity of tile pipe at varying grades. Discharge is given in cubic feet per second. Convert this into gallons by multiplying by 7.50

In the absence of any regulation as to size of house connections to main sewer, the minimum size of such connection shall be 6" tile pipe; unless a larger size is required for drainage after careful calculation. See "Size of Main House Drain," page 663.

II. **When Main Sewer is Above Level of Basement Floor Grade:** In this case all drainage from floor drains or fixtures in basement must be run to a sump basin and elevated by means of a pump. If no water closets or urinals are to be installed in basement the pump will be described as a **bilge** pump. If water closets and urinals are to be provided in basement, the pump will be described as a **sewage ejector.**

Obviously all waste and soil lines that may be drained by gravity, such as all drainage from floors above the basement shall be run into a horizontal line and this carried under ceiling of basement and thence through the wall connecting to the main sewer at such distance below grade as necessary to properly drain the system. The discharge from Bilge Pump or Sewage Ejector shall be connected into the horizontal line under ceiling of basement at such point inside of building as may be convenient.

If a Bilge Pump is installed—the basin for a single pump should be as follows: For pump from 10 to 30 gallons per minute, basin to be 30" diameter: for a pump from 50 to 100 gallons per minute, basin to be 36" diameter and for a pump from 125 to 200 gallons per minute, basin to be 42" diameter. For two or duplex pumps—basin to be 48" diameter for pumps from 100 to 125 gallons per minute and 60" in diameter for pumps of 150 to 200 gallons per minute capacity. All basins should be 36" deeper than lowest inlet entering the same.

If a Sewage Ejector is installed, the basin for a single ejector shall be as follows: For an ejector from 50 to 75 gallons per minute —basin to be 36" in diameter; for an ejector of from 100 to 200 gallons per minute, the basin should be 42" diameter and for an ejector of 250 to 350 gallons per minute, the basin should be 48" in diameter. For two or duplex ejectors, the basin to be 48" in diameter for ejectors of from 50 to 100 gallons per minute and 60" in diameter for ejectors of from 125 to 350 gallons per minute. All basins should be 48" deeper than the lowest inlet entering the same.

The best motive power for Bilge pumps or Sewage ejectors is a direct connected vertical type electric motor—the operation of which is automatically controlled by means of a float switch.

Wherever possible, both Bilge pumps and Sewage ejectors should be installed in duplicate sets. With duplex pumps the automatic control is arranged so that the same will start one pump when the water level has raised, holding the second pump in reserve, and starting the second pump when the first is not capable of handling all the water. Both pumps will then operate until normal condition has been restored. The automatic control should be provided with a four-pole transfer switch so connected up that by throwing over switch, each pump will operate at alternate periods, holding the other as reserve and in this way, equalize the wear on the pumps.

Always ascertain and specify the correct electric current and provide for service wires to within 5 feet of pump to be furnished by contractor for Electrical Work. If current is Direct give the voltage and if current is Alternating give voltage cycles and phase.

The motors for pumps are usually mounted on a cast iron or steel cover which forms a support for motors, contact apparatus etc. The basins may be of cast iron, steel, brick or concrete. If of the latter materials basins must be thoroughly waterproofed.

A swinging check valve, cast iron body, brass mounted must be placed in the horizontal discharge pipe between pump and sewer.

Blow-off drainage from boilers cannot be run directly into bilge pump or sewage ejector basins—but must always discharge into a cast iron or steel blow-off basin or muffler tank. From this basin the drainage may then be run to bilge or sewage ejector basins, if it is impossible to drain the same by gravity.

Boiler Blow-Off Basins:

These are usually included under the heading of "Heating Work." The contractor for this work makes all connections between same, boiler blow-offs, drips, etc. When directly connected to the house sewage line the plumbing contractor makes such connection as also the venting of blow-off basins through roof. Attention in this connection is called to the requirements of the Chicago Ordinance prohibiting the discharge

from basins being made into tile sewers within any building, furthermore, that the water discharged into a sewer shall not exceed 120° "F." It is necessary therefore to use cast iron pipe and in order to prevent leaks of joints, therefor cast iron hub and spigot pipe should be made with iron cement instead of lead—or flanged pipe used with asbestos graphite gaskets.

of roof surface. For medium sized roofs 1 sq. in. in sectional area of the leader for each 200 sq. ft. of roof surface. For large roofs, 1 sq. inch in sectional area of the leader for each 250 sq. ft. of roof surface.

Judgment must be used in arranging downspouts so as to equalize the square feet of drainage as nearly as possible.

Diameter.	Slope, or Head Divided by Length of Pipe.							
	1 in 40	1 in 70	1 in 100	1 in 200	1 in 300	1 in 400	1 in 500	1 in 600
5 in.	.456	.344	.288	.204	.166	.144	.137	.118
6 in.	.762	.576	.482	.341	.278	.241	.230	.197
8 in.	1.70	1.29	1.08	.765	.624	.54	.516	.441
9 in.	2.37	1.79	1.50	1.06	.868	.75	.717	.613
Slope	1 in 60	1 in 80	1 in 100	1 in 200	1 in 300	1 in 400	1 in 500	1 in 600
10 in.	2.59	2.24	2.01	1.42	1.16	1.00	.90	.82
12 in.	4.32	3.74	3.35	2.37	1.93	1.67	1.5	1.37
Slope	1 in 100	1 in 200	1 in 300	1 in 400	1 in 500	1 in 600	1 in 700	1 in 800
15 in.	6.18	4.37	3.57	3.09	2.77	2.52	2.34	2.19

Table I.

The following Table (II) may be of service to determine the proper size of basin to be provided:

Table II.

For Boiler of 25 to 75 H. P. use Basin 36" diameter by 42" deep.

For Boiler of 100 to 200 H. P. use Basin 42" diameter by 60" deep.

For Boiler of 250 to 400 H. P. use Basin 60" diameter by 72" deep.

For more than 400 H. P. use two or more basins—using the above as multiples according to horse power of boiler.

A¾" or 1" cold water supply with control valve should be made to each blow-off basin and such connection will be found of service in cooling excessively hot blowoff water and help to condense steam vapors. Vents through roof must in no case be less than 4" for small basins and 6" for larger basins.

Downspouts and Downspout Drains:

In many localities the drainage from downspouts must be connected into a "Storm Water Sewer"—and not to the "Sanitary or house sewer." In either case arrangement of downspouts and drainage from same may be the same.

The best material to use for vertical inside downspouts is extra heavy cast iron pipe and fittings of proper size. All outside sheet metal downspouts should be connected into cast iron pipe and fittings above grade and cast iron pipe be run to proper depth below grade and connected to tile pipe by means of a cast iron quarter-bend.

Before making connection to roof—downspouts should be increased one size and the roof connection should be made to allow for expansion and contraction by means of a copper or lead sleeve. Roof fittings and strainers should be of cast iron and well flashed with copper or lead.

To determine the proper size for downspouts the following may be of service.

A rainfall of 1-inch in depth on an area of 100 square feet will give a run off of 62 gallons.

Downspouts proportioned as follows have been found in practice to give satisfactory results. For small roofs, 1 sq. inch in sectional area of the leader for each 150 sq. ft.

Outside downspouts should be avoided, especially in cold climates, as they are constantly giving trouble on account of freezing and therefore cause damage to roofs and walls.

Where roofs are covered with gravel or in localities where high winds are likely to cover roof with debris, etc., the downspouts should be provided with cast iron gravel basins or running traps with cleanouts. Gravel basins or traps must always be used when connecting downspout drains to sanitary sewers, where ordinances do not require such downspout drains to be run into outside catch basin as required by the Chicago ordinance.

Size of Main House Drain:

The size of the main house drain when serving as a combination drain (sanitary and rain water) may for all practical purposes be determined by the total surface area covered by the building or buildings and paved surfaces to be drained, by the following table, which is based on cast iron pipe. If vitrified tile sewer pipe is used the diameter of pipe as given must be increased one size for same area of drainage.

Square Feet of Drainage Area.

Diameter	Fall ⅛ in. per foot	Fall ¼ in. per foot	Fall ½ in. per foot
4 inch	1.500	1.800	2.500
6 "	3.000	5.000	7.500
8 "	6.000	9.100	13.600
10 "	9.000	14.000	20.000

Back Water Valves:

Whenever the grade or size of sewer in street is such that there is a possibility of the same backing up—the house sewer must be provided with a cast iron back water valve of approved type and this valve should be placed in a manhole or otherwise located so as to be accessible for inspection or repair. It is desirable to use a back water valve having a flushing connection so that the line may be flushed.

Where water closets or urinals are located in basements and connected to horizontal house sewers which are likely to back up during heavy rains it is advisable to place a double gate valve on the branch to such

CRANE CO.

GENERAL OFFICES: 836 S. MICHIGAN AVE., CHICAGO, ILLINOIS
NEW YORK: 4730 29TH STREET, LONG ISLAND CITY

VALVES, FITTINGS, PIPE, PLUMBING AND HEATING, PUMPS

Nation Wide Service through 134 Branches and more than 500 Wholesalers.

Crane Service to the Architect

PLUMBING

Kitchen Planning

Crane Co. has made comprehensive studies of kitchen planning, has designed sinks and cabinets to meet the most exacting requirements. Cooperation with architects includes the making of detailed blueprints, based on your sketches and dimensions, together with colored perspectives if desired.

Modern Bathrooms

Supplying fine quality bathroom fixtures, Crane Co. is equipped to execute room plans wherein these fixtures find their proper settings. This bathroom planning service is available, on request, to any architect, and consists of blueprints and perspectives incorporating any or all of vour suggestions.

The Laundry

Crane laundry equipment — from the trays to the water softener — is presented with a laundry planning service to back it up. Plans and perspectives based on your roughs executed according to well-studied laundry technique.

Well-Water Systems

Crane Shallow and Deep Well Pumps— outstanding in their dependability, low maintenance and long life, form the central unit in efficient, low-cost water systems for the home beyond the city mains. Complete data is available, together with every cooperation from Crane engineers.

HEATING

High Efficiency Boilers

Crane offers a complete line of boilers for automatic heat, including Stoker-fired coal burning boilers, Sustained Heat oil burning boilers and burner unit. The Fin-type boiler designed to use any commercial oil burner and the Basmor gas-fired boilers. The Crane Sectional and Round Boilers for hand firing are also available.

Radiation

Crane radiation is made in all styles and sizes, including wall-hung and legless. Invisible shields that direct the heat outwardly into the room are a patented feature. Shields protect walls and minimize ceiling and floor temperature differences.

Crane Convectors

Crane convectors offer a highly efficient type of concealed radiation. Enclosures are made in various styles fully recessed, partially recessed, free standing, wall-hung, and plastered-in. The removable plastic grille eliminates the utility door that spoils the appearance of the convector enclosure.

Crane Valves and Fittings

Crane valves and fittings are the very heart of any plumbing or heating system. They are rugged and extremely well proportioned. Careful designing and control of product from the raw material to the finished item on the stock room shelf assures the highest of quality on any job. Specify Crane for all roughing-in requirements.

fixture in addition to the back water valve on the main drain so as to prevent sewage from backing up through these fixtures in event of the back water valve not operating properly

Flush Tanks:

Whenever the sewer in street to which connection must be made forms what is known as a "dead end" it is desirable to provide a flush tank which when filled to a proper height with clean water, will automatically discharge the contents into the sewer and thereby keep the sewer free and prevent obstructions that might otherwise occur. These flush tanks may be of two types—as illustrated herewith. Type A being suitable for flushing more than one dead end; Type B may be used if the "dead end" will be continued at some later time—in which case the flush tank may be converted into a standard manhole by taking off the cap at end of siphon and removing the latter.

Soil Pipe System:

In the order of preference for soil, waste and vent systems, we would rate: A, Brass pipe and fittings; B, cast iron pipe, calked or threaded type, with cast iron fittings; C, genuine wrought iron pipe with cast iron fittings.

The very best and most durable material for soil, waste and vent systems is full iron pipe size annealed brass pipe with red metal cast brass fittings. However, the cost of such an installation is usually greater than most owners care to invest.

Cast iron extra heavy soil pipe and fittings are the most permanent and best for soil, waste and vent systems and should be used wherever possible. While the ordinances of some cities require the use of wrought iron pipe where buildings are over seven stories in height, there is no reason why this exception should be made, as extra heavy cast iron soil pipe and fittings have been used in buildings 16-stories in height and the joints double calked as described on page 667.

The new PERMO-LOCKT hub on cast iron soil pipe, adopted as a standard by all manufacturers of cast iron soil pipe, together with the new type expansion joint, makes it possible to use cast iron pipe and fittings throughout in a building of any height with the assurance of having the best and most permanent piping system that could be installed and nothing superior to it except all brass pipe and fittings.

Threaded cast iron pipe with cast iron fittings is now being manufactured and if properly made, should prove most desirable for use where durability is a consideration.

Considering such ordinances as require the use of wrought iron pipe for soil, waste and vent systems, we would say, that unfortunately these ordinances are not specific in stating that when wrought iron pipe is used —it shall be Genuine Wrought Iron, hence most installations are made with commercial steel pipe, which in the opinion of the writer should never be used. In many cases where the question of cost of genuine wrought as compared to steel has been an issue—Architects and Engineers have expressed the opinion that genuine wrought pipe was not worth the difference in cost. This opinion is not shared by the writer but as stated above, I believe, that extra heavy cast iron soil pipe is the most logical material to use when all facts are taken into consideration.

If Genuine Wrought iron pipe is used for soil, waste and vent piping—all vent extensions up thru roof should be terminated with extra heavy cast iron soil pipe—the length of which should not be less than 10 feet, and more, if possible, from below roof to top of pipe.

Simplicity in arrangement of soil, waste and vent stacks is desirable and it is extremely desirable to make diagrams of the system that will be of aid to the plumbing contractor as well as of being of service to the other contractors on the work. In order to be of service these diagrams must be accurately drawn and amplified by details where necessary.

The importance of a plumbing plan carefully laid out has unfortunately not been properly recognized. At the present time the cost of material is such that the Architect who is going to give his client the service for which he is paid—must more than ever consider every item that will form a part of the work.

The structural parts of a building are carefully analyzed, weights of steel columns, girders, etc., proportioned to the loads they

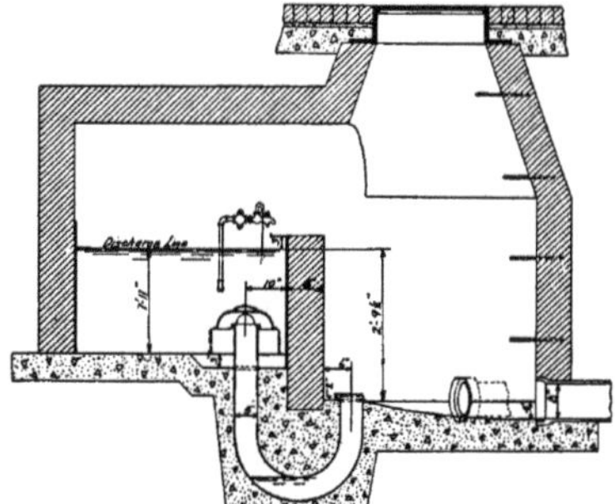

SECTION OF FLUSH TANK · TYPE "A".

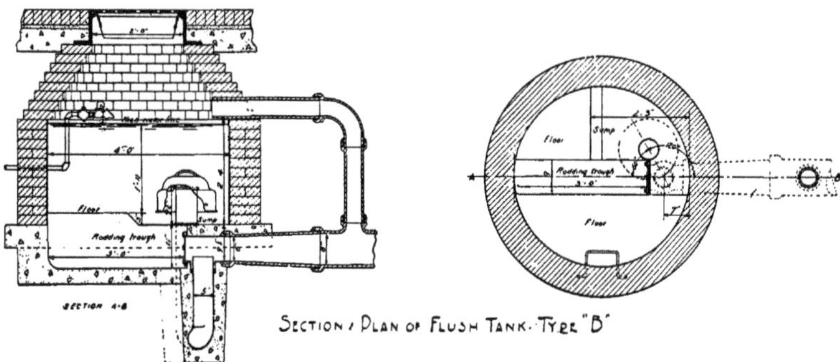

SECTION / PLAN OF FLUSH TANK · TYPE "B"

must carry, and all this work carefully detailed—and still the plumbing work is very rarely even laid out beyond a mere indication of the main run of soil or sewer lines—on the basement or foundation plan.

Specifications very often contain a clause requiring the successful bidder to submit a piping plan for the Architect's approval before commencing work. They might just as properly contain clauses asking the successful bidders to submit details for the elevations of the buildings, etc., etc. It is the Architect's duty to secure the best proposition possible for his client and therefore the plumbing work should be drawn—detailed and specified in such a manner that all bidders on the work may estimate on the same fixed basis and not permit them to submit figures based upon their ideas and conception of what may be required for the work. Such methods are very unsatisfactory and can only result in misunderstanding and most frequently in absolute failure at the expense of the client.

Whenever wrought iron pipe and cast iron drainage fittings are used, either asphalted in and out or galvanized—the stacks should be placed in pipe shafts so that the piping may be inspected and sections replaced when necessary without disturbing walls and partitions. All vents through roof should be of extra heavy cast iron soil pipe for a distance of not less than 10 feet below. Never place wrought iron pipe under basement floors. All such drainage pipe must be of extra heavy cast iron soil pipe and fittings.

When the building covers considerable area—it is desirable to use cast iron or waterproof concrete catch basins on the main lines and at intersections so as to permit of rodding the lines. In place of catch basins—large cleanouts may be used—which must always be the same size of pipe up to 6". Such cleanouts should be placed in manholes with cast iron covers large enough so that the lines can be rodded properly. Cleanouts must be placed at the foot of all stacks and wherever a change in direction of a horizontal line occurs. Cleanouts for best work should be of the heavy brass bell ferrule type with brass trap screw. With ferrules of iron the brass trap screw rusts in so that it is difficult to remove the same.

Changes in direction of horizontal lines should always be made on as full a sweep as possible, using Y-branches and 45° bends.

Connection between vertical stacks and horizontal drains in basement must always be made by means of Y-branches and 45° bends. Connection between horizontal lines on upper floors may be made by means of sanitary tees—although Y-branches are better.

All horizontal soil and waste lines should have a fall of ¼" to the foot toward outlets where possible.

All horizontal vent lines must be pitched so that water of condensation will drain freely into soil and waste lines or stacks, and foot of all vent stacks must be connected into a main soil or waste line or stack.

Reventing of each plumbing fixture is generally required. The Chicago ordinance prescribes this; other localities permit circuit venting and hence, the Architect must necessarily familiarize himself with the requirements of ordinances that may be in force in the locality in which his building is to be erected.

All main vent stacks must be extended up through roof. On pitched roofs, the vents may extend above roof 6 to 12 inches, on flat roofs 18 inches to 2 feet will be better in order to be safe in case of heavy fall of snow and to avoid dirt entering same.

In the Eastern, Central and North Western States it is necessary to increase all vent stacks at least one size up to 6 inch before passing through roof. The minimum size vent through roof should be 4 inch. All extensions through roof must be cast iron. Increasing stacks makes it possible to turn down lead or copper flashing into the pipe and leaves the extension free to provide for expansion and contraction. While caps or vent cowls should never be placed on top of vent stacks, it is desirable to use a strainer of cast iron of a removable type. Galvanized wire strainers are worthless. See Drawing.

Lead wastes are infrequently used in modern practice so we will only briefly mention them. When lead waste piping is used—it should be of a weight known as "medium" and when connected to wrought iron piping the connection must be made by means of extra heavy brass soldering nipples and a good heavy wiped joint. When connected to cast iron pipe—extra heavy brass bell ferrules must be used, wiped to the lead pipe and calked into the cast iron pipe.

JOINTING OF PIPE must be carefully done. For cast iron soil pipe—the following is a good method.

All joints of cast iron soil pipe shall be made with oakum and pure pig lead, bedded with hammer and calking iron. A gasket of well packed oakum shall be placed at the bottom of the hub extending above the rim of the spigot to prevent the escape of lead. The hub to be filled at one pouring and the lead calked with such force as to make the joint absolutely water tight under a pressure of at least 10 lbs. per square inch. All joints shall be filled at one pouring; if it fails to run full, it shall be dug out and repoured. Lead shall not be covered with paint, putty or otherwise.

Twelve ounces of lead should be allowed for each inch of diameter of pipe or fitting on which joint is made.

For buildings over six stories in height the cast iron soil pipe joints shall be double calked in the following manner: The oakum shall be well braided and before being placed in position shall be oiled and then well calked; then fill in the hub to within ⅝" of the top with molten soft pure lead and thoroughly calk. After the lead has been uniformly calked, fill in with molten lead to the top of huh and thoroughly calk—so as to make an absolutely perfect joint. All joints showing leaks under testing shall be dug out, repoured and double calked as above.

With cast iron pipe double calked as above, installations have been made in buildings 16 stories in height in which the entire system is still in excellent condition after a period of 27 years.

Joints between lead and cast iron pipe to be made by means of brass ferrules wiped to the lead pipe and calked into hub of cast iron fittings. Joints between lead and wrought iron pipe to be made by means of soldering nipples with hexagon nuts. Joints between wrought iron pipe and fittings to be screwed home into couplings or fittings without the use of any red lead or other compound.

No steam or cast bushed fittings to be used on any drainage or vent work.

Joints of tile pipe shall be made with neat Portland cement. A cleaner to be run through every length of pipe as it is laid so that no mortar used in jointing will adhere to the interior of the pipe. The connection between cast iron and tile pipe shall be made with a collar of concrete 6 inches thick and extending not less than 8 inches on each side of joint. See illustration.

All soil, waste and vent piping shall be tested. Ordinances usually prescribe the manner of testing which may be by means of water, air, peppermint or smoke on new work.

(The following illustrations show several methods for reventing plumbing fixtures in accordance with the Chicago practice and also by what is known as the "Circuit Venting" system.)

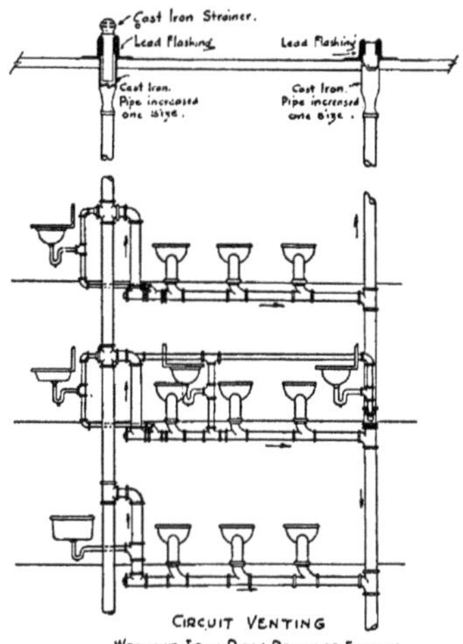

CIRCUIT VENTING
WROUGHT IRON PIPE & DRAINAGE FITTINGS.

For good work both water and peppermint tests should be made and if it is desired to be absolutely certain that integrant traps of water closets, etc., are perfect a smoke test may be made after fixtures are set.

In alteration work a peppermint test must always be made.

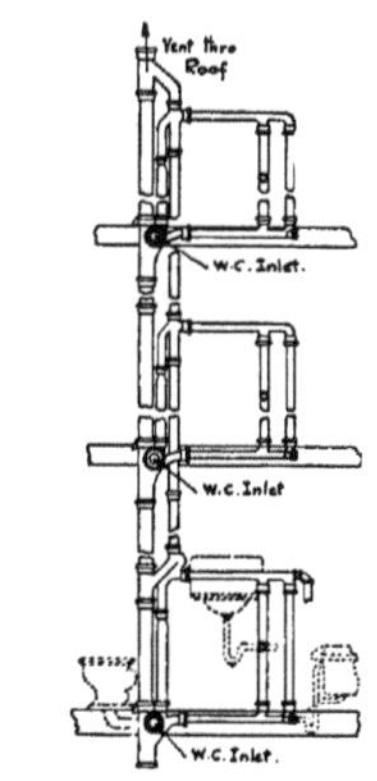

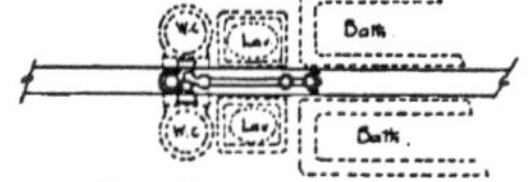

BATH ROOMS.(DOUBLE) ON ONE STACK. USING F & W. C.I. FITTINGS.

THE WATER SUPPLY

There are so many failures in the water supply system of buildings that it is evident that little study is given the problem which is one of most vital importance.

In order to provide an adequate supply of water for the particular building it is necessary to analyze the actual requirements based on a per capita consumption per day—and another factor that enters into the problem is the pressure under which the water will be delivered.

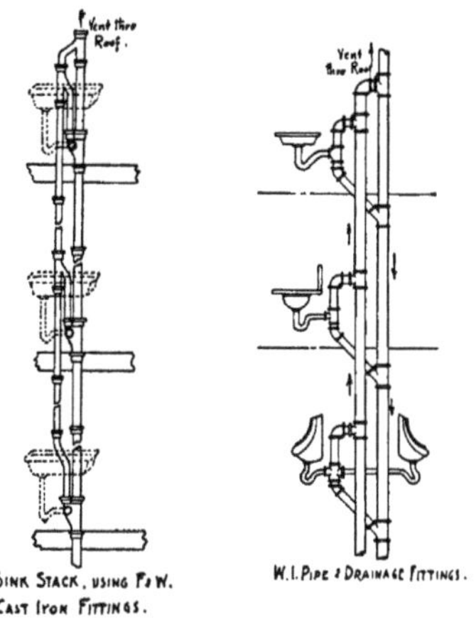

SINK STACK, USING F & W. CAST IRON FITTINGS.

W.I. PIPE & DRAINAGE FITTINGS.

Per capita requirements may be determined by the following tables, which are the minimum:

Schools (not boarding) 50 gallons per capita per day.

Industrial Plants & Factory Buildings—50 gallons per capita per day.

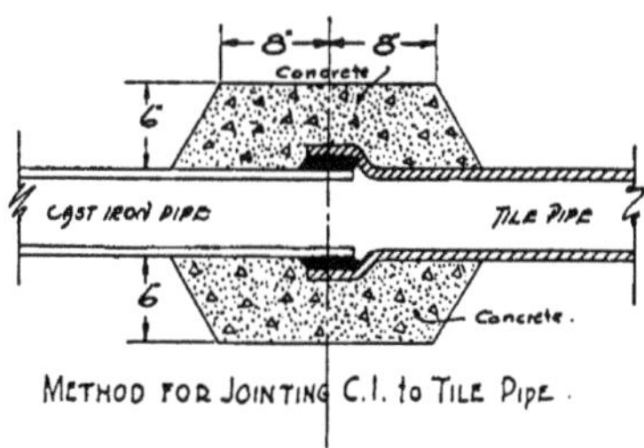

METHOD FOR JOINTING C.I. to TILE PIPE.

This does not include water that may be required directly in connection with plant operation in various manufacturing processes.

Hotels, Hospitals, Asylums, Sanitariums—150 to 200 gallons per capita per day.

Homes for the Aged, Orphan Asylums, Boarding Schools—Dormitories—100 gallons per capita per day.

To the above must be added water for sprinkling lawns, etc which must be based on the flow in gallons per minute of each ¾" lawn sprinkler installed—allowing for a

period of 3 to 4 hours for each sprinkler as a fair average.

Having determined the total quantity required for 24 hours—the next thing to determine is the proper pressure required for the work and in working this out the following must be considered:

If the average pressure is not sufficient to deliver water on the top floor of the building under at least 20 lbs. maintained pressure, it is advisable to provide a pumping system to increase the pressure so as to maintain an average of at least 20 lbs., on the top floor.

The following tables may be used to advantage in determining the sizes of main and branch supplies for buildings:

Equalizing Table of Areas of Taps

PIPE SIZES, INCHES	½	¾	1	1¼	1½	2	2½	3	4	5
½	1	1.7	2.8	4.9	6.6	11.	15.6	24.	32.	65.
¾		1.	1.6	2.6	3.8	6.2	8.9	13.8	23.	37.
1			1.	1.7	2.3	3.8	5.5	8.5	14.	23.
1¼				1.	1 3	2.2	3.1	4.9	8.	13.
1½					1.	1.6	2,3	3.6	6.2	9.7
2						1	1.4	2.2	3.8	5.3
2½							1.	1.3	2.6	4.1
3								1.	1 7	2.7
4									1	1.6
5										1.

Equalizing Table of Delivering Capacities of Pipes

DIAMETER, INCHES	¾	1	1¼	1½	2	2½	3	4	5	6
½	2 27	4.88	8.49	15.8	31.7	52.9	96.9	205.	377.	620.
¾		2.05	3 43	6.97	14.0	23.3	42.5	90.4	166.	273.
1			1 62	3.45	6.82	11.4	20.9	44.1	81.1	133.
1¼				1.69	2.67	5.94	11.6	23.7	47.4	78.5
1½					1.26	3.34	6.13	13.0	23.8	39.2
2						1.67	3 06	6.47	11.9	19.6
2½							1.83	3.87	7.12	11.7
3								2.12	3.89	6.39
4									1.84	3.02
5										1.65

Gallons per Minute Delivered From Circular Openings at Mains Under Various Net Pressures

HEAD, IN FEET	Pounds Pressure	DIAMETER OF OPENING, INCHES								
		¼	⅜	½	⅝	¾	1	1¼	1½	2
10	4.33					33	56	91	131	224
20	8.66	5	12	21	32	46	82	123	185	328
30	13.09					57	101	158	226	404
40	17.32	7 5	16	30	46	66	112	182	262	466
50	21.65					73	130	206	299	520
60	25.95	9	20	36	58	80	143	223	329	572
70	30.28					85	154	239	348	616
80	34.65	10	23	41	64	92	164	258	370	656
90	38.98					97	173	271	391	692
100	43.31	11.	26	46	72	104	184	288	415	736
110	47.64					109	192	300	432	768
120	51.98	13.	28	50	79	114	202	316	454	808
130	56.31					118	209	325	471	836
140	60.61	13.5	31	55	81	122	217	336	491	868
150	64.97	14.	32	57	87	126	226	353	509	904

Where the water supply from City mains cannot be relied upon as sufficient in volume or pressure to supply all fixtures in the building it will be necessary to provide for reserve storage to insure a constant supply, and there are two kinds of systems to be considered—First the one most commonly known, a tank on the roof, and the other and more recent—a compression tank system with a closed pressure tank in the basement. The roof tank system is obsolete and not recommended—for the reason that in order to maintain a pressure of 20 lbs. on the top floor it would have to be elevated 50 feet above the floor to give this result. Furthermore such tanks require special provision to be made for their support, must be enclosed and generally considered from a standpoint of efficiency vs. expenditure, are out of question at the present time.

The best system is a compression tank pumping system—which we shall briefly describe. These systems may be divided in two kinds—one where the pressure of the water is so low that all must be pumped and the other where it is only necessary to increase the pressure for the upper floors—in which case the system is known as the "booster" type.

In the first type the water may be delivered from a well, cistern or city main and depending upon the source of supply a pump designed for that special work must be used. Wherever possible, when pump is within suction lift of the water (20 feet) a centrifugal or turbine type pump with direct connected motor is the best to use. These pumps are of greater efficiency, less noisy and are more economical in operation than piston pumps.

In order to determine the proper size of pump to install we refer to the following table—which should be checked up with the per capita allowance per day previously mentioned.

To apply the above—First ascertain the number of fixtures pump is to supply—be sure to include every kind of fixture. In case any fixtures are supplied direct from city main these should be deducted. Second—Multiply the number of fixtures by the proper decimal that may apply according to the class of building.

Stores & Shops .... .75
Office Buildings .... .75
Factories 1.00
Apartment Buildings 5
Hotels 8
Hospitals 1.00
Schools 8

The table is based upon an equal number of males and females and the figures represent the gallons per minute per fixture. If the major portion of occupants are females increase pump capacity 25 per cent.

Where more than 150 fixtures are to be supplied pump capacity may be reduced 15 to 25 per cent.

Where actual water requirements have been determined (by meter or otherwise) furnish a pumping unit capable of discharging three times the actual quantity used.

Example—The total number of fixtures to be supplied by pump in an office building is 120. =120×.75=90. Therefore 90 gallons per minute which pump must discharge. Now to determine the head—The water must be elevated 100 feet and develop a pressure of 20 lbs. The actual head therefore will be 150 feet and to this must be added the distance of suction lift, if any, and allowance for loss of head by friction in pipe. If suction lift is 20 feet—this added to 150 makes a total of 170 and allowance for friction, 10 per cent, makes a total head of 187 feet against which the pump would have to work. The problem worked out in this manner and reference to standard catalogues of pump manufacturers will enable anyone to select the proper equipment.

When the system is of the second type or "booster" system—the head against which pump will work is determined by the following method:

Pump location to highest fixture....	100 feet
Range from minimum to maximum pressure	100 "
	200 "
Deduct City pressure 25 lbs. in feet —60	60 "
Pump required for a total head of..	140 feet

Compression tanks should be installed of such size that the cycles of pump operation do not exceed three to four per hour. To insure this condition the tank should have a storage capacity of 25 to 30 times the capacity of pump per minute. To illustrate for a pump of 90 gallons per minute:—30 × 90=2700 gallons per tank—1/3 to 1/2 of the storage capacity of tank should be filled with air—at maximum working pressure.

Where the city pressure is not constant and less than 20 pounds, it is advisable to install a surge tank to which the suction end of the pump is connected. The suction line to be provided with a gate valve and two horizontal check valves. The supply to the tank should not be less than 2" and the supply controlled by a float valve of approved type. The storage capacity of the surge tank should be at least ten times greater than the delivery capacity of the pump in gallons per minute.

The surge tank may be of cypress or steel and provided with a removable cover and also a 2" drain connection or larger, valved and connected to sewer.

Where the city water pressure exceeds 20 pounds and is constant, the surge tank may be omitted if the pump is of the centrifugal or turbine type, and the suction pipe may then be connected to the main service direct. However, in this case, the cross sectional area of the main service pipe must be at least 50 per cent greater than the area of the suction pipe of the pump.

Suction pipe connections must be provided with two check valves and a gate valve.

For large installations duplicate pumps should be installed.

Service Connection to Building:

For water service connection to buildings under 2" in size extra strong lead pipe may be used with corporation stops and goosenecks as required by regulations of the Water Department.

It is now possible to obtain cast iron pipe, 1¼", 1½" and 2" diameter, which is made in 5-foot lengths, hub and spigot pattern or threaded and the use of either one is highly recommended for service connections instead of lead pipe. Wrought iron or steel pipe should never be used for service connections under ground.

For service connections of larger diameter than 2" cast iron water pipe in 12-foot lengths, hub and spigot pattern of proper class or weight to suit pressure should be used.

When cast iron pipe is brought into the building and up through floor, the same should terminate in a flanged end fitting about 12" above floor.

Heavy pattern stop cocks or gate valves provided with heavy tee handle operating rods should be placed on the service connection and provided with cast iron service boxes with cast iron covers. On the inside of the building the main service must be provided with gate valves and arranged for meter connection if required by the Water Department. Provision must be made for supply connections to inside fire standpipes as may be required.

From this point on the supply piping should be of the following material:

Water Supply Piping:

1. Red brass pipe, iron pipe size, properly annealed seamless tubing with red metal cast brass fittings for such work where the utmost durability is desired.

In some cases where it may be desired to effect some saving in cost, the hot water supply and return piping are specified to be of red brass and the cold water supply piping of genuine wrought iron pipe galvanized.

It is suggested that all specifications for work of the best class be drawn covering red brass pipe for the water supply lines, both cold and hot, or at least for the hot water supply and return lines, and then, if desired, an alternate bid may be asked for on the supply piping of genuine wrought iron pipe with galvanized malleable iron beaded fittings.

This will enable the Architect to take advantage of a reduction in cost if the owner desires to make such a saving, and the Architect cannot be criticized for not having specified the best material for the work.

2. Next to red brass pipe the best material to use is genuine wrought iron pipe galvanized.

3. Where cost is the sole consideration and durability of not vital importance, commercial steel pipe, galvanized, may be used.

The following suggestion is made to specification writers on the subject of Genuine Wrought Iron and Steel Pipe.

Most frequently the specification states that the supply piping shall be of wrought iron pipe and this is the cause of much misunderstanding and frequently the installation of material that is not wanted.

When the phrase, "wrought iron pipe," is used, it is commonly taken for granted by plumbing contractors that the grade of pipe known as "Commercial Steel Pipe," either black or galvanized, will be satisfactory. Perhaps in many cases it will be, but it frequently happens that the Architect intended that genuine wrought iron pipe was to be used. In order to clarify this matter it is suggested that the specifications designate whether the pipe throughout shall be genuine, wrought iron or commercial steel pipe. In case of the former, the words should be added "with the name of the manufacturer stamped on each length."

Flanges and Unions:

All pipe up to 2½" should be provided with galvanized malleable iron unions with brass seats and all pipe over 2½" to be provided with flanged unions having gaskets of asbestos graphite packing $\frac{1}{16}$" thick for best work and rainbow packing for average work.

Valves and Stops:

Valves should be heavy type brass double

gate valves up to 2½" and iron body bronze mounted flanged end for larger sizes. Valves should be of the "rising stem" type for the reason that with this type it can be quickly observed if the valve is "open" or "closed."

Globe valves of approved pattern with heavy brass body and soft rubber discs for cold water lines and composition discs for hot water lines may be used instead of gate valves on pipe from ½" to 1½". For larger pipe, gate valves should be used.

Fittings and Nipples:

Fittings for genuine wrought iron or steel water supply piping should be galvanized malleable iron beaded fittings. Plain fittings must never be used. If the pressure exceeds 75 pounds, extra heavy fittings should be used.

Nipples whether on genuine wrought iron or commercial steel pipe lines should in all cases be of genuine wrought iron. The nipple is the weakest part of any line, and for the very best work, nipples known as close or short pattern should be made of "extra strong" pipe. Genuine wrought iron nipples so stamped may now be obtained from manufacturers and makes it possible for the superintendent to assure himself that the on ractor is furnishing the proper article. t

THE HOT WATER SUPPLY for the building should be determined upon the actual requirements to suit the conditions of each case.

For instance—in the case of a hotel with 100 bath rooms—each containing lavatory and bath tub or shower—the demand for hot water is at a peak load—from 6:30 to 8:00 A. M. and 4:30 to 7:00 P. M. with lesser demands at noon and later at night.

To provide for such service a minimum of 30 to 40 gallons should be allowed for each bath room per hour—this with 100 rooms would mean a heater having a capacity of 3000 to 4000 gallons per hour to which must be added the quantity that will be required for kitchens, laundry, etc.

Generally speaking the following table may be used to determine size of hot water supply systems:

Schools (not boarding):

5—gallons per pupil per day for water used in lavatories,
6—gallons per minute for each shower
or
25—gallons for each pupil using the shower.

Hospital:

50—gallons per day for each person and add 50% of total for kitchen —laundry and general service.

Hotels:

50—gallons per day for each bath room and add 50% of total for general service.

If there is a Turkish bath in connection with the hotel add 100 gallons for each bather—based upon the capacity per hour of the establishment.

Apartments: Allow 100 gallons per day for each apartment having not more than 2 baths, for each additional bath add 25 gallons and 25% of the total for general service.

Factories: Allow 10 gallons for each employe per day for each wash basin and 25 gallons for each employe using showers.

Boarding Schools—Asylums—Homes, etc.: Allow 40 gallons per day for each person. For showers 25 gallons for each user and add 50% of the total for general service.

For smaller installations a hot water storage tank with steam coils for winter service and hot water heater for summer service makes a satisfactory installation. The tank should always be provided with a thermostatic control to prevent overheating the water. Tanks with coils should always have a manhole at one end.

In cases where the heating system is a **vapor** system, the water should be heated by means of a hot water heater the year around, as the pressure of the steam is too low to effectively heat the water by means of steam coils in the tank.

Where showers are used it is desirable to place a thermostatic hot water line control valve in the hot water supply main in order to prevent scalding. It is good practice to separate the system in Hotels, Hospitals, etc., so that the water supplied to bath tubs, lavatories and showers is controlled in this manner. It not only prevents possible scalding but saves fuel and increases the life of valves, faucets, etc., which excessively hot water materially shortens.

In larger installations—especially where both exhaust and live steam (high or low pressure) are available; the hot water system should be arranged in two units; the first a storage tank of proper size, called the primary heater, in which the water is heated by exhaust steam—from this heater it passes to the secondary heater which is provided with coils supplied by live steam under thermostatic control. The latter heater brings the water up to the desired degree of temperature at which the control is set.

Another and most economical type of heater is the instantaneous type—heated by low or high pressure—controlled by an automatic thermostatic device and using only such quantity of steam as necessary to heat the water actually used—to the temperature for which the control is set. This type of heater is very efficient and economical and is especially adapted to large installations as Hotels, Hospitals, Factories and wherever there may be a large variation in the demand for hot water throughout the day or night.

In order to ensure proper results, hot water systems must be perfect in circulation —wherever possible the **overhead** type system should be used with a riser to the top floor—horizontal supply mains and drop supplies to the fixtures on floors below with circulating return in basement. Hot water riser should have an air vent trap at highest point.

Pressure of hot and cold water systems should always be the same.

In some cases circulating pumps are necessary. These should always be of the centrifugal type with low speed motors and if direct current is available, motors should be provided with a variable speed control.

In conclusion of the suggestions for water supply system—I would say that in my experience most mistakes have been made in having the piping system too small and this is especially true in the case of hot water tanks and heaters.

A heater too small for the service will waste more fuel than one too large.

ARRANGEMENT OF TOILET ROOMS AND PLUMBING FIXTURES.

Few Architects realize how much the cost of the plumbing and heating on a building is governed by the design and location of toilet rooms. Many buildings are up several stories before the location of pipe chases or shafts are decided upon and many botched up piping jobs are the result of this neglect.

This again brings up the great need of proper plumbing plans and diagrams—showing the proper size and location of the piping and permitting the general contractor to provide chases in walls—leave openings in floors and provide pipe shafts of proper size for the work.

In residences with wood studs the partition carrying soil pipe must have at least 6" studs and a still better arrangement is to have a hollow space and use 4 or 6" studs flat wise and framed once or twice in their height as this saves cutting of studs for horizontal vent pipes.

If partitions are hollow tile, 6" thick tile should be used. Thin partitions of gypsum materials make very unsatisfactory parti-

tions for concealment of piping, as no secure anchorage can be had in same for bolts to fasten hangers or brackets for fixtures; furthermore, condensation on pipes dissolves sulphuric acid in gypsum and induces quick corrosion of metal.

With buildings of fireproof construction in which the floors are of reinforced concrete

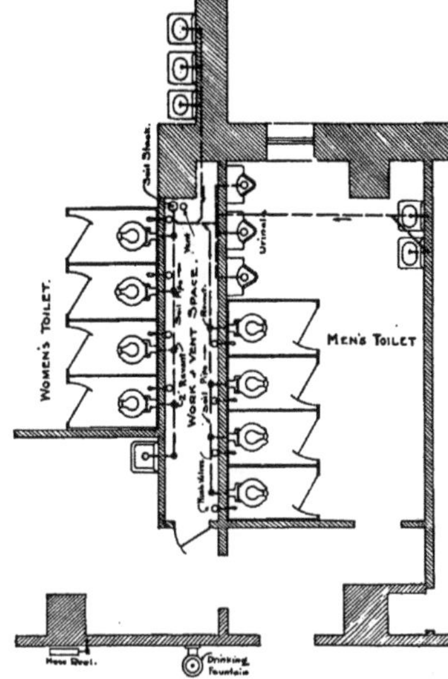

Toilet Rooms with Work-Vent Space between Same.

the location of bath and toilet rooms must receive careful study.

There are three schemes that may be used.

The first, a pipe shaft 2'-6" to 3' in width extending up through the building—in which all piping may be placed and fixtures all provided with wastes and supply connections to wall. (See illustration.) This arrangement is very desirable for Hospitals, Schools, Hotels, Office Buildings, etc.; it makes an ideal arrangement and is economical in cost of installation and maintenance. All pipe being exposed it is easily gotten at in case of repairs.

The second is to raise the floor of toilet rooms 7" to allow for piping being concealed in floor. This is sometimes objectionable and in the case of Hospitals, Homes and Institutions should not be done.

The third is to run the piping under the ceiling of room below—either exposed or concealing the same by furring down the ceiling.

In planning toilet rooms it is most important to ascertain the exact size of the various fixtures that are to be installed—so that these will be placed properly and to the best possible advantage.

This is especially necessary in the case of bath tubs and shower stalls. If recessed tubs are used, the exact length overall, distance the ends and back will extend into wall must be considered as there is always a difference between the nominal size of bath tub and their actual overall length, the end at which the waste and supply fixtures are to come should be shown and a paneled door of proper size provided so that the fittings can be properly installed and accessible in case of repairs. When recess tubs are used—it is always desirable to tile around the top of tub, as this makes a more permanent installation than a finish of hard plaster.

Shower stalls should never be less than 3'-0" x 3'-0" inside for a comfortable stall. 3'-2" x 3'-2" is the standard size adopted by plumbing manufacturers and should be used wherever possible. Stalls should be at least 6'-6" high. Solid porcelain receptors, grooved to receive marble partitions are the best and are absolutely leakproof. If marble floor slabs are used they must not be less than 2" thick and should be grooved all around to receive marble partitions.

The placing of sheet lead flashing underneath marble shower slab or tile for shower stalls on upper floors has been discontinued for the reason that very often the weight of the stall above same cracked the sheet lead

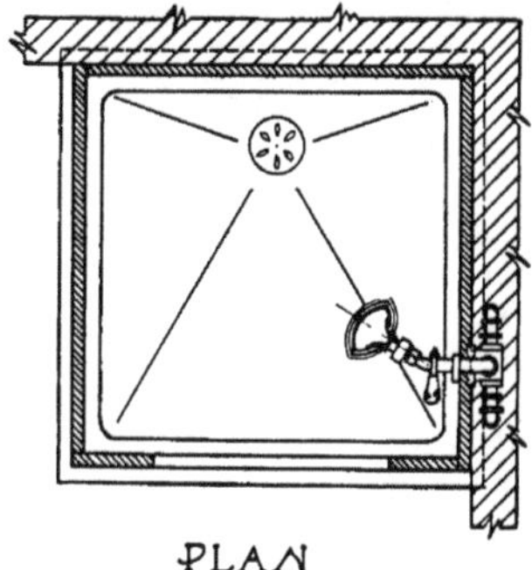

PLAN

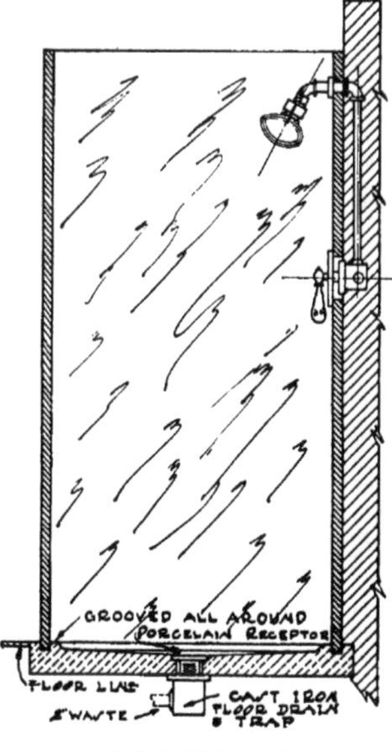

SECTION

Wherever There's A DWELLING *There's A* **MORTON CABINET** *to Fit The* STYLE *and* PURSE

"We're Specifying Morton Cabinets." Today,—as in many years of the past that brief phrase has marked the conclusion of investigation, comparison, and analysis of design, construction, and cost in the bathroom cabinet field.

MODEL R-26

THERE'S DE LUXE . . . trim, smart, modern,—yet restrainedly dignified. Morton De Luxe enhances and accents the finest bath rooms. Among the smartly designed models there is always one or more which strike just the right note even in the most individual bath room designs. De Luxe Cabinets have cushioned door . . . steel welded in one unit . . . "Mortelite" enamel finish . . . mirrors of finest plate glass with copper processed backs. Hinges are piano type,—chromium plated. Removable bulb-edged glass shelves. Removable razor blade box and aluminum tooth brush holder are conveniences. A 5-year guarantee to each cabinet mirror.

STANDARD

THEN STANDARD . . . a wide variety of sound, smart designs. As in De Luxe,—cushioned door, welded one-unit cabinet, "Mortelite" baked enamel finish, highest quality plate glass mirror with copper processed back. Stainless steel ratchet type clip holds mirror in place . . . bulb-edged glass shelves . . . piano type hinges and door stop. Removable razor blade box, aluminum tooth brush holder. Models with tubular lights carry mountings in chromium finish brass. The opalescent glass light tubes are enclosed top and bottom. Again,—a 5-year guarantee.

ECONOMY

SERVICE

ECONOMY . . . Sturdily built but cheaper. Welded in one unit. "Mortelite" baked enamel. Crystal glass mirrors with copper processed backs held by ratchet clips. Razor blade drop and concealed hinges. Bulb-edged glass shelves.

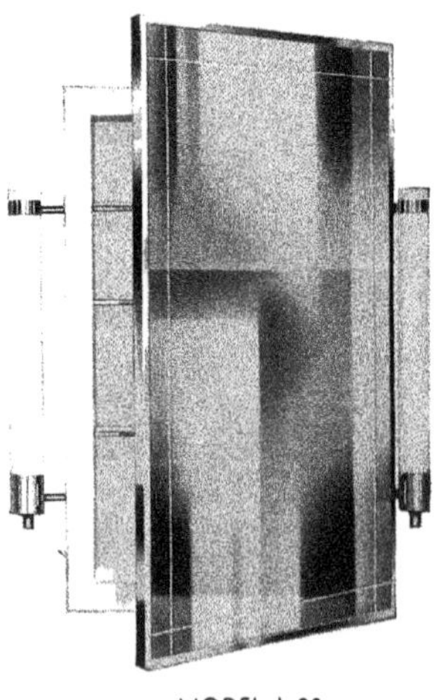

MODEL L-20

SERVICE . . . Low priced to meet the demand of the small home, apartment, or institution. Nevertheless, attractive in design. One-unit welded. "Mortelite" enamel. Plate glass mirrors with copper processed backs. Bulb-edged glass shelves, rust proof hinges.

MORTON MANUFACTURING COMPANY

5105-43 WEST LAKE STREET — CHICAGO, ILLINOIS

so that the installation of same as a means to prevent leaks—was a useless expense.

The best material to use for water proofing under marble or tile shower slabs is to build up three or four thicknesses of genuine asphalted felt well lapped and swabbed with asphaltum and the edges of the felt turned up at least 6" high at side walls. This is an inexpensive method and far more satisfactory than sheet lead.

A word of caution in connection with the kind of material to be used for shower stalls. Marble, slate, vitreous tile and salt-glazed brick are recommended as suitable, but the use of steel for shower stalls is not recommended unless stainless steel of approved quality or Monel metal, both of proper gauge, with welded or brazed joints is used.

It is desirable to place the controlling valves to shower head on one side of the stall near the entrance (see illustration page 447), so as to permit the water being turned on and tempered without wetting the bather. When a stop valve is placed in the supply to the shower head, it will be necessary to provide the hot and cold water supplies with check valves to prevent the bypass of water from either side in event that the valves on inlet of shower are not entirely closed. When there is no valve between the inlet valves and the shower head, check valves are not absolutely necessary.

All shower heads should be placed 6 feet above floor for adults and 5' 6" for school showers provided with an adjustable ball joint by means of which the angle of the shower head may be changed as desired. Shower heads arranged in this manner give better results and will not wet the bather's head unless he so desires.

When thermostatic or anti-scalding shower valves are used, it is always desirable to place on the hot and cold water supply line for each, a loose key compression shut off by means of which the supply can be controlled, which is necessary if the pressure is very high or the pressures of hot and cold water are not equal.

When plain compression type control valves for showers are used in place of thermostatic or anti-scalding valves—a thermostatic hot water control valve should be placed on the hot water supply line to the showers and set so as to prevent the hot water exceeding 110° F. in temperature.

The placing of plumbing fixtures against outside walls should be avoided. It is very unsatisfactory. Even if the supplies are carefully covered there is always danger of freezing. The custom of placing bath tubs under outside windows is most objectionable. This has been commonly done in apartment house work. A little study of grouping would have produced better results.

In public toilet rooms the arrangement of water closet stalls must be well considered. Where a number of these are to be installed the size of the stalls must be determined. The adopted standard width is 2'-6" centers for schools—they should not be less—but may be more. For adults the stalls should be 2'-10". Three (3) feet is the greatest width that should be used. To make them wider would be waste of space. The depth inside should not be less than 4'-6" with doors swinging in. This depth will allow the standard width—2 foot door to well clear the front of the closet bowl.

In factory, etc., and school work, especially primary grades, it is better to omit doors entirely and in this case the stalls need not b more than 3 feet, or at the most 3'-6" in depth.

If possible all flush tanks, piping, etc., should be concealed in a work space in rear of closet stalls. The wall of work space being formed by the backs of partitions or a built up wall as desired. Frequently this same work space is also utilized as a vent space, providing the back of each stall with a vent opening, protected by a ventilating hood or register face. This makes a most desirable arrangement for ventilating large toilet rooms—especially adapted for schools, asylums and all public toilet rooms.

The water closet stalls may be of marble, slate or steel according to the class of work. Steel partitions are very satisfactory and excellent for school and factory work.

The bottom of all partitions should be 12 inches above the floor. When marble is used the pilasters should be 1½ or 2 inches thick and grooved to receive the partitions. The backs should be cut out to receive the partitions and a top rail of marble corresponding in thickness to the pilasters and 3½ or 4" high extend along the entire front. The bottom of rail should not be less than 6'-6" high for schools and 7 feet for public toilet rooms. This arrangement does away entirely with brass floor and top standards and

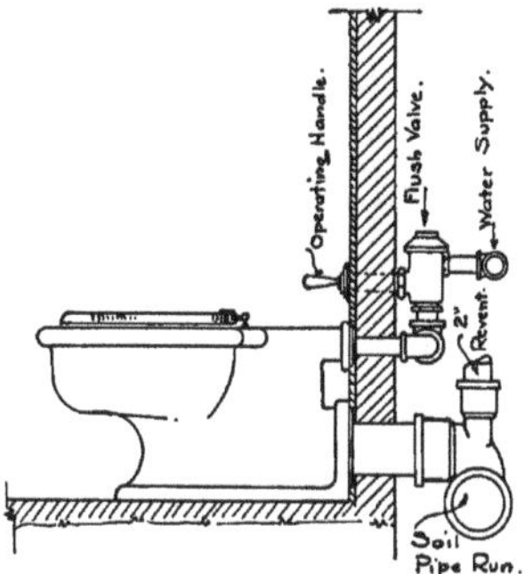

WATER CLOSET- WALL OUTLET.

all metal angles—very desirable for the reason that nickel plated brass work becomes tarnished very quickly and is rarely given the care it requires to keep the same in good condition.

If wood doors are used they should preferably be of the type known as "sanitary", perfectly flush without panels. The standard size is 2 feet wide, 5 feet high and 1⅛" thick. They should be provided with an adjustable N. P. box spring hinge and blank with check, door latches and stops and should always swing in, with spring set to hold the door open when not in use.

While on the subject of water closet stalls a word of caution regarding the floor is apropos. It frequently happens, especially in school and factory work, that the floors of toilet rooms are pitched toward a floor drain and whenever this is done the contractor doing the flooring work should be cautioned to keep that portion of the floor on which the water closets are to set perfectly level and establish his break line at least 3 inches forward of the front of the base of the water closet bowls. Unless this is done the plumber when setting the bowls will level them up with cement in order to obtain an even bearing and the cement under the base of the bowls either causes them to crack on account of unequal expansion and contraction or because of improper support throughout the entire base, the unequal strain on the ware will cause cracks.

For connecting water closets with floor outlets to soil pipe or fittings only cast iron bends of an approved type should be used—with a gasket of asbestos, graphited.

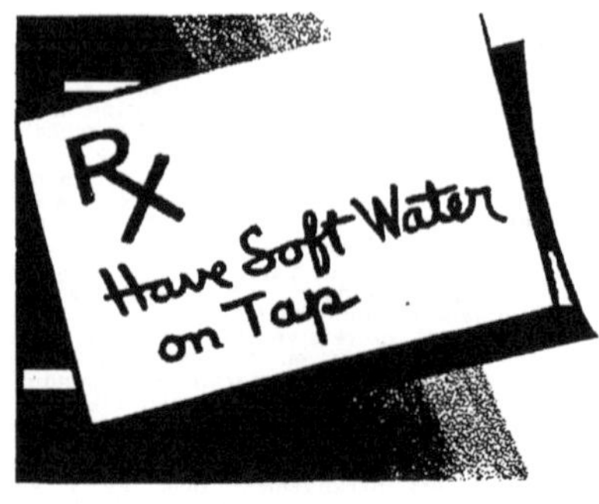

TRADE MARK REGISTERED

WATER SOFTENER

MODEL DS

Direct or internally salted type—highly efficient for small homes and bungalows and moderately hard water. "Easy" regeneration because of low height and equipped with permanently attached cast metal salt funnel. Made in five sizes—priced from $65.00. Bulletin DS-10 fully describes.

MODEL S. V. A.

Single Valve semi-automatic regeneration—avoids frequent handling of salt as salt tank has sufficient capacity for several months service, and salting is, by means of brine fed by an automatic ejector—regeneration is personally timed. Made in nine sizes—priced from $85.00. Bulletin S.V.A. 10 fully describes.

MODEL T. A.

Timed—"Measured Regeneration"—no handling of salt—no waiting—no confusing valves—just set the "RAINBOW TIMER" and all is done—20 seconds of attention at infrequent intervals produces continuous soft water service. Made in seven sizes—priced from $150.00. Bulletin T.A. 10 fully describes.

MODEL F. A.

Also "Measured Regeneration" without any personal attention—regenerates "on time" every 24 hours at 3 A.M. The ultimate in soft water service—the finest the market affords. Made in seven sizes—priced from $190.00. Bulletin F.A. 10 fully describes.

Specifications for installation are based on a diagnosis of each job—

- Water analysis
- Source of water
- How used
- Volume required
- Fixtures to be served
- Persons to be served
- Size of structure

All are considered to determine the type and size of RAINBOW Water Softener best suited to meet requirements, including the installation of RAINBOW Filters for iron, taste or odor removal, where necessary.

We will be pleased to have you submit any of your clients' water treating problems to us for consideration—without obligation.

ZEOLITE ENGINEERING CO.

INCORPORATED 1924

ZEOLITE WATER SOFTENERS · PRESSURE FILTERS

4640 RAVENSWOOD AVENUE CHICAGO

Connection between waste outlets of water closets with wall outlets and soil pipe should be made by means of heavy combination lead and iron ferrules, one end caulked into hub of cast iron fitting, the other wiped to a brass closet flange and asbestos graphite gasket.

This method of connection does not cover the wall hung type closets now being used of which each special type is provided with a special designed form of attachment to the soil pipe or stack.

Now as to the type of water closets to be used. There are today practically only two styles—one known as a siphon jet bowl, the other a washdown with jet. There are of course a large number of various special type bowls in the market but they are modifications of the above types.

The siphon jet bowl is the best to use on account of its more quiet action in flushing and also for the reason that the interior of the bowl presents less fouling surface, owing to the larger water surface.

The greater the cross-sectional area of the siphon limb the better the operation of the bowl. The minimum diameter of the siphon limb should be 2½" and 3" is better. The more uniform the passage is the less danger of stoppage. **All bowls should be tested out under water before shipment by the manufacturer—for two reasons: one to determine whether the ware is free from cracks—called "dunts" by the potteries, the other to be certain that the construction of the bowl is perfect.**

In many localities the water contains incrusting ingredients that may cause clogging up of the jet tubes in time. Such conditions may be remedied by emptying the water contained in the bowl and pouring a pint or more of "Commercial" Muriatic acid into the bowl. **The acid will dissolve the solids in the jet opening in about ½ to ¾ of an hour.**

However, where the water is extremely bad—it is advisable to use the washdown type of bowl with jet, which is not as apt to become stopped up as the jet openings are larger than in the siphon jet type and the tube has no pocket in which deposits can accumulate.

Where it is necessary to practice economy in the selection of fixtures—it is advisable to use washdown water closets with jets. For Schools and Factories this style is generally used.

There is another type of closet used today which is a composite of the siphon jet and washdown bowls. This bowl is known as the "reversed trap type" and when correctly designed and properly made, makes a very satisfactory closet. It has less fouling surface than the washdown bowl and is siphonic in action.

The conditions that are to be met in each case must necessarily determine the particular kind of closet that should be used. Also whether the bowls should have extended lips, floor or wall outlets, have low down tank, or flush valves or flushed automatically by seat operating valves. No fixed rule may be prescribed for such selection, which can only be made according to requirements of the work itself.

In the selection of water closets consideration must be given as to the manner in which the closets are to be flushed. Water closets with high tanks or low down tanks require a ½" supply connection, whereas these fixtures if operated by means of flush valves—require 1 to 1¼" supply connections to each flush valve. Water closets with automatic seat operating valves require ½" supply connections as a rule.

Where there is more than one water closet in a row or battery, the main supplies for such battery must be of a size that will adequately supply all fixtures. Reference to the following table will be of service:

Table of Branch Supplies for Water Closets.

The following table will be of service to determine the proper size of branch supplies for water closets from 1 to 12 fixtures in a battery. The size of pipe is based on a pressure of from 20 to 40 pounds.

For Automatic Seat Operating Water Closets or Water Closets with Low-Down or High Tanks. Inlets ½".

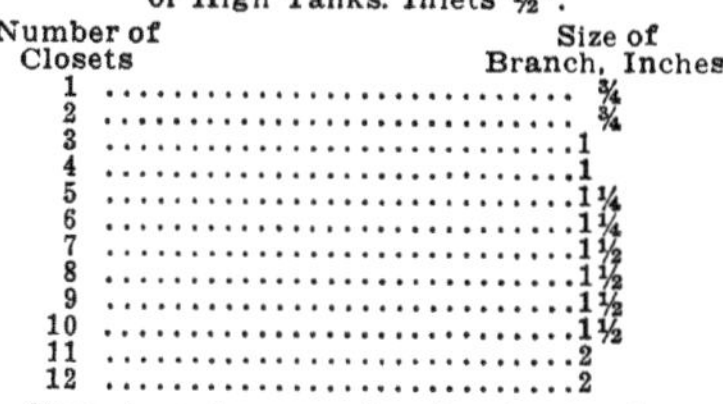

Number of Closets	Size of Branch, Inches
1	¾
2	¾
3	1
4	1
5	1¼
6	1¼
7	1½
8	1½
9	1½
10	1½
11	2
12	2

Each branch connection to closet valve or tank shall be ½ inch.

For Water Closets with Flush Valves Having 1¼" Inlets.

Number of Closets	Size of Branch, Inches
1	1¼
2	1½
3	2
4	2½
5	2½
6	2½
7	2½
8	2½
9	3
10	3
11	3
12	3

Each branch connection to flush valve shall be 1¼ or 1 inch, according to style of valve used.

In order to prevent water hammer in pipe lines, it is desirable that the supply connection for each flush valve be provided with an air chamber full size of the branch supply and at least 2' in length. In case of a battery of closets, the air chamber may be placed at the extreme end of the horizontal supply branch, and in this case should be at least 2" in diameter and 2' in length for a battery of six to ten. If there should be a double battery of closets placed back to back, the supply branch for each battery should be connected together and the air chamber placed at this point. In that case it would be well to provide an air chamber larger in diameter and at least 4' long.

Refer to table of "Delivering Capacities of Pipes" on Page 671 for sizes of branches where inlets are other than ½" or 1¼".

The water pressure must also be carefully considered for flush valves and automatic seat operating valve closets. For the former the minimum should be 15 lbs., and for the latter 20 lbs., at each bowl.

Consumption of water is another item to be considered. Tank closets will use 6 to 8 gallons per flush: those with flush valves from 6 to 8 gallons according to the pressure and automatic seat operating closets will only use 2½ to 3 gallons per flush.

Now regarding Urinals—At present there are three types.

First: The vitreous or solid porcelain 18" wide, low type and solid porcelain 18" and 24" wide, "full height," with lipped extension base. These urinals are usually set into the floor with lipped base flush with finished floor.

Second: The vitreous pedestal type, either washdown or siphon jet, and third: the vitreous or enamel iron wall hung, washdown or siphon jet type.

The first and third types may be obtained with local vents, a very desirable feature which serves the purpose of removing objectionable odors, when connected to a properly designed ventilating system.

Urinals should preferably be provided with automatic flush tanks. Individual flush valves are not desirable as urinals so equipped are never properly flushed.

When setting solid porcelain urinals into the floor a depth of 5" is required to bring the top of the drip receptor flush with the finished floor. Care should be taken to set these in accordance with instructions of the manufacturers. They must never be solidly set in a cement grout; an inch or more of dry sand should be put under same and a strip of expansion joint composition placed on the front edge and exposed sides so the concrete sub-base of floor will not adhere. The finished tile, terrazzo or cement may be run up against the expansion strip.

Regarding the other fixtures such as bath tubs, lavatories, sinks, slop sinks, etc., space will not permit going into details. The catalogues of manufacturers generally give all information necessary regarding same.

Obviously, the main factor determining the selection of fixtures should be quality, which implies the longest possible service and therefore the cheapest in the end. Fixtures in order to prove satisfactory must possess strength and the greatest resistance possible to effects of alkaline and mineral waters and other liquids or material likely to cause stains, or prove destructive to the glazed or enameled surface with which they come in contact. It is always the duty of the architect to make such recommendations to his client as will absolve him from all blame where defects develop in material, that has been used against his recommendation.

The class of the work in question should determine the character of the material to be used throughout, and naturally this will equally apply to the plumbing fixtures.

Vitreous ware of the best quality, is conceded the best and most durable material to use. In addition to water closet bowls and urinals of certain types, it is furnished for lavatories, drinking fountains, slop sinks and kitchen sinks in certain sizes.

Solid porcelain can be obtained for larger fixtures, such as bath tubs, stall urinals, slop sinks and kitchen sinks. Careful selection is necessary in order to obtain the best and while the danger of crazing of the surface glaze has been greatly reduced by improvements in methods of manufacture, it still remains a point to consider in making a selection.

If enameled iron sinks are used these should be of the grade known as "Acid resisting enamel."

Monel metal and certain brands of 18.8 stainless steel of tested non-corrosive merit—are excellent materials for sinks, drain boards, and table tops and especially well adapted for modern "built-in" fixtures and for kitchens in hospitals, hotels and institutions, these materials are superior to anything on the market today. However the use of untested stainless steel is cautioned against and only material of proven merit insofar as corrosive action is concerned, proper gauge in thickness and perfect construction of seams should be used. No material should be specified unless it has been proven by analysis and laboratory tests to be satisfactory for the purpose for which it is to be used.

By all means provide a clause in your specifications that all fixtures must be stamped with the name of the manufacturer and each piece properly labeled as to quality and showing individual inspection.

Brass Goods:

In order that the Architect may be assured of obtaining durable material great care should be used in the selection of the brass goods which include faucets, bibbs, stops, and supply piping for lavatories, bath tubs, showers, sinks, etc.

For first class work the following clause should be inserted in the specification.

"All brass work shall be red metal brass of a composition in accordance with the Navy Department standard—which is 85% copper, 5% tin, 5% zinc and 5% lead. All tubing such as flush pipes, etc., shall not be less than No. 14 gauge and all supply pipes shall be full iron pipe size, annealed, red brass. All nickel plated work shall be of the highest quality and subjected to the nickeling process for a period of not less than one hour. All faucets, valves and bibbs shall be provided with stems having movable loose discs. Discs to be of special hard fibre and not so called composition and all discs must have edges encased by a brass protecting rim.

Chromium plated metal is now being extensively used in place of nickel plated work. For the very best and most permanent untarnishable material "nickel silver" remains without equal.

"For all concealed valves or stops the operating part must be removable from face of wall and the discs the same as above and the seats of removable, renewable, type."

While on the subject of stops a few words are apropos. The supply connections to every fixture should be provided with stops either exposed or concealed type. The initial cost of stops is very small when compared to the cost of repairing damage to floors, plaster, ceiling and decoration caused by leaks.

Cost today, more than ever, is an important factor in considering the quality of plumbing fixtures that should be used. However, it would be very poor judgment to sacrifice quality of material in any line on account of cost. The work of the Architect is not for today, but for tomorrow, and he who builds well in all things will profit more than one who builds poorly; and hence, now more than ever skill in design and knowledge of materials and their proper use will be required of the Architect to secure results.

Nothing will cause as much annoyance and require as constant repairs as a poorly designed and cheap installation of plumbing. Repair bills are a constant reminder to the owner of mistakes made by the Architect, who failed to give in full the service for which he was paid.

CALL ALLEN FOR INTERIOR FIRE PROTECTION EQUIPMENT

Pioneers Since 1887 in the Improvement of Fire Protection Equipment, and Allen is STILL LEADING THE FIELD. The Name ALLEN means APPROVED, MODERN Interior Fire Protection at Competitive Prices. Leading Architects and Engineers Insist on Installing Allen's Approved Fire Protection Equipment as it Assures Their Clients of the Lowest Insurance Rates.

Chicago Standard Fire Hose Unit—Allen's Plate 1.

Allen's 'SAFETY' Concealed (Flush Type) Siamese is the modern fire department connection — pleasingly simple in design but rugged enough for the tallest building—and it's approved by all fire department engineers. It's SAFE for the pedestrian as this siamese extends only $4\frac{1}{4}$ inches from the wall. Made with sillcock as illustrated or without sillcock. Allen's Siamese are made of either red bronze or white nickel bronze.

Refer to Sweet's Catalog File for further information, or write for Catalog 245—Allen On Interior Fire Protection Equipment. This book contains complete data, specifications and working drawings on interior fire protection equipment.

ESTABLISHED 1887

W. D. ALLEN MANUFACTURING CO.

566 WEST LAKE STREET, CHICAGO, ILLINOIS **RANdolph 8181**

SOILWASTE AND VENT PIPES IN PLUMBING SYSTEMS

By THOMAS J. CLAFFY

This article is written with the object of presenting information in a convenient form relative to pipe sizes in plumbing systems, for the benefit of those who are interested. More detailed information may be obtained in the report of the Committee, hereafter referred to.

It has long been a recognized fact that definite information was lacking on which to base proper pipe sizes for soil, waste, drain, and rain water pipes within buildings. Architects and Engineers especially have been aware of the lack of uniformity in design and method of installing fixtures and piping systems in plumbing.

The war hastened a concentration of forces to bring about standardization in this respect. Early in 1921, Secretary of Commerce Herbert Hoover set in motion the work of a Committee appointed to bring about a standardization of building practices. Sub-committees were appointed, and the work of bringing about more uniformity of design and practice in plumbing was allotted to a sub-committee composed as follows:

George C. Whipple, Chairman,
Professor of Sanitary Engineering,
Harvard University.

Harry Y. Carson, C. E.,
Research Engineer, American Cast Iron Pipe & Foundry Co.,
Birmingham, Alabama.

William C. Groeniger,
Consulting Sanitary Engineer,
Columbus, Ohio.

Thomas F. Hanley,
National Association Master Plumbers,
Contracting Engineer, Chicago, Illinois.

A. E. Hansen,
Hydraulic and Sanitary Engineer,
New York City, N. Y.

James A. Messer,
President, James A. Messer Company,
Washington, D. C.

Albert L. Webster,
Consulting Engineer,
New York City, N. Y.

William J. Spencer,
Secretary-Treasurer, Building Trades Council,
American Federation of Labor, Washington, D. C.

Meetings were held at Washington, and after being duly organized, the Committee decided that its first duty was to define "plumbing." The discussion leading up to this is well worth reading. The definition is:

"Plumbing is the art of installing in buildings the pipes, fixtures, and other apparatus for bringing in the water supply and removing liquid and water-carried wastes."

The Committee's report says that—

"The air in sewers and drains often contains gases resulting from the decomposition of excreta, soap, fats, and other wastes, together with gases from mineral oils which may come from garages, streets, and industrial establishments. Illuminating gas may also find its way into sewers through leakage. Among these gases may be found methane, sulphuretted hydrogen, and carbonic oxide. In large amounts these gases are poisonous to the human system, and there are physiological objections to breathing them even in small quantities. Hence, the air of sewers or drains should be kept from entering buildings intended for human habitation or occupancy by the use of proper plumbing installations and by suitable ventilation of the rooms or compartments in which the plumbing fixtures are located. The smell of these gases and other emanations from decomposing organic matter is naturally repugnant to human beings. It not only offends the sensibilities, but may produce shallow breathing, headache and even nausea.

"In addition to the above facts, it is important to consider the bacteriological aspects of sewer and drain air, a subject upon which there has been some misunderstanding. In recent years bacteriologists have made studies which have thrown light upon this subject. They have shown by experiment that while sewage often contains disease-producing bacteria derived from human excreta and body wastes, these bacteria are rarely found in the air which escapes from sewers and drains. Hence, it has been argued by some that escaping sewer air has no influence on health. The committee does not agree with this conclusion. Health may be influenced by factors which do not cause specific diseases, for there are chemical and physiological as well as bacteriological factors involved. The investigations thus far made by bacteriologists should be considered to be merely a beginning of larger and more complete investigations, which will doubtless be made as the science of bacteriology advances. The committee is of the opinion, therefore, that until further light on this somewhat obscure subject has been obtained, the escape of sewer air from the house-drainage system, at frequent intervals or in considerable quantities, threatens the health of the building's occupants.

"The temporary losses of water seal in traps, which rarely occur and which are immediately replaced, do not involve any great danger to the health of the occupants, * * * * * but where a loss of seal is likely to be of frequent occurrence and not readily replaced, or where breaks in the system admit sewer air continually to a building, the health of the occupants is subject to the dangers heretofore described.

"For the above mentioned reasons, regulations governing the installation of plumbing have been established by law in many places. These regulations have been potent in improving living conditions throughout the country; in fact, they have even set the standards for those places where plumbing is not under public control.

"The committee believes that good plumbing is a matter which concerns health. Government has the right to protect the people's health, * * * * *"

This is the basis upon which is built all laws, rules, and regulations concerning plumbing installations within buildings.

The National Plumbing Code, prepared and submitted in the report of this Committee, is well worth a place in every architect's library. It is designed so as to apply in every part of the United States, and is national in its scope. The primary object was to standardize plumbing in small residences. That this has been well done is obvious.

In the report of this sub-committee as submitted for the year ending December, 1929, certain revisions are made in the tables of pipe sizes which are more in harmony with present practices.

The members of this committee are:

*George C. Whipple, Chairman.
Harry Y. Carson.
William C. Groeniger.
August E. Hansen.
J. L. Murphy.
John M. Gries, Chief of the Division of Building and Housing.
George N. Thompson, Secretary.
R. B. Hunter, In charge of Tests.

*Deceased.

Experiments.

Numerous practical experiments were carried out by the Bureau of Standards at Washington and by the Department of Sanitary Engineering at Harvard University, which aided this Committee in forming its conclusions. These experiments confirmed the findings of those whose experiences in tall building construction have been previously related.

The tables of pipe sizes are on a fixture unit basis, which is the most convenient form to use.

Table of Pipe Sizes.

"Fixture Unit: The following table based on the rate of discharge from a lavatory as the unit shall be employed to determine fixture equivalents:

	Fixture Units
One lavatory or wash basin...........	1
One kitchen sink	1½
One bathtub	2
One laundry tray......................	3
One combination fixture................	3
One urinal	3
One shower bath.......................	3
One floor drain........................	3
One slop sink.........................	4
One water-closet	6

"One bathroom group consisting of one water-closet, one lavatory, one bathtub, and overhead shower, or one water-closet, one lavatory and one shower compartment, shall equal 8 fixtures."

"One hundred and eighty square feet of roof or drained area in horizontal projection shall count as one fixture unit."

Vents, Required Sizes.

The required size of main vents or vent stacks shall be determined from the size of the soil or waste stack vented, the total number of fixture units drained into it, and the developed length of the vent, in accordance with the following table, interpolating when necessary between permissible lengths of vent given in the table:

Maximum Permissible Length of Units (in feet) for Soil and Waste Stacks.

Diameters of soil or waste stack (inches)	Number of Fixture Units	DIAMETER OF VENT (In Inches)									
		1¼	1½	2	2½	3	4	5	6	8	10
1¼................	1	45	.....	.....	.....	.....	.....	.....	.....	.....	.....
1½................	Up to 8	35	60	.....	.....	.....	.....	.....	.....	.....	.....
2.................	Up to 18	30	50	90	.....	.....	.....	.....	.....	.....	.....
2½................	Up to 36	25	45	75	105	.....	.....	.....	.....	.....	.....
3.................	12	.....	34	120	180	212	.....	.....	.....	.....	.....
3.................	18	.....	18	70	180	212	.....	.....	.....	.....	.....
3.................	24	.....	12	50	130	212	.....	.....	.....	.....	.....
3.................	36	.....	8	35	93	212	.....	.....	.....	.....	.....
3.................	48	.....	7	32	80	212	.....	.....	.....	.....	.....
3.................	72	.....	6	25	65	212	.....	.....	.....	.....	.....
4.................	24	.....	.....	25	110	200	300	340	.....	.....	.....
4.................	48	.....	.....	16	65	115	300	340	.....	.....	.....
4.................	96	.....	.....	12	45	84	300	340	.....	.....	.....
4.................	144	.....	.....	9	36	72	300	340	.....	.....	.....
4.................	192	.....	.....	8	30	64	282	340	.....	.....	.....
4.................	264	.....	.....	7	20	56	245	340	.....	.....	.....
4.................	384	.....	.....	5	18	47	206	340	.....	.....	.....
5.................	72	.....	.....	.....	40	65	250	390	440	.....	.....
5.................	144	.....	.....	.....	30	47	180	390	440	.....	.....
5.................	288	.....	.....	.....	20	32	124	390	440	.....	.....
5.................	432	.....	.....	.....	16	24	94	320	440	.....	.....
5.................	720	.....	.....	.....	10	16	70	225	440	.....	.....
5.................	1,020	.....	.....	.....	8	13	58	180	440	.....	.....
6.................	144	.....	.....	.....	.....	27	108	340	510	.....	.....
6.................	288	.....	.....	.....	.....	15	70	220	510	630	.....
6.................	576	.....	.....	.....	.....	10	43	150	425	630	.....
6.................	864	.....	.....	.....	.....	7	33	125	320	630	.....
6.................	1,296	.....	.....	.....	.....	6	25	92	240	630	.....
6.................	2,070	.....	.....	.....	.....	4	21	75	186	630	.....
8.................	320	.....	.....	.....	.....	.....	42	144	400	750	900
8.................	640	.....	.....	.....	.....	.....	30	86	260	750	900
8.................	960	.....	.....	.....	.....	.....	22	60	190	750	900
8.................	1,600	.....	.....	.....	.....	.....	16	40	120	525	900
8.................	2,500	.....	.....	.....	.....	.....	12	28	90	370	900
8.................	4,160	.....	.....	.....	.....	.....	7	22	62	252	840
8.................	5,400	.....	.....	.....	.....	.....	5	17	52	212	705

Soil and Waste Stacks.

"Every building in which plumbing fixtures are installed shall have a soil or waste stack, or stacks, extending full size through the roof. Soil and waste stacks shall be as direct as possible and free from sharp bends and turns. The required size of a soil or waste stack shall be determined from the distribution and total of all fixture units connected to the stack in accordance with the following table, except that no water-closets shall discharge into a stack less than 3 inches in diameter."

MAXIMUM FIXTURE UNITS ON ONE STACK

DIAMETER (inches)	With "Sanitary T" Inlets		With all 45° Y or "Combination Y and One-eighth Bend" Inlets		Maximum length, including extension as vent
	In one branch interval[1]	On any one stack	In one branch interval[1]	On any one stack	Feet
1¼	1	1	1	1	50
1½	2	8	4	12	65
2	9	16	15	36	85
3	24	48	45	72	212
4	144	256	240	384	300
5	324	680	540	1,020	390
6	672	1,380	1,122	2,070	510
8	2,088	3,600	3,480	5,400	750

[1]The term "branch interval" shall be interpreted to mean a vertical length of stack, not less than 8 feet, within which a branch or branches are connected, and the total fixture units on all branches connected to a stack within any 8-foot length shall not exceed the maximum permitted by the table in one "branch interval."

Branch and Individual Vents.

No vents shall be less than 1¼ inches in diameter. For 1¼ and 1½ inch wastes the vent shall be of the same diameter as the waste pipe, and in no case shall a branch or main vent have a diameter less than one-half that of the soil or waste pipe served, and in no case shall the length of a branch vent of given diameter exceed the maximum length permitted for the main vent serving the same size soil or vent stack.

Size of Drains, Sewers, and Horizontal Branches.

The required size of sanitary house drains, sanitary house sewers, and horizontal branches shall be determined on the basis of the total number of fixture units drained by them in accordance with the following table:

SANITARY SYSTEM ONLY

Diameter of Pipe (Inches)	Maximum Number of Fixture Units			Diameter of Pipe (Inches)	Maximum Number of Fixture Units		
	Slope ⅛ inch fall to 1 foot	Slope ¼ inch fall to 1 foot	Slope ½ inch fall to 1 foot		Slope ⅛ inch fall to 1 foot	Slope ¼ inch fall to 1 foot	Slope ½ inch fall to 1 foot
1¼	1	1	1	5	162	216	264
1½	2	2	3	6	300	450	600
2	5	[1]6	[1]8	8	990	1,392	2,220
3	[2]15	[2]18	[2]21	10	1,800	2,520	3,900
4	84	96	114	12	3,084	4,320	6,912

NOTE—[1]No water-closet shall discharge into a drainpipe less than 3 inches in diameter.
[2]Not more than two water-closets shall discharge into any 3-inch horizontal branch, house drain, or house sewer.

The required sizes of storm-water house drains and house sewers and other lateral storm drains shall be determined on the basis of the total drained area in horizontal projection in accordance with the following table:

STORM SYSTEMS ONLY

Diameter of Pipe (Inches)	Maximum drained roof area (square feet)[1]			Diameter of Pipe (Inches)	Maximum drained roof area (square feet)[1]		
	Slope ⅛ inch fall to 1 foot	Slope ¼ inch fall to 1 foot	Slope ½ inch fall to 1 foot		Slope ⅛ inch fall to 1 foot	Slope ¼ inch fall to 1 foot	Slope ½ inch fall to 1 foot
3	865	1,230	1,825	8	11,115	15,745	24,890
4	1,860	2,610	4,170	10	19,530	27,575	43,625
5	3,325	4,715	7,465	12	31,200	44,115	69,720
6	5,315	7,515	11,875	14	42,600	60,000	95,000

[1]The calculations in this table are based on a rate of rainfall of 4 inches per hour.

Combined Storm and Sanitary Sewer Systems.

Whenever a combined sewer system is employed, the required size of the house drain or house sewer shall be determined by multiplying the total number of fixture units carried by the drain or sewer by the conversion factor corresponding to the drained area and total fixture units, adding the product to the drained area and applying the sum to the preceding table for storm-water sewers section *115. No combined house drain or house sewer shall be less than 4 inches in diameter, and no combined house drain or house sewer shall be smaller in size than that required for the same number of fixture units or for the same roof area in separate systems.

*Note—Section 115—See following table.

CONVERSION FACTORS FOR COMBINED STORM AND SANITARY SYSTEM.

	NUMBER OF FIXTURE UNITS ON SANITARY SYSTEM															
	Up to 6	7 to 18	19 to 36	37 to 60	61 to 96	97 to 144	145 to 216	217 to 324	325 to 486	487 to 732	723 to 1,098	1,099 to 1,644	1,645 to 2,466	2,467 to 3,702	3,703 to 5,556	Over 5,556
p to 120	180	105	60	45	30	22	18	15	12	10	9.2	8.4	8.2	8.0	7.9	7.8
21 to 240	160	98	57	43	29	21	17.6	14.7	11.8	9.9	9.1	8.3	8.1	8.0	7.9	7.8
41 to 480	120	75	50	39	27	20	16.9	14.3	11.5	9.7	8.8	8.2	8.0	7.9	7.8	7.7
81 to 720	75	62	42	35	24	18	15.4	13.2	10.8	9.2	8.6	8.1	7.9	7.9	7.8	7.7
21 to 1,080	54	42	33	29	20	15	13.6	12.1	10.1	8.7	8.3	8.0	7.8	7.8	7.7	7.6
,081 to 1,620	30	18	16	15	12	11.5	11.1	10.4	9.8	8.4	8.1	7.9	7.7	7.7	7.6	7.5
,621 to 2,430	15	12	11	10.5	9.1	8.8	8.6	8.3	8.0	7.9	7.8	7.7	7.6	7.5	7.4	7.4
,431 to 3,645	7.5	7.2	7.0	6.9	6.6	6.5	6.4	6.3	6.2	6.3	6.4	6.4	6.8	7.0	7.1	7.2
,646 to 5,460	2.0	2.4	3	3.3	4.1	4.2	4.3	4.4	4.5	4.7	5.0	5.1	6.1	6.4	6.9	6.9
,461 to 8,190	0	2.0	2.1	2.2	2.3	2.4	2.5	2.6	2.8	3.2	3.7	4.6	5.0	5.6	6.2	6.4
,191 to 12,285	0	0	2.0	2.1	2.1	2.2	2.3	2.3	2.4	2.5	2.6	2.7	3.5	4.5	5.2	5.6
2,286 to 18,420	0	0	0	2.0	2.1	2.1	2.2	2.2	2.3	2.3	2.4	2.4	2.6	3.2	4.2	4.7
8,421 to 27,630	0	0	0	0	2.0	2.1	2.2	2.2	2.2	2.3	2.3	2.3	2.4	2.5	2.8	3.1
7,631 to 40,495	0	0	0	0	0	2.0	2.1	2.2	2.2	2.2	2.2	2.2	2.2	2.2	2.3	2.4
0,946 to 61,520	0	0	0	0	0	0	2.0	2.1	2.1	2.1	2.1	2.1	2.1	2.1	2.1	2.1
ver 61,520	0	0	0	0	0	0	0	2.0	2.0	2.0	2.0	2.0	2.0	2.0	2.0	2.0

In order to arrive at an understanding of vhat these tables mean, let us—

Assume that there are 12 complete bath-rooms, 12 kitchen sinks, 4 laundry tubs, 4 aundry water-closets and 4 laundry drains, the total number of fixture units would be:

16 water-closets	× 6	= 96
12 bathtubs	× 2	= 24
12 lavatories	× 0	= 12
12 sinks	× 1½	= 18
4 laundry tubs	× 3	= 12
4 floor drains	× 3	= 12
Total fixture units		=174

From the table of conversion factors for combined drainage systems, we find that for 7500 square feet of surface and 174 fixtures the factor is 2.5.

Following the rule we proceed:

174 × 2.5 = 435.0
7500 + 435 = 7935 sq. ft.

By referring to the storm-water drainage seetion we have:

7935 sq. ft. = 8″ sewer @ ⅛″ pitch
7935 sq. ft. = 8″ sewer @ ¼″ pitch
7935 sq. ft. = 6″ sewer @ ½″ pitch

If we keep in mind the fact that these calculations are based on a 4-inch rainfall per hour and that the heaviest rainfall here is but 2.3 inches for the same time, we are justified in revising these figures.

If it requires an 8-inch pipe to carry off a 4-inch rainfall, it is relatively easy to calculate that a rainfall of only two inches an hour will be carried off by a six-inch pipe at a pitch of ¼ inch to the foot.

There is, however, a generous margin of safety allowed in the figures set up by the Committee and it is considered good practice to err on the side of safety.

Chicago is built on a flat plain. It has a combined system of sewerage resulting in flooded basements and cellars where good engineering practice in the design of the house system has been ignored. There can be no permanent relief until the city has been provided with storm sewers or the drainage in existing buildings remodeled.

Common sense tells us that a storm-water drainage system must be designed and installed in the near future. It will prevent great annual loss in the destruction caused by flooding of basements and will mean economy and efficiency in the operation of sewage treatment plants.

NOTE.—See article on Plumbing Design in Tall Buildings, Page 517, 1922 Year Book.

Compare this table of pipe sizes with that in the new plumbing code for Chicago.

Study of the reference to a full size cross connection between soil or waste line and vent is well worth while in connection with the above tables.

Good practice would not permit the installation of a vertical waste carrier two to four hundred feet long without a break in it. At the angle fitting in such a break a full size connection between soil or waste line and vent should be made as suggested in the 1922 Year Book.

FIG. 1.

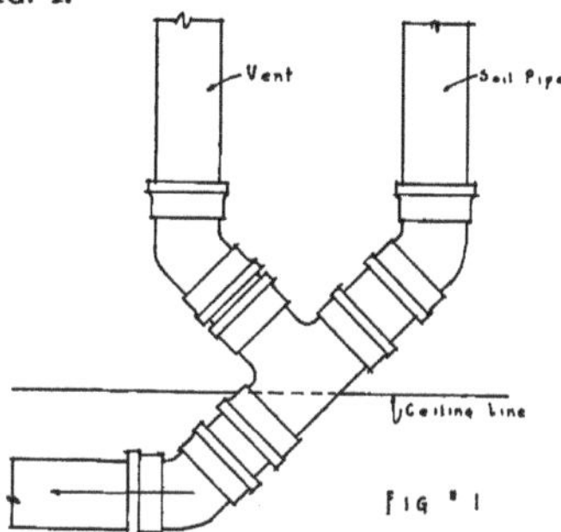

Method of connecting a main vent line into the bottom of a soil or waste line in a tall building to prevent excessive air compression.

The falling column of water hugs the bottom of the 45° angle extension and allows the air to separate from the water and freely escape up the vent line.

Experiments and actual installations prove this method to be correct.

Another method of connecting a main vent at the bottom of a soil or waste stack. This is a stock fitting for cast iron pipe.

Operation is the same as that in Fig. 1.

FIG. 2.

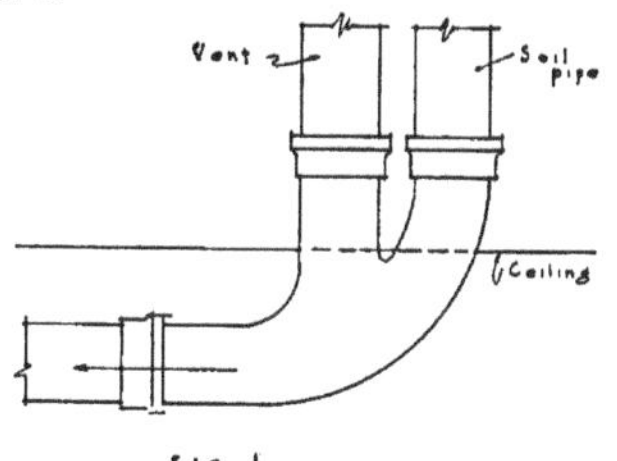

ARCHITECTURAL FINISHES OF VARNISH, ENAMEL AND LACQUER

By WAYNE R. FULLER, Technical Director

INTRODUCTION.

All finishes, whether produced with paint, varnish, enamel or lacquer, serve the dual purpose of protection and beautifying. In approaching any finishing problem it is first of all necessary to decide what is desired in respect to appearance, protection, and serviceability. The moment one begins to consider appearance he is confronted with two alternatives as to the general types of finish: it may be a clear, transparent finish that allows the natural grain and character of the surface to show through, or it may be an opaque finish that completely obscures the original surface, producing a uniform white, black or colored surface.

This article embraces a description of architectural finishes of varnish, oil enamel and lacquer. The word lacquer is here used in the broad sense of all nitrocellulose base finishes, clear and enamels. Interior walls are treated almost exclusively with opaque finishes and this type is also used frequently on interior trim, whether of wood or metal. When woodwork is to be given an enamel finish it is natural to choose a close grain wood having a smooth surface; among the wood most used for this purpose being poplar (whitewood), white pine, yellow pine and red gum. When woodwork is to be given a clear finish a wood having an attractive grain or figure should be selected. Among the woods most suitable for this purpose are oak, walnut, mahogany, chestnut, birch, maple, yellow pine and red gum.

VARNISH FINISHES.

GENERAL.

The major operations in producing a varnish finish are staining, filling and varnishing. A stain is used only if it is desired to change the natural color of the wood; otherwise, this operation is omitted. The primary function of a filler is to fill the pores of the wood in order that the complete finish will have a smooth, even surface. Fillers also have a secondary function of no small importance, to supplement the color effect produced by the stain. Usually the filler is about the same shade as the stain and simply reinforces it. When an only slightly stained effect is desired, it is sometimes obtained through the use of a colored filler without any stain, a common practice in finishing oak floors. Close grain woods like pine, red gum, maple, birch and whitewood (poplar) have no pores to be filled, hence filler is not used. The prevailing practice is to fill the open grain woods like oak, walnut, mahogany and chestnut, although at present there is quite a vogue for open pore or unfilled finishes on such woods. The varnishing usually consists in the application of two or three coats. On interior trim work a coat of sealer or first coater, usually shellac, is commonly applied just preceding the first coat of varnish. For interior work, either trim or floors, a high gloss is not as artistic or popular as a dull finish. This may be obtained by rubbing the final coat of gloss varnish or by using for this coat a varnish that dries dull.

STAINS.

While the fullest decorative possibilities of the use of stain have by no means been realized, it may be said with safety that great interest and better taste are more evident than ever before. Natural oak floors are giving way to floors that are colored by staining or filling in conformity to the general principle of decoration that base colors should be darker than the higher portions of the room. In some quarters the staining of hard wood floors is even treated as a discovery, although for many years this has been feasible and practiced to some extent.

The darker stains that formerly concealed much of the natural beauty of woods like walnut and mahogany are being replaced by lighter stains to lend greater emphasis to the inherent beauty of the wood itself. On panelled work high-lighting (i. e., wiping off part of the stain and filler on portions to produce contrast) has become a rather common procedure. The more widespread interest in matters achitectural has led to greater demand for period finishes concordant with the architecture. Authentic reproduction of the more important period effects has been greatly simplified for the architect and owner through the development, by the well-known varnish manufacturer, of a series of special stains matched directly to woodwork of the respective periods—among them an aged oak panel from an English room about 600 years old; the adzed oak beams of the Old Ship Meeting House, Bingham, Mass.; the aged pine panelling of a bed-chamber from an old house at Hampton, New Hampshire.

Since the architect and owner select a stain chiefly for its color value, they are vitally concerned in obtaining the effect that has been selected on the actual job. This introduces the questions of the wood and proper application. The shade of the stain is only one of the four factors that determine the effect resulting from stain, the others being: (1) the color of the particular piece of wood, (2) the porosity and texture of the wood, (3) the manner of applying the stain. When a filler is used, its color and manner of application also contribute to the final effect, the painter and architect assume that the stain is defective and give scarcely a thought to the other possibilities. Most stain difficulties are due to one or more of the other factors and they should be carefully investigated.

Since it is impossible to put two distinct colors into a single can, no stain will give the same color on woods of even slightly different tone. This means that if the wood differs in color on various parts of a job this variation will be reflected in the final finish, unless the painter is skillful enough to compensate for these differences in the process of application. It also means that dry, porous wood, having a tendency to absorb more stain, will be darker in unless the painter makes proper allowance in application. Occasionally painters and architects request a stain that will give the same color on the hard grain and soft grain of a wood or that will equalize the color of a wood that is darker in parts. It is inherently impossible for the stain manufacturer to comply with such a request; however, the same end can be partially accomplished through judicious application of the stain.

The Kind of Wood Effects Color of Stain.

One of the most prolific sources of disappointment is that the wood in the sample panel from which the stain is selected is quite different in tone and porosity from that used on the job; consequently, although the stain is the same the effect on the work is different. Various lots of white oak, red oak, birch, pine or almost any other wood differ substantially in color and porosity. To further complicate the situation, some finishers habitually apply stains heavier than others. Even with the same stain and the same wood, two painters are unlikely to produce exactly the same effect unless they have a sample panel as a guide. Experience has shown that for greatest assurance of securing the desired color, the stain and filler selected on the

basis of the manufacturer's sample panel should be applied by the painter on a typical, fair size piece of wood from the project. This sample panel should receive the varnish coats called for in the specification as well as the stain and filler. After this panel has been accepted by the painter and the architect, it should serve as the painter's guide and he should be able to practically duplicate it. The panel may be divided into sections showing the successive operations and, if desired, may be split, the architect retaining half of it as a check. Such a procedure leaves little room for excuses or difference of opinion.

Sources of Color.

Stains are divided into a number of classes which differ radically in their properties. There are smaller but appreciable differences in the stains of a given type as made by various manufacturers. Broadly speaking, the actual coloring materials in wood stains are coal tar dyes. Two exceptions to this rule are worthy of note. Gilsonite, a hardened natural pitch, is frequently used in benzol stains, pigment-oil stains and in some varnish stains, also there are some vegetable stains. (For illustration one derived from walnut hulls.)

Coal tar is the liquid that condenses in the heating of soft coal in sealed ovens to make coke and also in the destructive distillation of soft coal to make coal gas. This liquid upon separation yields a number of distinct substances, among them benzol, toluol, naphthalene and anthracene. Each of these compounds serves as the starting point from which, by a series of complicated chemical reactions, a large number of dyes are produced. Certain of these dyes have been found suitable for use in wood stains and most of the stain sold is made by merely dissolving one or more dyes in a proper liquid. Wood stains are classified according to the kind of solvent employed, the common solvents being benzol, water and alcohol. This classification does not include the pigment-oil stains, which must be considered by themselves; neither does it include varnish stains, which we will regard as color varnish rather than stain.

Oil Stains.

The benzol type of stain, easily first in importance, is variously known as an oil stain, penetrating stain, penetrating oil stain or benzol stain. The name "penetrating stain" is misleading, since these stains have no greater penetration on the average than do water stains. One of the other coal tar solvents, toluol, solvent naphtha, xylol or creosote oil, may be substituted for benzol. Benzol type stains are brilliant, are available in a wide but not complete range of colors, are reasonably but not extremely fast to light, do not change color in the package, dry quickly, do not raise the grain of wood. The fastness to light, while satisfactory for most interior requirements, is not such that the stains should be used for exterior work. Gray shades are hardly practicable, because of deficiency in lightfastness. These stains present no difficulty in application and due to solvent action of coal tar solvents on resins they penetrate summer growth better than other stains, minimizing contract between summer wood and spring wood. If varnish is applied directly over some of these stains, they bleed into it badly and seriously retard its drying. On the other hand, these stains as a class bleed into shellac only slightly and do not seriously affect its drying or other properties. If varnish is then applied over the shellac there is little "picking up" of the stain by the varnish, which dries properly and produces a satisfactory finish. These facts dictate the use of shellac or an equivalent first coater over a benzol type stain on trim before applying varnish. Trouble is frequently experienced in enameling over an old varnish finish that includes a benzol stain. It is difficult to seal in the benzol type stain, particularly mahogany, so that it will not discolor the enamel. A coat of shellac is frequently used for this purpose and while helpful in certain cases it does not always overcome the bleeding. Aluminum enamel, although frequently recommended for this purpose, is less effective than shellac. The frequent or long continued use of stain containing benzol endangers the health of the painter, since benzol is toxic. This danger may be eliminated by replacing benzol with one of the other coal tar solvents.

Water or Acid Stains.

Next in importance come the stains in which the solvent is water, known as acid stains or water stains. They get the name acid stain from the fact that when used for textiles an acid is added to the dye bath to exhaust the color on to the fiber. They are no more acid than any other type of stain in chemical reaction and the name is misleading in connection with their use on wood. These stains possess the desirable degree of richness, afford the most complete choice of colors and are superior to benzol stains in lightfastness. However, they are subject to gradual color change in the package and should be used within a reasonable time after manufacture. Due to their corrosive action on tin cans, they are at times marketed in earthenware jugs or colored glass bottles, which introduces the hazards of breakage in shipment and freezing in cold weather. It is now possible to use cans by employing a special can with a protective coating on the interior. Water stains are very easy to apply and penetrate deeply except on sappy wood. Sound practice requires that they be allowed to dry over night. Probably the main factor preventing the wider use of these stains is that the water raises the grain of the wood, especially in the case of the soft woods. This restricts their field to hard woods. The best practice is to sponge the wood with water, allow to dry and sand down any raised grain before applying the stain. It is, of course, impossible to do much sanding after the stain is applied or the stain will be sanded through. Water stains are practically non-bleeding under both varnish and shellac.

Spirit Stains.

Alcohol stains are also referred to as spirit stains. Their use is quite limited because of marked fading on exposure to light and difficulty in applying uniformly. They are very brilliant, penetrate poorly, do not raise the grain appreciably, dry quickly, bleed shellac badly and bleed varnish practically none.

Waterless Water Stains.

During the past few years there has appeared on the market a type of stain that is supposed to combine the desirable features of a benzol stain and a water stain. This type has been described as a waterless water stain, meaning that the dyes used are the same as in water stain but that the solvent is not water. These stains are usually made by first dissolving or dispersing the dye in triethanolamine or in one or more of the glycol derivatives: then thinning this concentrated material to the desired strength with denatured alcohol or a mixture of alcohol and benzol. Unfortunately, the more lightfast among the water soluble dyes can not be dissolved in this way: the lightfastness of these stains is not equal to the better water stains. The darker shades have a "painty" appearance and do not dry satisfactorily. Stains of this type have the further disadvantage of being subject to color change in the package.

Pigment-Oil Stains.

Radically different from the stains so far discussed, the pigment-oil stains have as the main coloring agent a finely ground, insoluble pigment instead of a soluble dye. The pigment may be supplemented by coal tar dyes of the benzol soluble type and gilsonite is used in some shades, as in the case of

benzol stains. The liquid consists of linseed oil, japan drier and volatile thinner. Thus, these stains are essentially paints reduced to a thin consistency for better penetration and clearness. As a class these stains are less brilliant than water stains and they are not practical in some light shades (e.g., silver gray), or in dark, rich shades (e.g., dark mahogany), due to partially obscuring the wood and imparting a "painty" appearance. Their main application is on the soft woods and they are not in general suitable for hard woods. They are superior to all other stains in lightfastness, do not raise the grain of wood, require over night drying, have the minimum tendency to bleed varnish and shellac.

FILLERS.

Filling is usually accomplished by the use of a material known as paste filler. As indicated by the name, this product comes in paste consistency and before using it is thinned with turpentine to the proper consistency for brushing. It may be obtained in colors corresponding to the various stains, e.g., light oak, dark oak, walnut, mahogany, and also uncolored for use in natural finishes on light colored woods. It is not necessary to furnish paste filler in as many shades as stain, since, for example, one mahogany paste filler will serve for mahogany stains varying considerably in shade. The color effect with a given filler depends largely upon the porosity of the wood and how heavy the filler is applied. After brushing on the filler it is allowed to dry or "set" until the painter determines by touch that it has reached the proper stage. The excess material is then removed by wiping crosswise of the grain with rags or cheesecloth. This operation should be performed so that all the filler is removed from the level surface of the wood without wiping it out of the pores. If the filler is wiped off too soon or if the work is done hastily and carelessly, inadequate filling is bound to result. It is important that the filler be allowed to dry thoroughly, which requires at least twenty-four hours for old-fashioned filler. Modern filler dries in four hours to recoat with shellac or varnish, although over night is required to recoat with lacquer. The surface should then be sanded well be fore applying shellac or varnish.

Fine furniture and musical instruments are sometimes double filled, the first coat being allowed to dry twenty-four hours before applying the second. Much of the Philippine mahogany employed for trim contains numerous worm holes. With this grade of Philippine mahogany double filling is required for reasonable assurance of filling the holes. With this exception, Philippine mahogany takes the same specification as other open grain hard woods. Silver gray effects present an exception to the general rule that fillers are the approximate shade of the stain. In this case the effect is obtained by applying white paste filler over a gray stain.

The basis of good paste fillers of the slow drying type is silex, linseed oil and japan drier. In the modern, quick drying fillers, the liquid is a special varnish. Silex is silica, which means that chemically it is the same as sea sand. It is made by powdering quartz and comes as a white powder consisting of very fine, sharp, colorless crystals. A natural filler contains only silex besides the linseed oil and japan drier. A colored filler contains in addition the color pigments required to impart the particular color.

If a filler contains an insufficient amount of non-volatile matter in the liquid it will be too porous. This may not be apparent at the time the work is finished but will result in a gradual sinking of the varnish film into the filler in the pores. The pores then appear incompletely filled and the whole finish has a skimpy appearance. This condition, known as shrinkage, is sometimes incorrectly ascribed to the varnish. If a colored filler contains insufficient color pigment or the color pigment is soluble in the succeeding coats, the filler may turn gray, a defect described as a "gray pore". Occasionally the use of an unsuitable liquid in a filler causes development of a hazy or cloudy appearance in the finishing coats.

Sealer Coats (Shellac, etc.).

Following the filling, the next operation on standing trim, such as base boards, doors and windows, is a coat of a suitable first coater, usually shellac. The primary object of this coat is to seal the surface and produce a smooth, non-absorbent foundation for the succeeding coats of varnish. If a benzol type stain is used, shellac is also required to seal in the stain. The thought naturally arises: Why use shellac when the stain is not of the benzol type? Why not apply an extra coat of varnish instead? Because shellac affords definite advantages in practicabilfty and economy for the painter. It dries ready for sanding in only a few hours and sands to a clean, smooth surface with much less work than varnish would require. Notwithstanding the rather thin body of shellac, it seals remarkably well and holds out succeeding coats of varnish as well as a full coat of varnish. Substitute shellac and adulterated shellac should not be used, as they are deficient in toughness. Occasionally painters thin shellac so greatly that much of its value is lost. Shellac varnish is usually made by cutting 4 to 5 pounds of gum shellac in a gallon of denatured alcohol. Such a shellac varnish should be thinned for use with not more than one part of alcohol to two parts of shellac.

It is inadvisable to use shellac on floors or exterior woodwork; on these surfaces the first coat of varnish should be applied directly on the stain or filler, as the case may be. The reason for this is that shellac reduces the durability of a varnish finish on floors and wherever there is much exposure to weather. Shellac is brittle in comparison with a high grade varnish. While this makes no practical difference on ordinary trim it becomes of serious moment wherever a high degree of elasticity is required for long service, as with floors, window sills and exterior work. Shellac is readily affected by water and turns white, even though protected by a waterproof varnish. Consequently, shellac should never be used as a first coater when a high degree of waterproofness is demanded of the finish. Shellac is also quite susceptible to heat, evidenced first as softening and in severe cases as blistering. If shellac is used as the first coater on a table top the varnish finish is much more liable to printing from hot dishes. A varnish that stands the boiling water test without damage when used alone, blisters badly under the same test if the first coat of varnish is replaced by a coat of shellac. Because of the rapid drying and convenience of shellac, some painters are prone to use it everywhere and to recommend it for purposes for which it is poorly suited. A considerable proportion of the complaints on varnish are traceable to the injudicious use of shellac in connection with the varnish.

Gum shellac is obtained principally from India, being formed on the twigs of trees by a small insect. After being gathered and subjected to a somewhat crude refining process, the material in molten form is poured in a thin film on wood cylinders. Upon hardening it is removed as thin flakes—the orange shellac of commerce. The bleached shellac used in white shellac varnish is made by treating the orange shellac with chlorine gas. Refined bleached shellac is made by extracting the wax from bleached shellac. It should never be used for ordinary, architectural purposes, since it is extremely brittle and lacking in durability. Since there are a number of standard grades of shellac, some of which are adulterated with rosin, the mere specification of shellac varnish does not assure a product of satisfactory quality. It is necessary to specify **pure** shellac varnish in order to require the use of rosin-free shellac. It is also important to specify the "cut" (num-

ber of pounds of gum shellac in a gallon of denatured alcohol) and amount of thinning permitted by the painter.

Liquid Fillers.

Liquid fillers are sometimes used in place of shellac. These fillers are made by incorporating a colorless pigment or extender in a varnish. They were formerly used extensively, but fell into disrepute because of the cheap, brittle nature of the varnish commouly used in them. When made with a durable, quick drying varnish and a well chosen extender, a liquid filler is fully as satisfactory as shellac for use on soft woods. In fact, it possesses certain advantages over shellac; waterproofness and a degree of elasticity that permits its use on floors and exposed surfaces. It is not wiped off like a paste filler and must be allowed to dry over night at least before it is ready to sand and recoat. It holds out the varnish coats about the same shellac and neither retards nor increases the tendency of stains to bleed into the varnish coats. If used over dark, rich stains it has a tendency to make them appear cloudy and dull. There is wide variation in the properties and qualities of the liquid fillers on the market. In making a choice, preference should be given to one that is elastic, yet dries hard and sands easily. Clearness of tone on the work should also receive careful attention.

The extenders most used in liquid fillers are asbestine, China clay and talc. A quick drying waterproof spar varnish makes the best vehicle.

Varnishes.

The final operation, varnishing, is intended to protect the surface, make it sanitary and washable and to bring out more fully the color of the stain and beauty of the wood. Three coats of varnish are required for a first class, long wearing floor finish. For a trim finish of the conventional, full bodied type, two coats of vanish should be applied over a coat of shellac or other sealer coater. Dull finish is now taken for granted on all particular interior work. At one time the only means of obtaining a dull finish was by rubbing of the final coat of gloss varnish with pumice and oil and this is still recommended for maximum results; but the modern, economical method is by substituting for the final coat of gloss varnish, a varnish that dries dull or so-called flat varnish. A full bodied trim finish would harmonize poorly with the open pore or antique effect now so popular, which is best attained by the use of a single coat of flat varnish directly over the shellac. For exterior doors and other woodwork at least two coats of a durable spar varnish should be applied without the use of a first coater. Adequate protection for a long period of time requires the use of at least two coats of varnish and three coats are preferable. The varnish may be applied direct to the bare wood or over stain or filler.

Oil Varnish.

Oil varnishes are made by cooking together under carefully controlled conditions the gums or resins, the drying oils and metallic driers; then thinning to the proper consistency for application with a volatile thinner.

The primary function of the oil is to make the varnish elastic, tough and weather-resistant. The chief varnish oil was formerly linseed oil, obtained from flax seed. However, in modern architectural varnishes tung oil or China wood oil obtained from the nuts of the Chinese tung tree, is used almost exclusively. This oil, while equally elastic and durable, dries much faster and harder and is very much more waterproof. Other oils may also be employed for special purposes. The object of the resin is to make the varnish dry faster and harder, rub better, have better body and gloss. Resins reduce the elasticity and weather-resistance. In the old style linseed oil varnishes the resins were fossil gums, frequently in combination with rosin. Fossil resins are also used in modern varnishes for certain purposes, especially furniture and rubbing varnishes. These resins have been largely replaced in modern China wood oil varnishes with various synthetic resins, which are superior to the fossil resins in many properties.

Fossil Resins

Fossil resins were formed by the gradual hardening (fossilizing) of the sap that exuded from trees that lived centuries ago. The gum is usually found in lumps a few to several feet beneath the surface of the ground and is obtained by shallow mining methods. Most of the fossil resins come from Africa, New Zealand and the East Indies. The hardest and best resins are Zanzibar animi, from Zanzibar on the eastern coast of Africa, and kauri, which comes from New Zealand. Zanzibar is no longer available in commercial quantities but kauri is still used to a considerable extent. Of the commercial gums, pontianak, from Borneo, ranks next to kauri in quality. The two most largely used resins are Manilla Copal from the East Indies and Congo Copal from the west coast of Africa. Each kind of gum is assorted into a number of grades based largely on color. The harder resins impart to a varnish better rubbing and greater resistance to "printing". In order to make the fossil resins soluble in the oil and thinner it is necessary to melt them and hold at 600 deg. to 650 deg. F. until approximately one-fourth the weight is driven off. The oil and drier are then incorporated with the molten resin.

Synthetic Resin.

The first synthetic resin to attain widespread use was ester gum, today the outstanding varnish resin in quantity used and importance. Ester gum, more accurately rosin ester, is manufactured by chemically combining rosin and glycerine. In the course of the reaction the rosin and glycerine disappear as separate materials and there results an entirely new substance of unique properties. This resin dissolves readily in oil and thinner when heated, gives a much lighter color than a fossil resin, is more waterproof, is more durable and being practically neutral produces a varnish that mixes with the basic pigments without thickening. Modern varnishes of the waterproof, rather quick drying type are based on the use of China wood oil with ester gum or rosin (or both). When good rubbing and polishing properties are required, a fossil resin must be used in substantial proportion for its greater hardness. Copal ester, made by combining Congo or one of the other fossil resins with glycerine, is used to a small extent. Recently other synthetic resins of radically different nature have been developed. These newer resins are responsible for the so-called four-hour varnishes and will be reserved for discussion in connection with those materials.

Metallic Driers.

Metallic driers are incorporated in varnish to accelerate the drying, which otherwise would be quite slow. The drying of a varnish consists of two stages. The first is the evaporation of the volatile thinner, which takes place rapidly and produces "setting". The second stage is the actual drying of the oil in the varnish by oxidation and polymerization. The oxygen of the air combines chemically with the oil in the varnish. This reaction is hastened by the presence in solution of certain metallic elements, although the latter are not themselves changed by the reaction. The metallic driers are various compounds of lead, manganese and cobalt. Most varnishes contain both lead and manganese. When an especially pale color is required, cobalt is usually employed. The driers are used in very small amount.

Solvents for Varnish

When the cooking together of the oils, resins and driers is completed there results a heavy varnish base, which is entirely too thick for application. A volatile solvent or thinner must therefore be added to reduce it to the proper consistency and to impart the required working and flowing properties. The standard varnish solvents are turpentine and mineral spirits. The latter is a petroleum product similar to gasoline but more highly refined and evaporating more slowly. A high grade mineral spirits is equal to turpentine as a thinner for most modern varnishes. In the early stages of drying the solvent, whether turpentine or mineral spirits, evaporates from the film. Any trace that may remain is too slight to have an appreciable effect on the properties of the film. It is immaterial what evaporates, as long as it keeps the various constituents in solution up to the time of evaporation and imparts the desired working properties. In these respects a typical varnish made with a properly selected mineral spirits can not be distinguished from the same formula made with turpentine. There is one noteworthy exception to this general statement. Turpentine is a stronger fossil resin solvent than mineral spirits and is required in formulas containing the harder, less soluble resins such as kauri. It is also used in part in many of the more expensive architectural varnishes, in deference to the painter's preference. However, it is little used in industrial varnishes, which are purchased more strictly on the basis of performance and economy.

Qualities of Varnish.

The qualities of a varnish are determined by the choice of raw materials, their relative amounts and the process of manufacture. Oil makes a varnish soft, elastic and resistant to the action of the weather. Resin makes a varnish hard, brittle, high in luster, readily deteriorated by the action of weather. The relative amount of oil and resin is the most important single factor in fixing the properties of a varnish. A varnish containing a large proportion of oil is called a long-oil varnish, while one containing a small proportion is referred to as a short-oil varnish. If to these we add a third division, the medium-oil varnishes, we have a classification that corresponds to the three major types of varnish from the standpoint of use. The long-oil group constitute the exterior varnishes, which may be separated into spar varnish and finishing varnish for automobiles, railway coaches, etc. These varnishes must be long in oil for exterior durability and do not require as rapid drying or as great hardness as interior varnishes. The medium-oil group is best represented by floor varnish. It is important that a floor varnish dry quickly so as to permit use of the floor within a short period and that after drying it be hard enough to withstand the scuffing and scraping to which it is subjected. At the same time it must be fairly elastic and tough in order to withstand the pounding heels and moving of furniture without cracking or scaling. A typical floor varnish is too short to have exterior durability and too long to rub with the ease and clearness of tone demanded of a rubbing varnish. The short-oil group are the rubbing and polishing varnishes. They must contain a large proportion of resin and a relatively small proportion of oil in order to secure satisfactory rubbing and polishing properties. While rather brittle, they render fair services on furniture, pianos and cabinet work. Interior trim varnishes are more difficult to classify, since they range from medium-oil to short-oil, depending on where the emphasis is placed by the individual manufacturer. There is a salutary trend toward trim varnishes that are more elastic and durable, even though this involves some sacrifice in rubbing properties. Such a varnish will give much better service on window sills and wherever exposed to strong sunlight. When a flat varnish is applied as the final coat on trim the rubbing properties of the gloss varnish becomes inconsequential and a medium-oil product should certainly be chosen.

The question of the exact kind of resin used in a varnish is usually a matter of great interest to the layman; he assigns to it a position out of all proportion to its true significance. In the old days of linseed oil-fossil resin varnish, this concern would have been largely warranted. With linseed oil it was important that the resin be extremely hard to compensate the softness of the oil. Moreover, the only resin other than fossil resin that was ever utilized was rosin, which had no particular merit in conjunction with linseed oil, produced a soft, "cheesy" film and was used mainly for lower cost. The general adoption of China wood oil had a profound effect upon the resin situation. China wood oil dries to such a hard film that an extremely hard resin was of no advantage except for rubbing and polishing varnishes. For many purposes the less expensive fossil resins were just as suitable as the harder, more expensive ones. Moreover, the use of rosin ester and rosin was found to overcome certain practical difficulties in the use of China wood oil and to greatly extend the possibility of utilizing this valuable, new oil. It was discovered that even rosin when used in reasonable proportion in conjunction with China wood oil afforded some definite advantages and produced a hard, tough, durable film, instead of the poor film that had characterized its use with linseed oil. Thus rosin ceased to be a mere substitute and adulterant.

The properties of a varnish depend in no small degree on the exact manner of combining and cooking the constituents. Many combinations that give the most desirable results when handled skillfully are impractical in less expert hands. It is impossible to predict what the properties of a varnish will be from the composition alone. At one time the development of new varnish formulas was in the hands of "practical" varnish makers. Today it is performed by trained chemists, who work out the precise procedure of cooking as well as the composition.

After the varnish is cooked, the kettle is drawn from the fire and it is allowed to partially cool, the thinner is then incorporated and finally it is clarified by passing through either a filter press or a centrifugal clarifier. The finished varnish is then pumped into large tanks for storage and aging. Aging for a period of a few months permits the very fine suspended matter that even a filter press will not remove, to settle out and thus slightly improves the clarity and brilliance. In the case of a press-filtered varnish this change is too slight to be discernable in the results when the varnish is used. There is no other change that can be ascertained by the most careful practical tests. All the important properties, such as working, drying, body, waterproofness and durability, remain unaltered. Consequently, aging dose not appreciably influence the quality of varnish. The elimination of aging as a factor for consideration tends to concentrate attention on the performance or physical properties of the varnish, which is the only thing that directly concerns the user.

The discussion of varnish raw materials and their uses has probably made it evident that there is no such thing as an all-around, perfect varnish. A perfect varnish would incorporate all the desirable features in the maximum degree, manifestly impossible when many of the qualities are inconsistent with one another. In practice it is necessary to balance one property against another and seek the compromise that is best adapted to the particular purpose. A general purpose varnish represents an attempt to reach a compromise that is most nearly suitable for all purposes. Obviously it can not serve

any purpose as well as a varnish designed for that particular use. Most general purpose varnishes fall between a floor varnish and a spar varnish in hardness, elasticity and other properties.

Testing of Varnish.

In order to determine the quality of a varnish or its suitability for a given purpose, it is necessary to consider a large number of qualities. Among the most important of these are working, flowing, drying, freedom from flatting or checking if exposed to drafts or gas fumes while drying, hardness, color, body, gloss, waterproofness and elasticity. It is comparatively easy to make a varnish that is outstanding in one or a few properties; the real problem is to make a product that embodies in a high degree the qualities that are most important for the specific use, yet sufficiently well balanced in other properties to be practical. To determine the performance that can be expected of a varnish under all conditions requires extensive practical experience with it or exhaustive tests by a trained varnish chemist. These tests cover all the properties enumerated above and frequently others in addition. Since the chemical analysis of varnishes is of little value, most of the tests are of a typical nature. The one chemical test that is frequently run is the percentage of non-volatile. This test is for the purpose of showing the proportion of volatile or thinner to the non-volatile or film forming materials—oils, resins and driers. In specifications for interior and spar varnish the usual requirement is a minimum of 45 per cent non-volatile. The layman is prone to assume that heavy consistency in a varnish indicates good body on the work, while it is really a minor factor. Without changing the composition of a varnish the manufacturer can increase the consistency at will by simply cooking slightly longer. Another mistake frequently made by the painter is in trying to judge the quality of a varnish by the odor, which has no bearing on the other qualities of a varnish. When it is desired to determine the elasticity of an exterior varnish more accurately than can be done by scratching the dried film with a pen knife, the kauri reduction test is used. In this test a solution of kauri gum, which is very hard and brittle, is added to the varnish in various proportions. These mixtures are then poured on tin and dried under specified conditions. The tins are bent over a one-eighth inch rod and note made of the smallest amount of kauri that causes the film to crack. The more kauri required to cause cracking the more elastic the varnish. When a kauri reduction test is included in a spar varnish specification, the common requirement is a kauri reduction value of at least 50 per cent, although the better spar varnishes have a value of 70 to 75 per cent. If China wood oil varnishes are undercooked, the oil has a tendency to dry flat should the varnish be exposed to a draft during the drying or to give a frosted or checked condition should fumes from combustion of coal, coke, gas or oil be present in the air while the varnish is drying. Special tests are required to preclude trouble from this source.

A Technical Specification for Varnish Is

Since the architect and painter are not in a position to conduct elaborate, conclusive tests they must look to the experience, knowledge and facilities of the varnish manufacturer. So far it has been found impossible to write a technical specification that assures a varnish of the best quality for a specific use. Many of the important qualities are of such nature that they can not be measured and specified exactly. A practical comparison of trim varnishes A and B may demonstrate that A has considerably better body and rubbing properties, yet no specification has been devised that will pass A and reject B. Varnishes are being improved so rapidly that before a specification calling for the best available varnish can be drawn up, some manufacturer is able to offer a distinctly better product. The Federal Government finds it necessary to buy its varnishes on technical specifications, for reasons that are obvious. As a result it must accept materials distinctly inferior to the best on the market. To take a concrete illustration, there is a Federal Goverment Specification for Water-Resisting Spar Varnish. The varnish required under this specification has no more than two-thirds the durabiliy of the better, widely sold brands of spar varnish. Recently the use of new synthetic resins has permitted the development of spar varnishes having approximately three times the durability of the varnish required by the Federal specification. This type of varnish is being marketed and is practical, yet it fails to pass the Federal specification on a trivial technicality. There is also a Federal specification for Interior Varnish. The varnish required is entirely too elastic for satisfactory rubbing, although a varnish having better rubbing qualities would have adequate durability for interior work.

Flat Varnishes.

Dull finish varnishes are made by adding a flatting agent to a gloss varnish. They appear flat or dull because the surface is rough, the irregularities being too minute for the eye to detect but causing the light to be reflected diffusely. Rubbing a gloss varnish makes it dull because the surface is filled with very fine scratches, which likewise produce diffuse reflection. The flatting agents are usually "soaps", the most common being aluminum stearate. The flatting agents tend to detract from certain desirable properties. Since they make a varnish heavy and somewhat jelly-like in consistency, it is necessary to use a larger amount of thinner, which means skimpy body on the work. Even with the increased thinner, it is difficult to obtain satisfactory flowing and freedom from brush marks. Some flat varnishes impart a cloudy or "milky" appearance when applied over dark or rich stains such as mahogany and walnut; at best, flat varnish does not produce quite as clear a tone as a rubbed varnish on dark woods. It is characteristic of flat varnishes that when scratched or rubbed heavily they show a grayish or whitish streak. Some products are more susceptible to this than others and in general the flatter a varnish is the more readily it marks. This property is not apparent on natural or light effects. This point is of little consequence on floors or most trim, but becomes more prominent in connection with dark furniture or doors. Flat varnishes are not as waterproof as clear varnishes, although the better products are satisfactory in this respect. In a flat varnish finish only the final coat is flat varnish. This keeps the clouding at a minimum and in the case of a bodied finish the use of a gloss varnish for the first two coats produces better body.

Color Varnishes.

Color varnishes or varnish stains are employed when it is desired to stain and varnish in the same operation. On new work or when refinishing in the same color, one or two coats of the color varnish are applied direct to the surface. When changing the color of an old finish, a dense-hiding ground color is first applied, followed by one or two coats of the color varnish. Two coats of the color varnish produce a much darker color than one coat. The results obtained with color varnish are by no means equal to

those secured when a separate stain is used, as there is a tendency for brush marks and laps to show and the color is less brilliant. On new work this condition can be improved by applying a coat of clear varnish next to the wood, followed by the color varnish. Naturally the main use of varnish stains is by amateurs for refinishing or where requirements are not very exacting.

Varnish stains are made by incorporating soluble dyes or insoluble pigments in a clear varnish. Most of the varnish stains on the market are of the dye type. These have relatively brilliant color and do not settle, but are subject to fading when exposed to sunlight. The varnish stains made with pigments or pigments and dyes combined are slightly less brilliant and may settle in the container, but are considerably more fast to light. The clear varnish used preferably should be a high grade floor varnish or varnish of similar nature. Some of the varnish stains are made with very cheap, brittle varnish.

Quick Drying or Four-hour Varnishes.

The discussion up to this point has been based on the type of oil varnish that has been on the market for years and which is still largely employed by the architect and painter. Within the past few years there has been an altogether new development: quick drying or so-called four-hour varnishes. These varnishes are frequently described as synthetic resin varnishes; true, they are, but so is every varnish containing ester gum, and that comprises a large proportion of the older varnishes. The distinctive feature of the quick drying varnishes is that they contain new types of synthetic resins, the essential constituents of the most common kind being carbolic acid and formaldehyde. These resins have the valuable property of causing more rapid drying of the oil. The effect seems to be largely one of polymerization and jellation, the varnish setting quickly and drying on the surface quickly, but not hardening through at a corresponding rate. In the past, unusual speed of drying has always been obtained by making a varnish short and hence lacking in durability. When a carbolic acid-formaldehyde resin is used, speed of drying may be obtained without shortness and does not indicate poor durability. Some of the synthetic resins are durable in themselves and consequently even short oil varnishes made with them may be durable. Resins of this durable type permit of making quick drying spar varnishes that are substantially more durable than the best of the slow drying spar varnishes. The quick drying principle may be applied to all types of varnish—floor, trim, spar, flat varnish, etc. Quick drying varnishes seldom dry as rapidly as implied by the name and claims. The quicker products for interior work dry in four hours so that a floor could be walked on lightly and the varnish could be recoated, providing sanding is dispensed with. In four hours these varnishes are not hard enough to sand, which the best practice requires before recoating. Moreover, on a floor they are not hard enough for heavy use. On a chair or furniture, "printing" would likely result. Quick drying varnishes become thoroughly hard only slightly faster than the usual type varnish of the same length.

New Type Glyptal Resin Varnishes.

The most recently developed type of synthetic resin varnish is made with a "glyptal" resin. The basic constituents of "glyptal" resins are phthalic acid (or anhydride) and glycerine, although invariably other materials enter into their composition. Varnishes of this type are in an early stage of development and exploitation, hence it is too early to foresee their value and no detailed discussion is required. One point seems to have been established: that it is possible to make varnishes of this type having remarkable exterior durability.

The practical advantages of more rapid drying and greater exterior durability are too obvious to require comment. On the other hand, it must be realized that as a class, quick drying varnishes have certain disadvantages: less body on the work, greater tendency to wrinkle if a heavy coat is applied, greater tendency to skin, greater tendency to gas check, darker color and greater tendency to yellow on the work. This statement should not be interpreted to mean that all quick drying varnishes embody all of the unfavorable features. Since quick drying varnishes are still in the stage of development, there is a greater variation in the quality of the products of different manufacturers than occurs in the slow drying varnishes, which suggests caution on the part of the painter and architect. At the same time the progress already made justifies the belief that quick drying varnishes are fundamentally sound and will displace the slow drying type.

Spirit Varnishes.

The word varnish ordinarily suggests an oil varnish or oleoresinous varnish. However, it has a broader meaning and unless qualified, includes the straight gum varnishes, made by simply dissolving a gum or resin in a suitable solvent. The most important of these are the spirit varnishes, in which the solvent is denatured alcohol. Shellac, the main spirit varnish, has already been discussed. Most shellac substitutes are crop Manilla Copal cut in denatured alcohol. Gloss oil is a straight gum varnish made by dissolving limed rosin in mineral spirits. Damar varnish is likewise a straight gum varnish and is made by dissolving damar gum from the East Indies in turpentine or mineral spirits. It is quite brittle and is used only for its extremely pale color and freedom from yellowing. Damar varnish should be used only in those exceptional cases where the required paleness can not be obtained otherwise.

ENAMEL FINISHES.

General.

It will simplify the discussion of enamel finishes to reserve wall and floor finishes for separate treatment and confine the first part of the discussion to interior trim. The major operations in producing an enamel trim finish are application of primer, undercoating and enamel. The primer, usually applied as soon as the trim is placed, is intended to seal the wood against marked changes in moisture content during construction, also to knit to the wood and furnish the proper bond or adhesion for the succeding coats. This coat likewise holds out the succeeding coats and contributes slightly toward the hiding required in the complete finish. Wood trim should always be back primed before being nailed in place, preferably at the mill. The primary function of the undercoating is to hide the surface solidly and produce the approximate shade desired. The undercoating also fills or levels up the surface and sands easily to a smooth foundation for the enamel coats. The enamel further enhances the smoothness and body of the finish, supplies any additional hiding required, produces a non-porous, washable surface having the exact color and degree of gloss desired. The enamel may have either a high gloss or an eggshell finish.

WOOD PRIMERS.

If the wood contains knots or sap streaks, the pitch should be sealed in with shellac or aluminum paint before applying the primer; otherwise, the final finish is likely to show a yellow or brown discoloration. The best primer is made by thinning four parts of

We Anticipate with Pleasure Your Visit to our Studios

T. C. GLEICH CO.

FRANKLIN MARLING, JR.—Successor

PAINTING • DECORATING

OUR WORK IS OUR HOBBY

3945 SHERIDAN ROAD
Phone Buckingham 4320-1
CHICAGO

a high grade undercoating with one part boiled linseed oil. This primer dries ready for recoating in twenty-four hours. Lead and oil, also, is widely used for this purpose. It satisfactory, although not equal to the undercoating and oil mixture in hiding, freedom from brush marks, drying and sanding properties. Following is a good formula for mixing lead and oil primer:

White Lead Paste...........	100 lbs.
Raw Linseed Oil............	4 gal.
Turpentine	2 gal.
Liquid Paint Drier...........	1/8 gal.

Any high grade, ready mixed outside house paint is at least equally suitable. Either of these primers should be allowed plenty of time for drying, at least two days. Aluminum paint has been advocated for this purpose. It has excellent priming and sealing properties but is difficult to hide with the succeeding coats and detracts from the smoothness of the finish. When economy compels the use of the minimum number of coats on interior work, it is essential that the first coat be undercoating, and the hiding may be improved by reducing the amount of boiled linseed oil to one part oil and eight parts undercoating.

Metal Priming.

Metal trim is usually primed at the factory. If any rust spots develop before finishing they should be sanded clean and touched up with red lead and oil. The work should be cleaned thoroughly with gasoline before finishing. Any metal surfaces that may lack primer should be given a coat of a good metal primer, and the standard product for this purpose is red lead and oil. Other suitable primers are blue lead and oil and iron oxide metal primer. White lead, also, can be recommended when the light color is advantageous. When economy is essential, a good undercoating may be applied direct to iron or steel.

From this point on, the finishing of wood and metal is the same.

Undercoatings.

The final result in an enamel finish is influenced by the undercoating almost, if not quite, as much as the enamel. Frequently the undercoating receives insufficient attention. It is not unusual for a fine, expensive enamel to be applied over a cheap, inferior undercoating or even over an ordinary flat wall paint. This practice defeats the very object in using a high grade enamel. A cheap undercoating is certain to leave a "ropy" surface, full of brush marks, and thus counteract the excellent leveling properties that are so carefully incorporated in a good enamel. A cheap undercoating is likely to be porous, causing the enamel to work hard and to sink in, thus detracting from the full body of the enamel. The poor grinding of an inferior undercoating causes the enamel finish to be gritty, and if the undercoating is too short the durability will also be reduced. The only way to obtain a quality enamel finish is by applying a first class enamel over an undercoating of corresponding quality. Painters as well as amateurs are prone to judge an undercoating by the way it works, hides and looks by itself; the correct way to judge it is **after** the enamel has been applied. It is only then that superior leveling, non-porosity and fineness of grinding are fully revealed.

For a first class job two coats of undercoating are required, but only one coat is used on a large portion of the work. Probably the majority of undercoatings are put out in the proper consistency for brushing. In this case they should not be thinned, except that in using the last part of a container it may be necessary to add thinner to replace the loss by evaporation. The unwarranted use of thinner for easier working and higher spreading rate means deficient body and hiding. Under normal drying conditions a modern undercoating should dry in twenty-four hours ready to sand or recoat. A smooth, clean enamel finish requires that the last coat of undercoating be thoroughly sanded, not merely given a light scuffing. The dust from sanding should be dusted off and just before enameling the surface should be gone over with a tack rag. For a colored finish the undercoating should be tinted with oil colors to the approximate shade of the enamel.

The most important properties that require consideration in the choice of an undercoating are ease of application, spreading rate, hiding power, leveling out of brush marks, drying rate, ease of sanding, fineness of grinding, degree of settling and caking in the can, durability and holding out of the enamel. One undercoating may give a much higher spreading rate than another and price comparisons are misleading unless proper allowance is made for this. The relative settling and caking of two undercoatings can not be determined by comparing cans bought at random, since one may be much older than the other. The only reliable procedure is to stir both cans thoroughly and then examine at intervals and note the difference. There is one test that is extremely valuable in examining undercoatings and which brings out a number of properties. This test uses a one foot by two feet panel of poplar or other close grain wood. The panel is first primed with lead and oil and after this coat has dried a broad flat black stripe is applied crosswise at the center of the panel. The panel is divided lengthwise in halves and the two undercoatings being compared are brushed on the respective halves. If there is a great difference in the ease of working it will be apparent; otherwise, the test will not cover this point. After the first coats have dried twenty-four hours they are inspected for hardness of drying, hiding, leveling and fineness of grinding. Differences (except drying) are most apparent over the black stripe. The halves are then recoated and after twenty-four hours again examined. One end of the panel is now sanded and note made of the ease of sanding and freedom from gumming the paper. The entire panel is finally given a single coat of a high grade gloss white enamel. After this has dried hard a final inspection is made to determine how the two undercoatings affect the appearance of the complete finish as to leveling, body, etc. The object of omitting the sanding one one end of the panel is to show the result when the undercoating receives practically no sanding, a rather common occurrence.

Composition of Under Coaters.

Undercoatings are made by incorporating pigments in a vehicle, consisting usually of bodied linseed oil or varnish and a thinner. The amount of pigment is normally in the neighborhood of sixty-three per cent by weight. The most extensively used white pigment is lithopone, which is a chemical or manufactured pigment containing thirty per cent zinc sulphide and seventy per cent barium sulphate. Lithopone is used because it hides better than white lead or zinc oxide, imparts better leveling properties, is less expensive and can be used with varnish vehicles that would "liver" with the other pigments mentioned. Other white pigments that are used in undercoatings are: zinc oxide, a fume pigment made by roasting zinc bearing ore or pellets of zinc metal in a furnace and collecting the fine white powder of zinc oxide; Titanox B, a chemical pigment consisting of twenty-five per cent titanium oxide and seventy-five per cent barium sulphate; Titanox C, a chemical pigment consisting of thirty per cent titanium oxide and seventy per cent barium sulphate; double-strength lithopone, containing fifty-five per cent zinc sulphide and forty-five per cent barium sulphate; titanated lithopone, which is regular lithopone having fifteen per cent titanium oxide incorporated with it. With the exception of zinc oxide, these pigments have even greater hiding power than lithopone. The choice of pigments depends upon the qualities desired and

also upon the knowledge and judgment of the formulator.

The pigments just mentioned all possess hiding power when incorporated in a film of drying oil or varnish. Most white undercoatings contain in addition one or more pigments that lack hiding power or are transparent. These pigments are known as inerts or extenders. They usually constitute only ten to twenty-five per cent of the total pigment and improve the body, improve the sanding properties, reduce settling and lower cost. The most widely used extender is asbestine, a naturally occurring magnesium silicate in the form of fine, needle-like crystals. It has the property of reducing settling and caking. Among the other extenders are whiting (calcium carbonate), powdered dolomite (calcium-magnesium carbonate), china clay, silica and talc, the latter a magnesium silicate of slightly different composition than asbestine and in the form of scale-like crystals.

To make a modern type undercoating which dries in twenty-four hours for sanding and recoating requires the use of a considerable proportion, at least, of China wood oil varnish in the vehicle. Some raw or heat bodied linseed oil may be used in conjunction with the varnish. The vehicle must, of course, be elastic in order to produce a durable undercoating.

The pigment is first incorporated with part of the vehicle by grinding on a flat stone mill, a steel cylindrical roller mill or a pebble mill. In modern plants the flat stone mill has given way to the more economical roller mills and pebble mills. The pebble mill is most economical and for this type of product also given maximum quality. When the grinding operation is completed the remainder of the vehicle is added and the material strained.

ENAMELS.

Two coats of enamel are required for a first class, full bodied finish. The first coat requires one to three days for drying, depending upon the kind of enamel and drying conditions. Since eggshell enamel has less body than gloss enamel, the first coat in an eggshell finish should be gloss enamel. A good eggshell enamel is so satisfactory that there is little advantage in going to the additional expense of producing a rubbed enamel finish by rubbing a gloss enamel. When a rubbed finish is desired the last coat should be allowed to dry quite hard. usually three days or longer. Pumice and water are recommended for rubbing, not pumice and oil as with varnish. When economy compels a four-coat enamel job the second coat of undercoating and the first coat of enamel should be replaced by a single coat made by mixing equal parts of undercoating and enamel. For a three-coat job the first coat should be undercoating; the second coat a mixture of equal parts undercoating and enamel; and the third coat enamel. Lack of skill or care is likely to be reflected in a skimpy job, brush marks or runs and sags. Enamel, like undercoating, should be applied as it comes from the can, unless there are directions to the contrary.

The properties that determine the quality of an enamel are ease of application, spreading rate, drying, leveling, body, fineness, gloss, hiding, initial color, degree of "after yellowing", durability, caking in container. Some enamels have a good gloss initially but lose it after a few weeks, hence final determination of this property should be made after an enamel has dried about a month. Enamels as a class have considerably less hiding than undercoatings, in consequence of the lower pigment content that must be used in order to obtain the higher gloss. Fortunately, it is not necessary that an enamel have great hiding, since the undercoating is relied on to produce solid covering. These statements contemplate an enamel of the conventional type for high grade white enamels, i. e., made with zinc oxide as the pigment. Much better hiding may be obtained by the use of other pigments. The initial color of an enamel should not be a "yellow white" or a "blue white", but as nearly as possible a neutral white. All enamel vehicles and most pigments have a slightly yellow color and this is neutralized by slight bluing. In the case of lithopone enamels the blue is frequently omitted, since the pigment is blued in the process of manufacture. Excessive bluing is likely to be an attempt to mask a color that is excessively yellow and the result is always lack of brightness. Moreover, a highly blued enamel does not produce clean tints. An enamel may have good initial color but "yellow" to a serious extent after application, particularly if it receives only subdued light. The method of testing on a panel that was described for undercoating is also applicable to enamel, when slightly altered by applying the black stripe over the undercoating.

From the standpoint of composition there are two major differences between an enamel and an undercoating. In the first place, an enamel has a lower percentage of pigment. In the second place, the vehicle of an enamel contains a larger proportion of non-volatile, i. e., oils and resins, and a smaller proportion of volatile or thinner. (In these respects an eggshell enamel falls between a gloss enamel and an undercoating). There are also less fundamental differences in the kind of constituents, as will be brought out in what follows:

Up to the present time most high grade architectural enamels have been zinc oxide-linseed oil enamels. The pigment has consisted of straight French process zinc oxide. Zinc oxide that is made in a single stage by roasting the ore is shown as American process; when it is made in two stages by first smelting the ore and then roasting the metallic zinc it is known as French process. The latter is purer, finer and has a white color. Ordinary raw or boiled linseed oil would be entirely unsuited for use in enamels, being deficient in body, gloss, leveling, color and yellowing. For enamels the oil is first refined by treatment with alkali and then "bodied" by heating for a prolonged period to a high temperature, usually 550 deg. to 600 deg. F. This heavy oil is thinned to the proper consistency with turpentine or mineral spirits and driers added. It then constitutes the vehicle for a straight oil enamel. The pigment is ground in this vehicle on a stone mill or roller mill and the resulting paste thinned to the correct consistency with additional vehicle. The details of this process are by no means as simple as the outline might indicate. Since the same kind of zinc oxide is available to all enamel manufacturers, the secret of superiority in an enamel lies primarily in the details of the preparation of the vehicle. A well-made enamel of this type is suitable for all interior and exterior purposes. The cheaper white enamels, such as gloss mill whites, are usually made with lithopone, or lithopone and not to exceed twenty-five per cent zinc oxide, as the pigment. They are not sufficiently weather-resistant for exterior use but if a long oil vehicle is used possess excellent interior durability. They have distinctly better hiding than a straight zinc oxide enamel. In the past few years a number of other pigments have come into more or less use in enamels: double strength lithopone, titanated lithopone, zinc sulphide, titanox, tianium oxide and antimony oxide. These pigments all possess excellent hiding and some are suitable for exterior as well as interior use. It is no longer safe to assume that every high grade white enamel does or should contain straight zinc oxide as the pigment.

Gloss enamels seldom if ever contain an extender. Eggshell enamels may contain a small proportion of an extender for its flatting effect.

With respect to the type of vehicle enamels may be divided into straight oil, varnish and spirit enamels. Obviously, a straight oil ena-

Palm Olive Building, Chicago Holabird & Root, Architects

mel is one in which the vehicle contains only oil, thinner and drier—no resin. A varnish enamel is one in which the vehicle is a varnish, in other words, contains resin as well as oil. A spirit enamel is one in which the vehicle contains no oil, i. e., a spirit varnish. Varnish enamels may be further divided into long oil and short oil. A long oil enamel is one that is made with a long oil varnish or with straight oil; a short oil enamel is one made with a short oil varnish. With straight zinc oxide enamels these classifications are of considerable practical importance, since it is necessary to use a straight oil vehicle with zinc oxide in order to make an enamel of maximum exterior durability. However, with pigments like Titanox and titanium oxide they are of less significance, since these pigments permit of making a varnish enamel possessing excellent durability.

The subject is further complicated by the possibility of using varnishes made with new synthetic resins that impart unique properties. The quick drying or so-called four-hour enamels are made by using a quick drying varnish as the vehicle. Many quick drying varnishes would thicken if mixed with zinc oxide and with such varnishes it is necessary to use the more inert pigments such as lithopone, titanated lithopone. Titanox and titanium oxide. Aside from thickening, the use of these pigments is advantageous because of their greater hiding power, which is doubly important in a quick drying enamel with its relatively thin film.

Enamels are manufactured in tints or colors as well as in white. Painters usually find it more convenient to make their own light tints by adding colors in oil to the white enamel, but a dark colored enamel of good quality could not be made in this way. A large number of color pigments are required in the manufacture of colored enamels and for best results these must be carefully selected and employed with understanding. Color pigments may be divided into four classes:

1—The earth colors or natural mineral pigments. These require only mining, washing, drying and sometimes calcining. Most important among them are the various iron oxides (Indian Red, Persian Gulf Oxide, Spanish Oxide, Prince's Metallic), raw umber, burnt umber, yellow ochre, raw sienna and burnt sienna. In general, these pigments are fast to light but not sufficiently bright and brilliant for use in high grade colored enamels. Their main uses are for tinting, in primers, surfacers and barn paints. Burnt umber is extensively used in brown paints and enamels, while raw sienna is the most widely used pigment for tinting white paints and enamels to ivory and cream shades.

2—Fire process pigments. These are made by roasting or combining various products in furnaces at high temperatures. The main ones are red lead, ultramarine blue, some English vermilion, Venetian red and some other iron oxides, bone black or drop black. Red lead is an oxide of lead which contains more oxygen than litharge. It finds little use in enamels, being of greatest value in anticorrosive metal primers and in metal paints. Ultramarine blue is a bright, transparent blue that is poor in lightfastness. Bone black, drop black and ivory black are synonymous terms referring to a black pigment that is made by calcining and grinding bones. It has a jet black color, produces a pure gray tint without the blue hue of lampblack and in light tinted enamels has the minimum tendency to cause streaking. It is very little used in black enamels because of relatively poor hiding and a tendency to settle.

3—Fume pigments. Aside from zinc oxide, the only important examples are lampblack, gas black and blue lead. Both blacks are made by incomplete combustion with the formation of finely-divided free carbon: lampblack being made by burning paraffin oil and gas black (carbon black) from natural gas. In the manufacture of high grade enamels, carbon black is used almost exclusively, because of its superior color and great hiding power. Lampblack is probably the most commonly used black pigment for tinting paints both by the painter and the paint manufacturer. It imparts a blue tone that makes it impractical for some shades. Blue lead is primarily a metal paint pigment that serves the same purpose as red lead. It is basic lead sulphate containing some metallic lead and carbon, which impart the characteristic color.

4—Chemical pigments. These are made by precipitation from solution and include both inorganic and organic pigments. The inorganie pigments comprise the most important blues, yellows and greens. The main blue is Prussian Blue, a complex iron cyanide. Prussian Blue is employed whenever a blue pigment is required, except for some especially bright, rich shades that can only be obtained with ultramarine blue. The main yellow and orange pigment is lead chromate. Most of the green pigments are made by mixing Prussian Blue and chrome yellow. Three precipitated organic pigments, coal tar derivatives, constitute the chief reds used in enamels; paranitraniline toner (para red), toluidine toner and lithol red lake. The most important maroon pigments are alizarine or madder lake and Bordeaux Lake, likewise coal tar derivatives. A lake is a pigment made by precipitating an organic dye on an inorganic base, frequently aluminum hydroxide, a barium salt or a calcium salt. Madder Lake is the more lightfast and expensive of the two maroons mentioned, while Bordeaux Lake has the better hiding.

FLOOR ENAMELS.

Enamel finishes are commonly used for concrete floors and frequently for soft wood floors and porch floors. The straight oil enamels that give such excellent results on standing work are not satisfactory for this purpose, drying too slowly and remaining too soft to stand traffic. The better quick drying, decorative enamel lines are ideally suited to the work, having just the right combination of hardness and elasticity. Similar properties can be incorporated in special floor enamel lines, which include only the colors that are most popular in this field. The durability of enamels recommended for floors varies greatly among different products. Any good floor enamel is satisfactory for use on porch floors but when maximum results are desired on a porch deck or boat deck it is advisable to use a porch and deck paint. Concrete basement floors present an extremely difficult finishing problem, due to seepage of moisture and water through many floors and the not uncommon presence of soap and grease, particularly around laundry tubs. The application of enamel or paint is attended with considerable uncertainty, even though the floor appears dry and clean. An enamel is less subject to softening, blistering and peeling than a paint, but even the best enamel does not afford assurance of satisfactory results.

WALL FINISHES.

General.

The painting of walls is such an involved subject that a comprehensive treatment of it is impossible in an article of this character. Accordingly, the discussion will be limited to those types of oil paints and enamels generally used for high grade work, omitting industrial or mill white paints, water paints and plastic paints.

Since most walls in residential buildings, clubs, schools, institutions, etc., are of plaster, the discussion will center on plaster walls. Paint technologists have written at length on the varying composition and properties of plaster, but they have been unable to devise any feasible plan for adjusting paint formulas to various plaster conditions. If plaster develops cracking, "popping" or sur-

face disintegration, these, obviously, are conditions that painting practice cannot prevent or correct. Likewise, if plaster contains excessive moisture when painted or moisture gets back of the plaster from some source, peeling is likely to occur regardless of the paint used or painting practice followed. The only point, in which plaster differs from other surfaces from a practical standpoint, is that it is more porous.

Primers.

The most important functions of a plaster primer are twofold. First, it must seal the surface and secondly, it must penetrate the surface. Probably the most common painting defect encountered on plaster walls is variation in the color of the final coat, producing a spotted appearance, which the painter describes as "flashing" or "hot spots". Frequently this is ascribed to "burning" by free alkali in the plaster. In fact, it is usually nothing but unequal absorption, due to failure of the primer to seal the surface perfectly. A good primer should have sufficient sealing that a single application over average plaster will prevent absorption of the next coat. On an exceptionally poor plaster job it may be necessary to touch up spots that show striking in. Assuming that a primer has adequate sealing properties, a reasonable degree of penetration is necessary to assure proper bonding and adhesion. The usual method of testing for satisfactory penetration is by applying a coat of the primer on a sheet of newspaper and noting whether the liquid comes through on the back of the paper.

The second requirement of a plaster primer is that it should have fair resistance to moisture and alkali. The word "fair" is used advisedly, because there is no advantage in a greater degree of resistance. A complete two or three coat wall finish does not permit moisture to pass through readily. If the painting is done before the plaster is dry or water gains access to the plaster through defective construction or a leaking water pipe, the entrapped water will force its way out and blistering or peeling is the natural result. Exceptional waterproofness of the primer is of little if any avail against this trouble. Peeling may occur on one job and not develop on another job where conditions appear to be the same. This inconstancy of results has led more than once to expectations and claims that had to be abandoned with broader experience. Any primer will stick on some jobs before the plaster is thoroughly dry but the writer is not familiar with any product that affords reasonable assurance of sticking on a damp wall.

When walls contain only a fair excess of moisture, failure of the paint usually takes the form of blistering or peeling. A wall in this condition is seldom sufficiently alkaline to cause chemical disintegration of the paint film; in other words, to produce true "burning". Occasionally, however, there is so much water in a newly painted wall that a fairly strong alkaline solution is formed; this solution saponifies the paint vehicle; and discolored areas or brown stains develop. When the condition is this severe no oil or varnish paint would withstand it for long, since all oils and varnishes are subject to saponification by alkali.

A Test for Alkaline Wall-surface Reaction.

The problem of moist and alkaline walls is one that is unlikely to be solved by paint formulation; the real solution is to provide favorable drying conditions, ventilation and heat, and allow plenty of time for the walls to dry. When walls have dried less than sixty days or drying conditions have been adverse, their condition for painting may be judged roughly by testing with a one per cent solution of phenolphthalein in ninety-nine per cent ethyl alcohol. Immediate development of a dark pink color serves as a warning that painting should be postponed. If the pink color develops slowly, there is unlikely to be difficulty. When it is necessary to paint a wall that is in doubtful condition, the surface alkali may be neutralized by brushing on a solution of two pounds of zinc sulphate in a gallon of water. At least one day should then be allowed for the added water to evaporate.

In evaluating wall primers we must not lose sight of a number of other properties that apply to practically all paint products, such as ease of application, leveling, hiding and drying. Brushing properties are very important because of the large surfaces to be coated and the tendency for the porous surface to cause "pulling" under the brush. At one time the standard wall primer was lead and oil. It proved deficient in sealing properties, was slow drying, had poor leveling and was poor in resistance to moisture and alkali. Using a lead and oil primer, it was impossible to obtain a satisfactory job in two coats. The modern wall primers are made with a vehicle of processed linseed oil or processed oil and varnish. The dense hiding pigments such as lithopone and Titanox C receive preference, and zinc oxide may also be used. Some primers of this type surface set in about five hours, although they are soft underneath and make the finishing coat pull under the brush unless allowed to dry over night. The primer should be tinted to the approximate shade of the finishing paint to follow.

FINISHING PAINTS.

On the type of work under consideration walls are usually finished in a flat effect or in a stippled flat or semi-gloss effect. When a plain flat effect is preferred a so-called flat wall paint is used. Flat wall paints have practically replaced lead and oil for finishing walls, due to superior hiding, leveling, drying and hardness. Only one coat of flat wall paint is required over a smooth plaster wall primed with a good primer, although an additional coat is frequently applied for increased body and smoothness. The difference between an ordinary flat wall paint and a superior product lies principally in better leveling, and a more washable surface. An ordinary grade of flat wall paint is likely to show distinct brush marks, which are leveled out by the better product. Many flat wall paints have a surface that is practically dead flat or that is rough, due to poor grinding. In either case, the paint is not completely washable. A good product should have a slight sheen, more apparent when viewed at an angle, and should be so finely ground as to present an absolutely smooth surface. A paint of this kind can be washed with a grit oap and water to remove finger prints, grease, pencil marks, etc., without damage to the surface.

The hiding pigment most widely used in flat wall paint is lithopone. Zinc oxide and Titanox C are also employed. Most flat wall paints contain a small proportion of extender, usually asbestine or whiting. The most common vehicle is heat bodied linseed oil thinned with mineral spirits, and varnish may be added for improved drying and hardening.

Stippled finish originated when lead and oil was the standard wall paint and was primarily a device for eliminating the brush marks that are inherent in lead and oil. With the development of flowing flat wall paints, stippled finish has declined, although still common. Lead and oil requires the application of two coats besides primer for a stippled job and has been partially superseded by special stippling paints that hide densely in one coat over primer. These paints are similar in construction to flat wall paints, except that they contain a substantial quantity of asbestine or other pigments that reduce the flowing quality and prevent the stipple from flowing out. The better products dry to a hard, eggshell or semi-gloss surface that is entirely washable. A stippled finish is advantageous when the plastering is not

smoothly done, as trowel marks and other imperfections are obliterated. For this reason and because of its washability, a semi-gloss stippled finish is popular for schools and other institutions where the conditions of service are severe.

The question of washability of wall paints is so simple that it is difficult to understand the air of mystery that beclouds it. Broadly speaking, the washability of a paint varies in direct relation to its gloss. Dead flat paints can not be washed successfully; as the gloss increases washability improves. A flat paint having a slight, satin sheen can be washed quite successfully; stains from fingerprints, grease and pencils being removed readily and ink stains with some difficulty. Paints having an eggshell gloss or higher permit ready removal of all kinds of stains, including ink. Semi-gloss or full gloss paints differ from eggshell gloss only in that less effort is required. The foregoing statements assume hard drying paints and do not include tincture of iodine or other stains from solutions containing alcohol or other solvents that penetrate deeply into the film.

Enameled Walls.

Gloss or eggshell enamels are not infrequently applied on the walls of bathrooms, kitchens, laundries, etc. The enamel should be of the same type that is used on woodwork. (It is impossible to obtain a satisfactory job in less than three coats and it is seldom that cost considerations allow more than three coats.) A good specification is one coat of wall primer, one coat of enamel undercoating and one coat of enamel. If it is inconvenient to use a wall primer, the first coat may consist of two parts undercoating, one part liquid filler and one part boiled linseed oil, thinned with turpentine to the proper consistency.

Concrete Walls.

When a wall of concrete, cement or stucco is to be finished, a first coat should have exceptional resistance to the alkali that is often present. This has led to the development of special first coaters known as cement coatings. When a flat finish is desired it is only necessary to apply one coat of cement coating and one coat of regular flat wall paint. (High grade cement coatings are recommended for exterior as well as interior work, and make an excellent flat finish or enamel undercoating for exterior concrete or brick surfaces.) (An interior gloss or enamel finish on concrete may be produced with one coat of cement coating, followed by two coats of enamel.)

LACQUER FINISHES.

General.

Nitrocellulose lacquers are of primary interest to manufacturers of automobiles, furniture and many other articles, but of secondary interest to the architect and painter. The extreme speed with which lacquer dries is much more important to the production manufacturer than to the painter and the former is in a much better position to employ the technical methods of application that insure maximum results, comfort and safety. A few years ago lacquer, in brushing form, promised to become a staple product for household use but it has given way to quick drying varnish and enamel. A substantial number of large architectural projects have had the interior woodwork finished with lacquer and it continues to enjoy some favor in this field. This article would be incomplete, therefore, if it did not treat lacquer, at least, briefly.

Clear Lacquer.

The architectural use of lacquer has been principally for clear finish on trim woods such as walnut, mahogany and oak. The staining and filling are much the same as for varnish. Water stain is the only kind that is altogether satisfactory. Filler should preferably be a hard drying type and should have a longer drying period than is required for varnish. Lacquer is especially well adapted to the antique, open pore effects. The type of clear lacquer that is ordinarily used on furniture, radio cabinets, etc., will not give satisfactory service on window sills and baseboards; instead a slower hardening, more elastic and durable type of lacquer should be chosen. Maximum durability is obtained by applying three coats of the clear lacquer directly over the stain and filler; results are also quite good if the first coat is replaced by a suitable lacquer base sanding sealer. Never use shellac or shellac and mixing lacquer. The clear lacquer may be a product that dries with a dull, rubbed effect or may have a full gloss, in which case it is usually rubbed to a dull finish. When trim is finished after erection by brushing, varnish has superior working properties and a more pleasant odor. Most of the large lacquer jobs have had the trim finished at the mill by spraying. Lacquer has less body than varnish but dries faster and is less susceptible to marring. When a lacquer of the type mentioned above is selected, there is little choice in ultimate durability. Lacquer contains more elements of uncertainty than does varnish.

The architectural use of pigmented lacquer has been largely, if not exclusively, by spraying. The main field for interior opaque finishes is on walls and lacquer enamel has been used on the walls of at least a few important jobs. Spraying is the only feasible method of applying lacquer on walls, and in rooms of ordinary size spraying involves working in an atmosphere of dense, disagreeable spray mist and fumes. On a wall the greater speed of drying of lacquer is of scant practical advantage. Possessing less body than a wall paint, lacquer has less power to conceal imperfections in the plaster. Lacquer produces a hard, mar resistant, washable finish, but these qualities can also be obtained in a carefully-chosen wall paint. All things considered, it seems improbable that lacquer will obtain wide use for walls.

Lacquer Composition.

What is lacquer? Essentially, it is a solution of nitrocellulose, resins and plasticizers in a suitable solvent mixture. The nitrocellulose is the distinctive ingredient that differentiates lacquer from varnish and enamel. It is to lacquer what the oil is to a varnish, making it tough and durable. Nitrocellulose is made by treating cotton, which is one form of cellulose, with a mixture of nitric and sulphuric acids. Nitrogen and oxygen are combined chemically with the cellulose to form a new substance—cellulose nitrate, or soluble nitrocellulose. The plasticizer is a material of a soft nature that is added to overcome the tendency of the nitrocellulose to contract and crack. Typical plasticizers are castor oil, raw or blown, dibutyl phthalate and tricresyl phosphate. The resin improves the body, gloss, hardness and rubbing properties. Among the common resins are rosin ester, damar and synthetic resins of both the "glyptal" (glyceryl phthalate) and carbolic acid-formaldehyde types. As in the case of varnish, most resins decrease durability if used in large proportion. This is not rue, however, of some of the glyceryl phthalate resins. There are a large number of lacquer solvents, affording the formulator for opportunity to base this selection on the requirements of the particular lacquer. A lacquer formula usually contains several solvents and representatives from several classes. Firstly, there are the true or active solvents, which have the capacity to dissolve nitrocellulose. Among these are ethyl acetate, butyl acetate, acetone, ethyl lactate, ethylene glycol mono ethyl ether (Cellosolve) and derivatives of

same. Secondly, there are the diluents, which are not themselves solvents of nitrocellulose but may be solvents for the resins. Most lacquers contain as diluents one or more alcohols, principally ethyl and butyl alcohol, and one or more coat tar hydrocarbons, principally toluene and xylol. Some lacquers contain paraffin hydrocarbons, i.e., a special petroleum distillate similar to benzine.

One of the fundamental differences between lacquer and varnish is in the nature of the drying process. In a typical lacquer drying involves only evaporation of the solvent, hence is quite rapid; while in a varnish drying consists in evaporation of the solvent and then drying of the oil by oxidation and polymerization.

The process of manufacturing clear lacquer is relatively simple. Each of the various types of nitrocellulose is dissolved separately by mixing with solvents in a large mixer. No heating is required. This gives a series of clear, viscous solutions. In like manner each resin employed is put into a separate solution. These nitrocellulose and resin solutions are then blended according to the particular formula and the proper addition of platicizer and thinner made. A pigmented lacquer or lacquer enamel is made by incorporating pigments with a clear lacquer. The pigment is ground in part of the clear lacquer, usually in a pebble mill, and the paste thinned with additional clear lacquer and thinner.

CASEIN PAINTS

In general it follows that nearly one hundred gallons of paint are used for interior surfaces for each single gallon used for exterior work. Interior paint being used primarily for decorative purposes and not for protective purposes and with interior surfaces being of plaster or masonry casein paints are highly suitable for the decoration of such surfaces.

Casein paints as they are known consist of casein used as a vehicle with strong coloring or high hiding pigments. Water is merely used as a thinner and does not act as a vehicle.

Some of the desirable properties of casein paint are the high light reflection, pleasing optical effects and truo color values. The casein film is substantially colorless and has none of the yellowing or darkening effects or changes found in other vehicle films. Casein has a great hiding and adhesive strength and a smaller quantity is required to hold the pigment than in oil or gum vehicles. Comparatively, casein absorbs very little light and does not change the color or character of the light when this light is reflected from a surface painted with casein paint.

The great holding power of casein makes for an extremely thin film surrounding the pigment particles and the film itself is porous and clear so that a beam of light reaches the particle of pigment without having passed through any substantial film of vehicle. The light is reflected directly from the surface of the pigment giving a true color reflection of the pigment and without having any light absorbed by the film of the vehicle.

This is the reason casein paints not only theoretically but practically reflect from five to fifteen percent more light in white paint or in any equivalent color than other paints. White casein paints reflect the true color of the pigment and do not obscure it with a yellow or tinted film of the vehicle.

The yellow in the film of vehicle is frequently counteracted or bleached in white oil paints by the addition of small amounts of ultra-marine. This yellow and blue causes a further greying of the paint which still further detract from the true color of the pigment giving it a cast. With casein paint requiring no such bleaching and with light reflected more directly from a pigment particle with a minimum of passage through the vehicle film means that the paint film has a higher percentage of light reflection and is more nearly that of the pigment used. This is the reason for the freshness and clarity of tint in casein paints.

Certain other qualities of casein paints are: Its porousness of surface which permits painting over wet plaster or concrete. The breathing of the surface and the drying process is not impeded and casein may be safely used either as a temporary treatment or as an under coating surface for the further application of casein paint or as an undercoat for any other paint whether it be oil paint, enamel or lacquer.

The flatness or lack of high gloss is a very desirable feature of casein paints. The casein binder having less light reflective qualities than other types of vehicles is responsible for this quality.

Casein paints have excellent sizing and hiding qualities. They are also useful in masking out certain blemishes and in the painting of asphalt coated surfaces casein paint does not dissolve the asphalt and stops or prevents the bleeding.

(Continued from page 535)

leave the forms in place until directed by the Architect to remove same. No concrete shall be poured when the temperature is below 34° Fahr., unless provision has been made for heating and protecting the work.

C. Tile shall be hard burned, free from damage, imperfections and properly scored on all exterior surfaces. Joints in tile between concrete joists shall be staggered in adjacent rows by starting alternately with half and whole tile so that joints in tile will not come opposite each other in alternate rows.

D. Reinforcing Steel shall be deformed bars of hard grade steel offering a mechanical bond with the concrete. Reinforcing shall be free from mill or rust scales and shall be imbedded not less than one diameter nor less than ¾" away from the exterior face of all concrete. Sizes of reinforcing shall be determined by the span and load to be carried as indicated on structural drawings.

(a) **Reinforcing** shall be supported with chairs and wired or otherwise tied in position so as to hold position during the process of placing concrete, substantially as shown by detail drawings.

(b) **Temperature and Expansion Reinforcement** shall be placed on top of tile running continuously in direction opposite to the direction of the length of joists consisting of not less than ⅜" bars placed not to exceed 16" o.c. and lapped not less than 8" at all splices, or in lieu of ⅜" bars approved fabric may be substituted of approximately equal weight and not less than 4" mesh.

E. Wheeling Planks shall be supported above the tile and reinforcing in such a manner as not to cause breakage of tile by vibration when wheeling over same or not to cause displacement of reinforcing.

F. Concrete used in floor construction shall consist of 7½ gal. water to one cu. ft. best American Portland Cement, to two cu. ft. clean sharp sand, torpedo or equal which will pass a ¼" mesh, and 3½ cu. ft. crushed stone or gravel of such a size as will pass through a ½" mesh and be retained on a ¼" mesh.

(a) **Concrete** as placed shall be well spaded and worked around reinforcing after pouring so as to make a solid, dense concrete which shall be perfectly bonded with the reinforcing and entirely free from voids.

(b) **The Placing of Concrete** shall be a continuous operation and the full depth of floor shall be poured at one time.

(c) **All Tile** should be wet before concrete is poured to effect good bonding. Soaking is not permissible. In pouring the ribs the concrete shall not be dumped into the joist, but on the previously placed concrete and worked forward, allowing the mortar to flow ahead in the joist.

(d) **Cessation-of-the-Work Joints** wherever required to be made shall be made exactly in the center of spans by setting vertical stop off dams and wherever cessation-of-the-work joints are made there shall be inserted in same metal stubs or dowels of sufficient number and area to develop at least three-fourths of the complete computed compressive strength of the concrete at the joint without reliance on butted joints in concrete, in excess of ¼ its computed strength. These stubs shall extend into the concrete on each side of the bond not less than 4". Under no circumstances shall stop-off joints be left with an incline from the vertical.

Illinois Society of Architects

1906, 134 N. La Salle Street, Chicago

The following is a list of the publications of the Society; further information regarding same may be obtained from the Financial Secretary.

FORM NO. 21, "INVITATION TO BID"—Letter size, 8½x11 in., two-page document, in packages of fifty at **75c,** broken packages, two for **5c.**

FORM NO. 22, "PROPOSAL"—Letter size, 8½x11 in., two-page documents, in packages of fifty, at **75c,** broken packages, two for **5c.**

FORM NO. 23, "ARTICLES OF AGREEMENT"—Letter size, 8½x11 in., two-page document, in packages of fifty, at **75c,** broken packages, two for **5c.**

FORM NO. 24, "BOND"—Legal size, 8x13 in., one page document, put up in packages of twenty-five, at **25c** per package, broken packages, three for **5c.**

FORM NO. 25, "GENERAL CONDITIONS OF THE CONTRACT"—Intended to be bound at the side with the specifications, letter size, 8½x11 in., ten-page document, put up in packages of fifty at **$2.50,** broken packages, three for **25c.**

FORM 26, CONTRACT BETWEEN ARCHITECT AND OWNER. Price, 5c each, in packages of fifty, **$1.25.**

FORM 1, BLANK CERTIFICATE BOOKS—Carbon copy, from 3¾x8½ in., price, **50c.** Two for **5c.**

FORM 4, CONTRACT BETWEEN THE OWNER AND CONTRACTOR—(Old Form.) Price, two for **5c.** Put up in packages of 50 for **$1.00.**

FORM E, CONTRACTOR'S LONG FORM STATEMENT—As required by lien law. Price, two for **5c.**

FORM 13, CONTRACTOR'S SHORT FORM STATEMENT—Price, **1c** each.

CODES OF PRACTICE AND SCHEDULE OF CHARGES—8½x11 in. Price, five for **10c.**

These documents may be secured at the Financial Secretary's office, suite 1906, 160 N. La Salle St., telephone Cent. 4214. We have no delivery service. The prices quoted above are about the cost of production. An extra charge will be made for mailing or expressing same. Terms strictly cash, in advance, with the order; except that members of the Society may have same charged to their account.

GLASS AND GLAZING

Glass making is one of the oldest known crafts, and its origin is lost in antiquity. However, it is only in comparatively modern times that glass has been recognized as a highly useful product, and has been made for purposes other than ornamentation. Although blown glass had been manufactured for some time previously, it was not until the 17th century that a process of manufacturing cast glass was developed in France from which the present method of making plate glass has been evolved. Present day large scale production of plate and window glass has been made possible by mechanical improvements perfected within very recent years. Blown glass was made on a small scale in the settlement of Jamestown, Va., about 1608, but it was not until 1860 that the first cast plate was produced in this country. The intervening seventy years between then and now were a long uphill struggle against foreign competition and technical difficulties, and the greatest improvements in manufacturing methods have been made within the last two or three decades. Present day glass manufacturing is one of the most modern of industries.

PLATE GLASS.

Manufacture.—Plate Glass is the finest flat glass made, and its superior quality is derived from its painstaking process of manufacture. The distinctive quality of Plate Glass is its perfectly flat parallel surfaces which insure clear, undistorted vision. As will be seen, the most difficult and laborious part of plate glass manufacturing—the grinding and polishing process—is directed toward the attainment of this result. The ingredients of Plate Glass are practically the same as those of Window Glass, the difference in the finished products being due to the more elaborate process of casting, grinding and polishing plate glass in contrast to the comparatively simple method of producing window glass by drawing it vertically upward in a flat sheet from a tank of molten glass. The principal ingredients of the melting batch are silica (white sand), soda, lime, and salt cake, to which are added proportionately small amounts of arsenic, charcoal, and cullet (broken glass).

At the present time, plate glass is made by one of two methods; it is either melted in a pot and cast, or it is melted in a huge tank from which it flows out through rolls. Both these processes are described briefly below:

Pot Making.—The pots of fire-clay in which the materials are melted together and fused, play such an important part in the successful manufacture of cast plate glass that the subject of pot making deserves special notice. The fact that it requires several years to make a pot, which lasts in service but a limited time, entails a heavy expense in the manufacture of cast plate glass.

The clays used in making the pots are exposed to the weather for a period of from one to two years, so that they will become thoroughly seasoned. They are then ground and screened and the finely sifted raw clay is mixed with coarse burned clay and water. The addition of burned clay reduces the liability of shrinking and cracking. The mixture is then kneaded and stored for about six months to ripen.

The pot itself is made by hand by highly skilled workmen, as the slightest defect, such as an air cavity, would cause the pot to crack in the furnace. The finished pot is then stored for from six months to a year for its final seasoning. Each pot holds a ton and a half of glass, and weighs 3,000 pounds. The average pot will cast approximately 600 sq. ft. of glass. The pots must be able to withstand a temperature of 3,000 degrees Fahrenheit as well as sudden changes in temperature. The life of the average pot, even after the greatest care has been taken in its manufacture, is only twenty days.

Melting and Casting.—Plate glass is melted in large gas heated furnaces which hold from twelve to twenty pots. A temperature of from 2,500 to 3,000 degrees Fahrenheit, for a period of 24 hours, is required to complete the melting. During this process the pot must be filled three times to insure a full pot of molten glass, as the intense heat causes a shrinkage of the raw materials. When the melting has reached an exact point, the pot is lifted from the furnace by huge electric tongs and taken to the casting table.

Until the last few years, plate glass was cast on a large rectangular metal table, and this method is still employed in making unusually large or thick plates. The contents of the pot were emptied on the table and flattened by a steel roller weighing twenty-five tons, making an enormous molten sheet about half an inch thick. This sheet was then shoved into an annealing oven, known as the "lehr," to prevent sudden cooling and consequent cracking. The lehr consists of a series of ovens, each succeeding one being slightly lower in temperature, through which the cast plates are shoven to insure gradual cooling.

The present method of casting plate glass, developed very recently, is slightly different from this procedure. The molten contents of the pot are now cast between two moving rolls which produce a sheet more uniform in thickness than the old process. This sheet immediately enters the lehr and is cooled gradually until it emerges at the other end, a large sheet of rough glass.

Tank Glass.—One of the most radical mechanical improvements in the manufacture of plate glass is the tank or continuous process. This method eliminates the necessity for pots and does away with the casting process. The raw materials are melted in a large tank built of refractory blocks, which is kept at a constant high temperature. At regular intervals additional quantities of raw materials are added to keep the tank full. A constant stream of molten glass flows from one end of the tank through rollers into the lehr. This continually moving ribbon of glass, passes slowly through the lehr and is cut off in the proper lengths at the cool end. In spite of the apparent simplicity of this method compared with the more laborious pot method, it is a triumph of modern engineering skill.

Grinding and Polishing.—The two processes of making plate glass described above both produce about the same kind of material —large plates of rough glass having no resemblance to the fine, highly polished plate glass to which the layman is accustomed. Although the rough glass is flawless and has the same inner transparency as the finished product, it remains for the grinding and polishing process to bring out its hidden quality and make it distinctively plate glass.

The rough plates are placed on flat topped steel cars and set in plaster of paris to keep them from slipping. Formerly the plates were placed on large revolving tables, but grinding and polishing is now a continuous process and literally, train loads of rough glass are constantly passing through the grinding and polishing machines. The cars are coupled together so that a continuous ribbon of glass is being ground and polished at the same time.

Grinding is done by means of massive ironshod runners which revolve over the surface of the glass. Water and sand are fed under the runners as they revolve, and under this powerful abrasive action, the surface is ground with absolute uniformity until all the

irregularities in the rough glass are removed. As the ribbon of glass passes further along under the battery of grinding machines, finer sands are substituted for the coarser grades until, during the final stages, a still finer abrasive—emery—is used in several degrees of fineness. The glass is then washed and the cars continue on to the polishing machines.

Polishing is similar to the grinding process, but instead of iron runners, the polishing machines are equipped with buffing disks of felt. Red oxide of iron, commonly called rouge, the finest known abrasive, is fed under the buffing disks as they revolve. This polishing process gives plate glass its beautiful brilliant surface.

After one side of the plate has been ground and polished, it is reversed and the process is repeated on the other side. Grinding and polishing reduces the plates to their proper thickness and it involves considerable loss of material as a large amount of the glass is ground away. The finished plates are inspected carefully for defects and cut to proper sizes.

Sizes.—Polished plate glass in regular 1/4" thickness, which has a tolerance from 7/32" to 9/32", can be made in very large sizes over 200 square feet in area. Plates as large as 144" x 240" (or 12 feet by 20 feet) containing 240 square feet have been made. However, such extraordinary sizes are very difficult to make, expensive, and dangerous to clean and handle. They are especially made to order, entail considerable delay in replacment when broken, and require special flat car shipment and special facilities for unloading and handling, as well as the most expert and skilled glaziers in setting. It is advisable, therefore, to confine the sizes of the plates required within the limits of 122" x 194". Prompt and economical deliveries of plates within these limitations may be secured from distributors' stocks. It is customary to measure glass sizes in inches, specifying the width first, then the height.

Thickness.—Although polished plate glass is manufactured in thicknesses ranging from 7/64" to 1 1/4", the standard produce is known as 1/8", 3/16" and 1/4" plate and is allowed to run from 7/32 to 9/32 of an inch thick. The other thicknesses are made specially and at an increased cost. The sash or rabbet for regular plate glass glazing should he made to accommodate glass full 5/16 of an inch thick.

1/8" PLATE GLASS.

This thickness of plate glass was developed to meet the demand for a fine plate glass of the weight and thickness to permit its use in standard sash with standard sash weights.

1/8" plate glass meets these requirements. It has all the visual advantages of heavier plate glass—high polish on both surfaces, freedom from distortion, and the ability to transmit with perfect clarity all objects seen through it.

This plate glass is glazed in standard 1 3/8" sash with ordinary sash weights.

Physical Characteristics of 1/8" Plate Glass.—1/8" Plate Glass is available in the same qualities as heavier plate glass.

Its range of thickness is from 6/64" to 9/64", and approximate weight per square feet—1.75 lbs.

Its strength is in direct proportion to the square of its thickness. Maximum size obtainable—72" x 123".

HEAVY PLATE GLASS.

Plate Glass thicker than the standard product is known as Heavy Plate Glass.

This Glass retains the qualities of the usual thickness of Plate Glass—clearness, visibility and polished surfaces—as well as being impervious to moisture, weather, cleaning chemicals, pencil marks and other disfiguring agents. Its strength, which is in proportion to the square of its thickness, is a feature that stands as a reason for many of its uses.

Among the home uses for Heavy Plate Glass are: book shelves, decorative panels and partitions, shower bath enclosures, end tables, modernistic furniture, table tops, clocks, paper weights, book ends.

In Commercial and Public buildings some of its uses are: bank fixtures, booths, Theatre Marquises, valances, lighting fixtures, radio acoustic chambers, refrigenator doors, soda fountain counters, counter tops, glass flooring and valances.

Other uses for Heavy Plate are: glass bottom boats, aquariums, ships, bullet resisting glass core portlights, submarine lighting of swimming pools, etc.

Physical Characteristics of Heavy Plate Glass.—It is available in Selected Commercial Quality in thicknesses ranging in 1/8" progression from 3/8" up to and including 1 1/4".

The weight per square ft. for 3/8" thickness is approximately 4.93 lbs.

The weight per square ft. for 1 1/4" thickness is approximately 16.45 lbs.

Maximum size—72" x 160".

Qualities.—Plate glass is made in three qualities ranging from the finest mirror or silvering quality to regular glazing quality. Quality is determined by the number and seriousness of defects in a plate, and the type of polish given to the surfaces. Regular glazing quality may contain some defects which in no way impair the value, beauty, or durability of the glass for ordinary use, such as small seeds or bubbles, short finish, ream or surface scratches, which are accepted as contingent with the regular run of plate glass, and even an open bubble or shot-hole (not clear through both surfaces) is passed in glazing quality, providing the plate is comparatively free from other defects and of good color and finish.

Weight.—Plate glass 1/4" thickness (7/32" to 9/32") weighs approximately 3 pounds per square foot bare, and its weight boxed for shipment may be roughly computed at 5 pounds per square foot. An easy method of figuring the approximate shipping weight of plate glass is as follows:

Extend the glass at 3 pounds per square foot. The weight of the box equals the square foot area of a plate of the greatest width and length of those packed therein, multiplied by 10. Thus:

1 plate 36" x 96" }
1 plate 60" x 84" } =59 sq. ft. x 6 =177 lbs.
Size of box—60" x 96"=40 sq. ft. x 10=400 lbs.

Total weight of shipment=577 lbs.

Edgework.—The value of plate glass for furniture tops, desks and tables, show-cases, shelves and numerous other purposes is generally recognized. The glass covering offers a clean, sanitary surface, and in addition to protecting and preserving the furniture, the highly polished surface of the glass enhances its appearance. For most of these purposes the edges of the plates must be polished—a process requiring considerable skill. The edges of plate glass may be polished to almost any specifications—they may be rounded, made square, or beveled to any desired width. The process of grinding and polishing the edges, or rounding of corners, curves or pattern lines may be roughly divided into five operations as follows:

Roughing is done on a large cast-iron wheel about 30" in diameter, having a corrugated surface which revolves rapidly in a horizontal plane. Either sand or carborundum mixed with water is used as an abrasive. The edge of the plate is brought in contact with the wheel until the proper amount of glass is ground off. Curved plates and those cut to pattern require great skill on the part of the operator.

Blue Green. The backing for any color of glass may be silver, gold or gunmetal.

Copper-Back Mirrors.—A new process has lately been introduced in mirror manufacturing known as copper-backing, which produces a mirror very much superior to and much more durable than ordinary patent-back mirrors. Copper-back mirrors are made the same way as any fine plate glass mirrors, except that, in addition, a layer of solid copper is deposited on the silver coating by an electrolytic process. This provides an impervious insulation, protecting the sensitive silvering from dampness and chemical reaction, and making the backing of the mirror more resistant against handling abuse. The delicate film of silver is completely protected by a sheathing of copper.

Copper-back mirrors are made in sizes up to 70" x 140".

Shocks.—The common sheet-mirror or looking-glass, used principally for the reflection of light rather than detailed images, is known to the trade as a "shock-mirror," and is made from ordinary window glass.

Installation.—Standard patent-back mirrors are susceptible to the effects of extreme cold or heat and moisture, and should be mounted with proper protection against dampness. An air space should be left between the mirror and wall, and care should be taken to avoid damp walls or plaster which has not properly dried out, before installing mirrors.

In glazing French doors with mirrors, or on Colonial work where small mullion glazing is specified, it is essential to have the panels on an absolutely uniform line, and rabbets of accurate depth, as the mirrors will otherwise reflect at different angles, resulting in distortion. A good effect may be obtained by using a large size mirror for a background with a false mullion over all.

BLUE AND FLESH TINTED PLATE GLASS.

Flesh Tinted Plate Glass.—As suggested by the name—a color tinted plate glass of a shade approximating the skin color commonly found in a Caucasian person. The color is slight in surface section, but considerably stronger in transverse section. Used in mirrors, Flesh Tinted Plate Glass produces reflections which minimize blues and violets, and emphasize flesh colors, thus offering flattering images.

Blue Plate Glass.—This is a plate glass of rich blue color, ideal for decorative use in modern homes and buildings. Blue Plate Glass is unusually attractive when fabricated into mirrors.

Flesh Tinted Mirrors or Blue Mirrors can be used with great success in home decoration, adding color, warmth and sparkle to all types of rooms. They are exceedingly ornamental in bars, stores and public and commercial buildings of all types. In addition, they both serve very well as unusual table tops, desk tops, book shelves, sill covers, etc.

Physical Characteristics of Flesh Tinted Plate Glass and Blue Plate Glass.—They are available in 13/64" thickness—of selected quality plate glass. Their weights are approximately 2⅝ lbs. per square foot—and strength equivalent to that of clear plate glass of equal thickness. Maximum sizes available—72" x 123".

WATER WHITE PLATE GLASS.

This is a plate glass that is true water white and colorless both in surface and in transverse section.

Water white plate glass has been developed primarily for use in multiple-glazing as in refrigerator cases, and in double-glazing of windows for purposes of insulating and air conditioning. This is the only type of glass so

PRESSED PRISM PLATE GLASS CO.

General Offices 33 North La Salle Street • Chicago

MANUFACTURERS OF

"Imperial" Ornamental Plate Glass

Styles O-1, O-2, O-3, O-4, O-5
Rubanite O-6, Texture O-7, and Other Special Patterns

"Imperial" Ornamental Plate Glass is made in one standard plate glass quality and furnished in standard plate glass thickness $\frac{1}{4}$ to $\frac{5}{16}$ in. The maximum size is 72x84 inches.

We have illustrated below three specimen patterns in full size. Samples of these and others in stock or in course of manufacture will be furnished on request. Correspondence regarding new or special designs is solicited.

The difference between "Imperial" and other Ornamental Glasses lies in the special process by which the brilliant patterns of "Imperial" are made. This process combines the essential features of polished plate glass manufacture with a die-pressure method of forming the characteristic ornamental prismatic surfaces.

These Illustrations Are Only Typical—Send for Other Samples

"Imperial" Style O-1 "Imperial" Style O-2 "Imperial" Rubanite O-6

far developed which does not effloresce or "bloom" when hermetically sealed with an air space between two sheets.

Water white plate glass has a transmission value for all the colors of the spectrum that is very nearly uniform, (88% to 92%) with a consequent transmission of the violet and blue light rays much higher than that of ordinary plate glass. As a result it transmits faithfully the natural colors of objects seen through it without changing the relative intensities of the colors, no matter how delicate the differentiation of tone and shade may be.

It will be found that this glass is excellent also for use in display cases of all kinds, and in mirrors where it is desired to obtain as nearly true reflections as possible, such as in dress shop and beauty shop interiors, etc. Jewelry stores have found water white plate glass very effective in the true to color display of their merchandise.

Physical Characters of Water White Plate Glass.—It is available in silvering quality in 7/64" thickness—weight per square foot, 1.44 lbs.

It is available in glazing quality in ¼" thickness—weight per square foot, 3.25 lbs.

Its strength is that of ordinary plate glass of equal thickness.

Maximum size obtainable—123" x 216".

TEMPERED PLATE GLASS.

This is polished plate glass, specially processed by heat and chilling so that it will support a weight 4 times as great as before tempering. It will also bend 4 times as far without breaking and its resistance to impact is 7 to 8 times greater. Plate glass can be tempered in thickness of from ¼" up to and including 1¼".

Varying surface temperatures do not affect this tempered plate glass as it can withstand, without breaking, a temperature of 650° F. on one surface, while the other is at ordinary room temperature. Its resistance to shock and impact is as strong at 15 degrees below zero F. as at ordinary temperature.

When tempered plate glass does break, it does not shatter into sharp fragments like ordinary glass, but crumbles into small, comparatively blunt-edged fragments.

Uses of Tempered Plate Glass are almost limitless where strength and safety are important considerations. It proves extraordinarily satisfactory when used for entrance doors and doors in lobbies of buildings and offices, for aquariums, cell doors, deck lights, fire screens, flooring, gas cooker doors, glass bottom boats gauge guards, kitchen equipment, laboratory equipment, partitions, portlights, road traffic signs, shelves, show cases, sight glasses, table and dresser tops, underwater lighting, etc.

Physical Characteristics of Tempered Plate Glass.—Its quality, thickness and weight are that of the glass before tempering.

Its strength is approximately 4 times that of glass of equal thickness before tempering.

Maximum size in which it may be obtained—48" x 108".

LAMINATED GLASS.

Laminated glass, commonly known as "Safety Glass," is standard equipment on all new automobiles. For this purpose, two sheets of either plate or window glass, with a transparent flexible plastic between, are welded into a single unit. In this form, the tendency of pieces of the glass to fly, when it is broken, is greatly lessened.

In multiple laminations of Plate Glass, its resistance to impact is increased to the extent that this type of glass is used for bullet resisting purposes.

In the fabrication of Bullet-Resisting Laminated Plate Glass, the same principles apply as in the lesser thicknesses of Safety Glass, with however from 3 to 5 sheets of plate glass being laminated into units from 1⅛" to 2" in thickness. Specifically, a heavy plate glass core is used in the center, with considerably thinner sheets of Plate Glass, being laminated to either side, using clear, flexible and strong plastic between.

The protective qualities of Bullet-Resisting Laminated Plate Glass are—for the commonly used 1⅛" thickness—that it will withstand from 4 to 10 shots from a .45 or .38 caliber revolver without penetration. Shots from revolvers of .32 caliber have practically no effect on this glass. It will, in fact, withstand scattered shots from a Thompson Sub-machine gun.

The 1½" thickness of Bullet-Resisting Laminated Plate Glass has been designed to offer protection against the new Smith & Wesson Magnum revolver which is the hardest-hitting sidearm so far developed.

In thicknesses of 2", this glass offers protection against shots from a 30-30 standard sheriff's weapon, and will withstand a shot from the terrifically powerful 30-06 Army Springfield, which has such force that it will penetrate seventy-five ⅞" pine boards or ½" thick boiler plate.

Banks and other institutions dealing with large sums of money or other valuables are users of the combined protection and visibility of Bullet-Resisting Laminated Plate Glass. Armored cars and trucks also make use of these qualities.

This type of Multiple Laminated Plate Glass can also be obtained in thicknesses of from ½" to 1⅛", which are highly suitable for safety shields in work-shops and laboratories.

Physical Characteristics of Bullet-Resisting Safety Plate Glass.—Weight per sq. ft. for the ½" thickness is 6.91 lbs., and for the 2" thickness is 27.11 lbs.

Maximum size 45" x 84".

Permissible tolerance in thickness—plus or minus ⅛".

HEAT ABSORBING PLATE GLASS.

This is a specially processed plate glass which absorbs heat. Thus, while it admits 70% to 75% of the sun's total light, it transmits less than 43% of the total heat.

When windows, skylights, etc., are glazed with heat absorbing plate glass, the solar heat entering that building through such openings is greatly reduced. Persons sitting adjacent to these windows are far more comfortable, and the glare resulting from high light intensity is considerably lessened.

Specifically, heat absorbing glass may be employed to advantage in southern and western exposures of all types of buildings, whether schools, residences, factories, hotels or office buildings, and will result when so used in greater bodily and visual comfort for building occupants. Used in glazing textile factories or warehouses, this glass prevents fading and bleaching of delicate colored fabrics from exposure to sunlight.

Physical Characteristics of Heat Absorbing Plate Glass.—It is available in glazing quality in ¼" thickness—wt. per sq. ft. 3.29 lbs.

Its strength is that of ordinary plate glass of equal thickness.

Maximum size obtainable—72" x 123".

X-RAY LEAD GLASS.

This is a glass developed primarily to protect X-Ray operators against continuous exposure to X-Rays, at the same time allowing clear vision of the apparatus and patient.

In single thickness it affords protection against X-Ray tubes operating under an impressed voltage of 100 K. W.

Its "lead coefficient"—.30—multiplied by the thickness of the glass gives the equivalent sheet lead thickness.

Physical Characteristics of X-Ray Lead Glass.—X-Ray Lead Glass is available in thicknesses of from 5.35 mm. to 7.35 mm.,

with maximum size of 40" x 72". Approximate wt. per sq. ft. is 5½ lbs.

The strength of this glass is approximately ⅔ that of ordinary plate glass of equal thickness.

Its color is golden yellow.

DOCUMENT PLATE GLASS.

This is a clear, canary-yellow plate glass, specially processed to minimize the harmful effects of light rays which cause paper and ink to fade and discolor.

It is particularly valuable in protecting rare documents in museums, libraries, etc.

Physical Characteristics of Document Plate Glass.—It is available in ¼" thickness, maximum size 40" x 72"—with approximate wt. per sq. ft. 5½ lbs.

The strength of this glass is approximately ⅔ that of plate glass of equal thickness.

BENT GLASS.

All kinds of flat glass can be bent—including window glass, laminated glass, plate and structural glass. In the process, flat sheet of manufactured glass is placed over a mold made in the desired shape. The glass is then heated until it softens and sinks, taking the shape of the mold. It is then carefully annealed.

Plate glass is the best type of transparent glass for bending, its polished surface retaining full brilliance and undistorted transparency.

The maximum size which can be bent is 8'4" x 12'0". Patterns or templates of the desired curve are necessary to insure an accurate mold and satisfactory results. It is not recommended that glass be bent to a curve exceeding a half circle or to acute bends approaching right angles, due to the risk of breakage or injury to the polished surfaces.

WINDOW GLASS.

Manufacture.—The modern method of manufacturing window glass by drawing it flat in a continuous sheet is a radical improvement over the old method of blowing and flattening cylinders by hand. The method in brief is as follows:

The raw ingredients (or batch) are melted in a large tank built of refactory blocks. From this tank the glass is drawn vertically upward as a continuous flat sheet, through annealing ovens to be cut off in finished sheets of the proper length. Although this process sounds quite simple, it has required years of experience in window glass manufacturing and a vast amount of technical research to perfect. The results of this new process have revolutionized the window glass industry. There are no more the uneven thickness, surface burns and stringy, wavy effects so long associated with window glass. Instead, flatness and brilliancy are achieved which never before were believed possible in fire finished glass. As either surface is equal in quality and the glass is no longer bowed, either surface may be glazed outward—thus eliminating a precaution which was necessary in glazing the old blown window glass.

Sizes.—Window glass in all thicknesses, except the unusually thin picture glass, is made in sizes up to 60" x 80". Picture glass is made in sizes having a maximum length of 40".

Thickness and Weight.—The thickness of window glass is generally measured by the number of lights contained in one inch, and a small variation is allowed either way. Picture glass measured approximately 13 lights to the inch and weighs about 16 ounces per square foot. Single strength measures from 10½ to 11 lights to the inch and weighs from 21 to 22 ounces per square foot. Double strength measures about 8 lights to the inch and weighs approximately 26 ounces per square foot. The "39 ounce" grade of heavy window glass runs approximately $\frac{3}{16}$" in thickness. Heavy window glass in the $\frac{7}{32}$" thickness weighs about 45 ounces per square foot.

Qualities.—There are three general classifications for grading the quality of regular single and double strength window glass as follows:

A.A.—The best quality obtainable in window glass. Higher than a commercial necessity.

A.—The highest grade for special commercial uses. Contains no defects to perceptibly interfere with straight vision.

B.—Glass free from glaring defects—but containing such imperfections as prevent its being graded as "A.A." quality or "A." quality.

Picture glass is made in superfine, select and commercial qualities which correspond approximately to AA, A, and B qualities as above, respectively. Heavy window glass is obtainable in two qualities, select and factory run.

Packing.—Window glass is packed in regular sizes approximately 50 square feet to the box up to the 100 united inch bracket (obtained by adding the width and length), and 100 square feet to the box in sizes over 100 united inches.

Shipping Weight.—Single strength in factory packages weighs from 65 to 75 pounds to the box (shipping weight.) Double strength in factory packages weighs from 85 to 110 pounds to the box, 50 foot boxes (shipping weight). Double strength in 100 foot cases weighs approximately 225 pounds.

WIRE GLASS.

The use of metal frames, metal window sash and fire-proof construction has increased the demand for wire glass, until the production of this material amounts to millions of square feet annually. Not only does this glass minimize the fire hazard, but its unyielding qualities even when cracked, make it the logical glass for skylights, elevator shafts, and stairwells, where these features are an important consideration.

Manufacture.—Wire glass is made by three methods: (1) Shuman process—by rolling a sheet of glass, laying the wire mesh upon it while the glass is still plastic, pressing the wire netting into the glass, and by a coincident process smoothing the surfaces. (2) Appert or Schmertz process—by rolling a thin sheet of glass and laying the wire mesh upon it and simultaneously pouring and rolling a second sheet of glass on top, embedding the wire. (3) Continuous or solid process—by mechanically crimping the wire netting and placing it on the casting table and pouring and rolling the glass over it to produce a sheet of wire glass. Various types of obscure and figured glasses (described in the next section) are made both plain and with wired construction.

Sizes and Thicknesses.—Wire glass is made in sheets as large as 50" x 130". The standard thickness is ¼", which has been approved by the National Board of Fire Underwriters. Other thicknesses are $\frac{3}{16}$" and ⅛" as well as ⅜" or heavier for special purposes.

Underwriters' Requirements.—It is necessary to follow certain rules and regulations in the making of fire-proof windows and construction, as provided by the National Fire Protection Association, and a copy of the requirements of the National Board of Fire Underwriters may be obtained from any member of the National Glass Distributors' Association.

Extracts from "Regulations of the National Board of Fire Underwriters for the Protection of Openings in Walls and Partitions Against Fire"

Effective October 15, 1930.

5212 **"Size of glass for fire windows**—Area of wired glass between supports shall not exceed 720 square inches, and the longer dimension of the glass shall not exceed 54 inches."

5214 **"Glazing Fire Windows**—(a) Windows shall be glazed with standard ¼-inch wired glass fitting the provided glass opening as closely as possible. The clearance between the edges of the glass lights and the metal forming the glass opening usually varies between 1/16 and ⅛ inch depending on the size of the glass. Glass shall be held in place and be mechanically secure without depending on the putty which is for weatherproofing only."

There are also restrictions and regulations governing the depth of rabbet, bearing of glass and style of metal frames and sash to meet the demands of fire retardant construction and to permit reglazing.

OBSCURE GLASS AND GLASS WITH PATTERNED SURFACES

Several kinds of glass are now manufactured which will admit light without permitting vision. These glasses are particularly suitable for partitions, doors and transoms in private offices, corridors in buildings, windows which face other windows or an unwanted view, as well as for other uses of a purely ornamental character.

Ground or Sandblasted Glass is made by blowing sand against either or both surfaces of the glass with compressed air, which gives it a frosted appearance. The sandblasting may also be done on only part of the plate or in designs and lettering.

Single and Double Process Chipped Glass is obtained by coating the surface of the glass with strong glue which, on drying, tears off thin flakes of glass and produces a delicate tracery pattern. When this process is repeated, in making double process chipped glass, a still more effective design results. Chipping may also be done only on part of the plate or in accordance with a pattern.

Figured Glass is a new type of obscure glass having an unusually fine texture of surface. Its ease of cleaning, excellent diffusive properties, and attractive appearance, make its uses practically limitless. This glass is made in both 3/16" and ¼" thickness in sizes up to 60" x 144". Both thicknesses are made with the tapestry finish on both sides, and the ¼" thickness is also made with one side polished.

Rolled Figured Glass is a cast glass which, instead of being ground and polished, is impressed on one side with a more or less ornamental pattern. These patterns are essentially prismatic, and as such serve to diffuse and distribute light. This type of glass is made in a wide variety of designs which, while admitting all the light, afford privacy and have excellent decorative possibilities. As mentioned above, some types of this glass are made with wired construction. Also several types are made with the smooth side ground and polished.

Embossed Glass and Etched Glass are produced by treating the surface with hydrofluoric acid. Embossed glass is translucent without being transparent, and has a delicate satin-like finish, similar to fine sandblasted glass. Etched glass is made in the same way in an endless variety of patterns.

Prismatic and Diffusing Glass.—Prism glass and diffusing glass is made in sheets by the rolled process and in smaller units by the pressed process. In sheets this glass is obtainable in sizes up to 60 inches high by about 120 inches long in the plain, and in slightly smaller sheets in the wired.

In pressed glass the units usually vary from 4 inches to 6 inches either in square or some geometrical shape, and these units must be built in metal frames to required sizes to fit openings.

Thickness varies in both sheet and pressed glass from approximately 3/16 to ¼ inch and shipping weight may be computed at from approximately 3 pounds per square foot for plain glass to 5 pounds per square foot for wired or metal glazed units.

There are several standard makes of prism glass as well as a very few standard and recommended designs of obscuring and diffusing glasses obtainable and it is especially desirable that careful consideration be given the selection and specification of a design of prism or diffusing glass best suited to give the desired results, depending upon the location and the conditions surrounding the openings in which the glass is required.

GLASS BLOCKS.

History.—Opaque Structural Glasses made their first appearance about 1910. Since then, their use has grown steadily, and they are now in common demand in the building industry for surface veneering. About 1929 a new form of structural glass was introduced from Europe. This was a practical common hollow block unit of glass. After many refinements and experiments with blocks of various kinds such as solid blocks and dished blocks of pressed glass, the present type of sealed, hollow, partially evacuated construction blocks was developed.

Manufacture. — Essentially, manufacture consists of mechanically pressing the two square halves and then sealing these together so as to make strong, hollow weather-tight blocks. The modern machine method of glass pressing is used. The glass is melted in large continuous gas-fired tanks, and fed in a molten stream through funnel-shaped, refractory lined feeders where the quantity is controlled by a reciprocating plunger, and automatic cut-off shears. From here the gobs of molten glass drop, guided by means of a chute, into polished cast iron moulds, six to twelve of which are mounted on a table revolving beneath the feeder and indexing with it. The glass is then pressed immediately to shape by a water-cooled cast iron plunger which registers with each mould on the table beneath it. After passing through successive indexing stages on the table, where the glass is cooled by blasts of air to the point so that it can be handled, the mould mechanically releases and the ware is removed. It is then placed on a conveyor belt and transferred to the next operation.

In this operation the two halves of the blocks are sealed together. This can be done in either one of two ways. One consists of reheating locally the four edges of the two half blocks simultaneously to fusing temperature, and pressing them together to form an integral glass sealed unit. The other method is to dip the four edges of the two halves comprising the block into molten aluminum, immediately followed by sealing in a press as before.

In the former method, the glass halves are placed in the chucks of a mechanically operated machine which rotates them in stages over and beneath stationary burners which come between the halves and heat the sealing edges. On the subsequent stages of the machine these halves are pressed together and held the very short interval necessary to set the glass. The chucks then automatically release and a mechanical "take-off" removes the finished block which is immediately placed in a lehr for annealing.

After the blocks are annealed they are coated on the four edges or mortar surfaces,

PYRAMID *Snap-On* METAL MOULDINGS

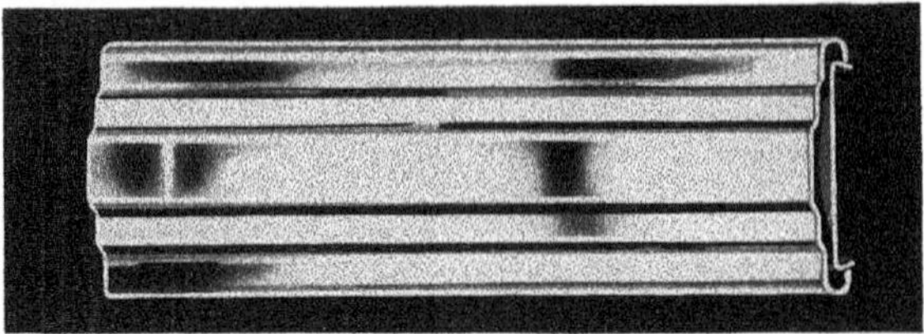

Pyramid Metal Mouldings afford a distinctly modern decorative trim. They are easy to install. The SNAP-ON feature conceals all nails or screws.

Pyramid Stainless Steel Mouldings are *not plated.* There is no coating to flake or wear off. They are made of Stainless Steel throughout—the ONLY Chromium metal that does not rust, tarnish or corrode.

The SNAP-ON feature of Pyramid Metal Mouldings insures economical application for new work or remodeling. Often this feature permits a labor saving of more than fifty per cent.

Installation is made after all other work is finished without regard to joints, thickness, or kind of wall covering material.

The many standard styles and sizes of Pyramid Metal Mouldings afford a wide variety of patterns to meet nearly every architectural need. Flat Top, Half Round, Corrugated Top, Half Round and Flat Top Ripple Corrugated. Special designs can be made to order.

Pyramid Mouldings are also available in Brass, Bronze, Copper or Aluminum. Bright Mirror or Satin Velvet Finish may be had on all metals. Catalog showing many standard patterns and many typical installations will be sent upon request.

Pyramid SNAP-ON Mouldings are easily applied. A little care will insure a neat attractive installation.

All SNAP-ON Mouldings have two members; a *track* or inner member and a *cover* or outer member.

HOW TO APPLY THE TRACK

Figure 1 Figure 2

1st. Fasten the track to the background to which the moulding is to be applied. Drive in the nails or screws as shown in *Figure 1. Be sure* they do not distort the *track* as in *Figure 2.* The nail heads should be flush with the edges of the track so they cannot dent the *cover* and mar the top surface.

2nd. On wood, wallboard or similar material, use cement coated nails of suitable length. On plaster walls, use shoemaker nails, driving them against a lath so the soft point will bend and form a key in the plaster. On *soft* wallboard, *always* use screws. They may also be used in paster walls.

HOW TO SNAP-ON THE COVER

Mouldings 1″ wide or less

Figure 3

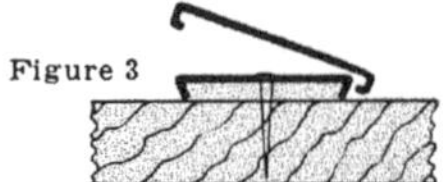

Place one edge of the *cover* under one edge of the *track.* On a horizontal application, the lower edge should be placed first. See *Figure 3.* Then apply pressure with the heel of the hand, pushing toward the opposite side of the *track* and the *cover* will readily SNAP-ON. Follow this method along the moulding *one foot* at a time.

No force is necessary. Use only your hands. *Do not* use tools or hammer.

Full data showing methods of cutting, mitering, bending and installation on other materials will be sent upon request.

PYRAMID METALS COMPANY

1335 North Wells Street **Chicago, Illinois**

with special water and alkaline resistant materials which permit the mortar to bond to the glass. They are then packed for shipment.

By a change in design of either or both the plungers and the moulds, a number of different sizes and face patterns are made. Blocks are now made in four sizes: 4¾" x 7¾"; 5¾" x 5¾"; 7¾" x 7¾" and 11¾" x 11¾"; standard thickness is 3⅞".

Characteristics.—Glass Blocks have greater heat insulation efficiency than single glazed window light construction and in sound insulating properties are comparable to other forms of masonry construction, being decidedly superior to single glazed window sash. Also there is considerably less solar heat transmission than with the use of glazed sash. In addition they are translucent without being transparent and there is less condensation than is usually encountered in sash construction where wide fluctuations of temperature and humidity occur.

Packing.—Glass Blocks are packed in cardboard containers in the following quantities: 6" block—12 to the carton; 8" block—8 to the carton; 12" block—3 to the carton. Corner blocks are packed 8 to the carton in the 6" size, and 4 to the carton in the 8" size.

Shipping Weight.—Cartons of glass blocks weigh approximately as follows: 6" block, 40 lbs.; 8" block, 48 lbs.; 12" block, 48 lbs. Corner blocks of 6" size weigh 27 lbs., and the 8" size weigh 24 lbs.

APPLICATION.

Heads and Jambs.—Provision for relative movements of the panel and the adjacent construction shall be in the form of approved pre-moulded water-proof expansion joint strips or water-proofed fibrous packings. A space at least ½" in depth shall be provided for expansion joint materials. In the case of interior partitions it is permissible to use mastics, soft fibrous packings, or heavy roofing felt, in such places as around door bucks and window frames. At least ¼" space must be provided at partition heads to allow for possible deflection of ceilings, etc.

For most satisfactory results glass block panels should be built in openings which have been provided with chases of the proper width and depth. Where panels are installed with wall anchors, these shall be so designed and placed as not to interfere with the free motion of the panel in the wall opening.

Sills.—Sills shall be of approved "slip construction" permitting greatest possible freedom of relative movement between the panel and the supporting structure. In no case shall "full chase" or channel sills be used.

Wall Ties.—In exterior panels wall ties shall be used in horizontal mortar joints as follows:

5¾" x 5¾" size—every four courses.
7¾" x 7¾" size—every three courses.
11¾" x 11¾" size—every course.

Wall ties shall run continuously with ends lapped at least 6", and shall not be anchored to the surrounding construction.

Wall ties shall be galvanized, and may be of the expanded metal type, perforated metal strips, or a double wire mesh formed of two parallel wires (.150" thick) with electrically welded cross wires at regular intervals.

Size.—Panels shall not be built in units exceeding 144 square feet of exposed surface area, and shall not have any dimension greater than 20 feet without provision for intermediate supports. All division supports shall provide for expansion and anchorage of panels in accordance with above specifications.

INSTRUCTIONS FOR SETTING.

Workmanship. — The glass-sealed Glass Blocks are designed to be laid up in square bond and with ¼" visible mortar joints. So laid, the average spacing center to center of the joints will be 6-inch, 8-inch or 12-inches.

For satisfactory appearance and performance careful workmanship is essential. All joints must be completely filled with mortar. Mortar joint beds must not be furrowed. To assist the mason in determining that joints are well filled the edges of glass-sealed Blocks are translucent—permitting visual inspection after each block is placed in position.

Blocks may be laid up in separate lifts of a limited number of courses each in order to avoid compaction. Where such interruptions in the work occur, the exposed horizontal and vertical edges of the blocks must not be covered with mortar until laying is resumed.

It is desirable that exposed mortar joints be kept wet for at least twenty-four hours after placing. Wet burlap hung over the panels will provide for this.

While mortar is still plastic and before the end of each day's work all excess mortar shall be cleaned off the faces of the blocks and the joints raked out to a depth necessary to expose the corners of the blocks as sharp clean lines, and the joints immediately tooled slightly concave and smooth.

All edges, both inside and outside, of glass block panels shall have provision for a caulking recess. This recess shall have a depth at least equal to its width and shall be filled with an approved mastic caulking compound after all mortar and other foreign material have been cleaned out.

Mortar Mix.—The proper mortar for glass block construction consists of one part Portland Cement, one part lime (putty or hydrated lime), and from four to six parts sand, measured by dry rodded volumes. The mortar must be well mixed with only enough water to produce a workable or fatty mix. Consistency should be as stiff as will permit good working to reduce shrinkage.

Re-tempered mortar that has taken initial set and mortar should not be used.

Mortar that has been placed but not covered with blocks for more than thirty minutes must be removed and replaced.

Setting accelerators should not be used, and freezing must be prevented without the use of anti-freeze compounds.

Cement.—Use a standard brand conforming with the specifications of the American Society for Testing Materials, A.S.T.M. Designation C9-30.

Lime.—Lime should be an approved brand, high calcium, well slaked Quick Lime, Hydrated Lime, or Mason's Hydrate. These materials shall conform respectively with the A.S.T.M. specifications C5-26 and C6-31.

Sand.—The sand used shall be free from silt, clay and loam in excess of 3% by weight as determined by decantation. Not more than 5% by weight shall pass a No. 100 mesh sieve and 100% shall pass through a No. 8 mesh sieve.

STRUCTURAL GLASS.

Structural Glass is an opaque, flat-surface, structural material, which is strong and durable. One or both of its surfaces may be ground and polished. It is impervious to moisture, stains, chemicals, pencil marks, grime, etc. It will not absorb odors of any kind. Cleaning of its polished surface is easily done by wiping with a damp cloth.

Structural Glass is available in a large variety of colors including Jade, Ivory, Gray, White, Forest Green, Black, Beige, Wine, Blue, and Orange.

This glass has a widespread use as a veneering material for store fronts. In this capacity, Structural Glass offers both practical and decorative features that make it

HOOKER

For glass, glazing and supplies, Hooker has been the first thought of architects and contractors for over 80 years. Far from being immersed in traditions or content with achievements of the past, Hooker Glass and Paint Manufacturing Company are constantly originating, investigating and adapting new developments in materials and methods in glass and paint products. Our stocks of plate, flat drawn, wire and rough glass are unusually complete; our facilities for cutting, beveling and silvering are unexcelled. Complete glazing service, featuring Kawneer Metal Store Fronts, L/O/F Vitrolite Structural Glass and Owens-Illinois Glass Blocks, is available at Chicago, Milwaukee, Davenport, Peoria, Muncie and Kalamazoo.

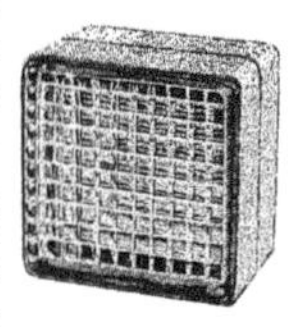

Hooker "Suede Back" mirrors constitute an outstanding development of the century. "Suede Back" provides a moisture-resisting, air-tight shield for the vulnerable silvering, making further treatment unnecessary. "Suede Back" gives maximum protection and longer life—at no extra cost. Complete installations

& PAINT

of any size or type will be made by our own trained crew.

Protective and decorative finishes of the highest type are offered under the KING label. The KING label is your assurance of quality and uniformity in paint and varnish products of all kinds made by Hooker's own modern factories.

KING 2-COAT PAINT SYSTEM provides the highest quality protective finish for the exteriors of all types of buildings, especially suitable for common brick and for cypress, southern pine or similar soft wood exteriors. The System features a special Foundation Coat, which seals porous surfaces and stops absorption, giving a smooth, waterproof foundation for the Finish Coat—at the same time gives the Finish Coat greater covering power.

KING VARNISHES for exterior and interior use are products of the Hooker manufacturing plants, and are unexcelled in quality and results—completely abreast of the newer developments in synthetic resins, gums and oils. KING VARNISHES provide a finish suitable for every application or type of use, and yielding the utmost satisfaction in appearance and durability.

GLASS

King Enamels include a product to meet every requirement. The latest achievement of the Hooker Laboratories in enamels is LUNA WHITE, utilizing synthetic gums and high-strength titanium pigments. As a result, a permanent white is obtained, combined with exceptional hiding power, light reflecting, high gloss and extreme durability. "It washes like a China dish."

King Paint and Varnish products cover the entire field of painting and decorating for protection and beauty for every part of every building. We welcome inquiries and solicit the opportunity of working with architects and builders in the solution of all painting and glazing problems.

MFG. CO.

659 W. WASHINGTON BLVD. **CHICAGO, ILL.**

extraordinarily suitable. Architects find that it lends itself to almost unlimited variations in modern storefront styling and architecture.

As a wall covering for Bathrooms and Kitchens Structural Glass is especially desirable in that it will not crack, craze, absorb moisture, etc., and can be used in numerous color combinations to create decorative effects that are unusual and attractive.

The sanitation and cleanliness obtained by using Structural Glass, particularly recommend it for toilet and shower compartments and the rooms surrounding. Other uses, both practical and decorative, include: counters, wainscoting, bulkheads, bases, table tops, deal plates, window sills, shelving, book-ends, trays, laboratory equipment, bank desks, switch-boards, refrigerator linings, scale tops or platforms, etc.

The installation of Structural Glass is handled similarly to marble. It may be installed over any hard, firm wall surface, but an allowance should be made for a space of ⅜" behind the glass for setting. Structural Glass is installed by means of a mastic which bonds permanently with the glass and the wall—yet allows for setting, shrinkage and expansion.

Although Structural Glass is manufactured in sizes up to 72" x 130", the largest size recommended for exterior use is 10 sq. ft. and for above second story height, 6 sq. ft.

Structural Glass is available in varying thicknesses ranging from ¼" to 1¼", with its use in each thickness depending on the nature of the installation.

The following listing will serve to correlate some of its uses with the thickness recommended:

Obscure Glazing .... ¼" thickness
Ceilings—wainscots ....11/32" thickness
Store Fronts—Strips,
  Caps, Bases .... 7/16" thickness
Bulkheads .... ¾" thickness
Laminated Partitions
  Solid Partitions
  Door and Window trim
  Deal Plates .... ⅞" thickness
Counter Tops
  Toilet Lintels
  Toilet Stiles
  Shower Seats .... 1¼" thickness

Physical Characteristics of Structural Glass. — The weight of Structural Glass ranges from 3.29 lbs. for 1 sq. ft. of ¼" thickness to 16.45 lbs. for 1 sq. ft. of 1¼" thickness.

In strength, this Glass is equal or superior to other materials commonly used for similar purposes.

GLAZING.

It is especially desirable that all glass to be specified for a building be placed under one section in the architect's specifications under the heading—"Glass and Glazing."

Accuracy is a necessity. Use a standard rule, true to gauge, specify the size plainly. For instance 56 inches might be confused if written 5' 6", and cut 66 inches—as 5 feet 6 inches. Always specify width first. In measuring, it is advisable to allow a little play and measure inside the rabbet. See that rabbet is made to accommodate glass of the thickness ordered; i. e. order glass of proper thicknes to fit rabbet. Measure the opening and see if all sides are squared, especially if metal work is to be glazed, it is essential to have perfect fit, and in large sizes it is not uncommon to find a warped frame, or not exactly square, slightly different at one side as compared with the other.

Be specific—it is better to give an abundance of information rather than leave anything indefinite, or to be taken for granted. Mistakes will follow carelessness and corrections involve loss of time and expense.

Large lights of plate glass should rest on two pads of felt, leather, lead, oakum or soft wood blocks, one near each end, not against bare metal, or at a single bearing-point which might cause breakage through settling of building, vibration, etc. The soft wood blocks or lead strips are to be preferred.

Do not fasten or bind glazing-mouldings too tight, as it is necessary to allow for expansion and contraction, vibration and re-adjustment of construction.

Use pure putty. Have sash-rabbet well oiled or painted so that putty will adhere. Give fresh putty glazing time to set before handling or hanging sash. Don't try to back-putty glass with corrugated or figured surface, as the putty cannot be removed from the ridges of the glass.

Steel sash glazing requires special putty for metal rabbets.

Caution.—When glass of any kind has been delivered to a building packed in cases or with paper between the sheets, it is advisable to store the glass under cover in a dry place and unpack it to avoid stains which come from drying out of damp hay, straw, paper, or other packing materials.

Glaze prism-glass with ribs inside—flat surface outside. Regular glazing is done with uncolored putty. If colored putty is desired, it should be specified accordingly. Glass is not bedded-in-putty or back-puttied unless specially ordered or specified.

METAL STORE FRONT CONSTRUCTION.

The use of metal corner-bars, division bars, sash and sills and the all-glass show-case or show-window has become so universal that few old-fashioned stores remain and all modern construction is marked by the absence of bulky posts or ponderous frames.

There are several standard makes of metal store-front construction, corner-bars, dividing bars, metal sash and sill, which fasten or secure the glass with a metal locking or clamping member and provide for drainage ventilation and illumination if desired.

It is well to give attention to the necessity for substantial strength in the retaining members and to using metal bars and construction of sufficient weight to insure strength and rigidity, and to avoid the use of the metal covered mouldings as a substitute for the all metal construction.

It is advisable to send working drawings or detailed plans of store fronts—and the utmost care should be exercised in furnishing accurate dimensions when ordering, so that a true fit of metal may be assured and proper allowance made for bearing contact or play of glass.

The architect should make definite specifications as to the material desired, giving names or numbers of bars, sill covering jamb bars, jamb covering, transom bars, transom covering and style of metal finish.

STANDARD RULES OF THE MEASUREMENT OF PLASTERING

Adopted by the Employing Plasterers' Association of Chicago.

LATH AND PLASTERING

to be measured by the superficial yard, rom floor to ceiling for walls, and from vall to wall for ceiling.

In rooms containing one or more horiontal angles between the floor and ceiling ine, the ceiling to be measured from wall to wall, as though all walls were vertical, for contents of ceiling, and from floor to highest point of ceiling for height of wall.

OPENINGS.

Openings in plastering to be measured between grounds. No deductions to be made for openings of two feet or less in width. One-half of contents to be deducted for openings two feet or more in width. The contents on all store front openings to be deducted, and the contractor to be allowed one foot six inches for each jamb by the height.

All beams or girders projecting below ceiling line to have one foot in width by total length added for each internal and external angle.

No openings to be deducted from "solid" or "hollow" metal lath and plaster partitions nor for openings in suspended ceilings containing less than 100 square feet, where furring is carried around such openings by plasterer. No openings to be deducted from cement wainscot or base.

CORNER BEADS, ARCHES, ETC.

All corner angles of more or less than 90 degrees, beads, "bullnoses," quirks, rule joints, and moldings, to be measured by the lineal foot on their longest extension, and one foot for each stop or miter.

CORNICES.

Length of cornices to be measured on walls. Plain cornices of one foot girth or less to be measured on walls by the lineal foot. Plain cornices exceeding one foot girth to be measured by the superficial foot. Add one lineal foot to girth for each stop or miter. Enriched cornices (cast work), by the lineal foot for each enrichment.

Arches, corbels, brackets, rings, center pieces, pilasters, columns, capitals, bases, rosettes, bosses, pendants and niches by the piece. Ceiling or frieze plates over eight inches wide by the square foot.

COLUMNS.

All columns to be measured by the lineal foot for plain plastered columns.

CEMENT WAINSCOTING AND BASE.

All cement wainscot to be measured by the square foot, and cement base by the lineal foot.

GROUNDS.

All grounds for various classes of work to be as follows, unless expressly specified to the contrary:

Grounds for 3-coat lath work..... 1 inch
Grounds for 3-coat metal lath work. ⅝ inch
Grounds for 3-coat metal lath work, on ½-inch iron furring.........1⅛ inch
Grounds for 3-coat metal lath work, on 1-inch iron furring..........1⅝ inch
Grounds for hard mortar metal lath work .... ⅝ inch
Grounds for hard mortar metal lath work, on ½-inch iron furring...1⅛ inch
Grounds for 2-coat work on brick or tile .... ⅝ inch
Grounds for hard mortar on brick or tile ....⅝ inch
Grounds for hard mortar lath work.. 1 inch
Grounds for plaster board........... 1 inch

Where metal lath is spoken of it applies to all wire or metal lath.

The Employing Plasterers' Association of Chicago solicit the co-operation and support of Architects and others in the Association's efforts to set the highest standard possible for plastering.

In many of the branches of building construction, efforts are tending towards the use of better material and workmanship, no material or finish for a building combines so fully the essentials for fire protection and sanitation at so low a cost to the owner as does plastering, and no other material that enters so largely into the construction of a building presents so large an area of visible surface as does plastering. The cost of plastering represents only a small percentage of the total cost of a building.

It is a necessary base for the most expensive decorations and in itself provides the requisites necessary for a finish interior. The association believes that so important an element in the construction and finish of a building is worthy of being well done, and that the best workmanship and material if specified and called for will more than compensate owners and architects in their requirements for such grade of work. The Employing Plasterers' Association of Chicago respectfully submits the following outline specification for lath and plaster work; all trade names of material have been omitted. Architects will find a list of standard materials in the Hand Book and elsewhere.

TENTATIVE OUTLINE SPECIFICATION FOR LATH AND PLASTER WORK.

Sand. All sand to be clean, sharp sand.

Lime. All lime to be fresh burned lump lime or an approved quality of Hydrated lime.

Lath. All wood lath to be No. 1 white pine 1½" lath free from sap and bark and even edged.

Nails. To be 3 penny fine 16 gauge wire nail.

Wire Lath. To be No. 18 Washburn and Moen gauge .0475⅜" mesh painted or No. 24 gauge metal lath painted with ribs not less than ⅛" wide, lath cut from sheet metal shall weigh not less than 3.4 lbs. per square yard.

Metal Lath. To conform to standards of the Metal Lath Manufacturers Assn. in the size and gauge specified.

Paper Backed Wire Fabric. To be not lighter than 16 W.E.N. gauge zinc coated wire, width not to exceed 2"x2" mesh with 26 U. S. Gauge V ribs or 12 W. & M. gauge zinc coated wire stiffners spaced not to exceed 4" apart and absorptive paper backing securely attached to metal in such a manner as to provide full imbedment of at least $\frac{1}{8}$" of plaster for at least one-half of total length of strands and one-half of total weight of metal.

Stucco. To be fresh.

Hair. To be well whipped cattle hair.

Fibre. To be long vegetable fibre.

Portland Cement. To be a brand that shall meet the requirements of the standard specifications for Portland Cement of the American Society for testing materials as revised to date by said Society.

Hard Plaster. To be an approved straight gypsum plaster.

Metal Corner Beads. To be a bead not less than 24 gauge galvanized.

Lathing. All wood lath to be nailed to each stud joist or bearing with joints broken not over seven lath to a break, no diagonal nor vertical lathing allowed, a full ⅜" key to be left for lime mortar and not less than a full ¼" for hard plaster.

Lime Mortar. To be composed of clean coarse sand, fresh lump lime and hair and fibre in proper proportions and to be well slaked and protected.

Putty. Lime putty to be run off in a tight putty box, thoroughly tempered and screened through a fine putty screen.

Hard Finish. To be composed of cold run lime putty, fresh plaster of paris and sand to be well troweled to a smooth even surface, free from blisters, checks and other imperfections.

Sand Finish. All float sand finish to be composed of lime putty and sand to be water floated with a float to an even granular or sand surface.

Scratch Coat. All scratch coating to be well laid on and surface covered with a full coat which is to be scratched with wire scratcher to be well under cut for the brown coat, all lime mortar scratch coating to be dry before applying the brown coat.

Brown Coat. All brown coating to be well applied, allowing only sufficient space for the finish coat, brown coat to be rodded and screeded with all angles straight and true, all hard plaster to be mixed in accordance with the directions of the manufacturer and no hard mortar to be floated with water nor shall any "dead" material be retempered or used.

Wire or Metal Lath. Shall be lapped at each joint or seam and shall be stapled every six inches with blued or galvanized staples.

Band Iron Furring. The following shall be furred with ¼", ½", ¾" or 1" corrugated band iron furring, such furring to be stapled to bearings and the wire or metal lath to be stapled over such band iron furring.

Suspended Ceilings. To be constructed with 1½" or 2" flat bars, angles or channels as may be called for, such principals shall be spaced 4' 0" on centers, hung with flat bar or not less ¼" rod hangers every 4' 0" securely fastened with approved clips to the structural framing or through the floor construction, in the event these hangers go through the floor construction they shall be provided with 6" channels or flat bar anchors, no hanger shall be supported from the bottom flange of the tile arch. The flat bar, angle or channel runners shall be cross furred 12" on centers with ¾" steel channels, securely secured to the principals with rod clips, entire construction to be lathed with No. 18 W. M. gauge ⅜" mesh painted wire lath or No. 24 U. S. Gov. standard gauge metal lath, lath to have lapped edges at each joining and to be tied to the channel furring every 6" with 18 gauge galvanized tie wire.

Furring. All false beam or cornice furring to be constructed of ¾" channel or 1" flat bar brackets not over 2' 0" apart lined out with intermediate furring supports and anchored or toggle bolted into the construction to be made to conform to the design so as to allow for a minimum of plaster, such brackets to be covered with 18 gauge wire or 24 U. S. Gov. gauge metal painted lath secured with 18 gauge galvanized tie wire, such furring to conform to the latest and best practice as to durability of construction.

Cornice Work. All moulded beams and cornices will be screeded and run in place wit moulds, with true lines and accurate mithes.

Ornamental Work. All ornamental work to be modeled by artistic modelers who will be approved by the architects. Models to be submitted for approval and no casts to be made until such models have been approved, all patterns to be gotten out by skilled mechanics with true and accurate lines.

Casts. All casts to be well made, the contractor to supply a sufficient number to meet the requirements of the job, all casts to be made in line, well and truly undercut and free from warps and other irregularities supplying all necessary shrinkers and stretchers.

Rough Casting. Lath the exterior of the house with 18 gauge wire or 24 U. S. Gov. metal painted lath stapled over 1" band iron furring scratch coat with mortar composed of 2 vols. of coarse, sharp sand 1 vol. of approved Portland cement, to which mixture add 15% of rich lime mortar, thoroughly scratched and undercut when this coat was "set," brown with mortar composed of 3 vols. sharp sand to 1 vol. Portland cement rod and straighten all surfaces and when this coat has "set" rough cast with mortar composed of 3 vols. of sharp sand or pebbles to 2 vols. Portland cement dashed on surface with a scoop or paddle to an even artistic finish.

Exterior Plastering on Wood Lath. Lath the exterior with No. 1 soft pine one-inch lath, nailed to each stud furring or bearing with not less than a 3 penny nail with full open ⅜" key space and not over seven lath to a break, plaster with 3 coats of cement plaster as called for under exterior plaster on metal lath, note the use of "hard plasters" so called are not recommended for exterior plastering.

Concrete Walls and Columns. All work on concrete walls and columns shall have such concrete well brushed with steel brushes and such concrete shall then be covered with a light coat of an approved bond cement as a bonding coat for the finish coat.

Concrete Ceilings. Shall first be washed with a solution of muriatic acid and such ceilings shall then be plastered as above.

Painted Walls. Walls that are to be coated with waterproofing shall first be scratch coated, then browned and finished.

Patching of Plaster. All patching of plaster damaged by other mechanics shall be paid for at the uniform scale of prices adopted by the Employing Plasterers' Association of Chicago, which scale of prices is set forth in the Hand Book.

Workmen's Compensation. This contractor shall insure his workmen under the provisions of the Workmen's Compensation Laws of the State of Illinois. This contractor shall also insure his liability for injury or death to "the public."

Scaffold. This contractor shall supply all necessary tools, scaffold and other appliances necessary to fulfill the requirements of the job, all scaffolding to be erected and maintained in accordance with the laws of the State relating to scaffolds.

Requirements. By Building Code in buildings of ordinary construction. At least three coats of plaster on all wood lath to 1 inch grounds.

By Union. All plain and ornamental plaster to the same contractor, the base coat of Portland cement under encaustic tile, cement base when installed independent of the floor or if more than 6" in height. All plastering regardless of the nature of the structure or of the material used.

RECOMMENDATIONS.

It is the experience of the industry that many defects that arise in the finish of homes and other structures, are due to improper and faulty construction of the underlying frame and basic construction, and not in lathing and plastering, which has a record of durability, worth and use extending over many centuries. For the purpose of assisting architects and others, the following important points are stressed, with respect to underlying construction. Proper and securely nailed cross bridging is vital to prevent deflection in studding and joist. It should be specified and secured upon every job. Wood cross furring should be securely nailed to joist. Green or partly seasoned lumber greatly minimizes the stability of nailing, due to shrinkage and deflection of such cross furring away from the joist.

Unsupported long spans of ceiling joist and splicing of ceiling joist are prolific causes of failure and should not be permitted. To secure the necessary minimum thickness of lath

and plaster, grounds of actual size and not lumber sizes should be specified and used. Jamb linings, etc., should also be detailed to meet such minimum thickness of lath and plaster. All plaster should be mixed in accordance with the manufacturer's directions, and should not be adulterated by over sanding, or the use of loamy or improper sand. Minimum thickness of plastering on Brick and Tile should be ⅝ of an inch, on wood lath; gypsum lath or fibrous lath ½" over the face of such lath, and on metal or wire lath not less than ⅝" over the face of such lath. Wood lath and Gypsum lath come ⅜" in thickness and the usual run of Fibrous lath ½" in thickness.

In the use of metal lath on ceilings of wood construction, all of such metal lath should be fastened in place at least every 6" to bearings with "A" not less than 6d nails driven to not less penetration than one inch into the joist, or "B" one inch blued or galvanized staples full driven, or "C" No. 12 gauge 1¼" blued nails with 7/16" heads. Such nails should be so driven as not to break the mesh of the lath. Wood lath should be nailed with 3d 16 gauge blued nails 16" apart with all ends nailed. Wood lath should be laid with broken joints in every 7e course and should have full ¼" key space for Gypsum plaster and ⅜" key space for limes mortar. Gypsum and Fibrous lath should be nailed with No. 12 gauge 1¼" blued nails with 7/16" heads spaced not over 4½" apart to bearings. Where metal lath is used over old wood lath ceilings, nails should be No. 11 gauge 1½" blued nails with 7/16" heads, spaced not over 6" apart. Where metal lath is applied over old wood lath and plaster, the nailing should be with No. 11 gauge 2" blued nails with 7/16" heads, spaced not over 6" apart. To secure a proper job of plastering, "body" coats are necessary and not a flimsy veneer of over sanded plaster.

The tentative outline specifications for lathing and plastering as recommended as an aid to architects is continued in the Hand Book. The several chapters for **Minimum** Standards for Lathing and Plastering will be available upon their approval by the above sponsors.

The Gypsum and Metal Lath specification Chapters have been submitted and they are now available. The Plastering Industry in all of its integral parts— manufacturer, contractor and labor—most highly value the good will, cooperation and support of the Architectural profession. It respectfully and unanimously solicits a continuation of their valuable cooperation in the industry's efforts to maintain proper and necessary standards of the work of the industry.

The use of soft pine lath, specify No. 1 white pine lath nailed to each stud, joist or bearing with 3 d. fine 16 gauge wire nails, with joints broken at least once in each seventh course or lath.

For better residence work specify one inch lath as above.

Wire or metal lath, specify No. 18 Washburn and Moen gauge wire lath ⅜" mesh, painted, or No. 24 U. S. Gov. standard metal lath painted, for better class work specify wire lath woven from galvanized strand or metal lath galvanized.

The use of wire or metal lath plastered insures slow burning construction, helps to prevent settlement cracks and bonds and ties all parts of the structure together. Its use is called for in almost every building, particularly on basement ceilings to prevent or retard fire on ceilings with long span joist construction on store ceilings and under other space subject to heavy use or abuse. Its use should also be general in all better class building, in rated buildings its use throughout entitles it to better classification for insurance.

The Association recommends the use of three coat plastering. This will insure a far better class of work, a better bonding together of buildings of ordinary construction due to the use of a greater body of material. The application of the second base coat enabling one to straighten out rod and line work. Specify three-coat dry work, first coat to be a scratch coat well scratched and under cut. When dry, apply a brown coat, this brown coat to be screened and rodded and when dry apply a finish coat.

The following suggestions are offered for guidance:

Sand. The use of clean, course, sharp sand is essential for good plastering.

Metal Lath. Should be laid with lapped edges or joinings and should be stapled to bearings every 6". No suspended ceilings should be supported from the bottom or soffit of tile.

Portland cement base coat behind encaustic tile, Opalite or kindred material should be specified under "Plastering" with one rodded coat scratched on tile or brick or a scratch and rodded brown coat scratched on metal or wire lath. We do not recommend Portland cement direct to gypsum partition or gypsum furrings.

Damp proofed, waterproofed or painted walls and ceilings are required to be given 3 coats. If a finish coat is desired, it should be so specified. All lathing plain and ornamental plastering should be specified under one heading in order to avoid divided responsibility for final results.

JURISDICTION CLAIMS.

By Plasterers' Union, any and all plastering regardless of the nature of the material, or of the structure to which it is applied, including Scagliola made under the "New Process" so called.

By Lathers' Union, all lathing, metal corner beads and all light iron furring designed, specified or used primarily as a support for lath and plaster, including "Hi Rib."

By Hodcarriers and Building Laborers' Union, all scaffolding erected for the use of plasterers.

PATCHING OF PLASTERING AFTER OTHER TRADES.

Patching of plastering after other mechanics shall not be done as a part of the contract price, and shall be paid for at the following scale of prices which have been adopted by and are recommended by the Employing Plasterers' Association of Chicago.

In accordance with wage agreements effective, and present prices of materials, the following scale of prices for patching of plastering after other mechanics and for work done upon a time and material basis, is respectfully submitted.

"The prices herein include cost of insurance of men under the provisions of the Workmen's Compensation Laws of the State of Illinois," and Federal Social Security taxes as at present in force.

Foreman Plasterer	$2.60	per hour
Plasterers	2.35	per hour
Foreman Laborer	2.60	per hour
Laborers	2.35	per hour
Plasterer Laborer	1.50	per hour
Mortar	3.50	per bbl.
Putty	4.00	per bbl.
Hydrated Lime	.80	per bag
Neat Hard Plaster	1.40	per bag
Stucco	1.40	per bag
Metal or Wire Lath	.45	per yard
1½" Pine Lath	.40	per bunch

Prices of labor as listed are for the periods expiring May 31, 1939.

Owing to abnormal conditions material prices are subject to change without notice, and labor scale will be proportionately increased where bonuses are required to be paid in order to get men.

18 gauge ⅜" mesh painted wire lath or 24 gauge expanded metal painted	.30	per yard

Where seven or more men are employed in one gang on same kind of work, foreman's time will be charged continuous while work is going on; where less than seven men are employed in one gang on same kind of work, foreman's time shall be counted one hour for each seven hours of men aggregate time employed on this work, unless foreman's time is required constantly, when he shall be so paid.

EXTRACTS CHICAGO BUILDING CODE ON PLASTER.

ACOUSTIC MATERIALS.

2109.01 General: Combustible acoustic material shall be applied directly to a non-combustible surface or to a ceiling of metal lath and plaster or to the members of heavy timber construction.

2110.03 PARTITIONS.

Item 3. Ordinary and Wood Frame Construction. In buildings of ordinary and wood frame construction solid or hollow partitions composed partially of combustible materials shall be permitted. Partitions of wood studs shall be lathed with wood or metal lath, gypsum, compressed fibre or similar plaster base which shall receive a coat or coats of lime, gypsum or cement plaster not less than one-half (½) inch in thickness, or, in lieu of said plaster base and plaster, other material of equal thickness and having equal fire-resistive and sanitation values may be used.

3201.03 Fire-Resistive Materials.

(g) **Metal Lath.** Metal lath of either wire mesh or expanded metal, to be used as a fire-resistive plaster base, shall weigh not less than two and five-tenths (2.5) pounds per square yard when used on vertical surfaces and not less than three (3) pounds per square yard when used on horizontal or sloping surfaces. Mesh lath shall not be considered as a fire-resistive material except when used as a plaster base in the same plant and integral with the plaster.

(h) **Plaster.** Wherever plaster is required or permitted as a fire-resistive material in this chapter, it shall consist of not less than two (2) coats having a total thickness of not less than one-half (½) inch applied directly to concrete or masonry and three-fourths (¾) inch applied to metal lath, measured from the face of the lath.

3202.01 WALLS AND PARTITIONS.

(d) **Two (2) Hour Fire-Resistive Walls.** Two (2) hour fire-resistive walls or partitions shall include any of the walls permitted under paragraphs (b) and (c) or walls of the following construction:

Brick, solid wall, three and three-fourths (3¾) inches thick, plastered both sides.

Cast-in-Place Gypsum poured four (4) inches thick, or hollow units three (3) inches thick, plastered both sides.

Hollow concrete units, eight (8) inches thick; or four (4) inches thick, plastered both sides.

Hollow clay tile, three (3) cells in wall thickness, eight (8) inches thick; or two (2) cells in wall thickness, eight (8) inches thick, plastered one (1) side; or two (2) cells in wall thickness, six (6) inches thick, plastered both sides.

Solid plaster partition on metal studs and metal lath two (2) inches thick.

(e) **One (1) Hour Fire-Resistive Walls.** One (1) hour fire-resistive walls or partitions shall include any of the walls permitted under paragraphs (b), (c) or (d) or walls or partitions of the following construction:

Brick, three and three-fourths (3¾) inches thick, or two and one-fourth (2¼) inches thick, plastered both sides.

Gypsum, solid or hollow units, three (3) inches thick, or solid units two (2) inches thick, plastered both sides.

Hollow concrete units, four (4) inches thick; or three (3) inches thick, plastered both sides.

Hollow clay tile, three and three-fourths (3¾) inches thick; or load-bearing tile three (3) inches thick, plastered both sides.

Hollow partition of either non-combustible or combustible studding, with metal lath and three-fourths (¾) inch of plaster; or gypsum lath and one-half (½) inch of plaster on both sides. Joints of gypsum lath shall be covered with strips of metal lath at least three (3) inches wide.

3202.02 FLOORS, ROOFS and CEILINGS.

(b) **Four (4) Hour Fire-Resistive Systems.** Four (4) hour fire-resistive floors or roof systems, shall include:

Hollow tile arches, ten (10) inches thick, plastered on the underside and with a solid or loose floor fill of non-combustible material two (2) inches thick.

(c) **Three (3) Hour Fire-Resistive Systems.** Three (3) hour fire-resistive floor or roof systems shall include any system permitted under paragraph (b) or the following:

Hollow tile arches, plastered, eight (8) inches thick for flat arches, six (6) inches thick for segmental arches, with a floor fill of solid or loose non-combustible material, two (2) inches thick.

(d) **Two (2) Hour Fire-Resistive Systems.** Two (2) hour fire-resistive floor or roof systems shall include any system permitted under paragraphs (b) and (c) or the following:

Steel joists with reinforced concrete slabs two and one-half (2½) inches thick protected on the underside by a ceiling of metal lath and plaster seven-eighths (⅞) inch thick.

Steel joists with reinforced gypsum slabs for roofs only, two and one-half (2½) inches thick, protected on the underside by a ceiling of metal lath and plaster seven-eighths (⅞) inch thick.

(e) **One (1) Hour Fire-Resistive Systems.** One (1) hour fire-resistive floor or roof systems shall include any systems permitted under paragraphs (b), (c) and (d) or the following:

Steel joists with reinforced concrete slabs two and one-half (2½) inches thick, protected on the underside by a ceiling of metal lath and plaster three-fourths (¾) inch thick.

Steel joists with reinforced gypsum slabs, for roofs only, two and one-half (2½) inches thick, protected on the underside by a ceiling of metal lath and plaster three-fourths (¾) inch thick.

Steel or wood joists, fire-stopped, with double board floor having insulating layer between boards and protected on the underside by a ceiling of three-fourths (¾) inch of plaster on metal lath or one-half (½) inch of plaster on gypsum lath. Joints of gypsum lath shall be covered with metal lath at least three (3) inches wide.

(f) **Reduction of Slab Thickness.** For the purpose of fire-resistive value only, in any floor or roof construction not required to have a protective ceiling on the underside thereof, the required slab thickness may be reduced one-half (½) inch if a protective ceiling of metal lath and plaster is applied thereto.

For Plastered Surfaces as Fire-Protective Covering for Metal Members:

See **Sec. 3202.03 Building Code** this volume.

MISCELLANEOUS AND USEFUL INFORMATION CONCERNING BUILDING ENGINEERING, TRADES AND MATERIALS.

The following pages contain tables, formulae, and miscellaneous information intended to be of assistance to architects in the preparation of plans, specifications, estimates, and the general supervision of the construction work. In order to make the classification simple and to follow a uniform system this matter is classified according to the Dewey System, see page 786, and the file or classification numbers are printed in small type at the head of each piece of matter falling under a different classification. As far as possible the names of authorities quoted are given but in some cases this has been impossible.

RULES AND FORMULAS FOR THE DESIGN OF SIMPLE WOOD BEAMS OR JOISTS.

When a beam is to be designed its length and the loads to which it is to be subjected are known, thus the maximum bending moment may be found.

The **allowable-working-strength** is assumed in accordance with engineering practice and must not be more than allowed by building laws, locally applicable. This **allowable-working-strength** is usually stated in municipal codes as a fixed number of pounds per square inch of cross sectional area, for each kind of material. This might just as well be stated in tons or any other unit of weight per square foot or any other unit of area, it being only important that whatever unit of dimension is used that the same unit shall be used both for areas, lengths and breadths.

Breadth-of-the-beam times **the-square-of-the-depth** divided by six equals **Bending-Moment** divided by **allowable-working-strength** per unit of area corresponding with unit of length used for stating the length and breadth of beam.

Bending-Moment (for beams uniformly loaded) equals **weight-to-be-supported-per-unit-of-length** times **the-square-of-the-total-number-of-units-of-length** divided by **eight.**

For a simple beam loaded with a single weight, the **maximum-Bending-Moment** (which is to be used in formula) equals **the-entire-load** times **[(the-length-of-the-beam)** minus **(the-distance-of-the-load-from-the-left-hand-end)]** times **the-distance-of-the-load-from-the-left-hand-end-of-the-beam** divided by **the-length-of-the-beam.**

If the load be movable **the-distance-of-load-from-left-hand-end** will be variable and the **maximum-moment** will be developed when the load is at the middle where the **maximum-Bending-Moment** is equal to **one-fourth-the-load** times **the-length-of-the-beam.** Placing the entire load on a beam at its center therefore produces the maximum strain that it is possible to produce on such beam by any position of such load.

APPLICATION OF ABOVE PRINCIPLES.

M=maximum bending moment.

S=the tensile or compressive unit stress per square inch allowable by building code or engineering practice for the material selected (See Section 539, Chicago Municipal Code, using the smallest value where there is a difference between compression and tension strength.)

l = length in inches of beam between supports.

b = breadth in inches of the beam.

d = depth in inches of the beam.

w = weight in pounds on beam including the weight of the beam itself per each inch of length.

W = total weight in pounds on beam = l w.

FOR UNIFORM LOADING.

$b = \frac{3\,w\,l^2}{4\,d^2\,S} = \frac{3\,W\,l}{4\,d^2\,S}$ = breadth of beam. $d = \sqrt{\frac{3\,w\,l^2}{4\,b\,S}} = \sqrt{\frac{3\,W\,l}{4\,b\,S}}$ = depth of beam

To find **b** it is necessary to assume a value for **d**. Also to find **d** it is necessary to assume a value for **b**. In case it is found that the value by formula is too large or too small for practical use, then assumed value must be changed so as to bring the computed value to a practical size.

Ultimate and Safe Strength of Masonry in Pounds Per Square Inch.

Material	Compression			Safe Bearing	Modulus of Elasticity		Shear			Tension			Weight per Cubic Foot	
	Ultimate		Safe		Ultimate		Ultimate		Safe	Ultimate		Safe		
	From	To	Av.		From	To	From	To	Av.	From	To	Av.	From	To
Hard Brick Work in P. C.	2000	3000	200	275	1,500,000	2 500,000	......	......	......	100	200	20	130	150
Common " P. C.	1500	2500	175	250	1,500,000	2,500,000	150	300	20	100	200	20	110	130
" " N C.	1000	2000	150	200	1,000,000	1,500,000	......	......	......	50	100	10	110	130
" " L. M.	800	1600	100	150	500,000	1,000,000	......	......	......	20	40	5	110	130
" " P.C.& LM.	1000	2000	150	200	1,000,000	1,500,000	......	......	......	50	100	10	110	130
Old Brick Work in P. C.	2000	3000	200	275	2,000,000	3,000,000	......	......	......	120	250	25	110	130
" " " N. C.	1500	2500	175	250	1,500,000	2,000,000	......	......	......	70	120	15	110	130
" " " L. M.	1000	2000	150	200	1,000,000	1,500,000	......	......	......	25	50	7	110	130
Brick Piers in P. C......	1500	2500	175	250	1,500,000	2,500,000	......	......	......	100	200	20	110	130
" " L. M......	800	1600	100	150	500,000	1,000,000	......	......	......	20	40	5	110	130
Rubble Work in P. C...	1000	2000	150	200	1,500,000	2,500,000	......	......	......	70	150	20	130	150
Coursed Rubble in P. C..	1500	2500	175	250	2,000,000	3,000,000	......	......	......	100	200	20	140	160
Neat P. C............	2000	4000	200	300	1,500,000	3,000,000	1200	2400	300	400	800	70	80	90
Neat N. C............	1000	3000	175	250	1,000,000	2,000,000	700	1500	125	200	400	30	60	70
P. C. Mortar 1 : 3......	1500	2500	175	250	1,000,000	2,000,000	200	400	35	200	400	30	120	130
N. C. Mortar 1 : 2......	800	1500	150	200	800 000	1,500,000	150	300	25	100	200	20	120	130
Lime Mortar...........	200	400	100	150	500,000	800,000	50	100	10	20	40	5	90	110
P. C. Stone Concrete 1:2:4	1500	3500	400	500	1,500,000	3,500,000	800	1200	125	200	400	40	140	150
N C. " " 1:2:5	1000	2000	200	300	1,000,000	2,000,000	500	1000	80	150	300	25	140	150
P. C. Cinder 1:2:5	800	1600	150	200	500,000	1,000,000	70	120	10	100	150	20	100	110
Granite..............	12000	20000	400	600	3,000,000	6,000,000	1200	2400	300	1200	2400	200	160	180
Limestone...........	6000	12000	350	500	2,000,000	5,000,000	1000	2000	175	1000	2000	175	150	170
Sandstone............	5000	10000	300	400	1,000,000	3,000,000	800	1600	125	800	1600	125	140	160
Brick and Tile.........	2000	5000	200	300	1,000,000	3,000,000	500	1000	80	500	1000	80	120	140

ULTIMATE AND SAFE STRENGTH OF IRON AND STEEL IN POUNDS PER SQUARE INCH

Material	Compression			Safe Bearing	Shear			Modulus of Elasticity		Weight per Cu. Ft.
	Ultimate		Safe Average		Ultimate		Safe	Ultimate		
	From	To			From	To	Average	From	To	
Hard Steel	36,000	40,000	18,000	26,000	45,000	55,000	12,000	28,000,000	31,000,000	490
Medium Steel	33,000	38,000	16,000	24,000	50,000	60,000	12,000	"	"	"
Steel Pins	33,000	38,000	16,000	24,000	50,000	60,000	12,000	"	"	"
Shop Rivets	24,000	29,000	16,000	24,000	50,000	60,000	12,000	"	"	"
Field Rivets	24,000	29,000	12,000	20,000	50,000	60,000	10,000	"	"	"
Cast Steel	60,000	90,000	12,000	26,000	50,000	60,000	12,000	29,000,000	32,000,000	"
Cast Iron	60,000	90,000	10,000	15,000	15,000	25,000	2,000	12,000,000	18,000,000	450

Material	Extreme Fiber Stress			Tension			Elastic Limit		Modulus of Resilience
	Ultimate		Safe Average	Ultimate		Safe Average	Ultimate		
	From	To		From	To		From	To	
Hard Steel	50,000	70,000	18,000	65,000	75,000	18,000	35,000	45,000	35
Medium Steel	40,000	60,000	16,000	60,000	70,000	16,000	30,000	40,000	35
Steel Pins	40,000	60,000	24,000	60,000	70,000	16,000	30,000	40,000	
Shop Rivets	40,000	60,000	24,000	48,000	58,000		24,000	30,000	
Field Rivets	40,000	60,000	18,000	46,000	54,000		24,000	30,000	
Cast Steel	60,000	90,000	16,000			18,000	35,000	50,000	
Cast Iron	30,000	40,000	3,500			3,000	10,000	20,000	1.2

BUILDING CODE REQUIREMENTS FOR LIVE LOAD IN VARIOUS CITIES

In Pounds Per Square Foot

	Baltimore, 1926	Boston (Tentative), 1938	Buffalo, 1930	Chicago, 1938	Cincinnati, 33	Indianapolis, 1925	Minneapolis, 1934	New Orleans, 1927	New York, 1937	Pacific Coast Bldg. Officials Conf., 1937	Peoria, 1937	Philadelphia, 1929	Rockford, 1930	St. Louis, 1930	San Francisco, 1937	Seattle, 1937	U. S. Dept. of Com., 1925	Washington, D. C., 1930
Apartments	40	40	50	40	40	...	40	40	40	40	40	40	50	50	40	40	40	40
Assembly Halls—Fixed Seats	...	75	...	75	50	60	80	80	75	50	50	60	...	...	75	75	50	70
Assembly Halls—Movable Seats	100	100	100	100	100	100	100	...	...	100	100	100	100	150	125	100	100	100
Dwellings	40	40	40	40	40	40	40	40	40	40	40	40	50	50	40	40	40	40
Hospitals—Bed and Living Rooms	40	40	50	40	40	40	40	40	40	40	40	40	50	50	40	40	40	40
Hotels	40	40	50	40	40	40	40	60	40	40	40	40	50	50	40	40	40	40
Heavy Manufacturing	120	250	120	100	150	120	100	150	120	125	125	200	150	150	250	125	...	...
Light Manufacturing	120	75	120	100	75	100	100	100	75	75	75	120	100	150	100	125	75	150
Heavy Warehouse	125	250	150	100	250	120	125	200	120	250	150	200	200	150	250	125	250	200
Offices	50	50	50	50	50	50	60	70	50	50	50	50	50	60	40	50	50	70
Schools—Class Rooms	50	50	50	50	50	50	60	50	60	40	40	50	75	75	75	50	50	40
Roofs—Slope less than 20°	30	30	..	25	25	30	30	30	30	20	25	30	30	30	30	40	30	30
Corridors and Stairways	...	100	...	100	100	100	100	100	100	100	75	100	100	...	...	...	100	100

First floors and special rooms, such as restaurants, places of public assembly, ballrooms, etc., in apartments, hotels, schools, hospitals and other buildings frequently require larger loads than shown in this table. Description of purpose such as warehouses or light and heavy manufacturing is faulty and require special loads other than shown.

Metric Tables.

	Approximate Equivalent.		Accurate Equivalent.
1 inch ....[length]	2½	cubic centimeters	2.539
1 centimeter	0.4	inch	0.393
1 yard	1	meter	0.914
1 meter (39.37 inches)	1	yard	1.093
1 foot	30	centimeters	30.479
1 kilometer (1,000 meters)	⅝	mile	0.621
1 mile	1½	kilometers	1.600
1 gramme ....[weight]	15½	grains	15.432
1 grain	0.064	gramme	0.064
1 kilogramme (1,000 grammes)	2.2	pounds avoirdupois	2.204
1 pound avoirdupois	½	kilogramme	0.453
1 ounce avoirdupois (437½ grains)	28 1/3	grammes	28.349
1 ounce troy, or apothecary (480 grains)	31	grammes	31.103
1 cubic centimeter ....[bulk]	1.06	cubic inch	1.060
1 cubic inch	16 1/3	cubic centimeters	16.386
1 liter (1,000 cubic centimeters)	1	U. S. standard quart	0.946
1 United States quart	1	liter	1.057
1 fluid ounce	29½	cubic centimeters	29.570
1 hectare (10,000 square meters)...[surface]	2½	acres	2.471
1 acre	0.4	hectare	0.40

In the nickel five-cent piece of our coinage is a key to the tables of linear measures and weights. The diameter of this coin is two centimeters, and its weight is five grammes. Five of them placed in a row will give the length of the decimeter, and two of them will weigh a decagram. As the kiloliter is a cubic meter, the key to the measure of length is also the key to the measure of capacity.

Handy Table.

Diameter of a circle × 3.1416 = circumference.
Radius of a circle × 6.283185 = circumference.
Square of the diameter of a circle × 0.7854 = area.
Square of the circumference of a circle × 0.07958 = area.
Half the circumference of a circle × half its diameter = area.
Circumference of a circle × 0.159155 = radius.
Square root of the area of a circle × 0.56419 = radius.
Circumference of a circle × 0.31831 = diameter.
Square root of the area of a circle × 1.12838 = diameter.
Diameter of a circle × 0.86 = side of inscribed equilateral triangle.
Diameter of a circle × 0.7071 = side of an inscribed square.
Circumference of a circle × 0.225 = side of an inscribed square.
Circumference of a circle × 0.282 = side of an equal square.
Diameter of a circle × 0.8862 = side of an equal square.
Base of a triangle × ½ the altitude = area.
Multiplying both diameters and .7854 together = area of an ellipse.
Surface of a sphere × 1/6 of its diameter = solidity.
Circumference of a sphere × its diameter = surface.
Square of the diameter of a sphere × 3.1416 = surface.
Square of the circumference of a sphere × 0.3183 = surface.
Cube of the diameter of a sphere × 0.5236 = solidity.
Cube of the radius of a sphere × 4.1888 = solidity.
Cube of the circumference of a sphere × 0.016887 = solidity.
Square root of the surface of a sphere × 0.56419 = diameter.
Square root of the surface of a sphere × 1.772454 = circumference.
Cube root of the solidity of a sphere × 1.2407 = diameter.
Cube root of the solidity of a sphere × 3.8978 = circumference.
Radius of a sphere × 1.1547 = side of inscribed cube.
Square root of (⅓ of the square of) the diameter of a sphere = side of inscribed cube.
Area of its base × ⅓ of its altitude = solidity of a on or pyramid, whether round, square, or triangu- lar.e
Area of one of its sides × 6 = surface of a cube.
Altitude of trapezoid × ½ the sum of its parallel sides = area.

Square root of ($\frac{1}{3}$ of the square of) the diameter of a sphere = side of inscribed cube.

Area of its base × $\frac{1}{3}$ of its altitude = solidity of a cone or pyramid, whether round, square, or triangular.

Area of one of its sides × 6 = surface of a cube.

Altitude of trapezoid × $\frac{1}{2}$ the sum of its parallel sides = area.

TABLE OF SQUARE ROOTS.

No.	Sq. Root.	No.	Sq. Root.	No.	Sq. Root.	No.	Sq. Root
25	5.	650	25.46	1400	37.42	2600	50.99
50	7.071	700	26.46	1450	38.08	2700	51.96
75	8.66	750	27.39	1500	38.73	2800	52.91
100	10.00	800	28.28	1550	39.37	2900	53.85
125	11.18	850	29.15	1600	40.00	3000	54.77
150	12.25	900	30.00	1650	40.62	3200	56.57
175	13.23	950	30.82	1700	41.23	3400	58.30
200	14.14	1000	31.62	1800	42.43	3600	60.00
250	15.81	1050	32.40	1900	43.59	3800	61.64
300	17.32	1100	33.16	2000	44.72	4000	63.24
350	18.70	1150	33.91	2100	45.82	4200	64.80
400	20.00	1200	34.64	2200	46.90	4400	66.32
450	21.21	1250	35.36	2300	47.95	4600	67.82
500	22.36	1300	36.06	2400	48.99	4800	69.28
550	23.45	1350	36.74	2500	50.00	5000	70.72
600	24.49						

Expansion of Water (Dalton).

Temperature.	Expansion.	Temperature.	Expansion.	Temperature.	Expansion.
22°	1.0009	72°	1.0018	152°	1.01934
32	1	92	1.00477	172	1.02575
*46	1	112	1.0088	192	1.03265
52	1.00021	132	1.01367	212	1.0466

*Greatest density at 39.1° Fahr.

Capacity of Bins and Boxes.

A box 24 inches long by 16 inches wide and 28 inches deep will contain a barrel, or three bushels; 24 by 16 inches and 14 inches deep contains half a barrel; 16 inches square and $8\frac{2}{5}$ inches deep will contain one bushel; 16 by $8\frac{2}{5}$ inches and 8 inches deep will contain half a bushel; 8 by $8\frac{2}{5}$ inches and 8 inches deep will contain one peck; 8 inches square and $4\frac{1}{5}$ inches deep will contain one gallon; 7 by 4 inches and $4\frac{4}{5}$ inches deep will contain half a gallon; 4 inches square and $4\frac{1}{5}$ inches deep will contain one quart; 4 feet long, 3 feet 5 inches wide and 2 feet 8 inches deep will contain one ton of coal, or 36 cubic feet.

Dimensions of a Barrel.—Diameter of head, 17 inches; bung, 19 inches; length, 28 inches; volume, 7,680 cubic inches.

Table Showing the Pressure of Water at Different Elevations.

Feet Head	Equals Pressure per Square Inch	Feet Head	Equals Pressure per Square Inch	Feet Head.	Equals Pressure per Square Inch	Feet Head	Equals Pressure per Square Inch.	Feet Head.	Equals Pressure per Square Inch	Feet Head	Equals Pressure per Square Inch
1	43	65	28.15	130	56.31	195	84 47	260	112 62	350	151.61
5	2 16	70	30.32	135	58.48	200	86 63	265	114 79	360	155.94
10	4 33	75	32 48	140	60.64	205	88 80	270	116.96	370	160.27
15	6 49	80	34 65	145	62 81	210	90 96	275	119.12	380	164.61
20	8 66	85	36.82	150	64.97	215	93 14	280	121 29	390	168 94
25	10.82	90	38 98	155	67.14	220	95.30	285	123.45	400	173.27
30	12 99	95	41.15	160	69.31	225	97 49	290	125 62	500	216 58
35	15.16	100	43 31	165	71.47	230	99 63	295	127.78	600	259.90
40	17.32	105	45.48	170	73.64	235	101 79	300	129.95	700	303 22
45	19 49	110	47 64	175	75.80	240	103 96	310	134.28	800	346.54
50	21.65	115	49.81	180	77 97	245	106 13	320	138 62	900	389 86
55	23.82	120	51.98	185	80.14	250	108 29	330	142 95	1.000	433.18
60	25 99	125	54 15	190	82.30	255	110 46	340	147.28	. . .	. . . .

For an exhaustive discussion of live loads in buildings send for "Report of Building Code Committee," Nov. 1, 1924, U. S. Dept. of Commerce entitled "Minimum Live Loads Allowable for Use in Design of Buildings." This report gives tables tabulating almost every kind of building occupancy.

Weights of Materials.

Dry Woods.

	Lbs. Board ft.	Lbs. Cubic ft.		Lbs. Board ft.	Lbs. Cubic ft
Apple	4.1	49.	Iron Wood	6.	71.
Ash, American white	3.9	47.	Larch	3.	35.
Birch	3.9	45.	Lignum Vitæ	6.9	83.
Beech	3.7	43.	Mahogany, Honduras	2.9	35.
Boxwood	5.	60.	Mahogany, Spanish	4.4	53.
Cedar, American	2.9	35.	Maple	4.1	49.
Cedar, W. Indian	3.9	47.	Maple, soft	3.5	42.
Cedar, Lebanon	2.5	30.	Oak, live	4.9	59.3
Cherry	3.5	42.	Oak, red	3.9	45.
Chestnut	3.4	41.	Oak, white	4.3	52.
Cork	1.3	15.	Pine, Southern	3.7	45.
Elm	2.9	35.	Pine, white	2.1	25.
Ebony	6.3	76.1	Pine, yellow	2.8	34.3
Hemlock	2.1	25.	Spruce	2.1	25.
Hickory	4.4	53.	Sycamore	3.1	37.
Hornbeam	2.9	47.	Walnut	3.2	38.

Building Materials—Stacked.

	Lbs. per cubic ft.		Lbs. per cubic ft.
Brick—pressed	150	Glass—window	157
" common	125	Granite	170
" soft	100	Lime—quick	53
Cement—Portland	100	Plaster of Paris	70
Cement—Rosedale	56	Sand	90-106
Cinders—dry	72	Sandstone	151
Cinders—packed	90	Shale	162
Earth—dry, shaken	82- 92	Slate	175
Earth—rammed	92-100	Trap rock	187

Masonry.

	Lbs. per cubic ft.		Lbs. per cubic ft.
Brick—pressed or paving	140	Granite	160
Brick—hard, common	120	Mortar and plaster	120
Brick—soft	100	Rubble—limestone, common	140
Brick—hollow	90	Rubble—limestone, cut face	150
Concrete—stone	150	Rubble—sandstone, common	140
Concrete—cinder	96	Rubble—sandstone, cut face	150

Standard Load-Bearing Wall Tile.

End construction:	Number of cells	Weight, each lbs.	Side construction:	Number of cells	Weight, each lbs.
3¾ by 12 by 12	3	20	3¾ by 5 by 12	1	9
6 by 12 by 12	6	30	8 by 5 by 12	2	16
8 by 12 by 12	6	36	8 by 5 by 12 ("L" shaped)	..	16
10 by 12 by 12	6	42	8 by 6¼ by 12 ("T" shaped)	4	16
12 by 12 by 12	6	48	8 by 7¾ by 12 (square)	6	24
			8 by 10¼ by 12 ("H" shaped)	7	32

Standard Partition Tile.

	Number of cells	Weight, each lbs.		Number of cells	Weight, each lbs.
3 by 12 by 12	3	15	8 by 12 by 12	4	30
4 by 12 by 12	3	16	10 by 12 by 12	4	36
6 by 12 by 12	3	22	12 by 12 by 12	4	40

Standard Split Furring Tile.

	Number of cells	Weight, each lbs.
2 by 12 by 12	..	9

Standard Book Tile.

	Lbs. per sq. ft.
3 by 12 by 18 to 24	18

Building Materials—In Construction.

Roofing.

	Lbs. per square ft.		Lbs. per square ft
Copper—sheet	0.75 to 1.25	Shingles—wood 16″	2
Felt and gravel	8 to 10	Singles—wood 16″	2
Iron—corrugated	1 to 3.75	Slate—average	10
Iron—galvanized	1 to 3	Tile—fancy, laid in mortar	25 to 30
Iron—sheet, black, painted	1.5	Tile—plain, average	12
Ready composition roofing	1 to 1.5	Tin and paint	1
Sheet lead	4 to 8	Zinc	1 to 2

Floors.

	Lbs. per sq. ft.		Lbs. per sq. ft.
Flat arches (tile) 3″ thick	17	Flat arches (tile) 12″ thick	39
" " " 4″ "	18	" " " 14″ "	43
" " " 6″ "	25	" " " 16″ "	49
" " " 8″ "	31	Book tile 2″ thick	15
" " " 10″ "	35	" " 3″ "	17
Brick arches 4″ thick and concrete	70	Beam tile	15

Table for Weights of Yellow Pine Joists, Studs and Rafters on the Assumption That One Board Foot of Y. P. Weighs 2.8 Pounds.

Spacing	Size	Weight per Sq. Foot	Size	Weight Per Sq. Foot	Size	Weight
12"	2"x4"	1.87	2"x6"	2.8	2"x8"	3.74
14"	"	1.60	"	2.4	"	3.20
16"	"	1.40	"	2.1	"	2.80
18"	"	1.25		1.87	"	2.50
20"	"	1.12	"	1.68		2.24
22"	"	1.02	"	1.53	"	2.04
12"	2"x10"	4.68	2"x12"	5.61	2"x14"	6.55
14"	"	4.00	"	4.80	"	5.60
16"	"	3.50	"	4.20	"	4.90
18"	"	3.13	"	3.75		4.38
20"		2.80		3.36		3.92
22"		2.55		3.06		3.57

Partitions.

	Lbs. per sq. ft.
Gypsum partition blocks 3" thick....	10
" " " 4" "	12
" " " 5" "	14
" " " 6" "	16
Plaster on brick, tile or concrete...	5
Partition tile 3" thick............	17
" " 4" "	18
" " 6" "	25
" " 8" "	31
" " 10" "	35

Ceiling.

	Lbs. per sq. ft.
Lath and plaster 2 coats...........	9
Lath and plaster 3 coats...........	10
Suspended ceiling	10

Sheathing, Flooring, etc.

	Lbs. per sq. ft.
Pine, Hemlock, Spruce, Poplar, Redwood, per inch thick..............	3
Chestnut, Maple	4

Weight per Square Foot of Sheet Lead.

1/62 inch thick	2 lbs.
3/64 " "	2½ "
1/25 " "	3 "
1/16 " "	4 "
1/14 " "	5 "
1/12 " "	6 "
1/10 inch thick	7 lbs.
⅛ " "	8 "
5/32 " "	10 "
3/16 " "	12 "
7/32 " "	14 "
¼ " "	16 "

Miscellaneous Items.

While the following items vary considerably in weight, the values given below are fair averages and may be used for preliminary computations.

	Lbs. per sq. ft.
Iron stair construction	50
Concrete stair construction	150
Wood stair construction	20
Sidewalk lights in concrete........	30
Reinforcement of concrete	6
Steel joists per sq. ft. of floor.....	6
Steel girders per sq. ft. of floor......	4

Contents of Storage Warehouses.

Material.	Weight per Cu. ft.	Allowable Height of Pile in ft.
Groceries Etc.		
Beans—in bags	40	8
Canned goods—cases	58	6
Coffee—roasted, in bags...	33	8
Coffee—green, in bags....	39	8
Flour—in barrels	40	5
Molasses—in barrels......	48	5
Rice—in bags	58	6
Sal Soda—in barrels......	46	5
Salt—in bags	70	5
Soap powder—in cases.....	38	8
Starch—in barrels	25	6
Sugar—in barrels	43	5
Sugar—in cases	51	6
Tea—in chests	25	8
Wines and Liquors, in bbls.	38	6
Dry Goods, Cotton, Wool, Etc.		
Burlap—in bales	43	6
Coir Yarn, in bales.......	33	8
Cotton — in bales, compressed	18	8
Cotton Bleached Goods — in cases	28	8
Cotton Flannel—in cases..	12	8
Cotton Sheeting—in cases.	23	8
Cotton Yarn—in cases....	25	8
Excelsior—compressed	19	8
Hemp—Manila, compressed	30	8
Linen Goods—in cases....	30	8
Wool—in bales, not compressed	13	8
Wool—worsteds, in cases..	27	8
Hardware, Etc.		
Sheet tin—in boxes.......	278	2
Wire—insulated copper, in coils	63	5
Wire—galvanized iron, in coils	74	4.5
Wire—magnet, on spools..	75	6
Drugs, Paints, Oils, Etc.		
Glycerine—in cases.......	52	6
Linseed oil—in bbls.......	36	6
Logwood extract—in boxes	70	5
Rosin—in bbls............	48	6
Shellac—gum	38	6
Soda — Caustic, in iron drums	88	3.33
Soda—Silicate, in bbls....	53	6
Sulphuric Acid	60	1.66
White Lead Paste—in cans	174	3.5
White Lead—dry	86	4.75
Red Lead and Litharge Putty—dry	132	3.75
Miscellaneous.		
Glass and Chinaware — in cases	40	5
Hides and Leather — in bales	20	8
Paper — newspaper and strawboard	35	6
Paper—writing and calendared	60	6
Rope—in coils	32	6

NOMENCLATURE OF DRAWINGS

We present in the following pages a collation of symbols for plan nomenclature, which we hope will be the means of bringing about a more uniform practice. In addition to the convenience, which will result from uniform practice to those compelled to examine, estimate from or execute plans from different offices; it will be found that the proficiency of draftsmen will not be so seriously affected on changing from office to office if practice becomes uniform.

General symbols presented have been collated from various sources. To assist memory those symbols have been selected which are suggestive in their make up.

For illustration all lines indicating water pipes have a periodic double indentation suggestive of a "w"; gas lines a periodic embryo "G", etc.

Lighting symbols are those adopted by the American Institute of Architects and the National Electrical Contractors' Association, except that 50 watts is taken as the standard for one light unit instead of 16 c. p.

Structural iron standard symbols; the Osborn systems are so generally understood and used that it hardly seems necessary to publish same. (See Cambria pocket book, 1906 edition, p. 309.)

GENERAL SYMBOLS

	In color system use
Earth	Black
Cinders	Green
Concrete	Brown
Stone	Blue
Brick	Red
Structural tile	Brown
Composition wall blocks	Blue
Architectural terra cotta	Brown
Plaster	Blue
Structural iron	Green
Sheet metal	Green
Floor tile, tile and mosaics	Brown
Marble (in elevation)	Blue
Marble (in section)	Blue
Terrazo	Black
Wood in section with grain (soft wood)	Yellow
Wood in section with grain (hard wood)	Brown
Wood in section across grain (soft wood)	Yellow
Wood in section across grain (hard wood)	Brown
Cork	Brown
Glass	Blue

Rubble

Dressed ashlar

Rubble stone

Rock faced ashlar

Dimension stone

Any stone dressed

Ashlar stone

Not described; small numeral refers to details and specifications

7 SIZE HERE — Column: Small numeral indicates No. of particular column

25 — Door: Small numeral indicates No. of particular door

50 W — Window: Small numeral indicates No. of particular window

N 3 — Indicates designating No. of a room or space.

+7-6 — Elevation of point; small numerals indicate elevation above zero point.

PIPING SYMBOLS

	In color system
Cold water	Blue
Hot water	Red
Hot water return	Red
Filtered or drinking water	Blue
Gas piping	Green
Air piping	Green
Compressed air piping	Green
Vacuum cleaning	Green

SEWERAGE AND DRAINAGE

Iron sewer pipe	Green
Sanitary iron sewer pipe	Green
Tile sewer	Red
Sanitary Tile Sewer	Red
Drainage tile	Brown
S.P. Soil pipe	Green
W.P. Waste pipe	Green
D.S. Down spout	Green
V.R. Vent riser	Green
Floor drain	Brown
2 Bracket: Prefix with "F" if for fuel	Blue
3 Ceiling: Prefix with "F" if for fuel	Blue
7 Floor outlet: Prefix with "F" if for fuel	Blue
4/2 Combined gas and electric; lower figure indicates No. of gas tips; upper figure Indicates No. of 50 watt electric lamps	Blue

STANDARD SYMBOLS FOR WIRING PLANS

As recommended and adopted by The Association of Electragists, International, The American Institute of Architects, and the American Institute of Electrical Engineers and approved by The American Engineering Standards Committee on March 6, 1924.

- Ceiling Outlet
- Ceiling Outlet (Gas and Electric)
- Ceiling Lamp Receptacle — Specification to Describe Type Such as Key, Keyless or Pull Chain
- Ceiling Outlet for Extensions
- Ceiling Fan Outlet
- Pull Switch (PS)
- Drop Cord (D)
- Wall Bracket
- Wall Bracket (Gas and Electric)
- Wall Outlet for Extensions
- Wall Fan Outlet
- Wall Lamp Receptacle — Specification to Describe Type Such as Key, Keyless or Pull Chain
- Single Convenience Outlet
- Double Convenience Outlet
- Junction Box (J)
- Special Purpose Outlet — Lighting, Heating and Power as Described in Specification
- Special Purpose Outlet — Lighting, Heating and Power as Described in Specification
- Special Purpose Outlet — Lighting, Heating and Power as Described in Specification
- Exit Light
- Floor Outlet
- Floor Elbow
- Floor Tee
- S^1 Local Switch—Single Pole
- S^2 Local Switch—Double Pole
- S^3 Local Switch—3 Way
- S^4 Local Switch—4 Way
- S^D Automatic Door Switch
- S^K Key Push Button Switch
- S^E Electrolier Switch
- S^P Push Button Switch and Pilot

- S^R Remote Control Push Button Switch
- Push Button
- Buzzer
- Bell
- Annunciator
- Interior Telephone
- Public Telephone
- Clock (Secondary)
- Clock (Master)
- Time Stamp
- Electric Door Opener
- Local Fire Alarm Gong (F)
- City Fire Alarm Station
- Local Fire Alarm Station (F)
- Fire Alarm Central Station
- Speaking Tube
- Nurse's Signal Plug (N)
- Maid's Plug (M)
- Horn Outlet
- District Messenger Call
- Watchman Station (W)
- Watchman Central Station Detector (W)
- Public Telephone—P B X Switchboard (PBX)
- Interconnection Telephone Central Switchboard (IX)
- Interconnection Cabinet

- Telephone Cabinet
- Telegraph Cabinet
- Special Outlet for Signal System — As described in specification
- Battery
- Tank Switch (T.S)
- Motor
- Motor Controller (M.C)
- Lighting Panel
- Power Panel
- Heating Panel
- Pull Box
- Cable Supporting Box
- Meter
- Transformer
- Branch Circuit, Run Concealed Under Floor Above
- Branch Circuit, Run Exposed
- Branch Circuit, Run Concealed Under Floor
- Signal Wires in Conduit — Concealed Under Floor
- Signal Wires in Conduit — Concealed Under Floor Above
- Tap Circuits Indicated by 2 Number 14 Conductors in ½" Conduit
- 3 Number 14 Conductors in ½" Conduit
- 4 Number 14 Conductors in ¾" Conduit Unless Marked ½"
- 5 Number 14 Conductors in ¾" Conduit
- 6 Number 14 Conductors in 1" Conduit Unless Marked ¾"
- 7 Number 14 Conductors in 1" Conduit
- 8 Number 14 Conductors in 1" Conduit

NOTE—If larger conductors than number 14 are used, use the same symbols and mark the conductor and conduit size on the run.

- Feeder Run Concealed Under Floor Above
- Feeder Run Exposed
- Feeder Run Concealed Under Floor
- Pole Line

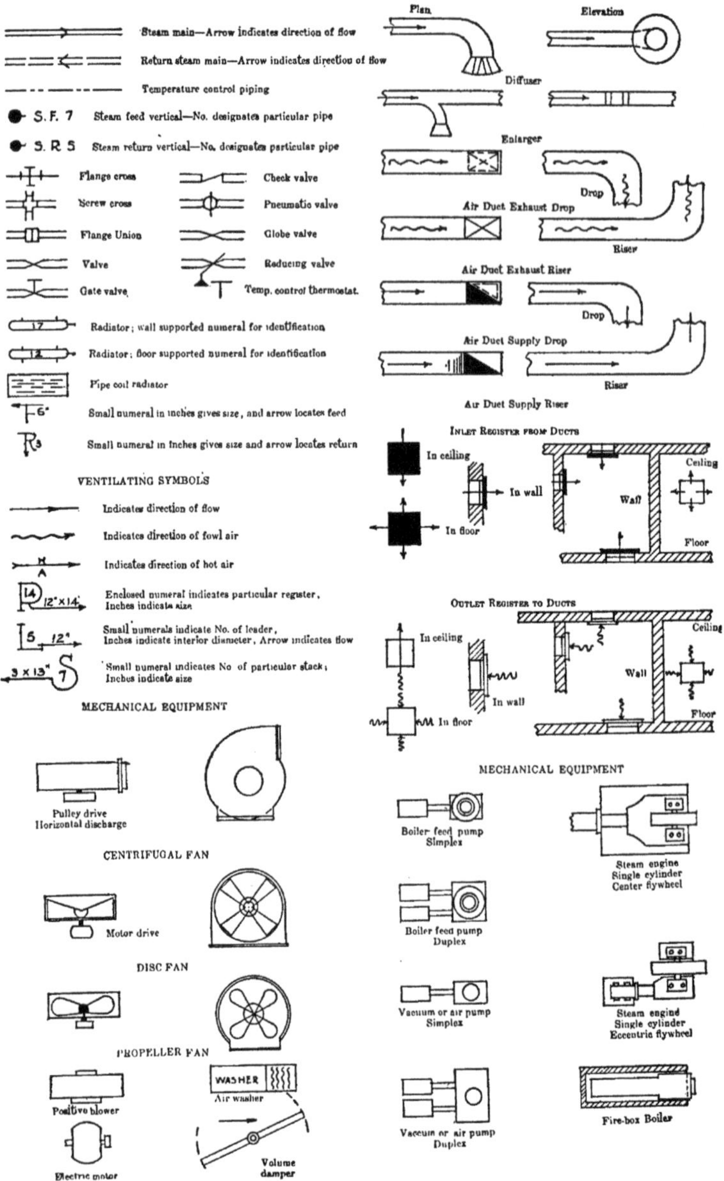
Steam main—Arrow indicates direction of flow
Return steam main—Arrow indicates direction of flow
Temperature control piping
S. F. 7
Steam feed vertical—No. designates particular pipe
S. R 5
Steam return vertical—No. designates particular pipe
Flange cross
Check valve
Screw cross
Pneumatic valve
Flange Union
Globe valve
Valve
Reducing valve
Gate valve
Temp. control thermostat
17
Radiator; wall supported numeral for identification
12
Radiator; floor supported numeral for identification
Pipe coil radiator
F 6"
Small numeral in inches gives size, and arrow locates feed
R3
Small numeral in inches gives size and arrow locates return
VENTILATING SYMBOLS
Indicates direction of flow
Indicates direction of fowl air
H
A
Indicates direction of hot air
14
12"×14
Enclosed numeral indicates particular register, Inches indicate size
5
12"
Small numerals indicate No. of leader, Inches indicate interior diameter, Arrow indicates flow
3 x 13"
7
Small numeral indicates No of particular stack; Inches indicate size
MECHANICAL EQUIPMENT
Pulley drive
Horizontal discharge
CENTRIFUGAL FAN
Motor drive
DISC FAN
PROPELLER FAN
Positive blower
WASHER
Air washer
Electric motor
Volume damper
Plan
Elevation
Diffuser
Enlarger
Drop
Air Duct Exhaust Drop
Riser
Air Duct Exhaust Riser
Drop
Air Duct Supply Drop
Riser
Air Duct Supply Riser
Inlet Register from Ducts
In ceiling
In wall
In floor
Ceiling
Wall
Floor
Outlet Register to Ducts
In ceiling
In wall
In floor
Ceiling
Wall
Floor
MECHANICAL EQUIPMENT
Boiler feed pump
Simplex
Steam engine
Single cylinder
Center flywheel
Boiler feed pump
Duplex
Vacuum or air pump
Simplex
Steam engine
Single cylinder
Eccentric flywheel
Vacuum or air pump
Duplex
Fire-box Boiler

For Welding Symbols see page 529.

TABLE OF TREADS AND RISERS

No. of Risers	6 Inch Rise. ft. in.	6¼ Inch Rise. ft. in.	6½ Inch Rise. ft. in.	6¾ Inch Rise. ft. in.	7 Inch Rise. ft. in.	7⅛ Inch Rise. ft. in.	7¼ Inch Rise. ft. in.	7⅜ Inch Rise. ft. in.	7½ Inch Rise. ft. in.	7⅝ Inch Rise. ft. in.	7¾ Inch Rise. ft. in.
1	6	6¼	6½	6¾	7	7⅛	7¼	7⅜	7½	7⅝	7¾
2	1 0	1 0½	1 1	1 1½	1 2	1 2¼	1 2½	1 2¾	1 3	1 3¼	1 3½
3	1 6	1 6¾	1 7½	1 8¼	1 9	1 9⅜	1 9¾	1 10⅛	1 10½	1 10⅞	1 11¼
4	2 0	2 1	2 2	2 3	2 4	2 4½	2 5	2 5½	2 6	2 6½	2 7
5	2 6	2 7¼	2 8½	2 9¾	2 11	2 11⅝	3 0¼	3 0⅞	3 1½	3 2⅛	3 2¾
6	3 0	3 1½	3 3	3 4½	3 6	3 6¾	3 7½	3 8¼	3 9	3 9¾	3 10½
7	3 6	3 7¾	3 9½	3 11¼	4 1	4 1⅞	4 2¾	4 3⅝	4 4½	4 5⅜	4 6¼
8	4 0	4 2	4 4	4 6	4 8	4 9	4 10	4 11	5 0	5 1	5 2
9	4 6	4 8¼	4 10½	5 0¾	5 3	5 4⅛	5 5¼	5 6⅜	5 7½	5 8⅝	5 9¾
10	5 0	5 2½	5 5	5 7½	5 10	5 11¼	6 0½	6 1¾	6 3	6 4¼	6 5½
11	5 6	5 8¾	5 11½	6 2¼	6 5	6 6⅜	6 7¾	6 9⅛	6 10½	6 11⅞	7 1¼
12	6 0	6 3	6 6	6 9	7 0	7 1½	7 3	7 4½	7 6	7 7½	7 9
13	6 6	6 9¼	7 0½	7 3¾	7 7	7 8⅝	7 10¼	7 11⅞	8 1½	8 3⅛	8 4¾
14	7 0	7 3½	7 7	7 10½	8 2	8 3¾	8 5½	8 7¼	8 9	8 10¾	9 0½
15	7 6	7 9¾	8 1½	8 5¼	8 9	8 10⅞	9 0¾	9 2⅝	9 4½	9 6⅜	9 8¼
16	8 0	8 4	8 8	9 0	9 4	9 6	9 8	9 10	10 0	10 2	10 4
17	8 6	8 10¼	9 2½	9 6¾	9 11	10 1⅛	10 3¼	10 5⅜	10 7½	10 9⅝	10 11¾
18	9 0	9 4½	9 9	10 1½	10 6	10 8¼	10 10½	11 0¾	11 3	11 5¼	11 7½
19	9 6	9 10¾	10 3½	10 8¼	11 1	11 3⅜	11 5¾	11 8⅛	11 10½	12 0⅞	12 3¼
20	10 0	10 5	10 10	11 3	11 8	11 10½	12 1	12 3½	12 6	12 8½	12 11
21	10 6	10 11¼	11 4½	11 9¾	12 3	12 5⅝	12 8¼	12 10⅞	13 1½	13 4⅛	13 6¾
22	11 0	11 5½	11 11	12 4½	12 10	13 0¾	13 3½	13 6¼	13 9	13 11¾	14 2½
23	11 6	11 11¾	12 5½	12 11¼	13 5	13 7⅞	13 10¾	14 1⅝	14 4½	14 7⅜	14 10¼
24	12 0	12 6	13 0	13 6	14 0	14 3	14 6	14 9	15 0	15 3	15 6
25	12 6	13 0¼	13 6½	14 0¾	14 7	14 10⅛	15 1¼	15 4⅜	15 7½	15 10⅝	16 1¾
26	13 0	13 6½	14 1	14 7½	15 2	15 5¼	15 8½	15 11¾	16 3	16 6¼	16 9½
27	13 6	14 0¾	14 7½	15 2¼	15 9	16 0⅜	16 3¾	16 7⅛	16 10½	17 1⅞	17 5¼
28	14 0	14 7	15 2	15 9	16 4	16 7½	16 11	17 2½	17 6	17 9½	18 1
29	14 6	15 1¼	15 8½	16 3¾	16 11	17 2⅝	17 - 6¼	17 9⅞	18 1½	18 5⅛	18 8¾
30	15 0	15 7½	16 3	16 10½	17 6	17 9¾	18 1½	18 5¼	18 9	19 0¾	19 4½

No. of Risers	7⅞ Inch Rise. ft. in.	8 Inch Rise. ft. in.	8¼ Inch Rise. ft. in.	8½ Inch Rise. ft. in.	9 Inch Rise. ft. in.	9½ Inch Rise. ft. in.	10 Inch Rise. ft. in.	10½ Inch Rise. ft. in.	11 Inch Rise. ft. in.	13 Inch Rise. ft. in.	14 Inch Rise. ft. in.
1	7⅞	8	8¼	8½	9	9½	10	10½	11	1 1	1 2
2	1 3¾	1 4	1 4½	1 5	1 6	1 7	1 8	1 9	1 10	2 2	2 4
3	1 11⅝	2 0	2 0¾	2 1½	2 3	2 4½	2 6	2 7½	2 9	3 3	3 6
4	2 7½	2 8	2 9	2 10	3 0	3 2	3 4	3 6	3 8	4 4	4 8
5	3 3⅜	3 4	3 5¼	3 6½	3 9	3 11½	4 2	4 4½	4 7	5 5	5 10
6	3 11¼	4 0	4 1½	4 3	4 6	4 9	5 0	5 3	5 6	6 6	7 0
7	4 7⅛	4 8	4 9¾	4 11½	5 3	5 6½	5 10	6 1½	6 5	7 7	8 2
8	5 3	5 4	5 6	5 8	6 0	6 4	6 8	7 0	7 4	8 8	9 4
9	5 10⅞	6 0	6 2¼	6 4½	6 9	7 1½	7 6	7 10½	8 3	9 9	10 6
10	6 6¾	6 8	6 10½	7 1	7 6	7 11	8 4	8 9	9 2	10 10	11 8
11	7 2⅝	7 4	7 6¾	7 9½	8 3	8 8½	9 2	9 7½	10 1	11 11	12 10
12	7 10½	8 0	8 3	8 6	9 0	9 6	10 0	10 6	11 0	13 0	14 0
13	8 6⅜	8 8	8 11¼	9 2½	9 9	10 3½	10 10	11 4½	11 11	14 1	15 2
14	9 2¼	9 4	9 7½	9 11	10 6	11 1	11 8	12 3	12 10	15 2	16 4
15	9 10⅛	10 0	10 3¾	10 7½	11 3	11 10½	12 6	13 1½	13 9	16 3	17 6
16	10 6	10 8	11 0	11 4	12 0	12 8	13 4	14 0	14 8	17 4	18 8
17	11 1⅞	11 4	11 8¼	12 0½	12 9	13 5½	14 2	14 10½	15 7	18 5	19 10
18	11 9¾	12 0	12 4½	12 9	13 6	14 3	15 0	15 9	16 6	19 6	21 0
19	12 5⅝	12 8	13 0¾	13 5½	14 3	15 0½	15 10	16 7½	17 5	20 7	22 2
20	13 1½	13 4	13 9	14 2	15 0	15 10	16 8	17 6	18 4	21 8	23 4
21	13 9⅜	14 0	14 5¼	14 10½	15 9	16 7½	17 6	18 4½	19 3	22 9	24 6
22	14 5¼	14 8	15 1½	15 7	16 6	17 5	18 4	19 3	20 2	23 10	25 8
23	15 1⅛	15 4	15 9¾	16 3½	17 3	18 2½	19 2	20 1½	21 1	24 11	26 10
24	15 9	16 0	16 6	17 0	18 0	19 0	20 0	21 0	22 0	26 0	28 0
25	16 4⅞	16 8	17 2¼	17 8½	18 9	19 9½	20 10	21 10½	22 11	27 1	29 2
26	17 0¾	17 4	17 10½	18 5	19 6	20 7	21 8	22 9	23 10	28 2	30 4
27	17 8⅝	18 0	18 6¾	19 1½	20 3	21 4½	22 6	23 7½	24 9	29 3	31 6
28	18 4½	18 8	19 3	19 10	21 0	22 2	23 4	24 6	25 8	30 4	32 8
29	19 0⅜	19 4	19 11¼	20 6½	21 9	22 11½	24 2	25 4½	26 7	31 5	33 10
30	19 8¼	20 0	20 7½	21 3	22 6	23 9	25 0	26 3	27 6	32 6	35 0

RULE FOR CACULATING PROPORTIONED WIDTH AND HEIGHT OF TREADS AND RISERS OF STAIRS.
Subtract the width of tread from 25 in. and the result will be twice the height of the riser. Thus: if the tread is 10 in. wide, then the height of riser proportionate to a 10-inch tread. This is exclusive of nosings.

BOWLING ALLEY DETAIL

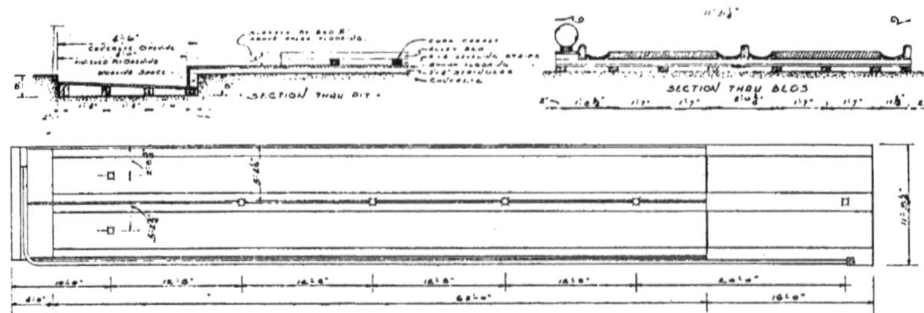

Size of the Billiard Room and Bowling Alleys.

Table	Outside dimensions	Room space required
1 ...2½'x 5'	2' 9" x 4' 10"	10' x12'
2 ...3' x 6'	3' 4" x 5' 11¼"	11' x14'
3 ...3½'x 7'	3' 11" x 7' 1"	12' x15'
4 ...4' x 8'	4' 7" x 8' 5"	14' 2"x18'
5 ...4½'x 9'	4' 11½"x 9' 1½"	14' 6"x18' 9"
6 ...5' x10'	5' 5½"x10' 1½"	15' x20'
7 ...6' x12'	6' 8" x12' 6"	16' x22'

1 is essentially a children's table, 2 and 3 sizes are provided to meet restricted space conditions, all sizes are adaptable to home use, sizes 4, 5 and 6 especially, 5 and 6 are the dominant commercial or club sizes, 7 English style standard cue length 57".

For dimensions required to use two or more tables of any size or sizes furnished, see manufacturer. For a single pair of 2 Regulation bowling alleys.

The length from back wall to the front of the approach, should never be less than 82 ft. This allows for pit and swinging cushion 4 ft. for alleys (to foul line) 63 ft. and for approach 15 ft. Width is 11'x5¼" but can be reduced if necessary. Space for players' seats for spectors should be in addition to the lengths and widths given. All drawings show concrete foundation construction which is necessary for basement installation, first floors where there is no basement.

Passenger Automobile Sizes.

To accommodate any passenger car in a residential garage the sizes of the largest cars are as follows:

Car—7 Passenger 1939 Model	Length Bumper to Bumper	Rear Tread
Packard 12	230 25/32"	62½"
Cadillac 16	222"	62½"
Chrysler Imperial	224⅞"	63"
Lincoln	223½"	60"

The rear tread measurement is from the center of one rear tire to the center of the other. The largest tires for passenger cars are 7.5". From 4" to 5" should be added for fenders on each side. Adding the 7.5" for the balance of the tire width and 10" for two fenders to the maximum tread of 63" for a Chrysler Imperial would be 80.5" wide for the car with no allowances for driving.

Heavy Trucks.

Length, 15 ft. to 26 ft.
Width, 6 ft. 0 in.
Height, 10 ft. 0 in.
Width on floor between wheel pockets, 48 in. Length of wheel pocket, 34 in.
Smallest practical door, 9 ft. 0 in. wide by 11 ft. 0 in. high; for largest trucks, 13 ft. 6 in. high.
Doors to alley should not be less than 12 ft. wide and should be set not less than 28 ft. from opposite side of alley.

Moving Vans.

Length, 13 ft. to 16 ft. 6 in.
Width, 7 ft. to 8 ft. 2 in.
Height, 10 ft. to 12 ft.
Smallest practical door 10 ft. 0 in. wide by 13 ft. 6 in. high.

CLEARANCE UNDER OLD ELEVATED RAILWAY STRUCTURES AND TROLLEY WIRES, 12 FT. 0 IN.

Clearance required by the city for steam roads, 13 ft. 6 in.

Architects will be perfectly safe in making the maximum limit of door heights for any sort of vehicle 13 ft. 6 in., standard subway height, as no vehicle can be used commercially on the streets of Chicago that will not clear steam road viaducts. They might go around elevated viaducts, but they can not go around steam road viaducts and there is a probability that any future elevated viaducts would be raised to the city standard height of 13 ft. 6 in.

FURNITURE DIMENSIONS. FILE 8270

Chairs—Height of seat, 18"; depth of seat, 19"; top of back, 38"; arms, 9" above seat.
Lounge—6' long, 30" wide.
Tables—Writing, height, 2'-5"; sideboards, height, 3'-0"; general height, 2'-6".
Note—The smallest size practical for knee holes, 2' high by 1'-8" wide.
Beds—Single, width, 3' to 4'; ¾ bed, width, 4'; double bed, width, 4'-6" to 5'-0", length 6'-6" to 6'-8"; standard double bed, 4'-6" x 6'-6"; footboards, 2'-6" to 3'-6" high; headboards, 5' to 6'-6".
Bureaus—Common, width, 3'-5" or 4'; depth, 1'-6" or 1'-8"; height, 2'-6" or 3'.
Commodes—Top, 1'-6" square and 2'-6" high.
Chiffoniers—3' wide, 1'-8" deep, 4'-4" high.
Cheval Glasses—Height, 6'-4" or 5'-0" or 5'-2"; width, 3'-2" or 2'-6" or 1'-8".
Washstands—Length, 3'-0"; width, 1'-6"; height, 2'-7".
Wardrobes—Length, 4'-6"—3'-0"; depth, 2'-0"—1'-5"; height, 8'-0".
Sideboards—Length, 5' to 6'; depth, 2'-2".
Pianos—Upright, length, 4'-10" to 5'-6"; height, 4'-4" to 4'-9"; depth, 2'-4". Square, length, 6'-8"; depth, 3'-4".
Billiard Tables—4'-8", 4½" x 9, 5' x 10. Must have 16' x 20' space.
Wardrobe Shelves—5'-10" high.
Coat Hooks—5'-6" high.
Flour Barrel—28" to 30" high and 20" to 21" dia.

DATA ON BUILDINGS WITH SIDINGS.

Clearance from face of building to center of track, 7'-0".
Height of loading decks:
For shipping, 4'-0".
For receiving, 3'-0".
Clearance from center of track to edges of loading decks:
Upper edge, 7'-0".
Lower edge, 5'-0".

Cocktail Bars.

The minimum sizes as recommended for bars and back bars are shown above. The space between the work board and back counter is of especial importance. In circular or odd shaped bars it is advised to allow greater space. Where the length of the bar is limited it is also well to provide sufficient space between work board and back counter for the passing of two bartenders. The top frame may be of any height even extending to the ceiling.

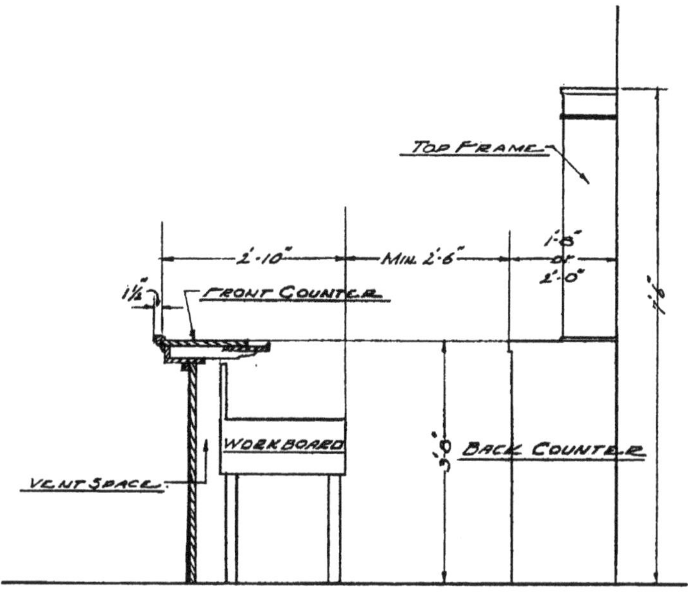

Electric Organs.

The Organ Console will fit in a space 5' square. Total weight of Console, including Pedal Clavier and Bench, is approximately 560 lbs. Tone Cabinet should be located some distance from the console to prevent a magnetic field setting up a hum and allowing for a short delay in the return of the sound to the organist. The Tone Cabinet should be located in a Reverberation Chamber so the sound will reflect from the walls and not direct from the cabinet which would be direct radiation from the Speaker Cones.

Where a Reverberation Chamber cannot be used special Tone Cabinets are available with self-contained reverberation qualities. The Tone Cabinet should be located above listeners where possible as on a balcony. Amplifiers are part of the Tone Cabinet.

Where a Reverberation Chamber is used Walls should be of a solid construction and hard plaster. Where desired, two chambers may be used. Reverberation increases in direct proportion to the volume of the room. Assuming solid construction and hard plaster walls and ceiling, the chambers should not be less than an eight foot cube (512 cu. ft.).

Electrical Equipment and wiring required are sufficient power of 115 volts 60 cycles A. C. current.

In the case of D. C. current a Rotary Converter with a frequency regulator. The frequency must be within a half cycle either way of 60 and must be constant. Special cable are used between the Console and the Tone Cabinet or Cabinets as well as Control Switches.

Size of Swimming Tank.

Swimming tanks that can be used for swimming contests must be exactly 20 yards in interior length, no less. (A tank ½ inch short would be ruled out of contest.) Eight yards wide is best, although 7 yards will pass; 4 feet deep at shallowest point and 8 feet deep at deepest point, which deepest point should be about 12 feet from end where springboard is placed. Depth at springboard end should be six feet. Interior of tank, both sides and bottom, should be white, and there should be three black lines on the bottom extending parallel with sides, and dividing the tank into four equal alleys; there should be a line across tank on bottom and up sides at exactly 2 yards from each end, measured horizontally, making lines exactly 16 yards apart horizontally.

SIZES OF FREIGHT CARS AND LEGAL RAILWAY CLEARANCES

FOR ILLINOIS

(Note: Designers should understand that there is a variation in legal Railway clearances in the various States. Particularly note Minnesota, Missouri, New Mexico, Texas and Wisconsin.)

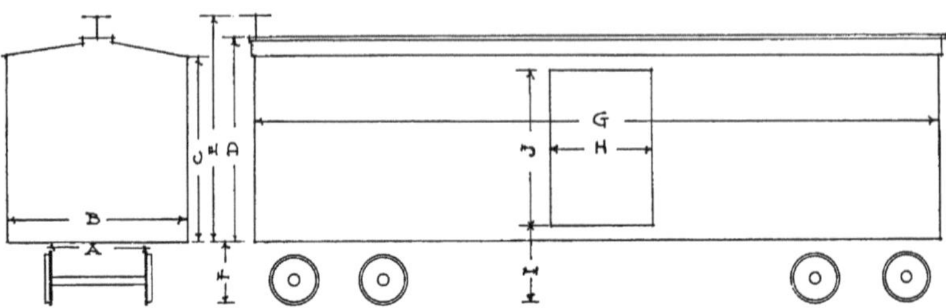

Type	Larger Freight Cars	A	B	C	D	E	F	G	H	I	J	K
Auto	C. & N. W. Ry.	56½"	9' 6"	11'4"	12' 3"	12'7"	34"	42'0"	120"	44"	10'0"	30' 8"
Auto	C. & N. W. Ry.	56½"	10' 5"	11'2"	12' 1"	—	35"	51'9"	174"	44"	9'10"	40' 9"
Box	C. & N. W. Ry.	56½"	10' 5"	11'1"	12' 0"	12'6"	32"	42'2"	72"	44"	9' 5"	31' 2"
Box	C. & N. W. Ry.	56½"	9' 4"	10'1"	10'11"	11'6"	32"	42'2"	72"	43"	8' 7"	31' 2"
Box	C. & N. W. Ry.	56½"	9' 4"	9'7"	10' 5"	—	32"	42'1"	72"	43"	8' 2"	31' 2"
Box	I. C. Ry.	56½"	9'11¼"	10'8"	11' 5"	12'0"	3'2¼"	44'6⅝"	6'3"	3'8"	7' 6"	—
Box	C. B. & Q. Ry.	56½"	9' 5"	—	—	—	—	42'3"	6'0"	3'7½"	9' 5"	30' 8½"
Auto	C. B. & Q. Ry.	56½"	10' 5⅝"	—	—	—	—	52'9½"	14'6¼"	3'7"	9' 6"	41' 3"
Stock	C. B. & Q. Ry.	56½"	9' 0⅞"	—	—	—	—	42'9½"	5'0"	3'7"	8'8⅞"	31' 3"
Gondola	C. B. & Q. Ry.	56½"	10' 0"	—	—	—	—	40'0"	—	46"	4' 3"	28'10"
Ballast	C. B. & Q. Ry.	56½"	10' 2"	7'2"	—	—	3'6"	34'0"	—	—	—	24' 5"

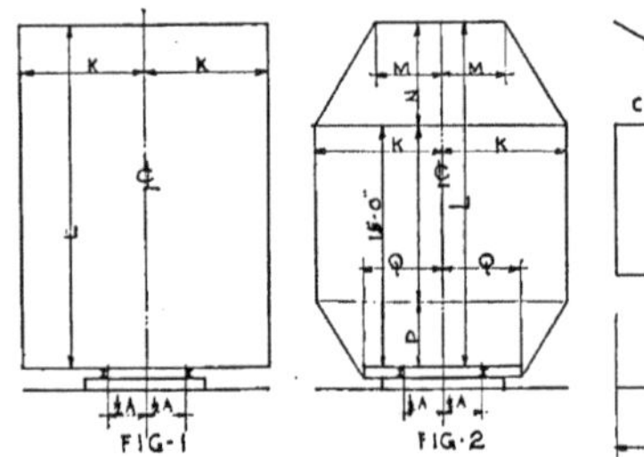

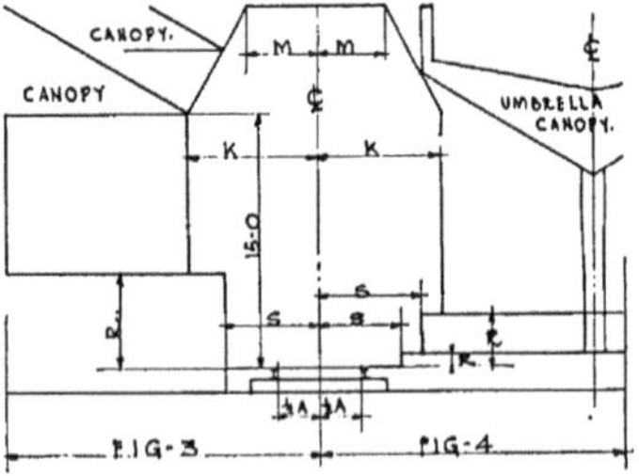

Fig. 1. Buildings and Miscellaneous Structures Adjacent to Main Tracks, L = 21'6", K = 8'0"; Adjacent to Subsidiary Passenger Tracks L = 21'6", K = 7'6"; Tracks entering buildings L = car clearance (in no case less than 15'6" and if to clear engines not less than 17'0"), K = 7'0".

Structures Adjacent to Subsidiary Freight Tracks except as otherwise specified. Tracks outside buildings L = 21'6", K = 8'0"; Tracks entering buildings L = practical car clearance 15'6", engine clearance 17'0", recommended in all cases where possible; K = 7'0".

Fig. 2. Bridges Supporting Main Tracks or Subsidiary Freight Tracks clearance shall be as follows: L = 21'3", M = 4'2", P = 4'0", Q = 5'0". Bridges spanning Main Tracks or Subsidiary Freight Tracks Fig. 1. L = 21'6", K = 8'0".

Fig. 3. High Freight Platforms R = not to exceed 3'8", Platforms used exclusively for refrigerator cars 3'3", S = 5'8", and for all Platforms higher than 3'8" S = not less than 8'0" for Main Line Tracks, and 7'6" for Subsidiary Tracks.

Fig. 4. High Passenger Platforms on Exclusive Passenger Tracks may have R = height of car floor above rail and clearance, according to special State restrictions. Low Passenger Platform R = 8" with S = to not less than 5'1", and very Low Passenger Platform R = 4" with S = to not less than 4'6".

MASONRY, PLASTERING AND FIREPROOFING.

Weight of Brickwork

Placing the weight of brickwork at 112 lb. per cubic foot, the weights per superficial foot for different walls are:

9 inch wall........................ 84 lb.
13 inch wall........................121 lb.
18 inch wall........................168 lb.
22 inch wall........................205 lb.
26 inch wall........................243 lb.

Measurement of Old Brick

Uncleaned rough from building dumped from 8 to 10 bricks per cubic foot, or average of 111 cubic feet to the M.

Uncleaned stacked on outside and interior of stack filled promiscuously 10-12 per cubic foot, or average of 91 cubic feet to the M.

Cleaned and closely stacked, 16 to 18 bricks per cubic foot, or actual average of 59 cubic feet to M. (Usually sold at 60 cubic feet to M to allow for waste and poor piling.)

Cleaned stacked on outside and interior filled promiscuously, 12 to 14 per cubic foot, or actual average of 77 cubic feet to M. (When sold from pile measure customary to count 80 cubic feet to M, to allow for waste and bats.)

Measurement of New Brickwork

The Chicago Masons and Builders' Association have arbitrarily assumed that a cubic foot of wall contains 22½ common brick, or 7½ brick to the superficial foot of 4-inch wall and 15 brick to the superficial foot of 8-inch wall. These figures of the Masons' and Builders' Association are frequently used for the appraisal of party walls, etc., but if so used, the price per M for work in wall should be reduced accordingly.

The actual number of Chicago common brick required for a cubic foot of solid wall varies from 17½ to 19½, and masons in purchasing brick usually reserve 18 brick per cubic foot of solid wall; and when so doing, rarely find an excess or shortage at the end of construction. When the walls are divided into many small piers, requiring much cutting, and consequently much waste, it is best to figure 20 brick to the cubic foot.

On account of the wide variance of practice on the part of masons in estimating, architects, when calling for estimates on brick work by the thousand, will avoid useless controversy by stipulating that quantity of brick will be determined by superficial wall measurement according to the following rule, which is very nearly correct, as Chicago brick now run. Divide the total number of superficial feet of wall surface of a given thickness by 160, and multiply the result by the number of brick widths the wall is thick, and the result will equal the number of thousands of brick contained. A four-inch wall will contain 6¾ brick to the superficial foot, or 1,000 brick to 160 square feet.

Miscellaneous Masonry Data

One hundred yards of plastering will require fourteen hundred laths, four and a half bushels of lime, four-fifths of a load of sand, nine pounds of hair and five pounds of nails, for two-coat work.

A load of mortar measures a cubic yard, requires a cubic yard of sand and nine bushels of lime, and will fill thirty hods.

A bricklayer's hod measuring one foot four inches by nine inches, equals 1,296 cubic inches in capacity, and contains twenty bricks.

A single load of sand or other materials equals a cubic yard.

Cement Mortars

Recent developments in building codes and construction practice have shown a marked tendency toward the more extended use of portland cement mortars.

Where greatest strength is required cement and sand mortar is almost invariably recommended; for a mortar with easy working and good weathering qualities, and with sufficient strength for all ordinary purposes, equal parts of cement and lime, with six parts of sand, is the formula generally adopted. The following paragraphs suggest good practice in specifying various types of mortar:

Mortar

Portland cement mortar used in laying up masonry shall be mixed in the proportion of one part of portland cement to not more than three parts of sand, measured by volume. Hydrated lime or lime putty may be added to an amount not exceeding 15 per cent, by volume, of the portland cement used.

Cement-and-lime mortar shall be mixed in the proportion of one part of portland cement and one part of lime to not more than 6 parts of sand, all by volume.

Lime or natural cement mortar shall be mixed in the proportion of one part of lime or cement to not more than 3 parts of sand measured by volume.

Whenever cement-and-lime mortar or natural cement mortar is used instead of portland cement mortar, the allowable working stresses on the masonry shall be reduced to 80 per cent of that allowed with portland cement mortar.

Whenever lime mortar is used instead of portland cement mortar, the maximum allowable working stress shall be reduced to 50 per cent of that specified for masonry laid up with portland cement mortar.

Where masonry is to be highly stressed, mortar is often tested. The following table shows what the strength of the various types of mortar should be, where tests are deemed advisable:

Strengths of Mortars Under Typical Field Condition*

Aggregates, by volume	Water Percent[1]	Compressive Strength at 28 days. Average of five 2-inch cubes or cylinders stored in air. Lbs. per sq. in.
1:3 Portland cement and sand	22	500
1:1:6 Portland cement, hydrated lime and sand.......	25	200
1:1:4 Portland cement, hydrated lime and sand.......	25	300
1:3 Hydrated lime and sand..	30	30

*See page 30, "Recommended Minimum Requirements for Masonry Wall Construction," Report of Building Code Committee, U. S. Department of Commerce.

[1]Percentages in terms of total weight of dry materials. The water proportions given are those ordinarily used for mortar for laying brick; not those necessarily resulting in the greatest mortar strength.

The cleanliness of the sand used has an important effect upon the strength of mortar or concrete. Excessive amounts of silt, clay, loam or organic matter are harmful. The influence of these impurities is somewhat irregular and therefore clean sand should be insisted upon. Care should also be taken not to permit the use of sand which has become mixed with soil at the bottom of storage piles.

The following data will assist in estimating the quantities of material needed for brick work when portland cement mortar is used: Ordinarily approximately 13.8 cu. ft. of mortar is required to lay 1,000 brick. 2.4 barrels of cement and 1.06 cu. yds. of sand are required to produce 1 cu. yd. of cement mortar. It is customary to assume that one bag of portland cement equals 1 cu. ft. and that a bag of hydrated lime equals about 1¼ cu. ft. In proportioning mortar, if the sand is thoroughly dry, a small reduction in the standard amount of sand used is desirable.

OVERLAYING CONSTRUCTION SHEET, SHINGLE AND COMPOSITION COVERING.

The average width of a shingle is four inches. Hence, when shingles are laid four inches to the weather each shingle averages 16 square inches, and 900 are required for a square of roofing (100 square feet). If 4½ inches to the weather, 800; 5 inches, 720; 5½ inches, 655; 6 inches, 600.

Slating.

FILE 695.2

Slating is estimated by the "square," which is the quantity required to cover 100 square feet. The slates are usually laid so that the third laps the first three inches.

Number of Slates per Square.

Size in Inches.	Pieces per Square.	Size in Inches.	Pieces per Square.	Size in Inches.	Pieces per Square.
6 × 12	533	8 × 16	277	12 × 20	141
7 × 12	457	9 × 16	246	14 × 20	121
8 × 12	400	10 × 16	221	11 × 20	137
9 × 12	355	9 × 18	213	12 × 22	126
7 × 14	374	10 × 18	192	14 × 22	108
8 × 14	327	12 × 18	160	12 × 24	114
9 × 14	291	10 × 20	169	14 × 24	98
10 × 14	261	11 × 20	154	16 × 24	86

The weight of slate per cubic foot is about 174 pounds, or per square foot of various thicknesses as follows:

Thickness in inches........	⅛	3/16	¼	⅜	½
Weight in pounds..........	1.81	2.71	3.62	5.43	7.25

The weight per square foot of roof tiling, set in iron or between wood rafters ready for slating, is about 12 pounds.

Tin Roofs.

Tin roofs should be laid with cleats.

There are two kinds of tin—"bright tin," the coating of which is all tin, that is, the tin proper; and "tern," "leaded," or "roofing" tin, the coating of which is a composition, part tin and part lead. This last will not rust any quicker, but the sulphur in soft coal smoke eats through the "leaded" coating sooner than through the "tinned."

Sizes of tin, 10 by 14 and 14 by 20, and two grades of thickness—IC light, and IX, heavy. For a steep roof (one-sixth pitch or over) the IC 14 by 20 tin ("leaded" if high up where little smoke will get to it; "bright" if low down), put on with a standing groove, and with the cross seams put together with a double lock, makes as good a roof as can be made. For flat roofs IX 10 x 14 "light" is best, laid with cleats, but the others make good roofs and any of them will last twenty-five years at least, if painted periodically.

Number of Square Feet a Box of Roofing Tin Will Cover.—For flat seam roofing, using ½-inch locks, a box of "14 by 20" size will cover about 192 square feet, and for standing seam, using ⅜-inch locks and turning 1¼ and 1½ inch edges, making 1-inch standing seams, it will lay about 168 square feet.

For flat seam roofing, using ½-inch locks, a box of "28 by 20" size will cover about 399 square feet, and for standing seam, using ⅜-inch locks and turning 1¼ and 1½ inch edges, making 1-inch standing seams, it will lay about 365 square feet.

Every box of roofing plates (IC or IX "14 by 20" or "28 by 20" sizes) contains 112 sheets.

For roofs and gutters use seven-pound lead; for hips and ridges, six-pound; for flashings, four-pound.

Gutters should have a fall of at least one inch in ten feet.

No sheet lead should be laid in greater length than ten or twelve feet without a dip to allow for expansion.

Joints to lead pipes require a pound of solder for every inch in diameter.

SANITARY EQUIPMENT

INCLUDING PLUMBING AND HEATING

Capacity of Cisterns.

FILE 697.3

For a circular cistern, square the diameter and multiply by .7854, for the area; multiply this by 1,728 and divide by 231, for number of gallons of one foot in depth; for a square cistern, multiply length by breadth, and proceed as above.

CIRCULAR CISTERN.

5 feet in diameter holds 4.66 bbls.
6 feet in diameter holds 6.71 bbls.
7 feet in diameter holds 9.13 bbls.
8 feet in diameter holds 11.93 bbls.
9 feet in diameter holds 15.10 bbls.
10 feet in diameter holds 18.65 bbls.

SQUARE CISTERN.

5 feet by 5 feet holds 5.92 bbls.
6 feet by 6 feet holds 8.54 bbls.
7 feet by 7 feet holds 11.63 bbls.
8 feet by 8 feet holds 15.19 bbls.
9 feet by 9 feet holds 19.39 bbls.
10 feet by 10 feet holds 23.74 bbls.

Dimension of Genuine Wrought Iron and Steel Pipe.

Nominal size	External diam.	Internal diameter		Wall thickness		Weight per ft.		Th'rds per inch	Test pressure	Couplings	
		Wr't iron	Steel	Wr't iron	Steel	Plain ends	Th'rds and c'plgs.			Nom. o.d.	Nom. length
Standard Weight Pipe—Black and Galvanized											
¼	.540	.360	.364	.090	.088	.42	.42	18	700	.685	1
⅜	.675	.489	.493	.093	.091	.56	.56	18	700	.848	1⅛
½	.840	.617	.622	.111	.109	.85	.85	14	700	1.024	1⅜
¾	1.050	.819	.824	.115	.113	1.13	1.13	14	700	1.281	1⅝
1	1.315	1.043	1.049	.136	.133	1.67	1.68	11½	700	1.576	1⅞
1¼	1.660	1.374	1.380	.143	.140	2.27	2.28	11½	1000	1.950	2⅛
1½	1.900	1.604	1.610	.148	.145	2.71	2.73	11½	1000	2.218	2⅜
2	2.375	2.060	2.067	.158	.154	3.65	3.67	11½	1000	2.760	2⅝
† 2½	2.875	2.460	2.469	.208	.203	5.79	5.81	8	1000	3.276	2⅞
† 3	3.500	3.059	3.068	.221	.216	7.57	7.61	8	1000	3.948	3⅛
3½	4.000	3.538	3.548	.231	.226	9.10	9.20	8	1000	4.591	3⅝
4	4.500	4.016	4.026	.242	.237	10.79	10.88	8	1000	5.091	3⅝
5	5.563	5.036	5.047	.263	.258	14.61	14.81	8	1000	6.296	4⅛
6	6.625	6.053	6.065	.286	.280	18.97	19.18	8	1000	7.358	4⅛
8	8.625	8.059	8.071	.283	.277	24.69	25.00	8	800	9.420	4⅝
8	8.625	7.967	7.981	.329	.322	28.55	28.80	8	1000	9.420	4⅝
10	10.750	10.181	10.192	.284	.279	31.20	32.00	8	600	11.721	6⅛
10	10.750	10.124	10.136	.313	.307	34.24	35.00	8	700	11.721	6⅛
10	10.750	10.005	10.020	.372	.365	40.48	41.13	8	900	11.721	6⅛
12	12.750	12.077	12.090	.336	.330	43.77	45.00	8	600	13.958	6⅛
12	12.750	11.985	12.000	.382	.375	49.56	50.70	8	800	13.958	6⅛

Grade Per Mile.

The following table will show the grade per mile:
An inclination of

1 foot in 15 is 352 feet per mile.
1 foot in 20 is 264 feet per mile.
1 foot in 25 is 211 feet per mile.
1 foot in 30 is 176 feet per mile.
1 foot in 35 is 151 feet per mile.
1 foot in 40 is 132 feet per mile.
1 foot in 50 is 106 feet per mile.
1 foot in 100 is 53 feet per mile.
1 foot in 125 is 42 feet per mile.

To find quantity of water elevated in one minute running at 100 feet of piston speed per minute: Square the diameter of the water cylinder in inches and multiply by 4. Example: Capacity of a 5-inch cylinder is desired. The square of the diameter (5 inches) in 25, which, multiplied by 4, gives 100, the number of gallons per minute (approximately).

Lifts used in Filing Stations in place of pits for greasing and under car work for passenger cars are 18' long and 7' wide and require a radius of 10' 3½" and a minimum head room of 12' 6".

The following is quoted from General Rules of the Division of Fire Prevention, State of Illinois. Revision of Sept. 1, 1938.

SERVICE STATIONS.

1. **Definition:** A service station is any place of business where gasoline, or any highly volatile fuels for motor vehicles or internal combustion engines, are sold or offered for sale at retail.

(a) This definition shall include also the private storage and dispensing of such products for the same purposes as those served by a service station, whether the storage is maintained for the use or benefit of the owner, lessee, agents or employes of either, or of any others.

2. **Storage Underground and Limited:** Storage shall be underground and the combined capacity of all storage tanks shall not exceed 12,000 gallons.

(As amended Sept. 1, 1938, to increase underground storage limit from 6,000 to 12,000 gallons.)

3. **Setting of Tanks:** Tanks shall be buried so that their tops will be not less than two feet below the surface of the ground or beneath 12 inches of earth and a slab of concrete not less than five inches in thickness and capable of sustaining a load of 250 pounds per square foot; slab shall be set on a firm, well tamped earth foundation and shall extend at least one foot beyond the outline of tanks in all directions.

(a) Tops of tanks shall be below the level of any piping to which tanks may be connected.

(b) Tanks shall be so located that no heavy trucks or other vehicles pass over them unless they are adequately protected by a reinforced concrete slab.

(c) Where soil conditions require, a firm foundation shall be provided.

(d) Tanks shall not be installed under any building or structure.

5. **Material and Construction of Tanks:** Tanks shall bear label of Underwriters' Laboratories or meet equivalent specifications.

(a) Tanks shall be thoroughly coated on the outside with tar, asphaltum or other suitable rust-resisting material.

(b) Tanks shall not be surrounded or covered by cinders or other material of corrosive effect. If the soil contains corrosive material, special protection shall be provided.

6. **Venting of Tanks:** Each tank shall be provided with a vent pipe, connected with the top of the tank and carried up to the outer air. Pipe shall be arranged for proper drainage to storage tank and its lower end shall not extend through top of tank for a distance of more than one inch; it shall have no traps or pockets.

(a) Upper end of pipe shall be provided with a goose-neck or T attachment, or weatherproof hood.

(b) Vent pipe shall be of sufficient cross-sectional area to permit escape of air and gas during the filling operation and in no case less than one inch in diameter. If a power pump is used in filling storage tank, and a tight connection is made to the fill pipe, the vent pipe shall not be smaller than the fill pipe.

(c) Vent pipe shall terminate outside of building not less than 12 feet above top of fill pipe, not less than four feet, measured vertically and horizontally, from any window or other building opening, and not less than 15 feet measured horizontally from any opening into the basement, cellar or pit of any building, and in a location which will not permit pocketing of gas. If a tight connection is made in the filling line, the terminus of vent pipe shall be carried to a point one foot above the level of the highest reservoir from which tank may be filled.

(d) Vent pipes from two or more tanks of the same class of liquid may be connected to one upright or main header. Area of header shall equal the combined area of the pipes connected to it. Connection to the header shall be not less than one foot above the level of the top of the highest reservoir from which tanks may be filled.

7. **Fill Pipes:** Fill pipes shall be carried to a location outside of any building, as remote as possible from any doorway or other opening into any building and in no case closer than five feet from any such opening.

(a) Location shall be in a place where there is minimum danger of breakage from trucks or other vehicles.

(b) Each fill pipe shall be closed by a screw cap or other tight fitting cap, preferably of a type which can be locked. Cap should be locked at all times when filling or gauging process is not going on.

9. **Pumps:** Liquids shall be withdrawn from tanks by means of approved pumps, equipped with metallic lined hose and non-ferrous discharge nozzle.

(a) No pump shall be located within a building.

(b) Curb pumps, or pumps located in any portion of a public street, are held to be unlawful by the Illinois Supreme Court. (People vs. Wolper, 350 Ill. 461.)

(c) Wiring of electric pumps and all electrical equipment in connection therewith shall conform to Article 32, National Electrical Code.

(d) Devices which discharge by gravity shall be so designed that it is impossible to retain in the gauging compartment materially more than 10 gallons of liquid, and so that it is not possible to lock the device without draining the liquid.

(e) Systems which employ continuous air pressure on storage tank in connection with gauging or vending device are prohibited.

(f) The use of aboveground storage tanks in connection with gauging or vending devices is prohibited.

10. **Piping:** Piping shall conform to the requirements set out in the rules under General Storage.

11. **Building:** No basement or excavation shall be permitted under any service station building.

(a) Floor level shall be above grade so as to prevent flow of liquids or vapors into building.

(b) Floor shall preferably be of concrete.

12. **No Inflammable Liquids Within Building; Exception:**

(a) No gasoline, naphtha or other liquids of Class I shall be kept inside of service station.

(b) No alcohol or other flammable antifreeze solutions shall be kept inside of service station except in original sealed containers. No transfer of such liquids from these receptacles to other receptacles shall be made inside the service station.

(As amended Sept. 1, 1938, to permit storage of flammable anti-freeze solutions inside building in manner provided.)

13. **Greasing Pits:** Every greasing pit installed within a building, or enclosed by three or more walls, shall be ventilated by a vent duct not less than six inches in diameter (or equivalent cross-sectional area if a noncircular duct is used.) Duct shall start within four inches of the floor and shall be extended on an upward diagonal or by an easy bend over to sidewall, thence straight up through roof to a height sufficient to draw off gasoline vapors which may accumulate in bottom of pit. Abrupt bends must be avoided and all joints must be tight. Floor of pit should pitch slightly toward corner where

duct is located, to facilitate flow of gases to duct.

(a) **Gasoline or naphtha shall never be used to clean out pit, whether pit is located within a building or enclosure, or outside in the open.**

(b) No sewer connection shall be permitted from any greasing pit.

(c) If electrically lighted, globes shall be of vapor-proof construction and wiring in conduit.

14. **Wash and Greasing Rooms:** If sewer connection is permitted by the city, an adequate grease trap shall be provided to intercept greases and oils. Trap shall be cleaned out at least every 30 days.

17. **Fire Extinguishers:** Each service station shall be equipped with at least one approved chemical fire extinguisher suitable for oil or gasoline fires.

19. **Approval of Plans:** Drawings or blue prints made to scale shall be submitted in triplicate to the State Fire Marshal and shall be approved by him before any new construction, addition or remodeling is undertaken. Drawings shall carry the name of the person, firm or company proposing the installation, the location with reference to city, village or town, and shall in addition show the following:

(As amended Sept. 1, 1938, to clarify meaning.)

(a) The plot to be utilized and its immediate surroundings on all sides; all property lines to be designated and adjacent streets and highways to be named.

(b) The complete installation as proposed, including tanks and their capacities, pumps, buildings, drives, and all equipment.

(c) Clearance from tanks to property lines as per Rule 4 (c).

(d) Type of construction of service station building or buildings, with a clear showing that there will be no basement, cellar or excavation under any portion.

(e) Location of basements, cellars or pits of other buildings on the property or on adjacent property, and location of tanks with reference thereto as provided in Rule 4(a). If a building has no basement, cellar or pit, make note to that effect.

(f) Location of sewers, manholes, catch-basins, cesspools, septic tanks, wells or cisterns (whether on the property, on adjacent property or in adjoining streets, highways or alleys), and location of tanks with reference thereto as provided in Rule 4(b). If there is no sewer, manhole or catch basin in a street or alley, or no sewer, cesspool, septic tank, well or cistern on a property, make notation to that effect in proper place.

(g) Location of vent pipe outlets as per Rule 6(c) and location of fill pipes as per Rule 7.

(h) Ventilation of greasing pit as per Rule 13, if greasing pit is located within a building or an enclosure.

(i) Drawings shall be accompanied by an application for approval made out in triplicate on blanks furnished by the State Fire Marshal.

Quantity of Brickwork in Barrel Drains and Wells.

Diameter in Clear.	Thickness of Brickwork.	Superficial Feet of Brick-work in One Linear Yard.	Number of Bricks Required for One Linear Yard
1 foot, 0 inches	0 feet, 4½ inches	16 feet, 6 inches	115
1 " 6 "	0 " 4½ "	21 " 2 "	148
2 " 0 "	0 " 4½ "	25 " 10 "	181
2 " 0 "	0 " 9 "	33 " 0 "	462
2 " 6 "	0 " 9 "	37 " 8 "	528
2 " 6 "	1 " 1 "	43 " 2 "	906
3 " 0 "	0 " 9 "	42 " 6 "	594
3 " 0 "	1 " 1 "	47 " 10 "	1004
3 " 6 "	0 " 9 "	47 " 1 "	659
3 " 6 "	1 " 1 "	52 " 7 "	1104
4 " 0 "	0 " 9 "	51 " 10 "	725
4 " 0 "	1 " 1 "	57 " 3 "	1203
5 " 0 "	0 " 9 "	61 " 3 "	857
5 " 0 "	1 " 1 "	66 " 9 "	1402
6 " 0 "	1 " 1 "	76 " 1 "	1597
7 " 0 "	1 " 1 "	85 " 6 "	1795

Tests for Pure Water.

Color: Fill a clean long bottle of colorless glass with the water; look through it at some black object. It should look colorless and free from suspended matter. A muddy or turbid appearance indicates soluble organic matter or solid matter in suspension. Odor: Fill the bottle half full, cork it, and leave it in a warm place for a few hours. If when uncorked it has a smell the least repulsive, it should be rejected for domestic use. Taste: If water at any time, even after heating, has a disagreeable taste, it should be rejected.

A simple semi-chemical test is known as the "Heisch test." Fill a clean pint bottle three-fourths full of the water; add a half-teaspoonful of clean granulated or crushed loaf sugar; stop the bottle with glass or a clean cork and let it stand in a light and moderately warm room for forty-eight hours. If the water becomes cloudy, or milky, it is unfit for domestic use.

APPROXIMATIONS OF RADIATION

By Samuel R. Lewis.

The computations for figuring heaters depend on accurate data as to the conductivity of the building and on the temperatures on each side of the wall. The figures should be worked out for each case by a competent engineer.

Steam Radiators

For the average room in a fairly well built house which is to be heated continuously, an approximation of the amount of steam radiation may be obtained by dividing the square feet of glass surface by 2 and the square feet of outside wall, not deducting the glass, by 13; the sum of these two being the sq. ft. of direct cast iron steam heating surface.

Hot Water Radiators

If hot water at 140-180 degrees is to be used the approximate steam radiation would be multiplied by 1.66.

Boilers

The boiler for a steam or water heating system should be selected on the basis of its guaranteed efficiency when burning some specified fuel at a given rate of combustion. For rough approximations with small boilers, the rating of the boiler should be about double the actual radiation. This addition allows for getting started, for possible unfortunate chimney, inefficient clinker removal, old leaking house, long periods between firing, etc.

One square foot of steam radiation requires from 600 lbs. to 800 lbs. of steam per heating season, or from 70 to 90 lbs. of bituminous coal per season, or about 1,200 cu. ft. of manufactured gas per season, or 4 gallons of average fuel oil per season.

One heat unit will warm .238 lbs. of air one degree at 70 degrees or about 55 cu. ft. of air.

One pound of ice in melting absorbs 144 heat units.

Hospitals and hotels use about 85 gallons of hot water per 24 hours per hot water fixture, plus 1 gallon per piece per day for laundries and plus 3 gallons per meal per person for kitchens.

The following is a fair average over the heating year of the percentages of fuel used during each month in Chicago:

Jan.....21%	April....9%	Nov.....13%
Feb.....18%	May 4%	Dec.....17%
March..14%	Oct......4%	Total..100%

Transmission of Heat by Various Substances.

FILE 697.4

Window glass being	1,000
Oak or Walnut	66
White Pine	80
Pitch Pine	100
Lath and Plaster	75 to 100
Brick (rough)	200 to 250
Brick Whitewashed	200
Granite or Slate	250
Sheet Iron	1,030 to 1,110

Table Showing Amount of Glass Surface which may be Heated by 1 Square Foot of Radiating Surface in Good Buildings.

	Hot Water.			Steam.	
Temperature of radiating surface (radiators) Fahr	160°	180°	200°	227° 5 Lbs.	240° 10 Lbs.
	Square Feet of Glass to 1 Square Foot Radiator Surface.				
Temperature above surrounding air 90°	1.9	2.3	2.8	3.3	3.8
" " " " 80°	2.3	2.9	3.5	4.0	4.6
" " " " 70°	3.0	3.6	4.2	5.0	5.7
" " " " 60°	4.0	4.6	5.25	6.0	7.0
" " " " 50°	5.0	6.0	6.8	8.0	9.0
" " " " 40°	6.9	8.0	8.2	10.0	11.5

Formulae for Figuring Radiation for Factories.

A formula for figuring radiation which is used by some of the best heating engineers in determining the amount of radiation for factory buildings is as follows: $\frac{G}{33} + \frac{W}{10.9} + \frac{V}{171} =$ sq. ft. of radiation in which, G = Glass Area.
W = Net Wall Area.
V = Volumn of air in the Room.

SIZE OF STANDARD FLUE LINING ON SALE ON THIS MARKET.

Outside size.	Inside size.	Inside area.
4¼ x 8½ in.	3⅛ x 7¼ in.	22.6 sq. in.
8½ x 8½ in.	7 x 7 in.	49 sq. in.
13 x 13 in.	11⅝ x 11⅝ in.	135 sq. in.
4½ x 13 in.	3⅛ x 11⅝ in.	36.5 sq. in.
8½ x 13 in.	6⅝ x 11⅝ in.	77 sq. in.
13 x 18 in.	11½ x 16¾ in.	193 sq. in.
8½ x 18 in.	6⅞ x 16½ in.	114 sq. in.
18 x 18 in.	15¾ x 15¾ in.	247 sq. in.
21 x 21 in.	**19½ x 19½ in.**	
24 x 24 in.	**21½ x 21½ in.**	

GENERAL RULE FOR BRICK STACKS.

Diameter of base should not be less than 1/10 of height if square, or round, 1/12 of height. Batter of stacks 3/100 of an inch to the foot in height. Thickness of brick work should be not less than one brick from top to 25 feet below same, changing to 1½ brick from 25 feet to 50 feet below top, increasing ½ brick in thickness for each succeeding 25 feet, measuring from the top downward.

Fireplace Flue Areas.

For three-story building, area at top of smoke chamber should be 1/12 of area of fireplace opening.

Two-story building area at top of smoke chamber should be 1/10 of area of fireplace opening.

One-story building area at top of smoke chamber should be ⅛ area of fireplace opening.

Throat of fireplace should never be less than 3 in. or more than 4½ in. by the width of fireplace opening.

Front edge of arch should never be thicker than one-half brick, approximately 4 in.

Splay of sides of flue from throat opening up to flue lining should be 2 in. to the foot. The raise from soffit or lintel, or from highest point or soffit to arch should be 6 in.

PRELIMINARY ELEVATOR PROPORTIONING

During the preparation of preliminary studies for a building it is always necessary to decide on tentative elevator requirements for the different types of buildings required to be designed. The following table of capacities, speeds and type of control for the various classes of buildings represents good general practice, but must be varied to meet special conditions:

	Capacity in Lbs.	Recommended Speed for Floors as shown					Control
		3	4–7	8–12	13–20	21–30	
1. Department Store	2,500– 4,000	100	200	300	400	500	Automatic Landing
2. Single Line Store	1,500– 2,500	100	300	400			Car Switch Voltage Control
3. Loft Building	2,000– 4,000	100	150	250			Car Switch
4. Public Building	2,500– 3,500	100	250	400	500		Car Switch with Automatic Landing
5. Hospitals (passenger)	1,500– 2,500	100	200	300	400		Dual (both Car Switch and Push Button)
6. Hospitals (service)	1,500– 2,500	60	100	150	300		Push Button Control
7. Factory (freight)	4,000–12,000	60	100	150			Car Switch (with landing device if trucking service)
8. Hotel (passenger)	2,000– 3,000	100	300	600	600	700	Automatic Landing Device
9. Hotel (service)	1,500– 2,500	100	200	400	500	600	Car Switch
10. Apartment Hotel (passenger)	1,500– 2,500	100	300	400	500	600	Car Switch Push Button or Dual
11. Garage (auto lift)	4,000–10,000	60	200	300	500	500	Car Switch
12. Office Building (local service)	2,000– 3,000	100	300	400	600	700	Car Switch with Voltage Control
13. High Office Building	1,500– 2,500	Local Run 10 floors, 600 F.P.M.			Express Run 10 floors, 600 F.P.M. 20 floors, 700 F.P.M. 30 floors, 800 F.P.M. 40 floors, 800 F.P.M. 50 floors, 900 F.P.M.		Signal Control

Each building is a transportation problem in itself to be solved only when the height of building, area per floor, type of tenants, visiting public and peak traffic periods are fully considered. Tables and figures given here can be taken only as generalities. However, there are certain fundamental considerations and rule of thumb calculations that aid the architects materially in the early drafts of a building.

In general, 25,000 square feet of rental area per elevator will give average service. Tower sections of high buildings are exceptions due to the unproportionate distance of travel. A traffic engineer of a reliable elevator company should early be consulted for tower buildings.

All students of elevator traffic agree that for buildings of 18 stories or more where the elevators are divided into local and express banks, more cars of smaller capacity give quicker emptying time and shorter interval of departure than do fewer large cars. While precedence has established a near standard of 2,500-pound capacity, a study of the traffic problems will show that with cars of 2,000-pound capacity and possibly one additional car bank will improve the handling of the building's population because fewer passengers means fewer stops, and so quicker round trip time.

The A. S. M. E. prescribes 75 pounds per square foot of effective cab area in calculating the carrying capacity of passenger cars. This is based on 150 pounds per person and 2 square feet per person.

There are two major classifications of control as related to electrical energy supply, rheostatic and voltage control. Rheostatic control has the full voltage of the power company's supply brought to the elevator control board. Acceleration and deceleration are obtained by short circuiting portions of the current through resistance grids where the current is dissipated as heat. These steps vary from two to five depending on the running speed of the elevators and are controlled through contacts in the car operator's switch.

Voltage control embodies a constant speed motor generator set for each elevator hoisting motor. Just the amount of current needed to properly accelerate the car is fed to the elevator motor in much the same way a locomotive engineer gradually increases the steam throttle opening. Voltage control gives infinitely smooth acceleration and deceleration and easy riding qualities which increases the life of equipment and permits economy of operation because no electricity is wasted through grid circuits.

The classifications of control as related to operation are: Push Button, Car Switch, and Signal Control.

Push button control is a general classification for automatic elevators. There are variations, such as Constant Pressure Push Button, Momentary Pressure Push Button, and Collective Control, any one of which may be using either rheostatic or voltage control power. Push button control finds its use in apartments or buildings where the elevators are rarely used. In either case the automatic feature is applied to dispense with an operator. It is inherent then, that proper application can be made only when every one who will use the elevator is familiar with its operation. Thus residential tenants, as in apartment buildings, may be schooled to manipulate the elevator, or the employees of a firm having elevators rarely used.

Car switch control requires the services of an operator to drive the car. This type of control is essential where there is a transitory traffic as in office and public buildings. Where large numbers of people must be handled for peak periods the push button control becomes inadequate and an operator with car switch control becomes a necessity.

Signal control is a combination of the better characteristics of both Push Button and Car Switch controls. It is the latest engineering accomplishment of the industry and embodies the automatic features of the Push Button elevator with its accurate landing at floor levels, but driven by an operator to supply that human element essential where masses are to be handled.

Merely high speed elevators do not mean fast service as seen by the following time analysis of an elevator cycle:

Loading time at main floor.
Time for closing doors and gates.
Accelerating time.
Running time at full speed.
Decelerating time.
Time for opening doors.
Unloading and loading time at all stops.
Time for closing doors and gates.

THE ORDERS AND THEIR APPLICATION

By ALFRED W. S. CROSS, M. A., F. R. I. B. A., and ALAN E. MUNBY, M. A.

Introduction.

So many scholarly works upon the Orders are in existence, that some explanation seems to be called for in introducing another series of articles upon a subject that is, to all appearances, already well worn.

Notwithstanding the consensus of opinion as to the general proportions that ought to be followed in their delineation, an opinion based upon the rules laid down by the architects of an early period of the Renaissance, a surprising divergence from the precepts and practices of these old masters of their art is to be found in many buildings of our own time.

The writers are only aware of the existence of one book which seems to meet the usual office requirements, and that is a work entitled: "Rules for Drawing the Several Parts of Architecture," by James Gibbs, published in 1732; a book that has never been reprinted and copies of which are not now readily obtainable. The object aimed at, and successfully attained, is an illustration and description of an example of each Order, not "after Gibbs," but representing one of a good average type of design so proportioned that the dimensions of the various parts bear simple and easily discernible ratios one to another.

An attempt has been made to co-ordinate the leading features of the book by re-drawing some of the illustrations, retaining the useful dimensions shown thereon and entirely re-writing the description of the plates, with the introduction of some general principles likely to be of value to the draughtsman and student, for which purpose the opinions of standard writers, particularly those of Sir William Chambers, have been freely incorporated.

Before attempting such a condensation of the material in the book it was thought desirable to ascertain how far the generalizations adopted by Gibbs really represent the proportions used by acknowledged authorities. For this purpose the average ratio of the diameter of the column to the height of the entablature, as being a relation which essentially affects the whole proportion of the Order, was obtained by measuring a number of recognized examples, and it may be of interest to give the results, as an indication of the actual value of the dimensions used.

The result renders it evident that the general proportions of the Orders as recommended for adoption by this architect are fully worthy of confidence.

Hence, it would obviously seem preferable to master a few main dimensions, and, having thus inculcated a general sense of proportion, to rely upon gaining familiarity with the plates by constant use, when the proportions of the smaller members of the compositions will become naturally assimilated. The Composite Order is given in Gibbs' book, but, owing to its similarity to the Corinthian and to the absence of a consensus of opinion as to its dimensions, it has not been included in the present wor t.

No encroachments have been shown on any of the Orders to avoid distracting attention from the dimensions. With the exception of the whole of the Tuscan Order and of the frieze of the Ionic Order there are few members, apart from mere fillets, which have not been enriched, by some form of ornament, in one or another example, the Doric naturally the least and the Corinthian the most. In the latter Order, in fact, even the cyma and corona of the cornice, in addition to the frieze, ogees and beads, are often ornamented, but, apart from the question of expense, it is undesirable to carry such elaboration too far, as when placed in close contact with each other, especially when a distant view is alone possible, one moulding will often rob another of its effect, and, indeed, the value of richness of detail is more often than not lost in this manner.

The enrichment of columns beyond ordinary flutings is generally to be deprecated, while the application of ornament to bases and pedestals is seldom either requisite or desirable.

However great may be the utility of drawings dealing with the Orders, it should never be forgotten that they are merely a means to an end, that end being an executed building. Those whose work is confined to a drawing board develop a strong tendency to consider their compositions solely from an elevational and artistic draughtsman's point of view, and every opportunity should be taken of checking this habit and of cultivating the art of thinking "in the round." The study of perspective of buildings, and, best of all, the preparation of models of portions of a proposed building, an occupation which often results in the discovery of latent defects of design, are alike of the greatest educational value to the student of architecture.

TABLE SHOWING THE APPROXIMATE RATIO BETWEEN THE LOWER DIAMETER OF THE COLUMN AND THE HEIGHT OF THE ENTABLATURE.

Tuscan.		Doric.		Ionic.		Corinthian.	
Alberti	1:1.5	Alberti	1:2.0	Alberti	1:1.4 / 1:1.7	Alberti	1:1.8
Palladio	1:1.8	Palladio	1:1.9			Palladio	1:2.0
Scamozzi	1:1.9	Scamozzi	1:2.1	Palladio	1:2.0	Scamozzi	2:2.0
Vignola	1:1.8	Vignola	1:2.0	Scamozzi	1:1.8	Vignola	1:2.5
—		Parthenon	1:2.0	Vignola	1:2.3	Pantheon	1:2.3
—		Baths, Diocletian	1:2.0	Fortuna (Rome)	1:2.3	Jupiter Stator	1:2.5
—		Temple Pæstum	1:1.7	Baths, Diocletian	1:1.9	Jupiter Tonans	1:2.2
—		Apollo, Delos	1:1.8	Minerva, Athens	1:2.3	Temple Antonius	1:2.3
St. Paul's Convent Garden	1:1.8	Bow Church, Portico	1:1.9	Iilius, Athens	1:2.3		
				Banqueting Hall	1:2.0	Hamrden Court	1:2.2
Average	1:1.76	Average	1:1.93	Average	1:2.00	Average	1:2.00
Gibbs	1:1.75	Gibbs	1:2.00	Gibbs	1:1.82	Gibbs	1:2.00

The above examples have not been selected with any intention of justifying the proportions adopted by Gibbs, but are merely cited as those which readily occurred to the mind, or of which the dimensions could be easily obtained.

THE SETTING UP OF AN ORDER.

(To be studied in connection with Plates I., II., III., IV. and V.)

The sequence followed in setting up an Order will be found to influence, to some extent, the rapidity and facility with which it can be accomplished. An outline of the method of procedure may, therefore, prove useful.

Usually the height of the Order is fixed by circumstances, as, for example, when it is to be applied to a given story of a building.

The total height having been settled, draw the limiting horizontal lines and then set out the vertical centre lines of the columns, thus dividing the frontage to be treated into bays appropriate to the exigencies of the design and having due regard to the correct intercolumniation of the Order adopted. If a pedestal is to be placed under the column, cut off one-fifth of the total height for it, and cut off one-fifth or one-sixth of the remainder (measured from the top limiting horizontal line) for the vertical height of the entablature; the intervening space gives the height of the column, including its cap and base. If no pedestal is to be used, divide the whole of the given height into five or six parts, cut off one of these parts, from the top, for the entablature, and the remainder gives the height of the column.

The Column. Since some of the dimensions of the entablature are in terms of the diameter of the column, the latter should be next developed. The term "diameter of the column" refers always to its greatest diameter—namely, that of the shaft just above the lower cincture. This dimension is one-seventh to one-tenth of the height between the soffit of the entablature and the top of the pedestal, or lower limit of the Order in the absence of a pedestal. If the centre lines of the piers do not represent the centres of the columns, as, for instance, when coupled columns are used, the centre line of one of the columns must now be decided upon and the diameter of the Order symmetrically disposed horizontally across it. A semi-diameter is then cut off, from the bottom of the column, for the height of the base, and it should be noticed that this—except in the Tuscan and alternative Doric Orders—does not include the fillet at the base of the shaft, the members above the upper torus being reckoned as part of the shaft, as are also the astragal and fillet below the necking of the capital of the column. The plinth and lower torus of the base project one-third and the upper torus one-fifth of a semi-diameter beyond the lower circumference of the shaft. The leading lines for the base having thus been obtained, cut off by a horizontal line the height of the capital from the top of the column, and (except in the Ionic Order) again below it, a height equal to one-sixth of a semi-diameter for the astragal and fillet below the necking.

The semi-diameter of the shaft at one-third of its height from the bottom is then divided into five or six parts, and four or five of these parts are taken as a semi-diameter at the top, below the astragal. The shaft may now be completed, the entasis being usually made to start from the greater diameter, one-third up the shaft, below which point it is a true cylinder until the cincture at the base is reached. This is the best method to adopt in the case of small scale drawings. Where large detailed drawings are in question the diameter may be alternatively divided at the base of the shaft instead of at one-third of the height, and the entasis extended throughout the whole length. The completion of the shaft enables the projection of the capital to be marked off, and also that of the astragal and fillet, which is equal to their combined height.

The Entablature. The development of the entablature can now be proceeded with, the architrave, frieze and cornice being ruled off horizontally and the members of each inserted (see dimensions). The projections for a returned end or section are obtained from the upper diameter of the shaft. The lowest member of the architrave, and also the frieze, lie vertically over the circumference of this upper end of the shaft. The projection of the cornice beyond the frieze line is equal to its height, except in the Doric Order, in which the projection is one-third more than its height of one diameter. Further rules dealing with minor projections and the position of the modillions, dentils, etc., will be supplied by a study of the plates and tabulated dimensions.

Pedestal. Finally, the pedestal, if any, should be divided vertically into four parts; the lower part is ruled off for the height of the plinth, one-third of the second part for the height of the base, and one-half of the top part for that of the cap. The projection of the die is equal to that of the base of the column, and the plinth and the cap of the pedestal extends beyond this for a distance equal to the height of the base of the pedestal previously obtained.

The above dimensions will all be found in the subjoined table, which represents an endeavour to bring together, in a form suitable for reference, sufficient information to make any glaring disproportion impossible.

A few of the minor divisions are only approximations; they will, however, be found to be sufficiently accurate for any but large detail drawings, in which it is not desirable to destroy all individuality by rigorous mechanical rules.

On the left hand will be found the dimension required and, in the intermediate column the fraction for each Order of the previously ascertained unit given in the right-hand column.

Plate I.

Plate I. represents the four Orders drawn to a common vertical height.

The pedestal may or may not be required and, if used, it is to be regarded as an addition to the Order, the relative dimensions of the parts of which are not altered by its removal or introduction.

The diameter of the column (by which is meant the diameter of the shaft following its lower cincture) is the ruling dimension from which most of the others are obtained, and the smaller circumference of the top of the shaft always coincides with the frieze line from which all the projections of the entablature are set out.

In judging the value of such projections it should be borne in mind that in execution the higher vertical faces of the composition will usually be much foreshortened to the observer and that there will be a consequent increase in the comparative value of neighboring projections.

A perusal of the table will indicate those dimensions which all the Orders have in common, but for convenience of reference they are further summarized thus:

Height of Pedestal, ⅕ total height of Order.

PLATE 1.

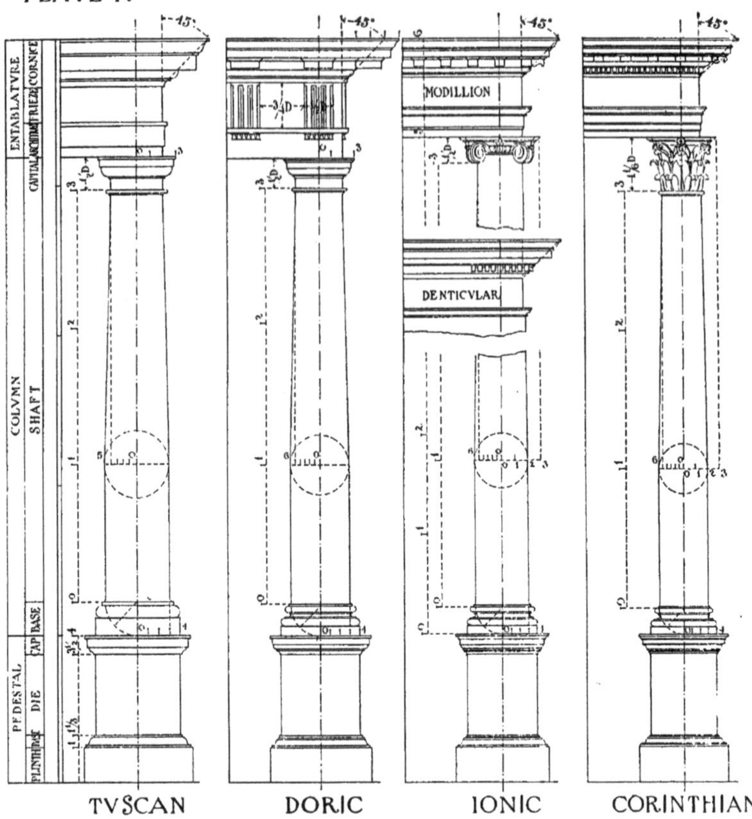

Height of Plinth, ¼ height of Pedestal.
Height of Pedestal Base, ⅓ height of Pedestal Plinth.
Height of Pedestal Cap, ½ height of Pedestal Plinth.
Projection of Cap and Plinth, ⅓ height of Pedestal Plinth.
Projection of Corona over Die, ¾ projection of Pedestal Cap.
Height of Column Base, ½ dameter of Column.
Projection of Base over Shaft, ⅓ semi-diameter of Column.

Pilasters. The general proportions allotted to the columns of the Orders apply also to pilasters, which may be regarded as columns square on plan, but almost universally deeply engaged. The projection of pilasters must be regulated by circumstances. If impost mouldings or other projections stop upon them, as on the inner wall of an arcade, these projections must be sufficient to take the mouldings, and if they line with engaged columns crowned by an entablature, they must have a projection similar to the columns, and therefore in such cases never less than a semi-diameter. Apart from these considerations, the projection should be about one-fourth of the diameter. Pilasters may be fluted or plain; if the former, the flutes should be, as far as possible, the same size as those of the adjoining columns, and always an odd number.

* * *

On plain faces 7 flutes (occasionally 9) are used, and therefore in the above case 4 flutes (or 5) would be employed on each side of the re-entering angle. The returned sides of pilasters should never be fluted unless the projection is as much as half of a diameter. The diameter assigned to a pilaster will be that of a column (if any) used in conjunction with it. The shaft may or may not be diminished.

If the pilaster stand alone it is best formed with the same top and bottom diameter, but if a column stand in front of it then it should be diminished to the same extent as the column. Entasis is not usually given to pilasters.

Unless columns and pilasters are monoliths the shafts should be built up of three drums and not two, as a central joint, unless exceptionally well executed, has a very disagreeable appearance.

"Practical Notes for Architectural Draughtsmen : The Orders and their Application."

Tabulated Dimensions of the Orders. Arranged Progressively as Required for Use.

	Dimension required.	Tuscan.	Doric.	Ionic.	Corinthian.	Dimension = 1.
	No Pedestal — Height of Entablature	$\frac{1}{5}$	$\frac{1}{5}$	$\frac{1}{6}$	$\frac{1}{6}$	Total height of Order.
	With Pedestal — Height of Pedestal	$\frac{1}{5}$	$\frac{1}{5}$	$\frac{1}{5}$	$\frac{1}{5}$	" " "
	With Pedestal — Height of Entablature	$\frac{1}{5}$	$\frac{1}{5}$	$\frac{1}{6}$	$\frac{1}{6}$	Height of Order less Pedestal.
	THE COLUMN.					
Vertical.	Diameter of Shaft	$\frac{1}{7}$	$\frac{1}{8}$	$\frac{1}{9}$	$\frac{1}{10}$	Height of Order less Entablature and Pedestal.
	Height of Base	$\frac{1}{2}$	$\frac{1}{2}$	$\frac{1}{2}$	$\frac{1}{2}$	Diameter of Shaft.
	" Base Plinth	$\frac{1}{2}$	$\frac{1}{3}$	$\frac{1}{3}$	$\frac{1}{3}$	Height of Base.
	" Lower Torus	$\frac{3}{4}$	$\frac{3}{8}$	$\frac{3}{8}$	$\frac{3}{10}$	" " less Plinth.
	" Upper Torus	—	$\frac{5}{6}$	$\frac{5}{6}$	$\frac{5}{6}$	" lower Torus.
	" Upper Torus and fillet under	—	1	1	1	" " "
	" Capital	$\frac{1}{2}$	$\frac{1}{2}$	$\frac{1}{2}$	$\frac{7}{6}$	Diameter of Shaft.
	" Necking	$\frac{1}{3}$	$\frac{1}{3}$	To eye $\frac{1}{3}$ Eye to abacus	To 1st leaf about $\frac{1}{3}$	Height of Capital (Corinthian less Abacus).
	" Top of Neck to top of Ovolo	$\frac{1}{3}$	$\frac{1}{3}$	$\frac{2}{3}$	To 2nd leaf about $\frac{1}{3}$	" " " " "
	" Abacus	$\frac{1}{3}$	$\frac{1}{3}$	$\frac{1}{3}$	2nd leaf to abacus about $\frac{1}{3}$	" " " " "
	" Astragal and fillet	$\frac{1}{6}$	$\frac{1}{6}$	—	$\frac{1}{6}$	Semi-diameter of Shaft.
	" Fillet below Astragal	$\frac{1}{3}$	$\frac{1}{3}$	—	$\frac{1}{3}$	Height of Astragal and Fillet.
Horizontal.	Projection of Base beyond Diameter	$\frac{1}{3}$	$\frac{1}{3}$	$\frac{1}{3}$	$\frac{1}{3}$	Semi-diameter of Shaft.
	" Upper Torus	—	$\frac{1}{5}$	$\frac{1}{5}$	$\frac{1}{5}$	" "
	Diminution of Shaft at Top	$\frac{1}{5}$	$\frac{1}{6}$	$\frac{1}{6}$	$\frac{1}{6}$	" "
	Projection of Cap over Shaft at Top	$\frac{1}{2}$	$\frac{1}{2}$	—	—	Semi-diameter of Shaft at Top.
	" Cap over Shaft at Base	—	—	$\frac{1}{2}$	$\frac{1}{2}$	Semi-diameter of Shaft.
	" Bead at top of Shaft	$\frac{1}{8}$	$\frac{1}{8}$	—	$\frac{1}{6}$	" "
	THE ENTABLATURE.					
Vertical.	Height of Architrave	$\frac{2}{7}$	$\frac{2}{8}$	$\frac{3}{10}$	$\frac{3}{10}$	Height of Entablature.
	" Frieze	$\frac{2}{7}$	$\frac{3}{8}$	$\frac{3}{10}$	$\frac{3}{10}$	" "
	" Cornice	$\frac{3}{7}$	$\frac{3}{8}$	$\frac{4}{10}$	$\frac{4}{10}$	" "
	" Fillet and Cyma	$\frac{1}{3}$	$\frac{2}{9}$	Both fillets $\frac{1}{4}$	Both fillets $\frac{1}{5}$	" Cornice.
	" Corona and Fillet over	$\frac{1}{3}$	$\frac{2}{9}$	$\frac{1}{4}$	$\frac{1}{5}$	" "
	" Base of Corona to top of Ovolo	—	$\frac{1}{4}$	$\frac{1}{4}$	$\frac{1}{5}$	" "
	" Top of Ovolo to Frieze	$\frac{1}{3}$	$\frac{2}{9}$	$\frac{1}{4}$	$\frac{1}{5}$	" "
Horizontal.	Total projection of Architrave over top diam. of Shaft	$\frac{1}{7}$	$\frac{1}{8}$	$\frac{1}{9}$	$\frac{1}{10}$	" Architrave.
	" top face of Architrave " "	$\frac{1}{5}$	$\frac{2}{3}$ (guttae)	$\frac{1}{5}$	$\frac{2}{5}$	Total projection of Architrave.
	" of Cornice over Frieze	1	$\frac{4}{5}$	1	1	Height of Cornice.
	Inset of Corona from top of Cornice	1	$\frac{5}{6}$	1	1	" Fillet and Cyma of Cornice.
	" Length of Modillions (or Mutules)	—	$\frac{1}{3}$	$\frac{1}{4}$	(and ogee) $\frac{1}{3}$	Diameter of Column.
	$\frac{1}{12}$ Breadth of Modillions (or Dentils)	—	(triglyph) $\frac{1}{2}$	M. $\frac{1}{6}$ D. $\frac{1}{12}$	$\frac{1}{6}$	" "
	$\frac{1}{24}$ Space between Modillions (or Dentils)	—	(metope) $\frac{3}{4}$	M. $\frac{1}{3}$ D. $\frac{1}{24}$	—	" "
	THE PEDESTAL.					
Vertical.	Height of Plinth	$\frac{1}{4}$	$\frac{1}{4}$	$\frac{1}{4}$	$\frac{1}{4}$	Height of Pedestal.
	" Base	$\frac{1}{3}$	$\frac{1}{3}$	$\frac{1}{3}$	$\frac{1}{3}$	" Plinth.
	" Cyma of Base	$\frac{2}{3}$	$\frac{1}{2}$	$\frac{1}{2}$	(and fillet) $\frac{1}{2}$	" Base.
	" Fillet below Cyma	1	1	1	1	Fillet above Cyma.
	" Cap	$\frac{1}{2}$	$\frac{1}{2}$	$\frac{1}{2}$	$\frac{1}{2}$	Height of Plinth.
	" Base of Corona to top of Cap	$\frac{2}{3}$	$\frac{1}{2}$	$\frac{1}{2}$	$\frac{1}{3}$	" Cap.
Hori-zontal.	Projection of Cap and Plinth over Die	1	1	1	1	" Base of Pedestal.
	" Corona of Cap over Die	$\frac{3}{4}$	$\frac{3}{4}$	$\frac{3}{4}$	$\frac{3}{4}$	Projection of Cap over Die.

Note.—The "Diameter" is always the greatest diameter of the drum of the Column. M. refers to the Modillion Cornice. D. to the alternative Dentil Cornice.

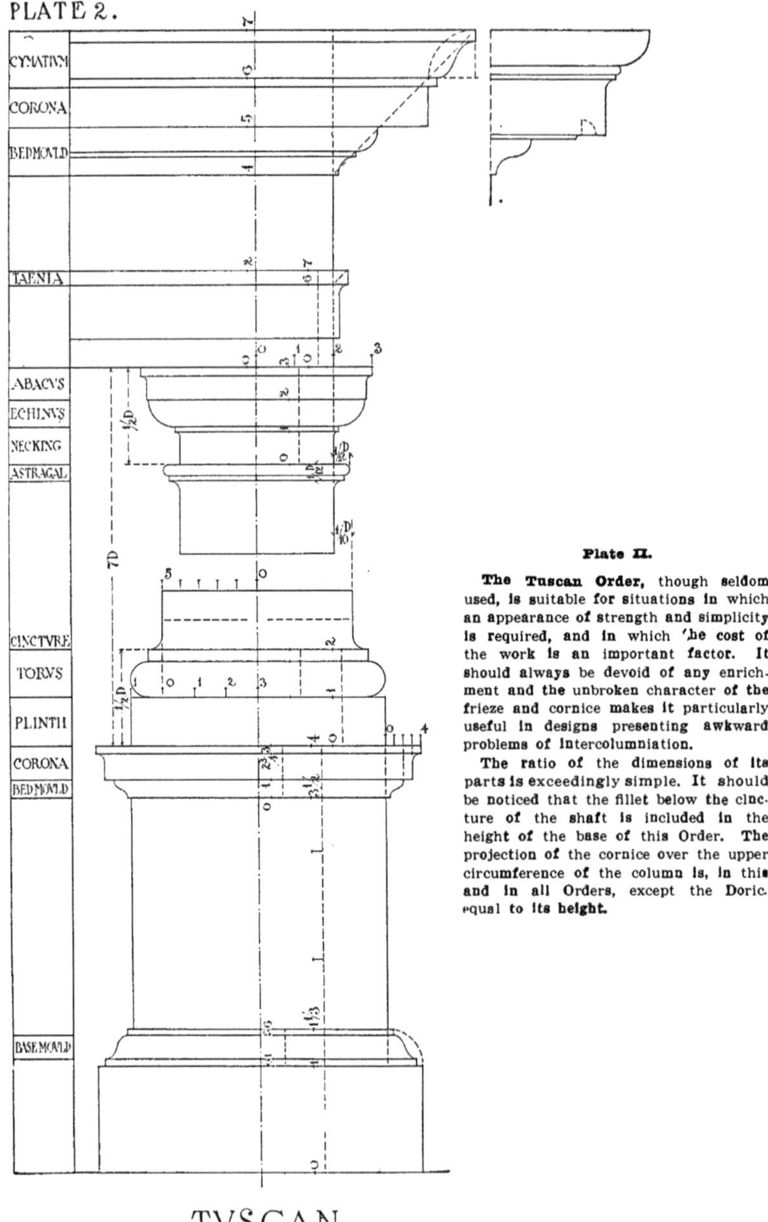

Plate II.

The Tuscan Order, though seldom used, is suitable for situations in which an appearance of strength and simplicity is required, and in which the cost of the work is an important factor. It should always be devoid of any enrichment and the unbroken character of the frieze and cornice makes it particularly useful in designs presenting awkward problems of intercolumniation.

The ratio of the dimensions of its parts is exceedingly simple. It should be noticed that the fillet below the cincture of the shaft is included in the height of the base of this Order. The projection of the cornice over the upper circumference of the column is, in this and in all Orders, except the Doric, equal to its height.

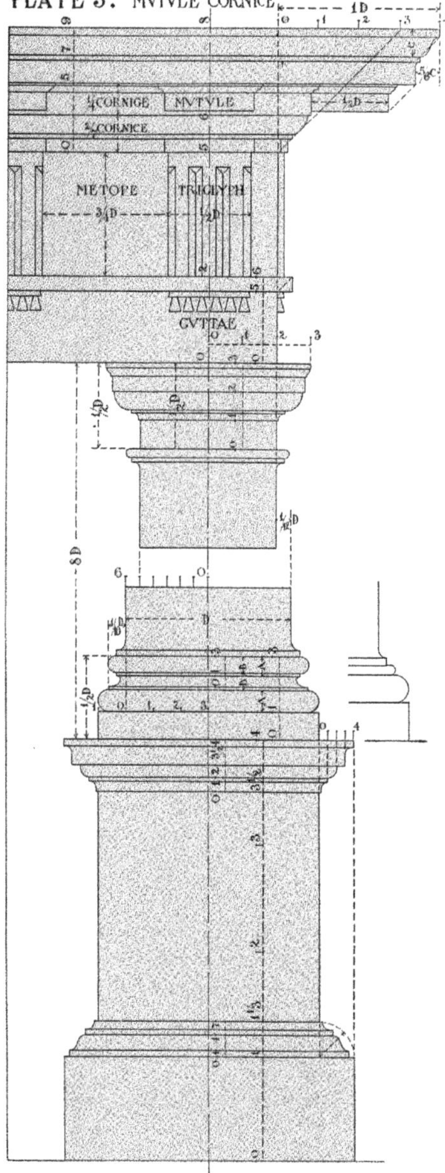

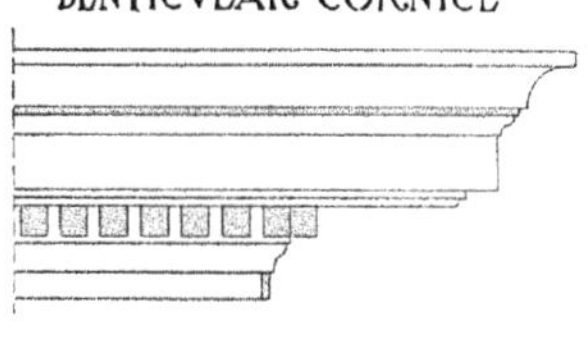

DORIC

Plate III.

The Doric Order is always effective when used in lower storeys, arcades, and door and window openings, but owing to the triglyphs upon the frieze, which must fall centrally over the columns, it is the most difficult to deal with when spacing is in question.

The dimensions of the cornice do not lend themselves to any simple ratio and its projection is always greater than that adopted for the other Orders. The 45° line from the top of the frieze at once gives the bed mould of the mutule course, and one-third of the height of the cornice added to the top projection of this guiding line gives the total projection, while the mutules are one-half a diameter in side elevation. Some considerable modifications of the Order, as here represented, will be found to exist in many recognised examples. Occasionally the mutules are dispensed with, and their bed mould is cut to form a dentil course, as in the Theatre of Marcellus. The cyma crowning the cornice is often replaced by a cavetto, while the Doric base (shown alternatively on the plate) sometimes replaces the more graceful attic base. When this base is used, the upper fillet should be included in the height of the base, as in the Tuscan Order.

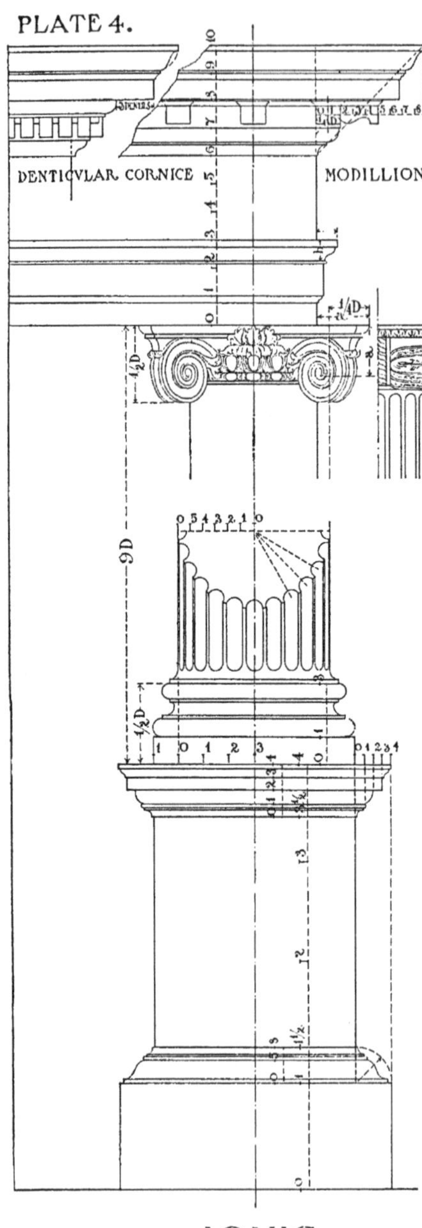

Plate IV.

The Ionic Order shows smaller variations from the pure Classic examples than any other, and its proportions are fairly simple.

Two styles of cornices are, however, used, the modillion and the dentil cornice, and although the method adopted by Gibbs of giving prominence to the former has been followed, it should be stated that the latter is more generally found in old examples, whilst the former is preferred by Palladio.

Represented side by side upon the plate the extent of the variation is easily discernible. A modillion or dentil should always be bisected by the centre line of the column and the spacing determined by the distance of this line from the frieze, as set out upon the drawing. The frieze is always plain and in larger works it is, preferably, kept flat. In smaller compositions, however, when narrow or when used over doors and windows a pulvinated frieze may be adopted with good effect.

The earlier alternative form of the Ionic capital in which the faces of the volutes are parallel to the plane of the elevation (not shown upon the drawings) may, of course, be substituted for the capital with angle volutes at 45°, though the latter has usually a much more graceful effect, particularly in small compositions. Of course, the geometrical method for setting out the volutes cannot be used in drawing such capitals in ordinary elevation. It should be noticed that the height of the capital in this Order is measured from the soffit of the volutes.

The centre of the eye is one-third of the height of the capital from its bottom and is in elevation placed just outside the top circumference of the shaft, while the horizontal fillet at the top of the shaft is immediately below the eye.

When the column is fluted the width of the fillets should be one-fourth to one-third that of the flutes. The flutes generally number twenty or twenty-four; in the latter case the simple method of setting them out on plan, as shown on the drawing, will be found of service.

The attic base is always used with the Ionic Order.

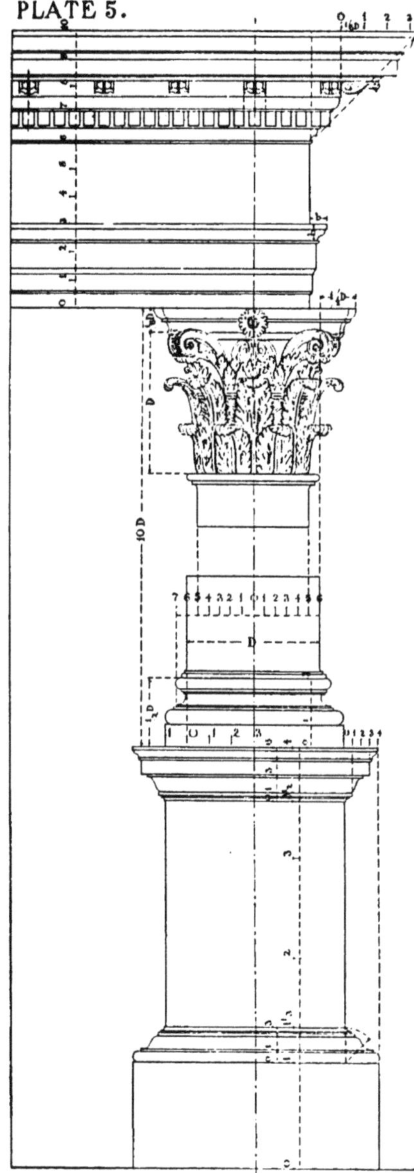

Plate V.

The Corinthian Order has been represented with considerable variations from the original type.

The Ionic entablature was often used by the ancients, supported by Corinthian columns, and the Corinthian cornice itself, though here represented with a dentil band, is often found without one. No general rule appears to exist for spacing the modillions or for their dimensions, the ratio of the width of the modillion to the space between two of them varying from 1 : 1½ to 1 : 2½, and again the number of the dentils between the modillions varies from 2 to 5 in different examples.

Both features should be symmetrically placed with reference to one another and to the centre line of the column, a point often neglected. To secure this result the following method is recommended :—Draw a modillion one-sixth of the diameter of the column in width, arranged symmetrically over the centre line of the column. Place another with its outside edge three and a half times its width within the total projection of the cornice, and thus obtain the spacing between the blocks. Divide the distance between two modillion centres into 15 parts, give two to a dentil, to be placed symmetrically under a modillion, and one to each space between the dentils, which will be found to bring the inside edge of the last dentil before the return, on the frieze line.

The form and projection of the leaves of the capital are largely matters of individual taste, but the general method of their arrangement will be evident after examining the drawing. It may, however, be noted that the eye of the volute is just outside the lower circumference of the shaft, and that the tiers of leaves divide the capital below the abacus into three approximate equal horizontal sections.

The column may or may not be fluted as in the Ionic Order.

The attic base, as used in the Ionic Order, is very generally employed—in fact, it is often preferable to adopt it, omitting the additional mouldings shown, for the sake of variety, on the drawing

PLATE 6.

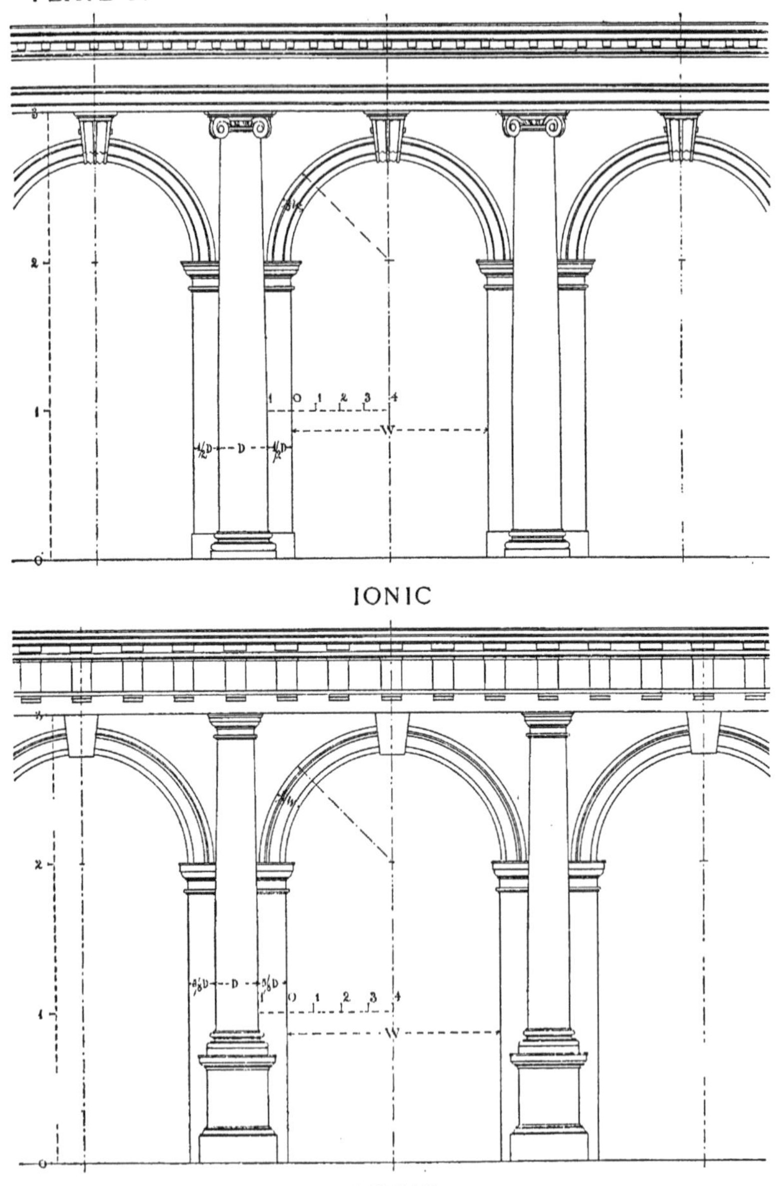

Plate VI.

The relations and dimensions given in this and similar subsequent plates must, therefore, be looked upon as necessarily somewhat elastic. At the same time, such dimensions as are given should not be disregarded, but considered in the light of proportions to be attained as far as the exigencies of the plan will admit.

The spacing of arcading dealt with in this plate should be governed by the height of the space to be treated, and it will be found that the best effects are obtained when the widths of the openings approximate to half of their height, and when the total width of the piers lies between one-half and two-thirds of that of the opening.

The spacing must also be considered in reference to the Order employed, so that when triglyphs, or modillions, are placed centrally over the columns their proper spacing may be interfered with as little as possible. It will thus be seen that a relation exists between the diameter of the column, the width of the pilaster, and the width of the opening. Again, the diameter of the column relatively to the opening will be influenced by the presence, or absence, of a pedestal to the Order. The summary shown, collected from Gibbs's work, giving the dimensions to be aimed at in order to comply with the above relations, will be found useful:

The height of the impost should always be about two-thirds of the height from the ground to the soffit of the architrave of the Order, whether a pedestal is in use or not.

Diameter of Column = 1.

	Tuscan.		Doric.		Ionic.		Corinthian.	
	No Ped.	With Ped.	No Ped.	With Ped.	No Ped.	With Ped.	No Ped.	With Ped.
Width of bay centre to center	6	7	6¼	7½	6	7½	6 5-12	8⅛
Width of one pilaster	½	⅔	½	⅝	½	⅝	9/16	7-10
Width of opening	4	4⅔	4¼	5¼	4	5¼	4⅓	5⅗

The archivolt or moulding running round the arch should be the same width as the pilaster (less any necessary clearance for the mouldings)—that is, about one-eighth of the width of the opening, which should also be the height of the impost cap to the bottom of the necking. Further details as to the members will be found on Plate VII.

PLATE 7.

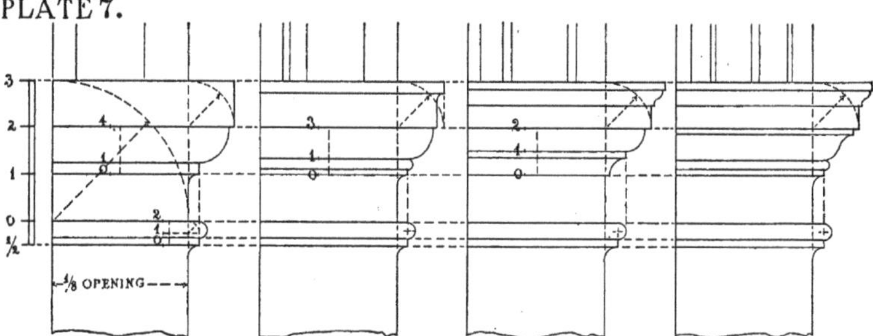

IMPOSTS AND ARCH MOVLDS

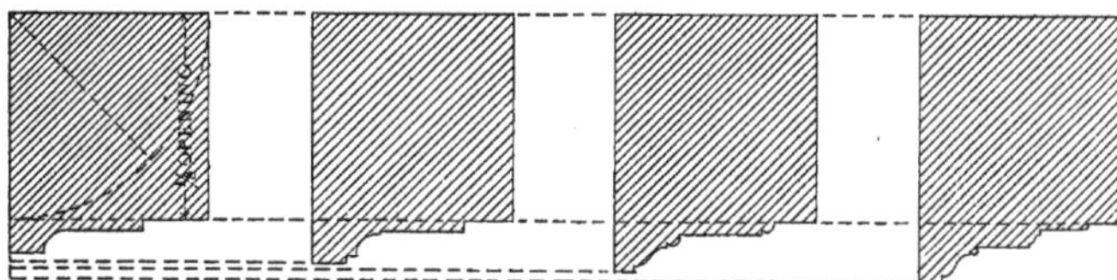

TVSCAN DORIC IONIC CORINTHIAN

Plate VII.

Impost Mouldings.

Details are here given of impost mouldings, with their archivolts, suitable for the different Orders. The divisions of the imposts are all simple and similar in each example, the height of the corona and of its mouldings above, if any, being equal to the height of the mouldings below, which, again, are equal to the necking. The bead and fillet below the necking are one-sixth of the height of the impost, the bead being double the height of the fillet. The projection of the impost beyond the line of the pilaster is equal to the height of the corona and member over in the first two Orders, while the projection of the corona itself is equal to this height in the last two.

The pilaster is square on plan, and, therefore, the plan of the archivolt is represented by this square upon which the mouldings are placed. An examination of these mouldings will show that they resemble the architraves given for their respective Orders, and their forms admit of similar variations. It will be noticed that the innermost face is always in the plane of the face of the pilaster, while the projection of the moulding at the extrados increases from about one-quarter the width of the whole archivolt in the Tuscan to one-third in the Corinthian Order.

SUBJECT INDEX.

System of Classification for Filing Data, Drawings, Plates, Catalogues, Etc., in Architects' and Contractors' Offices.

INTRODUCTION.

The decimal system of classification was devised and elaborated by Mr. Melvil Dewey, formerly director of the New York State Library. This system was intended primarily for the use of librarians in the classification and arrangement of books and pamphets, but it was soon found that the system also furnished a simple and effective means of classifying, indexing and filing literary matter of all kinds. Engineers have found it useful for indexing technical data and information, catalogs, reports, card systems, drawings, etc., and it has been found equally useful by manufacturing and business concerns.

The scheme and a considerable amount of the subject matter which follows has been obtained from the original publication of Mr. Dewey, but the outline on "Building," 690 to 699, has been compiled new by the Editor. The purpose of rewriting being to bring the Index more nearly in accord with the trade groups and divisions of modern practice. The index on "Ancient," "Mediaeval" and "Modern Architecture," 722 to 724 inclusive, has been completely revised in order to bring the same down to date in accord with the latest discoveries in the realm of the history of Architecture. For the revision of this material we are deeply indebted to Prof. Rexford Newcomb, Professor of Architecture of the University of Illinois. For its co-ordination with Dewey to Miss Winifred Fehrenkamp, Librarian of the Ricker Library of Architecture, also of the University of Illinois.

EXPLANATION OF THE DECIMAL SYSTEM.

The essential characteristic of the Dewey System is its method of division and subdivision. The entire field of knowledge is divided into nine chief classes numbered by the digits from 1 to 9. Matter of too general a nature to be included in any of these classes is put into a tenth class and indicated by 0. The following are the primary classes of the Dewey System:

0 **GENERAL WORKS**
1 **PHILOSOPHY**
2 **RELIGION**
3 **SOCIOLOGY**
4 **PHILOLOGY**
5 **NATURAL SCIENCE**
6 **USEFUL ARTS**
7 **FINE ARTS**
8 **LITERATURE**
9 **HISTORY**

Each of these classes is again divided into nine divisions, with a tenth division for general matter, and each division is separated into nine sections. The sections are again sub-divided and the process may be carried as far as desired.

It is thought that this system will be especially valuable, to architects for classifying drawings, catalogs, reports and technical data. Our space is too limited to publish the complete work, nor is it desirable. Should any one be sufficiently interested to go into the matter thoroughly, they should have Mr. Dewey's complete text on the subject. We are particularly concerned as practitioners of the profession of architecture with divisions 6 and 7, "Useful Arts" and "Fine Arts," comprising the following subject numbers:

600 USEFUL ARTS
610 **MEDICINE**
620 **ENGINEERING**
630 **AGRICULTURE**
640 **DOMESTIC ECONOMY**
650 **COMMUNICATION AND COMMERCE**
660 **CHEMICAL TECHNOLOGY**
670 **MANUFACTURES**
680 **MECHANIC TRADES**
690 **BUILDING**

Omitting all sub-divisions of this topic, with the exception of 690 "**Building,**" we publish the sub-divisions of same. As distinguished from "**Architectural Construction,**" "**Building**" has to do more particularly with the processes of construction and matters pertaining to trades and materials involved in the construction of buildings, should be more properly classified under "Building", while matters as to types and component architectural parts are more properly classified under **Architectural Construction.**

690 BUILDING — Materials and Trades.

690.0 **GENERAL.**
.01 **History.**
.011 History of Materials.
.012 History of the Art of Building.
.013 Biography of Architects.
.014 Biography of Builders.
.015 Biography of Craftsmen.
.02 **Organization of Construction.**
.03 **Finance of Building.**
.03-A Thru Building & Loan Associations.
.03-B Thru Cooperative Ownership.
.03-C Thru Bond Issue.
.03-D Thru Straight Loan.
.03-E Thru First and Junior Bonds.
.04 **Operation of Buildings.**
.05 **General Works on the Occupation and Art of Building.**
.50 Encyclopaedia.
.051 Manuals.
.052 Handbooks.
.053 Receipts.
.054 Periodicals.
.055 Society Proceedings.
.056 Trade Unions, Guilds, Etc.
.057 Contractor's Associations.
.058 Material Dealer's Associations.
.059 Insurance.

690.1 **EDUCATION OF PERSONNEL CONCERNED IN BUILDING.**
.11 **Education of Designers.**
.12 **Education of Supervisors.**
.13 **Education of Managers.**
.14 **Education of Craftsmen.**

690.2 **BUILDING MATERIAL IN THE ABSTRACT.**
(All special material should be classified under the appropriate trade.)

690.3 **PLANS FOR BUILDINGS.**
.30 **Incidents to the Preparation of Drawings.**
.301 Drafting Room Supplies.
.302 Drafting Methods.
.303 Cost Accounting.
.31 **Preliminary Studies.**
.32 **General Drawings.**
.33 **Scale Details.**
.34 **Full Size Details.**

690.4 **SPECIFICATIONS FOR BUILDINGS.**
.40 **Matter Pertaining to All Trades.**
.40-A General Conditions of the Contract.
.40-B Form of Agreement.
.40-C Form of Bid.

.40-D Form of Advertisement.
.40-E Form of Invitation to Bid.
.41 Earth Working and Transportation Trades. (See File 691.)
.41-A Preparation of Site.
.41-B Wrecking.
.41-C Shoring and House Moving.
.41-D Excavating.
.41-E Caisson and Special Foundations.
.41-F Construction Plan.
.41-G Maintenance Contract.
.41-I Grading and Filling.
.41-J Preparation of Soil, Sodding and Seeding.
.41-K Planting.
.41-Z Miscellaneous Labor not Otherwise Classified.

.42 Mortar Using Trades. (See File 692.)
.42-A Masonry Materials.
.42-B Foundation Work.
.42-C Concrete Work.
.42-D Stone Work.
.42-E Brick Work.
.42-F Fireproofing, Furring and Partitions.
.42-G Architectural Terra Cotta.
.42-H Paving.
.42-I Smoke Stacks of Masonry.
.42-J Plastic Reinforcement, Lathing and Furring.
.42-K Plastering.
.42-L Models, Clay and Plaster.
.42-M Plastic Insulation, Pipe Covering, Etc.
.42-N Marble and Substitutes (Including Slate, Structural Glass Terrazzo-Slabs, Etc.)
.42-O Tile and Substitutes.
.42-P Terrazzo Blocks.
.42-Z Miscellaneous Mortar Using Trades not Classified.

690.43 Wood-Working Trades and Hardware. (See File 693.)
.43-A Wood-Working Materials and Methods.
.43-B Carpentry.
.43-C Rough Carpentry Hardware.
.43-D Finish Hardware.
.43-E Revolving Doors.
.43-F Special Doors, Folding, Rolling, Etc.
.43-G Screens, Wood Frame, for Insects.
.43-H Wood Registers, Screens, Etc.
.43-I Mantels, Etc., of Wood.
.43-J Wood Specialties Show-Cases, Cabinets, Etc.
.43-K Seating for Assembly Pews, Opera Chairs, Etc.
.43-L Wood Platform Furniture, Pulpits, Lecturn Sedilia, Altars, and Altar Furniture.
.43-M Portable Furniture of Wood, Chairs, Etc.
.43-N Domestic Furniture.
.43-Z Miscellaneous Woodworking Trades not Otherwise Classified.

690.44 Heavy Metal Trades (employing metal heavier than No. 10 gauge). (See File 694.)
.44-A Metal Materials and Methods.
.44-B Structural Metal (over No. 10 gauge).
.44-C Miscellaneous Metal.
.44-D Ornamental Metal (over No. 10 gauge).
.44-E Vaults, Safes, Vault Doors, Etc.
.44-F Solid Metal Sash.
.44-G Heavy Metal Doors and Shutters.
.44-H Fire Escapes.
.44-I Stairs, Metal.
.44-J Fences, Metal.
.44-Z Miscellaneous Heavy Metal Trades not Otherwise Classified.

690.45 Sheet Metal Trades (employing metal of No. 10 gauge or less. See File 695).
.45-A Sheet-Metal Materials and Methods.
.45-B Ordinary Sheet-Metal.
.45-C Slate and Tile Roofing.

690

.45-D Ventilating Ducts, Fans, Stacks and Furnaces, Etc.
.45-E Hollow Metal Windows.
.45-F Metal Clad Wood Doors.
.45-G Enamel Sheet-metal Ceilings.
.45-H Art Sheet-metal Trim and Doors.
.45-I Enamel Sheet-metal Cabinets.
.45-J Enamel Sheet-metal Lockers.
.45-K Enamel Sheet-metal Radiator Covers and Seats.
.45-L Enamel Sheet-metal Toilet Partitions.
.45-M Metal Furniture.
.45-N Sheet-metal Utensils.
.45-O Drawn Sheet Metal Store Fronts, Etc.
.45-Z Miscellaneous Sheet Metal Trades not Otherwise Classified.

690.46 Brush, Broom and Swab-Using Trades (See File 696.)
.46-A Brush Trade Materials and Methods.
.46-B Water-proofing Membrane and Mastic or other Viscous Compositions mopped, broomed or swabbed in place.
.46-C Composition Roofing.
.46-D Plain Painting and Varnishing.
.46-E Decorations (Plain, Painted or Water Color).
.46-F Hangings, Fabrics, etc.
.46-G Upholstery.
.46-H Window Shades.
.46-I Mastic Tile and Sheet Floor Covering.
.46-J Rubber Tile and Sheet Floor Covering.
.46-K Cork Tile and Sheet Floor Covering.
.46-L Carpets, Linoleums, Etc., Floor Covering.
.46-M Plain Glass and Glazing.
.46-N Art Glass and Glazing.
.46-Z Miscellaneous Brush Trades not Otherwise Classified.

690.47 Pipe Trades. (See File 697.)
.47-A Pipe Trades Materials and Methods.
.47-B Sanitary Plant.
.47-B-1 Sewerage and Drainage.
.47-B-2 Sewerage and Bilge Pumps.
.47-B-3 Sewerage Disposal.
.47-B-4 Plumbing.
.47-B-5 Tanks and Towers for Water Supply, Stand Pipes.
.47-B-6 Gas Fitting.
.47-B-7 Gas Stoves, Etc.
.47-C Sprinkler Fitting.
.47-C-1 Storage Tanks and Towers.
.47-C-2 Pressure Tanks, Etc.
.47-C-3 Pumps.
.47-D Boiler Plant.
.47-D-1 Steel Stacks and Breeching.
.47-D-2 Tanks for Water Storage.
.47-D-3 Tanks for Oil Storage.
.47-D-4 Super Steam Heaters.
.47-D-5 Tube Blowers.
.47-D-6 Tube Cleaners.
.47-D-7 Furnaces.
.47-D-8 Stokers.
.47-D-9 Coal Handling Equipment.
.47-D-10 Ash Handling Equipment.
.47-D-11 Pulverized Coal Burners and Pulverizers.
.47-D-12 Oil Burners.
.47-D-13 Gas Burners.
.47-D-14 Draft Inducer Blowers.
.47-D-15 Soot Burners.
.47-D-16 Fuel Economizers.
.47-D-17 Smoke Indicators.
.47-D-18 Feed Water Heaters.
.47-D-19 Boiler Feed Pumps.
.47-D-20 Service Pumps.
.47-D-21 Fire Pumps.
.47-D-22 Governors for Pumps, Etc.
.47-D-23 Water Softeners.
.47-D-24 Lubricators.
.47-D-25 Injectors for Compound.
.47-D-26 Injectors for Water.
.47-D-27 Feed Water Regulators.
.47-D-28 Draft Regulators.
.47-D-29 Flow Meters.

690
.47-D-30 Draught Gauges.
.47-D-31 CO_2 Recorders.
.47-E Steam and Hot Water Fitting.
.47-E-1 Vacuum Pumps.
.47-E-2 Vacuum Valves.
.47-E-3 Miscellaneous Specialties.
.47-F Steam Power Plant.
.47-F-1 Engines.
.47-F-2 Compressors.
.47-G Vacuum Cleaning Plant.
.47-H Mechanical Refrigeration.
.47-H-1 Tanks.
.47-H-2 Compressors.
.47-H-3 Cooler Towers.
.47-I Mechanical Ventilation.
.47-I-1 Heating Units.
.47-I-2 Cooling Units.
.47-I-3 Air Washers.
.47-I-4 Fans and Engines.
.47-Z Miscellaneous Pipe Trades not Otherwise Classified.

690.48 **Wire and Conduit Trades** (See File 698).
.48-A Wire Trades Materials and Methods.
.48-B Electrical Conduit and Wiring.
.48-C Lighting Fixtures.
.48-D Electrical Power Work.
.48-E Electric Signs.
.48-F Private Telephone System.
.48-G Clock System.
.48-H Signal Clock System.
.48-I Fire Alarm System.
.48-J Burglar Alarm System.
.48-K Projecting Machines.
.48-Z Miscellaneous Electrical Trades not Otherwise Classified.

690.49 **Machinery and Miscellaneous Trades** (See File 699).
.49-A Machinery and Miscellaneous Materials and Methods.
.49-B Elevators.
.49-B-1 Passenger Elevators.
.49-B-2 Freight Elevators.
.49-B-3 Dumbwaiters.
.49-C Conveying Machines.
.49-D Mechanical Cleaners.
.49-E General Machinery.
.49-F Foundry Equipment.
.49-G Insulation, Pipe Covering, Etc. (See File 690.42-M).
.49-H Refrigerators, Coolers and Freezers.
.49-H-1 Ice Boxes.
.49-H-2 Electric Refrigeration.
.49-H-3 Gas Refrigeration.
.49-I Laundry Equipment.
.49-J Kitchen Equipment.
.49-K Laboratory Equipment.
.49-L Gymnasium Equipment.
.49-Z Miscellaneous Equipment not Otherwise Classified.

690.5 **ESTIMATES FOR BUILDINGS.**
.5-A Cube System.
.5-B Area System.
.5-C By Trades.
.5-D By Quantity Survey.

690.6 **CONTRACTS AND GENERAL CONDITIONS.**

690.7 **SUPERVISION OF CONSTRUCTION AND ACCOUNTS.**

690.8 **PROFESSIONAL SERVICES.**
.80-A Remuneration, Fees, Commissions.
.80-B Duties, Relationships, Etc.
.80-C Responsibility, Etc.
.80-D License or Registration.
.81 Architect.
.82 Structural Engineer.
.83 Mechanical Engineer.
.84 Sanitary Engineer and Surveyors.
.85 Electrical Engineer.
.86 Illuminating Engineer.
.87 Clerk of the Works, Draftsmen, Stenographers and Employees.
.88 Building Construction Manager.
.89 Specialists not otherwise Classified.

690.9 **LAWS AND RULES CONTROLLING BUILDING.**
.91 State or General Laws.
.92 Municipal Ordinances, Rules, Etc.
.93 Trade Rules.

690
.94 Findings, National Joint Board of Jurisdictional Awards.
.95 Lien Laws.
.96 Underwriters' Rules.
.97 Public Service Company's Rules.
.98 Liabilities of:
.981 Architects.
.982 Contractor.
.983 Workman.
.984 Owner.
.985 Bondsman.
.986 Liability Insurance Co.
.987 Adjoining Property Owner.
.988 Public.
.989 Any Other Responsibilities.

691 EARTH-WORKING, TRANSPORTATION AND TEAMING TRADES.

691.0 **TOOLS, UTENSILS, APPARATUS, ETC.**
.01 Shovels, Picks, Drills, Bars, Wheelbarrows, Etc.
.02 Plows, Scrapers, Trucks, Carts, Wagons, Teams, Tractors.
.03 Excavating, Trench and Mining Machinery.
.04 Hoists, Cranes, Pile Drivers, Conveyors, Hoisting Engines, Etc.
.05 Dummy Railroad Equipment. Tracks, Cars, Etc.
.06 Soil Testing Apparatus.
.07 Shoring, Sheet Piling, Piling, Caissons, Scaffolding, Etc.
.071 Wood.
.072 Metal.
.073 Concrete.
.08 Blasting Powder and Apparatus.
.09 Rock Crushers.

691.1 **MATERIALS TO BE REMOVED.**
.11 Common Earth, Clay, Sand, Gravel. Hard-pan, Conglomerate Rock, Etc.
.12 Trees, Shrubs, Etc.
.13 Rubbish, Etc.
.14 Buildings, Vaults, Pipes, Cisterns, Etc.

691.2 **DISPOSAL OF MATERIALS.**
.21 Stacking.
.22 Cartage.
.23 Dumps.

691.3 **UTILIZATION OF MATERIALS.**
.31 Sand and Gravel Stored for Mortar.
.32 Black Earth for Top Fill.
.33 Crushed Rock for Aggregate.
.34 Cleaning and Stacking Building Material for Use in New Building.
.35 Re-Planting and Protection of Trees and Shrubs.

691.4 **FILLING & GRADING MATERIAL.**
.5 FERTILIZER, SOIL TREATMENT.
.6 NURSERY STOCK, SODDING AND SEEDING.
.7 DRAINAGE MATERIAL.
.8 FROST PROTECTION.
.9

692 MORTAR-USING TRADES — (Inc. Masonry, Plastering, Tile and Marble Setting and the preparation for same).

692.0 **MASONRY APPARATUS.**
.01 Mixing Boxes, Platforms, Etc.
.02 Tools, Hose, Heaters, Etc.
.03 Mixers for Mortar and Concrete.
.04 Scaffolding, Horses, Planks, Etc.
.05 Forms.
.06 Erection Apparatus, Hoists, Cranes, etc.
.07 Shutes and Conveyors.

692.1 **MATERIALS FOR MASONRY.**
.11 Liquids, Water, Anti-freezing, Etc.
.12 Aggregate (a) Sand, (b) Stone Screenings, (c) Gravel, (d) Crushed Stone, (e) Crushed Slag, (f) Cinders, (g) Haydite, Etc.

692
.13 **Cementing Materials for Masonry.**
.131 Limes.
.132 Hydraulic Cements, (a) Natural, (b) Portland, (c) Miscellaneous.
.133 Gypsums, (a) Plaster of Paris, (b) Keene's Cement, (c) Miscellaneous.
.134 Magnesites.
.135 Asphaltic Cements.
.136 Composite Cements.
.137 Other Cements, Unclassified.
.138 Mortar Color.
.14 **Solids for Masonry.**
.141 Stone.
.142 Brick, (a) Adobe, (b) Burned Clay, (c) Sand Lime.
.143 Structural Partition and Load-bearing Tile.
.144 Terra Cotta, (a) Coping, (b) Ornamental Flue Lining, etc.
.145 Cement Blocks.
.146 Composite Blocks.
.147 Marble, Soapstone, Structural Slate and Glass Substitutes.
.148 Tile, Paving and Wall.
.149 Terrazzo Blocks and Slabs.
.15 **Mason's Hardware.**
.151 Anchors, Ties, Wall Boxes, Plates, Inserts, Scoopers, Sleeves, Etc.
.152 Thimbles, Ash and Coal Chutes, Clean-out Doors, Dampers, Grate Bars, Chimney Cap, Vent Gratings, Etc.
.153 Vault Lights, Sidewalk Doors, Etc.
.154 Screeds, Metal Expansion Joints.
.16 **Reinforcing for Masonry.**
.161 Bar Reinforcement.
.162 Fabric.
.163 Metal Lath.
.164 Wood-lath, Plaster Board.
.165 Fiber, Hair, etc.

692.2 **STONE CONSTRUCTION.**
.21 **Preservatives Treatment.**
.22 **Bond, Anchorage, Ties, Lewises, Etc.**
.23 **Cutting and Dressing of Stone, Stereotomy, Drips, Weathering, Etc.**
.24 **Setting, Joints, Mortar, Bedding, Etc.**
.25 **Cleaning and Pointing.**

692.3 **BRICK CONSTRUCTION.**
.31 **Preservative Treatment.**
.32 **Common Brick Work.**
.33 **Fire Brick Work.**
.34 **Face Brick Work.**
.35 **Laying Joints, Mortar, Etc.**
.36 **Chases, Fire-Stops, Corbels, Etc**
.37 **Bonds, Anchors, Etc.**
.38 **Cleaning and Pointing, Etc.**
.39 **Special Brick Work.**

692.4 **TERRA COTTA CONSTRUCTION.**
.41 **Preservative Construction.**
.42 **Bonding, Anchorage, Ties, Etc.**
.43 **Structural Tile Walls.**
.44 **Structural Tile Floors.**
.45 **Ornamental or Decorative Terra Cotta.**
.46 **Laying Joints, Mortar, Etc.**
.47 **Fitting Around Structural Parts.**
.48 **Centers, Supports, Protection.**
.49 **Cleaning, Pointing and Repairing.**
.5 **FIREPROOF CONSTRUCTION.**
.51 **Hollow Clay Tile, (a) Hard, (b) Porous.**
.52 **Gypsum Tile.**
.53 **Concrete.**
.54 **Tying, Fitting, Securing.**
.55 **Combination Construction.**
.56 **Centers, Forms, Etc.** (See 693.41 for Wood and 695 for Sheet-Metal.)
.59 **Patching, Repairing.**

692.6 **CONCRETE CONSTRUCTION.**
.61 **Massive, Caissons, Footings, Retaining Walls, Etc.**
.62 **High Duty Concrete.**
.63 **Hollow Concrete Building Blocks.**
.64 **Ornamental Concrete.**
.65 **Concrete Supported on the Ground, Paving of Walks, Floors, Drives, Etc.**
.66 **Waterproof Concrete.**
.67 **Reinforced Concrete.**

692
.671 Reinforcing Systems, Arranged Alphabetically.
.672 Forms and Centers. (See 693.41 for Wood; also 695 for Sheet-Metal.)
.673 Tests and Inspection.
.674 Data for Experiments.
.675 Formulae, (a) Vault Construction.
.676 Special Applications.

692.7 **DECORATIVE AND SANITARY WALL AND FLOOR SURFACING.**
.71 **Marble, Soapstone and Slate.**
.72 **Structural Glass.**
.73 **Terrazzo.**
.74 **Tile Mosaic,** (a) Ceramic, (b) Marble, (c) Glass.
.75 **Tile,** (a) Quarry, (b) Encaustic, (c) Marble, (e) Ornamental, (f) Composition Non-Slip, (g) Slate Flagging, (h) Rubber Tile.
.76 **Sanitary Composition Floors.**

692.8 **WATER-PROOFING AND HARDENERS.**
.81 **Integral Waterproofing** (for brush applied mastic and painting, waterproofing, see File 696).
.82 **Hardeners** (a) Surface, (b) Ad mixed.
.83 **Mortar Colors, Workability Mixtures**

692.9 **PLASTER TRADES.**
.91 **Interior Plaster.**
.91(a) Common Lime Plaster.
.91(b) Gypsum Plaster.
.91(c) Magnesite.
.91(d) Portland Cement Plaster.
.91(e) Lathing.
.91(f) Special Plasters.
.92 **Exterior Plaster.**
.93 **Modeling and Ornamental Plaster.**

693 WOOD WORKING TRADES.

.0 **APPARATUS, INCIDENTAL TOOLS, ETC.**
.01 **Mechanic's Tools.**
.02 **Wood-working Power Machinery,** (a) Saws, (b) Planers, (c) Stickers, (d) Sand-papering Machines, (e) Scraping Machines.
.03 **Kilns, Dryers.**
.04 **Scaffolding, Ladders, Horses and Benches.**

693.1 **MATERIALS.**
.11 **Lumber.**
.111 Timber, larger than 6"x6".
.112 Common Lumber.
.112 (a) Boards, Furring and Grounds.
.112 (b) Piece Stuff, Joists and Scantling.
.112 (c) Shingles, Wood and Composition.
.113 Finish Lumber.
.113 (a) Hardwood.
.113 (b) Soft Wood.
.113 (c) Flooring.
.114 Mill Stock.
.115 Veneers.
.116 Composition.
.117 Insulation Papers and Felts.
.12 **Glues.**
.13 **Rough Hardware.**
.131 (a) Nails, (b) Spikes, (c) Brads, (d) Hangers, Track, Etc.
.132 (a) Bolts, (b) Rods, (c) Anchors, Ties, (d) Screws, Etc.
.133 Rivets.
.134 (a) Washers, (b) Flitch Plates, (c) Splice Plates.
.135 Mill Construction Hardware, (a) Stirrups, (b) Hanger, (c) Column Caps, (d) Ties, (e) Box and Wall Anchors, (f) Bearing Plates, Etc.
.136 Double Hung Sash Hardware, (a) Pulleys, (b) Cords, (c) Chain, (d) Weights, (e) Spring Balances.
.137 Window Cleaning Hardware.
.138
.14 **Finish Hardware.**
.141 Hanging Hardware, (a) Butts, (b) Hinges, (c) Pivots, Etc.

693
.142 Controlling Hardware, (a) Bumpers, (b) Strikes, (c) Holders, (d) Hooks, (e) Stays, (f) Adjusters, Etc.
.143 Fastening Hardware, (a) Old Fashion Latches, (b) Spring Latches, (c) Catches, (d) Fasts, (e) Thumb Bolts, (f) Locks, Etc.
.144 Trimming Hardware, (a) Pulls, (b) Knobs, (c) Spindles, (d) Roses, (e) Escutcheons.
.145 Protection Hardware, (a) Kick Plates, (b) Push Plates, (c) Direction Plates or Signs, (d) Push Bars, Etc.
146 Operating Hardware, (a) Closers and Checks, (b) Springs, (c) Weights and Pulleys, (d) Window Poles, Etc.
.147 Weathering Hardware, (a) Weather Strips, (b) Thresholds, (c) Special Drips, (d) Metal Astrigals, (e) Casement Operators, Etc.
.148 Automatic and Panic Hardware.
.149 Miscellaneous Hardware not otherwise classified, (a) Wardrobe Hardware, (b) Showcase Hardware, (c) Toilet-room Hardware, (d) Ladder Hardware, (e) Castors, (f) Cabinet Hardware, (g) Gymnasium Apparatus, (h) Mail Boxes and Chutes, (i) Clothes Chutes.

693.2 **ORDINARY CONSTRUCTION.**
.21 **Balloon Construction for Frame Buildings.**
.22 **Joist Construction for Masonry Buildings.**
.23 **Trusses, etc.**

693.3 **HEAVY TIMBER CONSTRUCTION.**
.31 **Heavy Post and Timber Construction for Frame Buildings.**
.32 **Mill Construction for Masonry Buildings.**

693.4 **AUXILIARY WOOD CONSTRUCTION FOR FIREPROOF BUILDINGS.**
.41 **Centers, Forms, Protective Covering, Scaffolding, Etc.**
.42 **Grounds, Attachment Strips, Etc.**

693.5 **JOINERY AND MILL WORK.**
.51 **Frames and Sash.**
.511 Box Frames, Double Hung Sash.
.512 Casement Frames, Sash Opening In.
.513 Casement Frames, Sash Opening Out.
.514 Frames for Sash Hinged at Bottom, Swinging In at Top.
.515 Frames for Sash Hinged at Top, Swinging In at Bottom.
.516 Frames for Sash Hinged at Top, Swinging Out at Bottom.
.517 Frames for Horizontal Pivoted Sash
.518 Frames for Vertical Pivoted Sash.
.52 **Wood Interior Trim.**
.53 **Wood Floors.**
.54 **Blinds.**
.55 **Doors.**
.551 Ordinary Panel and Sanitary Doors.
.552 Special Revolving Doors.
.553 Folding, Accordion Doors.
.554 Rolling Doors.
.56 **Screens.**

693.57 **Mouldings.**

693.58 **Flooring Wood.**
.59 **Columns.**

693.6 **STAIR BUILDING.**

693.7 **ORNAMENTAL JOINERY.**
.71 **CABINET WORK, (a) Mantels, (b) Sideboards, (c) Cases, (d) Space Savers, (e) Panel Partitions.**
.72 **WOOD FURNITURE.**

693.8 **WOOD CARVING, WOOD LETTERS.**
.9 **MISCELLANEOUS.**

694 HEAVY METAL TRADES — (Employing Metal heavier than No. 10 gauge).

.0 **TOOLS, UTENSILS, APPARATUS, ETC.**

694
.01 **Job Machinery.**
.02 **Job Tools, Hammers, Sledges, Punches, Tongs, Reamers, Riveters, Forges, Etc.**
.03 **Derricks, Cable, Hoisting Machinery.**
.1 **MATERIALS USED IN THE METAL TRADES.**
.11 **Iron Products.**
.111 Cast-Iron.
.112 Wrought Iron.
.113 Steel.
.114 Alloys, (a) Copper Bearing Steel, (b) Nickel Steel, (c) Sheradized Steel.
.12 **Copper.**
.13 **Brass.**
.14 **Bronze.**
.15 **Aluminum.**
.16 **Miscellaneous Structural Metals.**

694.2 **STRUCTURAL METAL CONSTRUCTION.**
.21 **Fabrication.**
.211 Shop Drawings.
.22 **Framing.**
.221 Bases, Bearing Plates, Etc.
.222 Columns and Struts.
.223 Caps, Connections, Gussets, Etc.
.224 Girders, Beams, Etc.
.225 Suspenders, Tie-Rods, Chains Etc
.23 **Preservatives.**
.231 Paint. (See 696.)
.232 Galvanizing.
.233 Other Methods.

694.3 **MISCELLANEOUS METAL.**
.31 **Fire Escapes.**
.32
.33

694.4 **HEAVY METAL DOORS, SHUTTERS, ETC.**
.41 **Underwriters' Doors.**
.42 **Sidewalk Doors, Floor Plates.**
.43 **Shutters.**

694.5 **ORNAMENTAL METAL.**
.51 **Stairs, Thresholds.**
.52 **Enclosures, Guards, Grills, Fences, Gates, Flag Poles, Etc.**
.53 **Elevator Enclosures and Cages.**
.54 **Fireplace Trimming.**
.541 Andirons, Tongs, Pokers, Spark-screens, etc.
.542 Grate Frames, Dampers, Grates, Etc.
.543 Furniture.

694.6 **SOLID METAL SASH.**

694.7 **VAULT DOORS, SAFES, VAULTS, ETC.**
.71 **Vault Doors.**
.72 **Safes.**
.73 **Vaults and Bank Equipment.**

694.8 **Tablets Memorials, Signs, Bulletins, Etc.**
.9 **Bells and Miscellaneous.**

695 SHEET-METAL TRADES — (Employing Metal of No. 10 gauge or less).

695.0 **TOOLS, UTENSILS AND APPARATUS (used by the Sheet-Metal Trades).**
.01 **Brakes, Shears, Mallets, Hammers Etc.**
.02 **Welding Machines.**
.03 **Soldering Apparatus.**
.04 **Plating Apparatus.**

695.1 **SHEET-METAL MATERIALS.**
.11 **Sheet Iron.**
.111 Tin or Tin Coated Sheet Iron.
.112 Galvanized Iron.
.12 **Sheet Copper.**
.121 Planished Copper.
.13 **Zinc Sheet.**
.14 **Brass Sheet.**
.15 **Bronze Sheets.**
.16 **Other Sheet Metals.**
.17 **Solders, Fluxes, Etc.**
.18 **Hardware.**
.181 Rivets and Bolts.

695

.182 Nails, Tacks and Screws.
.183 Incidental Hardware.
.19 **Miscellaneous.**

695.2 **ORDINARY SHEET-METAL CONSTRUCTION.**

.21 **Roofs.**
.211 Tin.
.212 Galvanized Iron.
.213 Copper.
.214 Slate Shingles.
.215 Composition.
.216 Tile Shingles.
.217 Cement Tile.
.22 **Cornices, Etc.**
.23 **Flashing, Gutters, Valleys, Down-Spouts and Conductor Heads, Roofing, Etc.**
.24 **Sky-lights, Ventilator Heads, Etc.**
.251 **Furnace Work, Casings, Ducts and Stacks, Etc.**
.252 **Ventilation Ducts.**
.253 **Chutes, Etc.**

695.3 **FIRE RESISTING DOORS AND WINDOWS.**

.31 **Underwriters' Tin-Clad Doors.**
.32 **Underwriters Sheet-Metal Sash.**
.33 **Rolling Steel Shutters and Doors.**

695.4 **CEILINGS, STAMPED SHEET-METAL.**

695.5 **DRAWN SHEET-METAL.**

.51 **Store Front Bars.**
.52 **Showcase Bars, Etc.**
.53 **Copper Casements.**

695.6 **TRIM AND DOORS OF SHEET METAL.**

695.7 **FURNITURE OF SHEET METAL.**

695.8 **UTENSILS OF SHEET METAL.**

695.9 **Steel Joists, Forms, Etc.**

696 BRUSH, BROOM AND SWAB-USING TRADES.

.0 **BRUSH TRADE, TOOLS AND APPARATUS.**

.01 **Kettles, Buckets, Ladles, Swabs and Other Roofers' and Waterproofers' Tools.**
.02 **Brushes, Cans, Knives, Etc.**
.03 **Ladders, Scaffolding, Hoists, Etc.**
.04 **Drop Cloths.**
.05 **Grinders.**
.06 **Spraying Machines.**

696.1 **BRUSH TRADE MATERIALS AND METHODS.**

.11 **Roofing and Waterproofing Materials.**
.111 Felt.
.112 Paper.
.113 Gravel, Slag, Crushed Stone, Paving Tile, Etc.
.114 Tar and Asphalt.
.115 Creosote, Dips and Stains.
.12 **Painters' Materials.**
.121 Binders, (a) Oil, (b) Casein, (c) Dryers.
.122 Pigments, (a) White Lead, (b) Red Lead, (c) Zinc, (d) Graphite, (e) Whiting, (f) Lime, (g) Other Pigments.
.123 Colors, (a) Vegetable, (b) Mineral.
.124 Solvents, (a) Turpentine, (b) Benzine, (c) Alcohol, (d) Other Solvents.
.125 Wood Finishing Materials, (a) Stains, (b) Fillers, (c) Shellacs, (d) Varnishes, (e) Enamels, (f) Waxes, (g) Other Materials, (h) Lacquers.
.126 Prepared Paints.
.13 **Water Paints.**
.131 Binder, (a) Casein, (b) Glue, (c) Other Binders.
.132 Pigments, (a) Lime, (b) China Clay, (c) Whiting.
.133 Colors.
.14 **Wall Papers.**
.15 **Hangings and Coverings.**
.151 Fabrics.
.152 Leather, (a) Genuine, (b) Imitation.
.16 **Hanging Hardware Poles, Etc.**
.17 **Upholstery, (a) Tacks, (b) Feathers, (c) Hair, (d) Moss, (e) Ticking, (f) Cord, (g) Other Materials.**

696

.18 **Glazing Material.**
.181 Glass, (a) Common Glass, (b) Plate Glass, (c) Ornamental Glass, (d) Wire-glass, (e) Prismatic Glass, (f) Colored Glass, (g) Glass Substitutes.
.182 Putties.
.183 Tacks.
.184 Leading Bars, (a) Lead, (b) Zinc, (c) Copper, (d) Ventilators.
.19 **Other Materials.**

696.2 **WATER-PROOFING WORK.**

.21 **Brushed on Construction.**
.22 **Membrane.**
.23 **Calking.**

696.3 **COMPOSITION ROOFING WORK.**

.31 **Tar and Gravel Roofing.**
.32 **Asphaltum Composition Roofing.**
.33 **Promenade Deck Roofing.**
.34 **Mastic Floors.**
.35 **Composition Flashing.**

696.4 **PAINTING WORK.**

.5 **WOOD FINISHING WORK.**
.6 **GENERAL DECORATIONS.**
.61 **Ordinary Water Color Tinting.**
.62 **Fresco Painting, Stenciling, Etc.**
.63 **Mural Decorations.**
.7 **UPHOLSTERY.**
.8 **HANGINGS.**
.81 **Ordinary Window Shades, Awnings.**

696.82 **Lace Curtains.**
.83 **Draperies, Decorative Screens, Etc.**
.84 **Carpets, Rugs and Linoleums, Rubber Tile.**
.85 **Tents.**

696.9 **GLAZING.**

.91 **Common Glazing.**
.92 **Art Glass Glazing.**

697 PIPE TRADES.

.0 **TOOLS, UTENSILS AND APPARATUS.**
.01 **Mechanic's Chest Tools, Furnaces, Etc.**
.02 **Power Pipe Cutter, Benders, Dies, Etc.**
.03 **Scaffolding Ladders, Etc.**

697.1 **MATERIALS.**

.11 **Metals,** (a) Wrought Iron, (b) Steel, (c) Lead, (d) Brass, (e) White Metal.
.12 **Pipe.**
.121 Wrought Iron, (a) Black, (b) Galvanized.
.122 Steel, (a) Black, (b) Galvanized.
.123 Cast Iron, Duriron.
.124 Brass, Bronze and Copper.
.125 White Metal.
.126 Block Tin.
.127 Lead Lined Iron.
.128 Tin Lined Iron.
.129 Tile Pipe.
.13 **Pipe Fittings.**
.131 Screw Connections.
.132 Flange Connections.
.133 Union Connections, Expansion Joints.
.134 Caulked Connections.
.135 Valves, (a) Shut-off, (b) Gate, (c) Disk, (d) Other Valves, (e) Air Vents.
.136 Pipe Hangers, Supports, Etc.
.137 Under-Ground Conduit.
.14 **Tanks.**
.141 Hot Water.
.142 Cold Water, (a) Wood, (b) Metal.
.143 Oil Tanks.
.144 Gas Tanks.
.15 **Boiler.**
.151 Steel Water Tube.
.152 Steel Flue Tube.
.153 Cast Iron Sectional.
.16 **Stoves.**
.161 **Coal.**
.162 **Gas.**
.163 **Oil.**

.17 **Furnaces, Grates and Stokers for Coal and Oil. (a) Ordinary, (b) Smokeless, (c) Dutch-oven, (d) Oil Burning, (e) Mechanical Feed.**
.18 **Brass Goods.**
.19 **Pottery.**

697.2 **SEWERAGE, and Drainage** (See 692 for Masonry Sewers.)

697.21 **Drainage.**
.22 **Sewerage.**
.23 **Sewerage and Bilge Pumps.**
.24 **Sewerage Disposal Equipment.**
.3 **PLUMBING TRADE.**

697.31 **Plumbing Fixtures.**
.311 Roughing-in, (a) Durham System, (b) Cast-Iron Caulked Joint System.
.312 Water Supply, (a) Pumps, (b) Tanks, (c) Hose and Fire Apparatus, (d) Filters, (e) Sterilizers, (f) Ice Machinery, (g) Stills, Etc. (h) Domestic Heater, (i) Softeners, (j) Meters.
.313 Garbage and Sewage Disposal, (a) Bilge Pumps, Incinerators.
.314 Fixtures for Plumbing, (a) Floor Drains, (b) Cesspools, (c) Sinks, (d) Slop Sinks, (e) Laundry Wash Trays, (f) Lavatories, (g) Bath-tubs, (h) Showers, (i) Water Closets, (j) Urinals, (k) Bath and Toilet Room Trimmings, Paper-Holders, Towel Racks, Tumbler Holders, Soap Dishes, Etc.
.315 Laundry Machinery.
.316 Kitchen Machinery.

697.4 **GAS FITTING.**
.41 **Meters.**
.42 **Fixtures.**
.43 **Gas-water Heaters.**
.44 **Clothes Dryers.**
.45 **Gas Stoves.**

697.5 **MECHANICAL CLEANING.**
.6 **SPRINKLER FITTING.**
.60 Erecting Apparatus.
.61 Sprinkler-fitting Devices.
.62 Storage Tanks and Towers.
.63 Pressure Tanks, etc.
.64 Sprinkler Equipment Pumps.
.7 **HEATING, STEAM AND HOT WATER AND VENTILATION.**
.71 **One-Pipe Gravity.**
.72 **Two-Pipe Gravity.**
.73 **Vapor Two-Pipe.** (Systems arranged alphabetically.)
.74 **Vacuum.** (Systems arranged alphabetically.)
.75 **Radiation,** (a) Direct, (b) Direct-Indirect, (c) Indirect, (d) Hangers.
.76 **Boiler Plant.**
76-1 Steel Stacks and Breeching.
76-2 Tanks for Water Storage.
76-3 Tanks for Oil Storage.
76-4 Super Steam Heaters.
76-5 Tube Blowers.
76-6 Tube Cleaners.
76-7 Furnaces.
76-8 Stokers.
76-9 Coal Handling Equipment.
76-10 Ash Handling Equipment.
76-11 Pulverized Coal Burners and Pulverizers.
76-12 Oil Burners.
76-13 Gas Burners.
76-14 Draft Inducer Blowers.
76-15 Soot Burners.
76-16 Fuel Economizers.
76-17 Smoke Indicators.
76-18 Feed Water Heaters.
76-19 Boiler Feed Pumps.
76-20 Service Pumps.
76-21 Fire Pumps.
76-22 Governors for Pumps, Etc.
76-23 Water Softeners.
76-24 Lubricators.
76-25 Injectors for Compound.
76-26 Injectors for Water.
76-27 Feed Water Regulators.
76-28 Draft Regulators.
76-29 Flow Meters.

.76-30 Draught Gauges.
.76-31 CO_2 Recorders.
.77 **Mechanical Refrigeration.**
.771 Tanks.
.772 Compressors.
.773 Cooling Towers.
.78 **Mechanical Ventilation, Air Washers and Filters.**

697.8 **STEAM-POWER WORK, PUMPS, ETC.**
.81 **Engines.**
.82 **Compression.**
.83 **Pumps.**
.9 **OTHER PIPE TRADES.**

698 WIRE AND CONDUIT TRADES—Electrical Work of All Kinds.

.0 **TOOLS, UTENSILS AND APPARATUS.**

698.1 **MATERIALS FOR WIRE TRADES**
.11 **Conduit.**
.111 Pipe.
.112 Flexible Greenfield, Etc.
.113 Moulding, (a) Wood, (b) Metal.
.114 Tile and Porcelain.
.115 Knob and Tube Substitute.
.12 **Insulation.**
.13 **Wire,** (a) Gauges, (b) Kinds.
.14 **Switchboards. Miscellaneous Devices.**
.141 Switchboards.
.142 Switches, Switch Plates, Etc.
.143 Cut-out Cabinets, Fuses, Etc.
.144 Transformers.
.145 Receptacle Sockets, Plugs.
.146 Door Openers.
.147 Batteries.
.148 Meters, Instruments.
.15 **Lighting Fixtures, (a) Sockets, (b) General Fittings, (c) Pendents, (d) Brackets, (e) Indirect (f) Semi Indirect, (g) Special Reflectors, (h) Signs.**
.16 **Telephones, Speaking Tubes, Bells. Etc.**
.161 Private Telephones.
.162 Signal System, Alarms, Etc.
.163 Speaking Tube.
.164 Letter Boxes, Etc.
.17 **Motors and Generators.**
.18 **Lightning Rods.**
.19 **Miscellaneous, (a) Stoves, (b) Fans, (c) Time Systems, (d) Door Operators, (e) Electric Fire-Places.**

698.2 **GENERAL HOUSE WIRING FOR ILLUMINATING AND MINOR POWER WORK.**

698.3 **TELEPHONE WORK.**

698.4 **ELECTRIC POWER WORK.**

698.5 **CENTRAL STATION WORK.**

698.6 **OTHER ELECTRICAL WORK.**

699 MACHINERY TRADES AND MISCELLANEOUS BUILDING ITEMS—(Not Otherwise Classified).

699.0 **GENERAL MATTERS PERTAINING TO THE PREPARATION AND ERECTION OF MACHINERY.**

699.1 **MATERIALS.**

699.2 **ELEVATORS.**
.21 Passenger.
.22 Freight.
.23 Dumb Waiters.

699.3 **CONVEYING MACHINES.**
.31 **Belt Conveyors.**
.32 **Chain Conveyors.**
.33 **Pneumatic Tube Conveyors.**

699.4 **FOUNDRY EQUIPMENT.**

699.5 **GENERAL MACHINERY.**

699.6 **INSULATION, PIPE COVERING, ETC.** (See 692 for Plastic Pipe Covering.)

699.7 **REFRIGERATORS, COOLERS AND FREEZERS.**
699.8 **Kitchen, Laundry, Laboratory Equipment.**
.81 **Laundry Equipment.**
.82 **Kitchen Equipment.**
.83 **Laboratory Equipment.**
.84 **Gymnasium Equipment.**
699.9 **MISCELLANEOUS TRADES NOT OTHERWISE CLASSIFIED. ORGANS, CHIMES.**

700 FINE ARTS.

701 **PHILOSOPHY. THEORIES. UTILITY. AESTHETICS.**
702 **COMPENDS. OUTLINES.**
703 **DICTIONARIES. CYCLOPEDIAS.**
704 **ESSAYS. LECTURES. ADDRESSES.**
705 **PERIODICALS. MAGAZINES. REVIEWS.**
706 **SOCIETIES. TRANSACTIONS. REPORTS, ETC.**
707 **EDUCATION. STUDY AND TEACHING OF ART.**
708 **ART GALLERIES AND MUSEUMS.**
709 **HISTORY OF ART IN GENERAL** Divided like 930-999.

710 LANDSCAPE GARDENING.

711 **PUBLIC PARKS.**
712 **PRIVATE GROUNDS. LAWNS.**
713 **WALKS. DRIVES. BRIDGES.**
714 **WATER. FOUNTAINS. LAKES.**
715 **TREES. HEDGES. SHRUBS.** See also 634.9, Forestry; 582, Botany.
716 **PLANTS. FLOWERS.** .1, Plants; .2, Flowers; .3, Conservatories; .4, Window gardens; .5, Ferneries.
717 **ARBORS. SEATS. OUTLOOKS.**
718 **MONUMENTS. MAUSOLEUMS.**
719 **CEMETERIES.** See also 393.1, Earth burial; 614.61, Public health.

722 ANCIENT AND ORIENTAL ARCHITECTURE (Pagan).

722.0 **PRIMITIVE.**
.04 **Europe.**
.05 **Asia.**
.06 **Africa.**
.07 **America.**
722.1 **EGYPTIAN** (PERIOD DIVISION).
.11 **Ancient and Middle Empire** (4000-2000 B. C.)
.12 **Shepherd Kings** (2000-1600 B. C.)
.13 **Theban New Empire** (1600-1250 B. C.)
.14 **The Decadence** (1150-622 B. C.)
.15 **Restoration** (Saite Period 663-525 B. C.)
.16 **Ptolemaic Period** (332-30 B. C.)
722.2 **EASTERN ASIATIC.**
.21 **Chinese.**
.22 **Japanese.**
.23 **Korean.**
722.3 **WESTERN ASIATIC.**
.31 **Chaldean.**
.311 Summerian.
.312 Akkadian.
.32 **Assyrian.**
.33 **Persian.**
.331 Ancient.
.332 Sassanian.
.34 **Hittite.**
.35 **Phoenician.**
.36 **Jewish.**
.37 **Cypriote.**
.38 **Lycian.**
722.4 **CENTRAL ASIATIC** (INDIAN).
.41 **Hindu.**
.42 **Buddhist.**
.43 **Jaina.**
.44 **Dravidian.**
.45 **Chalukyan.**
.46 **Burmese.**
722.5 **ANCIENT AMERICAN.**
.51 **Mexico and Central America.**
.511 Toltec.
.512 Aztec.
.513 Maya.
.52 **South America.**
.521 Inca (Peruvian and others).
722.6 **PRE-CLASSIC.**
.61 **Aegean** (Grecian Archipelago)
.611 Cretan or Minoan.
.612 Mycenean.
.613 Trojan.
.62 **Etruscan** (Italy).

Classic.

722.7 **GREEK.**
.71 **Greece.**
.711 Archaic.
.712 Periclean
.713 Hellenistic.
.72 **Asia Minor.**
.73 **Italy.**
.74 **Sicily.**
.75 **Africa.**
722.8 **ROMAN.**
.81 **Italy.**
.811 Etruscan influence, 493-40 B. C.
.812 **Graeco-Roman influence**, 212-27 B
.813 Augustan, 27 B. C.-, 4 A. D.
.814 Imperial, 14 A. D.-313 A. D.
.82 **Greece.**
.83 **Germany and Austria.**
.84 **France.**
.85 **Spain.**
.86 **Asia.**
.87 **Africa.**
.88 **Britain.**
(Arbitrary numbers are used to designate Greek and Roman provinces).

723 MEDIAEVAL ARCHITECTURE.

723.1 **EARLY CHRISTIAN.**
.14 **Italy**
.15 **Syria.**
.16 **Egypt and North Africa** (Coptic).
723.2 **BYZANTINE.**
.245 **Italy.**
.247 **Russia.**
.2495 **Greece.**
.2496 **Constantinople and Vicinity.**
.2497 **Balkans.**
723.3 **MOHAMMEDAN OR SARACENIC.**
.346 **Spain** (Moorish).
.353 **Arabia.**
.3539 **Syria.**
.354 **India.**
.355 **Persia.**
.356 **Turkey.**
.362 **Egypt.**
.363 **North Africa** (Moorish).
723.4 **ROMANESQUE.**
.441 **Ireland and Scotland** (Celtic).
.442 **England.**
.4421 Saxon.
.4422 Norman.
.443 **Germanic Countries** (Rhenish).
.444 **France.**
.4441 Provencal.
.4442 Aquitainian.
.4443 Auvergnat.
.4445 Cluniac or Burgundian.
.4446 Norman.
.445 **Italy.**
.4451 Lombard.
.4452 Central.
.4453 Siculo-Arabic.
.446 **Spain and Portugal** (Iberian Peninsula).
.447 **Russia.**
.448 **Scandinavia.**
.4481 Norway.
.4482 Sweden.
.4483 Denmark.
.449 **Minor Countries.**
723.5 **GOTHIC.**
.541 **Ireland and Scotland.**
.542 **England.**
.5421 Early English, 1175-1307 (13th C.)
.5422 Decorated, 1307-1377 (14th C.)
.5423 Perpendicular, 1377-1485 (15 C.)

723
.5424 Tudor, 1485-1558 (Early 16th C.) or

.5421 { Lancet. / Geometrical (13th and 14th C.) / Curvilinear. }
.5422 Rectilinear (15th C.)
.5423 Tudor (Early 15th C)
.543 **Germanic Countries.**
.544 **France.**
.5441 Primary or Gothique (13th C.)
.5442 Secondary or Rayonnant (14th C.)
.5443 Tertiary or Flamboyant (15th C.)
.545 **Italy.**
.546 **Spain and Portugal** (Iberian Peninsula).
.547 **Russia.**
.548 **Scandinavian Countries.**
.549 **Minor Countries.**

724 MODERN.

724.1 **RENAISSANCE.**
.141 **Ireland and Scotland.**
.142 **England.**
.1421 Elizabethan.
.1422 Jacobean.
.1423 Queen Ann.
.1424 Georgian.
.143 **Germanic Countries.**
.144 **France.**
.1441 Valois Period.
{ Transition. / Francis I. / Advanced Renaissance. }
.1442 Bourbon.
{ Henri IV. / Louis XIV. / Rococo. }
.1443 Louis XVI.
.1444 Empire.
.145 **Italy.**
.1451 Early.
.1452 High.
.1453 Baroque.
.146 **Iberian Peninsula** (Spain and Portugal).
.147 **Russia.**
.148 **Scandinavian Countries.**
.149 **Minor Countries.**
.172 **Spanish Colonial** (Mission Mexico.)
.1721 California.
.1722 New Mexico.
.1723 Arizona.
.1724 Texas.
.1725 Florida (Arbitrary numbers given for geographical divisions).
.173 **American Colonial** (Including Georgian).
.1731 English (1607-1720).
.17311 New England.
.17312 New York and Middle States.
.17313 Maryland, Virginia, S. Carolina (Georgian Period).
.1723 Dutch (1720-1800).
.17321 New York.
.17322 Pennsylvania.
.1733 Swedish.
.17331 Pennsylvania.
.17332 New Jersey.
.1734 French.
.17341 Canada.
.17342 Mississippi Valley.

724.2 **CLASSIC REVIVAL** (Roman and Greek).
.241 **Ireland and Scotland.**
.242 **England.**
.243 **Germany.**
.244 **France.**
.245 **Italy.**
.246 **Spain.**
.247 **Russia.**
.248 **Scandinavian Peninsula.**
.249 **Minor Countries.**
.271 **United States** (1800-1850).

The Classic Revival began with the revival of Roman, but in Germany, England and America the Greek Revival eclipsed the Roman. Many public buildings in the United States were built during the Greek Revival.

724.3 **GOTHIC REVIVAL.**
.341 **Ireland and Scotland,** etc.
.373 **United States.**

724.4 **ROMANESQUE REVIVAL.**
.441 **Ireland and Scotland,** etc.
.473 **United States.**

724.5 **MODERNIST SCHOOL.**
.541 **Ireland and Scotland.**
.542 **England.**
.543 **Germany and Austria** (Secession Movement).
.544 **France** (Art Nouveau).
.545 **Italy,** etc.
.573 **United States** (Functionalism).
.6
.7
.8

724.9 **CONTEMPORARY TYPES.**
.941 **Ireland and Scotland,** etc.
.973 **United States.**
.9731 French Renaissance Vogue.
.9732 Neo-Romanesque Vogue.
.9733 Neo-Classic Vogue.
.9734 Neo-Gothic Vogue.
.9735 Spanish Vogue.
.9736 Italian Vogue.

DEWEY'S GEOGRAPHICAL TABLE.

These numbers are added to the various styles in order to designate geographical distribution.

4 EUROPE.
41 Ireland and Scotland.
42 England.
43 Germany.
431 Austria.
44 France.
45 Italy.
46 Spain.
47 Russia.
48 Scandinavian Countries.
49 Minor Countries.
5 ASIA.
6 AFRICA.
7 NORTH AMERICA.
71 Canada.
72 Mexico.
73 United States.

725 **PUBLIC BUILDINGS.**
.1 **Administrative. Governmental.**
.11 Capitols. Houses of Parliament.
.12 Ministries of War, State, etc.
.13 City and Town Halls. Bureaus. Public Offices. City Plans.
.14 Custom Houses. Bonded Warehouses. Excise Offices.

725
.15 Court Houses. Record Offices.
16 Post Offices, General and Special.
.17 Official Residences. Palaces of Rulers.
.18 Barracks. Armories. Police Stations.
.19 Engine Houses. Fire Alarm Stations.

725.2 **Business and Commercial.**
.21 Stores, Wholesale and Retail.
.22 Mixed Store, Office, and Apartment Buildings.
.23 Office Buildings. Telegraph. Insurance. Loft.
.24 Banks. Safe Deposit. Savings.
.25 Exchanges. Boards of Trade.
.26 Markets.
.27 Cattle Markets. Stock Yards.
.28 Abattoirs, Meat Packing Plants, etc.
.29 Other Business Buildings.

725.3 **Transportation and Storage.**
.31 Railway Passenger Stations.
.32 Railway Freight Houses.
.33 Railway Shops, Round Houses, Car Houses, Tanks, Stores.
.34 Dock Buildings. Wharf Boats and Houses.
.35 1, Warehouses; 2, Cold Storage; 3, Safe Deposit Storage.
.36 Elevators, Grain.
.37
.38
.39 Other.

725.4 **Manufactories.**
.41 Textile Factories or Mills. Wool, Cotton, Silk.
.42 Breweries. Malteries. Distilleries.
.43 Foundries. Machine Shops. Iron and Steel Works.
.44 Wood-working Mills, Furniture, Piano and Organ Factories.
.45 Carriage and Car Factories.
46 Paper Mills.
.47 Mills for Flour, Meal, Feed, etc.
.48 Pottery, Glass, Terra Cotta, Brick Works.
.49 Other Manufactories.

725.5 **Hospitals and Asylums.** See also 725.6. Reformatories.
.51 Sick and Wounded. Eye and Ear. Incurables. Lying-in.
.52 Insane.
.53 Idiotic. Feeble-minded.
.54 Blind. Deaf and Dumb.
.55 Paupers. Almshouses.
.56 Aged.
.57 Children. Orphans.
.58 Foundling.
.59 Soldiers' Homes.

725.6 **Prisons and Reformatories.**
.61 State Prisons. Penitentiaries.
.62 Jails. Cell Houses.
.63 Reformatories for Adults. Houses of Correction.
.64 Reform Schools.
.65 Inebriate Asylums.

725.7 **Refreshment. Baths. Parks.**
.71 Cafés. Restaurants.
.72 Saloons.
.73 Baths: Warm, Medicated, Turkish, Russian.
.74 Swimming Baths.
.75 Buildings for Watering Places, Spas, etc.
.76 Buildings for Parks and Streets, Public Comfort Stations.

725.8 **Recreation.**
.81 Music Halls Auditoriums.
.82 Theatres. Opera Houses.
.83 Halls for Lectures, Readings, etc.
.84 Bowling Alleys. Billiard Saloons.
.85 Gymnasiums. Turn Halls.
.86 Skating Rinks. Bicycle Rinks.
.87 Boat Houses. Bath Houses.
.88 Riding Halls and Schools.
.89 Shooting Galleries.

725.9 **Other Public Buildings.**
.91 Exhibition Halls.
.92 Temporary Halls. Tabernacles. Wigwams.
.93 Workingmen's Clubs and Institutes.
.94 Town Squares.
.95 Summer Recuperating Camps.

726 **ECCLESIASTICAL AND RELIGIOUS.**
.1 **Temples.**
.2 **Mosques.**
.3 **Synagogues.**
.4 **Chapels. Sunday-school Buildings.**
.5 **Churches.**
.51 Frame.
.52 Brick or Stone.
.521 Small Audt., seating less than 600.
.522 Large Audt., seating more than 600.
.6 **Cathedrals.**
.7 **Monasteries. Convents. Abbeys.**
.8 **Mortuary. Cemetery Chapels. Receiving Vaults. Tombs.**
.9 **Others, Y. M. C. A., etc.**

727 **EDUCATIONAL AND SCIENTIFIC.**
.1 **Schools.**
.11 Ward and Grammar.
.12 High Schools.
Study and Recitation Rooms. Not including dormitory or boarding.

727
.2 **Academies. Seminaries. Boarding Schools.**
.3 **Colleges. Universities.**
.4 **Professional and Technical Schools. Law, Theology, etc.**
.5 **Laboratories: Physical, Chemical.** See 542.1. **Biological, etc. Zoological and Botanic Gardens.** See also 590.7 and 580.7.
.6 **.1, Museums. .2, Herbariums.** See 580.7.
.7 **.1, Art Galleries. .2, Studios.**
.8 **Libraries.** See 022, Library Buildings.
.9 **Other. Learned Societies, etc.**

728 **RESIDENCES.**
.1 **Tenement Houses.**
.11 City Homes of Poor.
.12 Country Homes of Poor.
.13 Cités Ouvrieres.

728.2 **Collective Dwellings.**
.21 **Flats; one family to the floor.**
.211 Small Flats less than 8 rooms.
.212 Large Flats, 8 rooms or more.
.22 Apartment Houses; more than one family to floor.
.221 Five Suites or Less.
.222 Six Suites or More.
.2221 Elevator Service.
.2222 No Elevator Service.

728.3 **City Houses. Mansions. Palaces.**
.31 Between party-walls. Stone.
.32 Between party-walls. Brick.
.33 Between party-walls. Partly wood.
.34 Semi-detached, including end houses in city blocks. Stone.
.35 Semi-detached, including end houses in city blocks. Brick.
.36 Semi-detached, including end houses in city blocks. Partly wood.
.37 Detached. Stone.
.38 Detached. Brick.
.39 Detached. Partly wood.

728.4 **Club Houses. Buildings for Secret Societies.**
.5 **Hotels.**
.51 City Hotels.
.52 Summer Resorts.
.53 Country Inns.

728.6 **Village and Country Homes.**
.61 Village Dwellings. On small lots.
.62 Stone.
.63 Brick.
.64 Concrete or stucco.
.65 Part masonry, part wood.
.66 All wood, 1, less than 7 rooms; 2, 7-12 rm; 3, 13 rm or over.
.67 Farm Houses.
.68 Laborers' Cottages. 1, Frame, 2 Masonry.

728.7 **Seaside and Mountain Cottages Chalets.**

728.8 **Country Seats.**
.81 Castles.
.82 Chateux.
.83 Manor Houses.
.84 Villas
.85 Log Houses.
.86 Bungalows.

728.9 **Out-Buildings.**
.91 Porters' Lodges.
92 Servants' Quarters.
93 Kitchens and Laundries.
.94 .1, Stables. .2, Carriage Houses. .3, Garages.
.95 Barns, Granaries.
.96 Dairies.
.97 Ice Houses.
.98 Conservatories. Green Houses. Graperies.
.99 Other.

729 **ARCHITECTURAL DESIGN AND DECORATION.**

.1 **The Elevation.**

.11 Composition; .12, Distribution; .13. Proportion; .14, Light and Shade. .15, Perspective effect; .15, .16, .17, .18, .19.
For projection of shadows and graphics of light and shadow see 515.63 and 515.7.

729.2 **The Plan.**

.21 Elements required; .22, Distribution; .23, Proportion; .24, .25, .26, .27, .28, .29.

729.3 **Elementary Forms.** For construction of these forms see 721.

.31 Walls. Mouldings. Cornices. .32, Piers, Columns, Pilasters, Pedestals and the Orders. Colonnades. .33, Arches and Arcades. .34, Ceilings, Vaults and Domes. .35, Roof. Spires. Dormers. .36, Towers. .37, Gables and Pediments. .38, Doors and Windows. Bays. Oriels. .39, Stairs and Balustrades. See also 515.83, Stereotomy; 604.8, Building.

.4 **Painted Decoration.**

729.5 **Decoration in Relief.**

.6 **Incrustation and Veneering.**

.7 **Mosaic and Marble.**

.71 Mosaic Ceilings; .72, Mosaic Walls; .73, Mosaic Floors; .74, Other Mosaic designs; .75, .76, .77, .78, .79.

.8 **Stained Glass Design.** For technical processes see 666.1; for history see 748.

729.9 **Architectural Accessories and Fixed Furniture.**

.91 Altars, Pulpits, Tribunes, Dais Thrones (Ecclesiastical).

.92 Seating for Public Buildings.

.921 Benches; 2, Settees; 3, Portable Chairs and Opera Chairs.

.93 Domestic Chairs, Tables, Couches, Stools, Beds, etc.

.94 Buffets.

.95 Mantels. Overmantels. Andirons.

.96 Steel Furniture.

.97 Window Shades.

.98 .1, Organs. .2, Pianos.

.99 Lighting Fixtures.

730 SCULPTURE.

731 **MATERIALS AND METHODS.**

732 **ANCIENT.**

733 **GREEK AND ROMAN.**

734 **MEDIEVAL.**

735 **MODERN.**

736 **CARVING. SEALS. DIES. GEMS. CAMEOS.**

737 **NUMISMATICS. COINS. MEDALS.**

738 **POTTERY. PORCELAIN.**

739 **BRONZES. BRASSES. BRIC-A-BRAC.**

INDEX TO MISCELLANEOUS AND USEFUL INFORMATION

(With Volume and Page Numbers and Relative Index)

A.

B.

G.

H.

I.

J.

L.

M.

T.

U.

V.

W.

Y.

Z.

CLASSIFIED LIST OF ADVERTISERS

Advertisers are classified with a view to furnish Architects and others a ready reference list of firms engaged in the Building Business. Besides the Index to Advertisements on page 803, the number of the pages on which the Advertisements appear follows directly after each name Classified in this list. It is requested that those using it will kindly mention this publication in their correspondence.

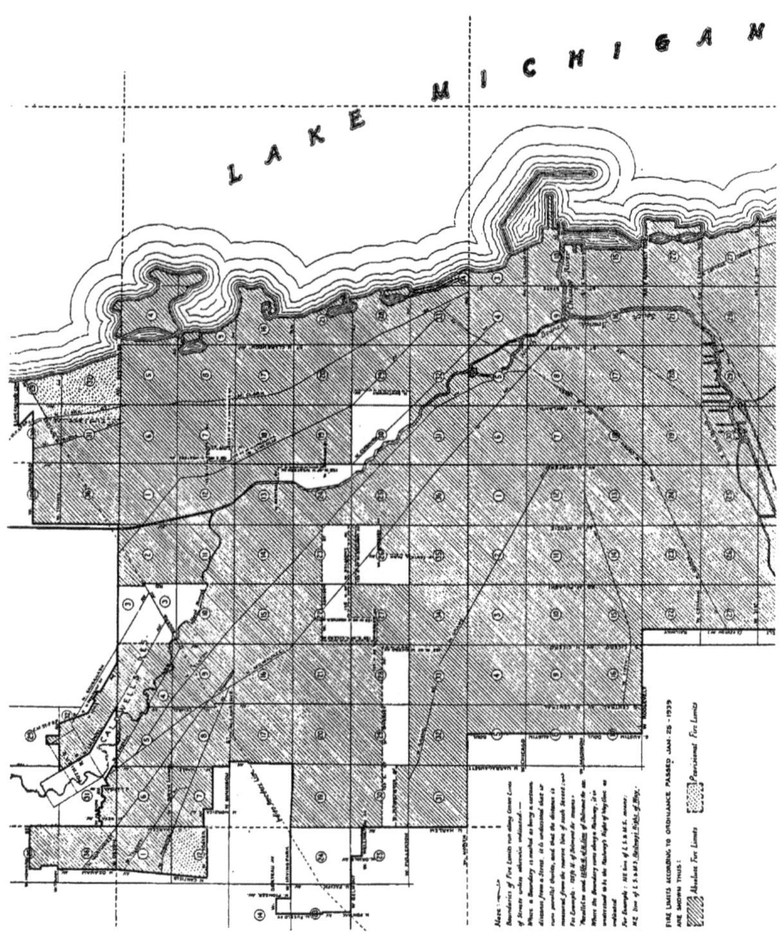

Fire Limits Map of Chicago, Passed January 25, 1939.

(Continued next page)

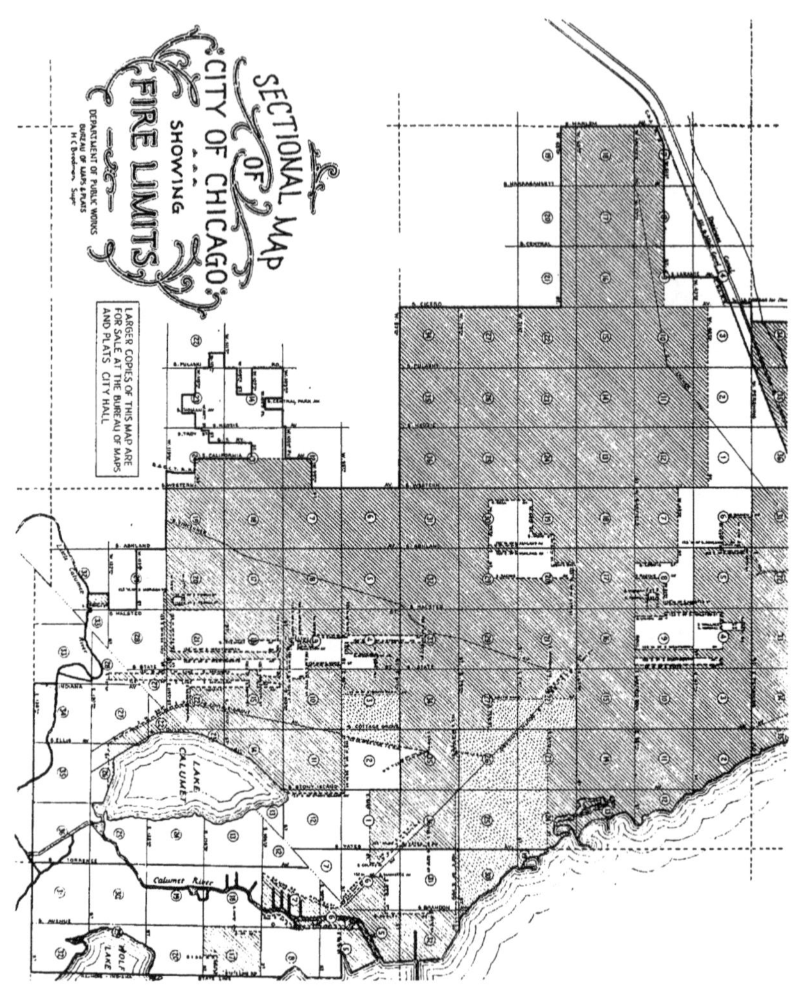
SECTIONAL MAP
OF
CITY OF CHICAGO
SHOWING
FIRE LIMITS
DEPARTMENT OF PUBLIC WORKS
BUREAU OF MAPS & PLATS
LARGER COPIES OF THIS MAP ARE
FOR SALE AT THE BUREAU OF MAPS
AND PLATS CITY HALL
LAKE CALUMET
Calumet River
WOLF LAKE

INDEX TO ADVERTISERS

TABLE OF CONTENTS

Lightning Source UK Ltd.
Milton Keynes UK
UKHW011836110219
337100UK00011B/1295/P